Annual Review of
Nuclear and Particle Science

Production Editor: Martha C. Benco
Managing Editor: Jennifer L. Jongsma
Bibliographic Quality Control: Mary A. Glass
Electronic Content Coordinator: Suzanne K. Moses
Illustration Editor: Douglas Beckner

Annual Review of Nuclear and Particle Science

Volume 60, 2010

Barry R. Holstein, *Editor*
University of Massachusetts

Wich C. Haxton, *Associate Editor*
University of Washington

Abolhassan Jawahery, *Associate Editor*
University of Maryland

www.annualreviews.org • science@annualreviews.org • 650-493-4400

Annual Reviews
4139 El Camino Way • P.O. Box 10139 • Palo Alto, California 94303-0139

Annual Reviews
Palo Alto, California, USA

International Standard Serial Number: 0163-8998
International Standard Book Number: 978-0-8243-1560-3
Library of Congress Catalog Card Number: 53-995

⊗ The paper used in this publication meets the minimum requirements of American National Standards for Information Sciences—Permanence of Paper for Printed Library Materials, ANSI Z39.48-1992.

TYPESET BY APTARA
PRINTED AND BOUND BY FRIESENS CORPORATION, ALTONA, MANITOBA, CANADA

Annual Review of Nuclear and Particle Science

Volume 60, 2010

Contents

Indexes

Errata

An online log of corrections to *Annual Review of Nuclear and Particle Science* articles may be found at http://nucl.annualreviews.org/errata.shtml

Related Articles

From the ***Annual Review of Astronomy and Astrophysics***, Volume 47 (2009)

Progenitors of Core-Collapse Supernovae
Stephen J. Smartt

Gravitational Waves from Merging Compact Binaries
Scott A. Hughes

The Chemical Composition of the Sun
Martin Asplund, Nicolas Grevesse, A. Jacques Sauval, and Patrick Scott

Gamma-Ray Bursts in the Swift Era
Neil Gehrels, Enrico Ramirez-Ruiz, and D.B. Fox

Transverse Charge Densities

Gerald A. Miller

Department of Physics, University of Washington, Seattle, Washington 98195-1560; email: miller@phys.washington.edu

Annu. Rev. Nucl. Part. Sci. 2010. 60:1–25

First published online as a Review in Advance on May 3, 2010

The *Annual Review of Nuclear and Particle Science* is online at nucl.annualreviews.org

This article's doi: 10.1146/annurev.nucl.012809.104508

0163-8998/10/1123-0001$20.00

Key Words

electromagnetic form factors, transverse momentum distributions, generalized parton distributions

Abstract

Electromagnetic form factors have long been used to probe the underlying charge and magnetization densities of hadrons and nuclei. Traditional three-dimensional Fourier transform methods are not rigorously applicable for systems with constituents that move relativistically. The use of the transverse charge density is a new, rigorously defined way to analyze electromagnetic form factors of hadrons. This review is concerned with the following issues: what a transverse charge density is, how one is extracted from elastic scattering data, the existing results, its relationship with other observable quantities, and future prospects.

Contents

1. INTRODUCTION

A truly impressive level of experimental technique, effort, and ingenuity has been applied to measuring the electromagnetic form factors of the proton, neutron (nucleon), and pion (1–4). These quantities are probability amplitudes that a given hadron can absorb a specific amount of momentum and remain in the ground state, and therefore they should supply information about charge and magnetization densities. The textbook interpretation of these form factors is that their Fourier transforms are measurements of the charge and magnetization densities. This interpretation is deeply entrenched in the thinking of nuclear and particle physicists and continues to guide intuition, as it has since the days of the Nobel prize–winning work of Hofstadter (5). Nevertheless, the relativistic motion of the constituents of the system renders the textbook interpretation incorrect (6).

The preceding statement leads to several questions: Is the statement correct? If so, how fast do the constituents have to move to violate the nonrelativistic interpretation? Why is it that the motion of the constituents and not that of the entire system that is relevant? The answers to these questions are probably displayed within the existing literature. However, obtaining general clear answers has been sufficiently difficult that posing even the first question of this paragraph would not lead to a unanimous answer. Moreover, there is another, more important question. If the nonrelativistic approach is not correct, then what is the correct procedure to determine model-independent information regarding hadronic charge and magnetization densities? How should we interpret the beautifully precise electromagnetic form factor data being produced at Jefferson Laboratory, Bates, Bonn, and Mainz?

The aim of this review is to present and clarify answers to these questions. The most immediate question is addressed here. The only way to determine model-independent information about the charge distribution is to study the transverse charge density $\rho(b)$ (7), which gives the density of a hadron moving at infinite momentum at a transverse (to the direction of rapid motion) distance b from the transverse center of momentum. It is only in the infinite momentum frame (IMF) that one can define such a center. Information about the charge density in terms of longitudinal coordinates is not yet available in a meaningful way.

1.1. Electromagnetic Form Factors Are Not Three-Dimensional Fourier Transforms of Charge Densities

Physicists are trained to believe that physics is independent of the inertial reference frame. In nonrelativistic quantum mechanics, this invariance allows us to express the wave function as a product of a plane-wave factor (that describes the motion of the center of mass) with a function that depends only on internal relative coordinates. Thus, for two spinless particles

$$\Psi(\vec{r}_1, \vec{r}_2) = e^{i\vec{P}\cdot\vec{R}}\phi(\vec{r}), \qquad 1.$$

where $\vec{P}$ is the total momentum, $\vec{R}$ is the position of the com, and $\vec{r} = \vec{r}_1 - \vec{r}_2$. Then the nonrelativistic form factor $F_{NR}(|\vec{q}\,|)$, which is the probability amplitude for the system to absorb a momentum $\vec{q}$ and remain in its initial state, is given by the integral

$$F_{NR}(|\vec{q}\,|) = \int d^3r|\phi(\vec{r})|^2 e^{i\vec{q}\cdot\vec{r}/2}, \qquad 2.$$

if the masses of the two constituents are the same. One sees that the square of a wave function or probability density is being probed.

However, the separation appearing in Equation 1 is not generally valid. For a similar but relativistic system, the Minkowski-space, Bethe-Salpeter wave function $\Phi(k, P)$ (where k is the relative four-momentum and P is the total four-momentum) can be written in a compact form, if the interaction kernel is given by a set of Feynman graphs (8, 9), by using the Nakanishi integral representation

$$\Phi(k; P) = -\frac{i}{\sqrt{4\pi}}\int_{-1}^{1} dz \int_0^\infty d\gamma \frac{g(\gamma, z)}{[\gamma + m^2 - \frac{1}{4}M^2 - k^2 - P\cdot kz - i\varepsilon]^3}, \qquad 3.$$

where m is the mass of each constituent scalar particle and M is the hadronic mass. The weight function $g(\gamma, z)$ itself is not singular, but the singularities of the Bethe-Salpeter amplitude are reproduced by using Equation 3 (10). The key feature of Equation 3 is that the covariant wave function depends explicitly on the total four-momentum P. Another way of presenting the contents of Equation 3 involves the relativistic boost operator used to obtain the wave function in a moving frame from one in which the hadron is at rest. The Nakanishi representation allows one to perform the boost merely by changing the value of P.

The explicit dependence on P dramatically influences our understanding of form factors because the initial and final hadrons each have a different momentum and therefore different wave functions. The presence of differing wave functions of the initial and final nucleons invalidates a naïve probability or density interpretation.

Assuming that only one of the scalar constituents carries charge, the electromagnetic form factor can be obtained by evaluating the impulse approximation triangle diagram, with the result (10)

$$(P + P')^\mu F(Q^2) = i\int \frac{d^4k}{(2\pi)^4}(P + P' + k)^\mu (k^2 - m^2)\Phi\left(\frac{1}{2}P - k, P\right)\Phi\left(\frac{1}{2}P' - k, P'\right), \qquad 4.$$

where $P' = P + q$.

The form factor of Equation 4 seems to differ markedly from the usual nonrelativistic expression of Equation 2, expressed in terms of momentum-space wave functions $\tilde{\phi}$:

$$F^{NR}(|\vec{q}|) = \int \frac{d^3k}{(2\pi)^3}\tilde{\phi}^*(\vec{k} + \vec{q}/2)\tilde{\phi}(\vec{k}). \qquad 5.$$

It is comforting that in the nonrelativistic limit, defined by the replacement

$$x = \frac{k^0 + k^3}{P^0 + P^3} \rightarrow \frac{1}{2} + \frac{k^3}{2m}, \quad M - 2m \ll 2m, \qquad 6.$$

the IMF light-front wave function becomes identical to the rest-frame instant-form wave function (11). This means (6) that if the condition in Equation 6 is true for a given model, then Equation 5 (and therefore Equation 2) can be obtained from Equation 4. In this case, one may extract $\rho(\mathbf{r}) \equiv |\tilde{\phi}(\mathbf{r})|^2$ from a three-dimensional Fourier transform of F_{NR}. However, the conditions in Equation 6 are not expected to be obtained for hadrons, although they are valid for nuclei (6). Furthermore, if the charged particles of a given system have different masses, one obtains (in the nonrelativistic limit for all constituents)

$$F^{NR}(\vec{q}) = \int d^3r \sum_i \rho_i(\vec{r})e^{-i\vec{q}\cdot\vec{r}m_i/M}, \qquad 7.$$

where ρ_i and m_i are the charge density and mass of the ith constituent, respectively, and one can not obtain the charge density from the form factor.

1.2. What Is a Transverse Charge Density?

A proper determination of a charge density requires the measurement of a matrix element of a density operator. We show here that measurements of the hadronic form factor directly involve the three-dimensional charge density of partons in the IMF, $\hat{\rho}_\infty(x^-, \mathbf{b})$. Before discussing this quantity, we provide a brief introduction to light-cone coordinates.

In the IMF, the time coordinate $ct = x^0/\sqrt{2}$ is expressed in a frame moving along the negative z direction, with a velocity nearly that of light, using the Lorentz transformation with the variable $x^+ = (x^0 + x^3)/\sqrt{2}$, in which the usual γ factor is absorbed by a Jacobian of an integral over volume (12). The x^+ variable is canonically conjugate to the minus component of the momentum operator $p^- \equiv (p^0 - p^3)/\sqrt{2}$. The longitudinal spatial variable is $x^- = (x^0 - x^3)/\sqrt{2}$, and its conjugate momentum is $p^+ = (p^0 + p^3)/\sqrt{2}$. It is this plus component of momentum that is associated with the usual Bjorken variable. The transverse coordinates x and y are written as $\mathbf{b}$ with the conjugate momentum $\mathbf{p}$. In the literature, these transverse coordinates are sometimes written with a subscript $\perp$, but here we use boldface to denote the transverse components of position and momentum vectors. Also, we employ a light-front time quantization that sets x^+ and the plus component of all spatial variables to zero. This means that x^- can be considered as the longitudinal variable $-\sqrt{2}x^3$. One extremely useful aspect of using these variables is that Lorentz transformations to frames moving with different transverse velocities do not depend on interactions. These transformations form the kinematic subgroup of the Poincaré group, so boosts in the transverse direction are accomplished by using the transverse component of the position operator (as in the nonrelativistic theory).

In the IMF, the electromagnetic charge density J^0 becomes J^+, and

$$\hat{\rho}_\infty(x^-, \mathbf{b}) = J^+(x^-, \mathbf{b}) = \sum_q e_q \bar{q}(x^-, \mathbf{b})\gamma^+ q(x^-, \mathbf{b}) = \sum_q e_q \sqrt{2} q_+^\dagger(x^-, \mathbf{b}) q_+(x^-, \mathbf{b}), \qquad 8.$$

where $q_+(x^\mu) = \gamma^0\gamma^+ q(x^\mu)/\sqrt{2}$, the independent part of the quark field operator $q(x^\mu)$. The time variable x^+ is set to zero. Note the appearance of the absolute square of quark field operators, which is the signature of a true density.

We are concerned with the relationship between charge density and the electromagnetic form factor $F(Q^2)$, determined from the current density via

$$F(Q^2) = \frac{\langle p^+, \mathbf{p}'|J^+(0)|p^+, \mathbf{p}\rangle}{2p^+}, \qquad 9.$$

where the normalization is $\langle p'^+, \mathbf{p}'|p^+, \mathbf{p}\rangle = 2p^+(2\pi)^3\delta(p'^+ - p^+)\delta^{(2)}(\mathbf{p}' - \mathbf{p})$. Spin or helicity dependence is ignored in the present subsection. We take the momentum transfer $q_\alpha = p'_\alpha - p_\alpha$ to be space-like, with the square of the space-like four-momentum transfer $q^2 = -Q^2$, and use the Drell-Yan frame with ($q^+ = 0,\ Q^2 = \mathbf{q}^2$). No longitudinal momentum is transferred, so the initial and final states are related only by kinematic transformations. Moreover, with this condition the current operator links Fock-state components with the same number of constituents. The matrix element in Equation 9 involves the combination of creation and destruction operators: $b^\dagger b - d^\dagger d$ for each flavor of quark so that the valence charge density is probed. The form factor F is independent of renormalization scale because the vector current $\bar{q}\gamma^\mu q$ is conserved (13).

The spatial structure of a spinless hadron can be examined by using states that are transversely localized (7, 13, 14) through a linear superposition,

$$|p^+, \mathbf{R} = \mathbf{0}\rangle \equiv \mathcal{N} \int \frac{d^2p}{(2\pi)^2}|p^+, \mathbf{p}\rangle, \qquad 10.$$

where $\mathcal{N}$ is a normalization factor satisfying $|\mathcal{N}|^2 \int \frac{d^2\mathbf{p}_\perp}{(2\pi)^2} = 1$. Wave packet representations can be used to avoid states normalized to δ functions (15, 16), but this leads to the same results as using Equation 10. Considering that $2p^+p^- - \mathbf{p}^2 = m_\pi^2 > 0$, (where $p^- > 0$), one finds that p^+ must be very large because the range of integration over $\mathbf{p}$ is large. Using such an ultralarge or infinite value of p^+ maintains the interpretation of a pion moving in the IMF with well-defined longitudinal momentum (15). It is in just such an infinite momentum that the parton interpretation of a hadron is valid. Setting the transverse center of momentum to zero (Equation 10) allows a meaningful transverse distance $\mathbf{b}$. A Fock-space parton representation of the position of the transverse center of momentum, $\mathbf{R}$, provides a useful relation between longitudinal momentum fractions and transverse positions (14),

$$\mathbf{R} = \sum_i x_i \mathbf{b}_i, \qquad 11.$$

where the sum is over the finite number of constituents in a given component.

Next, we relate the charge density

$$\rho_\infty(x^-, \mathbf{b}) = \frac{\langle p^+, \mathbf{R} = \mathbf{0}|\hat{\rho}_\infty(x^-, \mathbf{b})|p^+, \mathbf{R} = \mathbf{0}\rangle}{\langle p^+, \mathbf{R} = \mathbf{0}|p^+, \mathbf{R} = \mathbf{0}\rangle} \qquad 12.$$

to $F(Q^2)$. In the Drell-Yan frame, no momentum is transferred in the plus direction, so information regarding the x^- dependence of the density is not accessible. Therefore we integrate over x^-, using the relationship $q_+^\dagger(x^-, \mathbf{b})q_+(x^-, \mathbf{b}) = e^{i\hat{p}^+x^-}e^{-i\hat{\mathbf{p}}\cdot\mathbf{b}}q_+^\dagger(0)q_+(0)e^{i\hat{\mathbf{p}}\cdot\mathbf{b}}e^{-i\hat{p}^+x^-}$ to find

$$\rho(b) \equiv \int dx^- \rho_\infty(x^-, \mathbf{b}) = \langle p^+, \mathbf{R} = \mathbf{0}|\hat{\rho}_\infty(0, \mathbf{b})|p^+, \mathbf{R} = \mathbf{0}\rangle/(2p^+). \qquad 13.$$

The use of Equation 9 and the expansion in Equation 10 lead to the simplification of the right-hand side of the above equation:

$$\rho(b) = \int \frac{d^2q}{(2\pi)^2} F(Q^2 = \mathbf{q}^2) e^{-i\mathbf{q}\cdot\mathbf{b}}, \qquad 14.$$

where $\rho(b)$ is termed the transverse charge density, thereby giving the charge density at a transverse position b, irrespective of the value of the longitudinal position or momentum. The use of a three-dimensional coordinate-space density $\hat{\rho}_\infty(x^-, \mathbf{b})$ to obtain the transverse density appeared in Reference 17. Previous derivations (7, 13, 14, 18) used a density operator involving transverse position and longitudinal momentum variables; see Section 2.

If the nonrelativistic limit Equation 6 is valid, one finds that (6)

$$\rho^{NR}(b) = \int_{-\infty}^{\infty} dz \rho(r), \qquad 15.$$

where $\rho(r)$ is the square of the wave function obtained from Equation 2. This equation, as well as the integral over the longitudinal coordinate that appears in Equation 13, illustrates the nature of the transverse density as a reduction of the three-dimensional density. Moreover, these integrals appear in the Glauber theory (19) of high-energy scattering, in which scattering amplitudes are expressed in terms of transverse densities. A recent example can be found in Reference 20.

1.3. Simple Example

We use a simple example (6) in which a scalar particle Ψ is modeled as a bound state of two different scalar particles, ξ and χ, with a point-like coupling such that the interaction Lagrangian is $g\Psi\xi\chi$. The Bethe-Salpeter wave function can be obtained from Equation 3 by replacing $g(\gamma, z)/\sqrt{4\pi}$ with the simple coupling constant g. The explicit Bethe-Salpeter wave function can be obtained by straightforward integration, and it is proportional to the product of Klein-Gordan propagators. For this model, the evaluation of the form factor in Equation 4 can be performed in three different ways: One can (*a*) simply evaluate the Feynman integral, (*b*) work in the IMF with $P^3 \to \infty$ (21), or (*c*) proceed by first integrating over k^-. The resulting form factor is the same in all three cases (6) and can be expressed in terms of a three-dimensional integral, which involves a light-front wave function $\psi(x, \boldsymbol{\kappa})$:

$$\psi(x, \boldsymbol{\kappa}) \equiv g\left[M^2 - \frac{\boldsymbol{\kappa}^2 + m_1^2}{x} - \frac{\boldsymbol{\kappa}^2 + m_2^2}{1-x}\right]^{-1}. \qquad 16.$$

In this case,

$$F(Q^2) = \frac{1}{2(2\pi)^3} \int d^2\boldsymbol{\kappa} \int_0^1 \frac{dx}{x(1-x)} \psi^*(x, \boldsymbol{\kappa} + (1-x)\mathbf{q}) \psi(x, \boldsymbol{\kappa}). \qquad 17.$$

The relative variables are $(x, \boldsymbol{\kappa})$, where $x = k^+/P^+$, and $\boldsymbol{\kappa} = \boldsymbol{\kappa}_1 - \boldsymbol{\kappa}_2$. The function $\psi(x, \boldsymbol{\kappa})$ can be obtained from the Φ of Equation 3 by integrating over k^- (11, 22, 23).

One can obtain a relation between the form factor and a coordinate-space density by expressing the light-front wave function $\psi(x, \boldsymbol{\kappa})$ in terms of the canonically conjugate spatial variable $\mathbf{B} = \mathbf{b} - \mathbf{b}_2$, where $\mathbf{b}$ is the transverse position operator of the charged constituent. Then $\tilde{\psi}(x, \mathbf{B}) = \int \frac{d^2\boldsymbol{\kappa}}{(2\pi)^2} e^{i\boldsymbol{\kappa}\cdot\mathbf{B}} \psi(x, \boldsymbol{\kappa})$, and

$$F(Q^2) = \frac{1}{4\pi} \int d^2\mathbf{B} \int_0^1 \frac{dx}{x(1-x)} |\tilde{\psi}(x, \mathbf{B})|^2 e^{-i\mathbf{B}(1-x)\cdot\mathbf{q}}. \qquad 18.$$

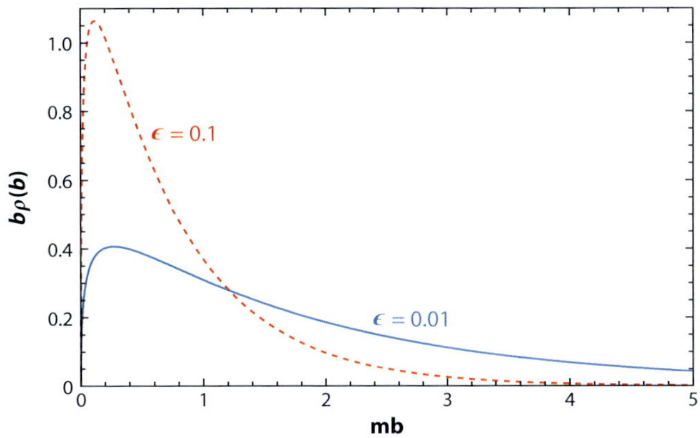

Figure 1

$b\rho(b)$ for the model of Equation 20. Results for two different values of the binding energy are displayed.

We express the form factor in terms of the position of the charged constituent by using the condition that the center of transverse momentum is set to zero (Equation 11 with two terms): $\mathbf{R} = 0 = x\mathbf{b} + (1-x)\mathbf{b}_2$ so that $\mathbf{B} = \mathbf{b}/(1-x)$. Expressing the form factor in terms of $\mathbf{b}$ gives

$$F(Q^2) = \frac{1}{4\pi}\int d^2\mathbf{b}\int_0^1 \frac{dx}{x(1-x)^3}\left|\tilde{\psi}\left(x, \frac{\mathbf{b}}{1-x}\right)\right|^2 e^{-i\mathbf{b}\cdot\mathbf{q}}. \qquad 19.$$

Comparing Equation 19 with Equation 14 allows us to identify the transverse density as

$$\rho(b) = \pi\int_0^1 \frac{dx}{x(1-x)^3}\left|\tilde{\psi}\left(x, \frac{\mathbf{b}}{1-x}\right)\right|^2, \qquad 20.$$

which for the present model is given by

$$\rho(b) = \frac{g^2}{2(2\pi)^3}\int_0^1 dx\frac{x}{(1-x)}K_0^2\left(\sqrt{m^2 - M^2x(1-x)}\frac{b}{1-x}\right). \qquad 21.$$

The mass M must be less than $2m$ for the hadron Ψ to be stable. We define a positive binding fraction ε, so that $M = 2m - \varepsilon m$ $(0 < \varepsilon < 2)$. Small values of ε correspond to the applicability of the nonrelativistic limit. The transverse densities of Equation 20 are shown in **Figure 1** for $\varepsilon = 0.01$ and 0.1. As expected from nonrelativistic intuition, the smaller value of the binding energy corresponds to a greater spatial extent.

2. ZOO OF NUCLEON DISTRIBUTIONS AND DENSITIES

The vast literature concerning the distribution functions that are used to describe nucleon structure includes generalized parton distributions (GPDs) (24–35) and transverse momentum distributions (TMDs) (36–42). The relationship among GPDs, transverse charge densities, and TMDs is discussed in the following paragraphs.

We begin by discussing GPDs and follow the discussion of Burkardt (14). Deep-inelastic scattering experiments allow the determination of parton distribution functions (PDFs), which give the probability that quarks carry a given fraction x of the nucleon longitudinal (+) momentum in the IMF. PDFs are the forward matrix element of a light-like correlation function,

$$q(x) = \langle P, S|\hat{O}_q(x, \mathbf{0})|P, S\rangle \qquad 22.$$

with

$$\hat{O}_q(x, \mathbf{0}) \equiv \int \frac{dx^-}{4\pi} \bar{q}\left(-\frac{x^-}{2}, \mathbf{0}\right) \gamma^+ q\left(\frac{x^-}{2}, \mathbf{0}\right) e^{ixP^+x^-}, \tag{23}$$

where $|P, S\rangle$ represents the wave function. Here we use the light-cone gauge $A^+ = 0$. The use of the canonical field expansion for the quark field operators explicitly shows that PDFs give the probability that the quarks carry a longitudinal momentum fraction x. We do not display the scale dependence of the PDFs in order to simplify the notation.

GPDs (24–35), which describe the scaling limit in real and virtual Compton scattering experiments, are defined by allowing the momenta and spins of the initial and final nucleons to differ:

$$\langle P', S'|\hat{O}_q(x, 0_\perp)|P, S\rangle \tag{24}$$

$$= \frac{1}{2\bar{p}^+}\bar{u}(P', S')\left(\gamma^+ H_q(x, \xi, t) + i\frac{\sigma^{+\nu}\Delta_\nu}{2M}E_q(x, \xi, t)\right)u(P, S), \tag{25}$$

where $\bar{p}^\mu = \frac{1}{2}(P^\mu + P'^\mu)$ is the mean momentum of the target, $\Delta^\mu = P'^\mu - P^\mu$ is the four-momentum transfer, and $t = \Delta^2$ is the invariant momentum transfer. The skewedness parameter $\xi = -\frac{\Delta^+}{2\bar{p}^+}$ represents the change in the longitudinal component of the nucleon momentum.

GPDs allow for a unified description of a number of hadronic properties. In the forward limit—$P' = P$—they reduce to conventional PDFs—$H_q(x, 0, 0) = q(x)$. Performing the integral over x causes the x^- coordinate to vanish so that the operator $\hat{O}_q$ is converted into a local current operator and results in the appearance of the usual Dirac and Pauli form factors of the nucleon,

$$\sum_q e_q \int dx H_q(x, \xi, t) = F_1(t) \tag{26}$$

and

$$\sum_q e_q \int dx E_q(x, \xi, t) = F_2(t). \tag{27}$$

The next step is to define impact parameter parton distributions (14). One localizes the nucleon in the transverse direction using the superposition of Equation 10 but with light-cone helicity states $|p^+, \mathbf{p}, \lambda\rangle = |P, S\rangle$ (43). For a transversely localized state $|p^+, \mathbf{R}_\perp = \mathbf{0}_\perp, \lambda\rangle$, Burkardt's impact parameter–dependent PDF is given by

$$q(x, \mathbf{b}) \equiv \langle p^+, \mathbf{R}_\perp = \mathbf{0}_\perp, \lambda|\hat{O}_q(x, \mathbf{b})|p^+, \mathbf{R}_\perp = \mathbf{0}_\perp, \lambda\rangle. \tag{28}$$

The operator $\hat{O}_q(x, \mathbf{b})$ is obtained by a translation of the operator appearing in Equation 23 in the transverse plane:

$$\hat{O}_q(x, \mathbf{b}) = \int \frac{dx^-}{4\pi} \bar{q}\left(-\frac{x^-}{2}, \mathbf{b}\right) \gamma^+ q\left(\frac{x^-}{2}, \mathbf{b}\right) e^{ixp^+x^-}. \tag{29}$$

The function $q(x, \mathbf{b})$ gives the probability density that a quark of momentum fraction x is located at a transverse position $\mathbf{b}$.

Burkardt (14) used the transversely localized states of the same helicity to show that $q(x, \mathbf{b})$ is the two-dimensional Fourier transform of the GPD H_q:

$$q(x, \mathbf{b}) = \int \frac{d^2q}{(2\pi)^2} e^{-i\mathbf{q}\cdot\mathbf{b}} H_q(x, t = -\mathbf{q}^2). \tag{30}$$

Integration over x using Equations 26 and 14 shows that

$$\rho(b) = \int dx \sum_q e_q\, q(x, b) = \int \frac{d^2q}{(2\pi)^2} F_1(t = -\mathbf{q}^2) e^{-i\,\mathbf{q}\cdot\mathbf{b}}. \qquad 31.$$

This means that the transverse density can be obtained either as an integral over x of $q(x, b)$ or as an integral over x^- of the three-dimensional coordinate-space density, as in Section 1.2. This equality of an integral over a momentum with one over a distance is an example of Parseval's theorem. In either case, model-independent information regarding the longitudinal coordinate is not available because the probe momentum is transverse; that is, $q^+ = 0$.

In the model of Section 1.3 (scalar hadron made of two scalar constituents), $q(x)$ can be obtained from the integrand of Equation 20 as

$$q(x, b) = \frac{\pi}{x(1-x)^3} \left| \tilde{\psi}\left(x, \frac{\mathbf{b}}{1-x}\right)\right|^2. \qquad 32.$$

For large values of x approaching unity, the transverse extent is very narrow because this corresponds to very large values of the relative transverse separation $\mathbf{B} = \mathbf{b}/(1-x)$ for finite values of $\mathbf{b}$. Thus, the transverse spatial extent explicitly depends on the value of x. Specifically, large values of x are associated with small values of b.

TMDs are another generalization of PDFs. The TMDs, which contain probability distributions regarding both the longitudinal momentum fraction x and the transverse momentum $\mathbf{k}$ carried by the quarks, are given by (36, 37, 40)

$$\Phi_q^\Gamma\left(x = \frac{k^+}{P^+}, \mathbf{k}\right) = \left\langle P, S \left| \int \frac{d\zeta^- d^2\boldsymbol{\zeta}}{2(2\pi)^3} e^{i(k^+\zeta^- - \mathbf{k}\cdot\boldsymbol{\zeta})} \bar{q}(0)\Gamma q(\zeta^-, \boldsymbol{\zeta}) \right| P, S\right\rangle, \qquad 33.$$

where the time separation $\zeta^+ = 0$. Here the quark field operators are the gauge-invariant operators in gauges for which the gauge potentials vanish at space-time infinity (44, 45). These operators are necessary for nonzero transverse spatial separations $\boldsymbol{\zeta}$. The operator Γ is a Dirac matrix that defines the quark density.

The initial and final states appearing in Equation 33 have the same momentum, which is the major difference between TMDs and GPDs. Another major difference is that GPDs involve densities in transverse coordinate space, whereas TMDs involve densities in transverse momentum space. However, one can use Wigner distributions (44) to show that GPDs and TMDs are obtained from different manipulations on the same operator. We define the reduced Wigner distribution as

$$W_q^\Gamma(\zeta^-, \boldsymbol{\zeta}, k^+, \mathbf{k}) = \frac{1}{4\pi} \int d\eta^- d^2\boldsymbol{\eta}\, e^{ik^+\eta^- - i\mathbf{k}\cdot\boldsymbol{\eta}} \bar{q}(\zeta^- - \eta^-/2, \boldsymbol{\zeta} - \boldsymbol{\eta}/2)\Gamma q(\zeta^- + \eta^-/2, \boldsymbol{\zeta} + \boldsymbol{\eta}/2). \qquad 34.$$

GPDs are obtained by taking the matrix element of $W_q^\Gamma(\zeta^- = 0, \boldsymbol{\zeta} = \mathbf{0}, k^+, \mathbf{k})$ between states of different momentum and then integrating over $\mathbf{k}$, as in

$$H_q(x, \xi, t) = \left\langle P', S \left| \int \frac{d^2\mathbf{k}}{(2\pi)^2} W_q^{\gamma^+}(\zeta^-, \boldsymbol{\zeta}, k^+, \mathbf{k}) \right| P, S\right\rangle, \qquad 35.$$

and TMDs are obtained by taking the matrix element of $W_q^\Gamma(\zeta^-, \boldsymbol{\zeta}, k^+, \mathbf{k})$ between states with identical momentum:

$$\Phi_q^\Gamma(x, \mathbf{k}) = \left\langle P, S \left| \int \frac{d\zeta^-}{(2\pi)^2} W_q^\Gamma(\zeta^-, \boldsymbol{\zeta}, k^+, \mathbf{k}) \right| P, S\right\rangle. \qquad 36.$$

The integral of a GPD over all values of x, as in Equations 26 and 27, converts the nonlocal bilinear appearing in Equation 29 into a local operator. More generally, the GPD is a correlation function of quarks at the same light-cone time $x^+ = 0$. The integral over x also gives $x^- = 0$, so both the time t and spatial z separations vanish. Thus, one can compare matrix elements involving the integral over x with related matrix elements computed using the equal-time formalism, $t = 0$. Moreover, performing the x integral allows a connection with lattice quantum chromodynamics (QCD) calculations. The lattice formulation is built using covariant Euclidean space. If one sets t and z to zero in a lattice calculation, one obtains a quantity that depends on transverse coordinates that are suitable for comparison with a relevant, transverse, experimentally measured quantity. This relation was pointed out in connection with transversity observables in Reference 46 and used in References 47–50.

A similar connection (51) between the light-front and equal-time formalisms occurs when using TMDs. A TMD (Equation 33) gives the probability that a quark has a three-momentum characterized by $(x, \mathbf{k})$. The relevant matrix element involves quarks separated at the same light-cone time, $\zeta^+ = 0$. Integration over x also sets ζ^- to zero, so one obtains a density evaluated by using quarks at the same time.

3. SINGULAR PIONIC TRANSVERSE CHARGE DENSITY

One must understand the pion to learn how QCD describes the interaction and existence of elementary particles. As a nearly massless excitation of the QCD vacuum with pseudoscalar quantum numbers, the pion plays a central role in nuclear and particle physics, as the carrier of the longest-ranged force between nucleons and as a harbinger of spontaneous symmetry breaking.

Computing the electromagnetic form factor of the pion, $F_\pi(Q^2)$, at asymptotically large values of Q^2 from first principles was one of the early challenges to the use of perturbative QCD (pQCD) in exclusive processes (52–55). Such calculations have been extended (56–59) by including effects of higher order in the strong coupling constant and higher-twist effects. The lowest-order results are approximately a factor of three smaller than existing data, and higher-order and higher-twist effects are not small at currently available values of Q^2 (59). It is widely believed that at large enough values of momentum transfer Q^2 the leading-order perturbative formula is valid, but such large values may be very large indeed (60). As a result, there is considerable experimental interest in studying the transition to the region in which pQCD can be applied. New measurements have been performed (61, 62), and more are planned (63). Here we review the first analysis (17) that provided a model-independent pionic transverse charge density.

Recent pion data (61, 62) provide a very accurate measurement of the longitudinal part of the electroproduction cross section and the related pion form factor up to a value of $Q^2 = 2.45\ \mathrm{GeV}^2$. The result is that existing data for the pion form factor are well represented by the monopole form

$$F_\pi(Q^2) = 1/(1 + R^2 Q^2/6), \qquad 37.$$

where $R^2 = 0.431\ \mathrm{fm}^2$. Equation 37 allows one to determine the transverse density from Equation 14, resulting in

$$\rho(b) = \frac{3K_0\left(\frac{\sqrt{6}b}{R}\right)}{\pi R^2}, \qquad 38.$$

where K_0 is a modified Bessel function of rank zero. For small values of b this function diverges as $\sim\log(b)$; thus, the transverse density is singular and infinite at the origin. This infinity cannot be "cured" by a renormalization procedure because the charge density under consideration is the matrix element of a valence quark operator between eigenstates of the full Hamiltonian. Furthermore, divergences of quark distribution functions that occur at small values of Bjorken x

do not occur here because the transverse charge density involves the difference between quark and antiquark densities. Many field theory models, derived even before QCD was established (64), obtain a form factor that corresponds to a divergent transverse density. Also, note that any model, such as vector meson dominance or holographic QCD (65–67), that yields a monopole form factor contains a central density with a logarithmic divergence. Thus, holographic QCD does not supply a representation of the soft component (60) of the pion wave function.

Intuition regarding a possible singularity in the central charge density may be gained from other examples. Suppose that the nonrelativistic limit, in which the quark masses are heavy, is applicable. Then the pion would be a pure $q\bar{q}$ object, and with the charge density represented by the Fourier transform of the form factor (Section 1.1). Using the form factor of Equation 37, the nonrelativistic three-dimensional density would be uniquely given by

$$\rho_{\mathrm{NR}}(r) = \frac{3}{2\pi\, r\, R^2} e^{\frac{-\sqrt{6}r}{R}}, \qquad 39.$$

where r is the distance relative to the pion center of mass. This density diverges at $r = 0$. If one takes $r = \sqrt{b^2 + z^2}$, as demanded by the rotational invariance of the nonrelativistic wave function, then one finds from Equation 15 that the nonrelativistic transverse charge density assumes the form of the $\rho(b)$ of Equation 38.

The divergence of the central transverse charge density encountered here may be a consequence of using a simple parameterization, presently untested by measurements at larger values of Q^2. Thus, we examine other approaches. Consider, for instance, pQCD, which provides a prediction (52–55) for asymptotically large values of Q^2 that

$$\lim_{Q^2\to\infty} F_\pi(Q^2) = 16\pi\alpha_s(Q^2) f_\pi^2/Q^2, \qquad 40.$$

where the pion decay constant $f_\pi = 93$ MeV, and in leading order

$$\alpha_s(Q^2) = \frac{4\pi}{(11 - \frac{2}{3}n_f)\ln\frac{Q^2}{\Lambda^2}}, \qquad 41.$$

where n_f is the number of quarks of mass smaller than Q and Λ is a parameter fixed by data. The $\log Q^2$ term in the denominator does not lead to nonsingular behavior of $\rho(b)$ for small values of b because $\rho(b) \sim \log\log\frac{1}{b}$, and the pQCD form factor corresponds to a singularity at short distance. This singularity would arise in any model form factor, even one based on sum rules (e.g., 60), that approaches pQCD asymptotically.

Chiral quark models (68) paint an alternative picture (69, 70), yielding transverse charge densities that behave singularly at the center as $\log b$ because the pion form factor yields a monopole form. Nevertheless, each and every observable quantity, including f_π and structure functions, is computed to be finite.

Gaussian models with GPDs $H(x, 0, Q^2)$ (see Equation 26) dominated by behavior near $x = 1$ present a set of examples that also yield a form factor with a $1/Q^2$ asymptotic behavior and have an impact parameter distribution that is well behaved at each value of x for all b. The key asymptotic features are captured in the simple formula (76, 77) $H(x, 0, Q^2)_{x\to 1} = (1 - x)^{n-1} e^{-\alpha(1-x)^n Q^2}$ with $n > 2$ so that $q(x, b)_{x\to 1} = \frac{1}{2\pi a(1-x)} e^{-b^2/(4a(1-x)^n)}$. This form shows that $q(x, b)$ is well behaved for all values of b and for each value of x, but the integral over x contains a logarithmic divergence.

Relativistic light-front constituent quark models (75, 78, 79) that describe existing form factor data have a nonsingular central charge density. The simplest way to derive these models (79) is by using the impulse approximation (evaluating the triangle diagram). One starts by evaluating the integral over the minus component of the loop momentum k^μ and then cutting off the remaining integral over $x = k^+/p^+$, $k_\perp$ by using a phenomenological wave function. The wave function of

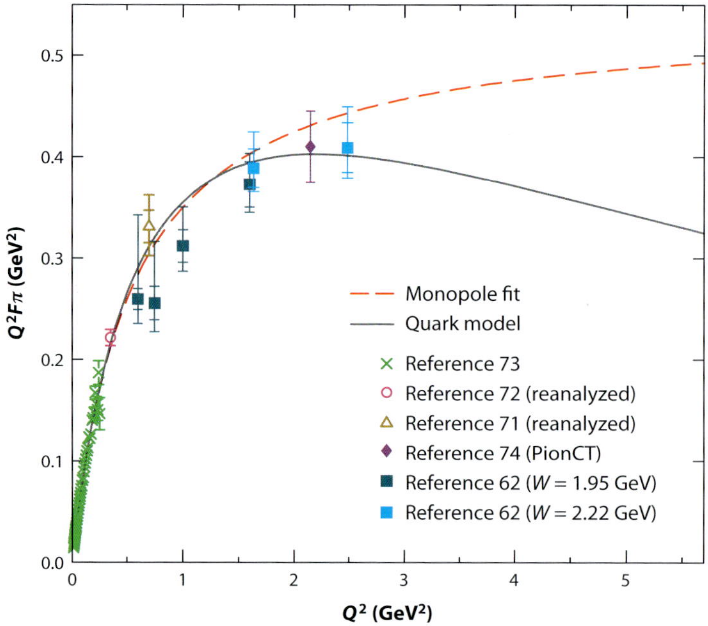

Figure 2

Pion form factor data as plotted in Reference 62. The data from References 71 and 72 have been reanalyzed using the method of Reference 62. The red dashed curve uses the monopole fit from Equation 37, and the gray solid line represents the constituent quark model of Reference 75. Reprinted from Reference 17 with permission from the American Physical Society.

Reference 75 is chosen to be a power-law form, and the resulting model describes the existing form factor data in the space-like region f_π and the transition form factor $f_{\pi\gamma}$, in which a virtual photon transforms a real pion into a real photon. The model form factor of Reference 75 and the monopole fit of Equation 37 are shown, along with the measured data, in **Figure 2**. Both the fit and the model provide a good fit to the data but present very different predictions for larger values of Q^2, for which measurements remain to be done.

The corresponding versions of $\rho(b)$ are shown in **Figure 3**. The singularity of Equation 38 is manifest as a rapidly rising function as b approaches zero, but the relativistic constituent quark model provides a smooth function $\rho(b)$. A planned experiment (63) aims to achieve results up to $Q^2 = 6$ GeV^2. This value of Q^2 should be large enough to resolve the differences between the model (75) and the monopole fit or, alteratively, to rule out both. However, this probably will not be sufficient to reach the regime in which pQCD might be valid (63). This picture (75) represents one useful phenomenology, but other interesting models exist, and computing the transverse density in those models would also be of interest.

4. NUCLEON TRANSVERSE CHARGE DENSITIES

Above, we emphasized that a proper determination of a nucleon charge density requires a connection to the square of a field operator, which can be achieved through the transverse charge density (7, 13, 14, 18, 80) $\rho(b)$ given by Equation 31. We review the first phenomenological analysis of existing data that determined $\rho(b)$ for the neutron and the proton (18). The neutron results contradict the ancient idea, obtained from both meson-cloud and gluon-exchange models (81–84),

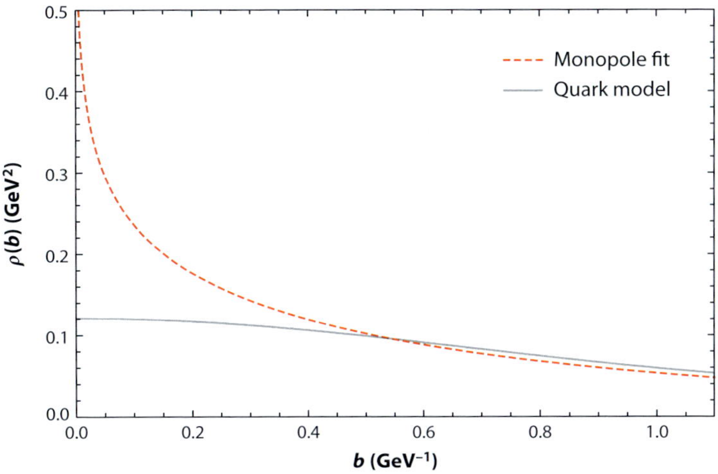

Figure 3

$\rho(b)$ corresponding to the two models shown in Figure 2. The red dashed curve represents the transverse density obtained from the monopole fit, and the gray solid line is obtained using the relativistic constituent quark model of Reference 75. Reprinted from Reference 17 with permission from the American Physical Society.

suggesting a positive value of the nonvanishing charge density at the center of the neutron. The meson-cloud picture is that the neutron sometimes undergoes a spontaneous quantum fluctuation into a heavy proton surrounded by a negatively charged pionic cloud, leaving a positive central density (81). The gluon-exchange mechanism (82–84) involves the repulsive nature of the one gluon–exchange interaction between two d quarks.

These well-motivated physical considerations concern statements about the three-dimensional Fourier transform of the neutron's electric form factor, $G_E^n(Q^2)$. However, such a transform is not the charge density. The only model-independent charge density is $\rho(b)$.

Recall the definitions of the form factors. The nucleon form factors, where $J^\mu(x)$ is the electromagnetic current operator in units of the proton charge, are given by

$$\langle p', \lambda' | J^\mu(0) | p, \lambda \rangle = \bar{u}(p', \lambda') \left(\gamma^\mu F_1(Q^2) + i \frac{\sigma^{\mu\alpha}}{2M} q_\alpha F_2(Q^2) \right) u(p, \lambda), \qquad 42.$$

where the momentum transfer $q_\alpha = p'_\alpha - p_\alpha$ is taken as space-like, with $q^+ = 0$, so that $Q^2 \equiv -q^2 = \mathbf{q}^2 > 0$. The nucleon polarization states are chosen to be those of definite light-cone helicities λ and λ' (43). The charge (Dirac) form factor is F_1, normalized such that $F_1(0)$ is the nucleon charge, and the magnetic (Pauli) form factor is F_2, normalized such that $F_2(0)$ is the anomalous magnetic moment. The Sachs form factors (85),

$$G_E(Q^2) \equiv F_1(Q^2) - \frac{Q^2}{4M^2} F_2(Q^2) \quad \text{and} \quad G_M(Q^2) \equiv F_1(Q^2) + F_2(Q^2), \qquad 43.$$

were introduced to provide an expression for the electron-nucleon cross section (in the one photon–exchange approximation) that depends on the quantities G_E^2 and G_M^2, but not on the product $G_E\, G_M$. In the Breit frame, in which $\mathbf{p} = -\mathbf{p}'$, G_E is the nucleon helicity-flip matrix element of J^0. Furthermore, the scattering of neutrons from the electron cloud of atoms measures the derivative $-dG_E(Q^2)/dQ^2$ at $Q^2 = 0$, which is widely interpreted as six times the mean-square charge radius of the neutron. However, any probability or density interpretation of G_E is spoiled

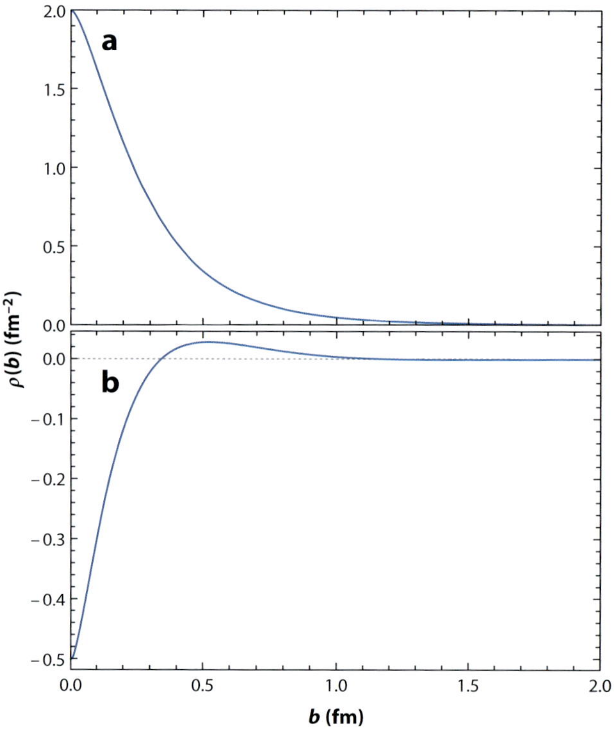

Figure 4

Nucleon $\rho(b)$. (*a*) Proton transverse charge density. (*b*) Neutron transverse charge density. These densities are obtained by using the parameterization of Reference 91.

by a nonzero value of Q^2, no matter how small, because the momentum difference between the initial and final states appears via the use of derivatives of momentum-conserving delta functions in the moments computed in Reference 85. Any attempt to analytically incorporate relativistic corrections in a p^2/m_q^2 type of expansion would be doomed by the presence of the quark mass m_q to be model dependent. This feature is explained more thoroughly in References 6 and 86.

We exploit Equation 31 by using measured form factors to determine $\rho(b)$. Recent parameterizations (87–91) of G_E and G_M are very useful, so we use Equation 43 to obtain F_1 in terms of G_E, G_M. Then $\rho(b)$ can be expressed as a simple integral of known functions,

$$\rho(b) = \int_0^\infty \frac{dQ\,Q}{2\pi} J_0(Qb) \frac{G_E(Q^2) + \tau G_M(Q^2)}{1+\tau}, \qquad 44.$$

where $\tau = \frac{Q^2}{4M^2}$ and J_0 is a cylindrical Bessel function.

A straightforward application of Equation 44 to the proton using the parameterizations of Reference 91 yields the results shown in **Figure 4*a***. The curves obtained by using the two different parameterizations overlap. Furthermore, there seems to be negligible sensitivity to form factors at very high values of Q^2 that are currently unmeasured. The density is peaked at low values of b but has a long positive tail, suggesting a long-ranged, positively charged pion cloud.

The neutron results are shown in **Figure 4*b***. The curves obtained by using the two different parameterizations seem to overlap. Surprisingly, the central neutron charge density is negative. The values of the integral of Equation 44 are somewhat sensitive to the regime $8 < Q^2 < 16\ \text{GeV}^2$,

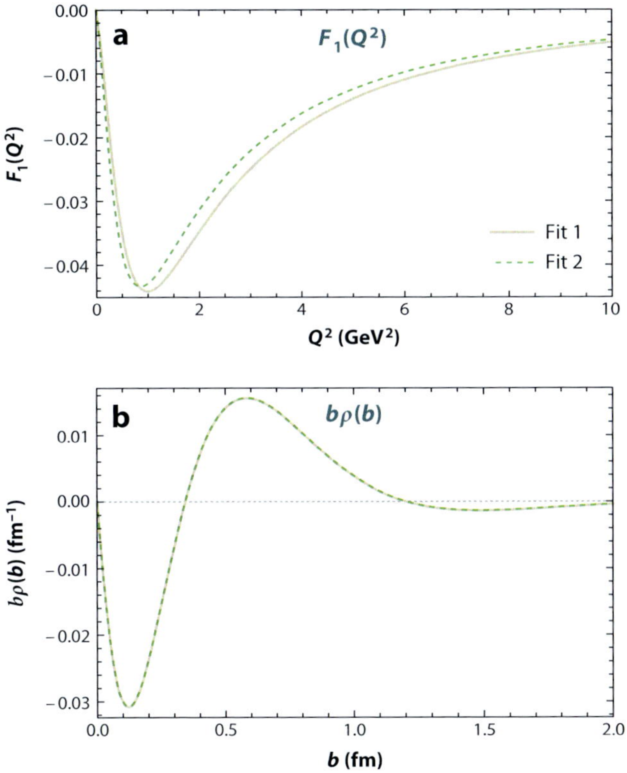

Figure 5

Neutron F_1 and $b\rho(b)$. (*a*) $F_1(Q^2)$. (*b*) $b\rho(b)$. The solid light brown curves are obtained using fit 1 of Reference 91, and the dashed green curves are obtained by using fit 2 of the same reference.

for which G_E is as yet unmeasured. Approximately 30% of the value of $\rho(0)$ arises from this region. That $\rho(b = 0) < 0$ was confirmed in References 80 and 92–94.

The negative central density deserves further explanation. See **Figure 5*a***, which shows F_1 for the neutron from two parameterizations of Reference 91. In both cases, F_1 is negative (because of the dominance of the G_M term of Equation 44) for all values of Q^2. This feature, along with taking $b = 0$ so that $J_0(Qb) = 1$ in Equation 44, immediately leads to the central negative result. The long-range structure of the charge density is captured by displaying the quantity $b\rho(b)$ in **Figure 5*b***. At very large distances from the center, $b\rho(b) < 0$, which suggests the existence of the long-ranged pion cloud. Thus, the neutron transverse charge density displays an unusual behavior, in which the positive charge density in the middle region is sandwiched by negative charge densities at the inner and outer reaches of the neutron. A simple model in which the neutron fluctuates into a proton and a π^- parameterized to reproduce the negative-definite nature of the neutron's F_1 (95) reproduces the negative transverse central density. In this case, the negative nature arises from pions that penetrate to the center. The change from the nominal positive value obtained from G_E can be understood as originating in the boost to the IMF (86).

One can gain information about the individual u and d quark densities by invoking charge symmetry [invariance under a rotation by π about the z (charge) axis in isospin space (96–99)] and by neglecting the effects of $s\bar{s}$ pairs (100). Model-independent information about nucleon structure is thereby obtained and shows, surprisingly, that the central density of the neutron is negative.

Future measurements of neutron electromagnetic form factors could render the present results more precise, or potentially modify them considerably. Obtaining a quantitative and intuitive understanding of these results presents a challenge to lattice QCD and to model builders.

4.1. Meaning of Central Density $b = 0$

The transverse coordinate $b = 0$ is the (transverse) center of the nucleon. Interestingly, in the IMF this position is also the center of the nucleon because in that frame the nucleon has no longitudinal extent. This can be seen by considering a light-cone wave function $\psi(x, \boldsymbol{\kappa})$. The coordinate canonically conjugate to $x = k^+/P^+$ is P^+x^- (101), so the coordinate-space wave function would depend on the product x^-P^+. Thus, when P^+ is infinite, the coordinate-space wave function vanishes unless $x^- = 0$. This means that in the IMF, the longitudinal density appears as a delta function and the position $b = 0$ is the true nucleon center (6).

4.2. Inclusive-Exclusive Connection Interpretation of the Negative Neutron Charge Density

GPDs contain information about the longitudinal momentum fraction x as well as the transverse position b. Information regarding these dependences is obtained from experiment via GPDs that reproduce both deep-inelastic scattering and elastic scattering data. Miller & Arrington (102) used this inclusive-exclusive connection to interpret the central neutron charge density and found that the center of the neutron is dominated by negatively charged d quarks. We review their argument in this section.

The quantities $q(x, b)$ are not measured directly but rather have been inferred from models that incorporate fits to parton distributions and electromagnetic nucleon form factors (103–106). Form factor sum rules (Equations 26 and 27) at zero skewness are used to model valence quark GPDs: $H_v^q \equiv H^q - H^{\bar{q}}$. This procedure yields the net contribution to the form factors from quarks and antiquarks. Possible effects of strangeness are neglected in such fits.

Each parameterization used (103–105) incorporates the Drell-Yan-West (107, 108) relationship between the behavior of the structure function $\nu W_2(x)$ near $x = 1$, measured in inclusive reactions, and the behavior of the electromagnetic form factor at large values of Q^2, measured in the exclusive elastic scattering process. Specifically, for a system of $n + 1$ valence quarks, described by a power-law wave function,

$$\lim_{x \to 1} \nu W_2(x) = (1 - x)^{2n-1} \to \lim_{Q^2 \to \infty} F_1(Q^2) = \frac{1}{Q^{2n}}. \qquad 45.$$

The value of n that defines the high-x behavior of the structure function also defines the high-Q^2 behavior of the form factor. This relation associates the behavior of large values of x with large momentum transfers, $Q^2 = \mathbf{q}^2$, which in turn corresponds to small values of b. There is a further connection between large values of x and small values of b that emerges from Equation 11 and was considered in the example given in Section 1.3 (and Equation 32). If a single quark carries momentum fraction x very near unity, only one term in the sum of Equation 11 survives; the requirement that this term vanish causes the corresponding transverse coordinate in turn to vanish. This interesting connection between different components of the position and momentum vectors is not associated with the uncertainty principle.

Given that the d quark dominates the high-x quark-distribution function of the neutron (e.g., 109), and given that the central charge density of the neutron is negative, it is natural to conclude that the central charge density of the neutron arises from the predominance of d quarks at the

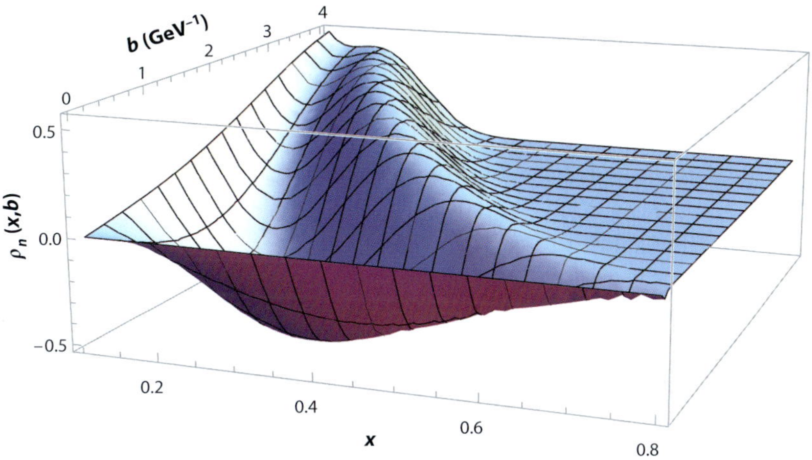

Figure 6

Neutron impact parameter charge distributions, $\rho_n(x, b)$, as a function of b.

center (**Figure 6**). One sees that for high values of x, b must be small to have a nonzero value of $\rho_n(x, b)$, and such nonzero values are negative.

4.3. Magnetization Density

We now use transverse densities to obtain the nucleon magnetization density in the IMF (110). This quantity is closely related to the transverse charge density for a polarized nucleon obtained by Carlson & Vanderhaeghen (80). Our starting point is the relation that $\boldsymbol{\mu} \cdot \mathbf{B}$ is the matrix element of $\vec{J} \cdot \vec{A}$ in a definite state, $|X\rangle$. This means that the magnetization density is $\frac{1}{2}(\vec{r} \cdot \vec{J})$.

Take the rest-frame magnetic field to be a constant vector in the 1 (or b_x) direction and the corresponding vector potential as $\mathbf{A} = Bb_y\hat{\mathbf{z}}$. Then consider the system in a frame in which the plus component of the momentum approaches infinity so that $\mathbf{J} \cdot \mathbf{A} \to J^+A^-$. The anomalous magnetic moment may be extracted by taking the matrix element of $\boldsymbol{\mu} \cdot \mathbf{B}$ in the state

$$|X\rangle \equiv \frac{1}{\sqrt{2}}[|p^+, \mathbf{R} = \mathbf{0}, +\rangle + |p^+, \mathbf{R} = \mathbf{0}, -\rangle], \qquad 46.$$

where $|p^+, \mathbf{R} = \mathbf{0}, +\rangle$ represents a transversely localized state of definite P^+ and light-cone helicity. The state $|X\rangle$ (14) may be interpreted as that of a transversely polarized target (111). The resulting anomalous magnetic moment μ_a is given by (110)

$$\mu_a = \left\langle X \left| \int d^2b\, b_y\, q_+^\dagger(0, \mathbf{b}) q_+(0, \mathbf{b}) \right| X \right\rangle, \qquad 47.$$

where $q(x^-, \mathbf{b})$ is a quark field operator and $q_+ = \gamma^0\gamma^+q$. This matrix element is the anomalous magnetic moment because the use of the transversely localized state $|X\rangle$ and the factor b_y suppresses the Dirac contribution (111).

Miller et al. (110) evaluated μ_a by using Burkardt's (14) impact parameter distribution and then integrating by parts. The result was

$$\mu_a = \frac{1}{2M} \int d^2b\, \rho_M(b), \qquad 48.$$

where

$$\rho_M(b) = \int \frac{d^2q}{(2\pi)^2} F_2(t = -\mathbf{q}^2) e^{-i\mathbf{q}\cdot\mathbf{b}} \qquad 49.$$

and the subscript M denotes the anomalous magnetic moment.

Equation 49 has an appealing simplicity. However, the directly obtained Equation 47 can also be evaluated immediately to find

$$\mu_a = \frac{-1}{2M} \int d^2b\, b_y \frac{\partial \rho_M(b)}{\partial b_y}, \qquad 50.$$

so the quantity $-b_y \frac{\partial \rho_M(b)}{\partial b_y} \equiv \tilde{\rho}_M(\mathbf{b})$ also has an interpretation as an anomalous magnetization density. The two integrals appearing in Equations 48 and 50 have the same value. Miller et al. (110) rejected the use of $\tilde{\rho}_M(\mathbf{b})$ as a magnetization density because of the appearance of an explicit direction y. However, this direction has a general interpretation as the transverse direction orthogonal to that of the transverse magnetic field. We evaluate $\tilde{\rho}_M(\mathbf{b})$ in terms of $F_2(Q^2)$ to find

$$\tilde{\rho}_M(\mathbf{b}) = \sin^2\phi\, b \int_0^\infty \frac{q^2\, dq}{2\pi} J_1(qb) F_2(q^2), \qquad 51.$$

where ϕ is the angle between the direction of $\mathbf{b}$ and that of the transverse magnetic field, which is also the direction of the nucleon polarization. Thus, the physical direction of the polarization or magnetic field provides a definite spatial direction. Indeed, the magnetization density $\tilde{\rho}_M(\mathbf{b})$ is largest in directions perpendicular to the direction of the nucleon polarization (or magnetic field) (**Figure 7**). The largest values occur for $\phi = \pi/2$, and the magnetization density peaks at approximately 0.5 fm. Furthermore, $\tilde{\rho}_M(\mathbf{b})$ vanishes if $b = 0$ or if $\phi = 0$. These features accord with the expectations of classical physics. A current in the z direction causes a magnetic dipole density $\sim\mathbf{r} \times \vec{J}$ in the x direction for positions $\mathbf{r}$ along the y direction. Therefore, we conclude that the quantity $\tilde{\rho}_M(\mathbf{b})$ is the preferred expression for the magnetization density.

Hoyer & Kurki (112) have computed the transverse density of the electron. Using their expression for the electron F_2 in Equation 50 leads to the correct result for the electron's anomalous magnetic moment, μ_a^e. Furthermore, the presence of the $\sin^2\phi$ term is consistent with Hoyer & Kurki's physical interpretation of the sign of μ_a^e as caused by currents in the positive (negative) z direction for positive (negative) values of y. Computing the cross product between $\mathbf{b}$ and those currents gives magnetization in the x direction for both positive and negative values of y.

To better understand this new magnetization density, consider the moments of both densities. Define the nth moment of ρ_M as

$$\langle b^{2n}\rangle_M \equiv \int d^2b b^{2n} \rho_M(b). \qquad 52.$$

These quantities are related to the nth derivative of $F_2(q^2)$. Similarly, the nth moment of $\tilde{\rho}_M$ is given by

$$\langle b^{2n}\rangle_{\tilde{M}} \equiv -\int d^2b b^{2n} b_y \frac{\partial \rho_M(b)}{\partial b_y}. \qquad 53.$$

Use of Equation 49 and integration by parts show that

$$\langle b^{2n}\rangle_{\tilde{M}} = (n+1)\langle b^{2n}\rangle_M. \qquad 54.$$

The $n = 0$ moment corresponds to the anomalous magnetic moment, which is the same for the two densities. The case $n = 1$, which defines the mean-square magnetic radius, is more interesting. Miller et al. (110) showed that $\langle b^2\rangle_M$ was slightly larger than $\langle b^2\rangle_{Cb}$, which is obtained from F_1. However, $\langle b^2\rangle_{\tilde{M}}$ is twice as large as $\langle b^2\rangle_M$ and therefore is much larger than $\langle b^2\rangle_{Cb}$! This clearly

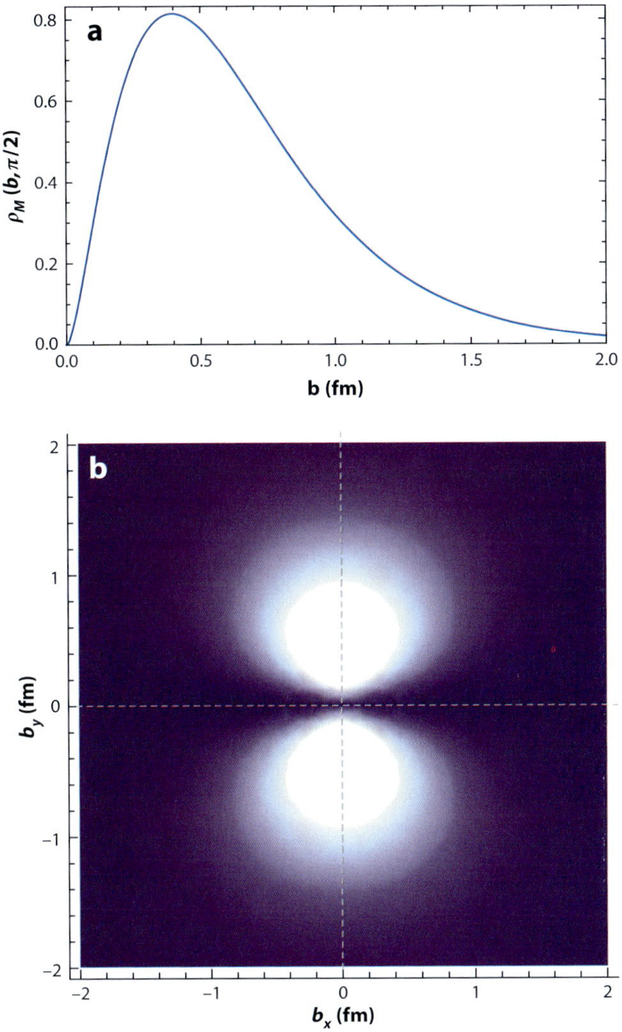

Figure 7

(*a*) $\tilde{\rho}_M\,[b = (\mathbf{b}, \phi = \pi/2)]$ as a function of **b**. (*b*) Density plot of $\tilde{\rho}_M\,(\mathbf{b})$. The horizontal axis is the direction of the applied magnetic field. The largest (smallest) values of $\tilde{\rho}_M$ are denoted by the brightest (darkest) areas. This panel is obtained by using a dipole parameterization for F2 of the proton.

shows that the proton's magnetization density extends much further than its charge density (110). This is surprising because it contradicts simple naïve intuition gained from the rapid fall of the ratio G_E/G_M with increasing momentum transfer (3), if one assumes that G_E (G_M) is related to the spatial extent of the charge (magnetization) density.

4.4. Noncylindrically Symmetric Transverse Charge Density and Shape of the Nucleon

Carlson & Vanderhaeghen (80) computed the transverse charge density $\rho_T^N\,(\mathbf{b})$ in a given state of transverse polarization For the case that the polarization is in the x direction, their result is

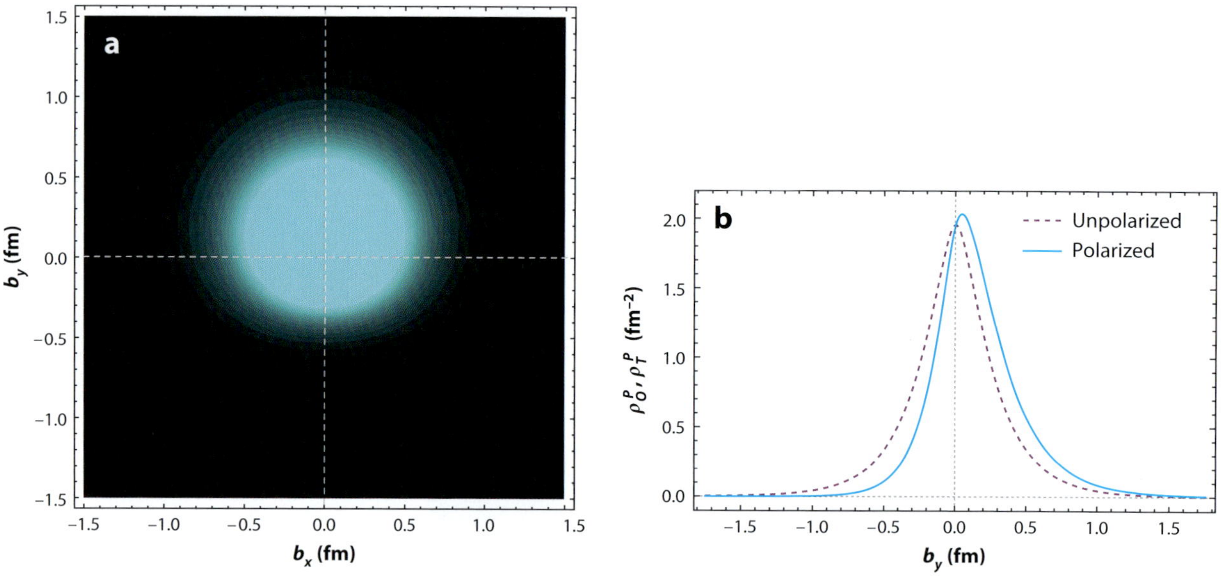

Figure 8

Quark transverse charge densities in the proton. (*a*) The density in the transverse plane for a proton polarized along the x axis. The light (dark) regions correspond with the largest (smallest) values of the density. (*b*) Comparison of the density along the y axis for an unpolarized proton (*dashed purple curve*) and for a proton polarized along the x axis (*solid light blue curve*). The momentum transfer q is the momentum added to the target proton. Modified from Reference 80 courtesy of M. Vanderhaeghen. The proton form factors are from Reference 90.

expressed as

$$\rho_T^N(\mathbf{b}) = \rho(b) + \sin\phi \int_0^\infty \frac{dQ}{2\pi}\frac{Q^2}{2M} J_1(bQ)F_2(Q^2), \tag{55}$$

where $\rho(b)$ is obtained from Equation 31. The sign of the second term of Equation 55 is obtained by using the definition that **q** is the momentum absorbed by the target nucleon. Carlson & Vanderhaeghen's results for the proton are shown in **Figure 8**.

Both the magnetization density $\tilde{\rho}_M(\mathbf{b})$ presented here and the transverse charge density of Reference 80 depend on the direction of **b**, indicating a nonspherical shape of the nucleon or the violation of cylindrical symmetry. These violations of cylindrical symmetry should not be confused with those present in earlier papers on the shape of the nucleon (51, 113, 114) that define a shape via the matrix elements of a spin-dependent density operator. The transverse charge densities that are the main subject of this article are matrix elements of currents that are local in transverse coordinate space and related to GPDs. In contrast, the shapes of References 51, 113, and 114, which are computed from momentum-space wave functions and probabilities, are matrix elements of quark densities for given values of transverse momenta and are related to TMDs. Also, note that the first example of Reference 113 is presented in coordinate space. The IMF version of the spin-dependent density operator is $q_+^\dagger(0, \mathbf{b})\frac{1}{2}(1 + \mathbf{n} \cdot \boldsymbol{\gamma})q_+(0, \mathbf{b})$, where $\mathbf{n}$ is an arbitrary transverse direction (49). Evaluating the matrix element of this quantity produces the spin-dependent density, which can be considered as the x^- integrated version of the coordinate-space results of Reference 113. The proton is nonspherical if the function $\tilde{A}''_{T10}$ of Reference 50 is nonvanishing, as recent lattice calculations demonstrate (115). One can also examine several different generalized densities $q_+^\dagger(0, \mathbf{b})\Gamma q_+(0, \mathbf{b})$, whose operators Γ are denoted in Reference 49.

5. TRANSVERSE TRANSITION CHARGE DENSITIES

There is much experimental interest in measuring transition form factors that represent probability amplitudes for a nucleon undergoing electron scattering to make a transition to a given baryon resonance. Substantial experimental efforts are under way at electron scattering laboratories such as Jefferson Laboratory in Virginia, ELSA in Bonn, and MAMI in Mainz. Moreover, there is considerable interest in using lattice techniques to make QCD calculations of these transition amplitudes.

Carlson & Vanderhaeghen (80) used empirical information regarding the $N \to \Delta$ transition form factors to map out the transition charge density which, in a transversely polarized N and Δ, contains monopole, dipole, and quadrupole patterns. The last of these corresponds to a deformation of the N and Δ transverse charge densities. Substantial deformations are observed. Lattice QCD calculations have been applied to these densities (47, 48). The Δ^+ charge density is elongated along the axis of the spin, and the quadrupole moment is larger than the value characterizing a point particle. This means the the Δ^+ is prolately deformed.

Tiator & Vanderhaeghen (116) used recent experimental data to analyze the electromagnetic transition from the nucleon to the $P_{11}(1{,}440)$ resonance, which is often considered the first radial excitation of the proton. These authors used the empirical transition form factors to find that the transition from the proton to the $P_{11}(1{,}440)$ is dominated by (*a*) up quarks in a central region approximately 0.5 fm wide and (*b*) down quarks in an outer band that extends up to approximately 1 fm. Tiator et al. (117) are extending the study of transition transverse densities to the S_{11} and D_{13} resonances.

6. NUCLEAR TRANSVERSE CHARGE DENSITIES

The same IMF formalism can also be applied to the analysis of nuclear form factors. The nonrelativistic approximation of Equation 6 is generally applicable for existing measurements of nuclear form factors (6). However, meson-exchange currents are known to be important for $Q^2 > 2\ \mathrm{GeV}^2$ (118). In this case, the charged constituents have different masses (see Equation 7), the pion moves relativistically, and the charge density is not a three-dimensional Fourier transform of the electromagnetic form factor. Moreover, Jefferson Laboratory plans to considerably extend the available range of momentum transfer for light nuclear form factors. Thus, transverse charge densities that supply model-independent information are of considerable relevance for nuclear physics.

Carlson & Vanderhaeghen (119) made the first computation of a nuclear transverse density in their analysis of the deuteron empirical transverse charge densities (120). The charge densities are characterized by monopole, dipole, and quadrupole patterns because the deuteron has a spin of unity. The authors (119) observed a dip in the center-of-charge density for helicity-zero deuterons. This central depression accords with standard nuclear force model calculations (121) and is a simple consequence of the deuteron D-state. We note that the deuteron is a nonrelativistic system as defined in Reference (6). Thus, these densities can, with a high degree of accuracy, be considered as an integral over z of the corresponding nonrelativistic density; see Equation 15. The dip is an important finding because it is model independent. Carlson & Vanderhaeghen (119) also found that transversely polarized deuterons show dipole and quadrupole structure in the charge densities. Their electric dipole and quadrupole moments depend only on the spin-1 particle's anomalous magnetic dipole moment and its anomalous electric quadrupole moment, which arise from its internal structure.

Nuclei are nonrelativistic systems (6), so for presently available data the principal distinction between transverse densities and the usual three-dimensional Fourier transform is that the former is an integral (Equation 15) over the latter. However, existing nuclear data extend only to approximately $Q^2 = 1\ \mathrm{GeV}^2$ (122–124). In this case, the motion of the nucleons within the nucleus is just barely relativistic. If, as expected, data are taken at higher values of Q^2, then relativistic effects can be expected to be important. One example is the nuclei with $A = 3$, mass M_3, and Sachs electric and magnetic form factors F_C and F_M (122). If $Q^2/4M_3^2$ is of order unity, the distinction between $F_1 = F_C - Q^2/4M_3^2 F_M$ and F_C becomes important, and the role of transverse densities may become very important.

7. OTHER APPLICATIONS

TMDs, which give the momentum density of hadrons, have also been studied (125, 126). The spatial distribution of the momentum component P^+ is related to Fourier transforms of gravitational form factors (125), which in turn are related to experimentally observed data. This is because gravitational form factors are matrix elements of the energy-momentum tensor obtainable as second Mellin moments of GPDs (25, 127, 128). The proton momentum density in the transverse plane is more compact than its charge density (125).

Any charge density deforms when subjected to a quasi-static external electric field; related measurable quantities are known as polarizabilities. Gorchtein et al. (129) extended the transverse charge density formalism to extract light-front quark charge densities related to polarizabilities and showed that the resulting induced polarizations can be extracted from proton-generalized polarizabilities (130, 131). The available data for the generalized electric polarizabilities of the proton yield a transverse density that shows an usual oscillatory pattern. A general expression for the transverse charge densities for any spin has been obtained (132).

8. SUMMARY AND FUTURE DIRECTIONS

Transverse charge densities are a new tool for analyzing electromagnetic form factors of systems composed of constituents that move relativistically. These quantities are hadronic charge densities as seen in a reference frame moving with infinite momentum. One of the main advantages of the IMF is that boosts in the transverse direction form a kinematic subgroup of the Poincaré group, so transformations to frames moving in a direction transverse to that of the infinite momentum are carried out via the transverse position operator. This is the same as the usual nonrelativistic Galilean transformation. Thus, the two-dimensional Fourier transformation of electromagnetic form factors provides a rigorous way to study charge and magnetization densities. Transverse charge densities involve matrix elements of local operators $[q_+^\dagger(t = 0, z = 0, \mathbf{b})\Gamma q_+(t = 0, z = 0, \mathbf{b})]$, which can be computed by using any of the light-front, equal-time, or lattice techniques.

The use of transverse charge densities has led to some very interesting findings. Examples include the negative nature of the central neutron charge density (18), caused by the dominance of d quarks at the center (102); the seemingly singular nature of the density at the center of the pion (17); the spatial extent of the proton's magnetization density being greater than that of its charge density (110); and the substantial deformation of the nucleon (113, 115) and the Δ baryon (80). The positive sign of the electron anomalous magnetic moment has been explained (112).

Transverse densities have been used to analyze other baryon transitions (116), nuclear charge densities (119), momentum densities (128), and generalized polarizabilities (129). Measurements of form factors at larger values of Q^2 than are presently available are needed to test the singularity of the pionic transverse density and the inner core of the neutron transverse density. Also, the use of transverse charge densities may become the chosen tool to analyze nuclear charge distributions once very high momentum transfer data become available. Transverse densities are relevant whenever matrix elements of an operator $[q_+^\dagger(t=0, z=0, \mathbf{b})\Gamma q_+(t=0, z=0, \mathbf{b})]$ appear. Thus, one may expect to see a host of interesting, informative, and important future applications of transverse charge densities.

DISCLOSURE STATEMENT

The author is not aware of any affiliations, memberships, funding, or financial holdings that might be perceived as affecting the objectivity of this review.

ACKNOWLEDGMENTS

My work is partially supported by the U.S. Department of Energy. I thank J. Arrington, A. Bernstein, M. Burkardt, C. Carlson, I. Cloët, P. Hoyer, E. Piasetzky, G. Ron, A. Radyushkin, J. Rinehimer, B. Tiburzi, M. Vanderhaeghen, and L. Zhu for useful discussions regarding transverse charge distributions.

LITERATURE CITED

1. Gao H-Y. *Int. J. Mod. Phys.* E12:1 (2003)
2. Hyde CE, de Jager K. *Annu. Rev. Nucl. Part. Sci.* 54:217 (2004)
3. Perdrisat CF, Punjabi V, Vanderhaeghen M. *Prog. Part. Nucl. Phys.* 59:694 (2007)
4. Arrington J, Roberts CD, Zanotti JM. *J. Phys. G* 34:S23 (2007)
5. Hofstadter R. *Rev. Mod. Phys.* 28:214 (1956)
6. Miller GA. *Phys. Rev. C* 80:045210 (2009)
7. Soper DE. *Phys. Rev. D* 15:1141 (1977)
8. Nakanishi N. *Prog. Theor. Phys. Suppl.* 43:1 (1969)
9. Nakanishi N. *Prog. Theor. Phys. Suppl.* 95:1 (1988)
10. Carbonell J, Karmanov VA, Mangin-Brinet M. *Eur. Phys. J. A* 39:53 (2009)
11. Miller GA, Tiburzi BC. *Phys. Rev. C* 81:035201 (2010)
12. Susskind L. *Phys. Rev.* 165:1547 (1968)
13. Diehl M. *Eur. Phys. J. C* 25:223 (2002)
14. Burkardt M. *Int. J. Mod. Phys. A* 18:173 (2003)
15. Burkardt M. *Phys. Rev. D* 62:071503 (2000)
16. Diehl M, Feldmann T, Jakob R, Kroll P. *Nucl. Phys. B* 596:33 (2001)
17. Miller GA. *Phys. Rev. C* 79:055204 (2009)
18. Miller GA. *Phys. Rev. Lett.* 99:112001 (2007)
19. Glauber RJ, Matthiae G. *Nucl. Phys. B* 21:135 (1970)
20. Luzum M, Romatschke P. *Phys. Rev. Lett.* 103:262302 (2009)
21. Gunion JF, Brodsky SJ, Blankenbecler R. *Phys. Rev. D* 8:287 (1973)
22. Lepage GP, Brodsky SJ. *Phys. Rev. D* 22:2157 (1980)
23. Carbonell J, Desplanques B, Karmanov VA, Mathiot JF. *Phys. Rep.* 300:215 (1998)
24. Mueller D, et al. *Fortschr. Phys.* 42:101 (1994)
25. Ji X-D. *Phys. Rev. D* 55:7114 (1997)
26. Radyushkin AV. *Phys. Rev. D* 56:5524 (1997)
27. Collins JC, Frankfurt L, Strikman M. *Phys. Rev. D* 56:2982 (1997)
28. Ji X-D. *J. Phys. G* 24:1181 (1998)
29. Radyushkin AV. In *At the Frontiers of Particle Physics: Handbook of QCD*, ed. M Shifman, p. 1037. Singapore: World Sci. (2001)
30. Goeke K, Polyakov MV, Vanderhaeghen M. *Prog. Part. Nucl. Phys.* 47:401 (2001)
31. Diehl M. *Phys. Rep.* 388:41 (2003)
32. Ji X-D. *Annu. Rev. Nucl. Part. Sci.* 54:413 (2004)

33. Belitsky AV, Radyushkin AV. *Phys. Rep.* 418:1 (2005)
34. Hagler P. *Phys. Lett.* B594:164 (2004)
35. Boffi S, Pasquini B. *Riv. Nuovo Cim.* 30:387 (2007)
36. Collins JC, Soper DE. *Nucl. Phys. B* 194:445 (1982)
37. Ralston JP, Soper DE. *Nucl. Phys. B* 152:109 (1979)
38. Belyaev VM, Ioffe BL. *Nucl. Phys. B* 313:647 (1989)
39. Anselmino M, Efremov A, Leader E. *Phys. Rep.* 261:1 (1995)
40. Mulders PJ, Tangerman RD. *Nucl. Phys. B* 461:197 (1996)
41. Chibisov B, Zhitnitsky AR. *Phys. Rev. D* 52:5273 (1995)
42. Pasquini B, Boffi S, Efremov AV, Schweitzer P. arXiv:0912.1761 [hep-ph] (2009)
43. Kogut JB, Soper DE. *Phys. Rev. D* 1:2901 (1970)
44. Ji X-D. *Phys. Rev. Lett.* 91:062001 (2003)
45. Belitsky AV, Ji X-D, Yuan F. *Nucl. Phys. B* 656:165 (2003)
46. Broniowski W, Prelovsek S, Santelj L, Arriola ER. arXiv:0911.4705 [hep-ph] (2009)
47. Alexandrou C, et al. *Phys. Rev. D* 79:014507 (2009)
48. Alexandrou C, et al. *Nucl. Phys. A* 825:115 (2009)
49. Diehl M, Hagler P. *Eur. Phys. J. C* 44:87 (2005)
50. Gockeler M, et al. (QCDSF Collab.) *Phys. Rev. Lett.* 98:222001 (2007)
51. Miller GA. *Phys. Rev. C* 76:065209 (2007)
52. Lepage GP, Brodsky SJ. *Phys. Lett.* B87:359 (1979)
53. Farrar GR, Jackson DR. *Phys. Rev. Lett.* 43:246 (1979)
54. Chernyak VL, Serbo VG, Zhitnitsky AR. *Sov. J. Nucl. Phys.* 31:552 (1980)
55. Efremov AV, Radyushkin AV. *Theor. Math. Phys.* 42:97 (1980)
56. Chernyak VL, Zhitnitsky AR. *Phys. Rep.* 112:173 (1984)
57. Li H-N, Sterman G. *Nucl. Phys. B* 381:129 (1992)
58. Braaten E, Tse S-M. *Phys. Rev. D* 35:2255 (1987)
59. Miller GA, Pasupathy J. *Z. Phys. A* 348:123 (1994)
60. Radyushkin AV. *Nucl. Phys. A* 532:141 (1991)
61. Blok HP, et al. *Phys. Rev. C* 78:045202 (2008)
62. Huber GM, et al. *Phys. Rev. C* 78:045203 (2008)
63. Huber GM, et al. *JLAB proposal E12-6-101.* **http://www.jlab.org/exp_prog/12GEV_EXP/E126101.html** (2006)
64. Goldberger ML, Soper DE, Guth AH. *Phys. Rev. D* 14:2633 (1976)
65. Brodsky SJ, de Teramond GF. *Phys. Rev. D* 77:056007 (2008)
66. Kwee HJ, Lebed RF. *J. High Energy Phys.* 01:027 (2008)
67. Grigoryan HR, Radyushkin AV. *Phys. Rev. D* 76:115007 (2007)
68. Broniowski W, Arriola ER, Golec-Biernat K. *AIP Conf. Proc.* 1030:286 (2008)
69. Ruiz Arriola E, Broniowski W. *Phys. Rev. D* 67:074021 (2003)
70. Broniowski W, Ruiz Arriola E. *Phys. Lett.* B574:57 (2003)
71. Brauel P, et al. *Z. Phys. C* 3:101 (1979)
72. Ackermann H, et al. *Nucl. Phys. B* 137;294 (1978)
73. Amendolia SR, et al. (NA7 Collab.) *Nucl. Phys.* B277:168 (1986)
74. Horn T, et al. *Phys. Rev. C* 78:058201 (2008)
75. Hwang C-W. *Phys. Rev. D* 64:034011 (2001)
76. Radyushkin AV. *Phys. Rev. D* 58:114008 (1998)
77. Burkardt M, Miller GA. *Phys. Rev. D* 74:034015 (2006)
78. Chung PL, Coester F, Polyzou WN. *Phys. Lett.* B205:545 (1988)
79. Frederico T, Miller GA. *Phys. Rev. D* 45:4207 (1992)
80. Carlson CE, Vanderhaeghen M. *Phys. Rev. Lett.* 100:032004 (2008)
81. Thomas AW, Theberge S, Miller GA. *Phys. Rev. D* 24:216 (1981)
82. Friar JL. *Part. Nucl.* 4:153 (1972)
83. Carlitz RD, Ellis SD, Savit R. *Phys. Lett.* B68:443 (1977)
84. Isgur N, Karl G, Sprung DWL. *Phys. Rev. D* 23:163 (1981)

85. Sachs RG. *Phys. Rev.* 126:2256 (1962)
86. Rinehimer JA, Miller GA. *Phys. Rev. C* 80:015201 (2009)
87. Bradford R, Bodek A, Budd HS, Arrington J. *Nucl. Phys. Proc. Suppl.* 159:127 (2006)
88. Kelly JJ. *Phys. Rev. C* 70:068202 (2004)
89. Arrington J. *Phys. Rev. C* 69:022201 (2004)
90. Arrington J, Melnitchouk W, Tjon JA. *Phys. Rev. C* 76:035205 (2007)
91. Alberico WM, Bilenky SM, Giunti C, Graczyk KM. *Phys. Rev. C* 79:065204 (2009)
92. Pasquini B, Boffi S. *Phys. Rev. D* 76:074011 (2007)
93. Boffi S, Pasquini B. *Mod. Phys. Lett. A* 24:2882 (2009)
94. Hwang DS, Kim DS, Kim J. *Phys. Lett.* B669:345 (2008)
95. Rinehimer JA, Miller GA. *Phys. Rev. C* 80:025206 (2009)
96. Miller GA, Thomas AW, Williams AG. *Phys. Rev. Lett.* 56:2567 (1986)
97. Miller GA, Nefkens BMK, Slaus I. *Phys. Rep.* 194:1 (1990)
98. Miller GA. *Phys. Rev. C* 57:1492 (1998)
99. Miller GA, Opper AK, Stephenson EJ. *Annu. Rev. Nucl. Part. Sci.* 56:253 (2006)
100. Acha A, et al. (HAPPEX Collab.) *Phys. Rev. Lett.* 98:032301 (2007)
101. Pirner HJ, Nurpeissov N. *Phys. Lett.* B595:379 (2004)
102. Miller GA, Arrington J. *Phys. Rev. C* 78:032201 (2008)
103. Guidal M, Polyakov MV, Radyushkin AV, Vanderhaeghen M. *Phys. Rev. D* 72:054013 (2005)
104. Diehl M, Feldmann T, Jakob R, Kroll P. *Eur. Phys. J. C* 39:1 (2005)
105. Ahmad S, Honkanen H, Liuti S, Taneja SK. *Phys. Rev. D* 75:094003 (2007)
106. Tiburzi BC, Detmold W, Miller GA. *Phys. Rev. D* 70:093008 (2004)
107. Drell SD, Yan T-M. *Phys. Rev. Lett.* 24:181 (1970)
108. West GB. *Phys. Rev. Lett.* 24:1206 (1970)
109. Pumplin J, et al. *J. High Energy Phys.* 07:012 (2002)
110. Miller GA, Piasetzky E, Ron G. *Phys. Rev. Lett.* 101:082002 (2008)
111. Burkardt M. *Phys. Rev. D* 72:094020 (2005)
112. Hoyer P, Kurki S. *Phys. Rev. D* 81:013002 (2010)
113. Miller GA. *Phys. Rev. C* 68:022201 (2003)
114. Kvinikhidze A, Miller GA. *Phys. Rev. C* 73:065203 (2006)
115. Schierholz G. (QCDSF Collab.) Presented at Workshop Transverse Partonic Struct. Hadrons, Yerevan, Armen. (2009)
116. Tiator L, Vanderhaeghen M. *Phys. Lett.* B672:344 (2009)
117. Tiator L, Drechsel D, Kamalov SS, Vanderhaeghen M. arXiv:0909.2335 [nucl-th] (2009)
118. Riska DO. In *Mesons In Nuclei*, vol. I, ed. M Rho, D Wilkinson, p. 755. Amsterdam: Elsevier (1979)
119. Carlson CE, Vanderhaeghen M. *Eur. Phys. J. A* 41:1 (2009)
120. Abbott D, et al. (JLAB t20 Collab.) *Eur. Phys. J. A* 7:421 (2000)
121. Forest JL, et al. *Phys. Rev. C* 54:646 (1996)
122. Amroun A, et al. *Nucl. Phys. A* 579:596 (1994)
123. Sick I. *Prog. Part. Nucl. Phys.* 47:245 (2001)
124. Sick I. *Phys. Rev. C* 77:041302 (2008)
125. Abidin Z, Carlson CE. *Phys. Rev. D* 78:071502 (2008)
126. Selyugin OV, Teryaev OV. *Phys. Rev. D* 79:033003 (2009)
127. Ji X-D. *Phys. Rev. Lett.* 78:610 (1997)
128. Abidin Z, Carlson CE. *Phys. Rev. D* 77:095007 (2008)
129. Gorchtein M, Lorce C, Pasquini B, Vanderhaeghen M. *Phys. Rev. Lett.* 104:112001 (2010)
130. Guichon PAM, Liu GQ, Thomas AW. *Nucl. Phys. A* 591:606 (1995)
131. Drechsel D, et al. *Phys. Rev. C* 57:941 (1998)
132. Lorce C. *Phys. Rev. D* 79:113011 (2009)

Reheating in Inflationary Cosmology: Theory and Applications

Rouzbeh Allahverdi,[1] Robert Brandenberger,[2] Francis-Yan Cyr-Racine,[3] and Anupam Mazumdar[4,5]

[1]Physics Department, University of New Mexico, Albuquerque, New Mexico 87131

[2]Physics Department, McGill University, Montréal, Québec H3A 2T8, Canada; email: rhb@physics.mcgill.ca

[3]Department of Physics and Astronomy, University of British Columbia, Vancouver, British Columbia V6T 1Z1, Canada

[4]Physics Department, Lancaster University, Lancaster LA1 4YB, United Kingdom

[5]Niels Bohr Institute, Copenhagen DK-2100, Denmark

Annu. Rev. Nucl. Part. Sci. 2010. 60:27–51

First published online as a Review in Advance on May 28, 2010

The *Annual Review of Nuclear and Particle Science* is online at nucl.annualreviews.org

This article's doi:
10.1146/annurev.nucl.012809.104511

0163-8998/10/1123-0027$20.00

Key Words

inflationary universe, reheating, early universe cosmology, preheating, parametric resonance

Abstract

Reheating is an important part of inflationary cosmology. It describes the production of Standard Model particles after the phase of accelerated expansion. We review the reheating process with a focus on an in-depth discussion of the preheating stage, which is characterized by exponential particle production due to a parametric resonance or tachyonic instability. We give a brief overview of the thermalization process after preheating and end with a survey of some applications to supersymmetric theories and to other issues in cosmology, such as baryogenesis, dark matter, and metric preheating.

Contents

1. INTRODUCTION

The inflationary model (1) has become the current paradigm of early universe cosmology. The first key aspect of the model is a phase of accelerated expansion of space, which can explain the overall homogeneity, spatial flatness, and large size of the current universe. Microscopic-scale quantum vacuum fluctuations that occur during the phase of acceleration are redshifted to currently observable scales and lead to a spectrum of cosmological fluctuations, which becomes scale invariant in the limit in which the expansion rate becomes constant in time (2).

Reheating at the end of the period of accelerated expansion is an important part of inflationary cosmology. Without reheating, inflation would leave behind a universe empty of matter. Reheating occurs through coupling of the inflaton field ϕ, which is the scalar field that generates the accelerated expansion of space, to Standard Model (SM) matter. Such couplings must be present at least via gravitational interactions. However, in many models of inflation there are couplings that take place through the matter sector of the theory directly.

Reheating was initially (3) analyzed using first-order perturbation theory and discussed in terms of the decay of an inflaton particle into SM matter particles. As first realized in Reference 4 (see also Reference 5), such a perturbative analysis may be misleading because it does not take into account the coherent nature of the inflaton field. A new view of reheating, which is based on the quantum mechanical production of matter particles in a classical background inflaton field, was then proposed (4).[1] As this analysis showed, it is likely that reheating involves a parametric

[1] See also References 6 and 7 for other approaches to the out-of-equilibrium dynamics of the inflaton field.

resonance instability. This proposal was studied more carefully in References 8 and 9 and analyzed in detail in Reference 10. The term preheating was coined (8) to describe the initial energy transfer from the inflaton field to matter particles. Typically, the state of matter after preheating is highly nonthermal and thus must be followed by a phase of thermalization.

The first goal of this review article is to present an introduction to the theory of preheating after inflation. In Section 2, we briefly review inflationary cosmology. In Section 3 we present a comprehensive analysis of preheating. The efficiency of preheating turns out to be rather model dependent. We first discuss standard preheating, which typically occurs in simple single-field inflation models such as those used in chaotic inflation. The efficiency of preheating can be much higher if a tachyonic direction develops, which is what occurs in certain small field inflation models and in so-called hybrid inflation, a model involving two scalar fields. Although preheating leads to a very rapid beginning of the process of energy transfer from the inflaton to SM matter, it typically does not drain most of the energy of the inflaton field. This occurs in a second stage that is characterized by the nonlinear interactions of the fluctuation modes, which have been highly excited by the preheating process. According to recent studies (11), this process is turbulent. The initial stage of turbulence (after which the bulk of the energy density is no longer in the inflaton field) is rapid, but the actual thermalization of the decay products takes much longer. These issues are briefly discussed in Section 4. Section 5 focuses on reheating in supersymmetric models. Finally, in Section 6 we give a brief overview of a number of applications of preheating in inflationary cosmology.

2. INFLATION MODELS AND INITIAL CONDITIONS FOR REHEATING

Cosmological inflation (1) is a phase of accelerated expansion of space. In the context of general relativity as the theory describing space and time, inflation requires scalar field matter. More precisely, the energy-momentum tensor of matter must be dominated by the almost constant potential energy density of the scalar field ϕ.

A scalar field is postulated to exist in the SM of particle physics: the Higgs field that is used to give elementary fermions their masses. To serve as a Higgs field, its potential energy must have a minimum at a nontrivial field value. The standard example is

$$V(\phi) = \frac{1}{4}\lambda(\phi^2 - \eta^2)^2, \tag{1}$$

where η is the vacuum expectation value (VEV) of ϕ. It is assumed that at high temperatures the symmetry is restored by finite temperature effects (see, e.g., Reference 12 for a review of field theory methods used in inflationary cosmology) and that $\phi = 0$. Once the temperature T falls below a critical value T_c, ϕ ceases to be trapped and begins to roll toward one of the lowest-energy states, $\phi = \pm\eta$. The SM Higgs must have a coupling constant λ, which is set by the gauge coupling constant and cannot be sufficiently small to yield a long time period of slow rolling of ϕ, which is required to obtain enough inflation [except possibly if ϕ is nonminimally coupled to gravity (13)]. Thus, scalar field–driven inflation requires us to go beyond the SM of particle physics. Once one makes this step, there are typically many candidate scalar fields that could be the inflaton, in particular in supersymmetric models.

For cosmological studies, the precise nature of the inflaton is often secondary, and hence simple toy models are used. So-called new inflation (14, 15) maintains the idea that ϕ starts out trapped near $\phi = 0$. Inflation takes place during the period in which ϕ is undergoing the symmetry-breaking phase transition and slowly rolling toward $\phi = \pm\eta$. The model was based on scalar field dynamics obtained by replacing the potential (Equation 1) by a symmetry-breaking potential of Coleman-Weinberg (16) form, in which the mass term at the field origin is set to zero and symmetry breaking is obtained through quantum corrections. However, new inflation models typically suffer from an initial condition problem (17).

Chaotic (or large field) inflation (18) is an alternative scenario. Inflation is triggered by a period of slow rolling of ϕ that is unrelated to a symmetry-breaking phase transition. The simplest example occurs in the toy model of a single scalar field with potential

$$V(\phi) = \frac{1}{2}m^2\phi^2, \tag{2.}$$

where m is the mass of ϕ [which is of the order 10^{-6} m_{pl} if the model is to yield the observed magnitude of cosmological fluctuations (2)]. Here, m_{pl} is the Planck mass defined via $m_{pl}^{-2} \equiv G$, where G is Newton's gravitational constant. It is assumed that ϕ begins at large field values and slowly rolls toward its vacuum state, $\phi = 0$. For inflation to be successful, two conditions must be satisfied. First, the energy density must be dominated by the potential energy term. Second, the acceleration term in the field equation

$$\ddot{\phi} + 3H\dot{\phi} = -V'(\phi), \tag{3.}$$

the Klein-Gordon equation in an expanding background, must be negligible compared with the other two terms. Here, H is the Hubble expansion rate and the prime symbol indicates the derivative with respect to ϕ. By using the Friedmann equation,

$$H^2 = \frac{8\pi G}{3}\left(\frac{1}{2}\dot{\phi}^2 + V(\phi)\right), \tag{4.}$$

we see that the slow-rolling conditions are only satisfied for super-Planckian field values $|\phi| > m_{pl}$. In Equations 3 and 4 we have taken the field configuration to be homogeneous. In the case of chaotic inflation, the homogeneous slow-roll trajectory is a local attractor in initial condition space (19), even in the presence of linear metric fluctuations (20); thus, this model is free from the initial condition problem of new inflation. In the context of "real" particle physics theories such as supersymmetric models, gravitational effects often steepen the potential for values of $|\phi|$ beyond the Planck mass and therefore prevent slow-roll inflation.

One way to avoid this problem but maintain the success of chaotic inflation is to add a second scalar field ψ to the sector of the theory responsible for inflation and to invoke a potential of the form

$$V(\phi, \psi) = \frac{1}{2}m^2\phi^2 + \frac{g^2}{2}\psi^2\phi^2 + \frac{1}{4}\lambda(\psi^2 - v^2)^2, \tag{5.}$$

where g and λ are dimensionless coupling constants and v is the VEV of ψ. For large values of $|\phi|$, the potential in ψ direction has a minimum at $\psi = 0$, whereas for small values of $|\phi|$, $\psi = 0$ becomes an unstable point. The reader can verify that in this model slow rolling of ϕ does not require super-Planckian field values. This two-field model is known as hybrid inflation (21).

Let us return to the toy model of chaotic inflation with the potential given in Equation 1. The slow-roll trajectory is given by

$$\dot{\phi} = -\frac{1}{2\sqrt{3\pi}}mm_{pl}, \tag{6.}$$

and the slow-roll conditions clearly break down at the field value

$$\phi_c = \frac{m_{pl}}{2\sqrt{3\pi}}. \tag{7.}$$

After the breakdown of slow rolling, ϕ commences damped oscillatory motion about $\phi = 0$, and the time-averaged equation of state is that of cold matter ($p = 0$, where p denotes pressure).

Asymptotically, for large times $mt \gg 1$ the solution approaches

$$\phi(t) \rightarrow \frac{m_{pl}}{\sqrt{3\pi}mt} \sin(mt). \tag{8.}$$

This scalar field configuration provides the classical background matter in the reheating phase.

3. INFLATON DECAY

3.1. Perturbative Decay

Reheating is a key part of inflationary cosmology. It describes the production of SM matter at the end of the period of accelerated expansion, when the energy density is stored overwhelmingly in the oscillations of ϕ. Historically, reheating was first treated perturbatively (3).

We assume that the inflaton ϕ is coupled to another scalar field χ. If we take the interaction Lagrangian to be

$$\mathcal{L}_{\text{int}} = -g\sigma\phi\chi^2, \tag{9.}$$

where g is a dimensionless coupling constant and σ is a mass scale, then the decay rate of the inflaton into χ particles is given by

$$\Gamma = \frac{g^2\sigma^2}{8\pi m}, \tag{10.}$$

where m is the inflaton mass. In the approach followed in Reference 3, the energy loss of the inflaton due to the production of χ particles was taken into account by adding a damping term to the inflaton equation of motion. In the case of a homogeneous inflaton field, the equation is

$$\ddot{\phi} + 3H\dot{\phi} + \Gamma\dot{\phi} = -V'(\phi). \tag{11.}$$

For a small coupling constant, the interaction rate Γ is typically much smaller than the Hubble parameter at the end of inflation. Thus, at the beginning of the phase of inflaton oscillations, the energy loss into particles is initially negligible compared with the energy loss due to the expansion of space. It is only when the Hubble expansion rate decreases to a value comparable to Γ that χ-particle production becomes effective. The energy density at the time when $H = \Gamma$ determines how much energy ends up in χ particles and thus determines the reheating temperature—the temperature of the SM fields after energy transfer:

$$T_R \sim (\Gamma m_{pl})^{1/2}. \tag{12.}$$

Because Γ is proportional to the square of the coupling constant g, which is generally very small, perturbative reheating is slow and produces a reheating temperature that can be very low compared with the energy scale at which inflation takes place.

There are two main problems with the perturbative decay analysis described above. First, even if the inflaton decay were perturbative, it is not justified to use the heuristic equation (11) because it violates the fluctuation-dissipation theorem: In systems with dissipation, there are always fluctuations, and these are missing in Equation 11. For an improved effective equation of motion see, for example, Reference 22. Second, and more importantly, the perturbative analysis does not take into account the coherent nature of the inflaton field. The inflaton field at the beginning of the period of oscillations is not a superposition of free asymptotic single-inflaton states but rather a coherently oscillating homogeneous field. The large amplitude of oscillation implies that it is well justified to treat the inflaton classically. However, the matter fields can be assumed to start off in their vacuum state (the redshifting during the period of inflation removes any matter particles present at the beginning of inflation). Thus, matter fields χ must be treated quantum mechanically. The improved approach to reheating that was initiated in Reference 4 (see also Reference 5) is to consider reheating as a quantum production of χ particles in a classical ϕ background.

3.2. Preheating

We present the preheating mechanism for the simple toy model with the interaction Lagrangian

$$\mathcal{L}_{\text{int}} = -\frac{1}{2}g^2\chi^2\phi^2, \tag{13}$$

where, as before, g is a dimensionless coupling constant.[2] In this subsection we neglect the expansion of space. Provided that the time period of preheating is small compared with the Hubble expansion time H^{-1}, this is a reasonable approximation. In the next subsection we explicitly include the expansion of space.

The quantum theory of χ-particle production in the external classical inflaton background begins by expanding the quantum field $\hat{\chi}$ into creation and annihilation operators $\hat{a}_k$ and $\hat{a}_k^\dagger$,

$$\hat{\chi}(t, \mathbf{x}) = \frac{1}{(2\pi)^{3/2}} \int d^3k(\chi_k^*(t)\hat{a}_k e^{i\mathbf{kx}} + \chi_k(t)\hat{a}_k^\dagger e^{-i\mathbf{kx}}), \tag{14}$$

where k is the momentum. If we assume that there are no nonlinearities in the χ sector of the theory, then the equation of motion for χ is linear and can be studied simply mode by mode in Fourier space. The mode functions then satisfy the equation

$$\ddot{\chi}_k + (k^2 + m_\chi^2 + g^2\Phi^2\sin^2(mt))\chi_k = 0, \tag{15}$$

where Φ is the amplitude of oscillation of ϕ. This is the Mathieu equation, which is conventionally written as

$$\chi_k'' + (A_k - 2q\cos 2z)\chi_k = 0, \tag{16}$$

where we have introduced the dimensionless time variable $z = mt$ and where the prime symbol now denotes the derivative with respect to z. Comparing the coefficients, we see that

$$A_k = \frac{k^2 + m_\chi^2}{m^2} + 2q, \text{ where } q = \frac{g^2\Phi^2}{4m^2}. \tag{17}$$

The growth of the mode function corresponds to particle production, as in the case of particle production in an external gravitational field (26). We return to this point in the next subsection. For now, let us state simply that exponential growth of the mode functions leads to an exponential growth of the number of χ particles and that the exponent of this growth is double the corresponding exponent of the mode functions.

It is well known that the Mathieu equation has instabilities for certain ranges of k and leads to exponential growth,

$$\chi_k \propto \exp(\mu_k z), \tag{18}$$

where μ_k is known as the Floquet exponent. For small values of q, such as $q \ll 1$, resonance occurs in a narrow instability band about $k = m$ (see **Figure 1**). Therefore, in this case we refer to narrow resonance (see Reference 27 for in-depth discussions of the Mathieu equation and its generalizations).

---

[2] Preheating in a conformally flat scalar field model was analyzed in References 23 and 24 and in a sine-Gordon potential in Reference 25.

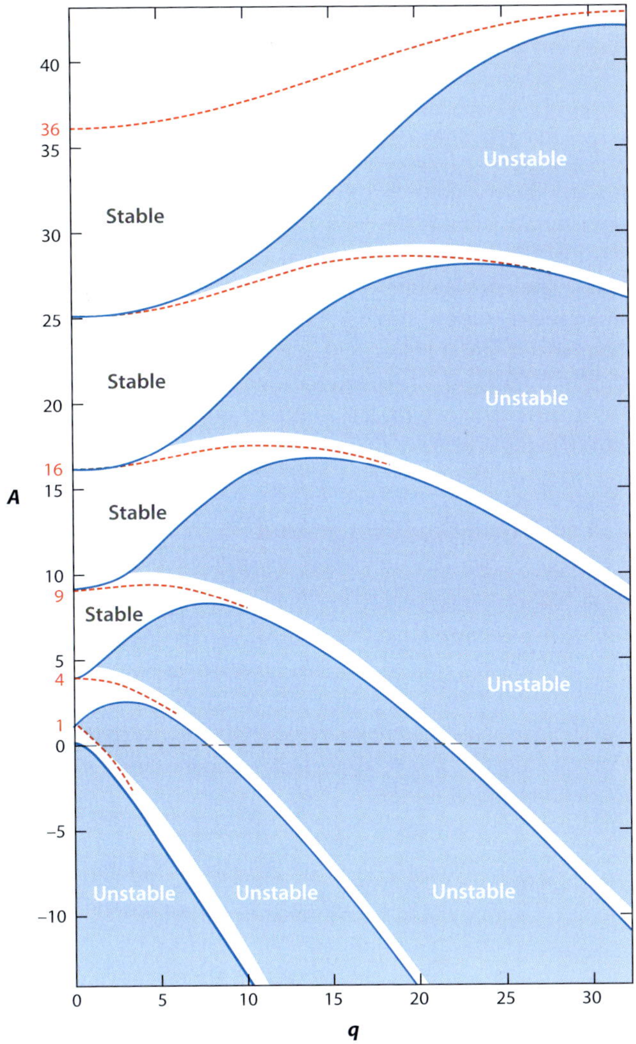

Figure 1

Instability bands of the Mathieu equation (from Reference 27). The horizontal axis is the parameter q from Equation 16; the vertical axis is the value of A. The shaded areas represent regions in parameter space in which there is a parametric resonance instability.

The resonance is much more efficient if $q \gg 1$ (8, 10). In this case, resonance occurs in broad bands. Specifically, the bands include all long-wavelength modes $k \to 0$. We then refer to broad parametric resonance. A condition for particle production is that the WKB (Wentzel-Kramers-Brillouin) approximation for the evolution of χ is violated. In the WKB approximation, we write $\chi_k \propto e^{\pm i \int \omega_k dt}$, which is valid as long as the adiabaticity condition

$$\frac{d\omega_k^2}{dt} \leq 2\omega_k^3 \qquad 19.$$

is satisfied. In the above, the effective frequency ω_k is given by

$$\omega_k = \sqrt{k^2 + m_\chi^2 + g^2\Phi(t)^2 \sin^2(mt)}. \qquad 20.$$

By inserting the effective frequency of Equation 20 into the condition of Equation 19 and following some algebra, we find that the adiabaticity condition is violated for momenta satisfying

$$k^2 \leq \frac{2}{3\sqrt{3}} gm\Phi - m_\chi^2. \qquad 21.$$

For modes with these values of k, the adiabaticity condition breaks down in each oscillation period when ϕ is close to zero. We conclude that the particle number does not increase smoothly, but rather in bursts, as was first studied in Reference 10.

So far, we have studied preheating in a toy model in which SM matter is modeled by a scalar field χ. However, in principle we are interested in the production of SM fermions. Such fermions could be produced after preheating into a scalar field χ, which then couples to fermions. In supersymmetric theories (discussed below), there are many channels in which this may happen. However, preheating into fermions is also effective, even though the occupation number of any fixed state cannot be greater than one (because of the Pauli exclusion principle). Preheating into fermions is discussed in detail elsewhere (28, 29).

3.3. Preheating in an Expanding Background

Provided that the Floquet exponent is not much smaller than unity, the parametric resonance instability leads to an energy transfer from the inflaton to matter particles, which is rapid on the scale of the Hubble time. Thus, an analysis neglecting the expansion of the universe is self-consistent. However, as discussed in detail in Reference 10, it is not too difficult to include the expansion of space.

First, we give a qualitative analysis of broad resonance in an expanding background characterized by the cosmological scale factor $a(t)$, which is increasing in time as given by the Friedmann equations. The equation of motion for χ is

$$\ddot{\chi}_k + 3H\,\dot{\chi}_k + \left(\frac{k^2}{a^2} + m_\chi^2 + g^2\Phi(t)^2 \sin^2(mt)\right)\chi_k = 0. \qquad 22.$$

The adiabaticity condition is now violated for momenta satisfying

$$\frac{k^2}{a^2} \leq \frac{2}{3\sqrt{3}} gm\Phi(t) - m_\chi^2. \qquad 23.$$

Note that the expansion of space makes broad resonance more effective because more k modes are redshifted into the instability band as time proceeds. Below, we explain that the improved analysis yields the same expression for the resonance band, except for the exact value of the numerical coefficient of the first term on the right-hand side. Broad parametric resonance ends when $q \leq 1/4$.

Let us move on to a quantitative analysis of this problem, building on the comprehensive work in Reference 10. It is convenient to eliminate the Hubble friction term in the equation of motion by rescaling the field variable. We consider the variable $X_k(t) = a^{3/2}(t)\chi_k(t)$, in terms of which the equation of motion (Equation 22) becomes

$$\ddot{X}_k + \omega_k^2 X_k = 0, \qquad 24.$$

where

$$\omega_k^2 = \frac{k^2}{a^2(t)} + m_\chi^2 + g^2\Phi^2(t)\sin^2(mt) - \frac{9}{4}H^2 - \frac{3}{2}\dot{H}. \qquad 25.$$

Note that, in the matter-dominated background that we are considering, the last two terms on the right-hand side cancel.

The equation of motion (Equation 24) represents a harmonic oscillator equation with a time-dependent frequency. The evolution of the solution is described by the WKB approximation (which entails the absence of particle production), unless the adiabaticity condition is violated. This occurs during short time intervals around the instances $t = t_j$, when $\phi = 0$, as discussed in the previous section. We label the intervals of adiabatic evolution by an integer j. In the jth interval (which lasts from t_{j-1} to t_j), the adiabatic evolution of X_k is given by

$$X_k^j(t) = \frac{\alpha_k^j}{\sqrt{2\omega_k}} e^{i\int \omega_k dt} + \frac{\beta_k^j}{\sqrt{2\omega_k}} e^{-i\int \omega_k dt}, \qquad 26.$$

where the coefficients α_k^j and β_k^j (known as the Bogoliubov coefficients) are constant and satisfy the normalization condition $|\alpha_k^j|^2 - |\beta_k^j|^2 = 1$ (derived from the Heisenberg uncertainty principle).

During the brief time periods during which ϕ is close to zero, we can use the approximation $\phi^2(t) \simeq \Phi^2 m^2 (t - t_j)^2$. By introducing the new time variable $\tau = g\Phi m(t - t_j)$ and a rescaled momentum $\kappa^2 = \frac{(k^2/a^2)+m_\chi^2}{g\Phi m}$, the equation of motion for X becomes

$$\frac{d^2 X_k}{d\tau^2} + (\kappa^2 + \tau^2) X_k = 0. \qquad 27.$$

This equation corresponds to scattering from a parabolic potential.

The nonadabatic evolution of X_k during the short intervals in which ϕ crosses the origin leads to a transformation of the coefficients of Bogoliubov type,

$$\begin{aligned} \begin{pmatrix} \alpha_k^{j+1} \\ \beta_k^{j+1} \end{pmatrix} &= \begin{pmatrix} \frac{1}{D_k} & \frac{R_k^*}{D_k^*} e^{-2i\theta_k^j} \\ \frac{R_k}{D_k} e^{2i\theta_k^j} & \frac{1}{D_k^*} \end{pmatrix} \begin{pmatrix} \alpha_k^j \\ \beta_k^j \end{pmatrix} \\ &= \begin{pmatrix} \sqrt{1+e^{-\pi\kappa^2}}\, e^{i\varphi_k} & i e^{-\frac{\pi}{2}\kappa^2 - 2i\theta_k^j} \\ -i e^{-\frac{\pi}{2}\kappa^2 + 2i\theta_k^j} & \sqrt{1+e^{-\pi\kappa^2}}\, e^{-i\varphi_k} \end{pmatrix} \begin{pmatrix} \alpha_k^j \\ \beta_k^j \end{pmatrix}, \end{aligned} \qquad 28.$$

where we define $\theta_k^j = \int_0^{t_j} dt\omega_k$ as the phase accumulated at t_j. The reflection and transmission coefficients R_k and D_k are given by

$$R_k = -\frac{i e^{i\varphi_k}}{\sqrt{1 + e^{\pi\kappa^2}}} \qquad 29.$$

and

$$D_k = \frac{e^{-i\varphi_k}}{\sqrt{1 + e^{-\pi\kappa^2}}}, \qquad 30.$$

where the phase φ_k is

$$\varphi_k = \arg\left\{\Gamma\left(\frac{1 + i\kappa^2}{2}\right)\right\} + \frac{\kappa^2}{2}\left(1 + \ln\frac{2}{\kappa^2}\right). \qquad 31.$$

Here, Γ stands for the complex gamma function. In accordance with the general theory of particle production in external fields (e.g., 26), the occupation number of the kth mode is

$$n_k = |\beta_k|^2. \qquad 32.$$

Making use of the Bogoliubov transformation (Equation 28), we obtain the recursion relation of the particle number

$$\begin{aligned} n_k^{j+1} &= e^{-\pi\kappa^2} + (1 + 2e^{-\pi\kappa^2}) n_k^j \\ &\quad + 2e^{-\frac{\pi}{2}\kappa^2} \sqrt{1 + e^{-\pi\kappa^2}} \sqrt{n_k^j(1 + n_k^j)} \sin\theta_{tot}^j, \end{aligned} \qquad 33.$$

where $\theta^j_{tot} = 2\theta^j_k - \varphi_k + \arg(\alpha^j_k) - \arg(\beta^j_k)$ per Equation 32 and $|\alpha^j_k|^2 = 1 + n^j_k$. From Equation 33, it follows that only modes with $\pi\kappa^2 \leq 1$ grow. Inserting the definition of κ into this condition, we obtain an expression for the resonance band that reproduces Equation 23, except for the numerical coefficient in the first term on the right-hand side of the equation, which is now $1/\pi$ instead of $2/(3\sqrt{3})$.

We can expand Equation 33 in the limit $n_k \gg 1$ and obtain

$$n^{j+1}_k = \left(1 + 2e^{-\pi\kappa^2} + 2e^{-\frac{\pi}{2}\kappa^2}\sqrt{1 + e^{-\pi\kappa^2}} \sin\theta^j_{tot}\right) n^j_k. \qquad 34.$$

The occupation number can grow only if the expression in the bracket is larger than one. For the fastest-growing mode, $(k = 0)$, $n^{j+1}_k > n^j_k$ if

$$-\frac{\pi}{4} < \theta^j_{tot} < \frac{5\pi}{4}, \qquad 35.$$

where we have taken $m_\chi = 0$. Now, because θ^j_{tot} has a rather complicated time dependency, it could be considered almost as a random variable. Thus, the range of phases (Equation 35) implies that the solution grows approximately 75% of the time.

We can define an effective Floquet exponent μ^j_k for the jth interval via

$$n^{j+1}_k = e^{2\pi\mu^j_k} n^j_k. \qquad 36.$$

By comparing this equation with Equation 34, we obtain

$$\mu^j_k = \frac{1}{2\pi} \ln(1 + 2e^{-\pi\kappa^2} + 2e^{-\frac{\pi}{2}\kappa^2}\sqrt{1 + e^{-\pi\kappa^2}} \sin\theta^j_{tot}). \qquad 37.$$

The so-called average Floquet exponent μ_k is determined by

$$\mu_k = \frac{\pi}{m\Delta t} \sum_j \mu^j_k, \qquad 38.$$

where Δt is the total duration of the resonance. By making use of Equation 37, we obtain

$$\mu_k \approx \frac{1}{2\pi} \ln 3 - \mathcal{O}(\kappa^2). \qquad 39.$$

That the exponent is of the order unity implies that broad parametric resonance in an expanding background is very efficient and can convert a substantial fraction of the inflaton energy density into matter in a time interval that is small compared with the Hubble time.

The total number density of χ particles is obtained by integrating over all values of k,

$$\begin{aligned} n_\chi(t) &= \frac{1}{(2\pi a)^3} \int_0^\infty d^3k n_k(t) \\ &= \frac{1}{2\pi^2 a^3} \int_0^\infty dk k^2 |\beta^0_k|^2 e^{2m\mu_k t}, \end{aligned} \qquad 40.$$

which can be estimated as (30)

$$n_\chi(t) \approx \frac{(gm\Phi_0)^{3/2} |\beta^0_m|^2 e^{2m\mu t}}{16\pi^3 a^3 \sqrt{\frac{m\mu t}{2\pi} + 1}}. \qquad 41.$$

This equation determines how fast the energy is drained from the inflaton field and, thus, the time interval that it takes before preheating is completed.

3.4. Termination of Preheating

In the above analysis of preheating, we neglected the back-reaction of the produced χ particles on the dynamics of the preheating process. The back-reaction arises at different places. For instance, the presence of χ particles changes the effective mass of the inflaton oscillations. Below, we use the rough criterion that this back-reaction effect be negligible as long as the change Δm_ϕ^2 in the square mass of the inflaton is smaller than m^2.

In the Hartree approximation, the change in the inflaton mass due to χ particles is given by

$$\Delta m_\phi^2 = g^2 \langle \chi^2 \rangle, \tag{42.}$$

where the pointed brackets indicate the quantum expectation value. The expectation value of χ^2 is given by

$$\langle \chi^2 \rangle = \frac{1}{2\pi^2 a^3} \int_0^\infty dk k^2 |X_k(t)|^2. \tag{43.}$$

By inserting the expansion of the field modes X_k in terms of the Bogoliubov coefficients, we find that

$$\Delta m_\chi^2(t) \simeq \frac{g n_\chi(t)}{|\phi(t)|}. \tag{44.}$$

This expression becomes ill defined when ϕ crosses zero. However, the equation of motion is still well defined at these points. To estimate the strength of back-reaction, we replace ϕ by its amplitude Φ. Thus, the condition under which this back-reaction effect is negligible is

$$n_\chi(t) \leq \frac{m^2 \Phi(t)}{g}. \tag{45.}$$

This is an implicit equation for the time t_1 when back-reaction can no longer be neglected. Note that the expression for the number density of χ particles was derived in Equation 41 and that Φ scales as a function of time as t^{-1}. The exponential growth of n_χ makes it clear that, up to logarithmic factors, the time interval of preheating is

$$\delta t \sim (\mu m)^{-1}. \tag{46.}$$

Thus, because $H \ll m$, this time interval is short compared with the Hubble expansion time, unless the Floquet exponent μ is suppressed.

Another condition must be satisfied so as to neglect the back-reaction of the χ particles on the preheating dynamics, namely that the energy in the χ particles be subdominant. By making use of the estimate $\rho_\chi \sim \langle (\nabla \chi)^2 \rangle \simeq k^2 \langle \chi^2 \rangle$ and inserting the value of $\langle \chi^2 \rangle$ at the time t_1 determined above, we see that ρ_χ is smaller than the potential energy of the inflaton field at the time t_1 as long as the value q at the time t_1 is larger than one, that is, if $q(t_1) > 1$. This condition is approximately the same as the condition for the effectiveness of broad resonance.

In many models, there is another mechanism that shuts off the resonance before either of the two conditions mentioned above becomes satisfied. Numerical studies (31–34) have shown that the scattering of χ particles off the inflaton condensate limits the value of the χ modes to a value lower than that which would be obtained from arguments such as the first one given above, which makes use of the Hartree approximation. For small values of q, the resonant period may be completely absent.

If $q(t_1) \gg 1$, then broad parametric resonance ends while most of the energy is still stored in the inflaton condensate. The further decay must then be analyzed in another fashion, for instance perturbatively or by using numerical simulations. Specifically, the time interval of matter-dominated dynamics after inflation could last a long time, which would lead to a low matter temperature after the matter field excitations have thermalized. However, if $q(t_1) \sim 1$, then the matter-dominated phase ends with the end of preheating. We still need to study how long it takes for the decay products to thermalize. The topic of thermalization after preheating is discussed in the next section. Before moving on to the study of thermalization after preheating, we discuss some variants of preheating in which the inflaton decay is much more efficient than in the chaotic inflation toy model described above.

3.5. Tachyonic Preheating

In the chaotic inflation model we have studied up to this point, the effective frequency of the χ oscillations is always positive. If it were negative, then we would obviously get an exponential instability. The simplest way to obtain this instability is simply to change the sign of the coupling term (Equation 13) in the interaction Lagrangian. This can be done without giving up stability of the model if we add quartic potential terms that dominate at large field values but are unimportant during preheating. This model has so-called negative coupling resonance (35). Similar negative coupling instabilities can also occur in models with cubic interaction terms (36–38).

Another simple model in which the effective frequency is negative for a certain time interval is the symmetry-breaking potential (Equation 1). For small field values, the effective mass of the fluctuations of ϕ is negative; therefore, a tachyonic resonance occurs (39, 40). For small field values, the equation for the fluctuations ϕ_k of ϕ is

$$\ddot{\phi}_k + (k^2 - m^2)\phi_k = 0. \qquad 47.$$

Thus, modes with $k < m$ grow with an exponent that approaches $\mu_k = 1$ in the limit $k \to 0$. Given initial vacuum amplitudes for the modes ϕ_k at the initial time $t = 0$ of the resonance, the field dispersion at a later time t is given by

$$\langle \delta\phi^2 \rangle = \int_0^m \frac{k\,dk}{4\pi} e^{2t\sqrt{m^2-k^2}}. \qquad 48.$$

The growth of the fluctuation modes terminates once the dispersion becomes comparable to the symmetry-breaking scale. Tachyonic preheating also occurs in hybrid inflation models such as the one shown in Equation 2. In this case, it is the fluctuations of ψ that have tachyonic form and that grow exponentially (39). Note that reheating in hybrid inflation was first studied in Reference 41 using the tools of broad parametric resonance. Fermion production in this context was discussed in Reference 42.

Another preheating mechanism that is more effective than the broad resonance process described above arises if the χ particles in the model of Equation 9 are coupled linearly to fermions such that the χ particles created when $\phi \sim 0$ decay after half a ϕ oscillation, thereby preventing the χ particles from slowing the decay of ϕ. This mechanism is termed instant preheating (43).

4. THERMALIZATION

4.1. Perturbative Considerations

Neither the perturbative decay of a Bose inflaton condensate (discussed in Section 3.1) nor the preheating mechanism (discussed later in Section 3) produces a thermal spectrum of decay products. For many questions in cosmology, it is not sufficient to know when inflation has terminated—it is crucial to know at what temperature the universe first takes on a thermal distribution. Examples are applications of reheating to baryogenesis and to nucleosynthesis constraints. In this section, we discuss perturbative thermalization and then summarize the results of recent nonperturbative studies of inflaton decay. We begin with perturbative considerations.

In a full thermal equilibrium, the energy density ρ and the number density n of relativistic particles scale as $\rho \sim T^4$ and $n \sim T^3$, where T is the temperature of the thermal bath. Thus, in full equilibrium the average particle energy is given by $\langle E \rangle_{\text{eq}} = (\rho/n)$, which obeys the scaling $\langle E \rangle_{\text{eq}} \sim \rho^{1/4} \sim T$.

However, if the inflaton decays perturbatively, then immediately after the inflaton decay has completed the energy density of the universe is given by

$$\rho \approx 3(\Gamma m_{pl})^2 \quad \text{and} \quad \langle E \rangle \approx m \gg \rho^{1/4}, \tag{49}$$

where m is the inflaton mass. Then, from conservation of energy, the number density of decay products is

$$n \approx \left(\frac{\rho}{m}\right) \ll \rho^{3/4}. \tag{50}$$

Therefore, perturbative inflaton decay results in a dilute plasma that contains a small number of very energetic particles.

Reaching full equilibrium requires redistributing the energy among different particles (kinetic equilibrium), as well as increasing the total number of particles, chemical equilibrium. Therefore, both number-conserving and number-violating reactions must be involved.

The most important processes for kinetic equilibration are $2 \to 2$ scatterings with gauge boson exchange in the t-channel. (Scalar exchange in t-channel diagrams is usually suppressed; also, vertices that arise from a Yukawa coupling are helicity suppressed.) The cross section for these scatterings is $\sigma_{2\to 2} \sim \alpha^2 |t|^{-1}$, where α is a gauge fine structure constant and the variable t is related to the exchanged energy ΔE and momentum $\Delta \mathbf{p}$ through $t = \Delta E^2 - |\Delta \mathbf{p}|^2$. Due to an infrared singularity, these scatterings are very efficient even in a dilute plasma (44, 45).

Chemical equilibrium is achieved by changing the number of particles in the reheat plasma. From Equation 50 it follows that, to reach full equilibrium, the total number of particles must increase by a factor of n_{eq}/n, where $n \approx \rho/m$ and the equilibrium value is $n_{\text{eq}} \sim \rho^{3/4}$. This can be a very large number, for instance, $n_{\text{eq}}/n \sim \mathcal{O}(10^3)$. It has been recognized (44, 46; see also 47, 48) that the most relevant processes are $2 \to 3$ scatterings with gauge boson exchange in the t-channel. The cross section for emitting a gauge boson whose energy is $E \sim (\Gamma m_{pl})^{1/2}$ (where, as above, Γ is the inflaton decay rate), from the scattering of two fermions (up to a logarithmic bremsstrahlung factor), is $\sigma_{2\to 3} \sim \alpha^3 (\Gamma m_{pl})^{-1}$. When these scatterings become efficient, the number of particles increases very rapidly (49).[3] As a result, full thermal equilibrium becomes established shortly thereafter.

On the basis of the above analysis, one can use the rate for the above inelastic scatterings as the thermalization rate Γ_{th} of the universe. This rate, at the time at which the inflaton decay completes, can be found by using Equations 49 and 50:

$$\Gamma_{\text{th}} \sim \alpha^3 \left(\frac{m_{pl}}{m}\right) \Gamma. \tag{51}$$

For typical values of $\alpha \sim 10^{-2} - 10^{-1}$ and $m \leq 10^{-5} m_{pl}$, we find that $\Gamma_{\text{th}} \geq \Gamma$. Therefore, the universe reaches full thermal equilibrium immediately after the completion of perturbative inflaton decay.

[3] Decays processes (considered in Reference 48) are helpful, but generally they cannot increase the number of particles to the required level.

4.2. Nonperturbative Considerations

The purely perturbative considerations of the previous subsection are subject to the same criticisms as the original perturbative analysis of the initial stages of reheating. Thus, we must turn to nonperturbative analyses. Some analytical approaches have been pioneered (6, 7); numerical studies, however, have proved more powerful.

Because the occupation numbers of the excited modes are typically very high after the initial stages of preheating, a classical field theory analysis should be justified. Initial numerical studies were pioneered in References 31–34 and 50. Two numerical packages to perform such simulations are publicly available (51, 52), as are detailed numerical simulations of tachyonic preheating (40). Here, we focus on numerical studies of preheating in models with narrow or broad parametric resonance (11).

The resonant phase of the reheating process described in Section 3 produces either (in the case of narrow resonance) field fluctuations in a narrow interval about the resonant frequency (which is set by the mass of the inflaton field) or (in the case of broad or tachyonic resonance) field fluctuations at all wave numbers smaller than the critical value (23), whose magnitude is set by the inflaton mass and amplitude.

Once the occupation numbers of the resonant modes become sufficiently large, rescattering of the fluctuations begins. As first studied in References 31–34, this terminates the phase of exponential growth of the occupation numbers. In the case of narrow resonance, new peaks in the spectrum of the number density $n(k)$ develop at harmonic frequencies. Soon, the spectrum shows excitations in a continuum band that reaches to $k = 0$ and has an ultraviolet (UV) cutoff whose value increases with increasing time.

The evolution of the field fluctuations evolves to a regime of turbulent scaling driven by the remnant oscillations of the inflaton condensate (11). The resulting distribution of fluctuations is characterized by the spectrum

$$n(k) \sim k^{-3/2}, \tag{52}$$

which is nonthermal [for a thermal distribution, we would have $n(k) \sim k^{-1}$].

The evolution during the phase of turbulence has been shown (11) to be self-similar in the sense that as a function of time the spectrum scales as

$$n(k, \tau) = \tau^{-q} n_0(k\tau^{-p}), \tag{53}$$

where τ is a rescaled time ($\tau = t/t_0$, where t_0 is the time when the turbulent scaling regime begins), n_0 gives the initial distribution of the particles, and p and q are positive rational numbers whose values are determined in numerical simulations. Equation 53 describes the overall growth in the number of fluctuation quanta and at the same time gives the increase of the UV cutoff frequency as a function of time.

The phase of turbulence ends once most of the energy has been drained from the inflaton field. At this time, quantum processes take over and lead to the thermalization of the spectrum. In the case of an $\mathcal{O}(N)$ scalar field model, the preheating and thermalization were considered analytically in the large-N approximation in Reference 53 and applied to the thermalization of fermions and gauge fields in hybrid inflation in Reference 54 (see also Reference 55). Note that the time interval of the resonant preheating phase is given by the inflaton mass m and thus is very short on the Hubble timescale. The period during which the initial rescatterings take place and which ends when the turbulent scaling distribution becomes established is longer than the period of the initial resonance, but not by a large factor (11). Thus, the time t_0 is of the same order of magnitude as the time when inflation ends. However, the period of driven turbulence is very long, especially if all of the coupling constants in the field theory model are small. If c_r is the fraction of the energy density ρ_I at the end of inflation, which remains in the inflaton at the beginning of the phase of turbulence, then a rough estimate of the time $\tau_{\rm th}$ of thermalization is (11)

$$\tau_{\rm th} \sim \left(\frac{(c_r \rho_I)^{1/4}}{m} \right)^{1/p}. \qquad 54.$$

A typical value of p is 1/7. Given the normalization of m from the observed magnitude of the cosmological fluctuations, we find $\tau_{\rm th} \sim c_r^{7/4} 10^{21}$. Therefore, the temperature at which thermal equilibrium finally becomes established is low. The resulting value of the reheating temperature roughly agrees with what is obtained from the perturbative arguments given in the previous subsection.

5. REHEATING IN SUPERSYMMETRIC MODELS

Here we consider reheating in supersymmetric models, which are an important application of the theory of reheating described above. Supersymmetry (SUSY) introduces new degrees of freedom, new parameters, and a large number of scalar fields that may acquire large VEVs during inflation. These elements can affect the various aspects of reheating discussed in Sections 3 and 4. In this section we demonstrate some of the possible effects in the context of a well-motivated SUSY model.

5.1. Inflaton Couplings to Matter Fields

The minimal supersymmetric SM (MSSM) is a well-motivated extension of the SM (for reviews see, e.g., References 56 and 57). The new fields in the MSSM are (*a*) scalar partners of leptons and quarks, known as sleptons and squarks, respectively; and (*b*) fermionic partners of gauge and Higgs fields, known as gauginos and Higgsinos, respectively. The MSSM superpotential is given by

$$W_{\rm MSSM} = h_u Q H_u u + h_d Q H_d d + h_e L H_d e + \mu H_u H_d, \qquad 55.$$

where H_u, H_d, Q, L, u, d, and e are chiral superfields representing the two Higgs fields (and their Higgsino partners), left-handed (s)quark doublets, right-handed up- and down-type (s)quarks, left-handed (s)lepton doublets, and right-handed (s)leptons, respectively. The dimensionless Yukawa couplings h_u, h_d, and h_e are 3×3 matrices in the flavor space, and we have omitted the gauge and flavor indices. The last term is the μ term, which is a SUSY version of the SM Higgs boson mass.

Now we consider inflaton couplings to the MSSM fields. For a gauge singlet inflaton, the only renormalizable coupling occurs through the superpotential term $2g\Phi H_u H_d$, where Φ is the inflaton superfield (for details, see Reference 58). Taking into account the inflaton superpotential mass term $(m/2)\,\Phi\Phi$, the renormalizable part of the scalar potential that is relevant for the inflaton decay into MSSM scalars is given by

$$V \supset \frac{1}{2} m^2 \phi^2 + g^2 \phi^2 \chi_1^2 + g^2 \phi^2 \chi_2^2 + \frac{1}{\sqrt{2}} g m \phi \chi_1^2 - \frac{1}{\sqrt{2}} g m \phi \chi_2^2, \qquad 56.$$

where $\chi_{1,2}$ denotes the scalar component of $(H_u \pm H_d)/\sqrt{2}$ superfields, and we have considered only the real parts of the inflaton, ϕ, and $\chi_{1,2}$ fields.

Even in the simplest SUSY setup, the scalar potential is more involved than in the non-SUSY case given in Equation 13, which in turn can alter the picture of preheating presented in Section 3 (45, 58). An interesting feature of Equation 56 is that both the cubic $\phi\chi^2$ and quartic $\phi^2\chi^2$

interactions appear and SUSY naturally relates their strengths.[4] Also, the inflaton coupling to fermionic partners of $\chi_{1,2}$ follows naturally from SUSY. The prospects for fermionic preheating are therefore the same as those for the bosonic case.

5.2. Supersymmetric Flat Directions

A key property of SUSY theories is the presence of flat directions in field space along which the potential identically vanishes (in the limit of unbroken SUSY). Such scalar fields (which are complex) can therefore obtain large VEVs along these special directions at no energy cost. These flat directions, which can be interpreted as a degeneracy of the vacuum state of SUSY theories, arise because of cancellations between fields of opposite charges in the D-term potential. A powerful tool for finding the flat directions has been developed in References 59–61 (for reviews, see References 62–64). Flat directions are classified by gauge-invariant monomials $\prod_{i=1}^{n} X_i$, where X_i are chiral superfields of the model. This ensures that the D-term part of the potential vanishes[5] along the direction $\langle\chi_1\rangle = \cdots = \langle\chi_n\rangle = \varphi$ (χ_i are scalar components of X_i), which corresponds to a two-dimensional subspace represented by a complex field φ. A flat direction VEV spontaneously breaks gauge symmetries and gives (SUSY-conserving) masses to the gauge bosons and gauginos similar to the Higgs mechanism in electroweak symmetry breaking (45, 58, 60, 65, 66). The induced masses for gauge/gaugino fields are $\sim\alpha^{1/2}|\varphi|$ (recall that α is a gauge fine structure constant). Similarly, a flat direction VEV induces (SUSY-conserving) masses $\sim h|\varphi|$ for those fields that have superpotential couplings to φ (h is a Yukawa coupling). Therefore, all fields that are coupled to a flat direction obtain very large masses.

The flat directions are massless if SUSY is exact, but they are lifted when SUSY is broken (which is assumed to happen at a scale of the order of teraelectronvolts), as a result of which they obtain a mass of $m_\varphi \sim \mathcal{O}(\text{TeV})$. Provided that $m_\varphi \ll H_{\text{inf}}$, where H_{inf} is the Hubble expansion rate during inflation, the flat direction can acquire a large VEV by the virtue of quantum jumps during inflation (62, 63). This can dramatically alter the postinflationary history of the universe, as we discuss below.[6]

5.3. Perturbative Decay

Consider a flat direction φ that has Yukawa couplings to the inflaton decay products χ. This arises, for example, for MSSM flat directions that are made of squark and/or slepton fields in which χ is a MSSM Higgs field; see Equation 56 (58). This results in the following term in the scalar potential:

$$V \supset h^2|\varphi|^2\chi^2, \qquad 57.$$

where h denotes a Yukawa coupling. Note that the first generation of leptons and quarks has a Yukawa coupling $\sim\mathcal{O}(10^{-5})$, whereas the rest of the SM Yukawa couplings are $>10^{-4}$. Because $|\varphi|$ is virtually frozen while $m_\varphi < H < H_{\text{inf}}$, it is only when $H \simeq m_\varphi$ that the flat direction starts its oscillations. Because the field is complex, an elliptical trajectory with an $\mathcal{O}(1)$ eccentricity typically results (60). Therefore, $|\varphi|$ redshifts as $|\varphi| \propto H^{-1}$.

---

[4]Note that the cubic term is required for a complete decay of the inflaton field.

[5]Because the total SM charge of a gauge-invariant monomial is zero by definition, the D-term potential involving only the fields used to build the monomial also vanishes because it is proportional to the sum of the charges.

[6]The development of large VEVs requires that the flat directions not obtain positive Hubble-induced supergravity corrections during inflation. This problem can be avoided, for example, by considering nonminimal Kahler potentials.

Although the flat direction has a large amplitude, the induced mass of the inflaton decay products obtained via Equation 57 leads to the inflaton decay being kinematically forbidden as long as $h|\varphi| \geq m/2$. There are thus two criteria for perturbative inflation decay. The first condition is that the decay of the inflaton into χ particles must be kinematically allowed; this becomes possible once the induced χ mass drops below the inflaton mass m. Taking into account the fact that once H falls below the value m_φ the field amplitude of φ decreases linearly in H, we find that the kinematic decay becomes possible once $H < (\frac{m}{h\varphi_0})m_\varphi$, where φ_0 is the initial VEV of the flat direction. The second condition is that $H < \Gamma$, where Γ is the rate for perturbative decay of the inflaton. Thus, the inflaton cannot decay until the Hubble rate has decreased to a value $H_{\rm dec}$

$$H_{\rm dec} = \min\left[\left(\frac{m}{h\varphi_0}\right)m_\varphi, \Gamma\right]. \qquad 58.$$

If φ_0 is sufficiently large, then $H_{\rm dec} \ll \Gamma$. This occurs if $\varphi_0 > h^{-1}\frac{m_\varphi m}{\Gamma}$. Flat directions can therefore significantly delay inflaton decay on purely kinematical grounds.

5.4. Nonperturbative Decay

To understand the preheating dynamics in the presence of flat directions, we consider the governing potential that is obtained from Equations 56 and 57:

$$V = \frac{1}{2}m^2\phi^2 + g^2\phi^2\chi^2 + \frac{g}{\sqrt{2}}m\phi\chi^2 + h^2|\varphi|^2\chi^2. \qquad 59.$$

As mentioned in the previous section, we generically have $h > 10^{-4}$, and g can be as large as $\sim\mathcal{O}(1)$. After mode decomposition of the field χ, the energy of the mode with momentum k, denoted by χ_k, is given by

$$\omega_k = (k^2 + 2g^2\phi^2 + \sqrt{2}gm\phi + 2h^2|\varphi|^2)^{1/2}. \qquad 60.$$

Let us freeze the expansion of the universe first. Including the expansion does not change our conclusions.

We now show that there is also a kinematic blocking of preheating if the initial value of the flat direction fields is large. We consider the most efficient case for preheating, namely large field inflation; that is, $\langle\phi\rangle > m_{pl}$. Note that for $g > 10^{-6}$, the inflaton induces a large mass $g\langle\phi\rangle > H_{\rm inf}$ for χ during inflation. As a result, χ quickly settles down to the minimum, even if it is initially displaced, and remains there. Therefore, φ does not receive any mass corrections from its coupling to χ during inflation.

As discussed in Section 5.3, in the interval $m_\varphi \leq H \leq m$, the flat direction VEV slides very slowly because of the underdamped motion due to a large Hubble friction term—it is effectively frozen. Nonperturbative production of χ quanta occurs if there is a nonadiabatic time variation in the energy, that is, if $d\omega_k/dt \geq \omega_k^2$. The inflaton oscillations result in a time-varying contribution to ω_k, whereas the flat direction coupling to χ yields a virtually constant piece. This constant piece weakens the nonadiabaticity condition. Indeed, time variation of ω_k is adiabatic at all times; in other words, $d\omega_k/dt < \omega_k^2$, provided that $h^2|\varphi|^2 > g\Phi m$, where $\Phi \sim \mathcal{O}(m_{pl})$ is the amplitude of the inflaton oscillations. Thus, there is no resonant production of χ quanta if

$$\varphi_0 > h^{-1}(gm_{pl}m)^{1/2}. \qquad 61.$$

Similar arguments lead to a kinematical blocking of fermionic preheating, as the symmetry between bosons and fermions implies similar equations for the momentum excitations; see Equation 60.

5.5. Thermalization

The flat direction VEV spontaneously breaks the SM gauge group. The gauge fields of the broken symmetries then acquire a SUSY-conserving mass of the order of $\alpha^{1/2}|\varphi|$. This mass provides a physical infrared cutoff for scattering diagrams with gauge boson exchange in the t-channel. Thus, the cross section for inelastic scatterings is now given by $\sigma_{2\to 3} \sim \alpha^2 |\varphi|^{-2}$. For large values of $|\varphi|$, the scattering rate is suppressed, which results in a delayed thermalization. The universe reaches thermal equilibrium when the Hubble expansion rate is (45)

$$H_{\rm th} = \min\left[10\alpha^2\left(\frac{m_{pl}}{\varphi_0}\right)^2\frac{m_\varphi^2}{m},\quad 10\alpha^2\left(\frac{m_{pl}}{\varphi_0}\right)^2\left(\frac{\Gamma}{m_\varphi}\right)^{1/2}\frac{m_\varphi^2}{m},\Gamma\right], \tag{62.}$$

where Γ is the rate for perturbative inflaton decay. This yields the following expression for the reheat temperature:

$$T_R \sim (H_{\rm th} m_{pl})^{1/2}. \tag{63.}$$

For very large values of φ_0, thermalization is considerably delayed (e.g., $H_{\rm th} \ll \Gamma$), and hence $T_R \ll (\Gamma m_{pl})^{1/2}$. This occurs for

$$\varphi_0 > 3\alpha\frac{m_\varphi m_{pl}}{(m\Gamma)^{1/2}}. \tag{64.}$$

Above, we have assumed that the flat direction condensate does not decay nonperturbatively. One may think that nonperturbative effects could also result in a fast decay of flat directions similar to preheating (36). However, there is a crucial difference between a rotating condensate and a radially oscillating condensate. The F-term couplings do not lead to resonant particle production from a rotating condensate (67, 68). D-term couplings of the flat direction can result in nonperturbative particle production (69). However, unlike the case of a radially oscillating field, the produced quanta have momenta that are less than the mass of the condensate in this case. The reason is that a rotating condensate has a $U(1)$ global charge[7] that, in the case of MSSM flat directions, is identified with the baryon and/or lepton number (59, 60, 70). Now, it is easy to show that the charge per particle in the rotating condensate is of $\mathcal{O}(1)$. Also, note that the baryon and lepton numbers of MSSM fields, whether in the flat direction condensate or produced from the flat direction rotation, are $\pm 1/3$ and ± 1, respectively. Thus, preservation of the baryon and lepton numbers by the D-term interactions, along with the conservation of energy density, implies that for an elliptic trajectory with $\mathcal{O}(1)$ eccentricity the number of produced quanta cannot be much smaller than the number of zero-mode quanta in the rotating condensate (66). Therefore, the nonperturbative decay of a rotating flat direction does not change the thermalization scenario described above.

Thermalization is delayed due to inducing large masses to the gauge and gaugino fields from the VEV of the flat direction. The induced mass by the plasma is given by $m_{eff}^2 \sim \alpha n/\langle E\rangle$, where $\langle E\rangle$ and n are the average energy and number density of quanta in the plasma, respectively. Because they cannot be very different from those in the condensate, the induced masses are also comparable.

6. CONSEQUENCES OF REHEATING/PREHEATING

6.1. Nonthermal Particle Creation

As discussed above, reheating and preheating lead to nonthermal particle production. In cosmology it is usually assumed that all particles start out in thermal equilibrium at the beginning of the

[7] This charge corresponds to the angular momentum of the rotating condensate in field space.

Standard Cosmology phase. However, reheating begins with out-of-equilibrium decay of the inflaton oscillations, and as we discussed, decay products may not reach full thermal equilibrium immediately. During the transition from inflation to Standard Cosmology, various nonthermal processes take place, and the assumption that all particles are in thermal equilibrium clearly breaks down. In the following subsections, we briefly mention a few applications of nonthermal particle production.

6.1.1. Baryogenesis and leptogenesis. The first application is to baryo- and leptogenesis. One of the several possible mechanisms to explain the observed asymmetry between baryons and antibaryons is to make use of out-of-equilibrium decay of superheavy Higgs and gauge particles (71). If reheating were purely perturbative, particles as heavy as the inflaton could be created either in inflaton decay (72) or from scatterings of inflaton decay products (46, 73).

Preheating, however, provides a mechanism to produce a large population of superheavy scalar particles much heavier than the inflaton. In Reference 74 this was studied making use of the same chaotic inflation Lagrangian which we have used in Section 3, where the χ scalars are the superheavy Higgs or gauge fields. However, a full numerical study (75) showed that large self-interactions may terminate the resonance before it becomes effective.

Another way to generate the observed baryon-to-entropy ratio is via leptogenesis (76), a scenario in which an asymmetry in the lepton number is produced and is then partially converted into baryon asymmetry via SM sphalerons (77). Preheating after inflation is a way to generate the initial lepton asymmetry. For example (78), preheating can produce a large number density of supermassive right-handed neutrinos in a model in which the inflaton couples to these neutrinos ψ via the standard fermionic preheating interaction term

$$\mathcal{L}_I = g\phi\bar{\psi}\psi. \qquad 65.$$

If hybrid inflation occurs at a scale close to the electroweak scale, then the nonthermal production of particles may provide the out-of-equilibrium condition that is necessary to achieve electroweak baryogenesis (79).

6.1.2. Dark matter. Another application of nonthermal particle creation during reheating is to excite dark matter. It is usually assumed that the dark matter particles are thermally distributed. This assumption is implicit in most current analyses of the prospects for dark matter detection in direct and indirect experiments. However, if the dark matter particles couple to the inflaton, then nonthermal production of dark matter during reheating is to be expected. If the dark matter particles have sufficiently strong interactions that allow them to thermalize during reheating, then the signatures of the initial nonthermal distribution are washed out. However, if the interactions do not permit thermalization after inflation, then the predictions concerning the dark matter distribution are quite different.

The production of out-of-equilibrium dark matter during preheating was put forward in Reference 78 and studied in detail in Reference 73. In the latter reference, the superheavy dark matter particles produced during reheating were termed Wimpzillas. Masses of Wimpzillas comparable to the Grand Unified Theory scale were considered (also see Reference 80 for a discussion of a purely gravitational production mechanism for Wimpzillas). The dark matter abundance that can be obtained by the preheating channel is very model dependent, whereas direct gravitational particle production produces dark matter of the required abundance for particle masses of $M_X \sim g^{1/2} 10^{15}$ GeV (80).

6.1.3. Moduli and gravitino production. Preheating could also produce dangerous and unwanted particles, such as particles with gravitationally suppressed couplings and weak scale masses that arise in many theories beyond the SM. Overproduction of these particles, if they are stable, could overclose the universe or, if they are unstable, ruin the success of big bang nucleosynthesis. Here we consider moduli and gravitino production during preheating.

The existence of bosonic and fermionic moduli fields is common in SUSY and superstring theories. Moduli (bosonic modulus, χ, and fermionic modulus, ψ) are typically coupled to the inflaton via nonrenormalizable interaction terms such as

$$\mathcal{L} \sim \phi^4 \frac{\chi}{m_{pl}} \text{ (bosonic)} \quad \text{and} \quad \mathcal{L} \sim \frac{\phi^2}{m_{pl}} \bar{\psi}\psi \text{ (fermionic)}. \qquad 66.$$

The production of moduli fields in chaotic and hybrid inflation reheating has been analyzed (81). Moduli fields can be parametrically amplified (their amplitude remains smaller than that of the inflaton fluctuations).

Another important example is the gravitino, the spin-$\frac{3}{2}$ partner of the graviton. Gravitinos are produced thermally from scatterings of light particles in the thermal bath. The number density of gravitinos thus produced can be obtained by solving the Boltzmann equation,

$$\dot{n}_X + 3Hn_X \simeq \langle \sigma v \rangle n_l^2, \qquad 67.$$

where n_X is the number density of the gravitinos; σ is the production cross section that scales as m_{pl}^{-2}; and $v \sim c$ is the relative velocity of scatterers l, whose number density is n_l. The resulting abundance is (82, 83)

$$\frac{n_X}{s} \sim 10^{-2} \frac{T_R}{m_{pl}}, \qquad 68.$$

where s is the entropy density and T_R is the reheat temperature of the universe. Big bang nucleosynthesis gives rise to an absolute upper bound of $(n_X/s) < 10^{-12}$ (the exact number depends on the gravitino mass and its decay modes), which in turn leads to an upper bound of $T_R < 10^9$ GeV (e.g., 84, 85).

Gravitino production during preheating has been studied (86). The gravitino equation of motion is the Rarita-Schwinger equation. Conformal invariance is broken during the reheating phase of inflationary cosmology. The presence of the oscillating inflaton field leads to a periodically varying correction to the effective gravitino mass that results in an instability, similar to the way an instability arises for spin-0 and spin-$\frac{1}{2}$ particle modes. The exact strength of the instability depends sensitively on the precise SUSY inflationary model one is considering.

Gravitinos with a helicity $\pm\frac{1}{2}$ component usually contain the Goldstino component—the inflatino (superpartner of the inflaton), whose interactions are not suppressed by m_{pl}. One would naturally expect them to be created in large abundance. However, in realistic scenarios, in which the scale of inflation is much higher than the scale of SUSY breaking [e.g., $H_{\text{inf}} \gg \mathcal{O}(100\,\text{GeV})$], it was argued (87) and later explicitly shown (88) that the helicity $\pm\frac{1}{2}$ states produced during preheating decay mainly in the form of inflatinos along with the inflaton.

6.2. Metric Preheating

Reheating may also affect metric fluctuations. Below, we discuss several ways in which this may occur.

6.2.1. Entropy fluctuations. The oscillating inflaton field can lead to the preheating of any fields to which it couples. The metric itself is no exception. Metric fluctuations come in three types (for details, see a review of the theory of cosmological perturbations, such as Reference 89)—scalar modes, vector modes, and tensor modes. In an expanding universe the vector modes are negligible because they decay. The tensor modes represent gravitational waves. The scalar modes are the cosmological perturbations, which are sourced by matter fluctuations.

It is possible (e.g., 89) to choose a coordinate system in which the metric including scalar metric fluctuations is diagonal:

$$ds^2 = a^2(\eta)[(1 + 2\Phi)d\eta^2 - (1 - 2\Psi)\gamma_{ij}dx^i dx^j], \qquad 69.$$

where Φ and Ψ are the two scalar metric fluctuation potentials. They are functions of space and time. In the absence of anisotropic stress (e.g., for scalar field matter) the two potentials are equal. In Equation 69, $a(\eta)$ is the scale factor of the background cosmology and η is conformal time defined via $dt = a d\eta$. Also, γ_{ij} is the background metric of the spatial sections after factoring out the cosmological expansion.

In the case of scalar field matter, the equation of motion for the metric fluctuation variable Φ reads (89)

$$\Phi'' + 2\left(\mathcal{H} - \frac{\phi_0''}{\phi_0'}\right)\Phi - \nabla^2\Phi + 2\left(\mathcal{H}' - H\frac{\phi_0''}{\phi_0'}\right)\Phi = 0, \qquad 70.$$

where the prime symbol indicates the derivative with respect to η and $\mathcal{H}$ is the Hubble expansion rate in conformal time. Also, ϕ_0 denotes the background value of the scalar field. Note that in this equation we assume that the perturbations are purely adiabatic (the relative density fluctuations in each matter component are the same). If entropy fluctuations were present, there would be a source term on the right-hand side of Equation 70 that is proportional to the entropy fluctuation.

Equation 70 suggests that the oscillations of the background inflaton field could induce parametric resonance of the metric fluctuations (90). The description of preheating in Section 3 might lead us to expect the long-wavelength modes to be the most sensitive to this instability. However, a careful study (91; see also References 92 and 93) showed that there is no instability of adiabatic metric perturbations during reheating. This can be most easily observed by focusing on a new variable ζ, the curvature fluctuation on constant density hypersurfaces:

$$\zeta \equiv \frac{2}{3}\frac{H^{-1}\dot{\Phi} + \Phi}{1 + w} + \Phi, \qquad 71.$$

where $w = p/\rho$ is the equation-of-state parameter. In the case of adiabatic fluctuations, ζ, on scales larger than the Hubble radius H^{-1}, satisfies the equation

$$\dot{\zeta}(1 + w) = 0. \qquad 72.$$

Although during the inflaton oscillations $w = -1$ is reached, ζ does not change.

However, for entropy fluctuations the result is rather different (94, 95). Provided that the entropy field (i.e., the χ field in the chaotic inflation preheating discussion of Section 3 or the ψ field in the hybrid inflation model) undergoes a parametric or tachyonic instability and thus leads to an exponential growth of the entropy fluctuation δS, then the curvature fluctuation ζ inherits this exponential growth. This follows because the entropy fluctuations act as a source for ζ. Equation 72 is replaced with

$$\dot{\zeta} = \frac{\dot{p}}{p + \rho}\delta S. \qquad 73.$$

This resonant growth of entropy fluctuations is important only in models in which the entropy fluctuations are not suppressed during inflation. Some recent examples have been studied (96, 97). The source for this instability need not be the oscillations of the inflaton field. In SUSY models, the decay of flat directions can also induce this instability (98).

6.2.2. Gravity waves. The equation of motion for gravitational waves is similar to that of scalar metric fluctuations (Equation 70), except that there is no coupling between the oscillating scalar field and the wave amplitude. Thus, there is no direct preheating of gravitational waves.

Nevertheless, gravitational waves can be produced by secondary processes. For example, they can be produced from the interaction of the classical matter waves produced during preheating (99). This is a rescattering effect. The induced gravitational-wave spectrum is not scale invariant but has a pronounced peak whose frequency is determined by the scale of inflation. For an inflation energy scale of 10^{15} GeV, the peak of the spectrum is at approximately 10^8 Hz. In hybrid inflation models, the scale of the peak can be in the kilohertz range relevant for Advanced LIGO (100). The amplitude of the peak is several orders of magnitude higher than the scale-invariant background of gravity waves produced directly during inflation.

Recently, several groups (101–104) have performed improved analyses of gravitational-wave production during preheating. The formalisms of References 101, 103, and 104 are quite general and were applied to, for instance, a $\lambda\phi^4$ model of inflation (104). Dufaux et al. (105) applied the formalism for the generation of gravitational waves to hybrid inflation, and Garcia-Bellido and colleagues (102) considered gravitational-wave production by the collision of bubbles formed during the tachyonic resonance in hybrid inflation models. The tachyonic instability leads to the formation of bubbles, and the collision of the bubbles is the primary source of gravitational waves. In the presence of fermionic couplings (see Equation 65), the inflaton can fragment to form non-topological solitons (106). Because the fragmentation of the inflaton condensate is inhomogeneous and anisotropic, it leads to large production of gravity waves (107).

7. DISCUSSION AND CONCLUSIONS

We present an overview of theory and application of reheating in inflationary cosmology. We emphasize the preheating mechanisms that in many models lead to rapid energy transfer between the inflaton and regular matter.

Our discussion of applications of reheating is superficial due to lack of space. We do not discuss, for instance, topological defect production during preheating (108–110); magnetic field generation (111, 112); induced non-Gaussianities (e.g., 113–116); preheating in theories with nonstandard kinetic terms (e.g., 117) or extra dimensions (possible excitation of Kaluza-Klein modes); applications to multifield inflation models (118) and to reheating in brane inflation models (119); and effects of noise on reheating (120, 121), which led to a new proof (122) of Anderson localization.

Importantly, resonant phenomena that are operative during reheating can also play a role in other areas of cosmology in which there are oscillating scalar fields. One example is in the context of the MSSM, where oscillating moduli fields can lead to resonances (98, 123). The study of resonant effects in early universe cosmology is a rich area of research that has hardly been explored. Many of the lessons learned in the context of inflationary reheating have more general applicability.

DISCLOSURE STATEMENT

The authors are not aware of any affiliations, memberships, funding, or financial holdings that might be perceived as affecting the objectivity of this review.

LITERATURE CITED

1. Guth AH. *Phys. Rev. D* 23:347 (1981)
2. Mukhanov VF, Chibisov GV. *JETP Lett.* 33:532 (1981)

3. Abbott LF, Farhi E, Wise MB. *Phys. Lett.* B117:29 (1982); Dolgov AD, Linde AD. *Phys. Lett.* B116:329 (1982); Albrecht AJ, Steinhardt PJ, Turner MS, Wilczek F. *Phys. Rev. Lett.* 48:1437 (1982)
4. Traschen JH, Brandenberger RH. *Phys. Rev. D* 42:2491 (1990)
5. Dolgov AD, Kirilova DP. *Sov. J. Nucl. Phys.* 51:172 (1990)
6. Boyanovsky D, et al. *Phys. Rev. D* 52:6805 (1995); Boyanovsky D, de Vega HJ, Holman R, Salgado JFJ. *Phys. Rev. D* 54:7570 (1996); Cormier D, Heitmann K, Mazumdar A. *Phys. Rev. D* 65:083521 (2002)
7. Baacke J, Heitmann K, Patzold C. *Phys. Rev. D* 55:7815 (1997)
8. Kofman L, Linde AD, Starobinsky AA. *Phys. Rev. Lett.* 73:3195 (1994)
9. Shtanov Y, Traschen JH, Brandenberger RH. *Phys. Rev. D* 51:5438 (1995)
10. Kofman L, Linde AD, Starobinsky AA. *Phys. Rev. D* 56:3258 (1997)
11. Micha R, Tkachev II. *Phys. Rev. Lett.* 90:121301 (2003); Micha R, Tkachev II. *Phys. Rev. D* 70:043538 (2004)
12. Brandenberger RH. *Rev. Mod. Phys.* 57:1 (1985)
13. Bezrukov FL, Shaposhnikov M. *Phys. Lett.* B659:703 (2008)
14. Linde AD. *Phys. Lett.* B108:389 (1982)
15. Albrecht AJ, Steinhardt PJ. *Phys. Rev. Lett.* 48:1220 (1982)
16. Coleman SR, Weinberg EJ. *Phys. Rev. D* 7:1888 (1973)
17. Goldwirth DS, Piran T. *Phys. Rep.* 214:223 (1992); Albrecht AJ, Brandenberger RH. *Phys. Rev. D* 31:1225 (1985)
18. Linde AD. *Phys. Lett.* B129:177 (1983)
19. Brandenberger RH, Kung JH. *Phys. Rev. D* 42:1008 (1990)
20. Feldman HA, Brandenberger RH. *Phys. Lett.* B227:359 (1989)
21. Linde AD. *Phys. Rev. D* 49:748 (1994)
22. Gleiser M, Ramos RO. *Phys. Rev. D* 50:2441 (1994)
23. Kaiser DI. *Phys. Rev. D* 56:706 (1997); Kaiser DI. *Phys. Rev. D* 57:702 (1998)
24. Greene PB, Kofman L, Linde AD, Starobinsky AA. *Phys. Rev. D* 56:6175 (1997)
25. Greene PB, Kofman L, Starobinsky AA. *Nucl. Phys. B* 543:423 (1999)
26. Birrell ND, Davies PCW. *Quantum Fields In Curved Space*. Cambridge, UK: Univ. Cambridge Press. 340 pp. (1982)
27. McLachlan NW. *Theory and Applications of Mathieu Functions*. Clarendon, UK: Oxford Univ. Press (1947)
28. Greene PB, Kofman L. *Phys. Lett.* B448:6 (1999); Greene PB, Kofman L. *Phys. Rev. D* 62:123516 (2000)
29. Baacke J, Heitmann K, Patzold C. *Phys. Rev. D* 58:125013 (1998)
30. Cyr-Racine F-Y. *Study of the growth of entropy perturbations in MSSM flat directions decay*. MSc thesis. McGill Univ. 107 pp. (2008)
31. Khlebnikov SY, Tkachev II. *Phys. Rev. Lett.* 77:219 (1996)
32. Khlebnikov SY, Tkachev II. *Phys. Lett.* B390:80 (1997)
33. Khlebnikov SY, Tkachev II. *Phys. Rev. Lett.* 79:1607 (1997)
34. Prokopec T, Roos TG. *Phys. Rev. D* 55:3768 (1997)
35. Greene BR, Prokopec T, Roos TG. *Phys. Rev. D* 56:6484 (1997)
36. Shuhmaher N, Brandenberger RH. *Phys. Rev. D* 73:043519 (2006)
37. Dufaux J-F, et al. *J. Cosmol. Astropart. Phys.* 0607:006 (2006)
38. Abolhasani AA, Firouzjahi H, Sheikh-Jabbari MM. arXiv:0912.1021 [hep-th] (2009)
39. Felder GN, et al. *Phys. Rev. Lett.* 87:011601 (2001)
40. Felder GN, Kofman L, Linde AD. *Phys. Rev. D* 64:123517 (2001)
41. Garcia-Bellido J, Linde AD. *Phys. Rev. D* 57:6075 (1998)
42. Garcia-Bellido J, Mollerach S, Roulet E. *J. High Energy Phys.* 0002:034 (2000)
43. Felder GN, Kofman L, Linde AD. *Phys. Rev. D* 59:123523 (1999)
44. Davidson S, Sarkar S. *J. High Energy Phys.* 0011:012 (2000)
45. Allahverdi R, Mazumdar A. *J. Cosmol. Astropart. Phys.* 0610:008 (2006)
46. Allahverdi R, Drees M. *Phys. Rev. D* 66:063513 (2002)
47. Jaikumar P, Mazumdar A. *Nucl. Phys. B* 683:264 (2004)
48. Allahverdi R. *Phys. Rev. D* 62:063509 (2000)

49. Enqvist K, Sirkka J. *Phys. Lett.* B314:298 (1993)
50. Felder GN, Kofman L. *Phys. Rev. D* 63:103503 (2001)
51. Felder GN, Tkachev I. arXiv:hep-ph/0011159 (2000)
52. Frolov AV. *J. Cosmol. Astropart. Phys.* 0811:009 (2008)
53. Berges J, Serreau J. *Phys. Rev. Lett.* 91:111601 (2003); Giraud A, Serreau J. *Nucl. Phys. A* 820:215C (2009)
54. Berges J, Borsanyi S, Serreau J. *Nucl. Phys. B* 660:51 (2003); Skullerud JI, Smit J, Tranberg A. *J. High Energy Phys.* 0308:043 (2003)
55. Tranberg A. *J. High Energy Phys.* 0811:037 (2008)
56. Nilles HP. *Phys. Rep.* 110:1 (1984)
57. Haber HE, Kane GL. *Phys. Rep.* 117:75 (1985)
58. Allahverdi R, Mazumdar A. *Phys. Rev. D* 76:103526 (2007)
59. Dine M, Randall L, Thomas SD. *Phys. Rev. Lett.* 75:398 (1995)
60. Dine M, Randall L, Thomas SD. *Nucl. Phys. B* 458:291 (1996)
61. Gherghetta T, Kolda SF, Martin SP. *Nucl. Phys. B* 468:37 (1996)
62. Enqvist K, Mazumdar A. *Phys. Rep.* 380:99 (2003)
63. Dine M, Kusenko A. *Rev. Mod. Phys.* 76:1 (2004)
64. Basboll A. arXiv:0910.0244 [hep-ph] (2009)
65. Allahverdi R, Mazumdar A. *J. Cosmol. Astropart. Phys.* 0708:023 (2007)
66. Allahverdi R, Mazumdar A. *Phys. Rev. D* 78:043511 (2008)
67. Allahverdi R, Shaw RAH, Campbell BA. *Phys. Lett.* B473:246 (2000)
68. Postma M, Mazumdar A. *J. Cosmol. Astropart. Phys.* 0401:005 (2004)
69. Olive KA, Peloso M. *Phys. Rev. D* 74:103514 (2006); Basboll A, Maybury D, Riva F, West SM. *Phys. Rev. D* 76:065005 (2007); Gumrukcuoglu AE, Olive KA, Peloso M, Sexton M. *Phys. Rev. D* 78:063512 (2008)
70. Affleck I, Dine M. *Nucl. Phys. B* 249:361 (1985)
71. Yoshimura M. *Phys. Rev. Lett.* 41:281 (1978); Yoshimura M. Erratum. *Phys. Rev. Lett.* 42:746 (1979); Ignatiev AY, Krasnikov NV, Kuzmin VA, Tavkhelidze AN. *Phys. Lett.* B76:436 (1978); Weinberg S. *Phys. Rev. Lett.* 43:1566 (1979)
72. Allahverdi R, Drees M. *Phys. Rev. Lett.* 89:091302 (2002)
73. Chung DJH, Kolb EW, Riotto A. *Phys. Rev. Lett.* 81:4048 (1998)
74. Kolb EW, Linde AD, Riotto A. *Phys. Rev. Lett.* 77:4290 (1996)
75. Kolb EW, Riotto A, Tkachev II. *Phys. Lett.* B423:348 (1998)
76. Fukugita M, Yanagida T. *Phys. Lett.* B174:45 (1986)
77. Kuzmin VA, Rubakov VA, Shaposhnikov ME. *Phys. Lett.* B155:36 (1985)
78. Giudice GF, Peloso M, Riotto A, Tkachev I. *J. High Energy Phys.* 9908:014 (1999)
79. Krauss LM, Trodden M. *Phys. Rev. Lett.* 83:1502 (1999); Garcia-Bellido J, Grigoriev DY, Kusenko A, Shaposhnikov ME. *Phys. Rev. D* 60:123504 (1999)
80. Chung DJH, Kolb EW, Riotto A. *Phys. Rev. D* 59:023501 (1999); Chung DJH, Kolb EW, Riotto A. *Phys. Rev. D* 60:063504 (1999)
81. Giudice GF, Riotto A, Tkachev II. *J. High Energy Phys.* 0106:020 (2001)
82. Ellis JR, Kim JE, Nanopoulos DV. *Phys. Lett.* B145:181 (1984)
83. Bolz M, Brandenburg A, Buchmuller W. *Nucl. Phys. B* 606:518 (2001); Bolz M, Brandenburg A, Buchmuller W. Erratum. *Nucl. Phys. B* 790:336 (2008)
84. Kawasaki M, Kohri K, Moroi T. *Phys. Rev. D* 71:083502 (2005)
85. Cyburt RH, Ellis JR, Fields BD, Olive KA. *Phys. Rev. D* 67:103521 (2003)
86. Maroto AL, Mazumdar A. *Phys. Rev. Lett.* 84:1655 (2000); Kallosh R, Kofman L, Linde AD, Van Proeyen A. *Phys. Rev. D* 61:103503 (2000); Giudice GF, Riotto A, Tkachev I. *J. High Energy Phys.* 9911:036 (1999); Kallosh R, Kofman L, Linde AD, Van Proeyen A. *Class. Quantum Gravity* 17:4269 (2000); Kallosh R, Kofman L, Linde AD, Van Proeyen A. Erratum. *Class. Quantum Gravity* 21:5017 (2004)
87. Allahverdi R, Bastero-Gil M, Mazumdar A. *Phys. Rev. D* 64:023516 (2001)
88. Nilles HP, Peloso M, Sorbo L. *Phys. Rev. Lett.* 87:051302 (2001); Nilles HP, Peloso M, Sorbo L. *J. High Energy Phys.* 0104:004 (2001)
89. Mukhanov VF, Feldman HA, Brandenberger RH. *Phys. Rep.* 215:203 (1992)

90. Bassett BA, Kaiser DI, Maartens R. *Phys. Lett.* B455:84 (1999)
91. Finelli F, Brandenberger RH. *Phys. Rev. Lett.* 82:1362 (1999)
92. Huang DH, Lin WB, Zhang XM. *Phys. Rev. D* 62:087302 (2000)
93. Afshordi N, Brandenberger RH. *Phys. Rev. D* 63:123505 (2001)
94. Bassett BA, Viniegra F. *Phys. Rev. D* 62:043507 (2000)
95. Finelli F, Brandenberger RH. *Phys. Rev. D* 62:083502 (2000)
96. Brandenberger RH, Frey AR, Lorenz LC. *Int. J. Mod. Phys. A* 24:4327 (2009)
97. Brandenberger RH, Dasgupta K, Davis AC. *Phys. Rev. D* 78:083502 (2008)
98. Cyr-Racine F-Y, Brandenberger RH. *J. Cosmol. Astropart. Phys.* 0902:022 (2009)
99. Khlebnikov SY, Tkachev II. *Phys. Rev. D* 56:653 (1997)
100. Garcia-Bellido J. arXiv:hep-ph/9804205 (1998)
101. Easther R, Lim EA. *J. Cosmol. Astropart. Phys.* 0604:010 (2006); Easther R, Giblin JT, Lim EA. *Phys. Rev. Lett.* 99:221301 (2007)
102. Garcia-Bellido J, Figueroa DG. *Phys. Rev. Lett.* 98:061302 (2007); Garcia-Bellido J, Figueroa DG, Sastre A. *Phys. Rev. D* 77:043517 (2008)
103. Dufaux JF, et al. *Phys. Rev. D* 76:123517 (2007)
104. Price LR, Siemens X. *Phys. Rev. D* 78:063541 (2008)
105. Dufaux JF, Felder GN, Kofman L, Navros O. *J. Cosmol. Astropart. Phys.* 0903:001 (2009)
106. Enqvist K, Kasuya S, Mazumdar A. *Phys. Rev. Lett.* 89:091301 (2002)
107. Kusenko A, Mazumdar A. *Phys. Rev. Lett.* 101:211301 (2008)
108. Khlebnikov S, Kofman L, Linde AD, Tkachev I. *Phys. Rev. Lett.* 81:2012 (1998)
109. Tkachev I, Khlebnikov S, Kofman L, Linde AD. *Phys. Lett.* B440:262 (1998)
110. Parry MF, Sornborger AT. *Phys. Rev. D* 60:103504 (1999)
111. Calzetta EA, Kandus A. *Phys. Rev. D* 65:063004 (2002); Davis A-C, Dimopoulos K, Prokopec T, Tornkvist O. *Phys. Lett.* B501:165 (2001); Boyanovsky D, Simionato M, de Vega HJ. *Phys. Rev. D* 67:023502 (2003); Mazumdar A, Stoica H. *Phys. Rev. Lett.* 102:091601 (2009)
112. Diaz-Gil A, Garcia-Bellido J, Garcia Perez M, Gonzalez-Arroyo A. *Phys. Rev. Lett.* 100:241301 (2008)
113. Enqvist K, et al. *Phys. Rev. Lett.* 94:161301 (2005)
114. Barnaby N, Cline JM. *Phys. Rev. D* 73:106012 (2006); Barnaby N, Cline JM. *Phys. Rev. D* 75:086004 (2007)
115. Kohri K, Lyth DH, Valenzuela-Toledo CA. arXiv:0904.0793 [hep-ph] (2009)
116. Chambers A, Rajantie A. *Phys. Rev. Lett.* 100:041302 (2008); Chambers A, Rajantie A. Erratum. *Phys. Rev. Lett.* 101:149903 (2008)
117. Lachapelle J, Brandenberger RH. *J. Cosmol. Astropart. Phys.* 0904:020 (2009)
118. Battefeld D, Kawai S. *Phys. Rev. D* 77:123507 (2008)
119. Barnaby N, Burgess CP, Cline JM. *J. Cosmol. Astropart. Phys.* 0504:007 (2005)
120. Zanchin V, Maia AJ, Craig W, Brandenberger RH. *Phys. Rev. D* 57:4651 (1998)
121. Zanchin V, Maia AJ, Craig W, Brandenberger RH. *Phys. Rev. D* 60:023505 (1999)
122. Brandenberger RH, Craig W. arXiv:0805.4217 [hep-th] (2008)
123. Allahverdi R, Enqvist K, Garcia-Bellido J, Mazumdar A. *Phys. Rev. Lett.* 97:191304 (2006); Allahverdi R, et al. *J. Cosmol. Astropart. Phys.* 0706:019 (2007)

LUNA: Nuclear Astrophysics Deep Underground

Carlo Broggini,[1] Daniel Bemmerer,[2] Alessandra Guglielmetti,[3] and Roberto Menegazzo[1]

[1] Istituto Nazionale di Fisica Nucleare, Sezione di Padova, Padova 35131, Italy; email: broggini@pd.infn.it, roberto.menegazzo@pd.infn.it

[2] Forschungszentrum Dresden-Rossendorf, Dresden 01314, Germany; email: d.bemmerer@fzd.de

[3] Università degli Studi di Milano and Istituto Nazionale di Fisica Nucleare, Sezione di Milano, Milano 20126, Italy; email: alessandra.guglielmetti@mi.infn.it

Annu. Rev. Nucl. Part. Sci. 2010. 60:53–73

First published online as a Review in Advance on May 20, 2010

The *Annual Review of Nuclear and Particle Science* is online at nucl.annualreviews.org

This article's doi: 10.1146/annurev.nucl.012809.104526

0163-8998/10/1123-0053$20.00

Key Words

underground measurements, solar hydrogen burning, big bang nucleosynthesis

Abstract

Nuclear astrophysics strives for a comprehensive picture of the nuclear reactions responsible for synthesizing chemical elements and for powering the stellar evolution engine. Deep underground in the Gran Sasso National Laboratory, the cross sections of the key reactions of the proton-proton chain and of the carbon-nitrogen-oxygen cycle have been measured right down to the energies of astrophysical interest. The salient features of underground nuclear astrophysics are summarized here. We review the main results obtained by LUNA during the past 20 years and discuss their influence on our understanding of the properties of the neutrino, the Sun, and the universe itself. Future directions of underground nuclear astrophysics toward the study both of helium and carbon burning and of stellar neutron sources in stars are outlined.

Contents

1. INTRODUCTION

The stars that fascinate us are not perfect and everlasting bodies, as believed by ancient philosophers. Gravity triggers the birth of a star, which then functions as a more or less turbulent chemical factory (1) and finally dies out in a quiet or violent way, depending on its initial mass (2). Only hydrogen, helium, and lithium are synthesized in the first few minutes after the big bang. All the other elements of the periodic table are produced in the thermonuclear reactions that take place within stars, in which the nuclear roots of life are embedded (3, 4). The aim of nuclear astrophysics is to obtain a comprehensive picture of all these reactions, which allow the transmutation of the chemical elements and provide the energy to run the engine of stellar evolution (5).

Knowledge of the reaction cross section at stellar energies lies at the heart of nuclear astrophysics. At these energies, the cross sections are extremely small. Such smallness makes the star lifetime of the length we observe, but it also makes impossible the direct measurement in the laboratory. The rate of the reactions, characterized by a typical energy release of a few megaelectronvolts—down to a few events per year—is not high enough to stand out from the background. The Laboratory for Underground Nuclear Astrophysics (LUNA), located in the Gran Sasso National Laboratory, began 20 years ago to run nuclear physics experiments in an extremely low-background environment. Just as the timbre of a note played on a piano cannot be distinguished in a crowded room, whereas it can in a music hall, the LUNA physicists are tuning their accelerators and detectors in the "music hall" located deep inside the mountain. This enables them to distinguish the tiny signal from nucleosynthesis reactions, thereby reproducing in the laboratory what occurs naturally inside the stars.

In this review, we describe the main features of thermonuclear reactions at very low energy, the characteristics of the background attainable at Gran Sasso, and the experimental equipment employed by LUNA. We then give an overview of hydrogen burning in stars and discuss the main results from LUNA, emphasizing their impact on models of the Sun and properties of the neutrino. Finally, we outline the next steps of underground nuclear astrophysics, particularly the study of helium and carbon burning and stellar neutron sources.

2. THERMONUCLEAR REACTIONS

Thermonuclear reactions between nuclei occur inside a star in a relatively narrow energy window, typically at an energy much lower than the height of the barrier arising from the Coulomb repulsion between nuclei. Consequently, reactions take place because of quantum mechanical tunneling, which leads to a very small (but not vanishing) probability that the incoming nucleus will penetrate the Coulomb barrier and reach its reaction partner.

Because of the tunneling, at these energies the reaction cross section $\sigma(E)$ drops almost exponentially with decreasing energy E,

$$\sigma(E) = \frac{S(E)}{E} \exp(-2\pi\eta), \qquad 1.$$

where $S(E)$ is the so-called astrophysical S-factor and $2\pi\eta = 31.29 Z_1 Z_2(\mu/E)^{1/2}$. Z_1 and Z_2 are the electric charges of the nuclei, μ is the reduced mass (in atomic mass units), and E is the energy (in kiloelectronvolts) in the center-of-mass system (2, 3). For most thermonuclear reactions, the astrophysical S-factor varies slowly with energy and contains all the nuclear physics information.

The reaction rate in the hot plasma of a star, with temperatures in the range of tens to hundreds of millions of degrees Kelvin, is obtained by weighting the reaction cross section $\sigma(E)$ with the energy distribution of the colliding nuclei: a Maxwell-Boltzmann $\phi(E)$ peaked at energies of 1 to 10 keV. The product of $\sigma(E)$ and $\phi(E)$ identifies the energy window in which reactions occur in the star: the Gamow peak. At lower energies the cross section is too small, whereas at higher energies the nuclei in the tail of the Maxwell-Boltzmann are too few. The energy release from the reaction, that is, the Q-value, corresponds to the mass difference between the entrance and exit channels.

To obtain the precise nuclear physics data required by modern astrophysics, one should measure the reaction cross section directly at the relevant stellar energy. For solar reactions, the cross-section values at the Gamow peak range from picobarns to femtobarns and even below (1b $= 10^{-24}$ cm^2). Although their ultralow count rates try the patience of physicists, these extremely small cross sections are a blessing to mankind because they account for the longevity of hydrogen burning in stars like our Sun.

However, the low count rates pose another experimental problem: In direct laboratory measurements at the Earth's surface, the signal-to-background ratio is too small. Much of this background is also associated with the cosmic rays that bombard Earth. So, instead of a direct measurement at astrophysical energies, data must be obtained at higher energies, where the signal-to-background ratio is more favorable, then extrapolated to the energies of interest. However, any extrapolation is fraught with uncertainty. For example, narrow resonances at low energy or even resonances below the reaction threshold may influence the cross section at the Gamow peak. Those effects cannot always be accounted for by an extrapolation.

Another effect can be studied at low energies: electron screening (6–9). In the laboratory, the electron cloud surrounding the target nucleus partially screens the nucleus's positive electric charge. This screening reduces the height of the Coulomb barrier, thus increasing the tunneling probability and eventually the reaction cross section. The screening effect has to be taken into account in order to derive the cross section for bare nuclei, which is the input data to nucleosynthesis network codes. These codes, in turn, must take into account the screening by electrons in the stellar plasma (6). Despite the impressive-sounding temperatures of many millions of degrees Kelvin in stellar interiors, the study of stars requires nuclear physics experiments performed at very low energy, measuring exceedingly small cross sections.

3. UNDERGROUND NUCLEAR ASTROPHYSICS

For stable stellar hydrogen burning, relevant temperatures range from 20 to 100×10^6 K, which correspond to Gamow peak energies of 10 to 50 keV, depending on the precise reaction of interest. The challenge of taking measurements at the Gamow peak involves suppressing the laboratory- and beam-induced background as well as enhancing the signal by boosting the beam intensity, target density, and detection efficiency.

3.1. Laboratory Background

The laboratory γ-ray background has two main sources: natural and cosmic ray–induced radioactivity. The decay of radioisotopes from natural decay chains produces a high density of γ-ray lines in the region below 2.6 MeV, where the highest-energy γ rays due to natural radioactivity are found. **Figure 1** illustrates the different steps in background reduction that are necessary for nuclear astrophysics experiments.

There is a sizable difference in background between, for instance, a naked detector and a commercially available graded shield (10) that includes 10 cm lead. This difference is due to the suppression of the soft component of cosmic rays and γ rays from nearby radioisotopes. It is possible to further reduce background by using a shallow underground laboratory at 100 mwe (meters water equivalent) and applying a more sophisticated shield (11). At that point, the remaining background would be dominated by muon-induced events.

Such muon-induced events stem from (*a*) the decay of the radioactive nuclei due to muon spallation and the capture of stopping negative muons, (*b*) the inelastic scattering and capture of the muon-induced neutrons, and (*c*) the energy loss of muons passing through the detector. Although the prompt muon effects can be reduced by surrounding the detector with a veto counter, the muon-induced radioactivity may have a relatively long lifetime (1–10 s), so rejection of such events by vetoing is not practical. Instead, the remaining background can be overcome by performing the experiments in a deep underground laboratory (12, 13), where the muon flux is greatly reduced [e.g., by six orders of magnitude (14) at the 3,800-mwe-deep Gran Sasso National Laboratory].

At higher γ-ray energies, specifically $E_\gamma > 2.6$ MeV, these considerations become more transparent. In this energy region, no significant improvement in background can be reached by applying an additional lead shield. Use of a shallow underground laboratory helps somewhat, and a muon veto reduces the background. Deep underground, the muon flux becomes negligible, and the veto does not further reduce the background counting rate. A further reduction below the very low γ-ray background observed at LUNA (15) can in principle be achieved by shielding the setup against neutrons. The neutron flux at Gran Sasso, which is three orders of magnitude lower than at the Earth's surface originates primarily from (α,n) reactions in the surrounding rock (16).

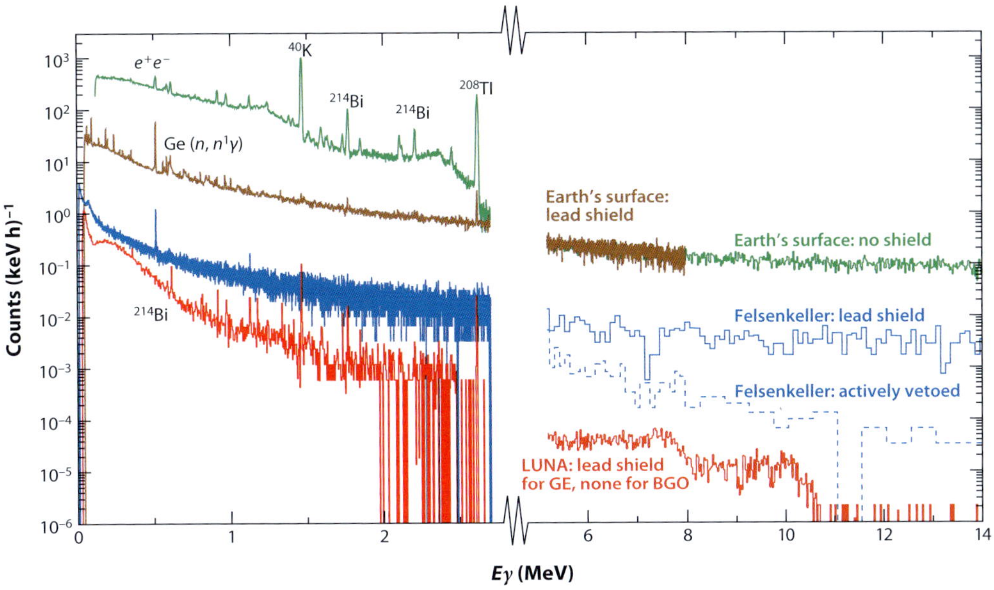

Figure 1

Laboratory γ-ray background measured with (*a*) a 100%–relative efficiency germanium detector at $E_\gamma < 2.7$ MeV and (*b*) a bismuth germanate (BGO) detector (scaled for equal volume with the germanium) for $5.2\,\mathrm{MeV} < E_\gamma < 14\,\mathrm{MeV}$. Green, the Earth's surface with no shield; brown, the Earth's surface with lead shield; blue, 110-mwe (meters water equivalent) underground Felsenkeller laboratory (Dresden) with lead shield (11); blue dashed line, actively vetoed spectrum in the 110-mwe underground Felsenkeller lab; red, 3,800-mwe LUNA lab with lead shield for the germanium (12) and no lead shield for the BGO (15).

Note that nuclear astrophysics is only one of the fields explored deep underground. The salient features of underground physics have been reviewed (17). Major underground activities are the detection of neutrinos generated by hydrogen burning in the Sun (18, 19), radioactivity in the Earth (20), and cosmic rays in the Earth's atmosphere (21) and the search for rare processes such as neutrinoless double-β decay (22), proton decay (23), and dark matter interactions (24). Recently, a compendium of major facilities was published (25).

3.2. Beam-Induced Background

In any experiment that uses an ion beam, the beam interacts with nuclei other than the target to be studied. Such nuclei may be found in the beam-transport system (especially in beam-limiting apertures, but also in drift tubes, magnets, and residual gas), or they may be part of the target itself. Such interactions can give rise to γ-ray background.

Such ion beam–induced background is independent of the underground depth and must be dealt with by (*a*) reducing the inventory of materials hit by the ion beam and (*b*) taking appropriate precautions to eliminate worrisome components. For low-energy nuclear astrophysics experiments, these tasks are greatly facilitated by two factors. First, contaminants with an atomic number greater than the target to be studied generally have a lower interaction probability than the target itself because of their higher Coulomb barrier. Many experiments study light nuclei of $Z \leq 8$, so common materials such as aluminum or steel generally do not give background. Second, the otherwise common activation of beam-line components is not an issue with low-energy protons or α beams. In cases where the background still interferes, it must be carefully identified, localized (e.g., 15), and whenever possible eliminated.

3.3. LUNA at Gran Sasso

There are several techniques to measure cross sections at the Gamow peak. So far, all of them have been pursued in only one laboratory: LUNA.

For nuclear reactions in which charged particles are emitted, it is usually possible to make an in-beam measurement at the surface of the Earth, except when coincident background due to cosmic rays is a problem. If so, only an underground measurement is feasible, as has been demonstrated for the $^3He(^3He,2p)\,^4He$ reaction (26, 27).

For nuclear reactions in which only γ rays are emitted, there are in principle two approaches. In-beam γ-ray spectrometry deep underground has been used to study $^2H(p,\gamma)^3He$ (28), $^{14}N(p,\gamma)^{15}O$ (29–33), $^3He(^4He,\gamma)^7Be$ (34, 35), and $^{15}N(p,\gamma)^{16}O$ (36) reactions. The in-beam approach requires experimenters to account for several uncertainties, such as the angular distribution of the emitted γ rays. This weakness can be overcome in selected cases where the created nucleus is radioactive, thus allowing an independent cross-check. An activation study with deep underground activity counting has been performed in the study of $^3He(^4He,\gamma)^7Be$ (37, 38). All these approaches benefit from the reduced background deep underground.

4. EXPERIMENTAL APPARATUS AT LUNA

The measurement of the cross section and the determination of the astrophysical *S*-factor for thermonuclear reactions require an experimental apparatus composed of an accelerator, a target, and a detection system.

4.1. Accelerators

Two different accelerators have been used at LUNA: a compact 50-kV "homemade" machine (39) and a commercial 400-kV one (40). Common features of the two accelerators are their high beam current, long-term stability, and precise beam-energy determination. The first feature maximizes the reaction rate, the second provides the long time typically needed for a cross-section measurement, and the third is important because of the exponential energy dependency of the cross section.

The 50-kV machine was designed and built at Ruhr Universität in Bochum, Germany, and then moved to Gran Sasso. It consists of a duoplasmatron ion source, an ion-beam extraction and acceleration system, and a double-focusing 90° analyzing magnet. Its compact shape optimizes the beam transmission. This machine can deliver beams of protons, $^3He^+$, and $^4He^+$ of 300 to 500 μA at energies between 10 and 50 keV, with an energy spread of less than 20 eV and a long-term stability (39). This capability allowed investigators to study (*a*) two fundamental reactions of the proton-proton (*pp*) chain, $^3He(^3He,2p)^4He$ and $^2H(p,\gamma)^3He$, at solar Gamow peak energies and (*b*) the screening effect in the $^2H(^3He,p)^4He$ reaction.

Even though it produced outstanding results (27, 28, 41), this accelerator should be considered as a pilot project toward a higher-energy facility, namely the 400-kV accelerator (**Figure 2**) (40). This electrostatic accelerator is embedded in a tank filled with an insulating mixture of N_2/CO_2 gas at 20 bar. The high voltage is generated by an inline Cockcroft-Walton power supply located inside the tank. The radio frequency ion source directly mounted on the accelerator tube can provide beams of 1 mA hydrogen and 500 μA He^+ over a continuous operating time of 40 days. The ions

Figure 2

The LUNA setup. The two beam lines are in the foreground, and the accelerator is in the background. The beam line on the left is dedicated to the measurements with solid targets, whereas the one on the right hosts the windowless gas target. The setup for the study of ^{3}He(^{4}He,γ)^{7}Be is shown during installation; the shield is only partially mounted.

can be sent into one of two different, parallel beam lines, thereby allowing the installation of two different target setups. In the energy range between 150 and 400 kV, the accelerator can provide up to 500 μA of protons and 250 μA of αs at the target stations, yielding 0.3-keV accuracy on the beam energy, 100-eV energy spread, and 5–eV h^{-1} long-term stability. Finally, the accelerator is controlled by programmable logic controller–based computers, which allow for safe operation over long periods without an operator present.

4.2. Targets and Ancillary Measurements

LUNA measurements have been performed with both solid and gas targets. Compared with gas targets, solid targets may contain a larger number of atoms per square centimeter: Typical areal densities range from 2×10^{17} to 2×10^{18}. They allow for the measurement of the cross-section angular dependency because the beam-hit position on the target can be precisely determined. Gas targets are more stable to beam bombardment and may reach extreme purity, which is essential to minimize the beam-induced background. The areal densities used in LUNA are between 5×10^{16} and 1×10^{18}. For solid targets, LUNA uses well-known production techniques such as implantation, evaporation, and sputtering (30). For gas targets a specific setup, consisting of a windowless gas target (i.e., a differentially pumped target) and a recirculation system (38, 42), was designed and installed at Gran Sasso and is depicted in **Figure 3**.

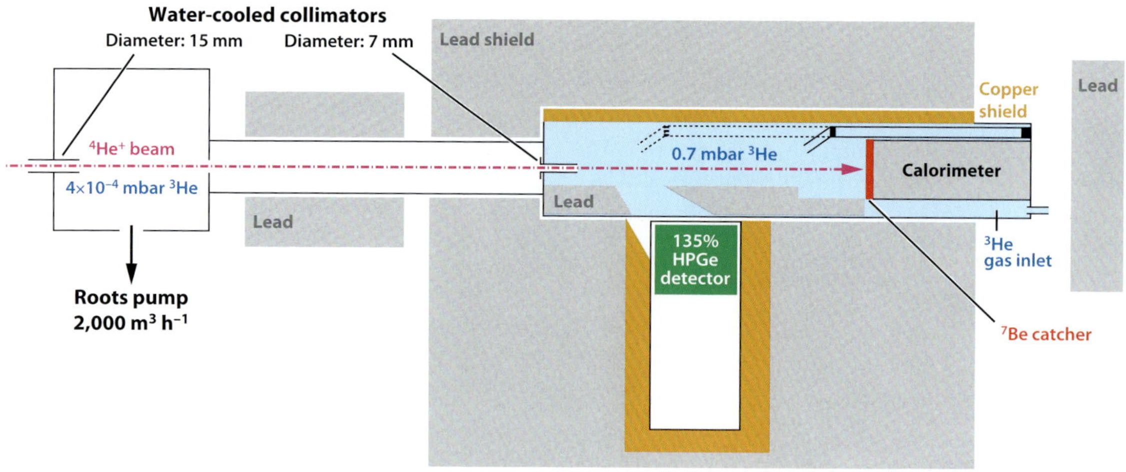

Figure 3

The setup for the study of $^3He(^4He,\gamma)^7Be$. The device to detect Rutherford scattering, the calorimeter, and the germanium detector are indicated.

A typical disadvantage of a gas target is the so-called beam-heating effect. This effect is due to the power deposition, which heats and thins the gas along the beam path. Different techniques are necessary to obtain the real number of atoms per square centimeter in the beam region. A technique based on Rutherford scattering has been employed to limit systematic uncertainties due to gas density to less than 2% (43). Alternatively, the target thickness along the beam can be obtained through the energy shift of a known resonance (32, 44).

Solid targets, in turn, may rapidly deteriorate under intense beams. Such deterioration may change the density profile, which must be known in determining the reaction rate. To monitor the target thickness and its stability under the beam, one may use the resonance scan technique. This approach consists of selecting a resonance of the reaction to be studied or of a parasitic reaction and measuring the yield profile as a function of the beam energy. The selected resonance should have an energy width smaller than or equal to the target thickness. The target thickness as well as the maximum yield can be continually monitored during the entire duration of the measurement.

For the cross-section measurement, it is essential to know the beam current on target. Whereas for a solid target the current is measured through a conventional Faraday cup, for a gas target this is not possible due to the ion charge–state changes (45). Therefore, LUNA has developed a beam calorimeter (42) with a constant temperature gradient between the hot side and the cold side. The power to the hot side is provided by the beam and by resistors embedded in the end cap of the calorimeter. The number of projectiles is then obtained by the difference in heating powers dissipated by the resistors with and without the beam, divided by the kinetic energy of the projectiles themselves at the calorimeter surface. If properly calibrated, the calorimeter can give a systematic uncertainty of the order of 1% on the beam current (42). Calorimeter calibration is an example of the measurements done in laboratories at the Earth's surface; it is an important step for the completion and integration of the results obtained underground.

4.3. Detectors

Apart from the $^3He(^3He,2p)^4He$ reaction, in which protons were detected through the use of commercial silicon detectors as single-stage or ΔE-E telescopes, all other LUNA measurements require the detection of γ rays. Here the choice of the most suitable detector depends on the desired physical information. The 4π bismuth germanate (BGO) summing crystal used at LUNA (42), a 28-cm-long cylinder with a 20-cm diameter and a 6-cm bore hole, can reach an efficiency as high as 70% for a 7-MeV γ ray, thus allowing the measurement of extremely low reaction yields. However, the BGO crystal's energy resolution is very poor and does not allow measurements of cascades and branching ratios to different levels because most of the γ-ray transitions are summed in a single peak. With a germanium detector, the efficiency dramatically decreases to the level of a few parts per mill, but the energy resolution is much better, allowing complex γ-ray cascades to be disentangled. Moreover, angular distribution measurements can be made by placing the detector at different angles with respect to the ion beam.

Nevertheless, summing effects can disturb the data evaluation for close geometry configurations. This can be addressed through the use of a composite detector such as a clover (46), wherein four small germanium detectors are enclosed inside the same cryostat, thereby reducing summing-in corrections.

5. HYDROGEN BURNING IN THE SUN

As the Sun's mass contracted from an initially large gas cloud, half of the gravitational energy released was radiated into the space and half was converted into the kinetic energy of hydrogen and helium nuclei, thereby increasing the temperature of the system (2). If gravitational energy had been the only energy source during this contraction and heating phase, then the Sun would have "died" at only 30 million years of age, as estimated by Lord Kelvin in the nineteenth century (47). Instead, at the central temperature of approximately 10 million degrees the kinetic energy of the hydrogen nuclei was high enough to penetrate the Coulomb barrier with significant probability and to switch on the hydrogen burning: $4^1\mathrm{H} \Rightarrow {}^4\mathrm{He} + 2e^+ + 2\nu_e$. This reaction releases useful energy and produces neutrinos detectable on Earth. Specifically, as the mass of the helium nucleus is lower than four times the proton mass, approximately 0.7% of the hydrogen rest mass was converted into energy in each of the transmutations.

Hydrogen fusion supplies the necessary energy to halt the contraction, and it provides all the energy required for the long life of the star. The Sun is a middle-aged main-sequence star whose shining arises from the burning of hydrogen fuel. The Sun began to burn hydrogen approximately 4.5 billion years ago. In another 5 billion years it will begin burning helium and will eventually turn into a celestial body that consists primarily of carbon and oxygen. In the central region of the Sun, which has a temperature of 15 million degrees and a density of $\sim$150 g cm^{-3}, hydrogen burning does not take place in one step but rather proceeds through a sequence of two-body reactions: the *pp* chain and the CNO cycle. The relative importance of the different reactions is determined by (*a*) the abundance of the nuclear species fusing together and (*b*) their fusion cross section at the Gamow peak energy.

5.1. Proton-Proton Chain

The first step and so-called bottleneck in the chain is the production of deuterium (48), which takes place through two different weak processes (**Figure 4**): (*a*) ${}^1\mathrm{H}(p,e^+\nu){}^2\mathrm{H}$, which gives rise to the so-called *pp* neutrinos, and (*b*) ${}^1\mathrm{H}(pe^-,\nu){}^2\mathrm{H}$, a source of *pep* neutrinos. The latter, a three-body process, is strongly suppressed (by a factor of approximately 400) compared with the former. Once produced, the deuterium quickly burns with hydrogen to synthesize ^{3}He. At this point, several possible scenarios emerge. The ${}^3\mathrm{He}(p,e^+\nu){}^4\mathrm{He}$ reaction has a negligible rate because it is a weak process that is further suppressed, compared with ${}^1\mathrm{H}(p,e^+\nu){}^2\mathrm{H}$, due to the atomic number

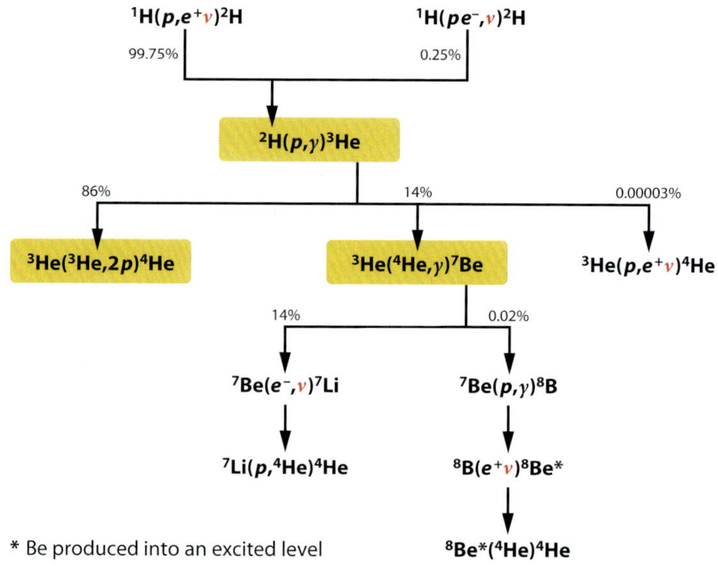

Figure 4

The proton-proton chain. The reactions studied by LUNA are highlighted in yellow.

of helium. The most probable fate of ^3He is to fuse with another ^3He nucleus to produce ^4He in the strong reaction $^3He(^3He,2p)^4He$. Approximately 14% of the time, the fusion takes place with the much more abundant ^4He through the electromagnetic process $^3He(^4He,\gamma)^7Be$. At this level, the chain branches again due to the competition between the electron-capture decay of ^7Be, $^7Be(e^-,\nu)^7Li$, and the fusion of ^7Be with hydrogen, $^7Be(p,\gamma)^8B$. The former is a weak process and is approximately 1,000-fold more probable than the latter scenario because it has no Coulomb barrier suppression. Once produced, ^7Li quickly fuses with hydrogen to produce ^8Be. which is extremely unstable and soon splits into two helium nuclei. $^7Be(p,\gamma)^8B$ seldom occurs, but this process is of crucial importance because it leads to the emission of high-energy neutrinos in the ^8B decay to ^8Be.

5.2. Carbon-Nitrogen-Oxygen Cycle

In the CNO cycle, the conversion of hydrogen into helium is achieved with the aid of carbon previously synthesized in older stars (3). Carbon works as a catalyst: It is not destroyed by the cycle, and it strongly affects the rate of the CNO cycle with its abundance. Because $^{15}N(p,\gamma)^{16}O$ has a cross section that is a factor of $\sim$2,000 lower than $^{15}N(p,\alpha)^{12}C$, the second CNO cycle is strongly suppressed compared with the first one. In the Sun, the CNO cycle accounts for only a small fraction of nuclear-energy production (less than 1%), and it is controlled by $^{14}N(p,\gamma)^{15}O$, the bottleneck reaction. Only at temperatures somewhat hotter than found at the Sun's center, 20 million degrees, would the production of energy through the CNO cycle equal that of the *pp* chain. Above this temperature, the CNO cycle dominates energy production, as the effects of the Coulomb barriers no longer strongly suppress the energy-production rate.

5.3. Neutrinos

Neutrinos are particles that interact weakly with matter. They travel at approximately the speed of light, and they reach the Earth 8 min after they are produced in the center of the Sun. Because we know that hydrogen fusion, $4^1H \Rightarrow {}^4He + 2e^+ + 2\nu_e$, produces 26.7 MeV of energy, and if we assume that the present luminosity of the Sun corresponds to the present nuclear energy–production rate (49) (it takes more than 10^4 years for electromagnetic energy to reach the surface of the Sun), then calculating neutrino flux is straightforward. We simply divide the solar luminosity, 3.85×10^{26} W, by the energy required to produce one neutrino, 13.35 MeV $\equiv 2.14 \times 10^{-12}$ J, to obtain a rate of 1.80×10^{38} neutrinos s^{-1}. This corresponds to a flux of approximately 60 billion neutrinos $cm^{-2}\ s^{-1}$ on the Earth. To calculate the energy spectrum of the solar neutrinos, we must know the cross section of the different reactions of the *pp* chain and the CNO cycle: They give rise to neutrinos of different energies.

In particular, *pp* neutrinos have a flux of 6.04×10^{10} $cm^{-2}\ s^{-1}$ (50), corresponding to 92% of the total neutrino flux. Their continuous spectrum has an end-point energy of 0.42 MeV, which makes the detection of these neutrinos extremely difficult. ^7Be neutrinos are produced with two different energies: 0.86 MeV (89.7% branching ratio) and 0.38 MeV. The 0.86-MeV ^7Be neutrinos are the second largest component of the spectrum, amounting to 7% of the total with a flux of 4.55×10^9 $cm^{-2}\ s^{-1}$. ^8B neutrinos have a much lower flux, 4.72×10^6 $cm^{-2}\ s^{-1}$; fewer than 1 neutrino in 10,000 comes from the ^8B decay. However, their relatively high end-point energy (~15 MeV) makes their detection much less difficult; unsurprisingly, ^8B neutrinos have been the best-studied neutrinos from the Sun. The CNO cycle produces neutrinos with end-point energies of 1.20 MeV (^{13}N) and 1.73 MeV (^{15}O). The latest results of the Standard Solar Model (50) predict that CNO neutrinos make a 0.5% contribution to the total flux.

In 1964 both Bahcall and Davis (51) proposed the detection of solar neutrinos in order to study the inner region of the Sun and thereby directly verify the hypothesis of nuclear-energy generation in stars. Nearly 40 years of experimental and theoretical work have shown that the source of the energy radiated by the Sun is hydrogen fusion in the solar interior. Also, the study of solar neutrinos has revealed something important about the nature of the neutrino itself: It oscillates. An electron neutrino produced inside the Sun may become a muon or tau neutrino by the time it reaches the Earth.

6. HYDROGEN BURNING STUDIED AT LUNA

LUNA started as a pilot project in 1991. In the following sections, we discuss LUNA's contributions over the past 20 years to our current understanding of the Sun and neutrinos.

6.1. $^2H(p,\gamma)^3He$: The Energy Source of the Protostar

Inside the Sun, the $^2H(p,\gamma)^3He$ reaction controls the equilibrium abundance of deuterium. The $^2H(p,\gamma)^3He$ reaction controls the life of protostars before they enter the main-sequence phase. Protostar models predict that a star forms by accretion of interstellar material onto a small, contracting core. Until the temperature is below 10^6 K, the main source of energy is the gravitational contraction. When the temperature approaches 10^6 K, the first nuclear "fire" is ignited inside the star: The primordial deuterium is converted into ^3He via $^2H(p,\gamma)^3He$, providing 5.5 MeV for each reaction. The total amount of nuclear energy generated by this deuterium burning is comparable to the entire gravitational-binding energy of the star. The onset of deuterium burning slows down the contraction, increases the lifetime of the star, and freezes its observational properties until the original deuterium is fully consumed. Reliable knowledge of the rate of $^2H(p,\gamma)^3He$ down to a few kiloelectronvolts (the Gamow peak in a protostar) is a fundamental prerequisite for the protostellar models.

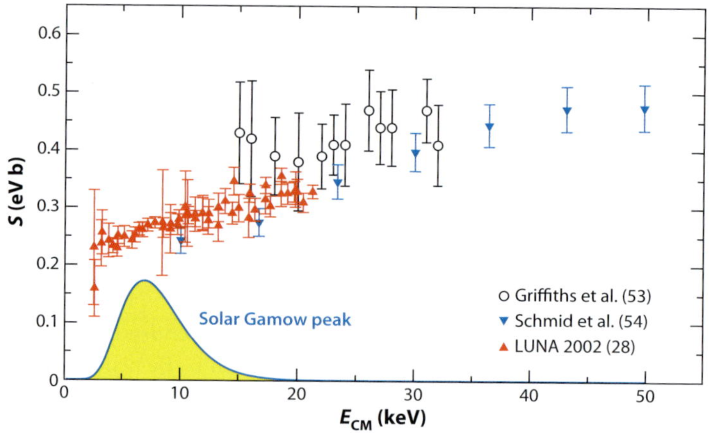

Figure 5

The $^2H(p,\gamma)^3He$ astrophysical factor $S(E)$ with the total error.

$^2H(p,\gamma)^3He$ is also a cornerstone in the big bang nucleosynthesis (BBN). Because of the deuterium bottleneck (52), that is, the photodisintegration of deuterium, the formation of ^3He is delayed until the temperature drops to approximately 8×10^8 K. Again, knowledge of the cross section at low energies is required.

The $^2H(p,\gamma)^3He$ cross-section measurement was performed at LUNA with the 50-kV accelerator connected to a differentially pumped gas-target system designed to fit the large BGO γ-ray detector (42). The BGO, placed around the deuterium target, detected the 5.5-MeV γ ray with 70% efficiency. The LUNA results (28), together with two previous measurements (53, 54) of the astrophysical factor $S(E)$ at low energy, are given in **Figure 5**. Their agreement with the theoretical calculations is excellent (55).

6.2. $^3He(^3He,2p)^4He$: In Search of the Resonance

LUNA's initial activity was focused on the $^3He(^3He,2p)^4He$ cross-section measurement within the solar Gamow peak (15–27 keV). This is a key reaction of the *pp* chain. A resonance at the thermal energy of the Sun was suggested long ago (56, 57) to explain the observed ^8B solar neutrino flux. Such a resonance would decrease the relative contribution of the alternative reaction $^3He(^4He,\gamma)^7Be$, which generates the branch responsible for ^7Be and ^8B neutrino production in the Sun.

LUNA's final setup comprises eight 1-mm-thick silicon detectors with an area of 5×5 cm^2 placed around the beam inside the windowless target chamber, which is filled with ^3He at a pressure of 0.5 mbar. The simultaneous detection of two protons unambiguously identified a $^3He(^3He,2p)^4He$ fusion reaction (Q-value: 12.86 MeV). **Figure 6** shows the results from LUNA (26, 27), together with higher-energy measurements (58–60) of the astrophysical factor $S(E)$.

For the first time, a nuclear reaction has been measured in the laboratory at the energy occurring in a star (27). Its cross section varies by more than two orders of magnitude in the measured energy range. At the lowest energy of 16.5 keV, it has the value of 0.02 pb, which corresponds to a rate of approximately two events per month—rather low even for the so-called silent experiments of underground physics. No narrow resonance has been found within the solar Gamow peak; as a consequence, the astrophysical solution of the ^8B and ^7Be solar neutrino problem based on its existence has been ruled out.

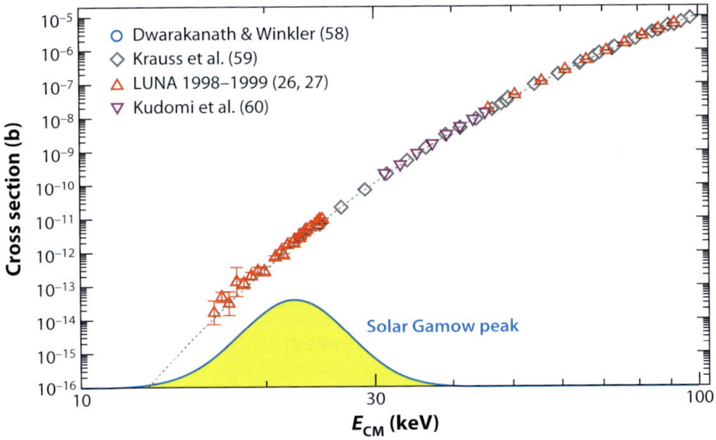

Figure 6

Cross section of the $^3He(^3He,2p)^4He$ reaction. Data from LUNA (26, 27) and from other groups (58–60). The dashed gray line is the extrapolation based on the measured $S(E)$-factor (27).

6.3. $^3He(^4He,\gamma)^7Be$: Solar Neutrino Oscillations

$^3He(^4He,\gamma)^7Be$ (Q-value: 1.586 MeV) is the key reaction for the production of ^7Be and ^8B neutrinos in the Sun because these neutrinos' flux depends almost linearly on its cross section. Unless a recoil separator is used (61), the cross section can be determined either from the detection of the prompt γ rays (62–68) or from the counting of the produced ^7Be nuclei (61, 66, 69–71). The latter approach requires detection of the 478-keV γ due to the excited ^7Li populated in the decay of ^7Be (half-life: 53.22 days). Both methods have been used to determine the cross section in the energy range $E_{CM} \geq 107$ keV, but the $S_{3,4}$ extracted from the measurements of the induced ^7Be activity was 13% higher than that obtained from the detection of the prompt γ rays (72).

An underground experiment was performed with the $^4He^+$ beam from the 400-kV accelerator in conjunction with a windowless gas target filled with ^3He at 0.7 mbar. The beam entered the target chamber and was stopped on the calorimeter (**Figure 3**). The ^7Be nuclei produced by the reaction inside the ^3He gas target were implanted into the calorimeter cap, which was removed after the irradiation and placed in front of a germanium detector for the measurement of the ^7Be activity.

In the first phase of this experiment, the $^3He(^4He,\gamma)^7Be$ cross section was obtained from the activation data alone (37, 38), with a total uncertainty of $\sim$4%. In the second phase, a new high-accuracy measurement that simultaneously used prompt and activation methods was performed down to the center-of-mass energy of 93 keV. The prompt-capture γ ray was detected by a 135% germanium, placed in close geometry with the target, and heavily shielded. The spectrum, taken at a beam energy of 250 keV, is shown in **Figure 7**. The astrophysical factors obtained with the two methods (34) are the same within the quoted experimental error (**Figure 8**). Similar conclusions have been reached in a new simultaneous activation and prompt experiment (73), which covered the E_{CM} energy range from 330 to 1,230 keV.

The energy dependency of the cross section seems to be theoretically well determined at low energy. If the normalization is the only free parameter, we can rescale the fit from Reference 74 to our data, yielding $S_{3,4}(0) = 0.560 \pm 0.017$ keV b. Thanks to our small error, the total uncertainty on the ^8B solar neutrino flux decreases from 12% to 10%, whereas the one on the ^7Be flux decreases from 9.4% to 5.5% (34). The ^7Be flux is now theoretically predicted with an error as small as the

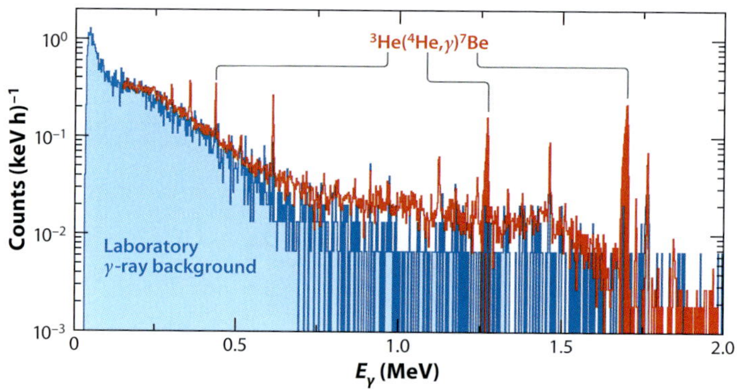

Figure 7

^{3}He(^{4}He,γ)^{7}Be spectrum at a beam energy of 250 keV (*red*) and the laboratory background (*blue*).

experimental one, which should soon be achieved by Borexino (75). Such small errors will allow us to precisely study the typical signature of neutrino oscillations in matter, namely the energy dependency of the oscillation probability.

The energy window covered by LUNA is above the Gamow peak of the Sun but well within the Gamow peak of the BBN. Our precise results clearly rule out the ^{3}He(^{4}He,γ)^{7}Be cross section as a possible source of the discrepancy between the predicted primordial ^{7}Li abundance (76) and the much lower observed value (77, 78).

6.4. ^{14}N(p,γ)^{15}O: The Composition of the Sun and the Age of the Universe

^{14}N(p,γ)^{15}O (Q-value: 7.297 MeV) is the slowest reaction of the CNO cycle; it controls the cycle's energy-production rate. Moreover, this reaction provides the ^{13}N and ^{15}O solar neutrino flux, which depends almost linearly on it, and it determines the age of globular clusters, which consist of 10^4 to 10^6 gravitationally bound stars that are the oldest stars in their galaxies. The luminosity of the turnoff point in the Hertzsprung-Russell diagram of a globular cluster, namely the point at which the main sequence turns toward cooler and brighter stars, is used to determine

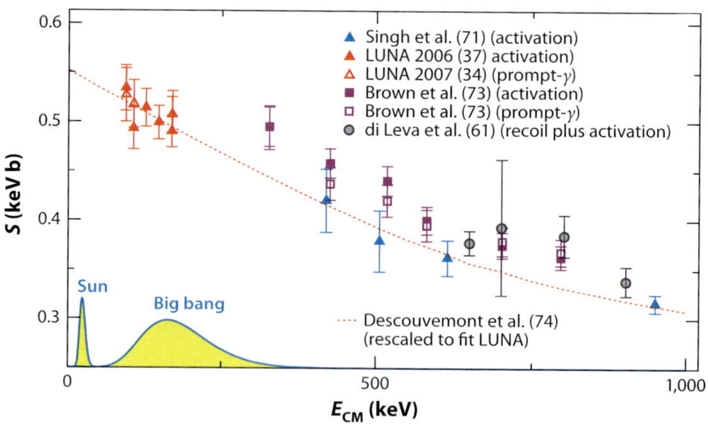

Figure 8

Astrophysical $S(E)$-factor for ^{3}He(^{4}He,γ)^{7}Be. The results from the modern, high-precision experiments are shown with their total error.

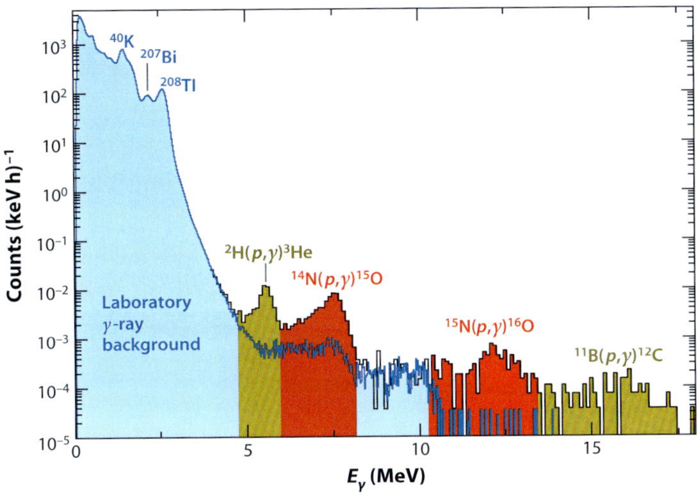

Figure 9

Bismuth germanate spectrum taken with a 100-keV proton beam on nitrogen of natural isotopic abundance. Red, peaks from $^{14}N(p,\gamma)^{15}O$ and $^{15}N(p,\gamma)^{16}O$; dark yellow, beam-induced background; blue, laboratory background. In spite of the small isotopic abundance of ^{15}N (0.366% only), the peak arising from $^{15}N(p,\gamma)^{16}O$ can be easily seen thanks to the much-reduced background.

the age of the cluster and to derive a lower limit on the age of the universe (79). A star at the turnoff point burns hydrogen in the shell through the CNO cycle; this is why the $^{14}N(p,\gamma)^{15}O$ cross section plays an important role in age determination.

In the first phase of the LUNA study, data were obtained down to 119 keV through the use of solid targets of TiN and a 126% germanium detector. The five different radiative-capture transitions that contribute to the $^{14}N(p,\gamma)^{15}O$ cross section at low energy were measured. The total cross section was then studied down to very low energy in the second phase of the experiment via the 4π BGO summing detector placed around a windowless gas target filled with nitrogen at a pressure of 1 mbar. The BGO spectrum at a beam energy of 100 keV is shown in **Figure 9**. At the lowest center-of-mass energy, 70 keV, a cross section of 0.24 pb was measured, yielding an event rate of 11 counts per day from the reaction.

The results obtained first with the germanium detector (29, 30) and then with the BGO setup (31) were approximately twofold lower than the existing extrapolation (72, 80, 81) from previous data (82, 83) at very low energy (**Figure 10**) and are in agreement with results from indirect methods (84–89). As a consequence, the CNO neutrino yield in the Sun is decreased approximately twofold, and the age of the globular clusters is increased by 0.7 to 1 billion years (89a) up to 14 billion years. The lower cross section also affects stars that are much more evolved than the Sun; in particular, the dredging up of carbon to the surface of asymptotic giant branch stars (90) is more efficient (91).

The main conclusion from the LUNA data has been confirmed by an independent study at higher energy (92). However, there is a 15% difference between the total S-factors extrapolated by the two experiments at the Gamow peak of the Sun. This difference arises from the extrapolation of the capture to the ground state in ^{15}O, a transition strongly affected by interference effects between several resonances and the direct-capture mechanism.

To provide precise data for the ground-state capture, a third phase of the $^{14}N(p,\gamma)^{15}O$ study was performed with a composite germanium detector in the beam-energy region immediately

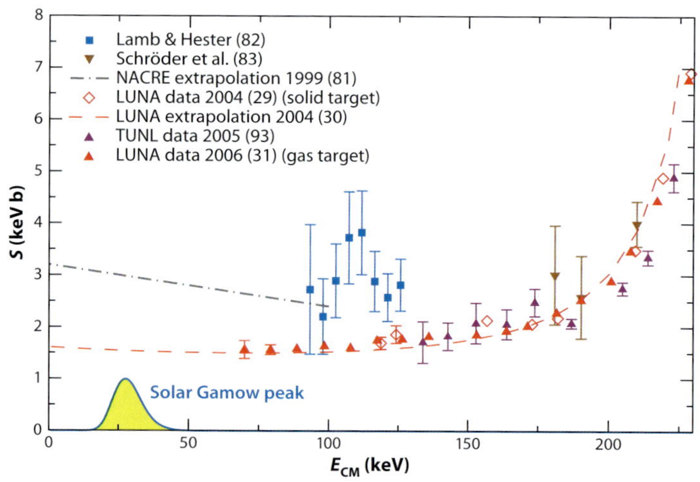

Figure 10

Astrophysical $S(E)$-factor of the $^{14}N(p,\gamma)^{15}O$ reaction. The errors are statistical only (the systematic ones are similar).

above the 259-keV resonance, where precise data effectively constrain a fit for the ground-state transition in the R-matrix framework (93). The total error on the S-factor was reduced to 8%: $S_{1,14}(0) = 1.57 \pm 0.13$ keV b (33). This is significant because, having finally solved the solar neutrino problem, we are now facing the problem of solar composition: There is a conflict between helioseismology and the new metal abundances that emerged from improved modeling of the photosphere (50, 94). Thanks to the relatively small error, it will soon be possible to measure the metallicity of the core of the Sun (i.e., the elements other than hydrogen and helium) by comparing the detected CNO neutrino flux with the predicted flux. The CNO neutrino flux in the low-metallicity scenario is ~35% lower than in the high-metallicity scenario. Use of these scenarios will allow us to test whether the early Sun was chemically homogeneous (95), which is a key assumption of the Standard Solar Model.

6.5. Ongoing Measurements

The solar phase of LUNA has almost reached its end. A new and rich program of nuclear astrophysics, mainly devoted to the MgAl and NeNa cycles, began at the 400-kV facility about two years ago with the measurement of $^{25}Mg(p,\gamma)^{26}Al$ (96). The MgAl and NeNa cycles are important for second-generation stars with central temperatures and masses that are higher than those of our Sun (3, 5). Due to their higher Coulomb barriers, these cycles are relatively unimportant for energy generation, although they are essential for the nucleosynthesis of elements with mass number higher than 20. Low-energy resonances (or the low-energy part of the direct capture) that are inaccessible in a laboratory at the Earth's surface could be measurable underground. Some reactions have already been measured aboveground, but an underground reinvestigation could substantially improve our knowledge of the related reaction rate in the different astrophysical scenarios responsible, for example, for the texture of the isotopes that are filling the universe.

LUNA is now measuring $^{15}N(p,\gamma)^{16}O$ with enriched ^{15}N targets. $^{15}N(p,\gamma)^{16}O$ is the leak reaction from the first to the second CNO cycle. The results already obtained with nitrogen of natural isotopic composition (0.366% ^{15}N) (36) extend to energies that are lower than ever measured before and provide a cross section at nova energies that is approximately twofold lower than previously believed.

The measurement in preparation, $^{2}H(^{4}He,\gamma)^{6}Li$, is not connected to the hydrogen-burning cycles but rather is a key reaction of the BBN. This reaction determines the amount of primordial ^{6}Li in the universe. Recently, the ^{6}Li isotope was detected in a number of metal-poor stars (97, 98), and its quantity is two to three orders of magnitude higher than what was expected from the BBN (76).

7. OUTLOOK

Measurements of cross sections directly at the energy of astrophysical interest have been successfully used to elucidate the reactions governing stellar hydrogen burning. To keep pace with the rapid progress of observational astronomy and astrophysical modeling, investigators should also explore the nuclear reactions that (*a*) govern helium and carbon burning and (*b*) produce the neutrons that give rise to the so-called astrophysical s-process. Owing to the different nature of these reactions, new techniques and new equipment are necessary for this task.

The $^{12}C(\alpha,\gamma)^{16}O$ reaction is often referred to as the Holy Grail of nuclear astrophysics (99–104). It plays a fundamental role in the evolution of stars during the helium-burning phase and determines the abundance ratio between carbon and oxygen, the two most important elements in the development of life. This abundance ratio influences the nucleosynthesis of elements up to the iron peak for massive stars, the cooling timescale of white dwarfs, and the properties of thermonuclear as well as core-collapse supernovae.

$^{12}C+^{12}C$ fusion reactions catalyze carbon burning. Their rate determines the evolution of massive stars, up to either a modest end as a white dwarf or a fiery death as a core-collapse supernova (113, 114). The fusion reactions also affect the ignition conditions and timescales of thermonuclear supernovae. The $^{13}C(\alpha,n)^{16}O$ and $^{22}Ne(\alpha,n)^{25}Mg$ reactions provide the neutrons for the buildup of the s-process isotopes (100, 111, 115–117). Most elements that are heavier than iron are produced through this process, which involves a series of subsequent neutron captures and β decays.

To study these and other exciting reactions, as listed in **Table 1**, a new, few-megavolt accelerator should be installed in a deep underground laboratory (118). It should also be able to accelerate not only protons and α particles, but also ions up to carbon or even oxygen so that carbon burning can be studied. As for the beam current, for all these ions an intensity of $\sim$1 mA would be required for the necessary sensitivity.

Table 1 Nuclear reactions of astrophysical interest recommended for study at future underground accelerator facilities

Type	Reaction	Q-value (MeV)	E_{Gamow} (keV)	Reference(s)
(α,γ)	$^{2}H(\alpha,\gamma)^{6}Li$	1.5	100–500	12
	$^{12}C(\alpha,\gamma)^{16}O$	7.2	300	104–106
	$^{14}N(\alpha,\gamma)^{18}F$	4.4	200–700	107
	$^{15}N(\alpha,\gamma)^{19}F$	4.0	500	108
^{12}C-induced	$^{12}C(^{12}C,\alpha)^{20}Ne$	4.6	1,500	12, 109
	$^{12}C(^{12}C,p)^{23}Na$	2.2	1,500	12, 109
(α,n)	$^{13}C(\alpha,n)^{16}O$	2.2	200	110, 111
	$^{22}Ne(\alpha,n)^{25}Mg$	–0.5	500	112

Such a project is presently being planned at several different sites in Europe and North America. It would be desirable to install a new underground accelerator at the site of the present LUNA machine, Gran Sasso National Laboratory, because of the facility's excellent infrastructure and proven low background. However, due to the limited underground laboratory space at Gran Sasso, other sites must be considered.

One such site is the Canfranc Laboratory in Spain (119), which is less well shielded (2,400 mwe) than Gran Sasso and therefore has a higher remaining muon flux. However, at the LUNA site the background is dominated by neutrons, not muons (15), and the neutron flux at the two sites is comparable. Access to this site is by horizontal road tunnel, which would facilitate the installation of complex, maintenance-intensive equipment such as an accelerator.

In the United States, plans to construct an underground facility at the site of the previous Homestake experiment include two accelerators with connected beam lines (120). One machine should have similar functions as LUNA's 400-kV accelerator, but with greatly increased beam intensity, whereas the second accelerator should be in the megavolt range and should study most of the reactions listed in **Table 1**.

Two projects in Boulby (United Kingdom, 2,800 mwe) and Praid (Romania, 900 mwe) (121, 122) are also in the planning stages. In these facilities, the laboratory should be placed in a salt matrix deep underground, where the levels of uranium and thorium and therefore neutron flux are generally much lower.

Regardless of the outcome of the ongoing siting discussion, both astrophysics and nuclear physics will greatly benefit from the new precision that will be enabled by a future, higher-voltage accelerator underground. A better understanding of nuclear burning in stars, obtained via direct new data, will allow us to model stellar scenarios that are now understood only in general terms. Also, improved experimental data for nuclear fusion reactions near or below the Coulomb barrier will present a fresh challenge for theoretical modeling of these reactions.

8. CONCLUSIONS

Twenty years ago, LUNA started underground nuclear astrophysics experiments in the core of Gran Sasso mountain, below 1,400 m of dolomite rock. The extremely low background has allowed physicists to perform experiments with count rates as low as a few events per year. For the first time, the important reactions that are responsible for hydrogen burning in the Sun have been studied down to the relevant stellar energies. As a consequence, 50 years after the first pioneering cross-section measurements were made, nuclear physics is no longer an important error source for the solar model, and solar neutrinos can now be exploited to probe the deep interior of the Sun. When applied to astrophysical scenarios different from the Sun, LUNA results increase the limit on the age of the universe up to 14 billion years.

Thanks to its underground environment, LUNA has spurred great improvement in our comprehension of hydrogen burning. In the next two decades, underground nuclear astrophysics will try to reach similar results in the study of the helium and carbon burning and of the neutron sources in the stars.

DISCLOSURE STATEMENT

The authors are not aware of any affiliations, memberships, funding, or financial holdings that might be perceived as affecting the objectivity of this review.

ACKNOWLEDGMENTS

It is a pleasure to thank our colleagues at LUNA, without whom the results reviewed in this paper could have not been obtained.

LITERATURE CITED

1. Eddington A. *Nature* 16:14 (1920)
2. Clayton DD. *Principles of Stellar Evolution and Nucleosynthesis*. Chicago: Univ. Chicago Press (1984)
3. Rolfs C, Rodney W. *Cauldrons in the Cosmos*. Chicago: Univ. Chicago Press (1988)
4. Clayton DD. *Handbook of Isotopes in the Cosmos: Hydrogen to Gallium*. Cambridge, UK: Cambridge Univ. Press (2003)
5. Iliadis C. *Nuclear Physics of Stars*. Weinheim, Ger.: Wiley (2007)
6. Salpeter EE. *Aust. J. Phys.* 7:373 (1954)
7. Assenbaum H, Langanke K, Rolfs C. *Z. Phys. A* 327:461 (1987)
8. Raiola F, et al. *Eur. Phys. J. A* 19:283 (2004)
9. Huke A, et al. *Phys. Rev. C* 78:015803 (2008)
10. Gilmore G. *Practical Gamma-Ray Spectrometry*. New York: Wiley. 2nd ed. (2008)
11. Köhler M, et al. *Appl. Radiat. Isot.* 67:736 (2009)
12. Caciolli A, et al. *Eur. Phys. J. A* 39:179 (2009)
13. Laubenstein M, et al. *Appl. Radiat. Isot.* 61:167 (2004)
14. Ahlen SP, et al. *Phys. Lett.* B249:149 (1990)
15. Bemmerer D, et al. *Eur. Phys. J. A* 24:313 (2005)
16. Wulandari H, Jochum J, Rau W, von Feilitzsch F. *Astropart. Phys.* 22:313 (2004)
17. Formaggio JA, Martoff C. *Annu. Rev. Nucl. Part. Sci.* 54:361 (2004)
18. Bethe H. *Phys. Rev.* 55:103 (1939)
19. von Weizsäcker CF. *Phys. Z.* 39:633 (1938)
20. Araki T, et al. *Nature* 436:499 (2005)
21. Gaisser TK, Honda M. *Annu. Rev. Nucl. Part. Sci.* 52:153 (2002)
22. Elliott SR, Vogel P. *Annu. Rev. Nucl. Part. Sci.* 52:115 (2002)
23. Perkins DH. *Annu. Rev. Nucl. Part. Sci.* 34:1 (1984)
24. Gaitskell RJ. *Annu. Rev. Nucl. Part. Sci.* 54:315 (2004)
25. Bettini A. *J. Phys. Conf. Ser.* 120:082001 (2008)
26. Junker M, et al. *Phys. Rev. C* 57:2700 (1998)
27. Bonetti R, et al. *Phys. Rev. Lett.* 82:5205 (1999)
28. Casella C, et al. *Nucl. Phys. A* 706:203 (2002)
29. Formicola A, et al. *Phys. Lett.* B591:61 (2004)
30. Imbriani G, et al. *Eur. Phys. J. A* 25:455 (2005)
31. Lemut A, et al. *Phys. Lett.* B634:483 (2006)
32. Bemmerer D, et al. *Nucl. Phys. A* 779:297 (2006)
33. Marta M, et al. *Phys. Rev. C* 78:022802(R) (2008)
34. Confortola F, et al. *Phys. Rev. C* 75:065803 (2007)
35. Costantini H, et al. *Nucl. Phys. A* 814:144 (2008)
36. Bemmerer D, et al. *J. Phys. G* 36:0452002 (2009)
37. Bemmerer D, et al. *Phys. Rev. Lett.* 97:122502 (2006)
38. Gyürky G, et al. *Phys. Rev. C* 75:035805 (2007)
39. Greife U, et al. *Nucl. Instrum. Methods A* 350:327 (1994)
40. Formicola A, et al. *Nucl. Instrum. Methods A* 507:609 (2003)
41. Costantini H, et al. *Phys. Lett.* B482:43 (2000)
42. Casella C, et al. *Nucl. Instrum. Methods A* 489:160 (2002)
43. Marta M, et al. *Nucl. Instrum. Methods A* 569:727 (2006)
44. Görres J, et al. *Nucl. Instrum. Methods* 177:295 (1980)
45. Allison SK. *Rev. Mod. Phys.* 30:1138 (1958)

46. Elekes Z, et al. *Nucl. Instrum. Methods A* 503:580 (2003)
47. Bahcall JN. arXiv:astro-ph/0009259 (2000)
48. Bahcall JN. *Neutrino Astrophysics*. Cambridge, UK/New York: Cambridge Univ. Press (1989)
49. Castellani V, et al. *Phys. Rep.* 281:309 (1997)
50. Peña-Garay C, Serenelli A. arXiv:0811.2424 [astro-ph] (2008)
51. Bahcall JN. *Phys. Rev. Lett.* 12:300 (1964); Davis R. *Phys. Rev. Lett.* 12:303 (1964)
52. Weinberg S. *Gravitation and Cosmology*. New York: Wiley (1972)
53. Griffiths GM, Lal M, Scarfe CD. *Can. J. Phys.* 41:724 (1963)
54. Schmid GJ, et al. *Phys. Rev. Lett.* 76:3088 (1996)
55. Marcucci L, Nollett K, Schiavilla R, Wiringa R. *Nucl. Phys. A* 777:111 (2006)
56. Fowler WA. *Nature* 238:24 (1972)
57. Fetisov VN, Kopysov YS. *Phys. Lett.* B 40:602 (1972)
58. Dwarakanath MR, Winkler H. *Phys. Rev. C* 4:1532 (1971)
59. Krauss A, Becker HW, Trautvetter HP, Rolfs C. *Nucl. Phys. A* 467:273 (1987)
60. Kudomi N, et al. *Phys. Rev. C* 69:015802 (2004)
61. di Leva A, et al. *Phys. Rev. Lett.* 102:232502 (2009)
62. Holmgren HD, Johnston RL. *Phys. Rev.* 113:1556 (1959)
63. Parker P, Kavanagh R. *Phys. Rev.* 131:2578 (1963)
64. Nagatani K, Dwarakanath M, Ashery D. *Nucl. Phys. A* 128:325 (1969)
65. Kräwinkel H, et al. *Z. Phys. A* 304:307 (1982)
66. Osborne J, et al. *Phys. Rev. Lett.* 48:1664 (1982)
67. Alexander T, Ball G, Lennard W, Geissel H. *Nucl. Phys. A* 427:526 (1984)
68. Hilgemeier M, et al. *Z. Phys. A* 329:243 (1988)
69. Robertson R, et al. *Phys. Rev. C* 27:11 (1983)
70. Volk H, Kräwinkel H, Santo R, Wallek L. *Z. Phys. A* 310:91 (1983)
71. Singh BN, Hass M, Nir-El Y, Haquin G. *Phys. Rev. Lett.* 93:262503 (2004)
72. Adelberger E, et al. *Rev. Mod. Phys.* 70:1265 (1998)
73. Brown TAD, et al. *Phys. Rev. C* 76:055801 (2007)
74. Descouvemont P, et al. *At. Data Nucl. Data Tables* 88:203 (2004)
75. Arpesella C, et al. *Phys. Rev. Lett.* 101:091302 (2008)
76. Coc A, et al. *Astrophys. J.* 600:544 (2004)
77. Ryan S, et al. *Astrophys. J.* 530:L57 (2000)
78. Bonifacio P, et al. *Astron. Astrophys.* 390:91 (2002)
79. Krauss L, Chaboyer B. *Science* 299:65 (2003)
80. Caughlan G, Fowler W. *At. Data Nucl. Data Tables* 40:283 (1988)
81. Angulo C, et al. *Nucl. Phys. A* 656:3 (1999)
82. Lamb W, Hester R. *Phys. Rev.* 108:1304 (1957)
83. Schröder U, et al. *Nucl. Phys. A* 467:240 (1987)
84. Bertone PF, et al. *Phys. Rev. Lett.* 87:152501 (2001)
85. Bertone PF, et al. *Phys. Rev. C* 66:055804 (2002)
86. Mukhamedzhanov A, et al. *Phys. Rev. C* 67:065804 (2003)
87. Nelson SO, et al. *Phys. Rev. C* 68:065804 (2003)
88. Yamada K, et al. *Phys. Lett.* B579:265 (2004)
89. Schürmann D, et al. *Phys. Rev. C* 77:055803 (2008)
89a. Imbrani G, et al. *Astran. Astrophys.* 420:625 (2004)
90. Herwig F. *Annu. Rev. Astron. Astrophys.* 43:435 (2005)
91. Herwig F, Austin SM. *Astrophys. J.* 613:L73 (2004)
92. Runkle RC, et al. *Phys. Rev. Lett.* 94:082503 (2005)
93. Lane AM, Thomas RG. *Rev. Mod. Phys.* 30:257 (1958)
94. Haxton WC, Serenelli AM. *Astrophys. J.* 687:678 (2008)
95. Haxton W. *J. Phys. Conf. Ser.* 173:012014 (2009)
96. Costantini H, et al. *Rep. Prog. Phys.* 72:086301 (2009)
97. Smith V, Lambert D, Nissen P. *Astrophys. J.* 408:262 (1993)

98. Asplund M, et al. *Astrophys. J.* 644:229 (2006)
99. Weaver TA, Woosley SE. *Phys. Rep.* 227:65 (1993)
100. Wallerstein G, et al. *Rev. Mod. Phys.* 69:995 (1997)
101. Buchmann L, Barnes C. *Nucl. Phys. A* 777:254 (2006)
102. Woosley SE, Heger A. *Phys. Rep.* 442:269 (2007)
103. Tur C, Heger A, Austin SM. *Astrophys. J.* 671:821 (2007)
104. Tang XD, et al. *Phys. Rev. Lett.* 99:052502 (2007)
105. Schürmann D, et al. *Eur. Phys. J. A* 26:301 (2005)
106. Matei C, et al. *Phys. Rev. Lett.* 97:242503 (2006)
107. Görres J, et al. *Phys. Rev. C* 62:055801 (2000)
108. Wilmes S, et al. *Phys. Rev. C* 66:065802 (2002)
109. Spillane T, et al. *Phys. Rev. Lett.* 98:122501 (2007)
110. Harissopulos S, et al. *Phys. Rev. C* 72:062801 (2005)
111. Heil M, et al. *Phys. Rev. C* 78:025803 (2008)
112. Jaeger M, et al. *Phys. Rev. Lett.* 87:202501 (2001)
113. Woosley SE, Heger A, Weaver TA. *Rev. Mod. Phys.* 74:1015 (2002)
114. Hillebrandt W, Niemeyer JC. *Annu. Rev. Astron. Astrophys.* 38:191 (2000)
115. Cameron AGW. *Publ. Astron. Soc. Pac.* 69:201 (1957)
116. Burbidge EM, Burbidge GR, Fowler WA, Hoyle F. *Rev. Mod. Phys.* 29:547 (1957)
117. Straniero O, Gallino R, Cristallo S. *Nucl. Phys. A* 777:311 (2006)
118. Nucl. Phys. Eur. Collab. Comm. *Roadmap for construction of nuclear physics research infrastructures in Europe.* **http://www.nupecc.org/pub/NuPECC_Roadmap.pdf** (2005)
119. José J, et al., eds. *Proc. Workshop Nucl. Astrophys. Oppor. Undergr. Lab., Canfranc, Spain.* **http://www.fnuc.es/workshop/LSC_Program.htm** (2009)
120. US Dep. Energy/Natl. Sci. Found. Nucl. Sci. Advis. Comm. arXiV:0809.3137 [nucl-ex] (2008)
121. Strieder F. *J. Phys. G* 35:014009 (2008)
122. Bordeanu C, et al. *J. Phys. G* 35:014011 (2008)

The Final Merger of Black-Hole Binaries

Joan Centrella,[1] John G. Baker,[1] Bernard J. Kelly,[1,2,3] and James R. van Meter[1,2,3]

[1]Gravitational Astrophysics Laboratory and [2]Center for Research and Exploration in Space Science and Technology, NASA Goddard Space Flight Center, Greenbelt, Maryland 20771; email: Joan.M.Centrella@nasa.gov, John.G.Baker@nasa.gov, Bernard.J.Kelly@nasa.gov, James.R.vanMeter@nasa.gov

[3]Department of Physics, University of Maryland, Baltimore County, Baltimore, Maryland 21250

Annu. Rev. Nucl. Part. Sci. 2010. 60:75–100

First published online as a Review in Advance on May 20, 2010

The *Annual Review of Nuclear and Particle Science* is online at nucl.annualreviews.org

This article's doi:
10.1146/annurev.nucl.010909.083246

0163-8998/10/1123-0075$20.00

Key Words

black holes, gravitational waves, numerical relativity

Abstract

Recent breakthroughs in the field of numerical relativity have led to dramatic progress in understanding the predictions of general relativity for the dynamical interactions of two black holes in the regime of very strong gravitational fields. Such black-hole binaries are important astrophysical systems and are a key target of current and developing gravitational-wave detectors. The waveform signature of strong gravitational radiation emitted as the black holes fall together and merge provides a clear, observable record of the process. After decades of slow progress, these mergers and the gravitational-wave signals they generate can now be routinely calculated using the methods of numerical relativity. We review recent advances in understanding the predicted physics of events and the consequent radiation and discuss some of the impacts of this new knowledge in various areas of astrophysics.

Contents

1. INTRODUCTION

1.1. Black Holes

It has been nearly a century since Albert Einstein's profound physical insight revealed our current description of gravitational physics, general relativity. Among the theory's extraordinary consequences was the predicted existence of black holes: nonlinearly self-gravitating, stable objects in which gravitational forces have completely overcome all other physical interactions. Although once considered a mathematical curiosity, general relativity's description of black holes now provides the best explanation for a wide class of astrophysical objects whose gravitational potential wells power many of the most energetic astronomical phenomena observed. These phenomena range from stellar black holes powering X-ray sources in the neighboring regions of our galaxy to supermassive monsters with masses of $\sim(10^6 - 10^9)M_\odot$, where $M_\odot$ is the mass of the Sun, and have far-reaching astrophysical consequences.

1.2. Gravitational-Wave Observations

In the coming decade, anticipated observations of gravitational waves from black-hole binaries will open a new window onto the universe. Interpreting such observations will richly engage aspects of our theoretical understanding of strong-field gravity that have never before been confronted with empirical observations.

The final coalescence of binaries consisting of two comparable-mass black holes, with mass ratios $q = M_1/M_2 \sim (1 - 10)$, is expected to be one of the strongest astrophysical sources of energy in the form of gravitational radiation. As the emission of gravitational waves removes energy and momentum from the binary, the orbits shrink and the black holes eventually merge into a single black hole, producing an intense burst of radiation. Given the advent of ground-based gravitational-wave detectors such as LIGO and VIRGO [which will detect mergers of black holes with masses in the range of $\sim(10 - 100)\, M_\odot$], and given that planning for the space-based LISA [which will observe mergers of massive black holes, with masses of $\sim(10^4 - 10^6)\, M_\odot$] is under way, knowledge of black-hole binary gravitational waveforms is urgent.

Inspiral: early binary evolution, when the holes are separated widely enough to be treated as point particles. In this stage, the orbital period is much shorter than the timescale on which the orbital parameters change, and the holes spiral together on quasi-circular orbits

Merger: strong-field dynamical regime in which the black holes plunge together and merge into a single, highly distorted black hole, surrounded by a common horizon

Ringdown: post-merger stage, where the remnant black hole sheds distortions in nonaxisymmetric modes through gravitational-wave emission and settles down into a quiescent rotating black hole

Post-Newtonian (PN) approximation: expansion of the full equations of general relativity in powers of $\varepsilon \sim v^2/c^2 \sim GM/(Rc^2)$, where v is the characteristic velocity of the source, M is its mass, and R is its characteristic size

1.3. Black-Hole Binary Coalescence

Black-hole binary coalescence can be considered in three stages. During the inspiral, the holes are separated widely enough to be treated as point particles. In this stage, the orbital period is much shorter than the timescale on which the orbital parameters change, and the holes spiral together on quasi-circular orbits. When the holes get so close together that they can no longer be approximated as point particles, they enter the merger phase. In this strong-field, dynamical regime, the black holes plunge together and merge into a single, highly distorted black hole, surrounded by a common horizon. Finally, there is the ringdown stage, where the remnant black hole sheds these distortions through gravitational-wave emission and settles down into a quiescent rotating black hole.

1.4. Gravitational Waves from the Coalescence

Knowledge of the gravitational waveforms from these three stages of black-hole coalescence is important for gravitational-wave detection and data analysis, as well as for astrophysical applications. The inspiral can be calculated analytically by using the post-Newtonian (PN) approximation. The inspiral gravitational waveform is a "chirp"—a sinusoid increasing in both frequency and amplitude. The ringdown can also be calculated analytically via techniques of black-hole perturbation theory, and the resulting gravitational waveforms are damped sinusoids. However, the merger stage can only be understood using full numerical simulations of the Einstein equations in three dimensions, and the resulting gravitational waveforms were unknown—until recently.

The final merger of comparable-mass black holes is a powerful source of gravitational radiation. The gravitational waveforms emitted by a black-hole binary with mass ratio q scale with the total mass M of the binary. If we set $c = 1$ and $G = 1$, we can express both length and timescale in terms of the total mass: $1M \sim 5 \times 10^{-6}(M/M_\odot)\,\mathrm{s} \sim 1.5(M/M_\odot)\,\mathrm{km}$. The strong-field merger produces a burst of gravitational radiation that lasts $\sim 100\, M$ and has a luminosity $\sim 10^{23} L_\odot$, which is greater than the combined luminosities of all the stars in the visible universe. For stellar black-hole binaries ($M \sim 10\, M_\odot$), this luminosity lasts for ~ 5 ms, and for massive black-hole binaries ($M \sim 10^6\, M_\odot$), it lasts for ~ 10 min.

1.5. Calculating the Merger

Understanding the final merger of a black-hole binary requires solving the full Einstein equations using the methods of numerical relativity. This endeavor has many challenging aspects, ranging from providing astrophysically relevant initial data to understanding the structure of the Einstein equations, and the solution eluded researchers for many years. Recently, a series of dramatic breakthroughs has ignited the field, making robust, stable, and accurate simulations of binary mergers possible for the first time. These models are improving our understanding of strong-field dynamics, thereby influencing gravitational-wave searches and other areas of astrophysics. In this article we provide an overview of these exciting developments, highlighting the key physical results.

2. STEPS TOWARD A ROBUST BLACK-HOLE BINARY MODEL

Computing the strong-field merger of two comparable-mass black holes has a long history, dating back more than 40 years. Overall, progress was generally slow and incremental, requiring the interplay among new ideas in black-hole modeling, mathematical investigations into the structure of the Einstein equations, the development of effective gauge conditions, advances in computational techniques, and the availability of high-performance computing resources. Here we provide a brief review of key developments in this arena.

2.1. Overview of Numerical Relativity Issues

Solving Einstein's equations on a computer typically requires slicing four-dimensional space-time into a stack of three-dimensional space-like hypersurfaces, each labeled by time t (1–3). The Einstein equations then divide into two sets: constraint equations and evolution equations. The constraints are a set of time-independent elliptic equations that must hold on each slice. Specifically, the constraints are solved first to obtain valid initial data for a black-hole binary simulation. Second, the data are propagated forward in time using the evolution equations.

2.1.1. Gauge freedom. General relativity has four space-time coordinate degrees of freedom, which are associated with four freely specifiable coordinates or gauge functions that govern the future development of both time and the spatial coordinates. These gauge functions are generally taken to be the lapse function α, which gives the lapse of proper time $(\alpha \cdot \delta t)$ between neighboring slices, and the shift vector β^i, which determines how the spatial coordinates develop from one slice to the next. Appropriate choices for the lapse and shift are crucial for successful black-hole evolutions (4, 5).

2.1.2. Formulation of the evolution equations. For many years, a primary challenge of numerical relativity was simply to decide which equations to solve (3). There is no unique formulation of the Einstein equations; rather, there are many choices regarding, for example, which variables to use, whether to write the equations as first order or second order in time, and which coordinate conditions to impose. These choices are not arbitrary because some formulations turn out to be more numerically "friendly" than others. In other words, although they are analytically equivalent when constraints and auxiliary variable definitions are assumed exact, some formulations may be unstable in practice. For example, the evolution equations may admit rapidly growing solutions that violate the constraints. In this case, although the initial data may contain only tiny errors, the subsequent evolution may produce violations of Einstein's equations that "blow up." There may also be pathological numerical solutions that do not represent solutions to the analytic evolution equations at all but rather are supported by the discrete numerical grid structure and depend on the details of the finite-differencing operators.

2.1.3. Numerical methods. Once a formulation has been chosen, the Einstein equations are solved numerically for various field variables on a grid of discrete points that represents the spacetime domain of interest. There are two general approaches for dealing with the spatial derivatives that appear in these equations. Finite differencing interpolates the derivative at a given point from the surrounding points according to a Taylor expansion in the grid spacing. Spectral methods assume a solution in the form of a summation of orthogonal functions; once the coefficients are obtained numerically, the derivatives can be found analytically. In both cases, time integration can be handled in a number of ways, most commonly by a Runge-Kutta algorithm. To date, most numerical relativity solutions have been carried out using finite differencing, with most current work employing a Cartesian grid in three spatial dimensions, although results from evolutions via spectral methods are now becoming available.

2.1.4. Initial conditions for black holes. To model astrophysical binaries, initial data must be generated for two black holes that move on quasi-circular inspiraling orbits and have a mass ratio q and some configuration of spins. One usually conceptualizes the initial state of the system with a particle-like description of the masses, positions, momenta, and spins of each black hole. Because general relativity is a field theory, such a description can be applied only as a first step. The model requires a full description for the initial field configuration that satisfies general relativity's four initial-value constraints and that somehow corresponds to the system we have described in these particle-like terms.

2.1.5. Evolving the binary. This system is evolved for several orbits and then through plunge, merger, and ringdown, for a total duration on the order of several hundred or more M. To obtain the gravitational waveforms, the radiation must be extracted from the simulation at large enough distances from the source to be in the wave zone. Because the length scales of the black holes are $\sim M$ and the wavelengths of the emitted radiation are $\sim(10-100)\ M$, some sort of variable resolution such as adaptive mesh refinement must be used to handle the very large computational domains needed.

2.2. The Lazarus Approach: Hybrid Black-Hole Merger Waveforms

By the late 1990s, numerical relativity had developed to the point that brief simulations of black-hole binaries in three spatial dimensions plus time were possible. These techniques were sufficient for evolving promptly merging "grazing collisions" of black holes (6, 7). However, the simulations were not indefinitely stable; rather, they typically crashed after $\sim(10-30)\ M$, well before any significant portion of a binary orbit could be evolved.

2.2.1. Hybrid simulations. In this arena, the Lazarus method emerged as a novel approach to obtaining black-hole binary waveforms, combining short numerical relativity simulations with black-hole perturbation methods (8). Because numerical relativity codes could then evolve only for a brief period, and because perturbation theory could approximate the late-time dynamics of the distorted remnant black hole, the Lazarus Project sought to apply numerical relativity to evolve the strong-field approach to merger, thereby providing a hybrid model for a significant part of the problem. Beginning from quasi-circular initial configurations near the start of the final merger, the black holes were evolved via numerical relativity for $\sim 15\ M$ until just before the simulation became inaccurate. Then, via a complicated interface, data from late in the numerical simulations were interpreted as initial data describing the perturbed final black hole and the emerging radiation.

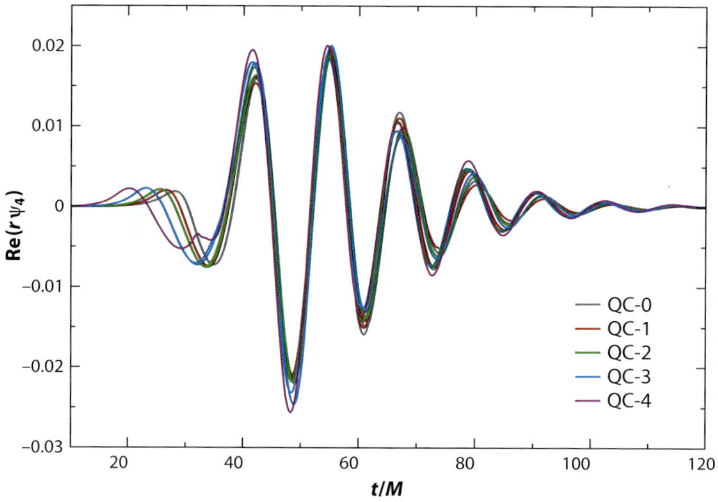

Figure 1

Waveforms from Lazarus models of equal-mass, nonspinning black-hole mergers. Shown is the $\ell = 2, m = 2$ component of $\mathrm{Re}(r\Psi_4)$, which corresponds to one of the two polarization states of the emitted gravitational radiation. Curves are plotted for ten simulations from five different initial black-hole separations (designated QC-0 through QC-4), each with two alternative times of transition to perturbative evolution. Reprinted from Reference 9. Copyright 2002, the American Physical Society (**http://link.aps.org/abstract/PRD/v65/e124012**).

Finally, black-hole perturbation theory techniques were applied to evolve this data and derive the full waveforms.

2.2.2. First results. The Lazarus simulations gave the first indication of what might be expected for the terminal burst of radiation from coalescing black-hole binaries. **Figure 1** shows that the dominant $\ell = 2, m = 2$ spherical harmonic component (of spin weight -2) has a brief burst of radiation smoothly joining the damped sinusoidal signal of the ringdown. The waveforms were remarkable for their simplicity, as they had predominantly circular polarization (in the $\ell = 2$, $m = 2$ mode) and a steady evolution of polarization frequency and amplitude. Because these characteristics were robust under variations in the model, it was conjectured that the waveforms from a binary starting from a wide separation late in the inspiral would be well characterized by these simple features. Subsequently, the Lazarus method was applied to the study of mergers of more generic black-hole binaries (10, 11).

2.3. Toward Evolving a Black-Hole Binary Orbit

Concurrently with the Lazarus investigations, further progress was being made toward full numerical relativity simulations of black-hole binary mergers, including new decompositions of the Einstein equations and new ways of controlling the evolution of the numerical coordinates.

2.3.1. The BSSN system. One major milestone was the development of a conformal formulation of the Einstein equations known as the Baumgarte-Shapiro-Shibata-Nakamura (BSSN) system (12), which overcame some instability problems associated with an earlier formulation (1, 13). In this approach, the set of evolution equations is written with first-order time derivatives and second-order spatial derivatives, and it is strongly hyperbolic (14). Stable time evolution was accomplished using a coordinate condition that evolves the lapse function dynamically, causing the slices to avoid crashing into the black-hole singularities (7, 15). However, the simulations were still limited to durations $\lesssim(30-40)M$ of stable evolution by (*a*) a failure of the spatial coordinate system known as grid stretching, in which the coordinates tend to fall into the black holes, and (*b*) instabilities related to how the black holes were handled numerically.

2.3.2. Controlling the spatial coordinates. Eliminating grid stretching required developing appropriate techniques for dynamically controlling the spatial coordinates, which are governed by the evolution of the shift vector. The first long-lasting evolutions of distorted black holes relied on a class of hyperbolic shift evolution schemes, known as Γ-freezing conditions, which were inspired by the BSSN formulation. With this approach, a single distorted black hole could be evolved indefinitely (e.g., for thousands of *M*) with reasonable accuracy (16). These studies allowed the full numerical determination of gravitational waveforms from black-hole ringdowns, as the distorted black hole settled down to a physically and numerically stable final state; the approach also provided a foundation for future advances in black-hole binary simulations (5).

2.3.3. Problems with moving black holes. However, another aspect of spatial gauge conditions remained a critical issue for long-lasting black-hole binary simulations. As general relativity allows arbitrary coordinate systems, many groups adopted coordinate conditions that did not allow black holes to move through the computational domain. Use of these conditions simplified the problem of dealing with the black-hole singularities, which were handled either by excising the black hole interiors (within the horizons) from the computational domain (17) or by representing the black holes as punctures (18). Although progress had been made toward implementing moving excision regions (19), it was computationally much simpler to demand that the excision region remain fixed on the grid (20). Similarly, the puncture treatment was formulated so that the black hole was fixed on the grid and the singularity was factored out and handled analytically in a time-independent manner. However, for binary configurations in which the black holes physically move, the cost of keeping the black holes fixed in coordinate space was paid by twisting and stretching of the dynamical fields, which eventually led to large computational errors.

For inspiraling binary configurations, one potential solution was to attempt to untwist the geometric-field data by imposing comoving coordinates via a shift vector that could be dynamically adjusted during the evolution of the binary to minimize motion of the black-hole horizons (21). With this approach, Brügmann et al. (22) achieved a significant milestone: the first simulation of a full orbital cycle of a black-hole binary system prior to merger. In this work, the authors applied a dynamically adjusted corotating frame that tracked the measured position of the black holes' apparent horizons. This process allowed a simulation that remained accurate for somewhat longer than the $\gtrsim 100$-M duration of a complete orbit. However, problems occurred with the large characteristic speeds that resulted near the outer boundaries, limiting the domains of these simulations and preventing gravitational waveforms from being measured effectively. The coordinate control also required fine-tuning that eventually failed, causing the simulation to crash as the black holes finally approached merger.

2.4. Robust Black-Hole Binary Simulations

The year 2005 was a watershed in the numerical simulation of black-hole binary coalescence. In that year, the final stumbling blocks in binary evolutions were overcome with a set of simulations that approached almost every aspect of the evolution problem—mathematical formulation, initial data, grid structure, and singularity handling—in different, but equally robust, ways.

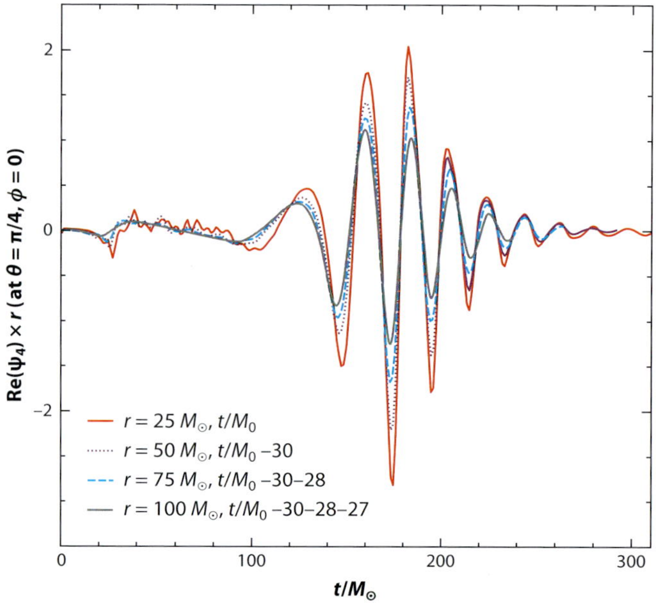

Figure 2

The first gravitational waveform for binary of equal-mass black holes evolving through a single plunge orbit, merger, and ringdown, computed by Pretorius (23). The waveforms were extracted at four radii from the source, then shifted in time to overlap for comparison. Reprinted with permission from Reference 23. Copyright 2005, the American Physical Society (**http://link.aps.org/abstract/PRL/v95/e121101**).

2.4.1. First orbit and merger simulation. In early 2005, Pretorius (23) shocked the numerical relativity community by announcing the first complete, robust simulations of an equal-mass black-hole merger. After completing $\sim$1 orbit, the black holes plunged and merged to form a single distorted black hole that then rang down. Pretorius extracted the gravitational waves to obtain the first inspiraling merger waveform directly from numerical relativity (**Figure 2**).

Pretorius (24) employed several techniques that were very different from most other numerical approaches to the black-hole binary problem. Rather than using the BSSN formulation, he applied a generalized harmonic formalism (25), directly integrating the space-time metric with evolution equations of second order in both space and time. These equations were implemented numerically using adaptive mesh refinement to allow high resolution around the black holes while maintaining a large computational domain. Pretorius utilized spatial coordinates compactified to draw spatial infinity into the computational domain, with a choice of gauge evolution strongly tied to his evolution formalism. Following Reference 26, he added terms to the evolution equations to specifically damp away any violations of the constraints. In his simulations, the black holes were excised and moved freely across the computational grid. Pretorius's success with such novel methods quickly raised questions as to whether the struggling, more widely pursued BSSN-based puncture approach might be off course.

2.4.2. Moving punctures. Later in 2005, however, a new robust method based on the BSSN formulation was announced. The moving puncture method was discovered simultaneously and independently by groups at the University of Texas at Brownsville (UTB) (27) and NASA's Goddard Space Flight Center (GSFC) (28). In this approach, the black holes are represented as punctures but are not constrained to remain fixed on the coordinate grid. Rather, they are allowed to move

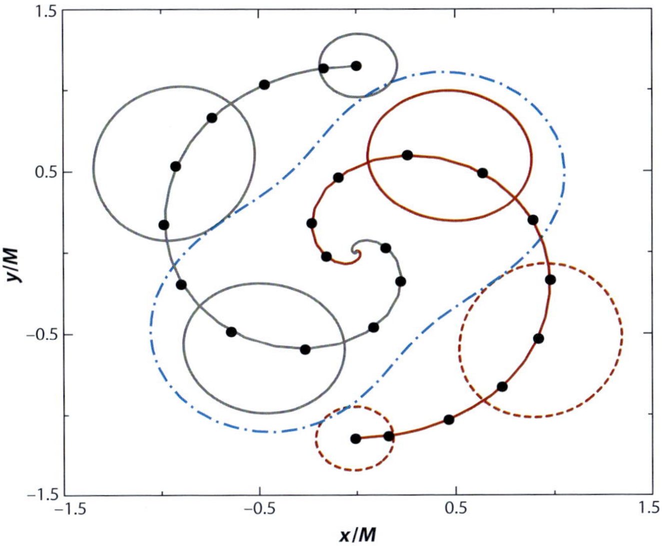

Figure 3

Puncture trajectories from the merger of an equal-mass, nonspinning black-hole binary calculated by the University of Texas at Brownsville group (27). The apparent horizons of the two premerger holes (*solid gray and dashed red lines*) expand due to coordinate effects. Also shown is the first detected common horizon (*blue dot-dashed line*); this horizon designates the point of merger and takes a peanut shape before it settles down. Reprinted with permission from Reference 27. Copyright 2006, the American Physical Society (**http://link.aps.org/abstract/PRL/v96/e111101**).

freely through the grid by means of novel coordinate conditions. **Figure 3** shows the trajectories of the puncture black holes as computed by the UTB group; the formation of a common horizon marks the point of merger. The moving puncture method eliminates the analytic representation of the puncture singularities in favor of an approximate numerical treatment within the black-hole horizons.

The UTB and GSFC groups had discovered and applied similar methods to the same problem: evolving an equal-mass nonspinning black-hole binary from near the final orbit, through merger and ringdown, and studying the gravitational waves. The first generation of merger waveforms from Pretorius, GSFC, and UTB showed the same simple burst of radiation ending in a damped-sinusoidal ringdown, and the waveforms were qualitatively consistent with the Lazarus project results discussed above and shown in **Figure 1**.

The discovery of the moving puncture method ignited the field of black-hole binary evolutions. Because the approach was based on commonly used methods, most researchers in the field quickly achieved accurate and stable evolutions using their existing codes, with the adoption of simple coordinate conditions (29). Suddenly, the game was on, and nearly all research groups were participating.

3. THE PHYSICS OF BLACK-HOLE BINARIES

With the advent of successful numerical evolutions of binaries that inspiral and merge, the numerical relativity community's focus changed to investigating the physics of binary mergers. This research has been enabled by continued technical improvements in areas ranging from initial data prescriptions to more accurate numerical methods.

3.1. Merger Dynamics and Waveforms

The astrophysical black-hole mergers that are the targets of current and future gravitational-wave detectors are expected to reach the merger stage after having radiated away any initial eccentricity (30) and proceeded through a long quasi-circular inspiral. All equal-mass, nonspinning binary merger simulations that begin from such orbits in the inspiral should produce the same gravitational waveform, subject only to rescaling with the total mass of the system. For many years, concerns about the accuracy and realism of black-hole binary initial data sets, including the effects of spurious gravitational radiation and eccentricity (31–33), had been raised. Now that numerical relativity can simulate the final merger, the next step has been to run models with enough orbits before the plunge and merger to obtain complete and reproducible waveforms starting from the late inspiral.

3.1.1. Equal-mass, nonspinning black holes. The GSFC group produced the first representation of the definitive waveform for the final stages of a merger of equal-mass, nonspinning black holes (34). The authors carried out a series of four simulations where the holes started from quasi-circular initial conditions at increasingly larger separations. In these runs, the holes completed $\sim$1.8, 2.5, 3.6, and 4.2 orbits before a common horizon was formed. To compare the results of these models, the authors chose the moment of peak gravitational-radiation amplitude as the fiducial time $t = 0$.

The orbital dynamics of the binaries can be examined by tracking the black-hole centers, which are given by the location of the punctures. The black-hole trajectories for each run were oriented so that they superposed at the fiducial $t = 0$. In the early stages of each run, the tracks clearly showed the effects of eccentricity in the initial conditions, where the amount of initial eccentricity decreased for more widely separated holes. As the holes spiraled together deeper into the strong-field regime, the tracks locked onto a universal trajectory independent of their initial conditions that continued for the last orbit, plunge, and merger.

Figure 4 reveals the corresponding universal gravitational waveform. Here, the dominant $\ell = 2, m = 2$ quadrupolar component for each run is shown, shifted in time so that the peak radiation amplitude occurs at $t = 0$. Starting from $t = -50\,M$, the waveforms show nearly perfect agreement, differing from one another at the level of 1%. The signals for the preceding few orbits agree at the level of $\sim$10%, except for a brief burst of spurious radiation at the beginning of each run.

Note that the merger waveform shows a remarkably simple shape, making a smooth transition from the inspiral chirp to the damped sinusoid of the ringdown. As the merger begins, both the wave amplitude and the frequency increase, albeit faster than in the inspiral. The amplitude reaches a peak and then decreases, dropping exponentially through the ringdown. The frequency increases monotonically to a maximum value that remains constant during the ringdown.

Of course, the black-hole binary merger dynamics and waveforms can be altered by the presence of large amounts of eccentricity (35) and spurious gravitational radiation (36) near the time of the plunge. However, the robustness of the merger to modest deviations from astrophysical initial conditions has allowed the study of many cases of interest using relatively short simulations, starting just a few orbits before the start of the plunge.

3.1.2. Unequal-mass, nonspinning black holes. Astrophysical black-hole binaries are unlikely to have exactly equal masses. Currently, numerical relativists can evolve systems with mass ratios of up to $q = 10$ (37). Starting from quasi-circular orbits, the simulations show that the merger phase for nonspinning, unequal-mass black-hole binaries is robust to modest deviations from these initial conditions and that it produces a generally simple waveform shape.

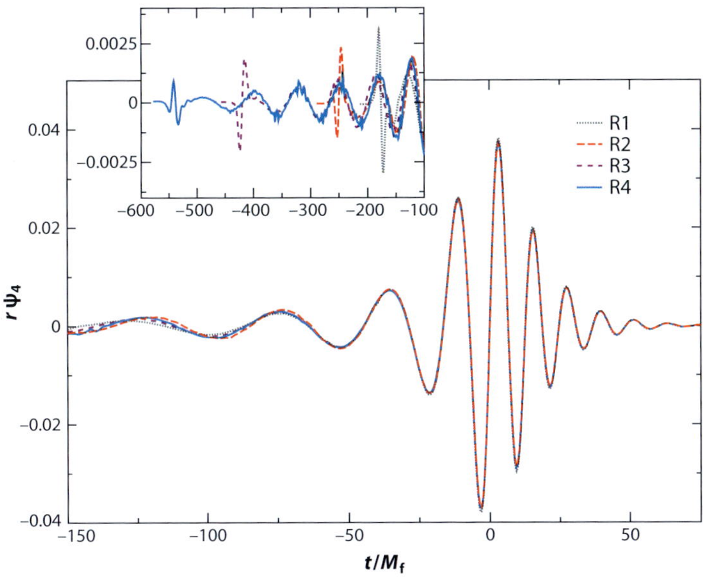

Figure 4

The universal waveform for equal-mass, nonspinning black holes calculated by the Goddard Space Flight Center group (34). Four simulations (labeled R1 through R4) were carried out, with increasingly larger initial separations between the black holes. The main panel shows how the waveforms converge to a universal shape by $t = 50\ M$, and the inset shows how each waveform begins. These waveforms are shifted in time so that the peak radiation amplitude occurs at $t = 0$. Reprinted from Reference 34. Copyright 2006, the American Physical Society (**http://link.aps.org/abstract/PRD/v73/e104002**).

An important tool for analyzing these mergers is a decomposition into spin-weighted spherical harmonic modes. Berti et al. (38) analyzed a set of unequal-mass, nonspinning mergers with mass ratios ranging from $q = 1$ to $q = 4$. Studying the multipolar distribution of the radiation, the authors found that the subdominant modes ($\ell > 2$) become more important in that they carry a larger fraction of the energy as q increases. Specifically, for $q > 2$, the $\ell = 3$ mode typically carries $\sim$10% of the radiated energy. Also, as expected from symmetry considerations, the odd-m modes are suppressed in the equal-mass limit.

Baker et al. (39) carried out a complementary study of the radiation from nonspinning mergers with mass ratios in the range $1 \leq q \leq 6$. The multipolar decomposition clearly shows that the hallmark simplicity of the waveform persists for $q > 1$ and extends to each of the spherical harmonic components $\ell \geq 2$; this property extends to the $q = 10$ case (37). In the full mode-summed waveform, this simple shape is also observed when viewed from the system's orbital axis, where the quadrupole mode dominates. A somewhat more complex appearance arises when the system is viewed off-axis, where higher multipoles contribute more strongly to the waveform at various angles.

Throughout the entire coalescence, each of the spherical harmonic waveform components is circularly polarized, with steadily varying phase and amplitude (39). For each mass ratio q, the different (ℓ, m) components share the same rotational phase and frequency. During the inspiral this is expected because the waveform phase is equal to the rotational phase multiplied by the mode number m. However, for the $\ell = m$ modes this relationship also holds throughout the merger and into the ringdown.

These properties suggest a simple conceptual interpretation in which the radiation is generated by an implicit rotating source. In this picture, each (ℓ, m) mode is generated separately by the (ℓ, m) moment of some implicit source. The fixed relationship for the $\ell = m$ modes implies that these components of the source rotate rigidly through the entire coalescence. The $\ell \neq m$ components are less rigid and peel away from the main $\ell = m$ trend during the merger (39).

3.1.3. Mergers of black holes with spin. The mass ratio q is a one-dimensional cut into the parameter space of black-hole binaries. The remaining parameter space is dominated by the spin angular momentum of each hole. Because spin is a vector quantity, this adds six more dimensions. Simulations of spinning black holes first focused on systems whose spins were expected, on the basis of PN arguments, to have the least effect on the orbital motion. Binaries with aligned and antialigned spins are relatively easy to treat because there is no precession of the orbital plane or the individual spins. When one or both holes have a spin not parallel to the orbital axis, spin-orbit and spin-spin interactions cause precession that can complicate the evolution and the resulting waveforms (40).

The first merger evolutions of spinning black holes were carried out by Campanelli et al. (41). The authors simulated the mergers of two equal-mass, highly spinning black holes with $(a/M)_{1,2} = 0.757$, and both spin vectors either aligned or antialigned with the orbital angular momentum. Here, $a \leq M$ is the magnitude of the black-hole spin angular momentum per unit mass. Campanelli et al. also evolved a nonspinning, equal-mass binary for comparison. All three binaries had the same initial orbital angular frequency that corresponded to an orbital period of $\sim 125\,M$, and they merged to form a rotating remnant black hole with $(a/M)_{\mathrm{final}} < 1$. However, the aligned system needed noticeably more orbits to merge than the others. Such behavior is caused by the spin-orbit interaction, which produces an effective force between the black holes that is either attractive or repulsive for the antialigned or aligned cases, respectively. All three binaries generated remarkably similar gravitational waveforms that had a simple shape, although the aligned case displayed a longer wave train and the antialigned case a shorter one, compared to the nonspinning binary.

The first fully numerical evolutions of strongly precessing spinning systems were carried out by Campanelli et al. (42, 43), who observed both significant precession of the orbits and a spin flip, in which the final postmerger black hole spins in the direction opposite that of the two initial black holes. More recent work is beginning to probe the effects of spin precession on waveforms (44).

Studies of the most general (noneccentric) parameter sets are still in the early stages. To date, no systematic study of waveform shapes and polarizations has been carried out. A preliminary study of the multipolar structure of gravitational waves from several classes of binaries with equal spins has been performed by Berti et al. (45). These authors considered an equal-mass case with aligned spins, several $q = 4$ binaries with antialigned (down-down) spins, and three unequal-mass binaries with spins initially in the orbital plane and pointing in opposite directions. By examining the distribution of gravitational-wave energy emitted by various modes, they found that, as in the case of nonspinning mergers, odd-ℓ multipoles are suppressed for $q = 1$ and that, as q increases, more energy is radiated in higher-ℓ multipoles.

3.2. The Spin of the Final Black Hole

Because the state of the final black hole formed from the coalescence depends primarily on termination of the inspiral and the late burst of radiation in the merger, it can be accurately probed with relatively short simulations of only a few orbits. The merger of two equal-mass, nonspinning black holes produces a final black hole with a moderately high spin, $(a/M)_{\rm final} \sim 0.69$ (9, 23, 27, 28, 46). Because each black hole starts out with no spin, and because any tidal spin-up is negligible (47), the spin of the final black hole arises from the orbital angular momentum of the original binary. Simulations show that, for this simplest black-hole merger, the final spin is universal—that is, independent of the initial black-hole separation—for modest deviations from quasi-circular initial conditions.

For mergers of nonspinning black holes with unequal masses, the spin of the final black hole decreases as the mass ratio q increases. Simulations with mass ratios of up to $q = 10$ show that the final spin parameter scales as $(a/M)_{\rm final} \sim q/(1+q)^2$, where $(a/M)_{\rm final} \approx 0.48$ for $q = 4$ and $(a/M)_{\rm final} \approx 0.26$ for $q = 10$ (37–39).

For mergers of spinning black holes, the effects of spin-orbit and spin-spin coupling can become important in determining the spin of the final black hole. In the simplest cases, the spin vectors are parallel to the orbital angular momentum. Depending on the mass ratio and the black-hole spin, the merger can result in a final black hole with a larger (spun-up) or smaller (spun-down) spin than either of the progenitors (10, 41, 48). In certain cases, the merger can lead to a spin flip in which the final black hole spins in a direction opposite that of the initial holes; for example, it is possible to produce a final nonspinning black hole, $(a/M)_{\rm final} = 0$, from the merger of two spinning holes (45, 49). More general black-hole binaries, with misaligned spins, are further complicated by the effects of precession (42, 50, 51).

Several groups have attempted to produce expressions for the final spin vector by using analytic techniques or by fitting to results from numerical simulations (49, 52, 53); see Reference 54 for a review. Of particular interest is whether a black-hole merger can produce a maximally spinning hole $[(a/M)_{\rm final} = 1]$ or, indeed, whether it can exceed the Kerr limit $[(a/M)_{\rm final} > 1]$. Current research suggests that this is not possible, even for mergers resulting from hyperbolic encounters (55) and mergers of highly boosted black holes (56).

3.3. Recoil Kicks from Gravitational Radiation

A notable phenomenon arising from asymmetric binary systems is the merger recoil or kick, a net movement of the end-state black hole from the system's center of mass that is caused by the anisotropic emission of gravitational radiation during the coalescence. Using a quasi-Newtonian approximation, Fitchett (57) produced a useful formula for the kick velocity, with significant applications in astrophysics. Several authors also calculated the recoil analytically using PN approximations (58). However, because the dominant part of the effect builds up in the strong-field regime close to merger, full numerical relativity simulations are needed for accurate calculations of the kick.

3.3.1. Kicks from nonspinning black-hole mergers. In 2006, Herrmann et al. (59) became the first group to demonstrate recoil from a fully numerical binary merger, specifically plunging black-hole binaries with mass ratios as large as $q \approx 3.1$. This demonstration was soon followed by a full orbit and plunge simulation by Baker et al. (60), who found a recoil of between 86 and 97 km s^{-1} for a mass ratio of $q = 1.5$.

A more systematic study of recoil from mergers of unequal-mass binaries was produced in 2007 by Gonzalez et al. (61), who studied the merger of binary systems with mass ratios between $q = 1$ and $q = 4$. **Figure 5** shows the resulting range of recoil speeds, together with several earlier numerical and analytical estimates. The authors synthesized these data into a single recoil formula, a nonlinear correction to the Fitchett formula, yielding a maximum recoil of 175 km s^{-1}

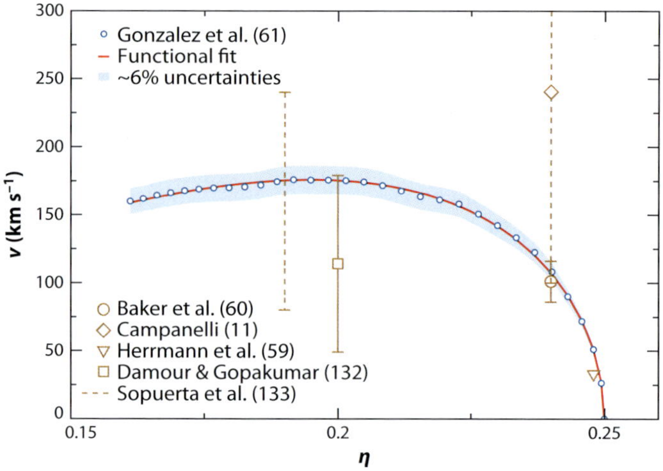

Figure 5

Recoil kicks for nonspinning black-hole mergers for mass ratios of $q \in \{1.0, 3.95\}$ or $\eta \in \{0.161, 0.25\}$; here, $\eta = q/(1+q)^2$. The open circles represent the results obtained from numerical relativity simulations by Gonzalez et al. (61). A functional fit is given by the continuous (*red*) line, and uncertainties of ~6% are indicated by the dashed (*blue*) lines above and below. For comparison, we also show earlier numerical results (*discrete symbols with error bars*) from Baker et al. (60), Campanelli (11), and Herrmann et al. (59), as well as perturbation theory predictions from Damour & Gopakumar (132) and Sopuerta et al. (133). Reprinted with permission from Reference 61. Copyright 2007, the American Physical Society (**http://link.aps.org/abstract/PRL/v98/e091101**).

for a mass ratio of $q \approx 3$. This formula was recently tested for the more extreme $q = 10$ mass ratio, with general agreement (37).

3.3.2. Spinning black-hole mergers and superkicks. Calculations of recoil in the much larger parameter space of spinning binaries began with the nonprecessing cases of holes with spins aligned (or antialigned) with the orbital angular momentum. Several studies of this region of parameter space (48, 62) revealed that the kick velocity has a quadratic dependence on initial spins, with a maximum kick of 448 km s^{-1} for extremal Kerr holes, $(a/M)_{1,2} = 1$. For these (anti)aligned cases—as well as for nonspinning, unequal-mass black-hole mergers—the direction of the kick velocity is always in the orbital plane.

Meanwhile, attention turned to more general black-hole spins. Campanelli et al. (43) speculated from PN arguments that optimal spin configurations could give rise to huge superkicks with velocities >1,000 km s^{-1}, out of the initial orbital plane. The first such superkick—with a velocity of approximately 2,500 km s^{-1}—was soon observed by Gonzalez et al. (63). Tichy et al. (50) argued that superkicks arise in mergers with general spin orientations, and greater insight into the mechanism of these kicks has been developed (64, 65).

With such a large parameter space to cover, construction of a general formula to describe kicks from arbitrarily spinning binaries may be useful. Baker et al. (66) used new and existing aligned-spin results to produce a single unifying formula for in-plane kicks; Campanelli et al. (43, 67) proposed an extension to this model, in which scaling of the superkick out-of-plane contribution is motivated by PN theory. The leading dependence in this formula on the angles between spins and linear momenta of the premerger holes has strong support from numerical simulations (64, 67, 68); however, the dominant scaling with mass ratio is still in dispute (69, 70).

3.4. Longer Waveforms: Modeling the Late Inspiral

Many key features of black-hole binary interactions can be usefully modeled with only a small handful of binary orbits before merger. Quantities such as radiated energy and momentum are bulked near the merger and are robust to the addition of one or two extra orbits (see Section 3.1.1). However, optimal observational studies of gravitational waveforms ultimately require theoretical predictions for the full waveform, starting at large separations during the inspiral.

Prior to the breakthroughs in numerical relativity, most information about the dynamics of compact binaries came from PN theory. This approach supplied the particle trajectories, energy flux, and—most importantly for detector scientists—the gravitational waveforms themselves. However, PN theory fails before the system merges, so the waveforms are necessarily incomplete. With the advent of numerical simulations that encompass many orbits, scientists finally have a way to develop complete information about the waveforms. Doing so requires longer simulations that reveal the last part of the binary inspiral and allow overlap with PN waveform predictions.

3.4.1. Low-eccentricity initial data. Simulations for these studies require careful attention to the initial configuration of the black holes. For a significant population of astrophysical binaries, the orbits are expected to have circularized through gravitational-wave emission over many orbits prior to merger. Numerical simulations try to mirror this expectation by selecting initial momenta consistent with near-zero eccentricity for the relatively small separations at which a full numerical relativity simulation becomes feasible.

For equal-mass, nonspinning binaries, two basic approaches have proven to be effective in reducing spurious eccentricity. The more direct method is to model the observed eccentricity, then adjust the momenta using a Newton-like step to zero it out; this approach has led to extremely low eccentricities (46, 71). The other approach, adopted by Husa et al. (72), is to use the PN Hamiltonian equations of motion to model the early evolution of the particle trajectories, starting from large ($\sim 50-100$-M) separations; the emission of radiation during this process naturally circularizes the orbit, and low-eccentricity momenta can be read off at the desired separation. This approach was also recently tested with spinning binaries (44).

The first long waveform was produced by the GSFC group (73, 74) for equal-mass, nonspinning black holes starting at $\sim$7 orbits or $\sim$14 gravitational-wave cycles before merger. Baker et al. (74) estimated their numerical errors in waveform phase as a function of frequency, finding that above a certain frequency numerical errors are smaller than internal errors in the PN sequence. Also, the numerical and PN waveform phases agree to within 1 rad of phase drift for somewhat more than ten gravitational-wave cycles preceding the last orbit before merger, comparable to numerical error estimates. Hannam et al. (75) improved on this work by comparing phase and amplitude between their low-eccentricity higher-resolution evolutions and PN waveforms.

3.4.2. Evolutions with spectral techniques. Numerical simulation codes based on pseudospectral differencing techniques are particularly well suited to long-waveform studies. The Caltech-Cornell group (76) adopted the constraint-damped generalized harmonic formalism used by Pretorius (25), an important component in developing a stable spectral evolution code. Also, like that of Pretorius, the Caltech-Cornell code handles black holes by excising the black-hole interiors. This group's spectral code also employs a numerical grid that tracks the black holes explicitly. Although there can be difficulties with the changing geometry as black holes merge, this approach allows efficient study of the long-lasting inspiral waveforms.

Recently this approach provided several of the longest and most accurate black-hole binary evolutions to date (46, 77). Specifically, the Caltech-Cornell researchers used their spectral code

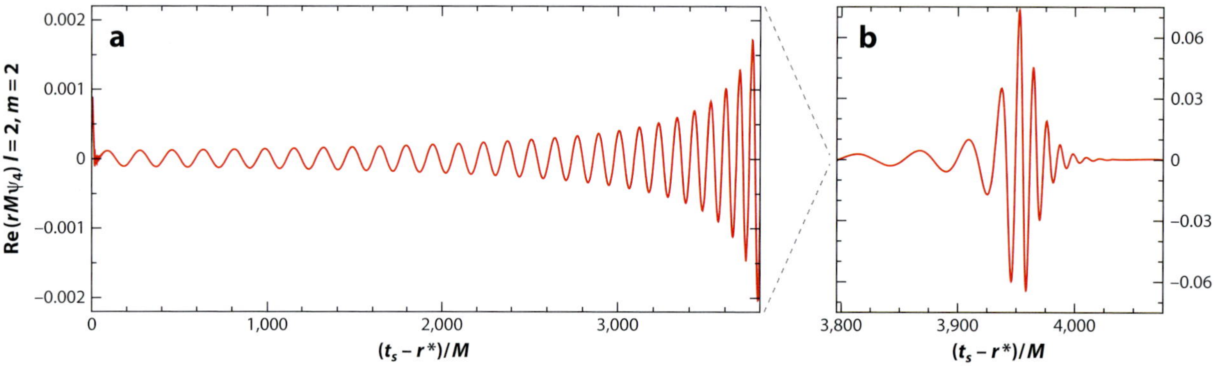

Figure 6

The longest gravitational waveform for an equal-mass, nonspinning black-hole binary merger computed by the Caltech-Cornell group (46). (*a*) The early stages of the waveform, during the inspiral. (*b*) The merger and ringdown portions of the waveform. Reprinted with permission from Reference 46. Copyright 2009, the American Physical Society (**http://link.aps.org/abstract/PRD/v79/e024003**).

to simulate an equal-mass, nonspinning binary that began 16 orbits and 32 gravitational-wave cycles before merger; see **Figure 6** (46). These much longer waveforms have been used to validate several competing PN models (71).

3.4.3. Comparing results. With so many premerger waveform cycles now available for the equal-mass case, the results can be cross-checked by comparing the complete waveform—inspiral, merger, and ringdown—between groups. Cross-checking is important to verify expectations that differences in methodology and residual numerical effects, such as unwanted eccentricity, are unimportant. A recent effort, dubbed the Samurai project, analyzed long waveforms from several groups in the light of detectability criteria for the LIGO and Virgo ground-based gravitational-wave detectors (78). The authors found that the available numerical relativity waveforms are indistinguishable in these detectors for signal-to-noise ratios (SNRs) $\lesssim$25 (78).

3.4.4. Long waveforms for more general black holes. Hannam et al. (79) have investigated the properties of highly spinning, orbit-aligned black holes, comparing their phase and amplitude with PN predictions over the last ten waveform cycles. In the best cases, the authors found phase agreement within 2.5 rad for 3.5 PN and amplitude agreement to within ~12% with restricted PN. This decline in agreement between numerical and PN results relative to the nonspinning case (75) may be attributable to the lower orders of accuracy currently available for spinning black holes in PN theory. The Caltech-Cornell group recently conducted long-lasting simulations of black holes with spins aligned and antialigned with the orbital angular momentum, calibrating a tunable PN waveform model to match the results (77, 80); see Section 5.1.

Generic mergers that involve nonaligned, and thus precessing, spins add four new degrees of freedom to the problem. A systematic understanding of waveforms generated by generic mergers requires considerably more study. Recent simulations have begun to explore generic examples (44).

4. APPLICATIONS IN ASTROPHYSICS

We have described various numerical relativity studies that have revealed some details of black-hole binary physics as implied by general relativity. Although more remains to be learned, this new understanding is already contributing to the planning and interpretation of astrophysical black-hole observations in which Einstein's theory is taken as the underlying model of gravitational physics.

4.1. Waveforms for Gravitational-Wave Observations

Experimental gravitational-wave detectors were first developed more than 40 years ago. Although advances in design have increased their sensitivity by many orders of magnitude, the extreme weakness of expected astrophysical signals (strain amplitudes $\delta L/L \sim 10^{-21}$) means that the observational challenge is still huge.

The output of any gravitational-wave detector is a data stream that must be combed through to find real signals. This search requires accurate "template" waveforms that represent our best picture of the radiation from expected sources; these templates can then be compared with the data stream through matched filtering. We refer the reader to a review on gravitational-wave astronomy for an overview of these techniques (81).

Crucially, the most important sources of gravitational radiation are expected to include the mergers of black-hole binaries. Before the advent of numerical relativity simulations of black-hole mergers, the only test waveforms available for use in data-analysis studies were based on PN theory. These waveforms were essentially only inspiral chirps and did not include the strong-field merger. The new, richer information now available from numerical relativity has revolutionized the data-analysis picture in several ways.

4.1.1. Detecting black-hole mergers. The availability of the plunge-merger-ringdown portion of the signal can greatly increase the SNR in the detector. Armed with a long numerical merger waveform of acceptable accuracy, we can extend it backward to cover an arbitrarily long inspiral by matching to a PN waveform. Such a hybrid waveform was first produced by the GSFC group for the equal-mass, nonspinning case (74). Using this hybrid, we can investigate the total achievable SNR, and the related distance reach, for current and future detectors. **Figure 7** demonstrates the gain in SNR achieved by including the merger portion of the waveform for the ground-based Advanced LIGO detector. Contours of SNR as function of redshift z and total binary mass M for the LISA detector are shown in **Figure 8**.

Full numerical waveforms, and the longer hybrid waveforms generated from them, can also be used to improve existing data-analysis techniques and template sets. Because previously developed gravitational-wave data-analysis algorithms were not based on knowledge of the merger waveforms, an obvious first step is to test how successfully these techniques detect the merger signals predicted by numerical simulations. In 2009, the NINJA project (82) used direct injection of a broad range of short and long numerical merger waveforms into mock LIGO and Virgo data streams for this purpose. The result was the most realistic testing ground to date for disparate data-analysis methods, which included full and partial waveform template matching, as well as unmodeled burst searches. Further studies of detection algorithms with numerical relativity waveforms are under way.

4.1.2. Measuring black-hole binary parameters. The SNR is only a crude guide to the specific detector response to gravitational waveforms, however. The merger portion of the waveform, although short in duration, may contain important new information not present in the inspiral. As a simple example, we expect the time of merger itself to be well localized with the full waveform, whereas it is not well defined in the inspiral-only signal.

More generally, the different (ℓ, m) modes of a binary's full inspiral-merger-ringdown waveform scale differently with the time to merger. Modes that were not significant during inspiral

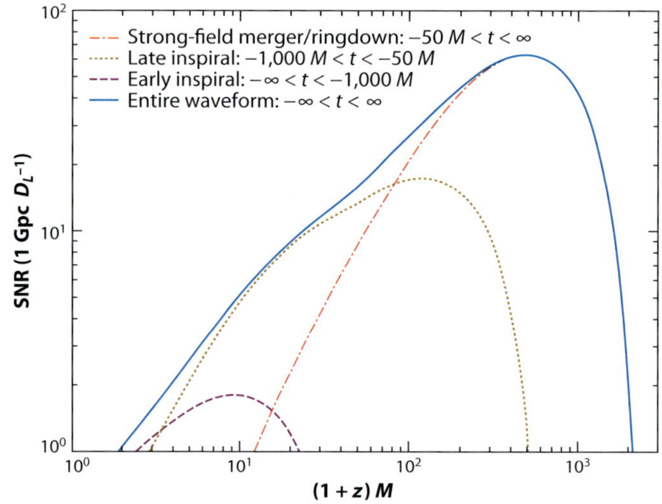

Figure 7

Gain in signal-to-noise ratio (SNR) by including the merger segment of the waveform for an equal-mass, nonspinning binary. The SNR for sources located at luminosity distance $D_L = 1$ Gpc is plotted versus (redshifted) mass for the Advanced LIGO detector. The purple dashed line shows the SNR calculated with post-Newtonian (PN) techniques for the early inspiral part of the waveform, $-\infty < t < -1000\,M$, where $t = 0$ is the time of maximum gravitational-radiation amplitude. The gold dotted line shows the SNR calculated with the late inspiral, $-1000\,M < t < -50\,M$, which is the transition region from PN to numerical relativity. The SNR for the strong-field merger and ringdown, $-50\,M < t < \infty$, was calculated with waveforms from a numerical relativity simulation (*red dash-dotted line*). The blue solid line shows the SNR from the entire waveform. Here, $t = 0$ marks the time of maximum gravitational radiation amplitude. Reprinted from Reference 74. Copyright 2007, the American Physical Society (**http://link.aps.org/abstract/PRD/v75/e124024**).

suddenly become more prominent in the merger, and the detailed information they carry becomes available to the observer (38, 39).

After gravitational waves from distant sources have been detected, we are most interested in identifying the physical parameters of the sources. Each of the seven intrinsic parameters of a black-hole binary (mass ratio and spin vectors) is generally imprinted on the gravitational-wave signal, along with some extrinsic parameters such as sky position and distance to the source. For sufficiently strong SNRs, expected for massive binary mergers observed by LISA, we can expect to be able to extract some of these parameters at high precision. Although these parameters can be partially disentangled using inspiral-only template information, the full merger waveform can help break parameter degeneracies and hence drive down uncertainties in several important physical parameters. Good localization of the source on the sky is especially important for the development of multimessenger astronomy. Recent parameter estimation studies of nonspinning mergers that include the merger waveforms indicate that LISA will be able to locate sky positions within a few arcminutes for binaries with $\sim 10^6\,M_\odot$ at cosmological distances (redshift $z = 1$) (83).

4.2. Consequences of Merger Recoil

As previously discussed, asymmetries in a black-hole binary system that arise from unequal masses and/or spins result in the anisotropic emission of gravitational radiation, ultimately imparting a recoil to the merged remnant black hole. Numerically, for certain configurations of black-hole

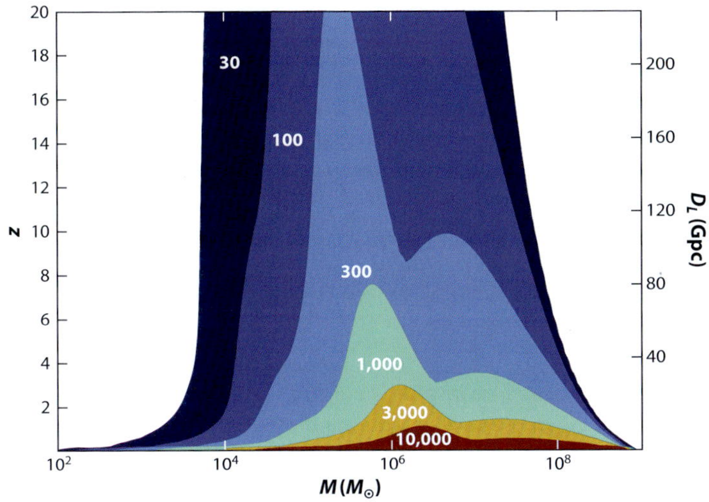

Figure 8

Contours of signal-to-noise ratio as a function of redshift z and total binary mass M are shown for the LISA detector. These contours have been calculated with a hybrid, equal-mass, nonspinning waveform by using post-Newtonian techniques for the early inspiral, matching to a numerical relativity waveform in the late inspiral, and continuing with the numerical waveform through the merger and ringdown. Reprinted from Reference 74. Copyright 2007, the American Physical Society (**http://link.aps.org/abstract/PRD/v75/e124024**).

spins, the recoil velocity can exceed the escape velocity of many galaxies. Because it is important to determine the theoretical probability that a massive black hole might be ejected from its host galaxy, there have been some preliminary calculations based on simple distributions of spins and mass ratios, as well as analytic fits of numerical results giving the dependence of the recoil on mass and spin (69, 84). Although a consensus on the exact probability of galactic ejection has yet to be achieved, there is general agreement that it is nonnegligible.

Such rogue black holes may have already been observed in the form of two rapidly moving quasars (85). These particular quasars may have originated from the mergers of massive black holes during the coalescence of their host galaxies. However, a previous study of quasar data found no indications of such recoil events, suggesting that they are rare (86).

The effect of recoil may also be observed less directly in its effect on the growth rates of black holes. Black holes ejected into the sparse intergalactic medium are less likely to encounter and merge with other black holes (87). Even the growth of recoiled black holes remaining within their host galaxies may be affected, as the motion of the black hole can modify the rate at which it accretes matter (88).

4.3. Mass and Spin Evolution

The expected distribution of masses and spins of astrophysical black holes is another topic of considerable interest in astrophysics. In our current understanding, black holes grow from smaller "seeds" that existed early in the history of the universe through a combination of mergers and the accretion of gas (89). Generally, most of the mass growth is believed to come from accretion, with mergers providing a modest increase.

However, the gravitational radiation emitted during black-hole coalescence carries away energy, reducing the overall system mass by several percent. The bulk of this loss happens quickly, in the final plunge and merger stages of the coalescence. With this rapid mass loss, nearby matter in an accretion disk around the remnant black hole might react to the accompanying sudden change in gravitational potential and produce a visible change in its electromagnetic profile—a possible electromagnetic counterpart to the burst of gravitational radiation (90, 91).

For the final spin, the situation is considerably more interesting and may answer questions about double-jet X-shaped radio sources (92). As discussed in Section 3.2, the final spin from black holes merging in vacuum depends on the magnitudes of the initial spins and their orientations relative to the orbital axis. Lousto et al. (53) have recently used their extended mass and spin formulas in a spin-evolution study, obtaining asymptotic spin distributions for black-hole merger remnants, assuming no accretion. However, interaction of the merging holes with surrounding gas may serve to align the binary spins before merger, changing the picture somewhat (93). Generally, the effects of binary merger and accretion have to be studied together for a coherent picture to emerge (94).

5. OUTLOOK

5.1. Complete Analytic Waveform Models

Gravitational-wave observatories may be sensitive to hundreds or thousands of wave cycles. Analysis of the observed data requires comparisons with model signals representing the full variety of possible mergers. Computing so many orbits of premerger evolution by using numerical relativity would be computationally very expensive.

The PN approximation provides accurate representations of the dynamics and waveforms for the long-lasting inspiral portion of the coalescence, during which the black holes remain fairly well separated while their velocities remain relatively low. The most valuable waveform models for gravitational-wave data analysis must combine the efficiency of the PN approach as well as accurately represent the final merger portion of the waveform, which is only understood by numerical simulations. Use of such models requires a means of analytically encoding the merger signals.

Several approaches are currently being explored for constructing these complete signal models. Some are based on the analytic effective one-body model of binary coalescence (95–97). **Figure 9** shows a comparison between an effective one-body waveform for the merger of a nonspinning black-hole binary with a mass ratio of $q = 4$ with a numerically simulated waveform (96). For specific cases, such models have been tuned to high accuracy to agree with numerical results (98). Another approach, which models the phenomenological shape of hybrid waveforms in frequency space (99), has been extended for the dominant waveform modes from spinning, but nonprecessing, mergers (100). Further development of these models, and the production of a family of simulated waveforms spanning the parameter space, is a current focus of broad-based research collaborations.

5.2. Improved Numerical Methods

Although current numerical methodology seems to guarantee robustly stable evolutions of any physical black-hole binary configuration, there is room for improvement in our specification of astrophysically realistic initial conditions, as well as in the accuracy to which we can reasonably extract the observable results of these evolutions.

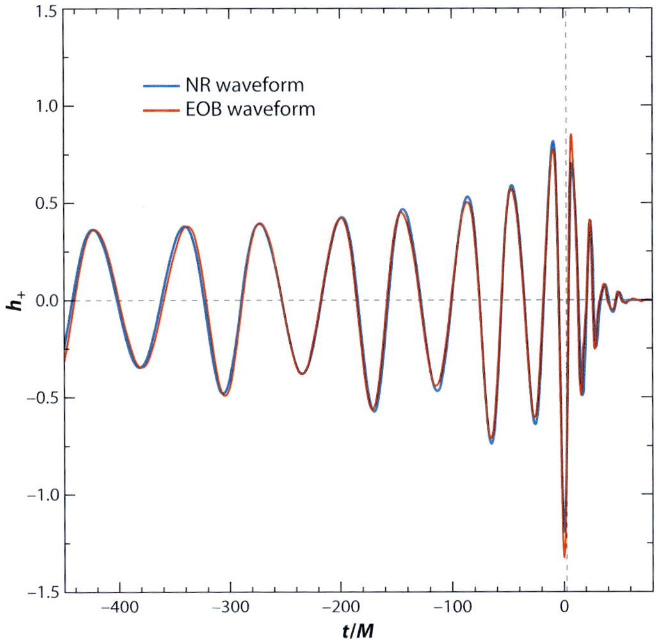

Figure 9

Comparison of merger waveforms for a $q = 4$ mass ratio, nonspinning black-hole binary calculated with both the analytic effective one-body (EOB) model and numerical relativity (NR). Reprinted with permission from Reference 96. Copyright 2007, the American Physical Society (**http://link.aps.org/abstract/PRD/v76/e104049**).

5.2.1. Initial data. The initial data models currently used as starting points for numerical simulations do not perfectly represent the intended astrophysical systems. Research continues to improve these models. For example, although some astrophysical black holes are expected to have near-extremal spins, common initial data cannot represent holes with $(a/M)_i \gtrsim 0.93$ (101, 102); novel methods are being developed to go beyond this limit (101). Also, current initial data models generally do not contain the physically appropriate radiation content for an inspiraling binary (32, 103), but rather harbor spurious radiation that is not astrophysical (104). Efforts to include initial radiation content more consistent with PN predictions are under way (105). Also, in the case of moving punctures, the initial coordinates typically undergo a rapid transition at the start of the evolution as the black holes relax into a more stable solution. Whereas the resulting transient pulse of gauge radiation does not alter the physics, it does contain fairly high frequencies that can be challenging to resolve, and it has motivated construction of analytic initial coordinates that are closer to the numerically evolved coordinates (106).

5.2.2. Evolution. More efficient and accurate methods of numerical evolution would make simulations of many binary orbits, particularly those with large mass ratios, more computationally practical. For representing spatial derivatives on a computational grid, the highest accuracy of finite differencing stencil yet achieved in the context of numerical relativity is eighth order in the grid spacing (although the accuracy of interpolation between refinement boundaries is not yet commensurate) (107). Spectral methods are generally more accurate but less robust than finite differencing, requiring fine-tuning for generic black-hole binaries (108); they are also unlikely to handle shocks in accreting matter (see Section 5.3). Alternatives to both finite differencing and spectral methods, such as finite-element methods, are also being investigated (109). Meanwhile, more efficient time-integration techniques that allow larger step sizes are being explored (110).

5.2.3. Wave extraction. Because the physical domain of a typical simulation is finite, gravitational radiation is usually extracted on a sphere of finite radius rather than at infinity. If multiple extraction surfaces are employed in a region of sufficiently high spatial resolution, then the radiation can be extrapolated to spatial infinity. A more accurate method, known as Cauchy-characteristic extraction, extracts the gravitational-wave data at a finite radius and then evolves it along null geodesics to future null infinity; this method is currently under development (111).

5.3. Including Gas and Magnetic Fields

In addition to being gravitationally "loud," black-hole mergers may also be electromagnetically visible. Massive black holes at the centers of galaxies are typically surrounded by gaseous accretion disks and magnetic fields. When the black holes merge, the dynamics of the gas and magnetic fields may produce electromagnetic signals, which are counterparts to the emitted gravitational radiation. For example, the inspiral may twist the fields, resulting in characteristic electromagnetic radiation (112, 113) as well as heating of the surrounding accretion disk (114). Also, the violent merger dynamics may induce shock waves in accreting matter, in turn generating electromagnetic radiation. The recoil of the merged remnant, in particular, may have such an effect on the accretion disk (90, 91, 115–120).

5.3.1. Multimessenger astronomy. Detection of electromagnetic counterparts of gravitational waves would be of great scientific value. Current models of complex merger physics (e.g., 121) could be directly tested. Einstein-Maxwell theory, which refers to the coupling of gravitational and electromagnetic fields on macroscopic scales, could be verified. For example, equality between the speed of gravity and the speed of light could be confirmed (112, 118). The field of astronomy would also benefit enormously, as the location and characteristics of gravitational-wave sources could be corroborated. Also, electromagnetically visible mergers could serve as standard candles, beacons by which to measure the accelerating expansion of the universe, while simultaneously playing the role of standard "sirens" emitting gravitational radiation. Cross-correlation of these signals could result in measurement of the cosmological dark energy to unprecedented accuracy (116, 118, 122–124).

5.3.2. Modeling matter. There has been some preliminary work on the dynamic effects of the space-time of a coalescing binary on surrounding matter. When the accretion disk was modeled as geodesic particles, large collision energies were found as the binary merged (125). Also, hydrodynamically simulating a gas cloud around the binary allowed the calculation of luminosity due to shocks (126). The generation of electromagnetic radiation by more direct means has also been simulated via the twisting of a magnetic field anchored in the accretion disk as it is frame-dragged by the binary (112, 113, 127). Future efforts will employ magnetohydrodynamic methods, in which challenges include adequately resolving shocks and turbulence in a dynamic space-time and accurately representing the divergence-free magnetic field on an adaptively refined grid.

6. CONCLUSION

We hope to have conveyed some of the excitement about recent progress in understanding black-hole binary physics, as well as the applications of this knowledge in astrophysics. These advances are the result of sustained efforts by a broad scientific community over many years. In a brief review, it is impossible to adequately represent all of the excellent work that has contributed to the current state of knowledge. We have provided only a few of the highlights as seen through the lens of our particular perspective.

We encourage interested readers to pursue the subject further. Resources are available on topics including numerical relativity techniques (3, 128), breakthroughs in black-hole merger simulations (129), black-hole simulations for gravitational-wave data analysis (130), and gravitational-wave science generally (81, 131).

DISCLOSURE STATEMENT

The authors are not aware of any affiliations, memberships, funding, or financial holdings that might be perceived as affecting the objectivity of this review.

ACKNOWLEDGMENTS

We acknowledge support from NASA grants 06-BEFS06-19 and 08-ATFP08-0126. B.J.K. was supported in part by an appointment to the NASA Postdoctoral Program at the Goddard Space Flight Center, administered by Oak Ridge Associated Universities through a contract with NASA.

LITERATURE CITED

1. Arnowitt R, Deser S, Misner CW. In *Gravitation: An Introduction to Current Research*, ed. L Witten, p. 227. New York: Wiley (1962)
2. Misner CW, Thorne KS, Wheeler JA. *Gravitation*. San Francisco: Freeman (1973)
3. Alcubierre M. *Introduction to 3+1 Numerical Relativity*. Oxford: Oxford Univ. Press (2008)
4. Smarr LL, York JW. *Phys. Rev. D* 17:2529 (1978)
5. Alcubierre M, et al. *Phys. Rev. D* 67:084023 (2003)
6. Brügmann B. *Int. J. Mod. Phys. D* 8:85 (1999)
7. Alcubierre M, et al. *Phys. Rev. Lett.* 87:271103 (2001)
8. Baker JG, Campanelli M. *Phys. Rev. D* 62:127501 (2000); Baker JG, Campanelli M, Lousto CO. *Phys. Rev. D* 65:044001 (2002)
9. Baker JG, Campanelli M, Lousto CO, Takahashi R. *Phys. Rev. D* 65:124012 (2002)
10. Baker JG, Campanelli M, Lousto CO, Takahashi R. *Phys. Rev. D* 69:027505 (2004)
11. Campanelli M. *Class. Quantum Gravity* 22:S387 (2005)
12. Shibata M, Nakamura T. *Phys. Rev. D* 52:5428 (1995); Baumgarte TW, Shapiro SL. *Phys. Rev. D* 59:024007 (1998)
13. York JW. In *Sources of Gravitational Radiation*, ed. LL Smarr, p. 83. Cambridge, UK: Cambridge Univ. Press (1979)
14. Nagy G, Ortiz OE, Reula OA. *Phys. Rev. D* 70:044012 (2004); Reula OA. *J. Hyperbolic Differ. Equ.* 1:251 (2004)
15. Anninos P, et al. *Phys. Rev. D* 52:2059 (1995)
16. Alcubierre M, et al. *Phys. Rev. D* 64:61501(R) (2001); Alcubierre M. *Class. Quantum Gravity* 20:607 (2003)
17. Thornburg J. *Class. Quantum Gravity* 4:1119 (1987)
18. Brandt SR, Brügmann B. *Phys. Rev. Lett.* 78:3606 (1997)
19. Shoemaker DM, et al. *Class. Quantum Gravity* 20:3729 (2003)
20. Alcubierre M, Brügmann B. *Phys. Rev. D* 63:104006 (2001)
21. Alcubierre M, et al. *Phys. Rev. D* 72:044004 (2005)
22. Brügmann B, Tichy W, Jansen N. *Phys. Rev. Lett.* 92:211101 (2004)
23. Pretorius F. *Phys. Rev. Lett.* 95:121101 (2005)

24. Pretorius F. *Class. Quantum Gravity* 23:S529 (2006)
25. Pretorius F. *Class. Quantum Gravity* 22:425 (2005)
26. Gundlach C, Martín-García JM, Calabrese G, Hinder I. *Class. Quantum Gravity* 22:3767 (2005)
27. Campanelli M, Lousto CO, Marronetti P, Zlochower Y. *Phys. Rev. Lett.* 96:111101 (2006)
28. Baker JG, et al. *Phys. Rev. Lett.* 96:111102 (2006)
29. van Meter JR, Baker JG, Koppitz M, Choi DI. *Phys. Rev. D* 73:124011 (2006)
30. Peters PC, Mathews J. *Phys. Rev.* 131:435 (1963); Peters PC. *Phys. Rev.* 136:B1224 (1964)
31. Lousto CO, Price RH. *Phys. Rev. D* 57:1073 (1998)
32. Damour T, Jaranowski P, Schäfer G. *Phys. Rev. D* 62:084011 (2000)
33. Pfeiffer HP, Cook GB, Teukolsky SA. *Phys. Rev. D* 66:024047 (2002)
34. Baker JG, et al. *Phys. Rev. D* 73:104002 (2006)
35. Hinder I, et al. *Phys. Rev. D* 77:081502(R) (2008)
36. Bode T, Shoemaker DM, Herrmann F, Hinder I. *Phys. Rev. D* 77:044027 (2008)
37. Gonzalez JA, Sperhake U, Brügmann B. *Phys. Rev. D* 79:124006 (2009)
38. Berti E, et al. *Phys. Rev. D* 76:064034 (2007)
39. Baker JG, et al. *Phys. Rev. D* 78:044046 (2008)
40. Apostolatos TA, Cutler C, Sussman GJ, Thorne KS. *Phys. Rev. D* 49:6274 (1994); Kidder LE. *Phys. Rev. D* 52:821 (1995)
41. Campanelli M, Lousto CO, Zlochower Y. *Phys. Rev. D* 74:041501(R) (2006)
42. Campanelli M, et al. *Phys. Rev. D* 75:064030 (2007)
43. Campanelli M, Lousto CO, Zlochower Y, Merritt D. *Astrophys. J.* 659:L5 (2007)
44. Campanelli M, Lousto CO, Nakano H, Zlochower Y. *Phys. Rev. D* 79:084010 (2009)
45. Berti E, et al. *Class. Quantum Gravity* 25:114035 (2008)
46. Scheel MA, et al. *Phys. Rev. D* 79:024003 (2009)
47. Campanelli M, Lousto CO, Zlochower Y. *Phys. Rev. D* 74:084023 (2006)
48. Pollney D, et al. *Phys. Rev. D* 76:124002 (2007)
49. Buonanno A, Kidder LE, Lehner L. *Phys. Rev. D* 77:026004 (2008)
50. Tichy W, Marronetti P. *Phys. Rev. D* 76:061502(R) (2007)
51. Dain S, Lousto CO, Zlochower Y. *Phys. Rev. D* 78:024039 (2008)
52. Kesden M. *Phys. Rev. D* 78:084030 (2008); Rezzolla L, et al. *Astrophys. J.* 679:1422 (2008); Rezzolla L, et al. *Astrophys. J.* 674:L29 (2008); Rezzolla L, et al. *Phys. Rev. D* 78:044002 (2008); Tichy W, Marronetti P. *Phys. Rev. D* 78:081501(R) (2008)
53. Lousto CO, Campanelli M, Zlochower Y. *Class. Quantum Gravity* 27:114006 (2010)
54. Rezzolla L. *Class. Quantum Gravity* 26:094023 (2009)
55. Healy J, Laguna P, Matzner RA, Shoemaker DM. *Phys. Rev. D* 81:081501 (2010)
56. Sperhake U, et al. *Phys. Rev. Lett.* 103:131102 (2009)
57. Fitchett MJ. *Mon. Not. R. Astron. Soc.* 203:1049 (1983)
58. Favata M, Hughes SA, Holz DE. *Astrophys. J.* 607:L5 (2004); Blanchet L, Qusailah MSS, Will CMW. *Astrophys. J.* 635:508 (2005); Damour T, Gopakumar A. *Phys. Rev. D* 73:124006 (2006); Wiseman AG. *Phys. Rev. D* 46:1517 (1992)
59. Herrmann F, Shoemaker DM, Laguna P. *Class. Quantum Gravity* 24:S33 (2007)
60. Baker JG, et al. *Astrophys. J.* 653:L93 (2006)
61. Gonzalez JA, et al. *Phys. Rev. Lett.* 98:091101 (2007)
62. Herrmann F, et al. *Astrophys. J.* 661:430 (2007); Koppitz M, et al. *Phys. Rev. Lett.* 99:041102 (2007)
63. Gonzalez JA, et al. *Phys. Rev. Lett.* 98:231101 (2007)
64. Brügmann B, et al. *Phys. Rev. D* 77:124047 (2008)
65. Schnittman JD, et al. *Phys. Rev. D* 77:044031 (2008)
66. Baker JG, et al. *Astrophys. J.* 668:1140 (2007)
67. Campanelli M, Lousto CO, Zlochower Y, Merritt D. *Phys. Rev. Lett.* 98:231102 (2007)
68. Lousto CO, Zlochower Y. *Phys. Rev. D* 77:044028 (2007)
69. Baker JG, et al. *Astrophys. J.* 682:L29 (2008)
70. Lousto CO, Zlochower Y. *Phys. Rev. D* 79:064018 (2009)
71. Boyle M, et al. *Phys. Rev. D* 76:124038 (2007)

72. Husa S, et al. *Phys. Rev. D* 77:044037 (2008)
73. Baker JG, et al. *Phys. Rev. Lett.* 99:181101 (2007)
74. Baker JG, et al. *Phys. Rev. D* 75:124024 (2007)
75. Hannam MD, et al. *Phys. Rev. D* 77:044020 (2007)
76. Lindblom L, et al. *Class. Quantum Gravity* 23:S447 (2006); Scheel MA, et al. *Phys. Rev. D* 74:104006 (2006)
77. Chu T, Pfeiffer HP, Scheel MA. *Phys. Rev. D* 80:124051 (2009)
78. Hannam MD, et al. *Phys. Rev. D* 79:084025 (2009)
79. Hannam MD, Husa S, Brügmann B, Gopakumar A. *Phys. Rev. D* 78:104007 (2008)
80. Pan Y, et al. arXiv:0912.3466 [gr-qc] (2009)
81. Camp JB, Cornish NJ. *Annu. Rev. Nucl. Part. Sci.* 54:525 (2004)
82. Aylott B, et al. *Class. Quantum Gravity* 26:165008 (2009)
83. McWilliams ST, Thorpe JI, Baker JG, Kelly BJ. *Phys. Rev. D* 81:064014 (2010)
84. Schnittman JD, Buonanno A. *Astrophys. J.* 662:L63 (2007)
85. Shields GA, et al. *Astrophys. J.* 707:936 (2009); Komossa S, Zhou H, Lu H. *Astrophys. J.* 678:L81 (2008)
86. Bonning EW, Shields GA, Salivander S. *Astrophys. J.* 666:L13 (2007)
87. Sesana A, Volonteri M, Haardt F. *Mon. Not. R. Astron. Soc.* 377:1711 (2007); Volonteri M. *Astrophys. J.* 663:L5 (2007)
88. Blecha L, Loeb A. *Mon. Not. R. Astron. Soc.* 390:1311 (2008)
89. Sesana A, Haardt F, Madau P, Volonteri M. *Astrophys. J.* 611:623 (2004)
90. Bode N, Phinney S. *Am. Phys. Soc. Meet. Abstr.* 1010 (2007)
91. O'Neill SM, et al. *Astrophys. J.* 700:859 (2009)
92. Merritt D, Ekers RD. *Science* 297:1310 (2002); Barausse E, Rezzolla L. *Astrophys. J.* 704:L40 (2009)
93. Bogdanovic T, Reynolds CS, Miller MC. *Astrophys. J.* 661:L147 (2007)
94. Berti E, Volonteri M. *Astrophys. J.* 684:822 (2008)
95. Pan Y, et al. *Phys. Rev. D* 77:024014 (2008)
96. Buonanno A, et al. *Phys. Rev. D* 76:104049 (2007)
97. Damour T, et al. *Phys. Rev. D* 78:044039 (2008)
98. Damour T, Nagar A. *Phys. Rev. D* 79:081503(R) (2009); Buonanno A, et al. *Phys. Rev. D* 79:124028 (2009)
99. Ajith P, et al. *Phys. Rev. D* 77:104017 (2008); Ajith P, et al. Erratum. *Phys. Rev. D* 79:129901(E) (2009)
100. Ajith P, et al. arXiv:0909.2867 [gr-qc] (2009)
101. Lovelace G, Owen R, Pfeiffer HP, Chu T. *Phys. Rev. D* 78:084017 (2008)
102. Dain S, Lousto CO, Takahashi R. *Phys. Rev. D* 65:104038 (2002)
103. Pfeiffer HP. *Initial data for black hole evolutions*. PhD thesis. Cornell Univ. 220 pp. (2005)
104. Hannam MD, et al. *Class. Quantum Gravity* 24:S15 (2007)
105. Kelly BJ, Tichy W, Campanelli M, Whiting BF. *Phys. Rev. D* 76:024008 (2007)
106. Hannam MD, Husa S, Murchadha NÓ. *Phys. Rev. D* 80:124007 (2009)
107. Lousto CO, Zlochower Y. *Phys. Rev. D* 77:024034 (2008) Pollney D, et al. arXiv:0910.3803 [gr-qc] (2009)
108. Szilágyi B, Lindblom L, Scheel MA. arXiv:0909.3557 [gr-qc] (2009)
109. Zumbusch G. *Class. Quantum Gravity* 26:175011 (2009)
110. Lau SR, Pfeiffer HP, Hesthaven JS. *Commun. Comput. Phys.* 6:1063 (2009)
111. Reisswig C, Bishop NT, Pollney D, Szilágyi B. *Class. Quantum Gravity* 27:075014 (2010); Babiuc MC, Bishop NT, Szilagyi B, Winicour J. *Phys. Rev. D* 79:084011 (2009)
112. Palenzuela C, et al. *Phys. Rev. Lett.* 103:081101 (2009)
113. Palenzuela C, Lehner L, Yoshida S. *Phys. Rev. D* 81:084007 (2010)
114. Reynolds CS, Garofalo D, Begelman MC. *Astrophys. J.* 651:1023 (2006)
115. Armitage PJ, Natarajan P. *Astrophys. J.* 567:L9 (2002); Milosavljevic M, Phinney ES. *Astrophys. J.* 622:L93 (2005); Dotti M, et al. *Mon. Not. R. Astron. Soc.* 372:869 (2006)
116. Kocsis B, Frei Z, Haiman Z, Menou K. *Astrophys. J.* 637:27 (2006)
117. Phinney ES. *Bull. APS April Meet.* 52:7002 (2007)
118. Kocsis B, Haiman Z, Menou K. *Astrophys. J.* 684:870 (2008)

119. Shields GA, Bonning EW. *Astrophys. J.* 682:758 (2008); Lippai Z, Frei Z, Haiman Z. *Astrophys. J.* 676:L5 (2008); Schnittman JD, Krolik JH. *Astrophys. J.* 684:835 (2008); Kocsis B, Loeb A. *Phys. Rev. Lett.* 101:041101 (2008); Haiman Z, et al. *Class. Quantum Gravity* 26:094032 (2009)
120. Haiman Z, Kocsis B, Menou K. *Astrophys. J.* 700:1952 (2009); Chang P, Strubbe LE, Menou K, Quataert E. arXiv:0906.0825 [astro-ph] (2009); Megevand M, et al. *Phys. Rev. D* 80:024012 (2009)
121. Balbus SA, Hawley JF. *Astrophys. J.* 376:214 (1991)
122. Lang RN, Hughes SA. *Class. Quantum Gravity* 26:094035 (2009); Kocsis B, Haiman Z, Menou K, Frei Z. *Phys. Rev. D* 76:022003 (2007); Jonsson J, Goobar A, Mortsell E. *Astrophys. J.* 658:52 (2007); Dalal N, Holz DE, Hughes SA, Jain B. *Phys. Rev. D* 74:063006 (2006)
123. Holz DE, Hughes SA. *Astrophys. J.* 629:15 (2005)
124. Arun KG, et al. *Class. Quantum Gravity* 26:094021 (2009)
125. van Meter JR, et al. *Astrophys. J.* 711:L89 (2010)
126. Bode T, et al. arXiv:0912.0087 [gr-qc] (2009)
127. Mösta P, et al. *Phys. Rev. D* 81:064017 (2010)
128. Bona C, Palenzuela-Luque C. *Lecture Notes in Physics*. Vol. 673: *Elements of Numerical Relativity*. Berlin/Heidelberg: Springer (2005)
129. Pretorius F. In *Physics of Relativistic Objects in Compact Binaries: from Birth to Coalescence*, ed. M Colpi, et al., p. 305. Heidelberg: Springer (2009)
130. Hannam MD. *Class. Quantum Gravity* 26:114001 (2009)
131. Sathyaprakash BS, Schutz BF. *Living Rev. Relativ.* 12:2. **http://www.livingreviews.org/lrr-2009-2** (2009)
132. Damour T, Gopakumar A. *Phys. Rev. D* 73:124006 (2006)
133. Sopuerta CF, Yunes N, Laguna P. *Phys. Rev. D* 74:124010 (2006)

Physics Accomplishments of HERA

C. Diaconu,[1] T. Haas,[2] M. Medinnis,[2] K. Rith,[2,3] and A. Wagner[2]

[1]CPPM, CNRS/IN2P3, Université de la Méditerranée, 13288 Marseille Cedex 9, France; email: diaconu@cppm.in2p3.fr

[2]DESY, 22603 Hamburg, Germany; email: tobias.haas@desy.de, michael.medinnis@desy.de, albrecht.wagner@desy.de

[3]Physikalisches Institut, Universität Erlangen-Nürnberg, 91058 Erlangen, Germany; email: klaus.rith@desy.de

Annu. Rev. Nucl. Part. Sci. 2010. 60:101–28

First published online as a Review in Advance on June 7, 2010

The *Annual Review of Nuclear and Particle Science* is online at nucl.annualreviews.org

This article's doi: 10.1146/annurev.nucl.012809.104458

0163-8998/10/1123-0101$20.00

Key Words

deep-inelastic scattering, proton structure functions, precision tests of QCD, search for new physics, spin-dependent structure functions, *B* meson production cross section

Abstract

The electron-proton collider HERA at DESY in Hamburg, Germany, delivered luminosity from 1992 to 2007. The center-of-mass energy was a factor of ten higher compared to previous lepton-nucleon scattering experiments. The research at HERA emphasized precision analyses of the proton structure at small x and high Q^2. In addition, precision tests of QCD provided significant new insights into the strong force. Diffractive scattering plays an important role in hard interactions. Many searches for new physics were performed at the electron-proton energy frontier. Additionally, two fixed-target experiments that used only the electron and proton beams of HERA, respectively, studied the spin structure of the nucleon and the production of strangeness, charm, and bottom in high-energy proton collisions. This review summarizes key results obtained by the four experiments at HERA: H1, ZEUS, HERMES, and HERA-B.

Contents

1. INTRODUCTION

Deep-inelastic lepton-nucleon scattering has always played a key role in exploring the structure of nucleons. HERA (Hadron-Elektron-Ringanlage, or Hadron-Electron Ring Accelerator), located at DESY in Hamburg (1), was the first machine in which leptons collided with protons in a storage ring. These collisions increased the center-of-mass energy by a factor of ten over previous fixed-target experiments. Data taking at HERA began in 1992 and ended in the summer of 2007.

HERA pioneered a new model of international cooperation. Partners from many countries financed one-quarter of the total cost through the construction of important components. Approximately 1,000 scientists from all over the world were involved in developing, constructing, and conducting the four major HERA experiments as well as in analyzing the data. HERA has provided a comprehensive and very precise overall picture of the proton and the forces acting within it. This review takes stock of the major results. After briefly summarizing the key features of the storage ring, we discuss the major results of the two collider experiments, H1 and ZEUS, and selected results from the two fixed-target experiments HERMES and HERA-B.

2. THE ACCELERATOR

The HERA accelerator had a circumference of 6.3 km. Inside the HERA tunnel, leptons (electrons or positrons) and protons were stored in two independent storage rings. Leptons and protons were preaccelerated in a chain of accelerators before being injected into the HERA ring, where they were accelerated to the nominal energies. The lepton ring was equipped with warm magnets, and the proton ring used superconducting magnets to provide the high magnetic field needed for proton energies in the teraelectronvolt range.

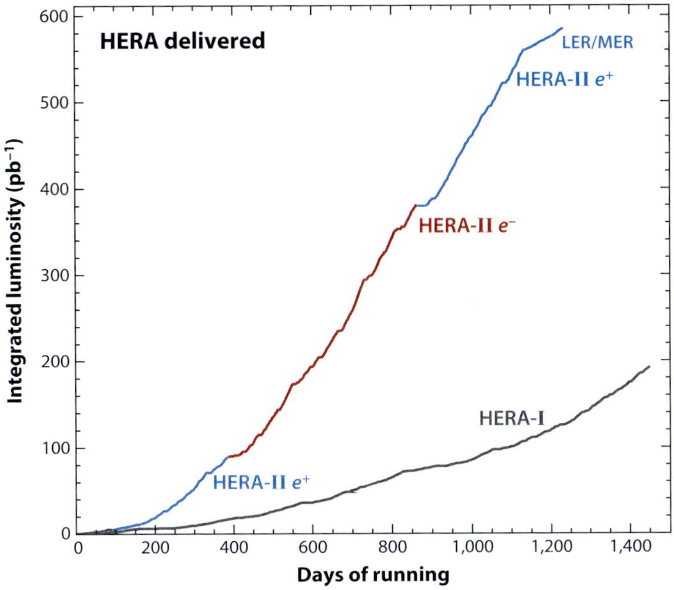

Figure 1

Luminosity delivered by HERA as a function of running days. Abbreviations: LER, low-energy running; MER, medium-energy running.

For HERA to be built and operated, substantial challenges had to be met. A few examples are the design of the superconducting magnets; the collision of different particle species; the large number of colliding bunches; and the short, 96-ns bunch-repetition time, which was also a major challenge for the experiments.

HERA had four interaction regions used by the experiments H1, ZEUS, HERMES, and HERA-B. Of the four experiments, H1 and ZEUS used the colliding lepton and proton beams, whereas HERMES used only the leptons and HERA-B only the protons.

HERA was operated with both electrons and positrons. From its startup in 1992 until 1997, HERA accelerated protons to 820 GeV and leptons to 27.6 GeV. It also delivered longitudinally polarized lepton beams to the HERMES experiment beginning in 1995. In 1998, the proton-beam energy was increased to 920 GeV, corresponding to a center-of-mass energy of 318 GeV. This operating period is referred to as HERA-I.

A significant luminosity upgrade program in 2000 resulted in a fourfold luminosity increase. At the same time, additional spin rotators were installed to provide longitudinal polarization to the colliding-beam experiments, H1 and ZEUS. This operating period is referred to as HERA-II. **Figure 1** shows the integrated collider luminosity delivered by HERA as a function of time. During the last few months of operation in 2007, the proton-beam energy was lowered to 460 GeV (medium energy) and 575 GeV (low energy) in order to measure one particular structure function, as explained in Section 4.1. The different run periods with the respective beam parameters are shown in **Table 1**. HERA operation ended in June 2007 after 15 years of data collection and after the design luminosity had been delivered.

The lepton beam in HERA became naturally transversely polarized by the emission of synchrotron radiation (2). The characteristic buildup time of polarization in HERA is approximately 40 min. Spin rotators on both sides of the experiments (3) changed the transverse polarization of the beam into a longitudinal one. The beam polarization was measured via independent

Table 1 Parameters of HERA running periods

Years	Lepton	E_{lep} (GeV)	E_{prot} (GeV)	E_{CMS} (GeV)	$\mathcal{L}$ (pb^{-1})	Period
1992–1994	e^-	27.6	820	300	2.2	HERA-I
1994–1997	e^+	27.6	820	300	70.9	HERA-I
1997–1998	e^-	27.6	920	318	25.2	HERA-I
1998–2000	e^+	27.6	920	318	94.9	HERA-I
2002–2004	e^+	27.6	920	318	89.7	HERA-II
2004–2006	e^-	27.6	920	318	290.9	HERA-II
2006–2007	e^+	27.6	920	318	180.6	HERA-II
2007 LER[a]	e^+	27.6	460	225	15.7	HERA-II
2007 MER	e^+	27.6	575	252	9.4	HERA-II

[a]Abbreviations: LER, low-energy running; MER, medium-energy running.

polarimeters for the transverse and longitudinal polarization (4, 5) by exploiting the spin-dependent cross section for Compton scattering of circularly polarized photons off leptons. During HERA-I operation, the polarization reached up to 65%. During HERA-II, due to larger beam-beam interactions, the degree of polarization was reduced to approximately 40%.

3. THE H1 AND ZEUS DETECTORS

The H1 and ZEUS detectors were large magnetic spectrometers with nearly hermetic coverage. They were designed following similar physics considerations but with different technical solutions, both for the calorimetric and the tracking measurements. They are described in detail elsewhere (6–9).

The main component of the H1 detector was the finely segmented liquid argon calorimeter (LAr) with resolutions, as measured with test beams, of $0.11\sqrt{E/\text{GeV}}$ and $0.50\sqrt{E/\text{GeV}}$ for electromagnetic and hadronic particles, respectively. The LAr was surrounded by a superconducting coil, which provided a solenoidal magnetic field of 1.16 T and an instrumented iron structure acting as both a shower tail catcher and a muon detector. In 1996 an upgrade was performed, during which the backward detectors were replaced by a backward drift chamber attached to a new lead fiber calorimeter with a high-resolution ($0.07\sqrt{E/\text{GeV}}$) electromagnetic section (EMC) followed by a hadronic section. In the upgraded configuration, the innermost central tracking detector of H1 also contained the central and backward silicon trackers. These were surrounded, sequentially, by the central inner proportional chamber and the central jet chamber (CJC), which was the main tracking device of H1. The CJC was complemented by drift chambers, and a further proportional chamber between the CJC chambers was used for triggering. Tracks in the forward direction were measured in the forward tracking detector, a set of drift chamber modules.

The main component of the ZEUS detector was the uranium–scintillator calorimeter. It was segmented into one EMC and either one or two hadronic sections. Under test-beam conditions, the energy resolutions were $0.18\sqrt{E/\text{GeV}}$ and $0.35\sqrt{E/\text{GeV}}$ for the EMC and hadronic sections, respectively. The timing resolution of the calorimeter was 1 ns for energy deposits greater than 4.5 GeV. Scintillator-tile presampler detectors were mounted in front of the calorimeter. Charged particles were tracked in the central tracking detector, which operated in a magnetic field of 1.4 T provided by a thin superconducting solenoid that was positioned inside the calorimeter. Planar drift chambers provided additional tracking in the forward and rear directions. For the HERA-II running period, a silicon strip microvertex detector was also installed.

Both H1 and ZEUS were also equipped with photon taggers, positioned at $\simeq 100$ m down the e-beam line, for a determination of the luminosity from Bethe-Heitler scattering, $ep \rightarrow ep\gamma$. The measurement accuracy of the luminosity was approximately 1–2% for each of the experiments.

4. PHYSICS RESULTS

A comprehensive review of collider physics results from HERA-I can be found in Reference 10. Here we emphasize recent results, particularly those from HERA-II.

4.1. Proton Structure

The principal processes of interest in the HERA physics program are the neutral-current (NC) and charged-current (CC) interactions of the electron or positron with the proton through the exchange of a neutral or charged virtual boson. These processes are characterized by either the scattered electron/positron or a neutrino in the final state.

The following variables are typically used to describe the kinematics: the virtuality of the exchanged boson, $-q^2 = Q^2$, where q is the four-momentum of the boson; the so-called Bjorken scaling variable, $x = Q^2/(2p \cdot q)$, where p is the four-momentum of the proton; the inelasticity, $y = p \cdot q / p \cdot k$, where k is the four-momentum of the incoming lepton; and the squared center-of-mass energy, $s = (p + k)^2$. They are related through $Q^2 = sxy$.

4.1.1. Proton structure formalism. The inclusive NC deep-inelastic $e^{\pm}p$ scattering cross section is given by a linear combination of structure functions. For unpolarized beams, it can be expressed as

$$\sigma^{\pm}_{r,\mathrm{NC}} = \frac{\mathrm{d}^2\sigma^{e^{\pm}p}_{\mathrm{NC}}}{dx\,\mathrm{d}Q^2} \cdot \frac{Q^4 x}{2\pi\alpha^2 Y_+} = \tilde{F}_2 \mp \frac{Y_-}{Y_+}\tilde{x}F_3 - \frac{y^2}{Y_+}\tilde{F}_L, \qquad 1.$$

where a propagator and a helicity factor are absorbed in the definition of the reduced cross section $\sigma^{\pm}_{r,\mathrm{NC}}$, and $Y_{\pm} = 1 \pm (1-y)^2$. The functions $\tilde{F}_2$ and $\tilde{F}_L$ include corrections to the purely electromagnetic structure functions F_2 and F_L due to Z^0 exchange. The function $x\tilde{F}_3$ arises completely from the Z^0 exchange. The structure functions depend only on the variables x and Q^2. More details can be found in Reference 11.

At low Q^2, the contribution of Z^0 exchange is negligible, and $\sigma_{r,\mathrm{NC}} = F_2 - y^2 F_L/Y_+$. The contribution of the term containing the structure function F_L is only significant for large values of y.

In the quark parton model, F_L is equal to zero (12), and F_2 is given as

$$F_2 = e_u^2(xU + x\bar{U}) + e_d^2(xD + x\bar{D}), \qquad 2.$$

where e_u and e_d denote the electric charge, in units of the elementary charge, of up- or down-type quarks. The parton distribution functions (PDFs) xU, xD, $x\bar{U}$, and $x\bar{D}$ denote the sums of up-type, down-type, and their antiquark distributions, respectively. Similar to the structure functions, they depend on x and Q^2. Below the b quark mass threshold, these sums are related to the quark distributions as follows:

$$xU = xu + xc, \quad x\bar{U} = x\bar{u} + x\bar{c}, \quad xD = xd + xs, \quad x\bar{D} = x\bar{d} + x\bar{s}, \qquad 3.$$

where xs and xc are the strange and charm quark distributions. Assuming symmetry between sea quarks and antiquarks, the valence quark distributions result from

$$xu_v = xU - x\bar{U}, \quad xd_v = xD - x\bar{D}. \qquad 4.$$

For inclusive unpolarized CC $e^{\pm}p$ scattering, the reduced cross section is defined as

$$\sigma^{\pm}_{r,\mathrm{CC}} = \frac{2\pi x}{G_F^2}\left[\frac{M_W^2 + Q^2}{M_W^2}\right]^2 \frac{\mathrm{d}^2\sigma^{e^{\pm}p}_{\mathrm{CC}}}{\mathrm{d}x\mathrm{d}Q^2}, \qquad 5.$$

and the corresponding CC structure functions are given by

$$\sigma^{\pm}_{r,CC} = \frac{Y_+}{2}W_2^{\pm} \mp \frac{Y_-}{2}xW_3^{\pm} - \frac{y^2}{2}W_L^{\pm}. \qquad 6.$$

The quantity G_F is the Fermi constant, and M_W is the mass of the heavy charged-vector bosons. In the quark parton model, $W_L^{\pm} = 0$, and the CC structure functions represent sums and differences of quark- and antiquark-type distributions, depending on the charge of the lepton beam, as

$$W_2^{+} = x\bar{U} + xD, \quad xW_3^{+} = xD - x\bar{U}, \quad W_2^{-} = xU + x\bar{D}, \quad xW_3^{-} = xU - x\bar{D}. \qquad 7.$$

From these equations it follows that

$$\sigma^{+}_{r,\mathrm{CC}} = x\bar{U} + (1-y)^2 xD, \quad \sigma^{-}_{r,\mathrm{CC}} = xU + (1-y)^2 x\bar{D}. \qquad 8.$$

Therefore, the NC and CC measurements may be used to determine the combined sea quark distribution functions, $x\bar{U}$ and $x\bar{D}$, and the valence quark distributions, xu_v and xd_v. A quantum chromodynamics (QCD) analysis in the DGLAP (Dokshitzer–Gribov–Lipatov–Altarelli–Parisi) formalism (13–17) also allows the gluon momentum distribution, xg, in the proton to be determined from scaling violations, namely the Q^2 evolution of the structure functions.

4.1.2. Inclusive deep-inelastic scattering cross sections. The HERA measurements of the inclusive deep-inelastic scattering (DIS) cross sections provide information on the structure of the proton in the ranges $10^{-6} \lesssim x \lesssim 0.5$ and $0.04\,\mathrm{GeV}^2 \lesssim Q^2 \lesssim 10^5\,\mathrm{GeV}^2$.

The H1 and ZEUS Collaborations achieved a significant reduction of their systematic uncertainties by combining 14 independent cross-section measurements performed with the HERA-I data for both the NC and CC processes (18). A significant part of the reduction in the uncertainties stems from a cross-calibration of the two detectors. In total, 1,402 individual data points are combined into 741 cross-section values on a common grid in the x, Q^2 plane. The uncertainties are extracted, taking 102 correlated sources into account.

In the so-called bulk region of the measurement, $10\,\mathrm{GeV}^2 \lesssim Q^2 \lesssim 100\,\mathrm{GeV}^2$, the level of the total uncertainty reaches $\approx$1%. An example is given in **Figure 2**, which shows the structure function xF_2 in three different bins of Q^2 from 3.5 GeV^2 to 650 GeV^2. At small Q^2, F_2 rises moderately toward low x. This rise becomes more and more pronounced as Q^2 increases. This rise is an effect of the gluon splitting into a quark-antiquark pair $g \to q\bar{q}$. Indeed, with growing Q^2, the time resolution of the probing photon improves, leading to an increasing number of interacting partons from gluon short time fluctuations into quarks. This rise is also qualitatively predicted by perturbative QCD within the so-called double-logarithmic approximation. It is well described by the QCD fit, as can be seen in **Figure 2**.

Figure 3 gives an overview of the data on the reduced cross section $\sigma^{+}_{r,NC}$ for NC $e^{+}p$ scattering as a function of Q^2 in different bins of x on a logarithmic scale. Also shown are data from fixed-target experiments. At low values of x, F_2 has a positive slope (positive scaling violations) originating from gluon splitting, whereas at high values of x, the slope becomes negative (negative scaling violations) due to gluon radiation. These effects arise because as Q^2 increases there is an increasing chance that a valence quark will radiate a gluon, which would effectively decrease the quark density at higher x while increasing it at lower x. In addition, at lower x there is an increased chance that a

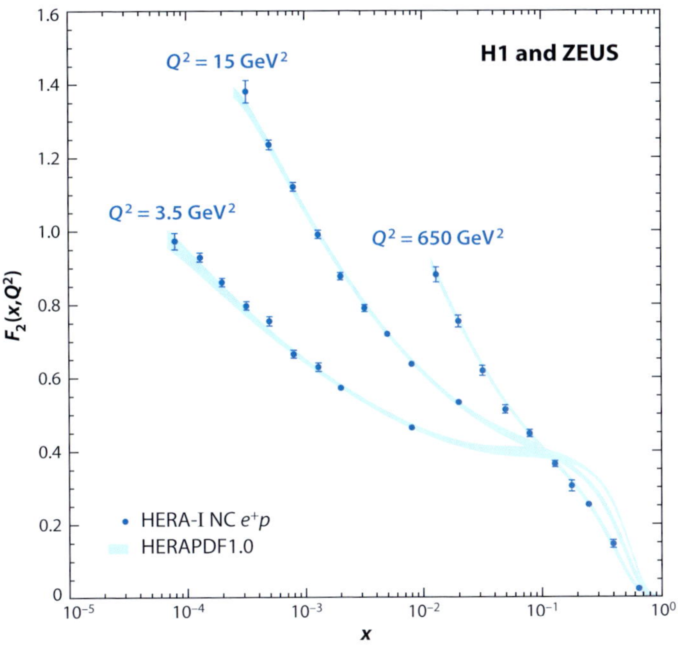

Figure 2

Structure function xF_2 of the proton measured by the H1 and ZEUS Collaborations at HERA in three bins of Q^2 as a function of x. Abbreviation: NC, neutral-current.

radiated gluon will split into a quark–antiquark pair, which would increase the low-x quark density. This figure demonstrates the size of the kinematic region covered as well as the good agreement with previous results.

These HERA data are the only precise source of information on the structure of the proton at values of x smaller than 10^{-3} and at values of Q^2 larger than a few hundred gigaelectronvolts squared. They are of particular interest at the Large Hadron Collider (LHC).

4.1.3. Electroweak effects at high Q^2. The center-of-mass energy of HERA was sufficiently large to access the region of phase space in which weak and electromagnetic effects are of comparable magnitude. This is clearly demonstrated in **Figure 4**, which shows the NC and CC $e^{\pm}p$ scattering cross sections as a function of Q^2 (19–21). At small values of Q^2, the NC process dominates because only electromagnetic effects contribute. When Q^2 is comparable to the mass squared of the Z^0 and $W^{\pm}$ bosons, the two cross sections are of similar magnitude. **Figure 4** also demonstrates that the CC cross section is much larger in e^-p than in e^+p scattering. In the former cross section, the exchanged W^- couples mostly to the u-valence quarks, which are approximately twice as abundant in the proton than the d-valence quarks.

A similar but smaller effect can be observed in the NC cross section, where the difference between the e^+p and e^-p cross sections is due to interference between photon and Z^0 exchange. From the difference one obtains the structure function xF_3 (Equation 1), which provides information about the valence quark distributions. These measurements currently also provide the most stringent constraint on the weak couplings of the u and d quarks.

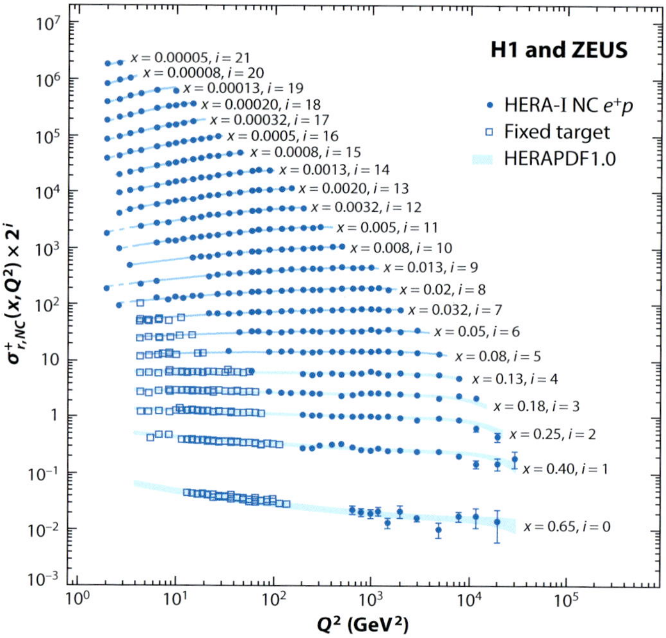

Figure 3

Structure function xF_2 of the proton measured by the H1 and ZEUS Collaborations at HERA as a function of Q^2 in bins of x. Abbreviation: NC, neutral-current.

During HERA-II operation, the lepton beam in H1 and ZEUS was longitudinally polarized, allowing tests of the helicity structure of electroweak interactions. The maximum polarization in this period reached values of around 40%. The most prominent effect of the lepton-beam polarization can be seen in the CC cross section. This cross section depends linearly on the polarization and should vanish for right-handed electrons or left-handed positrons. The three measurements of the CC cross sections made with unpolarized beams and with negatively and positively polarized beams lie on a straight line when plotted as a function of the beam polarization (20). The extrapolation of this line to polarizations of either $+1$ or -1 provides information about the possible existence of right-handed currents, which are excluded in the Standard Model (SM). No deviations from the SM expectation were observed, although the sensitivity of this measurement is not competitive with other methods (22, 23) that provide more stringent limits.

4.1.4. HERAPDF1.0. PDFs were obtained using a QCD fit to the combined H1 and ZEUS data from HERA-I (discussed in Section 4.1.2). The precision of these data combined with the understanding of the systematic uncertainties allows a rigorous extraction of the total error of the fit results, which also includes model and parameterization uncertainties. An example is presented in **Figure 5**, which shows u- and d-valence quarks as well as sea quark and gluon distributions at a scale of $Q^2 = 10{,}000$ GeV^2. The uncertainties on the sea quark and gluon distributions at values of $x < 10^{-2}$ amount to a few percent. The chosen scale is most relevant for the central production of any heavy object at the LHC, where the mass of the heavy object corresponds to the scale Q^2. The uncertainty on the PDF translates into a corresponding uncertainty for any theoretical prediction. For SM processes, such as the production of single $W^\pm$ or Z^0 bosons with subsequent leptonic decays, the total uncertainty stemming from knowledge of the PDFs is less than 3%.

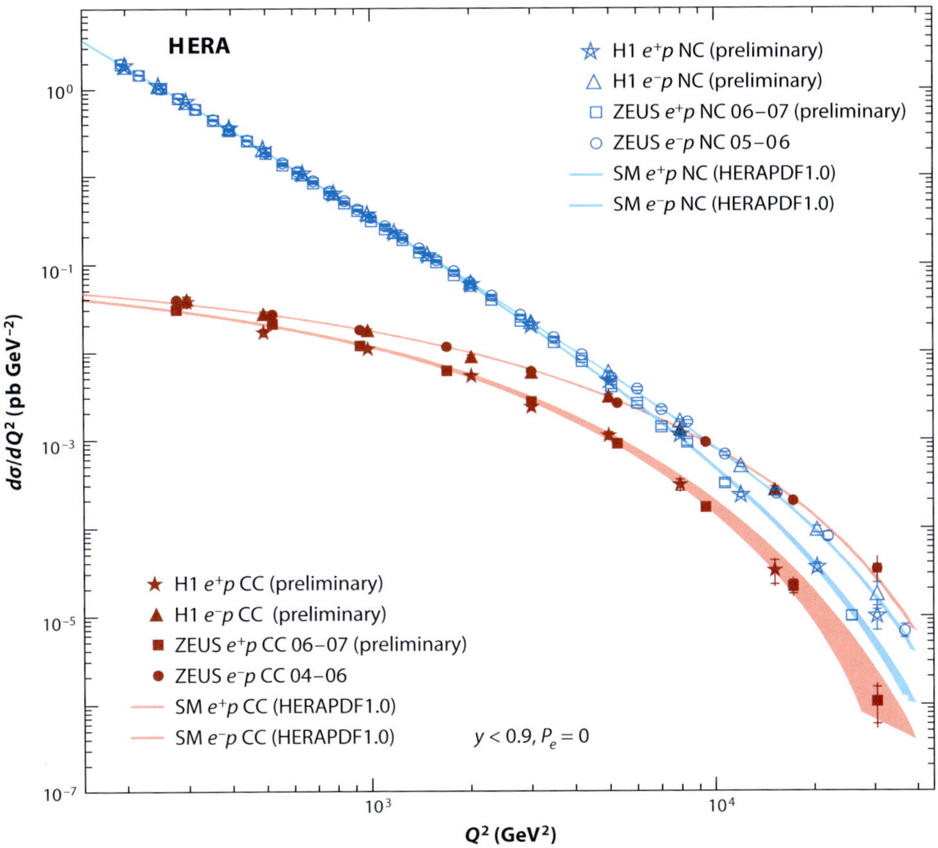

Figure 4

Cross section $\frac{d\sigma}{dQ^2}$ for deep-inelastic $e^{\pm}p$ neutral-current (NC) and charged-current (CC) scattering with positively and negatively polarized lepton beams. Abbreviation: SM, Standard Model.

4.2. Charm and Beauty Production

Heavy quarks were produced abundantly at HERA; charm constitutes roughly 20% and beauty approximately 5% of the total DIS cross section. Perturbative QCD should describe these cross sections adequately, given that the mass of the heavy quark provides a hard scale. Thus, these cross sections are an excellent testing ground for perturbative QCD and also provide a handle on the gluon density inside the proton.

However, it remains to be shown whether the charm quark contribution to the DIS cross section can be explained from gluon splitting or whether there might be an intrinsic, nonperturbative source of heavy quarks in the proton. To answer this question, one needs to overcome significant experimental and theoretical challenges. Heavy quarks can be tagged exclusively by identifying the respective mesons, or they can be identified inclusively by looking for displaced secondary vertices.

Significant progress compared with HERA-I has been made on both of these fronts. The HERA-II data sample provides large numbers of identified mesons, and the improved vertexing capabilities of both experiments yield yet another quantitative step in statistical precision (24–29). This fact is demonstrated in **Figure 6**, which presents the first combination of H1 and ZEUS

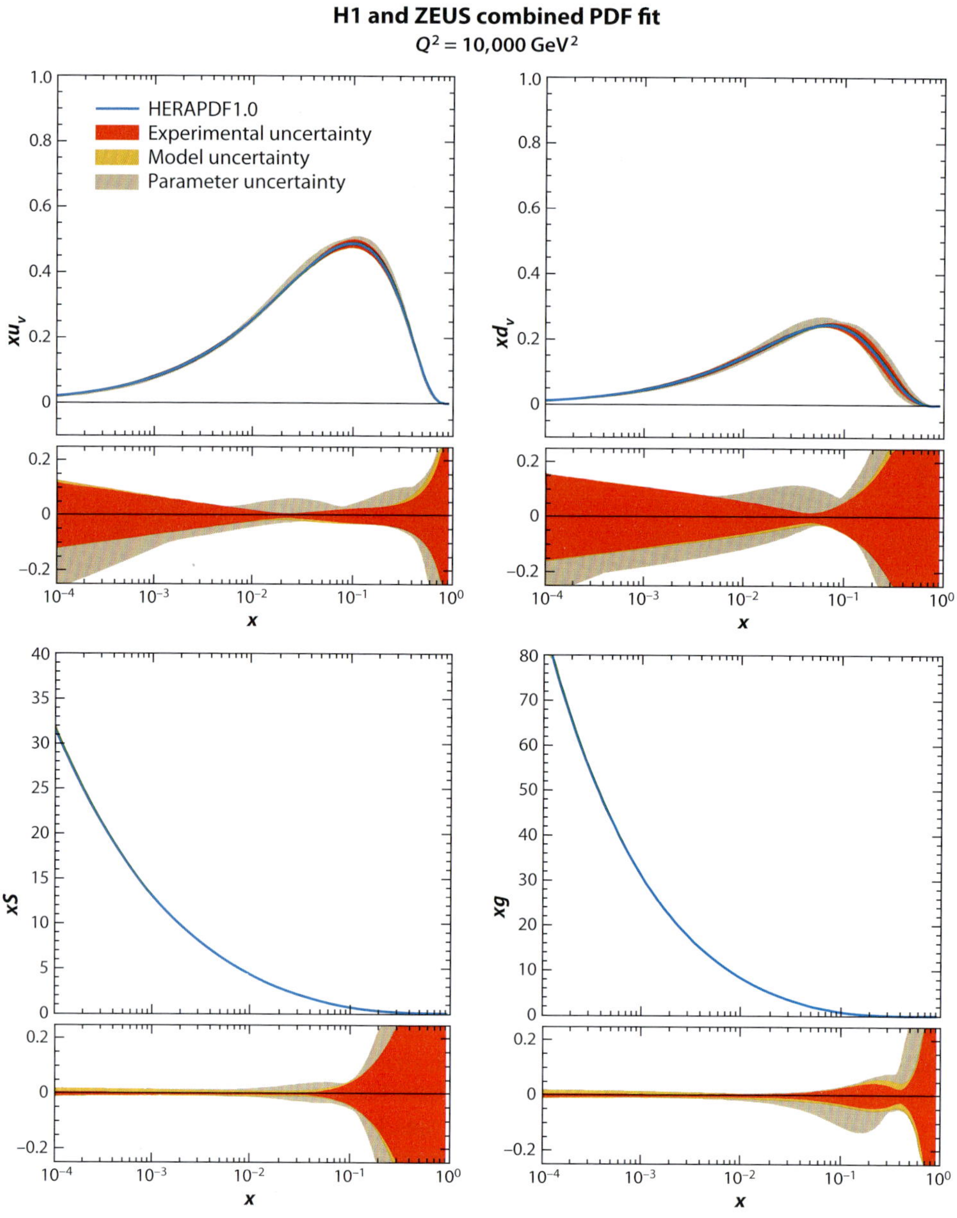

Figure 5

HERAPDF1.0 Parton distribution functions at a scale of $Q^2 = 10{,}000$ GeV^2 for u-valence quarks, d-valence quarks, sea quarks, and gluons as a function of the Bjorken scaling variable x.

results on the charm structure function, $xF_2^{c\bar{c}}$. The data are shown as a function of x in bins of Q^2. The data are compared to next-to-leading-order QCD predictions through use of the recent HERAPDF1.0 PDF. The data and the predictions are compatible at the levels of both the experimental and theoretical uncertainties, indicating that there is no sizable nonperturbative source of charm production. However, the uncertainties are sizable: The experimental uncertainties

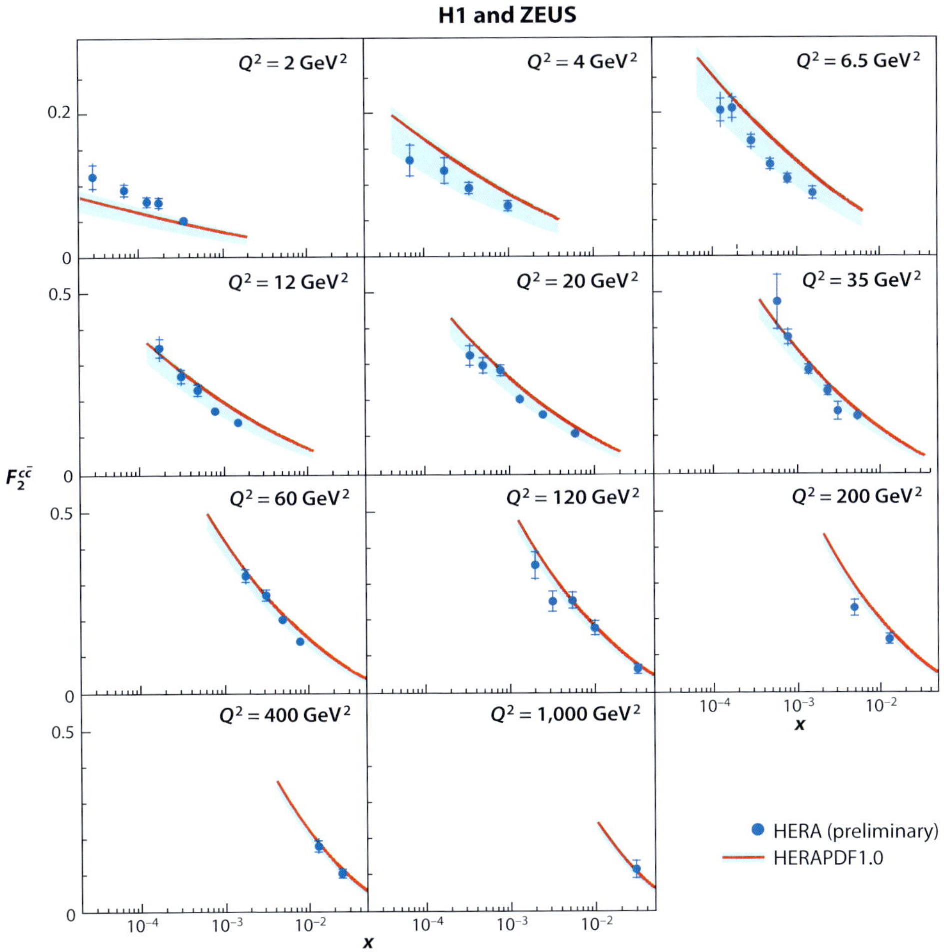

Figure 6

Combined H1 and ZEUS results on the structure function $F_2^{c\bar{c}}$ as a function of Bjorken x in bins of Q^2. The data are compared to the predictions based on the HERAPDF1.0 Parton distribution functions.

currently amount to approximately 10%. The major part of the theoretical uncertainties is from the detailed treatment of the next-to-leading-order calculation at the charm threshold. Significant improvements can be expected from ongoing experimental studies, from a better theoretical treatment, and from next-to-next-to-leading-order calculations.

4.3. Precise Tests of Quantum Chromodynamics

Proton structure and QCD can be investigated in more detail using measurements of the hadronic final states at HERA.

4.3.1. Particle production in deep-inelastic scattering. The production of charged particles depends on the details of the hadronization process, which governs the transformation of the produced partons into hadrons. Models of hadronization were confronted with data gathered from measurements of charged-particle multiplicities that were performed at HERA (30, 31).

Also, the asymmetry between positively and negatively charged particles has been measured. It is significant for particles carrying a large fraction of the corresponding quark momentum (32). As expected, there is a correlation between the charges found in the hadronic final state with those of the corresponding valence quarks in the protons. Thus, this measurement can further constrain the PDFs.

4.3.2. Jet production. Most of the time, DIS events contain a single jet, which corresponds to the scattered parton. In the Breit frame of reference, the struck quark is scattered along the virtual-photon direction. At leading order, it has no transverse momentum with respect to the direction of the virtual photon. Jets with large transverse energy E_T in the Breit frame result from the radiation of energetic gluons. As a consequence, the production rate of jets with high E_T is sensitive to the strong coupling constant, α_s.

Measurements of jet cross sections as functions of Q^2 and E_T have been made (33–35). The dependence of the cross sections on these variables allows testing of perturbative QCD at various scales in a single experiment, thus providing a unique handle on the strong interaction scale. This scale is inversely proportional to the effective distance at which the strong interaction is probed. The decrease in α_s with increasing scale, corresponding to a decreasing probed distance—often termed the running of α_s—demonstrates the asymptotic freedom of QCD. Studies of jet production in quasi-real photon-proton collisions (photoproduction regime) complement the studies in the DIS regime (36).

Results on jet production also help constrain the PDFs, particularly the gluon content of the proton. Fits combining both the inclusive DIS data and the jet data have successfully been performed by the ZEUS Collaboration (37). This combined approach enhances the robustness of the PDF and α_s determinations.

A summary of the measurements of the strong coupling obtained at HERA at the scale corresponding to the mass of the Z^0 boson. The total error is dominated by the theoretical uncertainty, which can be improved in the future by increasing the precision of the theoretical calculations (38, 39), and is shown in **Figure 7**.

4.4. Diffraction

The process of diffractive scattering has been known for a long time and has been extensively studied in hadron-hadron collisions. It was phenomenologically described in terms of Regge theory and considered to be a soft phenomenon that would not be present in hard interactions. Therefore, it came as a surprise when a sizable diffractive contribution, amounting to as much as 15% of the total γ^*p scattering cross section, was found in DIS at HERA through the observation of large-rapidity-gap events (40) inside the proton. In these processes, the protons escape intact even though hard hadronic interactions take place during the electron-proton collisions.

In the naïve picture of deep-inelastic interactions, such processes do not occur with a sizable rate because the proton is described as a gas of uncorrelated quarks and gluons. The photon knocks out one of the partons, leaving behind a colored state, and hadronization occurs between the struck parton and the proton remnant. The existence of a large diffractive component at high Q^2 shows that the structure of the proton is indeed not gas-like. Strong correlations between partons make the proton "grainy," so that the interactions take place between the photon fluctuating into a quark-antiquark pair and a colorless correlated state made of more than one parton inside the proton. In QCD this state comprises at least two gluons and is sometimes termed the Pomeron.

The Q^2 behavior of the inclusive diffractive cross section (41, 42) is described by QCD (**Figure 8**). It shows the reduced cross section or, roughly speaking, the diffractive structure

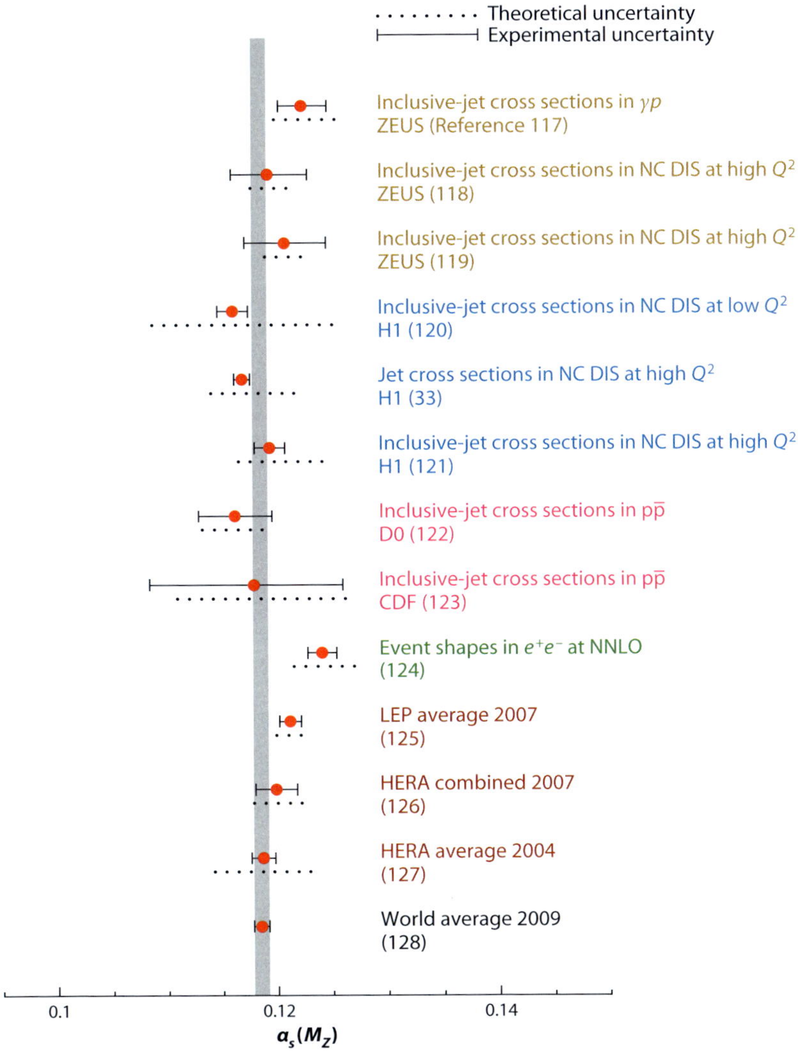

Figure 7

Different measurements of the strong coupling constant, at the mass of the Z^0 boson, compared with the world average. Abbreviations: DIS, deep-inelastic scattering; NC, neutral-current; NNLO, next-to-next-to-leading order.

function $F_2^{D(3)}$ for one value of the fractional momentum $x_{IP} = (M_X^2 + Q^2)/(W^2 + Q^2) = 0.01$ as a function of Q^2 in bins of the variable $\beta = x/x_{IP}$. The quantities M_X and W are the masses of the diffractively scattered system and of the total photon-proton system, respectively. The lines show the result of a QCD fit, which describes the inclusive data well.

Another important feature of diffractive scattering can be observed in quasi-real photon-proton γP scattering. **Figure 9** shows the total γP cross section together with the production cross sections for various vector mesons as a function of W. The total γP cross section rises slowly with W. For the vector mesons, this rise is faster and is well described by a power-like behavior whose power increases as the mass of the vector meson increases. This relationship indicates that, as the mass

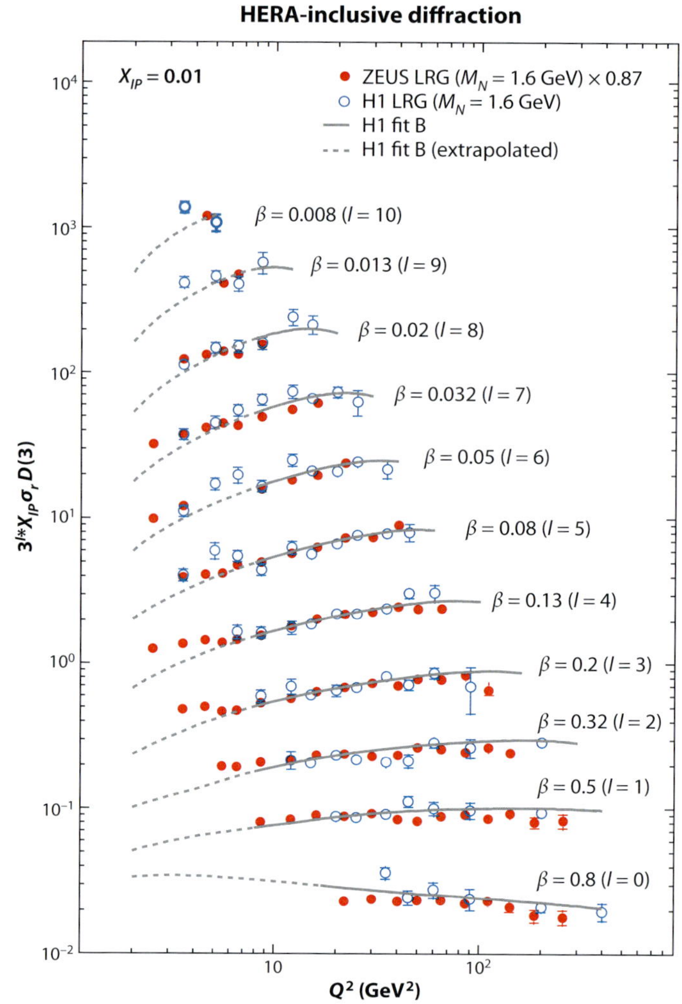

Figure 8

The diffractive structure function $F_2^{D(3)}$ as a function of Q^2 in different bins β at a value of $x_{IP} = 0.01$.

of the vector meson becomes large enough to provide a hard scale, these diffractive processes are indeed hard and described by perturbative QCD. The power of the rise with energy is close to two times that of inclusive DIS, indicating that it is roughly governed by the square of the gluon density. This finding has been confirmed by the agreement of the absolute size of the vector meson cross sections with the predictions via the gluon density extracted from inclusive measurements. These results pave the way for further studies at the LHC, or a future electron-ion collider, at which diffractive scattering will provide a very clean laboratory in which to test many aspects of the SM and also to search for new physics.

4.5. Searches for New Physics at the *ep* Energy Frontier

The HERA collider probed the structure of matter at the highest energy in *ep* collisions. It is complementary to two other colliders at the Fermi scale: LEP (e^+e^-) and the Tevatron ($p\bar{p}$).

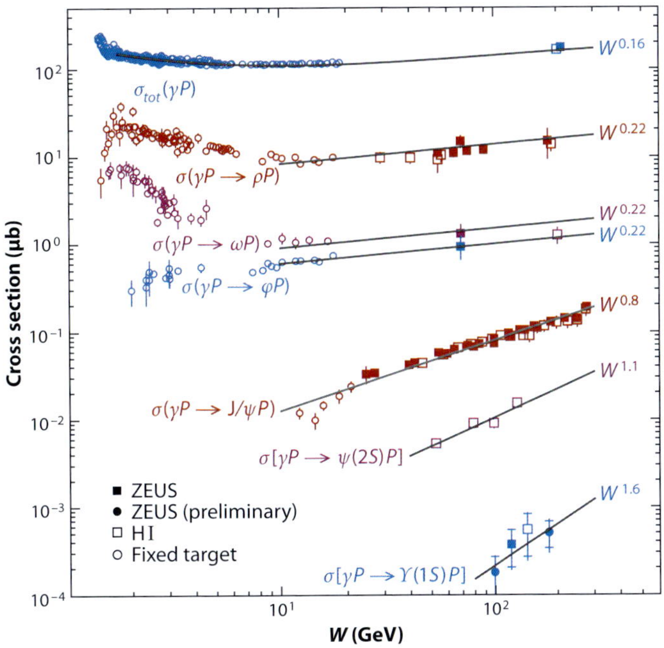

Figure 9

The total γP cross section and the cross sections for diffractive production of ρ, ω, ϕ, J/ψ, $\psi(2s)$, and $\Upsilon(1s)$ mesons as a function of the γP center-of-mass energy, W.

HERA data are particularly sensitive to new phenomena that can arise from the fusion of a beam lepton with a quark from the proton. Therefore, the measured DIS cross sections can be used to search for deviations from the predictions of the SM at large Q^2. The measurements are particularly sensitive to a possible finite size of the interacting quarks. If quarks are not point-like, the predicted SM cross section at largest Q^2 is distorted, following a form factor depending on a parameter corresponding to the effective quark radius (43). Because no deviation from the SM prediction has been found, an upper limit for the quark radius of approximately 0.7×10^{-18} m has been determined (**Figure 10**).

A new force that manifests itself at an effective scale Λ can modify the DIS cross sections at high Q^2 via contact interactions. Limits on Λ, the scale of the new physics, have been set at HERA in the 4.6–9-TeV range, depending on the chiral structure of the contact interactions (44). Specifically, the new interaction may be mediated by new bosons (leptoquarks) carrying both lepton and baryon numbers. If their masses are below the HERA center-of-mass energy, a resonant production is expected. The leptoquark decays lead to electron- or neutrino-jet final states, similar to DIS events. No such resonances have been found, and limits on the coupling of leptoquarks as a function of mass have been obtained (45). Searches for other signals of physics beyond the SM, namely excited fermions (46–48), supersymmetry (49), and a doubly charged Higgs (50), have been performed. Despite the unique reach of HERA into parts of the parameter space of these models, the searches revealed no significant deviation from the SM predictions.

Phenomena beyond the SM can also be identified via spectacular events that correspond to topologies for which the SM predicts small contributions. Events with energetic isolated leptons (electrons or muons) and missing transverse momentum have been measured at HERA (51, 52). Within the SM, such events originate predominantly from W boson production followed by

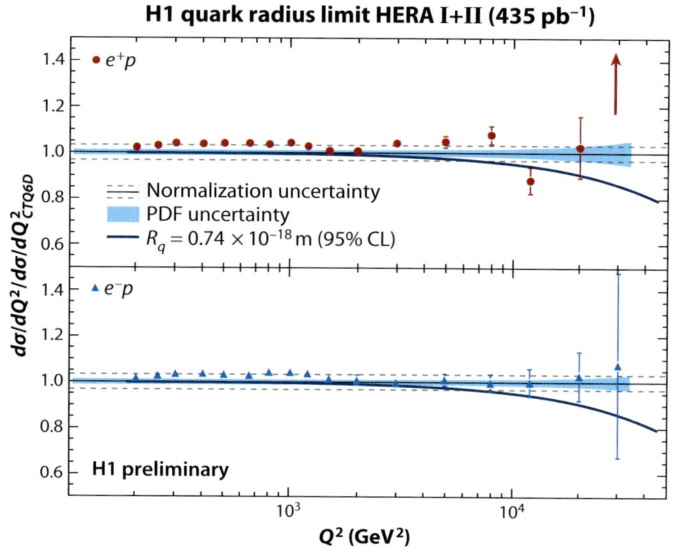

Figure 10

Ratio of the neutral-current cross section for e^+p and e^-p scattering, measured by the H1 experiment, divided by the expectation from the Standard Model. This ratio is used to extract a limit on a finite quark radius. Abbreviation: PDF, parton distribution function.

its leptonic decay, $ep \rightarrow eWX$, where $W \rightarrow \ell\nu$. The measurements of H1 and ZEUS have been combined (53), and a cross section for W production at HERA has been measured to be 1.06 ± 0.16 (stat.) $\pm$ 0.07 (sys.) pb, in agreement with the SM expectation of 1.26 ± 0.19 pb (54–56). Anomalous production of W bosons associated with a prominent hadronic jet may also be a signature for an anomalous production of single top quarks. Anomalous production of single top quarks is predicted in, for example, models with flavor-changing neutral currents via the coupling between the exchanged photon, a u quark from the proton, and the top quark. No signal was found, resulting in what is currently the most stringent limit on the magnetic coupling of the top quark (57).

Events with several high-P_T leptons at HERA have been observed and investigated in a combination of H1 and ZEUS analyses (58). Such events are produced mainly in photon-photon collisions, in which the interacting photons originate from the electron and the proton. All possible combinations of electrons and muons have been investigated. The results are shown in **Figure 11**. Events with high mass (i.e., above 100 GeV) have been observed by both experiments. To compare the dilepton and trilepton events, one must use the sum over the lepton transverse momentum $\sum P_T^\ell$. Seven di- and trilepton events have been observed in e^+p collisions with a scalar sum of the lepton transverse momenta above 100 GeV. Only 1.94 ± 0.17 events are expected. Such events have not been observed in e^-p collisions, for which 1.19 ± 0.12 is predicted. Total visible and differential dielectron and dimuon photoproduction cross sections have been extracted in a restricted phase space dominated by photon-photon collisions, and these cross sections are in good agreement with the SM prediction.

By using final-state particles or hadronic jets at large transverse momentum selected in the same polar angular range, a general search for deviations from the SM was performed by classifying events in classes of exclusive topologies (59). No significant indication of any new physics signal was

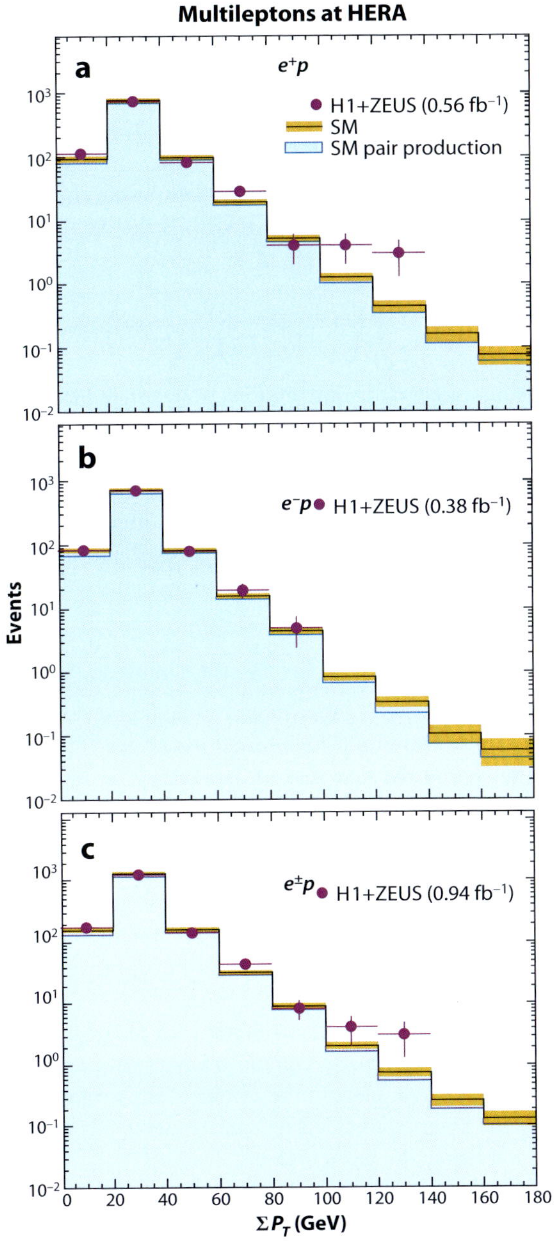

Figure 11

Multilepton production at HERA as a function of the sum of the transverse momenta of the leptons for (*a*) e^+p and (*b*) e^-p, scattering separately, and for (*c*) both processes together. The data are from a combined analysis of the H1 and ZEUS data. They correspond to a luminosity of 0.94 fb^{-1}. Abbreviation: SM, Standard Model.

observed. The results of this multichannel, model-independent analysis are in excellent agreement with the SM.

5. HERMES

5.1. Introduction

The HERMES experiment was proposed in early 1988 after the EMC Collaboration's (60) surprising observation that only a small fraction of the nucleon's spin could be attributed to the spins of the quarks. The primary scientific goal of HERMES is the detailed investigation of the spin structure of the nucleon and the determination of the partonic composition of its spin projection (61),

$$s_z = \frac{1}{2} = \left(\frac{1}{2}\Delta\Sigma + L_z^q\right) + (\Delta g + L_z^g). \qquad 9.$$

Here $\frac{1}{2}\Delta\Sigma$ (Δg) are the net integrated contributions of quark and antiquark (gluon) helicities in a reference frame in which the nucleon has very high (infinite) longitudinal momentum and in which L_z^q (L_z^g) is the z component of the orbital angular momentum of all quarks (gluons).

The proposal was based on two novel technologies: (*a*) the longitudinally polarized lepton beam in HERA and (*b*) a target cell internal to the lepton ring, where nuclear-polarized atoms from a high-intensity polarized source are present as pure atomic species without dilution from unpolarized target material. Data taking began in the early summer of 1995.

5.2. The Experiment

A detailed description of the HERMES experiment can be found elsewhere (62). The longitudinally polarized lepton beam scattered off the internal gas target. The 40-cm-long, open-ended, thin-walled target cell was fed in the first year of data taking by an optically pumped source of polarized ^{3}He atoms (63) and in the following years by an atomic-beam source (64) of hydrogen (H) or deuterium (D). The nuclear polarization of the atoms was flipped every 1 to 3 min. The average magnitude of the longitudinal target polarization with respect to the beam direction was approximately 0.46 ± 0.01 for ^{3}He and 0.85 ± 0.03 for both H and D. The transverse H polarization was 0.72 ± 0.06. The areal density of the polarized H (D) target was approximately 0.8×10^{14} (2×10^{14}) nucleons cm^{-2}, corresponding to a luminosity of approximately 1×10^{31} (2.4×10^{31}) nucleons cm^{-2} s^{-1} for an average beam current of 20 mA.

The HERMES detector was a forward spectrometer with a large number of tracking chambers and several particle-identification detectors. A dipole magnet surrounded the electron, and proton-beam pipes and a flux-exclusion iron plate in the center of the magnet gap shielded the electron and proton beams from the magnetic field. The spectrometer was constructed as two identical halves, mounted above and below the beam pipes.

Scattered leptons and hadrons produced in the inelastic reactions were detected and identified within a geometrical acceptance of $\pm$ 170 mrad horizontally and $\pm$ (40–140) mrad vertically. The trigger for scattered leptons was formed by a coincidence of signals from three hodoscope planes with signals from a radiation-hard lead-glass calorimeter (65). Lepton-hadron discrimination and particle identification were accomplished by using the information from four particle-identification detectors: the lead-glass calorimeter, a preshower detector located directly in front of the calorimeter, a transition-radiation detector, and a Cherenkov counter [primarily a dual-radiator ring-imaging Cherenkov counter (66)]. The system provided lepton identification with an average efficiency of 98 to 99% and a hadron contamination of less than 1%. The luminosity of the experiment was measured by detecting in coincidence either (*a*) electron (electron-positron) pairs from Møller (Bhabha) scattering of beam electrons (positrons) off the target-gas electrons or (*b*) photon pairs from e^+e^- annihilation (67).

5.3. Data Taking and Overview of Results

Data taking by HERMES can be subdivided into the three phases: 1995–2000, 2002–2005, and 2006–2007. During the first phase, data were taken mainly with longitudinally polarized ^{3}He (1995), H (1996–1997), and D (1998–2000) targets. These data yielded high-precision results on inclusive and semi-inclusive polarized DIS. They have confirmed and extended our understanding of the contribution of quark spins to the spin of the nucleon. During the second phase, measurements were performed with a transversely polarized H target that provided the first information about transverse momentum–dependent PDFs and fragmentation functions (FFs). During the third phase, measurements were made with high-density unpolarized H and D targets in conjunction with a dedicated recoil detector around the storage cell target to study in detail hard exclusive processes that help constrain generalized parton distributions (68–70). From 1995 onward, data were also collected with various other unpolarized gas targets (H_2, D_2, ^{3}He, ^{4}He, N_2, Ne, Kr, and Xe) to study the hadronization process (71).

The physics reach of the experiment extended well beyond the original scientific goal, and HERMES explored many details of hadron structure, hadron production, hadronic interactions, and especially hard exclusive reactions (72–81) with electromagnetic probes at center-of-mass energies of approximately 7 GeV. Many of the results published to date from the first and second phases are discussed in References 82 and 83. Some selected results are presented below. The data taken in the third phase with the recoil detector are still being analyzed.

5.4. Selected Results

As examples, we describe three results that illustrate the three data-taking configurations of HERMES: (*a*) the determination of the spin-dependent structure function $g_1(x)$ from measurements of inclusive DIS with both beam and target longitudinally polarized; (*b*) the measurement of azimuthal asymmetries in semi-inclusive DIS with a transversely polarized H target that provides information about transverse momentum–dependent PDFs and FFs; and (*c*) the extraction of the strange quark distribution function from kaon multiplicities in semi-inclusive DIS from a mostly unpolarized D target.

5.4.1. $\Delta\Sigma$ and the spin-dependent structure function g_1. Information about the net contribution of quark spins to the spin of the nucleon, $\frac{1}{2}\Delta\Sigma$, can be obtained from double-spin asymmetries (DSAs) in inclusive DIS of longitudinally polarized leptons from longitudinally polarized nucleons. The corresponding virtual-photon asymmetry A_1 is proportional to the ratio of the spin-dependent structure function g_1 and the unpolarized structure function F_1. At leading order in QCD, g_1 is defined as $g_1(x) = \frac{1}{2}\sum_q e_q^2(\Delta q + \Delta\bar{q})$, where $\Delta q = q^+ - q^-$ is the quark helicity distribution for massless quarks of flavor q in a longitudinally polarized nucleon in the so-called infinite-momentum frame. Here q^+ (q^-) is the number density of quarks with the same (opposite) helicity as that of the parent nucleon. At higher orders in QCD, g_1 is a convolution of quark, antiquark, and gluon helicity distributions (84). The first moments of these quantities at a fixed value of Q^2 are given by $\Gamma_1 = \int_0^1 g_1(x)dx$ and $\Delta q = \int_0^1 \Delta q(x)dx$, and $\Delta\Sigma$ is defined as $\Delta\Sigma = \sum_q(\Delta q + \Delta\bar{q})$. The HERMES results for $xg_1(x)$ for both the proton and the deuteron (85) are presented as a function of x in **Figure 12**. For $x < 0.04$, $g_1^d(x)$ becomes compatible

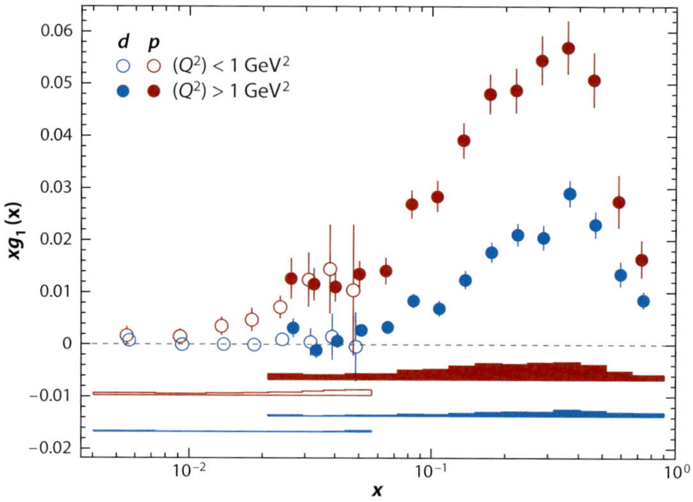

Figure 12

The spin-dependent structure function $xg_1(x)$ for the proton and the deuteron. The bands in the lower part of the graph represent systematic uncertainties.

with zero. Under the assumption that the first moment of $g_1^d(x)$ saturates for $x < 0.04$ [this is confirmed by data from the COMPASS Collaboration at lower values of x (86)], a value of $\Gamma_1^d = 0.00436 \pm 0.0012$ (stat.) ± 0.0033(sys.) has been extracted for $Q^2 = 5$ GeV^2. In the so-called modified minimal subtraction scheme (87), the gluon helicity distribution does not contribute to the first moment of g_1, and Γ_1^d is related to $\Delta\Sigma$ by

$$\Delta\Sigma = \frac{1}{\Delta C_S}\left(\frac{9\Gamma_1^d}{(1-\frac{3}{2}w_D)} - \frac{1}{4}a_8\Delta C_{NS}\right). \qquad 10.$$

Here $a_8 = 0.586$ is the octet axial charge determined from weak hyperon decay (88), $w_D = 0.05 \pm 0.01$ takes into account the D-state admixture in the deuteron-wave function, and C_{NS} (C_S) are QCD coefficients calculated to third (second) order in α_s. With these quantities, $\Delta\Sigma$ yields a value of $\Delta\Sigma = 0.330 \pm 0.011$(theor.) $\pm$ 0.025(exp.) $\pm$ 0.028(evol.) at $Q_0^2 = 5$ GeV^2. Consequently, the contribution of quark spins to the spin of the nucleon has been well determined to be approximately $\frac{1}{3}$; the rest must be attributed to gluons and orbital angular momenta. Furthermore, precise information on flavor-separated quark (antiquark) helicity distribution functions have been obtained from longitudinal DSAs for various identified hadrons in semi-inclusive polarized DIS (89). DSAs in the production of hadrons with high transverse momenta allowed access to the gluon helicity distribution function. From such measurements, HERMES extracted a rather small value of the average gluon polarization $\Delta g/g = 0.05 \pm 0.15$ in the range $0.1 < x < 0.4$ (90).

5.4.2. The strange quark distribution function. Experimental information about the PDF $s(x)$ of strange quarks (antiquarks) as a function of x is surprisingly scarce. Most of the experimental constraints are based on measurements of oppositely charged muon pairs in deep-inelastic neutrino and antineutrino scattering. In the absence of significant experimental constraints, most of the current global QCD fits of PDFs assume $s(x)$ [$\bar{s}(x)$] to be related to the distribution functions of light antiquarks by $s(x) = \bar{s}(x) = r[\bar{u}(x) + \bar{d}(x)]/2$, where $r \approx 0.3$–0.5 at some low factorization scale.

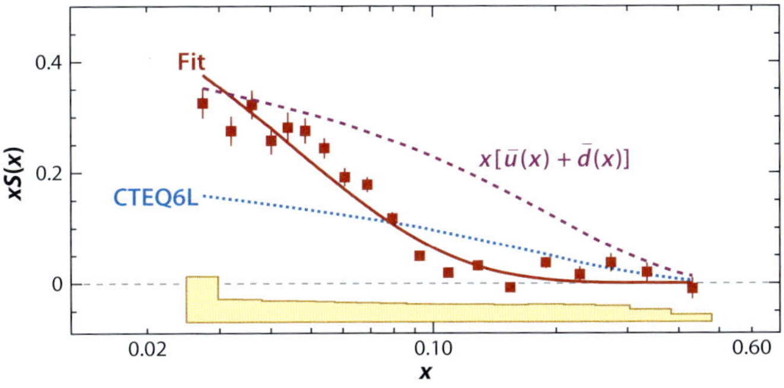

Figure 13

The strange parton distribution $xS(x)$ at $Q^2 = 2.5$ GeV^2 derived from the multiplicity of charged kaons in semi-inclusive deep-inelastic scattering from a deuterium target. The solid curve is a fit to the data. The shape is incompatible with $xS(x)$ from CTEQ6L as well as with the assumption that it is proportional to the light quark sea.

HERMES performed the first leading-order extraction of $S(x) = s(x) + \bar{s}(x)$ from the multiplicity of charged kaons in semi-inclusive DIS from a deuteron target (91). The measured kaon multiplicity cannot be reproduced by a fit by using the CTEQ6L (92) strange quark distribution functions and with the FFs of light and strange quarks into charged kaons as free parameters. The distribution $xS(x)$ extracted from these multiplicities is presented in **Figure 13**.

5.4.3. Transverse momentum–dependent distribution and fragmentation functions. A complete description of the partonic structure of the nucleon in leading twist requires three quark distribution functions that survive integration over intrinsic transverse momenta. These are the unpolarized quark distribution function $q(x)$, the quark helicity distribution function $\Delta q(x)$, and the chiral-odd transversity distribution function $\delta q(x)$ (93). There are five additional transverse momentum–dependent distribution functions that do not survive the integration (94). So far, they are essentially unexplored experimentally, and HERMES has performed pioneering measurements to access all of them. One example is the time-reversal-odd Sivers distribution function (95), $f_{1T,q}^{\perp}(x)$, which describes the distribution of unpolarized quarks in a transversely polarized nucleon and arises from a correlation between the momentum direction of the struck quark and the spin direction of its parent nucleon. It is of special interest as it can be related to the orbital angular momenta of quarks (96). One way to study this distribution function is via the azimuthal angular asymmetry in the distribution of hadrons produced in semi-inclusive DIS from a transversely polarized target. It gives rise to a $\sin(\phi - \phi_S)$ modulation in the azimuthal distribution, where ϕ (ϕ_S) is the azimuthal angle between the lepton scattering plane and the hadron production plane (the target spin axis). Its moment is proportional to a convolution of $f_{1T,q}^{\perp}(x)$ and the unpolarized FF $D_{1,q}(z)$, where z is the fraction of the virtual photon's energy carried by the hadron. Furthermore, the moment of the $\sin(\phi + \phi_S)$ modulation is proportional to a convolution of the transversity distribution function and the chiral-odd polarized Collins FF $H_{1,q}^{\perp}(z)$ (97). Results (98) for the Sivers moments for positive pions and kaons, obtained from data taken with the transversely polarized H target, are shown in **Figure 14**. The measured Sivers asymmetries are significantly positive, providing the first evidence for a time-reversal-odd PDF appearing in leptoproduction. Consequently one has to conclude that orbital angular momenta of quarks inside the nucleon are nonzero. However, at present it is not possible to quantitatively relate the magnitude of this asymmetry to the fraction of nucleon spin that can be attributed to orbital angular momenta of quarks.

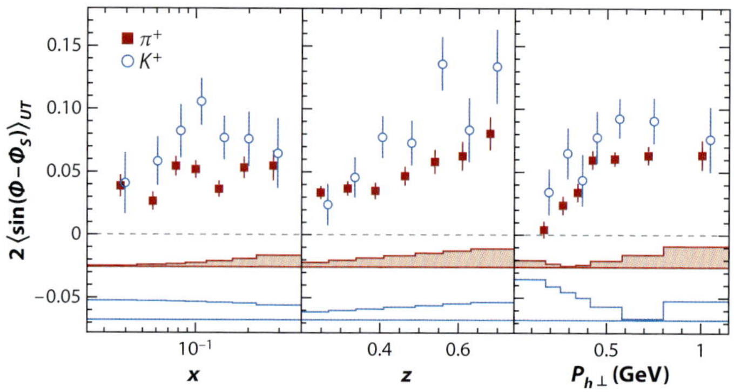

Figure 14

Results for the Sivers moment for positive pions and kaons obtained with a transversely polarized hydrogen target. The bands in the lower part of the graph represent systematic uncertainties.

6. HERA-B

6.1. Introduction

The HERA-B Collaboration proposed to measure *CP* violation in decays of *B* mesons to $J/\psi\ K^0$ at a time when the two e^+e^- *B* factory detectors, BaBar and BELLE, were being prepared. Although HERA-B was not designed to perform the detailed studies of *CP* violation that are possible at an e^+e^- machine, it was a viable and much more cost-effective alternative for making a first observation of *CP* violation in *B* meson decays. Achieving the needed integrated luminosity required a collision rate of 40 MHz, which in turn required the use of radiation-hard technologies and a complex, multilevel dilepton trigger and data-acquisition systems similar to those then under development for the LHC experiments. Thus, HERA-B also functioned as a platform for realistic tests of these new technologies prior to the start-up of the LHC.

The development of the trackers needed to cope with the intense particle fluxes at HERA proved more difficult than anticipated, and in 2000 it became apparent that HERA-B could not compete with the e^+e^- experiments. Nonetheless, the radiation issues had been overcome, and the experiment was nearing completion. The collaboration therefore requested and received continued support for a limited period to allow exploitation of the detector for a range of production studies with an emphasis on heavy-flavor production. The two principal goals were the determination of the total cross section for *b* quark production and the measurement of charmonium production and its dependency on the atomic number of the target nucleus.

HERA-B had been designed to measure all charged particles and photons produced in the central rapidity region by collisions of protons from the HERA proton ring, with the nuclei of target wires positioned in the halo of the beam. HERA-B consisted of a vertex detector, followed by a magnetic spectrometer with a forward acceptance of 15 to 220 mrad in the bending (horizontal) plane and 15 to 160 mrad in the nonbending (vertical) plane. These components were supplemented by electromagnetic calorimetry, a ring-imaging Cherenkov detector, and a muon identifier. At the first level of the three-level trigger system, lepton candidates were tracked at the 10-MHz HERA bunch-crossing rate. More elaborate tracking was performed at the second level in a large processor farm at an input rate of up to 40 kHz. Accepted events were fully reconstructed at the third level. More details on the detector and trigger system can be found in Reference 99 and references therein.

Data taking began in October 2002 and continued until March 2003. Most of the beam time was devoted to accumulating data with the dilepton trigger and carbon and tungsten targets for heavy-flavor studies. A total of 160 million dilepton triggers, including 250,000 reconstructed $J/\psi \rightarrow \mu^+\mu^-$ or $J/\psi \rightarrow e^+e^-$ decays, were recorded. Also, a 180 million–event sample was accumulated for studies of higher cross-section processes such as charm and strangeness production, with a minimum bias trigger on carbon, titanium, and tungsten targets.

6.2. Physics Results

The minimum bias data sample allowed for studies of strangeness (100) and charm (101) production and for a search for the production of pentaquark states (102). The dilepton sample was used to address several topics in the area of charm and beauty production and decay (101–109), including studies of charmonium (103–106) and bottomonium production (108) and the measurement of the total B meson production cross section (99, 107). As examples, the measurements of the B meson cross section and of the nuclear dependency of J/ψ production are briefly described below.

6.2.1. The total *B* meson production cross section. Two measurements of the B meson cross section were made. The first exploited the $B \rightarrow J/\psi + X$ decay (99), and the second (107) derived a cross section by extracting the number of events in which the decay chains of the produced B and $\bar{B}$ mesons included two oppositely charged muons, excluding the $B \rightarrow J/\psi + X$ decays when the produced J/ψ decays into two muons. For both measurements, systematic uncertainties due to triggering were minimized by deriving the product of luminosity and trigger efficiency from the large direct J/ψ signal and earlier measurements of the J/ψ total production cross section at the HERA-B energy (114).

In both cases, the vertex detector was crucial for separating B production from other processes. The J/ψ mesons from B decay were distinguished from direct J/ψ production mainly by requiring that the reconstructed J/ψ vertex be incompatible with the main event vertex. In the case of the double muonic decay technique, an impact parameter cut to the main vertex was an essential cut.

The two cross-section estimates were compatible and were thus combined to give a final estimate of $\sigma_{b\bar{b}} = 15.4 \pm 1.7_{\text{stat.}} \pm 1.2^{\text{uncorr.}}_{\text{sys.}} \pm 1.9^{\text{corr.}}_{\text{sys.}}$ nb nucleon^{-1}, where the parts of the systematic uncertainty that are uncorrelated and correlated (between the two measurements) are given separately. In **Figure 15**, the measurement is plotted along with two previous measurements from Fermilab (110, 111) and two predictions (112, 113). In the near-threshold region of the measurement, the QCD predictions must include not only leading-order and next-to-leading-order terms but also terms due to final-state radiation of soft and collinear partons (Sudakov effects). The HERA-B measurement and the predictions are in good agreement, although the uncertainties in the theoretical models are still large.

6.2.2. The nuclear dependence of J/ψ production. In 1986, Matsui & Satz (115) conjectured that a clean signature for the formation of a quark-gluon plasma in heavy-ion collisions would be the observation of a strong suppression of J/ψ production due to color-screening effects. This hypothesis was a strong motivator for several experimental studies of J/ψ production, both in heavy-ion collisions and also in collisions of protons on various nuclear targets (see Reference 116 and references therein). The proton-nucleus collision experiments were designed to measure the poorly understood suppression of J/ψ production in cold nuclear matter, that is, in the absence of a quark-gluon plasma. The E866 Collaboration measured J/ψ suppression for 820-GeV protons on two different nuclear targets over a wide range in Feynmann x ($x_F = p_z/p_{z\ \max}$, where p_z is the center-of-mass momentum of the produced J/ψ in the beam direction and $p_{z\ \max}$ is the maximum kinematically allowed value of p_z). In the year 2000, the E866 Collaboration reported a surprisingly

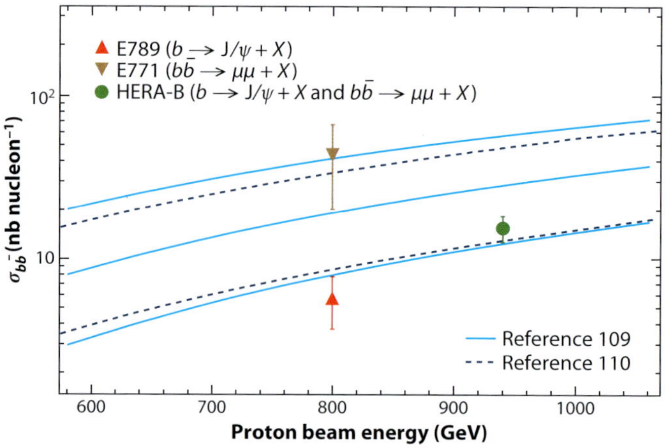

Figure 15

Three measurements [E771 (110), E789 (111), and HERA-B (107)] of the total $b\bar{b}$ production cross section as a function of proton energy for fixed-target collisions. The curves indicate upper and lower bounds of calculations by Bonciani et al. (112) and Kidonakis et al. (113).

large suppression as x_F approaches one (117). HERA-B extended the E866 measurements further into the unexplored negative x_F region.

The atomic mass number (A) dependency of a production process is often parameterized as a power law: $\sigma_{pA} = \sigma_{pN} \times A^{\alpha}$, where σ_{pA} is a production cross section for proton collisions with a nucleus of atomic mass number A, σ_{pN} is a proton-nucleon cross section of the same process, and α characterizes the nuclear dependency. HERA-B extracted α for J/ψ production by comparing J/ψ production on a carbon target to that of a tungsten target. The resulting x_F dependency of α is shown in **Figure 16**, together with the E866 (117) measurement and a lower-energy (450-GeV)

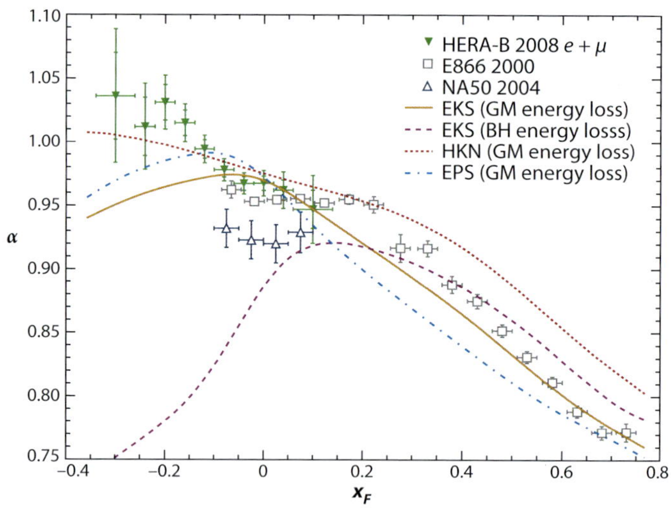

Figure 16

Measurements of the nuclear suppression parameter, α, as a function of x_F at a beam energy of 920 GeV by HERA-B (*filled green triangles*), at 820 GeV by E866 (*gray squares*), and at 450 GeV by NA50. The curves indicate model calculations by R. Vogt (119).

measurement by NA50 (118). In the region of overlap, the HERA-B and E866 measurements are in good agreement. The HERA-B measurements extend further into the negative x_F hemisphere and show that the suppression observed at positive x_F disappears and may instead become an enhancement. The results of a calculation by Vogt (119) for 920-GeV incident protons are also shown in the figure. The various curves correspond to different combinations of nuclear structure functions [EKS (120) or HKN (121)] and two different models of initial-state energy loss [BH (122) or GM (123)]. For more details, see Reference 119. The model variant that uses EKS and the GM energy-loss mechanism comes closest to describing the HERA-B data, although none of the model variants can simultaneously describe both the E886 and HERA-B data. The BH energy-loss mechanism is clearly ruled out. The lack of a good description of J/ψ production in the cold nuclear environment of proton-nucleus interactions precludes a common understanding of J/ψ production in both cold and hot nuclear matter.

7. SUMMARY

In its 15 years of operation, HERA helped uncover a wealth of different—and in part unexpected—aspects of the proton and of the strong and electroweak forces. The accelerator and the four experiments were a tremendous success and delivered a wealth of new information while exceeding the expectations in precision. HERA became a unique probe for physics at small x and provided detailed insight into the structure of the proton. Many results will be essential for the interpretation of the upcoming LHC measurements and for all other experiments in which precise knowledge of the proton structure is important.

The data analysis is not yet completed. However, it is already clear that the results will endure for a long time and will present a challenge to theorists who are reassessing their calculations, particularly in the field of QCD, in order to explain the measurements.

DISCLOSURE STATEMENT

The authors are not aware of any affiliations, memberships, funding, or financial holdings that might be perceived as affecting the objectivity of this review.

ACKNOWLEDGMENTS

Sincere thanks go to all those who have contributed to the success of HERA. In particular, Bjørn H. Wiik, Gustav-Adolf Voss, and Volker Soergel played a key role in the project's success during the planning and construction phases. HERA would not have existed without the substantial support of many institutes in many countries. The ideas and strong personal commitment of the people in the accelerator division and of all involved scientists were essential to creating the legacy of HERA.

LITERATURE CITED

1. Voss GA, Wiik BH. *Annu. Rev. Nucl. Part. Sci.* 44:413 (1994)
2. Sokholov AA, Ternov M. *Sov. Phys. Dokl.* 8:1203 (1964)
3. Buon J, Steffen K. *Nucl. Instrum. Methods A* 245:248 (1986)
4. Barber DP, et al. *Nucl. Instrum. Methods A* 329:79 (1993)
5. Beckmann M, et al. *Nucl. Instrum. Methods A* 479:334 (2002)
6. Abt I, et al. (H1 Collab.) *Nucl. Instrum. Methods A* 386:310 (1997)
7. Abt I, et al. (H1 Collab.) *Nucl. Instrum. Methods A* 386:348 (1997)

8. Appuhn RD, et al. *Nucl. Instrum. Methods A* 386:397 (1997)
9. Holm U, ed. (ZEUS Collab.) Status rep. **http://www-zeus.desy.de/bluebook/bluebook.html** (1993)
10. Klein M, Yoshida R. *Prog. Nucl. Part. Phys.* 61:334 (2008)
11. Klein M, Riemann TZ. *Z. Phys. C* 24:151 (1984)
12. Callan CG, Gross DJ. *Phys. Rev. Lett.* 22:156 (1969)
13. Altarelli G, Parisi G. *Nucl. Phys. B* 126:298 (1977)
14. Dokshitzer YL. *Sov. Phys. JETP* 46:641 (1977)
15. Gribov VN, Lipatov LN. *Sov. J. Nucl. Phys.* 15:438 (1972)
16. Gribov VN, Lipatov LN. *Sov. J. Nucl. Phys.* 15:675 (1972)
17. Lipatov LN. *Sov. J. Nucl. Phys.* 20:94 (1975)
18. Aaron FD, et al. (H1, ZEUS Collab.) arXiv:0911.0884 [hep-ex] (2009)
19. Chekanov S, et al. (ZEUS Collab.) *Eur. Phys. J. C* 62:625 (2009)
20. Chekanov S, et al. (ZEUS Collab.) *Phys. Lett.* B637:210 (2006)
21. Chekanov S, et al. (ZEUS Collab.) *Eur. Phys. J. C* 61:223 (2008)
22. Abulencia A, et al. (CDF Collab.) *Phys. Rev. D* 75:091101 (2007)
23. Abazov VM, et al. (DØ Collab.) *Phys. Rev. D* 72:011104(R) (2005)
24. Aktas A, et al. (H1 Collab.) *Eur. Phys. J. C* 51:271 (2007)
25. Aktas A, et al. (H1 Collab.) *Eur. Phys. J. C* 45:23 (2006)
26. Chekanov S, et al. (ZEUS Collab.) *Phys. Rev. D* 69:012004 (2004)
27. Breitweg J, et al. (ZEUS Collab.) *Eur. Phys. J. C* 12:35 (2000)
28. Chekanov S, et al. (ZEUS Collab.) *Eur. Phys. J. C* 65:65 (2010)
29. Chekanov S, et al. (ZEUS Collab.) *Eur. Phys. J. C* 63:171 (2009)
30. Aaron FD, et al. (H1 Collab.) *Phys. Lett.* B654:148 (2007)
31. Chekanov S, et al. (ZEUS Collab.) *J. High Energy Phys.* 08:077 (2009)
32. Aaron FD, et al. (H1 Collab.) *Phys. Lett.* B681:125 (2009)
33. Aaron FD, et al. (H1 Collab.) *Eur. Phys. J. C* 65:363 (2010)
34. Chekanov S, et al. (ZEUS Collab.) *Nucl. Phys. B* 765:1 (2007)
35. Aaron FD, et al. (H1 Collab.) arXiv:0911.5678 (2009)
36. Chekanov S, et al. (ZEUS Collab.) *Phys. Rev. D* 76:072011 (2007)
37. Chekanov S, et al. (ZEUS Collab.) *Eur. Phys. J. C* 42:1 (2005)
38. Daleo A, Gehrmann–De Ridder A, Gehrmann T, Luisoni G. *J. High Energy Phys.* 1001:118 (2010)
39. Gehrmann T, Glover EWN. *Phys. Lett.* B676:146 (2009)
40. Derrick M, et al. (ZEUS Collab.) *Phys. Lett.* B315:481 (1993)
41. Aaron FD, et al. (H1 Collab.) *Eur. Phys. J. C* 48:715 (2006)
42. Chekanov S, et al. (ZEUS Collab.) *Nucl. Phys. B* 816:1 (2009)
43. H1 Collab. *H1prelim-07-141.* **https://www-h1.desy.de/publications/H1preliminary.short_list.html** (2007)
44. ZEUS Collab. *ZEUS-prel-09-013.* **http://www-zeus.desy.de/physics/exo/ZEUS_PUBLIC/zeus-prel-09-013/ci_eps09.pdf** (2009)
45. H1 Collab. *H1prelim-07-164.* **https://www-h1.desy.de/publications/H1preliminary.short_list.html** (2007)
46. Aaron FD, et al. (H1 Collab.) *Phys. Lett.* B663:382 (2008)
47. Aaron FD, et al. (H1 Collab.) *Phys. Lett.* B666:131 (2008)
48. Aaron FD, et al. (H1 Collab.) *Phys. Lett.* B678:335 (2009)
49. H1 Collab. *H1prelim-09-061.* **https://www-h1.desy.de/publications/H1preliminary.short_list.html** (2009)
50. Aktas A, et al. (H1 Collab.) *Phys. Lett.* B638:432 (2006)
51. Aaron FD, et al. (H1 Collab.) *Eur. Phys. J. C* 64:251 (2009)
52. Chekanov S, et al. (ZEUS Collab.) *Phys. Lett.* B672:106 (2009)
53. Aaron FD, et al. (H1, ZEUS Collab.) *J. High Energy Phys.* 1003:035 (2010)
54. Baur U, Vermaseren JA, Zeppenfeld D. *Nucl. Phys. B* 375:3 (1992)
55. Diener DP, Schwanenberger C, Spira M. arXiv:hep-ex/0302040 (2003)

56. Diener DP, Schwanenberger C, Spira M. *Eur. Phys. J. C* 25:405 (2002); Nason P, Rückl, Spira M. *J. Phys. G* 25:1434 (1999); Spira M. arXiv:hep-ph/9905469 (1999)
57. Aaron FD, et al. (H1 Collab.) *Phys. Lett.* B678:450 (2009)
58. Aaron FD, et al. (H1, ZEUS Collab.) *J. High Energy Phys.* 10:013 (2009)
59. Aaron FD, et al. (H1, ZEUS Collab.) *Phys. Lett.* B674:257 (2009)
60. Ashman J, et al. (EMC Collab.) *Phys. Lett.* B 206:364 (1988)
61. Jaffe R, Manohar A. *Nucl. Phys. B* 337:509 (1990)
62. Ackerstaff K, et al. (HERMES Collab.) *Nucl. Instrum. Methods A* 417:230 (1998)
63. DeSchepper D, et al. *Nucl. Instrum. Methods A* 419:16 (1998)
64. Airapetian A, et al. (HERMES Collab.) *Nucl. Instrum. Methods A* 540:68 (2005)
65. Avakian H, et al. *Nucl. Instrum. Methods A* 417:69 (1998)
66. Akopov N, et al. *Nucl. Instrum. Methods A* 479:511 (2002)
67. Benisch TH, et al. *Nucl. Instrum. Methods A* 471:314 (2001)
68. Müller D, et al. *Fortschr. Phys.* 42:101 (1994)
69. Radyushkin AV. *Phys. Rev. Lett.* 78:610 (1996)
70. Ji X. *Phys. Lett.* B206:364 (1997)
71. Airapetian A, et al. (HERMES Collab.) *Nucl. Phys. B* 780:1 (2007)
72. Airapetian A, et al. (HERMES Collab.) *Phys. Rev. Lett.* 87:182001 (2001)
73. Airapetian A, et al. (HERMES Collab.) *Phys. Rev. D* 75:011103 (2007)
74. Airapetian A, et al. (HERMES Collab.) *J. High Energy Phys.* 0806:066 (2008)
75. Airapetian A, et al. (HERMES Collab.) *J. High Energy Phys.* 11:083 (2009)
76. Airapetian A, et al. (HERMES Collab.) *Phys. Rev. C* 81:035202 (2010)
77. Airapetian A, et al. (HERMES Collab.) *Phys. Rev. C* 81:035202 (2010)
78. Airapetian A, et al. (HERMES Collab.) *Eur. Phys. J. C* 62:659 (2009)
79. Airapetian A, et al. (HERMES Collab.) *Phys. Lett.* B659:486 (2008)
80. Airapetian A, et al. (HERMES Collab.) *Phys. Lett.* B679:100 (2009)
81. Airapetian A, et al. (HERMES Collab.) *Phys. Lett.* B682:345 (2010)
82. Rith K. *Prog. Nucl. Part. Phys.* 49:245 (2002)
83. Burkardt M, et al. *Rep. Prog. Phys.* 73:016201 (2010)
84. Altarelli G, et al. *Nucl. Phys. B* 496:337 (1997)
85. Airapetian A, et al. (HERMES Collab.) *Phys. Rev. D* 75:012007 (2007)
86. Alexakhin VY, et al. (COMPASS Collab.) *Phys. Lett.* B647:8 (2007)
87. Bardeen W. *Phys. Rev. D* 18:3398 (1978)
88. Amsler C, et al. *Phys. Lett.* B667:1 (2008)
89. Airapetian A, et al. (HERMES Collab.) *Phys. Rev. D* 71:012003 (2005)
90. Airapetian A, et al. (HERMES Collab.). arXiv:1002.3921 [hep-ex] (2010)
91. Airapetian A, et al. (HERMES Collab.) *Phys. Lett.* B666:446 (2008)
92. Pumplin J, et al. *J. High Energy Phys.* 0207:012 (2002)
93. Barone V, Ratcliffe PG. *Transverse Spin Physics*. Singapore: World Sci. 294 pp. (2003)
94. Mulders PJ, Tangerman RD. *Nucl. Phys. B* 461:197 (1996); Mulders PJ, Tangerman RD. Erratum. *Nucl. Phys. B* 484:538 (1997)
95. Sivers DW. *Phys. Rev. D* 41:83 (1990)
96. Burkardt M. *Phys. Rev. D* 66:114005 (2002)
97. Airapetian A, et al. (HERMES Collab.) *Phys. Rev. Lett.* 94:012002 (2005)
98. Airapetian A, et al. (HERMES Collab.) *Phys. Rev. Lett.* 103:152002 (2009)
99. Abt I, et al. (HERA-B Collab.) *Phys. Rev. D* 73:052005 (2006)
100. Abt I, et al. (HERA-B Collab.) *Phys. Lett.* B638:415 (2006); Abt I, et al. (HERA-B Collab.) *Eur. Phys. J. C* 50:315 (2007); Abt I, et al. (HERA-B Collab.) *Eur. Phys. J. C* 61:207 (2009)
101. Abt I, et al. (HERA-B Collab.) *Eur. Phys. J. C* 52:531 (2007); Abt I, et al. (HERA-B Collab.) *Phys. Lett.* B638:407 (2006)
102. Abt I, et al. (HERA-B Collab.) *Phys. Rev. Lett.* 93:212003 (2004)
103. Abt I, et al. (HERA-B Collab.) *Eur. Phys. J. C* 49:545 (2007)
104. Abt I, et al. (HERA-B Collab.) *Phys. Rev. D* 79:012001 (2009)

105. Abt I, et al. (HERA-B Collab.) *Eur. Phys. J. C* 60:525 (2009)
106. Abt I, et al. (HERA-B Collab.) *Eur. Phys. J. C* 60:517 (2009)
107. Abt I, et al. (HERA-B Collab.) *Phys. Lett.* B650:103 (2007)
108. Abt I, et al. (HERA-B Collab.) *Phys. Lett.* B638:13 (2006)
109. Abt I, et al. (HERA-B Collab.) *Phys. Lett.* B596:173 (2004)
110. Alexopoulis T, et al. (E771 Collab.) *Phys. Rev. Lett.* 82:41 (1999)
111. Jansen DM, et al. (E789 Collab.) *Phys. Rev. Lett.* 74:3118 (1995)
112. Bonciani R, et al. *Nucl. Phys. B* 529:424 (1998)
113. Kidonakis N, Vogt R. *Eur. Phys. J. C* 36:201 (2004)
114. Maltoni F, et al. *Phys. Lett.* B638:202 (2006)
115. Matsui T, Satz H. *Phys. Lett.* B178:416 (1986)
116. Lourenço C, Vogt R, Wöhri H. *J. High Energy Phys.* 0902:014 (2009)
117. Leitch MJ, et al. (E866 Collab.) *Phys. Rev. Lett.* 84:3256 (2000)
118. Alessandro B, et al. (NA50 Collab.) *Eur. Phys. J. C* 33:31 (2004)
119. Vogt R. *Phys. Rev. C* 61:035203 (2000); Vogt R. *Phys. Rev. C* 71:054902 (2005)
120. Eskola KJ, Kolhinin VJ, Salgado CA. *Eur. Phys. J. C* 9:61 (1999)
121. Hirai M, Kumano S, Nagai TH. *Phys. Rev. C* 70:044905 (2004)
122. Brodsky S, Hoyer P. *Phys. Lett.* B298:165 (1993)
123. Gavin S, Milan J. *Phys. Rev. Lett* 68:1834 (1992)
124. ZEUS Collab. *ZEUS-prel-08-008* (2008)
125. ZEUS Collab. *ZEUS-prel-09-006* (2009)
126. Chekanov S, et al. (ZEUS Collab.) *Phys. Lett.* B649:12 (2007)
127. H1 Collab. *H1prelim-08-032* (2008)
128. Aktas A, et al. *Phys. Lett.* B653:134 (2007)
129. Abazov VM, et al. (DØ Collab.) *J. High Energy Phys.* 1003:035 (2010)
130. Affolder T, et al. (CDF Collab.) *Phys. Rev. Lett.* 88:042001 (2002)
131. Dissertori G, et al. *J. High Energy Phys.* 0802:040 (2008)
132. Kluth S. arXiv:0709.0173 [hep-ex] (2007)
133. H1 Collab. *H1prelim-07-132* (2007); ZEUS Collab. *ZEUS-prel-07-025* (2007)
134. Glasman C. *AIP Conf. Proc.* 792:689 (2005)
135. Bethke S. *Eur. Phys. J. C* 64:689 (2009)

In Search of Extraterrestrial High-Energy Neutrinos

Luis A. Anchordoqui[1] and Teresa Montaruli[2]

[1]Department of Physics, University of Wisconsin, Milwaukee, Wisconsin 53201; email: anchordo@uwm.edu

[2]Department of Physics, University of Wisconsin, Madison, Wisconsin 53706; email: tmontaruli@icecube.wisc.edu

Annu. Rev. Nucl. Part. Sci. 2010. 60:129–62

First published online as a Review in Advance on June 7, 2010

The *Annual Review of Nuclear and Particle Science* is online at nucl.annualreviews.org

This article's doi:
10.1146/annurev.nucl.012809.104551

0163-8998/10/1123-0129$20.00

Key Words

neutrino telescopes, cosmic rays, dark matter

Abstract

We review the search for astrophysical neutrinos. We begin by summarizing the various theoretical predictions that correlate the expected neutrino flux with data from other messengers, specifically γ rays and ultrahigh-energy cosmic rays. We then review the status and results of neutrino telescopes both in operation and decommissioned as well as the methods used for data analysis and background discrimination. Particular attention is devoted to the challenge enforced by the highly uncertain atmospheric muon and neutrino backgrounds in relation to searches of diffuse neutrino fluxes. Next, we examine the impact of existing limits on neutrino fluxes on studies of the chemical composition of cosmic rays. After that, we show not only that neutrinos have the potential to discover astrophysical sources, but also that the huge statistics of atmospheric muons can be a powerful tool. We end by discussing the prospects for indirect detection of dark matter with neutrino telescopes.

Contents

1. INTRODUCTION

It has long been recognized that high-energy protons produced in cosmic-ray accelerators generate an observable flux of cosmic neutrinos, mainly through charged-pion production in collisions with the ambient gas or with radiation fields (1). As such, the emission of neutrinos necessarily implies the existence of relativistic baryons and, consequently, the acceleration of cosmic rays. Upcoming catalogs of cosmic-ray and neutrino sources would thus in principle be identical were it not for the different properties of the messengers.

Neutrinos can escape from the innermost regions of galaxies, where light and other kinds of electromagnetic radiation are blocked by matter, and hence can be tracers of processes that remain hidden from traditional photon astronomy. Furthermore, neutrinos carry with them information about the site and circumstances of their production. Undeflected and unabsorbed, they can point back to very distant sources, resolving the origin of the highest-energy cosmic rays and the underlying acceleration mechanism. Unlike γ rays, which can be produced via synchrotron radiation or inverse Compton scattering, neutrinos provide incontrovertible evidence for the acceleration of baryonic cosmic rays.

Megaelectronvolt neutrino astronomy has been possible for approximately 45 years (2). Thus far, two sources of nonatmospheric neutrinos have been identified: the Sun (3) and SN1987A (4). During the next decade, we will be able to observe the universe by using high-energy "neutrino light," which will illuminate a wondrous new view of the universe.

Candidate sources of high-energy neutrinos are of both galactic and extragalactic varieties; the latter are expected to dominate at the highest energies, just as for the parent cosmic rays. Energy-balance considerations suggest that the diffuse galactic flux of neutrinos predominantly originates in supernova (SN) remnants (5). Dense molecular clouds, often found in star-forming regions where SNe explode, are particularly efficient at converting protons into pions that decay into γ rays and neutrinos (6). Other possibilities for galactic sources include microquasars (7), γ-ray binaries (8), and pulsar wind nebulae (9). By far, the most likely extragalactic sources are active galactic nuclei (AGN) (12) and γ-ray bursts (GRBs) (13). (For a review see, e.g., Reference 10, and for neutrino flux estimates see Reference 11.) In addition, the so-called cosmogenic neutrinos are guaranteed to exist, although their amount remains highly uncertain. These neutrinos originate from the decay of charged pions produced in the interactions of protons with the isotropic photon background (14–16).

In terms of conventional astronomy, neutrino astronomers observe the neutrino sky through the atmosphere. This is both a curse and a blessing: The background of neutrinos produced by cosmic rays in interactions with air nuclei provides a beam essential for calibrating the detectors and demonstrating the neutrino measurement technique. It also provides an opportunity to probe neutrino standard oscillations and those arising from new physics, such as violation of Lorentz invariance (17). Especially unique is the energy range of the atmospheric neutrino beam between 10^4 GeV $\lesssim E_\nu \lesssim 10^8$ GeV, which is not within reach of accelerators, that may reveal the existence of neutrinos from prompt decays of charmed mesons and baryons (18). Over the next decade, a data set of the order of 1 million atmospheric neutrinos will be collected. The statistics will be so large that mapping of the Earth's interior will be possible via neutrino tomography (19).

Although neutrino astrophysics is the central motivation to build neutrino telescopes (NTs), ultrahigh-energy cosmic neutrinos are also unique probes of new physics, as their interactions are uncluttered by the strong and electromagnetic forces and can reach center-of-mass energies of $\sqrt{s} \sim 245$ TeV (20). Rates for new physics processes, however, are difficult to test because the flux of cosmic neutrinos is virtually unknown. It is possible to mitigate this difficulty by using multiple observables that allow one to decouple the effects of the flux and cross section (21). Large NTs may also open a new perspective for indirect searches of dark matter by detecting neutrinos produced via annihilation of weakly interactive massive particles (WIMPs) trapped within the Sun or at the center of the Earth (22).

The layout of this paper is as follows: In Section 2 we provide an overview of the connection between cosmic-ray observations and expected neutrino fluxes. In Section 3 we review the status of NT construction, operation, and data analysis. In Section 4 we analyze the upper limits on neutrino fluxes set by the different experiments to assess their scientific impact. Finally, in Section 5 we provide a summary and discuss future prospects.

2. COSMIC-RAY $\rightleftharpoons$ NEUTRINO CONNECTION

The cosmic-ray flux falls as an approximate power law in energy, $J_{\rm CR}(E) \propto E^{-\alpha}$, where $\alpha \simeq 2.7$ from approximately 10 GeV up to the "knee" in the spectrum at $E \sim 3 \times 10^6$ GeV, where it steepens to $\alpha \simeq 3.1$. It then steepens further to $\alpha \simeq 3.25$ at $E \sim 4 \times 10^8$ GeV (the so-called dip) and flattens back to $\alpha \simeq 2.69$ at $E \sim 3 \times 10^9$ GeV (the so-called ankle) (23). The spectrum extends up to at least $E \sim 10^{11}$ GeV, with a steep spectrum ($\alpha \simeq 4.2$) beyond $E = 5 \times 10^{10}$ GeV (24), in accord with the prediction by Greisen, Zatsepin, and Kuz'min (GZK) (25).

The energy density associated with the flux of galactic cosmic rays is $\rho_E \sim 10^{-12}$ erg cm^{-3} (26). This is approximately the value of the corresponding energy density $B^2/8\pi$ of the microgauss magnetic field in the Galaxy. Given that the average containment time of the cosmic rays in our Galaxy is $\sim 3 \times 10^6$ years, the power needed to maintain this energy density is $\sim 10^{-26}$ erg cm^{-3} s^{-1}. For a nominal volume of the galactic disk of 10^{67} cm^3, this requires an accelerator delivering power of approximately 10^{41} erg s^{-1}. This is 10% of the power produced by SNe, which releases 10^{51} erg every 30 years. This coincidence is the basis for the idea that shocks produced by SNe expanding into the interstellar medium are the origin of the galactic cosmic rays.

The conversion of 10^{50} erg of energy into particle acceleration is believed to occur by diffusive shock acceleration in the young (10^3–10^4 years) SN remnant expanding into the interstellar medium. If high-energy cosmic rays are indeed associated with the remnant, they will interact with hydrogen atoms in the interstellar medium to produce pions that decay into roughly equal numbers of photons and neutrinos: the former via $\pi^0 \to \gamma\gamma$ and the latter via $\pi^+ \to e^+ \nu_e \nu_\mu \bar{\nu}_\mu$ (and the conjugate process). These may provide us with indirect evidence for cosmic-ray acceleration.

The observation of these pionic γ rays has been one of the motivations for neutrino astronomy and, to a large extent, for γ-ray astronomy.

Whereas the details are complex and predictions can be treacherous, a simple estimate of the γ-ray flux associated with an SN remnant can be made following Aharonian et al. (27). The emissivity in pionic γ rays is simply proportional to the density of cosmic rays for energy in excess of 1 TeV, $n_{\rm CR}(> 1\,{\rm TeV}) \simeq 4 \times 10^{-14}\,{\rm cm}^{-3}$, and to the target density of hydrogen atoms, $n_{\rm H}$. The proportionality factor is determined by particle physics.

The emissivity (number/volume/time/energy) of neutral pions resulting from an isotropic distribution of highly relativistic particles following a power-law energy spectrum $dn_{\rm CR}(E_N)/dE_N \propto E_N^{-\alpha}$ is given by

$$Q_\pi(E_\pi) = c\, n_{\rm H} \int_{E_N^{\rm th}(E_\pi)}^{E_N^{\rm max}} \frac{dn_{CR}}{dE_N}(E_N)\frac{d\sigma_A}{dE_\pi}(E_\pi, E_N)dE_N, \qquad 1.$$

where $E_N^{\rm th}(E_\pi)$ is the minimum energy per nucleon required to produce a pion with energy E_π and $d\sigma_A(E_\pi, E_N)/dE_\pi$ is the differential cross section for the production of a pion with energy E_π in the lab frame due to the collision of a nucleus A of energy per nucleon E_N with a hydrogen atom at rest. The differential cross section can be parameterized by

$$\frac{d\sigma_A}{dE_\pi}(E_\pi, E_N) \simeq \frac{\sigma_0^A}{E_\pi} x F_\pi(x, E_N), \qquad 2.$$

where $x \equiv E_\pi/E_N$, $\sigma_0^A = A^{2/3}\sigma_0$ (with $\sigma_0 \simeq 40$ mb) provides a scaling of the cross section with the atomic mass number (26) and $F_\pi(x, E_N)$ is a fragmentation function. Taking into account that inelastic hadronic collisions lead to roughly equal numbers of π^0, π^+, and π^- mesons, as well as the mild variation with pseudorapidity, the parameterization (28)

$$F_\pi(x, E_N) = 4\beta B_\pi x^{\beta-1}\left(\frac{1-x^\beta}{1+r\,x^\beta(1-x^\beta)}\right)^4\left(\frac{1}{1-x^\beta} + \frac{r(1-2x^\beta)}{1+rx^\beta(1-x^\beta)}\right) \times \left(1 - \frac{m_\pi}{xE_N}\right)^{1/2} \qquad 3.$$

is found (29) to be consistent at 1σ level with data collected at the Tevatron by the CDF detector (30): $B_\pi = a + 0.25$, $\beta = 0.98/\sqrt{a}$, $r = 2.6/\sqrt{a}$, $a = 3.67 + 0.83L + 0.075L^2$, and $L = \ln(E_N/{\rm TeV})$. Substitution of Equation 2 into Equation 1 yields

$$Q_\pi(E_\pi) \simeq Z_{A\pi}(\alpha) Q_A^{A{\rm H}}(E_\pi), \qquad 4.$$

where

$$Q_A^{A{\rm H}}(E_N) = \sigma_0^A c\, n_{\rm H}\frac{dn_{\rm CR}}{dE_N}(E_N) \qquad 5.$$

and the spectrum-weighted moment of the inclusive cross section or so-called Z-factor is given by

$$Z_{A\pi}(\alpha) \equiv \int_0^1 x^{\alpha-1} F(x, E_N)dx. \qquad 6.$$

On average, the photons carry one-half of the energy of the pion, and thus the γ-ray emissivity is

$$Q_\gamma^{Ap}(E_\gamma) = 2\int_{E_\pi^{\rm min}(E_\gamma)}^{E_\pi^{\rm max}} \frac{Q_\pi^{Ap}(E_\pi)}{(E_\pi^2 - m_\pi^2)^{1/2}} dE_\pi \simeq Z_{\pi\gamma}(\alpha) Q_\pi^{Ap}(E_\gamma), \qquad 7.$$

where $E_\pi^{\min}(E_\gamma) = E_\gamma + m_\pi^2/(4E_\gamma)$ and $Z_{\pi\gamma}(\alpha) = 2/\alpha$. Assuming proton dominance and an emission spectrum $\propto E_N^{-2}$, the γ-ray emissivity above 1 TeV becomes

$$Q_\gamma(>1\,\mathrm{TeV}) = c\,\langle E_\pi/E_N\rangle \sigma_0\, n_\mathrm{H}\, n_\mathrm{CR}(>1\,\mathrm{TeV}) \simeq 10^{-29}\left(\frac{n_\mathrm{H}}{\mathrm{cm}^{-3}}\right)\frac{\mathrm{photons}}{\mathrm{cm}^3\,\mathrm{s}}, \quad 8.$$

where $\langle E_\pi/E_N\rangle \sim 0.2$ is the average energy of the secondary pions relative to the cosmic-ray protons. For different spectral indices, the quantity $\langle E_\pi/E_N\rangle$ is generalized to the spectrum-weighted moments for pion production by nucleons given in Equation 6.

The density of protons from an SN converting a total kinetic energy W of 10^{50} erg to proton acceleration is approximately given by ρ/W, where we assume that the density in the remnant is not very different from the ambient energy density $\rho_E \sim 10^{-12}$ erg cm^{-3} of galactic cosmic rays. This approximation is valid for young remnants in their Sedov phase (31). The total luminosity in γ rays is then

$$\mathcal{L}_\gamma(>1\,\mathrm{TeV}) = Q_\gamma \frac{W}{\rho_E} \simeq 10^{33}\,\mathrm{photons\,s^{-1}}. \quad 9.$$

The expected rate of teraelectronvolt photons from an SN at a distance $d \sim 1$ kpc is

$$\frac{d\mathcal{N}_\gamma}{d(\ln E_\gamma)}(>1\,\mathrm{TeV}) = \frac{\mathcal{L}_\gamma}{4\pi d^2} \simeq 10^{-11}\left(\frac{\mathrm{photons}}{\mathrm{cm}^2\,\mathrm{s}}\right)\left(\frac{W}{10^{50}\,\mathrm{erg}}\right)\left(\frac{n_\mathrm{H}}{\mathrm{cm}^{-3}}\right)\left(\frac{\mathrm{kpc}}{\mathrm{d}}\right)^2. \quad 10.$$

When the absorption effects are neglected, the all-flavor neutrino and photon fluxes are roughly equal, so we anticipate an event rate of three detected neutrinos per decade of energy per square kilometer year—a result readily obtained from the relation

$$\frac{d\mathcal{N}_\nu}{d(\ln E_\nu)}(>1\,\mathrm{TeV}) \simeq 10^{-11}\left(\frac{\mathrm{neutrinos}}{\mathrm{cm}^2\,\mathrm{s}}\right)\left(\frac{\mathrm{area}}{1\,\mathrm{km}^2}\right)\left(\frac{\mathrm{time}}{1\,\mathrm{year}}\right)P_\nu, \quad 11.$$

where the last factor indicates the average probability for neutrino detection. For the teraelectronvolt energy considered here, $P_\nu \sim 10^{-6}$ (1).

This estimate may be somewhat optimistic because we assume that the neutrino energy reaches 100 TeV with an E_ν^{-2} spectrum.[1] By contrast, if the galactic cosmic-ray spectrum extends at least up to the knee ($E \sim 3{,}000$ TeV), then some of the sources must produce ~100-TeV secondaries.

Photodisintegration of high-energy nuclei, followed by prompt photoemission from the excited daughter nuclei, also produces teraelectronvolt γ rays and antineutrinos. In this chain reaction, the nuclei act in analogy to Einstein's relativistic moving mirror to "double-boost" electronvolt starlight to teraelectronvolt energies for a Lorentz boost factor $>10^6$ (33). The teraelectronvolt antineutrino counterpart is produced through β decay of the emitted neutrons (34). Effective nucleus photodisintegration is expected in stellar associations harboring a large population of massive and early-type stars that can provide the required Lyman-α emission. Directional beams of galactic antineutrinos and γ rays may thus emerge as fluctuations over the diffuse intensity.

A plethora of explanations have been proposed to address the production mechanism of cosmic rays beyond the dip (35). In the absence of a single model that is consistent with all data, the origin of these particles remains a mystery. Clues to solve the mystery are not immediately forthcoming from the data, particularly because various experiments report mutually inconsistent results. In

[1]Such a spectral index is typical of Fermi acceleration mechanisms (32).

recent years, a somewhat confusing picture regarding the nature of these cosmic rays has been emerging. The HiRes data have been interpreted as a change in cosmic-ray composition, from heavy nuclei to protons, at $E \sim 4 \times 10^8$ GeV (36). This is an order of magnitude lower in energy than the previous crossover deduced from the Fly's Eye data (37).

The end point of the galactic flux is expected to be dominated by iron, as the large charge Ze of heavy nuclei reduces their Larmor radius (containment scales linearly with Z) and facilitates their acceleration to highest energy (again scaling linearly with Z). The dominance of nuclei in the high-energy region of the galactic flux carries the implication that any changeover to protons represents the onset of dominance by an extragalactic component. The inference from this HiRes data is therefore that the extragalactic flux is already beginning to dominate the galactic flux at $E \sim 4 \times 10^8$ GeV. Such a transition would appear to require considerable fine-tuning of the shape and normalization of the two spectra. It has been argued, however, that the spectral shape required for the extragalactic component can develop naturally during the propagation of extragalactic protons in the cosmic microwave background over cosmological distances (38). At $E \sim 3 \times 10^8$ GeV, the energy losses due to e^+e^- pair production and cosmic expansion are roughly equal and thus produce a steepening of an initially featureless power-law injection spectrum. Significantly, the onset of the extragalactic component would be well below $E_{\rm GZK} \sim 5 \times 10^{10}$ GeV, the threshold energy for resonant $p\gamma_{\rm CMB} \to \Delta^+ \to N\pi$ energy loss on the cosmic microwave background, and thus samples sources even at large redshift.

It is helpful to envision the cosmic-ray engines as machines in which protons are accelerated and (possibly) permanently confined by the magnetic fields of the acceleration region. The production of neutrons and pions and subsequent decay produce neutrinos, γ rays, and cosmic rays. If the neutrino-emitting source also produces high-energy cosmic rays, then pion production must be the principal agent for the high-energy cutoff on the proton spectrum. Conversely, because the protons must undergo sufficient acceleration, inelastic pion production needs to be small below the cutoff energy. Consequently, the plasma must be optically thin. Because the interaction time for protons is greatly increased over that of neutrons as a result of magnetic confinement, the neutrons escape before interacting and, upon decay, give rise to the observed cosmic-ray flux. The foregoing can be summarized as three conditions on the characteristic nucleon interaction timescale $\tau_{\rm int}$, the neutron decay lifetime τ_n, the characteristic cycle time of confinement $\tau_{\rm cycle}$, and the total proton confinement time $\tau_{\rm conf}$: (*a*) $\tau_{\rm int} \gg \tau_{\rm cycle}$, (*b*) $\tau_n > \tau_{\rm cycle}$, and (*c*) $\tau_{\rm int} \ll \tau_{\rm conf}$. The first condition ensures that the protons attain sufficient energy. The first and second conditions also allow the neutrons to escape the source before decaying. The third condition permits sufficient interaction to produce neutrons and neutrinos. These three conditions together define an optically thin source (39). A desirable property to reproduce the almost structureless energy spectrum is that a single type of source will produce cosmic rays with a smooth spectrum across a wide range of energy.

The cosmic-ray flux above the ankle is often summarized as one 3×10^{10}–GeV particle per kilometer square per year per steradian. This can be translated into an energy flux (41)

$$E\{EJ_{\rm CR}\} = \frac{3 \times 10^{10}\ \text{GeV}}{(10^{10}\ \text{cm}^2)(3 \times 10^7\ \text{s})\ \text{sr}} = 10^{-7}\ \text{GeV cm}^{-2}\ \text{s}^{-1}\ \text{sr}^{-1}. \qquad 12.$$

From this equation, we can derive the energy density ρ_E in ultrahigh-energy cosmic rays using flux = velocity × density, or

$$4\pi \int dE\{EJ_{\rm CR}\} = c\rho_E. \qquad 13.$$

This leads to

$$\rho_E = \frac{4\pi}{c}\int_{E_{\min}}^{E_{\max}} \frac{10^{-7}}{E}\,\mathrm{d}E\,\frac{\mathrm{GeV}}{\mathrm{cm}^3} \simeq 3\times 10^{-19}\,\frac{\mathrm{TeV}}{\mathrm{cm}^3}, \tag{14}$$

taking the extreme energies of the accelerator(s) to be $E_{\max}/E_{\min} \simeq 10^3$. The power required for a population of sources to generate this energy density over the Hubble time of 10^{10} years is

$$E^2\frac{\mathrm{d}\dot{N}_{\mathrm{CR}}}{\mathrm{d}E}\bigg|_{E_{\min}} = \frac{\dot{\varepsilon}_{\mathrm{CR}}^{[10^{19},10^{21}]}}{\ln(10^{21}/10^{19})} \approx 10^{44}\,\mathrm{erg\,Mpc^{-3}\,year^{-1}}, \tag{15}$$

where an energy spectrum $\propto E^{-2}$ has been assumed (40).

This works out to approximately $3\times10^{39}\,\mathrm{erg\,s^{-1}}$ per galaxy, $3\times10^{42}\,\mathrm{erg\,s^{-1}}$ per cluster of galaxies, $\sim 2\times10^{44}\,\mathrm{erg\,s^{-1}}$ per active galaxy, or $\sim 2\times10^{52}$ erg per cosmological γ-ray burst (41). The coincidence between these numbers and the observed output in electromagnetic energy of these sources explains why they have emerged as the leading candidates for cosmic-ray accelerators. The coincidence is consistent with the relationship between cosmic rays and photons built into the previously discussed optically thin source.

The energy density of neutrinos produced through $p\gamma$ interactions of these protons can be directly tied to the injection rate of cosmic rays,

$$E_\nu^2\frac{\mathrm{d}N_\nu}{\mathrm{d}E_\nu} \approx \frac{3}{8}\varepsilon_\pi\, t_{\mathrm{H}}\, E^2\frac{\mathrm{d}\dot{N}_{\mathrm{CR}}}{\mathrm{d}E}, \tag{16}$$

where t_{H} is the Hubble time and ε_π is the fraction of the energy injected into protons lost in photopion interactions. (The factor of $\frac{3}{8}$ comes from the fact that, close to threshold, roughly half the pions produced are neutral, thus not generating neutrinos, and one-quarter of the energy of charged-pion decays goes to electrons rather than neutrinos.)[2] The Waxman-Bahcall bound is defined by the condition $\varepsilon_\pi = 1$:

$$\begin{aligned}[E_\nu^2\Phi_\nu]_{WB} &\approx \left(\frac{3}{8}\right)\xi_Z\,\varepsilon_\pi\, t_{\mathrm{H}}\frac{c}{4\pi}E^2\frac{\mathrm{d}\dot{N}_{CR}}{\mathrm{d}E}\\ &\approx 2.3\times10^{-8}\,\varepsilon_\pi\,\xi_Z\,\mathrm{GeV\,cm^{-2}\,s^{-1}\,sr^{-1}},\end{aligned} \tag{17}$$

where the parameter ξ_z accounts for the effects of source evolution with redshift and is expected to be approximately three (42). For interactions with the ambient gas (i.e., pp rather than $p\gamma$ collisions), the average fraction of the total pion energy carried by charged pions is approximately two-thirds, compared with one-half in the photopion channel. In this case, the upper bound given in Equation 17 is enhanced by 33% (43).

The actual value of the neutrino flux depends on what fraction of the proton energy is converted to charged pions (which then decay to neutrinos). To quantify this, we follow Waxman-Bahcall and define ε_π as the ratio of charged-pion energy to the emerging nucleon energy at the source. For resonant photoproduction, the inelasticity is kinematically determined by requiring equal boosts for the decay products of the Δ^+ (44), giving $\varepsilon_\pi = E_{\pi^+}/E_n \approx 0.28$, where E_{π^+} and E_n are the emerging charged-pion and neutron energies, respectively. For $pp \to NN +$ pions, where N indicates a final-state nucleon, the inelasticity is ≈ 0.6 (45). This then implies that the energy carried away by charged pions is approximately equal to the emerging nucleon energy, yielding (with our definition) $\varepsilon_\pi \approx 1$.

[2]The average neutrino energy from the direct pion decay is $\langle E_{\nu_\mu}\rangle^\pi = (1-r)\,E_\pi/2 \simeq 0.22\,E_\pi$, and that of the muon is $\langle E_\mu\rangle^\pi = (1+r)\,E_\pi/2 \simeq 0.78\,E_\pi$, where r is the ratio of muon to the pion mass squared. In muon decay, because the ν_μ has approximately one-third of its parent energy, the average muon neutrino energy is $\langle E_{\nu_\mu}\rangle^\mu = (1+r)E_\pi/6 = 0.26\,E_\pi$.

At production, if all muons decay, the neutrino flux consists of equal fractions of ν_e, ν_μ, and $\bar{\nu}_\mu$. Originally, the Waxman-Bahcall bound was presented for the sum of ν_μ and $\bar{\nu}_\mu$ (neglecting ν_e), motivated by the fact that only muon neutrinos are detectable as track events in NTs. Because oscillations in the neutrino sector mix the different species, we chose instead to discuss the sum of all neutrino flavors. When the effects of oscillations are accounted for, nearly equal numbers of the three neutrino flavors are expected on Earth (46). (For a detailed calculation of the flavor ratios see, for instance, Reference 47.)

If the injected cosmic rays include nuclei heavier than protons, then the neutrino flux expected from the cosmic-ray sources may be modified. Nuclei undergoing acceleration can produce pions, just as protons do, through interactions with the ambient gas, so the Waxman-Bahcall argument would be unchanged in this case. However, if interactions with radiation fields dominate over interactions with matter, the neutrino flux would be suppressed if the cosmic rays are heavy nuclei. This is because the photodisintegration of nuclei dominates over pion production at all but the very highest energies. Defining κ as the fraction of nuclei heavier than protons in the observed cosmic-ray spectrum, the resulting neutrino flux is then given by

$$E_\nu^2 J_\nu \approx (1-\kappa)[E_\nu^2 \Phi_\nu]_{\rm WB}. \qquad 18.$$

Unfortunately, current observations do not allow a conclusive determination of κ. For $1.6 \leq \alpha \leq 2.1$, the data can be well reproduced if the emitted cosmic rays consist entirely of nuclei with masses in the intermediate (carbon, nitrogen, or oxygen) to heavy (iron, silicon) range, but a mixture of protons (97%) and heavier species (3%) is also acceptable (16, 48).

The diffuse neutrino flux has an additional component originating in the energy losses of ultrahigh-energy cosmic rays en route to Earth. Ultrahigh-energy protons above the GZK cutoff interact with the cosmic microwave and infrared backgrounds as they propagate over cosmological distances. These interactions generate pions and neutrons, which decay to produce neutrinos (14). The accumulation of these neutrinos over cosmological time is known as the cosmogenic neutrino flux. Ultrahigh-energy nuclei also interact with the cosmic microwave and infrared backgrounds, undergoing photodisintegration. The disassociated nucleons then interact with the cosmic microwave and infrared backgrounds to produce cosmogenic neutrinos (49). In the limit that the cosmic backgrounds are opaque to cosmic-ray nuclei, full disintegration occurs and the resulting cosmogenic neutrino spectrum is not dramatically different from that predicted in the all-proton case (assuming the cosmic-ray spectrum extends to high enough energies to produce protons above the GZK cutoff). In contrast, if a significant fraction of cosmic-ray nuclei remain intact, the resulting flux of cosmogenic neutrinos can be considerably suppressed. Of course, the predicted neutrino flux also depends on the spectrum, the source evolution, and the nature of the emitted cosmic rays (14–16). Representative spectra of cosmogenic neutrinos are shown in **Figure 1** for pure Fe and bimodal Fe + p mass compositions consistent with data (16). The estimates shown in **Figure 1** are framed in the context of the usual concordance cosmology (50) of a flat universe dominated by a cosmological constant with $\Omega_\Lambda \sim 0.7$; the rest is cold dark matter (CDM) with $\Omega_{\rm m} \sim 0.3$. (A rigorous compilation of the Λ CDM parameters is provided in Section 4.5.) The Hubble parameter is given by $H^2(z) = H_0^2[\Omega_{\rm m}(1+z)^3 + \Omega_\Lambda]$, normalized to its value today of 70 km s^{-1} Mpc^{-1}. The time dependency of the redshift can be expressed via $dz = -dt\,(1+z)H$. The cosmological evolution of the source density per comoving volume is parameterized as

$$\mathcal{L}_i(z, E) = \mathcal{H}(z)\mathcal{L}_i(0, E), \qquad 19.$$

where $\mathcal{H}(z)$ is the cosmological evolution of the cosmic-ray sources, which is taken here to follow the luminosity density evolution of quasi-stellar objects (QSOs): $\mathcal{H}_{\rm QSO}(z) = (1+z)^3$ for $z < 1.9$, $\mathcal{H}_{\rm QSO}(z) = (1+1.9)^3$ for $1.9 < z < 2.7$, and $\mathcal{H}(z) = (1+1.9)^3 \exp\{(2.7-z)/2.7\}$ for $z > 2.7$ (15).

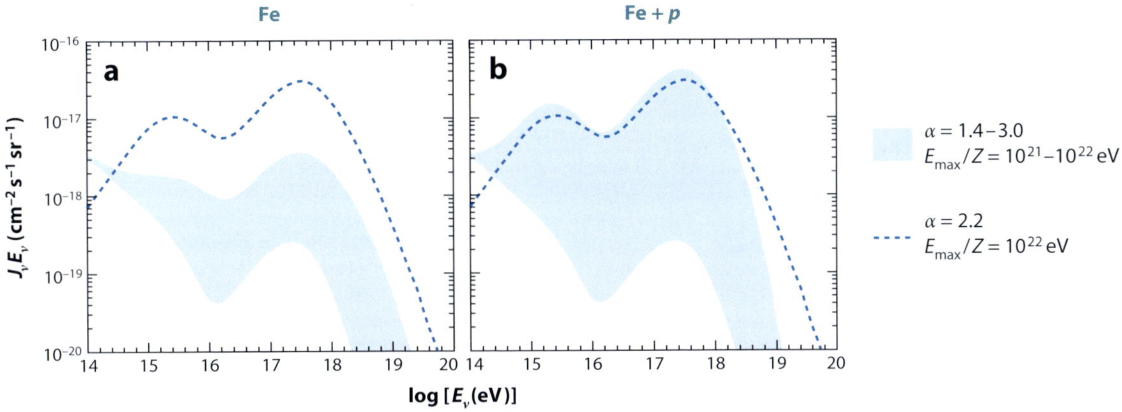

Figure 1

The light blue areas indicate the range of the cosmogenic neutrino spectrum for $\alpha = 1.4–3.0$ and $E_{max}/Z = 10^{21}–10^{22}$ eV for (*a*) pure iron and (*b*) iron and proton. In each frame, we show for comparison the prediction for an all-proton spectrum with $\alpha = 2.2$ and $E_{max} = 10^{22}$ eV (*dark blue dashed curves*) (16).

For such a cosmological evolution, which is similar to that describing the star-formation rate, $\xi_Z \approx 3$ (42).

In summary, the case for doing high-energy neutrino astronomy is compelling. The challenge is to deliver the technology to build a neutrino detector with the largest possible effective area and the best possible angular and energy resolution sensitive to these cosmic neutrino fluxes. We discuss this challenge next.

3. NEUTRINO TELESCOPES

3.1. Detection Principle

It took approximately 35 years between Markov's (51) conception of the NT detection principle and the operation of the first full detector. Although NTs use the well-established technique of photomultiplier tubes (PMTs), building them presents challenges. The difficulties in terms of construction arise from (*a*) the necessity of instrumenting extremely large regions, mainly because of the smallness of the neutrino-nucleon cross section and the steeply decreasing energy spectrum of neutrino fluxes of atmospheric and astrophysical origin; (*b*) the required darkness and transparency of the medium to detect the faint Cherenkov light produced by ultrarelativistic charged particles; and (*c*) the need for filtering neutrino events out of the more abundant atmospheric muons. Detectors are located under 1 to 4 km of water or ice, where for $E_\nu \gtrsim 100$ GeV, the downward-going background of atmospheric muons is approximately five to six orders of magnitude larger than the atmospheric neutrino flux coming from all directions (see **Figure 2**).

NTs are tridimensional matrices of photodetectors made up of strings of optical modules (OMs) that house large photocathode PMTs and protect them from both the water-column pressure (when they are installed in seas or in lakes) and the pressure during ice refreezing (for Antarctic detectors). In some cases OMs also include front-end electronics. The NT energy threshold depends on the vertical spacing between PMTs and the horizontal distance of strings; typical values range from 50 GeV to 1 TeV. The threshold is not sharp, but the detection efficiency increases with energy. Neutrinos interact with the medium nuclei in and around the detector through charged-current ($\nu_{lepton} + X \rightarrow$ lepton $+ X$) and neutral-current ($\nu_{lepton} + X \rightarrow \nu_{lepton} + X$)

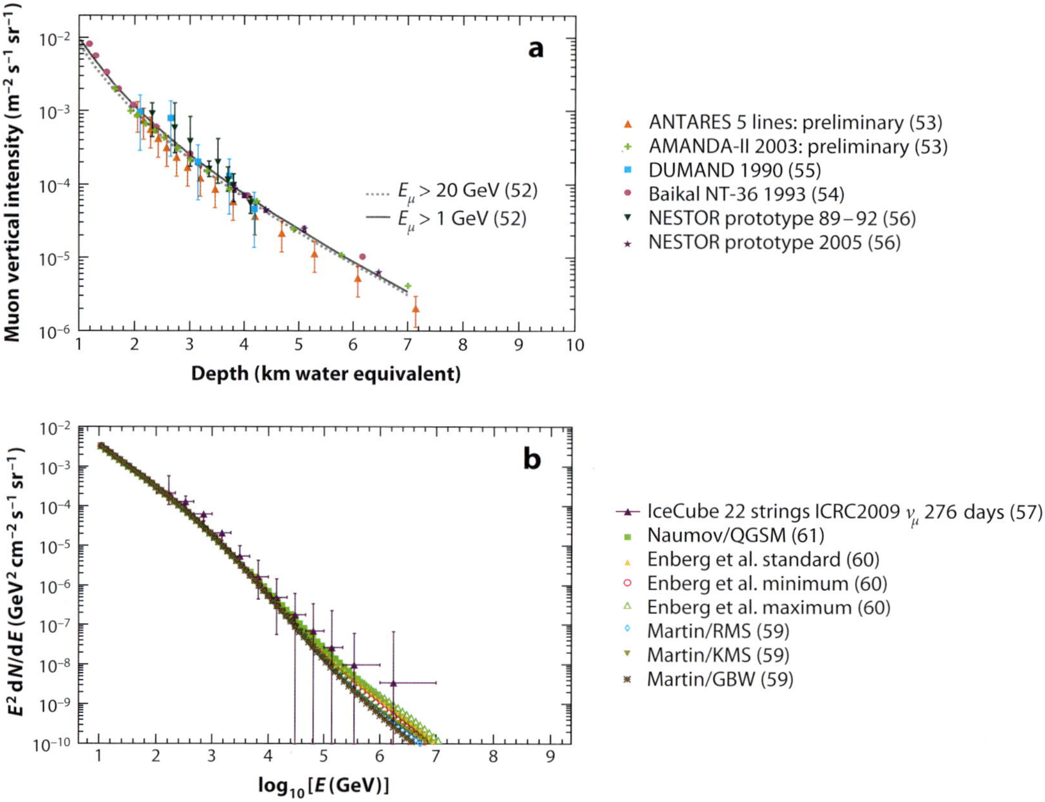

Figure 2

(*a*) Atmospheric muon vertical intensity measured by underwater and ice arrays as a function of depth compared with calculations (52). No correction is applied for the slightly different muon propagation properties between water and ice. Data are from ANTARES and AMANDA-II (53), Baikal NT-36 (54), the DUMAND prototype string (55), and NESTOR (for two different prototypes) (56). (*b*) Preliminary atmospheric unfolded neutrino spectrum measured with 22 strings of IceCube (57) compared with predictions for $\nu_\mu + \bar{\nu}_\mu/3$. [The factor 1/3 accounts for the lower charged-current (CC) interaction cross section of antineutrinos up to approximately 10^5 GeV compared with neutrinos.] Data are accompanied by large error bars. A preliminary evaluation of systematic errors, which are mainly due to the ice purity–level dependency with depth, is included in the error bars. The energy bin size was selected to match the energy resolution of the experiment. [Because the energy resolution is 0.3 on the $\log_{10}(E_\mu)$, there are three bins per decade.] The size of the last two bins is larger and is determined by the desire to accommodate bins with vanishing contents. Conventional neutrino fluxes are calculated as in Equation 58. Prompt models are calculated in the framework of perturbative quantum chromodynamics in Equation 60 (std represents optimal parameters, whereas min and max indicate the range of variation of the model parameters) and in Equation 59 for different structure functions. The quark-gluon string model (QGSM) (61) is also shown.

interactions. Charged-current deep-inelastic collisions above 10 GeV dominate over all other processes with the exception of $\bar{\nu}_e e$ interactions because of the intermediate-boson resonance formed in the neighborhood of $E_\nu^{\rm res} = M_W^2/2m_e \approx 6.3 \times 10^6$ GeV, which is generally referred to as the Glashow resonance (62). The resonant reactions $\bar{\nu}_e e^- \to W^- \to \bar{\nu}_\mu \mu^-$ and $\bar{\nu}_e e^- \to W^- \to$ hadrons may offer a detectable signal if the $\bar{\nu}_e$ flux originates in *pp* collisions (43).[3] The

[3] In *pp* collisions, the nearly isotopically neutral mix of pions upon decay creates a neutrino population in the ratio $N_{\nu_\mu} = N_{\bar{\nu}_\mu} = 2N_{\nu_e} = 2N_{\bar{\nu}_e}$. In contrast, photopion interactions leave the isotopically asymmetric process $p\gamma \to \Delta^+ \to \pi^+ n$ as the dominant source of neutrinos. Then, at production, $N_{\nu_\mu} = N_{\bar{\nu}_\mu} = N_{\nu_e} \gg N_{\bar{\nu}_e}$.

neutrino-nucleon cross section rises almost linearly with energy (63). This implies that (*a*) the interaction probability increases with energy, indicating that NTs are optimal detectors at high energies, and (*b*) neutrinos begin to be absorbed into the Earth. The shadowing effect of the Earth depends not only on the neutrino energy but also on the incident zenith angle (64). Near 10^5 GeV, the interaction length of muon and electron neutrinos is approximately the Earth's diameter. Thus, almost 80% (40%) of the neutrinos with $\cos\theta = -1\,(-0.7)$ are absorbed, where θ is the zenith angle. This indicates that at very high energy, where the astrophysical signals are expected to emerge from the atmospheric muon and neutrino backgrounds, detectors should have a large acceptance. There is a completely different behavior for τ neutrinos. In propagating through the Earth, they interact, producing τ leptons that decay back into a ν_τ of lower energy in what is referred to as a regeneration effect (65). The two leptonic channels, with a branching fraction of approximately 17%, also produce ν_e and ν_μ. The relevance of this phenomenon in terms of event rates at detectors depends on the neutrino spectrum.

NTs detect ultrarelativistic ($\beta \sim 1$) secondary particles produced in the interaction of the incoming neutrino with an atomic nucleus. These particles, which typically travel faster than the speed of light in the detector medium, emit directional Cherenkov light at an angle $\theta_C = \cos^{-1}[1/\beta n] \sim 41°$, where n is the refraction index of the medium. Between 300 and 600 nm, the region where PMTs are mostly sensitive, approximately 3.2×10^4 Cherenkov photons per meter are emitted by a bare muon track in the ice. Once the OM acceptance and PMT quantum efficiency are accounted for, only 6% of these photons, on average, contribute to the signal. The muon neutrino event topology is distinct from that of electron and τ neutrinos. Secondary muons travel long distances and can be reconstructed with an angular resolution better than 1° at high energy. Specifically, the kinematic angle between the muon and the neutrino in the deep-inelastic regime ($\geq$10 GeV) decreases with neutrino energy as $\sqrt{\langle\theta^2_{\nu\mu}\rangle} \propto \sqrt{m_p/E_\nu}$ (26), where m_p is the proton mass. For $E_\nu \geq 10$ TeV, the angle becomes negligible with respect to the intrinsic detector resolution. NTs measure the leading edge time of the PMT signal and, in some cases, the fully digitized waveform and the integrated charge. This information is used to reconstruct tracks for muons that travel long distances across the detector and determine their energy as well. Charged-current interactions of electron and τ neutrinos and all-flavor neutral-current interactions produce cascade-like events. The resulting hadronic and electromagnetic showers are small with respect to the scale of the separation between OMs; thus, they are almost a point-like source of light that propagates with some anisotropy that remembers the direction of the incoming neutrino. This prevents the achievement of an angular resolution for neutrinos comparable to the muon channel.

Muons lose energy according to

$$\frac{\mathrm{d}E_\mu}{\mathrm{d}x} = -a - b\,E_\mu, \qquad 20.$$

where $a = 2.0 \times 10^{-6}\,\mathrm{TeV\,cm^2g^{-1}}$ and is the continuous ionization term. The stochastic energy loss term bE, with $b = 4.2 \times 10^{-6}\,\mathrm{cm^2g^{-1}}$ (50), dominates at energies $\geq$1 TeV in water or ice. The distance a muon travels before its energy drops below some energy threshold E_μ^{th}, known as the muon range, is then given by

$$\lambda_\mu = \frac{1}{b}\ln\left[\frac{a + b\,E_\mu}{a + b\,E_\mu^{\mathrm{th}}}\right]. \qquad 21.$$

For muons with $E_\mu > 1$ TeV propagating in water or ice, stochastic losses such as bremsstrahlung, pair production, and photonuclear processes ($b\,E_\mu$ term) dominate over the ionization constant a term. In the first kilometer, a muon of approximately 100 TeV typically loses energy in two to three showers carrying more than 10% of its initial energy. Near the end of its range, the muon

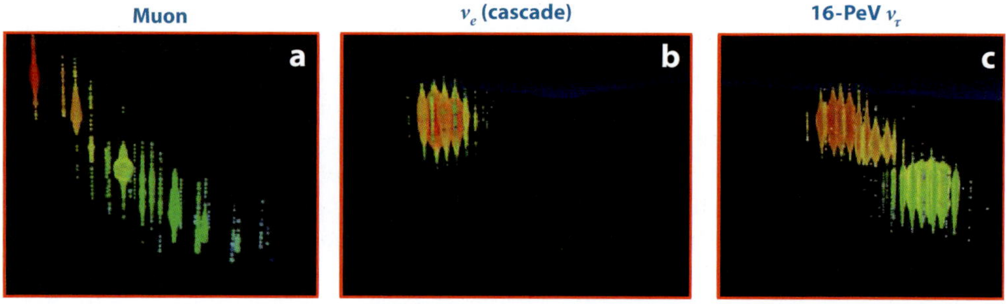

Figure 3

Simulated events in the IceCube detector, visualized using the IceCube event display, showing the three typical topologies discussed in Section 3. (*a*) A muon event of 100 TeV. (*b*) A cascade event induced by a 100-TeV ν_e. (*c*) A double-bang event induced by a 16-PeV ν_τ. The shading represents the time sequence of the hits. The size of the dots corresponds to the number of photoelectrons detected by the individual photomultipliers.

becomes a minimum ionizing particle emitting light that creates single-photoelectron signals at a distance of just over 10 m from the track. Above 1 TeV, the muon energy can be reconstructed with an energy resolution in $\log_{10} E_\mu$ of approximately 0.3. Above a certain energy, saturation effects of the PMT electronics and the size of the detector limit the energy reconstruction ability. In fact, because many of the events are not fully contained at very high energy, part of the Cherenkov cone can miss the PMTs so that part of the charge that is proportional to the muon energy is lost. The parent neutrino energy can be inferred from simulation of the $\nu(\bar{\nu}) + N \rightarrow \mu^-(\mu^+)$ interactions. For $E_{\nu,\bar{\nu}} > 10^5$ GeV, the average value of the energy taken by the lepton is $\sim$0.75 (63). In most cases, the well-reconstructed cascade events develop inside the detector; hence, the neutrino energy resolution is largely improved, up to 0.1 to 0.2 in $\log_{10} E_\mu$. Tau neutrinos produce similar events to electron neutrinos, up to approximately 10^6 GeV, because NTs have not yet demonstrated the ability to discriminate between electromagnetic and hadronic showers. However, for $E_\nu \sim 2 \times 10^6$ GeV, τ neutrinos deserve special attention: For such an energy, the τ lepton range is approximately 100 m. Consequently, both the $\nu_\tau N$ interaction and the shower (hadronic or electromagnetic) from the τ decay can be separated (66). These are called double-bang events and are in principle a background-free topology. Nevertheless, the expected event rates are limited by the small energy window in which these events are detectable. Indeed, this topology is limited in energy to the range of a few petaelectronvolts for the τ track to be long enough, up to approximately 100 PeV, to contain the two cascades in the instrumented volume (the τ range is 1 km at approximately 200 PeV). The three different neutrino topologies are shown in **Figure 3**.

Atmospheric muons and neutrinos constitute a troublesome background for searches of neutrinos produced in astrophysical sources. Moreover, the predictions for such a background are quite uncertain at energies $\geq$1 TeV. Therefore, background discrimination at NTs requires different techniques depending on the analysis. However, all these analyses use the fact that neutrinos can cross the entire Earth, although with a shadowing effect that depends on energy and nadir angle. By selecting up-going muons induced by neutrinos, researchers reduce the residual background of atmospheric muons to those events that are down-going but are misreconstructed as up-going. For cascade events, background discrimination is accomplished through containment cuts because neutrinos have their vertex in the instrumented region, whereas external tracks are produced mostly by atmospheric muons.

For point-source searches, the directional information is the crucial cipher. Being undeflected by intergalactic and galactic magnetic fields, neutrinos preserve directionality from their own sources and tend to cluster around it, whereas the background distribution is random (see Section 4.1). The muon channel provides the "golden sample" pointing back to primary sources. This is because the angular resolution is better than 1°, whereas cascade-like events have poor resolution of approximately 10° to 30°, depending on the medium.

For diffuse flux searches, the energy measurement is determinant because the signal spectrum is expected to be harder than the neutrino background (95) from atmospheric cascades initiated by cosmic rays. When protons and nuclei enter the atmosphere, they collide with air molecules and produce various kinds of secondary particles, which in turn interact with or decay or propagate to the ground, depending on their intrinsic properties and energies. Above 1 GeV, the most abundant particles at sea level are neutrinos and muons (50). The properties of these particles, which constitute the main background for astrophysical searches, are described in Section 4.2.

3.2. Detectors in Operation and Decommissioned

The DUMAND and the Baikal Collaborations pioneered underwater NT technologies in the deep ocean near Hawaii (67) and in Lake Baikal in Siberia (68), respectively. In November 1987, the DUMAND Collaboration measured the muon vertical intensity at depths ranging between 2 and 4 km, with a prototype line at a depth of 4.8 km approximately 30 km off the Hawaiian shore (55). The project was canceled in 1995. In 1993 a first configuration of 36 PMTs on three strings (NT-36) was installed in Lake Baikal at the shallow depth of 1.1 km located 3.6 km offshore (54). The experiment was later upgraded to larger configurations (NT-72-96-144). In 700 days of effective live time, 320 million muon events were detected, and the first atmospheric neutrino measured by an NT was selected in this sample and reported in 1996. The configuration of the experiment (NT-200) with 192 OMs was put into operation in April 1998. It consists of an umbrella-like structure of eight (72-m-long) strings, with up-looking and down-looking OMs (containing 37-cm-diameter PMTs developed for the project, known as QUASAR-370). Three external strings at 100 m from the center of NT200, each with 12 pairs of OMs, were added in April 2005 to increase the cascade sensitivity at very high energies (NT200+). The Baikal Collaboration is working toward the construction of a gigaton-volume detector with approximately 2,100 to 2,500 OMs (arranged over 95 to 100 strings separated by distances of the order of 80 to 120 m) grouped in clusters to form independent subarrays (69). A prototype string has been acquiring data since April 2008.

The NESTOR Collaboration began surveys of an area close to the Peloponnese coast at a mean depth of approximately 4 km in the 1990s. They measured the muon flux with a hexagonal prototype floor with 12 up-looking and down-looking PMTs (56), and they studied by simulation the possibility of deploying a telescope made of towers with 12 of these hexagonal floors (32 m in diameter) vertically spaced by 20 to 30 m. The NESTOR Collaboration is operating a deep-sea station off the shore of Pylos at a depth exceeding 3.5 km connected to the onshore laboratory by an electro-optical cable.

In 1987 Halzen (70) developed the idea of "ice-fishing" neutrinos. The Antarctic muon and neutrino detector array (AMANDA) was located below the surface of the Antarctic ice sheet at the geographic South Pole. During 1993 and 1994, in an exploratory phase, the four-string AMANDA-A array was deployed and instrumented with 80 PMTs spaced at 10-m intervals from 810 to 1,000 m. (The scattering length at that depth is too short to allow degree-level reconstruction.) A deeper array of ten strings, referred to as AMANDA-B10, was deployed during the austral summers between 1995 and 1997 to depths between 1,500 and 2,000 m. The instrumented volume of AMANDA-B10 formed a cylinder with a diameter of 120 m, overlooked by 302 OMs (Hamamatsu R5912-2, 20-cm diameter) (71). During December 1997 and January 2000, the detector was expanded by adding nine outer strings of PMTs. The composite AMANDA-II array of 19 strings and 677 OMs forming two concentric cylinders with a larger diameter of 200 m became operative in 2000 and continued until 2009. The OMs were connected to cables supplying the 8-inch PMT power and transmitting analog signals. Some strings used fiber-optic cables reducing cross-talk effects, but most used coaxial twisted-pair cables. One of the strings pioneered the digitization technique, with data acquisition for waveform digitization in each module communicating to the surface, designed as a prototype for the IceCube digital OMs (DOMs) and data transmission.

The IceCube NT, currently being deployed near the Amundsen-Scott station, comprises a cubic kilometer of ultraclear ice approximately 1 mile below the South Pole surface, instrumented with long strings of sensitive photon detectors that record light produced when neutrinos interact in the ice. The deep array is complemented by IceTop, a surface air shower detector consisting of a set of 160 frozen water tanks. With its surface array, IceCube becomes a three-dimensional air shower array for studies of cosmic rays up to the exaelectronvolt range. It is also useful for calibrating the NT, and it serves as a partial veto for the downward events at high energy. IceCube strings are placed 124 m apart on a hexagonal grid. Each string is instrumented (between 1.5 and 2.5 km below the surface) with 60 DOMs vertically separated by approximately 17 m. The DOMs enclose 10-inch Hamamatsu PMTs with a quantum efficiency of $\sim$20% at approximately 400 nm. Six additional strings with new high-efficiency ($\sim$25–30% at 400 nm) PMTs to be deployed in between seven standard IceCube strings will constitute the deep core. This change will improve the sensitivity of the detector in the region $\lesssim$1 TeV, important for dark matter searches and neutrino oscillations. The outer IceCube strings around DeepCore and the upper layers of DOMs will be used as a veto shield against the background of atmospheric muons, allowing the identification of down-going neutrino-induced muons when the neutrino vertex is in the inner detector. Construction of the observatory will be completed in the austral summer of 2010 to 2011. The drilling of strings has been so successful that as many as 20 strings can be installed in a season (from mid-November to mid-February, when airplane flights are possible to and from the South Pole). The basic element of IceCube, the DOM, is made up of a glass spherical pressure vessel 33 cm in diameter, containing a 25-cm-diameter PMT, a high-voltage supply, light-emitting diodes for energy and time calibrations, and an electronic motherboard. The motherboard digitizes and time-stamps PMT signals (72). The performance of the first string (deployed in the 2005–2006 season), its calibration and test methods, and the data acquisition and installation of the detector are described in Reference 73.

The completion of the undersea neutrino observatory ANTARES, located approximately 2.5 km below the sea surface offshore of Toulon in southern France, took several years. Since 1996, the Collaboration has had sea campaigns comprising more than 60 line deployments to test water transmission, light optical background, biodeposition, and sedimentation (74). The permanent electro-optical cable (approximately 40 km in length) transmitting data and power between the shore station in La Seyne-sur-Mer and the junction box offshore has been operating since November 2002. Between 2003 and 2005, various prototype lines and a mini-instrumentation line for environmental parameter measurements were deployed (75). In 2006, the first two lines of the detector were deployed, and two additional lines were installed in 2007. In May 2008, ANTARES was completed with a total of 12 lines. Lines were connected initially by a submarine and then by an unmanned remote-operated vehicle. The ANTARES observatory comprises the 12 mooring lines, a line specifically dedicated to marine environmental monitoring, a seismometer, and a biocamera for bioluminescence studies. The lines are anchored to the seabed at a depth of 2,475 m and are held vertical by buoys. These buoys are free floating so each line moves in response to the sea current, with movements of a few meters for typical values of 5 cm s^{-1}. An acoustic positioning system, comprising transponders and receivers, gives a real-time measurement of the position of the OMs with a precision better than 10 cm, typically every 2 min. Tiltmeters and compasses provide their orientation. Seventy-five OMs along each line between ~2,400 and 2,000 m are grouped in triplets on stories. The three PMTs [Hamamatsu 10-inch (76)] look downward at 45° from the vertical to prevent transparency loss due to sedimentation and optimize detection for up-going tracks. PMTs are housed inside pressure-resistant glass spheres comprising two halves closed by applying an underpressure of 200 to 300 mbar. The setup (including the PMT, the glass sphere, the silicon gel for optical coupling between the glass and the photocathode, and the μ metal cage for shielding the Earth's magnetic field) is referred to as the OM (77). Stories also include titanium containers housing the front-end electronics, with a pair of ASIC chips per PMT used for signal processing and digitization. This provides the time stamp and the amplitude of the PMT signal. Each of the OMs contains a pulsed LED for calibration of the relative variations of PMT transit time, and a system of LED and laser optical beacons allows the relative time calibration of different OMs. An internal clock system, which is synchronized by GPS to Universal Time with a precision of ~100 ns, distributes the 20-MHz clock signal from the shore. Time calibrations allow a precision at the level of 0.5 ns, ensuring the receipt of an angular resolution at the level of 0.3° muons above 10 TeV for point-source searches (78). All data above a threshold of approximately one-third of a photoelectron pulse are sent to shore for further online filtering. This requires coincidences between PMTs on the same story-rejecting hits not compatible with the propagation of light in water between hit PMTs (79).

Despite the pros and cons of each, ice and water are proving to be suitable transmission media. Photon-propagation properties are characterized by a much longer scattering length $\lambda_{\rm scatt}$ in seawater compared with ice. The effective scattering length, defined in terms of the average scattering angle of the photons, θ, reads $\lambda_{\rm eff} = \lambda_{\rm scatt}/(1 - \langle\cos\theta\rangle)$. In seawater, $\lambda_{\rm eff}$ is approximately 100–200 m and is much larger than in ice owing to a more forward peaked angular distribution. By contrast, the absorption length, which determines the distance photons can propagate before being absorbed, is much larger in ice (~100–200 m) than in water (~50–70 m).

Seawater is a less quiet environment than ice. Additionally, bioluminescence and β decay of ^{40}K can produce a baseline of optical background in the sea of the order of 30 to 100 kHz and megahertz bursts that cause a random dead time for OMs. No current or bioluminescence is present in ice. Once water developed in the drilling process is frozen, the risk of short circuits due to water leakages in the OMs is largely reduced, and PMTs can be switched on. By contrast, once ice freezes, the components are not recoverable, whereas entire instrumented lines can be recovered from water for repairs. The average background rate in IceCube PMTs is ~280 Hz, which depends mainly on the radioactivity from the OM and PMT materials. The lower optical noise in ice makes IceCube a very sensitive detector for SN-collapse searches.[4]

[4]Approximately 99% of the binding energy of a Type II SN is expected to be released in neutrinos. These SNe originate in the gravitational collapse of massive giant stars, heavier than 8 solar masses, into a neutron star. Current upper limits on the number of supernova explosions in our Galaxy, from scintillator and water Cherenkov detectors, are getting closer to the predicted rates of approximately two per century (85). In the early phase (~10 ms), known as deleptonization or neutronization, electron neutrinos are emitted in the electron-capture process ($e^- + p \rightarrow n + \nu_e$); all subsequent (~10 s) neutrino flavors are produced in reactions such as $e^+ + e^- \rightarrow \nu + \bar{\nu}$ in the thermalization phase. The passage of a large flux of megaelectronvolt-energy neutrinos during a period of seconds will be detected as an excess of single counting rates in all individual OMs. Each OM will be triggered by positrons from the inverse-β reaction: $\bar{\nu}_e + p \rightarrow e^+ + n$, where the visible positron energy is given by $E_{\rm vis} = E_{\bar{\nu}_e} - Q + m_e = E_{\bar{\nu}_e} - 0.789$ MeV (86). IceCube, with 5,160 OMs, can observe the 10-ms pulse as an increased rate over the average background rate (280 Hz) in the PMTs. For example, an SN at a distance of 7.5 kpc produces a rate of approximately 10^6 Hz over this time, yielding a 34σ effect. An SN in the Large Magellanic Cloud would produce a 5σ effect.

The neutrino effective area, the sensitive area "seen" by neutrinos producing detectable muons when entering the Earth, is a useful parameter to determine event rates and the performance of a detector for different analyses. The expected event rate,

$$N_\mu = \int\int\int \mathrm{d}E_\nu\, \mathrm{d}\cos\theta_\nu\, \mathrm{d}\phi_\nu\, A_\nu^{\text{eff}}(E_\nu,\theta_\nu,\phi_\nu) J_\nu(E_\nu,\theta_\nu,\phi_\nu), \qquad 22.$$

of a model predicting a neutrino spectrum J_ν depends on the neutrino-nucleon cross section $\sigma(E_\nu)$, the shadowing of the Earth $P_{\text{Earth}}(E_\nu,\theta_\nu)$, the track-reconstruction quality cuts, and the selection criteria for background rejection. The shadowing factor depends on column depth, namely the integral of the Earth density profile $\rho(\theta_\nu,\ell)$ for a given direction of the neutrino θ_ν over the distance in the Earth ℓ: $P_{\text{Earth}}(E_\nu,\theta_\nu) = e^{-N_A\times\sigma(E_\nu)\times\int_\ell \rho(\theta_\nu,\ell)d\ell}$, where N_A is the Avogadro number. The neutrino effective area is a parameter calculable through simulations and is formally given by

$$A_\nu^{\text{eff}}(E_\nu,\Omega_\nu) = V_{\text{eff}}(E_\nu,\theta_\nu,\phi_\nu)\times(\rho N_A)\times\sigma(E_\nu)\times P_{\text{Earth}}(E_\nu,\theta_\nu), \qquad 23.$$

where (ρN_A) is the number of nucleons in the target material assumed isoscalar and $V_{\text{eff}}(E_\nu,\theta_\nu,\phi_\nu) = V_{\text{gen}}(E_\nu,\theta_\nu,\phi_\nu)\times N_{\text{sel}}(E_\nu,\theta_\nu,\phi_\nu)/N_{\text{gen}}(E_\nu,\theta_\nu,\phi_\nu)$ is the effective volume of the interactions generated in a volume V_{gen} of the neutrinos that produce detectable muons. The effective volume depends on the energy of the neutrinos that determines how far muons can travel from the vertex (muon range) and accounts for the selection efficiency of an analysis (the efficiency factor $N_{\text{sel}}/N_{\text{gen}}$).

The point spread function is the percentage of muons that are reconstructed inside an angular half-cone width from the neutrino or source direction. It depends strongly on the medium properties and on the reconstruction algorithms. In the final configuration of IceCube, the angular resolution at energies between 10 and 100 TeV will be such that approximately 50% of the reconstructed muons from the direction of a neutrino point source will be inside an angle of 0.5°. Seawater has longer effective scattering lengths than Antarctic ice; hence, the resolution in ANTARES or in a future cubic-kilometer detector in the Mediterranean (84) will be of the order of 0.2° to 0.3°. The effective area and the point spread function of the different NTs are compared in **Figure 4**.

Cosmic-ray experiments, like those of the Pierre Auger Observatory, provide a complementary technique for ultrahigh-energy neutrino detection by looking for deeply developing, large zenith angle (>60°) air showers (87). At these large angles, hadronic air showers have traversed the equivalent of several times the depth of the vertical atmosphere, and their electromagnetic component has been extinguished far away from the detector. Only very high energy core–produced muons survive past two equivalent vertical atmospheres. Therefore, the shape of a hadronic (background in this case) shower front is very flat and very prompt in time. In contrast, a neutrino shower appears generally equivalent to a normal shower, which makes straightforward the process of distinguishing neutrino-induced events from background hadronic showers. Moreover, very low ν_τ fluxes could be detected efficiently by Auger detectors by looking at the interaction in the Earth crust of quasi-horizontal ν_τ inducing a horizontal cascade at the detector (88).

4. EXPERIMENTAL RESULTS AND SCIENTIFIC IMPACT

The main goal of NTs is to detect astrophysical neutrinos with $E_\nu \geq 100$ GeV. As we discussed in Section 3, muons constitute the "golden sample" for neutrino astronomy because the achievable angular resolution is at subdegree level. At energies ≥10 TeV, the limiting angle is mainly due to the transit time spread of the PMTs and to the scattering of light.

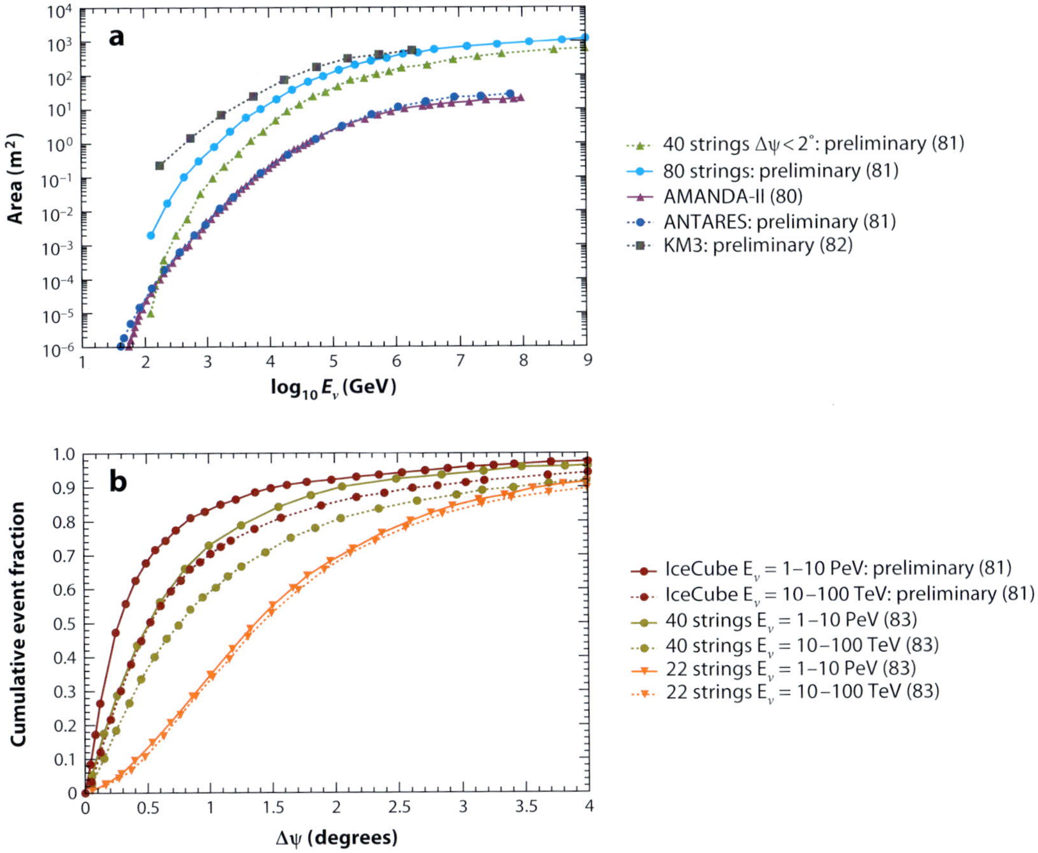

Figure 4

(*a*) Effective area for muon neutrinos, integrated over the lower hemisphere, as a function of the neutrino energy. The selected cuts ensure a pointing capability suitable for point-source searches. The area for 40 strings of IceCube is obtained with the requirement that tracks are reconstructed inside 2° from a source. The full IceCube area is obtained using the same cuts and is therefore preliminary. The AMANDA-II area is from the final analysis in Reference 80. The ANTARES effective area is obtained with preliminary selection cuts based on simulation (81). The KM3NeT area is for a possible configuration still under debate (82). (*b*) Cumulative distribution of reconstructed muon events inside a cone of semiangle $\Delta\Psi$ (point spread function) from the neutrino direction. The different curves indicate growing configurations of IceCube for two neutrino energy bins and the point-source analysis cuts (83). The IceCube curves are preliminary because they are based on cuts optimized for the 40-string analysis.

Although imaging atmospheric Cherenkov telescopes can prove their point spread function on copious sources of γs, such as the Crab Nebula, no astrophysical high-energy neutrino source has yet been shown to exist. Hence, the "standard candle" for NTs must be an "antisource," such as the Moon shadow. The Moon disk, with a diameter of approximately 0.5°, blocks primary cosmic rays, thus producing a deficit in the muon flux. IceCube, using 3 months of data of the 40-string configuration, begins to find evidence at 5σ level of this deficit, which is proof of the absolute timing and pointing capability of the detector (89). A detector with an ice platform on top (such as IceCube) is able to test, for a limited solid angle, the angular resolution using events in coincidence with the extensive air shower array IceTop. Sea detectors, such as ANTARES, can instead use either a scintillator array carried by a boat or the reconstruction of the shape of the nearby continental crust using horizontal muons (90).

4.1. Steady and Flaring Point-Source Searches

Searches of point-like sources have various advantages when compared with searches of diffuse fluxes. For example, in the former, one can use the direction, energy, and eventually the time distributions to reject the background of atmospheric neutrinos. As a matter of fact, the signal is expected both to distribute around the source according to the detector point spread function and to have a harder energy spectrum than the background. Flaring sources as well as GRBs are particularly interesting because, if the neutrinos are emitted in coincidence with the γ flare, it is possible to reduce dramatically the background of neutrinos over a short time window. Because the background of the experiment is flat in right ascension (which is directly connected to the hour of the day), many "equivalent experiments" with only background can be reproduced by scrambling the right ascension of events and keeping fixed other information, such as the declination and the energy proxy. This has the advantage that data, rather than simulation, can be used for estimating the background, so that resulting significances of the signal and the background are not affected by a possible imperfect understanding of the detector or the detector medium. Moreover, for searches of flares or GRBs, data outside the short time window can be used to reproduce equivalent experimental samples where no signal is present. These techniques are similar to the off-on source techniques applied in photon astronomy, where the background outside the source pixel region is used to evaluate the significance of the observed source. In the many neutrino astronomy point-source searches (91), binned methods that would look for clusters of events on top of the background from specific directions or in the entire sky were applied: The search bin was optimized depending on the expected signal shape (typically assumed to be $\propto E_\nu^{-2}$). However, to avoid missing signal events at the border of bins, grids or bins had to be moved, which penalized the search with several trial factors. Many experiments (92) use likelihood methods (93; J. Braun, manuscript in preparation) that can be up to 40% more sensitive depending on the amount of information used. These methods use probability distribution functions that maximize the discrimination between the signal and background. The signal is characterized with respect to the background by the source directional feature and by the harder spectrum expected from astrophysical sources compared with that of atmospheric neutrinos. For time-dependent sources, for instance, periodical binaries such as microquasars or flares from blazars, periodicity assumptions, or light curves from γ-ray, X-ray, and optical telescopes can be used.

From IceCube's 22-string configuration, researchers collected 5,114 events from the Northern Hemisphere in 276 days (80). No signal has been found, and upper limits have been set for sources in the Northern Hemisphere. For illustrative purposes we note that, for Geminga, the Crab Nebula, Markarian 421, M87, and Cygnus OB2, the upper limits of the 90%-CL interval for a flux $\Phi \leq \Phi_{90} \times 10^{-12}\,\mathrm{TeV}^{-1}\,\mathrm{cm}^{-2}\,\mathrm{s}^{-1}(E_\nu/\mathrm{TeV})^{-2}$ are $\Phi_{90} = 9.67$, 10.35, 14.35, 7.91, and 15.28, respectively. In addition, a hot spot was found with a posttrial probability of 1.34%. The excess, however, was not confirmed using the 6-month data of the 40-string configuration (that by then encompassed in sensitivity the previous configuration by a factor of approximately two), so the hot spot has been considered a background fluctuation (83). This 40-string analysis selects up-going neutrino events from the Northern Hemisphere at a rate of approximately 40 per day. In this region of the sky, the experiment is sensitive mainly to events in the 10–100-TeV region for a $\propto E_\nu^{-2}$ spectrum. Very recently, by using a scrambled background of very high energy atmospheric muons, the IceCube Collaboration extended the field of view for point-source searches to the Southern Hemisphere. In this area of the sky, the sensitive region, for a spectrum $\propto E_\nu^{-2}$, moves to higher energies (i.e., the petaelectronvolt-to-exaelectronvolt range) (94). As an illustration, the resulting 90%-CL upper limits $\Phi \leq \Phi_{90} \times 10^{-9}\,\mathrm{GeV}^{-1}\,\mathrm{cm}^{-2}\,\mathrm{s}^{-1}(E_\nu/\mathrm{GeV})^{-2}$, using 22 strings of IceCube, are $\Phi_{90} = 558.2$, 245.0, and 27.1 for Centaurus A, Sagittarius A*, and 3C279, respectively. (The

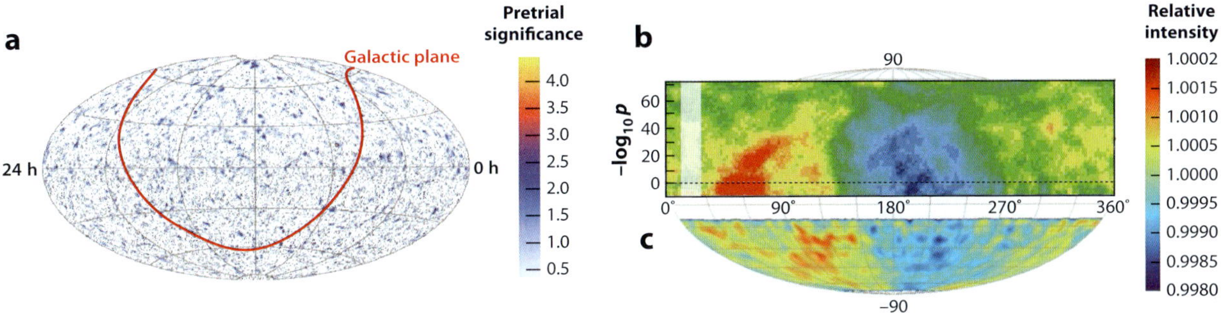

Figure 5

(*a*) Muon-event sky map (z axis color code indicates pretrial significance) for 6 months of data taken with the 40-string IceCube array, from July 2008 to December 2008. Of the 17,777 black dots on the sky map, 6,797 are up-going neutrino candidates (zenith >90°) from the Northern Hemisphere. The color shading indicates the significance of the data, and the curved red line represents the galactic plane (83). (*b*) Celestial cosmic-ray intensity map for Tibet AS γ data taken from 2001 to 2005. The vertical color bin width for the relative intensity calculated within each declination band is 2.5×10^{-4} (107). The highest and lowest intensities are observed in the region of the tail-in and loss-cone in the direction of the helio-tail, and on the right the excess is in the direction of Cygnus. (*c*) Downward-going muon sky map (expressed as relative intensity ≡ number of events in an angular bin in right ascension and declination divided by the total number of events in the declination band) observed by 22 strings of the IceCube neutrino telescope with data taken between April 2006 and April 2007 (106).

associated energy intervals that contain 90% of the signal are $\Delta E_\nu/\text{GeV} = 2 \times 10^6 - 8 \times 10^8$, $1 \times 10^6 - 9 \times 10^8$, and $5 \times 10^4 - 1 \times 10^8$.) For the full data sample of 40 strings, the IceCube discovery reach (5σ) for $E_\nu^2 \Phi_{90}$ (averaged over right ascension and in the declination region between 0° and 90°) is expected to be between 1 and 5×10^{-11} cm^{-2} s^{-1}. This is near to the exclusion flux for 22 strings and is a factor of approximately three higher than the exclusion limits for the same configuration. With the 40-strings result, flux limits for galactic sources begin to be in the region of interest: $E_\nu^2 \Phi \sim 10^{-12} - 10^{-11}$ TeV cm^{-2} s^{-1} (see Section 2). According to estimates based on γ-ray observations, it would be possible to achieve 5σ significance for Milagro Pevatrons with ~10 years of IceCube data collection (6). The 40-string sky map is shown in **Figure 5*a***.

4.2. Atmospheric Muons and Neutrinos

Atmospheric muons and neutrinos detected by NTs above some tens of gigaelectronvolts are produced in cosmic ray–induced showers and come mainly from pions and kaons. At lower energies (≤10 GeV), muon and electron neutrinos are also generated by muon decays. They were extensively studied in the past two decades because of the so-called atmospheric neutrino problem, that is, a deficit of muon-induced neutrino events that was explained by muon neutrino oscillations into τ neutrinos (for reviews see, e.g., References 95 and 96). At energies above 100 GeV, where NTs operate, atmospheric neutrino oscillation studies are not a primary target. Nonetheless, although DeepCore and ANTARES photocathode coverage are modest, they will collect a very large data set of atmospheric neutrinos, thereby allowing the possibility of observing neutrino oscillations. The main reasons for using NTs to measure atmospheric neutrinos are (*a*) to calibrate the detector, thus demonstrating their ability to discriminate neutrino events from the huge background of atmospheric muons; (*b*) to understand hadronic interactions in the forward region at energies not yet accessible by colliders; and (*c*) to study new physics scenarios.

The dominant production channels of atmospheric and muon neutrinos detected in NTs are the decay chains of mesons created in hadronic cascades. The pion decay chain

($\pi^+ \rightarrow \mu^+\nu_\mu \rightarrow e^+\nu_e\nu_\mu\bar{\nu}_\mu$ and the conjugate process) dominates the atmospheric neutrino production at approximately gigaelectronvolt energies. At energies of some gigaelectronvolts, the muon decay length begins to be comparable to the atmospheric altitude where most muons and neutrinos are produced (10–20 km). Hence, depending on the incoming zenith angle, muons may stop decaying. At energies of $\sim$100 GeV, kaon decay becomes increasingly important with respect to pion decay. In this energy range, the atmospheric neutrino spectrum can be parameterized as (26)

$$\frac{\mathrm{d}N_{\nu_\mu}}{\mathrm{d}\ln E_\nu} = A_{\mathrm{tot}} E_\nu^{\alpha} \left[\frac{A_\nu}{1 + B_p E_\nu (\cos\theta)^* / \varepsilon_\pi} + 0.635 \times \frac{B_\nu}{1 + B_k E_\nu (\cos\theta)^* / \varepsilon_k} \right], \qquad 24.$$

where 0.635 is the branching ratio of $K^\pm$ decays; $\varepsilon_\pi = 115$ GeV and $\varepsilon_K = 850$ GeV are the critical energies, respectively, for pion and kaon decay; $\alpha \sim 1.7$ is the spectral index of the primary spectrum, A_{tot}; A_ν and B_ν are normalization constants; and B_p and B_k are constant terms with energy. The term $(\cos\theta)^*$ is equal to $\cos\theta$ for zenith angles such that $\cos\theta \lesssim 0.3$, and for horizontal directions ($\cos\theta \geq 0.3$), it has a larger value that accounts for the curvature of the Earth. The $1/(E_\nu \cos\theta)$ dependency is characteristic of atmospheric processes where there is competition between interaction and decay, because the decay probability increases proportionately to the interaction one has for longer horizontal paths in the atmosphere. The muon flux in this energy region has a similar dependency, with the important difference of the energy fraction taken by muons in two-body pion and kaon decays. We remind the reader that, on average, in pion decay the neutrino takes one-quarter of the energy of the pion while the muon takes the rest. This implies that, above $\sim$1 TeV, muon production is dominated by pion decay and neutrino production by kaon decay (approximately 85% of muons are produced via pion decay, whereas only 25% of neutrinos come from the decay of charged pions). The electron neutrino formula is more complicated, with two similar terms due to $K^\pm$ and K_L decays from which ν_e are produced directly, and a term due to muon decay into electron neutrinos is determined by $\pi^\pm$, $K^\pm$, and K_L decays. In this energy window, the flavor ratios are $w_e : w_\mu : w_\tau \approx 1/20 : 19/20 : 0$ (97), and the energy spectra are a function of the zenith angle of the atmospheric cascades. As already observed, mesons in inclined showers spend more time in rarefied atmosphere where they are more likely to decay rather than interact. As a result, the spectra of highly inclined neutrinos are harder than those of nearly vertical neutrinos and resemble their primary spectrum. Above $\sim 10^5$ GeV, kaons are also significantly attenuated before they decay, and the "prompt" component, arising mainly from very short-lived charmed mesons ($D^\pm$, D^0, D_s, and Λ_c), dominates the spectrum (18). Such a prompt neutrino flux is isotropic with flavor ratios of 12/25:12/25:1/25. Because prompt neutrinos have a hard spectrum resembling the primary one (below the knee, the differential cosmic-ray spectrum goes as $\sim E^{-2.7}$), they provide a persisting background for astrophysical neutrinos that follow a power law $\sim E_\nu^{-2}$.

Various Monte Carlo high-precision calculations have been developed to study atmospheric neutrino oscillations (58, 95, 98). For energies below 10 GeV, high precision needs to account for the angle of secondaries with respect to primaries (99), the solar modulation effects, and the Earth geomagnetic field deflections. At higher energies, these effects become negligible, and one-dimensional calculations that neglect the angle of secondaries with respect to primaries are suitable. Uncertainties in atmospheric neutrino fluxes are of the order of 15% in the region of $\sim$10 GeV, rising to $\sim$20 to 30% at $\sim$100 GeV -1 TeV (100). These are mainly due to the rather poor knowledge of the primary cosmic-ray flux and the hadronic interaction models. A precise estimate in the prompt region is not possible because charm production in the atmosphere is not well constrained by collider measurements (101). The various calculations differ by up to two orders of magnitude (18) (see **Figure 2** and References 59–61 for some examples). The experiments LHCf and TOTEM, dedicated to measuring particles emitted in the very forward region (102), will improve our understanding of hadronic models at laboratory energies of approximately 10^{17} eV.

A cubic-kilometer NT measures of the order of kilohertz muon events, the so-called second calibration beam. With such high statistics, high-precision verifications become possible; for instance, how well the detector details and the effects of the light propagation in the radiator are simulated. Nonetheless, hadronic interaction models used in the atmospheric shower simulations are affected by theoretical uncertainties, not only experimental ones. Some of these models are similar to those used to simulate atmospheric neutrinos. This represents a drawback because the two calibration beams should be used to understand the same detector independent of theoretical uncertainties. Another complication that affects shower models is the difficulty of investigating errors arising from imperfect knowledge of high-energy primary cosmic-ray composition and hadronic models with those resulting from the fact that simulations do not account for the seasonal variation of the atmosphere. These simulations are very time consuming because rates of events are very high and high-energy showers develop many particles in the atmosphere. Hadronic interaction models that foresee a larger K^+ production (such as SIBYLL 2.1) show better agreement than others (e.g., QGSJET II) in the high $x = E_{\mathrm{secondary}}/E_{\mathrm{primary}}$ region (103). The preference for such hadronic models is also indicated by the muon ratio measurement of MINOS (104) that shows a rise between 0.3 and 1 TeV, which is consistent with an increasing contribution to the muon charge ratio from kaons.

An NT's high statistics measurement of the atmospheric neutrino flux above $\sim$100 GeV is an important benchmark for all these calculations. A fit of the atmospheric muon neutrino flux (see Equation 24) was recently performed with AMANDA-II data to determine variations with respect to existing models (105). The allowed region in the parameter space normalization versus spectral index was derived relative to the Bartol flux (98), using the angular distribution and an energy proxy for 5,511 neutrino events collected in 1,387 days. The best-fit point indicates that, at 640 GeV, the data prefer a higher normalization by approximately 10% and a slightly harder spectrum by approximately 0.056. Using a sample of 4,492 up-going neutrino-induced muon events collected in 275.5 days, with a contamination of 5% from misreconstructed atmospheric muons, the IceCube Collaboration unfolded the atmospheric neutrino spectrum for the 22-string configuration (57). This spans a wider energy range up to petaelectronvolt energies and is compared with some conventional (from π and K decays) and prompt neutrino models in **Figure 2*b***. Systematic errors are so large at $E_\nu > 10$ TeV that it is not yet possible to disentangle an eventual contribution from prompt neutrinos (originating in the decay of charmed mesons and baryons) from an astrophysical component (60).

NTs can collect a huge statistics of muons, of the order of billions. Thus, they are sensitive to anisotropies of amplitudes $<10^{-3}$. For example, in **Figure 5*b*** we show the map in equatorial coordinates of approximately 5 billion atmospheric muons detected with the 22 strings of IceCube. These muons are produced by cosmic rays with a median energy of 14 TeV and have a median angular resolution of 3°; they also show an anisotropy in their arrival direction of amplitude $(6.4 \pm 0.2) \times 10^{-4}$. The observed anisotropy persists with high statistical significance to energy in excess of 100 TeV with a smaller but measurable amplitude. This observation is not compatible with the Compton-Getting effect (109), which produces an energy-independent dipole anisotropy in the arrival direction of cosmic rays as a result of the relative motion of the Solar System with respect to the rest frame of galactic cosmic rays. This explanation was already excluded by the Tibet Array (107), which in the absence of an anisotropy at 300 TeV concluded that galactic cosmic rays corotate with the Galaxy. Other experiments observed anisotropies in the Northern Hemisphere at the scale of 10° to 30° (108): In particular, Milagro applied a cut that discriminates hadrons with respect to γs. The observed anisotropy with IceCube in the opposite hemisphere persists at energies >100 TeV, thus making it difficult to explain such anisotropy as an effect of the heliosphere or the heliospheric tail. Interestingly, it has been speculated that the effect may be due to the propagation of cosmic rays produced by nearby (~100-pc) SNRs and pulsars (110). Given this, the presence of Geminga between the anisotropy regions in the Northern Hemisphere is noteworthy. In the Southern Hemisphere, the Vela pulsar is located near the highest-intensity regions observed by IceCube. However, such regions are so broad that we can speculate only about responsible local sources in the absence of a more precise calculation, including the intensity of galactic magnetic fields. This in turn could be correlated to observations at lower energies of an excess in the positron-to-electron fraction with respect to secondary production models (111) that have also been interpreted as propagation of cosmic rays from local sources such as SNRs and pulsars (112). The sites of the acceleration of cosmic rays might also not be isolated, but may instead involve superbubbles (overlapping SNRs) (113).

4.3. Searches for Diffuse Cosmic Neutrino Fluxes

Diffuse fluxes from extragalactic sources present more promising material for research compared with single-source fluxes because of the larger event rates, with a signal that may extend up to the highest energies where some of the extragalactic sources (e.g., as AGN and GRBs) can contribute. Nonetheless, these searches have to rely more on simulations than point-source searches do, and hence they are subject to larger systematic errors. The basic concept is that because of the harder spectrum the astrophysical signal should show up at high energies above the atmospheric neutrino background. However, the theoretical uncertainties on high-energy atmospheric neutrino fluxes (see Section 4.2) and the experimental errors in the high-energy region where the statistics is low limit the sensitivity and reliability of this search. For large statistics, the uncertainty in the prompt neutrino flux can be mitigated by triggering on τ neutrinos, which have the double-bang characteristic signature in the petaelectronvolt energy range (**Figure 3**). Specifically, the detection of a double-bang event would confirm neutrino oscillations while establishing the discovery of high-energy cosmic neutrinos because τ neutrino production in the atmosphere is negligible. **Figure 6** presents a collection of experimental limits compared with representative models for classes of sources such as AGN, GRBs, and cosmogenic neutrinos. These upper limits can in turn be used to constrain the origin and nature of ultrahigh-energy cosmic rays.

4.4. Cosmic-Ray Proton Fraction

Consider the working hypothesis introduced in Section 2 for optically thin sources: The relative number and energy of the neutrinos and neutrons depend only on kinematics, which implies approximate equipartition of the decaying pion's energy between the neutrinos and the electron, and the relative radiation density in the source. On average, each interaction will produce η neutrinos per neutron with relative energy $\in$ per neutrino; that is, $\eta = \langle N_\nu \rangle / \langle N_n \rangle$ and $\in = \langle E_\nu \rangle / \langle E_n \rangle$. The neutrino emissivity of flavor i is then given by

$$\frac{\Delta E_{\nu_i}}{N_{\nu_i}} \mathcal{L}_{\nu_i}(z, E_{\nu_i}) = \frac{\Delta E_n}{N_n} \mathcal{L}_n(z, E_n). \qquad 25.$$

Assuming flavor universality as well as $\in \simeq E_\nu / E_n \simeq \Delta E_\nu / E_n \simeq \Delta E_\nu / \Delta E_n$ and $\eta \simeq N_{\text{all}\,\nu} / N_n$, one arrives at the neutrino source luminosity (per comoving volume)

$$\mathcal{L}_{\text{all}\,\nu}(z, E_\nu) \simeq \frac{\eta}{\in} \mathcal{L}_n(z, E_\nu / \in). \qquad 26.$$

For the (hypothetical) source where pion production proceeds exclusively via resonant $p\gamma \to \Delta^+$ scattering, the neutrino multiplicity and the relative energy are fixed: $\eta = 3$ and $\in = 0.07$.

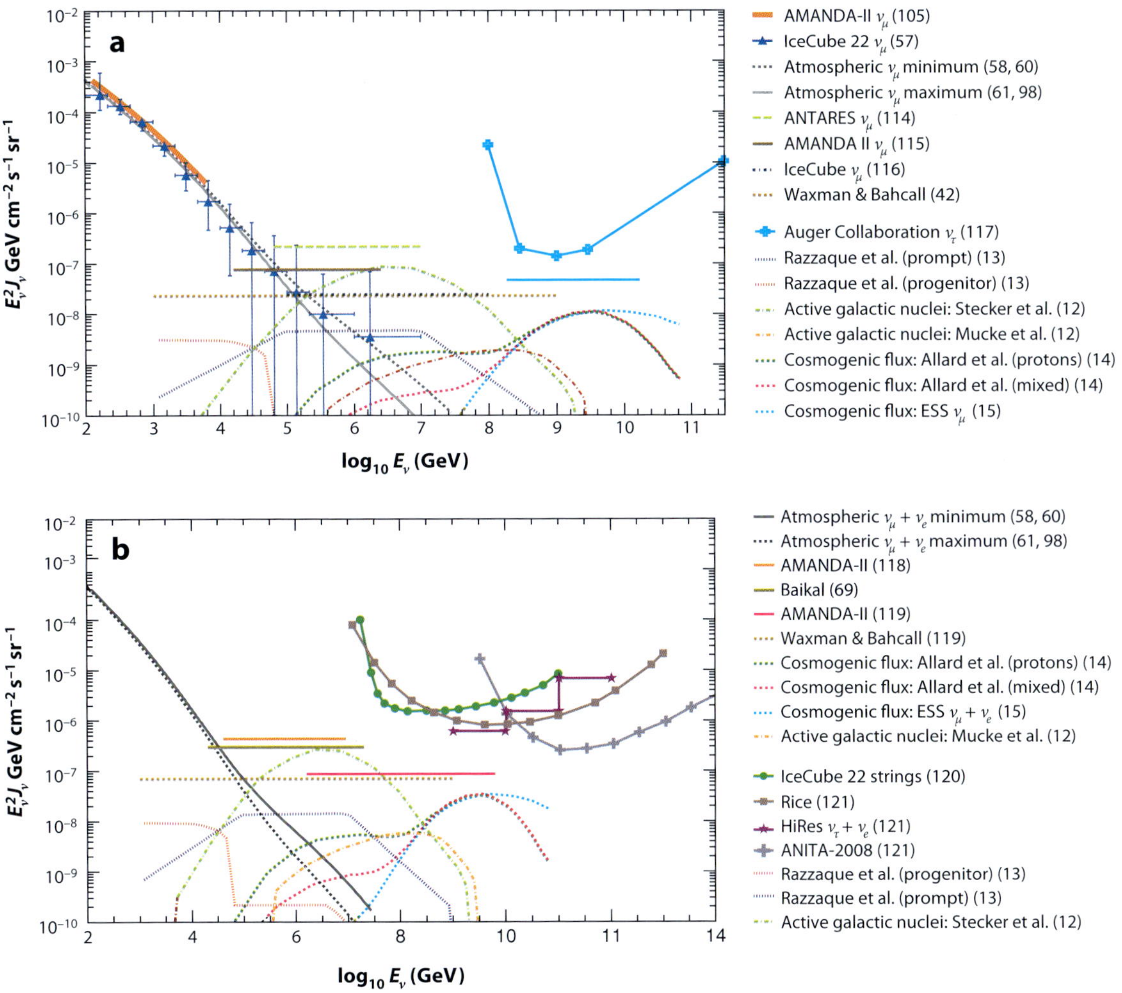

Figure 6

(*a*) The blue triangles indicate the atmospheric ($\nu_\mu + \bar{\nu}_\mu$) flux averaged over the lower hemisphere as measured by 22 strings of IceCube (57), and the falling dotted gray line indicates AMANDA-II measurements (105). For comparison, predictions of the atmospheric neutrino flux (conventional plus prompt) are indicated by the falling gray solid lines (58, 60, 61, 98). The horizontal lines represent different 90%-CL limits and sensitivities on E_ν^{-2} fluxes. From top to bottom: ANTARES estimated sensitivity for 1 year (*light green dashed line*) (114), AMANDA-II muon neutrino limit for 804 days (115), and IceCube estimated sensitivity for 1 year (116). For comparison, also shown are the Waxman-Bahcall upper limit corrected for oscillations (42), neutrino flux predictions from active galactic nuclei (12) and prompt and precursor emission from γ-ray bursts (13), and various estimates of the nearly guaranteed cosmogenic flux (14, 15). The ultrahigh-energy upper limit on the τ neutrino flux (differential and $\propto E_\nu^{-2}$) reported by the Pierre Auger Collaboration is also shown (117). (*b*) Same as in panel *a* but for experiments measuring more than one flavor, such as measuring cascades induced by all-flavor neutrinos. Atmospheric $\nu_\mu + \nu_e$ fluxes averaged over the lower hemisphere are shown as in panel *a*. The horizontal solid lines are 90%-CL limits for a flux $\propto E_\nu^{-2}$. From top to bottom: AMANDA-II all-flavor cascade limit for 1,001 days (118), Baikal cascade limit for 1,038 days (69), and all-flavor ultrahigh-energy limit (preliminary) for 507 days of AMANDA-II (119). The differential in energy upper limits (90% CL) are for the 22 strings of IceCube (120) and several ultrahigh-energy neutrino–detection experiments (121), including those with radio detectors that look for the coherent Cherenkov radiation produced by the excess of electrons in neutrino-induced showers in dense media (e.g., ice and the lunar regolith). For further details, see Reference 85.

Note that the relation (26) derived for optically thin sources can be regarded as a lower limit on the neutrino luminosity as long as energy-loss processes in the source are negligible (122). Therefore, this conservative expectation can be translated into an upper limit on the extragalactic proton fraction in ultrahigh-energy cosmic rays, exploiting, for instance, the experimental upper limit on the diffuse neutrino flux from IceCube or the preliminary AMANDA data (119).

For this procedure, one introduces test functions of the neutron source luminosity of the form $\mathcal{L}_n^{\text{test}}(0, E) = \mathcal{L}_0(E/E_{\max})^{-1} \exp(-E/E_{\max})$, with an exponential energy cutoff $E_{\max}$ that varies between 10^8 GeV and 10^{12} GeV with a logarithmic step-size of $\Delta \log_{10} E = 0.25$ (122). The source function per unit redshift is given in Equation 19. The source luminosity per comoving volume is assumed to follow that of the star-formation rate (SFR): $\mathcal{H}_{\text{SFR}}(z) = (1+z)^{3.4}$ for $z < 1$, $\mathcal{H}_{\text{SFR}}(z) = N_1(1+z)^{-0.3}$ for $1 < z < 4$, and $\mathcal{H}_{\text{SFR}}(z) = N_1 N_4 (1+z)^{-3.5}$ for $z > 4$, with appropriate normalization factors $N_1 = 2^{3.7}$ and $N_4 = 5^{3.2}$ (123). Each neutron test luminosity is related to a neutrino luminosity by the ratio of Equation 26. After propagation, the accumulated contribution of extragalactic and cosmogenic neutrinos is normalized to the neutrino limit. The results are shown in **Figure 7**. The AMANDA preliminary limit on the diffuse neutrino flux (119) severely constrains the extragalactic proton contribution in cosmic rays at energies below 10^9 GeV. Following the same line of thought, one can verify that if the evolution of cosmic-ray sources follows that of AGN (124), the upper limit on the diffuse flux of τ neutrinos from the Pierre Auger Observatory (117) marginally constrains the proton fraction at the end of the spectrum (125).

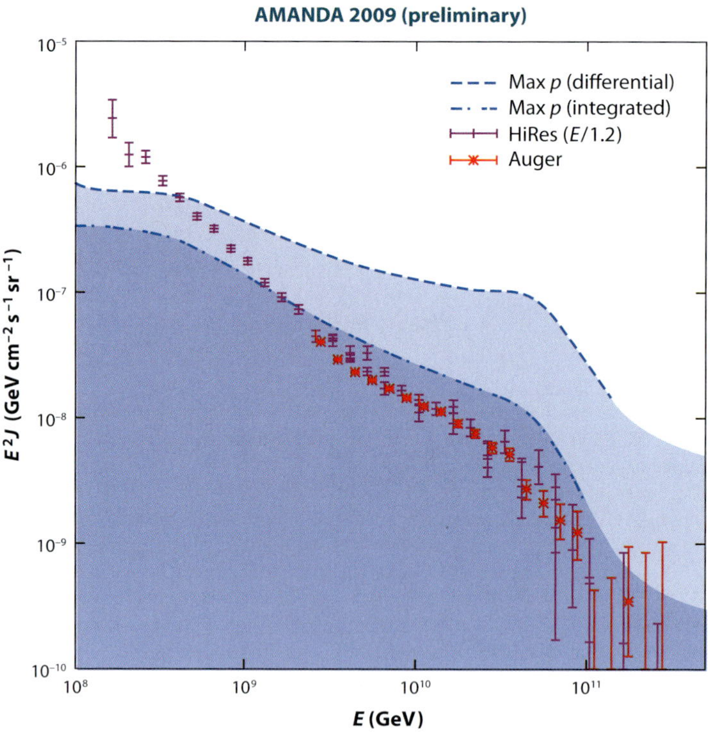

Figure 7

Integrated and differential upper limits on the proton contribution in ultrahigh-energy cosmic rays derived from the AMANDA bound on diffuse neutrinos (see Reference 119; based on Reference 122). The HiRes data have been scaled by approximately 20% to match the Auger data.

4.5. Dark Matter Searches

Over the past few years, a flood of high-quality data (126) from the Supernova Cosmology Project, the Supernova Search Team, the Wilkinson Microwave Anisotropy Probe, the Two-Degree Field Galaxy Redshift Survey, and the Sloan Digital Sky Survey have pinned down cosmological parameters to percent-level precision, establishing a new paradigm of cosmology. A surprisingly good fit to the data is provided by a simple geometrically flat Friedman-Robertson-Walker universe, in which 30% of the energy density is in the form of nonrelativistic matter ($\Omega_m = 0.30 \pm 0.04$) and 70% in the form of a new, unknown dark energy component with strongly negative pressure ($\Omega_\Lambda = 0.70 \pm 0.04$). The matter budget has only three free parameters: the Hubble parameter $h = 0.70^{+0.04}_{-0.03}$, the matter density $\Omega_m h^2 = 0.138 \pm 0.012$, and the density in baryons $\Omega_b h^2 = 0.0230^{+0.0013}_{-0.0012}$.[5] This implies that the structure of the universe is dictated by the physics of still-undiscovered CDM ($\Omega_{CDM} h^2 = 0.115 \pm 0.012$) and that the galaxies we see today are the remnants of relatively small overdensities in the nearly uniform distribution of matter in the very early universe.

The simplest model for CDM consists of WIMPs (128). The most popular CDM candidates are (*a*) the lightest supersymmetric particle (such as the well-motivated neutralino χ) that is stabilized in *R*-parity-conserving models of supersymmetry (SUSY) (129) and (*b*) the lightest Kaluza-Klein state appearing in models of universal extra dimensions (130). Generic WIMPs were once in thermal equilibrium, but they decoupled while remaining strongly nonrelativistic. The relic abundance of SUSY WIMPs can be found by integrating the Boltzmann equation (131),

$$\frac{\mathrm{d}n}{\mathrm{d}t} + 3Hn = -\langle \sigma v \rangle (n^2 - n_{eq}^2), \qquad 27.$$

where n is the present number density of SUSY WIMPs, H is the expansion rate of the universe at temperature T, $n_{eq} = g(m_\chi T/2\pi)^{3/2} e^{-m_\chi/T}$ is the equilibrium number density, g is the number of internal degrees of freedom of the WIMP, $\langle \sigma v \rangle$ is the thermally averaged annihilation cross section, and m_χ is the neutralino mass. Note that in the very early universe, when $n \simeq n_{eq}$, the right-hand side of Equation 27 is small and the evolution of the density is dominated by Hubble expansion. As the temperature falls below m_χ, however, the equilibrium number density becomes suppressed and the annihilation rate increases, rapidly reducing the number density. Finally, when the number density falls sufficiently, the rate of depletion due to expansion becomes greater than the annihilation rate, $H \geq n\langle \sigma v \rangle$, and the neutralinos freeze out of thermal equilibrium. The freeze-out temperature T_f depends logarithmically on $\langle \sigma v \rangle$, but for models with teraelectronvolt-scale SUSY breaking, one finds that $T_f/m_\chi \sim 0.05$. SUSY WIMPs are, by far, the favored candidate for CDM, because for masses of order 100 GeV to 10 TeV, a present density of $\Omega_\chi h^2 \sim 0.1$ is produced fairly naturally (132).

Many approaches to detect dark matter have been developed. Such endeavors include direct detection experiments, which aim to observe the scattering of dark matter particles with the target material of the detector (133), and indirect detection experiments, which are designed to search for the products of WIMP annihilation into γ rays, antimatter, and neutrinos (134).

The annihilation of WIMPs into photons typically proceeds through a complex set of processes, and the final spectrum contains a detailed imprint of WIMP annihilation that can, in principle, reveal features such as the WIMP spin and/or other particles in the dark sector (135). On the one hand, tree-level annihilation of WIMPs into quarks or leptons (or heavier states that

[5]The third component is consistent with the estimate from big bang nucleosynthesis on the basis of measurements of deuterium in high-redshift absorption systems, $\Omega_b h^2 = 0.020 \pm 0.002$ (127).

decay into them) yields a continuum of photon energies, with an upper cutoff approximately at the WIMP mass. On the other hand, loop-level annihilation into a photon and X results in a line at energy $E_\gamma = m_\chi[1 - M_X^2/(4m_\chi^2)]$, where M_X is the X boson in the final state (in SUSY models, X is either another photon or a Z). The line emission typically has smaller magnitude than continuum emission, but it provides a clean signal that helps discriminate against backgrounds. The differential flux of photons arising from CDM annihilation observed in a given direction making an angle ψ with the direction of the galactic center (GC) is given in Reference 135:

$$\phi^\gamma(\psi, E_\gamma) = \int \bar{J}\frac{1}{2}\frac{D_\odot}{4\pi}\frac{\rho_\odot^2}{m_\chi^2}\sum_f \langle\sigma v\rangle_f \frac{dN_f}{dE_\gamma} d\Omega, \qquad 28.$$

where $\bar{J} = (1/\Delta\Omega)\int_{\Delta\Omega} J(\psi)d\Omega$ denotes the average of J over the solid angle $\Delta\Omega$ (corresponding to the angular resolution of the instrument) normalized to the local density: $J(\psi) = (D_\odot\rho_\odot^2)^{-1}\int_{\ell=0}^{\infty}\rho^2[r(\ell,\psi)]d\ell$. Furthermore, the coordinate ℓ runs along the line of sight, which in turn makes an angle ψ with respect to the direction of the GC (i.e., $r^2 = \ell^2 + D_\odot^2 - 2\ell D_\odot \cos\psi$); the subindex f denotes the annihilation channels with one or more photons in the final state and dN_f/dE_γ is the (normalized) photon spectrum per annihilation; and $\rho(\vec{x})$, $\rho_\odot = 0.3\,\text{GeV cm}^{-3}$, and $D_\odot \simeq 8.5$ kpc, respectively, denote the CDM density at a generic location $\vec{x}$ with respect to the GC, its value at the Solar System location, and the distance of the Sun from the GC. Indeed, the GC has long been considered to be among the most promising targets for detection of CDM annihilation, particularly if the halo profile of the Milky Way is cusped in its inner volume (136). This has been complicated, however, because H.E.S.S. observations disfavored CDM emission from the GC and revealed instead a very high energy source (HESS J1745-290) (137). To discover CDM photon emission, the contribution of the many point sources near the GC (138) needs to be subtracted. Imaging atmospheric Cherenkov telescopes or satellite and balloon-borne experiments also look for γs from high–CDM density regions relatively close to the Earth ($\lesssim$100 kpc), such as dwarf spheroidal satellite galaxies of the Milky Way, clusters of galaxies, and intermediate-mass black holes (85).

NTs indirectly search for the presence of dark matter by taking advantage of the Sun's ability to capture large numbers of WIMPs over time. Over billions of years, a sufficiently large number of WIMPs can accumulate in the Sun's core to allow for their efficient annihilation. Such annihilations produce a wide range of particles, most of which are quickly absorbed into the solar medium. Neutrinos, on the other hand, may escape the Sun and be detected in experiments on Earth. Specifically, the time evolution of the WIMP population in the Sun is controlled by

$$\dot{N} = C_\odot - A_\odot N^2, \qquad 29.$$

where

$$C_\odot \approx 3\times 10^{18}\,\text{s}^{-1}\left(\frac{\rho_{\text{local}}}{0.3\,\text{GeV cm}^{-3}}\right)\left(\frac{270\,\text{km s}^{-1}}{\bar{v}_{\text{local}}}\right)^3\left(\frac{\sigma_{\text{eff}}}{10^{-6}\,\text{pb}}\right)\left(\frac{1\,\text{TeV}}{m_\chi}\right)^2 \qquad 30.$$

is the Sun's rate of capture (139); ρ_{local} is the local dark matter density; $\bar{v}_{\text{local}}$ is the local root mean-square velocity of halo dark matter particles, $A_\odot = \langle\sigma v\rangle/V_{\text{eff}}$; $V_{\text{eff}} = 1.8\times 10^{26}(\text{TeV}/m_\chi)^{3/2}\,\text{cm}^3$ is the effective volume of the core Sun determined by matching the core temperature with the gravitational potential energy of a single WIMP at the core radius (140); and $A_\odot$ is related to the annihilation rate by

$$\Gamma = \frac{1}{2}A_\odot N^2. \qquad 31.$$

The rate at which WIMPs are captured in the Sun depends on the nature of the interaction the WIMP undergoes with nucleons in the Sun. These elastic scattering processes are often broken into two classifications: (*a*) spin-dependent interactions in which cross sections, σ_{SD}, increase with the spin of the target nuclei and (*b*) spin-independent interactions in which cross sections, σ_{SI}, increase as the square of the total number of nucleons in the target. Thus, the effective WIMP-on-nucleus (hydrogen and helium) elastic scattering cross section has two contributions: $\sigma_{\rm eff} = \sigma_{SD} + \sigma_{ID}$. The present WIMP annihilation rate is found by solving Equation 29 for the annihilation rate at a given time

$$\Gamma = \frac{1}{2} C_\odot \tanh^2\left(\frac{t}{\tau_{eq}}\right), \quad 32.$$

where $\tau_{\rm eq} = 1/\sqrt{C_\odot A_\odot}$ is the timescale required to reach equilibrium between capture and annihilation. When $\tau_{\rm eq}$ becomes comparable to or larger than the age of the Solar System ($t_\odot \simeq 4.5$ billion years), the system has not yet reached equilibrium and the annihilation rate is strongly suppressed, $\Gamma \approx \frac{1}{2} C_\odot (\sqrt{C_\odot A_\odot} t_\odot)^2$, whereas when $(C_\odot A_\odot) t_\odot \gg 1$, the annihilation rate is saturated at $\Gamma = \frac{1}{2} C_\odot$.

As they annihilate, WIMPs can generate neutrinos through a wide range of channels. Annihilations to heavy quarks, τ leptons, gauge bosons, and Higgs bosons can generate neutrinos in the subsequent decay (22). The total flux of neutrinos emitted by the Sun as a result of WIMP annihilation is then

$$\left.\frac{\mathrm{d}F_\odot^{\nu_\mu}}{\mathrm{d}E_\nu}\right|_{\rm source} = C_\odot F_{\rm Eq} \sum_j \left(\frac{\mathrm{d}N_{\nu_\mu}}{\mathrm{d}E_\nu}\right)_j e^{-E_\nu/150\,\mathrm{GeV}}, \quad 33.$$

where $F_{\rm Eq}$ is the nonequilibrium suppression factor ($\approx$1 for capture-annihilation equilibrium) and $(\mathrm{d}N_{\nu_\mu}/\mathrm{d}E_\nu)_j$ is the ν_μ flux produced by the j decay channel per WIMP annihilation. Note that neutrinos produced near the center of the Sun interact with the solar medium, yielding a depletion of the emission spectrum by a factor of $\sim e^{-E_\nu/150\,\mathrm{GeV}}$ (22). Finally, the ν_μ flux reaching the Earth is

$$\phi_\odot^{\nu_\mu} = \frac{1}{4\pi \mathrm{d}^2} \left.\frac{\mathrm{d}F_\odot^{\nu_\mu}}{\mathrm{d}E_\nu}\right|_{\rm source}, \quad 34.$$

where $\mathrm{d} \approx 1.5 \times 10^8$ km is the Earth-to-Sun distance. All in all, the possibility that neutrino-detection experiments will detect dark matter critically depends on the capture rate of WIMPs in the Sun, which in turn depends on the elastic scattering cross section of these particles.

Whereas γ-ray experiments set limits on $\langle \sigma v \rangle$ or on the γ-ray flux, NTs measure the muon-induced flux by neutrinos and hence constrain this flux as a function of m_χ. A collection of various limits on the muon flux is shown in **Figure 8**. The conversion from muon flux to cross-section limits is not model independent: In **Figure 8** it was done using DarkSUSY (150). Equilibrium is assumed between capture and annihilation rates in the Sun, so that the annihilation rate is proportional to the spin-dependent and -independent cross sections. A limit on σ_{SD} is found by setting to zero the spin-independent cross section. This procedure is indispensable to show complementarity between searches and to combine results from different techniques. As an illustration, in **Figure 8** we compare the limits on σ_{SD} with those obtained from direct detection experiments including CDMS (141), COUPP, and KIMS (149).

The described indirect searches for CDM are complementary approaches to the direct searches and the possible direct production at the LHC. Although indirect detection channels are subject to large astrophysical uncertainties, they explore different regions of the parameter space and a very wide range of candidate particle masses.

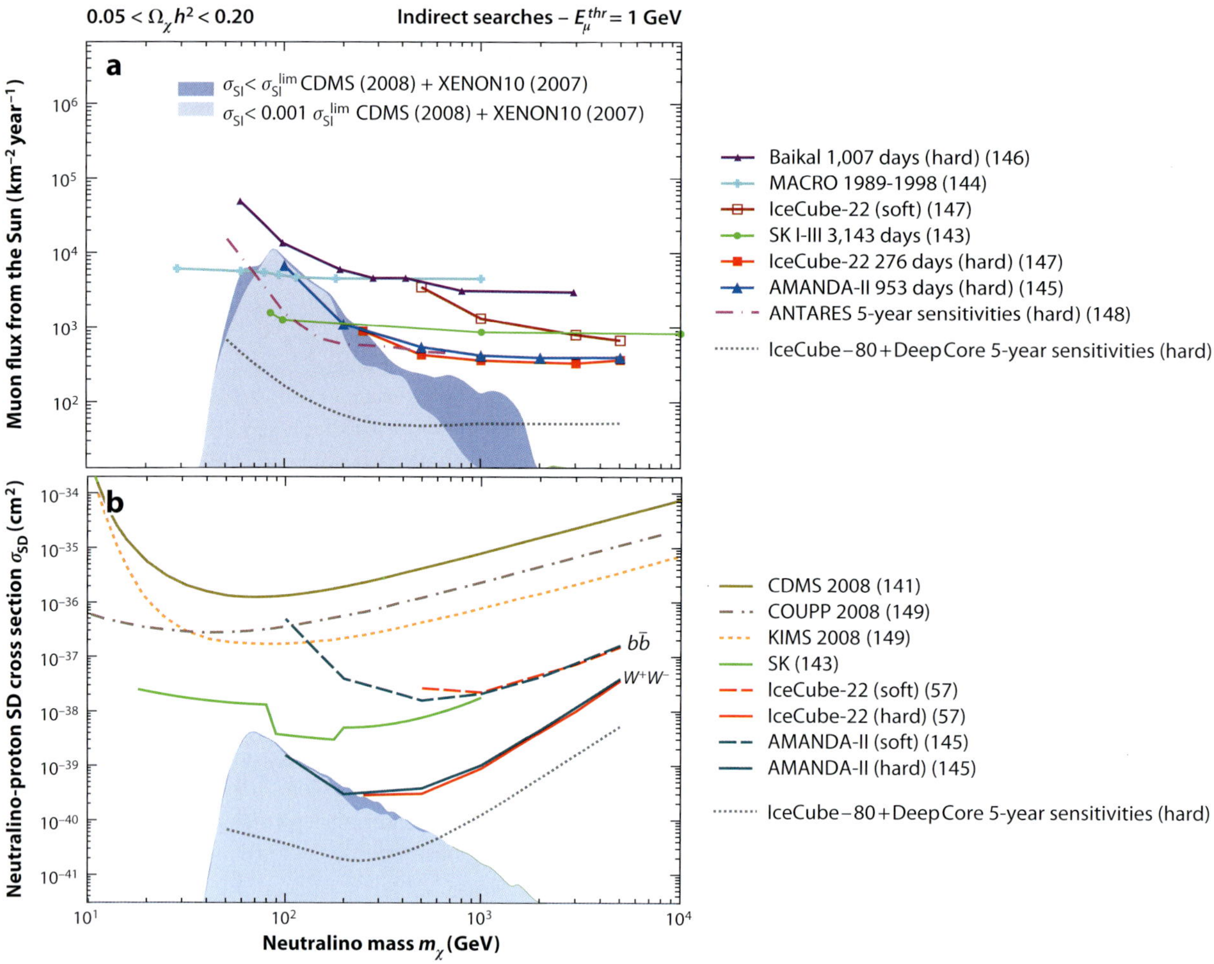

Figure 8

(*a*) Upper limits (90% CL) on the muon fluxes from neutralino annihilation in the Sun as a function of neutralino mass for soft ($b\bar{b}$) and hard (W^+W^-) channels (85, 147). The lighter blue region indicates the supersymmetry parameter space compatible with direct detection limits on the spin-independent cross section from CDMS (141) and XENON10 (142). The darker blue area represents the projected sensitivity for the same region of the parameter space, assuming that direct detection limits are 100 times better. Experimental limits that include a correction for detector threshold within the common assumption of $E_{\nu,\mathrm{thr}} = 1$ GeV are shown for Super-Kamiokande (SK) (143), MACRO (144), AMANDA-II (145), Baikal (146), and 22 strings of IceCube for the hard and soft channels (147). For comparison, the projected sensitivity of ANTARES after 5 years of data taking is also shown (148). (*b*) 90%-CL upper limits from AMANDA-II and 22 strings of IceCube on the spin-dependent cross section for soft and hard channels (147). Shaded blue regions follow the same conventions used in panel *a*. For comparison, we also show direct search limits on σ_{SD} from CDMS (141), COUPP (149), and KIMS.

5. CONCLUSIONS AND FUTURE PROSPECTS

The hunt for neutrinos has just begun with the advent of suitably sized detectors to measure the expected small event rates. Given their dimensions, these experiments are very challenging to build. IceCube, the first cubic-kilometer detector, is expected to be completed on schedule in 2011. If the predictions from the measured γ-ray fluxes are indeed correct, then the first detection of sources, such as the Milagro Pevatrons with hard spectra up to a few tens of teraelectronvolts, may become possible after a few years of data accumulation. Capable of accessing sources that are opaque in photons, NTs may broaden our reach to the unknown similar to the way many other astronomical instruments have discovered unexpected signals from the universe. The Mediterranean community is operating ANTARES, the first completely underwater detector, at the scale of the IceCube's precursor AMANDA-II. Other research and development programs to build a cubic-kilometer detector in the Mediterranean sensitive to the GC are ongoing. In Italy in July 2007, a ~100-km electro-optical cable was installed offshore of Capo Passero (at approximately 3.5 km below sea surface) by the NEMO Collaboration. A prototype tower of the proposed cubic-kilometer array will be connected in the near future. ANTARES, NEMO, and NESTOR joined their efforts to form the Consortium KM3NeT to design the cubic-kilometer detector in the Mediterranean. High-quantum efficiency PMTs are being studied for this detector. Most proposals aim at increasing the photocathode area per OM and the isotropy of light detection, minimizing the transit time spread, maximizing the collection efficiency and the peak-to-valley ratio for a single-photoelectron signal (connected to the capability of counting photoelectrons in one event), and minimizing costs and power consumption.

Whether in the ice or in the sea, building a detector a factor of ten more sensitive than IceCube is not an easy task. A new detector in the sea, with a cost similar to that of IceCube, could not achieve such a goal with the current technology, despite the fact that in the sea the angular resolution for point-source searches can be better by a factor of two to three compared with IceCube. It is also hard to obtain a uniform increment in area at all energies while avoiding cost divergence. A significant increase in the area at very large energies implies larger separations between instrumented lines; thus, this penalizes the 1–100-TeV region, which is important for galactic sources.

Projects are under way to build detectors with resolution of the order of 100 times that of IceCube using techniques that exploit the advantage of attenuation lengths that are a factor of approximately ten greater than the optical one in the same medium. Particularly interesting is the radio technique in ice. The acoustic technique in seawater or ice also seems encouraging. None of these techniques, however, is background free. They also have the disadvantage of higher energy thresholds compared with those of the NTs discussed in this review. Such thresholds reduce the degree of overlap with the optical technique, preventing effective cross calibrations that facilitate understanding the background. Moreover, thresholds of the order of 10^{18} eV are suitable for extragalactic neutrinos but not for galactic sources. Detectors combining the radio and acoustic techniques are also desirable. However, for cross calibrations, the radio technique seems currently more promising due to the low cost of antennas, the lower power consumption versus that of acoustic devices, and the larger attenuation length in the ice.

In summary, after 20 years of careful work by many research groups around the world, we are in possession of a tantalizing body of data, more than sufficient to stimulate our curiosity but not yet sufficient to pin down high-energy neutrino sources. The upcoming high-quality observations from the giant-aperture NTs under construction will generate a data sample of unprecedented size, ushering us into a golden age of neutrino astronomy.

DISCLOSURE STATEMENT

The authors are not aware of any affiliations, memberships, funding, or financial holdings that might be perceived as affecting the objectivity of this review.

ACKNOWLEDGMENTS

We are grateful to Markus Ahlers, Francis Halzen, Luciano Moscoso, Paolo Desiati, Ellen Zweibel, Christian Spiering, Tom Gaisser, Todor Stanev, Tom Weiler, Francesco Arneodo, Haim Goldberg, Subir Sarkar, Diego Torres, and Evelyn Malkus for comments and discussions. L.A.A. is supported by the U.S. National Science Foundation under grant PHY-0757598 and the University of Wisconsin-Milwaukee Research Growth Initiative. T.M. is supported by the U.S. National Science Foundation Office of Polar Programs; U.S. National Science Foundation Physics Division; University of Wisconsin-Madison Alumni Research Foundation; and INFN, Sezione di Bari, Italy.

LITERATURE CITED

1. Reines F. *Annu. Rev. Nucl. Part. Sci.* 10:1 (1960); Greisen K. *Annu. Rev. Nucl. Part. Sci.* 10:63 (1960); Gaisser TK, Halzen F, Stanev T. *Phys. Rep.* 258:173 (1995). Erratum. 271:355 (1996); Learned JG, Mannheim K. *Annu. Rev. Nucl. Part. Sci.* 50:679 (2000); Halzen F, Hooper D. *Rep. Prog. Phys.* 65:1025 (2002); Lipari P. *Nucl. Instrum. Methods A* 567:405 (2006)
2. Bahcall JN, Fowler WA, Iben I, Sears RL. *Astrophys. J.* 137:344 (1963)
3. Cleveland BT, et al. *Astrophys. J.* 496:505 (1998); Hirata KS, et al. *Phys. Rev. Lett.* 63:16 (1989); Fukuda Y, et al. *Phys. Rev. Lett.* 81:1158 (1998); Fukuda Y, et al. Erratum. *Phys. Rev. Lett.* 81:4279 (1998); Ahmad QR, et al. *Phys. Rev. Lett.* 89:011301 (2002)
4. Hirata K, et al. *Phys. Rev. Lett.* 58:1490 (1987); Bionta RM, et al. *Phys. Rev. Lett.* 58:1494 (1987)
5. Erlykin AD, Wolfendale AW. *J. Phys. G* 31:1475 (2005); Alvarez-Muniz J, Halzen F. *Astrophys. J.* 576:L33 (2002); Morlino G, Blasi P, Amato E. *Astropart. Phys.* 31:376 (2009); Villante FL, Vissani F. *Phys. Rev. D* 78:103007 (2008)
6. Anchordoqui L, Halzen F, Montaruli T, O'Murchadha A. *Phys. Rev. D* 76:067301 (2007); Anchordoqui L, Halzen F, Montaruli T, O'Murchadha A. Erratum. *Phys. Rev. D* 77:069906 (2008); Beacom JF, Kistler MD. *Phys. Rev. D* 75:083001 (2007); Halzen F, O'Murchadha A. *Phys. Rev. D* 76:123003 (2007); Halzen F, Kappes A, O'Murchadha A. *Phys. Rev. D* 78:063004 (2008)
7. Levinson A, Waxman E. *Phys. Rev. Lett.* 87:171101 (2001); Distefano C, Guetta D, Waxman E, Levinson A. *Astrophys. J.* 575:378 (2002)
8. Aharonian FA, Anchordoqui LA, Khangulyan D, Montaruli T. *J. Phys. Conf. Ser.* 39:408 (2006); Torres DF, Halzen F. *Astropart. Phys.* 27:500 (2007)
9. Amato E, Guetta D, Blasi P. *Astron. Astrophys.* 402:827 (2003); Bednarek W. *Mon. Not. R. Astron. Soc.* 345:847 (2003)
10. Bednarek W, Burgio GF, Montaruli T. *New Astron. Rev.* 49:1 (2005); Becker JK. *Phys. Rep.* 458:173 (2008)
11. Kistler MD, Beacom JF. *Phys. Rev. D* 74:063007 (2006); Kappes A, Hinton J, Stegmann C, Aharonian FA. *Astrophys. J.* 656:870 (2007); Kappes A, Hinton J, Stegmann C, Aharonian FA. Erratum. *Astrophys. J.* 661:1348 (2007)
12. Stecker FW, Done C, Salamon MH, Sommers P. *Phys. Rev. Lett.* 66:2697 (1991). Erratum. 69:2738 (1992); Stecker FW. *Phys. Rev. D* 72:107301 (2005); Atoyan A, Dermer CD. *Phys. Rev. Lett.* 87:221102 (2001); Múcke A, et al. *Astropart. Phys.* 18:593 (2003); Anchordoqui LA, Goldberg H, Halzen F, Weiler TJ. *Phys. Lett.* B600:202 (2004)
13. Waxman E, Bahcall JN. *Phys. Rev. Lett.* 78:2292 (1997); Alvarez-Muniz J, Halzen F, Hooper DW. *Phys. Rev. D* 62:093015 (2000); Razzaque S, Meszaros P, Waxman E. *Phys. Rev. Lett.* 90:241103 (2003); Dermer CD, Atoyan A. *Phys. Rev. Lett.* 91:071102 (2003); Razzaque S, Meszaros P, Waxman E. *Phys. Rev. D* 68:083001 (2003); Becker JK, Halzen F, O'Murchadha A, Olivo M. arXiv:0911.2202 [astro-ph] (2009)
14. Berezinsky VS, Zatsepin GT. *Phys. Lett. B* 28:423 (1969); Stecker FW. *Astrophys. J.* 228:919 (1979); Hill CT, Schramm DN. *Phys. Lett.* B131:247 (1983); Fodor Z, Katz SD, Ringwald A, Tu H. *J. Cosmol. Astropart. Phys.* 0311:015 (2003); Allard D, et al. *J. Cosmol. Astropart. Phys.* 0609:005 (2006)

15. Engel R, Seckel D, Stanev T. *Phys. Rev.* D64:093010 (2001)
16. Anchordoqui LA, et al. *Phys. Rev. D* 76:123008 (2007)
17. Coleman SR, Glashow SL. *Phys. Rev. D* 59:116008 (1999); Fogli GL, Lisi E, Marrone A, Scioscia G. *Phys. Rev. D* 60:053006 (1999); Gonzalez-Garcia MC, Halzen F, Maltoni M. *Phys. Rev. D* 71:093010 (2005); Battistoni G, et al. *Phys. Lett.* B615:14 (2005)
18. Zas E, Halzen F, Vazquez RA. *Astropart. Phys.* 1:297 (1993); Gondolo P, Ingelman G, Thunman M. *Astropart. Phys.* 5:309 (1996); Costa CGS. *Astropart. Phys.* 16:193 (2001); Beacom JF, Candia J. *J. Cosmol. Astropart. Phys.* 0411:009 (2004)
19. Gonzalez-Garcia MC, Halzen F, Maltoni M, Tanaka HKM. *Phys. Rev. Lett.* 100:061802 (2008)
20. Anchordoqui L, Halzen F. *Ann. Phys.* 321:2660 (2006)
21. Anchordoqui LA, Feng JL, Goldberg H, Shapere AD. *Phys. Rev. D* 65:124027 (2002); Kusenko A, Weiler TJ. *Phys. Rev. Lett.* 88:161101 (2002)
22. Jungman G, Kamionkowski M. *Phys. Rev. D* 51:328 (1995); Crotty P. *Phys. Rev. D* 66:063504 (2002); Cirelli M, et al. *Nucl. Phys. B* 727:99 (2005); Cirelli M, et al. Erratum. *Nucl. Phys. B* 790:338 (2008); Lehnert R, Weiler TJ. *Phys. Rev. D* 77:125004 (2008); Halzen F, Hooper D. *Phys. Rev. D* 73:123507 (2006)
23. Anchordoqui L, Paul T, Reucroft S, Swain J. *Int. J. Mod. Phys. A* 18:2229 (2003); Nagano M, Watson AA. *Rev. Mod. Phys.* 72:689 (2000)
24. Abbasi R, et al. *Phys. Rev. Lett.* 100:101101 (2008); Abraham J, et al. *Phys. Rev. Lett.* 101:061101 (2008)
25. Greisen K. *Phys. Rev. Lett.* 16:748 (1966); Zatsepin GT, Kuzmin VA. *JETP Lett.* 4:78 (1966)
26. Gaisser TK. *Cosmic Rays and Particle Physics*. Cambridge: Cambridge Univ. Press (1990)
27. Aharonian FA, Drury LO, Voelk HJ. *Astron. Astrophys.* 285:645 (1994)
28. Kelner SR, Aharonian FA, Bugayov VV. *Phys. Rev. D* 74:034018 (2006); Kelner SR, Aharonian FA, Bugayov VV. Erratum. *Phys. Rev. D* 79:039901 (2009)
29. Anchordoqui LA, et al. *Phys. Rev. D* 80:103004 (2009)
30. Abe F, et al. *Phys. Rev. D* 41:2330 (1990)
31. Torres DF, et al. *Phys. Rep.* 382:303 (2003)
32. Fermi E. *Phys. Rev.* 75:1169 (1949)
33. Anchordoqui LA, et al. *Phys. Rev. Lett.* 98:121101 (2007)
34. Anchordoqui LA, Goldberg H, Halzen F, Weiler TJ. *Phys. Lett.* B593:42 (2004)
35. Torres DF, Anchordoqui LA. *Rep. Prog. Phys.* 67:1663 (2004); Olinto AV. *Phys. Rep.* 333:329 (2000)
36. Bergman DR. *Nucl. Phys. Proc. Suppl.* 136:40 (2004)
37. Bird DJ, et al. *Phys. Rev. Lett.* 71:3401 (1993)
38. Berezinsky V, Gazizov AZ, Grigorieva SI. *Phys. Rev. D* 74:043005 (2006)
39. Ahlers M, et al. *Phys. Rev. D* 72:023001 (2005)
40. Waxman E. *Astrophys. J.* 452:L1 (1995)
41. Gaisser TK. arXiv:astro-ph/9707283 (1997)
42. Waxman E, Bahcall JN. *Phys. Rev. D* 59:023002 (1999)
43. Anchordoqui LA, Goldberg H, Halzen F, Weiler TJ. *Phys. Lett.* B621:18 (2005)
44. Stecker FW. *Phys. Rev. Lett.* 21:1016 (1968)
45. Alvarez-Muniz J, et al. *Phys. Rev. D* 66:033011 (2002)
46. Learned JG, Pakvasa S. *Astropart. Phys.* 3:267 (1995)
47. Kashti T, Waxman E. *Phys. Rev. Lett.* 95:181101 (2005); Lipari P, Lusignoli M, Meloni D. *Phys. Rev. D* 75:123005 (2007); Pakvasa S, Rodejohann W, Weiler TJ. *J. High Energy Phys.* 0802:005 (2008)
48. Anchordoqui LA, Hooper D, Sarkar S, Taylor AM. *Astropart. Phys.* 29:1 (2008)
49. Hooper D, Taylor A, Sarkar S. *Astropart. Phys.* 23:11 (2005); Ave M, et al. *Astropart. Phys.* 23:19 (2005); Hooper D, Sarkar S, Taylor AM. *Phys. Rev. D* 77:103007 (2008)
50. Amsler C, et al. *Phys. Lett.* B667:1 (2008)
51. Markov MA. *Proc. 10th Int. Conf. High Energy Phys., Rochester*, ed. ECG Sudarshan, et al., p. 578 (1960); Markov MA, Zheleznykh IM. *Nucl. Phys.* 27:385 (1961)
52. Sinegovskaya TS, Sinegovsky SI. *Phys. Rev. D* 63:096004 (2001); Carminati G, Margiotta A, Spurio M. *Nucl. Instrum. Methods A* 602:95 (2009)

53. Bazzotti M. *Proc. 31st Int. Cosmic Ray Conf.,HE 2.1 0340* (2009); Desiati P. *Proc. 28th Int. Cosmic Ray Conf.*, p. 1373 (2003)
54. Belolaptikov IA, et al. *Astropart. Phys.* 7:263 (1997)
55. Babson J, et al. *Phys. Rev. D* 42:3613 (1990)
56. Aggouras G, et al. *Astropart. Phys.* 23:377 (2005)
57. Chirkin D (IceCube Collab.) *HE 2.2 1418* (2009)
58. Honda M, et al. *Phys. Rev. D* 75:043006 (2007)
59. Martin AD, Ryskin MG, Stasto AM. *Acta Phys. Polon. B* 34:3273 (2003)
60. Enberg R, Reno MH, Sarcevic I. *Phys. Rev. D* 78:043005 (2008)
61. Fiorentini G, Naumov VA, Villante FL. *Phys. Lett.* B510:173 (2001); Bugaev EV, Naumov VA, Sinegovsky SI, Zaslavskaya ES. *Nuovo Cim.* C12:41 (1989); Bugaev EV. *Phys. Rev. D* 58:054001 (1998)
62. Glashow SL. *Phys. Rev.* 118:316 (1960)
63. Gandhi R, Quigg C, Reno MH, Sarcevic I. *Astropart. Phys.* 5:81 (1996); Gandhi R, Quigg C, Reno MH, Sarcevic I. *Phys. Rev. D* 58:093009 (1998); Anchordoqui LA, Cooper-Sarkar AM, Hooper D, Sarkar S. *Phys. Rev. D* 74:043008 (2006); Cooper-Sarkar A, Sarkar S. *J. High Energy Phys.* 0801:075 (2008)
64. L'Abbate A, Montaruli T, Sokalski I. *Astropart. Phys.* 23:57 (2005)
65. Halzen F, Saltzberg D. *Phys. Rev. Lett.* 81:4305 (1998)
66. Bugaev E, Montaruli T, Shlepin Y, Sokalski I. *Astropart. Phys.* 21:491 (2004)
67. DUMAND Proposal. *Rep. HDC-3-88* (1980)
68. Bezrukov LB, et al. *Proc. Int. Conf. Neutrino-84, Singapore*, p. 550. Hackensack, NJ: World Sci. Publ. (1984)
69. Kuzmichev L, et al. (Baikal Collab.) Presented at 31st *Intl. Cosm. Ray Conf., HE2.5 1091* (2009)
70. Halzen F, et al. Presented at *4th Int. Symp. Neutrino Telesc.*, Venice, Italy (1992)
71. Andres E, et al. *Astropart. Phys.* 13:1 (2000); Andres E, et al. *Nature* 410:441 (2001)
72. Abbasi R, et al. *Nucl. Instrum. Methods A* 601:294 (2009)
73. Achterberg A, et al. *Astropart. Phys.* 26:155 (2006)
74. Amram P, et al. *Astropart. Phys.* 19:253 (2003); Amram P, et al. *Astropart. Phys.* 13:127 (2000); Aguilar JA, et al. *Astropart. Phys.* 23:131 (2005)
75. Aguilars JA, et al. *Astropart. Phys.* 26:314 (2006); Aguilars JA, et al. *Nucl. Instrum. Methods A* 581:695 (2007)
76. Aguilars JA, et al. *Nucl. Instrum. Methods A* 555:132 (2005)
77. Aguilars JA, et al. *Nucl. Instrum. Methods A* 484:369 (2002)
78. Aguilars JA, et al. *Nucl. Instrum. Methods A* 578:498 (2007)
79. Aguilars JA, et al. *Nucl. Instrum. Methods A* 570:107 (2007)
80. Abbasi R, et al. *Phys. Rev. D* 79:062001 (2009)
81. Montaruli T. *Nucl. Phys. Proc. Suppl.* 190:101 (2009)
82. Sapienza P, Coniglione R, Distefano C, Migneco E. *Nucl. Instrum. Methods A* 602:101 (2009)
83. Dumm J, et al. (IceCube Collab.) *OG 2.5 0653* (2009)
84. Katz UF. *Nucl. Instrum. Methods A* 567:457 (2006)
85. Montaruli T. arXiv:0910.4364 [hep-ph] (2009)
86. Halzen F, Jacobsen JE, Zas E. *Phys. Rev. D* 53:7359 (1996)
87. Berezinsky VS, Smirnov AY. *Phys. Lett.* B 48:269 (1974); Capelle KS, Cronin JW, Parente G, Zas E. *Astropart. Phys.* 8:321 (1998); Anchordoqui L, et al. *Ann. Phys.* 314:145 (2004)
88. Bertou X, et al. *Astropart. Phys.* 17:183 (2002); Feng JL, Fisher P, Wilczek F, Yu TM. *Phys. Rev. Lett.* 88:161102 (2002); Fargion D. *Astrophys. J.* 570:909 (2002)
89. Boersma D, Gladstone L, Karle A. (IceCube Collab.) *OG 2.5.1173* (2009)
90. Borriello E, et al. *Nucl. Phys. Proc. Suppl.* 190:150 (2009)
91. Ambrosio M, et al. *Astrophys. J.* 546:1038 (2001); Oyama Y, et al. *Phys. Rev. D* 39:1481 (1989); Berger C, et al. *Z. Phys. C* 48:221 (1990); Abe K, et al. *Astrophys. J.* 652:198 (2006); Desai S, et al. *Astropart. Phys.* 29:42 (2008); Achterberg A, et al. *Phys. Rev.* D 75:102001 (2007)
92. Abbasi R, et al. *Astrophys. J.* 701:L47 (2009); Thrane E, et al. *Astrophys. J.* 704:503 (2009)
93. Braun J, et al. *Astropart. Phys.* 29:299 (2008); Aguilar JA, Hernandez-Rey JJ, *Astropart. Phys.* 29:117 (2008); Alexandreas DE, et al. *Nucl. Instrum. Methods A* 328:570 (1993)

94. Abbasi R, et al. *Phys. Rev. Lett.* 103:221102 (2009)
95. Gaisser TK, Honda M. *Annu. Rev. Nucl. Part. Sci.* 52:153 (2002)
96. Gonzalez-Garcia MC, Maltoni M. *Phys. Rep.* 460:1 (2008)
97. Lipari P. *Astropart. Phys.* 1:195 (1993)
98. Barr GD, et al. *Phys. Rev. D* 70:023006 (2004)
99. Battistoni G, et al. *Astropart. Phys.* 12:315 (2000)
100. Barr GD, Gaisser TK, Robbins S, Stanev T. *Phys. Rev. D* 74:094009 (2006)
101. Berghaus P, Montaruli T, Ranft J. *J. Cosmol. Astropart. Phys.* 0806:003 (2008)
102. D'Alessandro R, et al. *Nucl. Phys. Proc. Suppl.* 190:52 (2009); Anelli G, et al. *J. Instrum.* 3:S08007 (2008)
103. Ahn EJ. *Phys. Rev. D* 80:094003 (2009); Ostapchenko S. *Am. Inst. Phys. Conf. Proc.* 928:118 (2007); Kochanov AA, Sinegovskaya TS, Sinegovsky SI. arXiv:0906.0671 [astro-ph] (2009); Berghaus P, et al. *J. Phys. Conf. Ser.* 136:042019 (2008)
104. Adamson P, et al. *Phys. Rev. D* 76:052003 (2007)
105. Abbasi R, et al. *Phys. Rev. D* 79:102005 (2009)
106. Abbasi R, et al. arXiv:0907.0498 [astro-ph] (2009)
107. Amenomori M. *Science* 314:439 (2006)
108. Abdo AA, et al. *Astrophys. J.* 698:2121 (2009); Abdo AA, et al. *Phys. Rev. Lett.* 101:221101 (2008); Guillian G, et al. *Phys. Rev. D* 75:062003 (2007); Zhang JL, et al. Presented at *31st Intl. Cosm. Ray Conf., HE 1.1 0814* (2009); Vernetto S, et al. (IceCube Collab.) *OG 2.3 0399* (2009)
109. Compton AH, Getting IA. *Phys. Rev.* 47:817 (1935)
110. Salvati M, Sacco B. arXiv:0802.2181 (2008); Drury L, Aharonian F. *Astropart. Phys.* 29:420 (2008); Erlykin AD, Wolfendale AW. arXiv:0906.3949 [hep-ph] (2009)
111. Adriani O, et al. *Nature* 458:607 (2009); Chang J, et al. *Nature* 456:362 (2008); Aharonian F, et al. *Phys. Rev. Lett.* 101:261104 (2008); Abdo AA, et al. *Phys. Rev. Lett.* 102:181101 (2009)
112. Hooper D, Blasi P, Serpico PD. *J. Cosmol. Astropart. Phys.* 0901:025 (2009); Yuksel H, Kistler MD, Stanev T *Phys. Rev. Lett.* 103:051101 (2009); Blasi P. *Phys. Rev. Lett.* 103:051104 (2009); Mertsch P, Sarkar S. *Phys. Rev. Lett.* 103:081104 (2009); Ahlers M, Mertsch P, Sarkar S. arXiv:0909.4060 [astro-ph] (2009)
113. Butt Y. *Nature* 460:701 (2009)
114. Coyle P (ANTARES Collab.) arXiv:1002.0754 [astro-ph] (2009)
115. Achterberg A, et al. *Phys. Rev.* D 76:042008 (2007); Achterberg A, et al. Erratum. *Phys. Rev. D* 77:089904 (2008)
116. Ahrens J, et al. *Astropart. Phys.* 20:507 (2004)
117. Abraham J, et al. arXiv:0906.2347 [astro-ph] (2009); Abraham J, et al. *Phys. Rev. D* 79:102001 (2009); Abraham J, et al. *Phys. Rev. Lett.* 100:211101 (2008)
118. Abbasi R, Desiati P (IceCube Collab.) arXiv:0907.0498 [astro.ph] (2009)
119. Silvestri A, et al. Presented at *31st Intl. Cosm. Ray Conf., OG 2.5 0549* (2009)
120. Mase K, Ishihara A, Yoshida S (IceCube Collab.) *HE 1.4 0861* (2009)
121. Abbasi R, et al. arXiv:0803.0554 [astro-ph] (2008); Gorham PW, et al. *Phys. Rev. Lett.* 103:051103 (2009); James CW, et al. arXiv:0907.4332 [astro-ph] (2009); Kravchenko I, et al. *Phys. Rev. D* 73:082002 (2006)
122. Ahlers M, Anchordoqui LA, Sarkar S. *Phys. Rev. D* 79:083009 (2009)
123. Hopkins AM, Beacom JF. *Astrophys. J.* 651:142 (2006)
124. Hasinger G, Miyaji T, Schmidt M. *Astron. Astrophys.* 441:417 (2005); Stanev T. arXiv:0808.1045 [astro-ph] (2008)
125. Anchordoqui LA, et al. arXiv:0907.5208 [astro-ph] (2009)
126. Riess AG, et al. *Astron. J.* 116:1009 (1998); Perlmutter S, et al. *Astrophys. J.* 517:565 (1999); Komatsu E, et al. *Astrophys. J. Suppl.* 180:330 (2009); Colless M, et al. arXiv:astro-ph/0306581 (2003); Tegmark M, et al. *Phys. Rev.* D 69:103501 (2004); Reid BA, et al. arXiv:0907.1659 [astro-ph] (2009)
127. Burles S, Nollett KM, Turner MS. *Astrophys. J.* 552:L1 (2001); Cyburt RH, Fields BD, Olive KA. *Phys. Lett.* B567:227 (2003)
128. Bergstrom L. *Rep. Prog. Phys.* 63:793 (2000); Bertone G, Hooper D, Silk J. *Phys. Rep.* 405:279 (2005); Feng JL. *J. Phys. G* 32:R1 (2006)

129. Goldberg H. *Phys. Rev. Lett.* 50:1419 (1983); Ellis JR, et al. *Nucl. Phys. B* 238:453 (1984)
130. Servant G, Tait TMP. *Nucl. Phys. B* 650:391 (2003); Cheng HC, Feng JL, Matchev KT. *Phys. Rev. Lett.* 89:211301 (2002)
131. Scherrer RJ, Turner MS. *Phys. Rev. D* 33:1585 (1986); Scherrer RJ, Turner MS. Erratum. *Phys. Rev. D* 34:3263 (1986)
132. Griest K, Kamionkowski M, Turner MS. *Phys. Rev. D* 41:3565 (1990); Drees M, Nojiri MM. *Phys. Rev. D* 47:376 (1993); Birkedal-Hansen A, Jeong E. *J. High Energy Phys.* 0302:047 (2003)
133. Goodman MW, Witten E. *Phys. Rev. D* 31:3059 (1985)
134. Silk J, Olive KA, Srednicki M. *Phys. Rev. Lett.* 55:257 (1985); Srednicki M, Olive KA, Silk J. *Nucl. Phys. B* 279:804 (1987); Ng KW, Olive KA, Srednicki M. *Phys. Lett.* B188:138 (1987); Bergstrom L, Edsjo J, Gondolo P. *Phys. Rev. D* 58:103519 (1998); Feng JL, Matchev KT, Wilczek F. *Phys. Rev. D* 63:045024 (2001); Barger VD, Halzen F, Hooper D, Kao C. *Phys. Rev. D* 65:075022 (2002)
135. Rudaz S, Stecker FW. *Astrophys. J.* 368:406 (1991); Bergstrom L, Ullio P, Buckley JH. *Astropart. Phys.* 9:137 (1998); Dodelson S, Hooper D, Serpico PD. *Phys. Rev. D* 77:063512 (2008); Bertone G, et al. *Phys. Rev. D* 80:023512 (2009); Goodenough L, Hooper D. arXiv:0910.2998 [hep-ph] (2009); Anchordoqui L, et al. arXiv:0912.0517 [hep-ph] (2009)
136. Navarro JF, Frenk CS, White SDM. *Astrophys. J.* 490:493 (1997)
137. Aharonian FA, et al. *Phys. Rev. Lett.* 97:221102 (2006); Aharonian FA, et al. Erratum. *Phys. Rev. Lett.* 97:249901 (2006)
138. Aharonian F, et al. *Nature* 439:695 (2006)
139. Gould A. *Astrophys. J.* 388:338 (1992); Jungman G, Kamionkowski M, Griest K. *Phys. Rep.* 267:195 (1996)
140. Griest K, Seckel D. *Nucl. Phys. B* 283:681 (1987); Griest K, Seckel D. Erratum. *Nucl. Phys. B* 296:1034 (1988); Gould A. *Astrophys. J.* 321:560 (1987)
141. Ahmed Z, et al. *Phys. Rev. Lett.* 102:011301 (2009)
142. Angle J, et al. *Phys. Rev. Lett.* 100:021303 (2008)
143. Tanaka T (Super-Kamiokande Collab.) Presented at *31st Intl. Cosm. Ray Conf., HE 2.3 0622* (2009); Desai S, et al. *Phys. Rev. D* 70:083523 (2004); Desai S, et al. Erratum. *Phys. Rev. D* 70:109901 (2004)
144. Ambrosio M, et al. *Phys. Rev. D* 60:082002 (1999)
145. Braun J, Hubert D (IceCube Collab.) *HE 2.3 0834* (2009)
146. Avrorin A, et al. (IceCube Collab.) *HE 2.3 1165* (2009)
147. Abbasi R, et al. *Phys. Rev. Lett.* 102:201302 (2009)
148. Lim GMA (ANTARES Collab.) *HE 2.3 0031* (2009)
149. Behnke E, et al. *Science* 319:933 (2008); Lee HS, et al. *Phys. Rev. Lett.* 99:091301 (2007)
150. Gondolo P, et al. *J. Cosmol. Astropart. Phys.* 0407:008 (2004)

The Construction and Anticipated Science of SNOLAB

F. Duncan,[1] A.J. Noble,[2] and D. Sinclair[3]

[1]SNOLAB, Lively, Ontario P3Y 1N2, Canada and Department of Physics, Laurentian University, Sudbury, Ontario P3E 2C6, Canada; email: fduncan@snolab.ca

[2]Department of Physics, Queen's University at Kingston, Kingston, Ontario K7L 3N6, Canada; email: potato@snolab.ca

[3]Department of Physics, Carleton University, Ottawa, Ontario K1S 5B6, Canada and TRIUMF, Vancouver, British Columbia V6T 2A3, Canada; email: Sinclair@physics.carleton.ca

Annu. Rev. Nucl. Part. Sci. 2010. 60:163–80

First published online as a Review in Advance on June 7, 2010

The *Annual Review of Nuclear and Particle Science* is online at nucl.annualreviews.org

This article's doi:
10.1146/annurev.nucl.012809.104513

0163-8998/10/1123-0163$20.00

Key Words

astroparticle physics, deep underground laboratory, dark matter, double-β decay, neutrinos, supernovae, seismology

Abstract

A new laboratory for astroparticle physics has been created in the Creighton Mine near Sudbury, Ontario, Canada. The laboratory is located at a great depth to provide shielding from the cosmic-ray backgrounds, and it has been constructed as a clean room to protect experiments from radioactive dust contamination. In this review, we discuss the motivation for the establishment of the laboratory and the design and construction process. We also introduce the initial suite of experiments.

Contents

1. INTRODUCTION

A new laboratory has been constructed at the Creighton Mine (owned and operated by Vale Inco) near Sudbury, Ontario, Canada to house experiments in astroparticle physics and other areas of science that can benefit from its infrastructure. The measurements that can be performed in this unique underground environment address some of the most important and fundamental science topics imaginable. What are the dark matter particles left over from the big bang that make up a substantial fraction of the universe and influence its evolution? What are the detailed properties of these unknown particles, and how do they fit into or modify the basic laws of elementary particle physics? What are the detailed properties of our Sun and its energy-generation processes? How do stars collapse into supernovae, generating a substantial fraction of the elements in the universe? What are the masses and transformation properties of neutrinos, and how do they fit into the basic laws of physics?

The new laboratory, known as SNOLAB, builds on the infrastructure created for the SNO experiment, which successfully resolved the solar neutrino problem by demonstrating that neutrinos, which originate in the Sun of one flavor, change to other flavors while propagating to the Earth (1–3). The proposed new facility joins a developing network of underground science facilities around the world, each with its own natural advantages. The niche created by SNOLAB is to be the deepest and cleanest in the world, making it the place of choice for the most sensitive of experiments to be performed.

Only by going deep underground to escape the cosmic rays from outer space and by creating an ultraclean environment to avoid natural radioactivity can one observe the elusive dark matter particles and neutrinos. These conditions are also required to perform other sensitive measurements of rare physics and astrophysics processes. Taking these unique particle telescopes deep underground is analogous to mounting optical telescopes on the tops of remote mountains to study astrophysical sources of light. In both cases, researchers seek to create the very best conditions to study rare processes in the universe and to push the fundamental frontiers of science.

The concept for the new laboratory developed around the idea of a single campus located at a depth of 2 km, where several experiments could share a common infrastructure. The entire campus would be built as a clean room, and much of the design has been driven by the need to establish this clean environment in a mine, which is naturally very dirty. Cleanliness is essential because dust and other common contaminants are radioactive to some degree, and all of the work that goes into producing ultraclean materials is wasted if the detectors become contaminated by exposure through handling. Accomplishing the required cleanliness requires not only a substantial air-filtration system but also facilities for cleaning all people and materials that enter the laboratory. This cleanliness barrier is a common infrastructure for all experiments.

Other major infrastructure provided by SNOLAB includes services such as power distribution, cooling, and ultrapure water. SNOLAB also supports the experimental program with a staff of highly trained technical personnel, engineers and scientists responsible for the operation of the facility and the integration of experiments. A new office building on the surface also provides infrastructure support for experiments in the form of clean room laboratory space, low-background counting facilities, and computing resources.

The design of the laboratory was based on input from a series of international workshops held to examine the most promising of the future experiments. In response to a request for letters of interest, approximately 20 experiments were suggested. These proposals were reviewed by an international experiment advisory committee of experts in experimental and theoretical astroparticle physics, and the most promising were encouraged to proceed with development into full projects. As anticipated, the experiments were mainly in the areas of searches for weakly interacting massive particle (WIMP) dark matter; searches for neutrinoless double-β decay; and studies of low-energy neutrinos from the Sun, the Earth, and supernovae. A requirements matrix was produced for these projects, and the detailed design of the laboratory was developed to meet these requirements.

The original concept for the underground campus involved excavating a single large hall into which many experiments could be placed. This idea evolved to excavating several smaller spaces so that there could be greater isolation between experiments. There were several reasons for adopting this approach, even though it restricts the flexibility of the laboratory in assigning space. Because these experiments are inherently extremely sensitive, several of them are influenced by acoustic or electric noise from nearby activity. Thus, the exact layout of equipment and any local traffic were important components of the experimental design. These factors required that experiments have control over their space, which is more easily managed when the experiment spaces are separated.

A second and more compelling consideration was concern about the safety of experiments. Many of the proposed detectors require large volumes of noble gases or liquids (because they can be made extremely pure). Although these detectors will be designed, through the use of proper engineering procedures, to be intrinsically safe, it is desirable to isolate such experiments to minimize the impact of any failure. The result of this process is the laboratory shown in **Figure 1**. The facility includes the original SNO cavern (the world's largest cavern at such a depth), the Cube Hall for general-purpose large detectors, the Ladder Labs for smaller experiments, and the Cryopit for experiments that require either large volumes of noble gases or liquids or other materials that present safety concerns. Other spaces are available for prototyping projects and for ancillary equipment in support of experiments such as cryogenic and purification plants.

2. THE FACILITY

The SNOLAB facility consists of an underground campus and a surface facility to support the underground operations. As an expansion of the original facility excavated for the SNO experiment

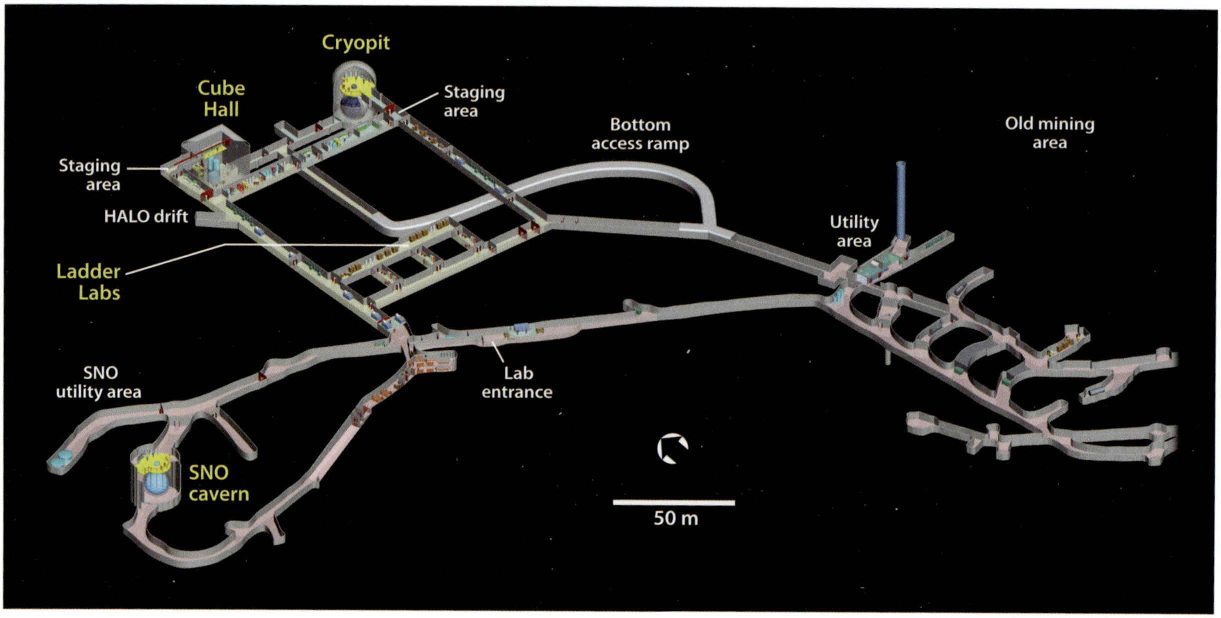

Figure 1

SNOLAB underground facility. Reprinted with permission from Hatch Acres.

(4), the underground campus is located on the 6,800-ft level of Creighton Mine. SNOLAB coexists with Vale Inco's mining activities at Creighton, the bulk of which now occur deeper in the mine than the 6,800-ft level. All of the mine hoists, ventilation, power, and water infrastructure are operated by the mining company. The laboratory is situated in a hanging wall of norite granite rock several hundred meters from the ore body and any mining interest. The surface profile of the mine site is flat, and the 2,070-m overburden gives approximately 6,000 mwe (meters water equivalent) shielding, resulting in a cosmic-ray muon flux of 0.27 μm^{-2} day^{-1}. Access to the underground campus is via the number Nine Shaft, which is located approximately 1.5 km from the laboratory entrance.

The location of the underground campus, the means of access to the facility, and the environmental conditions of the mine provide constraints on the design of the facility and its operation. As shown in **Figure 1**, the campus consists of (*a*) a series of large excavations and connecting tunnels (termed drifts) that are maintained as a clean room and (*b*) an area that is operated "dirty," meaning that it is exposed to the normal mine environment. The dirty areas include a utility area containing the laboratory's cooling (chiller) system, make-up air handler, power substation, (future) emergency generator, and sewage-treatment plant. The clean room area contains the experiment areas; their ancillary equipment and support areas; the laboratory service areas (including areas for air handling, power distribution, and material handling); and personnel facilities including a lunchroom, shower and change rooms, washrooms, and a refuge station. The total excavation of the laboratory, including clean and dirty spaces, is 46,600 m^3, of which 37,200 m^3 are operated as a CLASS 2000 (or better) clean room. This clean room space includes the experimental areas, summarized in **Table 1**, as well as the ancillary spaces used to support the experiments, service areas, and personnel and material-handling areas.

Table 1 Clean room laboratory spaces[a]

Area	Dimensions	Area	Volume
SNO cavern	24 m (diameter) × 30 m (h)	250 m^2	9,400 m^3
Ladder Labs	32 m (l) × 6 m (w) × 5.5 m (h)	190 m^2	960 m^3
	23 m (l) × 7.5 m (w) × 7.6 m (h)	170 m^2	1,100 m^3
Cube Hall	18.3 m (l) × 15 m (w) × 19.7 m (h)	280 m^2	5,600 m^3
Cryopit	15 m (diameter) × 19.7 m (h)	180 m^2	3,900 m^3
Total		1,320 m^2	29,370 m^3

[a]These do not include support and service areas.

2.1. Design Considerations

The foremost consideration in the design of the SNOLAB underground facility was geotechnical. Geotechnical evaluation and design were performed by Itasca Consulting Canada. Itasca was then retained to evaluate the ground conditions during the subsequent excavation. The new excavations were located away from the mining activities to keep the change from the ambient rock stress due to mining below 5%. The possibility of mining-induced seismicity was also considered, leading to the requirement that the laboratory be able to withstand an event, situated 160 m away (the distance to the mined-ore body), of 4.1 M on the Nuttli scale. In 2006, a 4.1-M event was observed, and although there was no significant damage to the SNOLAB facilities, the maximum plausible event was subsequently upgraded to 4.3, which corresponds to an estimated ground motion at the laboratory of 800 mm s^{-1} peak particle velocity at a frequency of 5 Hz. Forcing functions for this hypothetical event are available to facilitate the design of experiments. Exploratory core drilling was performed to determine the rock conditions in the areas intended for excavation. Geotechnical simulations of the excavations were performed to optimize the cavity shapes and necessary structural support. Along with the 2-m resin-grouted rock bolts and the 10-cm mesh-steel screen applied to all excavated surfaces, the larger excavations (Cube Hall and Cryopit) were fitted with extensive ground support consisting of cables grouted either 7 or 10 m into the rock on a 1.5-m grid. A 10-cm layer of shotcrete (spray-on concrete) was applied to all spaces intended to be part of the clean room area. Areas in which shotcrete was not applied used galvanized screen to enhance longevity. As part of the facility monitoring, extensometers were installed around the larger new excavations. These complemented similar monitors that were installed as part of SNO, along with a microseismic system consisting of an array of acoustic sensors. These systems are monitored to ensure there is no undetected ground motion.

The location of SNOLAB on the 6,800-ft level of Creighton Mine presents a number of unusual environmental considerations. SNOLAB was constructed as a monolithic clean room because the norite rock from which the laboratory is excavated contains 1.00 ± 0.13% K, 1.11 ± 0.13 ppm U, and 5.56 ± 0.52 ppm Th, along with these elements' decay products such as ^{222}Rn. For many of the SNOLAB experiments, these radioisotopes constitute significant backgrounds; therefore, great effort is expended to control so-called mine dust in the facility. All clean room areas are painted to encapsulate dust from concrete floor and shotcrete wall linings. In addition to the radiological contamination from mine dust, ^{222}Rn gas is a problem: Its air concentration averages 130 Bq m^{-3} but varies depending on location, air flow, and time of year.

Another consideration is that the air pressure at this depth of the mine is approximately 25% higher than on the surface. A significant factor in the design of any pressure or vacuum system underground is that not only is the ambient pressure higher than on the surface, but there can also be pressure swings of as much as 3% over a period of a few minutes due to changes in the forced air ventilation of the mine. The nominal rock temperature at this depth is 42°C. Following the initial cooldown of the rock after excavation, the nominal temperature leads to a steady-state heat load in the laboratory of approximately 100 kW.

An issue that became apparent during SNO operations was the presence of low concentrations (a few parts per billion) of H_2S in the mine air, which was presumed to evolve from the sulfide-rich ore body. The presence of H_2S led to observable corrosion of copper and caused problems with electronic circuit boards. The problem was partially mitigated during SNO operations through the installation of NaOH-impregnated charcoal filters in one of the recirculating air handlers inside the clean space; however, because the filters did not treat the make-up air to the laboratory, they were only partially effective. Incorporated into the design for SNOLAB is a filter bank in the main air intake that is intended to remove the H_2S before it enters the laboratory. The effectiveness of this new strategy is being evaluated.

2.2. Laboratory Layout

The layout of the underground campus was dictated by the requirements of a representative set of low-background astroparticle physics experiments, combined with structural considerations based on geotechnical information. The main experimental areas of the facility are the original SNO cavern, the Cube Hall, the Cryopit, and the Ladder Labs. The dimensions of these spaces are summarized in **Table 1**. Although most of the laboratory is situated at a single elevation in the mine, the largest excavations (SNO cavern, Cube Hall, Cryopit) are depressed below this level; ramps were used during construction to access their bases. **Figure 2** shows a cross section of the Cube Hall and the Cryopit.

2.2.1. SNO cavern. A barrel-shaped cavity measuring 22 m in maximum diameter and 30 m in height is the largest excavation at SNOLAB; it accounts for one-quarter of the excavated volume and was originally built to house the SNO detector. An 8-mm-thick polyurethane liner was sprayed onto the shotcrete walls of the cavern, which was then filled with water to shield the 18-m-diameter SNO detector from radioactivity in the rock. The liner waterproofs the cavity and prevents the ingress of ^{222}Rn from the surrounding rock. Entry into the SNO cavern is from the main level of the laboratory onto a deck supported 24 m above the floor of the cavern. The SNO detector is suspended below this deck. When the cavern is drained of water, one can access the area below the deck via bosun's chair.

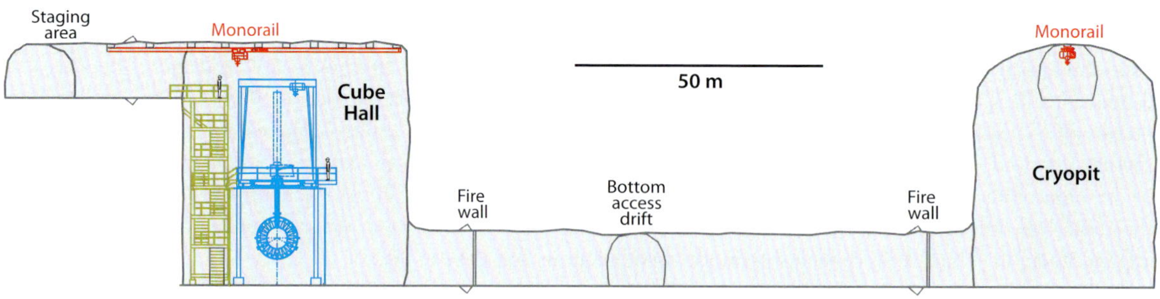

Figure 2

Cross section of the Cube Hall and the Cryopit.

2.2.2. Cube Hall. Able to hold a cube-shaped object approximately 15 m in size, the Cube Hall is SNOLAB's second largest excavation. It is 18 m long and 15 m wide at its base and has curved walls, yielding a maximum width of 18 m. The total floor-to-ceiling height (termed the back height) is 19 m, and access to the hall is through a staging area 15 m above the floor of the hall. A stair tower provides access to the floor. At the time of excavation, a 10-t monorail was installed along the highest point of the back in the hall. Although it is possible to flood the hall with water to provide a single large water shield, the initial experimental program leaves the hall dry.

2.2.3. Cryopit. The Cryopit is a straight-walled cylindrical excavation with a dome-shaped back measuring 15 m in diameter and 19 m in total height. As in the Cube Hall, clean entry to the Cryopit is 15 m above the floor. The Cryopit is designed to house a large, 10–12-m scale detector that is envisioned to be in a water shield that fills the cavity. As the name Cryopit implies, one possible candidate for the detector is a large (tens of tonnes) cryogenic liquid detector. To facilitate such a detector, the Cryopit was isolated from the rest of the laboratory and provisions were made during excavation to allow the installation of high-pressure bulkheads to seal the Cryopit from the other inhabitable laboratory spaces. This seal, combined with an exhaust system to vent boil-off gases to the mine's return air system, would help to mitigate a failure of the cryogenic containment. A 5-t monorail crane extends from the staging area to the center of the Cryopit. A ladder provides access from the staging area to the floor of the pit for construction; the ladder will be removed as the pit is prepared for water fill.

2.2.4. Ladder Labs. The Ladder Labs are a network of tunnels that derive their name from their resemblance to sections of a ladder. The intent is to house medium-scale (4–5-m) experiments in one of the rails of the ladder while placing control rooms and ancillary equipment in the rungs. The second rail provides service access to the Ladder Labs. It is expected that the Ladder Labs will be able to house three to four medium-scale experiments and experiment prototypes.

In addition to the spaces allocated for the detectors, there are adjacent spaces intended for control rooms, staging materials, and ancillary equipment such as pumps, cryogenic refrigerators, purification systems, and so forth. The Cube Hall and Cryopit share a common ramp to their bottom level, which was used for excavation of the spaces. This level has been outfitted with an airlock and is used only to transport large equipment that would not readily fit through the main entry to the laboratory. The SNO cavern had a similar ramp that was used during construction, but it was subsequently partially back-filled with concrete to ensure containment of water within the cavern.

The outfitting of the new experimental spaces was deliberately kept minimal, as the design of the spaces preceded the allocation of experiments to them. Where possible, the permanent infrastructure was designed to accommodate future modifications as the experiments' needs are identified. For example, walls may be installed in the Ladder Labs to isolate experiments.

Integral to the concept of operating all the experiment spaces within a common clean room is the incorporation of the material-handling and personnel areas. Most of the materials entering the laboratory are brought underground on rail cars that operate on a 2-ft gauge railway running from the mine shaft to the laboratory. These rail cars are loaded at the SNOLAB surface facility and are then transported underground in the cage (the elevator car that travels in the shaft), which has dimensions of 3.7 m $\times$ 1.5 m $\times$ 2.6 m. All materials entering the laboratory must either be transported in a clean fashion from the surface facility or be cleaned prior to entering the underground laboratory. Cleaning is done in a dedicated "car wash" facility. The rail cars are rolled into the car wash, and the materials are cleaned before entering the clean space. If the materials are too delicate to be washed underground, they are transported in sealed blue boxes from the surface. On entry to the car wash, the exterior of the blue box is cleaned and opened, and the materials are then unloaded.

Similar to materials, personnel must also be "cleaned" prior to entering the laboratory. Persons entering the laboratory must first shower and change into clean room clothing in the segregated male or female change and shower areas. They then enter the personnel area of the laboratory, which contains a lunchroom, washrooms, a meeting room, and a laundry facility, which alleviates the need to ship the clean room clothing to the surface for cleaning. The personnel area also acts as a refuge station. In the event that there is a fire or other emergency in the mine, SNOLAB personnel will enter the refuge station and isolate it from the mine ventilation, using breathing air piped in from the surface. Other support spaces for the experiments include storage space in the clean room area and a low-background counting facility. The low-background facilities at SNOLAB, which are in operation at the time of writing, include a high-purity germanium counter for γ counting, radon assay capability (6), Lucas cells, a large $\beta\alpha$ coincidence counting system for measurements of thorium and uranium daughters, electrostatic counters, a variety of water-sampling methods (which use MnOx and HTiO as ion exchange), and neutron counters (7, 8). Funding has been provided to augment these components with additional high-purity germanium counters, large-area α counters, and emanation chambers. Possible future additions for the underground clean room area include a chemistry laboratory and a machine shop.

2.3. Services

Power to the laboratory comes from the three-phase, 13.8-kV power-distribution system in the mine and is stepped down to three-phase, 600-V power for distribution within the laboratory via two mine power centers (substations) that provide 500 kVA and 1,500 kVA, respectively. The 600-V power is distributed within the laboratory through four motor control centers to the experiment and service areas. A future addition to power distribution will be a 150-kW diesel generator, which will provide backup power for vital communications and experiment process systems during prolonged power outages.

Waste heat is removed from the laboratory via a chilled water system circulating 10°C water throughout the laboratory to air handlers and to experiment process systems. The chiller consists of five parallel turbine units and a secondary water loop that transfers heat to 2,800 m^3 min^{-1} of mine ventilation air. The ventilation air exhausts into a sealed return air raise, which in turn exhausts to the main exhaust air system for the mine. The chiller can remove 1 MW of heat from the laboratory.

The clean room ventilation system for SNOLAB consists of 13 recirculating air handlers that cool and filter the air within the laboratory; the air is filtered through high-efficiency particulate-absorbing (HEPA) filters. Make-up air from the mine passes through a series of successively finer filters, including HEPA filters. Nominally, there are ten air changes per hour with 10% make-up air for all areas except for the largest excavations (Cube Hall and Cryopit), where there are five air changes per hour. Although most of the ducting in the laboratory is rigid, cloth ducting designed for clean room use has been stalled in the experiment spaces. This product has many advantages: It is relatively inexpensive, provides a more uniform air distribution, and is easily taken down for cleaning. The 13 air handlers are used to configure different pressure zones within the laboratory, with the experiment areas at the highest pressure cascading through the service areas to personnel and transition areas to the ambient mine air pressure. Pressure balancing is achieved through a combination of motorized dampers and variable speed drives on all the new air handlers (the original SNO air handlers are fixed speed). Typically, one can maintain 25-Pa pressure differences between pressure zones by modulating the fraction of make-up air and exhaust air from different zones. To compensate for rapid changes in the mine air pressure, the main make up-air handler can deliver up to 170% of its nominal flow.

Utility water to the laboratory comes from the normal mine water distribution. The normal iron piping on the SNOLAB level has been replaced with high-density polyethylene piping, which proved during SNO operations to be very successful at preventing corrosion. After modest filtration, utility water is distributed in the laboratory as so-called industrial water (meaning that although it is chlorinated, regulatory testing is not performed to determine its potability) for domestic use, including showers.

This water is also the supply for the SNO ultrapure water plant, which has been maintained as a general facility for future experiments. The plant consists of a first, conventional stage with charcoal filters and water softeners followed by a large reverse-osmosis system that can process ~170 liter min^{-1}. The subsequent polishing stages include ion-exchange beds, a UV sterilization system, and a large vacuum degasser. These stages are followed by a high-purity reverse-osmosis system. The end product is 18.3 MΩ cm water with ultralow levels of uranium and thorium daughters, including radon, at flows in excess of 150 liter min^{-1}. Also, a distributed nitrogen cover-gas system with extremely low levels of radon is available. The source of this gas is from boil-off of liquid nitrogen, and the current capacity is some tens of liters per minute.

The 6,800-ft level of Creighton mine is extremely dry. Thus, dewatering considerations deal only with water introduced by the laboratory and are primarily limited by the general mine-dewatering system. SNOLAB is budgeted at 150 liter min^{-1} for dewatering, although this capacity can be exceeded for short periods by advance arrangement with mine operations. A small membrane bioreactor sewage-treatment plant has been installed to process sewage and so-called gray water from showers and wash basins in the laboratory. Although the sewage-treatment plant requires regular maintenance, its effluent is of appropriate quality for discharge to the general mine-dewatering system.

Communication between the surface and the underground facility is provided by a fiber-optic link consisting of 12 multimode fibers that provide a bandwidth of 200 MB s^{-1}; however, the bandwidth will be increased to multigigabytes per second following the completion of a planned link consisting of six single-mode fibers. Communication between SNOLAB and elsewhere is provided by a 1–GB s^{-1} fiber-optic link to the Ontario Research and Innovation Optical Network, with a 100–MB s^{-1} very local area network–to–commodity network. Although most of the telephony in the laboratory is voice-over-internet protocol, there is also an analog phone that provides a redundancy as well as direct contact with the Creighton Mine control room.

The laboratory infrastructure is controlled and monitored through a conventional building-automation system based on the BACNet protocol that is used in many commercial building applications. This system allows (*a*) remote control of the major laboratory systems, including the chiller and air handlers; and (*b*) monitoring of environmental parameters such as temperature, pressure, and gases such as O_2, CO, and NO. The latter two gases are by-products of mining. The fire alarm is a hybrid of an older zone-style system in the original SNO areas and an addressable head system in the new areas. An unusual feature is the use of the local area network to remotely monitor the fire alarm system from the surface and to interface with Creighton Mine's general fire alarm system.

2.4. Construction

The construction of SNOLAB was divided into two phases (C1 and C2). In C1, the caverns and tunnels were excavated, and shotcrete was applied to the walls and concrete floors. In C2, the laboratory was outfitted with structural steel and walls, and the equipment and electrical and plumbing systems were installed. The construction was split into phases because it would have been difficult to find a single contractor with expertise in both mining and outfitting. Carleton University was responsible for the overall construction, and project management and design were contracted to Hatch Engineering. Excavation was performed by J.S. Redpath between September 2004 and June 2008. Outfitting was contracted to Comstock of Canada Ltd. and took place after the majority of excavation was complete, from May 2007 to May 2009. In total, 87,000 t of rock were excavated for the SNOLAB expansion, in addition to the 47,000 t that had been removed for the original SNO facility. Approximately 40% of the rock from the SNOLAB expansion was disposed of by lifting it to surface as part of the mine waste rock–removal cycle. However, the remainder was crushed to make roadbed material, which was used by Creighton Mine in its mining development. The excavation used 8,300 t of shotcrete and concrete as well as 30,000 rock bolts, 8,500 sheets of ground-support screen, and 2,000 cable bolts.

In the original SNO excavation, most of the surfaces had a spray-on shotcrete lining on the walls, which resulted in a very rough surface that is difficult to paint and seal in the dust, as well as very difficult to clean. The exception was the SNO cavern, which had a second-pass, hand-troweled shotcrete surface over which a polyurethane liner was applied to make the cavern watertight and to prevent radon ingress from the rock. For the SNOLAB expansion, the second smoothed shotcrete coat was applied not only in the experiment areas but also in the service areas. This coating provides a smooth surface on which to paint and is greatly facilitating cleaning efforts in the laboratory.

Outfitting of the underground campus was in many respects similar to the construction of a surface building—except that many materials were constrained by the access underground, there were very few straight walls, and work had to be adjusted to accommodate the cage schedule. The larger pieces of equipment, such as the air handlers, sewage-treatment plant, and mine power center, were modularly constructed to fit in the cage. Some equipment was too large to travel on rail cars and still fit within the drift, so it was dragged from the shaft station to the construction site on so-called stone boats (large steel plates designed to fit inside the cage and to be towed down the drifts). Most of the ventilation was distributed through rigid spiral-wound ducting fabricated from galvanized-steel ribbon stock that was shipped underground. This distribution system had the double advantage of minimizing the number of rail cars required to get the material underground and facilitating the field fitting necessary to accommodate the irregular excavations. Some materials, such as the beams for platforms, were too long to fit in the cage, so workers brought them underground by "slinging" them under the cage.

A significant consideration in the construction of SNOLAB expansion was that the SNO experiment was still operational at the beginning of excavation. To minimize disruption of SNO, access for the majority of the excavation took place through an old mining area, with a so-called breakthrough to the access drift to SNO late in the excavation. Blasting was monitored and adjusted to keep seismicity and concussion below 200 mm s^{-1} peak particle velocity and 500 Pa, respectively, at the entrance to the SNO facility. This approach resulted in less than a week of downtime for the SNO experiment, most of which was for changing electrical services. Similarly, the year-long overlap between the excavation and outfitting contractors required careful coordination. Disruption to the outfitting from the excavation blasting (the primary concerns being the concussion and the resulting dust) was mitigated by the construction of temporary blast barricades that were removed after completion of the excavation. **Figure 3** shows the current state of the Ladder Labs and Cube Hall.

Although the excavation and outfitting of the underground spaces were performed by contractors, SNOLAB took direct responsibility for the final cleaning and transitioning to clean room operations of the underground facility. The transition to clean room operations required extensive

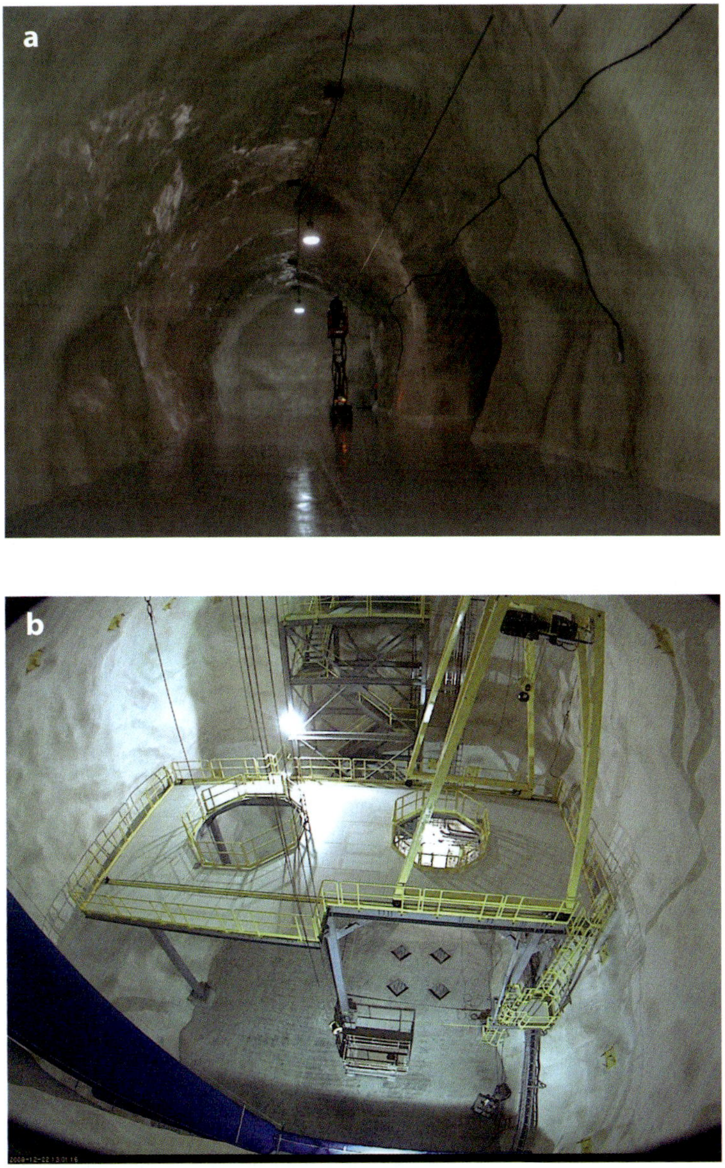

Figure 3

(*a*) The Ladder Labs. (*b*) The Cube Hall after installation of the stair tower and service platform for the DEAP-3600 and MiniCLEAN experiments. The DEAP-3600 experiment will be situated below the deck opening on the left. MiniCLEAN will be situated below the opening on the right.

cleaning of all surfaces and equipment; although pressure washing was done throughout all the new areas, more detailed cleaning had to be done by hand. The transition to clean room operations has been staged: The new personnel facilities became operational in February 2009 and the Ladder Labs in August 2009. As of this writing, following major infrastructure installation for the new experiments to be sited there, cleaning of the Cube Hall is currently under way and expected to be finished by spring 2010.

2.5. Clean Room Operation

Although SNOLAB operates as a clean room, the usual clean room rules are not followed. In contrast to a conventional clean room, most of the services (piping, electrical wiring, ventilation) are left exposed to facilitate maintenance and upgrades. Whereas the expanded SNOLAB areas have a smooth wall finish, the original SNO areas have a very rough wall finish. Nonetheless, monitoring shows that the experiment areas achieve a count of fewer than 2,000 particulates larger than 0.5 μm per cubic foot. Service areas such as access drifts and the lunchroom typically achieve a count better than 5,000. Achieving this level of cleanliness is attributed to a rigorous cleaning program. In addition to monitoring of the air particulates, dust deposition is measured with witness plates; the amount of mine dust deposited is determined by X-ray fluorescence measurement of iron, which makes up approximately 6% of the norite rock. This technique, originally developed for SNO, is being continued for SNOLAB and can achieve a sensitivity of 0.1 μg of mine dust per square centimeter (5).

2.6. Surface Facilities

To support the underground campus, SNOLAB has a surface facility located approximately 300 m from the mine shaft. It consists of a 3,100-m^2 building that was added to two 370-m^2 buildings that existed during SNO operations. The surface facility contains seating for 65 personnel in a combination of offices and open-format cubicles. It has a 100-seat auditorium as well as smaller meeting rooms. There are change rooms and showers for staff going underground. The facility has several dedicated control rooms from which experiments are monitored when there are no personnel underground. The surface facility also contains 440 m^2 of CLASS 1000 clean room laboratories to support the underground operations. These clean laboratories include a chemistry lab, a low-background counting facility for radioassays, and a final assembly room for preparing materials to go underground. Services include an ultrapure water supply; off gas header; and a distributed nitrogen gas supply derived from liquid nitrogen boil-off, which provides a low-radon cover gas. A small warehouse is used to stage materials underground. A 150-kW backup generator maintains computer systems, communications, and monitoring of underground systems in a power outage.

An important consideration when the expanded surface facility for SNOLAB was designed was to provide space to stage materials going underground. When the SNO experiment was constructed, there were surface facilities for the handling of heavy water, including a 70-t storage tank, and a dedicated rail line to transport specially designed tank cars from a surface transfer station to the mine shaft. These features are being maintained and expanded. The new surface clean rooms will allow preassembly of detector components under clean conditions, after which they will be loaded into sealed rail cars for shipping underground in a car wash similar in design to the one underground. The surface clean rooms will effectively become an extension of the underground clean room space. The goal is to maximize the clean room work that can be done on surface.

In addition to the facility located at Creighton Mine, SNOLAB maintains a small (230-m^2) laboratory at nearby Laurentian University. This laboratory is intended for work with radiological agents, including spikes that should not be kept at the SNOLAB site out of concern for contamination.

2.7. Operations Personnel

A very valuable resource provided by SNOLAB is a highly trained workforce experienced in the operation of a clean underground facility. Many of the staff at SNOLAB have been with the project since the early 1990s, when SNO began. There are approximately 45 permanent staff members, including skilled technicians responsible for the operation of the facility, for cleaning and maintenance, and for the installation of new equipment. These staff members are skilled in areas including chemistry, electrical systems, instrumentation, millwrighting, pipe fitting, welding, tramming, rigging, and process systems. There is also a strong engineering team responsible for the operation and maintenance of the facility and a design group assisting experiments with design to integrate them into the laboratory. A research scientist team, with student and postdoctoral support, helps build the scientific culture at the lab. These on-site scientists are major contributors to the experimental program and are largely responsible for managing the facility. They are supported by an administrative team that oversees finances, human resources, and purchasing.

3. THE SCIENCE

The program of science workshops that had guided the development of the laboratory was continued to review the selected experiments, to make recommendations to the experiments on how to proceed, and finally to select the initial suite of experiments. This review by the laboratory occurs in parallel with the review carried out by funding agencies and is an ongoing process. At the time of writing, five major projects are funded and in operation or construction. Three are designed to search for dark matter, one is a general-purpose detector with a primary focus in double-β decay but with additional scientific reach in low-energy solar neutrinos and in geoneutrinos, and one is monitoring for supernova neutrinos. Brief outlines of these experiments follow.

Several other experiments are in the pipeline. Typically they are projects for which some final development is required to demonstrate their feasibility or optimize their design. Some await final funding decisions. All these experiments seek to detect very rare processes in a background produced by naturally occurring radioactivity. Therefore, to be successful, such experiments must (*a*) be very large, (*b*) be made from materials containing ultralow levels of radioactivity, (*c*) be shielded from external sources of radiation, and (*d*) operate in this clean state for several years. The infrastructure and resources provided by SNOLAB provide the best available conditions in which to accomplish this goal.

3.1. Dark Matter Experiments

The evidence for the existence of dark matter is now irrefutable. Galactic rotation curves have long demanded the need for extra nonluminous matter in the Galaxy (9). The recent results from WMAP (Wilkinson Microwave Anisotropy Probe; 10) require a dark matter component of approximately 25%, and the striking observation of the Bullet cluster gives convincing support to the existence of unknown, nonbaryonic, nonluminous matter pervading the universe. In contrast, the normal baryonic matter contained in stellar objects and dust contributes only ~5% of the matter/energy budget of the universe.

Also, the existence of dark matter as a WIMP is highly motivated by the supersymmetric models of elementary particles. These models are based on a concept that for every particle we are familiar with, there is a corresponding partner: The normal bosons have fermion partners, and the familiar fermions have boson partners. The lightest supersymmetric partner should be stable and is thus a candidate for dark matter. There are many parameters in the model whose values are not known, but for a wide range of these parameters, the predicted mass and interaction cross sections should be within reach of the next generation of experiments.

Searches for WIMP dark matter are particularly sensitive to cosmic rays, and the great depth of SNOLAB makes it a very attractive location for such searches. The various detection techniques require separation of nuclear recoil events from the dominant background processes, which produce minimum ionizing particles in the detectors. However, such detectors cannot distinguish recoils from collisions between regular matter and WIMPs, as opposed to collisions of neutrons and the detector matter. The fast neutrons coming from the interaction of cosmic-ray muons with the rock are particularly difficult to shield against and are best eliminated by ranging out these muons, which can be done by going very deep underground. It is not surprising that more than half of the proposals for SNOLAB involved searches for WIMP dark matter. The importance of maximal depth for dark matter searches has been stressed by Mei & Hime (11) in a thorough study of the shielding needs of dark matter experiments.

The DEAP/CLEAN dark matter search program at SNOLAB is a Canada/U.S. collaboration that uses single-phase liquid argon scintillation detectors (12). In addition to the DEAP-1 detector currently operational in the old SNO area at SNOLAB, the Cube Hall will host two experiments, MiniCLEAN and DEAP-3600, which have fiducial masses of 150 kg and 1,000 kg, respectively. Whereas DEAP-3600 is a dedicated argon detector, MiniCLEAN can use either argon or neon as the target material. Particle interactions in the liquid target produce argon dimers in either singlet or triplet states, which deexcite by scintillation light in the short UV. The light, in turn, is wavelength shifted and detected by photomultiplier tubes (PMTs). Because nuclear recoils (such as from a WIMP interaction) preferentially excite the short-lived singlet states, electromagnetic backgrounds can be excluded by pulse shape discrimination by looking at the fraction of prompt versus late light detected in an event.

DEAP-1, a prototype detector with a 7-kg liquid argon fiducial mass, is used primarily to characterize the pulse shape discrimination and backgrounds and has been operational at SNOLAB since 2007. MiniCLEAN (**Figure 4**), which contains more than 500 kg of liquid argon and has a fiducial mass of 150 kg, uses a conventional vacuum insulated cryostat viewed by 92 PMTs

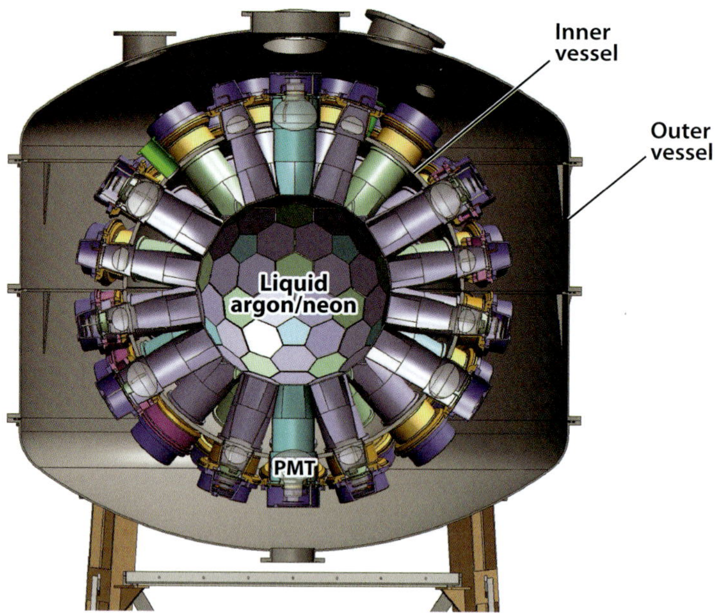

Figure 4

The MiniCLEAN liquid argon/neon dark matter detector. Abbreviation: PMT, photomultiplier tube.

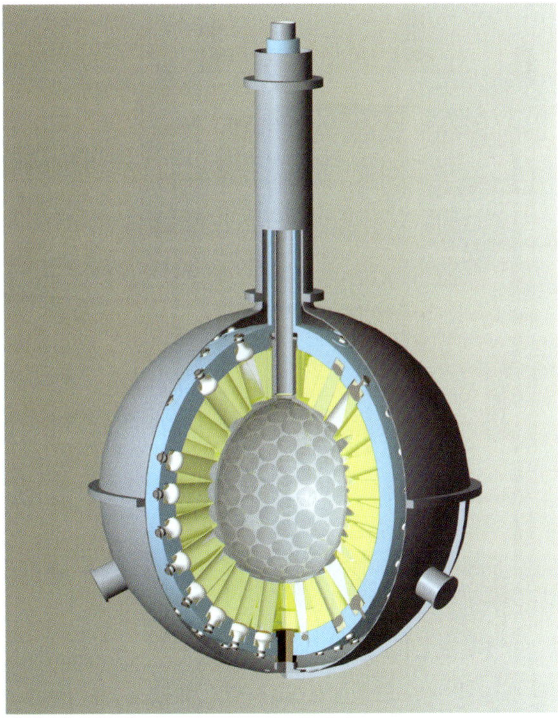

Figure 5

The DEAP-3600 liquid argon dark matter detector.

immersed in the liquid argon. Intended to reach a sensitivity to 100-GeV WIMPs of 2×10^{-45} cm^2, MiniCLEAN will be situated in a 6-m-diameter water shield in the Cube Hall. Installation is expected to begin in 2010. DEAP-3600 (**Figure 5**) is a 3,600-kg liquid argon detector with a 1,000-kg fiducial mass contained in an acrylic cryostat viewed by approximately 260 room-temperature PMTs. DEAP-3600 is intended to have a physics reach of 10^{-46} cm^2 for a 100-GeV WIMP—a factor of 500 more sensitive than current experimental dark matter searches. DEAP-3600 will be housed in an 8-m-diameter shield tank and is expected to begin installation in 2011.

PICASSO is a dark matter search experiment that exploits bubble chamber technology to search for evidence of spin-dependent WIMP-nucleon interactions. The active material consists of superheated liquid droplets of C_4F_{10} suspended in a radioactively clean gel matrix. Nuclear recoils deposit sufficient heat within a droplet to induce a phase transition that can be detected by sensitive piezoelectric transducers. By design, PICASSO is insensitive to γ and minimum ionizing particles. Background shielding to neutrons is accomplished by depth (located in SNOLAB), material selection, and local water shielding. PICASSO has also discovered a means of discriminating between nuclear recoils and α particles that significantly enhances the potential sensitivity of the experiment (13).

PICASSO has been running at SNOLAB for several years, gradually incrementing the total sensitive mass and lowering the backgrounds. PICASSO is particularly sensitive to spin-dependent interactions, and its most recent results are among the best limits in the world (14). **Figure 6** illustrates the features of a PICASSO detector.

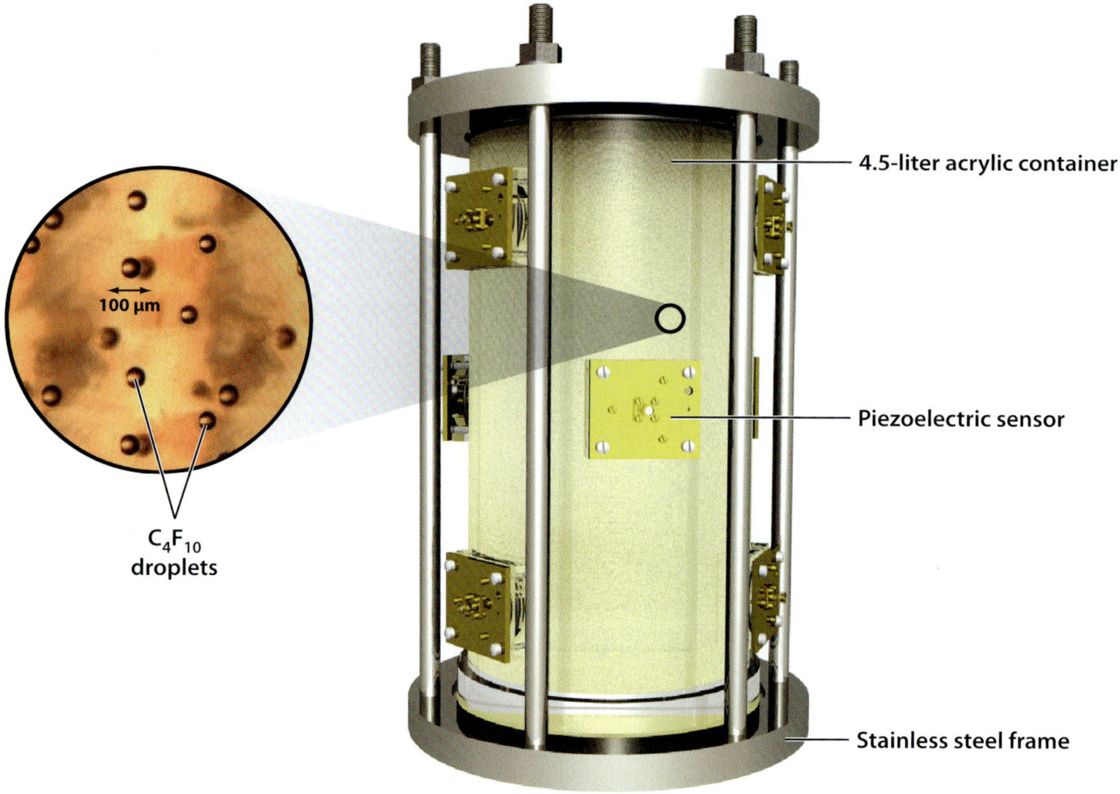

Figure 6

One of the 32 PICASSO detectors currently in operation at SNOLAB. The active C_4F_{10} droplets are held within the gel matrix inside the container. Nine piezoelectric sensors mounted on the outside walls listen for the telltale signals of an induced phase transition.

3.2. Supernova Watch

In a type II supernova, approximately 99% of the energy is released in a burst of neutrinos as the very hot neutron star cools. Measuring the properties of this burst can give information about the dynamics of the supernova collapse as well as the evolution of the neutrino flavor. The neutrino burst arrives hours before the optical explosion can be detected, so a neutrino detector can act as an alert to the astronomical community. The rate of detectable supernovae is very low, perhaps one event per 50 years, but the potential physics impact is enormous, as evidenced by the scientific output from the observation of only a few neutrinos observed in SN 1987a in experiments not optimized for supernova detection.

As a result, supernova detectors must be very reliable, low maintenance, and capable of long-term unattended operation. HALO (15) is such a detector. It uses approximately 80 tons of lead to produce neutrons from either a charged-current or neutral-current interaction between neutrinos. The neutrons are detected in embedded ^{3}He detectors from the neutral-current detector phase of SNO (16). For a supernova occurring within our Galaxy, the array would produce a burst of neutron events. SNO$^+$ (described in more detail below) will also be sensitive to supernova neutrinos. Both HALO and SNO$^+$ will be a part of the worldwide Supernova Early Warning System (SNEWS; 17), which will alert the community of an event in advance of the optical signal.

3.3. Search for Neutrinoless Double-Beta Decay (SNO^+)

Following the discovery of neutrino oscillations at Super-Kamiokande and SNO, we learned that neutrinos are strongly mixed and that they have mass. Although we can set limits on the mass from the properties of the oscillations, together with the bounds on neutrino mass from tritium β decay, the actual mass of the neutrino states remains unknown. One process that may reveal this mass is the observation of neutrinoless double-β decay. Not only would an observation of this process provide an indication of the absolute neutrino mass scale, it would show that neutrinos are Majorana particles—that is, they are their own antiparticles. Many of the models that attempt to explain the very small mass of neutrinos require them to have this property.

A project that will provide a sensitive search for double-β decay is SNO^+ (18). This experiment will adapt the SNO detector for use with a liquid scintillator, linear alkylbenzene, in place of the heavy water in the original SNO detector. The scintillator will be loaded with neodymium, an attractive candidate for the double-β decay process. Liquid scintillators suffer from poorer energy resolution than some proposed techniques for this study, but the proposed detector will achieve exceptional sensitivity both from the large mass of neodymium that can be incorporated in the scintillator and from the excellent shielding against backgrounds. External backgrounds will be controlled by several meters of pure water surrounding the acrylic vessel. Contributions from the vessel itself will be reduced by self-shielding within the massive scintillator. Self-shielding is possible because the SNO cavern is the largest in the world that is designed for human occupancy at this depth, and the SNO^+ project exploits this feature to create a uniquely sensitive detector. The detector could also be used to measure low-energy neutrinos from the Sun from the fundamental *pep* process (19) and to measure the geoneutrino flux (20) to fix the amount of uranium and thorium in the Earth's crust.

3.4. Beyond Astroparticle Physics

Although most of the proposed experiments fall within the area of astroparticle physics, the existence of a clean, deep, well-instrumented laboratory would facilitate experiments in other branches of physics. For example, unique information on seismic waves may be obtained if a full three-dimensional array of sensors could be used in place of the normal surface array. A test with a few stations has been carried out at SNOLAB, and data of interest on nearby seismic events have already been obtained and published (21).

4. CONCLUSIONS

The construction of the laboratory is now complete, and initial construction of a new suite of experiments is under way. An exciting future in astroparticle physics is already unfolding with initial results from PICASSO and DEAP-1. A new generation of ultrasensitive experiments searching for WIMP dark matter and neutrinoless double-β decay are under construction. These experiments have the potential to change the way we understand the universe. They are made possible by a new, very deep, and very clean laboratory for science.

DISCLOSURE STATEMENT

The authors are not aware of any affiliations, memberships, funding, or financial holdings that might be perceived as affecting the objectivity of this review.

LITERATURE CITED

1. Aharmim B, et al. (SNO Collab.) *Phys. Rev. C* 75:045502 (2007)
2. Aharmim B, et al. (SNO Collab.) *Phys. Rev. C* 72:055502 (2005)
3. Aharmim B, et al. (SNO Collab.) *Phys. Rev. Lett.* 101:111301 (2008)
4. Boger J, et al. (SNO Collab.) *Nucl. Instrum. Methods A* 449:172 (2000)
5. Hallman ED. (SNO Collab.) *AIP Conf. Proc.* 785:163 (2004)
6. Blevis I, et al. *Nucl. Instrum. Methods A* 517:139 (2004)
7. Anderson TC, et al. *Nucl. Instrum. Methods A* 501:386 (2003)
8. Anderson TC, et al. *Nucl. Instrum. Methods A* 501:399 (2003)
9. Zwicky F. *Astrophys. J.* 86:217 (1937)
10. Bennett LL, et al. (WMAP Collab.) *Astrophys. J. Suppl. Ser.* 148:1 (2003)
11. Mei DM, Hime A. *Phys. Rev. D* 73:053004 (2006)
12. Boulay MG, Hime A. arXiv:astro-ph/0411358v1 (2004)
13. Aubin F, et al. *New J. Phys.* 10:103017 (2008)
14. Archambault S, et al. *Phys. Lett.* B682:185 (2009)
15. Virtue CJ. *J. Phys. Conf. Ser.* 136:042077 (2008)
16. Amsbaugh JF, et al. *Nucl. Instrum. Methods A* 579:1054 (2007)
17. Antonioli P, et al. *New J. Phys.* 6:114 (2004)
18. Chen MC. (SNO^+ Collab.) arXiv:0810.3694 [hep-ex] (2008)
19. Chen MC. *J. Phys. Conf. Ser.* 120:052001 (2008)
20. Chen MC. *Earth Moon Planets* 99:221 (2006)
21. Atkinson GM, et al. *Seismol. Res. Lett.* 79:119 (2008)

Multiparton Scattering Amplitudes via On-Shell Methods

Carola F. Berger[1] and Darren Forde[2,3]

[1]Center for Theoretical Physics, Massachusetts Institute of Technology, Cambridge, Massachusetts 02139; email: cfberger@mit.edu

[2]Theory Division, CERN, CH-1211 Geneva 23, Switzerland

[3]NIKHEF Theory Group, NL-1098 XG Amsterdam, The Netherlands; email: darren.forde@cern.ch

Annu. Rev. Nucl. Part. Sci. 2010. 60:181–205

First published online as a Review in Advance on July 1, 2010

The *Annual Review of Nuclear and Particle Science* is online at nucl.annualreviews.org

This article's doi: 10.1146/annurev.nucl.012809.104538

0163-8998/10/1123-0181$20.00

Key Words

QCD, precision calculations, next-to-leading order, generalized unitarity, recursion relations

Abstract

We present an overview of recent developments, based on on-shell techniques, in the calculation of multiparton scattering amplitudes at one loop that are relevant for phenomenological studies at hadron colliders. These new on-shell methods make efficient use of the physical properties of the hard scattering, such as unitarity and factorization.

Contents

1. INTRODUCTION

With the recent first collisions at the Large Hadron Collider (LHC), we are entering a new era of discovery in particle physics. The LHC has now replaced the Tevatron at Fermilab as the highest-energy particle accelerator, colliding protons at very high energies in search of new physics beyond the Standard Model. These high-energy collisions are a fertile environment for the production of high-multiplicity events, that is, events with many particles in the final state that have very complex signatures in the detectors. If we are to take full advantage of the discovery potential at high-energy hadron colliders such as the Tevatron and the LHC so as to disentangle the Standard Model from new physics signals, we need to have a precise understanding of the physics that will occur there. This necessitates computations of the Standard Model background, especially of quantum chromodynamics (QCD) processes, to at least next-to-leading order (NLO) in the perturbative series. Leading-order (LO) computations are in general not able to correctly predict the shapes of kinematical distributions.

For a good understanding of the physics at hadron colliders, a great variety of different processes needs to be computed to NLO accuracy (e.g., 1). This necessitates automation of these computations. The basic ingredients in an NLO computation are matrix elements that describe real emission processes, matrix elements for the one-loop virtual corrections, and integration of these over the final state phase space. Historically, the bottleneck in the automation of NLO computations has been the one-loop virtual contributions. Over the past few years, rapid progress has been made in the development of new techniques for these one-loop computations. These advancements have been motivated both by a greater desire to understand the structure of scattering amplitudes as well as by the need for improved efficiency and automation in the computation of one-loop matrix elements. The ability to automate and "mass-produce" amplitudes requires approaches that are both numerically stable and straightforward to implement as an algorithm.

The standard approach to performing this class of computations has relied heavily on Feynman diagram techniques. There have been many impressive results with this approach (see Reference 1 and references therein; cross sections for six-point processes that have been computed via Feynman diagrammatic methods can be found in, e.g., Reference 2). However, Feynman diagrams suffer from two problems. First, there is a factorial growth in the number of terms as the multiplicity of partons in the process increases. Second, each Feynman diagram is gauge dependent. This means that there are large cancellations between terms that combine to give the gauge-independent amplitudes. It is these two problems that make automated approaches using Feynman diagrams difficult as the number of partons increases.

Consequentially, the focus of much recent progress has been to sidestep these issues. On-shell recursion and unitarity methods work with gauge-independent amplitudes as building blocks instead of Feynman diagrams. It is these new techniques that we focus on in this review. An earlier review of these techniques has been presented (3). However, many of the details given therein have been superseded by even more efficient techniques, which we present in the following sections.

Generally, as we explain in more detail below, a one-loop amplitude can be decomposed into a set of scalar box, triangle, bubble, and tadpole integrals, that is, integrals with four-, three-, two-, or one-loop propagators, respectively. These scalar integrals contain all the logarithmic and polylogarithmic dependence of the amplitude and are multiplied by rational coefficients. There are also purely rational terms in the amplitude.

The original unitarity approach (4) applied two-particle cuts in four dimensions and was used to produce many results (5). All the terms containing (poly)logarithms and associated constants could be computed via two-particle cuts. Terms with only a rational dependence on momentum invariants, however, could not be computed in this way, so separate techniques were required (5). A systematic approach for the computation of these rational terms without the use of Feynman diagrams did not appear until a few years ago. The starting points for these developments were the recursion relations, developed at tree level by Britto et al. (6). An earlier version of a tree-level recursion relation was the off-shell Berends-Giele recursion (7). The proof of the Britto et al. on-shell recursion relations relies only on the factorization properties of the amplitudes and on Cauchy's theorem. They could therefore be adapted to the more complicated problem of computing the rational parts of loop amplitudes (8, 9). On-shell recursion for the rational parts of loop amplitudes was first used in an analytic context (9) and then further adapted into a numerical procedure (10).

At the same time, improvements to the original unitarity approach were also occurring. Brandhuber et al. (11) used two-particle cuts with maximally helicity violating[1] vertices to compute certain sets of one-loop amplitudes. The application by Bern at al. (5) and Britto et al. (12) of generalized unitarity to the computation of box coefficients highlighted the benefits of examining not the one-loop integral as a whole, but rather its integrand. The work of Ossola, Papadopoulos & Pittau (OPP; 13) and their colleagues (14) followed in this vein. Upon separation of the integrand into a standard set of basis terms, the problem could be reduced to solving for their coefficients numerically. Forde (15) focused on the simple analytic extraction of triangle and bubble coefficients and used the known analytic behavior of the integrand to straightforwardly extract the bubble and triangle coefficients. A numerical adaptation of this procedure suitable for automation was then presented in Reference 10.

The investigation of the one-loop integral itself has also presented new directions to explore. Britto et al. (16) showed that the integral, when written in a canonical form, can be directly integrated. This procedure was developed further to produce new results (17) and also to provide a general analytic structure for the coefficients of one-loop amplitudes (18, 19). Taking further advantage of the analytic properties of the two-particle cut amplitude, Mastrolia (20) applied

[1]An amplitude written in terms of the helicity of its external legs is maximally helicity violating if all but two of its external legs are of the opposite helicity. At tree level, all amplitudes with all legs of the same helicity and with all but one of the same helicity vanish.

Stokes' theorem and a generalized residue theorem to compute bubble coefficients via direct integration.

Extending the four-dimensional cut techniques to D dimensions (in dimensional regularization) has also provided a second, very fruitful approach for the computation of the rational terms. In D dimensions, the rational terms develop branch cuts and so are accessible via unitarity cuts. Giele et al. (21) elucidated the extra structures present in D dimensions beyond those of the original four-dimensional OPP integrand. Thus, the authors computed the rational terms. Following up on this work and on an earlier result relating masses to D-dimensional unitarity (22), Badger (23) presented an alternative computational approach where the $(D-4)$-dimensional terms are treated as an additional mass in the loop. A different approach advanced by Papadopoulos and colleagues (24) involves splitting the rational computation into two pieces. One part is computed from the one-loop integrand and is an extension to the original OPP approach, and the second part comes from a reduced form of Feynman diagrams.

These approaches have been implemented in several automated tools for the computation of one-loop amplitudes: BLACKHAT (10), CUTTOOLS/ONELOOP (25, 26), ROCKET (27), and others (28). These programs, combined with tools for the real emission part (29, 30), have yielded a host of new results at NLO (31–36).

Below we review the main ideas of all these developments. Due to lack of space, we regret that we can present neither all details nor an exhaustive list of all results, and we refer the reader to the cited literature and references therein.

We begin with a review of our notation and the general structure of multiparton scattering amplitudes and explain how to construct such amplitudes recursively at tree level. We use the spinor formalism, and readers unfamiliar with spinor techniques and their application to the calculation of tree-level and one-loop amplitudes may wish to consult, for example, References 37–39. In Section 3 we discuss the use of (generalized) unitarity to construct one-loop amplitudes from tree-level amplitudes, both the cut and the rational part. In Section 4 we explain how to alternatively employ a recursive approach to obtain the purely rational nonlogarithmic terms of amplitudes that cannot be constructed via unitarity in four dimensions. We conclude with a summary and give an outlook on the expected progress in the near future.

2. STRUCTURE OF AMPLITUDES

Below we briefly review our notation, which closely follows Reference 38 (see also Reference 37). We then discuss the general structure of tree and one-loop amplitudes in renormalizable gauge theories to set the stage for the subsequent sections, in which we discuss new methods for the computation of these amplitudes.

2.1. Notation and Color Decomposition

In the following we consider amplitudes where all color[2] and coupling information have been stripped off. We can express any amplitude in terms of some basic color (and coupling) factors, which are multiplied by color-ordered subamplitudes, also known as primitive amplitudes. These primitive amplitudes, which are defined for specific cyclic orderings of the external partons, carry all the kinematic information but no explicit color indices. The full amplitude is then assembled from the primitive amplitudes by dressing them with appropriate color factors. There exist different, but

[2]Color here and below refers to any group theory factors.

equivalent, ways of grouping the primitive amplitudes together into so-called partial amplitudes, and we refer the reader to the literature for further information (37, 38, 40, 41, and references therein). In what follows, we discuss the computation of the primitive amplitudes, concentrating on the description of the evaluation of the kinematic part. As shown recently (42), the algorithms that we present below can be extended to amplitudes with color information.

We express the primitive amplitudes in terms of spinor inner products,[3]

$$\langle j\, l\rangle = \langle j^-|l^+\rangle = \bar{u}_-(k_j)u_+(k_l), \quad [j\, l] = \langle j^+|l^-\rangle = \bar{u}_+(k_j)u_-(k_l), \qquad 1.$$

where $u_\pm(k)$ are massless two-component (Weyl) spinors with momentum k and positive or negative chirality, respectively, which we also write as

$$(\lambda_i)_\alpha \equiv [u_+(k_i)]_\alpha\,, \quad (\tilde{\lambda}_i)_{\dot\alpha} \equiv [u_-(k_i)]_{\dot\alpha}\,. \qquad 2.$$

Massless four-momenta can be reconstructed from the spinors by

$$k_i^\mu(\sigma_\mu)_{\alpha\dot\alpha} = (\not{k}_i)_{\alpha\dot\alpha} = (\lambda_i)_\alpha(\tilde{\lambda}_i)_{\dot\alpha}. \qquad 3.$$

Spinor products can thus be used to construct the usual momentum dot products via

$$\langle i\, j\rangle[j\, i] = \frac{1}{2}\mathrm{Tr}[\not{k}_i\not{k}_j] = 2k_i\cdot k_j = s_{ij}. \qquad 4.$$

Furthermore, we use the following notation for sums of cyclically consecutive external momenta and their invariant masses,

$$K^\mu_{i\ldots j} \equiv k_i^\mu + k_{i+1}^\mu + \cdots + k_{j-1}^\mu + k_j^\mu \qquad 5.$$

and

$$s_{i\ldots j} \equiv K^2_{i\ldots j}, \qquad 6.$$

where all indices are to be understood mod n for n-particle amplitudes.

The above formalism can be extended to include massive spinors and vectors, using the well-known decomposition of any (not necessarily light-like) four-vector k into a sum of two light-like four-vectors:

$$k^\mu = k^{\flat\mu} + \frac{k^2}{2k\cdot q}q^\mu. \qquad 7.$$

Here q is a fixed light-like four-vector, fixing the axis of the spin for spinors, and $k^\flat$ is the associated projection of the massive vector k. Massive spinors can be constructed via

$$u_-(k,q) = \frac{1}{\langle k^\flat q\rangle}(\not{k}+m)|q^+\rangle = |k^{\flat-}\rangle + \frac{m}{\langle k^\flat q\rangle}|q^+\rangle, \qquad 8.$$

and

$$u_+(k,q) = \frac{1}{[k^\flat q]}(\not{k}+m)|q^-\rangle = |k^{\flat+}\rangle + \frac{m}{[k^\flat q]}|q^-\rangle, \qquad 9.$$

and the same can be done for the conjugate spinors (43). Here, the label $\pm$ indicates that the spinors $u_\pm$ are eigenstates of the projector $(1\pm\not{s}\gamma^5)$, with the spin vector $s = k/m - m/(k\cdot q)q$. The normalization is chosen to allow a smooth limit to the massless case.

[3]Note that we use the sign convention of most of the QCD literature; in the twistor literature a different sign convention for $[j\, l]$ is used (e.g., 6).

2.2. On-Shell Recursions at Tree Level

An efficient recursive technique for computing tree-level multiparton scattering amplitudes was developed more than 20 years ago (7) and adapted for numerical implementation in various computer codes (41, 44, 45). Berends-Giele recursion connects smaller off-shell currents together to produce amplitudes. More recently it was realized that, through the use of complex kinematics, amplitudes can be computed entirely using only smaller on-shell amplitudes. This leads to more compact analytic expressions not only at tree level but, for rational terms, also at loop level. Here we briefly review the on-shell recursion relations for tree-level amplitudes found and proved in Reference 6. A recursive approach at loop level is not quite so straightforward, as we discuss in Section 4.

At tree level, the on-shell recursion relations rely on general properties of complex functions as well as on factorization properties of scattering amplitudes. The proof (6) of the tree-level relations employs a parameter-dependent complex continuation $[j, l\rangle$, or shift, of two of the external massless spinors, j and l, in an n-point process,

$$[j, l\rangle : \tilde{\lambda}_j \to \tilde{\lambda}_j - z\tilde{\lambda}_l, \quad \lambda_l \to \lambda_l + z\lambda_j, \qquad 10.$$

where z is a complex number. The corresponding momenta are then continued in the complex plane as well, whereby they remain massless, $k_j^2(z) = 0 = k_l^2(z)$, and overall momentum conservation is maintained.

An on-shell amplitude A containing the momenta k_j and k_l also becomes dependent on the parameter z: $A(z)$. The physical amplitude is given by $A(0)$. When A is a tree amplitude or finite one-loop amplitude, $A(z)$ is a rational function of z. At tree level, $A(z)$ has only simple poles. These poles arise only from the shifted propagators of the amplitude. For example,

$$\frac{i}{K^2_{r\ldots l\ldots s}} \to \frac{i}{K^2_{r\ldots s} + z\langle j^- | \not{K}_{r\ldots s} | l^- \rangle}, \qquad 11.$$

if the set of legs $\{r, \ldots, s\}$ contains leg l, which is shifted according to Equation 10. In the vicinity of the location of the pole z_{rs}, the complex-continued amplitude is then schematically given by

$$\lim_{z \to z_{rs}} A(z) = \sum_h A_L^h(z) \frac{i}{K^2_{r\ldots s} + z\langle j^- | \not{K}_{r\ldots s} | l^- \rangle} A_R^{-h}(z), \qquad 12.$$

where $h = \pm 1$ labels the helicity of the intermediate state and L and R denote amplitudes with fewer legs, which the propagator Equation 11 connects. The number of poles z_{rs} in the complex plane is given by the number of ways the set of external legs can be partitioned such that the legs j and l always appear on opposite sides of the z-dependent propagator.

We can now use Cauchy's theorem,

$$\frac{1}{2\pi i} \oint_C \frac{dz}{z} A(z) = 0, \qquad 13.$$

where the contour C is taken around the circle at infinity and the integral vanishes if the complex-continued amplitude $A(z)$ vanishes as $z \to \infty$. Evaluating the integral as a sum of residues, we can then solve for the physical amplitude $A(0)$ to obtain

$$A(0) = -\sum_{\text{poles}\,\alpha} \operatorname*{Res}_{z=z_\alpha} \frac{A(z)}{z} = \sum_{r,s} \sum_h A_L^h(z = z_{rs}) \frac{i}{K^2_{r\cdots s}} A_R^{-h}(z = z_{rs}). \qquad 14.$$

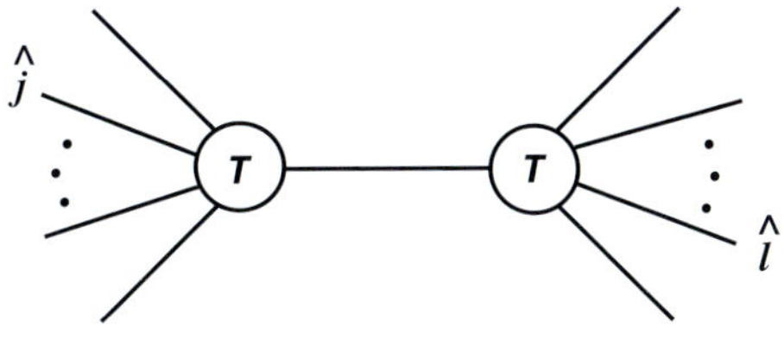

Figure 1

Schematic representation of a typical tree-recursive contribution to Equation 14. T represents an on-shell tree amplitude. The momenta $\hat{j}$ and $\hat{l}$ are complex-continued, on-shell momenta.

The on-shell amplitudes with fewer legs, A_L and A_R, are evaluated in kinematics that have been shifted by Equation 10 with $z = z_{rs}$, where Equation 11 has a pole,

$$z_{rs} = -\frac{K^2_{r\ldots s}}{\langle j^- | \not{K}_{r\ldots s} | l^- \rangle}. \tag{15}$$

Below, such shifted, on-shell momenta are denoted by $k(z = z_{rs}) \equiv \hat{k}$. A typical contribution to the sums in Equation 14 is illustrated in **Figure 1**.

We have thus succeeded in expressing the n-point amplitude A in terms of sums over on-shell, but complex-continued, amplitudes with fewer legs, which are connected by scalar propagators. These recursion relations can be extended to massive QCD and other theories (46, 47). Moreover, for certain helicity configurations, this recursion relation can be solved explicitly, leading to new all-multiplicity expressions for these amplitudes (48).

The basic ingredients needed to obtain such a recursion relation are (*a*) complex momenta and analysis, which are necessary to make three-point vertices nonvanishing; (*b*) factorizability, which is responsible for the simple pole structure; and (*c*) the vanishing of the boundary contribution as $z \to \infty$. At tree level in QCD, one can always find complex continuations where this boundary condition vanishes. However, as we discuss in Section 4 below, this is not the case at the one-loop level, and a recursive approach becomes considerably more complicated. Other theories (such as scalar ϕ^4 theory) already have nonvanishing $z \to \infty$ behavior at the tree level, which spoils the recursive approach. Studies of the origin of these boundary terms and their relation to the Lagrangian have been performed (49). Furthermore, additional amplitude structures related to on-shell recursion relations and twistor space have been uncovered (50).

2.3. Structure of One-Loop Amplitudes

We now turn to the discussion of gauge-theory one-loop amplitudes, the main subject of this review. Hereafter, we denote one-loop amplitudes explicitly with a superscript by $A^{1\text{-loop}}$ and tree amplitudes without a superscript simply by A.

By use of reduction techniques (51–53), any m-point scalar integral, $m > 4$, can be reduced to scalar integrals with at most four propagators. That is, any one-loop n-point amplitude $A_n^{1\text{-loop}}$ can be decomposed into a basis B_4 of scalar box, triangle, bubble, and tadpole integrals, with rational coefficients in four dimensions. In D dimensions, for example when working in dimensional regularization in which $D = 4 - 2\varepsilon$, then the basis B_D is extended to include a scalar pentagon. The coefficients of the D-dimensional-basis scalar integrals can be decomposed into purely four-dimensional coefficients after expanding in ε. Purely rational terms are generated when terms that are higher order in ε in the coefficients are multiplied by the poles in the integrals:

$$A_n^{1\text{-loop}} = \sum_{j\in B_D} c_j^D \mathcal{I}_j^D = \sum_{j\in B_4} c_j^{D=4} \mathcal{I}_j^D + \mathcal{R}_n. \tag{16}$$

Illustrative examples of integrals o f the basis B are shown in **Figure 2**.

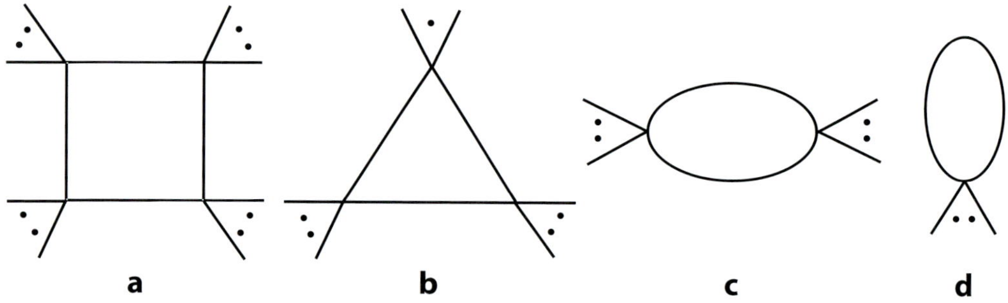

Figure 2

Representative examples of integrals appearing in Equation 16. (*a*) A box (four-point) integral. (*b*) A triangle (three-point) integral. (*c*) A bubble (two-point) integral. (*d*) A tadpole integral. Each corner can have one or more external momenta attached to it. The tadpole vanishes when all internal propagators are massless.

We explain in the next two sections how to obtain these coefficients and rational terms in efficient ways. The scalar integrals contain infrared and ultraviolet divergences that are regulated via dimensional regularization and that depend logarithmically or polylogarithmically on momentum invariants. The integrals appearing in Equation 16 are known and tabulated in, for example, References 4 and 54. In order to compute one-loop matrix elements the task is therefore reduced to the determination of the coefficients; it is not necessary to perform any integrals.

For amplitudes with only massless particles, the tadpoles vanish. In $\mathcal{N} = 4$ supersymmetric theories, as counting of powers of loop momenta in one-loop integrals reveals, only box integrals contribute with (D–4)-dimensional coefficients (i.e., free of ε terms); in $\mathcal{N} = 1$ supersymmetry, bubble, triangle, and box integrals contribute with four-dimensional coefficients (4). That is, theories with unbroken supersymmetries do not contain purely rational terms that are not associated with any of the integrals in the basis. One-loop supersymmetric amplitudes can therefore be completely reconstructed from unitarity cuts, as we discuss in the next section.

3. EXTRACTION OF INTEGRAL COEFFICIENTS VIA UNITARITY

The problem of computing Equation 16 has been reduced to determining the coefficients c_j, which multiply the known basis integral functions in the most efficient manner possible. The nature of Equation 16 suggests the use of unitarity cuts to isolate particular integral coefficients. At the most basic level, a unitary cut effectively replaces a propagator with an on-shell delta function; specifically, we "cut" the propagator with the replacement

$$\frac{1}{p^2 - m^2 + i\epsilon} \to \delta^{(+)}(p^2 - m^2). \qquad 17.$$

In the original unitarity approach (4), only two propagators were cut, but more recent developments have highlighted the benefits of applying multiple cuts (5, 12). The application of multiple cuts is known as generalized unitarity and has become the foundation of the most recent developments in the literature.

Our starting point is the form of Equation 16 decomposed in the B_4 basis; we postpone until Section 3.3 the discussion of the general D-dimensional case using B_D. The purely rational terms are independent of any possible cuts in four dimensions, and therefore only the remaining "cut-constructible" pieces are accessible via a unitarity technique.

We apply a number of cuts to the expression of the one-loop amplitude and match this expression to that of the basis decomposition (Equation 16) with the same set of cuts applied. Doing so allows us to directly relate the cut expression to the basis integral coefficients. This procedure is repeated with as many different sets of cuts as are needed to compute all the basis coefficients. Rather than actually applying the cuts to the full expression for the one-loop amplitude, computed for example with Feynman diagrams, we construct the cut expression simply by multiplying appropriate on-shell tree amplitudes together. This allows us to take advantage of efficient, compact forms of tree amplitudes produced via recursion relations, such as those of Section 2.2.

Below, we describe two unitarity approaches for the extraction of integral coefficients. They both utilize knowledge of the integrand, $\tilde{A}(l)$, of the one-loop amplitude, $A_n^{\text{1-loop}} = \int dl \tilde{A}_n(l)$, to derive the basis integral coefficients. The first, described in Section 3.1, is based on the examination of the behavior of the loop integrand and the loop momenta in the complex plane. The second, described in Section 3.2, known as the OPP method, relies on computing the coefficients of the general structure of the loop integrand itself.

3.1. Generalized Unitarity in Four Dimensions

In general we want to isolate as few basis integral coefficients as possible with each cut that we consider. It is easy to verify that a quadruple cut can be used to isolate, on the basis side, a single box coefficient. There are not enough propagators in the bubble and triangle integrals to accommodate so many cuts. The set of cuts we require to compute all box coefficients corresponds simply to all possible boxes that could be present.

3.1.1. Boxes. A very straightforward way to extract a specific box coefficient from a quadruple cut expression was proposed by Britto et al. (12). The momentum circulating inside a loop without cuts is off shell and can therefore be parameterized in terms of four free components. Applying a cut to one of the propagators in the expression of the one-loop amplitude reduces the number of free components in this loop momentum by one. Then, third and fourth cuts reduce the number of free components to zero. The loop momentum for the box is then completely frozen by the four delta-function constraints. The desired scalar box coefficient can now be read off from the resulting rational expression,

$$\int \frac{d^4 l}{(2\pi)^4} \frac{d(l)}{(l^2+i\epsilon)(l_1^2+i\epsilon)(l_2^2+i\epsilon)(l_3^2+i\epsilon)}\bigg|_{l_i^2+i\epsilon \to \delta^+(l_i^2)} \to \frac{1}{2}\sum_{\text{solutions}} d(l_{\text{solution}}). \qquad 18.$$

The coefficient is a product of four tree amplitudes that sit at the four corners of the cut box, as illustrated in **Figure 3**. The momenta flowing into the trees satisfy the four cut constraints. Generally, there are two possible solutions to the constraints parameterizing the box. For example, with at least one massless leg (leg 1) we have (10, 55)

$$l_\pm^\mu = \frac{\langle 1^\mp | \not{K}_2 \not{K}_3 \not{K}_4 \gamma^\mu | 1^\pm \rangle}{2\langle 1^\mp | \not{K}_2 \not{K}_4 | 1^\pm \rangle}. \qquad 19.$$

Inserting each solution into the product of four trees at each corner and then summing the two results gives the complete box coefficient:

$$d_0 = \frac{1}{2}\sum_{a=\pm} d_a, \quad d_a = A_1(l_a)A_2(l_a)A_3(l_a)A_4(l_a). \qquad 20.$$

Further discussion on the use of both solutions and alternative approaches can be found in Reference 56.

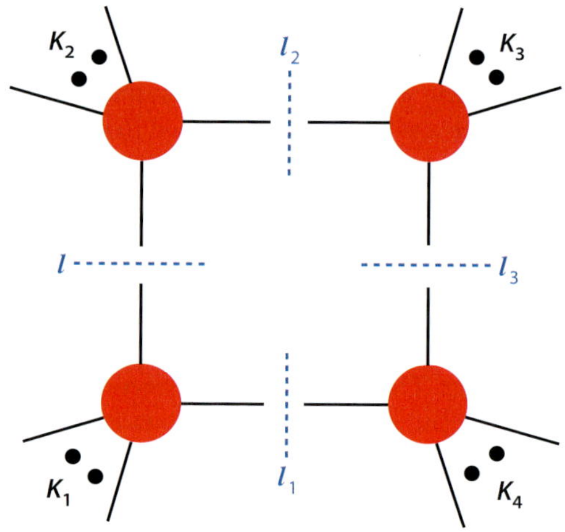

Figure 3

A quadruple-cut one-loop integral isolating the single box term $d_0(K_1^2, K_2^2, K_3^2, K_4^2)$.

3.1.2. Triangles. To compute triangle coefficients, we first apply a triple cut to our one-loop amplitude. Unlike the box case, this cut isolates not merely a single triangle but also any boxes that also contain the same triple cut. Furthermore, we are left with a loop integral with a single free component. Our cut expression therefore still depends on the loop integration. Extracting the triangle coefficient isolated by the triple cut therefore requires two steps. First, we must remove the boxes polluting the triple-cut expression. Second, we must relate the triangle integral, which depends on the remaining free loop parameter, to the scalar triangle basis integral.

As was first proposed in Reference 15, both issues can be solved simultaneously by examining the analytic behavior of the triple-cut expression in the free parameter t of the loop momentum. We choose the following specific parameterization of the loop momentum:

$$l^\mu = K_1^{\flat,\mu} + K_2^{\flat,\mu} + \frac{t}{2}\langle K_1^{\flat,-}|\gamma^\mu|K_2^{\flat,-}\rangle + \frac{1}{2t}\langle K_2^{\flat,-}|\gamma^\mu|K_1^{\flat,-}\rangle. \qquad 21.$$

The massless momenta $K_i^{\flat,\mu}$ are given by

$$K_1^{\flat,\mu} = \gamma\alpha\frac{\gamma K_1^\mu + S_1 K_2^\mu}{\gamma^2 - S_1 S_2}, \quad K_2^{\flat,\mu} = -\gamma\alpha'\frac{\gamma K_2^\mu + S_2 K_1^\mu}{\gamma^2 - S_1 S_2},$$
$$\gamma_\pm = -(K_1\cdot K_2) \pm \sqrt{\Delta}, \quad \Delta = -\det(K_i\cdot K_j) = (K_1\cdot K_2)^2 - K_1^2K_2^2,$$
$$\alpha = \frac{S_2(S_1-\gamma)}{(S_1S_2-\gamma^2)}, \quad \alpha' = \frac{S_1(S_2-\gamma)}{(S_1S_2-\gamma^2)}, \qquad 22.$$

where $S_i = K_i^2$.

By definition, any box terms containing our chosen triple cut also contains an additional propagator, $1/(l-K_4)^2$. Inserting the parameterization for l^μ of Equation 21 into this additional propagator leads to the development of two poles, $t_\pm$, in the propagator $(l-K_4)^2 \propto (1/t)(t-t_-)(t-t_+)$. We can take advantage of the occurrence of these poles to separate the box terms from the triangle pieces.

The numerators of these box poles are given by an effective quadruple cut generated by the extra pole, along with the original triple cut. As discussed above, the quadruple cut of a box actually corresponds to one of the two contributions in the construction of a box coefficient. Therefore, each box term can be written as a sum of two residue terms, $\sum_{i=\pm} d_i/[\chi_i(t-t_i)]$, where d_i is the residue corresponding to the box coefficient of the pole t_i and χ_i is a constant factor depending on the box in question.

Analytically, to remove these box terms we simply expand the parameterized triple-cut integrand expression around $t \to \infty$. The box terms behave as $1/t \to 0$ in this limit and thus drop out. Taking a parameter to infinity numerically is problematic, so instead we use a different approach in the computer code BLACKHAT (10). Considering t as a complex parameter, the box terms appear as poles in the complex plane of t, whereas the triangle coefficient is at the origin of the t plane. To remove the box terms we simply systematically "clean" the complex plane by subtracting all box pole terms from our triple-cut expression. A final complication to both numerical and analytic approaches is the presence of the $1/t$ factor in the box propagator. To account for this, we add back the sum of all box terms evaluated at $t = 0$; that is, we add $(d_- - d_+)/[\chi_i(t_- - t_+)]$ for each box term. An alternative approach to this last step and further discussion on the analytic properties of the three-mass triangle can be found in References 15 and 57. Also, the application of on-shell recursion to the computation of certain triangle coefficients can be found in Reference 58.

After eliminating the box terms, we are left with a finite power series in t,

$$C(t) \equiv A_1(t)A_2(t)A_3(t) - \sum_{i=\pm} \frac{d_i}{\chi_i(t - t_i)} = \sum_{j=-n}^{n} c_j \int dt\, t^j, \qquad 23.$$

where the c_j are rational coefficients. The upper and lower limits $\pm n$ of the sum are determined by the theory in question; for example, $n = 3$ for renormalizable theories. To relate this sum of terms to the scalar triangle integral we first note that, for this particular parameterization, the integrals over any power of t in Equation 23 vanish, as proved in Reference 15. Then the only remaining term, $c_0 \int dt$, is in the form of a rational coefficient multiplying a scalar triangle integral. In an analytic formalism, the series expansion around $t \to \infty$ automatically isolates this term, so we can directly relate this sole remaining term to the desired coefficient of the basis triangle integral we have isolated with our triple cut. Numerically, our final step is to note that, because the power series in t terminates at a finite power, n, the full contour integral is equivalent to a discrete Fourier projection (10) with $2n + 1$ evaluation points. The triangle coefficient is therefore given by

$$c_0 = \frac{1}{2n+1} \sum_{j=-n}^{n} C\left(t_0 e^{2\pi i\, j/(2n+1)}\right). \qquad 24.$$

The arbitrary complex number t_0 is the radius of the numerical Fourier projection, as illustrated in **Figure 4**. For technical details, see Reference 10.

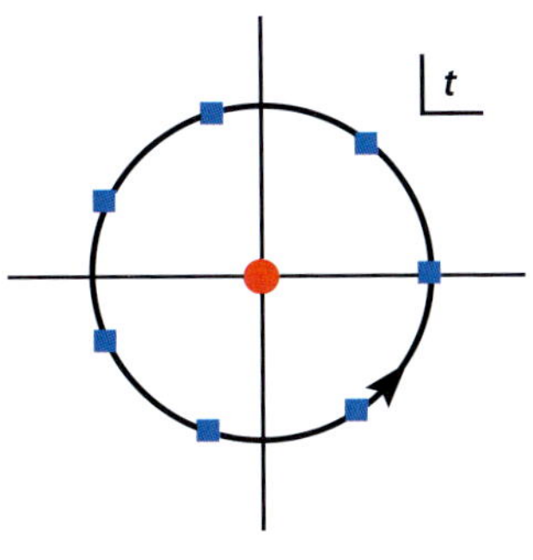

Figure 4

The points on the circle used by the discrete Fourier projection (Equation 24).

3.1.3. Bubbles. The computation of bubble coefficients proceeds along similar lines as above. A two-particle cut isolates a single bubble coefficient but also captures triangles and boxes that share the same cut. Again we use the differing analytic properties of the terms with additional propagators to separate the triangle and box terms from the bubble coefficient.

As before, we start from a specially chosen parameterization of the loop momentum. A two-particle cut leaves two free parameters, y and z. We then choose to parameterize the two-particle cut, bubble, and loop momenta as

$$l_i^{\mu}(y,z) = \frac{1}{2}K_i^{\mu} + \left(y - \frac{1}{2}\right)(\tilde{K}_1^{\mu} - \chi^{\mu}) + \frac{z}{2}\langle\tilde{K}_0^{-}|\gamma^{\mu}|\chi^{-}\rangle + \frac{y(1-y)}{2z}\langle\chi^{-}|\gamma^{\mu}|\tilde{K}_1^{-}\rangle. \qquad 25.$$

Here χ, an arbitrary massless momentum, is used to define the massless momentum $\tilde{K}_1^{\mu} = K_1^{\mu} - \chi^{\mu}$, with its normalization chosen so that $K_1 \cdot \chi = K_1^2/2$.

The triangle and box terms sharing the two-particle cut contain at least one additional propagator $1/(l - K_2)^2$. The parameterization of Equation 25 then introduces poles in y or z. The residue of the poles in terms of y is given by, effectively, a triple-cut expression, which is a combination of the pole and the original two-particle cut. Contained inside this expression are a single triangle, isolated by the effective triple cut, and possibly box terms with the same cut.

Unfortunately, the more complicated structure of the momentum parameterization and the bubble integrand means that a simple extension of the triangle procedure is not so straightforward. This is because for the parameterization of Equation 25 the integrals over positive powers of y are nonzero and are given by

$$\int dy y^n = \frac{1}{n+1}\int dy. \qquad 26.$$

The integrals over powers of z still vanish. We must therefore alter our approach for extracting the bubble coefficient. As before, we discard any terms in the expansion around $z \to \infty$ that depend on z, but we must retain the coefficients of all powers of y (which are all guaranteed to be positive due to our parameterization choice). For a renormalizable theory, the maximum power is y^2. We can relate these terms to the scalar integral using Equation 26, which is independent of y and z: $\int dydz$. The sum of the resulting terms forms only part of the bubble coefficient.

The source of the remaining contribution is the triangle expression from the residues of the poles in z. This effective triple cut does not vanish in the double expansion in y and z. Therefore, we obtain an additional contribution, which needs to be subtracted from the part of the bubble coefficient computed in the previous paragraph. Further details on this subtraction term can be found in Reference 15.

To numerically extract the bubble coefficient, we must first clean the complex plane of all pole terms. Computing the residues of each pole involves simply computing the triple cut at the location of each pole. Once all pole terms have been removed, we are free to extract the bubble coefficient from the remaining nonpole terms using a double discrete Fourier projection, in both y and z. Instead of naïvely evaluating at as many points as there are coefficients in y and z we use the nature of our parameterization to reduce the number of points at which we need to evaluate z by one. The coefficient in a renormalizable theory is given by

$$b_0 = \frac{1}{20}\sum_{j=0}^{4}\left[B\left(y=0,\ z=t_0e^{2\pi ij/5}\right) + 3B\left(y=2/3, z=t_0e^{2\pi ij/5}\right)\right], \qquad 27.$$

where $B(y, z)$ denotes the two-particle cut from which the triangles and boxes that share this cut have been subtracted. The more general case is given in Reference 10. Alternative approaches to the computation of the bubble coefficients, such as Mastrolia's (20) use of Stokes' theorem, have been proposed.

3.1.4. Massive particles. The expressions given above are for amplitudes with purely massless particles. The addition of massive particles that do not circulate in the loop is straightforwardly accommodated within the above expressions, with no changes. Including massive particles inside the circulating loop requires further exposition. Two changes are required. First, the loop-momentum parameterization needs to be extended to include massive particles. Second, we must compute the coefficients of tadpole integrals in addition to the bubbles, triangles, and boxes. A detailed discussion of the extension of the procedure to the computation of massive particles as well as of the computation of the tadpole terms themselves is given in Reference 59. There is also the problem of the wave function renormalization with massive particles, which was addressed in Reference 60.

3.2. Extraction of Integral Coefficients at the Integrand Level

So far we have described methods that use analytic limits or a combination of the subtraction of poles on the complex plane and discrete Fourier projections. An alternative approach developed by OPP uses the knowledge of the general structure of the integrand $\tilde{A}(l)$ instead. The integrand is built up from a standard set of terms. These terms either vanish upon integration or correspond to one of the scalar integral basis functions of Equation 16. Computing the scalar basis integral coefficients then reduces to the problem of finding the coefficients of the OPP decomposition of the integrand.

In four dimensions, the integrand can be written as

$$\tilde{A}_n(l) = \sum_{1\le i_1<i_2<i_3<i_4\le n} \frac{d_{i_1i_2i_3i_4}(l)}{D_{i_1}D_{i_2}D_{i_3}D_{i_4}} + \sum_{1\le i_1<i_2<i_3\le n} \frac{c_{i_1i_2i_3}(l)}{D_{i_1}D_{i_2}D_{i_3}} + \sum_{1\le i_1<i_2\le n} \frac{b_{i_1i_2}(l)}{D_{i_1}D_{i_2}} + \sum_{1\le i_1\le n} \frac{a_{i_1}(l)}{D_{i_1}}, \qquad 28.$$

with the propagator $D_i = (l - K_i)^2 - m_i^2$, where the mass of the cut propagator with momentum l_i has mass m_i. Hereafter we include massive particles in the discussion. The form of the numerators in Equation 28 depends on the basis we choose for the loop momenta. We wish to choose this momentum basis so that each scalar integral basis coefficient of B_4 corresponds to a single term in the integrand decomposition and so that the integrals over the remaining structures vanish. In the form originally presented by OPP (13), a momentum parameterization for the box, triangle, and bubble that is very similar to Equations 19, 21, and 25 was used.

An alternative momentum parameterization, presented by Ellis et al. (61), is related to the van Neerven–Vermaseren basis (52). The generic form of a momentum in this basis is

$$l^\mu = \sum_{j=1}^{D_P} v_j^\mu + \sum_{j=1}^{D_T} \alpha_j n_j^\mu. \qquad 29.$$

This is a decomposition into two sets of basis vectors. The vectors v_j^μ span the physical space defined by the external legs K_i, and the vectors n_j^μ span the space transverse to this physical space. For a box, $D_T = 1$ and $D_P = 3$; for a triangle, $D_T = 2$ and $D_P = 2$; and for a bubble, $D_T = 3$ and $D_P = 1$. The basis vectors are chosen such that $n_i \cdot n_j = \delta_{ij}$, $n_i \cdot K_j = 0$, and $n_i \cdot v_j = 0$. The v_js are chosen such that any cut legs are on shell.

The numerators of the propagator terms are then arranged in the following way (61):

$$d_{i_1i_2i_3i_4}(l) = c^0_{i_1i_2i_3i_4} + c^1_{i_1i_2i_3i_4}t_1, \tag{30.}$$

$$c_{i_1i_2i_3}(l) = c^0_{i_1i_2i_3} + c^1_{i_1i_2i_3}t_1 + c^2_{i_1i_2i_3}t_2 + c^3_{i_1i_2i_3}(t_1^2 - t_2^2) + t_1t_2(c^4_{i_1i_2i_3} + c^5_{i_1i_2i_3}t_1 + c^6_{i_1i_2i_3}t_2), \tag{31.}$$

$$b_{i_1i_2}(l) = c^0_{i_1i_2} + c^1_{i_1i_2}t_1 + c^2_{i_1i_2}t_2 + c^3_{i_1i_2}t_3 + c^4_{i_1i_2}(t_1^2 - t_3^2) + c^5_{i_1i_2}(t_2^2 - t_3^2) + c^6_{i_1i_2}t_1t_2 + c^7_{i_1i_2}t_1t_3 + c^8_{i_1i_2}t_2t_3, \tag{32.}$$

and

$$a_{i_1}(l) = c^0_{i_1} + c^1_{i_1}t_1 + c^2_{i_1}t_2 + c^3_{i_1}t_3 + c^4_{i_1}t_4 \tag{33.}$$

for the tadpole, bubble, triangle, and box, $i = 1, 2, 3$, and 4, respectively. Here the $t_j = (n_j \cdot l)$ depend on the specific parameterization of the loop momentum, which is chosen differently for boxes, triangles, bubbles, and tadpoles. The coefficients we wish to compute are the c^0 terms, which correspond to the scalar basis integral coefficients. We do not discuss the computation of the tadpole terms here; methods to compute these can be found elsewhere (13, 19, 59–61). The remaining coefficients multiply t_j terms (or combinations of such terms), which vanish upon integration over l. As in Section 3.1, the choice of the representation for the loop momentum is crucial in this approach.

Trying to solve for the entire set of coefficients at once is clearly not the optimal approach. Using unitarity cuts, we can isolate individual terms of the integrand. The application of all possible cuts allows us to systematically solve for all of the coefficients sequentially. Isolating a single box term $d_{i_1i_2i_3i_4}$ with a quadruple cut works in the same way as in Section 3.1. Again we have two solutions for the completely frozen box loop momentum. For the momentum basis corresponding to the integrand decomposition of Equation 30, these solutions are

$$l^\mu_\pm = v^\mu_1 + v^\mu_2 + v^\mu_3 \pm \sqrt{m_0^2 - (v_1 + v_2 + v_3)^2}\, n^\mu_1. \tag{34.}$$

As above, the box coefficients in the OPP basis are given by products of four tree amplitudes,

$$c^0_{i_1i_2i_3i_4} = \frac{1}{2}\sum_{a=\pm} d_a, \quad d_a = A_1(l_a)A_2(l_a)A_3(l_a)A_4(l_a),$$
$$c^1_{i_1i_2i_3i_4} = \frac{1}{2}(d_+ - d_-), \tag{35.}$$

where $c^0_{i_1i_2i_3i_4}$ is the desired scalar box integral coefficient, d_0 (Equation 20). The other coefficient is needed for the computation of the triangle coefficients.

For the computation of the triangle coefficients, we apply a triple cut to isolate the triangle of interest. Much like before, this means that we also have box terms polluting the result. To solve this problem, we subtract the complete box contribution from the integrand. To find the triangle coefficients $c^k_{i_1i_2i_3}$, we do not evaluate $A_1(l)A_2(l)A_3(l)$, but instead

$$A_1(l)A_2(l)A_3(l) - \sum_{i_4}\frac{d_{i_1i_2i_3i_4}(l)}{D_{i_4}} \tag{36.}$$

at different choices for the loop momentum of the triangle. This process contrasts with the procedure followed in the previous section. There we subtracted the residue of the box pole from the triangle rather than subtracting the entire contribution of the box (Equation 20) from the integrand as above. The triangle-momentum parameterization corresponding to the integrand structure of Equation 31 contains one free parameter. which we label α:

$$l^\mu = v_1^\mu + v_2^\mu + \sqrt{m_0^2 - \alpha^2 - (v_1 + v_2)^2}\, n_1^\mu + \alpha\, n_2^\mu. \tag{37}$$

There are seven unknown coefficients in Equation 31, so we need to evaluate this at seven, in principle arbitrary, values of α to determine all coefficients. For increased numerical stability, however, they can be chosen to lie on a circle as in the previous section. The resulting set of linear equations can then be solved to find the full set of coefficients, which is needed in the computation of the bubble coefficient.

Finally, to compute the bubbles we apply a double cut, which again isolates a single bubble but also any triangle and box coefficients that share the same cut. Again we simply subtract the unwanted terms from the integrand to remove them. To find the coefficients $b^k_{i_1 i_2}$ we evaluate

$$A_1(l)A_2(l) - \sum_{i_3} \frac{c_{i_1 i_2 i_3}(l)}{D_{i_3}} - \frac{1}{2!} \sum_{i_3 i_4} \frac{d_{i_1 i_2 i_3 i_4}(l)}{D_{i_3} D_{i_4}} \tag{38}$$

at different values of the loop momentum of the bubble. Here the bubble loop momentum corresponding to the parameterization Equation 32 is

$$l^\mu = v_1^\mu + \sqrt{m_0^2 - \alpha_1^2 - \alpha_2^2 - v_1^2}\, n_1^\mu + \alpha_1 n_2^\mu + \alpha_2 n_3^\mu, \tag{39}$$

with two free parameters, α_1 and α_2. Solving for the nine coefficients in Equation 32 requires nine linear equations. These can be generated by choosing nine different values of l via the choice of the two free parameters. The full set of nine coefficients is required only if we wish to compute the tadpole coefficients, which appear only if massive particles are present in the loop. Further details can be found in References 60 and 61.

3.3. Rational Terms from *D*-Dimensional Generalized Unitarity

We have so far restricted ourselves to working with cut legs in four dimensions, keeping the rational terms out of reach. Considering cuts in D dimensions allows us to use the B_D integral basis of Equation 16, bringing the computation of all terms within our grasp. This is related to van Neerven's important observation (62) that dispersion relations for Feynman integrals converge in dimensional regularization. There are many different approaches to go beyond four-dimensional cuts, as was proposed in Reference 22. The additional $D-4$ components can be related to massive terms in four dimensions (22). In a related approach, the $D-4$ components can be converted to an additional integral that can be used to compute the full amplitude (63). Alternatively, we can directly extend the approaches of either Section 3.1 or Section 3.2, as we discuss below, beginning with the generalization of Section 3.2.

3.3.1. *D*-dimensional unitarity at the integrand level.

OPP have suggested a two-step computational procedure. Here the $D-4$ terms in the numerator of the integrand are computed using a separate set of Feynman diagrams. The corresponding Feynman rules have been derived for QCD and electroweak theories in a series of papers (24, 64). The $D-4$ contributions from the denominator of the integrand are found by computing coefficients of an extended OPP basis structure for the integrand (13, 24).

A direct extension of the OPP approach to D dimensions was derived by Giele et al. (21). This approach combines trees in higher dimensions with an extended integrand basis for the one-loop integrand, taking into account the additional structure in higher dimensions. Again, as in

Section 3.2, the choice of the loop momentum determines the form of the integrand structures. The form of the higher-dimensional loop momenta can be decomposed into a four-dimensional part $\bar{l}$ and an orthogonal $(D-4)$-dimensional part $\tilde{l}$; $l^\mu = \bar{l}^\mu + \tilde{l}^\mu$. The on-shell constraint then means that $l^2 = \bar{l}^2 + \tilde{l}^2$, so the four-dimensional component effectively becomes massive with mass $-\tilde{l}^2 = \mu^2$, where μ^2 is the scale of the higher-dimensional subspace.

Because the external momenta remain in four dimensions, there are a limited number of additional integrand structures that can be present, because the external momenta are orthogonal to the $D-4$ additional transverse dimensions. The numerator structures can depend only on the higher-dimensional terms through even powers of μ. Similarly, there can be at most one additional cut leg because μ^2 is fixed by the fifth constraint. Therefore we must include a pentagon in the integrand basis, so for a renormalizable theory the extended integrand structures are given by

$$\tilde{A}_n(l) = \sum_{1\le i_1<i_2<i_3<i_4<i_5\le n} \frac{\tilde{e}_{i_1i_2i_3i_4i_5}(l)}{D_{i_1}D_{i_2}D_{i_3}D_{i_4}D_{i_5}} + \sum_{1\le i_1<i_2<i_3<i_4\le n} \frac{\tilde{d}_{i_1i_2i_3i_4}(l)}{D_{i_1}D_{i_2}D_{i_3}D_{i_4}} + \sum_{1\le i_1<i_2<i_3\le n} \frac{\tilde{c}_{i_1i_2i_3}(l)}{D_{i_1}D_{i_2}D_{i_3}} + \sum_{1\le i_1<i_2\le n} \frac{\tilde{b}_{i_1i_2}(l)}{D_{i_1}D_{i_2}} + \sum_{1\le i_1\le n} \frac{\tilde{a}_{i_1}(l)}{D_{i_1}}. \quad 40.$$

Here the numerator coefficients are given by

$$\begin{aligned}
\tilde{e}_{i_1i_2i_3i_4i_5}(l) &= c^0_{i_1i_2i_3i_4i_5},\\
\tilde{d}_{i_1i_2i_3i_4}(l) &= d_{i_1i_2i_3i_4}(\bar{l}) + \mu^2(c^2_{i_1i_2i_3i_4} + t_1c^3_{i_1i_2i_3i_4}) + \mu^4c^4_{i_1i_2i_3i_4},\\
\tilde{c}_{i_1i_2i_3}(l) &= c_{i_1i_2i_3}(\bar{l}) + \mu^2\left(t_1c^8_{i_1i_2i_3} + t_2c^9_{i_1i_2i_3} + c^{10}_{i_1i_2i_3}\right),\\
\tilde{b}_{i_1i_2}(l) &= b_{i_1i_2}(\bar{l}) + \mu^2c^{10}_{i_1i_2},\\
\tilde{a}_{i_1}(l) &= a_{i_1}(\bar{l}),
\end{aligned} \quad 41.$$

where, again, $t_j = n_j \cdot l$.

The attentive reader may wonder how this basis can be used if we have a fractional number of additional dimensions, for example when $D = 4 - 2\varepsilon$. This problem can be sidestepped by noting that the dependence of the amplitudes on the $(D-4)$-dimensional subspace is linear. The fractional dimensional structure of the amplitude can be reconstructed by evaluating the amplitude at two different integer dimensions. By using $A_D^{\text{1-loop}} = A_0^{\text{1-loop}} + (D-4)A_1^{\text{1-loop}}$ the full D-dimensional amplitude can then be reconstructed (21), which allows us to freely choose the exact form of the regularization scheme. The integer dimension chosen when evaluating the trees that enter the calculations is arbitrary up to the constraint that the dimension has to be an even number if fermions are present in the loop.

The final step is to relate the new integrand structures of Equation 41, proportional to powers of μ, to their contribution in Equation 16. As before, terms proportional to t_j terms vanish, leaving us with terms proportional only to μ. Scalar integrals multiplied by powers of μ can be related to scalar integrals in higher dimensions. From Equation 41 we have the following different integrals:

$$\begin{aligned}
\int d^Dl \frac{\mu^2}{D_{i_1}D_{i_2}D_{i_3}D_{i_4}} &= -\frac{D-4}{2} I^{D+2}_{i_1i_2i_3i_4} \overset{D\to 4}{\to} 0\\
\int d^Dl \frac{\mu^4}{D_{i_1}D_{i_2}D_{i_3}D_{i_4}} &= \frac{(D-4)(D-2)}{4} I^{D+4}_{i_1i_2i_3i_4} \overset{D\to 4}{\to} -\frac{1}{6}\\
\int d^Dl \frac{\mu^2}{D_{i_1}D_{i_2}D_{i_3}} &= -\frac{(D-4)}{2} I^{D+2}_{i_1i_2i_3} \overset{D\to 4}{\to} -\frac{1}{2}\\
\int d^Dl \frac{\mu^2}{D_{i_1}D_{i_2}} &= -\frac{(D-4)}{2} I^{D+2}_{i_1i_2} \overset{D\to 4}{\to} \frac{m^2_{i_1}+m^2_{i_2}}{2} - \frac{1}{6}S_{i_1i_2}.
\end{aligned} \quad 42.$$

Here $S_{i_1 i_2}$ is the mass of the bubble labeled by i_1 and i_2. Examining the limits of these integrals as we return to four dimensions, we observe that the integrals are finite and correspond to purely rational numbers. Each new integrand factor therefore contributes to the rational term.

Computing the one-loop amplitude is now very similar to the cut-computation procedure described in Section 3.2. Starting with the pentagon, we find that the loop momenta are completely frozen by the four cuts on the four-dimensional component of l_5^μ and the constraint on $\mu^2 = -\tilde{l}^2$ from the fifth propagator $(\bar{l} + \tilde{l} - K_5)^2 = m_5^2$. The resulting expression is then inserted into the quintuple cut so that we find the pentagon coefficient using

$$c^0_{i_1 i_2 i_3 i_4 i_5} = A_1(l)A_2(l)A_3(l)A_4(l)A_5(l). \qquad 43.$$

The trees must be evaluated, as described above, at two different integer dimension choices for the internal loop momentum. The pentagon coefficient is proportional to $D - 4$, so it vanishes in the four-dimensional limit. We need it only for subtraction when computing the box terms, as we now explain.

The box is computed as above but with the pentagon subtracted:

$$A_1(l)A_2(l)A_3(l)A_4(l) - \sum_{i_5} \frac{e_{i_1 i_2 i_3 i_4 i_5}(l)}{D_{i_5}}. \qquad 44.$$

With three additional coefficients to be determined, compared with the four-dimensional case, we need three additional evaluation points in order to generate enough equations to solve for all coefficients. All coefficients are required for the computation of the triangle terms. Computing the triangle and bubble contributions to the rational terms follows in a similar manner as for the boxes; we direct the reader to Reference 21 for further details on the computation of all such contributions.

3.3.2. Four-dimensional generalized unitarity with a $(D-4)$-dimensional mass. The differing dependence of the rational terms of Equation 41 on μ^2 suggests that we could use the analytic structure of the integrand to compute the rational terms at the integral level. This approach of Badger (23) is similar in spirit to the extraction of the cut coefficients described in Section 3.1. Here the loop momentum is massive and four dimensional rather than massless and D dimensional.

To compute the rational contribution of the box, we begin from the quadruple-cut expression of a box. Now each cut leg has an additional mass μ^2. This cut isolates a single box coefficient as well as any pentagon terms that share the same cut. Schematically we have

$$A_1(l(\mu^2))A_2(l(\mu^2))A_3(l(\mu^2))A_4(l(\mu^2)) = r_1 + r_2\mu^2 + r_3\mu^4 + \sum_i \frac{e_i^0}{\chi_i(\mu^2 - \mu_i^2)}, \qquad 45.$$

where μ_i^2 is the pole in μ^2 for the pentagon i and χ_i is a constant factor depending on the pentagon in question. The additional propagator of the pentagon terms shows up as a $1/(\mu^2 - \mu_i^2)$ factor. We wish to separate these terms, as well as the other terms in the powers series in μ^2, from each other. Here we want only the coefficient r_3 of μ^4, which is the term that contributes to the purely rational part (cf. Equation 42). Simply expanding $A_1(l(\mu^2))A_2(l(\mu^2))A_3(l(\mu^2))A_4(l(\mu^2))/\mu^4$ around the limit $\mu^2 \to \infty$ gives us the coefficient directly.

Performing a triple cut isolates a single triangle as well as boxes and pentagons. These objects contain poles in both μ^2 and the unconstrained component of the loop momenta, t. We are interested only in the coefficient of μ^2, so series-expanding around both $t \to \infty$ and $\mu^2 \to \infty$ isolates this single term. For the bubbles, a similar procedure applies in three parameters; a complete description is given in Reference 23.

As in Section 3.1, taking an infinite limit numerically is difficult to do. Similar to the numerical approach to the computation of the cut-containing terms described in Section 3.1, we can adapt Badger's method, which is the approach implemented within BLACKHAT (10). Starting from Equation 45 we clean the complex μ plane by subtracting all pentagons at their poles. We then perform a discrete Fourier projection in μ to extract the rational contribution r_3. For the rational contribution of a triangle, we consider the triple-cut expression and then clean the complex μ plane by subtracting all terms with an additional propagator. Each such term has a pole in t, and its numerator is a quadruple cut. Here we subtract the quadruple-cut residue of the additional propagator and not the integrand box and pentagon terms, as we would in the OPP approaches. We therefore prevent any loss of numerical stability that would arise from cancellations between these box and pentagon terms because these pieces are never separated out in the quadruple cut. The computation of the rational contribution from the two-particle cut of a bubble follows along a similar line.

4. ON-SHELL RECURSION AT ONE LOOP

As above, an alternative approach to obtain the full amplitude is to use four-dimensional unitarity methods for the cut-constructible terms and to use recursion relations for the purely rational remainder. The recursive approach previously developed (8, 9) has already been reviewed extensively (3). Although this approach is very useful for analytical calculations and even allows one to obtain closed-form all-multiplicity results (65), it is not easy to cast into a form suitable for numerical implementation. This difficulty arises from the removal of spurious singularities via the introduction of overlap terms, which is difficult to perform in an automated fashion. The approach has subsequently been modified to allow efficient numerical implementation into the automated C^{++} library BLACKHAT (10). Here we discuss this modified approach for amplitudes with massless particles in the loop.

We begin by dividing the amplitude into cut-constructible and rational terms, as in Equation 16. The rational terms are defined by setting all scalar integrals to zero:

$$A_n^{1\text{-loop}} = \sum_{j \in B_4} c_j^{D=4} \mathcal{I}_j^D + \mathcal{R}_n, \qquad 46.$$

$$C_n \equiv \sum_{j \in B_4} c_j^{D=4} \mathcal{I}_j^D, \quad \mathcal{R}_n \equiv A_n^{1\text{-loop}}\big|_{\mathcal{I}_j^D \to 0}. \qquad 47.$$

Below, we assume that the cut-containing terms have been computed via the generalized unitarity methods described in the previous section.

The basic idea is to complex-continue $\mathcal{R}$ and use Cauchy's theorem to reconstruct the rational term from its poles in the complex plane, similarly to the tree-level approach introduced in Section 2.2. We add this rational term to the previously computed cut terms C to obtain the full physical amplitude at $z = 0$. However, two of the basic premises of the derivation in Section 2.2 do not generally hold at one loop for two reasons: (*a*) We cannot always find shifts (10) such that the amplitude vanishes as $z \to \infty$, and (*b*) the division of the amplitude into cut and rational parts introduces the presence of spurious, unphysical poles in the complex plane when considering the rational part separately. That is, the rational part has the following form upon complex continuation:

$$\mathcal{R}_n(z) = \mathcal{R}_n^P(z) + \mathcal{R}_n^S(z) + \mathcal{R}_n^{\text{large}\,z}. \qquad 48.$$

Here $\mathcal{R}_n^P$ denotes the contribution from physical poles, which are as at the tree-level simple poles, whereas the spurious poles can be simple or double poles. The contributions $\mathcal{R}_n^{\text{large}\,z}$ display

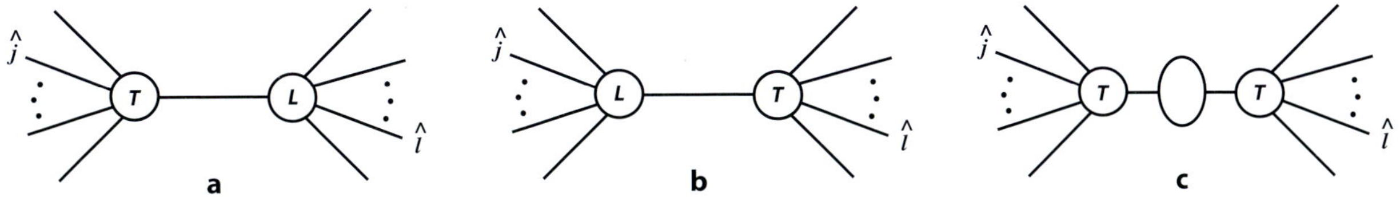

Figure 5

Schematic representation of one-loop recursive contributions to Equation 50. T and L refer to the tree and the rational part of loop vertices, respectively.

polynomial growth in z:

$$\mathcal{R}_n^P(z) = \sum_\alpha \frac{A_\alpha}{z - z_\alpha},$$
$$\mathcal{R}_n^S(z) = \sum_\beta \left(\frac{B_\beta}{(z - z_\beta)^2} + \frac{C_\beta}{z - z_\beta} \right),$$
$$\mathcal{R}_n^{\text{large } z} = \sum_{\sigma=0}^{\sigma_{\max}} D_\sigma z^\sigma . \qquad 49.$$

Here, A, B, C, and D are functions of the external momenta. We postpone discussion of the case in which $\mathcal{R}_n^P(z)$ has a more complicated structure to Section 4.2.

The physical contribution is recursively computed as in the tree-level case:

$$-\sum_{\text{poles}\,\alpha} \operatorname*{Res}_{z=z_\alpha} \frac{\mathcal{R}_n^P(z)}{z} = \sum_{r,s} \sum_h \left\{ A_L^h(z = z_{rs}) \frac{i}{K_{r\ldots s}^2} \mathcal{R}_R^{-h}(z = z_{rs}) + \mathcal{R}_L^h(z = z_{rs}) \frac{i}{K_{r\ldots s}^2} A_R^{-h}(z = z_{rs}) \right.$$
$$\left. + A_L^h(z = z_{rs}) \frac{i\mathcal{F}(K_{r\ldots s}^2)}{K_{r\ldots s}^2} A_R^{-h}(z = z_{rs}) \right\} . \qquad 50.$$

Here, we express the rational term as a sum of products of rational terms from lower-point amplitudes (defined according to Equation 47) with lower-point tree amplitudes. The last term with $\mathcal{F}$ corresponds to a one-loop correction to the propagator. Equation 50 is illustrated in **Figure 5**.

The full amplitude is found by combining all rational and cut contributions:

$$A_n^{1\text{-loop}}(0) = C_n(0) - \sum_{\text{poles}\,\alpha} \operatorname*{Res}_{z=z_\alpha} \frac{\mathcal{R}_n^P(z)}{z} + \mathcal{R}_n^S(0) + \mathcal{R}_n^{\text{large } z} . \qquad 51.$$

In the next section, we show how to obtain the last two contributions, starting with the spurious pole contribution.

4.1. Spurious Poles

The division into cut and rational parts introduces spurious singularities in each of these terms that, however, cancel in the full amplitude. These spurious singularities are already present in real kinematics. Cauchy's theorem requires us to sum over all poles, whether physical or unphysical. In Reference 9 this problem was remedied by adding additional rational terms to the cut part. These rational terms are constructed such that the cut and the rational parts individually do not contain spurious singularities upon continuation into the complex plane. However, these extra rational terms then contribute additional terms to the residues at the physical poles in Equation 50, which have to be subtracted from the rational part $\mathcal{R}$ in the recursive construction by so-called overlap terms in order to avoid double counting (9). This approach leads to compact expressions in analytical calculations, but it is not particularly amenable to numerical implementation.

An alternative way of dealing with the spurious singularities is to use the fact that we know that they cancel in the full amplitude. In other words, we can extract the spurious residues from the known cut parts,

$$\mathcal{R}_n^S(0) = -\sum_{\text{spur poles } \beta} \operatorname*{Res}_{z=z_\beta} \frac{\mathcal{R}_n^S(z)}{z} = +\sum_{\text{spur poles } \beta} \operatorname*{Res}_{z=z_\beta} \frac{C_n(z)}{z}, \tag{52.}$$

where $C_n(z)$ is the complex-continued cut part. The spurious poles in $C_n(z)$ come from the vanishing of complex-continued Gram determinants, $\Delta(z)$, associated with bubble, triangle, and box integrals. Because the spurious poles cancel between the rational and the cut parts, the spurious contribution to the residues from the cut part can only be rational. To compute the residue, we therefore series-expand the logarithms and polylogarithms around the location of the vanishing Gram determinants and obtain a series of rational functions. The spurious contribution is thus given by evaluating

$$\mathcal{R}_n^S(0) = \sum_{\Delta_m(z_\beta)=0} \operatorname*{Res}_{z=z_\beta} \left[\sum_j \frac{c_j^{D=4}(z)\, \mathcal{I}_j^D(z)\big|_{\text{rat}}}{z} \right] \equiv \sum_{\Delta_m(z_\beta)=0} \operatorname*{Res}_{z=z_\beta} \frac{E_n^\beta(z)}{z}, \tag{53.}$$

where the subscript rat indicates that we take the rational part of the series expansion of the integrals around the spurious poles. We have introduced the abbreviation E_n^β for these rational terms. The spurious poles z_β are located where the shifted Gram determinants vanish: $\Delta_m(z_\beta) = 0$, where $m = 2$, 3, and 4 for bubble, triangle, and box integrals, respectively. Note that poles in Gram determinants of box integrals generally also appear in the daughter triangle and bubble integrals.[4] Singularities in triangle integrals feed down into the daughter bubble coefficients, but they do not affect the parent box integrals.

The expansion of an integral around the location where its Gram determinant vanishes can be obtained (66) by using the dimension-shifting relation (53) iteratively,

$$\mathcal{I}_j^D = \frac{1}{2}\left(\sum_{i=1}^{j} c_i \mathcal{I}_{j-1}^D[i] + (j-1-D) c_0 \mathcal{I}_j^{D+2} \right), \tag{54.}$$

where $\mathcal{I}_{j-1}^D[i]$ denotes the lower-point integral obtained from $\mathcal{I}_j^D$ by removing the ith propagator. The coefficients c_i and c_0 are given by

$$c_i = \sum_{k=1}^{j} \left(Y^{-1}\right)_{ik}, \quad c_0 = \sum_{k=1}^{j} c_i \sim \Delta_j, \tag{55.}$$

where Y^{-1} denotes the inverse of the modified Cayley matrix. These modified Cayley matrices are listed in, for example, Reference 54. As indicated, the coefficient c_0 is proportional to the Gram determinant. We obtain the series expansion of $\mathcal{I}_j^D$ in terms of its Gram determinant,

$$\mathcal{I}_j^D\big|_{\text{rat}} = \sum_{k=0} r_k \Delta_j^k, \tag{56.}$$

with rational coefficients r_k that are found by using Equation 54 iteratively. The explicit expression for the rational expansion of the three-mass triangle is listed in Reference 10. Similar expansions

[4]By daughter integrals we mean integrals that can be found from the parent integral by collapsing one or more of the loop propagators.

for the remaining integrals can be obtained as described above and are listed in a forthcoming publication (C.F. Berger, Z. Bern, L.J. Dixon, F. Febres Cordero, D. Forde, H. Ita, D.A. Kosower & D. Maître, in preparation).

Numerically, we can evaluate Equation 53 by using a discrete Fourier sum, which here approximates only the contour integral. We evaluate the quantity given in the square bracket of Equation 53, $E_n^\beta(z)/z$, at m points equally spaced around a circle of radius δ_β in the z plane, centered on the pole location z_β:

$$\mathcal{R}_n^S(0) \approx \frac{1}{m} \sum_\beta \sum_{j=1}^{m} \delta_\beta e^{2i\pi j/m} \frac{E_n^\beta(z_\beta + \delta_\beta e^{2i\pi j/m})}{z_\beta + \delta_\beta e^{2i\pi j/m}}. \qquad 57.$$

The sum over β runs over all locations of spurious poles where Gram determinants vanish. For technical details of the numerical implementation, we refer the reader to Reference 10.

4.2. Contribution from Infinity

The remaining rational contribution $\mathcal{R}_n^{\text{large } z}$ is the boundary contribution in the contour integral as $z \to \infty$. Although it is generically possible to find complex continuations that have vanishing boundary contributions, these shifts usually have additional contributions that cannot be constructed recursively as in Equation 50. In other words, we have instead

$$\mathcal{R}_n^P(0) = \mathcal{R}_n^{P\,\text{recursive}} + \mathcal{R}_n^{P\,\text{nonstd}}, \quad \mathcal{R}_n^{P\,\text{recursive}} \equiv -\sum_{\text{poles }\beta} \operatorname*{Res}_{z=z_\beta} \frac{\mathcal{R}_n^P(z)}{z}. \qquad 58.$$

These nonstandard contributions (labeled by the superscript nonstd) arise in configurations in which two external momenta with the same helicity are on one side of the partition and all other legs are on the other side. The complex-factorization properties of these configurations are not yet fully understood. The sum over poles β in Equation 58 is only over the channels that factorize, namely those that do not display the aforementioned problematic behavior. Conversely, it is possible to find shifts that avoid these nonstandard channels, but only at the price of reintroducing a boundary contribution.

The solution to this problem (9) is to use two independent complex continuations to determine the boundary contribution. Let us denote the primary shift by $[j,l\rangle$ and the auxiliary shift by $[a,b\rangle$ in the notation of Equation 10. We then have two relations for the same amplitude, analogous to Equation 51,

$$A_n^{1\text{-loop}}(0) = \operatorname*{Inf}_{[j,l\rangle} A_n^{1\text{-loop}} + C_n(0) - \operatorname*{Inf}_{[j,l\rangle} C_n + \mathcal{R}_n^{P\,\text{recursive},[j,l\rangle} + \mathcal{R}_n^{S[j,l\rangle}(0) \qquad 59.$$

and

$$A_n^{1\text{-loop}}(0) = C_n(0) - \operatorname*{Inf}_{[a,b\rangle} C_n + \mathcal{R}_n^{P\,\text{recursive},[a,b\rangle} + \mathcal{R}_n^{P\,\text{nonstd},[a,b\rangle} + \mathcal{R}_n^{S[a,b\rangle}(0). \qquad 60.$$

We have indicated with additional superscripts which shift has been employed. $\mathcal{R}_n^S$ is evaluated according to Equation 53. We have also used

$$\mathcal{R}_n^{\text{large } z} = \operatorname*{Inf}_{[j,l\rangle} A_n^{1\text{-loop}} - \operatorname*{Inf}_{[j,l\rangle} C_n, \qquad 61.$$

where $\operatorname*{Inf}_{[j,l\rangle} A$ is the unknown large-z behavior of the full amplitude found from a Laurent expansion of $A_n^{1\text{-loop}}(z)$ around $z = \infty$ with the shift $[j,l\rangle$, and similarly for $C_n(z)$. Thus, both Equations 59 and 60 contain unknown terms. If we now apply the primary shift $[j,l\rangle$ to the auxiliary recursion

(Equation 60) and take the limit $z \to \infty$, we can extract the large-z behavior of the primary shift,

$$\operatorname*{Inf}_{[j,l\rangle} A_n^{1\text{-loop}} = \operatorname*{Inf}_{[j,l\rangle} C_n - \operatorname*{Inf}_{[j,l\rangle} \left(\operatorname*{Inf}_{[a,b\rangle} C_n \right) + \operatorname*{Inf}_{[j,l\rangle} \left(\mathcal{R}_n^{P\,\text{recursive}\,[a,b\rangle} \right) + \operatorname*{Inf}_{[j,l\rangle} \left(\mathcal{R}_n^{S\,[a,b\rangle}(0_{[a,b\rangle}) \right), \qquad 62.$$

where all terms on the right-hand side are now either known or recursively constructible if

$$\operatorname*{Inf}_{[j,l\rangle} \left(\mathcal{R}_n^{P\,\text{nonstd},[a,b\rangle} \right) = 0. \qquad 63.$$

Putting everything together, we find the full amplitude from Equation 59, along with Equations 50, 53, and 62, and the cut terms are constructed as described in the previous section.

5. CONCLUSIONS AND OUTLOOK

In this review, we have presented an overview of recent developments in the calculation of multiparton scattering amplitudes at the one-loop level. These developments are based on on-shell techniques that make efficient use of the physical properties of the hard scattering, such as unitarity and factorization. The basic ingredients in these new approaches are on-shell tree-level or lower-point one-loop amplitudes instead of Feynman diagrams, which allow us to sidestep many of the complications associated with the use of Feynman diagrams.

Furthermore, these new techniques allow for efficient algorithmic implementation and hence the construction of efficient, numerically stable, and fast computer codes such as BLACKHAT (10), CUTTOOLS/ONELOOP (25, 26), ROCKET (27), and others (28). With these new techniques and computer tools, a flurry of results relevant for the LHC have recently been computed (31–36), and we expect further rapid progress in the near future.

Nevertheless, much work remains to be done to bring one-loop calculations to the same level of automatization as tree-level computations. Such work should ideally begin with a Lagrangian and produce complete events including parton-shower and hadronization corrections. Further open issues include, for example, a better understanding of the complex-factorization properties of one-loop amplitudes and the generalization of the new techniques to higher loops in nonsupersymmetric theories. At two loops, such a generalization must necessarily begin with a complete set of all possible basis integrals that can appear. Such a complete set of two-loop master integrals is currently not yet available for a general gauge theory.

The past few years have seen an unprecedented progress in the development of techniques for the computation of multiparton one-loop scattering amplitudes, which are an essential ingredient in precision calculations for the LHC. These new methods have also been used to study the higher-loop structure of $\mathcal{N} = 4$ supersymmetric Yang-Mills theory and $\mathcal{N} = 8$ supergravity (67). Their basic ingredients are unitarity, factorization, and complex analysis—properties that are quite generic. It is thus not inconceivable that these new techniques will find further application beyond those presented or referenced in this review.

DISCLOSURE STATEMENT

The authors are not aware of any affiliations, memberships, funding, or financial holdings that might be perceived as affecting the objectivity of this review.

ACKNOWLEDGMENTS

We thank Zvi Bern, Lance Dixon, Fernando Febres Cordero, Tanju Gleisberg, Harald Ita, David Kosower, and Daniel Maître for fruitful collaboration. We also thank Zvi Bern and Lance Dixon for valuable comments. C.F.B. thanks the Aspen Center for Physics for hospitality. This work is supported in part by funds provided by the U.S. Department of Energy under cooperative research agreement DE-FC02-94ER40818.

LITERATURE CITED

1. Bern Z, et al. (NLO Multileg Work. Group) arXiv:0803.0494 [hep-ph] (2008)
2. Reiter T. arXiv:0903.0947 [hep-ph] (2009); Bredenstein A, Denner A, Dittmaier S, Pozzorini S. *Phys. Rev. Lett.* 103:012002 (2009); Binoth T, et al. arXiv:0910.4379 [hep-ph] (2009)
3. Bern Z, Dixon LJ, Kosower DA. *Ann. Phys.* 322:1587 (2007)
4. Bern Z, Dixon LJ, Dunbar DC, Kosower DA. *Nucl. Phys. B* 425:217 (1994); Bern Z, Dixon LJ, Dunbar DC, Kosower DA. *Nucl. Phys. B* 435:59 (1995)
5. Bern Z, Dixon LJ, Kosower DA. *Nucl. Phys. B* 513:3 (1998)
6. Britto R, Cachazo F, Feng B. *Nucl. Phys. B* 715:499 (2005); Britto R, Cachazo F, Feng B, Witten E. *Phys. Rev. Lett.* 94:181602 (2005)
7. Berends FA, Giele WT. *Nucl. Phys. B* 306:759 (1988)
8. Bern Z, Dixon LJ, Kosower DA. *Phys. Rev. D* 73:065013 (2006)
9. Berger CF, et al. *Phys. Rev. D* 74:036009 (2006)
10. Berger CF, et al. *Phys. Rev. D* 78:036003 (2008)
11. Brandhuber A, Spence BJ, Travaglini G. *Nucl. Phys. B* 706:150 (2005); Bedford J, Brandhuber A, Spence BJ, Travaglini G. *Nucl. Phys. B* 706:100 (2005); Bedford J, Brandhuber A, Spence BJ, Travaglini G. *Nucl. Phys. B* 712:59 (2005); Brandhuber A, Spence BJ, Travaglini G. *J. High Energy Phys.* 0702:088 (2007)
12. Britto R, Cachazo F, Feng B. *Nucl. Phys. B* 725:275 (2005)
13. Ossola G, Papadopoulos CG, Pittau R. *Nucl. Phys. B* 763:147 (2007)
14. Mastrolia P, Ossola G, Papadopoulos CG, Pittau R. *J. High Energy Phys.* 0806:030 (2008)
15. Forde D. *Phys. Rev. D* 75:125019 (2007)
16. Britto R, Cachazo F, Feng B. *Phys. Rev. D* 71:025012 (2005); Britto R, Buchbinder E, Cachazo F, Feng B. *Phys. Rev. D* 72:065012 (2005)
17. Britto R, Feng B, Mastrolia P. *Phys. Rev. D* 78:025031 (2008)
18. Britto R, Feng B. *J. High Energy Phys.* 0802:095 (2008); Britto R, Feng B, Yang G. *J. High Energy Phys.* 0809:089 (2008)
19. Britto R, Feng B. *Phys. Lett.* B681:376 (2009)
20. Mastrolia P. *Phys. Lett.* B678:246 (2009); Mastrolia P. *Lett. Math. Phys.* 91:199 (2010)
21. Giele WT, Kunszt Z, Melnikov K. *J. High Energy Phys.* 0804:049 (2008)
22. Bern Z, Morgan AG. *Nucl. Phys. B* 467:479 (1996)
23. Badger SD. *J. High Energy Phys.* 0901:049 (2009)
24. Ossola G, Papadopoulos CG, Pittau R. *J. High Energy Phys.* 0805:004 (2008); Draggiotis P, Garzelli MV, Papadopoulos CG, Pittau R. *J. High Energy Phys.* 0904:072 (2009)
25. Ossola G, Papadopoulos CG, Pittau R. *J. High Energy Phys.* 0803:042 (2008)
26. van Hameren A, Papadopoulos CG, Pittau R. *J. High Energy Phys.* 0909:106 (2009)
27. Ellis RK, et al. *J. High Energy Phys.* 0901:012 (2009)
28. Lazopoulos A. arXiv:0812.2998 [hep-ph] (2008); Winter JC, Giele WT. arXiv:0902.0094 [hep-ph] (2009); Kilian W, Kleinschmidt T. arXiv:0912.3495 [hep-ph] (2009)
29. Gleisberg T, Krauss F. *Eur. Phys. J. C* 53:501 (2008); Gleisberg T, et al. *J. High Energy Phys.* 0902:007 (2009)
30. Frederix R, Frixione S, Maltoni F, Stelzer T. *J. High Energy Phys.* 0910:003 (2009)
31. Berger CF, et al. *Phys. Rev. Lett.* 102:222001 (2009); Berger CF, et al. *Phys. Rev. D* 80:074036 (2009)
32. Ellis RK, Melnikov K, Zanderighi G. *Phys. Rev. D* 80:094002 (2009); Melnikov K, Zanderighi G. arXiv:0910.3671 [hep-ph] (2009)
33. Melnikov K, Schulze M. *J. High Energy Phys.* 0908:049 (2009)
34. Ossola G, Papadopoulos CG, Pittau R. *J. High Energy Phys.* 0707:085 (2007)
35. Binoth T, Ossola G, Papadopoulos CG, Pittau R. *J. High Energy Phys.* 0806:082 (2008)

36. Bevilacqua G, et al. *J. High Energy Phys.* 0909:109 (2009)
37. Mangano ML, Parke SJ. *Phys. Rep.* 200:301 (1991)
38. Dixon LJ. arXiv:hep-ph/9601359 (1996)
39. Bern Z, Dixon LJ, Kosower DA. *Annu. Rev. Nucl. Part. Sci.* 46:109 (1996)
40. Paton JE, Chan HM. *Nucl. Phys. B* 10:516 (1969); Cvitanovic P, Lauwers PG, Scharbach PN. *Nucl. Phys. B* 186:165 (1981); Berends FA, Giele W. *Nucl. Phys. B* 294:700 (1987); Mangano ML. *Nucl. Phys. B* 309:461 (1988); Zeppenfeld D. *Int. J. Mod. Phys. A* 3:2175 (1988); Bern Z, Kosower DA. *Nucl. Phys. B* 362:389 (1991); Del Duca V, Frizzo A, Maltoni F. *Nucl. Phys. B* 568:211 (2000); Del Duca V, Dixon LJ, Maltoni F. *Nucl. Phys. B* 571:51 (2000); Maltoni F, Paul K, Stelzer T, Willenbrock S. *Phys. Rev. D* 67:014026 (2003)
41. Kanaki A, Papadopoulos CG. *Comput. Phys. Commun.* 132:306 (2000)
42. Giele W, Kunszt Z, Winter J. arXiv:0911.1962 [hep-ph] (2009)
43. Kleiss R, Stirling WJ. *Nucl. Phys. B* 262:235 (1985)
44. Mangano ML, et al. *J. High Energy Phys.* 0307:001 (2003)
45. Gleisberg T, Höche S. *J. High Energy Phys.* 0812:039 (2008)
46. Schwinn C, Weinzierl S. *J. High Energy Phys.* 0704:072 (2007); Badger SD, Glover EWN, Khoze VV, Svrček P. *J. High Energy Phys.* 0507:025 (2005); Badger SD, Glover EWN, Khoze VV. *J. High Energy Phys.* 0601:066 (2006)
47. Bedford J, Brandhuber A, Spence BJ, Travaglini G. *Nucl. Phys. B* 721:98 (2005); Cachazo F, Svrček P. arXiv:hep-th/0502160 (2005); Bjerrum-Bohr NEJ, et al. *J. High Energy Phys.* 0601:009 (2006); Benincasa P, Boucher-Veronneau C, Cachazo F. *J. High Energy Phys.* 0711:057 (2007); Brandhuber A, McNamara S, Spence B, Travaglini G. *J. High Energy Phys.* 0703:029 (2007); Elvang H, Freedman DZ, Kiermaier M. *J. High Energy Phys.* 0904:009 (2009); Elvang H, Freedman DZ, Kiermaier M. *J. High Energy Phys.* 0906:068 (2009)
48. Britto R, et al. *Phys. Rev. D* 71:105017 (2005); Forde D, Kosower DA. *Phys. Rev. D* 73:065007 (2006); Drummond JM, Henn JM. *J. High Energy Phys.* 0904:018 (2009)
49. Draggiotis PD, Kleiss RHP, Lazopoulos A, Papadopoulos CG. *Eur. Phys. J. C* 46:741 (2006); Vaman D, Yao YP. *J. High Energy Phys.* 0604:030 (2006); Arkani-Hamed N, Kaplan J. *J. High Energy Phys.* 0804:076 (2008); Cheung C. *J. High Energy Phys.* 1003:098 (2010); Feng B, Wang J, Wang Y, Zhang Z. *J. High Energy Phys.* 1001:019 (2010)
50. Mason L, Skinner D. arXiv:0903.2083 [hep-th] (2009); Hodges A. arXiv:0905.1473 [hep-th] (2009); Arkani-Hamed N, Cachazo F, Cheung C, Kaplan J. *J. High Energy Phys.* 1003:020 (2010); Arkani-Hamed N, Cachazo F, Cheung C. arXiv:0909.0483 [hep-th] (2009); Mason L, Skinner D. *J. High Energy Phys.* 0911:045 (2009); Bullimore M, Mason L, Skinner D. *J. High Energy Phys.* 1003:070 (2010); Kaplan J. *J. High Energy Phys.* 1003:025 (2010); Arkani-Hamed N, Bourjaily J, Cachazo F, Trnka J. arXiv:0912.3249 [hep-th] (2009)
51. Brown LM, Feynman RP. *Phys. Rev.* 85:231 (1952); Melrose DB. *Nuovo Cim.* 40:181 (1965); Passarino G, Veltman MJG. *Nucl. Phys. B* 160:151 (1979); Stuart RG. *Comput. Phys. Commun.* 48:367 (1988); van Oldenborgh GJ, Vermaseren JAM. *Z. Phys. C* 46:425 (1990); Pittau R. *Comput. Phys. Commun.* 104:23 (1997); Pittau R. *Comput. Phys. Commun.* 111:48 (1998); Weinzierl S. *Phys. Lett.* B450:234 (1999)
52. van Neerven WL, Vermaseren JAM. *Phys. Lett.* B137:241 (1984)
53. Bern Z, Dixon LJ, Kosower DA. *Phys. Lett.* B302:299 (1993); Bern Z, Dixon LJ, Kosower DA. Erratum. *Phys. Lett.* B318:649 (1993); Bern Z, Dixon LJ, Kosower DA. *Nucl. Phys. B* 412:751 (1994)
54. Ellis RK, Zanderighi G. *J. High Energy Phys.* 0802:002 (2008)
55. Risager K. arXiv:0804.3310 [hep-th] (2008)
56. Cachazo F. arXiv:0803.1988 [hep-th] (2008)
57. Bjerrum-Bohr NEJ, Dunbar DC, Perkins WB. *J. High Energy Phys.* 0804:038 (2008)
58. Bern Z, Bjerrum-Bohr NEJ, Dunbar DC, Ita H. *Nucl. Phys. Proc. Suppl.* 157:120 (2006)
59. Kilgore WB. arXiv:0711.5015 [hep-ph] (2007)
60. Ellis RK, Giele WT, Kunszt Z, Melnikov K. *Nucl. Phys. B* 822:270 (2009)
61. Ellis RK, Giele WT, Kunszt Z. *J. High Energy Phys.* 0803:003 (2008)
62. van Neerven WL. *Nucl. Phys. B* 268:453 (1986)
63. Anastasiou C, et al. *J. High Energy Phys.* 0703:111 (2007)

64. Garzelli MV, Malamos I, Pittau R. *J. High Energy Phys.* 1001:040 (2010)
65. Forde D, Kosower DA. *Phys. Rev. D* 73:061701 (2006); Berger CF, et al. *Phys. Rev. D* 75:016006 (2007); Berger CF, Del Duca V, Dixon LJ. *Phys. Rev. D* 74:094021 (2006); Berger CF, Del Duca V, Dixon LJ. Erratum. *Phys. Rev. D* 76:099901 (2007); Badger SD, Glover EWN, Risager K. *J. High Energy Phys.* 0707:066 (2007)
66. Campbell JM, Glover EWN, Miller DJ. *Nucl. Phys. B* 498:397 (1997); Giele WT, Glover EWN. *J. High Energy Phys.* 0404:029 (2004); Giele W, Glover EWN, Zanderighi G. *Nucl. Phys. Proc. Suppl.* 135:275 (2004)
67. Bern Z, Dixon LJ, Smirnov VA. *Phys. Rev. D* 72:085001 (2005); Bern Z, et al. *Phys. Rev. D* 77:025010 (2008); Arkani-Hamed N, Cachazo F, Kaplan J. arXiv:0808.1446 [hep-th] (2008); Bern Z, et al. *Phys. Rev. Lett.* 103:081301 (2009)

Efimov States in Nuclear and Particle Physics

Hans-Werner Hammer[1] and Lucas Platter[2,3]

[1]Helmholtz-Institut für Strahlen- und Kernphysik (Theorie) and Bethe Center for Theoretical Physics, Universität Bonn, 53115 Bonn, Germany; email: hammer@hiskp.uni-bonn.de

[2]Institute for Nuclear Theory, University of Washington, Seattle, Washington 98195; email: lplatter@phys.washington.edu

[3]Department of Physics, The Ohio State University, Columbus, Ohio 43210

Annu. Rev. Nucl. Part. Sci. 2010. 60:207–36

The *Annual Review of Nuclear and Particle Science* is online at nucl.annualreviews.org

This article's doi:
10.1146/annurev.nucl.012809.104439

0163-8998/10/1123-0207$20.00

Key Words

universality, Efimov physics, few-body systems, discrete scale invariance, hyperspherical formalism, effective field theory

Abstract

Particles with resonant short-range interactions have universal properties that do not depend on the details of their structure or their interactions at short distances. In the three-body system, these properties include the existence of a geometric spectrum of three-body Efimov states and a discrete scaling symmetry, which leads to log-periodic dependence of observables on the scattering length. Similar universal properties appear in the four-body system and possibly higher-body systems as well. For example, universal four-body states have recently been predicted and observed in experiments. These phenomena are often referred to as Efimov physics. We review their theoretical description and discuss applications in different areas of physics with a special emphasis on nuclear and particle physics.

Contents

1. INTRODUCTION

The scattering of particles with sufficiently low kinetic energy is determined by the particles' S-wave scattering length a, assuming that their de Broglie wavelengths are large compared with the range of the interaction. Generically, the scattering length a is comparable in magnitude to the range ℓ of the interaction: $|a| \sim \ell$. In exceptional cases, the scattering length can be much larger in magnitude than the range: $|a| \gg \ell$. Such a large scattering length requires the fine-tuning of a parameter characterizing the interactions to the neighborhood of a critical value at which a diverges to $\pm\infty$. If the scattering length is large, the particles exhibit properties that depend on a but are insensitive to the range and other details of the short-range interaction. These properties are universal: They apply equally well to any nonrelativistic particle with short-range interactions that produce a large scattering length (1, 2).

For example, in the case of equal-mass particles with mass m and $a > 0$, there is a two-body bound state near the scattering threshold with binding energy $B_d = \hbar^2/(ma^2)$. The corrections to this formula are suppressed by powers of ℓ/a. This bound state corresponds to a pole of the two-particle scattering amplitude at $E = -B_d$. If the scattering length is negative, there is a universal virtual state that corresponds to a pole on the unphysical second sheet in the complex energy plane.

The key evidence for universal behavior in the three-body system was the discovery of the Efimov effect in 1970 (3). In the unitary limit $1/a \to 0$, the two-body bound state is exactly at the two-body scattering threshold $E = 0$. Efimov showed that in this limit there are infinitely many arbitrarily shallow three-body bound states whose binding energies $B_t^{(n)}$ have an accumulation point at $E = 0$. The Efimov effect is only one aspect of universal properties in the three-body system. It has universal properties not only in the unitary limit but whenever the scattering length is large compared with the range ℓ. The universal properties include a discrete scaling symmetry. We refer to universal aspects associated with this discrete scaling symmetry as Efimov physics.

Although they are well established theoretically, Efimov states are difficult to precisely identify in nature because (*a*) typical systems are not in the unitary limit and (*b*) the scattering length cannot be varied. Perhaps the simplest example in nuclear physics is the triton. The triton can be interpreted as the ground state of an Efimov spectrum in the *pnn* system with total spin $J = 1/2$. Because the ratio ℓ/a is only approximately 1/3, the whole spectrum contains only one state, but the low-energy properties of the triton can be described in this scenario. A promising system for observing several Efimov states is ^{4}He atoms, which have a scattering length that is more than a factor of ten larger than the range of the interaction. Calculations that use accurate potential models indicate that a system of three ^{4}He atoms has two three-body bound states or trimers. The ground-state trimer can be interpreted as an Efimov state, and it has been observed in experiments involving the scattering of cold jets of ^{4}He atoms from a diffraction grating (4). The excited trimer is universally believed to be an Efimov state, but it has not yet been observed.

The rapid development of the field of cold atom physics has opened up new opportunities for the experimental study of Efimov physics. These opportunities have been made possible by two separate technological developments. One is the technology for cooling atoms to the extremely low temperatures at which Efimov physics plays a crucial role. The other is the technology for controlling the interactions between atoms. By tuning the magnetic field to a Feshbach resonance, the scattering lengths of the atoms can be controlled experimentally and made arbitrarily large. Both developments were crucial in recent experiments that provided the first indirect evidence for the existence of Efimov states in ultracold atoms (5).

Overviews of Efimov physics in ultracold atomic gases can be found elsewhere (1, 2, 6). In this review, we focus on universal aspects and Efimov states in nuclear and particle physics. Even though the scattering length cannot be varied, there are many systems close to the unitary limit at which Efimov physics is relevant. They include few-nucleon systems, halo nuclei, and weakly bound hadronic molecules. These systems can be described in a universal effective field theory (EFT) that implements an expansion around the unitary limit. Three-body bound states can be interpreted as Efimov trimers.

In the next section, we review the physics of the Efimov effect, beginning with a brief account of the history. In the following sections, we discuss applications in nuclear and particle physics. We end with a summary and outlook.

2. PHYSICS OF THE EFIMOV EFFECT

2.1. History

The first hints on universal behavior in the three-body system came from the 1935 discovery of the Thomas collapse (7), which is closely related to the Efimov effect. Thomas (7) studied the zero-range limit for potentials with a single two-body bound state with fixed energy. Using a variational argument, he showed that the binding energy $B_t^{(0)}$ of the deepest three-body bound state diverges to infinity in this limit. Thus, the spectrum of three-body bound states is unbounded from below.

Further progress was made through the application of the zero-range limit to the three-nucleon system. An integral equation for S-wave neutron-deuteron scattering via zero-range interactions was derived in 1957 by Skorniakov & Ter-Martirosian (8). For the spin-quartet channel, this integral equation has no bound-state solutions and is well behaved. In the spin-doublet channel, however, it has solutions for arbitrary energy (9), including bound-state solutions. If the solution is fixed by requiring a specific three-body energy, the resulting equation still has a discrete spectrum that extends to minus infinity (10, 11), in agreement with the earlier result of Thomas. Although a prediction for the spin-doublet neutron-deuteron scattering length was obtained by use of the triton binding energy as input (12), most subsequent work focused on finite-range forces that avoid this pathology at high energies.

In 1970, Efimov (3) realized that one should focus on the physics at low energies, $E \ll \hbar^2/(m\ell^2)$, and not on the deepest states. In this limit, where zero-range forces are adequate, he arrived at some surprising results. He pointed out that when $|a|$ is sufficiently large compared with the range ℓ of the potential, there is a sequence of three-body bound states whose binding energies are spaced roughly geometrically in the interval between $\hbar^2/(m\ell^2)$ and $\hbar^2/(ma^2)$. As $|a|$ is increased, new bound states appear in the spectrum at critical values of a that differ by multiplicative factors of e^{π/s_0}, where s_0 depends on the statistics and the mass ratios of the particles. In the case of spin-doublet neutron-deuteron scattering and for three identical bosons, s_0 is the solution to the transcendental equation

$$s_0 \cosh \frac{\pi s_0}{2} = \frac{8}{\sqrt{3}} \sinh \frac{\pi s_0}{6}. \qquad 1.$$

Its numerical value is $s_0 \approx 1.00624$, so $e^{\pi/s_0} \approx 22.7$. As $|a|/\ell \to \infty$, the asymptotic number of three-body bound states is

$$N \to \frac{s_0}{\pi} \ln \frac{|a|}{\ell}. \qquad 2.$$

In the limit $a \to \pm\infty$, there are infinitely many three-body bound states with an accumulation point at the three-body scattering threshold with a geometric spectrum

$$B_t^{(n)} = (e^{-2\pi/s_0})^{n-n_*} \hbar^2 \kappa_*^2/m, \qquad 3.$$

where m is the mass of the particles and κ_* is the binding wave number of the branch of Efimov states labeled by n_*. The geometric spectrum in Equation 3 is the signature of a discrete scaling symmetry with scaling factor $e^{\pi/s_0} \approx 22.7$. It is independent of the mass or structure of the identical particles and independent of the form of their short-range interactions. The Efimov effect can also occur in other three-body systems if at least two of the three pairs have a large S-wave scattering length, but the numerical value of the asymptotic ratio may differ from 22.7.

A formal proof of the Efimov effect was subsequently given by Amado & Noble (13, 14). The Thomas and Efimov effects are closely related. The deepest three-body bound states found by Thomas's variational calculation can be identified with the deepest Efimov states (15). The mathematical connection of the Efimov effect to a limit cycle was discussed in Reference 16.

The universal properties in the three-body system with large scattering length are not restricted to the Efimov effect. The dependency of three-body observables on the scattering length or the energy is characterized by scaling-behavior modulo coefficients that are log-periodic functions of a (17, 18). This behavior is characteristic of a system with a discrete scaling symmetry.

In 1981, Efimov (19) proposed a new approach to the low-energy few-nucleon problem in nuclear physics that, in modern language, was based on perturbation theory around the unitary limit. Remarkably, this program works reasonably well in the three-nucleon system at momenta that are small compared with M_π. The Efimov effect requires that a boundary condition be imposed on the wave function at short distances. The boundary condition can be fixed by means of either the spin-doublet neutron-deuteron scattering length or the triton binding energy as input. If the deuteron binding energy is used as the two-body input and if the boundary condition is fixed by using the spin-doublet neutron-deuteron scattering length as input, the triton binding energy can be predicted with an accuracy of 6%. The accuracy of the predictions can be further improved by taking into account the effective range as a first-order perturbation (20). Thus, the triton can be identified as an Efimov state associated with the deuteron and the spin-singlet virtual state as a *pn* state with large scattering length (19).

In the three-nucleon system, this program was implemented within an EFT framework by Bedaque et al. (21–23). The authors (23) found that the renormalization of the EFT requires a $SU(4)$-symmetric three-body interaction with an ultraviolet (UV) limit cycle. The three-body force depends on a parameter Λ_* that is determined through a renormalization condition that plays the same role as Efimov's boundary condition. $SU(4)$ symmetry was introduced in 1937 by Wigner (24) as a generalization of the $SU(2) \times SU(2)$ spin-isospin symmetry, allowing for a mixing of spin and isospin degrees of freedom in symmetry transformations. It is satisfied to a high degree in the energy spectra of atomic nuclei. Exact Wigner symmetry requires the S-wave scattering lengths in the spin-triplet and spin-singlet channels to be equal. However, if both two-body scattering lengths are large, it is a very good approximation even if they are different, because the symmetry-breaking terms are proportional to the inverse scattering lengths (25).

This EFT is ideally suited to calculating corrections to the universal results in the scaling limit. Its application to few-nucleon physics is discussed in Section 3.

2.2. Hyperspherical Methods

Coordinate-space methods are a valuable tool for the analysis of the three-body problem with short-range interactions (17, 26, 27). In this subsection, we introduce the hyperspherical approach that has been used to obtain important results about the Efimov spectrum. The material of this subsection is based on the discussion of the hyperradial formalism in Reference 1. For three particles of equal mass, the Jacobi coordinates are defined as

$$\mathbf{r}_{ij} = \mathbf{r}_i - \mathbf{r}_j \quad \text{and} \quad \mathbf{r}_{k,ij} = \mathbf{r}_k - \frac{1}{2}(\mathbf{r}_i + \mathbf{r}_j), \tag{4.}$$

where the triple (ijk) is a cyclic permutation of the particle indices (28). The hyperradius R and hyperangle α_k are then defined by

$$R^2 = \frac{1}{3}(\mathbf{r}_{12}^2 + \mathbf{r}_{23}^2 + \mathbf{r}_{31}^2) = \frac{1}{2}\mathbf{r}_{ij}^2 + \frac{2}{3}\mathbf{r}_{k,ij}^2 \quad \text{and} \quad \alpha_k = \arctan\left(\frac{\sqrt{3}|\mathbf{r}_{ij}|}{2|\mathbf{r}_{k,ij}|}\right), \tag{5.}$$

respectively. In the center-of-mass system, the Schrödinger equation in hyperspherical coordinates is given by

$$\left(T_R + T_{\alpha_k} + \frac{\Lambda_{k,ij}^2}{2mR^2} + V(R, \Omega)\right)\Psi(R, \alpha, \Omega) = E\Psi(R, \alpha, \Omega), \tag{6.}$$

where

$$T_R = \frac{\hbar^2}{2m}R^{-5/2}\left(-\frac{\partial}{\partial R^2} + \frac{15}{4R^2}\right)R^{5/2}, \tag{7.}$$

$$T_\alpha = \frac{\hbar^2}{2mR^2}\frac{1}{\sin 2\alpha}\left(-\frac{\partial^2}{\partial\alpha^2} - 4\right)\sin 2\alpha, \tag{8.}$$

and

$$\Lambda_{k,ij}^2 = \frac{\mathbf{L}_{ij}^2}{\sin^2\alpha_k} + \frac{\mathbf{L}_{k,ij}^2}{\cos^2\alpha_k}, \tag{9.}$$

where $\Omega = (\theta_{ij}, \phi_{ij}, \theta_{k,ij}, \phi_{k,ij})$ and the Ls in Equation 9 are the usual angular-momentum operators with respect to these angles.

We assume that the potential V depends only on the magnitude of the interparticle separation:

$$V(\mathbf{r}_1, \mathbf{r}_2, r_3) = V(r_{12}) + V(r_{23}) + V(r_{31}). \tag{10.}$$

We now employ the usual Faddeev decomposition of ψ for three identical bosons and neglect subsystem angular momentum:

$$\Psi(R, \alpha, \Omega) = \psi(R, \alpha_1) + \psi(R, \alpha_2) + \psi(R, \alpha_3). \tag{11}$$

The solution of the corresponding Faddeev equation can then be expanded in a set of eigenfunctions of the hyperangular operator; specifically,

$$\psi(R, \alpha) = \frac{1}{R^{5/2}\sin(2\alpha)} \sum_n f_n(R)\phi_n(R, \alpha). \tag{12}$$

This expansion leads to separate differential equations for the hyperangular functions ϕ_n and the hyperradial functions f_n. Particularly for the hyperradial functions, we obtain

$$\begin{aligned} Ef_n(R) &= \left[\frac{\hbar^2}{2m}\left(-\frac{\partial^2}{\partial R^2} + \frac{15}{4R^2}\right) + V_n(R)\right] f_n(R) \\ &+ \sum_m \left[2P_{nm}(R)\frac{\partial}{\partial R} + Q_{nm}(R)\right] f_m(R), \end{aligned} \tag{13}$$

where the hyperradial potential $V_n(R)$ is defined by

$$V_n(R) = (\lambda_n(R) - 4)\frac{\hbar^2}{2mR^2}, \tag{14}$$

and $P_{nm}(R)$ and $Q_{nm}(R)$ are potentials that induce coupling between different hyperradial channels (1).

For hyperradii R, which are much larger than the range ℓ over which V is nonzero, the solution of the equation for the hyperangular function ϕ_n for large α is

$$\phi_n^{(\mathrm{high})}(\alpha) \approx \sin\left[\sqrt{\lambda_n}\left(\frac{\pi}{2} - \alpha\right)\right]. \tag{15}$$

However, for $R \gg \ell$ and small α, the solution for the hyperangular part can be written as

$$\phi_n^{(\mathrm{low})}(\alpha) = A\psi_0(\sqrt{2}R\alpha) - \frac{8\alpha}{\sqrt{3}}\sin\left(\sqrt{\lambda_n}\frac{\pi}{6}\right), \tag{16}$$

where A is a constant and ψ_0 is the zero-energy solution to a two-body Schrödinger equation with the two-body potential V,

$$\psi_k(r) = \frac{\sin[kr + \delta(k)]}{k} = \frac{\sin\delta(k)}{k}[\cos(kr) + \cot\delta\sin(kr)]. \tag{17}$$

As $k \to 0$, this expression yields $\psi_0(r) = r - a$, and we can use this asymptotic two-body wave function in Equation 16. This gives

$$\phi_n^{(\mathrm{low})}(\alpha) = A(\sqrt{2}R\alpha - a) - \frac{8\alpha}{\sqrt{3}}\sin\left(\sqrt{\lambda_n}\frac{\pi}{6}\right). \tag{18}$$

Because $V = 0$ in this region, this result must be consistent with Equation 15. This is achieved by (*a*) choosing

$$A = -\frac{1}{a}\sin\left[\sqrt{\lambda_n}\frac{\pi}{2}\right], \tag{19}$$

which ensures that $\phi_n(\alpha)$ is continuous across the boundary between low and high solutions at $\alpha \approx \ell/R$, and (*b*) the condition

$$\cos\left(\sqrt{\lambda_n}\frac{\pi}{2}\right) - \frac{8}{\sqrt{3\lambda_n}}\sin\left(\sqrt{\lambda_n}\frac{\pi}{6}\right) = \sqrt{\frac{2}{\lambda_n}}\sin\left(\sqrt{\lambda_n}\frac{\pi}{2}\right)\frac{R}{a} \tag{20}$$

on λ_n, which ensures that $\phi_n(\alpha)$ has a continuous first derivative as $\alpha \to \ell/R$. We note that if these equations are satisfied, then λ_n, and hence ϕ_n, is independent of R for $R \ll |a|$. Indeed, as long as Equations 19 and 20 are satisfied, Equation 15 is the result for ϕ for all α, such that $\alpha > \ell/R$. Solving Equation 20 in the limit $R \ll |a|$ we find the lowest eigenvalue

$$\lambda_0 = -s_0^2 \left(1 + 1.897\frac{R}{a}\right), \tag{21.}$$

where $s_0 = 1.00624\ldots$. This is the only negative eigenvalue, and therefore only this channel potential is attractive. If we now focus on the unitary limit, where $|a| \to \infty$, we have $\lambda_0 = -s_0^2$. Because it can also be shown that the coupling potentials P_{nm} and Q_{nm} vanish in this regime, the hyperradial equation (Equation 13) in the lowest channel becomes

$$\frac{\hbar^2}{2m}\left(-\frac{\partial^2}{\partial R^2} - \frac{s_0^2 + \frac{1}{4}}{R^2}\right) f_0(R) = E f_0(R). \tag{22.}$$

This equation holds for $R \gg \ell$. If we desire a solution for negative E, the requirement of normalizability for f_0 mandates that

$$f_0^{(0)}(R) = \sqrt{R} K_{is_0}(\sqrt{2}\kappa R), \tag{23.}$$

where the superscript (0) indicates that we are working in the unitary limit, and the subscript 0 refers to the solution for the hyperchannel corresponding to λ_0, which is the only one that supports bound states. The binding energy of these bound states is related to the κ of Equation 23 by

$$B_t \equiv \frac{\hbar^2 \kappa^2}{m}. \tag{24.}$$

Because the attractive $1/R^2$ potential produces a spectrum that is unbounded from below, some other short-distance physics is needed to stabilize the system. If the two-body potential is known, this short-distance physics is provided by the two-body potential V, which becomes operative for $R \sim \ell$. An alternative approach is to add an additional term to Equation 22 that summarizes the impact of the two-body V. We take this potential to be a surface δ function at a radius $1/\Lambda$ (29),

$$V_{SR}(R) = H_0(\Lambda)\Lambda^2 \delta\left(R - \frac{1}{\Lambda}\right), \tag{25.}$$

where H_0 is adjusted as a function of Λ such that the binding energy of a particular state, say $B_t^{(n_*)}$ (with a corresponding κ_*, given by Equation 24), is reproduced. Note that because V_{SR} is operative only at small hyperradii $R \sim 1/\Lambda$, it corresponds to a three-body force. (See Reference 30 for a realization of this in a momentum-space formalism.) In physical terms, we anticipate $\ell \sim 1/\Lambda$ because we know that once we consider hyperradii of order $1/\Lambda$, the potential V begins to affect the solutions.

Given that our focus is on predictions of the theory that are independent of details of V, we can consider the extreme case and take the limit $\ell \to 0$. In this limit, the form of K_{is_0} as $R \to 0$ guarantees that, once H_0 is fixed to give a bound state at $B_t^{(n_*)}$, the other binding energies in this hyperradial eigenchannel form a geometric spectrum. Specifically, $B_t^{(n)} = \hbar^2\kappa_n^2/m$, where

$$\kappa_n = \left(e^{-\pi/s_0}\right)^{n-n_*} \kappa_*. \tag{26.}$$

Here n_* is the index of the bound state corresponding to κ_*. Equation 26 holds for all κ_n, such that $\kappa_n \ll \Lambda$. (Note that the subscript on κ now denotes the index of the bound state in adiabatic channel zero.) The continuous scale invariance of the $1/R^2$ potential has been broken down to a discrete scale invariance by the imposition of particular short-distance physics on the problem through the short-distance potential of Equation 25 (29).

The discrete scale invariance of the three-body wave function, which is exact in the limit of infinite scattering length and zero range, also has implications for finite range. A perturbative calculation of the effect of a finite effective range on the bound-state spectrum showed that the spectrum remains unchanged (31) and that corrections to binding energies are of order $(r_0/a)^2$, where r_0 is the effective range of the interaction.

2.3. Efimov Spectrum

The hyperspherical methods discussed above can also be used to obtain the binding energy spectrum at finite scattering length. The short-distance boundary condition that was used to fix the binding energy in the unitary limit also determines the bound-state spectrum at finite scattering length. The binding momentum κ_* introduced above can therefore be considered a convenient parameter that determines the value of all universal few-body observables of the corresponding universality class.

The exact discrete scaling symmetry observed in the limit of infinite scattering length also exists if κ_* is kept fixed and if a and other variables such as the energy are rescaled:

$$\kappa_* \rightarrow \kappa_*, \quad a \rightarrow \mathcal{S}_0^m a, \quad E \rightarrow \mathcal{S}_0^{-2m} E. \tag{27}$$

Observables such as binding energy and cross sections scale with integer powers of $\mathcal{S}_0 = \exp(\pi/s_0)$ under this symmetry. For example, the binding energy of an Efimov state (trimer), which is a function of a and κ_*, scales as

$$B_t^{(n)}(\mathcal{S}_0^m a, \kappa_*) = \mathcal{S}_0^{-2m} B_t^{(n-m)}(a, \kappa_*). \tag{28}$$

This implies for positive scattering length that

$$B_t^{(n)}(a, \kappa_*) = F_n[2s_0 \ln(a\kappa_*)] \frac{\hbar^2 \kappa_*^2}{m}. \tag{29}$$

The function F_n parameterizes the scattering length dependency of all Efimov trimers exactly in the limit of vanishing range. The function F_n satisfies

$$F_n(x + 2m\pi) = (e^{-2\pi/s_0})^m F_{n-m}(x). \tag{30}$$

The scattering length dependency of the bound-state spectrum is shown in **Figure 1**. We plot the quantity $K \equiv \mathrm{sgn}(E)(m|E|)^{1/2}/\hbar$ against the inverse scattering length (note that the axes have been rescaled such that the scaling factor is approximately 2.2 instead of 22.7). For bound states, K corresponds to the binding momentum. Only a few of the infinitely many Efimov branches are shown. A given physical system has a fixed scattering length value and is denoted by the vertical dashed line. Changing the parameter κ_* by a factor $\mathcal{S}_0$ corresponds to multiplying each branch of trimers with this factor without changing their shapes. One important result is that three-body bound states exist for positive and negative scattering lengths. This is remarkable for the latter case because the two-body subsystem is unbound for $a < 0$. At a negative scattering length that we denote with a'_*, a bound state with a given κ_* has zero binding energy. As the scattering length is increased, the binding energy increases until it crosses the atom-dimer threshold at the positive scattering length a_*. The quantities a_* and a'_* can also be used to quantify a universality class of Efimov states.

2.4. Universal Properties

Other calculable observables also display the discrete scaling symmetry discussed above for the bound-state spectrum. The atom-dimer cross section fulfills, for example, the constraint

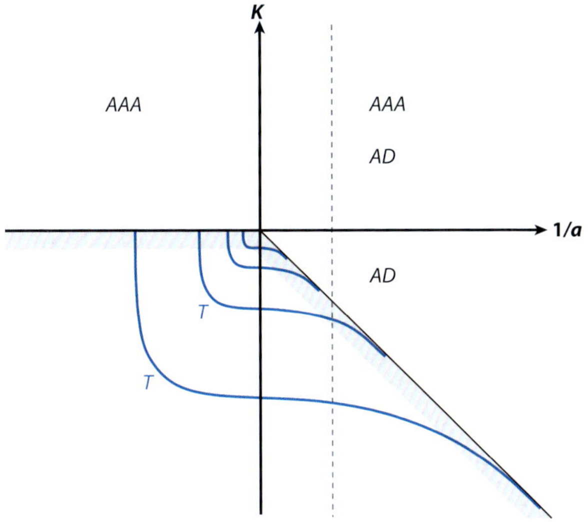

Figure 1

The Efimov plot for the three-body problem. We show $K \equiv \mathrm{sgn}(E)(m|E|)^{1/2}/\hbar$ versus the inverse scattering length. The allowed regions for three-atom scattering states and atom-dimer scattering states are labeled *AAA* and *AD*, respectively. The blue lines labeled *T* are two of the infinitely many branches of Efimov states. The gray shading indicates the threshold for scattering states. States along the vertical dashed line have a fixed scattering length. Adapted from Reference 1 with permission.

$$\sigma_{AD}(\mathcal{S}_0^{-2m}E; \mathcal{S}_0^{m}a, \kappa_*) = \mathcal{S}_0^{2m}\sigma_{AD}(E; a, \kappa_*) \qquad 31.$$

under rescaling. At $E = 0$, the cross section is related to the atom-dimer scattering length $\sigma_{AD} = 4\pi|a_{AD}|^2$. This relationship implies that the atom-dimer scattering length can be written as

$$a_{AD} = f[2s_0\ln(a\kappa_*)]a, \qquad 32.$$

where $f(x)$ is a periodic function with period 2π.

An observable that has been crucial for the experimental detection of Efimov physics in ultracold atoms is the three-body recombination rate. In ultracold gases, atoms can undergo inelastic three-body collisions in which a two-body bound state is formed. The dimer and remaining atom gain kinetic energy in this process and can leave the atomic trap. Such processes lead to a measurable loss of particles in the atomic trap. For negative scattering length, atoms can only recombine into deep dimers that have binding energy of order $\hbar^2/(mR^2)$. For positive scattering length, atoms can recombine into (*a*) shallow dimers with binding energy $\hbar^2/(ma^2)$ and (*b*) deep dimers. The recombination rate that is a measure for the loss rate of atoms scales as $\hbar a^4/m$ multiplied by a log-periodic coefficient that is a function of a and κ_*. The recombination rate constant α can therefore be written as

$$\alpha(a) = g[2s_0\ln(a\kappa_*)]\frac{\hbar a^4}{m}. \qquad 33.$$

The analytic form of the function $g(x)$ is known and can be found in References 1 and 6. Here we focus on the qualitative features of positive scattering length (**Figure 2**). At positive scattering length, interference effects lead to log-periodically spaced minima in the recombination rate. At negative scattering length, free atoms can only recombine into deep dimers. This process is enhanced dramatically whenever an Efimov trimer is at threshold.

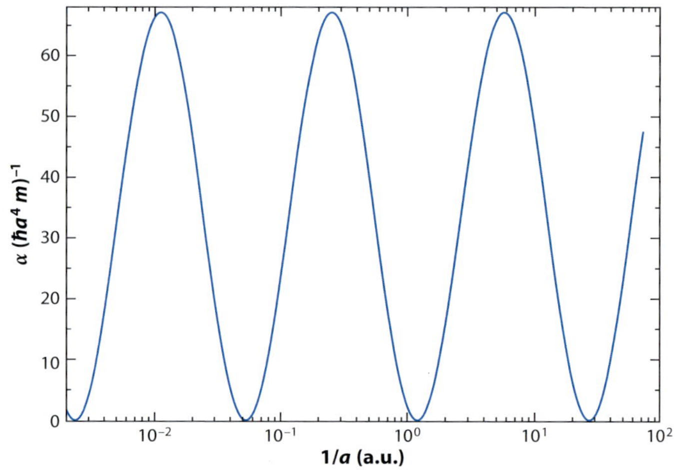

Figure 2

The recombination rate constant α for positive scattering length a in units of $\hbar a^4/m$ as a function of $1/a$ (in arbitrary units).

A different perspective on Efimov physics can be gained by keeping the two-body scattering length fixed and varying the three-body parameter. As a result, all three-body observables are correlated and lead to correlation lines when plotted against each other. One of these lines is the Phillips line (32): the correlation that results when the trimer binding energy is plotted against the atom-dimer scattering length. Such correlations have frequently been observed in nuclear physics, in which different phase-shift equivalent potentials are employed in few-body calculations.

Because one three-body parameter is required for a description of the three-body system with zero-range interactions, one must determine how many parameters are needed for calculations in the n-body system. A first step toward answering this question was taken in Reference 33. The authors of this work showed that the two-body scattering length and one three-body parameter are sufficient to make predictions for four-body observables. Results in a more detailed analysis (34) led to the conclusion that every trimer state is tied to two universal tetramer states with binding energies related to the binding energy of the next-shallower trimer. In the unitary limit $1/a = 0$, the relation between the binding energies is

$$B_4^{(0)} \approx 5\,B_t \quad \text{and} \quad B_4^{(1)} \approx 1.01\,B_t, \qquad 34.$$

where $B_4^{(0)}$ denotes the binding energy of the deeper of the two tetramer states and $B_4^{(1)}$ denotes the binding energy of the shallower of the two.

A recent calculation by von Stecher et al. (35) supports these findings and extends them to higher numerical accuracy. For the relation between universal three- and four-body bound states in the unitary limit, the authors found that

$$B_4^{(0)} \approx 4.57\;B_t \quad \text{and} \quad B_4^{(1)} \approx 1.01\;B_t, \qquad 35.$$

which is consistent with the results given in Equation 34 within the numerical accuracy.

The results obtained by Hammer & Platter (34) were presented in the form of an extended Efimov plot (**Figure 3**). Four-body states must have a binding energy larger than that of the deepest trimer state. An extended version of this four-body Efimov plot was also presented by von Stecher et al. (35). The authors of this study calculated more states with higher numerical accuracy and extended the calculation of the four-body states to the thresholds at which they

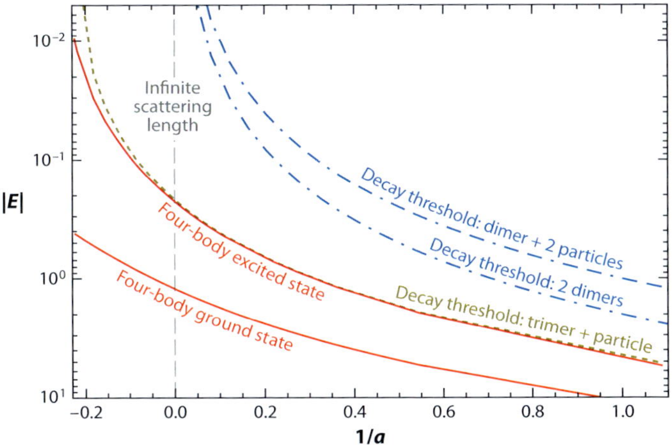

Figure 3

The extended Efimov plot for the four-body problem. Shown is the absolute value of energy E versus the inverse scattering length. Both quantities are given in arbitrary units. The lower (upper) solid red line denotes the four-body ground (excited) state. The dashed dark yellow line represents the threshold for decay into a ground-state trimer and a particle. The lower (upper) blue dashed-dotted line represents the threshold for decay into two dimers (a dimer and two particles). The vertical dashed line indicates the infinite scattering length.

become unstable. From these results, they extracted the negative values of the scattering lengths at which the binding energies of the tetramer states become zero and found that

$$a^*_{4,0} \approx 0.43 a'_* \quad \text{and} \quad a^*_{4,1} \approx 0.92 a'_*. \qquad 36.$$

These numbers uniquely specify the relative position of three- and four-body recombination resonances. This information was essential for Ferlaino et al.'s (36) subsequent observation of these states in ultracold atoms.

Calculations for larger numbers of particles by means of a model that incorporates the universal behavior of the three-body system were carried out by von Stecher (37). His findings indicate that there is at least one n-body state tied to each Efimov trimer, and he also found numerical evidence for a second excited five-body state.

2.5. Observation in Ultracold Atoms

The first experimental evidence for Efimov physics in ultracold atoms was presented by Kraemer et al. (5) in 2006. This group used ^{133}Cs atoms in the lowest hyperfine spin state. They observed resonant enhancement of the loss of atoms from three-body recombination that can be attributed to an Efimov trimer crossing the three-atom threshold. Kraemer et al. also observed a minimum in the three-body recombination rate that can be interpreted as an interference effect associated with Efimov physics. In a subsequent experiment with a mixture of ^{133}Cs atoms and dimers, Knoop et al. (38) observed a resonant enhancement in the loss of atoms and dimers. This loss feature can be explained by an Efimov trimer crossing the atom-dimer threshold (39). The most exciting recent developments in the field of Efimov physics involve universal tetramer states. Ferlaino et al. (36) observed two tetramers in an ultracold gas of ^{133}Cs atoms that confirm the results by Platter et al. (33), Hammer & Platter (34), and von Stecher et al. (35).

Recent experiments with other bosonic atoms have provided even stronger evidence of Efimov physics in the three- and four-body sectors. Zaccanti et al. (40) measured the three-body recombination rate and the atom-dimer loss rate in an ultracold gas of ^{39}K atoms. They observed two atom-dimer loss resonances and two minima in the three-body recombination rate at large positive values of the scattering length. The positions of the loss features are consistent with the universal predictions, with a discrete scaling factor of 22.7. Zaccanti et al. also observed loss features at large negative scattering lengths. Barontini et al. (41) obtained the first evidence of the Efimov effect in a heteronuclear mixture of ^{41}K and ^{87}Rb atoms. They observed three-atom loss resonances at large negative scattering lengths in both the K-Rb-Rb and K-K-Rb channels, for which the discrete scaling factors are 131 and 3.51×10^5, respectively. Gross et al. (42) measured the three-body recombination rate in an ultracold system of ^{7}Li atoms. They observed a three-atom loss resonance at a large negative scattering length and a three-body recombination minimum at a large positive scattering length. The positions of the loss features, which are in the same universal region on different sides of a Feshbach resonance, are consistent with the universal predictions, with a discrete scaling factor of 22.7. Pollack et al. (43) measured the three-body recombination in a system of ^{7}Li atoms in a hyperfine state different from the system considered by Gross et al. Pollack et al. (43) observed a total of 11 three- and four-body loss features. The features obey the universal relations on each side of the Feshbach resonance separately; however, a systematic deviation of 50% is found when features on different sides of the Feshbach resonance are compared.

Efimov physics has also been observed in three-component systems of the ^{6}Li atom. For the three lowest hyperfine states of ^{6}Li atoms, the three-pair scattering lengths approach a common large negative value at large magnetic fields, and all three have nearby Feshbach resonances at lower fields that can be used to vary the scattering lengths (44). The first experimental studies of many-body systems of ^{6}Li atoms in the three lowest hyperfine states have recently been carried out by Ottenstein et al. (45) and by Huckans et al. (46). These groups' measurements of the three-body recombination rate revealed a narrow loss feature and a broad loss feature in a region of low magnetic field. Theoretical calculations of the three-body recombination rate supported the interpretation that the narrow loss feature arises from an Efimov trimer crossing the three-atom threshold (47–49). Very recently, another narrow loss feature was discovered in a much higher region of the magnetic field by Williams et al. (50) and by Jochim and coworkers (51). Williams et al. used measurements of the three-body recombination rate in this region to determine the complex three-body parameter that governs Efimov physics in this system. This parameter, together with the three scattering lengths as functions of the magnetic field, determines the universal predictions for ^{6}Li atoms in this region of the magnetic field.

3. APPLICATIONS IN NUCLEAR PHYSICS

The properties of hadrons and nuclei are determined by quantum chromodynamics (QCD), a nonabelian gauge theory formulated in terms of quark and gluon degrees of freedom. At low energies, however, the appropriate degrees of freedom are the hadrons. Efimov physics and the unitary limit can serve as a useful starting point for EFTs that describe hadrons and nuclei at very low energies. For convenience, we work in natural units, where $\hbar = c = 1$.

In nuclear physics, there are a number of EFTs, all of which are useful for a certain range of systems (52–54). At very low energies, where Efimov physics plays a role, all interactions can be considered short range, and even the pions can be integrated out. This so-called pionless EFT is formulated in an expansion of the low-momentum scale M_{low} over the high-momentum scale M_{high}. It can be understood as an expansion around the limit of infinite scattering length or, equivalently, around near-threshold bound states. Its breakdown scale is set by a one-pion exchange, $M_{high} \sim M_\pi$, whereas $M_{low} \sim 1/a \sim k$. For momenta k of the order of the pion mass M_π, pion exchange becomes a long-range interaction and has to be treated explicitly. Doing so leads to the chiral EFT, whose breakdown scale M_{high} is set by the chiral symmetry–breaking scale Λ_χ. The pionless theory relies only on the large scattering length and is independent of the short-distance mechanism generating it. This theory is therefore ideally suited to unraveling universal phenomena driven by the large scattering length, such as limit-cycle physics (55, 56) and the Efimov effect (3). In this section, we focus on the aspects of nuclear EFTs related to Efimov physics. More complete overviews of the application of EFTs to nuclear phenomena in general are available (52–54, 57).

3.1. Few-Nucleon Systems

In the two-nucleon system, the pionless theory reproduces the well-known effective range expansion in the large scattering length limit. The renormalized S-wave scattering amplitude to next-to-leading order (NLO) in a given channel takes the form

$$T_2(k) = \frac{4\pi}{m}\frac{1}{-1/a - ik}\left[1 - \frac{r_0 k^2/2}{-1/a - ik} + \ldots\right], \qquad 37.$$

where k is the relative momentum of the nucleons and the dots indicate corrections of order $(M_{low}/M_{high})^2$ for typical momenta $k \sim M_{low}$. In the language of the renormalization group, Equation 37 corresponds to an expansion around the nontrivial fixed point for $1/a = 0$ (58, 59). The pionless EFT becomes very useful in the two-nucleon sector when external currents are considered, and it has been applied to a variety of electroweak processes. These calculations are reviewed in detail in References 52 and 53.

Here we focus on the three-nucleon system. It is convenient (but not mandatory) to write the theory using so-called dimeron auxiliary fields (60). We need two dimeron fields, one for each S-wave channel: (*a*) a field t_i with spin (isospin) 1 (0) representing two nucleons interacting in the $^3S_1$ channel (the deuteron) and (*b*) a field s_a with spin (isospin) 0 (1) representing two nucleons interacting in the $^1S_0$ channel (23). Specifically,

$$\begin{aligned}\mathcal{L} = {}& N^\dagger\left(i\partial_t + \frac{\vec{\nabla}^2}{2m}\right)N - t_i^\dagger\left(i\partial_t + \frac{\vec{\nabla}^2}{4m} - \Delta_t\right)t_i \\ & - s_a^\dagger\left(i\partial_t + \frac{\vec{\nabla}^2}{4m} - \Delta_s\right)s_a - \frac{g_t}{2}\left(t_i^\dagger N^T\tau_2\sigma_i\sigma_2 N + \text{h.c.}\right) \\ & - \frac{g_s}{2}\left(s_a^\dagger N^T\sigma_2\tau_a\tau_2 N + \text{h.c.}\right) - G_3 N^\dagger\left[g_t^2(t_i\sigma_i)^\dagger(t_j\sigma_j)\right. \\ & \left. + \frac{g_t g_s}{3}\left((t_i\sigma_i)^\dagger(s_a\tau_a) + \text{h.c.}\right) + g_s^2(s_a\tau_a)^\dagger(s_b\tau_b)\right]N + \ldots, \end{aligned} \qquad 38.$$

where i and j are spin; a and b are isospin indices; and g_t, g_s, Δ_t, Δ_s, and G_3 are the bare coupling constants. The Pauli matrices σ_i (τ_a) operate in spin (isospin) space. This Lagrangian goes beyond leading order (LO) and already includes the effective range terms. The coupling constants g_t, Δ_t, g_s, and Δ_s, are matched to the scattering lengths a_α and effective ranges $r_{0\alpha}$ in the two channels ($\alpha = s, t$). Alternatively, one can match to the position of the bound-state/virtual-state pole γ_α in the T-matrix instead of the scattering length, which often improves convergence (61).

The term proportional to G_3 constitutes a Wigner $SU(4)$ symmetric three-body interaction. It only contributes in the spin-doublet S-wave channel. When the auxiliary dimeron fields t_i and s_a are integrated out, an equivalent form containing only nucleon fields is obtained. At LO, when the

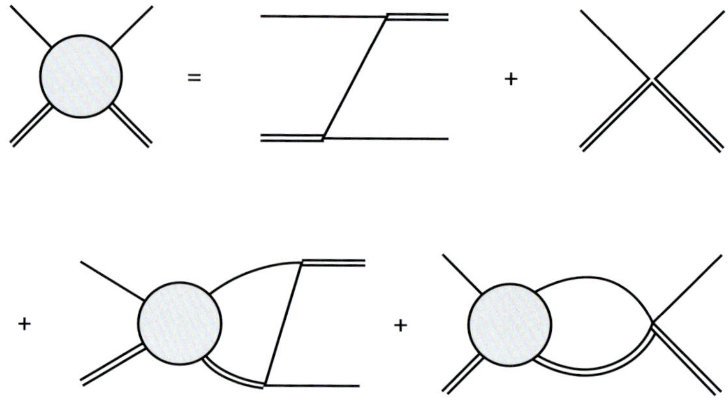

Figure 4

The integral equation for the particle-dimeron scattering amplitude. The single (double) line indicates the particle (dimeron) propagator. Adapted from Reference 1 with permission.

effective range corrections are neglected, the spatial and time derivatives acting on the dimeron fields are omitted and the field is static. The coupling constants g_α and Δ_α, $\alpha = s, t$ are then not independent, and only the combination g_α^2/Δ_α enters in observables. This combination can then be matched to the scattering length or pole position.

The simplest three-body process to consider is neutron-deuteron scattering below the breakup threshold. To focus on the main aspects of renormalization, we suppress all spin-isospin indices and complications from coupled channels in the three-nucleon problem. The resulting system is equivalent to a system of three spinless bosons with large scattering length. If the scattering length is positive, the particles form a two-body bound state analog to the deuteron, which we generically term a dimeron. The LO integral equation for particle-dimeron scattering is shown schematically in **Figure 4**. Projected on total orbital angular momentum $L = 0$, it takes the form

$$T_3(k, p; E) = \frac{16}{3a} M(k, p; E) + \frac{4}{\pi} \int_0^\Lambda dq\, q^2\, T_3(k, q; E) \times \frac{M(q, p; E)}{-1/a + \sqrt{3q^2/4 - mE - i\epsilon}}, \qquad 39.$$

where the inhomogeneous term reads

$$M(k, p; E) = \frac{1}{2kp} \ln\left(\frac{k^2 + kp + p^2 - mE}{k^2 - kp + p^2 - mE}\right) + \frac{H(\Lambda)}{\Lambda^2}. \qquad 40.$$

Here $H(\Lambda)$ is a running coupling constant that determines the strength of the three-body force $G_3(\Lambda) = 2mH(\Lambda)/\Lambda^2$, and Λ is a UV cutoff introduced to regularize the integral equation. Note that the three-body force is enhanced and enters at LO in this theory. The magnitude of the incoming (outgoing) relative momenta is k (p), and $E = 3k^2/(4m) - 1/(ma^2)$. The on-shell point corresponds to $k = p$, and the phase shift can be obtained via $k \cot \delta = 1/T_3(k, k; E) + ik$. For $H \equiv 0$ and $\Lambda \to \infty$, Equation 39 reduces to the STM (Skorniakov–Ter-Martirosian) equation (8). It is well known that the STM equation has no unique solution (9). The regularized STM equation has a unique solution for any given (finite) value of the UV cutoff Λ, but the solution strongly depends on the value of Λ. In the EFT framework, cutoff independence of the amplitude is achieved by an appropriate "running" of $H(\Lambda)$ (30, 62),

$$H(\Lambda) = \frac{\cos[s_0 \ln(\Lambda/\Lambda_*) + \arctan s_0]}{\cos[s_0 \ln(\Lambda/\Lambda_*) - \arctan s_0]}, \qquad 41.$$

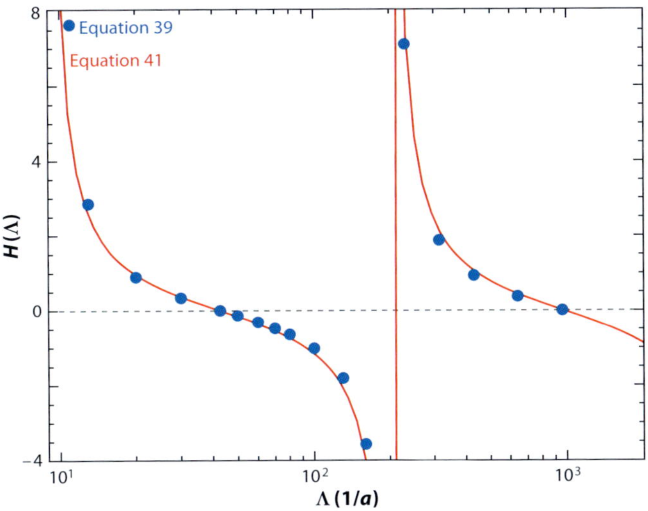

Figure 5

The three-body coupling H as a function of the cutoff Λ for a fixed value of the three-body parameter Λ_*. The solid red line shows the analytical expression from Equation 41, whereas the blue dots show results from the numerical solution of Equation 39.

where Λ_* is a dimensionful three-body parameter generated by dimensional transmutation. The dependence of the three-body coupling H on the cutoff Λ is shown in **Figure 5** for a fixed value of the three-body parameter Λ_*. A good agreement between the methods of Equations 39 and 41 is observed, indicating that the renormalization is well under control. Adjusting Λ_* to a single three-body observable allows us to determine all other low-energy properties of the three-body system.[1] Because $H(\Lambda)$ in Equation 41 vanishes for certain values of the cutoff Λ, it is possible to eliminate the explicit three-body force from the equations by working with a fixed cutoff that encodes the dependency on Λ_*. Doing so justifies tuning the cutoff Λ in the STM equation to reproduce a three-body datum and using the same cutoff to calculate other observables, as suggested by Kharchenko (63). Equivalently, a subtraction can be performed in the integral equation (64, 65). In all cases, one three-body input parameter is needed for the calculation of observables.

The discrete scaling symmetry of the Efimov spectrum is manifest in the running of the coupling $H(\Lambda)$. The spectrum of three-body bound states of this EFT is exactly the Efimov spectrum. The integral equations for the three-nucleon problem derived from the Lagrangian (Equation 38) are a generalization of Equation 39. (For their explicit form and derivation, see, e.g., Reference 66.)

For S-wave nucleon-deuteron scattering in the spin-quartet channel, only the spin-1 dimeron field contributes, and the integral equation becomes (8, 21, 22)

$$T_3^{(3/2)}(p, k; E) = -\frac{4\gamma_t}{3} K(p, k) - \frac{1}{\pi} \int_0^\infty dq\, q^2 D_t(q; E) K(p, q) T_3^{(3/2)}(q, k; E), \qquad 42.$$

where

$$K(p, k) = \frac{1}{pk} \ln\left(\frac{p^2 + pk + k^2 - mE}{p^2 - pk + k^2 - mE}\right), \qquad 43.$$

$D_t(q; E)$ is the full spin-1 dimeron propagator, and $\gamma_t \approx 45$ MeV is the deuteron pole momentum.

[1] Note that the choice of the three-body parameter Λ_* is not unique. For alternative definitions, see Reference 1.

This integral equation has a unique solution for $\Lambda \to \infty$, and there is no three-body force in the first few orders. An S-wave three-body force is forbidden by the Pauli principle in this channel because all nucleon spins must be aligned to obtain $J = 3/2$. The spin-quartet scattering phases $k \cot \delta^{(3/2)} = 1/T_3^{(3/2)}(k, k; E) + ik$ can therefore be predicted to high precision from two-body data alone.

In the spin-doublet channel, both dimeron fields as well as the three-body force in the Lagrangian (Equation 38) contribute (23). This contribution leads to a pair of coupled integral equations for the T-matrix. The renormalization of this equation is easily understood in the unitary limit, which corresponds to a Wigner $SU(4)$ symmetry of the theory (Equation 24). In the unitary limit, these two integral equations decouple. One of the two equations has the same structure as the equation for the bosonic problem (Equation 39), whereas the other one is similar to the equation in the quartet channel (Equation 42). Thus, one needs a new parameter that is not determined in the two-nucleon system in order to fix the (leading) low-energy behavior of the three-nucleon system in this channel. This parameter corresponds to the $SU(4)$ symmetric three-body force proportional to G_3 in the Lagrangian (Equation 38) (23). The three-body parameter gives a natural explanation of universal correlations between different three-body observables such as the Phillips line, a correlation between the triton binding energy and the spin-doublet neutron-deuteron scattering length (32). These correlations are driven purely by the large scattering length, independent of the mechanism responsible for it. If the spin-doublet neutron-deuteron scattering length is given, the triton binding energy is predicted. In this scenario, the triton emerges as an Efimov state. This scenario can be tested by using the effective theory to predict other three-body observables.

Higher-order corrections to the amplitude, including those due to two-nucleon effective range terms, can be included perturbatively. This was first done at NLO for the scattering length and triton binding energy (20) and for the energy dependency of the phase shifts (64). The authors of References 66 and 67 demonstrated that it is convenient to iterate certain higher-order range terms to extend the calculation to next-to-next-to-leading order (NNLO). Here a subleading three-body force was also included, as required by dimensional analysis. More recently, Platter & Phillips (68) showed, using the subtractive renormalization, that the leading three-body force is sufficient to achieve cutoff independence up to NNLO in the expansion in M_{low}/M_{high}. The results for the spin-doublet neutron-deuteron scattering phase shift at LO (23), NLO (64), and NNLO (69) are shown in **Figure 6**. There is excellent agreement with the available phase-shift analysis and a calculation that uses a phenomenological nucleon-nucleon interaction. From dimensional analysis, one would expect the subleading three-body force at NNLO. Whether there is a suppression of the subleading three-body force or simply a correlation between the leading and subleading contributions is not understood.

Three-nucleon channels with higher orbital angular momentum are similar to the spin quartet for S-waves, and three-body forces do not appear until very high orders (70). A general counting scheme for three-body forces based on the asymptotic behavior of the solutions of the LO STM equation was proposed by Griesshammer (71). A complementary approach to the few-nucleon problem is given by the renormalization group, in which the power counting is determined from the scaling of operators under the renormalization group transformation (72). This method leads to consistent results for the power counting (73–75).

Three-body calculations with external currents are still in their infancy. However, a few exploratory calculations have been carried out. Universal properties of the triton charge form factor have been investigated (76), and neutron-deuteron radiative capture has been calculated (77–79). The electromagnetic properties of the triton were also recently investigated (80, 81). This work raises the possibility of carrying out accurate calculations of electroweak reactions at very low energies for astrophysical processes.

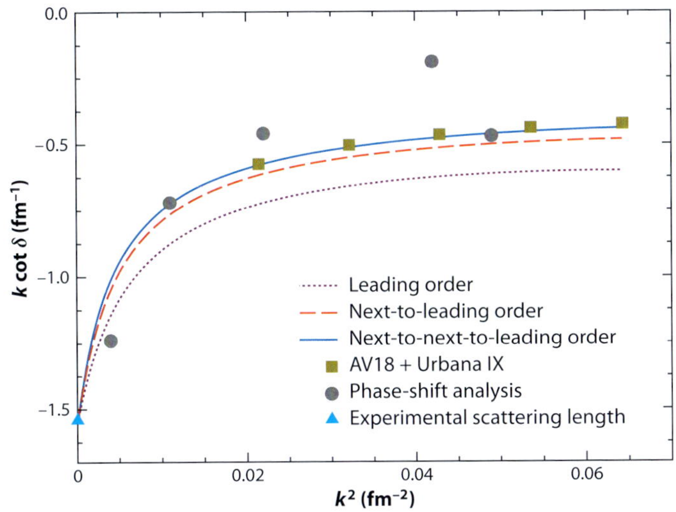

Figure 6

Phase shifts for neutron-deuteron scattering below the deuteron breakup at leading order (*purple dotted line*), next-to-leading order (*red dashed line*), and next-to-next-to-leading order (*solid blue line*). The dark yellow squares and gray circles represent the results of a phase-shift analysis and a calculation that used AV18 and the Urbana IX three-body force, respectively. The light blue triangle represents the experimental scattering length. Adapted from Reference 69 with permission.

The pionless approach has also been extended to the four-body sector (33, 82). To be able to apply the Yakubovsky equations, the authors used an equivalent effective quantum mechanics formulation. The study of the cutoff dependency of the four-body binding energies revealed that no four-body parameter is required for renormalization at LO. As a consequence, there are universal correlations in the four-body sector that are also driven by the large scattering length. The best-known example is the Tjon line: a correlation between the triton and α particle binding energies, B_t and B_α, respectively. Of course, higher-order corrections break the exact correlation and generate a band. In **Figure 7**, we show this band together with some calculations that make use of phenomenological potentials (83) and a chiral EFT potential with explicit pions (84, 85). All calculations with interactions that give a large scattering length must lie within the band. Different short-distance physics and/or cutoff dependency should only move the results along the band. This can be observed, for example, in the NLO results with the chiral potential indicated by the squares in **Figure 7** or in the few-body calculations with the low-momentum nucleon-nucleon potential $V_{\mathrm{low\ k}}$ (86). The $V_{\mathrm{low\ k}}$ potential is obtained from phenomenological nucleon-nucleon interactions by integrating out high-momentum modes above a cutoff Λ but leaving two-body observables (such as the large scattering lengths) unchanged. The results of few-body calculations with $V_{\mathrm{low\ k}}$ are not independent of Λ but lie close to the Tjon line (cf. figure 2 of Reference 86).

Another interesting development is the application of the resonating group model to solve the pionless EFT for three- and four-nucleon systems (87). This method allows for a straightforward inclusion of Coulomb effects. Kirscher et al. (87) extended previous calculations in the four-nucleon system to NLO and showed that the Tjon line correlation persists. Moreover, they calculated the correlation between the singlet S-wave ^{3}He-neutron scattering length and the triton binding energy. Preliminary results for the halo nucleus ^{6}He have been reported (88).

The pionless theory has also been applied within the no-core shell model framework. Here the expansion in a truncated harmonic oscillator basis is used as the UV regulator of the EFT. The

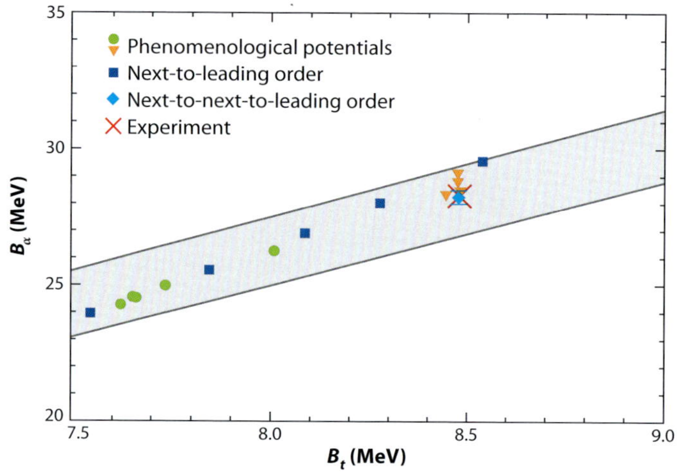

Figure 7

The Tjon line correlation as predicted by the pionless theory. The colored circles and triangles represent various calculations that use phenomenological potentials (83). The dark blue squares represent the results of a chiral effective field theory at next-to-leading order for different cutoffs, whereas the light blue diamond represents the next-to-next-to-leading order result (84, 85). The red cross indicates the experimental point. Adapted from Reference 82 with permission.

effective interaction is determined directly in the model space, where an exact diagonalization in a complete many-body basis is performed. The authors of Reference 89 calculated the 0^+ excited state of ^{4}He and the ^{6}Li ground state using the deuteron, triton, and α particle ground states as input. The first 0^+ excited state in ^{4}He was calculated within 10% of the experimental value, whereas the ^{6}Li ground state was approximately 70% of the experimental value, in agreement with the 30% error expected for the LO approximation. These results are promising and should be improved if range corrections are included. Finally, the spectrum of trapped three- and four-fermion systems was calculated with the same method (90). In this case, the harmonic potential is physical and not simply used as an UV regulator. For an update on this work, see Reference 91.

3.2. Quark Mass Dependence and Infrared Limit Cycle

In the following paragraphs, we discuss the possibility of an exact infrared limit cycle and the Efimov effect in a deformed version of QCD with quark masses slightly larger than their physical values. The quark mass dependency of the chiral nucleon-nucleon interaction was calculated to NLO in the chiral counting (92, 93). At this order, the quark mass dependency is synonymous with the pion mass dependency because of the Gell-Mann–Oakes–Renner relation: $M_\pi^2 = -(m_u + m_d)\langle 0|\bar{u}u|0\rangle / F_\pi^2$, where $\langle 0|\bar{u}u|0\rangle \approx (-290\,\text{MeV})^3$ is the quark condensate. In this section, we therefore use the term pion mass dependency instead of quark mass dependency, which is more convenient for our purpose. The pion mass dependency of the nucleon-nucleon scattering lengths in the $^3S_1$–$^3D_1$ and $^1S_0$ channels, as well as the deuteron binding energy, has been calculated 92–94.

In principle, the pion mass dependence of the chiral nucleon-nucleon potential is determined uniquely. However, the extrapolation away from the physical pion mass generates errors. The dominating sources are the constants $\bar{C}_{S,T}$ and $\bar{D}_{S,T}$, which give the corrections to the LO contact terms $\propto M_\pi^2$ and cannot be determined independently from fits to data at the physical pion mass. A smaller effect is due to the error in the low-energy constant $\bar{d}_{16}$, which governs the pion mass dependence of g_A. Both effects generate increasing uncertainties as one extrapolates away from the physical point.

In the calculation given in Reference 93, the size of the two constants $\bar{D}_S$ and $\bar{D}_T$ was constrained from naturalness arguments, assuming that $-3 \leq F_\pi^2 \Lambda_\chi^2 \bar{D}_{S,T} \leq 3$, where $\Lambda_\chi \simeq 1$ GeV is the chiral symmetry–breaking scale. These bounds are in agreement with resonance saturation estimates (95). The constant $\bar{d}_{16}$ was varied in the range $\bar{d}_{16} = -0.91 \ldots -1.76$ GeV^{-2} (96). This range was used to estimate the extrapolation errors of two-nucleon observables such as the deuteron binding energy and the spin-singlet and spin-triplet scattering lengths (93). In the chiral limit, the deuteron binding energy was found to be of natural size: $B_d \sim F_\pi^2/m \simeq 10$ MeV. Note, however, that if larger uncertainties in the low-energy constants $\bar{D}_S$ and $\bar{D}_T$ are assumed, one cannot make a definite statement about the binding of the deuteron in the chiral limit (92, 94). For pion masses above the physical value, however, all calculations show similar behavior.

In **Figure 8**, we show the inverse scattering lengths in the spin-triplet and spin-singlet channels from Reference 93, together with some recent lattice results from Reference 99. **Figure 8** also shows that a scenario in which both inverse scattering lengths vanish simultaneously at a critical pion mass of approximately 200 MeV is possible. For pion masses below the critical value, the spin-triplet scattering length is positive and the deuteron is bound. As the inverse spin-triplet scattering length decreases, the deuteron becomes more and more shallow and finally becomes unbound at the critical mass. Above the critical pion mass, the deuteron exists as a shallow virtual state. In the spin-singlet channel, the situation is reversed: The spin-singlet deuteron is a virtual state below the critical pion mass and becomes bound above. On the basis of this behavior, Braaten & Hammer (56) conjectured that one should be able to reach the critical point by varying the up and down quark masses m_u and m_d independently because the spin-triplet and spin-singlet channels have different isospin. In this case, the triton would display the Efimov effect, which corresponds to the occurrence of an infrared limit cycle in QCD. A complete investigation of

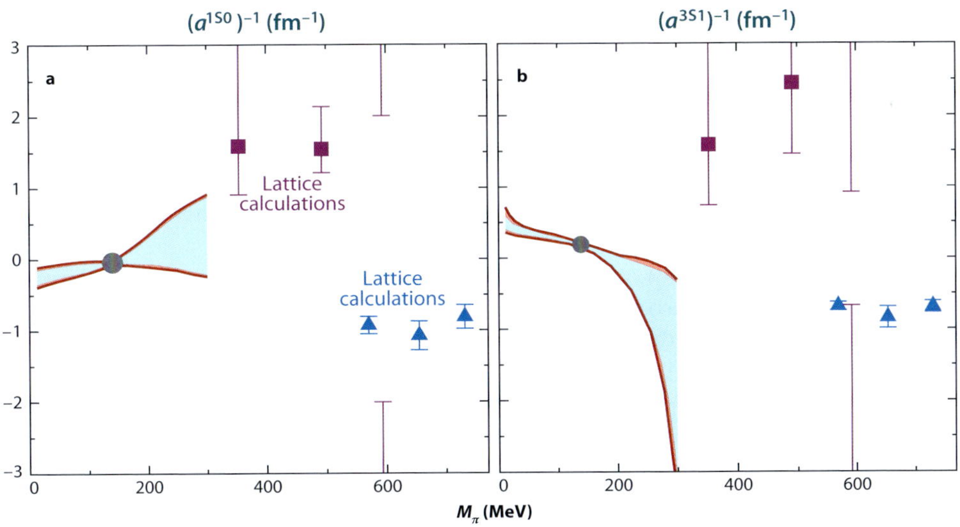

Figure 8

Inverse of the S-wave scattering lengths in the (*a*) spin-singlet and (*b*) spin-triplet nucleon-nucleon channels as a function of the pion mass M_π. The blue triangles represent the lattice calculations from References 97 and 98; the purple squares represent those from Reference 99. Adapted from Reference 54 with permission.

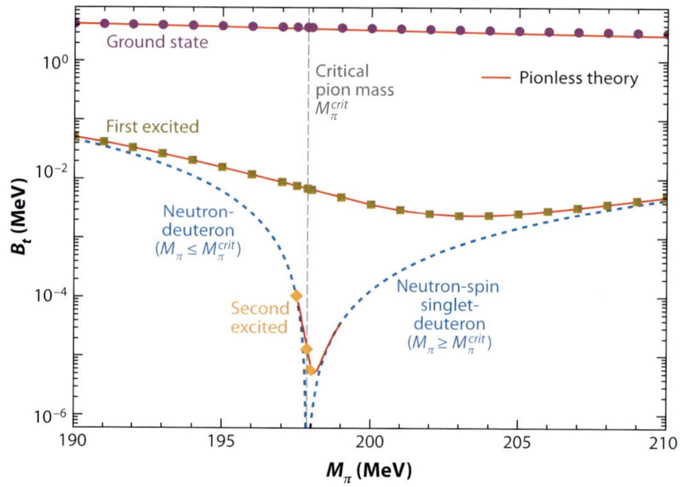

Figure 9

Binding energies B_t of the triton ground state and first two excited states as a function of M_π. The circles, squares, and diamonds represent the chiral effective field theory result, and the solid red lines represent calculations in the pionless theory. The vertical gray dotted line indicates the critical pion mass M_π^{crit}, and the dashed blue lines are the neutron-deuteron ($M_\pi \leq M_\pi^{crit}$) and neutron–spin singlet–deuteron ($M_\pi \geq M_\pi^{crit}$) bound-state thresholds. Adapted from Reference 101 with permission.

this issue requires the inclusion of isospin-breaking corrections and therefore higher orders in the chiral EFT. However, the universal properties of the limit cycle have been investigated by considering specific values of $\bar{D}_S$ and $\bar{D}_T$ that lie within the naturalness bound and that cause the spin-singlet and spin-triplet scattering lengths to become infinite at the same value of the pion mass.

The properties of the triton around the critical pion mass have been studied for one particular solution with a pion mass, M_π^{crit}, of 197.8577 MeV (100). From the solution of the Faddeev equations, the binding energies of the triton and the first two excited states in the vicinity of the limit cycle were calculated for this scenario in chiral EFT. The binding energies are given in **Figure 9**. The neutron-deuteron ($M_\pi \leq M_\pi^{crit}$) and neutron–spin singlet–deuteron ($M_\pi \geq M_\pi^{crit}$) thresholds are also shown. Directly at the critical mass, these thresholds coincide with the three-body threshold, and the triton has infinitely many excited states. **Figure 9** also shows the LO calculations in the pionless theory that use the pion mass dependence of the nucleon-nucleon scattering lengths and one triton state from chiral EFT as input. The chiral EFT results for the other triton states in the critical region are reproduced very well. The binding energy of the triton ground state varies only weakly over the whole range of pion masses and is approximately one-half of the physical value at the critical point. The excited states are strongly influenced by the thresholds and vary much more strongly.

These studies have been extended to NNLO in the pionless EFT and neutron-deuteron scattering observables (101). The higher-order corrections in the vicinity of the critical pion mass are small. This is illustrated in **Figure 10**.

A final answer to the question of whether an infrared limit cycle can be realized in QCD can be obtained only by solving QCD directly. Determining whether this solution can be achieved by appropriately tuning the quark masses in a lattice QCD simulation (102) would be particularly interesting. The first full lattice QCD calculation of nucleon-nucleon scattering has been reported (99), but statistical noise presented a serious challenge. A promising recent high-statistics study of

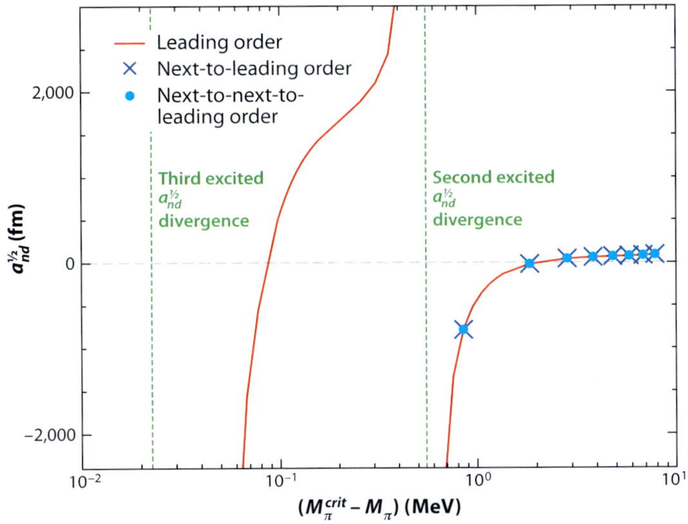

Figure 10

Doublet neutron-deuteron scattering length $a_{nd}^{1/2}$ in the critical region computed in the pionless effective field theory. The solid red line represents the leading-order result, and the light blue crosses and dark blue circles represent the next-to-leading order and next-to-next-to-leading order results, respectively. The vertical green dotted lines indicate the pion masses at which $a_{nd}^{1/2}$ diverges because the second and third excited states of the triton appear at the neutron-deuteron threshold. These singularities in $a_{nd}^{1/2}(M_\pi)$ are a clear signature that the limit cycle is approached in the critical region. Adapted from Reference 101 with permission.

three-baryon systems also presented initial results for a system with the quantum numbers of the triton, such that lattice QCD calculations of three-nucleon systems are now within sight (103). For a review of these activities, see Reference 104.

Such calculations require a detailed understanding of the modification of the Efimov spectrum in a cubic box. For identical bosons, there are significant finite volume shifts even for moderate box sizes (105). These shifts show universal scaling behavior that could be exploited to reduce the computational effort (106). The extension of these studies to the triton case is in progress.

3.3. Halo Nuclei

A special class of nuclear systems exhibiting universal behavior consists of halo nuclei (27, 107). A halo nucleus comprises a tightly bound core surrounded by one or more loosely bound valence nucleons. The valence nucleons are characterized by a very low separation energy compared with those in the core. As a consequence, the radius of the halo nucleus is large compared with the radius of the core. A trivial example is the deuteron, which can be considered a two-body halo nucleus. The root mean square radius of the deuteron is approximately three times larger than that of the constituent nucleons. Halo nuclei with two valence nucleons are particularly interesting examples of three-body systems. If none of the two-body subsystems is bound, the halo nuclei are termed Borromean halo nuclei. (This name is derived from the heraldic symbol of the Borromeo family of Italy, which consists of three rings interlocked in such a way that if any one of the rings is removed, the other two separate.) The most carefully studied Borromean halo nuclei are ^{6}He and ^{11}Li, which have two weakly bound valence neutrons (107). In the case of ^{6}He, the core is a ^{4}He nucleus, which is also known as the α particle. The two-neutron separation energy for ^{6}He is ~1 MeV, which is small compared with the ~28-MeV binding energy of the α particle. The $n\alpha$ system has no bound states, and the ^{6}He nucleus is therefore Borromean. There is, however, a strong P-wave resonance in the $J = 3/2$ channel of $n\alpha$ scattering, which is sometimes referred to as ^{5}He. This resonance is responsible for the binding of ^{6}He. Thus, ^{6}He can be interpreted as a bound state of an α particle and two neutrons, both of which are in $P_{3/2}$ configurations.

Because of the separation of scales in halo nuclei, they can be described by extensions of the pionless EFT. One can assume the core to be structureless and treat the nucleus as a few-body system of the core and the valence nucleons. Corrections from the structure of the core appear in higher orders and can be included in perturbation theory. Cluster models of halo nuclei then appear as LO approximations in this halo EFT. A new feature is the appearance of resonances as in the $n\alpha$ system, which leads to a more complicated singularity structure and renormalization compared with the few-nucleon system discussed above (108).

The first application of EFT methods to halo nuclei was carried out in References 108 and 109, in which the $n\alpha$ system (^{5}He) was considered. The authors found that for resonant P-wave interactions, both the scattering length and effective range have to be included nonperturbatively at LO. At threshold, however, only one combination of coupling constants is fine-tuned, and the EFT becomes perturbative. Because the $n\alpha$ interaction is resonant in the P-wave and not in the S-wave, the binding mechanism of ^{6}He is not the Efimov effect. However, this nucleus can serve as a laboratory in which to study the interplay of resonance structures in higher partial waves.

Three-body halo nuclei composed of a core and two valence neutrons are of particular interest due to these systems' potential to display the Efimov effect (3). Because the scattering length cannot be easily varied in halo nuclei, one looks for Efimov scaling between different states of the same nucleus. Such analyses assume that the halo ground state is an Efimov state.[2] They have previously been carried out in cluster models and the renormalized zero-range model (e.g., 110–112). A comprehensive study of S-wave halo nuclei in halo EFT was recently carried out (113). This work provided binding energy and structure calculations, including error estimates, for various halo nuclei. In a confirmation of earlier results by Fedorov et al. (110) and Amorim et al. (111), ^{20}C was found to be the only candidate nucleus for an excited Efimov state, assuming the ground state is also an Efimov state. This nucleus consists of a ^{18}C core with spin and parity quantum numbers $J^P = 0^+$ and two valence neutrons. The nucleus ^{19}C is expected to have a $\frac{1}{2}^+$ state near threshold, implying a shallow neutron-core bound state and therefore a large neutron-core scattering length. The value of the ^{19}C energy, however, is not known well enough to make a definite statement about the appearance of an excited state in ^{20}C. An excited state with a binding energy of ~65 keV is marginally consistent with the current experimental information.

The matter form factors and radii of halo nuclei can also be calculated in the halo EFT (113, 114). As an example, we show the various one- and two-body matter density form factors $\mathcal{F}_c$, $\mathcal{F}_n$, $\mathcal{F}_{nc}$, and $\mathcal{F}_{nn}$ with LO error bands for the ground state of ^{20}C as a function of the momentum transfer k^2 (from Reference 113) in **Figure 11**. The theory breaks down for momentum transfers on the order of the pion mass squared ($k^2 \approx 0.5$ fm^{-2}).

From the slope of the matter form factors, one can extract the corresponding radii:

$$\mathcal{F}(k^2) = 1 - \frac{1}{6}k^2\langle r^2\rangle + \ldots . \qquad 44.$$

Information about these radii has been extracted from experiments for some halo nuclei based on intensity interferometry and Dalitz plots (115). Within the error estimates, the extracted values

[2]It is also possible that only the excited state is an Efimov state and that the ground state is more compact. This scenario cannot be ruled out, but it is also less predictive.

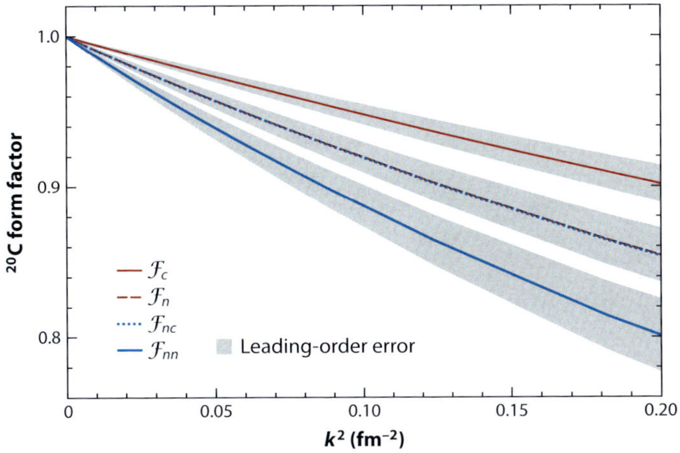

Figure 11

The one- and two-body matter density form factors $\mathcal{F}_c$, $\mathcal{F}_n$, $\mathcal{F}_{nc}$, and $\mathcal{F}_{nn}$ with leading-order error bands (*gray areas*) for the ground state of ^{20}C as a function of the momentum transfer k^2.

are in good agreement with the theoretical predictions from halo EFT (113). For the possible ^{20}C excited state, the halo EFT at LO predicts neutron-neutron and neutron-core radii of order 40 fm, whereas the ground-state radii are of order 2–3 fm. The theoretical errors are estimated to be of order 10%. Assuming a natural value for the effective range of the n-^{18}C interaction, $r_0 \approx 1/M_\pi$, NLO predictions for these radii have recently been obtained (116). The LO results were found to be stable under inclusion of the leading effective range corrections, and the typical errors could be reduced to approximately 1–2%.

Scattering observables offer a complementary window into Efimov physics in halo nuclei, and recent model studies have focused on this issue. Specifically, the trajectory of the possible ^{20}C excited state was extended into the scattering region to find a resonance in n-^{19}C scattering (117, 118). A detailed study of n-^{19}C scattering near an Efimov state was also carried out (119).

The simplest strange halo nucleus is the hypertriton, a three-body bound state composed of a proton, a neutron, and the Λ. The total binding energy is only approximately 2.4 MeV. The separation energy for the Λ, $E_\Lambda = 0.13$ MeV, is tiny compared with the binding energy of the deuteron, $B_d = 2.22$ MeV. The hypertriton can therefore also be considered a two-body halo nucleus. It has been studied in both two-body and three-body approaches (120–122). A study of the hypertriton in the halo EFT has been performed (123). The ΛN scattering lengths are not well known experimentally because the few scattering data are at relatively high energies. If the ΛN scattering lengths are large, the hypertriton is probably bound due to the Efimov effect. In this case, there would also be a correlation between the Λd scattering length $a_{\Lambda d}$ and the hypertriton binding energy B_t^Λ analog to the Phillips line in the neutron-deuteron system (121). In **Figure 12**, we show this Phillips line correlation for three values of the ΛN pole position γ_i (123). For small hypertriton energies B_t^Λ, the different Phillips lines coincide exactly (the physical hypertriton corresponds to $M\,B_t^\Lambda \approx 1.06\,\gamma_t^2$) and deviate from each other only at very large binding energies, at which the EFT breaks down. For all practical purposes, the Phillips line is therefore independent of γ_i. Whether the Efimov effect plays a role for the hypertriton is an open question. Most modern hyperon-nucleon potentials, however, favor a natural ΛN scattering length (54).

Another powerful method that can be used to investigate the Efimov effect in three-body halo nuclei at existing and future facilities with exotic beams (such as FAIR and FRIB) is Coulomb

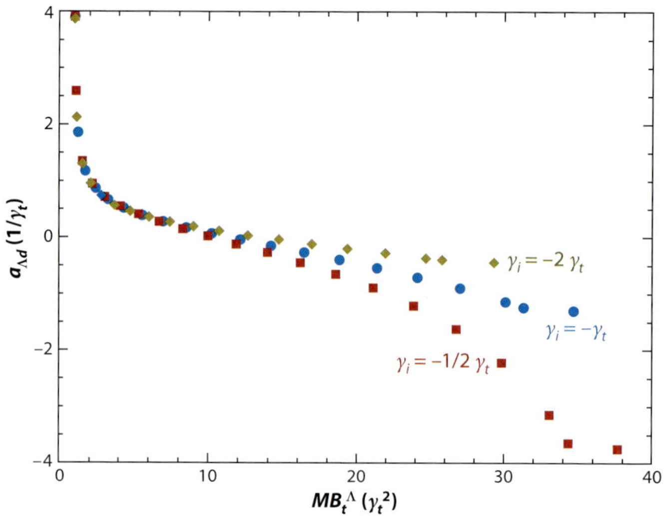

Figure 12

Phillips line in the hypertriton channel for different values of the ΛN pole position γ_i (all quantities are in units of the deuteron pole position $\gamma_t \approx 45$ MeV). Adapted from Reference 123 with permission.

excitation. In such experiments, a nuclear beam scatters off the Coulomb field of a heavy nucleus. Such processes can populate excited states of the projectile that subsequently decay, leading to its so-called Coulomb dissociation (124). The halo EFT offers a systematic framework for a full quantum-mechanical treatment and can be used to predict the signature of the Efimov effect in these reactions.

3.4. Three-Alpha System and Coulomb Interaction

The excited 0^+ state in ^{12}C is known as the Hoyle state. Its properties are important for stellar astrophysics because it determines the ratio of carbon to oxygen in stellar helium burning. Efimov suggested (3, 17) that the Hoyle state could be explained as a universal trimer of α particles. (For a more detailed discussion, see Reference 28.) In this case, the universal properties are modified by the long-range Coulomb interaction. The modified Efimov spectrum must be understood before any definite statement about the nature of the Hoyle state can be made.

Many recent studies have focused on the consistent inclusion of the Coulomb interaction in two-body halo nuclei such as the $p\alpha$ and $\alpha\alpha$ systems (125, 126). Specifically, the $\alpha\alpha$ system shows a surprising amount of fine-tuning between the strong and electromagnetic interactions. It can be understood in an expansion around the limit where, when the electromagnetic interactions are turned off, the ^8Be ground state is exactly at threshold and exhibits conformal invariance (126). In this scenario, the Hoyle state in ^{12}C would indeed appear as a remnant of an excited Efimov state. To better understand the modification of the Efimov spectrum and limit cycles by long-range interactions such as the Coulomb interaction, Hammer & Higa (127) investigated a one-dimensional inverse square potential supplemented with a Coulomb interaction. The results indicate that the counterterm required to renormalize the inverse square potential alone is sufficient to renormalize the full problem. However, the breaking of the discrete scale invariance through the Coulomb interaction leads to a modified bound-state spectrum. The shallow bound states are strongly influenced by the Coulomb interaction, whereas the deep bound states are dominated by the inverse square potential. These results support the conjecture that the Hoyle state is an Efimov state of α particles, but a full calculation of the 3α system—including Coulomb in the halo EFT—is missing. Calculations with the fermionic molecular dynamics model and electron scattering data, however, support a pronounced α cluster structure of the Hoyle state (128).

4. APPLICATIONS IN PARTICLE PHYSICS

4.1. Hadronic Molecules

In recent years, many new and possibly exotic charmonium states have been observed at the B factories at SLAC, at KEK in Japan, and at the CESR collider at Cornell. These observations have revived the field of charmonium spectroscopy (129–131). Several of the new states exist very close to scattering thresholds and can be interpreted as hadronic molecules. If they are sufficiently shallow, one may ask whether there are any three-body hadronic molecules bound by the Efimov effect.

A particularly interesting example is the $X(3872)$, discovered by the Belle Collaboration (132) in $B^{\pm} \to K^{\pm}\pi^{+}\pi^{-}J/\psi$ decays and quickly confirmed by CDF (133), DØ (134), and BaBar (135). The state has likely quantum numbers $J^{PC} = 1^{++}$ and is very close to the $D^{*0}\bar{D}^{0}$ threshold. As a consequence, the $X(3872)$ has a resonant S-wave coupling to the $D^{*0}\bar{D}^{0}$ system. An extensive program provides predictions for its decay modes based on the assumption that it is a $D^{*0}\bar{D}^{0}$ molecule with even C-parity:

$$(D^{*0}\bar{D}^{0})_{+} \equiv \frac{1}{\sqrt{2}}\left(D^{*0}\bar{D}^{0} + D^{0}\bar{D}^{*0}\right). \qquad 45.$$

This assumption naturally explains several puzzling features, such as the apparently different mass in the $J/\psi\pi^{+}\pi^{-}$ and $D^{*0}\bar{D}^{0}$ decay channels and the isospin-violating decays (136, 137). A status report with references to the original literature can be found in Reference 138.

Use of the latest measurements in the $J/\psi\pi^{+}\pi^{-}$ channel show the mass of the $X(3872)$ (139) to be $m_X = (3{,}871.55 \pm 0.20)$ MeV, which corresponds to an energy relative to the $D^{*0}\bar{D}^{0}$ threshold of

$$E_X = (-0.26 \pm 0.41)\,\text{MeV}. \qquad 46.$$

The central value corresponds to a $(D^{*0}\bar{D}^{0})_{+}$ bound state with binding energy $B_X = 0.26$ MeV [but a virtual state cannot be excluded from the current data in the $J/\psi\pi^{+}\pi^{-}$ and $D^{*0}\bar{D}^{0}$ channels (140–142)]. The $X(3872)$ is also very narrow; its width is smaller than 2.3 MeV.

Because the $X(3872)$ is so close to the $D^{*0}\bar{D}^{0}$ threshold, it has universal low-energy properties that depend only on its binding energy (143). Close to threshold, the coupling to charged D mesons can be neglected because the $D^{*+}\bar{D}^{-}$ threshold is ~8 MeV higher in energy. Therefore, the properties of the $X(3872)$ can be described in a universal EFT with contact interactions only. Unfortunately, there is no Efimov effect in this system (143), so universal bound states of the X and D^0 or D^{*0} mesons do not exist. The reason for this is that there are insufficient pairs with resonant interactions, as only the $D^{*0}\bar{D}^{0}$ and $D^{0}\bar{D}^{*0}$ interactions are resonant. However, it is possible to provide model-independent predictions for the scattering of D^0 and D^{*0} mesons and their antiparticles off the $X(3872)$. This scattering process is determined to LO by the $D^{*0}\bar{D}^{0}$ and $D^{0}\bar{D}^{*0}$ interactions only.

The corresponding cross sections as a function of the center-of-mass momentum k (139) are shown in **Figure 13**. The difference between the contribution of S-waves ($L = 0$) and the full cross section (including all partial waves up to $L = 6$) is negligible for momenta below the bound-state pole momentum γ. Our results are given in units of the scattering length and may be scaled

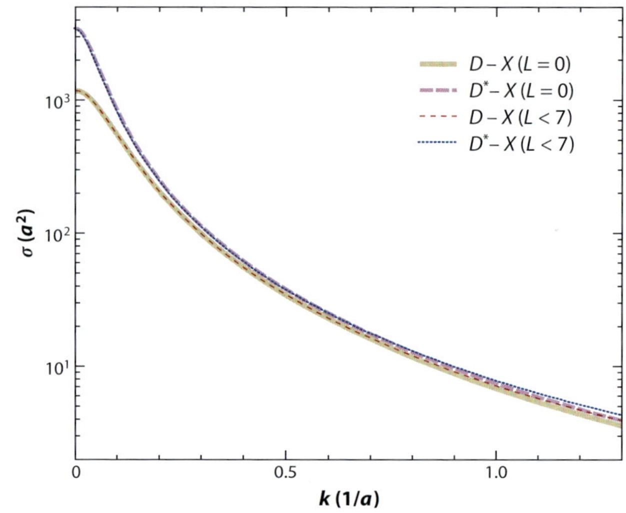

Figure 13

Total cross section for scattering of D^0 and D^{*0} mesons off the $X(3872)$ for S-waves ($L = 0$) and including higher partial waves with $L < 7$, in units of the scattering length a. The cross section is the same for the scattering of particles as it is for the scattering of antiparticles. Adapted from Reference 139 with permission.

to physical units once a is determined. At present, the error in the experimental value for E_X in Equation 43 implies a large error in the scattering length. Specifically, we obtain the range $\gamma = (0 \ldots 36)$ MeV for the pole momentum and the range $a = (5.5 \ldots \infty)$ fm for the scattering length with central values $\gamma = 22$ MeV and $a = 8.8$ fm. Using the central value of the scattering length, we obtain for the scale factor $a^2 = 0.78$ b. This factor can become infinite if the $X(3872)$ is directly at threshold, whereas the lower bound from the error in E_X would give a value of 0.3 b. Even in this case, the total cross section at threshold would be of the order of 300 b for D^0X scattering and 1,000 b for $D^{0*}X$ scattering. It may be possible to extract the scattering within the final-state interactions of B_c decays and/or other LHC events. Observation of enhanced final-state interactions would provide an independent confirmation of the nature of the $X(3872)$.

There may also be hadronic three-body molecules that are bound due to the Efimov effect, but currently no strong candidate states are known. This situation will be improved by new experiments at facilities such as FAIR and Belle II, which have a dedicated program to study exotic charmonium states.

5. SUMMARY AND OUTLOOK

Any few-body system with short-range interactions that has a two-body scattering length larger than the range of the underlying interaction displays universal properties and Efimov physics. This statement is independent of the typical length scale of the system, and atomic, nuclear, and particle physics can provide examples of universality. Although much progress has recently been made in experiments with ultracold atoms, the concept of Efimov physics was originally devised for the few-nucleon problem. As we have shown, it can serve as a starting point for a description of very low energy nuclear phenomena in an expansion around the unitary limit.

In this review, we have discussed the manifestation of Efimov physics with a strong emphasis on nuclear and particle physics. There are a number of such systems that display low-energy universality associated with Efimov physics. The most important example from nuclear physics is the triton. The nucleon-nucleon scattering length is large compared with the range of the internuclear interaction. Phase shift–equivalent nucleon-nucleon potentials therefore necessarily give results for the triton binding energy and neutron-deuteron scattering length that are correlated and lie on the Phillips line. The implications of universality on the four-nucleon system have also been explored. The Tjon line, a correlation between three-nucleon and four-nucleon binding energies, is a result of the large two-nucleon scattering length.

Halo nuclei could provide a further example of few-body universality. Specifically, two-neutron halos such as ^{20}C could display Efimov physics in the form of an excited three-body state due to the large core-neutron scattering length. Although initial studies of bound-state observables have become available, scattering calculations represent an exciting opportunity for future applications of the EFT approach. So-called $p - t$ reactions, in which a triton is formed in a collision of a proton and a two-neutron halo, will provide an important benchmark.

Several new charmonium states have recently been discovered close to scattering thresholds and can be interpreted as hadronic molecules. If they are sufficiently shallow, they have universal properties associated with large scattering length physics. The best-known example is the $X(3872)$, which may be interpreted as a $D^{*0}\bar{D}^0$ molecule with even C-parity. There may also be three-body hadronic molecules bound by the Efimov effect, but currently no strong candidates are known.

The separation of scales between scattering length and range facilitates the application of an EFT that reproduces at LO the results obtained by Efimov and is known as pionless EFT. Within this framework, corrections to the zero-range limit can be calculated systematically in a small parameter expansion in powers of $k\ell$ and ℓ/a. The number of few-body calculations that include higher-order corrections is growing. As a consequence, the expansion around the unitary limit provides a useful starting point for a controlled description of very low energy phenomena in nuclear and particle physics. The intricate consequences of Efimov physics are explicit in this framework, and universal correlations between observables arise naturally. Moreover, this theory is an ideal tool with which to unravel universal properties and establish connections between different fields of physics.

The constituents of nuclear few-body systems can have charge, unlike the neutral atoms used in experiments with ultracold gases. Electroweak observables thus provide additional information about few-body universality. However, they are also of interest by themselves, and pionless EFT guarantees a consistent framework for the calculation of observables with the minimal number of parameters because it is straightforward to consider external currents in any EFT. The first calculations for observables such as form factors and capture rates have been performed, but many more remain to be calculated. Of very great interest in this context are thermal capture rates in the four-nucleon sector that are relevant to big bang nucleosynthesis.

DISCLOSURE STATEMENT

The authors are not aware of any affiliations, memberships, funding, or financial holdings that might be perceived as affecting the objectivity of this review.

ACKNOWLEDGMENTS

We thank Eric Braaten and Vitaly Efimov for discussions. Our research was supported by the National Science Foundation under grant number PHY-0653312, by the UNEDF SciDAC Collaboration under U.S. Department of Energy grant numbers DE-FC02-07ER41457 and DE-FG02-00ER41132, by the Deutsche Forschungsgemeinschaft through SFB/TR16, and by the Bundesministerium für Bildung und Forschung under contract numbers 06BN411 and 06BN9006.

LITERATURE CITED

1. Braaten E, Hammer HW. *Phys. Rep.* 428:259 (2006)
2. Platter L. *Few Body Syst.* 46:139 (2009)
3. Efimov V. *Phys. Lett.* B33:563 (1970)
4. Schöllkopf W, Toennies J. *Science* 266:1345 (1994)
5. Kraemer T, et al. *Nature* 440:315 (2006)
6. Braaten E, Hammer HW. *Ann. Phys.* 322:120 (2007)
7. Thomas L. *Phys. Rev.* 47:903 (1935)
8. Skorniakov GV, Ter-Martirosian KA. *Sov. Phys. JETP* 4:648 (1957)
9. Danilov G. *Sov. Phys. JETP* 13:349 (1961)
10. Faddeev LD, Minlos RA. *Sov. Phys. JETP* 14:1315 (1962)
11. Minlos RA, Faddeev LD. *Sov. Phys. Dokl.* 6:1072 (1962)
12. Danilov G, Lebedev V. *Sov. Phys. JETP* 17:1015 (1963)
13. Amado RD, Noble JV. *Phys. Lett.* B35:25 (1971)
14. Amado RD, Noble JV. *Phys. Rev. D* 5:1992 (1972)
15. Adhikari SK, et al. *Phys. Rev. A* 37:3666 (1988)
16. Albeverio S, Hegh-Krohn R, Wu TT. *Phys. Lett.* A83:105 (1981)
17. Efimov V. *Sov. J. Nucl. Phys.* 12:589 (1971)
18. Efimov V. *Sov. J. Nucl. Phys.* 29:546 (1979)
19. Efimov V. *Nucl. Phys. A* 362:45 (1981)
20. Efimov V. *Phys. Rev. C* 44:2303 (1991)
21. Bedaque PF, van Kolck U. *Phys. Lett.* B428:221 (1998)
22. Bedaque PF, Hammer HW, van Kolck U. *Phys. Rev. C* 58:R641 (1998)
23. Bedaque PF, Hammer HW, van Kolck U. *Nucl. Phys. A* 676:357 (2000)
24. Wigner E. *Phys. Rev.* 51:106 (1937)
25. Mehen T, Stewart IW, Wise MB. *Phys. Rev. Lett.* 83:931 (1999)
26. Nielsen E, Fedorov D, Jensen A, Garrido E. *Phys. Rep.* 347:373 (2001)
27. Jensen AS, Riisager K, Fedorov DV, Garrido E. *Rev. Mod. Phys.* 76:215 (2004)
28. Efimov V. *Comm. Nucl. Part. Phys.* 19:271 (1990)
29. Braaten E, Phillips D. *Phys. Rev. A* 70:052111 (2004)
30. Bedaque PF, Hammer HW, van Kolck U. *Phys. Rev. Lett.* 82:463 (1999)
31. Platter L, Ji C, Phillips DR. *Phys. Rev. A* 79:022702 (2009)
32. Phillips A. *Nucl. Phys. A* 107:209 (1968)
33. Platter L, Hammer HW, Meissner U-G. *Phys. Rev. A* 70:052101 (2004)
34. Hammer HW, Platter L. *Eur. Phys. J. A* 32:113 (2007)
35. von Stecher J, D'Incao J, Greene C. *Nat. Phys.* 5:417 (2009)
36. Ferlaino F, et al. *Phys. Rev. Lett.* 102:140401 (2009)
37. von Stecher J. *J. Phys. B* 43:101002 (2010)
38. Knoop S, et al. *Nat. Phys.* 5:227 (2009)
39. Helfrich K, Hammer HW. *Europhys. Lett.* 86:53003 (2009)
40. Zaccanti M, et al. *Nat. Phys.* 5:586 (2009)
41. Barontini G, et al. *Phys. Rev. Lett.* 103:043201 (2009)
42. Gross N, Shotan Z, Kokkelmans S, Khaykovich L. *Phys. Rev. Lett.* 103:163202 (2009)
43. Pollack SE, Dries D, Hulet RG. *Science* 326:1683 (2009)
44. Bartenstein M, et al. *Phys. Rev. Lett.* 94:103201 (2005)
45. Ottenstein TB, et al. *Phys. Rev. Lett.* 101:203202 (2008)
46. Huckans JH, et al. *Phys. Rev. Lett.* 102:165302 (2009)
47. Braaten E, Hammer HW, Kang D, Platter L. *Phys. Rev. Lett.* 103:073202 (2009)

48. Naidon P, Ueda M. *Phys. Rev. Lett.* 103:073203 (2009)
49. Floerchinger S, Schmidt R, Wetterich C. *Phys. Rev. A* 79:053633 (2009)
50. Williams JR, et al. *Phys. Rev. Lett.* 103:130404 (2009)
51. Lompe T, et al. arXiv:1003.0600 [cond-mat] (2010)
52. Beane SR, et al. arXiv:nucl-th/0008064 (2000)
53. Bedaque PF, van Kolck U. *Annu. Rev. Nucl. Part. Sci.* 52:339 (2002)
54. Epelbaum E, Hammer HW, Meissner U-G. *Rev. Mod. Phys.* 81:1773 (2009)
55. Mohr RF, et al. *Ann. Phys.* 321:225 (2006)
56. Braaten E, Hammer HW. *Phys. Rev. Lett.* 91:102002 (2003)
57. Epelbaum E. *Prog. Part. Nucl. Phys.* 57:654 (2006)
58. Kaplan DB, Savage MJ, Wise MB. *Nucl. Phys. B* 534:329 (1998)
59. Birse MC, McGovern JA, Richardson KG. *Phys. Lett.* B464:169 (1999)
60. Kaplan DB. *Nucl. Phys. B* 494:471 (1997)
61. Phillips DR, Rupak G, Savage MJ. *Phys. Lett.* B473:209 (2000)
62. Bedaque PF, Hammer HW, van Kolck U. *Nucl. Phys. A* 646:444 (1999)
63. Kharchenko VF. *Sov. J. Nucl. Phys.* 16:173 (1973)
64. Hammer HW, Mehen T. *Nucl. Phys. A* 690:535 (2001)
65. Afnan IR, Phillips DR. *Phys. Rev. C* 69:034010 (2004)
66. Bedaque PF, Rupak G, Griesshammer HW, Hammer HW. *Nucl. Phys. A* 714:589 (2003)
67. Griesshammer HW. *Nucl. Phys. A* 744:192 (2004)
68. Platter L, Phillips DR. *Few Body Syst.* 40:35 (2006)
69. Platter L. *Phys. Rev. C* 74:037001 (2006)
70. Gabbiani F, Bedaque PF, Griesshammer HW. *Nucl. Phys. A* 675:601 (2000)
71. Griesshammer HW. *Nucl. Phys. A* 760:110 (2005)
72. Wilson KG. *Rev. Mod. Phys.* 55:583 (1983)
73. Barford T, Birse MC. *J. Phys. A* 38:697 (2005)
74. Birse MC. *Phys. Rev. C* 77:047001 (2008)
75. Ando S-I, Birse MC. *Phys. Rev. C* 78:024004 (2008)
76. Platter L, Hammer HW. *Nucl. Phys. A* 766:132 (2006)
77. Sadeghi H, Bayegan S. *Nucl. Phys. A* 753:291 (2005)
78. Sadeghi H, Bayegan S, Griesshammer HW. *Phys. Lett.* B643:263 (2006)
79. Sadeghi H. *Phys. Rev. C* 75:044002 (2007)
80. Sadeghi H. arXiv:0908.2052 [nucl-th] (2009)
81. Sadeghi H, Bayegan S. *Few Body Syst.* 47:167 (2010)
82. Platter L, Hammer HW, Meissner U-G. *Phys. Lett.* B607:254 (2005)
83. Nogga A, Kamada H, Gloeckle W. *Phys. Rev. Lett.* 85:944 (2000)
84. Epelbaum E, et al. *Phys. Rev. Lett.* 86:4787 (2001)
85. Epelbaum E, et al. *Phys. Rev. C* 66:064001 (2002)
86. Nogga A, Bogner SK, Schwenk A. *Phys. Rev. C* 70:061002 (2004)
87. Kirscher J, Griesshammer HW, Shukla D, Hofmann HM. *Eur. Phys. J. A* 44:239 (2010)
88. Kirscher J, Griesshammer HW, Shukla D, Hofmann HM. arXiv:0909.5606 [nucl-th] (2009)
89. Stetcu I, Barrett BR, van Kolck U. *Phys. Lett.* B653:358 (2007)
90. Stetcu I, Barrett BR, van Kolck U, Vary JP. *Phys. Rev. A* 76:063613 (2007)
91. Stetcu I, Rotureau J, Barrett BR, van Kolck U. *J. Phys. G* 37:064033 (2010)
92. Beane SR, Savage MJ. *Nucl. Phys. A* 717:91 (2003)
93. Epelbaum E, Meissner U-G, Gloeckle W. *Nucl. Phys. A* 714:535 (2003)
94. Beane SR, Bedaque PF, Savage MJ, van Kolck U. *Nucl. Phys. A* 700:377 (2002)
95. Epelbaum E, Meissner U-G, Gloeckle W, Elster C. *Phys. Rev. C* 65:044001 (2002)
96. Fettes N. *Pion-nucleon physics in chiral perturbation theory*. Ph.D. thesis. Univ. Bonn. 273 pp. (2000)
97. Fukugita M, et al. *Phys. Rev. Lett.* 73:2176 (1994)
98. Fukugita M, et al. *Phys. Rev. D* 52:3003 (1995)
99. Beane SR, Bedaque PF, Orginos K, Savage MJ. *Phys. Rev. Lett.* 97:012001 (2006)
100. Epelbaum E, Hammer HW, Meissner U-G, Nogga A. *Eur. Phys. J. C* 48:169 (2006)

101. Hammer HW, Phillips DR, Platter L. *Eur. Phys. J. A* 32:335 (2007)
102. Wilson KG. *Nucl. Phys. Proc. Suppl.* 140:3 (2005)
103. Beane SR, et al. *Phys. Rev. D* 80:074501 (2009)
104. Beane SR, Orginos K, Savage MJ. *Int. J. Mod. Phys.* E17:1157 (2008)
105. Kreuzer S, Hammer HW. *Phys. Lett.* B673:260 (2009)
106. Kreuzer S, Hammer HW. *Eur. Phys. J. A* 43:229 (2010)
107. Zhukov MV, et al. *Phys. Rep.* 231:151 (1993)
108. Bertulani CA, Hammer HW, van Kolck U. *Nucl. Phys. A* 712:37 (2002)
109. Bedaque PF, Hammer HW, van Kolck U. *Phys. Lett.* B569:159 (2003)
110. Federov DV, Jensen AS, Riisager K. *Phys. Rev. Lett.* 73:2817 (1994)
111. Amorim AEA, Frederico T, Tomio L. *Phys. Rev. C* 56:R2378 (1997)
112. Mazumdar I, Arora V, Bhasin VS. *Phys. Rev. C* 61:051303 (2000)
113. Canham DL, Hammer HW. *Eur. Phys. J. A* 37:367 (2008)
114. Yamashita MT, Tomio L, Frederico T. *Nucl. Phys. A* 735:40 (2004)
115. Marques FM, et al. *Phys. Rev. C* 64:061301 (2001)
116. Canham DL, Hammer HW. *Nucl. Phys. A* 836:275 (2010)
117. Yamashita MT, Frederico T, Tomio L. *Phys. Lett.* B660:339 (2008)
118. Mazumdar I, Rau ARP, Bhasin VS. *Phys. Rev. Lett.* 97:062503 (2006)
119. Yamashita MT, Frederico T, Tomio L. *Phys. Lett.* B670:49 (2008)
120. Cobis A, Jensen AS, Fedorov DV. *J. Phys. G* 23:401 (1997)
121. Fedorov DV, Jensen AS. *Nucl. Phys. A* 697:783 (2002)
122. Congleton JG. *J. Phys. G* 18:339 (1992)
123. Hammer HW. *Nucl. Phys. A* 705:173 (2002)
124. Bertulani CA, Baur G. *Phys. Rep.* 163:299 (1988)
125. Higa R. *Mod. Phys. Lett. A* 24:915 (2009)
126. Higa R, Hammer HW, van Kolck U. *Nucl. Phys. A* 809:171 (2008)
127. Hammer HW, Higa R. *Eur. Phys. J. A* 37:193 (2008)
128. Chernykh M, et al. *Phys. Rev. Lett.* 98:032501 (2007)
129. Eichten E, Godfrey S, Mahlke H, Rosner JL. *Rev. Mod. Phys.* 80:1161 (2008)
130. Voloshin MB. *Prog. Part. Nucl. Phys.* 61:455 (2008)
131. Godfrey S, Olsen SL. *Annu. Rev. Nucl. Part. Sci.* 58:51 (2008)
132. Choi SK, et al. (Belle Collab.) *Phys. Rev. Lett.* 91:262001 (2003)
133. Acosta DE, et al. (CDF II Collab.) *Phys. Rev. Lett.* 93:072001 (2004)
134. Abazov VM, et al. (DØ Collab.) *Phys. Rev. Lett.* 93:162002 (2004)
135. Aubert B, et al. (BaBar Collab.) *Phys. Rev. D* 71:071103 (2005)
136. Braaten E, Lu M. *Phys. Rev. D* 76:094028 (2007)
137. Braaten E, Lu M. *Phys. Rev. D* 77:014029 (2008)
138. Braaten E. arXiv:0808.2948 [hep-ph] (2008)
139. Canham DL, Hammer HW, Springer RP. *Phys. Rev. D* 80:014009 (2009)
140. Bugg DV. *Phys. Lett.* B598:8 (2004)
141. Hanhart C, Kalashnikova YS, Kudryavtsev AE, Nefediev AV. *Phys. Rev. D* 76:034007 (2007)
142. Voloshin MB. *Phys. Rev. D* 76:014007 (2007)
143. Braaten E, Kusunoki M. *Phys. Rev. D* 69:074005 (2004)

Particle Physics Implications of F-Theory

Jonathan J. Heckman

School of Natural Sciences, Institute for Advanced Study, Princeton, New Jersey 08540; email: jheckman@sns.ias.edu

Annu. Rev. Nucl. Part. Sci. 2010. 60:237–65

First published online as a Review in Advance on July 16, 2010

The *Annual Review of Nuclear and Particle Science* is online at nucl.annualreviews.org

This article's doi: 10.1146/annurev.nucl.012809.104532

0163-8998/10/1123-0237$20.00

Key Words

F-theory, Grand Unified Theories, String Phenomenology

Abstract

We review recent progress in realizing Grand Unified Theories (GUTs) in a strongly coupled formulation of type IIB string theory known as F-theory. This review's main emphasis is on the expected low-energy phenomenology of a minimal class of F-theory GUTs. We introduce the primary ingredients in such constructions, then present qualitative features of GUT models in this framework such as GUT breaking, doublet-triplet splitting, and proton decay. Next, we review proposals for realizing flavor hierarchies in the quark and lepton sectors. We discuss possible supersymmetry-breaking scenarios and their consequences for experiments, as well as geometrically minimal realizations of F-theory GUTs that incorporate most of these features.

Contents

1. INTRODUCTION

Of the many vacua of string theory, some are presumably compatible with experiment. Even so, not a single string-based model has been found that satisfies all known constraints. Extracting predictions from the vast list of possibilities requires narrowing the search to phenomenologically promising vacua. Once suitable assumptions are imposed, we can then ask whether the remaining vacua are compatible with observation and, moreover, what new phenomena to expect. Correlations between seemingly unrelated physical features then translate into predictions of the model.

For the purposes of particle physics, gravity is expected to play a significant role only at energies near the Planck scale, $M_{pl} \sim 10^{19}$ GeV. As a first step in identifying promising vacua, it is thus natural to focus on a limit in which the effects of gravity decouple from the rest of particle physics. More formally, this can be phrased as the existence of a local model in which the particle physics degrees of freedom localize on a space-time-filling brane that can be separated from the rest of the string compactification. This by itself is a rather mild condition that can be met in many string constructions (1), again providing a landscape of potentially interesting vacua.

As an additional criterion for selecting promising vacua, we also assume that low-energy supersymmetry will soon be found and that, moreover, this should be viewed as evidence for Grand Unified Theories (GUTs). Although circumstantial, the evidence for GUT-like structures based on $SU(4) \times SO(4)$, $SU(5)$, or $SO(10)$ gauge groups (2–4) includes the unification of the gauge coupling constants in the Minimal Supersymmetric Standard Model (MSSM) and the intriguing fact that the chiral matter unifies into complete GUT multiplets charged under $SU(5)$. See Reference 5 for a recent review of some aspects of GUTs.

Although the requirements of a local model and GUT-like structures can individually be met, combining them significantly narrows the available options in string-based models. A strongly coupled formulation of IIB string theory, known as F-theory (6–8), provides a potentially promising starting point for realizing GUTs in local models. For recent work on F-theory GUTs, see, for example, References 9 through 63. Here we investigate what types of F-theory GUT geometries realize elements of the Standard Model and, conversely, how the geometry constrains the available model-building options.

1.1. Grand Unified Theory–Like Structures

We now briefly review the aspects of GUTs we aim to realize in string theory. Throughout, we work in conventions that are compatible with $\mathcal{N} = 1$ supersymmetry, so that all fermions are taken to be left-handed Weyl spinors.

In the MSSM, there are three generations of chiral superfields given by Standard Model analogs of the quark doublets Q, antiups U, and antidowns D transforming in the respective $SU(3)_C \times SU(2)_L \times U(1)_Y$ representations

$$Q : (3, 2)_{1/6}, \quad U : (\overline{3}, 1)_{-2/3}, \quad \text{and} \quad D : (\overline{3}, 1)_{1/3}. \qquad 1.$$

There are also three generations of lepton doublets L and antileptons E in the representations

$$L : (1, 2)_{-1/2} \quad \text{and} \quad E : (1, 1)_1. \qquad 2.$$

Compared with the Standard Model, the Higgs sector is extended to include Higgs up and Higgs down chiral superfields in the representations

$$H_u : (1, 2)_{1/2} \quad \text{and} \quad H_d : (1, 2)_{-1/2}. \qquad 3.$$

The Higgs fields couple to the chiral matter through the superpotential terms

$$W_{MSSM} \supset \lambda^u_{ij} H_u Q^i U^j + \lambda^d_{ij} H_d Q^i D^j + \lambda^l_{ij} H_d L^i E^j, \qquad 4.$$

which respectively induce masses for the up-type, down-type, and charged leptons of the Standard Model.

A remarkable feature of the MSSM is that when extrapolated up from the weak scale, the gauge coupling constants unify to one-loop order at a common value of $\alpha_{GUT} = g^2/4\pi \sim 1/24$ at an energy scale of 2×10^{16} GeV. This suggests that the gauge groups of the Standard Model unify into a bigger group such as $SU(5)$. Indeed, the chiral matter of the MSSM also unifies into the $SU(5) \supset SU(3)_C \times SU(2)_L \times U(1)_Y$ representations

$$10_M \rightarrow Q \oplus U \oplus E \quad \text{and} \quad \overline{5}_M \rightarrow D \oplus L. \qquad 5.$$

The Higgs up and Higgs down fields of the MSSM fit in the 5_H and $\overline{5}_H$, and the interaction terms for the up- and down-type quarks are respectively controlled by the interaction terms

$$5_H \times 10_M \times 10_M \quad \text{and} \quad \overline{5}_H \times \overline{5}_M \times 10_M. \qquad 6.$$

This is only the first step in realizing a GUT model. For example, a realistic GUT must also contain mechanisms for breaking the GUT group, removing the Higgs triplets from the 5_H and $\overline{5}_H$, generating more realistic flavor textures, maintaining a sufficiently stable proton, and coupling to a supersymmetry-breaking sector.

1.2. Local Models

Our main focus in this article is on a class of string theory–based constructions known as local models (1). Because gravity is quite weak even at the GUT scale, the aim of a local model is to first reproduce details of the Standard Model, and worry about gravity later. Geometrically, a local model corresponds to a system in which gauge theory localizes on a $(4 + k)$-dimensional subspace of the ten-dimensional string theory. Starting from the ten-dimensional Einstein-Hilbert action and the $(4 + k)$-dimensional Yang-Mills action, the resulting gauge coupling constant and four-dimensional Newton's constant respectively depend on the internal volumes as

$$(g_{YM}^{4d})^2 \propto \mathrm{Vol}(\mathcal{M}_k)^{-1} \quad \text{and} \quad G_N^{4d} \propto \mathrm{Vol}(\mathcal{M}_6)^{-1}. \qquad 7.$$

Decoupling gravity corresponds to a limit in which the ratio of the characteristic radii $[\mathrm{Vol}(\mathcal{M}_k)]^{1/k}[\mathrm{Vol}(\mathcal{M}_6)]^{1/6}$ becomes parametrically small. In perturbative type II superstring theory, this suggests associating the closed-string sector with gravity and associating the open-string sector with gauge theory. In this case, the Standard Model degrees of freedom localize on Dirichlet (D)-branes. As a point of terminology, we refer to a Dp-brane as one that fills the temporal direction and p spatial directions. In F-theory, there are nonperturbative generalizations of D-branes that allow us extra model-building options. For reviews of model-building efforts with open strings, see Reference 64.

Perturbative open-string degrees of freedom provide nearly all of the qualitative ingredients necessary to realize the Standard Model. The transfer matrix of open strings between such D-branes defines the color-space degrees of freedom of the gauge theory. For example, n spacetime-filling D3-branes at the same point in $\mathbb{R}^6$ realizes four-dimensional $\mathcal{N} = 4$ $U(n)$ gauge theory. It is also possible to engineer SO and USp gauge theories by use of unoriented open strings. Open strings stretched between different stacks of D-branes correspond to matter fields charged under the fundamental of one gauge group and the antifundamental of the other. With unoriented strings it is also possible to generate matter charged in the two-index symmetric and antisymmetric representations of a gauge group. Note that all of the matter content of the Standard Model transforms in such representations. In string-perturbation theory, the gauge-invariant interaction terms always contract fundamental indices with antifundamental indices.

There are, however, drawbacks to local models based on purely perturbative degrees of freedom. Much of this tension stems from the absence of certain GUT-like structures. Although it is possible to engineer $SU(5)$ or, more precisely, $U(5)$ gauge theories with the matter content of a GUT, terms such as the $5_H \times 10_M \times 10_M$ interaction cannot be generated in string-perturbation theory because such interaction terms do not match up an equal number of fundamental and antifundamental indices. Other GUT structures based on $SO(10)$ or exceptional groups fare worse. The spinor 16 of $SO(10)$ cannot be realized in open-string perturbation theory, and E-type gauge group structures are also unavailable. All these drawbacks point to the need for extra nonperturbative ingredients.

1.3. Organization of This Review

The rest of this review is organized as follows. In Section 2, we introduce the primary building blocks of F-theory GUTs. In Section 3, we review how this class of models addresses some of the qualitative issues present in four-dimensional GUT models. In Section 4, we discuss supersymmetry-breaking scenarios. In Section 5, we discuss more detailed aspects of flavor physics for quarks and leptons, and in Section 6, we review some of the characteristics of minimal models based on a single point of E_8 unification. Section 7 presents our conclusions and possible directions for future investigation.

The intent of this article is to provide a point of entry to the literature. For earlier reviews with slightly different emphases, see References 20, 21, 37, and 55. Also, some important topics are treated only briefly, if at all. Most noticeably absent is a discussion of recent efforts to include the effects of gravity (24, 29, 32, 36, 41, 43, 46, 56, 59).

2. BUILDING BLOCKS OF F-THEORY GRAND UNIFIED THEORIES

2.1. Brief Review of F-Theory

A remarkable feature of type IIB string theory is that it is invariant under the duality group $SL(2, \mathbb{Z})$, which corresponds to all 2×2 matrices with integer entries with unit determinant. Under the action of the $SL(2, \mathbb{Z})$ duality group, the Neveu-Schwarz (NS) and Ramond-Ramond (RR) two-forms transform as a two-component doublet. Moreover, complexifying the string coupling g_s by combining it with the RR zero-form C_0, the axio-dilaton

$$\tau = C_0 + \frac{i}{g_s} \qquad 8.$$

transforms as $\tau \to (a\tau + b)/(c\tau + d)$ under integers a, b, c, and d such that $ad - bc = 1$.

Under the weak/strong duality map $\tau \to -1/\tau$ given by $a = d = 0$ and $b = -c = -1$, the fundamental string and D1-brane interchange roles. For more general duality group transformations, the fundamental string maps to a bound state of p fundamental strings and q D1-branes or a (p, q) string. Because fundamental strings end on D-branes, acting by the duality group provides us with new strongly coupled (p, q) analogs of the more familiar D-branes on which strings of the same (p, q) type can end. Branes of different (p, q) type can also form nonperturbative bound states.

The profile of the axio-dilaton is closely connected with the presence of seven-branes in the compactification. As an illustrative example, consider a D7-brane filling $\mathbb{R}^{7,1}$ and sitting at a point of the remaining two spatial directions. The one-form flux F_1 sourced by the D7-brane is locally defined by the relation $F_1 = dC_0$. Globally, we cannot write F_1 in this form. Indeed, passing around a circle surrounding the D7-brane, C_0 shifts as $C_0 \to C_0 + 1$. Similarly, n D7-branes source n units of F_1 flux, inducing $C_0 \to C_0 + n$. This corresponds to an $SL(2, \mathbb{Z})$ transformation with $a = d = 1$, $b = n$, and $c = 0$. We see that, acting by $SL(2, \mathbb{Z})$ transformations, different seven-branes affect the axio-dilaton in different ways.

F-theory (6) provides a geometric formulation of strongly coupled type IIB vacua that automatically keeps track of τ in the presence of seven-branes. The main idea is to specify the profile of τ by including two additional geometric directions in addition to the usual ten space-time dimensions of string theory. Let us stress that these two extra dimensions are on a somewhat different footing from that of the other ten. In this 12-dimensional geometry, the parameter τ is to be viewed as the shape of a two-torus, or elliptic curve (6). This two-torus can be visualized as a parallelogram in the complex plane $\mathbb{C}$, with vertices at 0, 1, τ, and $\tau + 1$ and parallel sides identified. The shape of this torus is invariant under $SL(2, \mathbb{Z})$ transformations $\tau \to (a\tau + b)/(c\tau + d)$, which makes manifest the action of the $SL(2, \mathbb{Z})$ duality group of IIB string theory. The presence of seven-branes affects the value of τ and is indicated in the geometry by allowing τ to attain different values at different points in the space-time.

For the purposes of model building, we are particularly interested in compactifications of F-theory that preserve four real supercharges in four dimensions. Retaining this amount of supersymmetry imposes the geometric condition that these supercharges be covariantly constant in the four complex–dimensional internal space. Much as in other string compactifications, this condition is met for F-theory compactified on a four complex–dimensional Calabi-Yau space. Unlike other string compactifications, the six physical internal directions define a complex three-dimensional manifold (threefold) B_3 that need not be Calabi-Yau.

The geometry of the complex four-dimensional Calabi-Yau manifold, or Calabi-Yau fourfold, automatically includes the effects of seven-branes. As we have seen, encircling a seven-brane leads to an $SL(2,\mathbb{Z})$ transformation of τ. Although this is an acceptable description away from the seven-brane, this means that on top of the seven-brane, the shape of the two-torus becomes singular and effectively pinches off. As a hypersurface in $\mathbb{C}^2$ with coordinates x and y, the elliptic curve can be modeled in Weierstrass form as

$$y^2 = x^3 + fx + g, \tag{9}$$

where f and g generally have nontrivial position dependency on B_3. The location of the seven-branes are then given by roots of the discriminant of the cubic in x:

$$\Delta = 4f^3 + 27g^2. \tag{10}$$

The equation $\Delta = 0$ corresponds to a one complex–dimensional equation in B_3 and so sweeps out a two complex–dimensional subspace. As a polynomial in the coordinates of the threefold base B_3, Δ may factorize into irreducible polynomials, so that $\Delta = \Delta_1 \cdots \Delta_n$. Each factor $\Delta_i = 0$ then defines the location of seven-branes in B_3.

2.2. Geometry and Gauge Theory

The locations of seven-branes are dictated by where the elliptic curve of an F-theory compactification pinches off. The precise ways that this pinching off can occur admits a classification in terms of the ADE Dynkin diagrams familiar from gauge theories, which is known as the Kodaira classification of singular fibers. Locally, we can describe the eight–real dimensional space of the Calabi-Yau fourfold in terms of two complex directions spanned by the seven-brane, and two additional complex directions given by an ADE-type singularity (see **Table 1**). As their name suggests, these singularities correspond to geometries in which a number of S^2s intersecting according to the ADE Dynkin diagram have collapsed to zero size.

In M-theory and type IIA string theory, it is known that compactifying on an ADE singularity realizes the corresponding ADE gauge group $SU(n)$, $SO(2n)$, or $E_{6,7,8}$ in the uncompactified directions, and the situation in F-theory is no different. For example, in M-theory, integrating the three-form potential C_{abc} over each S^2 of the Dynkin diagram yields a $U(1)$ gauge boson. M2-branes wrapped over the remaining S^2s are charged under these $U(1)$s and constitute the off-diagonal gauge bosons of the ADE gauge group. The duality between circle compactifications of IIB and IIA string theory lifts to F-theory and M-theory, allowing us to translate this derivation

Table 1 Dictionary for ADE singularities and seven-brane gauge groups

ADE type	Equation in $\mathbb{C}^3$	F-theory gauge group
A_n	$y^2 = x^2 + z^{n+1}$	$SU(n+1)$
D_n	$y^2 = x^2 z + z^{n-1}$	$SO(2n)$
E_6	$y^2 = x^3 + z^4$	E_6
E_7	$y^2 = x^3 + xz^3$	E_7
E_8	$y^2 = x^3 + z^5$	E_8

over to the F-theory side. Although this approach may be unsuitable for GUT phenomenology, it is also possible to engineer the non–simply laced groups $USp(2n)$, $SO(2n+1)$, F_4, and G_2.

Associated with a singular ADE fiber of an F-theory compactification is a corresponding seven-brane with that ADE gauge group. Just as in perturbative type IIB string theory, space-time-filling seven-branes that occupy the same subspace in the internal directions allow us to engineer higher-dimensional gauge theories. Although our ultimate interest lies in seven-branes that wrap four compact internal directions, let us first consider the maximally supersymmetric gauge theory of seven-branes on $\mathbb{R}^{7,1}$ with gauge group G_S engineered by an ADE singularity of type (by abuse of notation) G_S. To arrive at the eight-dimensional seven-brane theory in flat space, let us first consider ten-dimensional supersymmetric Yang-Mills theory on $\mathbb{R}^{9,1}$ with 16 real supercharges and gauge group G_S. The field content of this ten-dimensional theory consists of an adjoint-valued gauge boson A_I and an adjoint-valued gaugino, which transforms as a 16-component Majorana-Weyl spinor. Dimensionally reducing this ten-dimensional theory to eight dimensions yields the maximally supersymmetric eight-dimensional gauge theory. Focusing on the bosonic-field content, reduction to eight dimensions yields an eight-dimensional gauge boson transforming as a vector under $SO(7,1)$, given by those components of A_I with directions along the eight space-time directions. The two additional gauge bosons of A_I then combine to form a complex scalar Φ.

The eigenvalues of the complexified scalar Φ determine the positions of the seven-branes in the threefold base B_3. Because the order of the singularity determines the gauge group on the seven-brane, these vacuum expectation values (VEVs) must also deform the geometry of the ADE-type singularity. Consider, for example, a stack of n seven-branes located at $z = 0$ with gauge group $SU(n)$. This corresponds to an A_{n-1} singularity,

$$y^2 = x^2 + z^n. \qquad 11.$$

We can break the gauge group down to $U(1)^{n-1}$ through the VEV $\Phi = \mathrm{diag}(t_1, \ldots, t_n)$, where $t_1 + \cdots + t_n = 0$. In the geometry this translates to the deformation

$$y^2 = x^2 + \prod_{i=1}^{n} (z - t_i). \qquad 12.$$

Note that for generic values of z, each factor in this product behaves roughly as a constant. However, near the locus $z = t_i$, we encounter the location of one of the seven-branes. At this location, the center of mass of the brane system is kept fixed, and the n seven-branes are separated.

More general Higgsing patterns of gauge groups also translate to complex deformations or to the unfolding of a singularity. See Reference 65 for further discussion on the deformation theory of the ADE singularities. Such deformations of G_S are parameterized in the gauge theory by the Casimirs of Φ, which in the above example with $\Phi = \mathrm{diag}(t_1, \ldots, t_n)$ are given by $\mathrm{Tr}(\Phi^n)$, which can in turn be expressed in terms of the elementary symmetric polynomials in the t_i. Thus, the VEVs of the Casimirs specify a breaking pattern that dictates how the seven-branes are distributed in the geometry.

We can also consider more general VEVs for Φ with nontrivial position dependency on the world volume spanned by the original stack of seven-branes. This corresponds to rotating some of the seven-branes of the original stack so that they now occupy a different set of four internal directions of the geometry. Such seven-branes then intersect the original stack along two real dimensions. To illustrate this, consider a $U(2)$ gauge theory described by two coincident seven-branes. When $\Phi = \mathrm{diag}(\phi, -\phi)$, this breaks $U(2)$ down to $U(1)^2$. Letting s denote a local holomorphic coordinate of S, when $\phi \sim s$, we see that for nonzero s, the gauge group is broken, but at $s = 0$, the $U(2)$ symmetry is restored, corresponding to the meeting of the two seven-branes. Modes Ψ charged under both stacks of branes obey a wave equation obtained from the dimensional reduction of the covariant derivative of ten-dimensional Yang-Mills theory to eight dimensions. Although the full system of equations is somewhat more involved, at a schematic level the mode Ψ obeys

$$\frac{\partial \Psi}{\partial \bar{s}} - \phi \cdot \Psi = 0, \qquad 13.$$

so that $\Psi \sim \exp(-|s|^2)$ exhibits Gaussian falloff away from the locus of symmetry enhancement at $s = 0$. In other words, matter becomes trapped along the intersection of the seven-branes. For further discussion on matter localization in F-theory, see References 10 and 66.

Shifting back and forth between the gauge-theory description and geometric data allows us to treat some of the more subtle aspects of symmetry enhancements in an F-theory compactification in terms of eight-dimensional gauge theory. Indeed, although at generic points the singularity type over a complex surface is given by G_S, over complex one-dimensional curves this enhancement can jump to a higher rank $G_\Sigma \supset G_S$, and over points it can jump again to $G_p \supset G_\Sigma \supset G_S$. As in our discussion surrounding Equation 13, we can treat the system as a G_p gauge theory that has been Higgsed down to the lower singularity types. The position dependency of Φ then dictates how these matter curves meet in the geometry.

2.3. The Four-Dimensional Effective Theory

In eight-dimensional flat space $\mathbb{R}^{7,1} = \mathbb{R}^{3,1} \times \mathbb{C}^2$, the components of the eight-dimensional gauge boson split up into two four-component vectors, each of which transforms (*a*) as a vector under one factor of space-time and (*b*) as a scalar of the other factor. The remaining bosonic degrees of freedom correspond to the scalar Φ, which controls the position of the seven-brane. The theory of a space-time-filling seven-brane that wraps four small directions is quite similar because, in a sufficiently small four-dimensional patch of the internal geometry, we can again treat the theory as an eight-dimensional gauge theory on $\mathbb{R}^{3,1} \times \mathbb{C}^2$. The main subtlety is how to take this local description and patch it together more globally in a way consistent with supersymmetry in $\mathbb{R}^{3,1}$. Although a full discussion is beyond the scope of this review, the main point is that on compactifications that preserve four real supercharges, we can organize all of the field content into four-dimensional $\mathcal{N} = 1$ superfields, which transform as differential forms with a given number of holomorphic and antiholomorphic indices on the two complex–dimensional surface S wrapped by the seven-brane. In terms of four-dimensional $\mathcal{N} = 1$ supermultiplets labeled by points of the internal geometry, the fields of the seven-brane theory organize into an adjoint-valued vector multiplet $W_{(0,0)}$, a chiral multiplet $\mathbb{A}_{(0,1)}$ with one antiholomorphic index, and a chiral multiplet $\Phi_{(2,0)}$ with two holomorphic indices (9, 10).

Dimensionally reducing the eight-dimensional gauge theory to four dimensions, the volume of the two complex–dimensional surface determines the gauge coupling constant of the four-dimensional gauge theory through the relation

$$\alpha_{GUT}^{-1} = \frac{4\pi}{g_{YM}^2} \sim M_*^4 \mathrm{Vol}(S), \qquad 14.$$

Where M_* is a mass scale close to the string scale $M_* \sim 10^{17}$ GeV. Roughly speaking, the radius of the volume factor $\mathrm{Vol}(S)^{-1/4} \sim M_{GUT}$ sets the characteristic scale of unification.

Our discussion so far has focused on the theory of an isolated seven-brane. Achieving realistic phenomenology requires intersecting the GUT seven-brane with other seven-branes of the compactification. In perturbative type IIB setups, the pair-wise and triple intersections of stacks of D7-branes respectively yield bifundamental matter and Yukawa couplings in the four-dimensional theory. In F-theory, the analogous phenomenon is due to other seven-branes wrapping complex

Table 2 Geometric ingredients of F-theory Grand Unified Theories

Dimension	Ingredient
10	Gravity
8	Gauge theory
6	Chiral matter (plus flux)
4	Cubic interaction terms

surfaces S' with gauge groups $G_{S'}$, which intersect the GUT stack of seven-branes. In this context, the gauge theory localizes on complex two-dimensional surfaces S wrapped by seven-branes, chiral matter localizes on complex one-dimensional curves defined by the intersection of such surfaces $\Sigma = S \cap S'$, and cubic interaction terms such as Yukawa couplings localize at points defined by the triple intersection of matter curves, $p = \Sigma_1 \cap \Sigma_2 \cap \Sigma_3$ (see **Table 2**).

The intersection of seven-branes along a complex one-dimensional curve Σ corresponds to a further enhancement in the ADE type of the singularity, which we denote by $G_\Sigma \supset G_S \times G_{S'}$. As mentioned in Section 2.2, we can locally model this situation in terms of an eight-dimensional gauge theory with gauge group G_Σ, which is Higgsed down to $G_S \times G_{S'}$ (9, 10, 66, 67). Some of the off-diagonal components of the fields $\mathbb{A}_{(0,1)}$ and $\Phi_{(2,0)}$ then become trapped along the locus at which the gauge group is restored to all of G_Σ. The resulting theory then contains a six-dimensional hypermultiplet charged under both G_S and $G_{S'}$. The adjoint representation of G_Σ decomposes into irreducible representations of $G_S \times G_{S'}$, some of which are charged under both seven-branes,

$$ad(G_\Sigma) = (R, R') + (\overline{R}, \overline{R'}) + \cdots, \qquad 15.$$

where both R and R' are nontrivial. In four-dimensional $\mathcal{N} = 1$ language, this six-dimensional field corresponds to a collection of vector-like pairs in the $(R, R') \oplus (\overline{R}, \overline{R'})$ trapped along the curve.

This analysis of matter localization recovers the usual result from perturbative type IIB string theory. For example, the intersection of D7-branes with gauge groups $G_S = SU(n)$ and $G_{S'} = SU(m)$ contains six-dimensional matter in the $(n, \overline{m}) \oplus (\overline{n}, m)$ of $SU(n) \times SU(m)$, corresponding to open strings stretched between the two stacks of D7-branes.

In an $SU(5)$ GUT, matter in the 10 and matter in the $\overline{5}$ of $SU(5)$ are respectively given by enhancements in the singularity type to $SO(10)$ and $SU(6)$. F-theory vacua also extend the available matter content beyond perturbatively realized states. For example, although the spinor 16 of $SO(10)$ cannot be generated by open-string states with two ends, it is present in geometries in which $G_S = SO(10)$ enhances to E_6 because the decomposition of the adjoint of $E_6 \supset SO(10) \times U(1)$ contains

$$78 \to 45_0 \oplus 1_0 \oplus 16_{-3} \oplus \overline{16}_{+3}. \qquad 16.$$

The kinetic terms of the matter-field wave functions are also determined by the geometry. Because these fields localize on matter curves, in a holomorphic basis of chiral superfields the corresponding kinetic terms are of the form

$$M_*^2 \operatorname{Vol}(\Sigma) \cdot \int d^4\theta \, \Psi^\dagger \Psi. \qquad 17.$$

In particular, matter localized on distinct matter curves has diagonal kinetic terms. A canonical normalization of matter fields then corresponds to performing the rescaling $\Psi \to [M_*^2 \operatorname{Vol}(\Sigma)]^{-1/2} \Psi$.

Further enhancements in the singularity type at points of the geometry induce interaction terms between matter localized on curves. To illustrate this scenario, consider a local point enhancement to E_6 that is Higgsed down to $SO(10)$ and $SU(6)$ along complex curves and $SU(5)$ in the bulk of

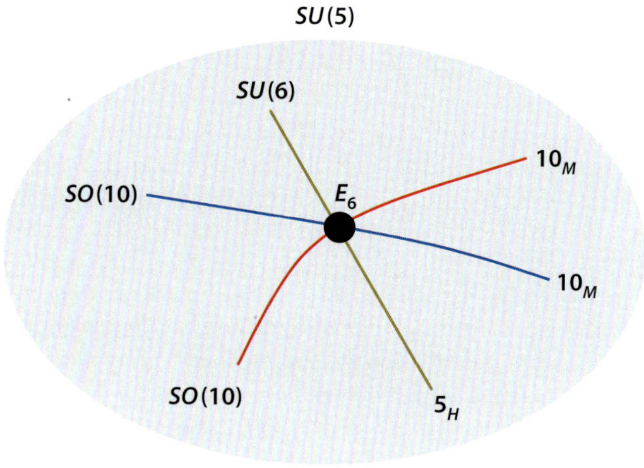

Figure 1

Depiction of an E_6 singularity enhancement. The gray-shaded ellipsoidal shape denotes the internal directions of the seven-brane, with gauge group $SU(5)$. Along complex one-dimensional curves, the singularity enhances to $SO(10)$ and $SU(6)$, where matter in the 10 and 5 of $SU(5)$, respectively, localizes. There is a further enhancement to E_6 at a point of the geometry at which the Yukawa interaction $5_H \times 10_M \times 10_M$ localizes.

the seven-brane. The adjoint representation of E_6 decomposes into irreducible representations of $SU(5) \times U(1)^2$ as

$$78 \to 24_{0,0} \oplus 1_{0,+2} \oplus 1_{0,0} \oplus 5_{+6,0} \oplus 10_{-3,+1} \oplus 10_{-3,-1} \oplus c.c. \qquad 18.$$

Returning to the discussion surrounding Equation 13, fields with different $U(1)^2$ charges localize on different curves. An interesting feature is that in the interaction term $5 \times 10 \times 10$, each field has a distinct $U(1)$ charge, so that these three matter fields automatically localize on three distinct matter curves. In other words, once two of the matter curves meet, a third appears to form the $5 \times 10 \times 10$ interaction. See **Figure 1** for a depiction of this geometry. Although this statement is basically correct, we refine it in Section 5, in which we discuss seven-brane monodromy.

More abstractly, in a neighborhood p of the enhancement to a singularity of type G_p, we can model the configuration of seven-branes in terms of an eight-dimensional theory with gauge group G_p Higgsed down to G_{Σ_i} along various matter curves and G_S elsewhere on the seven-brane. In the eight-dimensional theory, the internal gauge fields and $\Phi_{(2,0)}$ interact through the superpotential (9, 10)

$$W_{\text{parent}} = \int \text{Tr}(\overline{\partial}\mathbb{A}_{(0,1)} + \mathbb{A}_{(0,1)} \wedge \mathbb{A}_{(0,1)}) \wedge \Phi_{(2,0)}. \qquad 19.$$

Expanding $\mathbb{A}_{(0,1)} = \langle\mathbb{A}\rangle + \delta\mathbb{A}$ and $\Phi_{(2,0)} = \langle\Phi\rangle + \delta\Phi$ yields cubic superpotential terms coupling the $\delta\mathbb{A}$ and $\delta\Phi$s of the form

$$W_p = \lambda_{123}\Psi_1\Psi_2\Psi_3 \qquad 20.$$

for chiral superfields Ψ_i localized on curves of the internal geometry. Here the Yukawa coupling λ_{123} is given by the overlap of the Ψ_i wave functions in the internal geometry

$$\lambda_{123} = \int_S \Psi_1\Psi_2\Psi_3. \qquad 21.$$

In order for the coupling of Equation 20 to be generated, the product of the Ψs must form a gauge-invariant operator under the gauge groups of the seven-branes participating in the Yukawa interaction. However, the D-term

$$\int d^4\theta \frac{\Psi_1^\dagger \Psi_2 \Psi_3}{\Lambda_{UV}} \qquad 22.$$

may instead be gauge invariant. Integrating out Kaluza-Klein modes of mass Λ_{UV} localized on the Ψ_1 curve generates this higher-dimension operator (19). Note that the suppression scale Λ_{UV} is simply the Kaluza-Klein mass scale associated with harmonics on the matter curve, which in turn is set by the characteristic radius of the seven-brane theory to be near the GUT scale. This means that, if present, such terms dominate over similar M_{pl}-suppressed contributions.

Note that although these ingredients appear to provide the most promising route for model building, we may also consider models in which some of the matter content propagates in the world-volume of the seven-brane. In this more general context, there are additional interactions such as those between three bulk matter fields and those between one bulk mode and two matter fields on the same curve.

2.4. Seven-Brane Flux

The Standard Model has three chiral generations of matter. In an F-theory compactification, these data are encoded in how gauge-field flux in the internal directions of the seven-branes couples to the higher-dimensional fields of the seven-brane theory.

A background flux on a seven-brane takes values in a subgroup $\Gamma \subset G_S$. Much as in compactifications of the heterotic string, this breaks the gauge group in four dimensions down to the commutator subgroup of Γ in G_S. Note that a flux in an abelian subgroup such as $U(1)_Y$ hypercharge can then break $SU(5)$ down to $SU(3)_C \times SU(2)_L \times U(1)_Y$.

Matter fields localized at the pair-wise intersection of seven-branes with gauge groups G_S and $G_{S'}$ couple to both sets of background fluxes. These fields obey the six-dimensional Dirac equation

$$\not{D}_6 \Psi = (\not{D}_4 + \not{D}_\Sigma)\Psi = 0, \qquad 23.$$

where $\not{D}_4$ and $\not{D}_\Sigma$, respectively, denote the Dirac operator in $\mathbb{R}^{3,1}$ and the matter curve Σ. Because the coupling $\overline{\Psi}\not{D}_\Sigma\Psi$ of the higher-dimensional theory descends to a mass term of the four-dimensional theory, it follows that the massless modes correspond to zero modes of the Dirac operator along the matter curve.

Index theory now determines the number of zero modes. For abelian-field strengths F_S and $F_{S'}$,

$$\text{the number of chiral modes} = q_S \int_\Sigma \frac{F_S}{2\pi} + q_{S'} \int_\Sigma \frac{F_{S'}}{2\pi}, \qquad 24.$$

where the qs denote the charges of the fields under the respective gauge groups. By tuning the discrete parameters of a compactification, it is in principle possible to realize the spectrum of the Standard Model (14). Even so, as of this writing, a fully consistent compactification realizing precisely the matter curves and interaction terms of the Standard Model has yet to be constructed.

3. GRAND UNIFIED THEORY BREAKING AND ITS CONSEQUENCES

Having introduced the primary building blocks of an F-theory GUT, we now study some of their consequences for low-energy physics. To frame the following discussion, we often focus on the case of a minimal $SU(5)$, F-theory GUT with a seven-brane wrapping a complex surface S. Matter in the 10_M and $\overline{5}_M$ respectively localizes on enhancements to $SO(10)$ and $SU(6)$, with Higgs fields localized on additional $SU(6)$ enhancements. To reach the Standard Model, $SU(5)$ must be broken to $SU(3)_C \times SU(2)_L \times U(1)_Y$. This has consequences for the massless spectrum, proton decay, and gauge coupling unification.

3.1. Local Models and Grand Unified Theory–Breaking Fluxes

Breaking the GUT group down to the Standard Model gauge group turns out to be surprisingly restrictive in F-theory. In four-dimensional GUT models and higher-dimensional stringy models, the primary means to break the GUT group are (*a*) VEVs for adjoint-valued chiral superfields and (*b*) discrete Wilson lines. The local-model condition obstructs both of these options in an F-theory GUT (9, 10, 14, 17). In the seven-brane theory, the VEVs of adjoint-valued chiral superfields determine the location of the two complex–dimensional surface wrapped by the seven-brane S inside the threefold base B_3. Because gravity is decoupled in a local model, this seven-brane cannot explore B_3, so there are no such zero modes. Note that in some GUT models such as flipped $SU(5)$ GUTs, GUT breaking can be realized through the VEVs of fields in representations different from the adjoint. We briefly touch on this point below when we discuss F-theory GUT implementations of flipped $SU(5)$ GUTs.

Wilson line breaking is also problematic because it requires a nontrivial fundamental group $\pi_1(S) \neq 0$. If there exists a limit where B_3 remain of finite size while S contracts to a point, S is required to be a del Pezzo surface, namely a two complex–dimensional surface of positive curvature. For further discussion on del Pezzo surfaces geared toward physicists, see, for example, Reference 68. All such surfaces are simply connected, obstructing Wilson line breaking. Additional discussion on restrictions expected from the local-model condition can be found in References 29, 32, 52, and 59. In principle, this condition can be weakened to allow a milder version of decoupling (43).

GUT breaking by gauge-field fluxes provides an alternative to these options. Recall that nontrivial flux in a subgroup $\Gamma \subset G_S$ breaks the gauge group down to the commutator subgroup. Indeed, setting $\Gamma = U(1)_Y \subset SU(5)$ yields the commutator subgroup $SU(3)_C \times SU(2)_L \times U(1)_Y$.

It may appear surprising that this mechanism has not been exploited in earlier work on compactifications of superstring theory. As shown in Reference 69, this mechanism does not work in compactifications of the perturbative heterotic string due to a generalized Chern-Simons coupling to the NS two-form B in the ten-dimensional Lagrangian density

$$L_{Het} \supset (dB + A \wedge \langle F \rangle)^2, \qquad 25.$$

where $\langle F \rangle$ is the field strength in the internal directions of the compactification. Dualizing B to an axion generates a string-scale mass for A via the Stueckelberg mechanism. Remixing by other gauge bosons is possible, although it distorts gauge coupling unification (69; also see References 70 and 71 for recent discussion).

The analogous coupling in the seven-brane theory is between the RR four-form potential C_4, the gauge-field strengths in $\mathbb{R}^{3,1}$, and the internal directions

$$\int_{\mathbb{R}^{3,1} \times S} C_4 \wedge Tr(F_{\mathbb{R}^{3,1}} \wedge F_S). \qquad 26.$$

Assuming two of the legs of C_4 wedge with F_S, this leaves a two-form in $\mathbb{R}^{3,1}$, which is the analog of B in Equation 25.

The global topology of the complex threefold base B_3 can prevent the gauge field from developing a mass. The essential point is that the coupling of Equation 26 involves two-forms such as F_S defined purely on S, as well as a two-form given by two legs of C_4 defined on the entire threefold base B_3. Depending on how S embeds in B_3, the integral of the wedge product of C_4 and F_S can vanish. When this happens, there is no coupling of the gauge field to an axion, and the gauge field remains massless (14, 17). Preventing the gauge field from developing a string-scale mass imposes a mathematical condition on the relative cohomology of S and B_3, as well as a choice of flux F_S on S, and this condition can be met in explicit examples. In practical terms, what is required is that the two-cycle Poincaré dual to F_S in S lift to a trivial two-cycle in B_3. This is quite similar to the topological condition found in Reference 72, which prevents the $U(1)_Y$ hypercharge gauge boson from developing a mass in models based on D3-brane probes of singularities. An interesting feature of this result is that although string dualities connect certain heterotic and F-theoretic vacua, when this topological criterion is satisfied no heterotic dual exists (14, 17).

The presence of a background internal gauge-field flux in the hypercharge direction or hyperflux also affects the zero-mode spectrum of the theory. Consider again a minimal $SU(5)$ GUT model. The internal gauge bosons of $SU(5)$ in the $(3, 2)_{-5/6} \oplus (\overline{3}, 2)_{5/6}$ of $SU(3)_C \times SU(2)_L \times U(1)_Y$ are in exotic representations of the Standard Model gauge group, and they can be avoided only for an essentially unique choice of cohomology class for the hyperflux (14). An important subtlety for this type of GUT-breaking flux is the overall factor of five in the hypercharge of these off-diagonal elements. To avoid introducing exotics, it is necessary to allow seemingly fractional hyperfluxes (14). For all matter fields to couple to an integral number of net flux quanta, fluxes must be activated on the other seven-branes of the compactification (14, 17).

Increasing the rank of the GUT group only adds to the number of constraints that a background flux must satisfy. Breaking $SO(10)$ to $SU(3)_C \times SU(2)_L \times U(1)_Y \times U(1)$ by fluxes always generates exotics propagating in the bulk of the seven-brane (14). However, there do exist fluxes that break $SU(6)$ down to the Standard Model without introducing such bulk exotics (54).

It is also possible to construct hybrids that incorporate flux breaking to a four-dimensional model, then a further breaking of the GUT group down to the Standard Model. For example, in a flipped $SU(5) \times U(1)$ GUT, a VEV for matter in the $10_{-1} \oplus \overline{10}_{+1}$ leads to the Standard Model gauge group. Partial flux breaking of $SO(10)$ down to $SU(5) \times U(1)$ has been used as a starting point for the construction of flipped $SU(5)$ F-theory GUTs (14, 39, 40), which provides a natural embedding of a flipped GUT in a higher unified structure. It is also possible to consider flux breaking of $SO(10)$ down to $SU(3) \times U(1) \times SU(2)_L \times SU(2)_R$, as in Reference 22. Suitable VEVs for additional fields descending from vector-like pairs in the $16 \oplus \overline{16}$ of $SO(10)$ can then break the GUT group down further.

Fluxes also affect the spectrum of zero modes localized on matter curves. For example, in an $SU(5)$ GUT with matter in the $\overline{5}$, the decomposition of the $\overline{5}$ reveals that the triplets and doublets couple differently to the internal hyperflux

$$\overline{5} \rightarrow (\overline{3}, 1)_{+1/3} \oplus (1, 2)_{-1/2}. \qquad 27.$$

In other words, when the net hyperflux through a matter curve is nonzero, we do not retain full GUT multiplets in the zero-mode spectrum. An economical way to evade this problem is to demand that flux through chiral matter curves be zero (14). In less minimal $SO(10)$ models, flux from $U(1)_Y$ and $U(1)_{B-L}$ can also be chosen so that different components of a GUT multiplet localize on distinct curves (22).

Doublet-triplet splitting in the Higgs sector by fluxes occurs when the Higgs field localizes on a curve with nonzero hyperflux (14). Because H_u and H_d have opposite $U(1)_Y$ quantum numbers, to get a H_u and H_d zero mode these fields should localize on different matter curves (14).

3.2. Proton Decay

Proton decay is a signature of GUT models (see Reference 73 for a review) and has been studied in F-theory GUT models (14, 17, 48, 74). Dangerous dimension-four and -five operators that can induce rapid proton decay must be sufficiently suppressed to evade current experimental bounds. However, the coefficients of dimension-six operators provide a potentially exciting window into GUT physics.

The most phenomenologically problematic couplings are cubic F-terms such as $\overline{5}_M \times \overline{5}_M \times 10_M$. Such couplings are absent in the MSSM because of the existence of a $\mathbb{Z}_2^{matter}$ matter parity under which it is assumed that chiral matter is odd and that the Higgs fields are even. Geometrically, we can exclude such contributions by demanding the absence of such Yukawa enhancement points. The absence of such interactions is also compatible with symmetry considerations, either through embedding $\mathbb{Z}_2^{matter} \subset U(1)_{PQ} \subset E_8$ (e.g., 19) or by viewing it as a geometric action of the compactification (14, 38). E_7 singularity structures in F-theory GUTs also effectively forbid dimension-four proton decay operators (74).

Dimension-five operators such as

$$\frac{\eta_5}{M_{GUT}} \cdot \int d^2\theta \, QQQL, \qquad 28.$$

where η_5 is a numerical coefficient, induce rapid proton decay primarily through the channel $p \to K^+\overline{\nu}$. Current bounds on the lifetime of the proton require $\eta_5 < 10^{-10}$. In a four-dimensional GUT, doublet-triplet splitting can also inadvertently generate this operator. For example, when the Higgs up and Higgs down triplets couple to the MSSM matter fields through interactions such as

$$W_{GUT} \supset T_u QQ + T_d QL + M_{GUT} T_u T_d, \qquad 29.$$

heavy triplet exchange generates the operator $QQQL/M_{GUT}$. This problem is potentially worse in higher-dimensional models because a whole Kaluza-Klein tower of excitations becomes a concern.

Doublet-triplet splitting by hyperflux automatically addresses this issue by geometrically sequestering the Higgs up and down on distinct matter curves. This means that the Higgs triplets pair up with other Kaluza-Klein modes and that the offending dimension-five operator is not generated (14). In an intersecting seven-brane configuration, the absence of dangerous dimension-five operators can also be traced to the presence of additional anomalous symmetries under which the MSSM fields are charged, such as a $U(1)$ Peccei-Quinn (PQ) symmetry. As advocated elsewhere (19, 33, 42), this $U(1)$ naturally fits inside the unfolding of an E_8 singularity.

Dimension-six operators such as

$$\frac{\eta_6}{M_{GUT}} \cdot \int d^4\theta \, U^\dagger E^\dagger QQ, \qquad 30.$$

where η_6 is a numerical coefficient, induce proton decay primarily through the channel $p \to e^+\pi^0$, but at a level that impinges on the regions currently being probed by experiment. In a four-dimensional GUT, the coefficient η_6 approximately equals g_{GUT}^2 and is generated by the exchange of off-diagonal gauge bosons of the broken GUT group. A similar effect is present in F-theory GUTs from all of the off-diagonal elements of the broken gauge group. Moreover, because these modes actually fit inside an entire Kaluza-Klein tower of heavy states, the full contribution from such effects is summed up by a two-point function integrated over the internal directions wrapped by the seven-brane. Due to the presence of internal volume factors for the heavy modes participating in the corresponding Green's function, there is a relative parametric enhancement in some decay channels, in comparison to the four-dimensional decay rate, of a factor of $\log \alpha_{GUT}^{-1}$

(17, 21). Unfortunately, the overall ambiguity in the normalization of the internal Green's function means that extracting a precise coefficient is difficult to perform with present techniques (17).

3.3. Gauge Coupling Unification

Starting from the weak scale and evolving to higher energies, we define the GUT scale as the energy at which the $SU(2)_L$ and $U(1)_Y$ gauge couplings [normalized to embed in $SU(5)$] are equal. Unification occurs provided that the $SU(3)_C$ gauge coupling also meets at this same scale. At one-loop order, the MSSM couplings unify, and at two-loop order, the value of α_3 is roughly 4% lower than α_{GUT}, although the precise value depends on teraelectronvolt- and GUT-scale threshold corrections. Indeed, particular patterns of soft breaking masses can sometimes help with gauge coupling unification (75).

As in many GUT models, the GUT-breaking sector itself can shift the tree-level values of the gauge coupling constants. This shift is comparable in magnitude to the effects from subleading two-loop corrections to the evolution of the MSSM couplings. These contributions come with various signs, and importantly, there exist regimes of parameter space in F-theory GUTs in which gauge coupling unification can be retained (17, 25). Effects from the closed-string sector connected with hyperflux breaking also affect the running of couplings (44), although a complete discussion of such effects is beyond the scope of this article.

Consider the contribution from GUT-breaking fluxes. In the seven-brane theory, the gauge-field strength couples to itself through kinetic terms, as well as terms quartic in the field strength. The net contribution is given by the kinetic term for gauge fields of the eight-dimensional gauge theory and by the Chern-Simons-like coupling of the seven-brane to the background RR zero-form potential C_0,

$$M_*^4 \int_{\mathbb{R}^{3,1}\times S} \mathrm{Tr}(F \wedge {}_*F) + \int_{\mathbb{R}^{3,1}\times S} C_0 \wedge \mathrm{Tr}(F^4), \qquad 31.$$

where we have suppressed the order-one coefficients multiplying each term. Prior to GUT breaking, the GUT coupling is controlled by the volume wrapped by the seven-brane so that $\alpha_{GUT}^{-1} = M_*^4 \mathrm{Vol}(S)$. Activating an internal-field, gauge-field strength breaks the GUT group and shifts the relative values of the gauge coupling constants by

$$\alpha_i(M_{GUT})^{-1} \rightarrow \alpha_i(M_{GUT})^{-1} + k_i, \qquad 32.$$

where k_i is an order-one number that depends on the details of the instanton number for the flux, as well as on the breaking pattern. The percentage change in the value of the couplings is then on the order of $\alpha_{GUT} \sim 5\%$. These effects were studied for $U(1)_Y \subset SU(5)$ GUT-breaking fluxes (17), and a different embedding of abelian fluxes in $U(5)_{GUT}$ was studied in related IIB compactifications (25). In both cases, the sign of the correction tended to increase the mismatch in unification.

Incomplete GUT multiplets of heavy Kaluza-Klein modes also induce threshold corrections to unification (17, 25). Adding a single vector-like pair of Higgs triplets 10^{15}–10^{16} GeV allows unification to be retained (25). Summing over the entire tower of Kaluza-Klein states and including the contribution from the $\mathrm{Tr}(F^4)$ terms, the specific linear combination of one-loop determinants for the Laplacians of k-forms that enters the threshold correction can be recast as a quasi-topological invariant (17; M. Wijnholt, private communication). This invariant is known as the holomorphic Ray-Singer torsion (76),

$$\mathbb{T} \equiv \frac{1}{2} \sum_k (-1)^{k+1} \log \det{}' \Delta_k, \qquad 33.$$

where Δ_k denotes the Laplacian of bundle-valued k-forms, the one-loop determinants are evaluated in ζ-function regularization, and det′ denotes the determinant with all zero modes omitted. These threshold effects can in principle have either sign, thus preserving precision unification in a region of parameter space (17).

4. SUPERSYMMETRY BREAKING

So far, our discussion has focused on the dynamics of the model near the GUT scale. Making contact with observation requires a discussion of supersymmetry breaking, and in particular how all of the superpartners develop masses that are compatible with current experimental bounds. In keeping with the philosophy of local models, in this section we focus on supersymmetry-breaking scenarios that rely on dynamics in the vicinity of the seven-brane controlled by the VEV of a GUT singlet,

$$\langle X \rangle = x + \theta^2 F_X. \tag{34.}$$

Our aim is to realize scenarios in which the parameters F_X and x generate viable soft supersymmetry-breaking masses for the MSSM gauginos and scalars.

In addition to the soft masses, it is also necessary to generate weak-scale coefficients for the B_μ and μ terms,

$$L_{eff} \supset B\mu h_u h_d + \int d^2\theta \mu H_u H_d + \text{h.c.}, \tag{35.}$$

where h denotes the bosonic component of the superfield H. The values of these parameters, in addition to the soft supersymmetry-breaking terms in the Higgs sector, dictate electroweak symmetry breaking and consequently the value of the ratio of the Higgs up and Higgs down VEVs, $\tan\beta = v_u/v_d$.

In many supersymmetric models, especially those derived from string theory, there are typically many additional, nearly flat scalar directions that must also be stabilized. Here we assume that these moduli have been stabilized through high-scale dynamics, which do not interfere with our present discussion. See, for example, Reference 77 for recent discussion on combining certain moduli-stabilization scenarios with supersymmetry breaking in F-theory GUT models.

4.1. Moduli-Dominated Scenarios

Although it is natural to decouple the dynamics of many moduli in a local model, the overall volume modulus of the complex surface wrapped by the GUT brane remains dynamical, even after decoupling gravity. This suggests that the VEV of this volume modulus could be a dominant source of supersymmetry-breaking effects (78). Because this modulus couples to all of the MSSM superpartners with particular scaling dimensions, a VEV for it would generate a soft supersymmetry-breaking pattern of masses for the MSSM. The authors of Reference 12 studied the phenomenology of this scenario in greater detail and found that achieving appropriate electroweak symmetry breaking requires particular scaling dimensions for the fields of the MSSM. This type of scaling is most easily achieved in configurations in which all of the matter localizes on matter curves. In particular, in such scenarios the Yukawas originate from the triple intersection of seven-branes, which fits with the main emphasis of models presented above.

In this class of models, a bino-like lightest neutralino constitutes the lightest supersymmetric particle (LSP), providing a natural dark matter candidate for this scenario. Generating an appropriate relic abundance also requires a stau next-to-lightest supersymmetric particle (NLSP), with a mass close to that of the bino and a large value of $\tan\beta \sim 40$.

4.2. Gauge-Mediation Scenarios

In a limit where gravity is decoupled from gauge theory, and where the volume modulus is stabilized due to high-scale dynamics, gauge-mediated supersymmetry-breaking scenarios are quite natural (see Reference 79 for a review of gauge mediation). Such scenarios are quite attractive because transmission of supersymmetry breaking by gauge fields does not introduce new sources of flavor violation, which is potentially problematic in other approaches. Gauge-mediated supersymmetry scenarios in F-theory GUTs have been studied (16, 19). See, for example, References 80 and 81 for other work on realizing gauge mediation in string-based models.

In minimal gauge mediation, the supersymmetry-breaking GUT singlet X couples to vector-like pairs of messenger fields $Y \oplus Y'$ through the superpotential term

$$W \supset XYY'. \tag{36.}$$

Messenger and gauge loops then generate soft supersymmetry-breaking masses for the gauginos and scalars of the MSSM. The precise mass spectrum depends on the scale of supersymmetry breaking $\sqrt{F_X}$, the messenger scale x, the representation content and number of messengers, and the messenger-scale values for the μ and B_μ terms. As a rule of thumb, generating teraelectronvolt-scale colored superpartners requires $F_X/x \sim 10^5$ GeV.

This type of scenario has a clean geometric realization in F-theory GUTs (14). When Y and Y' localize on distinct matter curves, they meet a third curve, normal to the GUT seven-brane, on which X localizes. In principle, Y and Y' can correspond to matter in the $5 \oplus \overline{5}$ or the $10 \oplus \overline{10}$.

Gauge-mediated interactions generate soft masses but do not address how the μ-term of the MSSM is generated. We must avoid a GUT scale value for the μ-parameter in the MSSM superpotential

$$W_{MSSM} \supset \mu H_u H_d. \tag{37.}$$

This condition is met in models in which the Higgs up and down are sequestered on distinct matter curves because the requisite term is not invariant under the gauge symmetries of the other seven-branes intersecting the GUT stack (14, 19). In the low-energy effective theory, this scenario corresponds to a nearly exact anomalous $U(1)$ PQ symmetry. Such a PQ symmetry descends from another seven-brane of the compactification that intersects the GUT stack so that all matter fields are charged under $U(1)_{PQ}$. As a point of terminology, we refer to this seven-brane as the PQ seven-brane of the compactification.

In minimal setups, achieving a weak-scale μ-term determines the required scale of supersymmetry breaking. The higher-dimension operator

$$\int d^4\theta \frac{X^\dagger H_u H_d}{\Lambda_{UV}} \tag{38.}$$

induces $\mu \sim F_X/\Lambda_{UV}$. In the geometry, this operator is generated at points at which the X curve and Higgs curves meet (19). The scale $\Lambda_{UV} \sim 10^{15}$ GeV is close to the GUT scale (14, 19), which requires $F_X \sim 10^{17}$ GeV^2 to generate a weak-scale $\mu \sim 100$ GeV. Combined with the requirement $F_X/x \sim 10^5$ GeV, this implies $x \sim 10^{12}$ GeV (19).

The leading contribution to the $B\mu$-term, which is compatible with $U(1)_{PQ}$, descends from the operator

$$\int d^4\theta \frac{X^\dagger X X^\dagger H_u H_d}{\Lambda_{UV}^3}. \tag{39.}$$

This operator produces a value for $B\mu \sim (x/\Lambda_{UV}) \cdot \mu^2$, which is far smaller than μ^2. This is the primary contribution to the $B\mu$-term because of localization of the matter fields on Riemann surfaces. This yields the messenger-scale boundary condition $B\mu \sim 0$. In addition, the messenger-scale A-terms are also zero due to brane localization, so that additional potentially problematic CP-violating phases are not generated. Finally, at low energies, this scenario leads to large values of $\tan\beta \sim 20-35$.

Additional contributions to the soft masses arise from anomalous $U(1)$ gauge bosons that couple the supersymmetry-breaking sector to the visible sector (82). The effects of integrating out an anomalous $U(1)_{PQ}$ gauge boson were also studied (19); the authors showed that there are additional contributions to the soft breaking parameters from higher-dimension operators of the form

$$L_{eff} \supset -g_{PQ}^2 e_X e_\Psi \int d^4\theta \frac{X^\dagger X \Psi^\dagger \Psi}{M_{U(1)_{PQ}}^2}, \quad 40.$$

where e_X and e_Ψ denote the PQ charges of X and Ψ, g_{PQ} denotes the gauge coupling of the $U(1)_{PQ}$ gauge theory, and $M_{U(1)_{PQ}}$ is the mass of the $U(1)_{PQ}$ gauge boson. In principle, $M_{U(1)_{PQ}}$ can range from masses somewhat below the GUT scale to much higher values.

Combining the effects of minimal gauge mediation with the effects of heavy PQ gauge boson exchange induces a PQ deformation of the soft masses away from the usual minimal gauge-mediation spectrum

$$m_{soft}^2 = m_{mGMSB}^2 - q\,\Delta_{PQ}^2, \quad 41.$$

where $\Delta_{PQ} \sim g_{PQ}^2 F/M_{U(1)_{PQ}}$ and where $q \sim -e_X e_\Psi$ is determined by the product of the X and Ψ PQ charge of the field Ψ in question. The PQ charges of the MSSM chiral matter are opposite in sign to that of X. This lowers the mass of the scalars, such as the stau, but does not shift the masses of the gauginos. The actual value q depends on the neutrino sector of the model because compatibility with the other interaction terms of the MSSM uniquely determines a choice of PQ charge assignments compatible with the neutrino scenarios in Reference 33. See **Figure 2** for a plot of the characteristic mass spectrum of a Majorana neutrino scenario at minimal and maximal PQ deformation.

Generation of a hierarchically suppressed supersymmetry-breaking scale is more challenging, but it can be arranged through a Polonyi term,

$$W_{eff} \supset \exp[-\mathrm{Vol}(S_{PQ})]X, \quad 42.$$

which is induced by Euclidean D3-instantons wrapped on the PQ seven-brane (15, 16, 18, 19). See References 60, 62, and 63 for additional recent discussion on D3-instantons in F-theory. Here, realizing the precise scales $x \sim 10^{12}$ GeV, $F_X \sim 10^{17}$ GeV^2 requires a certain amount of tuning in the Kähler potentials of X and other axion-like fields that contribute to the quantum chromodynamics (QCD) axion (19). In principle, it is also possible to use the supergravity effective potential to generate an appropriate potential for X (16, 83), although doing so requires additional assumptions about the cosmological constant.

The supersymmetry-breaking sector also feeds into the physics of axions. The phase of X corresponds to a Goldstone mode of the nearly exact global $U(1)_{PQ}$ symmetry and constitutes, along with other axion-like fields, the QCD axion (19). Interestingly, the numerology demanded by gauge mediation is that $x \sim 10^{12}$ GeV. This value for x also corresponds to the value of the axion decay constant $x \sim f_a \sim 10^{12}$ GeV, which is a phenomenologically acceptable value. The axion supermultiplet also influences the cosmology of F-theory GUTs (26). The bosonic partner of the axion—the saxion—develops a weak-scale mass due to supersymmetry breaking. The fermionic component corresponds to the axino, which is also the spin-1/2 component of the gravitino. As the Goldstino of local supersymmetry breaking, the gravitino has a mass of $F_X/M_{pl} \sim 10-100$ MeV that is set by the supersymmetry-breaking scale. The saxion field oscillates in the early universe, eventually decaying prior to the start of big bang nucleosynthesis. When it decays, it releases a

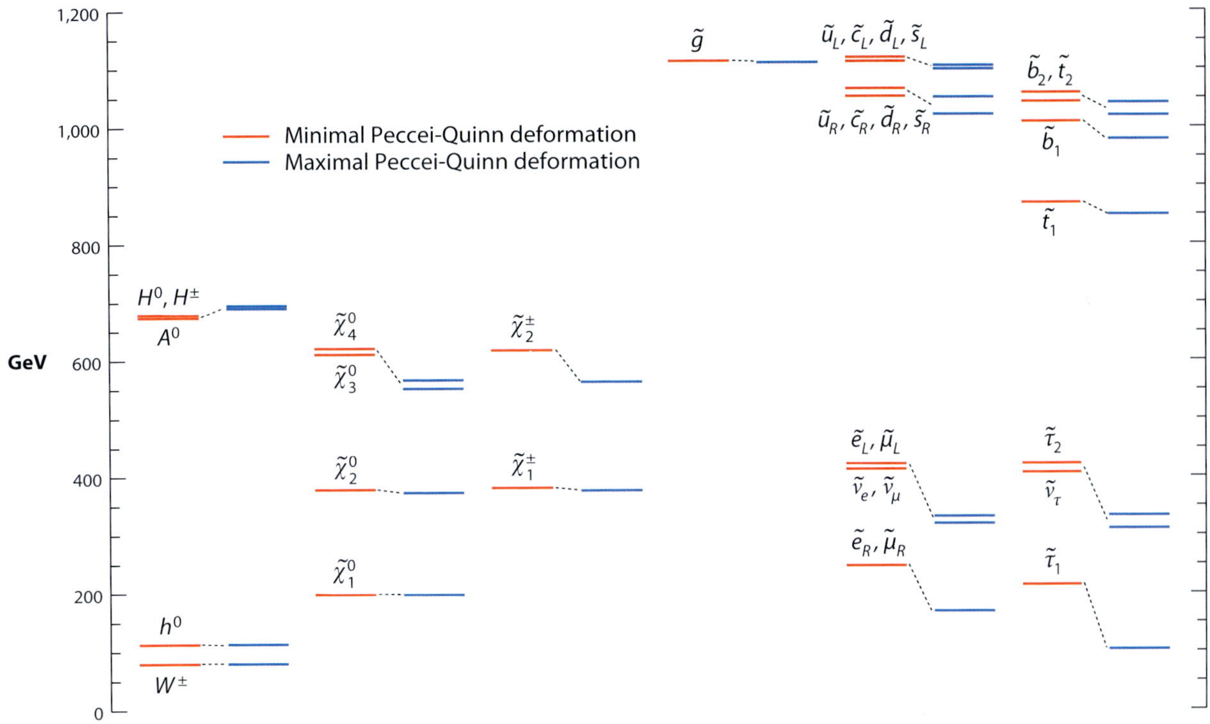

Figure 2

Plot (31, 61) of the sparticle mass spectrum (indicated by tildes) in an F-theory Grand Unified Theory with one vector-like pair of messengers in the $10 \oplus \overline{10}$ and $F_X/x = 5 \times 10^4$ GeV for minimal (*left, red columns*) and maximal (*right, blue columns*) Peccei-Quinn deformation. At $\Delta_{PQ}^{\min} = 0$, a bino-like lightest neutralino $\tilde{\chi}_1^0$ is the next-to-lightest supersymmetric particle. For moderate values of $\Delta_{PQ} \sim 100$ GeV, the NLSP transitions to the lightest stau $\tilde{\tau}_1$, which persists until a tachyon develops near $\Delta_{PQ}^{\max} \sim 200$ GeV.

significant amount of entropy. The diluted relic abundance of gravitinos makes the saxion field a viable dark matter candidate (26). In principle, axionic dark matter can also constitute a component of dark matter. Note that saxion decay also dilutes the relic abundance of other possible dark matter candidates (42).

5. FLAVOR IN F-THEORY

Proceeding down in energy scales, we have studied the physics at the GUT scale, and supersymmetry breaking. We now turn to realizations of flavor for the quarks and leptons of the Standard Model and its extension to massive neutrinos.

The masses and mixing angles of the quark sector exhibit striking and rather mysterious flavor hierarchies. For example, the masses of the u, c, and t quarks are 0.002, 1, and 170 GeV, respectively, and the magnitudes of the Cabibbo-Kobayashi-Maskawa (CKM) quark-mixing matrix elements are (84)

$$|V_{CKM}| \sim \begin{pmatrix} |V_{ud}| & |V_{us}| & |V_{ub}| \\ |V_{cd}| & |V_{cs}| & |V_{cb}| \\ |V_{td}| & |V_{ts}| & |V_{tb}| \end{pmatrix} \sim \begin{pmatrix} 1 & 0.2 & 0.004 \\ 0.2 & 1 & 0.04 \\ 0.008 & 0.04 & 1 \end{pmatrix}. \quad 43.$$

By contrast, neutrino-oscillation experiments indicate a far lower characteristic mass scale of order 0.05 eV, and compared to the CKM matrix, the leptonic mixing matrix V_{PMNS} is less hierarchical (84). Moreover, the magnitude of the (1, 3) element of the Pontecorvo-Maki-Nakagawa-Sakata (PMNS) matrix $|V^{1,3}_{PMNS}| = \sin\theta_{13}$ is still consistent with a value of zero and is bounded above by 0.2.

GUT structures naturally explain some aspects of flavor. For example, the masses of the b quark and τ lepton unify at the GUT scale (86), which fits with embedding their Yukawas in the interaction term $\overline{5}_H \times \overline{5}_M \times 10_M$. Geometrically, order-one coefficients for both $\overline{5}_H \times \overline{5}_M \times 10_M$ and $5_H \times 10_M \times 10_M$ seem the easiest to arrange, and this is the case we focus on here. Because the top and bottom quarks have different masses, order-one coefficients for the couplings $\overline{5}_H \times \overline{5}_M \times 10_M$ and $5_H \times 10_M \times 10_M$ are most compatible with large tan-β scenarios.

On the one hand, the limited representation content for zero modes in an F-theory GUT eliminates some common mechanisms in four-dimensional models (85) for correlating GUT breaking with flavor structure. On the other hand, these same geometric ingredients suggest new possibilities for both quarks and leptons. To frame our discussion, we again focus on minimal $SU(5)$ F-theory GUTs. We show that minimal geometries required to realize the MSSM superpotential also produce rank-one Yukawas. Subleading corrections to these Yukawas responsible for flavor hierarchies are then induced through violations of local symmetries, either from background ambient three-form fluxes in the case of quarks or from nonchiral massive modes of the compactification in the case of neutrinos.

5.1. Geometry of Flavor

The overlap of matter-field wave functions in the internal directions of the seven-brane determines the up-type Yukawa matrix

$$\lambda^u_{ij} = \int_S \psi^{H_u} \psi^Q_i \psi^U_j, \tag{44}$$

and there are similar expressions for the down-type and charged-lepton Yukawas. The maximal overlap between matter localized on different curves occurs where these curves geometrically intersect. To leading order, the up-type Yukawa matrix is given by a sum over all such intersection points:

$$\lambda^u_{ij} = \sum_p \psi^{H_u}(p)\psi^Q_i(p)\psi^U_j(p). \tag{45}$$

Note that each summand is an outer product of two flavor vectors, and so defines a rank-one 3×3 matrix. In other words, a single interaction point already generates the expected structure of one generation of weak-scale mass and two generations that, to first approximation, are massless (14, 23). In actual compactifications, there are additional enhancement points of the geometry (51, 52). Realizing a single heavy generation then suggests either that these points are tuned to be close together (14, 51) or that the MSSM fields participate in only one such point (14, 23).

The number of heavy generations depends on how many curves participate in a Yukawa interaction. First, consider a model in which the three generations of matter are distributed on up to three matter curves. The $5 \times 10 \times 10$ interaction requires a point of enhancement from $SU(5)$ to at least E_6. Returning to our discussion surrounding Equation 18, this interaction term couples two 10s, which would seem to localize on different matter curves because they have distinct charges under the abelian factor of $SU(5) \times U(1)^2 \subset E_6$. This coupling is phenomenologically problematic because the Yukawa matrix then has zeroes along the diagonal. Assuming roughly similar normalizations for the kinetic terms, this corresponds to a model with only one hierarchically light generation (14).

In actual compactifications of F-theory, the curves on which the two 10s localize are often merely different branches of a single complex one-dimensional curve (28). This is an example of a more general phenomenon known as seven-brane monodromy, which constitutes an important element in achieving realistic flavor structure in an F-theory GUT.

Seven-brane monodromy is best motivated by example. Consider the breaking pattern defined by unfolding an A_{n-1} singularity $y^2 = x^2 + z^n$ as

$$y^2 = x^2 + z^{n-2}(z^2 + 2\beta z + \gamma) = x^2 + z^{n-2}(z - t_+)(z - t_-), \qquad 46.$$

where $t_\pm = -\beta \pm \sqrt{\beta^2 - \gamma}$. This corresponds to an $SU(n-2)$ stack of seven-branes at $z = 0$ and another seven-brane at $z^2 + 2\beta z + \gamma = 0$. The branch cuts in $t_\pm$ identify the two seemingly different seven-branes at $z = t_+$ and $z = t_-$.

Per our discussion surrounding Equation 18, in terms of the eight-dimensional $SU(n)$ gauge theory locally defined at $z = t_\pm = 0$, the VEVs of the Casimirs of Φ can be packaged in terms of the coefficients in the unfolding of the singularity (10). When Φ is not diagonalizable, the corresponding matter curves exhibit branch cuts, reflected in the $t_\pm$.

We now define seven-brane monodromy in more general terms. Consider again a configuration in which the singularity type is G_S at generic points of the complex surface wrapped by the seven-brane, which enhances to G_p at a point of the geometry. The unfolding of this singularity is specified by coordinates t_i of the Cartan subalgebra of G_p, which are to be considered as the eigenvalues of Φ. The monodromy group of the seven-brane configuration G_{mono} permutes the t_is. In a breaking pattern of the form $G_p \supset G_S \times \Gamma$, G_{mono} is a subgroup of the Weyl group of Γ.

Returning to the example of the $5 \times 10 \times 10$ interaction, note that under the breaking pattern $E_6 \supset SU(6) \times SU(2) \supset SU(5) \times U(1) \times SU(2)$, the adjoint of E_6 decomposes as

$$78 \rightarrow (24, 1) \oplus (1, 3) \oplus (5_{-6}, 1) \oplus (\overline{5}_{+6}, 1) \oplus (10_{-3}, 2) \oplus (\overline{10}_{-3}, 2), \qquad 47.$$

so that the 10s transform as a doublet of $SU(2)$. Indeed, the Weyl group of $SU(2)$ is $\mathbb{Z}_2$, and monodromy acts by interchanging the components of the doublet. The geometry then consists of a single E_6 enhancement point at which one 10 curve and one 5 curve meet to form the $5 \times 10 \times 10$ interaction. Thus, monodromy in the seven-brane configuration allows us to achieve two hierarchically light generations (28).

5.2. Quark Yukawas

We now turn to the Yukawas of the lighter quark generations. A common way to generate flavor hierarchies is through the Froggatt-Nielsen mechanism (87). In this scenario, the matter fields of the Standard Model are charged under an additional horizontal symmetry, where different generations have distinct Froggatt-Nielsen charges. Spontaneous breaking of this horizontal symmetry generates additional corrections $\lambda_{ij} \sim \varepsilon^{a_i + b_j}$, producing hierarchical masses and mixing angles. This begs the question, however, as to how the parameters a_i and b_j are to be chosen, as well as how the parameter ε is fixed.

It is possible to engineer F-theory GUTs utilizing this four-dimensional mechanism (57). Even without introducing explicit GUT singlets, the ingredients of an F-theory compactification provide similar, although more stringy, variants on this theme. Matter localized on complex one–dimensional curves have similar charges induced by the action of the internal Lorentz group (23). More precisely, the matter wave functions satisfy $\overline{\partial}_A \psi = 0$ and so are locally holomorphic. Organizing the wave-function solutions according to their order of vanishing near a Yukawa point yields $\psi^i \sim z^{3-i} + O(z^{4-i})$ for $i = 1, 2, 3$ in a three-generation model. The action of the internal Lorentz group corresponds to a rephasing symmetry of z, providing a horizontal symmetry of the chiral generations. Background gauge-field and three-form fluxes of an F-theory compactification distort the profile of matter-field wave functions (23):

$$\psi \to \exp(\mathcal{M}_{i\bar{j}} z_i \bar{z}_{\bar{j}})\psi, \tag{48.}$$

where the z_i denote local holomorphic coordinates in a patch of the Yukawa point. This distortion of wave functions then suggests a possible mechanism for generating corrections to Yukawas. Note that by dimensional analysis, $\mathcal{M}$ scales as M_{GUT}^2.

Even so, the topological structure of a single nonzero entry turns out to be quite robust against physical perturbations. Gauge-field fluxes, for example, distort the profile of the physical matter-field wave functions but fail to distort the Yukawa structure (47, 49). Yukawa-distorting fluxes correspond to three-form fluxes such as $H = H_R + \tau H_{NS}$, which induce higher-dimension operator deformations of the seven-brane gauge theory. The effects of three-form H flux are locally captured in a patch of the Yukawa-enhancement point in terms of a noncommutative deformation of the seven-brane theory (47). In principle, instanton effects can also generate similar noncommutative deformations (53), the effects of which can also be captured in terms of an effective contribution from H fluxes (88).

Having different hypercharges, the different components of a GUT multiplet couple to the background fluxes differently, leading to different wave-function and Yukawa distortions for the various generations. This is important because whereas the bottom quark and τ lepton masses unify near the GUT scale, the masses of other particles, $m_c \neq m_\mu$ and $m_d \neq m_e$, do not. In other words, the geometry of the compactification preserves a leading-order notion of matter unification, which is violated by flux distortion of the wave functions (14, 23, 47).

Expanding in successive powers of $\bar{z}$ in $\mathcal{M}$, either by expanding in higher powers of $\mathcal{M}$ in the exponential of Equation 48 or by including higher-order derivative corrections, yields the structures (23)

$$\lambda_{DER} \sim \begin{pmatrix} \varepsilon^5 & \varepsilon^4 & \varepsilon^3 \\ \varepsilon^4 & \varepsilon^3 & \varepsilon^2 \\ \varepsilon^3 & \varepsilon^2 & 1 \end{pmatrix} \quad \text{and} \quad \lambda_{FLX} \sim \begin{pmatrix} \varepsilon^8 & \varepsilon^6 & \varepsilon^4 \\ \varepsilon^6 & \varepsilon^4 & \varepsilon^2 \\ \varepsilon^4 & \varepsilon^2 & 1 \end{pmatrix}, \tag{49.}$$

where each entry is multiplied by an order-one complex coefficient that we suppress. As the DER in its name suggests, the λ_{DER} expansion is given by expanding to first order in the flux and then expanding in successive gradients, or derivatives, of the flux. The λ_{FLX} expansion is given by expanding to leading order in the gradients of the flux, and then expanding in successive powers of the first gradient.

Because the matter-field wave functions couple differently to the background fluxes, the precise value of ε and the dominance of the FLX or DER expansion can depend on the matter field in question (23). Identifying up-type quark Yukawas with λ_{FLX} and down-type quarks with λ_{DER} yields hierarchical up-type quark masses

$$m_u : m_c : m_t \sim \varepsilon_u^8 : \varepsilon_u^4 : 1 \tag{50.}$$

and down-type quark masses

$$m_d : m_s : m_b \sim \varepsilon_d^5 : \varepsilon_d^3 : 1. \tag{51.}$$

Although this result is somewhat heuristic, it reproduces the observed masses and mixing angles surprisingly well (23). Numerically, achieving a match to the observed quark masses works best when $\varepsilon_u \sim 0.26$ and $\varepsilon_d \sim 0.27$. This is quite close to the value expected on the basis of the relative scaling of $\mathcal{M}$ to the string scale $\varepsilon \sim M_{GUT}^2/M_*^2 \sim \sqrt{\alpha_{GUT}} \sim 0.2$ (23, 47).

Achieving a hierarchical CKM matrix is more subtle. The CKM matrix measures the mismatch between the gauge and mass quark eigenstates. Because the profile of matter-field wave functions varies over the geometry of the seven-brane, achieving a roughly diagonal CKM matrix requires near alignment of the up- and down-type Yukawa points. Assuming that this further condition for point unification has been met, the resulting CKM matrix is then (23, 47)

$$V_{CKM}^{F-th} \sim \begin{pmatrix} 1 & \varepsilon & \varepsilon^3 \\ \varepsilon & 1 & \varepsilon^2 \\ \varepsilon^3 & \varepsilon^2 & 1 \end{pmatrix} \sim \begin{pmatrix} 1 & 0.2 & 0.008 \\ 0.2 & 1 & 0.04 \\ 0.008 & 0.04 & 1 \end{pmatrix} \qquad 52.$$

at the GUT scale. In principle, this matrix should be evolved to lower energy scales, but doing so multiplies various entries by order-one coefficients, which is already beyond the approximation scheme adopted here. Returning to Equation 43, it is surprising how well these crude numerical estimates match with observation.

From the perspective of Yukawa enhancements, putting E_6 and $SO(12)$ together can be accomplished through a higher unification to either E_7 or E_8. Including neutrinos pushes the singularity type at the Yukawa point up to E_8 (42). We return to this theme in Section 6 when we discuss minimal unification at a point of E_8.

5.3. Neutrino Models

The numerology of the seesaw mechanism $m_\nu \sim M_{weak}^2/\Lambda_{UV}$, for Λ_{UV} close to the GUT scale, strongly suggests a connection between GUTs and neutrinos. It is well known that Majorana and Dirac neutrino masses can respectively be generated by the higher-dimension operators

$$\int d^2\theta \frac{(H_u L)^2}{\Lambda_{UV}} \quad \text{and} \quad \int d^4\theta \frac{H_d^\dagger L N_R}{\Lambda_{UV}}. \qquad 53.$$

The Higgs VEVs $H_u \sim M_{weak}$ and $H_d \sim \theta^2 F_{H_d} \sim \theta^2 M_{weak}^2$ induce the mass terms

$$\int d^2\theta \frac{M_{weak}^2}{\Lambda_{UV}} N_L N_L \quad \text{and} \quad \int d^2\theta \frac{M_{weak}^2}{\Lambda_{UV}} N_L N_R. \qquad 54.$$

Right-handed singlets that interact with the MSSM fields can originate from matter localized on curves normal to the GUT stack (14, 33, 34), but in principle they can also originate from moduli fields (38). Right-handed neutrinos in the spinor 16 of an $SO(10) \supset$ flipped $SU(5)$ F-theory GUT are also possible (14, 39, 40).

The flavor structure of $SU(5)$ models with neutrinos localized on singlet curves has been better developed, so we focus on this case. In such scenarios, the interaction with the MSSM fields localizes at a point of enhancement from $SU(5)$ to at least $SU(7)$. Decomposing the adjoint of $SU(7)$ to $SU(5) \times U(1)^2$ yields

$$48 \rightarrow 24_{0,0} \oplus 1_{0,0} \oplus 1_{-2,0} \oplus 1_{2,0} \oplus 5_{+1,-7} \oplus 5_{-1,-7} \oplus \overline{5}_{-1,+7} \oplus \overline{5}_{+1,+7}, \qquad 55.$$

so that the MSSM fields correspond to a pairing of the 5 and $\overline{5}$ with a GUT singlet $1_{\pm 2,0}$, corresponding to the right-handed neutrino. Depending on the $U(1)$ charge assignments of the matter fields and the neutrino scenario in question, integrating out massive modes of the compactification would then generate the higher-dimension operators of Equation 53. For Majorana scenarios, seven-brane monodromy is especially important because it breaks $U(1)$ symmetries that would otherwise forbid Majorana mass terms.

The participation of massive nonchiral modes dilutes the flavor hierarchies in comparison to what is present in the quark and charged-lepton Yukawas, yielding neutrino masses

$$m_{\nu_1} : m_{\nu_2} : m_{\nu_3} \sim \varepsilon^2 : \varepsilon : 1, \qquad 56.$$

which is known as a normal mass-hierarchy scenario. Up to order-one factors, the predicted ratio of atmospheric and solar neutrino mass splittings is then (31)

$$\frac{\Delta m_{sol}^2}{\Delta m_{atm}^2} \sim \varepsilon^2 \sim \alpha_{GUT} \sim 0.04, \qquad 57.$$

which is numerically quite close to the experimental value of 0.03 (84). Again, note that this numerology works surprisingly well considering all of the crude approximations performed.

Large neutrino mixing angles are generically present in models in which the neutrino interaction point is geometrically separated from the lepton-interaction point. This finding is in accord with the observed large mixing angles θ_{atm} and θ_{sol}, but it leads to a certain amount of tension with the current upper bound on the mixing angle $\theta_{13} < 0.2$.

As in the quark sector, realizing a hierarchical mixing matrix naturally suggests unifying the charged-lepton and neutrino interaction points. Even when the neutrino and charged-lepton interaction points are close together, the Kaluza-Klein dilution of flavor hierarchies still generates large mixing angles with PMNS mixing matrix (33)

$$V_{PMNS} \sim \begin{pmatrix} V_{e1} & \varepsilon^{1/2} & \varepsilon \\ \varepsilon^{1/2} & V_{\mu 2} & \varepsilon^{1/2} \\ \varepsilon & \varepsilon^{1/2} & V_{\tau 3} \end{pmatrix}, \qquad 58.$$

where the Vs are constrained by unitarity of V_{PMNS}. Using the same parametric scaling, $\varepsilon \sim \sqrt{\alpha_{GUT}}$, as in the quark sector yields $\varepsilon^{1/2} \sim 0.45$, leading to order-one solar- and atmospheric-neutrino mixing. In addition, the model also predicts

$$\theta_{13} \sim \alpha_{GUT}^{1/2} \sim 0.2. \qquad 59.$$

In other words, the expectation is that θ_{13} is close to the current experimental bound. This expectation also appears to be a common feature of neutrino models with additional interaction points (34).

Majorana and Dirac neutrino scenarios can in principle be distinguished by neutrinoless double-beta decay experiments. The corresponding decay amplitude is generated only in Majorana scenarios and is proportional to a mass term $m_{\beta\beta}$, which in F-theory GUTs is on the order of 6 meV (33, 34). This value is quite small and may be within reach of experiments only on the horizon of a decade (89).

More options are available in less minimal scenarios. Anarchic neutrino mixing and mass matrices can be realized either by separating the charged-lepton and neutrino interaction points (33) or by including multiple neutrino enhancement points (34). Tuning the locations of the neutrino interaction points, neutrino models with an inverted mass hierarchy can also be realized (34).

6. MINIMAL E_8 UNIFICATION

In this article, we have demanded that the geometry of an F-theory GUT satisfy numerous phenomenological constraints. These include a list of matter curves where GUT multiplets localize and interaction terms for the MSSM superpotential, neutrino sector, and supersymmetry-breaking sector (for gauge-mediation models). From this perspective, each requirement would seem to add an additional degree of arbitrariness to the models we have considered. In addition, it is not altogether clear whether all of these ingredients can consistently combine in a single, unified framework.

The requirements of flavor physics that the quark and leptons exhibit hierarchies in the mixing matrices suggests that the E_6, $SO(12)$, and $SU(7)$ interaction points unify into a single point of E_8 enhancement (33, 42). This E_8 enhancement breaks down to $SU(6)$ and $SO(10)$ along the GUT matter curves, and it is also compatible with the supersymmetry-breaking sector introduced in Section 4. The matter curves and interaction terms are then determined by the breaking pattern for the adjoint of $E_8 \supset SU(5)_{GUT} \times SU(5)_{\perp}$,

$$248 \rightarrow (24, 1) \oplus (1, 24) \oplus (5, 10) \oplus (\overline{5}, \overline{10}) \oplus (\overline{5}, 10) \oplus (5, \overline{10}). \qquad 60.$$

Given the large set of independent directions in the Cartan of $SU(5)_{\perp}$, such an unfolding might appear to generate a large number of extraneous matter curves. Seven-brane monodromy, which already figures prominently in the physics of flavor, propagates to other sectors of the model, constraining the ways that additional matter can be added. In fact, it is possible to classify the possible breaking patterns of E_8 with seven-brane monodromy, subject to the following physical conditions: (*a*) hierarchical CKM and PMNS matrices; (*b*) a Kaluza-Klein seesaw for neutrinos; (*c*) $U(1)_{PQ}$ forbidding a bare μ-term; and (*d*) a μ-term from either $\int d^4\theta \frac{X^{\dagger} H_u H_d}{\Lambda_{UV}}$ or $\int d^2\theta S H_u H_d$, where X and S are to be considered GUT singlets that develop suitable VEVs to induce a μ-term. In addition to identifying many matter curves, the monodromy group also identifies the $U(1)$s in $SU(5)_{\perp}$. In Majorana neutrino scenarios, only $U(1)_{PQ}$ survives, and in Dirac neutrino scenarios, only $U(1)_{PQ}$ and $U(1)_{B-L}$ survive. See **Figure 3** for a depiction of a minimal E_8 model.

There are typically just enough matter curves to accommodate the Standard Model, and there is a very constrained messenger sector. In all Majorana neutrino scenarios, and all but one Dirac

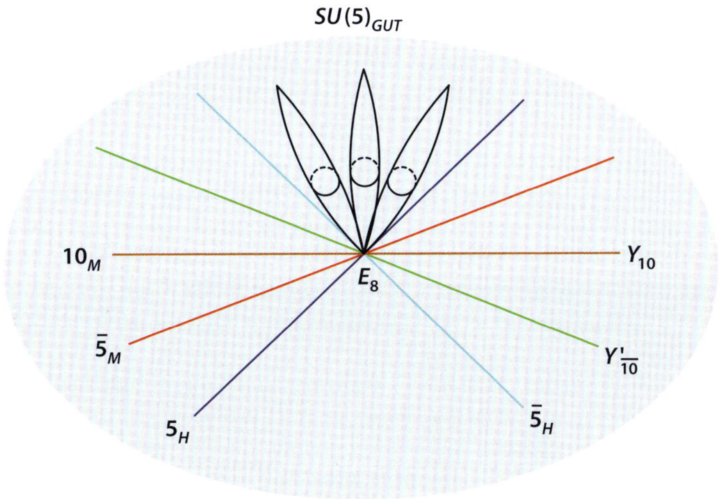

Figure 3

Depiction of a minimal model with E_8 point unification. In the proposed model, all of the interaction terms descend from a single E_8 point of enhancement. Here, matter curves inside the seven-brane are depicted by straight lines, and curves normal to the seven-brane are depicted by cigar-shaped tubes. In these minimal models, there is typically just enough room to accommodate the Minimal Supersymmetric Standard Model spectrum and a minimal messenger sector (*Y*s) in the $10 \oplus \overline{10}$. Abbreviation: GUT, Grand Unified Theory.

neutrino scenario, the messengers transform in vector-like pairs in the $10 \oplus \overline{10}$. Moreover, the matter curve for the 10_M also supports the messenger (42). In loop-exchange diagrams involving such messenger fields, increasing the number of messengers or, equivalently, the dimension of the representation tends to lower the masses of the scalar partners relative to the gauginos. In particular, once the effects of the PQ deformation are included, the mass of the lightest stau tends to be lowered relative to the mass of the bino (42).

The qualitative experimental signatures of this minimal gauge-mediation model are dictated by the scale of supersymmetry breaking and by the identity of the NLSP. The NLSP decays to its Standard Model counterpart and the gravitino with decay rate F_X^2/m_{NLSP}^5, leading to a lifetime on the order of 1 s to 1 h. This means that on timescales probed by colliders, the NLSP is quasi-stable, which is somewhat different from other gauge-mediation scenarios.

Next consider bino NLSP scenarios. In gauge-mediation scenarios with a low scale of supersymmetry breaking, the bino NLSP promptly decays inside the detector to a gravitino and photon. This signature is not generated in F-theory GUT scenarios with a bino NLSP, because the higher scale of supersymmetry breaking means that the bino decays only outside the detector. Rather, the bino leaves the detector as missing energy, much as in gravity-mediated supersymmetry breaking. Nevertheless, the different particle spectrum, especially for colored states, allows such scenarios to be distinguished from F-theory GUTs (31).

A characteristic feature of quasi-stable stau NLSP scenarios is that, once produced in a collider, the stau appears as a heavy charged particle, which registers in the detector as a so-called fake muon. Compared to the signatures of other supersymmetric models with a bino LSP or a quasi-stable NLSP, this particle has a clean signature with low Standard Model background. The expected collider phenomenology of such F-theory GUT scenarios at the CERN Large Hadron Collider has been studied (61).

7. CONCLUSIONS

F-theory GUTs provide a geometric framework for connecting string-scale physics to phenomena of the Standard Model. In this article we have introduced the primary ingredients that enter into such constructions, emphasizing the tight interplay between these ingredients in phenomenological models. Although the geometry is flexible enough to accommodate many aspects of the Standard Model, it is also rigid enough to favor particular phenomenological scenarios.

Minimal scenarios that make use only of ingredients found within a single E_8 point seem particularly rigid and predictive. What is particularly interesting is that within this specific class of models, there are various avenues by which they can be falsified and thus indirectly tested by the Large Hadron Collider and by other upcoming experiments for neutrinos, dark matter, and possibly proton decay.

The construction of explicit geometries realizing all of these ingredients remains an active area of investigation, and such considerations will probably provide an important guide for future model-building efforts. Consistently coupling such models to gravity remains an important avenue of investigation, potentially providing new constraints that cannot be seen in purely local models. Moreover, extracting the collider and flavor physics signatures of broader classes of such vacua may help to illuminate whether F-theory GUTs are realized in nature. Although any one aspect of a model could be viewed as suggestive, the fact that constraints from the geometry propagate to many aspects of an F-theory GUT makes the prospects of nontrivial correlations and the prospect of contact with distinct experiments especially exciting.

DISCLOSURE STATEMENT

The author is not aware of any affiliations, memberships, funding, or financial holdings that might be perceived as affecting the objectivity of this review.

ACKNOWLEDGMENTS

I thank C. Vafa for a very stimulating collaboration, which led to much of the work reviewed here. In addition, I thank my collaborators on related F-theory projects. I also thank J. Marsano, S. Raby, C. Vafa, B. Wecht, and E. Witten for helpful comments on the draft. My research is supported by National Science Foundation grant PHY-0503584.

LITERATURE CITED

1. Antoniadis I, Kiritsis E, Tomaras TN. *Phys. Lett.* B486:186 (2000); Aldazabel G, Ibáñez LE, Quevedo F, Uranga AM. *J. High Energy Phys.* 0008:002 (2000); Verlinde H, Wijnholt M. *J. High Energy Phys.* 0701:106 (2007)
2. Pati JC, Salam A. *Phys. Rev. D* 8:1240 (1973)
3. Georgi H, Glashow SL. *Phys. Rev. Lett.* 32:438 (1974)
4. Georgi H. *Particles and Fields, 1974: Proceedings of the APS Division of Particles and Fields*, ed CE Carlson, p. 575. New York: AIP (1975); Fritzsch H, Minkowski P. *Ann. Phys.* 93:193 (1975)
5. Raby S. Review on "Grand Unified Theories." In Amsler C, et al. *Phys. Lett.* B667:1 (2008)
6. Vafa C. *Nucl. Phys. B* 469:403 (1996)
7. Morrison DR, Vafa C. *Nucl. Phys. B* 473:74 (1996)
8. Morrison DR, Vafa C. *Nucl. Phys. B* 476:437 (1996)
9. Donagi R, Wijnholt M. arXiv:0802.2969 [hep-th] (2008)
10. Beasley C, Heckman JJ, Vafa C. *J. High Energy Phys.* 0901:058 (2009)
11. Hayashi H, et al. *Nucl. Phys. B* 806:224 (2009)
12. Aparicio L, Cerdeño DG, Ibáñez LE. *J. High Energy Phys.* 0807:099 (2008)
13. Buchbinder EI. *J. High Energy Phys.* 0809:134 (2008)
14. Beasley C, Heckman JJ, Vafa C. *J. High Energy Phys.* 0901:059 (2009)
15. Heckman JJ, et al. arXiv:0808.1286 [hep-th] (2008)
16. Marsano J, Saulina N, Schäfer-Nameki S. *Phys. Rev. D* 80:046006 (2009)
17. Donagi R, Wijnholt M. arXiv:0808.2223 [hep-th] (2008)
18. Marsano J, Saulina N, Schäfer-Nameki S. arXiv:0808.2450 [hep-th] (2008)
19. Heckman JJ, Vafa C. *J. High Energy Phys.* 0909:079 (2009)
20. Heckman JJ, Vafa C. arXiv:0809.3452 [hep-ph] (2008)
21. Wijnholt M. arXiv:0809.3878 [hep-th] (2008)
22. Font A, Ibáñez LE. *J. High Energy Phys.* 0902:016 (2009)
23. Heckman JJ, Vafa C. arXiv:0811.2417 [hep-th] (2008)
24. Blumenhagen R, Braun V, Grimm TW, Weigand T. *Nucl. Phys. B* 815:1 (2009)
25. Blumenhagen R. *Phys. Rev. Lett.* 102:071601 (2009)
26. Heckman JJ, Tavanfar A, Vafa C. arXiv:0812.3155 [hep-th] (2008)
27. Bourjaily JL. arXiv:0901.3785 [hep-th] (2009)
28. Hayashi H, Kawano T, Tatar R, Watari T. *Nucl. Phys. B* 823:47 (2009)
29. Andreas B, Curio G. arXiv:0902.4143 [hep-th] (2009)
30. Chen CM, Chung YC. *Nucl. Phys. B* 824:273 (2010)
31. Heckman JJ, Kane GL, Shao J, Vafa C. *J. High Energy Phys.* 0910:039 (2009)
32. Donagi R, Wijnholt M. arXiv:0904.1218 [hep-th] (2009)
33. Bouchard V, Heckman JJ, Seo J, Vafa C. arXiv:0904.1419 [hep-ph] (2009)
34. Randall L, Simmons-Duffin D. arXiv:0904.1584 [hep-ph] (2009)
35. Heckman JJ, Vafa C. arXiv:0904.3101 [hep-th] (2009)

36. Marsano J, Saulina N, Schäfer-Nameki S. *J. High Energy Phys.* 0908:030 (2009)
37. Bourjaily JL. arXiv:0905.0142 [hep-th] (2009)
38. Tatar R, Tsuchiya Y, Watari T. *Nucl. Phys. B* 823:1 (2009)
39. Jiang J, Li T, Nanopoulos DV, Xie D. arXiv:0905.3394 [hep-th] (2009)
40. Jiang J, Li T, Nanopoulos DV, Xie D. *Phys. Lett.* B677:322 (2009)
41. Blumenhagen R, Grimm TW, Jurke B, Weigand T. *J. High Energy Phys.* 0909:053 (2009)
42. Heckman JJ, Tavanfar A, Vafa C. arXiv:0906.0581 [hep-th] (2009)
43. Marsano J, Saulina N, Schäfer-Nameki S. *J. High Energy Phys.* 0908:046 (2009)
44. Conlon JP, Palti E. *Phys. Rev. D* 80:106004 (2009)
45. Font A, Ibáñez LE. arXiv:0907.4895 [hep-th] (2009)
46. Blumenhagen R, Grimm TW, Jurke B, Weigand T. arXiv:0908.1784 [hep-th] (2009)
47. Cecotti S, Cheng MCN, Heckman JJ, Vafa C. arXiv:0910.0477 [hep-th] (2009)
48. Li T, Nanopoulos DV, Walker JW. arXiv:0910.0860 [hep-ph] (2009)
49. Conlon JP, Palti E. arXiv:0910.2413 [hep-th] (2009)
50. Choi KS. arXiv:0910.2571 [hep-th] (2009)
51. Hayashi H, Kawano T, Tsuchiya Y, Watari T. arXiv:0910.2762 [hep-th] (2009)
52. Córdova C. arXiv:0910.2955 [hep-th] (2009)
53. Marchesano F, Martucci L. arXiv:0910.5496 [hep-th] (2009)
54. Chung YC. arXiv:0911.0427 [hep-th] (2009)
55. Vafa C. arXiv:0911.3008 [math-ph] (2009)
56. Marsano J, Saulina N, Schäfer-Nameki S. arXiv:0912.0272 [hep-th] (2009)
57. Dudas E, Palti E. arXiv:0912.0853 [hep-th] (2009)
58. Leontaris GK, Tracas ND. arXiv:0912.1557 [hep-ph] (2009)
59. Grimm TW, Krause S, Weigand T. arXiv:0912.3524 [hep-th] (2009)
60. Cvetič M, Garcia-Etxebarria I, Richter R. *AIP Conf. Proc.* 1200:246 (2010)
61. Heckman JJ, Shao J, Vafa C. arXiv:1001.4084 [hep-ph] (2010)
62. Blumenhagen R, Collinucci A, Jurke B. arXiv:1002.1894 [hep-th] (2010)
63. Cvetič M, Garcia-Etxebarria I, Halverson J. arXiv:1003.5337 [hep-th] (2010)
64. Blumenhagen R, Cvetič M, Langacker P, Shiu G. *Annu. Rev. Nucl. Part. Sci.* 55:71 (2005); Blumenhagen R, Körs B, Lüst D, Stieberger S. *Phys. Rep.* 445:1 (2007); Malyshev D, Verlinde H. *Nucl. Phys. Proc. Suppl.* 171:139 (2007)
65. Katz S, Morrison DR. *J. Algebr. Geom.* 1:449 (1992)
66. Katz SH, Vafa C. *Nucl. Phys. B* 497:146 (1997)
67. Bershadsky M, Johansen A, Pantev T, Sadov V. *Nucl. Phys. B* 505:165 (1997)
68. Iqbal A, Neitzke A, Vafa C. *Adv. Theor. Math. Phys.* 5:769 (2002)
69. Witten E. *Nucl. Phys. B* 268:79 (1986)
70. Blumenhagen R, Moster S, Weigand T. *Nucl. Phys. B* 751:186 (2006)
71. Tatar R, Watari T. *Nucl. Phys. B* 810:316 (2009)
72. Buican M, et al. *J. High Energy Phys.* 0701:107 (2007)
73. Raby S. arXiv:hep-ph/0211024 (2002)
74. Tatar R, Watari T. *Nucl. Phys. B* 747:212 (2006)
75. Raby S, Ratz M, Schmidt-Hoberg K. arXiv:0911.4249 [hep-ph] (2009)
76. Ray DB, Singer IM. *Ann. Math.* 98:154 (1974)
77. Blumenhagen R, et al. *J. High Energy Phys.* 0909:007 (2009)
78. Conlon JP, Quevedo F, Suruliz K. *J. High Energy Phys.* 0508:007 (2005); Conlon JP, Cremades D, Quevedo F. *J. High Energy Phys.* 0701:022 (2007); Conlon JP. *J. High Energy Phys.* 0803:025 (2008); Conlon JP, Quevedo F. *J. Cosmol. Astropart. Phys.* 0708:019 (2007); Conlon JP, Abdussalam SS, Quevedo F, Suruliz K. *J. High Energy Phys.* 0701:032 (2007)
79. Giudice GF, Rattazzi R. *Phys. Rep.* 322:419 (1999)
80. Floratos E, Kokorelis C. arXiv:hep-th/0607217 (2006)
81. Cvetič M, Weigand T. arXiv:0807.3953 [hep-th] (2008)
82. Arkani-Hamed N, Dine M, Martin SP. *Phys. Lett.* B431:329 (1998)
83. Ibe M, Kitano R. *J. High Energy Phys.* 0708:016 (2007)

84. Amsler C, et al. *Phys. Lett.* B667:1 (2008)
85. Georgi H, Jarlskog C. *Phys. Lett.* B86:297 (1979)
86. Chanowitz MS, Ellis JR, Gaillard MK. *Nucl. Phys. B* 128:506 (1977); Buras AJ, Ellis JR, Gaillard MK, Nanopoulos DV. *Nucl. Phys. B* 135:66 (1978); Einhorn MB, Jones DRT. *Nucl. Phys. B* 196:475 (1982); Ellis JR, Nanopoulos DV, Rudaz S. *Nucl. Phys. B* 202:43 (1982)
87. Froggatt CD, Nielsen HB. *Nucl. Phys. B* 147:277 (1979)
88. Baumann D, et al. arXiv:0912.4268 [hep-th] (2009)
89. O'Sullivan K. (EXO Collab.) *J. Phys. Conf. Ser.* 120:052056 (2008)

Jet Physics at the Tevatron

Anwar A. Bhatti[1] and Don Lincoln[2]

[1]Laboratory of Experimental High-Energy Physics, The Rockefeller University, New York, New York 10065; email: bhatti@fnal.gov

[2]Fermi National Accelerator Laboratory, Batavia, Illinois 60510; email: lincoln@fnal.gov

Annu. Rev. Nucl. Part. Sci. 2010. 60:267–97

First published online as a Review in Advance on July 16, 2010

The *Annual Review of Nuclear and Particle Science* is online at nucl.annualreviews.org

This article's doi:
10.1146/annurev.nucl.012809.104430

0163-8998/10/1123-0267$20.00

Key Words

QCD, jet cross section, dijet resonance search, compositeness, extra dimensions

Abstract

Jets have been used to verify the theory of quantum chromodynamics (QCD), measure the structure of the proton, and search for physics beyond the Standard Model. In this article, we review the current status of jet physics at the Tevatron, a $\sqrt{s} = 1.96$ TeV $p\bar{p}$ collider at the Fermi National Accelerator Laboratory. We report on recent measurements of the inclusive jet production cross section, dijet production measurements, and the results of searches for physics beyond the Standard Model using jets.

Contents

1. INTRODUCTION

The theory of quantum chromodynamics (QCD) (1) is currently the best description of the fundamental strong force. This theory describes the color interaction between quarks as being mediated by gluons, the vector bosons of the strong force. QCD has been successfully tested in e^+e^- (2), ep (3), pp (4), and $p\bar{p}$ (5) collisions.

One of the basic properties of QCD is that its coupling strength α_s decreases with the energy of the interaction, and at sufficiently high energies, QCD calculations can be performed using perturbation theory in powers of α_s (6). Currently, these perturbative QCD (pQCD) calculations are available at next-to-leading order (NLO) for many processes and, in some cases, at next-to-next-to-leading order (NNLO). Leading-order (LO) calculations, supplemented with parton shower calculations (7), are used in several Monte Carlo event generators (8, 9). Also, matrix element generators that match NLO calculations of rates for QCD processes with a parton shower Monte Carlo event generator are also available (10), but only a limited number of processes have been implemented.

The pQCD calculations result in a small number of partons in the final state, whereas experimenters observe jets of particles. These jets retain the kinematic properties (energy and momentum) of the parent partons (quarks or gluons). To facilitate comparison between data and calculation, researchers have devised jet-finding algorithms that are insensitive to the difficult-to-calculate low-energy phenomena that govern the transition from low-multiplicity partons to high-multiplicity particle final states. There are several jet-finding algorithms, and the details of the measurements are sensitive to the choice of algorithm.

Although events in which jets are created are used for a detailed understanding of the strong force, it is also possible that such events could reveal new physical phenomena, including quark substructure (compositeness), extra spatial dimensions, and new particles that decay into jets.

Because of their high energy, jets can probe very small distances. At the Tevatron, the highest-p_T jets can probe distances down to $\mathcal{O}(10^{-17})$ cm.

In 2001, the Fermilab Tevatron $p\bar{p}$ collider commenced Run II, with a collision energy of 1.96 TeV. By contrast, the 1992–1996 Run I had an energy of 1.8 TeV. Even this relatively small increase in energy leads to a substantial increase in jet production with large transverse momentum, p_T, by approximately a factor of three at $p_T = 500$ GeV/c. The beam intensity of Run II is much higher than that of Run I due to the addition of the Main Injector and the Recycler Ring to the Fermilab accelerator complex (11). Also, both the CDF (12) and DØ detectors (13) were upgraded. The results reported here utilize an order-of-magnitude-higher integrated luminosity than reported previously (5).

2. PERTURBATIVE QUANTUM CHROMODYNAMICS

The theory of QCD describes the behavior of those particles (quarks q and gluons g) that experience the strong force. It is broadly modeled on the theory of quantum electrodynamics (QED), which describes the interactions between electrically charged particles. However, unlike the electrically neutral photon of QED, the gluons (the force-mediating bosons of the strong interaction) carry the strong charge. This fact greatly increases the complexity in calculating the behavior of matter undergoing interactions via the strong force.

The mathematical techniques required to make these calculations can be found in textbooks (e.g., 14). Instead of giving an exhaustive description of those techniques here, we focus on those aspects of the calculations employed most frequently in the experimental analysis, thereby clarifying the phenomena that experimentalists investigate.

At high energies, the strong interactions between two hadrons can be factorized into three components: (*a*) the probability of finding the partons in the hadrons, (*b*) the interaction between quasi-free partons, and (*c*) fragmentation and hadronization of the final-state partons. The process is schematically shown in **Figure 1**. The cross section of the hadron-hadron scattering with four-momenta P_1 and P_2 can be written as (14)

$$\sigma(P_1, P_2) = \sum_{i,j} \int dx_1 \int dx_2\, f_i(x_1, \mu_F^2) f_j(x_2, \mu_F^2) \hat{\sigma}_{ij}[p_1, p_2, \alpha_s(\mu_R^2), Q^2/\mu_F^2, Q^2/\mu_R^2]. \quad 1.$$

The hard interaction between partons i and j is given by $\hat{\sigma}_{ij}$, where $p_1 = x_1 P_1$ and $p_2 = x_2 P_2$ are four-momenta of the two partons. This interaction is independent of the incoming hadrons' structure and can be calculated with pQCD. The function $f_i(x, \mu_F^2)$ is the probability of finding a parton i with momentum fraction x at the scale μ_F, and it is known as the parton distribution function (PDF). The sum i, j is over the partons in the two respective hadrons; α_s is the strong coupling constant; μ_R is the renormalization scale, which is used to remove nonphysical infinities inherent in fixed-order calculations; and Q is the characteristic scale of the interaction and is related to some physical scale in the interaction, such as the p_T of the leading jet.

This factorization of hadron-hadron interactions into a short-distance interaction (hard interaction) and a long-distance interaction (PDF) is done at an arbitrary energy scale μ_F. A parton emitted below the scale μ_F is considered to be part of the hadron structure and thus is described by the PDFs. The hadron structure is measured by many different experiments, especially lepton-hadron scattering experiments. The PDFs are determined by fitting these data and are parameterized at a scale μ_0. The QCD evolution equations (7), currently available at NNLO in perturbation theory, are used to evolve these PDFs to any different scale μ. The cross section is a convolution of the PDFs and the parton-level cross section, and to be consistent, both must be calculated to the same order in perturbation theory. A complete calculation, including all orders

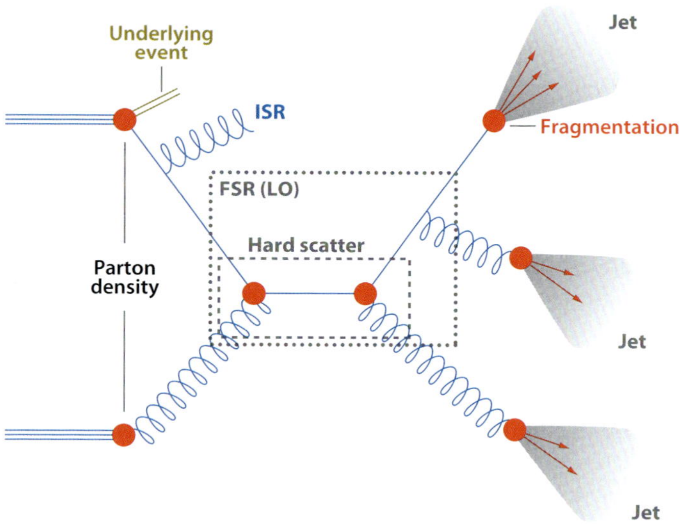

Figure 1

Stylized hadron-hadron collision, with relevant features labeled. Note that a leading-order (LO) calculation of the hard scatter (*dashed gray line*) will assign a jet to final-state radiation (FSR) that would be included in the hard scatter calculation by a next-to-leading-order calculation (*dotted gray line*). Abbreviation: ISR, initial-state radiation.

in the perturbation series, must be independent of μ_F, but a fixed-order calculation depends on this choice. To evaluate the sensitivity to μ_F, this scale is usually varied up and down by a factor of two. Scale dependencies are expected to decline with the addition of higher-order terms in the calculation.

In **Figure 2**, examples of LO and NLO Feynman diagrams for jet production are shown. At NLO, the jet cross section receives contributions from virtual corrections to the two-parton final state and from real corrections from the three-parton final state. Both contributions are divergent, but the sum is finite. An NLO pQCD calculation can describe up to three jets in the final state. For inclusive jet production, many different implementations of NLO pQCD calculations are available (15–18). All these programs use Monte Carlo integration techniques to calculate the real and virtual contribution to the cross sections. The EKS program (15) calculates the cross sections for user-predefined cuts on transverse momentum and rapidity ranges. JETRAD (16) and NLOJet++ (17) generate parton events with weights (both positive and negative), and thus full event kinematics are available to the user for jet clustering, detector acceptance calculations, and study of any other distribution. These programs require a huge amount of processing power to reach the desired precision. A large fraction of this processing power is consumed in evaluating the PDFs. In the FastNLO program (18), the convolution in Equation 1 is modified to a product of a perturbatively calculable piece $\sigma_{n,i,k,l,m}(\mu)$ that is calculated using NLOJet++, the strong coupling constant, and a function $F_a(x_1, x_1, \mu_F)$ that depends on only the PDFs and the factorization scale μ_F. The time-consuming piece σ is calculated only once. The function F_a is calculated on a grid of different values of x_1, x_2, and μ_F only once, and it is interpolated between those points while evaluating the full cross section. This procedure significantly speeds up the calculations.

Near threshold, the phase space for the emission of real gluons is limited, and large logarithmic corrections to the above cross-section calculation may arise from the incomplete cancellation of infrared (IR) divergences against the virtual gluon emission contributions. For jet production, these corrections are expected to contribute at very high x, where parton distributions fall very

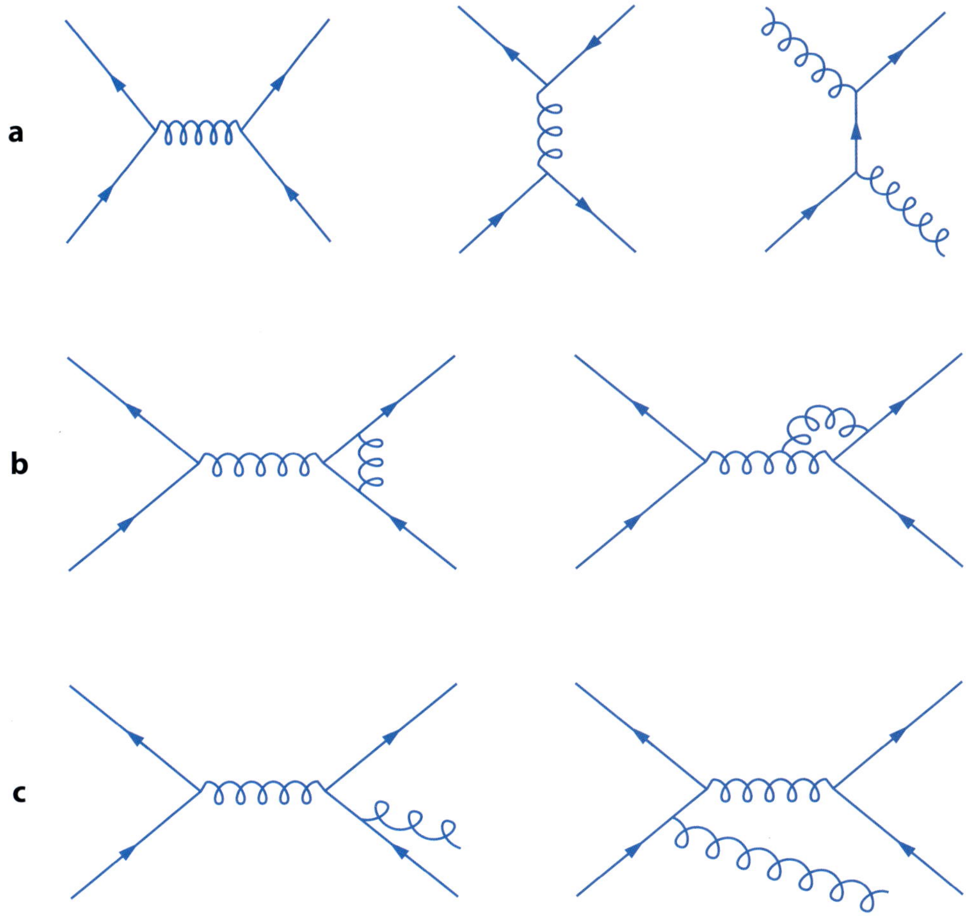

Figure 2

(*a*) Leading-order tree diagrams. (*b*) Next-to-leading-order loop diagrams. (*c*) Next-to-leading-order tree diagrams.

steeply. For the inclusive jet cross section, these threshold corrections have been calculated to NNLO at next-to-leading-logarithmic accuracy (19) and are found to be small. However, the corrected cross section shows a substantial reduction of the scale μ_F and μ_R dependence.

The partons radiate when they pass through a color field. In this type of radiation, two forms dominate: collinear and soft radiation. Collinear radiation is in the direction of the parent parton, whereas soft radiation is simply low-energy emission. Such soft and collinear radiation can be calculated in a leading-logarithmic approximation to all orders, and this algorithm is a crucial component of event generators. In this formalism, a gluon radiates another gluon or converts into a $q\bar{q}$ pair according to DGLAP (Dokshitzer-Gribov-Lipatov-Altarelli-Parisi) splitting functions (7) that depend on α_s and on a variable z, which describes the energy sharing between two daughter partons. Similarly, a quark or an antiquark can radiate a gluon that radiates further. This sequential radiation results in a shower of partons. The radiation process is continued until the parton virtual mass t is smaller than a mass scale t_0. This simple procedure is augmented by angular ordering; that is, each subsequent emission is required to have a smaller angle so as to simulate color coherence effects. Color coherence leads to suppression of soft gluon radiation in certain regions of phase space. In the final-state showers, the radiation is limited to a cone defined by color flow lines, and the emission angle at each branch point is smaller than the previous emission angle. Partons with a virtual mass $t \leq t_0 \sim 1$ GeV/c^2 are combined into hadrons by means of a phenomenological hadronization model. The hadronization models have been tuned to reproduce the jet structure observed at e^+e^- colliders. The showering process described above is used for radiation from outgoing partons, which is known as final-state radiation (FSR). For radiation from incoming partons, known as initial-state radiation (ISR), the event generators use so-called backward evolution. First, the momentum fractions x_1 and x_2 of partons participating in the hard interaction are determined. Second, the parton shower that preceded the hard interaction is subsequently reconstructed, evolving partons from the hard interaction scale Q backward in time toward smaller Q, at which the PDF f is evaluated. The color coherence in ISR is slightly more complicated but still follows the same angular ordering. Interference between ISR and FSR is implemented in HERWIG but not in PYTHIA. The independent variable t, through which the evolution of $\alpha_s(t)$ is parameterized, is not unique. In the PYTHIA showering algorithm, the squared mass m^2 of the branching parton is used as the evolution variable. HERWIG uses $t = m^2/[2z(1-z)]$, where $z = E_b/E_a$ is the ratio of daughter parton energy (E_b) to parent parton energy (E_a). In recent versions of PYTHIA, $t = p_{\mathrm{T}}$ of the branching parton is also available as the evolution variable.

Occasionally, the radiated parton is at a sufficiently large angle to the parent parton and carries enough energy that it leads to an identifiable jet. Because such jets are typically of lower p_{T}, they do not dominate the event kinematics. However, this lower-p_{T} radiation becomes important in studies of jet multiplicity. This part of ISR/FSR can either be considered as a component of the hard scatter $\hat{\sigma}_{ij}$ or can be treated independently. For instance, **Figure 1** shows one box surrounding the LO scatter and another box surrounding an NLO diagram. At NLO pQCD, one parton emitted from either an incoming or outgoing parton is part of the short-distance hard cross section $\hat{\sigma}_{ij}$. In LO event generators, these parton emissions are treated quasi-independently of the hard scatter and are part of the leading-log showering process.

The partons from the incoming hadrons that do not participate in the hardest scatter in an event also interact, but these interactions are normally soft. The particles produced in these multiparton interactions are, on average, isotropically distributed in the allowed $y - \phi$[1] space and can overlap with the jets produced in the hard interactions. Being soft, multiparton interactions are in the nonperturbative regime and thus are implemented in event generators using QCD-inspired phenomenological models. The parameters of these multiparton interaction models have been tuned to reproduce the transverse energy and multiplicity distributions of the particles observed far away from the hard jets in collider data (20, 21). The beam remnants, namely the partons that do not participate in either hard or multiparton interactions, go along the beam direction. However, they do carry color, and to become color singlets, they must exchange (soft) partons with the rest of the event. The particles produced in multiparton interactions and from the hadronization of beam remnants collectively constitute the underlying event.

As described above, pQCD predictions for jet production are available at NLO at the parton level only. These predictions cannot, in principle, be compared directly with the data, which are

[1]The important kinematic variables are the rapidity $y = \frac{1}{2} \ln \frac{E+p_z}{E-p_z}$; the azimuthal angle $\phi = \tan^{-1} \frac{p_y}{p_x}$; and the transverse momentum $p_T = \sqrt{p_x^2 + p_y^2}$, where a particle or jet has a four-momentum (E, p_x, p_y, and p_z) and the positive z is defined by the proton beam's direction, positive y is up, and the x direction is that required to make a right-handed coordinate system. In addition, the pseudorapidity $\eta = -\ln[\tan(\theta/2)]$ is sometimes used, where θ is the polar angle, measured with respect to the proton beam axis. The boost rapidity is defined as $y_{boost} = 0.5|y_1 + y_2|$.

available at the particle level because these parton-level calculations do not include hadronization effects and the contribution from the underlying event. However, event generators include hadronization, underlying-event energy, and ISR/FSR to all orders in the leading-log approximation, but the hard interaction is calculated at LO only. Thus, to compare data with QCD predictions, a hybrid scheme is generally used. The parton-level NLO calculations are correcting for the underlying-event and hadronization effects before they are compared with data. One can determine these corrections with Monte Carlo event generators by comparing the jets obtained by clustering the pQCD partons with the jets obtained by clustering the particles after the hadronization process. The parton jets are obtained from Monte Carlo events in which the underlying-event simulation (multiple parton scattering) has been turned off.

Over the past decade, there has been a great deal of progress in simulating high jet-multiplicity events using tree-level matrix elements. In ALPGEN (22), events with the exclusive parton multiplicities $n = 2, 3, 4$, and 5 are generated by using matrix elements from pQCD at the tree level. To include the effect of soft and collinear emission to all orders (albeit in the leading-log approximation), these events are passed through a showering program, such as PYTHIA. The phase spaces of the matrix elements and the parton showering program overlap. In particular, showering programs occasionally generate hard partons, which can lead to a state that has already been generated by the matrix element. To avoid this double counting, a matching criterion is used. For example, in ALPGEN for inclusive N jet production, the number of jets produced by clustering the partons that are produced by the matrix element must be the same as the number of those produced after the showering for $n \leq N$. The events that do not satisfy this condition are rejected. For the $n = N$ parton state, the showering algorithm is allowed to produce a higher jet-multiplicity state. The spectra from each different multiplicity are combined to form the full spectrum. SHERPA (23) employs a different matching procedure (24) in which parton showers above a cutoff k_T-like measure (Equation 4) are vetoed. Both ALPGEN and SHERPA have been extensively tested at the Tevatron in W/Z + jet production, but we do not discuss these studies here due to space constraints (25).

Although lepton-nucleon deep-inelastic scattering (DIS) experiments can precisely measure the quark content of the proton, this precision is not achievable for the gluon, especially at high x. At low x, the gluon distribution can be determined precisely using QCD scaling violation in DIS data. Studying the high-x gluon distribution functions requires data from hadron-hadron scattering. The effect of including Tevatron jet data in global fits to determine PDFs is described in Section 5.3.

3. JET CLUSTERING ALGORITHMS

Because partons carry the strong charge, they are not directly observable. A parton showers into many partons, which combine to form a large number of particles that travel in roughly the same directions as the initial parton. The kinematic properties of the initial parton can be inferred from the jet of collimated particles. For this inference, these particles or partons must be clustered into a jet by an algorithm. In pQCD, at NLO and higher orders, a jet algorithm is needed to define physics observables that are well defined, that is, that are soft/collinear safe. Jet algorithms are run on a few partons generated in pQCD calculations to construct such variables. Experimentally, the final-state particles are observed as tracks in the tracker systems or as towers of energy in the calorimeter. These tracks or towers must also be combined into a reconstructed jet so that they can be compared to the parton produced in the hard interaction. In the following sections, we collectively refer to (*a*) partons in a pQCD calculation, (*b*) partons or (*c*) particles produced in Monte Carlo event generators, (*d*) towers or (*e*) tracks, or (*f*) reconstructed particles observed in a detector as the objects that function as input for a clustering algorithm.

For a valid comparison between observations and theoretical predictions, the clustering algorithm must satisfy some basic criteria (26, 28). The algorithm must be safe against soft (IR) and collinear radiation, invariant to boosts along the beam direction, and insensitive to the nonperturbative hadronization effects. In an algorithm that is not safe against soft/collinear radiation, the virtual and real contributions in pQCD calculations do not cancel completely, and thus the predicted cross sections are ill defined. One should be able to run the same algorithm on the detector calorimeter towers or tracks, particles or the multiparton state from event generators, and partons in fixed-order pQCD calculations and obtain sensible results. Experimentally, a jet clustering algorithm should be insensitive to the energy from additional hadron-hadron collisions in the same bunch crossing that overlaps the energy from hard interactions, and it should not consume too many computer resources such as processing power. Finally, the algorithm must be completely specified.

Commonly used jet-finding algorithms can be divided into two categories: (*a*) cone clustering algorithms and (*b*) pair-wise recombination algorithms. With a few exceptions, only cone clustering algorithms have been used at hadron colliders. The cone clustering algorithms used prior to the Tevatron Run II were not IR/collinear safe (26, 27), and it was proposed to add an additional seed at the midpoint of stable cones. This made the new algorithm IR/collinear safe to NLO for the inclusive jet cross-section measurement. For other physics observables, it is either safe at LO only or unsafe at all orders (28). Various issues related to jet reconstruction are extensively discussed in a recent review (29).

3.1. Cone Clustering Algorithm

For jet studies in Run II, both the CDF and DØ Collaborations are using the Midpoint Algorithm, as laid out by the QCD Workshop recommendations (26), but the two implementations differ in some details. Below we describe the implementation of this algorithm by the CDF Collaboration.

The clustering process starts with a list of all objects to be clustered. In simulated events, all the particles or partons are included without a p_T threshold. However, in data, the calorimeter towers are required to have $p_T \geq 100$ MeV/c to minimize the effect of detector noise. From the initial list, a second list of seed objects is made with the requirement that the p_T of the objects exceed a fixed threshold of 1.0 GeV/c. At each seed location, the four-momentum of the cluster is determined by summing the four-momenta of all the objects within a distance $R = \sqrt{(y - y_c)^2 + (\phi - \phi_c)^2}$ from the seed (y_c, ϕ_c). The four-momenta are summed using the E-scheme (26)

$$(E, p_x, p_y, p_z) = \sum_i (E, p_x, p_y, p_z)_i, \qquad 2.$$

$$p_T = \sqrt{p_x^2 + p_y^2}, y_c = \frac{1}{2} \ln\left(\frac{E + p_z}{E - p_z}\right), \phi_c = \tan^{-1}(p_y/p_x). \qquad 3.$$

This yields massive jets with $m^2 = E^2 - p^2$ and is different from the Snowmass scheme (30) used in Run I, where the clustering centroid was defined as the E_T-weighted average of η and ϕ and jets were massless. Using the center of the cluster as a new seed location, the process is iterated until the center of the circle (y_c, ϕ_c) coincides with the position of the cluster four-momentum.

After all the stable cones have been identified, beginning with real seeds, an additional search for stable cones takes place in which the midpoints between the initial set of stable cones are used as seed locations. This cone-finding algorithm allows that the same object may be part of many cones. The shared objects are uniquely assigned to a single cone via a split-merge algorithm, as specified in Reference 26. If two stable cones share objects, the shared energy is compared to the energy of the lower-p_T cone. If the ratio of the shared transverse energy to the transverse energy of the lower-p_T cone is higher than an energy fraction f_{merge}, the two cones are merged. Otherwise, on the basis of proximity, the shared objects are assigned to the nearest cone. The two collaborations use different values of f_{merge}: CDF uses $f_{merge} = 0.75$, and DØ uses $f_{merge} = 0.50$. This split-merge procedure may lead to jets that are not circular in $y - \phi$ space. After all the objects above threshold have been uniquely assigned to a stable cone, the jet kinematics are determined through the same E-scheme.

A cone clustering algorithm can be made IR safe to all orders if a stable cone is evaluated at each point in $y - \phi$ space. Such an algorithm is very processing-power intensive even when the number of particles is modest, and thus its use is not practical beyond some parton-level pQCD calculations. Recently, a new seedless cone clustering algorithm that is IR and collinear safe to all orders in perturbation theory was proposed. The Seedless Infrared Safe Cone algorithm (31) uses the fact that a circle enclosing a set of particles can be moved around such that two of the particles lie on its circumference. Consequently, all stable circles can be reconstructed by considering all possible pairs of particles. After determining all the stable circles the algorithm merges and splits the stable circles to uniquely assign the particles to a single circle. This algorithm is fast and has been used at the Tevatron for comparison purposes only.

3.2. Pair-Wise Clustering Algorithm

The cone algorithm combines all the objects within a distance R from the seed. In contrast, the recombination algorithms combine pairs of objects based on some measure d_{ij} and are an attempt to undo the showering of partons. The k_T algorithm (32, 33) begins with a list of protojets given by four-momentum (E, p_x, p_y, and p_z). All the objects to be clustered are considered to be protojets. The transverse momentum p_T, rapidity y, and azimuthal angle ϕ of a protojet are calculated using Equation 3.

For each protojet i and the pair i, j, where $i \neq j$, d_i and d_{ij} are defined as

$$d_i = p_{T,i}^2;$$
$$d_{ij} = \min(p_{T,i}^{2p}, p_{T,j}^{2p})\frac{(y_i - y_j)^2 + (\phi_i - \phi_j)^2}{D^2}, \qquad 4.$$

where D is the parameter that controls the size of the jet. For the k_T algorithm, the parameter $p = 1$. The algorithm determines the minimum d_{min} of the d_i and all the d_{ij}. If $d_{min} = d_i$, the protojet is not mergeable and is promoted to a jet. Otherwise, the protojets i and j are merged into a single protojet with the four-momentum $(E_{ij}, \vec{p}_{ij}) = (E_i + E_j, \vec{p}_i + \vec{p}_j)$. The process is repeated until no protojets are left.

The k_T algorithm has been extensively used at e^+e^- and ep colliders. At hadron colliders, the environment is more challenging. The energy from multiparton interactions, beam remnants, and pileup can contribute to the jets and must be taken into account. The large particle multiplicity observed in hadron-hadron collisions requires substantial computer power to process an event. Thus, the use of the k_T algorithm has been limited at hadron colliders. The DØ Collaboration measured the inclusive jet cross section in Run I (34) using this algorithm. In the Tevatron Run II, the k_T algorithm was used only by the CDF Collaboration to measure the inclusive jet cross section (described in Section 5.2).

Recently, two additional recombination algorithms that use $p = 0$ (Cambridge-Aachen) (35) and $p = -1$ (anti-k_T) (36) in Equation 4 were proposed. The Cambridge-Aachen algorithm combines particles based only on their relative distance. The anti-k_T algorithm combines the highest-p_T objects in the events first. This process leads to circular jets, which like the cone jets have well-defined areas. Thus far, these algorithms have not been used at the Tevatron.

4. JET ENERGY SCALE DETERMINATION

At the Tevatron, jets are generally measured with a calorimeter, which is sensitive to both charged and neutral particles. Both CDF and DØ utilize sampling calorimeters, which measure only a small fraction of the energy of the particles. This observed energy is multiplied by a calibration constant so that it is equal to the sum of the energies of the incident particles. The calorimeter response is different for hadrons, photons, electrons, and muons. For hadrons, the response depends on the momentum and the type of the particles, whereas for photons and electrons it is almost momentum independent. Muons normally deposit a small amount of energy (~few GeV) in the calorimeter, which is almost independent of the muon momentum. Neutrinos escape without interacting and lead to an imbalance in the measured p_{T} in the event. The observed jet energy must be corrected for the calorimeter response and other detector effects. CDF and DØ employ different techniques to determine these jet energy scale corrections. The CDF Collaboration's technique (37) depends on an accurate modeling of the calorimeter response to single particles and on knowledge of the p_{T} spectrum of the particles in a jet, whereas the DØ technique is data driven and utilizes the fact that in photon-jet events $p_{\mathrm{T},Jet} = p_{\mathrm{T},\gamma}$ (38). These techniques are used to calibrate the central region of the calorimeter, where the tracking system is available to measure the charged-particle momentum and where the calorimeter response is uniform.

The calorimeter response was extended to other regions ($|\eta| < 0.1$, $|\eta| > 0.7$ for CDF and $|\eta| > 0.5$ for DØ) by using dijet balancing to scale the jet energy response in the other regions to the one in the central region. The energy from additional $p\bar{p}$ interactions in the same bunch crossing was subtracted; the correction is based on the number of reconstructed primary vertices in an event. For cone jets, this correction is determined from minimum bias events by summing the energy in towers in a cone of radius R placed randomly in the calorimeter. The procedure for k_{T} jets is described in Section 5.2.

4.1. Photon-Jet Balancing

In this technique, the jet energy is determined by scaling the measured jet p_{T} to the photon's p_{T} in photon-jet events. The photon energy is measured by the electromagnetic (EM) calorimeter, which is linear and has very good energy resolution. In the approximation of $2 \rightarrow 2$ scattering, the jet transverse energy p_{T} is equal to the photon p_{T}. The real situation is a little more complicated due to the presence of ISR, the energy not clustered in the jet, and contributions to the clustered jet from the multiple parton interactions. To be insensitive to these effects (especially ISR), DØ evaluates the missing p_{T} ($\not{E}_{\mathrm{T}}$) projection fraction along the photon direction using

$$R_{had} = 1 + \frac{\vec{\not{E}}_{\mathrm{T}} \cdot \vec{p}_{\mathrm{T},\gamma}}{p^2_{\mathrm{T},\gamma}}. \qquad 5.$$

The hadronic recoil correction factor, R_{had}, is the scale factor to the entire recoil system. By requiring that the jet be back to back with the photon, and in the absence of any additional jet(s) in the event, R_{had} is almost equal to the jet response. The derived response is expressed in the jet energy E' determined from the p_{T} of the photon and the position of the balancing jet using $E' \equiv p^{\gamma}_{\mathrm{T}}/\sin\theta^{\mathrm{jet}}$, as both p^{γ}_{T} and the direction of the jet are accurately measured. Thus, E' provides a better estimate of the jet energy than does the direct jet energy measurement by the calorimeter. It is preferred over jet p_{T} because the calorimeter response depends on the energy of the incident particles; therefore, the parameterization of calorimeter response in E' is more natural.

The EM calorimeter is calibrated using the electrons from Z boson decays such that the reconstructed Z boson mass is equal to the world average (39). The EM calorimeter response to electrons and photons is similar, but not the same, because photons begin their shower later than electrons. This difference is small and is evaluated using simulated events at $p_T = 100$ GeV/c. The estimated uncertainty on the photon energy scale is 0.5% at low E' and 0.8% at high E'. Using this procedure, the DØ Collaboration has achieved a 1% accuracy on the jet energy scale in photon-jet events. The current statistics of the $\gamma+$ jets sample limit the direct measurement of the jet energy corrections in the central region to $E' < 350$ GeV. The response is extrapolated to higher energies using Monte Carlo, which has been tuned to the data. The correction to a single jet with a given algorithm and size is deduced from R_{had} using simulated events.

The calorimeter response to jets depends on their flavor, as the particle spectrum and multiplicity for quark-initiated and gluon-initiated jets are different. The jet energy scale corrections determined from $\gamma+$ jet events are valid only for the flavor composition of $\gamma+$ jet events. Event topologies with different flavor composition have different jet energy scale corrections. Indeed, DØ tuned the single-pion response in their detector simulator to data and used PYTHIA to generate photon-quark, photon-gluon, and dijet events. The author of this study (38) found that the gluon jet response was 8% (2%) lower for jets with 20 GeV (500 GeV) of energy. In QCD jet production, the fraction of gluon-initiated jets changes with jet p_T, so the corrections were adjusted to account for this variation in flavor composition. With these additional corrections, the uncertainty on the jet energy scale was reduced to an unprecedentedly small value.

4.2. Jet Corrections Using Single-Particle Response

Another approach to determine the jet energy correction is based on a knowledge of the calorimeter response to each particle that makes up a jet. The CDF Collaboration measured the calorimeter response to charged hadrons and electrons using both $p\bar{p}$ collider and test-beam data. The calorimeter simulation was tuned to reproduce the measured response. The calorimeter response to a jet was determined by a convolution of the single-particle response with the type and momentum distribution of particles constituting a jet, as given by a fragmentation model. CDF used QCD dijet events with the PYTHIA fragmentation model to measure the default jet corrections. The PYTHIA fragmentation model agreed well with the particle p_T and multiplicity distributions in a jet measured in $p\bar{p}$ data. The HERWIG event generator was used to cross-check the PYTHIA fragmentation functions, and the results that were determined using two generators agreed well. In this procedure, the difference in calorimeter response to gluon-initiated and quark-initiated jets is automatically included. Although this procedure requires a detailed knowledge of the calorimeter response and a well-tuned simulation, it has the advantages that (*a*) the correction can be easily determined for any event topology over the entire kinematic range and (*b*) real and simulated data have the same corrections and thus can be treated on an equal footing.

5. INCLUSIVE JET CROSS SECTION

The inclusive jet cross-section measurement (40–45) has been used to test QCD and to search for physics beyond the Standard Model by searching for an excess of events at high p_T. During Run I at $\sqrt{s} = 1.8$ TeV, the search was limited by both theoretical and experimental systematic uncertainties, even with an integrated luminosity of 100 pb^{-1}. The uncertainty on the jet energy scale dominated the experimental uncertainty. NLO pQCD calculations (15–17) significantly reduced the dependence on the factorization and renormalization scales, and the remaining dependence was ~10%—almost independent of jet p_T for $p_T > 100$ GeV/c. The jet cross section is not very sensitive to nonperturbative hadronization effects. The underlying event contributes approximately 2 GeV/c of p_T to a jet and thus is significant only at low p_T. The main theoretical uncertainty arises from uncertainty in the PDFs, especially for high-x gluons. Unfortunately, inclusive jet production is the only process in which the high-x gluon distribution can be directly measured. The other possible process, photon-jet production, has a limited reach in x, and the associated theoretical uncertainties are large. The gluon distributions are also measured from lepton-proton scattering data through QCD scaling violations, but these measurements are also limited to low-x values. Because of these limitations, the Run II inclusive jet cross section has been used primarily to constrain the gluon content of the proton. The data at high y are particularly useful as they probe high x at lower Q values, where the contribution of physics beyond the Standard Model, if any, is negligible.

Due to the higher center-of-mass energy and much larger integrated luminosity, Run II jet measurements extend the jet spectrum to higher jet p_T, compared to the Run I measurements, by approximately 200 GeV/c. Both CDF and DØ implemented an improved jet clustering algorithm. The jet clustering algorithms used in Run II are IR/collinear safe at least to the order of the available pQCD calculations. An accurate determination of the calorimeter response from the $p\bar{p}$ data as well as a refinement in the techniques to determine the jet energy scale have led to reduced systematic uncertainty compared to Run I. In previous inclusive measurements, the hadronization effects were ignored, as they were much smaller than both the experimental and theoretical uncertainties. In Run II, both collaborations have evaluated the effect of hadronization and corrected the parton-level pQCD calculation. Also, the pQCD calculations are corrected for the energy from the underlying event, which is determined using tuned event generators. In contrast, in Run I, energy from the underlying event was removed from jets in data.

The CDF Collaboration measured the inclusive jet cross section using (*a*) the cone clustering (46) algorithm, where the cone size $R = 0.7$, and (*b*) the k_T clustering algorithm, where $D = 0.5$, 0.7, and 1.0 (47). The DØ Collaboration recently published the inclusive jet cross section (48) using the cone clustering algorithm, where the cone size $R = 0.7$. These three measurements are described below.

5.1. Measurement Using Cone Clustering Algorithm

The most common jet-finding algorithms at the Tevatron are variations of the cone clustering algorithm.

5.1.1. DØ Collaboration. The DØ Collaboration analyzed 0.7 fb^{-1} of data taken from 2004 to 2005 to measure the inclusive jet cross section for $p_T > 50$ GeV/c in six rapidity bins, $|\Delta y| = 0.4$ wide, over the range $0 < |y| < 2.4$. The data were collected by triggering on a jet passing a p_T threshold. Six triggers with p_T thresholds of 15, 25, 45, 65, 95, and 125 GeV/c were used to collect data. Due to high production rates, only predetermined fractions of the lower-threshold triggers were recorded. The efficiency for triggering on jet events was measured by using data collected with a muon trigger that did not rely on calorimeter activity. These different jet triggers were combined to form the full p_T spectrum, with each trigger contributing to a unique p_T range. Only those data for which the trigger efficiency was >98% were used. The events were required to have a reconstructed primary vertex, and the position of the $p\bar{p}$ interaction had to be within 50 cm of the detector center along the beam direction. These requirements ensured that the jets followed the projective geometry of the calorimeter, and thus their energy was accurately measured. The consequence of these requirements was a reduction of only $7.0 \pm 0.5\%$ in the integrated luminosity. The primary vertex was reconstructed using charged-particle tracks that were measured using silicon microstrip and scintillating fiber detectors located inside a 2-T solenoidal magnetic field (13).

The triggered data include events containing cosmic-ray interactions, beam halo, and detector noise. These contributions are mostly asymmetric and lead to a large imbalance in the momentum in the plane transverse to the beam direction, $\not{E}_T$. In contrast, for QCD jet production, $\not{E}_T$ is ideally zero, apart from a small neutrino contribution. In QCD events, $\not{E}_T$ arises mainly from fluctuations in calorimeter response and is much smaller than the total energy observed in the detector. Most of these background events are removed by requiring the ratio of $\not{E}_T$ to the transverse momentum of the leading jet to be small. Remaining backgrounds are removed by requiring that the shape of energy deposition in the calorimeters be consistent with the expected shape from a hadronic jet. The shape of energy deposition for a jet is very different from that of the energy deposited by a cosmic muon or a beam halo particle, as a jet consists of many particles. These shape requirements also remove photons and electrons. These requirements are highly efficient for the signal, and the remaining background is estimated to be <0.1%.

The measured p_T of each jet is corrected for calorimeter nonlinearity and energy lost in uninstrumented regions. These average jet-by-jet energy corrections do not correct the smearing (bin-to-bin migration) of jets due to the finite energy resolution. This smearing is determined using an iterative procedure. It is assumed that the particle-level physics (true) spectrum is described by the function

$$F(p_T, y) = N_0 \left(\frac{p_T}{100\,\text{GeV}/c}\right)^{\alpha} \left(1 - \frac{2p_T \cosh(y_{\min})}{\sqrt{s}}\right)^{\beta} \exp(-\gamma p_T), \qquad 6.$$

where $y_{\min}$ is the rapidity lower bin edge. This functional form is a good representation of the NLO pQCD prediction and fits the measured raw inclusive jet spectrum well. This true spectrum is smeared using the jet energy resolution function, which is determined using $p\bar{p}$ collider dijet data and simulated dijet events. The resulting smeared spectrum is compared with data by use of a χ^2 test. The process is iterated to determine the best parameters (N_0, α, β, γ) of the true function, $F(p_T, y)$. This true spectrum is used to correct the migrations between bins in p_T in the observed data. In the central region, the migration correction is a multiplicative factor that is 0.8 to 0.9 at low p_T and 0.7 at high p_T, with a strong dependence on y. The true spectrum $F(p_T, y)$ is measured separately for each rapidity bin. The jet rapidity is measured very precisely, and thus migration between rapidity bins is small. The y migration corrections are less than 2% in most bins, and 10% in the highest-p_T bin, where the spectrum is the steepest. The rapidity unsmearing is applied after the p_T unsmearing. After the jet energy scale and resolution smearing corrections, the observed data distribution is corrected to the particle level and is completely independent of the detector properties.

The observed inclusive particle jet spectrum compared with the NLO pQCD predictions, corrected for underlying-event and hadronization effects, is shown in **Figure 3**. The NLO pQCD predictions are calculated by using NLOJet++ and FastNLO (17, 18), which use α_s^3 matrix elements with $\mu_R = \mu_F = p_T^{\text{jet}}$. The PDFs from CTEQ6.5M (49), which include the Run I inclusive jet data, are used. The experimental uncertainty, dominated by the jet energy scale uncertainty, is 12% at $p_T = 50$ GeV/c and 17% at $p_T = 550$ GeV/c in the $|y| < 0.4$ bin. The uncertainty is higher in other y bins. These results are the most precise to date. The main theoretical uncertainties arise from the uncertainty on the PDFs and the missing higher-order terms in the perturbation series. As is customary, the effect of higher-order terms is evaluated by varying the renormalization and factorization scales. Fortunately, the change in cross section from varying these scales is almost independent of the p_T of the jets. The predicted cross section changed by ~10 to 15% when the scale was changed to $\mu_F = \mu_R = 2p_T$ or $\mu_F = \mu_R = p_T/2$. There is a good agreement between the data and the theoretical predictions over the whole p_T range, which spans 50 GeV/c to 550 GeV/c. Over this p_T range, the cross section falls by ten orders of magnitude.

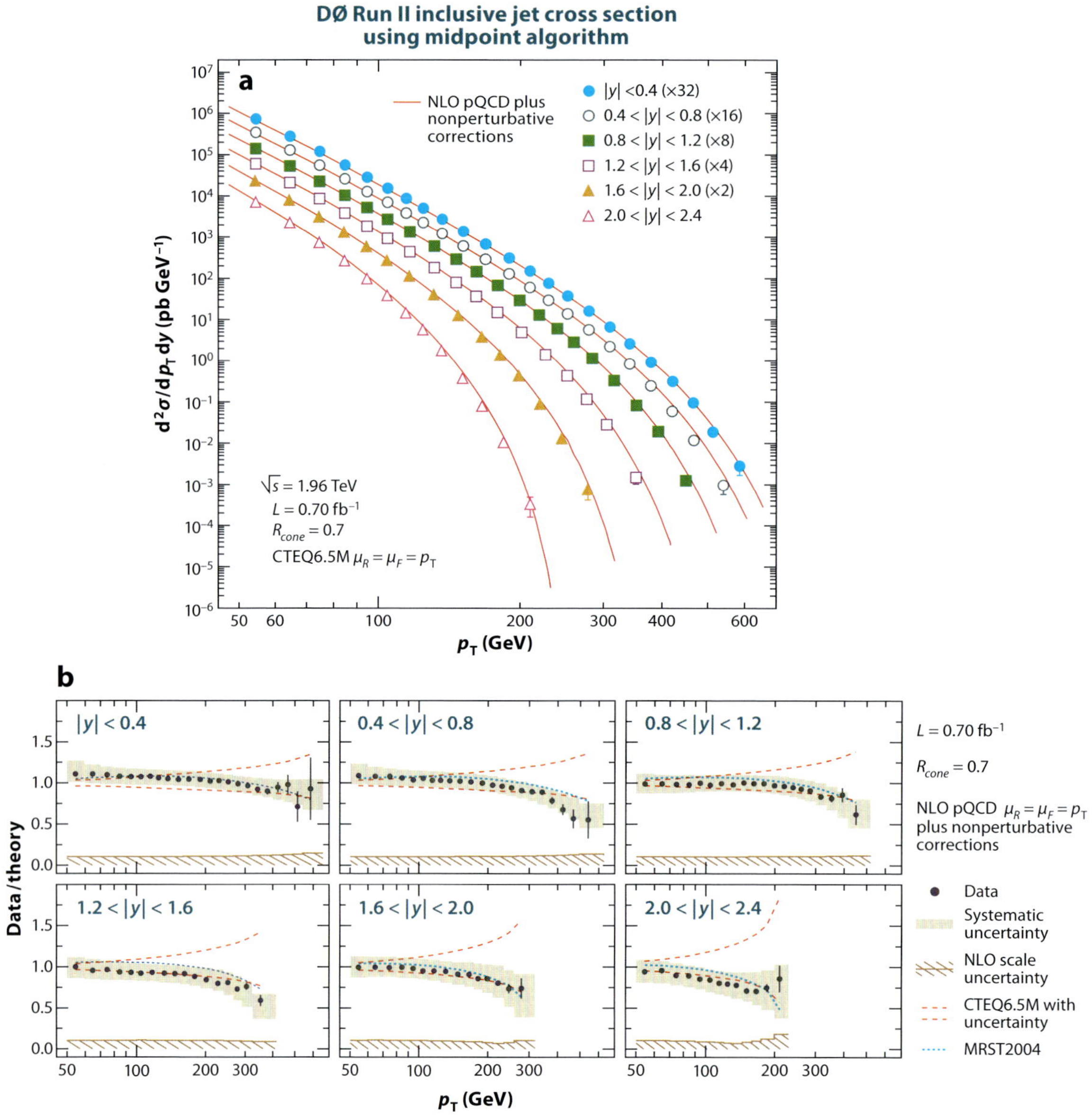

Figure 3

(*a*) DØ's observed inclusive jet differential cross sections corrected to the particle level in six rapidity regions compared to next-to-leading-order (NLO) quantum chromodynamics (QCD) predictions (48). The NLO QCD predictions are calculated with the CTEQ6.5M parton distribution functions. (*b*) Ratios of the measured cross sections over the NLO QCD predictions. The dashed blue curves show the NLO perturbative QCD (pQCD) prediction calculation that used MRST2004, divided by the same calculation that used CTEQ6.5M. The data agree with the theory quite well and have remarkably small systematic uncertainties.

The data prefer the lower bound of the theoretical prediction, favoring a smaller gluon content of the proton at high x. The theoretical uncertainty arising from the uncertainties in the PDFs is larger than the experimental uncertainties. These data, along with CDF inclusive jet data, have been used in the global fits to improve the precision of the gluon distribution function. The results of these fits are described in Section 5.3.

5.1.2. CDF Collaboration. The CDF Collaboration measured the inclusive jet cross section for a cone size of $R = 0.7$, using data corresponding to 1.13 fb^{-1} (46). The measurement spans five rapidity bins, $|y| < 0.1$, $0.1 < |y| < 0.7$, $0.7 < |y| < 1.1$, $1.1 < |y| < 1.6$, and $1.6 < |y| < 2.1$. These bins are matched to the CDF calorimeter structure (12) and thus are different from the binning used in DØ's analysis. The data were collected between 2002 and 2005 using jet triggers with four thresholds: 20, 50, 70, and 100 GeV/c. To prevent saturation of the data-acquisition system by jet triggers, only 1/808, 1/35, and 1/8 of lower-threshold triggers were recorded. An average correction is applied to the transverse energy of each jet to form the jet p_T spectrum, which was corrected for bin-to-bin migration of jets due to finite jet energy resolution. CDF used simulated events to evaluate the smearing corrections. The corrections depend on the shape of the true jet p_T spectrum and the jet energy resolution. A large sample of QCD jet events was generated using the PYTHIA event generator (8) and passed through the CDF detector simulation. The detector simulation was tuned to describe the single-particle response measured in $p\bar{p}$ collisions (37). These simulated data were analyzed with the same procedure as the one used for the real data to obtain the smeared spectrum. The bin-to-bin migration effect was determined by taking the ratio of the smeared jet spectrum and the particle jet spectrum. For this procedure to be valid, the smeared p_T spectrum of the simulated events must match the spectrum measured in data. The two spectra are very close, but not exactly the same. The simulated particle jet–smeared p_T spectrum was adjusted (reweighted) to force it to agree with the measured spectrum. Reweighting changes the unsmearing corrections by only a few percent. The unsmearing correction is <5% for $p_T^{jet} < 300$ GeV/c and increases to as much as 20% at $p_T = 500$ GeV/c.

The corrected jet p_T spectrum is compared to pQCD predictions evaluated with the FastNLO (18) program using the CTEQ6.1M PDFs (50). The renormalization and factorization scales (μ_R and μ_F) are chosen to be $p_T/2$—the same as those used in the global QCD analysis—to determine the PDFs (50). Use of $\mu_R = \mu_F = p_T^{jet}$ gives up to 10% smaller predictions in the cross section. The pQCD predictions are corrected for underlying-event and hadronization effects measured using the procedure described in Section 2. While clustering the partons produced by FastNLO, CDF used an ad hoc parameter, R_{sep}, which was introduced to mimic the split and merge procedure in iterative cone clustering algorithms (33). At order α_s^3, the final state can have up to three partons. Depending on their relative p_T and their separation in $y - \phi$ space, these partons can be clustered into two or three jets. Two partons are clustered into a single jet if they are within R from the jet centroid and within $R \times R_{sep}$ of each other. A value of $R_{sep} = 1.3$ is used in this calculation. A value of $R_{sep} = 2.0$ (i.e., the midpoint algorithm without R_{sep}) yields less than a 5% increase in cross section for NLO QCD predictions. As shown in **Figure 4**, the data are in good agreement with the theoretical predictions. The experimental uncertainty, dominated by the jet energy scale uncertainty, is comparable to the theoretical uncertainty, which in turn is dominated by the PDF uncertainty.

5.2. Measurement Using the k_T Clustering Algorithm

The k_T clustering algorithm, which combines objects in pairs to reconstruct a jet, is IR and collinear safe at all orders in perturbation. The CDF Collaboration (47) measured the inclusive jet cross

CDF Run II inclusive jet cross section using midpoint algorithm

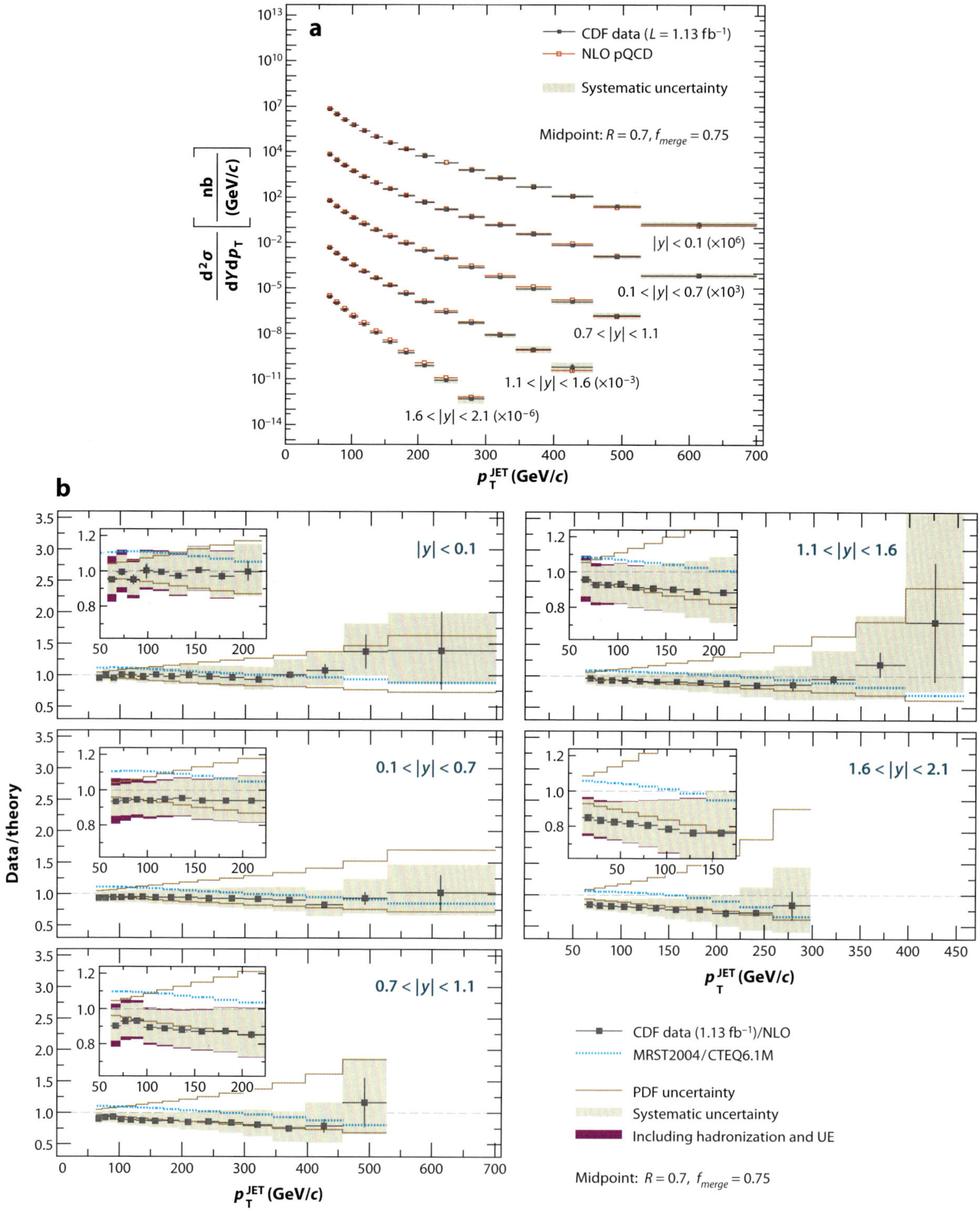

section by using the k_T clustering algorithm with $D = 0.4$, 0.7, and 1.0, using 1.0 fb^{-1} of data in the same rapidity region as in the cone-based analysis (46). The analysis procedure is similar to the one described in Section 5.1.2, except for the correction for multiple interactions, which are determined using a novel approach. The measured jet transverse momenta were corrected for this effect by removing a certain amount of transverse momentum, $\delta_{p_T}^{\rm mi} \times (N_V - 1)$, where N_V denotes the number of primary vertices in the event. The value of $\delta_{p_T}^{\rm mi}$ was determined by requiring the shape of the p_T spectrum at high instantaneous luminosity ($\mathcal{L}^{\rm inst}$) to be the same as the one at low $\mathcal{L}^{\rm inst}$. After the shape of two spectra were made the same, the data at low $\mathcal{L}^{\rm inst}$ and high $\mathcal{L}^{\rm inst}$ were combined. The study was carried out independently for each rapidity region, and the results were consistent with a common value of $\delta_{p_T}^{\rm mi} = 1.86 \pm 0.23$ GeV/c. The corresponding correction for the cone jets is 0.97 ± 0.29 GeV/c, which is measured by summing the p_T in a cone of $R = 0.7$ in minimum bias events.

In **Figure 5** the p_T spectra are compared with NLO QCD predictions using the CTEQ6.1M PDFs (50), where $\mu = 0.5 \times p_T^{\rm max\,jet}$ for $D = 0.7$. The theoretical predictions are calculated with the JETRAD (16) program. The data are in very good agreement with the QCD prediction except in the highest-rapidity bin ($1.6 < |y| < 2.1$), where the data are lower than the prediction but are well within experimental systematic uncertainties. The theoretical uncertainties are dominated by the PDF uncertainties and are comparable to or larger than the experimental uncertainties. The theoretical predictions using the MRST2004 PDFs are very close to those based on the CTEQ6.1M PDFs, except in the $1.6 < |y| < 2.1$ bin, where the MRST2004 cross section is smaller but still within the PDF uncertainty on the CTEQ6.1M prediction. The results for jet sizes $D = 0.5$ and $D = 1.0$ show similar behavior.

The jet p_T spectra measured using two different clustering algorithms are expected to be different and can be compared only via theoretical predictions. The ratios of data to theory from two analyses were compared; they are in very good agreement with each other except in the $0.7 < |y| < 1.1$ region, where the cross section measured using k_T is ~5% higher. In this y region, the CDF calorimeter coverage is not uniform, which leads to a large variation in calorimeter response and poor jet energy resolution. The two CDF analyses have similar experimental uncertainties. Thus, both the k_T and cone clustering algorithms can be successfully used at hadron colliders.

5.3. Determination of Gluon Distribution Function

The PDF $f_i(x, \mu)$, which is the probability of finding a parton with a type $i = g, q, \bar{q}$ with momentum fraction x and mass scale μ, must be experimentally determined. The PDFs for gluons, light quarks, and antiquarks ($u, d, s, \bar{u}, \bar{d}$, and $\bar{s}$) are normally determined from experimental data. For heavier quarks, namely c and b, they are normally dynamically generated through gluon splitting. Data from (*a*) $e^{\pm}p$ collisions at the ZEUS and H1 experiments (3); (*b*) νp, $\bar{\nu} p$, νn, and $\bar{\nu} n$ collisions at CCFR/NuTeV (52); (*c*) Drell-Yan (lepton pairs and W/Z bosons) production in pp (53) and $p\bar{p}$ collisions (54); (*d*) jet data from the Tevatron; and (*e*) many other experiments, especially low-energy experiments, are used to extract the PDFs using a global fit. These experiments are sensitive to different f_i. For example, $e^{\pm}p$ experiments are sensitive to the sum of the q and $\bar{q}$

Figure 4

(*a*) CDF's observed inclusive jet differential cross sections corrected to the particle level in five rapidity regions compared to next-to-leading-order (NLO) quantum chromodynamics (QCD) predictions (46). The NLO QCD predictions are calculated with the CTEQ6.1M parton distribution functions (PDFs). (*b*) Ratios of the measured cross sections over the NLO QCD predictions. The theory describes the data quite well. Abbreviations: pQCD, perturbative QCD; UE, underlying event.

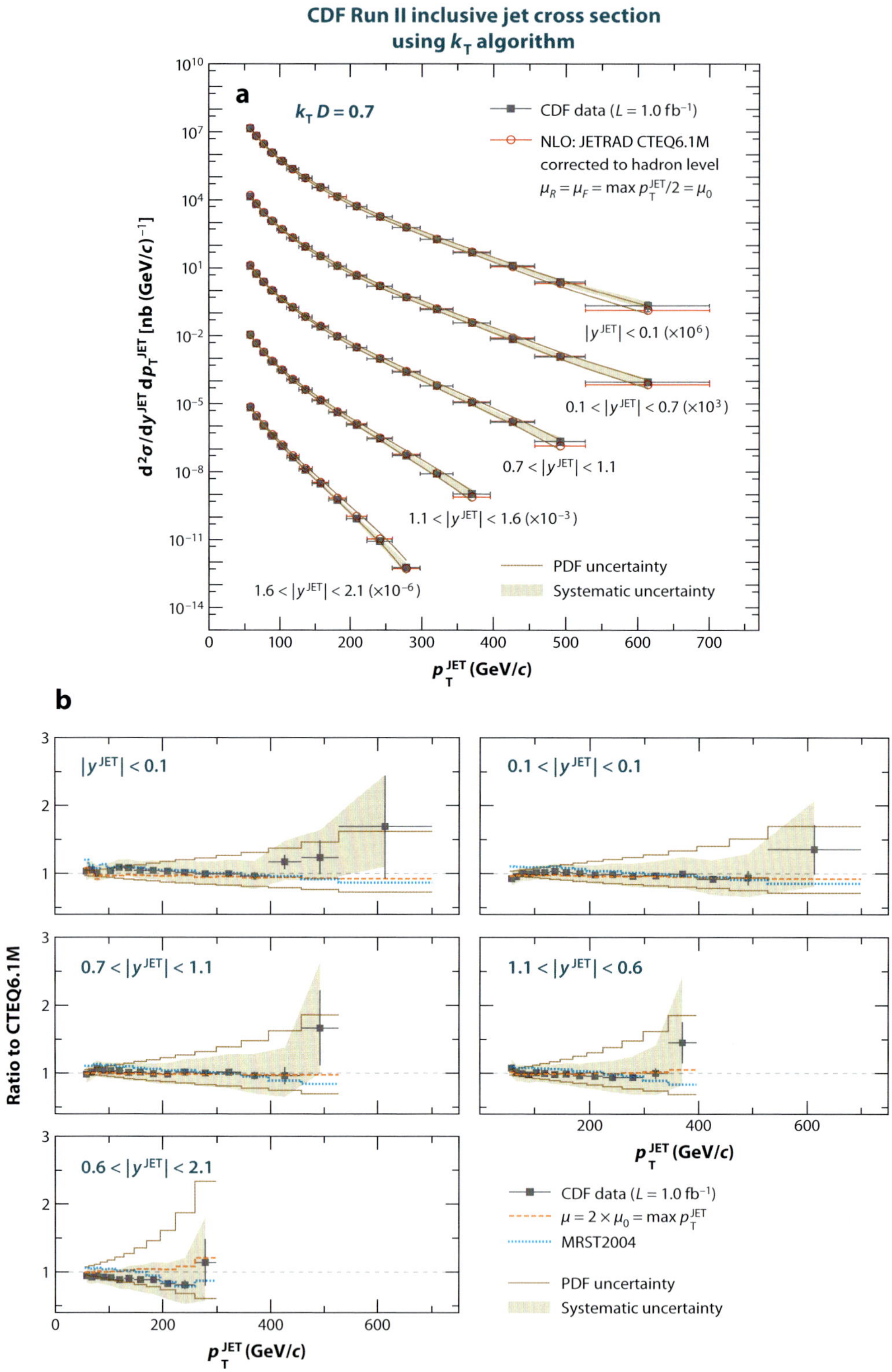

CDF Run II inclusive jet cross section
using k_T algorithm
a
k_T D = 0.7
CDF data (L = 1.0 fb^{-1})
NLO: JETRAD CTEQ6.1M
corrected to hadron level
$\mu_R = \mu_F = \max p_T^{JET}/2 = \mu_0$
$|y^{JET}| < 0.1$ (×10^6)
$0.1 < |y^{JET}| < 0.7$ (×10^3)
$0.7 < |y^{JET}| < 1.1$
$1.1 < |y^{JET}| < 1.6$ (×10^{-3})
$1.6 < |y^{JET}| < 2.1$ (×10^{-6})
PDF uncertainty
Systematic uncertainty
$d^2\sigma/dy^{JET}dp_T^{JET}$ [nb (GeV/c)$^{-1}$]
p_T^{JET} (GeV/c)
b
$|y^{JET}| < 0.1$
$0.1 < |y^{JET}| < 0.1$
$0.7 < |y^{JET}| < 1.1$
$1.1 < |y^{JET}| < 0.6$
$0.6 < |y^{JET}| < 2.1$
Ratio to CTEQ6.1M
p_T^{JET} (GeV/c)
CDF data (L = 1.0 fb^{-1})
$\mu = 2 \times \mu_0 = \max p_T^{JET}$
MRST2004
PDF uncertainty
Systematic uncertainty

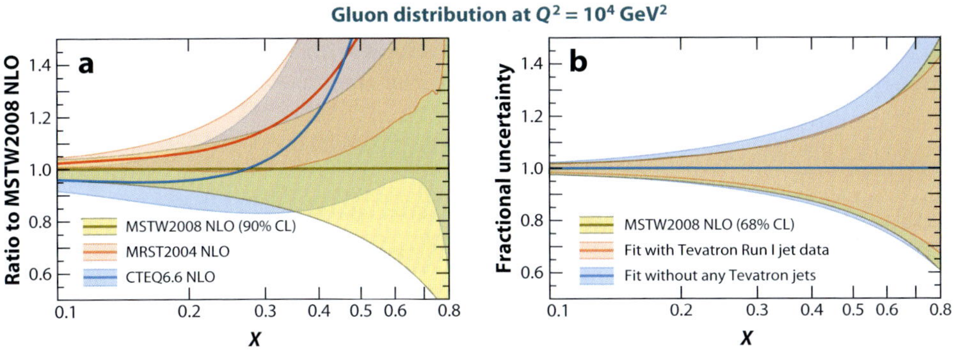

Figure 6

(*a*) Ratios of various gluon parton distribution functions (PDFs) to the MSTW2008 PDF. (*b*) The MSTW2008 gluon PDF uncertainties for variations on the inclusion or exclusion of Tevatron data (55). Abbreviation: NLO, next-to-leading order.

distributions weighted by $e^2_{q(\bar{q})}$ and cannot distinguish between quark and antiquark distributions. Neutrino and antineutrino data are used to differentiate between q and $\bar{q}$. The Tevatron jet data play a significant role in constraining the gluon distribution at high x. The gluon distribution at low x are determined mainly from the scaling violations in lepton-nucleon scattering data. Normally, the results of these global analyses are fit at some initial scale μ_0 using tens of parameters. The PDF can be evolved to any arbitrary scale μ using QCD evolution equations, which are available at NNLO approximation in perturbation theory.

Both the MSTW (55) and CTEQ (56) Collaborations have included Run II jet data in their global analysis to update the PDFs. The three Run II inclusive jet measurements are more accurate than Run I measurements, span a larger p_T range, and are consistent with each other (55). Because two CDF measurements used the same data, the MSTW Collaboration is using the k_T jet spectrum, whereas the CTEQ Collaboration is using the cone-based measurements. The Run I jet measurements do not play a significant role in the fit; thus, the MSTW Collaboration has dropped those data from the new fits. Comparisons between (*a*) the gluon distribution $g(x)$ determined in the new MSTW fit and (*b*) the gluon distributions from MRST2004 (57) and CTEQ6.6 (58) fits, along with the 90%-uncertainty band, are shown in **Figure 6*a*** for $\mu^2 = Q^2 = 10^4$ GeV^2. The new $g(x)$ is lower than previous fits for $x \geq 0.3$ but is within the still large systematic uncertainties. Because α_s and $g(x)$ always appear as a product, the values of α_s and $g(x)$ are strongly anticorrelated. The value of α_s in the three sets of PDFs is different, and thus $g(x)$ is also expected to be slightly different. The fractional uncertainty on the gluon distribution is shown in **Figure 6*b***. At $x = 0.4$ and $\mu^2 = 10^4$ GeV^2, the uncertainty reduces from 18%, when the jet data are excluded from the fit, to 12%, when jet data are included in the fit. This modest extra constraint will make the predictions more precise at the LHC in processes in which gluon-quark scattering dominates.

Figure 5

(*a*) CDF's inclusive jet differential cross sections measured using the k_T clustering algorithm in five rapidity regions. The data have been corrected to the particle jets and are compared to next-to-leading-order (NLO) quantum chromodynamics (QCD) predictions (47). The NLO QCD predictions are calculated with the CTEQ6.1M parton distribution functions (PDFs). (*b*) Ratios of the measured cross sections over the NLO QCD predictions. The perturbative NLO QCD calculations are in good agreement with the data.

5.4. Determination of the Strong Coupling Constant

The hard cross section for jet production ($\hat{\sigma}_{ij}$) in Equation 1 at high p_T can be expanded in powers of the strong coupling constant to the nth order in perturbation theory,

$$\hat{\sigma}_{ij} = \alpha_s^2 \sum_{m=0}^{n} c_{ij}^{(m)} \alpha_s^m, \qquad 7.$$

where the perturbative coefficients $c_{ij}^{(m)}$ are functions of the kinematic variables and the factorization and renormalization scales only. The coefficients $c_{ij}^{(m)}$ are available for $m = 0$ and $m = 1$. By use of the jet cross-section measurement, the strong coupling constant α_s can be determined with Equation 1, provided that the PDFs are known. This technique was used by the CDF Collaboration to measure α_s at different p_T values and show its running with the hard scattering scale in Run I (59). In Run II, the DØ Collaboration used the same principle but employed an improved technique to measure α_s (60) from the data used to measure the inclusive jet cross section. The hadronization and underlying-event corrections were determined using PYTHIA and applied to theoretical predictions. The hadronization (underlying-event) corrections vary from -15% ($+30\%$) to -3% ($+6\%$) for $p_T = 50$ GeV/c to 600 GeV/c (48).

The perturbative results are the sum of $\mathcal{O}(\alpha_s^3)$ pQCD calculations (17, 18), supplemented with $\mathcal{O}(\alpha_s^4)$ (two-loop) corrections for the threshold effects (19). The PDFs are taken from the MSTW2008 NNLO parameterization, and the renormalization and factorization scales are set to be equal to the p_T of the jet. The LO (NLO) MSTW2008 PDFs are available for 21 different values of $\alpha_s(Mz)$, ranging from 0.107 to 0.117 (0.110 to 0.130) in steps of 0.001. This α_s is used to evolve the PDFs.

Commonly available parton distributions from the MSTW and CTEQ groups include Tevatron jet data in the global fit. To avoid any correlation between the input PDFs and the extracted α_s, only 22 of the 110 available jet data points are used. These selected points contribute to the x region ($x \leq 0.25$), where PDFs in global fits are determined mainly by other experimental data and are not strongly influenced by Tevatron jet data. The jet data begin to affect $g(x)$ at $x \sim 0.2$. The change in $g(x)$ due to the inclusion of jet data is less than 5% for $x \lesssim 0.25$ (55).

The central $\alpha_s(Mz)$ result is obtained by minimizing the χ^2 with respect to $\alpha_s(M_Z)$ and by integrating over the nuisance parameters for the correlated uncertainties. The variation of $\alpha_s(p_T)$ versus p_T is shown in **Figure 7*a***. The running of α_s as a function of jet p_T follows the QCD evolution equation. The data points from the H1 and ZEUS experiments follow the same curve but have large uncertainties. The α_s, evolved to $\mu = M_Z$, is shown in **Figure 7*b***. The combined result of 22 selected points, $\alpha_s(M_Z) = 0.1161^{+0.0041}_{-0.0048}$, is consistent with the world average of 0.1184 ± 0.0007 (61), although the (mostly theoretical) uncertainties are large.

6. SEARCHES FOR PHYSICS BEYOND THE STANDARD MODEL USING JETS

Due to large theoretical and experimental uncertainties in the jet production rate, the high-p_T tail cannot be used to improve upon the current limits on new interactions (62). However, there are kinematic distributions from jet events that can be used to identify physics beyond the Standard Model. Below we describe these searches using the dijet mass spectrum and the dijet angular distributions. These searches are not very sensitive to the jet energy scale, PDFs, or renormalization or factorization scales.

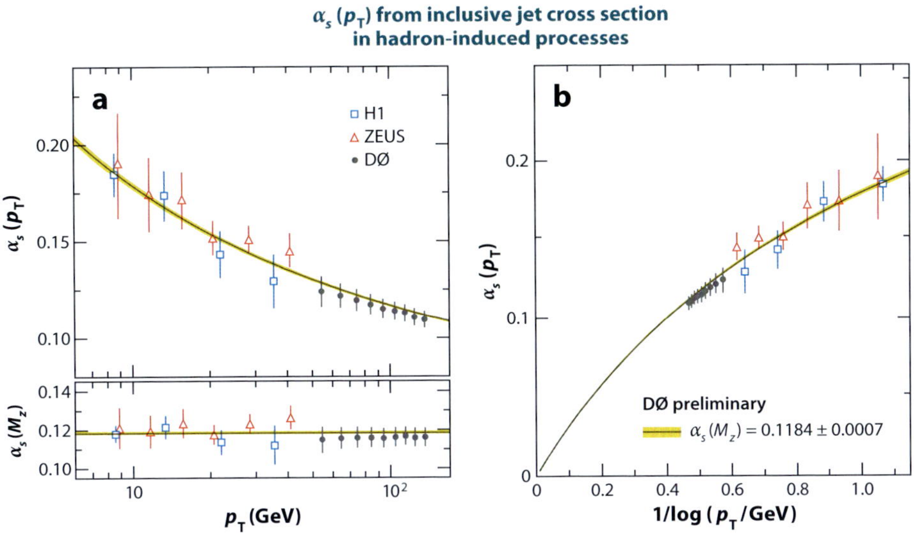

Figure 7

(*a*) Measurement of α_s as a function of the transverse momentum of the jet from several experiments, as compared to the expected variation of α_s. Here $\alpha_s(M_Z) = 0.1184 \pm 0.0007$ (60). (*b*) Plot of α_s as a function of $1/\log(p_T)$, which supports the idea of asymptotic freedom.

6.1. Dijet Mass Spectrum

Many new physics models predict particles that decay into two high-p_T jets. Such particles can be identified by the reconstructed mass of the dijet system, provided that their intrinsic mass width is narrow. Such models include excited quarks (63), axigluons (64), flavor-universal colorons (65), color-octet techni-ρ (66), Randall-Sundrum gravitons (67), heavy vector bosons (68), and diquarks in the string-inspired E_6 model (69). The excited quarks q^* decay into qg. The heavy vector bosons W' and Z' decay into $q\bar{q}'$ or $q\bar{q}$, respectively. The axigluon A decays into $q\bar{q}$, and the E_6 diquarks D and D^c decay into $\bar{q}\bar{q}$ and qq, respectively. Both the Randall-Sundrum graviton G^* and the color-octet techni-ρ ρ_{T8} decay into either a $q\bar{q}$ or gg pair, but their branching ratios are different.

All these models predict an intrinsic mass width that is much smaller than both the detector resolution and the mass-broadening effects arising from QCD radiation. These models can be divided into three categories depending on the decay channel: gg, gq, and $q\bar{q}$. The expected mass shapes for q^*, G^*, W', and Z' particles with a mass of 800 GeV/c^2 are shown in **Figure 8**. Because q^* and G^* decay into gluons, their widths are broader than the widths of W' and Z'. Gluons radiate more than quarks, resulting in a broader dijet mass distribution. These distributions are close and change the final limits by only 10 to 20%. These shapes can be used to search for resonance structure, independent of the model details.

To measure the dijet mass spectrum (70), the CDF Collaboration used the same data set as in the inclusive jet cross-section measurement described in Section 5.1.2 (46). The dijet mass is reconstructed from the two highest-p_T jets using

$$m_{jj} = \sqrt{(E_1 + E_2)^2 - (\vec{p}_1 + \vec{p}_2)^2}. \qquad 8.$$

Jets produced by new physics are expected to be produced more centrally than by Standard Model processes, and we expect a better signal-to-background ratio in the central region. Thus, only those events in which two leading jets have $|y| \leq 1.0$ are used. Moreover, the CDF calorimeter

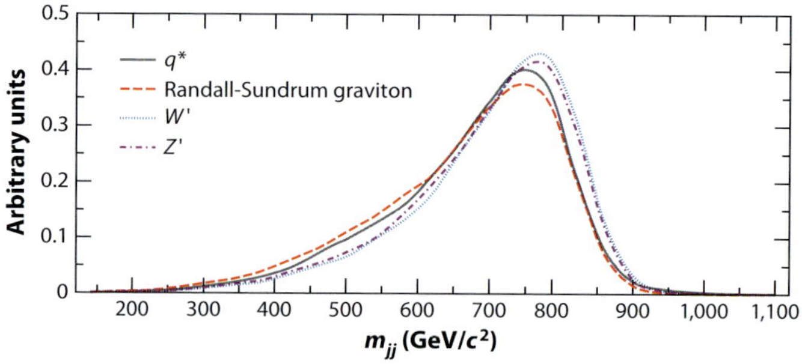

Figure 8

Expected dijet mass distributions for simulated signals for the following new physical models: $q^* \to qg$; Randall-Sundrum graviton $\to gg, q\bar{q}$; $W' \to q\bar{q}'$; and $Z' \to q\bar{q}$, where the particles have a mass of 800 GeV/c^2.

is best understood in this region. As shown in **Figure 9**, these data, after all corrections, are in good agreement with the NLO QCD calculations. The dijet mass spectrum before smearing corrections is shown in **Figure 10**. To guard against any theoretical biases, dijet resonance searches are performed without the use of smearing corrections. The measured mass spectrum is fit to a smooth background given by

$$\frac{d\sigma}{dm_{jj}} = p_0(1-x)^{p_1}/x^{p_2+p_3\ln(x)}, \quad x = m_{jj}/\sqrt{s}, \qquad 9.$$

where p_0, p_1, p_2, and p_3 are free parameters. The dijet mass spectra predicted by PYTHIA, HERWIG, and NLO pQCD can be described well by this functional form. The fit to the measured dijet mass spectrum is shown in **Figure 10*a***. The data are well described by this smooth function, with a χ^2

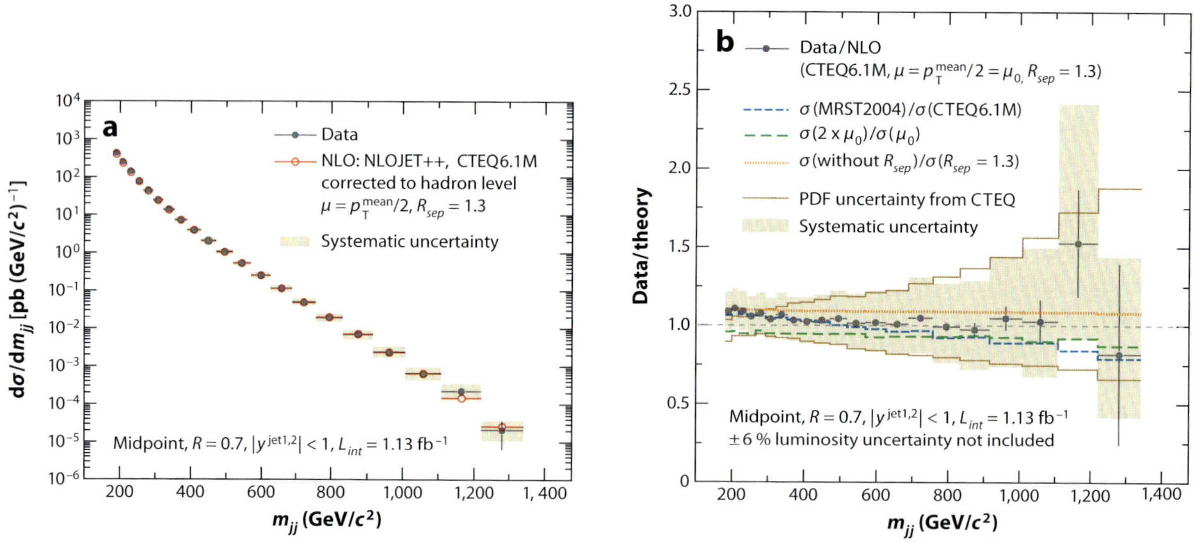

Figure 9

(*a*) CDF's measurement of the dijet mass cross section for events in which the two highest p_T have $|y| < 1.0$ (70), compared to next-to-leading-order (NLO) calculations by use of CTEQ6.1M. (*b*) Ratio of data to NLO theory. The experimental uncertainties are dominated by jet energy scale uncertainty and are comparable to the theoretical parton distribution function (PDF) uncertainty.

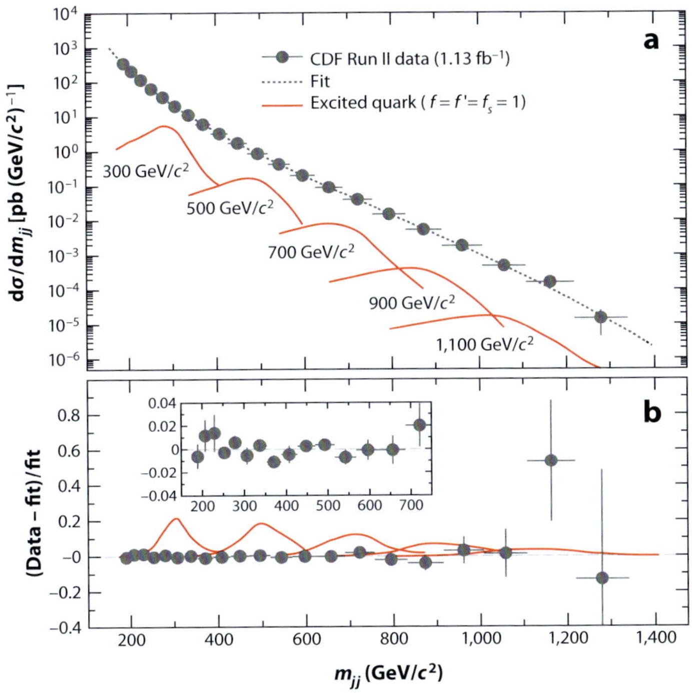

Figure 10

(*a*) CDF's measured dijet mass spectrum (70). The dashed gray curve shows the fit to Equation 9. Also shown are the predicted dijet mass distributions of the excited quark q^* for masses of 300, 500, 700, 900, and 1,100 GeV/c^2, respectively. (*b*) The fractional difference between the measured dijet mass difference and the fit (*points*) compared to q^* signals, divided by the fit to the measured dijet mass spectrum (*curves*).

of 16 for 17 degrees of freedom. The deviation from the smooth curve is shown in **Figure 10*b***. These data are used to determine the exclusion limits on the existence of new particles decaying into jets, as there is no evidence for the existence of any resonant structure. The experimental limits are determined for $\sigma^{sig} \equiv \sigma \times B \times A$, where σ is the theoretical new particle–production rate, B is the probability of the particle decaying into two jets, and A is the kinematic acceptance of the resulting particle jets to have $|y| < 1.0$.

The upper limits on σ^{sig} are evaluated using a likelihood function,

$$L = \prod_i \mu_i^{n_i} \exp(-\mu_i)/n_i!, \qquad 10.$$

where $\mu_i = n_i^{sig} + n_i^{QCD}$ is the predicted number of events in bin i. The QCD dijet background n_i^{QCD} is determined using Equation 9 by evaluating $\mathcal{L}_i \times \varepsilon_i \times \Delta m_{jj} \times d\sigma/dm_{jj}|_i$, where Δm_{jj}, ε_i, and $\mathcal{L}_i$, are the bin width, the trigger efficiency, and the integrated luminosity for bin i, respectively. The number of expected signal events n_i^{sig} is given by $\sigma^{sig} \times \mathcal{L}_i \times \varepsilon_i \times (n_i/n_{tot})$, where ε_i is the signal event–selection efficiency in the ith dijet mass bin and n_i/n_{tot} is the predicted signal fraction in bin i. For each value of σ^{sig}, the likelihood is maximized with respect to the four parameters in Equation 9. This profiled likelihood is integrated over Bayesian priors for the parameters describing the systematic uncertainties (71). A flat prior on σ^{sig} is used to extract a Bayesian upper limit on that parameter.

Table 1 The limits on the masses of particles decaying into dijets in various models[a]

Model	Parameters	Excluded region
q^*	$f = f' = f_s$	260–870
Axigluons		260–1,250
Coloron		290–630
E_6 diquark		260–1,100
ρ_{T8}		280–840
W'	Standard Model	280–840
Z'	Standard Model	320–740
G^*	$k/M_{Pl} = 0.1$	—

[a]Limits are in units of GeV/c^2 (70).

The mass exclusion limits for W', Z', q^*, and G^* are determined by comparing the limits obtained using their respective signal shapes and the predicted theoretical cross section. For other models, the limits obtained for these four signal shapes are used. The q^* signal shape (qg) is used for axigluons, the flavor-universal coloron, and the E_6 diquark, as these particles do not decay into modes that include gluons, and thus their signal shapes are expected to be narrower than that of q^*. For ρ_{T8}, the limits obtained for the G^* shape are used. Both the ρ_{T8} and the G^* decay into $q\bar{q}$ or gg, but the branching fraction of G^* into gg is higher. Therefore, in all the above cases, the obtained exclusion limits are conservative. The limits for these models are given in **Table 1**.

The dijet mass spectrum measured by the DØ Collaboration in six rapidity bins (72) is shown in **Figure 11**. The rapidity bin is labeled by the higher of the two jet rapidities. The data are compared to NLO pQCD predictions computed by NLOJet++ (17) using the MSTW2008 NLO PDFs (55) with scale $\mu_R = \mu_F = (p_{T,1} + p_{T,2})/2$ and are corrected for nonperturbative effects determined by using the PYTHIA event generator. The data and QCD predictions are in reasonable agreement for the $|y| < 0.4$ region, which is not surprising given that the MSTW2008 global fit includes the inclusive jet data (see Section 5.3). For higher $|y|$ bins, the data are below the theoretical predictions but are within 1σ of the total experimental systematic uncertainty. The DØ Collaboration is searching for new particles using these data.

6.2. Dijet Angular Distributions

The angle between the initial- and final-state partons in the center-of-momentum frame is sensitive to the spin of the exchanged or the intermediate particle and thus can be used to search for physics beyond the Standard Model. At hadron colliders, dijet production is dominated by the t channel exchange of a gluon, which is a massless vector boson, and the angular distribution has the familiar Rutherford scattering form

$$\frac{d\hat{\sigma}}{d\cos\theta^*} \sim \frac{1}{(1-\cos\theta^*)^2} = \frac{1}{\sin^4(\theta^*/2)}, \qquad 11.$$

where θ^* is the angle between the jet and the beam direction in the dijet center-of-momentum frame. The angular distribution of the new particles proposed in many new physics scenarios is

Figure 11

(*a*) DØ's dijet mass measurement for six rapidity bins (72) compared to next-to-leading order (NLO) using MSTW2008 PDFs. (*b*) Ratio of data to theory. The experimental systematic uncertainties are very small. The dashed red lines show the effect of using the CTEQ6.6M parton distribution functions (PDFs).

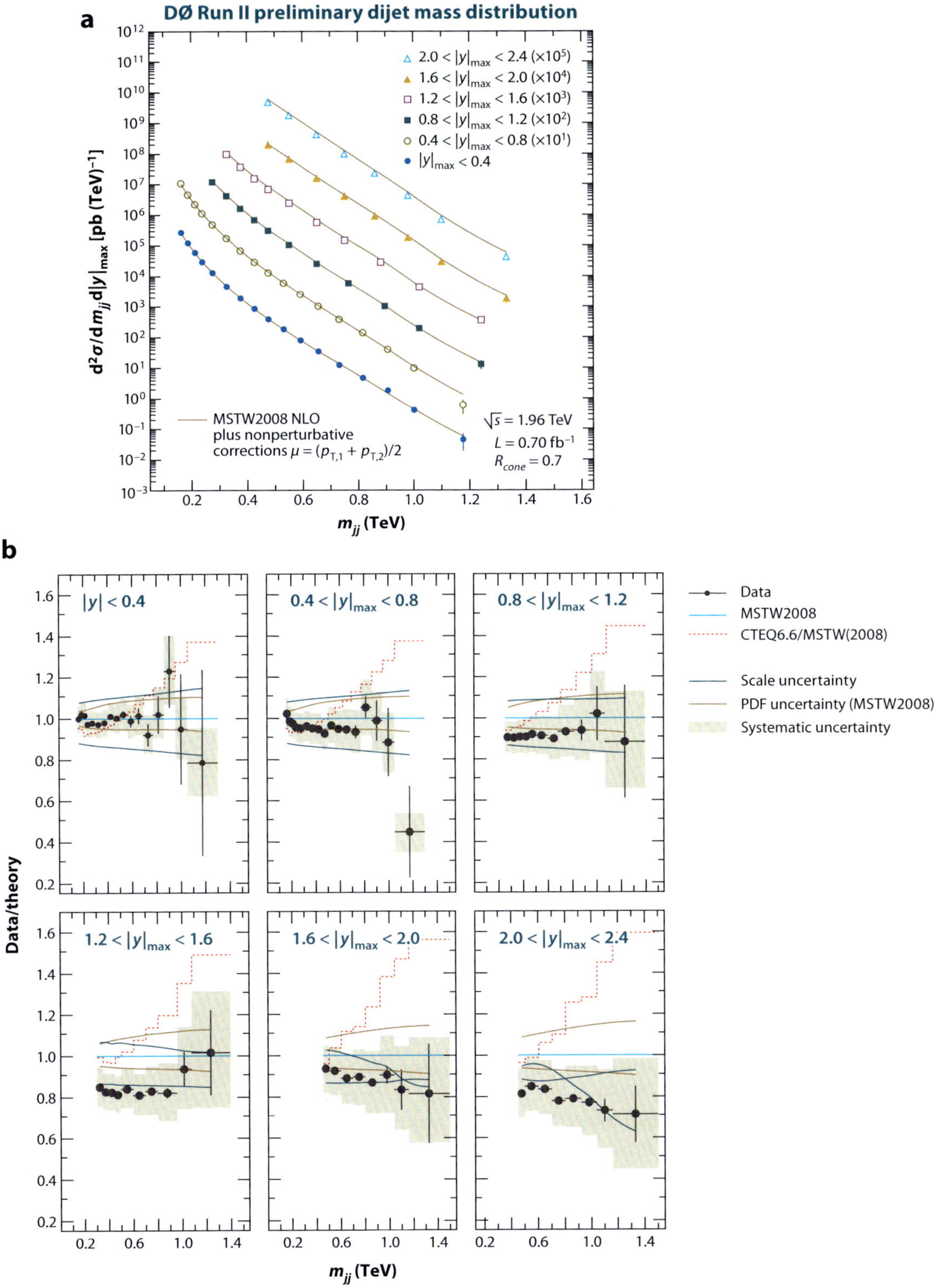
a
DØ Run II preliminary dijet mass distribution
2.0 < |y|max < 2.4 (×10^5)
1.6 < |y|max < 2.0 (×10^4)
1.2 < |y|max < 1.6 (×10^3)
0.8 < |y|max < 1.2 (×10^2)
0.4 < |y|max < 0.8 (×10^1)
|y|max < 0.4
MSTW2008 NLO
plus nonperturbative
corrections μ = (pT,1 + pT,2)/2
√s = 1.96 TeV
L = 0.70 fb^-1
Rcone = 0.7
d²σ/dmjj d|y|max [pb (TeV)^-1]
mjj (TeV)
b
|y| < 0.4
0.4 < |y|max < 0.8
0.8 < |y|max < 1.2
1.2 < |y|max < 1.6
1.6 < |y|max < 2.0
2.0 < |y|max < 2.4
Data
MSTW2008
CTEQ6.6/MSTW(2008)
Scale uncertainty
PDF uncertainty (MSTW2008)
Systematic uncertainty
Data/theory
mjj (TeV)

relatively flat in $\cos\theta^*$. For example, the angular distribution of spin-1 particles (W', Z', axigluon, coloron) decaying in fermions is $d\sigma/d\cos\theta^* \sim 1+\cos^2\theta^*$. Theories in which quarks are composite particles, but where the compositeness scale is much higher than the available energy, can be parameterized by an effective Lagrangian of the type (62, 73, 74)

$$\mathcal{L} = \eta \frac{g^2}{4\Lambda^2}(\bar{q}_i\gamma^\mu q_i)(\bar{q}_j\gamma_\mu q_j), \quad i = L, R; \; j = L, R,$$

where Λ is a parameter in the theory that controls the characteristic energy of the new interactions. The parameter η is ± 1 and determines the sign of interference between new interactions and the Standard Model interactions. The main effect of the substructure is to increase the proportion of centrally produced jets, which can be observed in the jet angular distributions (73).

The ADD LED (Arkani-Hamed–Dimopoulos–Dvali large extra dimensions) models (75, 76) that have been proposed to solve the hierarchy problem, namely the difference between the electroweak scale (~100 GeV) and the Planck scale M_{Pl} ($\sim 10^{19}$ GeV), assume the existence of extra spatial dimensions in which gravity is allowed to propagate. As a consequence, gravity appears weak in the three conventional spatial dimensions. The Planck scale, the number of extra dimensions n, their size R, and an effective Planck scale M_S are related by $M_{Pl} = M_S R^n$. Experimentally, M_S can be measured for different values of n. The Kaluza-Klein excitations of the graviton can be exchanged between partons and thus contribute to jet production, resulting in jets that are central. There are two different formalisms to describe LED models, GRW (Giudice-Rattazzi-Wells; 77) and HLZ (Han-Lykken-Zhang; 78). In the HLZ formalism, the subleading dependence on the number n of extra dimensions is also included.

In some models (79–81), extra dimensions are assumed to exist at the TeV^{-1} distance scale. In these models, Kaluza-Klein excitations of Standard Model bosons modify various production cross sections, and gluons can propagate through the extra dimensions, which changes the jet cross section. The strength of the interaction is given by the model parameter, the compactification scale M_C.

To search for new physics, instead of studying the $\cos\theta^*$ distribution directly it is convenient to use the χ_{dijet} distribution, which removes the Rutherford singularity: χ_{dijet} is defined as $\exp(y_1 - y_2)$, where y_1 and y_2 are rapidities (see footnote 1) of the two highest-p_T jets in an event. For $2 \to 2$ scattering of massless partons, the variable χ_{dijet} is related to the partonic center-of-momentum frame polar angle θ^* by $\chi_{dijet} = (1+\cos\theta^*)/(1-\cos\theta^*)$.

The CDF Collaboration studied the dijet angular distributions using 106 pb^{-1} of data from Run I (82). The data exclude at 95% CL a model of quark substructure in which only up and down quarks are composite and the contact interaction scale is $\Lambda^+_{ud} \leq 1.6$ TeV or $\Lambda^-_{ud} \leq 1.4$ TeV, where the subscript refers to the flavor of quarks assumed to be composite and the superscript $\pm$ refers to the sign of the interference. For a model in which all quarks are composite, the excluded regions are $\Lambda^+ \leq 1.8$ TeV and $\Lambda^- \leq 1.6$ TeV. In Run II, the DØ Collaboration measured the χ_{dijet} distribution in the $\chi_{dijet} \leq 16$ range in ten dijet mass m_{jj} bins covering the $0.25 < m_{jj} < 1.1$ TeV/c^2 range, using up to 0.7 fb^{-1} of data collected during 2004 and 2005 (83). The boost of the two-jet system is required to be $y_{\text{boost}} \equiv 0.5 \times |y_1 + y_2| \leq 1$. This requirement, combined with the dijet mass cut and the range of the χ_{dijet} distribution, restricts the highest allowable rapidity to $|y_{1,2}| < 2.4$, where the DØ detector performance is well understood. The measured distributions are corrected for detector effects using events generated with PYTHIA v6.419 (8) with Tune QW (21) and the MSTW2008 LO PDFs. This procedure corrects for the migration between dijet mass bins, as well as the shape of the χ_{dijet} distributions in each mass bin. The corrected normalized differential cross-section distributions ($1/\sigma_{dijet} \times d\sigma/d\chi_{dijet}$) at the particle level are shown in **Figure 12** for ten dijet mass bins. The NLO pQCD predictions are computed using

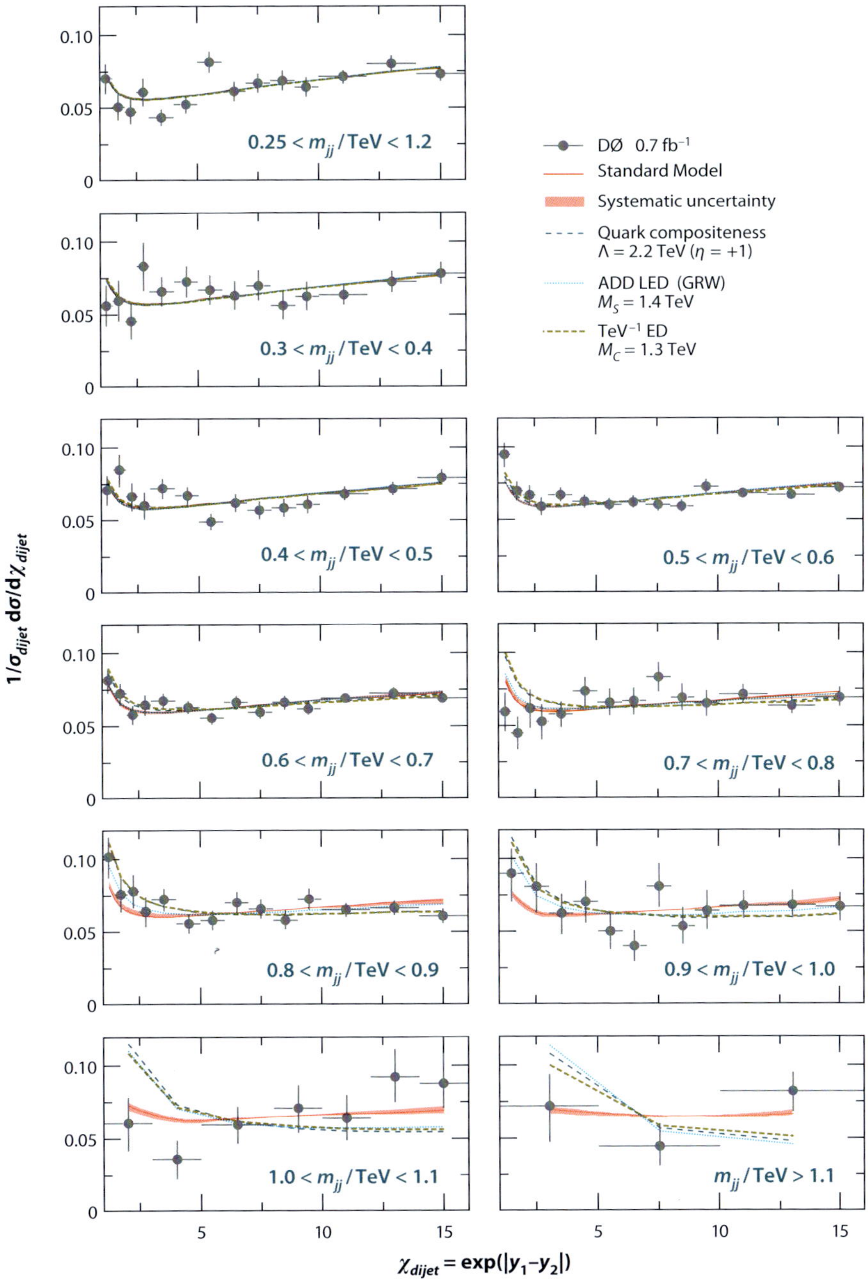

Figure 12

DØ's measurement of the dijet χ for ten dijet mass bins (83). The solid red line is the Standard Model prediction at next-to-leading order, and the shaded area shows the systematic uncertainty. The variously colored dashed lines show the predictions for various new physics models. Abbreviations: ADD LED, Arkani-Hamed–Dimopoulos–Dvali large extra dimensions; ED, extra dimensions; GRW, Giudice-Rattazzi-Wells.

Table 2 Expected and observed 95%-CL limits on various new physics models[a]

Model (parameter)	Expected (TeV)	Observed (TeV)
Quark compositeness (Λ)		
$\eta = +1$	2.76	2.84
$\eta = +-1$	2.75	2.82
TeV^{-1} ED (M_C)	1.60	1.55
ADD LED (M_S)		
GRW	1.47	1.59
HLZ $n = 3$	1.75	1.89
HLZ $n = 4$	1.47	1.59
HLZ $n = 5$	1.33	1.43
HLZ $n = 6$	1.24	1.34
HLZ $n = 7$	1.17	1.26

[a]Abbreviations: ADD, Arkani-Hamed–Dimopoulos–Dvali; ED, extra dimensions; GRW, Giudice-Rattazzi-Wells; HLZ, Han-Lykken-Zhang; LED, large extra dimensions.

FastNLO (18) based on NLOJet++ (17). These parton-level predictions are corrected for the hadronization and underlying-event contributions, which are evaluated using PYTHIA. The theoretical uncertainties on the Standard Model χ_{dijet} distributions arising from the uncertainty on the PDFs and the uncertainty on renormalization and factorization scales are less than 2% and 5%, respectively. These data are in good agreement with the Standard Model predictions and thus are used to set exclusion limits in the parameter space of quark compositeness, ADD LED, and TeV^{-1} models. Calculations for all these models are available only at LO, whereas pQCD calculations can be performed at NLO. For this analysis, the expected distributions for each new model are calculated at LO and then scaled by k-factors ($k = \sigma_{\text{NLO}}/\sigma_{\text{LO}}$) determined from pQCD calculations. The k-factors vary from 1.25 to 1.5. All these models predict a higher rate as $\chi_{dijet} \to 1$ and as m_{jj} increases. However, the magnitude of the excess is different for different models. A Bayesian procedure (45) is used to obtain 95%-CL limits on the mass scale parameters Λ, M_C, and M_S in the above models. The results in which the prior is chosen to be flat in the model cross section are given in **Table 2**. Other choices give similar but slightly higher limits (83).

The limits on M_C obtained in this analysis, although inferior to indirect limits from precision electroweak measurements (81), are the first direct search for TeV^{-1} extra dimensions at a particle collider. The limits on M_S in different formalisms of ADD LED are, on average, slightly higher than the recent DØ results obtained using dielectron and diphoton data (85). The quark compositeness limits are the most stringent limits to date.

7. CONCLUSIONS

Since the beginning of Run II, there has been a significant increase in the experimental data used in the jet analyses at the Tevatron. Due to the higher production cross section and the increase in available integrated luminosity, the inclusive jet cross-section measurement has been extended to transverse momenta of 600 GeV/c. The experimental uncertainty is still dominated by the uncertainty on the jet energy scale, which is +31/−26% in the $p_T = 457$–527 GeV/c bin in the $0.1 < |y| < 0.7$ region for CDF measurements and +16.0/−15.5% in the $p_T = 490$–540 GeV/c bin in the $|y| < 0.5$ region for DØ measurements. Both collaborations used a modern jet-finding algorithm, the midpoint algorithm, which is IR and collinear safe at NLO in pQCD for measurements of the inclusive jet cross section. Also, the treatment of the underlying-event energy and hadronization effects has improved over the techniques used in Run I. The experimental uncertainties are lower than the theoretical uncertainties, which are dominated by the uncertainties on the PDFs. These data have been used in global fits to determine the parton distributions and have decreased the uncertainty on the gluon distribution for the $x \geq 0.3$ region. The new gluon distribution is slightly lower than that determined by Run I jet cross-section measurements. This increase in the accuracy of gluon distributions will make the prediction of various processes more precise.

These jet data have been successfully used to search for physics beyond the Standard Model via jet kinematic distributions. The dijet mass spectrum has been used to expand the exclusion regions in parameters of excited quarks in quark compositeness models, E_6 diquarks, axigluons, and heavy vector W and Z bosons, as well as the techni-ρ in color-octet models. The 95%-CL lower mass limits range from 630 GeV/c^2 for colorons to 1.25 TeV/c^2 for axigluons. The dijet angular distribution has been used to extend the limits on the quark compositeness mass scale, the ADD LED, and the TeV^{-1} extra dimensions. The 95%-CL lower limit on the compositeness mass scale is 2.8 TeV, and the 95%-CL lower limit for the compactification mass scale in the TeV^{-1} model is 1.5 TeV. The limits on the ADD LED range from 1.9 TeV to 1.3 TeV, depending on the number of extra dimensions in the HLZ formalism. The 95%-CL lower limit in the GRW formalism is 1.6 TeV. In most cases, these are the best limits to date.

DISCLOSURE STATEMENT

The authors are not aware of any affiliations, memberships, funding, or financial holdings that might be perceived as affecting the objectivity of this review.

ACKNOWLEDGMENTS

We thank Andrew Beretvas, Kenichi Hatakeyama, Markus Wobisch, and Marek Zielinski for reading the manuscript and for insightful comments.

LITERATURE CITED

1. Gross DJ, Wilczek F. *Phys. Rev. D* 8:3633 (1973); Fritzsch H, Gell-Mann M, Leutwyler H. *Phys. Lett.* B47:365 (1973); Politzer HD. *Phys. Rev. Lett.* 30:1346 (1973)
2. Stenzel H. *Nucl. Phys. B* 152:23 (2006)
3. Gwenlan C. *Acta Phys. Polon. B* 35:377 (2004)
4. Fabjan C, McCubbin N. *Phys. Rep.* 403:165 (2004)
5. Blazey GC, Flaugher BL. *Annu. Rev. Nucl. Part. Sci.* 49:633 (1999)
6. Salam GP. arXiv:0906.1736v3 [hep-ph] (2009)
7. Gribov VN, Lipatov LN. *Sov. J. Nucl. Phys.* 15:438 (1972); Altarelli G, Parisi G. *Nucl. Phys. B* 126:298 (1977); Dokshitzer Y. *Sov. Phys. JETP* 46:641 (1977)
8. Sjöstrand T, et al. *J. High Energy Phys.* 0605:026 (2006)
9. Corcella G, et al. *J. High Energy Phys.* 0101:010 (2001)
10. Frixione S, Webber BR. *J. High Energy Phys.* 0206:29 (2002)
11. Bhat CM. *Int. J. Mod. Phys. A* 16S1C:1187 (2001)
12. Abulencia A, et al. *J. Phys. G* 34:2457 (2007)
13. Abazov VM, et al. *Nucl. Instrum. Methods A* 565:463 (2006)
14. Ellis RK, Stirling WJ, Webber BR. *QCD and Collider Physics*. Cambridge, UK: Cambridge Univ. Press. (1996)

15. Ellis SD, Kunszt Z, Soper DE. *Phys. Rev. Lett.* 64:2121 (1990)
16. Giele WT, Glover EWN, Kosower DA. *Nucl. Phys. B* 403:633 (1993)
17. Nagy Z. *Phys. Rev. D* 68:094002 (2003)
18. Kluge T, Rabbertz K, Wobisch M. arXiv:hep-ph/0609285 (2006)
19. Kidonakis N, Owens JF. *Phys. Rev. D* 63:054019 (2001)
20. Field R. Presented at Fermilab ME/MC Workshop, Fermilab, Batavia, Ill. **http://www.phys.ufl.edu/~rfield/cdf/FNAL_Workshop_10-4-02.pdf** (2002)
21. Field R, et al. (Tev4LHC QCD Work. Group) arXiv:hep-ph/0610012 (2006)
22. Mangano ML, et al. *J. High Energy Phys.* 0307:001 (2003)
23. Gleisberg T, et al. *J. High Energy Phys.* 0402:056 (2004)
24. Catana S, Krauss F, Kuhn R, Webber BR. *J. High Energy Phys.* 0111:063 (2001); Krauss F. *J. High Energy Phys.* 0208:015 (2002)
25. Ellis SD, et al. *Prog. Part. Nucl. Phys.* 60:484 (2008)
26. Blazey GC, et al. arXiv:hep-ex/0005012v2 (2000)
27. Kilgore W, Giele WT. *Phys. Rev. D* 55:7183 (1997)
28. Buttar C, et al. arXiv:0803.0678v1 [hep-ph] (2008)
29. Salam GP. arXiv:0906.1833v1 [hep-ph] (2009)
30. Huth JE, et al. *FERMILAB-CONF-90-249-E* (1990)
31. Salam GP, Soyez G. *J. High Energy Phys.* 0705:086 (2007)
32. Catani S, et al. *Nucl. Phys. B* 406:187 (1993)
33. Ellis SD, Soper DE. *Phys. Rev. D* 48:3160 (1993)
34. Abazov VM, et al. *Phys. Lett.* B525:211 (2002)
35. Dokshitzer Yu, Leder G, Moretti S, Webber B. *JHEP* 9708:001 (1997); Wobisch M, Wengler T. arXiv:hep-ph/9907280 (1999)
36. Cacciari M, Salam GP, Soyes G. *J. High Energy Phys.* 0804:063 (2008)
37. Bhatti A, et al. *Nucl. Instrum. Methods A* 556:375 (2006)
38. Voutilainen M. *FERMILAB-THESIS-2008-19* (2008)
39. Amsler C, et al. *Phys. Lett.* B667:1 (2008)
40. Arnison G, et al. *Phys. Lett.* B172:461 (1986)
41. Alitti J, et al. *Phys. Lett.* B257:232 (1991); Alitti J, et al. *Z. Phys. C* 49:17 (1991)
42. Abe F, et al. *Phys. Rev. Lett.* 70:1376 (1993)
43. Abe F, et al. *Phys. Rev. Lett.* 77:438 (1996)
44. Affolder T, et al. *Phys. Rev. D* 64:032001 (2001)
45. Abbott B, et al. *Phys. Rev. D* 64:032003 (2001)
46. Aaltonen T, et al. *Phys. Rev. D* 78:052006 (2008)
47. Abulencia A, et al. *Phys. Rev. D* 75:092006 (2007)
48. Abazov VM, et al. *Phys. Rev. Lett.* 101:062001 (2008)
49. Tung WK, et al. *J. High Energy Phys.* 02:053 (2007); Pumplin J, et al. *J. High Energy Phys.* 07:12 (2002)
50. Stump D, et al. *J. High Energy Phys.* 10:046 (2003)
51. Deleted in proof
52. Tzanov M. *Int. J. Mod. Phys. A* 20:3759 (2005)
53. Towell RS, et al. *Phys. Rev. D* 64:052002 (2001)
54. Abe F, et al. *Phys. Rev. Lett.* 81:5754 (1998); Aaltonen T, et al. *Phys. Rev. Lett.* 102:181801 (2009); Aaltonen T, et al. arXiv:0908.3914v3 [hep-ex] (2009); Abazov VM, et al. *Phys. Rev. D* 76:012003 (2007)
55. Martin AD, et al. *Eur. Phys. J. C* 63:189 (2009)
56. Pumplin J, et al. arXiv:0904.2424v2 [hep-ph] (2009)
57. Martin AD, Roberts RG, Stirling WJ, Thorn RS. *Phys. Lett.* B604:61 (2004)
58. Nadolsky PM, et al. *Phys. Rev. D* 78:013004 (2008)
59. Affolder T, et al. *Phys. Rev. Lett.* 88:042001 (2002)
60. Abazov VM, et al. arXiv:0911.2710 [hep-ex] (2009)
61. Bethke S. *Eur. Phys. J C* 64:689 (2009)
62. Eichten E, Lane K, Peskin M. *Phys. Rev. Lett.* 50:811 (1983)

63. Baur U, Hinchliffe I, Zeppenfeld D. *Int. J. Mod. Phys. A* 2:1285 (1987); Baur U, Spira M, Zerwas PM. *Phys. Rev. D* 42:815 (1990)
64. Frampton PH, Glashow SL. *Phys. Lett.* B190:157 (1987); Bagger J, Schmidt C, King S. *Phys. Rev. D* 37:1188 (1988)
65. Chivukula RS, Cohen AG, Simmons EH. *Phys. Lett.* B380:92 (1996); Simmons EH. *Phys. Rev. D* 55:1678 (1997)
66. Lane KD, Ramana MV. *Phys. Rev. D* 44:2678 (1991); Lane K, Mrenna S. *Phys. Rev. D* 67:115011 (2003)
67. Randall L, Sundrum R. *Phys. Rev. Lett.* 83:3370 (1999); Bijnens J, et al. *Phys. Lett.* B503:241 (2001)
68. Eichten E, Hinchliffe I, Lane K, Quigg C. *Rev. Mod. Phys.* 56:579 (1984); Eichten E, Hinchliffe I, Lane K, Quigg C. *Rev. Mod. Phys.* 58:1065 (1986)
69. Hewett JL, Rizzo TG. *Phys. Rep.* 183:193 (1989)
70. Aaltonen T, et al. *Phys. Rev. D* 72:112002 (2009)
71. Demortier L. Presented at Conf. Adv. Stat. Tech. Part. Phys., Durham, UK (2002)
72. DØ Collab. *DØ Note 5919-Conf.* **http://www-d0.fnal.gov/Run2Physics/WWW/results/prelim/QCD/Q14/Q14.pdf** (2009)
73. Eichten E, Lane K. arXiv:hep-ph/9609298v1 (1996)
74. Lane KD. arXiv:hep-ph/9605257 (1996)
75. Arkani-Hamed N, Dimopoulos S, Dvali G. *Phys. Lett.* B429:263 (1998)
76. Atwood D, Bar-Shalom S, Soni S. *Phys. Rev. D* 62:056008 (2000)
77. Giudice GF, Rattazzi R, Wells JD. *Nucl. Phys. B* 544:3 (1999)
78. Han T, Lykken JD, Zhang RJ. *Phys. Rev. D* 59:105006 (1999)
79. Dienes KR, Dudas E, Gherghetta T. *Nucl. Phys. B* 537:47 (1999)
80. Pomarol A, Quiros M. *Phys. Lett.* B438:255 (1998)
81. Cheung K, Landsberg G. *Phys. Rev. D* 65:076003 (2002)
82. Abe F, et al. *Phys. Rev. Lett.* 77:5336 (1996)
83. Abazov VM, et al. *Phys. Rev. Lett.* 103:191803 (2009)
84. Deleted in proof
85. Abazov VM, et al. *Phys. Rev. Lett.* 102:051601 (2009)

Beta Beams

Mats Lindroos[1,2] and Mauro Mezzetto[3]

[1]Accelerator Design, ESS Scandinavia, Lund, SE-221, Sweden

[2]Accelerators and Beams, CERN, Genève 23, CH-1211, Switzerland; email: mats.lindroos@cern.ch

[3]Istituto Nazionale di Fisica Nucleare, Sezione di Padova, Padova, I-35131, Italy; email: mezzetto@pd.infn.it

Annu. Rev. Nucl. Part. Sci. 2010. 60:299–325

First published online as a Review in Advance on July 16, 2010

The *Annual Review of Nuclear and Particle Science* is online at nucl.annualreviews.org

This article's doi: 10.1146/annurev.nucl.012809.104455

0163-8998/10/1123-0299$20.00

Key Words

accelerators, charge conjugation–parity (*CP*) violation, long-baseline neutrino experiments, neutrino oscillations, radioactive ions

Abstract

The ultimate goal of neutrino oscillation physics is the search for leptonic *CP* violation, which will require neutrino beams that are much more intense and pure than those used in present experiments. Beta beams are an attractive innovative possibility in this direction. Neutrinos are generated by the beta decays of radioactive nuclei and are accelerated at very high energies. The resulting neutrino beam consists of only one easily predictable flavor of neutrinos (ν_e or $\bar{\nu}_e$). A realistic beta beam design, which has already been demonstrated by the Eurisol Design Study, is based on CERN's PS and SPS accelerators. The beta beam concept has also been extended in several other directions, including high energy, high Q, electron capture, and low energy. Both the accelerator complex and the physics potential of a neutrino experiment are reviewed here. We emphasize the beta beam design based on the CERN PS and SPS, but we also discuss other possibilities.

Contents

1. NEUTRINO OSCILLATIONS

The discovery of neutrino oscillations (1) has established that neutrinos have mass and mix. The existence of neutrino masses is the first solid experimental fact that has required physics beyond the Standard Model.

Neutrino oscillations are consistently described by three families: ν_1, ν_2, and ν_3. These families have mass values of m_1, m_2, and m_3 that are connected to the flavor eigenstates ν_e, ν_μ, and ν_τ by a mixing matrix, U. The neutrino oscillation probability depends on three mixing angles, θ_{12}, θ_{23}, and θ_{13}; two mass differences, $\Delta m^2_{12} = m^2_2 - m^2_1$ and $\Delta m^2_{23} = m^2_3 - m^2_2$; and a CP phase, δ_{CP}. Additional phases are present in case the neutrinos are Majorana particles, but they do not influence neutrino flavor oscillations.

1.1. Future Discoveries in Neutrino Oscillations

Out of seven parameters, three have not yet been measured in neutrino oscillations:

1. The mixing angle θ_{13} is the key parameter of three-neutrino oscillations. It regulates at first order all the oscillation processes that could contribute to the measurement of sign(Δm^2_{31}) and δ_{CP}.

2. The neutrino mass hierarchy, the mass order of the three mass eigenstates, can be fixed by measuring the sign of Δm_{31}^2. Its value could be +1 (normal hierarchy), in which case ν_1 would be the lightest neutrino, or −1 (inverted hierarchy), for which the lightest would be ν_3. Its value is of great importance for double-beta decay experiments (2), and it could shed light on possible flavor symmetries.
3. The *CP* phase δ_{CP} is the holy grail of ultimate neutrino oscillation searches. The demonstration of leptonic *CP* violation (LCPV) and the knowledge of the value of this phase may help us understand the origin of observed baryon asymmetry in the universe, thereby providing a strong indication—although not proof—that leptogenesis is the explanation for this baryon asymmetry (3).

All these parameters can be measured via subleading $\nu_\mu \rightarrow \nu_e$ oscillations that represent the key process of any future discovery in neutrino oscillation physics.

1.1.1. Leptonic *CP* violation. The phenomenon of *CP* (or *T*) violation in neutrino oscillations manifests itself by a difference in the oscillation probabilities of, say, $P(\nu_\mu \rightarrow \nu_e)$ versus $P(\bar{\nu}_\mu \rightarrow \bar{\nu}_e)$ (*CP* violation) or $P(\nu_\mu \rightarrow \nu_e)$ versus $P(\nu_e \rightarrow \nu_\mu)$ (*T* violation). Extensive studies, such as those published by CERN (4), the European Network BENE (5), and the International Scoping Study (6), have been performed to establish the physics potential of future facilities in discovering leptonic *CP* violation (7). According to equation 1 of Reference 8, when matter effects are not negligible, the transition probability $\nu_e \rightarrow \nu_\mu$ ($\bar{\nu}_e \rightarrow \bar{\nu}_\mu$) at second order in perturbation theory in θ_{13}, $\Delta m_{12}^2/\Delta m_{23}^2$, $|\Delta m_{12}^2/a|$, and $\Delta m_{12}^2 L/E_\nu$ is

$$P^{\pm}(\nu_e \rightarrow \nu_\mu) = X_\pm \sin^2(2\theta_{13}) + Y_\pm \cos(\theta_{13})\sin(2\theta_{13})\cos\left(\pm\delta - \frac{\Delta m_{23}^2 L}{4E_\nu}\right) + Z, \qquad 1.$$

where ± refers to neutrinos and antineutrinos, respectively. The coefficients of the two equations are

$$\begin{cases} X_\pm = \sin^2(\theta_{23})\left(\dfrac{\Delta m_{23}^2}{|a - \Delta m_{23}^2|}\right)^2 \sin^2\left(\dfrac{|a - \Delta m_{23}^2|L}{4E_\nu}\right), \\[2ex] Y_\pm = \sin(2\theta_{12})\sin(2\theta_{23})\left(\dfrac{\Delta m_{12}^2}{a}\right)\left(\dfrac{\Delta m_{23}^2}{|a - \Delta m_{23}^2|}\right)\sin\left(\dfrac{aL}{4E_\nu}\right)\sin\left(\dfrac{|a - \Delta m_{23}^2|L}{4E_\nu}\right), \\[2ex] Z = \cos^2(\theta_{23})\sin^2(2\theta_{12})\left(\dfrac{\Delta m_{12}^2}{a}\right)^2 \sin^2\left(\dfrac{aL}{4E_\nu}\right), \end{cases} \qquad 2.$$

where $a(\mathrm{eV}^2) = \pm 2\sqrt{2}G_F n_e E_\nu = 7.6 \times 10^{-5}\rho(\mathrm{gcm}^{-3})E_\nu(\mathrm{GeV})$ changes sign by exchanging neutrinos with antineutrinos.

Searches for θ_{13} look for experimental evidence of ν_e appearance in excess of what is expected from the solar terms. The present limit on θ_{13}, which is driven mainly by the CHOOZ reactor experiment (9), is $\sin^2\theta_{13} \leq 0.035\ (0.056)$ at 90% CL (3σ) (10).

One of the interesting aspects of Equation 1 is the occurrence of matter effects that, unlike the straightforward θ_{13} term, depend on the sign of the mass-difference sign(Δm_{31}^2). These terms could allow extraction of the mass hierarchy, but they may also be a background to the *CP*-violating effect, from which they can be distinguished by the different neutrino energy dependence.

CP violation can be seen as interference between the solar and atmospheric oscillation for the same transition. Of experimental interest is the *CP*-violating asymmetry A_{CP},

$$A_{CP} = \frac{P(\nu_\mu \rightarrow \nu_e) - P(\bar{\nu}_\mu \rightarrow \bar{\nu}_e)}{P(\nu_\mu \rightarrow \nu_e) + P(\bar{\nu}_\mu \rightarrow \bar{\nu}_e)}, \qquad 3.$$

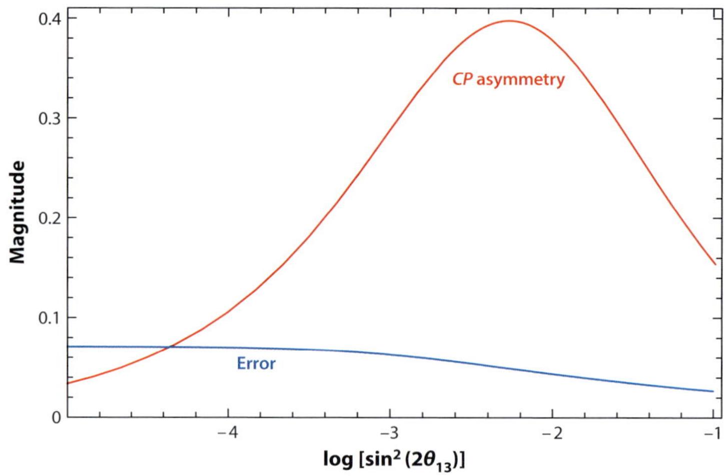

Figure 1

Magnitude of the *CP* asymmetry (*red curve*) at the first oscillation maximum for $\delta = 1$ as a function of the mixing angle $\sin^2 2\theta_{13}$. The blue error curve indicates the dependence of the statistical plus systematic error on such a measurement. The curves have been computed for the baseline beta beam option at the fixed energy $E_\nu = 0.4$ GeV, $L = 130$ km, statistical plus 2% systematic errors. Reproduced from Reference 11.

which is displayed in **Figure 1** as a function of θ_{13}, or the equivalent time-reversal asymmetry A_T.

1.1.2. The problem of degenerate solutions. The richness of the $\nu_\mu \to \nu_e$ transition is also its weakness: It is difficult to extract all the genuine parameters unambiguously. Due to the three-flavor structure of the oscillation probabilities, for a given experimental result several different disconnected regions of the multidimensional space of parameters could fit the experimental data, resulting in degenerate solutions.

Traditionally, these degeneracies are referred to as the intrinsic or $(\delta_{\rm CP}, \theta_{13})$ degeneracy (8), the hierarchy or sign(Δm^2_{31}) degeneracy (12), and the octant or θ_{23} degeneracy (13). These degenerate solutions lead to an eightfold ambiguity in θ_{13} and $\delta_{\rm CP}$ (14), thereby seriously obstructing the determination of θ_{13}, $\delta_{\rm CP}$, and sign(Δm^2_{31}).

1.2. Experimental Setups

Several different experimental setups have been envisaged for the searches of nonvanishing values of θ_{13} and the subsequent searches of leptonic *CP* violation; we describe these experiments in this section.

1.2.1. Conventional neutrino beams. Conventional neutrino beams are produced through the decay of π and K mesons that are generated by a high-energy proton beam striking small-Z, needle-shaped, segmented targets. Positive (negative) mesons are sign selected and focused (defocused) by large-acceptance magnetic lenses into a long, evacuated decay tunnel in which ν_μs ($\overline{\nu}_\mu$s) are generated. In the case of positive charge selection, the ν_μ beam has typically a few percent of $\bar{\nu}_\mu$ contamination (from the decay of the residual π^-, K^-, and K°) and ~1% of ν_e and $\bar{\nu}_e$ arising from three-body $K^\pm$, K_0, and μ decays.

The precision of the evaluation of the intrinsic ν_e contamination is limited by our knowledge of π and K production in the primary proton beam target, which requires a devoted hadroproduction experiment. Recently, the Harp experiment (15) measured both the K2K (16) and the MiniBooNE (17) targets, covering most of the useful pion phase space and successfully improving the description of the two beam lines.

Close detectors are used to directly measure beam neutrinos and backgrounds. For a discussion about close detectors and systematic errors in future LBL experiments, see Reference 18. Current long-baseline experiments with conventional neutrino beams can search for $\nu_\mu \rightarrow \nu_e$ transitions even if they are not optimized for such studies. For instance, the K2K experiment published an analysis about ν_e appearance (19) but did not improve upon the CHOOZ limit. MINOS at NuMI (20) has published preliminary results (21) showing a statistically insignificant excess of ν_e-like events in the data sample. The OPERA detector (22) at CNGS (23), which is also suitable for electron detection, can reach a 90%-CL sensitivity of $\sin^2 2\theta_{13} = 0.06$ ($\Delta m^2_{23} = 2.5 \times 10^{-3}$ eV^2) (24) for five years' exposure to the CNGS beam at a nominal intensity of 4.5×10^{19} protons on target per year.

1.2.2. Second-generation long-baseline experiments. The focus of second-generation LBL experiments will be the measurement of θ_{13} through the detection of subleading $\nu_\mu \rightarrow \nu_e$ oscillations. The T2K (Tokai to Kamioka) experiment (25) will aim neutrinos from the Tokai site of J-PARC (30 GeV, 0.75 MW) to the Super-Kamiokande detector 295 km away. The neutrino beam is situated at an off-axis angle of 2.5, ensuring a neutrino peak energy of approximately 0.6 GeV. The beam line is equipped with a set of dedicated on-axis (INGRID) and off-axis (ND280) near detectors at a distance of 280 m. It is expected that the sensitivity of the experiment in a five-year ν_μ run at the full J-PARC beam intensity will be of the order of $\sin^2 2\theta_{13} \leq 0.006$ (90% CL, computed for $\delta_{CP} = 0$).

The NOνA experiment, with an upgraded NuMI off-axis neutrino beam (26) ($E_\nu \sim 2$ GeV and a ν_e contamination lower than 0.5%), a totally active 15-kton liquid scintillator detector, and a baseline of 810 km (12 km off-axis), is currently under construction at FNAL. This experiment aims to explore $\nu_\mu \rightarrow \nu_e$ oscillations with a sensitivity ten times better than that of MINOS.

1.2.3. Reactor experiments. Another approach to searching for nonvanishing θ_{13} is to look at $\bar{\nu}_e$ disappearance through the use of nuclear reactors as neutrino sources. The Double Chooz (27) experiment will employ a near detector as well as a far detector in the same location as the former CHOOZ detector. The sensitivity after five years of data taking will be $\sin^2 2\theta_{13} = 0.025$ at 90% CL (27).

The Daya Bay project in China (28) could reach a $\sin^2 2\theta_{13}$ sensitivity around 0.01. A sketch of θ_{13} sensitivities as a function of time, following the schedule reported in the experimental proposals, is given in **Figure 2**.

1.2.4. Neutrino super beams. Consensus exists that even a global fit of T2K plus NOνA plus reactors will not be able to provide firm results (3σ or better) for leptonic *CP* violation (30) or sign(Δm^2_{31}) (30, 31), no matter what the value of θ_{13} is. A further generation of long-baseline neutrino experiments will be needed to address this important search. These experiments should be at least one order of magnitude more sensitive than T2K or NOνA, a condition equivalent to an increase of two orders of magnitude on neutrino statistics, and there should be a consequent, important reduction of systematic errors. To meet such a challenge, conventional neutrino beams must be pushed to their ultimate limits (neutrino super beams) (32), and gigantic (megaton-scale) neutrino detectors must be built.

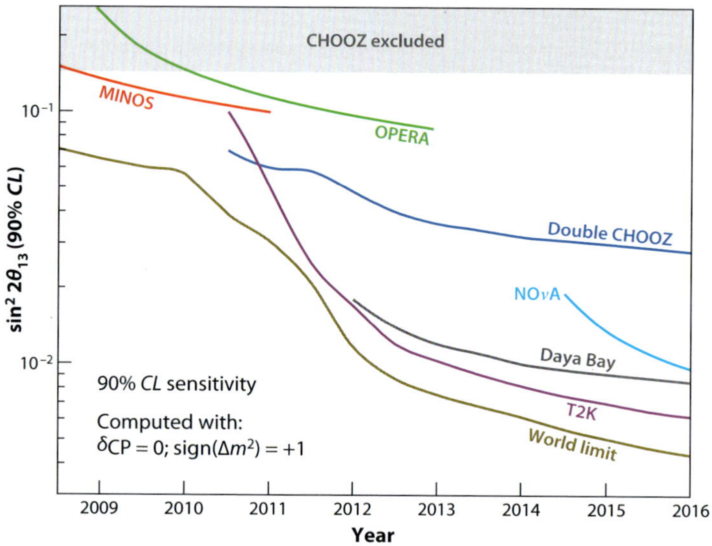

Figure 2

Evolution of experimental $\sin^2 2\theta_{13}$ sensitivities as function of time, computed assuming $\delta_{CP} = 0$. Experiments are assumed to provide results after the first year of data taking. Reproduced from Reference 29.

Phase II of the T2K experiment, often referred to as T2HK (where HK stands for Hyper-Kamiokande; 33), calls for (*a*) an increase of beam power up to the maximum feasible with the accelerator and target (4 MW of beam power); (*b*) antineutrino runs; and (*c*) a very large, 520-kton, water Cherenkov detector to be built close to Super-Kamiokande. A more advanced version of T2HK is the T2KK project (34), where half of the HK detector would be installed in Japan and the second half would be mounted in Korea, at a baseline of approximately 900 km, around the second oscillation maximum.

A wide-band beam has been proposed as the next phase of the FNAL main injector neutrino program following the end of the Tevatron program (35). A conventional wide-band neutrino would be sent to a megaton water Cherenkov (or liquid argon) detector at the Homestake mine at a baseline of 1,290 km. Wide-band beams possess the advantages of a higher on-axis flux and a broad energy spectrum. The latter allows the first and second oscillation nodes in the disappearance channel to be observed, providing a strong tool with which to solve the degeneracy problem (36). However, experiments served by wide-band beams must determine the incident neutrino energy with good resolution, as well as eliminate the background from the high-energy tail of the spectrum. The very long baseline decreases the event rate at the far detector in an experiment, at which the statistics are very important, and reduces the sensitivity of the experiment to θ_{13} and *CP* violation but increases its sensitivity to mass hierarchy.

1.2.5. CERN-SPL. Here we describe the CERN-SPL (Superconducting Proton Linac) superbeam in more detail, emphasizing its possible synergy with the CERN-Fréjus beta beam (see Section 3.6). In the CERN-SPL superbeam project (37) the planned 4-MW SPL would deliver a 3.5 GeV/*c* H^- beam on a Hg target to generate a neutrino beam with an average energy of

$\sim$0.3 GeV.[1] The ν_e contamination from K would be suppressed by threshold effects, and the resulting ν_e/ν_μ ratio ($\sim$0.4%) would be known within 2% error. The use of near and far detectors (the latter would be located at $L = 130$ km in the Fréjus area) would allow for both ν_μ disappearance and $\nu_\mu \rightarrow \nu_e$ appearance studies. The physics potential of the SPL superbeam, with a water Cherenkov far detector having a fiducial mass of 440 kton, has been extensively studied (39, 40). The most updated sensitivity estimations for this setup have been published in Reference 42 and are described in Section 3.6.

The MEMPHYS detector (43) is a megaton-class water Cherenkov designed to be located at Fréjus, 130 km from CERN; this detector would address both the nonaccelerator (nucleon decay, neutrinos from supernovae neutrinos from burst events or from relic explosions, solar and atmospheric neutrinos) domain and the accelerator (superbeam, beta beam) domain (44).

1.3. New Concepts on Neutrino Beams

The intrinsic limitations of conventional neutrino beams can be overcome if the neutrino parents are fully selected, collimated, and accelerated to a given energy. This can be attempted within the muon lifetime, leading to the neutrino factory (45), or within beta-decaying ion lifetimes, leading to the beta beam (46, 47).

This challenging approach may result in several important improvements to conventional neutrino beams: (*a*) The neutrino fluxes would be derived simply from the knowledge of the number of parents circulating in the decay ring and from their Lorentz boost factor γ; (*b*) the energy shape of the neutrino beam would be defined by two parameters, the end-point energy Q_β of the beta-decaying parent and its Lorentz boost factor γ; and (*c*) the intrinsic neutrino backgrounds would be suppressed (in the case of beta beams) or reduced to wrong-sign muons (golden channel in neutrino factories). The technological problems derive from the fact that the parents need to be unstable particles, so a fast, efficient acceleration scheme is required.

2. MACHINE ASPECTS

2.1. A Possible Beta Beam Facility

The first proposal for a beta beam facility (46, 48) suggested that the ions be produced in a thick target using a proton beam of 1 to 2 GeV, extracted as neutral atoms, and reionized and bunched in a high-frequency electron cyclotron resonance source. The first step of acceleration would be a linear accelerator, which would bring the ions to a kinetic energy of 150 MeV per nucleon. Thereafter, the beam would be injected into a rapid cycling synchrotron, in which the energy would be increased to 500 MeV per nucleon. After this first step, the beam would enter the existing CERN accelerator complex, from where first it would be accelerated in the Proton Synchrotron (PS) to its maximum energy, then transferred to the Super Proton Synchrotron (SPS), and finally ejected into a decay ring. The last step would be performed through a scheme that would permit the new ions to be merged with the ions already circulating in the decay ring so that ions that can still decay (and create neutrinos) would not be wasted by being ejected from the decay ring too soon. The details of each step in this scheme are discussed below, and an overview

[1] SPL is foreseen as one of the elements of a new injection chain for the SPS, in light of the LHC luminosity upgrades (38). In this context, a power of 0.4 MW would be sufficient. Extensions to 4 MW could be driven by the needs of a neutrino superbeam or a proton driver for a neutrino factory and/or a proton driver for EURISOL.

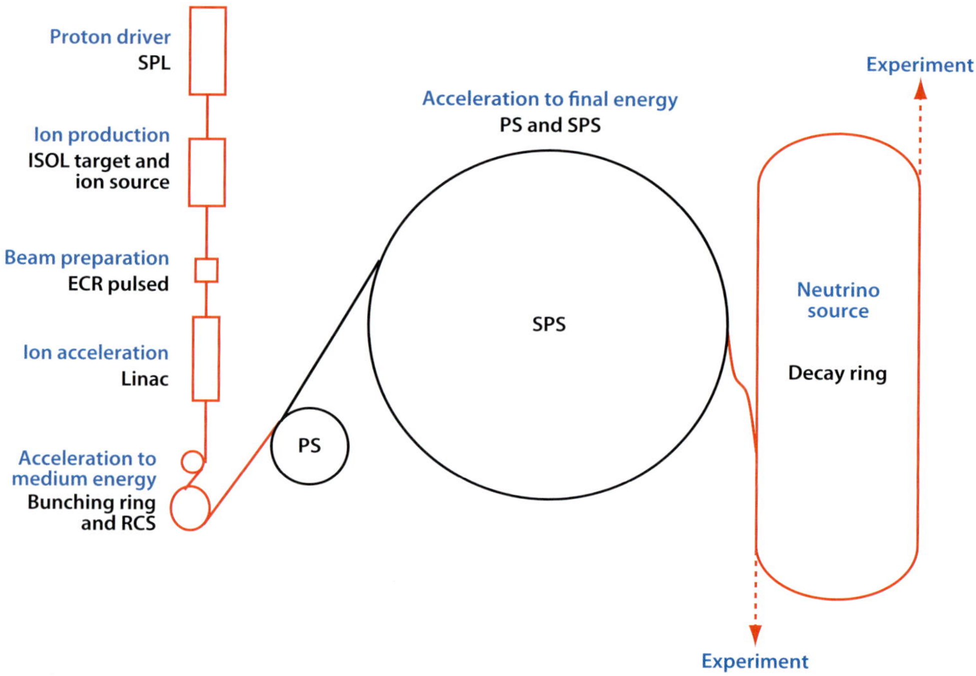

Figure 3

An overview of a possible beta beam facility at CERN, as discussed in Reference 48. Note that the existing accelerators, the Proton Synchrotron (PS) and the Super Proton Synchrotron (SPS), were proposed as part of the injector chain, which would represent major savings for the proposal. Abbreviations: ECR, electron cyclotron resonance; ISOL, isotope separation online; RCS, rapid cycling synchrotron.

of the facility is shown in **Figure 3**. The key ingredient of this setup would be the reuse of existing accelerators to reduce cost and gain time for the construction of a beta beam facility.

2.2. The Beta Beam Isotopes

The choice of parent ions has a great impact on the design of the facility and its final performances, as we discuss below.

2.2.1. Which isotope to use. The CERN 2002 beta beam study (48) established a list of suitable isotopes considering lifetime and contamination risk; this list is reprinted in **Table 1** (β^- emitters) and **Table 2** (β^+ emitters). For this first study, ^{6}He and ^{18}Ne were selected; their at-rest half-lives are on the order of 1 s, which corresponds to the typical cycling time of low-energy accelerators (too-long lifetimes lead to a large accumulation of charges). Also, they are easy to produce, are in the gas phase at room temperature, and have no dangerous daughter products.

2.2.2. Isotope production. Radioactive ions must be produced continuously because there is no way to stop them from decaying. Nuclear physicists employ two different methods for online production of exotic radioactive ions: the in-flight method and the isotope separation online (ISOL) method (50). The ISOL method uses a thick target in which a beam of particles is stopped. This method typically produces much-higher-intensity beams than the in-flight method does.

Table 1 Some possible isotopes that are β^- emitters[a]

Isotope	A/Z	$T_{1/2}$ (s)	Q_β ground state to ground state (MeV)	Q_β effective (MeV)	E_β average (MeV)	E_ν average (MeV)
^{6}He	3.0	0.80	3.5	3.5	1.57	1.94
^{8}He	4.0	0.11	10.7	9.1	4.35	4.80
^{8}Li	2.7	0.83	16.0	13.0	6.24	6.72
^{9}Li	3.0	0.17	13.6	11.9	5.73	6.20
^{11}Be	2.8	13.8	11.5	9.8	4.65	5.11
^{15}C	2.5	2.44	9.8	6.4	2.87	3.55
^{16}C	2.7	0.74	8.0	4.5	2.05	2.46
^{16}N	2.3	7.13	10.4	5.9	4.59	1.33
^{17}N	2.4	4.17	8.7	3.8	1.71	2.10
^{18}N	2.6	0.64	13.9	8.0	5.33	2.67
^{23}Ne	2.3	37.2	4.4	4.2	1.90	2.31
^{25}Ne	2.5	0.60	7.3	6.9	3.18	3.73
^{25}Na	2.3	59.1	3.8	3.4	1.51	1.90
^{26}Na	2.4	1.07	9.3	7.2	3.34	3.81

[a]Data from Reference 48.

To enhance diffusion and effusion out of the target, the target must be hot. In high-intensity facilities, the heat produced by the beam is more than sufficient, and the target designer must eliminate the excess heat so that the target is not destroyed. Several variations of the ISOL method for beta beams exist, and studies for the production of beta beam–relevant isotopes (see **Table 3**) have so far been performed for (*a*) the converter method (51, 52), (*b*) direct production with a low-energy projectile (53), and (*c*) the recently proposed production-ring concept (54, 55).

Table 2 Some possible isotopes that are β^+ emitters[a]

Isotope	A/Z	$T_{1/2}$ (s)	Q_β ground state to ground state (MeV)	Q_β effective (MeV)	E_β average (MeV)	E_ν average (MeV)
^{8}B	1.6	0.77	17.0	13.9	6.55	7.37
^{10}C	1.7	19.3	2.6	1.9	0.81	1.08
^{14}O	1.8	70.6	4.1	1.8	0.78	1.05
^{15}O	1.9	122	1.7	1.7	0.74	1.00
^{18}Ne	1.8	1.67	3.3	3.0	1.50	1.52
^{19}Ne	1.9	17.3	2.2	2.2	0.96	1.25
^{21}Na	1.9	22.4	2.5	2.5	1.10	1.41
^{33}Ar	1.8	0.17	10.6	8.2	3.97	4.19
^{24}Ar	1.9	0.84	5.0	5.0	2.29	2.67
^{35}Ar	1.9	1.77	4.9	4.9	2.27	2.65
^{37}K	1.9	1.22	5.1	5.1	2.35	2.72
^{80}Rb	2.2	34	4.7	4.5	2.04	2.48

[a]Data from Reference 48.

Table 3 Estimates for the production rate in the target of a few beta beam isotopes[a]

Isotope	Method	Rate within reach (ions s^{-1})	Reference[b]
^{18}Ne	ISOL at 1 GeV and 200 kW	$<8 \times 10^{11}$	48
^{6}He	ISOL converter at 1 GeV and 200 kW	$<5 \times 10^{13}$	48
^{18}Ne	Direct production through $^{16}O(^{3}He,n)^{18}Ne$	$<1 \times 10^{13}$	53
^{6}He	ISOL converter at 40-MeV deuterons and 80 kW	$<6 \times 10^{13}$	52
^{8}Li	Production ring through $^{7}Li(d,p)^{8}Li$	$<1 \times 10^{14}$	54

[a]The extraction efficiency from the target can vary considerably from a few percent up to 90%.

[b]The references refer to the method rather than to the production limits. Abbreviation: ISOL, isotope separation online.

The converter method uses a primary cooled target for the production of neutrons, which in turn produce the isotope of interest in a secondary target. Compared to direct irradiation, which uses protons, neutrons are less destructive and generate less heat at an equivalent production rate. The production of ^{6}He has been studied (51, 52) using the reaction $n + {}^{9}Be = {}^{6}He + {}^{4}He$, which has a large cross section of approximately 100 mb. The direct production method is the simplest way of creating a new isotope. One nucleus is accelerated and merged with another nucleus in a target at an energy that is high enough to overcome the Coulomb barrier but low enough not to destroy the newly formed nuclei through spallation or fission. This process has been studied for beta beams (53); to produce a sufficient number of ^{18}Ne isotopes for a beta beam facility via a MgO solid target, 120 mA of a primary ^{3}He beam at approximately 13 MeV of total energy are required.

The production-ring concept uses direct production, which requires that a very thin target be placed in the closed orbit of a ion-storage ring with continuous injection. The combination of energy loss in the transverse directions in the target with reacceleration results in net beam cooling (54, 55). A wedge-shaped gas target has been used in a dispersive region of the ring, which added longitudinal cooling (54). Furthermore, the proposed use of a gas target made it possible to handle a large amount of beam power. The produced ions were collected with a second target downstream. A fixed-field alternating-gradient accelerator with large longitudinal acceptance has been used to manage the beam without any longitudinal cooling (55). In both proposals the beam would be injected partially stripped, and the energy of the circulating ions would be kept high enough to assure that all the ions emerge fully stripped after the target. The production of ^{8}B and ^{8}Li with ^{3}He and deuterium as projectiles and a liquid lithium target (56) of enriched ^{6}Li or ^{7}Li was proposed in Reference 57, which also contains a full six-dimensional analysis of the cooling process.

2.3. Ion Transfer, Ionization, and Bunching

The ions produced in a thick ISOL target must be collected, ionized, and bunched before they can be accelerated. Bunching must occur before the ions can be injected into a synchrotron. An electron cyclotron resonance source can operate in either afterglow or preglow mode (58) for ionization and bunching of ions. Tests of preglow operation—in which neutral atoms are ionized and extracted in a pulse—for beta beams have been performed by the EURISOL Design Study (see **http://www.eurisol.org**), and further work by the EURONU Design Study (see **http://www.euronu.org**) is in progress.

2.4. Acceleration

Accelerating a heavy ion is more challenging than accelerating a proton because in the case of the heavy ion (*a*) the charge–per–mass unit (*Q*/*A*) ratio is smaller and (*b*) the so-called charge-exchange reactions are more severe. The smaller *Q*/*A* ratio makes acceleration and transverse and longitudinal focusing more demanding, and the greater charge-exchange reactions increase losses during acceleration and storage.

The beam is first accelerated with a linac, and the resulting semicontinuous beam is then injected into the first synchrotron over a large number of turns. The combination of the chosen working point (*Q*) and a deliberate shift of the central beam orbit with dipolar magnets allow one to inject over as many as 50 to 100 turns with only moderate (<30%) losses (59).

2.5. Beam Losses

Manipulation of the ion beam between the different accelerators—for instance, for stripping and injection—induces losses. Furthermore, the collision with residual gas in the accelerator itself and the decay of radioactive ions add to these losses. The lost ions can both induce radioactivity and cause vacuum degradation (60, 61). For the beta beam studied in Reference 48, the arc lattice was derived for ^{18}Ne and ^{6}He, and large-aperture (160-mm-diameter) dipoles were used with a length that allowed all the lost ions to be intercepted with absorbers after the dipole (62, 63). The choice of absorber length and material is important because the absorbers could serve simply as a point for the ion to break up in a hadron cascade that could overheat the following (superconducting) dipole (64). Two possible solutions are (*a*) to develop open midplane superconducting magnets (65) and (*b*) to use a thick liner inside the dipole that would distribute the decay products over a larger volume of the magnet mass and coil. In the straight sections, the decay products could be successfully kept in the machine and extracted to a beam dump before the arc (62).

2.6. Stacking and Storage

A bunched neutrino beam enables an efficient suppression of background in the detector (see Section 3.3.2). Because the neutrino beam will be bunched if the ion beam is bunched (**Figure 4**),

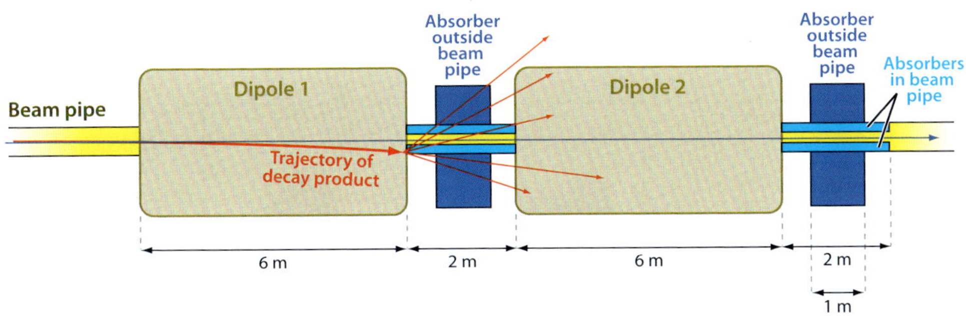

Figure 4

The neutrino beam from the beta beam facility will have a bunched structure that mirrors the bunch structure of the ion beam. The neutrino bunch originates from the ion bunch labeled Dipole 1. The neutrinos from the decay of ions in the bunch labeled Dipole 2 have not yet left the ring but are traveling along the straight section with the same velocity as the ions. Figure modified with permission from Elena Wildner, CERN.

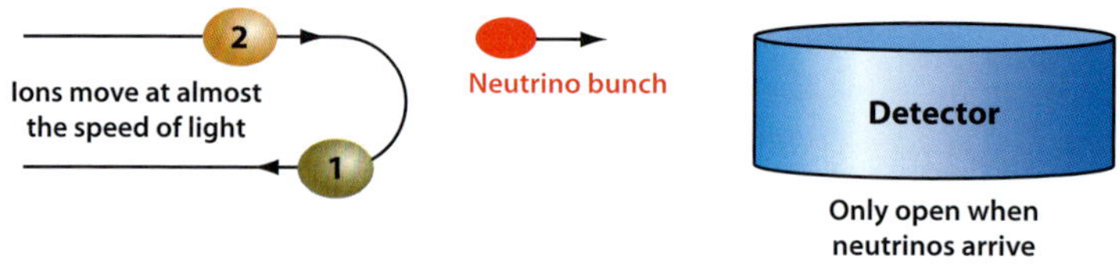

Figure 5

Sketch of the dipole and absorber arrangement in the arcs of the decay ring, which is designed to preserve the dipole from irradiation.

it is necessary to use some form of stacking scheme for the incoming ion bunches; beam cooling is not an option because the cooling rate is too long.

The beam loss from radioactive decay in the arcs of the decay ring cannot be avoided and the dipoles must be adapted in length so that the daughter products are lost in absorbers between the dipoles and not in the dipoles themselves (**Figure 5**).

The bunching (duty factor) can be expressed as

$$\text{duty factor} = \frac{N t_b}{T_{rev}}, \tag{4.}$$

where N is the number of filled bunches in the decay ring, t_b is the length of the individual bunch in seconds, and T_{rev} is the revolution time. For a beta beam facility operating at Lorenz $\gamma = 100$ that uses isotopes with an average neutrino energy spectrum of a few megaelectronvolts at rest, a duty factor no larger than 10^{-2} is required.

A longitudinal stacking scheme has been proposed (66) in which the radio-frequency (RF) systems are used to stack from the center of each bunch. For each injection cycle, the outer parts of the bunch are pushed outside the potential wall formed by the RF system and lost. This process has been simulated and tested in the CERN PS with very good results (**Figure 6**). In the proposed CERN beta beam facility (48), the merging could be performed up to 15 times for an A/Q of 3 and up to 20 times for an A/Q of 2, at which point the longitudinal focusing is better. The efficiency of the merging would be further limited by the phase stability of the RF system. Assuming realistic RF parameters and stability, an overall stacking efficiency of 80% could be reached for both ^{6}He and ^{18}Ne (68). The total beam power injected would be approximately 1 MJ for the reference facility, and up to 50% of that beam power would be lost from the bucket during stacking and

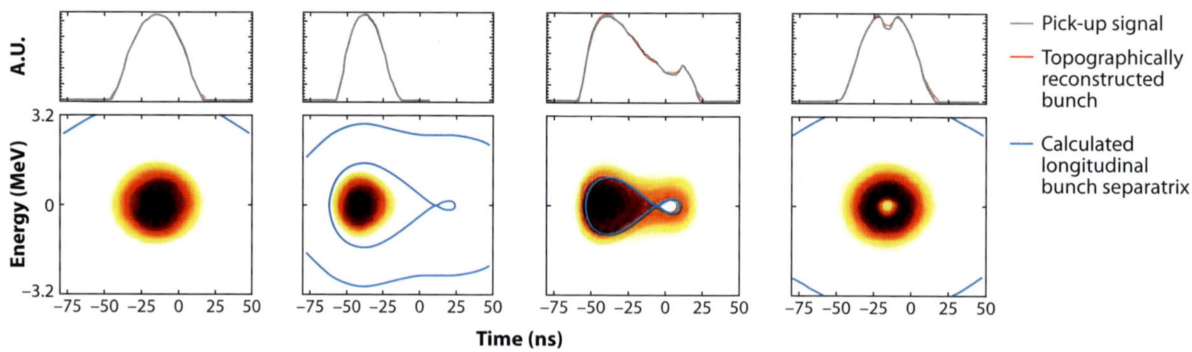

Figure 6

The stacking scheme tested in the CERN Proton Synchrotron (66), in which a high-intensity proton beam was merged with a small part of empty phase space with high efficiency. (*Top*) The line current for the bunches during the merging process. (*Bottom*) The tomographic (67) reconstruction of the bunches.

would have to be "scraped" away with longitudinal collimators. The decay ring (62) would have to be specially designed for this stacking scheme: It should have a large dispersion region for the injection and another large dispersion region for beam collimation.

A further consideration for the decay-ring design is the average (optical) beam divergence D in the straight sections. It can be estimated from the vertical and horizontal beam emittance ε and the Lorentz β that

$$D_{average} = \frac{\varepsilon_{h,v}}{\beta_{average}}.$$

2.7. Annual Rate of Neutrinos

The purpose of the beta beam facility is to produce a well-collimated beam of single-flavor electron (anti)neutrino beams. For an ideal beta beam facility, in which all accelerated ions can be stored in the decay ring until they produce neutrinos and are lost, the annual rate is equal to the number of ions injected.

If the decay ring can stack only a certain number of bunches, the upper intensity limit in the machine can be calculated from the truncated series of repeated injections in the decay ring. In **Figure 7** the actual fraction of stored ions, compared with the ideal case in which all injected ions are stored in the ring, is plotted as a function of the number of merges for both ^{6}He and ^{18}Ne. Note that the realistic case of only 15 merges for ^{6}He and 20 merges for ^{18}Ne corresponds to a stacking-scheme efficiency of only 54% and 26%, respectively. The annual rate at the end of one of the straight sections for a conventional year of 10^7 s, in which the relative length of the straight section (f) is compared with the circumference of the ring, can be written as

$$A = 10^7 f \frac{N}{T}\left(1 - e^{-\lambda n T}\right), \qquad 5.$$

where N is the number of ions per injection into the decay ring, T is the time interval between injections, λ is the decay constant, and n is the number of completely efficient merges. This is the maximum annual rate because it assumes perfect stacking from the center of the bunch. The decay constant for the radioactive decay producing neutrinos in the correct energy interval is not

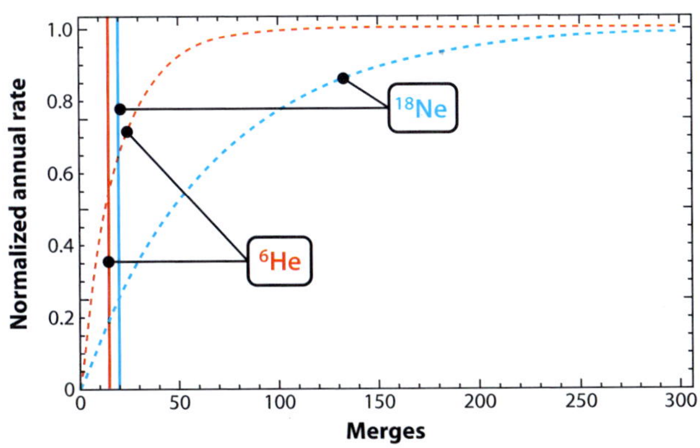

Figure 7

The fractions of ^{6}He ions (*red line*) and ^{18}Ne ions (*blue line*) stored in the decay ring are plotted as a function of the number of completely efficient merges. The point at which the curves eventually converge is the ideal case in which all ions are accumulated in the ring until they decay. The vertical red and blue lines for each ion type mark the stacking limits for the beta beam studied in Reference 48.

necessarily the only decay constant determining the decay rate of the ions in the decay ring. Other loss processes such as vacuum collisions may occur, resulting in a change of the ion's mass-to-charge ratio. For monochromatic neutrino beams arising from electron-capture decay, there is often a radioactive branch of competing $\beta+$ decay with a different decay constant. The annual rate of the proposed beta beam facility (48) is assumed to reach 1.1×10^{18} electron neutrinos per year from the decay of ^{18}Ne and 2.9×10^{18} electron antineutrinos per year from the decay of ^{6}He. To reach these ambitious goals, a production rate of 2×10^{13} ions per species is required. The overall efficiency from the ion source to the number of ions injected into the decay ring is assumed to reach 12% for ^{18}Ne and 25% for ^{6}He. Per the above-quoted efficiencies for the stacking scheme, the total efficiency from the ion source to ions decaying in one straight section (which is 36% of the total circumference) is 1% and 5% for ^{18}Ne and ^{6}He, respectively.

2.8. Possible Future Developments

If the annual rate, energy, or energy resolution of the neutrinos could be increased, the beta beam concept would have a larger scope. New particles can be injected only when the magnets in a synchrotron are ramped back to the field corresponding to the injection energy. The simplest way to make use of this lost production time is to accumulate the ions produced in this interval. A study for the CERN 2002 beta beam (69) proposed use of a low-energy storage ring that would yield a gain of up to a factor of five in the overall annual rate.

In principle, the two types of ions can be kept in the decay ring at the same time. The ions would be kept in different bunches sufficiently well separated in time. This scheme would (*a*) impose a fixed relationship between the γ of the two isotopes, (*b*) require very large aperture dipoles, (*c*) need double injectors for both ion types, and (*d*) require a highly complex RF system for the accumulation process.

A conceptual study performed at FNAL for a higher γ of 300 used the Tevatron as the preaccelerator (49). The opening angle of the kinematic focusing angle would change with $\gamma(\Theta \approx 1/\gamma)$; the isotope lifetime would be longer due to increased time dilatation; and the decay ring would have to be larger or the dipole magnets more powerful (see **Table 4**) to cope with the increased magnetic rigidity of the radioactive ions.

2.9. Acceleration of Partly Stripped Ions

Proton-rich isotopes decaying through electron capture can be used to create a monoenergetic neutrino beam (70, 71). The equivalent process on the neutron-rich side is bound beta decay. The

Table 4 Some decay-ring options for a different Lorenz γ of ^{6}He[a]

γ	Rigidity (Tm)	Ring length (m)[b]	Dipole field (T)[c]
100	935	4,197	3.1
150	1,403	6,296	4.7
200	1,870	8,395	6.2
350	3,273	14,691	10.9
500	4,676	20,987	15.6

[a]The decay-ring arcs are for all cases considered to be completely filled with dipoles.

[b]Assuming a fixed field of 5 T and a single straight section of 36% of the circumference.

[c]Assuming an arc radius of 300 m and a decay-ring length of 6,885 m.

branching rate for this process is generally very small, but it could be used for a monoenergetic antineutrino beam (72).

A definite requirement for electron-capture decay is that the nuclei be only partly stripped so that an electron is available for capture. At high energy, the likelihood of picking up an electron is vanishingly small. However, the likelihood of losing an electron, expressed in an equivalent half-life, is on the order of minutes in a ring with a modern ultrahigh-vacuum system. The modified annual rate for 10^7 species can be written as

$$\text{rate} = \frac{Nf}{T} \times \frac{\lambda_{ec}/\gamma_{top}}{\lambda_{ec}/\gamma_{top} + \lambda_{vac}} \times \left(1 - e^{-mT\left(\lambda_{ec}/\gamma_{top}+\lambda_{vac}\right)}\right) \times 10^7,$$

where N is the number of incoming ions into the decay ring per injection cycle, f is the fraction of the decay ring (i.e., a straight section pointing toward the detector), and T is the time between injection cycles.

The combination of the high charge state, the longer half-life, and the electron-stripping losses would—even with exceptionally short-lived and electron-capture decaying rare-earth isotopes (73)—require a large number of ions to be accelerated and stored in the decay ring to keep the annual rate high. For the cases studied in Reference 74, the tune shift in the CERN accelerators PS and SPS would have to peak well above 0.25 to keep the annual rate at 10^{18} electron neutrinos at the end of one straight section for a snowmass year.

3. CERN-FRÉJUS BETA BEAM PHYSICS POTENTIAL

3.1. General Principles

A beta beam is produced from the decay of a high-energy radioactive ion beam, which results in a pure ν_e or $\bar{\nu}_e$ beam. The flavor transitions that can in principle be studied in this facility are

$$\begin{aligned} &\nu_e \to \nu_\mu, \quad \nu_e \to \nu_e, \quad \nu_e \to \nu_\tau, \\ &\bar{\nu}_e \to \bar{\nu}_\mu, \quad \bar{\nu}_e \to \bar{\nu}_e, \text{ and } \quad \bar{\nu}_e \to \bar{\nu}_\tau. \end{aligned}$$

In the laboratory frame, the neutrino flux Φ^{lab} is given by

$$\left.\frac{d\Phi^{\text{lab}}}{dSdy}\right|_{\theta\simeq 0} \simeq \frac{N_\beta}{\pi L^2}\frac{\gamma^2}{g(y_e)} y^2(1-y)\sqrt{(1-y)^2 - y_e^2}, \qquad 6.$$

where N_β is the number of ion decays per unit time, Q_β is the end-point kinetic energy of the beta particle, γ is the relativistic Lorentz boost factor, m_e is the mass of the electron, dS is the element of the solid angle, L is the distance between the decay ring and the detector, $0 \leq y = \frac{E_\nu}{2\gamma Q_\beta} \leq 1 - y_e$, $y_e = m_e/Q_\beta$, and

$$g(y_e) \equiv \frac{1}{60}\left\{\sqrt{1-y_e^2}\,(2 - 9y_e^2 - 8y_e^4) + 15y_e^4 \log\left[\frac{y_e}{1-\sqrt{1-y_e^2}}\right]\right\}. \qquad 7.$$

The intensity and the energy shape of the neutrino beam are determined by only four quantities: N_β, Q_β, γ, and L. Once these parameters are fixed, the neutrino flux can be precisely calculated because the kinematics of beta decay is very well known (75).

There are some approximate scaling laws for different parameters (assuming that N_β is constant); for instance, the maximum γ to which a given accelerator can accelerate a ion is proportional to Z/A. For instance, if SPS can accelerate protons up to 450 GeV, ^{6}He ($Z/A = 2/6$) can be accelerated up to $\gamma = 150$. The neutrino flux Φ at a far detector placed at a distance L is

$$\Phi \propto \frac{\gamma^2}{L^2}$$

because the emission angle of the neutrino from the parent ion, in the laboratory frame, is proportional to γ^{-1}. Because the optimal distance L is defined by the oscillation Δm^2: $L \propto E_\nu / \Delta m^2$ and $E_\nu \propto \gamma Q_\beta$, the flux becomes

$$\Phi \propto \frac{(\Delta m^2)^2}{Q_\beta^2}.$$

Considering that the neutrino interaction rate I at the far detector is $I = \sigma \Phi$ and that the neutrino cross section σ goes as $\sigma \propto E_\nu$ (this scaling law becomes inaccurate for $E_\nu < 5$ GeV), a merit factor $\mathcal{M}$ can be derived:

$$\mathcal{M} \propto \frac{\gamma}{Q_\beta}. \qquad 8.$$

It follows that performances of a beta beam scale as the Lorentz boost factor γ and that they are inversely proportional to the end-point energy Q_β of the parent ions.

3.2. The CERN-Fréjus Configuration

The CERN beta beam can accelerate ^{6}He ions up to $\gamma = 150$ and ^{18}Ne ions up to $\gamma = 250$. Given the characteristics of the ^{6}He decay, this translates to mean neutrino energies of up to ~600 MeV, equivalent to a maximum baseline of 300 km. The only realistic candidate site for the excavation of a megaton-class detector fitting this request is the Fréjus site, at a distance of 130 km from CERN. For this distance, the optimal γ for ^{6}He is $\gamma \simeq 100$. Higher γ values would increase interaction rates in the detector but would not significantly increase the oscillated event interaction rates because the baseline would no longer fit the oscillation pattern. Furthermore, background rates would rise (see Section 3.3.2).

Smaller γ values would suppress background rates in the detector; indeed, for this reason $\gamma_{^6\mathrm{He}} = 66$ had been the initial choice for the CERN-Fréjus configuration (CFBB) (76, 77). Under this condition, however, the neutrino flux is smaller and a larger fraction of the ν_μ events created by oscillations produces a muon below the Cherenkov light-production threshold ($p_\mu > 120$ MeV/c).

CFBB is not designed to be the optimal configuration for a beta beam experiment. It is intended to be a realistic setup in which both the beam and the detector sites are chosen from among realistic conditions.

3.3. Data Analysis

The most sensitive processes in a beta beam experiment are $\nu_e \to \nu_\mu$ transitions (see Section 3.4). These processes introduce an experimental problem that has not yet been confronted: the detection of a small number of ν_μ events in a pure ν_e beam. The $\nu_e \to \nu_\mu$ transitions can be complemented by $\nu_e \to \nu_\mu$ transitions, in which a small deficit in ν_e spectrum is sought. The combination of the two processes demands a massive detector that can measure both electrons and muons with precision and high purity. The natural baseline option is the MEMPHYS detector.

3.3.1. Signals.

The neutrino flux in this setup is shown in **Figure 8**. In this energy range, almost all the neutrino charged-current interactions are quasi-elastic interactions, which are two-body configurations that are very favorable for a water Cherenkov detector because the neutrino energy

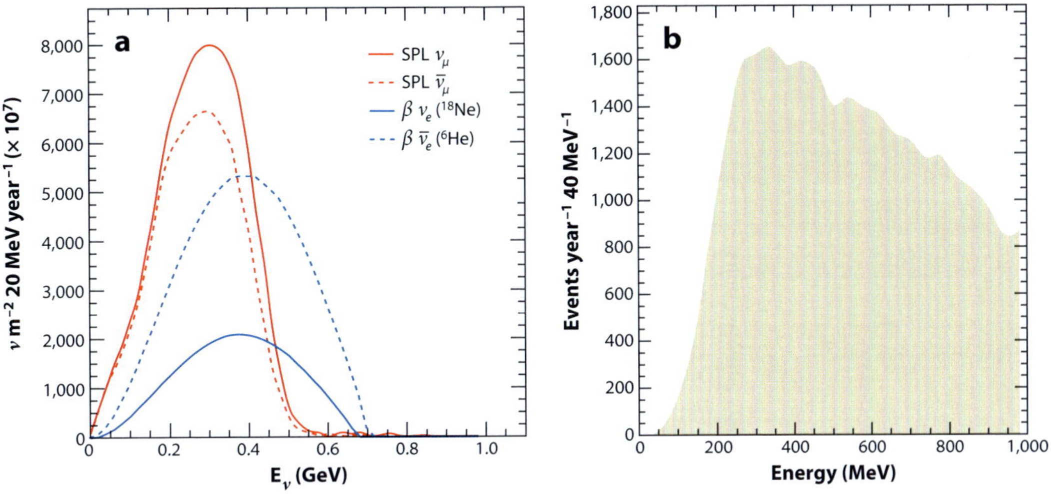

Figure 8

(*a*) Neutrino flux of a beta beam ($\gamma = 100$), together with that of the CERN SPL (Superconducting Proton Linac), at a distance of 130 km. (*b*) Rate of atmospheric $\nu_\mu + \bar{\nu}_\mu$ interactions in MEMPHYS, integrated in 1 year.

can be derived simply by measuring the momentum and the direction of the outgoing lepton. The precision in measuring the neutrino energy is shown in **Figure 9**. The energy of single-ring non-quasi-elastic events is underestimated because of the different kinematics. This effect is hardly visible in the low-energy bins, in which the non-quasi-elastic event fraction is small. The non-Gaussian features of energy reconstruction are taken into account via migration matrices that connect the true and reconstructed neutrino energies (42). Data reduction is shown in **Figure 10** for ^{18}Ne events.

3.3.2. Backgrounds. Although a beta beam is completely free of ν_μ ($\bar{\nu}_\mu$), backgrounds can be introduced through an imperfect performance of the neutrino detector. The experimental sensitivity requires that the electron-muon misidentification rate in the detector be kept below 10^{-4}. This rate can be guaranteed in a water Cherenkov detector, thanks to the particle-identification algorithms based on the very different topologies of muons' and electrons' rings. Furthermore, a muon can be positively identified through the detection of its decay in a Michel electron (the probability that a negative muon will be captured before decay is 22% in water).

The charged pions produced in resonant neutrino-nucleus interactions can also be misidentified as muons, thereby generating backgrounds. At CERN-Fréjus energies, the pion production is suppressed just below or close to the threshold (337 MeV/*c*) and because the detection of the outgoing pion requires that it have a momentum greater than 159 MeV/*c*. Furthermore, most of the negative pions are absorbed before the completion of the decay chain.

Atmospheric neutrinos are another source of background in a beta beam experiment. The spectrum of ν_μ and $\bar{\nu}_\mu$ (**Figure 8*b***) overlaps the spectrum of oscillated signals, providing a copious source of backgrounds. The direction of the outgoing muon is not strong enough to eliminate this background because both the quasi-elastic kinematics and the Fermi motion generate a loose correlation between the outgoing lepton and the incoming neutrino. At the energies of the $\gamma =$ 100 beta beam, the angular resolution is approximately 0.25 rad. Therefore, the only other way to avoid atmospheric neutrino background is to keep the time in which beam neutrinos arrive to the

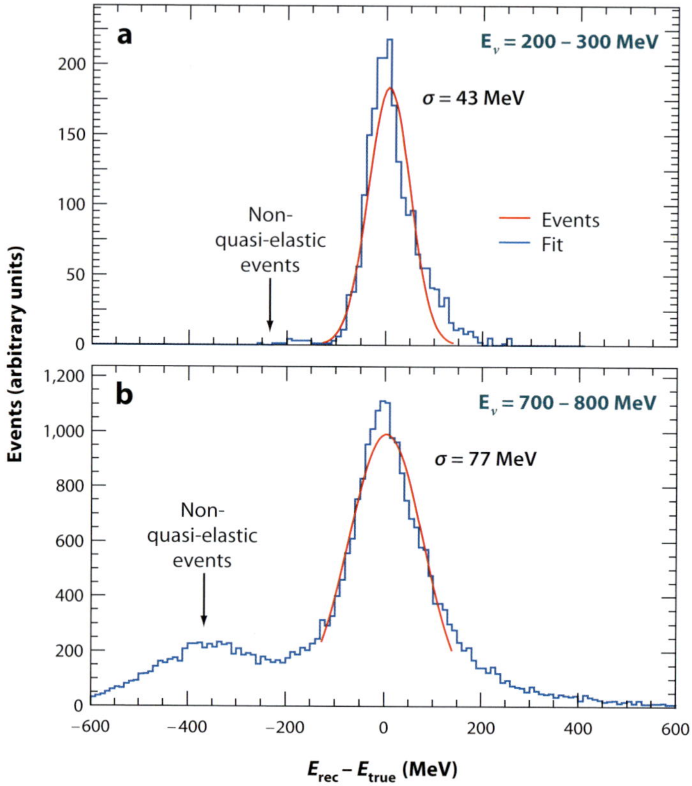

Figure 9

Energy resolution for ν_e interactions in the (*a*) 200–300-MeV and (*b*) 700–800-MeV energy ranges. The quantity displayed is the difference between the reconstructed and true neutrino energies.

detector very short. In other words, the duty cycle of the beta beam decay ring must be very short. As computed in References 78 and 79, a duty cycle of 10^{-2} is needed to keep the atmospheric neutrino background rate below the neutral-current pion background rate.

The fraction of background events with respect to the fully oscillated sample, following the analysis selection, is approximately 0.2%—well below the ~1% that is characteristic of super-beam experiments. These backgrounds (**Figure 10**) have a different spectral distribution from that of oscillated events, reducing their impact on oscillation analysis (see Section 3.4.1). For an independent computation of backgrounds in a beta beam setup, see Reference 80.

3.4. Oscillation Analysis

Most of the results described in this section are taken from Reference 42, and they assume a detector mass of 440 kton, a running time $(\nu + \bar{\nu})$ of 5 + 5 years, a beam intensity of $2.9(1.1) \times 10^{18}$ ^{6}He (^{18}Ne) decays per year, and 5% for the systematic errors.

3.4.1. θ_{13} searches. One can search for nonzero values of θ_{13} by exploiting $\nu_e \rightarrow \nu_\mu$ transitions; the discovery limits [computed with GLoBES software (82)] are shown in **Figure 11*a***. The main limiting factor for the beta beam performances is the statistics. The impact of systematics is very small; systematical errors dominate over statistical errors if they become $\sigma_{\rm bkgr} \gtrsim 6\%$.

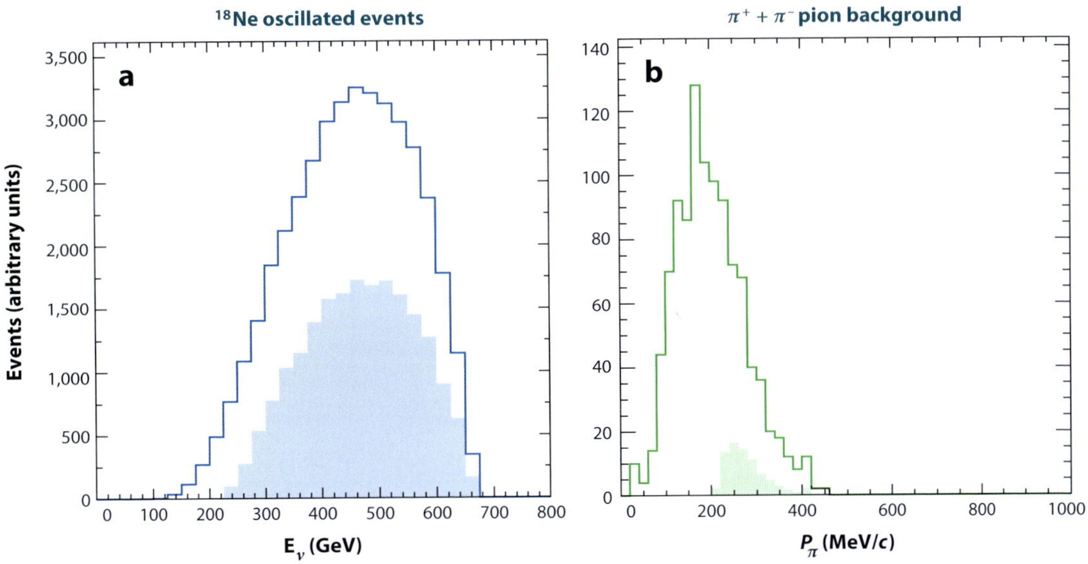

Figure 10

Event reduction for (*a*) ^{18}Ne oscillated events and (*b*) pion background, $\pi^+ + \pi^-$. White histograms shows event rates before any selection, and the shaded-color histograms show event rates after selection cuts that include particle identification and the detection of the Michel electron, which is the signature of a muon decay.

Also, $\nu_e \rightarrow \nu_e$ transitions contribute to θ_{13} sensitivity. They are marginal if the overall systematic error over the ν_e signal is approximately 2% (for the sake of comparison, reactor experiments plan to reach systematic errors of approximately 0.2% in $\bar{\nu}_e$ disappearance to reach sensitivities of $\sin^2 2\theta_{13} \simeq 0.01$). As computed in Reference 40, the CFBB experiment could reach sensitivities of $\sin^2 2\theta_{13} \leq 0.02$ (90% CL) to ν_e disappearance.

3.4.2. Leptonic *CP*-violation searches. Leptonic *CP*-violation searches are performed by comparing event rates and spectra in neutrino and antineutrino runs (see Section 1.1.1). In the specific setup of CFBB—where matter effects are negligible, meaning that no other process competes with LCPV to generate differences between neutrino and antineutrinos—the simple comparison of neutrino and antineutrino oscillation rates can provide evidence of LCPV independently from any neutrino oscillation model. The LCPV discovery potential is shown in **Figure 11*b***.

3.5. Combined Analyses with the Atmospheric Neutrinos

Beta beam and atmospheric neutrino data form a truly synergic combination: Together the two samples provide more information than would be expected from statistics alone (83). Beta beams have very limited capabilities in measuring $\text{sign}(\Delta m^2_{31})$ and resolving degeneracies. However, atmospheric neutrinos, even if measured with large statistics, cannot measure $\text{sign}(\Delta m^2_{31})$ in the absence of a measured value of θ_{13}, precisely what beta beams measure best.

The combination of atmospheric plus beta beam data (42) leads to a nontrivial sensitivity to the neutrino mass hierarchy, that is, to $\text{sign}(\Delta m^2_{31})$ as shown in **Figure 11*c***. For beta beam data

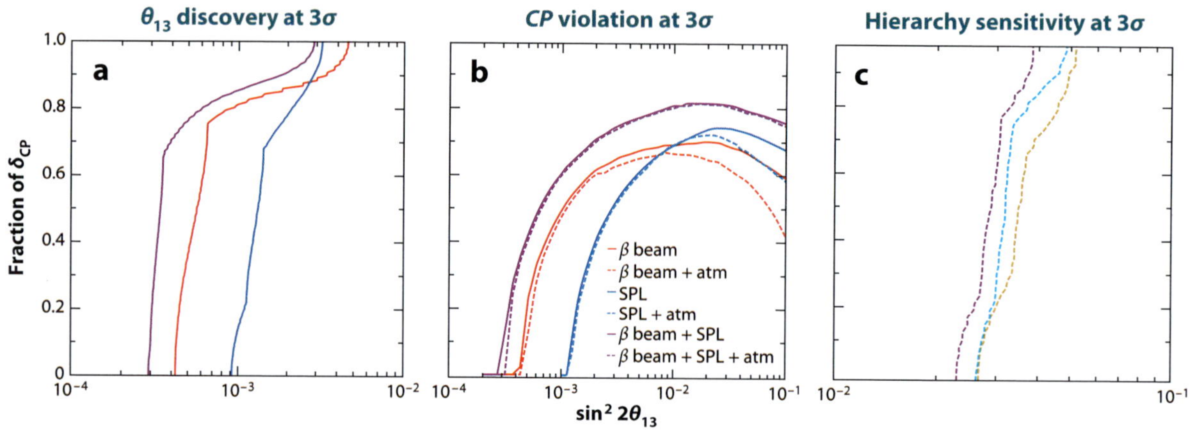

Figure 11

3σ sensitivities for the CERN-Fréjus beta beam configuration (*red lines*), the Superconducting Proton Linac superbeam (SPL; *blue lines*), and a combination of the two (*purple lines*). The solid and dashed curves represent the inclusion and noninclusion, respectively, of the atmospheric neutrino data (atm). Computed as a function of the fraction of all possible values of δ_{CP}. (*a*) Sensitivity to θ_{13}. (*b*) Sensitivity to leptonic *CP* violation. (*c*) Sensitivity to the mass hierarchy. Adapted from Reference 42, courtesy of T. Schwetz.

alone, there is almost no sensitivity in the CFBB setup because of the very small matter effects due to the relatively short baseline. However, if data from atmospheric neutrinos are included, the mass hierarchy can be identified at 3σ CL, provided that $\sin^2 2\theta_{13} \gtrsim 0.03-0.04$. Atmospheric data can also help in breaking degeneracies (42).

3.6. Combined Analyses with the Superconducting Proton Linac Superbeam

Soon after beta beams were first proposed (46), it was realized that neutrinos created by the SPL could be fired to the same detector (40, 41). The injector of a beta beam complex must be a 1–3-GeV linac, which is precisely the energy of the SPL. Furthermore radioactive ion production requires at most 0.2 MW, whereas SPL can deliver up to 4 MW of power.

Under these circumstances, a very intense superbeam (see Section 1.2.5) could run at the same time as a beta beam. The typical energy of a neutrino beam created by the SPL could nicely match the energy of a $\gamma = 100$ beta beam (**Figure 8**), so the two neutrino beams could share the same baseline and therefore the same detector. Combined SPL and beta beam performances are shown in **Figure 11**.

Combining a superbeam with a beta beam in the same experiment could provide an experimental environment with very unique characteristics: (*a*) The two beams could be used to separately study *CP* channels such as $\nu_\mu \to \nu_e$ versus $\bar{\nu}_\mu \to \bar{\nu}_e$ and $\nu_e \to \nu_\mu$ versus $\bar{\nu}_e \to \bar{\nu}_\mu$; (*b*) they could be mixed to study *T* transitions such as $\nu_\mu \to \nu_e$ versus $\nu_e \to \nu_\mu$ and $\bar{\nu}_\mu \to \bar{\nu}_e$ versus $\bar{\nu}_e \to \bar{\nu}_\mu$; and (*c*) they could be mixed to study *CPT* transitions such as $\nu_\mu \to \nu_e$ versus $\bar{\nu}_e \to \bar{\nu}_\mu$ and $\nu_e \to \nu_\mu$ versus $\bar{\nu}_\mu \to \bar{\nu}_e$. The addition of a superbeam to a beta beam could also complement some of the weak points of the beta beam, namely its lack of sensitivity to the atmospheric parameters θ_{23} and Δm^2_{23} and its lack of ν_μ events in the close detector, which is useful for calibrating beta beam signal efficiency and measuring the ν_e/ν_μ cross-section ratio.

In an SPL superbeam–plus–beta beam experiment, all the channels would be measured in the same detector with small background rates. This would be highly beneficial for systematic errors and would provide redundancy in the oscillation signals. The latter feature should not be underestimated in an experimental field that so far is completely unexplored.

4. PHYSICS POTENTIAL OF OTHER BETA BEAM SETTINGS

Several different new concepts to explore the full capabilities of a beta beam setup have been developed. These concepts aim to obtain higher neutrino fluxes and higher neutrino energies, to introduce innovative tools in neutrino oscillation searches, and to measure neutrino properties in the 10–100-MeV energy range.

4.1. High-Energy Beta Beams

High-energy beta beams (HEBB) were first discussed in Reference 81. In this paper, the proposed accelerator would be able to accelerate ^{6}He up to $\gamma = 350$ [2.3 times higher than the maximum $\gamma(^6\text{He})$ reachable at the SPS], a condition that could be fulfilled by a machine than can accelerate protons at 1 TeV. The same number of ion decays per year as for the CFBB has been considered.[2]

Two major upgrades to the accelerator scheme are needed for HEBB. Of course a new accelerator would be necessary. Proton accelerators that run at 1 TeV energy have been recently dismantled (e.g., HERA at DESY) or are going to be shut down (e.g., the Tevatron at FNAL). The LHC is a collider with a very slow acceleration cycle, which makes it unsuitable for the acceleration of the large number of radioactive ions required for a beta beam. A possible energy upgrade of the LHC would require a new higher-energy injector, SPS+ (38), which could be used for HEBB. Also, the decay ring would be heavily affected by a γ increase of the stored ions, as shown in **Table 4**. Important advantages of HEBB would be the greater statistics (cf. the beta beam merit factor; see Equation 8) and the possibility of increasing the baseline length to the point at which sensitivity to sign(Δm^2_{31}) becomes large.

The authors of Reference 84 studied the case of a $\gamma = 350$ beta beam fired to a water Cherenkov detector at approximately 700 km. As discussed in Section 3.3.1, increasing the average neutrino energy causes the fraction of well-reconstructed events to decrease until the point at which the flux increase provided by the higher γ is eliminated by the loss of quasi-elastic interaction events. This occurs for $\gamma \simeq 400$ (84).

Backgrounds from neutral-current events are much more apparent in HEBB than in CFBB, but they cluster at small energies. A simple lower cut in the visible energy keeps neutral-current backgrounds at a tolerable level (84). Also, atmospheric neutrinos integrated in the signal energy range increase much less than signal events compared with CFBB. This feature implies that in HEBB the bounds to the beta beam duty cycle derived from the atmospheric neutrino background rate are less severe, allowing for higher duty cycles.

A $\gamma = 350$ beta beam would definitely be better than CFBB, but only marginally better (as far as θ_{13} and LCPV sensitivities are concerned) than an SPS-based beta beam at the maximum γ ($\gamma = 150$ for ^{6}He and ^{18}Ne) and at the optimal baseline ($L = 300$ km) (84). For a comparison of beta beam setups at different γ, see Reference 85.

As emerges from the above discussion, a water Cherenkov detector would be somewhat limited in the energy range of HEBB, if only quasi-elastic events can be efficiently reconstructed. Different detector technologies that might overcome this problem have been considered for HEBB. For example, the case of a totally active scintillating detector (TASD), derived from the NOνA project,

[2]In a superbeam setup, this rate would be equivalent to a 2.3-fold power increase.

has been considered (86). The authors (86) studied the scaling of performances with the number of ion decays per year (either this number or its product with the γ factor could be assumed to be constant), showing that assumptions about the scaling law are very important for overall comparisons.

A different detector technology consists of an iron calorimeter whose sensitive elements, 2-cm-thick glass RPC planes with a 2-mm gas-filled gap, are interleaved with 4-cm-thick iron plates (87). This configuration has the advantage of providing a higher density than a TASD detector, such that a 40-kton detector could fit in a LNGS hall—a very attractive experimental situation. A full simulation of this detector has been performed, allowing for a robust sensitivity estimation. The fraction of neutral-current backgrounds with respect to the nonoscillated ν_e events is 8.8×10^{-3} at $\gamma = 580$, a much higher rate than the 10^{-3} rate assumed (but not computed) at $\gamma = 500$ for a TASD detector. The overall performances of this setup almost match those of the CERN-Fréjus scenario, again assuming a constant ion-decay rate. The combined sensitivity with atmospheric neutrinos of this setup has also been studied (88).

4.2. Beta Beams Based on ^{8}B and ^{8}Li Ions

^{8}B and ^{8}Li ions have a significantly higher Q_β value than do ^{6}He and ^{18}Ne ions, as can be derived from **Tables 1** and **2**. As discussed in Section 3.1, higher-Q_β ions can allow greater neutrino energies for the same γ:

$$E_\nu^{\max} = 2\gamma Q_\beta. \qquad 9.$$

Furthermore, the *Z*/*A* of the ^{8}B and ^{8}Li ions are higher than those of ^{6}He and ^{18}Ne. Taking into account the β^- emitters, ^{8}B/^{8}Li ions could produce a neutrino beam 4.74 times more energetic than a ^{6}He/^{18}Ne beam for the same accelerator energy and with a shorter decay-ring length. However, the merit factor of a ^{8}B/^{8}Li beam (see Section 3.1) is smaller than that of a ^{6}He/^{18}Ne beam because it is inversely proportional to Q_β and thus would produce smaller fluxes at the same neutrino energy. Approximately four times more ions are needed to allow a ^{8}B/^{8}Li beta beam to match the performances of a ^{6}He/^{18}Ne beta beam at the same γ (79).

As discussed in Section 2.2.2, an innovative way to produce ^{8}B/^{8}Li ions has been proposed (54). This procedure can, in principle, produce two to three orders of magnitude more radioactive ion fluxes. The feasibility and performances of this injection scheme will be studied in the context of the European Design Study EUROnu. The physics of a ^{8}B/^{8}Li beta beam based on the FNAL Main Injector and a massive liquid argon detector at Soudan has been discussed (89), and a setup based on the Tevatron and the detector at DUSEL has also been studied (90).

The authors of Reference 91 have studied the case of a mixed ^{8}B/^{8}Li and ^{6}He/^{18}Ne beta beam based on SPS. A 500-kton water Cherenkov detector with a baseline of approximately 700 km would receive the ^{8}B/^{8}Li beta beam at the first oscillation maximum and the ^{6}He/^{18}Ne beta beam at the second oscillation maximum. CFBB's rate of ion decays per year is assumed for ^{8}B and ^{8}Li. This setup has little or null advantage as far as θ_{13} and LCPV are concerned, whereas it outperforms CFBB as far as sign(Δm^2_{31}) sensitivity is concerned.

Another study (92) considered the case of a single ^{18}Ne exposure at $\gamma = 450$ (within the reach of the SPS+) and with a 50-kton iron detector placed at a baseline of 1,050 km (CERN-Boulby mine). The combination of high-energy, ^{8}B/^{8}Li-based beta beams would allow the so-called magic baseline, L_{magic}, to be covered.

The concept of a magic baseline (14, 93) derives from the observation that, for $\rho L = \sqrt{2}\pi / G_F Y_e$ (Y_e is the electron fraction inside the earth) in Equation 1, any δ_{CP} dependence disappears from $P_{e\mu}$, thereby allowing sign(Δm^2_{31}) effects to be measured without any degenerate solution. According to the Preliminary Reference Earth Model (94) earth matter density profile, $L_{\mathrm{magic}} \simeq 7{,}690$ km, so the resonance energy for matter effects would be

$$E_{\mathrm{res}} \equiv \frac{|\Delta m_{31}^2| \cos 2\theta_{13}}{2\sqrt{2} G_F N_e} \simeq 7\,\mathrm{GeV} \qquad 10.$$

for $|\Delta m_{31}^2| = 2.4 \times 10^{-3}\,\mathrm{eV}^2$ and $\sin^2 2\theta_{13} = 0.1$.

Note that close to matter resonance, the flux of oscillated events at the detector falls as a function of approximately $1/L$ (against the $1/L^2$ fall of vacuum oscillations), which means that longer baselines may be preferable.

Studies of beta beams at L_{magic} have been initiated (95) within the context of the India-based Neutrino Observatory (INO) (**http://www.imsc.res.in/~ino**), where a large magnetized iron calorimeter is set to come up. The CERN-INO distance, which is 7,152 km, approaches L_{magic}. The slope at which the decay ring should be built so as to point at a detector 7,000 km away is approximately 34.5°, so it would be very challenging to build. In any event, two detectors at different distances are needed because the detector at L_{magic} is insensitive to any LCPV effect by construction of this type of setup (96–98). The possibility of constructing a high-γ, high-Q beta beam has great potential (97, 99).

4.3. Monochromatic Neutrino Beams

Monochromatic neutrino beams based on the electron-capture process (ECB; see Section 2.9) are certainly an intriguing experimental setup, but for LCPV searches they have two major apparent limitations: There is no way to create antineutrino beams (a conceptual possibility for the production of monochromatic antineutrino beams is discussed in the following paragraph), and they do not provide any spectral information, which is very important for the solution of degeneracies. To overcome these limitations, researchers have introduced several interesting experimental strategies. For instance, studies of θ_{13} and LCPV in an ECB setup based on the ^{150}Dy ion (a lifetime of 3.1 min and $Q = 1{,}794$ keV) have been proposed (70, 100). In such studies, the beam would be run at two γs tuned to the first and second oscillation maxima. Performances of ECB in these configurations are very promising. Reference 101 proposes a more aggressive strategy that is based on $^{110}_{50}Sn$ isotopes ($Q = 267$ keV and a lifetime of 4.11 h) at $\gamma = 2{,}500$ and a baseline of 600 km.

A way to generate monochromatic antineutrino beams has also been described (72). This approach is based on the process of the bound-state beta decay (102), in which the electron is created in a previously unoccupied bound atomic state and the antineutrino is emitted at a fixed energy. Although there exist candidates such as $^{108}_{47}\mathrm{Ag}^{46+}$ with $\tau_{1/2} = 24.4$ s and neutrino energies of 1.90 and 1.67 MeV for the ECB and bound beta lines, respectively, the branching ratios for such processes are approximately 1%, making it very difficult to produce significant neutrino fluxes even conceptually.

4.4. Low-Energy Beta Beams

Beta beams are the ideal tool with which to measure neutrino cross sections because the neutrino beam flux can be predicted with high precision. This feature has been extensively discussed in the literature for neutrino energies of approximately 100 MeV, for which numerous interesting nonoscillation neutrino experiments are possible.

Construction of a low-energy facility in the 100-MeV range has been proposed (103) for (*a*) nuclear-structure studies and neutrino-nucleus interactions (103–107), (*b*) electroweak tests of the Standard Model (103, 108–110), and (*c*) studies of core-collapse supernova physics

(103, 111, 112). For a facility to operate in this energy range, the decay-ring characteristics and the detector locations would have to be reoptimized (103, 104, 113, 114).

Neutrino-nucleus interactions are of great interest in the fields of neutrino physics, nuclear physics, and astrophysics, among others. Specifically, physicists seek precise knowledge of (*a*) the neutrino detector response in neutrino experiments and in core-collapse supernova observatories that aim to detect the relic supernova neutrino background (44) by using neutrino interaction on argon (115) and carbon or oxygen (116) and (*b*) neutrinos from an (extra)galactic explosion (44). For instance, the $1n$ or $2n$ emission associated with charged-current events in a supernova lead-based observatory depends on the average electron neutrino energy, which encodes information on the still-unknown third neutrino-mixing angle, θ_{13} (117).

Studies of neutrino-nucleon reactions will improve our understanding of supernova dynamics (118, 119) and the yields of the r-process nucleosynthesis that could take place in such environments (120) and that contribute to the energy transfer (from accretion-disk neutrinos to nucleons) in γ-ray burst models (121, 122). Finally, an understanding of the subtleties of neutrino-nucleon interactions will be important for terrestrial observation of neutrino signals (123, 124). In addition to these astrophysical applications, a precise knowledge of the nuclear response of neutrinos will improve our knowledge of the nuclear isospin and spin-isospin responses, which in turn will aid the search for physics beyond the Standard Model through neutrinoless double-beta decay (106), among other efforts.

Several applications for fundamental interaction studies of low-energy beta beams have been described, including the measurement of the Weinberg angle at low momentum transfer (109); a conserved vector-current test with neutrino beams (110); the measurement of the neutrino magnetic moment (105); the measurement of coherent neutrino-nucleus elastic scattering (125); the sensitivity to extra neutral gauge bosons, leptoquarks, and R-parity-breaking interactions (126); and the search for sterile neutrinos (127). For a more detailed discussion of physics at low-energy beta beams, see Reference 128.

DISCLOSURE STATEMENT

The authors are not aware of any affiliations, memberships, funding, or financial holdings that might be perceived as affecting the objectivity of this review.

ACKNOWLEDGMENTS

Many thanks to all our colleagues in the worldwide neutrino physics community. A special thanks to all our colleagues in the beta beam task group in the European Community (EC)-supported BENE (within the Infrastructure Activity CARE) and EUROnu networks. Special thanks to Thomas Schwetz for his kind help in preparing some of the physics potential plots.

We wish to thank our home institutes, Lund University and CERN for M.L. and Istituto Nazionale di Fisica Nucleare for M.M., for supporting their work. M.L. also wishes to thank the Physics Department at the University of Liverpool, where he is a visiting professor, for its support.

We acknowledge financial support from the EC under the FP6 Research Infrastructure Action Structuring the European Research Area: EURISOL DS, project contract number 515768 RIDS. We also acknowledge financial support from the EC under the European Commission Framework Program 7 Design Study: EUROnu, project number 212372. The EC is not liable for any use that may be made of the information contained herein.

LITERATURE CITED

1. Fukuda Y, et al. (Super-Kamiokande Collab.) *Phys. Rev. Lett.* 81:1562 (1998)
2. Feruglio F, Strumia A, Vissani F. *Nucl. Phys. B* 637:345 (2002); Feruglio F, Strumia A, Vissani F. Addendum. *Nucl. Phys. B* 659:359 (2003)
3. Fukugita M, Yanagida T. *Phys. Lett.* B174:45 (1986)
4. Blondel A, et al. *CERN-2004-002, ECFA-04-230* (2004)
5. Baldini A, et al. (BENE Steer. Group) Midterm scientific report. **http://www.cap.bnl.gov/mumu/pubs/BeneMidtermReport.pdf** (2006)
6. Bandyopadhyay A, et al. (ISS Phys. Work. Group) *Rep. Prog. Phys.* 72:106201 (2009)
7. Guglielmi A, Mezzetto M, Migliozzi P, Terranova F. *Phys. Rep.* 434:47 (2006); Battiston R, Mezzetto M, Migliozzi P, Terranova F. arXiv:0912.3372 [hep-ex] (2009)
8. Burguet-Castell J, et al. *Nucl. Phys. B* 608:301 (2001)
9. Apollonio M, et al. (CHOOZ Collab.) *Eur. Phys. J. C* 27:331 (2003)
10. Schwetz T, Tortola M, Valle JWF. *New J. Phys.* 10:113011 (2008)
11. Blondel A, et al. *Acta Phys. Polon. B* 37:2077 (2006)
12. Minakata H, Nunokawa H. *J. High Energy Phys.* 0110:001 (2001)
13. Fogli GL, Lisi E. *Phys. Rev. D* 54:3667 (1996)
14. Barger V, Marfatia D, Whisnant K. *Phys. Rev. D* 65:073023 (2002)
15. Catanesi MG, et al. (HARP Collab.) *Nucl. Instrum. Methods A* 571:527 (2007)
16. Catanesi MG, et al. (HARP Collab.) *Nucl. Phys. B* 732:1 (2006)
17. Catanesi MG, et al. (HARP Collab.) *Eur. Phys. J. C* 52:29 (2007)
18. Huber P, Mezzetto M, Schwetz T. *J. High Energy Phys.* 0803:021 (2008)
19. Ahn MH, et al. (K2K Collab.) *Phys. Rev. Lett.* 93:051801 (2004)
20. Fermilab NuMI Group. NumI Facil. Techn. Des. Rep. *FNALNuMI-346* (1998)
21. Adamson P, et al. (MINOS Collab.) *Phys. Rev. Lett.* 103:261802 (2009)
22. Guler M, et al. (OPERA Collab.) *CERN-SPSC-P-318, LNGS-P25-00* (2000); Pessard H. (OPERA Collab.) arXiv:astro-ph/0504033 (2005); Guler M, et al. (OPERA Collab.) *CERN-SPSC-2000-028* (2000)
23. Acquistapace G, et al. *CERN 98-02, INFN/AE-98/05* (1998); Acquistapace G, et al. Addendum. *CERN-SL/99-034(DI), INFN/AE-99/05* (1999)
24. Komatsu M, Migliozzi P, Terranova F. *J. Phys. G* 29:443 (2003)
25. Itow Y, et al. arXiv:hep-ex/0106019 (2001)
26. Ayres DS, et al. (NOvA Collab.) arXiv:hep-ex/0503053 (2005)
27. Ardellier F, et al. (Double Chooz Collab.) arXiv:hep-ex/0606025 (2005)
28. Guo X, et al. (Daya Bay Collab.) arXiv:hep-ex/0701029 (2007)
29. Mezzetto M. *Proc. Int. Workshop Neutrino Telesc., 13th, Venice*, ed. M Baldo-Ceolin, p. 421. Padova: Padova Univ. Press (2009); Mezzetto M, Schwetz T. arXiv:1003.5800 [hep-ph] (2010)
30. Huber P, et al. *Nucl. Phys. Proc. Suppl.* 145:190 (2005); Huber P, Lindner M, Schwetz T, Winter W. *J. High Energy Phys.* 0911:044 (2009)
31. Mena O, Nunokawa H, Parke SJ. *Phys. Rev. D* 75:033002 (2007)
32. Richter B. arXiv:hep-ph/0008222 (2000)
33. Kobayashi T. *J. Phys. G* 29:1493 (2003)
34. Kajita T, Minakata H, Nakayama S, Nunokawa H. *Phys. Rev. D* 75:013006 (2007); Ishitsuka M, Kajita T, Minakata H, Nunokawa H. *Phys. Rev. D* 72:033003 (2005)
35. Diwan M, et al. arXiv:hep-ex/0608023 (2006)
36. Diwan MV, et al. *Phys. Rev. D* 68:012002 (2003)
37. Autin B, et al. *CERN-2000-012* (2000); Garoby R. *CERN-AB-2005-007* (2005)
38. Bruning O, et al. *CERN-LHC-PROJECT-REPORT-626*. **http://paf.web.cern.ch/paf/** (2002); Scandale W. *Nucl. Phys. Proc. Suppl.* 154:101 (2006)
39. Gomez-Cadenas JJ, et al. *Proc. Int. Workshop Neutrino Telesc., 9th, Venice*, ed. M Baldo-Ceolin, 2:463. Padova: Padova Univ. Press (2001); Blondel A, et al. *Nucl. Instrum. Methods A* 503:173 (2001); Mezzetto M. *J. Phys. G* 29:1771 (2003); Apollonio M, et al. arXiv:hep-ph/0210192 (2002); Campagne JE, Cazes A. *Eur. Phys. J. C* 45:643 (2006)

40. Mezzetto M. *Nucl. Phys. Proc. Suppl.* 149:179 (2005)
41. Mezzetto M. *Nucl. Phys. Proc. Suppl.* 143:309 (2005)
42. Campagne JE, Maltoni M, Mezzetto M, Schwetz T. *J. High Energy Phys.* 0704:003 (2007)
43. de Bellefon A, et al. arXiv:hep-ex/0607026 (2006)
44. Autiero D, et al. *J. Cosmol. Astropart. Phys.* 0711:011 (2007)
45. Geer S. *Phys. Rev. D* 57:6989 (1998); Geer S. Erratum. *Phys. Rev. D* 59:039903 (1999); Geer S. *Annu. Rev. Nucl. Part. Sci.* 59:347 (2009)
46. Zucchelli P. *Phys. Lett.* B532:166 (2002)
47. Lindroos M, Mezzetto M. *Beta Beams*. London: Imperial Coll. Lond. Press (2009)
48. Autin B, et al. *J. Phys. G* 29:1785 (2003)
49. Albright C, et al. APS Jt. Study Rep. Future Neutrino Phys. *FNAL-TM-2259* (2004)
50. Ravn HR, Allardyce BW. In *Treatise on Heavy-Ion Science*, ed. DA Bromley, 8:363. New York: Plenum (1989)
51. Thiollière N, et al. *CEA DAPNIA-06-274, EURISOL 03-25-2006-0004* (2006)
52. Hass M, et al. *J. Phys. G* 35:104042 (2008)
53. Loislet M, Mitrofanov S. Presented at *Beta Beam Town Meet., EURISOL, 6th, Helsinki* (2007)
54. Rubbia C, Ferrari A, Kadi Y, Vlachoudis V. *Nucl. Instrum. Methods A* 568:475 (2006)
55. Mori Y. *Nucl. Instrum. Methods A* 562:591 (2006)
56. Reed C, et al. Presented at Int. Conf. Radioact. Nucl. Beams, 7th, Cortina d'Ampezzo, Italy (2006)
57. Neuffer DV. *Fermi Natl. Lab. NFMCC-doc-516* (2007)
58. Sortais P, et al. *Rev. Sci. Instrum.* 75:1610 (2004)
59. Lachaize A, Tkatchenko A. *EURISOL/DS/12-25-2008-0012* (2008)
60. Omet C, et al. *New J. Phys.* 8:284 (2006)
61. Magistris M, Silari M. *CERN-TIS-2003-017-RP-TN* (2003)
62. Chancé A, Payet J. *Proc. Eur. Part. Accel. Conf., 10th, Edinburgh*, p. 1933. **https://oraweb.cern.ch/pls/epac06/toc.htm** (2006)
63. Chancé A, Payet J. *Proc. Eur. Part. Accel. Conf., 11th, Genoa*, p. 3104. **http://www.epac08.org/index.php?n=Main.ProceedingsDownload** (2008)
64. Jones FW, Wildner E. *Proc. Part. Accel. Conf. (PAC07), Albuquerque*, p. 3232. Albuquerque: IEEE (2007)
65. Gupta R, et al. *IEEE Trans. Appl. Supercond.* 14:259 (2004)
66. Benedikt M, Hancock S. *Nucl. Instrum. Methods A* 550:1 (2005)
67. Hancock S, Lindroos M, McIntosh E, Metcalf M. *Comput. Phys. Commun.* 118:61 (1999)
68. Hancock S. *AIP Conf. Proc.* 981:89 (2008)
69. Källberg A, Lindroos M. *EURISOL DS/TASK12/TN-05-04* (2005); Källberg A, Lindroos M, Simonsson A. *Proc. Eur. Part. Accel. Conf. 10th, Edinburgh*, p. 1693. **https://oraweb.cern.ch/pls/epac06/toc.htm** (2006)
70. Bernabeu J, Burguet-Castell J, Espinoza C, Lindroos M. *J. High Energy Phys.* 0512:014 (2005)
71. Sato J. *Phys. Rev. Lett.* 95:131804 (2005)
72. Fukumi A, et al. arXiv:hep-ex/0612047 (2006)
73. Nacher E. *Beta decay studies in the N = Z and the rare-earth regions using total absorption spectroscopy techniques*. PhD. thesis, Univ. Valencia. 158 pp. (2004)
74. Lindroos M, Bernabeu J, Burguet-Castell J, Espinoza C. *Proc. Sci.* HEP2005:365 (2006)
75. Masood SS, et al. *Phys. Rev. C* 76:045501 (2007)
76. Mezzetto M. *J. Phys. G* 29:1771 (2003)
77. Bouchez J, Lindroos M, Mezzetto M. *AIP Conf. Proc.* 721:37 (2004)
78. Mezzetto M. *Nucl. Phys. Proc. Suppl.* 155:214 (2006)
79. Fernandez-Martinez E. *Nucl. Phys. B* 33:96 (2010)
80. Ishihara C. *P.S NuFact* 08:44 (2008)
81. Burguet-Castell J, et al. *Nucl. Phys. B* 695:217 (2004)
82. Huber P, Lindner M, Winter W. *Comput. Phys. Commun.* 167:195 (2005); Huber P, et al. *Comput. Phys. Commun.* 177:432 (2007)
83. Huber P, Maltoni M, Schwetz T. *Phys. Rev. D* 71:053006 (2005); Gandhi R, et al. arXiv:hep-ph/0506145 (2005)

84. Burguet-Castell J, et al. *Nucl. Phys. B* 725:306 (2005)
85. Winter W. *Phys. Rev. D* 78:037101 (2008)
86. Huber P, et al. *Phys. Rev. D* 73:053002 (2006)
87. Terranova F, Marotta A, Migliozzi P, Spinetti M. *Eur. Phys. J. C* 38:69 (2004)
88. Donini A, et al. *Eur. Phys. J. C* 48:787 (2006)
89. Rubbia C. arXiv:hep-ph/0609235 (2006)
90. Agarwalla SK, Huber P. arXiv:0909.2257 [hep-ph] (2009)
91. Donini A, Fernandez-Martinez E. *Phys. Lett.* B 641:432 (2006)
92. Meloni D, et al. *J. High Energy Phys.* 0807:115 (2008)
93. Huber P, Winter W. *Phys. Rev. D* 68:037301 (2003); Smirnov AY. arXiv:hep-ph/0610198 (2006); Minakata H. *Proc. Int. Workshop Neutrino Telesc., 13th, Venice*, p. 163 (2007)
94. Dziewonski AM, Anderson DL. *Earth Phys. Planet. Interiors* 25:297 (1981)
95. Agarwalla SK, Choubey S, Goswami S, Raychaudhuri A. *Phys. Rev. D* 75:097302 (2007); Agarwalla SK, Choubey S, Raychaudhuri A. *Nucl. Phys. B* 771:1 (2007); Agarwalla SK, Choubey S, Raychaudhuri A. *Nucl. Phys. B* 798:124 (2008)
96. Coloma P, Donini A, Fernandez-Martinez E, Lopez-Pavon J. *J. High Energy Phys.* 0805:050 (2008)
97. Agarwalla SK, Choubey S, Raychaudhuri A, Winter W. *J. High Energy Phys.* 0806:090 (2008)
98. Choubey S, Coloma P, Donini A, Fernandez-Martinez E. *J. High Energy Phys.* 0912:020 (2009)
99. Agarwalla SK, Choubey S, Raychaudhuri A. *Nucl. Phys. B* 805:305 (2008)
100. Bernabeu J, Espinoza C. *Phys. Lett.* B664:285 (2008); Bernabeu J, et al. *J. High Energy Phys.* 0906:040 (2009)
101. Rolinec M, Sato J. *J. High Energy Phys.* 0708:079 (2007)
102. Bahcall JN. *Phys. Rev.* 194:495 (1961)
103. Volpe C. *J. Phys. G* 30:L1 (2004); Volpe C. *Nucl. Phys. Proc. Suppl.* 155:97 (2006)
104. Serreau J, Volpe C. *Phys. Rev. C* 70:055502 (2004); Volpe C. *Nucl. Phys. Proc. Suppl.* 143:43 (2005)
105. McLaughlin GC. *Phys. Rev. C* 70:045804 (2004)
106. Volpe C. *J. Phys. G* 31:903 (2005)
107. Lazauskas R, Volpe C. *Nucl. Phys. A* 792:219 (2007)
108. McLaughlin GC, Volpe C. *Phys. Lett.* B591:229 (2004)
109. Balantekin AB, de Jesus JH, Volpe C. *Phys. Lett.* B634:180 (2006)
110. Balantekin AB, de Jesus JH, Lazauskas R, Volpe C. *Phys. Rev. D* 73:073011 (2006)
111. Jachowicz N, McLaughlin GC. *Phys. Rev. Lett.* 96:172301 (2006)
112. Jachowicz N, McLaughlin GC, Volpe C. *Phys. Rev. C* 77:055501 (2008)
113. Lazauskas R, Balantekin AB, de Jesus JH, Volpe C. *Phys. Rev. D* 76:053006 (2007)
114. Amanik PS, McLaughlin GC. *Phys. Rev. C* 75:065502 (2007)
115. Cocco AG, et al. *J. Cosmol. Astropart. Phys.* 0412:002 (2004)
116. Volpe C, Welzel J. arXiv:0711.3237 [astro-ph] (2007)
117. Engel J, McLaughlin GC, Volpe C. *Phys. Rev. D* 67:013005 (2003)
118. Balantekin AB, Fuller GM. *J. Phys. G* 29:2513 (2003)
119. Horowitz CJ. *Phys. Rev. D* 65:043001 (2002)
120. Meyer BS, McLaughlin GC, Fuller GM. *Phys. Rev. C* 58:3696 (1998)
121. Ruffert M, Janka HT, Takahashi K, Schaefer G. *Astron. Astrophys.* 319:122 (1997)
122. Kneller JP, McLaughlin GC, Surman R. *J. Phys. G* 32:443 (2006)
123. Vogel P, Beacom JF. *Phys. Rev. D* 60:053003 (1999)
124. Beacom JF, Farr WM, Vogel P. *Phys. Rev. D* 66:033001 (2002)
125. Bueno A, Carmona MC, Lozano J, Navas S. *Phys. Rev. D* 74:033010 (2006)
126. Barranco J, Miranda OG, Rashba TI. *Phys. Rev. D* 76:073008 (2007)
127. Agarwalla SK, Huber P, Link JM. *J. High Energy Phys.* 1001:071 (2010)
128. Volpe C. *J. Phys. G* 34:R1 (2007); Volpe C. *Nucl. Phys. A* 752:38 (2005)

Precision Muon Capture

Peter Kammel[1] and Kuniharu Kubodera[2]

[1]Department of Physics, University of Illinois at Urbana-Champaign, Urbana, Illinois 61801; email: pkammel@illinois.edu

[2]Department of Physics and Astronomy, University of South Carolina, Columbia, South Carolina 29208; email: kubodera@sc.edu

Annu. Rev. Nucl. Part. Sci. 2010. 60:327–53

First published online as a Review in Advance on July 16, 2010

The *Annual Review of Nuclear and Particle Science* is online at nucl.annualreviews.org

This article's doi:
10.1146/annurev-nucl-100809-131946

0163-8998/10/1123-0327$20.00

Key Words

electroweak nucleon form factors, effective field theory, time-projection chamber, solar nuclear fusion

Abstract

A new generation of experiments is studying muon capture on hydrogen, deuterium, and ^{3}He with significantly improved precision using a novel active-target method. At the same time, chiral effective field theories (EFTs) allow the calculation of low-energy electroweak observables on protons and the simplest nuclei within a rigorous quantum chromodynamics (QCD)-based framework. This experimental and theoretical progress significantly enhances our understanding of the axial current. In the case of the nucleon, the unambiguous experimental determination of the pseudoscalar coupling g_P clarifies a long-standing puzzle and provides a basic test of QCD at low energies. In the case of deuterium and ^{3}He, the new precision data allow a comprehensive test of EFT calculations of weak few-body reactions and determine an essential low-energy constant relevant for related astrophysics reactions of fundamental interest.

Contents

1. INTRODUCTION

Muon capture on the proton,

$$\mu^- + p \rightarrow n + \nu_\mu, \tag{1.}$$

is a basic charged-current weak process. In the low-energy regime relevant to muon capture, the Standard Model electroweak interaction reduces to an effective Fermi interaction, $H_W = (G_F/\sqrt{2})V_{ud}L_\mu J^\mu$, between the leptonic current L_μ and the hadronic current J_μ. Here $G_F = 1.166367(5) \times 10^{-5}$ GeV^{-2} is the Fermi coupling constant, and $|V_{ud}| = 0.97418(27)$ is the CKM (Cabibbo-Kobayashi-Maskawa) matrix element (1; see **http://pdg.lbl.gov** for the 2009 partial update). The leptonic current is given simply by $L_\mu = \bar{\psi}_{\nu_\mu}\gamma_\mu(1-\gamma_5)\psi_\mu$. By contrast, although the hadronic current $J^\mu = V^\mu - A^\mu$, written in terms of the quark fields, has the simple structure $J^\mu = \bar{q}_d\gamma^\mu(1-\gamma^5)q_u$, it cannot be written explicitly in terms of hadronic fields because of the nonperturbative nature of the strong interactions in the low-energy regime. Meanwhile, the matrix elements of V_μ and A_μ between nucleon states [in the absence of the so-called second-class current (2)] can be parameterized in the general form

$$\begin{aligned} \langle n|V^\mu|p\rangle &= \bar{u}_n\left[F_V(q^2)\gamma^\mu + \frac{iF_M(q^2)}{2m_N}\sigma^{\mu\nu}q_\nu\right]u_p, \\ \langle n|A^\mu|p\rangle &= \bar{u}_n\left[F_A(q^2)\gamma^\mu\gamma^5 + \frac{F_P(q^2)}{m_\mu}q^\mu\gamma^5\right]u_p. \end{aligned} \tag{2.}$$

Here m_μ and m_N are the muon and nucleon masses, respectively, and the momentum transfer is defined by $q = p_n - p_p$, where p_p (p_n) is the proton (neutron) momentum.[1] $F_V(q^2)$ and $F_M(q^2)$ are

[1] For γ-matrices, spinors, and so on, we follow the notation of Bjorken & Drell (3).

the vector and weak-magnetic form factors, and $F_A(q^2)$ and $F_P(q^2)$ are the axial and pseudoscalar form factors. Because by definition such form factors subsume all strong-interaction effects, their study provides valuable information about the symmetries and dynamics of the underlying fundamental theory. Historically, the study of F_V and F_M played an important role in establishing the conserved vector current (CVC) hypothesis (4), which explained the conservation of the vector charge and was a precursor of electroweak unification. The study of F_A and F_P was closely related to the emergence of the idea of the partial conservation of the axial current (PCAC) and to its deeper significance as a consequence of chiral symmetry and its breaking. The fundamental idea of spontaneous symmetry breaking, which was proposed five decades ago and for which Nambu (5) received the Nobel prize in 2008, paved the way for explaining the generation of particle masses and for the development of chiral perturbation theory (χPT), which is a low-energy effective field theory (EFT) of the fundamental theory of quarks and gluons—quantum chromodynamics (QCD).

With the advent of QCD, the study of electroweak form factors has acquired new interest. One facet is that it should be possible to derive these form factors from a QCD-based, first-principle calculation; comparison of the results with data will test direct consequences of QCD. The second facet is that χPT allows us to relate various hadronic and electroweak observables to a limited number of parameters known as the low-energy constants (LECs), which enter into the χPT Lagrangian. The electroweak form factors can be used either to constrain LECs or, if the relevant LECs are already known, to verify the model-independent predictions of χPT against experimental data. When these predictions are sufficiently precise at the calculated order, their comparison with experiment sensitively tests the symmetry structure of the underlying fundamental theory.

These developments enhance the necessity of precise measurements of the weak-interaction form factors. Of the four form factors in Equation 2, F_P is by far the least well known. Muon capture on the proton is the most direct probe for obtaining information on F_P. Although muon capture on a complex nucleus is easier to measure because of the higher capture rates, the extraction of F_P is compounded with nuclear many-body effects. At this point, we distinguish two types of muon capture on the proton. Ordinary muon capture (OMC) is the process in Equation 1, whereas radiative muon capture (RMC) refers to $\mu^- p \rightarrow \gamma n \nu_\mu$. OMC allows one to deduce $g_P \equiv F_P(q_m^2)$, where $q_m^2 \equiv -0.88\, m_\mu^2$ is the momentum transfer relevant to OMC, and RMC probes $F_P(q^2)$ for a finite range of q^2 allowed by three-body kinematics. Experimental efforts to study OMC span more than 40 years, whereas the first measurement of RMC was achieved only a decade ago (7). Unfortunately, as shown in **Figure 1**, the situation before the recent MuCap experiment (8) was inconclusive, either because the experiments lacked precision or because the interpretation of data was clouded by uncertainties in muon atomic and molecular processes. Note that the RMC result in the original publication (7) indicated a 4.2σ deviation from the χPT prediction and that this deviation has not been fully explained despite intense theoretical scrutiny (9, 10). The MuCap experiment developed a novel active-target method based on a hydrogen time-projection chamber (TPC) and selected conditions in which muon atomic physics uncertainties were minimal. An initial data set has given a first precise OMC result with an unambiguous interpretation; the final analysis of the full data with ten-times-higher statistics is currently under way.

Equation 1 gives the leading-order contribution to muon capture in multinucleon systems. We note that, for simple few-nucleon systems, it is now possible to make an accurate comparison of theory and experiment, owing to (*a*) the recent development of new experimental techniques that underlie the high-precision measurement of g_P, among others, and (*b*) concurrent sophistication in EFT methods. In the case of muon capture on ^{3}He, the capture rate has already been measured to 0.3% precision using the TPC technique (11). The new MuSun experiment (12; also see **http://www.npl.uiuc.edu/exp/musun**) aims to measure the rate of muon capture

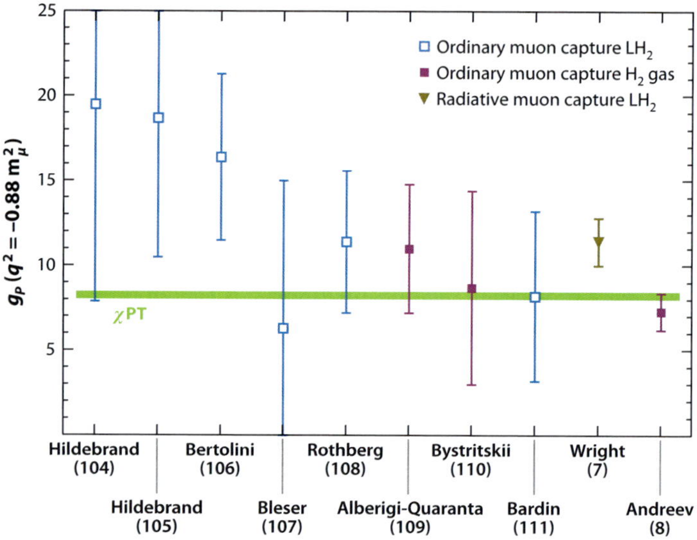

Figure 1

Pseudoscalar coupling g_P determined from muon-capture experiments, compared to the precise theoretical prediction. The present updated evaluation incorporates new data (6) on an important muon molecular rate. The recent MuCap result (8) is largely independent of molecular effects and supports the theory prediction [chiral perturbation theory (χPT)].

on the deuteron,

$$\mu^- + d \rightarrow n + n + \nu_\mu, \qquad 3.$$

to better than 1.5% precision. This measurement will impact our understanding of a number of weak processes of high astrophysical interest. Two prominent examples are *pp* fusion in the sun and νd reactions; the latter processes were observed in the Sudbury Neutrino Observatory (SNO) experiment (13) to determine the flavor composition of solar neutrinos. Recent EFT calculations (14, 15) have demonstrated that all these reactions involve the same axial two-body current term and are governed by a single LEC that parameterizes poorly constrained short-distance physics common to these processes. Muon capture on the deuteron offers a theoretically clean way to determine this LEC.

At present we are at the confluence of two exciting developments in the field of muon capture. For the first time, experiments on hydrogen and the lightest nuclei are approaching subpercent precision, corresponding to an order-of-magnitude improvement over the past decade. At the same time, χPT has been established as a systematic low-energy expansion of QCD and naturally covers the energy domain relevant to muon capture. It is a powerful framework for calculating weak processes on a single nucleon, the deuteron, and even few-nucleon systems. In this review we concentrate on muon capture in simple systems that are already or will soon become amenable to rigorous QCD-based calculations. The theoretical concepts and results are described in Section 2. The experimental challenges related to the interplay between weak processes and the rich atomic physics phenomena induced by negative muons in hydrogen isotopes are discussed in Section 3, together with an evaluation of the earlier capture experiments. The methodology, status, and results of the new precision experiments are described in Section 4. In Section 5 we summarize the current status of pertinent physics issues and discuss future prospects. This article is not meant to be a comprehensive review of muon capture in general, and for those aspects that have not been covered here we refer the reader to recent reviews (9, 10, 16).

2. THEORY OF MUON CAPTURE

2.1. Chiral Symmetry

Chiral symmetry is one of the basic symmetries of the QCD Lagrangian $\mathcal{L}_{\mathrm{QCD}}$ (for details, see, e.g., Reference 17). Consider the two-flavor case (u and d quarks), and let ψ represent an isospin doublet field, $\psi = (u, d)$. The quark fields can be decomposed into so-called left-handed and right-handed components via $\psi_L \equiv \frac{1}{2}(1+\gamma_5)\psi$ and $\psi_R \equiv \frac{1}{2}(1-\gamma_5)\psi$, respectively. Let Ω represent a rotation in isospin space. In the limit of massless quarks, $\mathcal{L}_{\mathrm{QCD}}$ is invariant not only with respect to an overall rotation,

$$\psi \to \psi' \equiv \Omega\psi, \qquad 4.$$

but also with respect to independent isospin rotations of the left-handed and right-handed quark sectors $\psi_L \to \psi'_L \equiv \Omega_L\psi_L$ and $\psi_R \to \psi'_R \equiv \Omega_R\psi_R$, where Ω_L and Ω_R represent two independent rotations in isospin space. This property, termed $SU(2)_L \times SU(2)_R$ symmetry, is equivalent to the combination of isospin symmetry and chiral symmetry. The former refers to invariance against the transformation in Equation 4, and the latter refers to invariance with respect to a transformation (known as a chiral transformation),

$$\psi \to \psi' \equiv \Omega\gamma_5\psi. \qquad 5.$$

The isospin transformation generates a CVC V_μ, $\partial_\mu V^\mu = 0$, whereas the chiral transformation generates an axial vector current A_μ, which is also conserved in the massless quark limit. A surprising fact is that, although $\mathcal{L}_{\mathrm{QCD}}$ possesses chiral symmetry, the ground state (vacuum) of $\mathcal{L}_{\mathrm{QCD}}$ does not respect this symmetry. This feature, known as spontaneous breaking of chiral symmetry, gives rise to massless pseudoscalar Nambu-Goldstone bosons, which can be identified with pions. Of course, the actual pions have finite masses, but it is the presence of finite (albeit small) quark masses that is responsible for the masses of the physical pions. The finite quark masses ($m_q \neq 0$) also lead to a slight breakdown of chiral symmetry, and as a result the conservation of the axial vector current is slightly violated: $\partial_\mu A^\mu = m_q \bar{q}_d \gamma^5 q_u$, which is the expression of PCAC at the quark level.

2.2. Partial Conservation of the Axial Current and F_P

Taking the divergence of Equation 2 yields

$$\langle n|\partial_\mu A^\mu|p\rangle = i\,\bar{u}_n\left[F_A(q^2)2m_N + \frac{F_P(q^2)}{m_\mu}q^2\right]\gamma^5 u_p. \qquad 6.$$

The observation that the quantum numbers of $\partial_\mu A^\mu(x)$ are the same as those of the pion motivated the PCAC hypothesis $\partial_\mu A^\mu(x) \propto \phi(x)$, where $\phi(x)$ is the pion field. The proportionality constant can be related to the pion decay constant $F_\pi = 92.4 \pm 0.3$ MeV, so that the PCAC relation reads $\partial_\mu A^\mu(x) = (1/\sqrt{2})F_\pi m_\pi^2 \phi(x)$. The matrix element of this equality between the nucleon states and Equation 6 leads to

$$F_A(q^2)2m_N + \frac{F_P(q^2)}{m_\mu}q^2 = 2F_\pi \frac{m_\pi^2}{-q^2+m_\pi^2} G_{\pi NN}(q^2), \qquad 7.$$

where $G_{\pi NN}(q^2)$ is the πNN vertex function. Applying this relation to the $q^2 = 0$ case gives $m_N F_A(0) = F_\pi G_{\pi NN}(0) = F_\pi g_{\pi NN}[G_{\pi NN}(0)/G_{\pi NN}(m_\pi^2)]$, where $g_{\pi NN} = G_{\pi NN}(m_\pi^2)$ is the physical πNN coupling constant. If $G_{\pi NN}(q^2)$ does not vary significantly between m_π^2 and zero, Equation 7 implies that $m_N F_A(0) = F_\pi g_{\pi NN}$, which is known as the Goldberger-Treiman relation. This relation is satisfied at the 2% level if we use $F_A(0) \equiv g_A = 1.2695 \pm 0.0029$ (1) from neutron β decay and a modern value of $g_{\pi NN} = 13.05 \pm 0.08$ (18–20). For $q^2 \neq 0$, Equation 7 leads to an expression for $F_P(q^2)$ in terms of $F_A(q^2)$ and $G_{\pi NN}(q^2)$. If we expand Equation 7 around $q^2 = 0$ and use the Goldberger-Treiman relation, we arrive at the Adler-Dothan formula (21). By taking only the dominating first term (i.e., the pion-pole term) in this expansion, the PCAC prediction can be summarized as

$$F_P(q^2) = \frac{2m_\mu m_N}{m_\pi^2 - q^2} F_A(0). \qquad 8.$$

At $q^2 = q_m^2$, Equation 8 gives $g_P \equiv F_P(q_m^2) = 6.77 F_A(0) = 8.58$.

2.3. Chiral Perturbation Theory

In describing low energy-momentum phenomena characterized by a scale Q, which is sufficiently smaller than the chiral scale $\Lambda_\chi \sim 1$ GeV (e.g., 17, p. 80), we can exclude from the QCD Lagrangian those degrees of freedom that belong to Λ_χ or higher, retaining only pions and nucleons as active hadronic degrees of freedom. The resulting EFT, χPT, is a low-energy EFT of QCD, and as such it retains all symmetries, including the (approximate) chiral symmetry, of QCD. The terms in the χPT Lagrangian, $\mathcal{L}_{\chi\mathrm{PT}}$, can be organized into a perturbative expansion in powers of Q/Λ_χ. The effective nature of EFT is reflected in the presence of LECs, which parameterize the high-energy physics that, in generating the low-energy EFT, has been integrated out. Insofar as all the LECs up to a specified chiral order n can be fixed either from theory or from fitting to experimental values of appropriate observables, $\mathcal{L}_{\chi PT}$ serves as a complete (and hence model-independent) Lagrangian to the given order of expansion. χPT can serve as a model-independent framework for describing low-energy nuclear phenomena (e.g., 22, 23). To this end, it is useful to treat the nucleon fields appearing in $\mathcal{L}_{\chi\mathrm{PT}}$ as nonrelativistic particles, suppressing the antinucleon degrees of freedom and retaining only the "large" components of the nucleon fields as explicit degrees of freedom. The resulting scheme is known as the heavy-baryon chiral perturbation theory (HBχPT).

Bernard et al. (24) used HBχPT to calculate F_P and arrived at

$$F_P(q^2) = \frac{2m_\mu g_{\pi NN} F_\pi}{m_\pi^2 - q^2} - \frac{1}{3} F_A(0) m_\mu\, m_N\, r_A^2, \qquad 9.$$

where r_A^2 is the nucleon axial radius defined as $F_A(q^2) = F_A(0)[1 + \frac{1}{6} r_A^2 q^2 + \cdots]$. Essentially this is the result derived earlier based on PCAC (21, 25). An important point, however, is that owing to the systematic nature of HBχPT we can be confident that the corrections must be of higher order in chiral expansion. The two-loop corrections to Equation 9 have been evaluated by Kaiser (26), and the results indicate that, so long as the relevant LECs are of natural size, such two-loop corrections are tiny. The result of a different HBχPT calculation by Fearing et al. (27) agrees with Equation 9, apart from the differences in the treatment of small corrections arising from the q^2 dependence of $G_{\pi NN}$. Similar results have also been obtained by Ando et al. (28) and by Bernard et al. (29). With the use of the experimental value of $r_A^2 = 0.44 \pm 0.02$ fm^2 (31) and the other well-known LECs on the right-hand side of Equation 9, HBχPT gives a robust and accurate prediction: $g_P = 8.26 \pm 0.23$ (10, 24). The original PCAC prediction (Equation 8) is understood to be the leading term in a chiral expansion, and corrections can be unambiguously understood in terms of higher-order contributions in this expansion.

2.4. Recent Lattice Results

A lattice QCD calculation should in principle allow the determination of all nucleon electroweak observables from first principles, so considerable effort has been invested in lattice QCD calculations of the nucleon weak-interaction form factors (32–41). Progress in such calculations is characterized by the use of a finer lattice spacing, a larger lattice size, an increased number of gauge configurations, and a lower (and hence closer to the physical) value of the u and d quark mass. In connection with the last point, χPT is often invoked to extrapolate the numerical results obtained with the currently manageable value of the quark mass to the case with the physical value. Earlier works (32–34) were based on the quenched approximation, which ignored the excitations of quark-antiquark pairs, but recent calculations tend to avoid this approximation. Also, for weak-interaction form factor calculations, it is important to use a formalism that preserves chiral symmetry, and to this end, the domain-wall fermion (DWF) formalism has been developed (42, 43).

An extensive lattice QCD calculation of the nucleon weak-interaction form factors has been carried out by the RBC and UKQCD Collaborations (41). This is an unquenched calculation based on the DWF formalism, which includes three flavors (u, d, and s quarks), and the light quark mass (u and d quark masses taken to be degenerate) is varied in a range that corresponds to the pion mass between 0.33 GeV and 0.67 GeV. The form factors have been calculated for momentum transfers $0.2 < q^2 < 0.75$ GeV^2, and the main results concerning the axial vector currents are as follows. By extrapolating $g_A \equiv F_A(0)/F_V(0)$ to the physical value of m_π with χPT, $g_A = 1.19(6)_{stat}(4)_{sys}$ is obtained, which is 7% smaller than the experimental value. The calculated form factor $F_A(q^2)$ shows the empirically observed dipole behavior, $F_A(q^2)/F_A(0) = (1-q^2/M_A^2)^{-2}$, but the theoretical value of M_A is 37% larger than the experimental $M_A = 1.03$ GeV. The q^2 dependence of the calculated $F_P(q^2)$ is consistent with the pion-pole dominance pattern; however, $\alpha \equiv \frac{(q^2-m_\pi^2)F_P(q^2)}{2m_N F_A(q^2)}$, which should be approximately one, is 10–20% smaller than unity. As for the μp-capture point value, an extrapolation of the calculated g_P down to the physical pion mass point gives a value of g_P that is consistent with experiment (8). For a precise test of (lattice) QCD concerning the weak-interaction form factors, a larger lattice size is needed, and the systematic errors have to be studied further as the quark mass is decreased toward its physical value.

2.5. Muon Capture on the Proton

We survey the current status of theoretical studies of the two types of muon capture on the proton: OMC and RMC. For OMC, once the form factors that appear in Equation 2 are determined, all observables—apart from radiative corrections and the finite–proton size effects (see below)—can be readily calculated. By contrast, for RMC, a standard diagram calculation becomes complicated because it involves a number of delicate issues, such as the off-shell extrapolation of the electroweak form factors and the implementation of gauge invariance. Meanwhile, χPT provides an unambiguous framework that automatically preserves gauge invariance as well as the symmetries of QCD.

2.5.1. Ordinary muon capture. The ground state of a μp atom has two hyperfine states: $F = 0$ (singlet) and $F = 1$ (triplet). The OMC rate from a hyperfine state F is given by (e.g., 28)

$$\Lambda_F = \frac{|\phi_\mu(0)|^2}{16\pi m_\mu m_p}\left(\frac{E_\nu}{E_\nu + \sqrt{m_n^2 + E_\nu^2}}\right)\frac{1}{2F+1}\sum_{F_z,h}|M(h;F,F_z)|^2, \qquad 10.$$

Table 1 Recent theoretical results of muon-capture rates from the triplet (Λ_t) and singlet states (Λ_s) of the μp atom and the quartet (Λ_q) and doublet (Λ_d) states of the μd atom, respectively[a,b]

Λ_t	Λ_s	$\Lambda_s(1 + \text{RC})$	F_A (0)	$F_P(q_m^2)$	Reference	Method
12.0(1)	688.4(3.8)		1.2670(35)	8.475(76)	46	Pheno
	692.3(3.4)	711.5(4.6)	1.2695(29)	8.2(2)	44	Pheno
12.9	687.4		1.2670	8.44(23)	29	χPT, Δ
11.9	695		1.267	8.47	28	χPT
Λ_q	**Λ_d**		**F_A (0)**	**$F_P(q_m^2)$**	**Reference**	**Method**
	397–400		1.262	8.46	47	SNPA
	416(7)		1.2605	8.53	48	SNPA
11.4	416(6)		1.2694	8.59	49	SNPA
	386		1.267	8.47	14	EFT*

[a]The pionless EFT calculation (15) cannot be expressed by a single value and is discussed in the text.

[b]Abbreviations: EFT*, hybrid effective field theory; Pheno, the standard diagrammatic calculation using the phenomenological form factors of Equation 2; RC, radiative corrections (calculated in Reference 44); SNPA, standard nuclear physics approach; χPT, chiral perturbation theory.

where $E_\nu = 99.15$ MeV is the neutrino energy and $\phi_\mu(0)$ is the value at the origin of the radial wave function for the μp ground state. The helicity amplitude $M(h; F, F_z)$ is specified by the final neutron helicity h, the initial hyperfine spin, and its z component. $M(h; F, F_z)$ contains the form factors in Equation 2 evaluated at $q^2 = q_m^2$. Because $\Lambda_s \equiv \Lambda_{F=0}$ is approximately 60 times larger than $\Lambda_t \equiv \Lambda_{F=1}$, most studies of OMC concentrate on Λ_s. High-precision calculations need to consider the radiative corrections (RCs) in Equation 10; see Section 2.7. χPT calculations of the amplitude for Equation 1 were carried out by Fearing et al. (27), Ando et al. (28), and Bernard et al. (29). These calculations begin with $\mathcal{L}_{\chi\text{PT}}$ written up to a specified chiral order and then evaluate the amplitude, including tree and loop diagrams. The form factors appearing in Equation 2 can be related to the LECs that characterize $\mathcal{L}_{\chi\text{PT}}$. But because we need to determine these LECs by using the same sources of information as used to constrain the form factors, the χPT calculations lead to the same results as a phenomenological treatment, as far as observables in OMC are concerned. However, once the LECs are determined, χPT provides a systematic and model-independent framework for calculating other processes, such as the RMC process discussed below. **Table 1** summarizes the results of recent calculations. For the vector form factors, the standard values of $F_V(q_m^2) = 0.976$ and $F_M(q_m^2) = 3.583$ are used, and the axial form factors used in the individual calculations are given explicitly to facilitate comparison. As for the μp atomic wave function, the point-charge expression $|\phi_\mu(0)|^2 = (\alpha m_\mu^r)^3/\pi$, where $m_\mu^r = m_\mu m_p/(m_\mu + m_p)$, is changed by a factor of 0.995 if the finite proton size (30) is considered. The phenomenological calculations introduce this correction, whereas it would be of higher order in the EFT expansion.

2.5.2. Radiative muon capture. RMC on the proton, $\mu^- p \to \gamma \nu_\mu n$, exhibits higher sensitivity to $F_P(q^2)$ than does OMC because the three-body final state allows the momentum transfer q^2 to come close to the pion pole. Because the standard diagram calculations of RMC have been thoroughly surveyed (9), we summarize only the χPT calculations.

χPT allows us to carry out a systematic calculation that preserves gauge invariance as well as the symmetries of QCD. Furthermore, it automatically incorporates the effects of the form factors (up to a specified chiral order). The first HBχPT calculation of RMC up to $\mathcal{O}(q^2)$ was performed by Meissner et al. (50). An extension of the calculation to $\mathcal{O}(q^3)$ was achieved by Ando & Min (51), who confirmed that the higher-order effects are small, thereby demonstrating the rapid convergence of chiral expansion. An $\mathcal{O}(q^2)$ calculation by Bernard et al. (29) included the Δ explicitly with the use of the small-scale expansion (52). All such calculations indicate that the RMC rate obtained by Wright et al. (7) is in conflict with the χPT results.

2.6. Weak Processes in Two-Nucleon Systems

In the single-nucleon sector, experimental and theoretical developments have reached a level at which all relevant LECs can be controlled with reasonable accuracy. This implies that, if the EFT-based approach can be extended to two-nucleon systems with comparable accuracy, and if sufficiently accurate data in two-nucleon systems are available to fix the relevant LECs, then we can make reliable model-independent predictions for all electroweak reactions in the two-nucleon system. The planned precision measurement of the μd-capture rate is expected to play an important role in this program. Before explaining recent theoretical developments, we briefly recapitulate the traditional method, known as the standard nuclear physics approach (SNPA), then describe the developments based on HBχPT.

2.6.1. Theoretical methods. In SNPA, we assume that an A-nucleon system is described by the Hamiltonian $H_{\mathrm{A}} = \sum_{i=1}^{\mathrm{A}} t_i + \sum_{i<j}^{\mathrm{A}} v_{ij}^{\mathrm{phen}} + \sum_{i<j<k}^{\mathrm{A}} v_{ijk}^{\mathrm{phen}}$, where t_i is the kinetic energy of the ith nucleon, v_{ij}^{phen} (v_{ijk}^{phen}) is a phenomenological potential involving two nucleons (three nucleons), and the nuclear wave function is obtained as an exact solution to the A-body Schrödinger equation. There is freedom in choosing possible forms of the short-range parts of phenomenological potentials. For v_{ij}^{phen}, there exist a number of high-precision phenomenological *NN* potentials that can reproduce all existing two-nucleon scattering data as well as the properties of the deuteron. The three-body interaction v_{ijk}^{phen} is constrained by observables in A = 3 nuclei. The electroweak transition operators in SNPA consist of (*a*) one-body terms $\sum_{i=1}^{\mathrm{A}} \mathcal{O}_i$, which can be determined from the coupling of the single nucleon to the electroweak current, and (*b*) two-body terms $\sum_{i<j}^{\mathrm{A}} \mathcal{O}_{ij}$, which are derived from one boson–exchange diagrams. SNPA has been extensively applied to the description of nuclear observables in light nuclei (53).

Although SNPA is believed to work with reasonable accuracy, it involves a degree of model dependence because of its phenomenological nature. In principle, HBχPT should allow us to treat multinucleon systems in a model-independent manner. In applying HBχPT to nuclear systems, however, we encounter the problem that a new scale enters that is related to the smallness of nuclear binding compared with m_π. As proposed by Weinberg (54), one can construct an EFT that is applicable to multinucleon systems by applying χPT only to irreducible diagrams, that is, diagrams that do not contain purely nucleonic intermediate states. The contributions of reducible diagrams are incorporated by solving the Schrödinger equation for a Hamiltonian in which the potential is identified with the contribution of the irreducible diagrams. For convenience, we refer to Weinberg's approach (54) and its subsequently developed variants as nuclear χPT. In calculating the matrix element of an electroweak nuclear transition in nuclear χPT, we need to evaluate $\mathcal{M}^{\mathrm{EFT}} = \langle \Psi_f^{\mathrm{EFT}} | \sum_i^{\mathrm{A}} \mathcal{O}_i^{\mathrm{EFT}} + \sum_{i<j}^{\mathrm{A}} \mathcal{O}_{ij}^{\mathrm{EFT}} | \Psi_i^{\mathrm{EFT}} \rangle$, where both the initial and final nuclear wave functions and the transition operators are derived from HBχPT. The transition operators can contain terms that involve three or more nucleons, but they belong to higher orders in the χPT expansion. If all the quantities appearing in $\mathcal{M}^{\mathrm{EFT}}$ are evaluated from $\mathcal{L}_{\chi\mathrm{PT}}$ with no unknown LECs, we have a model-independent estimate of $\mathcal{M}^{\mathrm{EFT}}$ (to the chiral expansion order under consideration). In considering the specific case of μd capture, two crucial points are that (*a*) for the low-energy Gamow-Teller (GT) transition that governs this process, the one-body transition operators, $\mathcal{O}_i^{\mathrm{EFT}}$s, are well known and (*b*) the two-body operators, $\mathcal{O}_{ij}^{\mathrm{EFT}}$s, contain only

a single unknown LEC. This LEC, denoted by $\hat{d}^R$, parameterizes the strength of a contact-type four-nucleon coupling to the axial current. Furthermore, as noted by Park et al. (55), the same LEC $\hat{d}^R$ also controls the two-body GT amplitudes for solar *pp* fusion, low-energy νd reactions, and tritium β decay. This fact implies that, once $\hat{d}^R$ is determined using any one of these reactions, we can make robust predictions for the remaining processes.

1. Hybrid EFT (EFT*). Above, we emphasized the formal merits of nuclear EFT. However, in actual calculations it is a major challenge to generate, strictly within the EFT framework, nuclear wave functions of an accuracy comparable to that of the SNPA wave functions. A way to avoid this difficulty is to replace Ψ^{EFT} with Ψ^{phen}, which is a phenomenological nuclear wave function obtained as an exact eigenstate of the phenomenological nuclear Hamiltonian H_A. Thus we use the transition operators derived from HBχPT and sandwich them between Ψ_i^{phen} and Ψ_f^{phen}. This hybrid approach, termed EFT*, can be applied to complex nuclei (A $=$ 3, 4 ...) with essentially the same accuracy and ease as to the A $=$ 2 system and enables us to determine $\hat{d}^R$ from observables pertaining to complex nuclei (if the A $=$ 2 observables are insufficient). A concrete example is described in Section 2.6.2. For discussion of the formal shortcomings of EFT* and the estimation of their practical consequences, see, for instance, Reference 56.
2. Pionless EFT. When the characteristic momentum p of a phenomenon is much smaller than m_π, one can eliminate (integrate out) pions from $\mathcal{L}_{\text{eff}}$ by choosing $\Lambda_{\text{cut}} \sim m_\pi$; one then has pionless EFT (57–59). All the nucleon-nucleon interactions and electroweak currents are described by point-like contact interactions with systematic expansion in powers of p/m_π. Up to next-to-next-to-leading order, the two-body GT transition is controlled by a single term (60, 61), whose unknown coupling constant is denoted by $L_{1,\text{A}}$. In pionless EFT based on the Kaplan-Savage-Wise (KSW) scheme (57, 62, 63), all the low-energy GT transition amplitudes in the A $=$ 2 systems (μd, *pp*, and νd) can be written model independently and analytically in terms of the low-energy scattering parameters (scattering lengths and effective ranges) and $L_{1,\text{A}}$. At present, however, one cannot take full advantage of this formal simplicity of pionless EFT, because no data pertaining to the A $=$ 2 systems that have sufficient accuracy to yield a significant constraint on $L_{1,\text{A}}$ (64, 65) are available. In the literature (61), $L_{1,\text{A}}$ is fixed from the theoretical result of the GT amplitude calculated by other methods (SNPA or EFT*), but with this procedure the attractive feature of formal self-consistency is compromised. The MuSun experiment is important in that it will allow the determination $L_{1,\text{A}}$ with much higher precision within the A $=$ 2 system.
3. Full nuclear χPT. It is of great current interest to formulate nuclear physics in a fully consistent χPT framework without resorting to phenomenological models. There have been intensive efforts to derive nuclear interactions (*NN* potential, *NNN* potential, etc.) based on HBχPT (66–68). So far, this approach has been limited mostly to *NN* scattering and nuclear rearrangement reactions that involve only strong interactions. An ab initio calculation of tritium β decay based on HBχPT has recently been reported (69), and Γ_β^t and the binding energy of the A $=$ 3 nucleus can provide a robust and stringent constraint on the LEC c_D, which in turn determines $\hat{d}^R$. The full EFT calculation of the μd-capture rate with the use of this $\hat{d}^R$ has yet to be performed.

2.6.2. Muon capture on the deuteron. Extensive SNPA-based calculations of μd capture (**Table 1**) were carried out by Tatara et al. (47) and by Adam et al. (48) 20 years ago. The primary conclusions of Tatara et al. are as follows: (*a*) The doublet-capture rate is $\Lambda_d = 397 - 400\,\text{s}^{-1}$; (*b*) the results are stable against the use of different realistic *NN* potentials available at that time;

and (*c*) more than 50% of the contributions to Λ_d come from higher partial-wave states ($L \geq 1$) of the final two-neutron relative motion. Very recently, Ricci et al. (49) carried out a detailed SNPA calculation for μd capture based on a hard-pion chiral Lagrangian. By use of the Nijmegen I potential (70) and a particular model for evaluating the Δ excitation–exchange current, these authors arrived at $\Gamma_{\mu d} = 416\,\mathrm{s}^{-1}$. Thus the results obtained within SNPA show some scatter, and although Ricci et al. present rather detailed discussion of its possible origins, further studies appear to be needed.

An EFT* calculation of the μd-capture rate was done by Ando et al. (14), who used $\hat{d}^R$ determined from Γ_β^t (55). A potential problem for the application of χPT is the kinematical region where the momentum transfer to the final two nucleons is rather large, and the convergence property of chiral expansion may be poor. However, the contribution to Λ_d from this region is very small (14), and it is safe to use EFT* for the total μd-capture rate. By construction, $\mathcal{L}_{\chi\mathrm{PT}}$ is valid only well below Λ_χ, whereas the use of the phenomenological nuclear Hamiltonian H_A introduces high-momentum components above Λ_χ. To remedy this problem, Ando et al. (14) introduced a momentum cutoff Λ_{NN} that eliminates high-momentum components in the two-nucleon relative wave functions. If the effects of neglected higher-order terms and deviations from the framework of EFT* itself are sizable, they would be expected to cause significant Λ_{NN} dependence. Therefore the stability of the results against Λ_{NN} may be taken as a measure of model independence of an EFT* calculation. Ando et al.'s calculation (14) resulted in $\Lambda_d = 386\ \mathrm{s}^{-1}$, with a high degree of stability against variations of Λ_{NN}.

A calculation of Λ_d in pionless EFT was carried out by Chen et al. (15). This approach is most reliable in a kinematic region in the three-body final state where the energy transfer to the two-nucleon system is small compared to m_π, with the neutrino carrying most of the energy. Thus the precision of the $L_{1,\mathrm{A}}$ extraction could be significantly improved—if the capture rate leading to this limited kinematic region could be determined experimentally—by separately measuring the total capture rate and the small contribution due to the high-energy region, which can be identified experimentally by detecting high-energy neutrons.

2.6.3. Related astrophysical neutrino reactions. As mentioned, the two-body currents responsible for the GT transition for μd capture, *pp* fusion, and νd reactions are all controlled by the single common LEC $\hat{d}^R$ (or $L_{1,\mathrm{A}}$ in pionless EFT). This implies that the determination of $\hat{d}^R/L_{1,\mathrm{A}}$ through a precision measurement of Λ_d in the MuSun experiment has an important impact on calculations of the *pp*-fusion rate and the νd cross sections. Existing calculations on these processes are predictive only if constrained by Γ_β^t derived from tritium decay. Given the fundamental nature of these reactions, it is highly desirable to determine $\hat{d}^R$ from an $\mathrm{A} = 2$ observable. The two-nucleon system (*a*) can be calculated within the self-consistent framework of full nuclear χPT, (*b*) avoids the complexity of the three-nucleon system, and (*c*) allows a systematic evaluation of the uncertainties involved.

The low-energy $pp \rightarrow de^+\nu_e$ process is the primary solar nuclear reaction, and hence a precise estimate of its rate is a crucial input for the further elaboration of solar models. A benchmark calculation of this reaction based on SNPA was carried out by Schiavilla et al. (71). Park et al. (55) performed an EFT* calculation of *pp* fusion up to next-to-next-to-leading order (NNLO), and after fitting $\hat{d}^R$ to Γ_β^t, they arrived at the *pp*-fusion *S*-factor (at zero energy): $S_{pp}(0) = 3.94 \times (1 \pm 0.008) \times 10^{-25}$ MeV b, where the errors are estimated from the dependence on the cutoff parameter Λ_{NN}, the uncertainty in the experimental value of Γ_β^t, and the possible contributions of higher-order chiral terms. The computation of the *pp*-fusion rate in pionless EFT was carried out to second order (72, 73) and then to fifth order (61); the results are consistent with those obtained in EFT* if $L_{1,\mathrm{A}}$ is treated as a free parameter.

In connection with the SNO solar neutrino experiments (13), reliable estimates of the cross sections for νd reactions are extremely important. Calculations based on SNPA were carried out by several groups (47, 74–77), and an evaluation in EFT* has been reported (78). Again, Γ_β^t was used to fix the poorly known strength of the $N\Delta$ axial vector coupling in SNPA and to fix $\hat{d}^R$ in EFT*. As long as the constraint from Γ_β^t was respected, both approaches led to practically identical results. Meanwhile, Butler & Chen (60) and Butler et al. (61) applied pionless EFT to these reactions by using the value of $L_{1,A}$ determined by fitting to the *pp*-fusion GT matrix element calculated in SNPA or EFT*. The νd cross sections obtained by these authors (60, 61) are in perfect agreement with those found in SNPA (77) and in EFT* (78). Mosconi et al. (79) compared the νd reaction cross sections for various models that differed in the implementation of what these authors termed nuclear PCAC, and they concluded that the results obtained in SNPA and EFT* had substantial uncertainties of 2 to 3%. Note, however, that this scatter of the unconstrained calculations will be significantly reduced once the strength of the two-body current is constrained by an appropriate observable, such as Γ_β^t.

At the level of ~1% precision, it becomes important to examine the uncertainty in the *nn* scattering length, a_{nn}. Extraction of a_{nn} from the analysis of the $nd \to nnp$ reaction still shows a few inconsistencies (80–82), so that the standard value $a_{nn} = -18.63 \pm 0.27$ (exp.) $\pm\, 0.30$ (theor.) fm (83) is based on the $\pi^- d \to \gamma nn$ reaction. The theoretical uncertainty in the analysis of this reaction can be drastically reduced with the use of EFT, provided that $\hat{d}^R$ is known with sufficient accuracy (84, 85).

2.7. Radiative Corrections

It is customary to decompose RCs of order α into so-called outer and inner corrections (86–88). The outer correction is a universal function of the electron energy and is independent of the details of hadronic physics, whereas the inner correction reflects short-distance physics and the hadronic structure. The best available estimates of RCs have been given by Marciano & Sirlin (88) for neutron β decay and by Czarnecki et al. (44) for μ capture. In these works, the inner corrections arising from γ − (weak boson) loop diagrams are divided into high-momentum and low-momentum parts. The former is evaluated in the current quark picture, whereas the latter is estimated with the use of the phenomenological weak form factors of the nucleon. For μp capture, the electroweak RC in the semileptonic reaction and in the QED corrections to the muonic atom wave function amounts to a significant increase of the capture rate by $2.8 \pm 0.4\%$ (44).

HBχPT offers a model-independent framework in which to calculate RCs, provided that all the relevant LECs are known. A first HBχPT calculation of RCs for neutron β decay was carried out to order α by Ando et al. (89). In this approach, the short-distance physics contributions can be condensed into two LECs, one relevant to the Fermi constant G_F and the other to the axial coupling constant g_A. If these LECs can be determined from empirical data, we would have a totally model-independent way to evaluate RCs to neutron β decay. A similar situation should exist for any low-energy weak-interaction processes on a single nucleon.

As for RCs in multinucleon systems, all existing works (apart from studies related to the $0^+ \to 0^+$ transitions) concentrate on the single-nucleon RC diagrams because the genuinely two-body-type RCs are expected to be small. Kurylov et al. (90) carried out extensive studies of RCs in the A = 2 systems based on the Marciano & Sirlin approach (88), and their results have been widely used in analyzing the A = 2 weak-interaction observables (also see Reference 91). It is nevertheless desirable to have an explicit evaluation of RCs from two-body diagrams, and an EFT-based estimation of RCs should be extended to the A = 2 systems.

3. MUON ATOMIC PHYSICS AND INTERPRETATION OF EXPERIMENTS

Negative muons stopped in matter rapidly form muonic atoms. From these atomic states, muons can either decay via the process $\mu^- \rightarrow e\bar{\nu}_e \nu_\mu$ with a rate close to the free muon decay rate—$\lambda_\mu^+ = 1/\tau_\mu^+ \approx 0.455 \times 10^6$ s^{-1}—or be captured by the nucleus, following the basic reaction of Equation 1. The capture rate is proportional to the overlap of the muon atomic wave function with the nucleus, which after summing over the available protons Z, leads to a strong increase of the capture rate in light nuclei roughly proportional to Z^4. The muon atomic physics of hydrogen isotopes is unique, as μp and μd atoms are neutral systems with a small muonic Bohr radius of $a_\mu = \frac{\hbar^2}{m_\mu e^2} = 2.56 \times 10^{-3}$ Å. They easily penetrate the electron shell of neighboring atoms, where they experience three-body Coulomb interactions with other hydrogen nuclei (92). Charge-exchange scattering (93) leads to transitions between the muon atomic hyperfine states with spin $F_\pm = I \pm 1/2$, where I is the proton spin 1/2 or the deuteron spin 1, respectively. Hyperfine transitions decisively change the observed muon-capture rate because the helicity structure of weak interaction suppresses capture from the F_+ state and favors capture from the F_- state (see **Table 1**). Thus the capture rates Λ_s from the singlet μp state and Λ_d from the doublet μd state are the primary experimentally accessible observables. Muon capture in hydrogen is a rare process, as Λ_s and Λ_d amount only to 1.6×10^{-3} and 0.9×10^{-3} fractions of the free muon decay rate λ_μ^+. In collision with other hydrogen nuclei, μp and μd atoms can also form muonic hydrogen molecules, $(pp\mu)^+$ and $(dd\mu)^+$, respectively, which resemble the H_2^+ hydrogen ion of ordinary molecular physics. In muonic hydrogen molecules, the internuclear distance is reduced to the scale of order a_μ, so that muon-catalyzed nuclear fusion proceeds spontaneously by tunneling through the repulsive Coulomb barrier. If chemical impurities with atomic number Z are present, the muon is transferred to these nuclei at a high rate because of the significantly larger muonic binding energies on higher Z elements.

Thus the study of the basic muonic capture processes (Equations 1 and 3) is naturally intertwined with muon atomic and molecular phenomena in hydrogen isotopes. Fortunately, much has been learned about these processes during muon-catalyzed fusion studies (92). Nevertheless, muon atomic physics poses several important challenges to muon-capture experiments:

1. The composition of atomic/molecular states from which muon capture occurs has to be precisely known; otherwise, the experimental results cannot be interpreted in terms of fundamental weak couplings.
2. Muon capture from any material other than hydrogen has to be avoided during the measuring period because capture rates in wall materials exceed those in hydrogen by several orders of magnitude.
3. The target hydrogen has to be ultrapure to eliminate muon transfer to trace impurities.

3.1. Muon Capture and Atomic Physics in Hydrogen

Many of the collisional processes in hydrogen isotopes scale simply with the hydrogen density, so it is conventional to normalize those rates to the liquid hydrogen number density, $n_0 = 4.25 \times 10^{22}$ atoms cm^{-3}, and express the target density ϕ relative to n_0. At relevant hydrogen densities $\phi \geq 0.01$, negative muons slow down, are captured in highly excited atomic orbits, and then cascade down to the 1s ground states of the atom—the triplet and singlet hyperfine states $(\mu p)_t$ (F = 1) and $(\mu p)_s$ (F = 0)—at a much faster pace than the subsequent reaction steps (**Figure 2**). Because the hyperfine splitting energy of 0.18 eV exceeds the available thermal energy, the upper triplet state irreversibly depopulates to the energetically lower singlet state with a calculated rate of $\approx 2 \times 10^{10}\,\phi$ s^{-1} (94). Subsequently, $(\mu p)_s$ atoms can form $pp\mu$ molecules with a rate of $\phi\lambda_{pp}$.

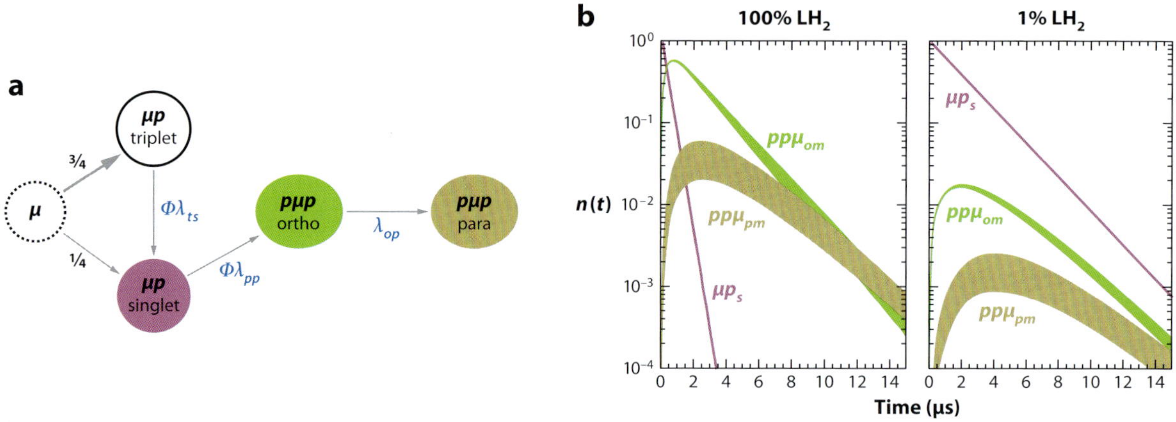

Figure 2

(*a*) Sequence of muon-induced atomic and molecular reactions in pure hydrogen. Muons can decay at the rate of λ_μ^+ or be captured with a capture rate characteristic of the atomic/molecular state they are occupying. (*b*) Populations $n(t)$ of μp singlet atoms and $pp\mu$ ortho- and paramolecules calculated according to the kinetics of panel *a*, using our best estimate for λ_{pp} and λ_{op}. The filled bands reflect the uncertainty in λ_{op}. At $\phi = 1$ the molecular states dominate, whereas at $\phi = 0.01$ the muon spends 95% of its life in the singlet μp state.

Formation of the excited $(pp\mu)_{om}$ orthostate with relative angular momentum $\mathrm{L} = 1$ between the two protons has a calculated rate of $1.8 \times 10^6\ \mathrm{s}^{-1}$, whereas the formation of the $\mathrm{L} = 0$ $pp\mu$ parastate is much slower ($7.5 \times 10^3\ \mathrm{s}^{-1}$) (95). The $pp\mu$ is stable against fusion, as the fusion rate via weak interactions is completely negligible. Because of the Fermi statistics of the two-proton system, in the $\mathrm{L} = 1$ state the two protons have a total spin of $I_{pp} = 1$, whereas in the $\mathrm{L} = 0$ state they have $I_{pp} = 0$. This spin statistics has two important consequences.

The OMC rates in the ortho- and paramolecule can be expressed as

$$\Lambda_{om} = 2\gamma_{om}\left(\frac{3}{4}\Lambda_s + \frac{1}{4}\Lambda_t\right), \quad \Lambda_{pm} = 2\gamma_{pm}\left(\frac{1}{4}\Lambda_s + \frac{3}{4}\Lambda_t\right). \qquad 11.$$

The molecular overlap factors are $2\gamma_{om} = 1.009 \pm 0.001$ and $2\gamma_{pm} = 1.143 \pm 0.001$; the relative contributions of Λ_s and Λ_t account for the population of singlet and triplet μp components in the total $pp\mu$ molecular spin function (96).

The transition rate λ_{op} between ortho- and parastates involves a proton spin flip, and thus enters only through the relativistic part of the molecular wave function. Once formed, $(pp\mu)_{om}$ is quickly embedded in electronic molecular clusters in collision with nearby H_2 molecules. The energy gained in the ortho-para transition is released by emission of an electron from these clusters, and $\lambda_{op} = 7.1 \pm 1.2 \times 10^4\ \mathrm{s}^{-1}$ is calculated on the basis of this mechanism (96). If isotopic or chemical impurities of concentration C_X are present in the hydrogen target, muons can transfer to deuterium at the rate of $\phi\, C_D 1.4 \times 10^{10}\ \mathrm{s}^{-1}$ (97) and to oxygen and nitrogen at typical rates of $\phi C_O\, 8.5 \times 10^{10}\ \mathrm{s}^{-1}$ (98) and $\phi C_N\ 3.4 \times 10^{10}\ \mathrm{s}^{-1}$ (99), respectively.

The experimental observables in μp capture are either the capture yield Y_n, obtained by integrating the neutron time distribution

$$\frac{dn_n}{dt} = \Lambda_s\, n_s(t) + \Lambda_{om} n_{om}(t) + \Lambda_{pm} n_{pm}(t), \qquad 12.$$

or the effective disappearance rate λ_μ^- of negative muons in hydrogen, derived from the decay electron time distribution

$$\frac{dn_e}{dt} = \lambda_\mu^+[n_s(t) + n_{om}(t) + n_{pm}(t)]. \qquad 13.$$

An expression analogous to Equation 12 can be written for the differential rate $d\Gamma/dk$ for the emission of a photon with energy k in the RMC process. Note that in RMC the triplet state has a much higher rate than the singlet state.

In the case of Equation 12, the experiment determines Y_n as the combination of capture rates from the $(\mu p)_s$, $(pp\mu)_{om}$, and $(pp\mu)_{pm}$ states. In the case of Equation 13, λ_μ^- is sensitive to the lifetimes in these three states, which are affected by their respective capture rates. As $\Lambda_s : \Lambda_{om} : \Lambda_{pm} \approx 1 : 3/4 : 1/4$, the populations $n(t)$ of the three relevant states must be accurately known to interpret a specific experimental condition. In the case of liquid targets $\phi = 1$, capture takes place predominantly from $pp\mu$ molecules, and both Y_n and λ_μ^- are highly sensitive to λ_{op} (see **Figure 2*b***). At low densities, capture from the singlet μp atom dominates; Y_n measures essentially Λ_s, and the negative muon disappearance rate equals the sum of the free muon decay rates and Λ_s, so that

$$\lambda_\mu^- \approx \lambda_\mu^+ + \Lambda_s. \qquad 14.$$

Experiments (100–102) used different techniques to measure λ_{pp} in liquid hydrogen or gas. The results are not completely consistent, and we have inflated the uncertainty by a factor $S = 1.76$, according to the Particle Data Group (PDG) prescription (see equations 1 and 2 of Reference 1), to determine our best experimental estimate: $\lambda_{pp} = 2.3 \pm 0.2 \times 10^6\,\mathrm{s}^{-1}$. By contrast, the only method applied to determine λ_{op} is the measurement of the neutron time distribution at $\phi = 1$, where for $\lambda_{pp}t \gg 1$, Equation 12 can be approximated as $dn_n/dt \propto \exp(-\lambda_\mu^+ t)[\Lambda_{pm} + (\Lambda_{om} - \Lambda_{pm})\exp(-\lambda_{op}t)]$. The first experiment (103) gave the result $\lambda_{op} = 4.1 \pm 1.4 \times 10^4\,\mathrm{s}^{-1}$, which has been used for evaluating capture experiments over the past 20 years, although it was always slightly inconsistent with the theoretical prediction (96). More recently, the experiment was repeated at TRIUMF (6) and yielded a very different value: $\lambda_{op} = 11.1 \pm 1.9 \times 10^4\,\mathrm{s}^{-1}$. Clearly, this is an unsatisfactory situation. Because there is no obvious reason to favor one result over the other, we again adopt the PDG convention to average the two results to obtain our best estimate, $\lambda_{op} = 6.6 \pm 3.4 \times 10^4\,\mathrm{s}^{-1}$, where the error is inflated by $S = 3$.

3.2. Review of Earlier Capture Experiments in Hydrogen

The process shown in Equation 1 was first observed in 1962 by Hildebrand (104), who used the bubble chamber technique (105, 106). The first counter experiment (107, 108) observed the time distribution of Equation 12 in highly purified liquid H_2 through the detection of neutrons with liquid scintillators; thus the so-called neutron method was established. Because the absolute efficiency of the neutron detectors has to be calculated by Monte Carlo methods, this neutron method is fundamentally limited to a few percent accuracy. Experiments (109, 110) used gas targets at low density to measure Λ_s in the μp singlet atom. The CERN experiment reported $\Lambda_s = 651 \pm 57\,\mathrm{s}^{-1}$, which is based on an innovative method that placed a system of proportional counters inside a hydrogen target at $\phi = 0.011$. A similar method at Dubna that made use of scintillators inside the hydrogen gas yielded $\Lambda_s = 686 \pm 88\,\mathrm{s}^{-1}$. Both techniques attempted to reduce muon stops and muonic atom diffusion onto the target walls, which become a very serious background with the limited hydrogen mass of gaseous compared to liquid targets. To overcome the uncertainties inherent in the neutron detector efficiency, an experiment at Saclay (111) derived the capture rate from the measured electron time distribution (Equation 13) in a liquid target. No knowledge of the detector efficiency is required in this lifetime method, but the difference between the decay rates λ_μ^- and λ_μ^+ has to be measured with high precision (cf. Equation 14). A capture rate of $435 \pm 17\,\mathrm{s}^{-1}$ was reported (see References 103 and 112 and the update in Reference 9). In 1998, the first measurement of the extremely rare RMC process was achieved at TRIUMF (7). It used a liquid target and a large acceptance pair spectrometer to observe 397 ± 20 photons above an energy threshold of 60 MeV. After background subtraction, 279 ± 26 signal events remained, and a branching ratio of $R_\gamma = 2.10 \pm 0.21 \times 10^{-8}$ was obtained for this energy cut.

All experimentally observed capture rates are weighted sums over the individual rates from the three states $(\mu p)_s$, $(pp\mu)_{pm}$, and $(pp\mu)_{om}$. Their relative proportions depend on experimental conditions, specifically the density ϕ and the measuring time interval $[t_1, t_2]$ after the muon stopped in the target. Experiments at low gas density permit a clear interpretation, as the muon resides mostly in $(\mu p)_s$. The most accurate OMC experiment (103) is very sensitive to the ortho-para transition because (*a*) it used a liquid target and (*b*) its effective start $t_1 \approx 2.5$ μs of the measuring period was relatively late due to the pulsed-beam structure, so that both $(pp\mu)_{om}$ and $(pp\mu)_{pm}$ significantly contributed.

To evaluate the overall consistency of the available results, we can either solve the kinetic equations according to **Figure 2** or use the sensitivities given in the original experimental publications to extract the most important physics parameter, the pseudoscalar coupling g_P. Such an analysis has been performed (103) and later updated (9). Our evaluation (**Figure 1**) differs from the previous one (9) in three ways: Our averaged $\lambda_{op} = 6.6 \pm 3.4 \times 10^4\,\mathrm{s}^{-1}$ includes the new λ_{op} data (6); the theoretical predictions on OMC have been updated with new RCs (44); and the recently improved muon lifetime gives $\tau_\mu = 2.197034(21) \times 10^{-6}$ s, which corresponds to $\lambda_\mu^+ = 0.4551591(43) \times 10^6\,\mathrm{s}^{-1}$ (1, 113, 114). Compared to the earlier results (9), g_P derived from the Saclay experiment acquires a significant additional uncertainty due to insufficient knowledge of the ortho-para proportions underlying that experiment. The RMC result has a weaker dependence on λ_{op}; the extracted g_P changes only modestly and remains above the theoretical prediction. A more detailed analysis performed in conjunction with the new MuCap result is presented in Section 5.

3.3. Muon Atomic Physics and Capture in Deuterium

Although the early stage of muon atomic processes in deuterium follows the description for hydrogen, several distinctive features appear after formation of the two hyperfine states, the quartet $(\mu d)_q$ ($F = 3/2$) and the doublet $(\mu d)_d$ ($F = 1/2$) (**Figure 3**). The hyperfine transition rate $\lambda_{qd} = 37 \times 10^6\,\mathrm{s}^{-1}$ (115) is much smaller than that in hydrogen, and at room temperature reverse upward

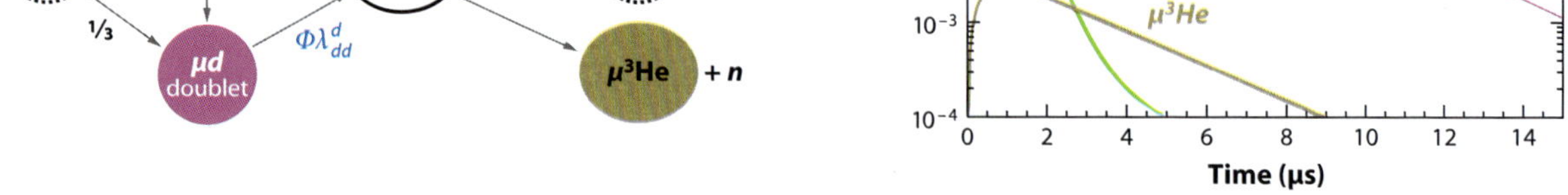

Figure 3

(*a*) Sequence of muon-induced atomic and molecular reactions in pure deuterium. (*b*) Populations $n(t)$ of the two μd hyperfine states and the $\mu\,^3$He atom calculated according to the kinetics for the conditions of the MuSun experiment. The density $\phi = 0.05$ leads to a sufficiently fast depopulation of the upper μd quartet states, so the muon spends 86% of its life in the μd doublet state. The widening of the lines reflects the small uncertainties introduced by assuming a conservative 3σ uncertainty in λ_{qd}.

transitions are possible because the μd hyperfine splitting is only 0.048 eV. As a consequence of the remarkable phenomenon of resonant formation of an extremely loosely bound $dd\mu$ state, the molecule formation rates from the two hyperfine states λ_{dd}^{q} and λ_{dd}^{d} exhibit a dramatically different temperature dependence (116). Finally, the deuterons in the $dd\mu$ undergo fusion in less than 1 ns into the two channels $dd \rightarrow {}^3$He (0.82 MeV) $+n$ (2.45 MeV) and $dd \rightarrow t$ (1.01 MeV) $+$ p (3.01 MeV). After fusion, the muon usually begins a new reaction chain; in only ~7% of the fusions it is captured in an atomic orbit of μ^3He and remains bound until it decays or undergoes nuclear muon capture on ^{3}He. The μd system has been intensively investigated (92), and the kinetic rates have been determined to percent-level precision (117).

The early μd-capture experiments were based on the above mentioned neutron method. Calibration of the neutron detector efficiency is difficult because a continuous spectrum of neutrons peaked around 1.5 MeV is emitted in its three-body final state. Moreover, 2.45-MeV neutrons from $dd\mu$ fusion create a significant background. The first generation of experiments tried to reduce this background by using hydrogen targets with a low deuterium concentration that function as effective deuterium targets. An experiment (118) with a liquid target measured $\Lambda_d = 369 \pm 96\,\mathrm{s}^{-1}$, and another experiment (119) at $\phi = 0.032$ reported $\Lambda_d = 445 \pm 60\,\mathrm{s}^{-1}$. At the low densities used in the second experiment, the μd spin population was essentially unknown at the time, and it was conjectured to be mostly doublet. The discovery of the strong temperature dependence of the $dd\mu$ formation made the μd hyperfine structure populations directly measurable in the fusion time distributions and indicated that the condition of the experiment (119) was closer to that expected for a statistical population (115, 117); therefore the quoted result should be corrected upward by as much as a factor of three, according to the statistical weight of $(\mu d)_d$. This contradictory situation stimulated a new generation of experiments based on pure deuterium targets. The lifetime method, which used a liquid deuterium target, was employed at Saclay (120) and avoided the background from fusion neutrons. The result was $\Lambda_d = 470 \pm 29\,\mathrm{s}^{-1}$. An experiment at PSI used the neutron method but reduced the gas density to $\phi = 0.04$ and the temperature to $\mathrm{T} = 34$ K to suppress fusion neutrons; the result was $\Lambda_d = 409 \pm 40\,\mathrm{s}^{-1}$ (121). In summary, the present experimental situation is insufficient to provide much guidance on the axial two-body current. The best experiments have uncertainties of 6.2 to 10% and are marginally consistent, and the most precise experiment (120) deviates by nearly 3σ from theory (14). If correct, such a discrepancy would have major ramifications for the related astrophysical processes discussed in Section 2.6.3.

4. NEW GENERATION OF MUON-CAPTURE EXPERIMENTS

4.1. Hydrogen Time-Projection Chambers

The central idea for the new muon-capture experiments is the use of active targets consisting of TPCs filled with ultrapure H_2, D_2, or ^{3}He at high pressures. This novel technique addresses the criteria listed in Section 3: (*a*) Low target densities, for which the interpretation of experiments is clear, can be realized; (*b*) the dangerous background from muons stopping or diffusing to the walls can be suppressed by reconstructing each incoming muon track in three dimensions in the TPC; and (*c*) the very demanding purity conditions in the TPC can be continuously verified in situ by monitoring nuclear recoils after muon stops, which are characteristic of muon capture on $Z > 1$ elements.

The muon-capture program evolved from a series of TPCs developed at PNPI for muon-catalyzed fusion studies and the μ^3He-capture experiment (11). These TPCs are gridded ionization chambers without gas gain. For experiments in which sufficient energy is deposited, compared to the 36.5-eV average energy required to produce an ion pair in hydrogen (122), the ionization

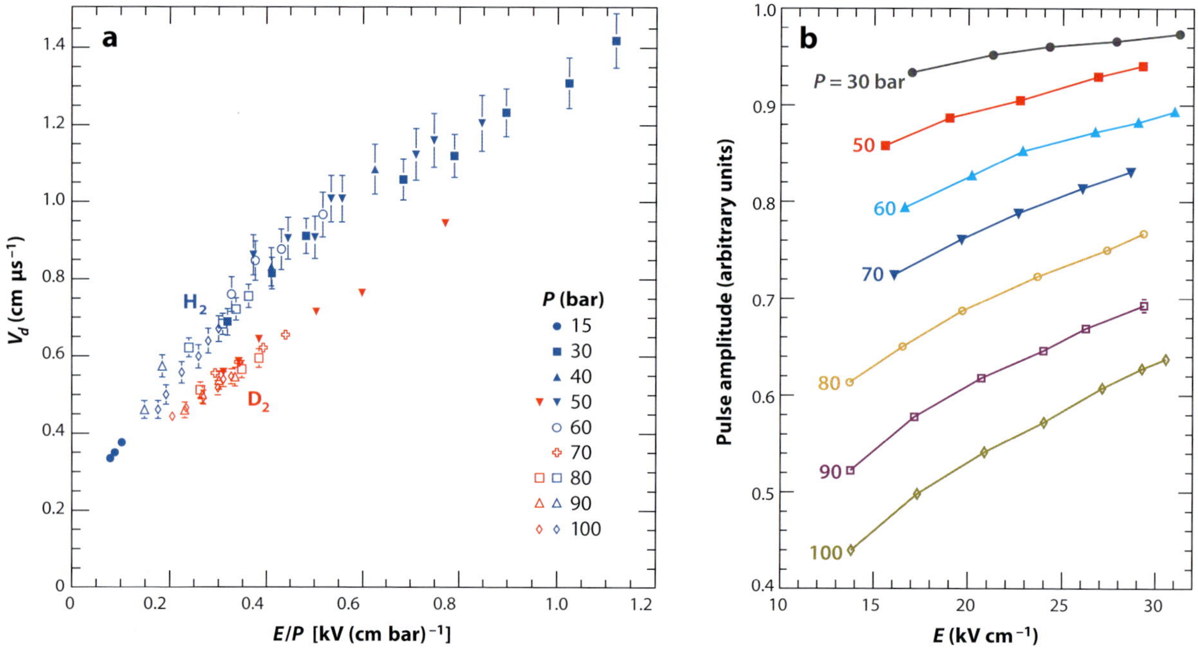

Figure 4

(*a*) Electron drift velocities V_d in protium and in deuterium versus reduced electric field E/P [kV (cm bar)$^{-1}$]. (*b*) Signal loss due to electron-ion recombination for 4.78-MeV α particles in D_2 as function of electric field E (kV cm^{-1}) for different pressures. Reproduced from Reference 117.

chambers can achieve good energy resolutions, as they are free of fluctuations from gas amplification. The chamber had a sensitive volume of 15 cm^3, and −40 kV voltage was applied on a cathode-grid distance of 14 mm and −3.5 kV on a grid-anode distance of 0.8 mm. The apparatus operated at hydrogen pressures of approximately 50 bar ($\phi = 0.05$) (117) and 120 bar of ^{3}He at room temperature (11). The energy resolution achieved was $\sigma = 30$ keV. As shown in **Figure 4**, hydrogen is a slow-drift gas, and high-drift fields are required to achieve a drift velocity of 0.5 cm µs^{-1} and to reduce signal loss due to initial recombination. The recombination loss depends on the primary ionization density; that is, it is much larger for α particles than for minimum ionizing particles.

The MuCap experiment needed hydrogen gas at $\phi = 0.01$ and 4-mm anode spacing. At these conditions, the wire chamber has to operate with gas amplification to clearly detect muons entering ($\Delta E/\text{anode} \approx 30$ keV) and stopping ($\Delta E/\text{anode} \approx 250$ keV) after 15 cm in the TPC volume. The worldwide experience with hydrogen proportional chambers is very limited; the basic parameters were established in a pioneering experiment (123). A main characteristic is the first Townsend coefficient, which can be expressed as $\alpha/p = A \exp(-bp/E)$, where $A = 3.83 \times 10^3$ (bar cm)$^{-1}$, $b = 104$ (kV cm^{-1} bar^{-1}), E is in kilovolts per centimeter, and p is in bar. The integral over α along the electron drift path towards the anode determines the average gas gain $G = \exp[\int_{r1}^{r2} \alpha(r)dr]$ (124, p. 37). At 10 bar ($\phi = 0.01$), high surface fields of $E_{sur} = 800$ kV cm^{-1} are required to obtain G up to 10^4. In an extensive research and development program (126, 127) at PNPI and PSI, the MuCap Collaboration built several prototypes to optimize the operating conditions and develop the technology for an ultraclean chamber for stable production data taking from 2004 to 2007. Gains G of 0.5 to 1×10^4 were obtained in the laboratory (126, 127), but because of long-term stability issues in the intense muon beam, a lower G of approximately 100 was chosen for the main data taking.

4.2. MuCap: Muon Capture on the Proton

The goal of the MuCap experiment (8) is a 1% measurement of the muon-capture rate Λ_s from the μp singlet state. Λ_s is determined with the lifetime method according to Equation 14, so that $\sigma(\Lambda_s)^2 = \sigma(\lambda_\mu^+)^2 + \sigma(\lambda_\mu^-)^2$. MuCap must determine the negative muon disappearance rate λ_μ^- in hydrogen to 15 ppm, while the $\tau_\mu^+ = 1/\lambda_\mu^+$ will be determined in the dedicated μ^+ lifetime experiments MuLan (113) and FAST (114) to better than 2 ppm. The low hydrogen density of $\phi = 0.011$ allows a direct measurement of Λ_s with only a small correction from $pp\mu$ capture (**Figure 2**).

The experiment was performed at the $\pi E3$ beam line of the Paul Scherrer Institute, with a high-intensity muon beam of 10^5 muons s^{-1} at 34 MeV/c. As shown in **Figure 5**, incident muons traverse beam counters (μSC and μPC) to enter the TPC, which has a sensitive volume of $15 \times 12 \times 28$ cm^3 and is placed in the hydrogen pressure vessel of 10 bar at ambient temperature. The ionization electrons produced by incoming muons travel downward with a drift velocity of 0.55 cm μs^{-1} in an applied field of 2 kV cm^{-1} toward a multiwire proportional chamber that contains perpendicular anode and cathode wires. The anode plane consists of 25-μm-diameter gold-plated tungsten wires with 4-mm spacing. The high voltage over the 3.5-mm-wide half-gap during the main data runs was 5.0 to 5.4 keV. The TPC was surrounded by two cylindrical wire chambers (ePC1 and ePC2) and by a hodoscope barrel (eSC) of plastic scintillators, which tracked the electrons to the muon decay vertex with 3π acceptance. Large-volume neutron detectors detected the 5.2-MeV capture neutrons as an auxiliary measurement of the capture process to verify the muon kinetics according to Equation 12. To cope with the high data rate, digital TPC data based on different energy thresholds were continuously recorded. For highly selected event topologies the full TPC was read out with 12-bit flash analog-to-digital converters (ADCs) (**Figure 6**).

The MuCap experiment derived the time between the muon entrance and the electron decay from fast scintillators (μSC and eSC), and additional key information was provided by the slower wire chambers. In the TPC, muon stops were completely reconstructed in three dimensions; thus the stability of the main experimental observable λ_μ^- could be systematically verified for

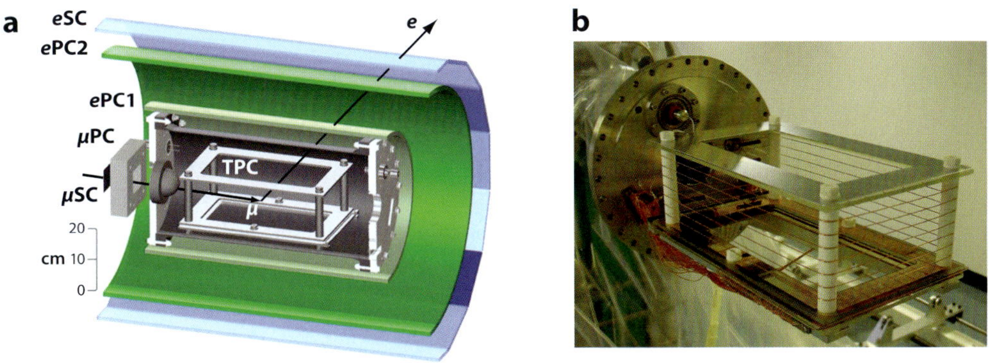

Figure 5

(*a*) Cutaway view of the MuCap setup (8). (*b*) Hydrogen time-projection chamber (TPC). Panel *a* reprinted from Reference 8 and panel *b* reprinted from Reference 125 with permission.

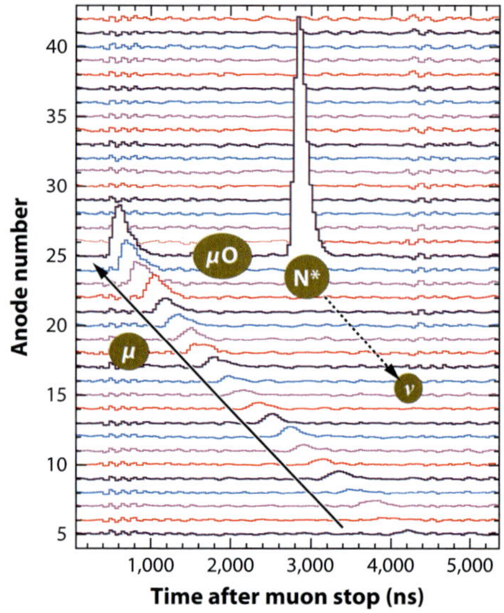

Figure 6

Example of a rare impurity-capture event in the time-projection chamber (TPC) occurring at the 10^{-5} level (129). An incoming muon stops at anode 25 and transfers to oxygen. After ≈ 2 μs, a nitrogen nucleus is observed on the stop anode, in the reaction $\mu^- + \mathrm{O} \rightarrow \mathrm{N}^* + \nu_\mu$. Muon decay or μp capture leaves only a clean muon track without additional signals in the TPC.

increasingly restrictive cuts on the stopping volume so as to suppress μp diffusion and identify rare large-angle μp Coulomb scattering, which are processes that might lead to muons reaching the walls. Systematic distortions due to muon combinatorics were avoided by requiring that only one muon be present during the measuring and TPC drift period of 24 μs. Although this condition traditionally had been enforced by rejecting muon pileup detected by the entrance counters, MuCap developed a muon-on-request scheme. It actively deflected the beam with the MuLan kicker (128) once a muon had entered the apparatus, thereby preventing significant pileup losses and increasing the data rate threefold. The muon and electron detectors operated independently to minimize instrumental correlations. Only the TPC had a small, but nonnegligible, probability of detecting delta electrons created by decay electrons. The muon-tracking algorithm was designed to be robust against associated time distortions.

The hydrogen gas had to be chemically pure at the few-parts-per-billion level to prevent muon transfer to impurities and the subsequent capture $\mu^- + Z \rightarrow (Z-1) + \nu_\mu$, which produces a yield Y_Z of nuclear recoils. The resulting correction to λ_μ^- is proportional to Y_Z. The final TPC was build from high-vacuum-compatible materials (125). The fully assembled detector structure could be heated up to 120°C; this process was part of a several-weeks-long cleaning/pumping procedure to reach sufficiently low outgasing levels. The TPC gas was continuously recycled through a cryopurification system (129), which achieved a purity better than 20 ppb during the main data runs. The TPC monitored the yield Y_Z with sub-parts-per-million precision (**Figure 6**). The calibration factors between Y_Z and the correction to λ_μ^- were determined through the use of dedicated fillings with known impurity admixtures, and λ_μ^- was corrected on the basis of the continuously monitored Y_Z. The hydrogen had to be isotopically pure to prevent muon transfer to deuterium. Due to a Ramsauer-Townsend minimum, the $\mu d + p$ cross sections are abnormally small, and $d\mu$ atoms can diffuse for distances up to a few centimeters (130), thereby potentially introducing time-dependent losses. MuCap developed an on-site isotopic cryogenic distillation column (131), which achieved record isotopic purity of $\leq$6 ppb, as verified by a high-sensitivity accelerator mass spectrometry method (132).

The MuCap data taking is complete; $N_\mu \approx 1.5 \times 10^{10}$ fully reconstructed μe events were recorded. An analysis of an intermediate data set of $N_\mu = 1.6 \times 10^9$ has been published (8). A muon-disappearance rate of $\lambda_\mu^- = 455\,851.4 \pm 12.5_{stat} \pm 8.5_{sys}$ s^{-1} was measured. After minor corrections for the reduced muon decay rate in the μp bound state and a small contribution from $pp\mu$ capture, the μ^+ decay rate λ_μ^+ (113) was subtracted according to Equation 14 to obtain a singlet-capture rate of $\Lambda_s = 725.0 \pm 13.7_{stat} \pm 10.7_{sys}$ s^{-1}. The analysis of the full data set is expected to improve the precision up to threefold.

4.3. MuSun: Muon Capture on the Deuteron

The goal of the MuSun experiment (12) is to measure the capture rate Λ_d from the μd doublet state to better than 1.5%. This rate can be calculated to high precision from model-independent EFT once the important but poorly constrained LEC termed $L_{1,\mathrm{A}}$ in pionless EFT and $\hat{d}^R$ in χPT is determined. The experiment is designed both to resolve the long-standing inconclusive experimental status of this process and to determine $L_{1,\mathrm{A}}$ and $\hat{d}^R$ in the two-nucleon system with greatly improved precision. The precision of the $L_{1,\mathrm{A}}$ determination can be estimated by error propagation from the expression $\Lambda_d = a + bL_{1,\mathrm{A}}$ (15). With the anticipated experimental error, but an exact theoretical prediction for Λ_d, $L_{1,\mathrm{A}}$ would be determined to better than 1.1 fm^3. A future precision calculation of Λ_d might achieve an uncertainty of 0.5% in the χPT expansion and a conservatively estimated additional error 0.7% induced by g_P and other small theoretical uncertainties. Although challenging, such a combined theoretical and experimental effort promises to calibrate $L_{1,\mathrm{A}}$ to approximately ± 1.25 fm^3, compared with the precision of approximately ± 6 fm^3 that is presently available from two-nucleon observables.

MuSun will use the lifetime technique and the TPC method to define clean muon stops in deuterium. However, the target conditions strikingly differ from those of MuCap, as a higher-density cryogenic TPC at $\phi = 0.05$ and $T = 30$ K is required to achieve conditions in which muon atomic processes lead to negligible uncertainties in the extraction of Λ_d. The populations for these selected conditions are shown in **Figure 3*b***. The μd doublet state dominates; the hyperfine transition rate of $\lambda_{qd} = 37.0 \pm 0.3 \times 10^6$ s^{-1} (117) is very well known, and the $dd\mu$ formation rate from this state is small (115, 117), so μ^3He production is slow. This analysis assumption can be verified in situ by simultaneously observing $dd\mu$ fusion events to monitor the μd hyperfine population.

The cryogenic TPC is currently under construction. It will be an ionization chamber with a Frisch grid; the sensitive pad area will be 12×10 cm^2, segmented into approximately 50 pads. Its height of 8 to 10 cm will determine the required drift voltage of up to 100 kV. This drift voltage will yield an electric field–to–pressure ratio of $E/P = 0.2$ kV (cm atm)$^{-1}$, which according to **Figure 4** will provide an acceptable recombination loss of 0.82 and a drift velocity of $V_d = 0.4$ cm μs^{-1}. Due to high gas density, $\phi = 0.05$, the design does not include gas amplification but rather relies on high resolution ($\sigma \approx 30$ keV) in the chamber to clearly identify muons and to detect them without systematic interference with the fusion signals (catalyzed with a probability of 5% per muon). To achieve such resolution, the cryogenic cold head will be vibrationally isolated from the chamber by an intermediate, liquid neon heat-transfer system, and the signals will be routed to custom-designed low-noise preamplifiers via low-capacity Kapton® flex cables. All signals on the TPC pads will be digitized with flash ADCs and continuously read out for a detailed analysis of complex event topologies generated by the overlap of muon and fusion tracks. Fusion events will be recorded both in the TPC and with fast neutron detectors. The decay electrons will be detected

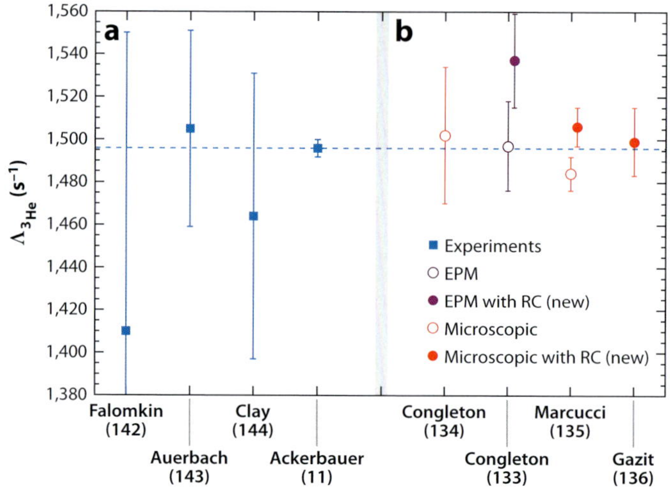

Figure 7

Results for the μ^3He-capture rate $\Lambda_{^3\mathrm{He}}$ from (*a*) experiments and (*b*) theory. The values of two original calculations (*open circles*) are revised with the recently calculated electroweak radiative corrections (RCs) to give the filled circles (according to Reference 44). Abbreviation: EPM, elementary particle method.

with the wire chamber/scintillator tracker developed for MuCap. Isotopic purity and muon atom diffusion will be less critical than in MuCap, but a high chemical purity at the 1-ppb level will be required. MuSun plans to achieve this level of purity by using only selected ultrahigh-vacuum materials in the TPC construction, using continuous gas circulation (129), and relying on the low partial pressure of impurities at the cryogenic environment.

4.4. Muon Capture on ^{3}He

Figure 7*a* presents the experimental data of the rate $\Lambda_{^3\mathrm{He}}$ for the process $\mu + {}^3\mathrm{He} \rightarrow t + \nu_\mu$, showing the impressive precision of $\Lambda_{^3\mathrm{He}} = 1496.0 \pm 4.0\,\mathrm{s}^{-1}$ achieved with the TPC technique (11), compared with earlier measurements (142–44). This experimental result can be compared with a number of elaborate calculations based on the elementary particle method (EPM) (133), as well as on SNPA and EFT* calculations (**Figure 7*b***) (133–136). The recent calculation of electroweak RCs has important consequences at this precision level and reduces the previous agreement between the data and the EPM results, as pointed out in Reference 44. The microscopic calculation (137) and a new EFT* calculation (136) agree with the experimental result. This agreement is robust against changes in the cutoff Λ_{NN} and against different choices of the NN potential.

5. RESULTS AND OUTLOOK

Figure 8 summarizes the current information on g_P based on the most accurate OMC result in liquid H_2 (103), the RMC measurement (7), and the new MuCap result on OMC in low-density gas (8). Because the muon molecular transition rate λ_{op} is required to extract g_P, each experiment defines an allowed region in the g_P versus λ_{op} plane. The problem in the results before MuCap is evident: If the small experimental value λ_{op}^{Ex1} (103) were adopted, then g_P from RMC would significantly exceed the precise theoretical prediction. By contrast, a large value λ_{op}^{Ex2} (6) could lower the RMC value of g_P close to the χPT value, but then g_P from OMC liquid H_2 would become exceedingly small. The MuCap result is a significant step toward the solution of this long-standing

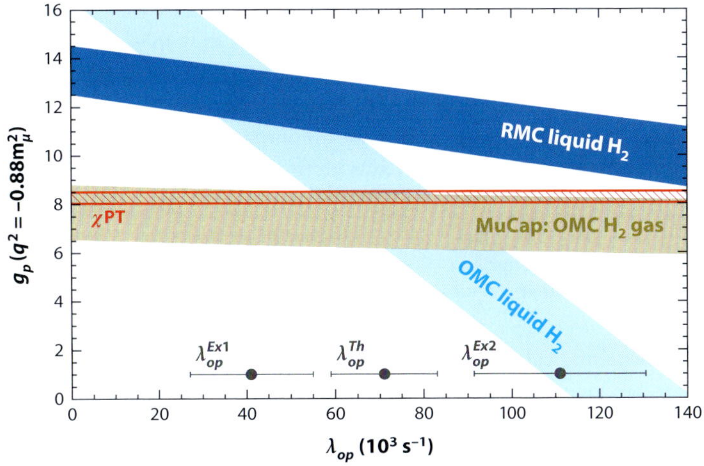

Figure 8

Experimental and theoretical determinations of g_P versus the ortho-para transition rate λ_{op} in the $pp\mu$ molecule. Previous experiments [ordinary muon capture (OMC) liquid H_2 (111); radiative muon capture (RMC) (7)] are inconclusive because their determination of g_P strongly depends on the poorly known ortho-para transition rate in muonic molecular hydrogen λ_{op}. The MuCap result for OMC gas (8) supports chiral perturbation theory (χPT) and is nearly independent of molecular effects. Adapted from Reference 8 with permission.

puzzle, as it largely avoids the molecular complications. Thus $g_P(q^2 = -0.88\, m_\mu^2) = 7.3 \pm 1.1$ can be extracted from the comparison of experiment with Standard Model theory, including electroweak RCs (44), where the error incorporates a conservative estimate of the contribution from muon atomic uncertainties. The result agrees with theory within one standard deviation and thus seems to exclude a dramatic breakdown of the χPT prediction for muon capture on the proton, as the RMC result was originally thought to suggest. Given the robustness of the theoretical prediction for g_P, such a discrepancy would be a serious problem for χPT and our basic understanding of chiral symmetry breaking. Our evaluation in **Figure 1**, which results in an OMC world average (excluding the early bubble chamber experiments) of $g_P = 10.1 \pm 2.3$ and $g_P = 7.8 \pm 1.0$, excluding and including the MuCap result, respectively, also indicates consistency with theory. Further supporting evidence comes from the recent EFT* analysis (136) of the high-precision experiment on μ^3He capture (11). Albeit derived from a more complex nuclear physics calculation, $g_P(q^2 = -0.954 m_\mu^2) = 8.13 \pm 0.6$ is determined with impressive precision and is in good agreement with theory. Final conclusions should await the forthcoming analysis of the full MuCap data set, which was obtained under cleaner conditions and with ten-times-higher statistics than the presently available result.

By design, the MuCap experiment is insensitive to λ_{op} and cannot resolve the basic ambiguity in the interpretation of experiments with liquid H_2 targets (**Figure 8**). Our evaluation used an averaged $\lambda_{op} = 6.6 \pm 3.4 \times 10^4\,s^{-1}$ with inflated errors to account for the inconsistent measurements of λ_{op}. With this assumption, we obtained $g_P = 11.6 \pm 1.24$ from RMC, higher than both the theoretical prediction and the MuCap result. Rather than use this averaging convention, it would be better to directly address the inconsistency shown in **Figure 8** experimentally by remeasuring RMC or OMC in liquid H_2 or significantly improving on the λ_{op} measurement. The latter possibility might naturally arise from the high-density TPC technique developed for MuSun if gas densities of $\phi = 0.1$–0.2 can be reached; in those densities, the capture neutron time spectrum becomes quite sensitive to λ_{op}.

If Λ_s can be determined at the level of 0.7%, other parameters that influence the extraction of g_P from Λ_s will require careful attention. An uncertainty of 0.46% to Λ_s is introduced by the uncertainty in the form factors in Equation 2, dominated by $F_A(0)$; another 0.4% error results from the estimated uncertainty in RCs. Also, the disagreement between the PDG average (1) and the recent measurement (138) of the neutron lifetime amounts to 0.7%, the same size as the expected uncertainty in Λ_s. Better consistency in neutron lifetime and asymmetry measurements is critical for semileptonic precision experiments and their potential in searches for new physics (46).

Muon capture on deuterium offers a rare opportunity to study the axial current in the $A = 2$ system, in particular its short-range two-body term, which is parameterized by an LEC in EFT. An accurate determination of this LEC is essential for reducing the theoretical uncertainty in the *pp*-fusion rate and the νd reaction cross sections. In the near future, muon capture will provide the best chance for an experimental calibration of these reactions from two-nucleon data, but significant work still needs to be done. On the experimental side, the new MuSun experiment (12) needs to achieve a very challenging precision of 1.5% in the capture rate Λ_d. Theory should provide even higher precision at the 0.5% level to link the relevant processes within the model-independent EFT framework. Thus far, EFT* and pionless EFT calculations have been performed, but a fully consistent χPT calculation including electroweak RCs will be required. This program would determine the LEC, which is currently known with 100% error from the two-nucleon system, to approximately 20%. The three-nucleon system provides high-quality experimental data on axial vector reactions, as the rates for both tritium β decay and muon capture on ^{3}He have been measured precisely. In recent years, EFT* calculations and even full EFT calculations have advanced to this level. The new precision data thus allow a comprehensive test of few-body methods against QCD-based EFT calculations on weak reactions in the two- and three-nucleon systems at the subpercent level. Verifying that QCD-based theory can calculate basic nuclear electroweak processes at this precision level will be a significant achievement and will be of practical relevance for the prediction of important astrophysical weak processes.

Finally, we note that hydrogen TPCs may find additional applications in nuclear and particle physics experiments because they afford active targets for nucleons and deuterium without any interfering background material. Hydrogen TPCs were initially developed for small-angle scattering (139) and photoproduction (123) and more recently for the study of exotic beam reactions (140; AT-TPC project at MSU). The muon-capture program has considerably expanded the experience of hydrogen TPCs in the density range of 0.01 to 0.10 relative to liquid hydrogen density, where the high-density chambers operate at cryogenic conditions to allow thin-walled detector configurations. Further efforts to obtain stable gas amplification with micropattern devices (141) could broaden the applications of TPCs. The technology of maintaining ultrahigh purity at the few-parts-per-billion level has been demonstrated, and the chambers can be scaled to differing experimental requirements.

DISCLOSURE STATEMENT

The authors are not aware of any affiliations, memberships, funding, or financial holdings that might be perceived as affecting the objectivity of this review.

ACKNOWLEDGMENTS

P.K. expresses his gratitude to all his collaborators in the experimental program at PSI. Without their dedicated efforts and ingenuity, the experimental progress reported in this review would not have been possible. K.K. is grateful to F. Myhrer and M. Oka for useful communications. The authors' work was supported by the U.S. National Science Foundation under grant number

PHY-0855569 (for P.K.) and grant number PHY-0758114 (for K.K.). P.K.'s current address is Department of Physics, University of Washington, Seattle, Washington 98195.

LITERATURE CITED

1. Amsler C, et al. *Phys. Lett.* B667:1 (2008)
2. Weinberg S. *Phys. Rev.* 112:1375 (1958); Grenacs L. *Annu. Rev. Nucl. Part. Sci.* 35:455 (1985); Severijns N, Beck M, Navaliat-Cuncic O. *Rev. Mod. Phys.* 78:991 (2006)
3. Bjorken JD, Drell SD. *Relativistic Quantum Mechanics*. New York: McGraw-Hill (1964)
4. Zel'dovich YB, Gershtein SS. *Sov. Phys. JETP* 2:576 (1957); Feynman RP, Gell-Mann M. *Phys. Rev.* 109:193 (1958)
5. Nambu Y. *Phys. Rev. Lett.* 4:380 (1960)
6. Clark JHD, et al. *Phys. Rev. Lett.* 96:073401 (2006)
7. Wright DH, et al. *Phys. Rev. C* 57:373 (1998)
8. Andreev VA, et al. *Phys. Rev. Lett.* 99:032002 (2007)
9. Gorringe T, Fearing HW. *Rev. Mod. Phys.* 76:31 (2004)
10. Bernard V, Elouadrhiri L, Meissner U-G. *J. Phys. G* 28:R1 (2002)
11. Ackerbauer P, et al. *Phys. Lett.* B417:224 (1998)
12. Andreev VA, et al. arXiv:1004.1754v1 [nucl-ex] (2010)
13. Aharmim B, et al. *Phys. Rev. Lett.* 101:111301 (2008)
14. Ando S, Park TS, Kubodera K, Myhrer F. *Phys. Lett.* B533:25 (2002)
15. Chen J-W, Inoue T, Ji X-D, Li Y-C. *Phys. Rev. C* 72:061001 (2005)
16. Measday DF. *Phys. Rep.* 354:243 (2001)
17. Georgi H. *Weak Interactions and Modern Particle Theory*. New York: Benjamin (1984)
18. Stoks V, Timmermans R, de Swart JJ. *Phys. Rev. C* 47:512 (1993)
19. Arndt RA, Stravsky II, Workman RL. *Phys. Rev. C* 52:2246 (1995)
20. de Swart JJ, Rentmeester MCM, Timmermans RGE. *Proc. Menu97* 13:96 (1997)
21. Adler SL, Dothan Y. *Phys. Rev.* 151:1267 (1966)
22. Bernard V, Kaiser N, Meissner U-G. *Int. J. Mod. Phys. E* 4:193 (1995)
23. Bernard V, Kaiser N, Meissner U-G. *Annu. Rev. Nucl. Part. Sci.* 57:33 (2007)
24. Bernard V, Kaiser N, Meissner U-G. *Phys. Rev. D* 50:6899 (1994)
25. Wolfenstein L. *High-Energy Physics and Nuclear Structure*. New York: Plenum (1970)
26. Kaiser N. *Phys. Rev. C* 67:027002 (2003)
27. Fearing HW, Lewis R, Mobed N, Scherer S. *Phys. Rev. D* 56:1783 (1997)
28. Ando S-I, Myhrer F, Kubodera K. *Phys. Rev. C* 63:015203 (2001)
29. Bernard V, Hemmert TR, Meissner U-G. *Nucl. Phys. A* 686:290 (2001)
30. Borie E. *Phys. Rev. A* 71:032508 (2005)
31. Liesenfeld A, et al. *Phys. Lett.* B468:20 (1999)
32. Sasaki S, et al. *Phys. Rev. D* 68:054509 (2003)
33. Tang A, Wilcox W, Lewis R. *Phys. Rev. D* 68:094503 (2003)
34. Boinepalli, et al. *Phys. Rev. D* 74:093005 (2006)
35. Edwards RG, et al. *Phys. Rev. Lett.* 96:052001 (2006)
36. Ali Khan A, et al. *Nucl. Phys. B (Proc. Suppl.)* 153:128 (2006)
37. Alexandrou C, et al. *Phys. Rev. D* 76:094511 (2007)
38. Sasaki S, Yamazaki T. *Phys. Rev. D* 78:014510 (2008)
39. Högler P, et al. *Phys. Rev. D* 77:094502 (2008)
40. Lin H-W, et al. *Phys. Rev. D* 78:014505 (2008)
41. Yamazaki T, et al. *Phys. Rev. D* 79:114505 (2009)
42. Kaplan DV. *Phys. Lett.* B288:342 (1992)
43. Furman V, Shamir Y. *Nucl. Phys. B* 439:54 (1995)
44. Czarnecki A, Marciano WJ, Sirlin A. *Phys. Rev. Lett.* 99:032003 (2007)
45. Deleted in proof

46. Govaerts J, Lucio-Martinez J-L. *Nucl. Phys. A* 678:110 (2000)
47. Tatara N, Kohyama Y, Kubodera K. *Phys. Rev. C* 42:1694 (1990)
48. Adam J, Truhlik E, Ciechanowicz S, Schmitt KM. *Nucl. Phys. A* 507:675 (1990)
49. Ricci P, Truhlik E, Mosconi B, Smejkal J. *Nucl. Phys. A* 837:110 (2010)
50. Meissner T, Myhrer F, Kubodera K. *Phys. Lett.* B416:36 (1998)
51. Ando S, Min D-P. *Phys. Lett.* B417:177 (1998)
52. Hemmert TR, Holstein BR, Kambor J. *J. Phys. G* 24:1831 (1998)
53. Carlson J, Schiavilla R. *Rev. Mod. Phys.* 70:743 (1998)
54. Weinberg S. *Phys. Lett.* B251:288 (1990); Weinberg S. *Nucl. Phys. B* 363:3 (1991); Weinberg S. *Phys. Lett.* B295:114 (1992)
55. Park TS, et al. *Phys. Rev. C* 67:055206 (2003)
56. Kubodera K, Park T-S. *Annu. Rev. Nucl. Part. Sci.* 54:19 (2004)
57. Kaplan DB, Savage MJ, Wise MB. *Nucl. Phys. B* 478:629 (1996)
58. Bedaque PF, Hammer H-W, van Kolck U. *Phys. Rev. Lett.* 82(3):463 (1999)
59. Chen J-W, Rupak G, Savage MJ. *Nucl. Phys. A* 653:386 (1999)
60. Butler M, Chen J-W. *Nucl. Phys. A* 675:575 (2000)
61. Butler M, Chen J-W, Kong X. *Phys. Rev. C* 63:035501 (2001)
62. Kaplan DB, Savage MJ, Wise MB. *Nucl. Phys. B* 534:329 (1998)
63. Beane SR, et al. In *At the Frontier of Particle Physics: Handbook of QCD*, vol. 1, ed. M Shifman, p. 133. Singapore: World Sci. (2001)
64. Butler M, Chen J-W, Vogel P. *Phys. Lett.* B549:26 (2002)
65. Chen J-W, Heeger KM, Robertson RGH. *Phys. Rev. C* 67:025801 (2003)
66. Epelbaum E. *Prog. Part. Nucl. Phys.* 57:654 (2006)
67. Epelbaum E, Hammer H-W, Meissner U-G. *Rev. Mod. Phys.* 81:1773 (2009)
68. Nogga A, Timmermans RGE, van Kolck U. *Phys. Rev. C* 72:054006 (2005)
69. Gazit D, Quaglioni S, Navratil P. *Phys. Rev. Lett.* 103:102502 (2009)
70. Stoks VGJ, et al. *Phys. Rev. C* 49:2950 (1994)
71. Schiavilla R, et al. *Phys. Rev. C* 58:1263 (1998)
72. Kong X, Ravndal F. *Phys. Rev. C* 64:044002 (2001)
73. Ando S, et al. *Phys. Lett.* B668:187 (2008)
74. Ying S, Haxton W, Henley EM. *Phys. Rev. D* 40:3211 (1989)
75. Kubodera K, Nozawa S. *Int. J. Mod. Phys. E* 3:101 (1994)
76. Nakamura S, Sato T, Gudkov V, Kubodera K. *Phys. Rev. C* 63:034617 (2001)
77. Nakamura S, et al. *Nucl. Phys. A* 707:561 (2002)
78. Ando S, et al. *Phys. Lett.* B555:49 (2003)
79. Mosconi B, Ricci P, Truhlik E, Vogel P. *Phys. Rev. C* 75:044610 (2007)
80. Huhn V, et al. *Phys. Rev. C* 63:014003 (2000)
81. van Witch W, Ruan X, Witala H. *Phys. Rev. C* 74:014001 (2006)
82. Gonzàlez Trotter DE, et al. *Phys. Rev. Lett.* 83:3788 (1999); Gonzàlez Trotter DE, et al. *Phys. Rev. C* 73:034001 (2006)
83. Chen Q, et al. *Phys. Rev. C* 77:054002 (2008)
84. Gardestig A, Phillips DR. *Phys. Rev. C* 73:014002 (2006); Gardestig A, Phillips DR. *Phys. Rev. C* 74:017001 (2006)
85. Gardestig A. *J. Phys. G* 36:053001 (2009)
86. Sirlin A. *Phys. Rev.* 164:1767 (1967)
87. Sirlin A. *Rev. Mod. Phys.* 50:573 (1978)
88. Marciano WJ, Sirlin A. *Phys. Rev. Lett.* 56:22 (1986)
89. Ando S, et al. *Phys. Lett.* B64:250 (2004)
90. Kurylov A, Ramsey-Musolf MJ, Vogel P. *Phys. Rev. C* 67:035502 (2003)
91. Fukugita M, Kubota T. *Phys. Rev. D* 72:071301 (2005)
92. Breunlich WH, Kammel P, Cohen JS, Leon M. *Annu. Rev. Nucl. Part. Sci.* 39:311 (1989)
93. Gershtein SS. *Sov. Phys. JETP* 7:318 (1958); Gershtein SS. *Sov. Phys. JETP* 9:927 (1959); Gershtein SS. *Sov. Phys. JETP* 13:488 (1961)

94. Cohen JS. *Phys. Rev. A* 43:4668 (1991)
95. Faifman MP. *Muon Catal. Fusion* 4:341 (1989)
96. Bakalov DD, Faifman MP, Ponomarev LI, Vinitsky SI. *Nucl. Phys. A* 384:302 (1982)
97. Cohen JS. *Phys. Rev. A* 43:3460 (1991); Adamczak A, et al. *At. Data Nucl. Tables* 62:255 (1996)
98. Werthmuller A, et al. *Hyperfine Interact.* 116:1 (1998)
99. Thalmann Y-A, et al. *Phys. Rev. A* 57:1713 (1998)
100. Bleser EJ, et al. *Phys. Rev.* 132:2679 (1963)
101. Conforto G, Rubbia C, Zavattini E, Focardi S. *Nuovo Cim.* 33:1001 (1964)
102. Bystricky VM, et al. *Sov. Phys. JETP* 43:606 (1976)
103. Bardin G, et al. *Phys. Lett.* B104:320 (1981)
104. Hildebrand RH. *Phys. Rev. Lett.* 8:34 (1962)
105. Hildebrand RH, Doede JH. *Proc. 1962 Int. Annu. Conf. High Energy Phys., Geneva*, p. 418 (1962)
106. Bertolini E, et al. *Proc. 1962 Int. Annu. Conf. High Energy Phys., Geneva*, p. 421 (1962)
107. Bleser E, et al. *Phys. Rev. Lett.* 8:288 (1962)
108. Rothberg JE, et al. *Phys. Rev.* 132:2664 (1963)
109. Quaranta AA, et al. *Phys. Rev.* 177:2118 (1969)
110. Bystritsky VM, et al. *Sov. Phys. JETP* 39:19 (1974)
111. Bardin G, et al. *Nucl. Phys. A* 352:365 (1981)
112. Martino J. In *Ettore Majorana International Science Series*, vol. 23: *Erice 1984: Fundamental Interactions in Low-energy Systems*, ed. G Fiorentini, P Dalpiaz, G Torelli, p. 43. New York: Plenum (1984)
113. Chitwood DB, et al. *Phys. Rev. Lett.* 99:032001 (2007)
114. Barczyk A, et al. *Phys. Lett.* B663:172 (2008)
115. Kammel P, et al. *Phys. Lett.* B112:319 (1982); Kammel P, et al. *Phys. Rev. A* 28:2611 (1983)
116. Zmeskal J, et al. *Phys. Rev. A* 42(3):1165 (1990)
117. Balin DV, et al. *Phys. Elem. Part. At. Nuclei* 42:In press (2011)
118. Wang I-T, et al. *Phys. Rev.* 139:0B1539 (1965)
119. Bertin A, Vitale A, Placci A, Zavattini E. *Phys. Rev. D* 8:3774 (1973)
120. Bardin G, et al. *Nucl. Phys. A* 453:591 (1986)
121. Cargnelli M, et al. *Proc. Yamada Conf. Nucl. Weak Process Nucl. Struct., 23rd*, p. 115 (1989)
122. Smirnov IB. *Nucl. Instrum. Methods A* 554:474 (2005)
123. Chapin TJ, et al. *Nucl. Instrum. Methods* 197:305 (1982); Chapin TJ, et al. *Nucl. Instrum. Methods* 225:550 (1984)
124. Sauli F. *Rep. CERN-77-09* (1977)
125. Petitjean C. (MuCap Collab.) *Nucl. Instrum. Methods A* 600:338 (2009)
126. Maev EM, et al. *Nucl. Instrum. Methods A* 478:158 (2002)
127. Maev EM, et al. *Nucl. Instrum. Methods A* 515:288 (2003)
128. Barnes MJ, Wait GD. *IEEE Trans. Plasma Sci.* 32:1932 (2004)
129. Ganzha VA, et al. *Nucl. Instrum. Methods A* 578:485 (2007)
130. Adamczak A. *Eur. Phys. J. D* 41:493 (2007)
131. Alekseev I, et al. (Deuterium Remov. Unit MuCap Exper.) Presented at NHA Annu. Hydrog. Conf., Sacramento. **http://nha.confex.com/nha/2008/techprogram/MEETING.HTM** (2008)
132. Dobeli M, et al. *Nucl. Instrum. Methods B* 266:1820 (2008)
133. Congelton JG, Fearing HW. *Nucl. Phys. A* 552:534 (1993)
134. Congleton JG, Truhlik E. *Phys. Rev. C* 53:956 (1996)
135. Marcucci LE, et al. *Phys. Rev. C* 66:054003 (2002)
136. Gazit D. *Nucl. Phys. A* 827:408c (2009)
137. Marcucci LE, et al. *Phys. Rev. C* 63:015801 (2001)
138. Serebrov AP, et al. *Phys. Rev. C* 78:035505 (2008)
139. Vorobyov AA, et al. *Nucl. Instrum. Methods* 119:509 (1974)
140. Demonchy CE, et al. *Nucl. Instrum. Methods A* 583:341 (2007)
141. Sauli F, Sharma A. *Annu. Rev. Nucl. Part. Sci.* 49:341 (1999)
142. Falomkin IV, et al. *Phys. Lett.* 3:229 (1963)
143. Auerbach LB, et al. *Phys. Rev.* 138:B127 (1965)
144. Clay DR, Keuffel JW, Wagner RL, Edelstein RM. *Phys. Rev.* 140:B586 (1965)

Flavor Physics Constraints for Physics Beyond the Standard Model

Gino Isidori,[1,2] Yosef Nir,[3] and Gilad Perez[3]

[1]INFN, Laboratori Nazionali di Frascati, 00044 Frascati, Italy; email: gino.isidori@lnf.infn.it

[2]Institute for Advanced Study, Technische Universität München, 80333 München, Germany

[3]Department of Particle Physics and Astrophysics, Weizmann Institute of Science, Rehovot 76100, Israel; email: yosef.nir@weizmann.ac.il, gilad.perez@weizmann.ac.il

Annu. Rev. Nucl. Part. Sci. 2010. 60:355–80

First published online as a Review in Advance on July 16, 2010

The *Annual Review of Nuclear and Particle Science* is online at nucl.annualreviews.org

This article's doi:
10.1146/annurev.nucl.012809.104534

0163-8998/10/1123-0355$20.00

Abstract

Over the past decade, much progress in experimentally measuring and theoretically understanding flavor physics has been achieved. Specifically, the accuracy of the determination of the CKM elements has been greatly improved, and a large number of (*a*) flavor-changing neutral-current processes involving $b \to d$, $b \to s$, and $c \to u$ transitions and (*b*) *CP*-violating asymmetries have been measured. No evidence for new physics has been established. Consequently, strong constraints on new physics at a high scale apply. In particular, the flavor structure of new physics at the teraelectronvolt scale is strongly constrained. We review these constraints and discuss future prospects to better understand the flavor structure of physics beyond the Standard Model.

Contents

1. INTRODUCTION

The Standard Model (SM) of particle physics provides a mathematical description of three fundamental forces—namely the strong, weak, and electromagnetic forces—that act among the elementary spin-half particles: quarks and leptons. The SM is based on symmetry principles. Specifically, the symmetry of the interactions is described by a gauge group, $SU(3)_C \times SU(2)_L \times U(1)_Y$. However, the ground state of the universe respects only the $SU(3)_C \times U(1)_{EM}$ subgroup. This structure of symmetries implies that the various interactions are mediated by spin-one force carriers: eight massless gluons for the strong interaction, two charged massive W bosons and a single neutral massive Z boson for the weak interaction, and a massless photon for the electromagnetic interaction.

For the gauge symmetry principle to be implemented, the quarks and the leptons must reside in representations of the groups. For example, with regard to the $U(1)_{EM}$ group, the representation means simply that each particle carries a well-defined electromagnetic charge. With regard to the $SU(3)_C$ group, quarks reside in triplets (3), whereas leptons reside in singlets (2). To allow for the spontaneous breaking of $SU(2)_L \times U(1)_Y$ to $U(1)_{EM}$, the SM includes a spin-zero particle: the Higgs boson. Quarks and leptons can interact via Higgs-mediated interactions that, unlike gauge interactions, are not ruled by symmetry principles. It is via these interactions that quarks and leptons acquire their different masses.

Clearly, the SM does not fully describe nature. It must be extended in order to account for the observed neutrino masses, dark matter, and the baryon asymmetry of the universe. The term new physics (NP) is used for extensions of the SM that correct these and other deficiencies of the SM. NP models may include additional elementary particles, additional forces, and larger symmetry groups. However, the success of the SM in describing the strong, weak, and electromagnetic interactions among the quarks and the leptons requires that NP models include the SM as a low-energy effective theory.

In particle physics, the term flavors is used to describe several copies of the same gauge representation, namely several fields that are assigned the same quantum charges. Within the unbroken $SU(3)_C \times U(1)_{EM}$ gauge group of the SM, there are four different types of particles (each of which comes in three flavors):

1. Up-type quarks in the $(3)_{+2/3}$ representation: u, c, and t.
2. Down-type quarks in the $(3)_{-1/3}$ representation: d, s, and b.
3. Charged leptons in the $(1)_{-1}$ representation: e, μ, and τ.
4. Neutrinos in the $(1)_0$ representation: ν_1, ν_2, and ν_3.

The term flavor physics refers to interactions that distinguish among flavors. By definition, gauge interactions—that is, interactions that are related to unbroken symmetries and are therefore mediated by massless gauge bosons—do not distinguish among flavors and do not constitute part of flavor physics. Within the SM, flavor physics refers to the weak and Yukawa interactions. With NP, there are likely to be additional flavored interactions.

The term flavor changing refers to processes in which the initial and final flavor numbers (that is, the number of particles of a certain flavor minus the number of antiparticles of the same flavor) are different. In flavor-changing charged-current processes, both up-type and down-type flavors and/or both charged-lepton and neutrino flavors are involved. Examples include (*a*) muon decay via $\mu \to e\bar{\nu}_i\nu_j$ and (*b*) $K^- \to \mu^-\bar{\nu}_j$, which corresponds, at the quark level, to $s\bar{u} \to \mu^-\bar{\nu}_j$. Within the SM, these processes are mediated by the W bosons and occur at tree level. (The term tree level refers to processes that can proceed via the exchange of a single W boson and in which, consequently, the amplitude is quadratic in the weak coupling.) In flavor-changing neutral-current (FCNC) processes, either up-type or down-type flavors, but not both, and/or either charged-lepton or neutrino flavors, but not both, are involved. Examples include (*a*) muon decay via $\mu \to e\gamma$ and (*b*) $K_L \to \mu^+\mu^-$, which corresponds, at the quark level, to $s\bar{d} \to \mu^+\mu^-$. Within the SM, these processes do not occur at tree level (and therefore their amplitude is at least quartic in couplings), and they are often highly suppressed. This situation makes FCNC particularly sensitive to NP: If NP does not have the same flavor-suppression factors as in the SM, then it could contribute to FCNC comparably to the SM even if it takes places at energy scales that are orders of magnitude higher than the weak scale. The fact that flavor physics is a very sensitive probe of high-energy physics is the main reason for experimental efforts to measure flavor parameters and theoretical efforts to interpret these data.

We emphasize that the suppression of FCNC within the SM is not a generic consequence of its basic symmetry structure but rather is a result of its particle content and of the peculiar numerical values of some of its parameters. It is therefore quite possible that the flavor features of NP are different and would lead to significant enhancements of various FCNC processes. The fact that no such enhancements have been observed is difficult to explain within well-motivated extensions of the SM with particles that are heavier than the weak-force carriers by one or two orders of magnitude. This situation is known as the NP flavor puzzle. A more quantitative and detailed discussion of this puzzle is given in Section 3. Specific examples of NP models in which strong enhancements of FCNC could occur include supersymmetry (Section 5) and models of extra dimensions (Section 6).

CP violation provides another reason for the interest in flavor physics. In the context of cosmology, *CP* violation is a necessary condition for baryogenesis, that is, the dynamical generation of the baryon asymmetry of the universe. However, within the SM, *CP* is violated only in flavor-changing interactions, and quantitatively, it is much too tiny to explain the observed baryon asymmetry. New sources of *CP* violation are therefore necessary, and they might be discovered through their effects on flavor-changing processes.

In this review, we focus on flavor physics in the quark sector. Flavor physics in the lepton sector, however, provides an equally powerful probe of new physics. Without new particles beyond the SM, the rate of FCNC processes in the charged-lepton sector, such as $\mu \to e\gamma$, is suppressed by the tiny neutrino masses to levels that are many orders of magnitude below present and foreseeable experimental sensitivity. Thus, any near-future observation of such a process will constitute an unambiguous signal of NP. We limit our discussion to the quark sector because this is the sector in which there has been much progress in recent years, attained mainly via measurements at the B factories.

2. THE STANDARD MODEL

The SM is defined as follows: The gauge symmetry is

$$G_{\rm SM} = SU(3)_{\rm C} \times SU(2)_{\rm L} \times U(1)_{\rm Y}. \qquad 1.$$

It is spontaneously broken by the vacuum expectation value (VEV) of a single Higgs scalar, $\phi(1,2)_{1/2}$ ($\langle\phi^0\rangle = v/\sqrt{2}$):

$$G_{\rm SM} \to SU(3)_{\rm C} \times U(1)_{\rm EM}. \qquad 2.$$

There are three fermion generations, each of which consists of five representations of $G_{\rm SM}$:

$$Q_{Li}(3,2)_{+1/6}, \quad U_{Ri}(3,1)_{+2/3}, \quad D_{Ri}(3,1)_{-1/3}, \quad L_{Li}(1,2)_{-1/2}, \quad \text{and} \quad E_{Ri}(1,1)_{-1}. \qquad 3.$$

The SM Lagrangian $\mathcal{L}_{\rm SM}$ is the most general renormalizable Lagrangian that is consistent with the gauge symmetry (Equation 1), the particle content (Equation 3), and the pattern of spontaneous symmetry breaking (Equation 2). It can be divided into three parts:

$$\mathcal{L}_{\rm SM} = \mathcal{L}_{\rm kinetic+gauge} + \mathcal{L}_{\rm Higgs} + \mathcal{L}_{\rm Yukawa}. \qquad 4.$$

The source of all flavor physics is in the Yukawa interactions,

$$-\mathcal{L}_{\rm Yukawa} = Y^d_{ij}\bar{Q}_{Li}\phi D_{Rj} + Y^u_{ij}\bar{Q}_{Li}\tilde{\phi} U_{Rj} + Y^e_{ij}\bar{L}_{Li}\phi E_{Rj} + \text{h.c.}, \qquad 5.$$

where $\tilde{\phi} = i\tau_2\phi^\dagger$. This part of the Lagrangian is generally flavor dependent (that is, $Y^f \not\propto 1$) and CP violating.

In the absence of the Yukawa matrices Y^d, Y^u, and Y^e, the SM has a large $U(3)^5$ global symmetry:

$$G_{\rm global}(Y^{u,d,e} = 0) = SU(3)^3_q \times SU(3)^2_\ell \times U(1)^5. \qquad 6.$$

The non-Abelian part of this symmetry is particularly relevant to flavor physics:

$$SU(3)^3_q = SU(3)_Q \times SU(3)_U \times SU(3)_D,$$
$$SU(3)^2_\ell = SU(3)_L \times SU(3)_E. \qquad 7.$$

What is important for our purposes is that $\mathcal{L}_{\rm kinetic+gauge} + \mathcal{L}_{\rm Higgs}$ respect the non-Abelian flavor symmetry $S(3)^3_q \times SU(3)^2_\ell$, under which

$$Q_L \to V_Q Q_L, \quad U_R \to V_U U_R, \quad D_R \to V_D D_R, \quad L_L \to V_L L_L, \quad \text{and} \quad E_R \to V_E E_R, \qquad 8.$$

where the V_i are unitary matrices. The Yukawa interactions (Equation 5) break the global symmetry

$$G_{\rm global}(Y^{u,d,e} \neq 0) = U(1)_B \times U(1)_e \times U(1)_\mu \times U(1)_\tau. \qquad 9.$$

[The gauged $U(1)_Y$ also remains a good symmetry.] Thus, the transformations of Equation 8 are not a symmetry of $\mathcal{L}_{\rm SM}$. Instead, they correspond to a change of the interaction basis. These observations also offer an alternative way of defining flavor physics: The term here refers to interactions that break the $SU(3)^5$ symmetry (Equation 8). Thus, the term flavor violation is often used to describe processes or parameters that break the symmetry.

Using Equation 8, one can choose an interaction basis in which the number of parameters is minimized. A useful example of such a basis for the quark Yukawa matrices is

$$Y^d = \lambda_d, \quad Y^u = V^\dagger \lambda_u, \tag{10}$$

where $\lambda_{d,u}$ are diagonal,

$$\lambda_d = \mathrm{diag}(y_d, y_s, y_b), \quad \lambda_u = \mathrm{diag}(y_u, y_c, y_t), \tag{11}$$

and where V is a unitary matrix that depends on three real angles and one complex phase. There are ten quark flavor parameters: nine real ones and a single phase. In the mass basis, six of the real parameters are the six quark masses, whereas the remaining three real parameters and one imaginary parameter appear in the Cabibbo-Kobayashi-Maskawa (CKM) matrix (2, 3) V, which describes the couplings of the charged weak-force carriers, the $W^\pm$ bosons, with quark-antiquark pairs.

Within the SM, all flavor-changing physics comes from the V matrix. There are various ways to choose the four parameters of V. One of the most convenient ways is given by the Wolfenstein parameterization, where the four mixing parameters are λ, A, ρ, and η; $\lambda = |V_{us}| = 0.23$ plays the role of an expansion parameter; and η represents the CP-violating phase (1, 4):

$$V \simeq \begin{pmatrix} 1-\frac{1}{2}\lambda^2-\frac{1}{8}\lambda^4 & \lambda & A\lambda^3(\rho-i\eta) \\ -\lambda+\frac{1}{2}A^2\lambda^5[1-2(\rho+i\eta)] & 1-\frac{1}{2}\lambda^2-\frac{1}{8}\lambda^4(1+4A^2) & A\lambda^2 \\ A\lambda^3\left[1-\left(1-\frac{1}{2}\lambda^2\right)(\rho+i\eta)\right] & -A\lambda^2+\frac{1}{2}A\lambda^4[1-2(\rho+i\eta)] & 1-\frac{1}{2}A^2\lambda^4 \end{pmatrix}. \tag{12}$$

One can assume that flavor-changing processes are fully described by the SM and can check the consistency of the various measurements with this assumption. The values of λ and A are known accurately (5) from, respectively, $K \to \pi \ell \nu$ and $b \to c \ell \nu$ decays:

$$\lambda = 0.2257 \pm 0.0010, \quad A = 0.814 \pm 0.022. \tag{13}$$

Then, one can express all the relevant observables as a function of the two remaining parameters, ρ and η, and check whether there is a range in the $\rho\eta$ plane that is consistent with all measurements. The list of observables includes the following:

1. The rates of inclusive and exclusive charmless semileptonic B decays depend on $|V_{ub}|^2 \propto \rho^2 + \eta^2$.
2. The CP asymmetry in $B \to \psi K_S$: $S_{B\to\psi K} = \sin 2\beta = \frac{2\eta(1-\rho)}{(1-\rho)^2+\eta^2}$.
3. The rates of various $B \to DK$ decays depend on the phase γ, where $e^{i\gamma} = \frac{\rho+i\eta}{\sqrt{\rho^2+\eta^2}}$.
4. The rates of various $B \to \pi\pi, \rho\pi, \rho\rho$ decays depend on the phase $\alpha = \pi - \beta - \gamma$.
5. The ratio between the mass splittings in the neutral B and B_s systems is sensitive to $|V_{td}/V_{ts}|^2 = \lambda^2[(1-\rho)^2 + \eta^2]$.
6. The CP violation in $K \to \pi\pi$ decays, ε_K, depends in a complicated way on ρ and η.

The resulting constraints are shown in **Figure 1**.

The consistency of the various constraints is impressive. Specifically, the following ranges for ρ and η can account for all the measurements (5):

$$\rho = +0.135^{+0.031}_{-0.016}, \quad \eta = +0.349 \pm 0.017. \tag{14}$$

Therefore, it is very likely that flavor violation and CP violation in flavor-changing processes are dominated by the CKM mechanism (7–9). One can go a step further and allow for arbitrary NP in all flavor-changing processes except for those that have contributions from SM tree diagrams.

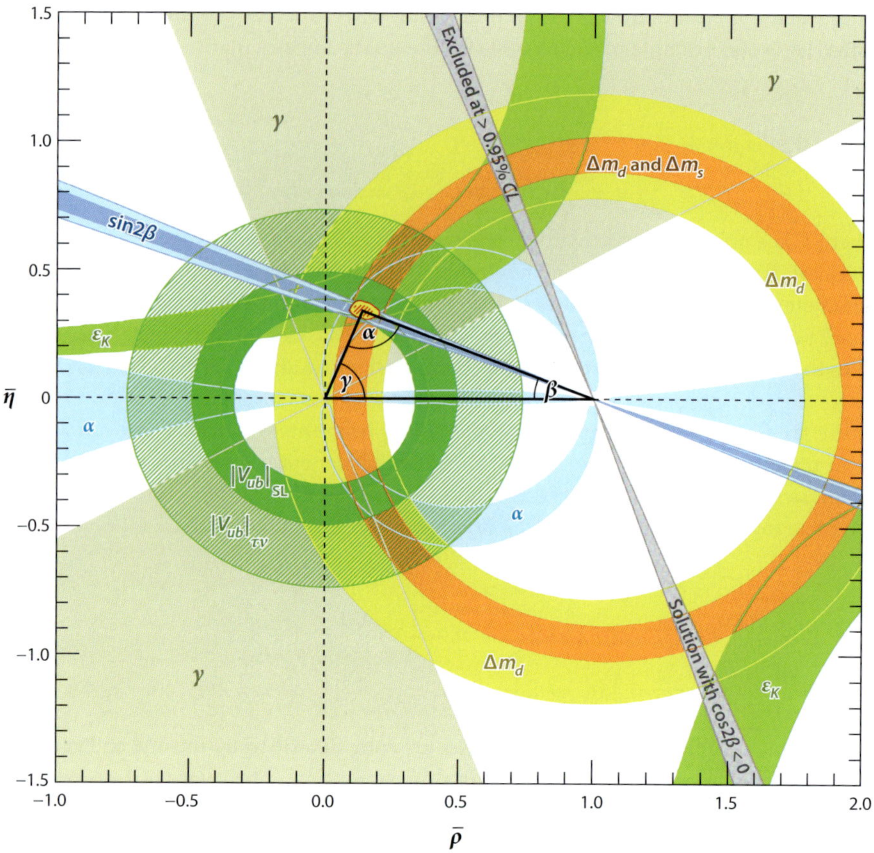

Figure 1

Allowed region in the $\rho\eta$ plane. Superimposed are the individual constraints from charmless semileptonic B decays ($|V_{ub}/V_{cb}|$); mass differences in the B^0 (Δm_d) and B_s (Δm_s) neutral meson systems; and CP violation in $K \to \pi\pi(\varepsilon_K)$, $B \to \psi K$ (sin2β), $B \to \pi\pi$, $\rho\pi$, $\rho\rho$ (α), and $B \to DK$ (γ) decays. Reprinted from Reference 6.

Then, one can quantitatively constrain the size of NP contributions to processes such as neutral meson mixing. We do so in the next section.

3. MODEL INDEPENDENT CONSTRAINTS

To describe NP effects in flavor physics, we can follow one of two main strategies: (*a*) We can build an explicit NP model, specifying the new matter fields and symmetries beyond the SM ones, or (*b*) we can analyze the NP effects by using a generic effective-theory approach or by integrating out the new heavy fields. The first approach is more predictive, but also more model dependent. We follow this approach in Sections 5 and 6 in two well-motivated SM extensions. Here and in the next section we follow the second strategy, which is less predictive but also more general.

Assuming the new degrees to be heavier than SM fields, we can integrate them out and describe NP effects by means of a generalization of the Fermi theory. The SM Lagrangian becomes the renormalizable part of a more general local Lagrangian, which includes an infinite tower of operators with dimension $d > 4$, constructed in terms of SM fields, suppressed by inverse powers of an effective scale $\Lambda > M_W$:

$$\mathcal{L}_{\rm eff} = \mathcal{L}_{\rm SM} + \sum \frac{c_i^{(d)}}{\Lambda^{(d-4)}} O_i^{(d)}(\text{SM fields}). \tag{15}$$

This general bottom-up approach allows us to analyze all realistic extensions of the SM in terms of a limited number of parameters (the coefficients of the higher-dimensional operators). The drawback of this method is the impossibility of establishing correlations of NP effects at low and high energies: The scale Λ defines the cutoff of the effective theory. However, correlations among different low-energy processes can still be established by implementing specific symmetry properties, such as the minimal flavor-violation (MFV) hypothesis (Section 4). Experimental tests of such correlations allow us to test and establish general features of the new theory, which hold independently of the dynamical details of the model. Specifically, B, D, and K decays are extremely useful in determining the flavor symmetry–breaking pattern of the NP model.

3.1. Bounds from $\Delta F = 2$ Down-Type Transitions

The starting point for this analysis is the observation that in several realistic NP models we can neglect nonstandard effects in all cases in which the corresponding effective operator is generated at tree level within the SM. This general assumption implies that the experimental determination of the CKM matrix via tree-level processes is free from the contamination of NP contributions. Using this determination, we can unambiguously predict meson-antimeson mixing and FCNC amplitudes within the SM and compare them with data, thereby constraining the couplings of the $\Delta F = 2$ operators in Equation 15. The term $\Delta F = 2$ operators refers to four-quark operators that change flavor numbers by two units, such as $\bar{s}d\bar{s}d$. Each $\Delta F = 2$ amplitude is then conveniently parameterized in terms of the shift induced in the modulo and the CP-violating phase, or the real and imaginary parts (10, 11):

$$\frac{\langle B_q|\mathcal{L}_{eff}|\bar{B}_q\rangle}{\langle B_q|\mathcal{L}_{\rm SM}|\bar{B}_q\rangle} = C_{B_q} e^{2i\phi_{B_q}}, \quad \frac{\mathrm{Re}[\langle K^0|\mathcal{L}_{\rm eff}|\overline{K^0}\rangle]}{\mathrm{Re}[\langle K^0|\mathcal{L}_{\rm SM}|\overline{K^0}\rangle]} = C_{\Delta m_K} \quad \xrightarrow{\mathrm{Re}\to\mathrm{Im}} C_{\varepsilon_K}. \tag{16}$$

An updated analysis of these constraints has been presented (**Figure 2**) (12). There are three main conclusions that can be drawn from this analysis.

First, in all three accessible short-distance amplitudes ($K^0 - \overline{K^0}$, $B_d - \bar{B}_d$, and $B_s - \bar{B}_s$), the magnitude of the NP amplitude cannot be greater than the SM short-distance contribution. The latter is suppressed both by the Glashow-Iliopoulos-Maiani (GIM) mechanism, namely the smallness of the mass differences between second- and first-generation quarks, and by the hierarchical structure of the CKM matrix, specifically the smallness of quark-mixing angles,

$$\mathcal{A}_{\rm SM}^{\Delta F=2} \approx \frac{G_F^2 m_t^2}{16\pi^2}\left(V_{ti}^* V_{tj}\right)^2 \times \langle \bar{M}|(\bar{Q}_{Li}\gamma^\mu Q_{Lj})^2|M\rangle \times F\left(\frac{M_W^2}{m_t^2}\right), \tag{17}$$

where F is a function of $\mathcal{O}(1)$ that gives the dependence on the masses of the top quark and the W boson, which are the virtual intermediate particles in this amplitude. As a result, NP models with teraelectronvolt-scale flavored degrees of freedom and $\mathcal{O}(1)$ effective flavor-mixing couplings are ruled out. To set explicit bounds, let us consider, for instance, the subset of left-handed $\Delta F = 2$ operators in the generic effective Lagrangian (Equation 15), namely

$$\Delta\mathcal{L}^{\Delta F=2} = \sum_{i\neq j} \frac{c_{ij}}{\Lambda^2}(\bar{Q}_{Li}\gamma^\mu Q_{Lj})^2, \tag{18}$$

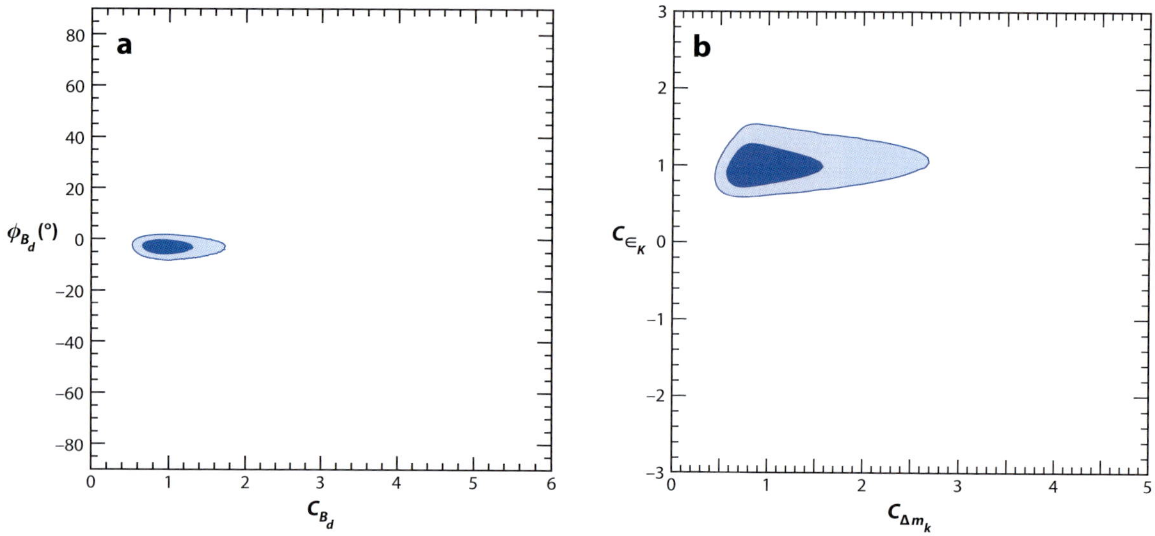

Figure 2

Constraints on the effective parameters encoding new physics effects in (*a*) $B_d - \bar{B}_d$ mixing and (*b*) $K^0 - \overline{K^0}$ mixing, as obtained by the UTfit Collaboration (12).

where the c_{ij} are dimensionless couplings. The condition $|\mathcal{A}_{\rm NP}^{\Delta F=2}| < |\mathcal{A}_{\rm SM}^{\Delta F=2}|$ implies that

$$\Lambda > \frac{4.4\,\text{TeV}}{|V_{ti}^* V_{tj}|/|c_{ij}|^{1/2}} \sim \begin{cases} 1.3 \times 10^4\,\text{TeV} \times |c_{sd}|^{1/2} \\ 5.1 \times 10^2\,\text{TeV} \times |c_{bd}|^{1/2} \\ 1.1 \times 10^1\,\text{TeV} \times |c_{bs}|^{1/2} \end{cases}. \qquad 19.$$

The strong bounds on Λ for generic c_{ij} of order one is a manifestation of what in many specific frameworks (supersymmetry, technicolor, etc.) is known as the flavor problem: If we insist that NP emerge in the teraelectronvolt region, we have to conclude that it possesses a highly nongeneric flavor structure.

Second, in the case of $B_d - \bar{B}_d$ and $K^0 - \overline{K^0}$ mixing, in which both *CP*-conserving and *CP*-violating observables are measured with excellent accuracy, there is still room for a sizable NP contribution (relative to the SM one), provided that it is to a large extent aligned in phase with the SM amplitude [$\mathcal{O}(0.01)$ for the K system and $\mathcal{O}(0.3)$ for the B_d system]. This is necessary because the theoretical errors in the observables used to constrain the phases, $S_{B_d\to\psi K}$ and ε_K, are smaller with respect to the theoretical uncertainties in Δm_{B_d} and Δm_K, which constrain the magnitude of the mixing amplitudes.

Third, in the case of $B_s - \bar{B}_s$ mixing, the precise determination of Δm_{B_s} does not allow large deviations in modulo with respect to the SM. The constraint is particularly severe if we consider the ratio $\Delta m_{B_d}/\Delta m_{B_s}$, in which hadronic uncertainties cancel to a large extent. However, the constraint on the *CP*-violating phase is quite poor. Present data from CDF (13) and DØ (14) indicate a large central value for the *CP*-violating phase, contrary to the SM expectation. However, the errors are still large, and the disagreement with the SM is at approximately the 2σ level. If the disagreement persists and becomes statistically significant, it would not only signal the presence of physics beyond the SM but would also rule out a whole subclass of MFV models (see Section 4).

Fourth, in D–$\bar{D}$ mixing, we cannot estimate the SM contribution from first principles; however, to a good accuracy this contribution is *CP* conserving. As a result, strong bounds on possible

Table 1 Bounds on representative dimension-six $\Delta F = 2$ operators[a]

Operator	Bounds on Λ (TeV)		Bounds on c_{ij} ($\Lambda = 1$ TeV)		Observables
	Re	Im	Re	Im	
$(\bar{s}_L\gamma^\mu d_L)^2$	9.8×10^2	1.6×10^4	9.0×10^{-7}	3.4×10^{-9}	Δm_K; ε_K
$(\bar{s}_R d_L)(\bar{s}_L d_R)$	1.8×10^4	3.2×10^5	6.9×10^{-9}	2.6×10^{-11}	Δm_K; ε_K
$(\bar{c}_L\gamma^\mu u_L)^2$	1.2×10^3	2.9×10^3	5.6×10^{-7}	1.0×10^{-7}	Δm_D; $\lvert q/p \rvert$; ϕ_D
$(\bar{c}_R u_L)(\bar{c}_L u_R)$	6.2×10^3	1.5×10^4	5.7×10^{-8}	1.1×10^{-8}	Δm_D; $\lvert q/p \rvert$; ϕ_D
$(\bar{b}_L\gamma^\mu d_L)^2$	5.1×10^2	9.3×10^2	3.3×10^{-6}	1.0×10^{-6}	Δm_{B_d}; $S_{B_d\to\psi K}$
$(\bar{b}_R d_L)(\bar{b}_L d_R)$	1.9×10^3	3.6×10^3	5.6×10^{-7}	1.7×10^{-7}	Δm_{B_d}; $S_{B_d\to\psi K}$
$(\bar{b}_L\gamma^\mu s_L)^2$	1.1×10^2	1.1×10^2	7.6×10^{-5}	7.6×10^{-5}	Δm_{B_s}
$(\bar{b}_R s_L)(\bar{b}_L s_R)$	3.7×10^2	3.7×10^2	1.3×10^{-5}	1.3×10^{-5}	Δm_{B_s}

[a]Bounds on Λ are quoted assuming an effective coupling $1/\Lambda$; alternatively, the bounds on the respective c_{ij}s are given assuming that $\Lambda = 1$ TeV. Observables related to *CP* violation are separated from the *CP*-conserving observables with semicolons. In the B_s system we quote only a bound on the modulo of the new physics amplitude derived from Δm_{B_s}. For the definition of the *CP*-violating observables in the D system, see Reference 15.

nonstandard *CP*-violating contributions can still be set. The resulting constraints are second only to those from ε_K, and unlike in the case of ε_K, they are controlled by experimental statistics and could be significantly improved in the near future.

A more detailed list of the bounds derived from $\Delta F = 2$ observables is shown in **Table 1**, where we give the bounds for two representative sets of dimension-six operators: (*a*) left-left operators (also present in the SM) and (*b*) operators with a different chirality that arise in specific SM extensions. The bounds on the latter are stronger, especially in the kaon case, because of the larger hadronic matrix elements. The constraints related to *CP* violation correspond to maximal phases and are subject to the requirement that the NP contributions be smaller than 30% (60%) of the SM contributions (9) in the B_d (K) system. Because the experimental status of *CP* violation in the B_s system is not yet settled, we require simply that the NP contributions be smaller than the observed value of Δm_{B_s} (for less naïve treatments see, e.g., References 6 and 12).

3.2. Correlations Between *K* and *D* Mixing

Two different features can provide flavor-related suppression factors: degeneracy and alignment. Generally, low-energy measurements can constrain only the product of these two suppression factors. However, an interesting exception occurs for the operators with a left-left current structure of the type shown in Equation 18, where there is an independent constraint on the level of degeneracy (16). Consider operators of the form

$$\frac{1}{\Lambda_{\mathrm{NP}}^2}[\bar{Q}_{Li}(X_Q)_{ij}\gamma_\mu Q_{Lj}][\bar{Q}_{Li}(X_Q)_{ij}\gamma^\mu Q_{Lj}], \qquad 20.$$

where X_Q is a hermitian matrix. Without loss of generality, we can choose to work in the basis defined in Equation 10:

$$Y^d = \lambda_d, \quad Y^u = V^\dagger \lambda_u, \quad X_Q = V_d^\dagger \lambda_Q V_d, \qquad 21.$$

where λ_Q is a diagonal real matrix and V_d is a unitary matrix that parameterizes the misalignment of the operator of Equation 20 with the down mass basis.

The experimental constraints that are most relevant to our study come from $K^0 - \overline{K^0}$ and $D^0 - \overline{D^0}$ mixing, which involve only the first- and second-generation quarks. When studying NP effects, ignoring the third generation is often a good approximation to the physics at hand. Indeed, even when the third generation does play a role, our two-generation analysis applies as long as there are no strong cancellations with contributions related to the third generation. In a two-generation framework, V depends on a single mixing angle (the Cabibbo angle θ_c), whereas V_d depends on a single angle and a single phase. To understand the various aspects of our analysis, however, it is useful to provisionally set the phase to zero and study only CP-conserving observables. We therefore have

$$\lambda_Q = \mathrm{diag}(\lambda_1, \lambda_2), \quad V = \begin{pmatrix} \cos\theta_c & \sin\theta_c \\ -\sin\theta_c & \cos\theta_c \end{pmatrix}, \quad V_d = \begin{pmatrix} \cos\theta_d & \sin\theta_d \\ -\sin\theta_d & \cos\theta_d \end{pmatrix}. \qquad 22.$$

It is convenient to define

$$\lambda_{12} = \frac{1}{2}(\lambda_1 + \lambda_2), \quad \delta_{12} = \frac{\lambda_1 - \lambda_2}{\lambda_1 + \lambda_2}, \quad \Lambda_{12} = \delta_{12}\lambda_{12}. \qquad 23.$$

Thus, λ_{12} parameterizes the overall flavor-diagonal suppression of X_Q (in particular, loop factors), δ_{12} parameterizes the suppression arising from approximate degeneracy between the eigenvalues of X_Q, and θ_d, and $\theta_c - \theta_d$ parameterize the suppression that comes from alignment with the down and up sectors, respectively.

The main point is that alignment can entirely suppress the contribution to either $K^0 - \overline{K^0}$ mixing ($\theta_d = 0$) or $D^0 - \overline{D^0}$ mixing ($\theta_d = \theta_c$), but not both. Thus, the flavor measurements give a constraint on Λ_{12} that reads (16)

$$\Lambda_{12} \leq 3.8 \times 10^{-3} \left(\frac{\Lambda_{\rm NP}}{1\,\mathrm{TeV}}\right). \qquad 24.$$

If we switch on the CP-violating phase γ in V_d, then for $0.03 \lesssim |\sin\gamma| \lesssim 0.98$ (16)

$$\Lambda_{12} \leq \frac{4.8 \times 10^{-4}}{\sqrt{\sin 2\gamma}} \left(\frac{\Lambda_{\rm NP}}{1\,\mathrm{TeV}}\right). \qquad 25.$$

With a loop suppression of order $\lambda_{12} \sim \alpha_2$, the degeneracy should be stronger than 0.02.

3.3. Top Physics

Although the present direct determination of top quark decay rates is at the $\mathcal{O}(1\%)$ level of accuracy (17), an orders-of-magnitude improvement is expected at the LHC. With 100 fb^{-1} of data, the LHC will be sensitive (at 95% CL) to branching ratios of $\mathcal{O}(10^{-5})$ in the $t \to u^i\, Z/\gamma$ channels (18), where $u^i = u, c$. Within the SM, these channels have branching ratios in the 10^{-13} range, so any experimental observation would be a clear sign of physics beyond the SM. The LHC sensitivity to additional $\Delta t = 1$ processes—such as $t \to c\, G, \phi$—is more limited (e.g., 19).

The impact of the projected LHC bounds on top flavor violation can be described in a model-independent manner. The leading contributions to the above processes are mediated via a set of dimension-six operators (20), which can be classified according to the SM global flavor symmetries (21). Focusing on the chirality of the quark fields, we can write these operators as $\mathcal{O}_{A^tB}$, where the indices A^t and B can assume the value L or R (A^t denotes top quark chirality). Bounds from B physics can give severe constraints on the operators with $B = L$ (21). The situation changes in the case of a flavor alignment of these effective operators, such that they are diagonal in the down-type mass basis. In this limit, no bounds can be derived on the $\mathcal{O}_{LR}$ and $\mathcal{O}_{RL}$ operators (22). However, the $\mathcal{O}_{LL}$ operators cannot be aligned simultaneously with the down and up quark mass bases. As a result, by combining bounds on $t \to u^i$ with bounds from $b \to s l^+ l^-$, we can get stringent constraints on the operator $\mathcal{O}^b_{LL} = \bar{Q}_i \gamma^\mu (X_Q)_{ij} Q_j (\phi^\dagger \overleftrightarrow{D}_\mu \phi)/\Lambda_t^2$. The flavor-nonuniversal part of the X_Q matrix can be written as $X_Q = a_i T_i$, such that the projection onto the $SU(3)$ generators has a unit norm ($\sum_i a_i^2 = 1$). We can then identify a unique combination that minimizes the combined bound from top and B constraints. Taking into account the projected LHC sensitivity on $t \to u^i$ decays, we can expect to reach bounds of $\mathcal{O}(600\,\text{GeV})$ on the scale Λ_t (22) in the absence of a NP signal.

4. MINIMAL FLAVOR VIOLATION

A very reasonable albeit pessimistic setup that avoids the NP flavor problem is the MFV hypothesis. Under this assumption, flavor-violating interactions are linked to the known structure of Yukawa couplings not just within but also beyond the SM. As a result, nonstandard contributions in FCNC transitions are suppressed to a level that is consistent with experiments, even for $\Lambda \sim$ few teraelectronvolts. One of the most interesting aspects of the MFV hypothesis is that it can naturally be implemented within the generic effective Lagrangian in Equation 15. Furthermore, SM extensions in which the flavor hierarchy is generated at a scale much higher than other dynamical scales tend to flow to the MFV class of models in the infrared (IR).

The MFV hypothesis consists of (*a*) a flavor symmetry and (*b*) a set of symmetry-breaking terms (23). The symmetry is the large global symmetry of the SM Lagrangian in the absence of the Yukawa couplings shown in Equation 7. Because this global symmetry, particularly the $SU(3)$ subgroups that control quark flavor-changing transitions, is already broken within the SM, we cannot promote it to be an exact symmetry of the NP model. Some breaking appears at the quantum level because of the SM Yukawa interactions. The most restrictive assumption we can make to consistently protect quark flavor mixing beyond the SM is to assume that Y^d and Y^u are the only sources of flavor-symmetry breaking that are also present in the NP model. To consistently implement and interpret this hypothesis, we can assume that $SU(3)_q^3$ is a good symmetry and promote $Y^{u,d}$ to be nondynamical fields (spurions) with nontrivial transformation properties under $SU(3)_q^3$:

$$Y^u \sim (3, \bar{3}, 1), \quad Y^d \sim (3, 1, \bar{3}). \tag{26}$$

If the symmetry breaking occurs at very high energy scales, at low energies we would be sensitive only to the background values of the Y, that is, to the ordinary SM Yukawa couplings. Employing effective-theory language, an effective theory satisfies the criterion of the MFV hypothesis in the quark sector[1] if all higher-dimensional operators, constructed from SM and Y fields, are invariant (formally) under the flavor group $SU(3)_q^3$ (23). Invariance under CP may or may not be imposed as well.

According to this criterion, one should in principle consider operators with arbitrary powers of the (dimensionless) Yukawa fields. However, a strong simplification arises from the observations that (*a*) all the eigenvalues of the Yukawa matrices—except for the top one and, possibly, the bottom one (see below)—are small and (*b*) the off-diagonal elements of the CKM matrix are very suppressed. Working from the basis presented in Equation 10, and neglecting the ratio of light quark masses over the top mass, we obtain

$$[Y^u (Y^u)^\dagger]^n_{i \neq j} \approx y_t^{2n} V_{ti}^* V_{tj}. \tag{27}$$

[1]The notion of MFV can also be extended to the lepton sector. However, in this case there is no unique way to define the minimal sources of flavor symmetry breaking if we want to keep track of nonvanishing neutrino masses (24, 25).

As a consequence, including high powers of the Yukawa matrices amounts only to a redefinition of the overall factor in Equation 27, and the leading $\Delta F = 2$ and $\Delta F = 1$ FCNC amplitudes get exactly the same CKM suppression as in the SM:

$$\mathcal{A}(d^i \to d^j)_{\mathrm{MFV}} = (V^*_{ti} V_{tj}) \mathcal{A}^{(\Delta F=1)}_{\mathrm{SM}} \left[1 + a_1 \frac{16\pi^2 M^2_W}{\Lambda^2}\right] \qquad 28.$$

and

$$\mathcal{A}(M_{ij} - \bar{M}_{ij})_{\mathrm{MFV}} = (V^*_{ti} V_{tj})^2 \mathcal{A}^{(\Delta F=2)}_{\mathrm{SM}} \left[1 + a_2 \frac{16\pi^2 M^2_W}{\Lambda^2}\right], \qquad 29.$$

where $\mathcal{A}^{(i)}_{\mathrm{SM}}$ are the SM loop amplitudes and a_i are $\mathcal{O}(1)$ real parameters. The a_i depend on the specific operator considered but are flavor independent, which implies the same relative correction in $s \to d$, $b \to d$, and $b \to s$ transitions of the same type.

Within the MFV framework, several of the constraints used to determine the CKM matrix (and especially the unitarity triangle) are not affected by NP (26). In this framework, NP effects are negligible not only in tree-level processes but also in a few clean observables that are sensitive to loop effects, such as the time-dependent *CP*-violating asymmetry in $B_d \to \psi K_{L,S}$. Indeed, the structure of the basic flavor-changing coupling in Equation 29 implies that the weak *CP*-violating phase of $B_d - \bar{B}_d$ mixing is $\arg[(V_{td} V^*_{tb})^2]$, exactly as in the SM. This construction provides a natural (a posteriori) justification as to why no NP effects have been observed in the quark sector: By construction, most of the clean observables measured at *B* factories are insensitive to NP effects in the MFV framework.

In **Table 2**, we list a few representative examples of the bounds on the higher-dimensional operators in the MFV framework. For simplicity, only leading spurion dependence is shown in the left-hand column. The built-in CKM suppression leads to bounds on the effective scale of NP, not far from the teraelectronvolt region. These bounds are very similar to the bounds on flavor-conserving operators derived by precision electroweak tests. This observation reinforces the conclusion that a deeper study of rare decays is definitely needed in order to clarify the flavor problem; the experimental precision on the clean FCNC observables required to obtain bounds more stringent than those derived from precision electroweak tests (and possibly discover NP) is typically in the 1–10% range.

Although MFV seems to be a natural solution to the flavor problem, we stress that (*a*) it is not a theory of flavor (there is no explanation for the observed hierarchical structure of the Yukawas) and (*b*) we are still far from proving the validity of this hypothesis from data [in effective-theory

Table 2 Bounds on the scale of new physics (at 95% CL) for representative $\Delta F = 1$ (27) and $\Delta F = 2$ (12) minimal flavor-violation operators[a] and the corresponding observables used to set the bounds

Operator	Bound on Λ	Observables
$H^\dagger \left(\bar{D}_R Y^{d\dagger} Y^u Y^{u\dagger} \sigma_{\mu\nu} Q_L\right) (e F_{\mu\nu})$	6.1 TeV	$B \to X_s\gamma$, $B \to X_s \ell^+\ell^-$
$\frac{1}{2}(\bar{Q}_L Y^u Y^{u\dagger} \gamma_\mu Q_L)^2 X^{X^X}_{iii}$	5.9 TeV	ε_K, Δm_{B_d}, Δm_{B_s}
$H^\dagger_D \left(\bar{D}_R Y^{d\dagger} Y^u Y^{u\dagger} \sigma_{\mu\nu} T^a Q_L\right) (g_s G^a_{\mu\nu})$	3.4 TeV	$B \to X_s\gamma$, $B \to X_s \ell^+\ell^-$
$\left(\bar{Q}_L Y^u Y^{u\dagger} \gamma_\mu Q_L\right) (\bar{E}_R \gamma_\mu E_R)$	2.7 TeV	$B \to X_s \ell^+\ell^-$, $B_s \to \mu^+\mu^-$
$i \left(\bar{Q}_L Y^u Y^{u\dagger} \gamma_\mu Q_L\right) H^\dagger_U D_\mu H_U$	2.3 TeV	$B \to X_s \ell^+\ell^-$, $B_s \to \mu^+\mu^-$
$\left(\bar{Q}_L Y^u Y^{u\dagger} \gamma_\mu Q_L\right) (\bar{L}_L \gamma_\mu L_L)$	1.7 TeV	$B \to X_s \ell^+\ell^-$, $B_s \to \mu^+\mu^-$
$\left(\bar{Q}_L Y^u Y^{u\dagger} \gamma_\mu Q_L\right) (e D_\mu F_{\mu\nu})$	1.5 TeV	$B \to X_s \ell^+\ell^-$

[a]Assuming effective coupling $\pm 1/\Lambda^2$.

Table 3 Some predictions derived in the minimal flavor-violation (MFV) framework

Observable	Experiment	MFV prediction[a]	Standard Model prediction
β_s from $\mathcal{A}_{CP}(B_s \to \psi\phi)$	[0.10, 1.44] at 95% CL	0.04(5)*	0.04(2)
$\mathcal{A}_{CP}(B \to X_s\gamma)$	<6% at 95% CL	<0.02*	<0.01
$\mathcal{B}(B_d \to \mu^+\mu^-)$	$<1.8 \times 10^{-8}$	$<1.2 \times 10^{-9}$	$1.3(3) \times 10^{-10}$
$\mathcal{B}(B \to X_s\tau^+\tau^-)$	—	$<5 \times 10^{-7}$	$1.6(5) \times 10^{-7}$
$\mathcal{B}(K_L \to \pi^0\nu\bar{\nu})$	$<2.6 \times 10^{-8}$ at 90% CL	$<2.9 \times 10^{-10}$	$2.9(5) \times 10^{-11}$

[a]Asterisks indicate that the predictions are not valid in the general MFV case at large tan β (see Section 4.2).

language, there is still room for sizable new sources of flavor-symmetry breaking in addition to the SM Yukawa couplings (28)]. A proof of the MFV hypothesis can be achieved only with positive evidence of physics beyond the SM that exhibits the flavor-universality pattern (same relative correction in $s \to d$, $b \to d$, and $b \to s$ transitions of the same type) predicted by the MFV assumption. Although this goal is difficult to achieve, the MFV framework is quite predictive and thus could easily be falsified. In **Table 3**, we list some clean MFV predictions that could be falsified by future experiments. Violations of these bounds would imply not only physics beyond the SM but also a clear signal of new sources of flavor-symmetry breaking beyond the Yukawa couplings.

The idea that the CKM matrix also rules the strength of FCNC transitions beyond the SM has become popular in recent literature and has been implemented and discussed in several papers. We stress that the CKM matrix represents only part of the problem: A key role in determining the structure of FCNCs is played by quark masses or by the Yukawa eigenvalues. In that respect, the MFV criterion provides the maximal protection of FCNCs (or the minimal violation of flavor symmetry) because the full structure of Yukawa matrices is preserved. At the same time, this criterion is based on a renormalization group–invariant symmetry argument, which can be implemented independently of any specific hypothesis about the dynamics of the NP framework. This model-independent structure does not hold in most of the alternative definitions of MFV models that can be found in the literature. For instance, the definition used in Reference 29, known as constrained MFV, requires that the effective FCNC operators playing a significant role within the SM also be the only relevant ones beyond the SM. This condition is realized only in weakly coupled, teraelectronvolt-scale theories that have only one light Higgs doublet, such as the MSSM with small tan β, where no large logs or sizable anomalous dimensions are present (30, 31). This condition does not hold in several other frameworks, such as Higgsless models, five-dimensional MFV models, and the MSSM with large tan β.

In NP models in which sizable anomalous dimensions are present, the expansion in powers of the Yukawa spurions cannot be truncated at the first nontrivial order. In this limit, denoted general MFV, higher-order terms in the third-generation Yukawa couplings need to be resummed. As shown in Equation 27, if only the top quark Yukawa coupling is of order one, this resummation has negligible consequences for flavor-violating processes. [To measure the effect requires that the accuracy on rare K (D) decays become strong enough to probe the contributions related to the charm (strange) quark Yukawa coupling (30).] However, significant differences from a linear expansion may arise if both top quark and bottom quark Yukawa couplings are order one, as occurs in, for instance, the large–tan β regime.

4.1. Minimal Flavor Violation at Large tan β

If the Yukawa Lagrangian contains more than a single Higgs field, we can still assume that the Yukawa couplings are the only irreducible breaking sources of $SU(3)_q^3$, and we can change their overall normalization. A particularly interesting scenario is the two-Higgs-doublet model in which the two Higgs fields, ϕ_U and ϕ_D, are coupled separately to up- and down-type quarks:

$$-\mathcal{L}_{\text{Yukawa}}^{\text{2HDM}} = Y_{ij}^{d}\,\bar{Q}_{Li}\phi_D D_{Rj} + Y_{ij}^{u}\,\bar{Q}_{Li}\phi_U U_{Rj} + Y_{ij}^{e}\,\bar{L}_{Li}\phi_D E_{Rj} + \text{h.c.} \qquad 30.$$

This Lagrangian is invariant under an extra $U(1)$ symmetry with respect to the one-Higgs Lagrangian in Equation 5: a symmetry under which the only charged fields are D_R and E_R (charge $+1$) and ϕ_D (charge -1). This symmetry, denoted $U(1)_{PQ}$, prevents tree-level FCNCs and implies that $Y^{u,d}$ are the only sources of the $SU(3)_q^3$ breaking that appears in the Yukawa interaction (similar to the one-Higgs-doublet scenario). Consistently with the MFV hypothesis, we can assume that $Y^{u,d}$ are the only relevant sources of the $SU(3)_q^3$ breaking that appears in all the low-energy effective operators. This assumption is sufficient to ensure that flavor mixing is still governed by the CKM matrix, and it naturally guarantees a good agreement with present data in the $\Delta F = 2$ sector. However, the extra symmetry of the Yukawa interaction allows us to change the overall normalization of $Y^{u,d}$ with interesting phenomenological consequences in specific rare modes. These effects are related only to the large value of bottom Yukawa and, indeed, can also be found in other NP frameworks where there is no extended Higgs sector but where the bottom Yukawa coupling is of order one (30).

Assuming the Lagrangian in Equation 30, the normalization of the Yukawa couplings is controlled by the ratio of the VEVs of the two Higgs fields or by the parameter

$$\tan\beta = \langle\phi_U\rangle/\langle\phi_D\rangle. \qquad 31.$$

For $\tan\beta \gg 1$, the smallness of the b quark and τ lepton masses can be attributed to the smallness of $1/(\tan\beta)$ rather than to the corresponding Yukawa couplings. As a result, for $\tan\beta \gg 1$ we can no longer neglect the down-type Yukawa coupling. Moreover, the $U(1)_{PQ}$ symmetry cannot be exact: It must be broken, at least in the scalar potential, to avoid the presence of a massless pseudoscalar Higgs. Even if the breaking of $U(1)_{PQ}$ and $SU(3)_q^3$ is decoupled, the presence of $U(1)_{PQ}$-breaking sources can have important implications for the structure of the Yukawa interaction, especially if $\tan\beta$ is large (23, 32–34). Indeed, we can consider new dimension-four operators, such as

$$\varepsilon\,\bar{Q}_L Y^d D_R \tilde{\phi}_U \quad \text{or} \quad \varepsilon\,\bar{Q}_L Y^u Y^{u\dagger} Y^d D_R \tilde{\phi}_U, \qquad 32.$$

where ε denotes a generic MFV-invariant $U(1)_{PQ}$-breaking source. Even if $\varepsilon \ll 1$, the product $\varepsilon \times \tan\beta$ can be $\mathcal{O}(1)$, which would induce large corrections to the down-type Yukawa sector:

$$\varepsilon\,\bar{Q}_L Y^d D_R \tilde{\phi}_U \xrightarrow{\text{vev}} \varepsilon\,\bar{Q}_L Y^d D_R \langle\tilde{\phi}_U\rangle = (\varepsilon \times \tan\beta)\bar{Q}_L Y^d D_R \langle\phi_D\rangle. \qquad 33.$$

Because the b quark Yukawa coupling becomes $\mathcal{O}(1)$, the large–tan β regime is particularly interesting for helicity-suppressed observables in B physics. One of the clearest phenomenological consequences is a suppression (typically in the 10–50% range) of the $B \to \ell\nu$ decay rate with respect to its SM expectation (35). Potentially measurable effects in the 10–30% range are also expected in $B \to X_s\gamma$ (36) and ΔM_{B_s} (37). Given the present measurements of $B \to \ell\nu$, $B \to X_s\gamma$, and ΔM_{B_s}, none of these effects seems to be favored by data. However, present errors are still sizable compared to the estimated NP effects.

The most striking signature could arise from the rare decays $B_{s,d} \to \ell^+\ell^-$, whose rates could be enhanced over the SM expectations by more than one order of magnitude (38). An enhancement of both $B_s \to \ell^+\ell^-$ and $B_d \to \ell^+\ell^-$, respecting the MFV relation $\Gamma(B_s \to \ell^+\ell^-)/\Gamma(B_d \to \ell^+\ell^-) \approx |V_{ts}/V_{td}|^2$, would be an unambiguous signature of MFV at large $\tan\beta$ (27).

4.2. Minimal Flavor Violation with Additional Flavor-Diagonal Phases

The breaking of the $SU(3)_q^3$ flavor group and the breaking of the discrete CP symmetry are not necessarily related, and we can add flavor-diagonal CP-violating phases to generic MFV models (30, 39, 40). Because of the experimental constraints on electric dipole moments (EDMs), which are generally sensitive to such flavor-diagonal phases (40), in this more general case the bounds on the scale of NP are substantially higher with respect to the minimal case, where the Yukawa couplings are assumed to be the only breaking sources of both symmetries (23)

If $\tan\beta$ is large, then the inclusion of flavor-diagonal phases also has interesting effects on flavor-changing processes. The main consequences, derived in a model-independent manner, can be summarized as follows (30): (*a*) Extra CP violation can arise only from flavor-diagonal CP-violating sources in the ultraviolet (UV) theory; (*b*) the extra CP phases in $B_s - \bar{B}_s$ mixing provide an upper bound on the amount of CP violation in $B_d - \bar{B}_d$ mixing; and (*c*) if operators containing right-handed light quarks are subdominant, then the extra CP violation is equal in the two systems and is negligible in transitions between the first two generations. Conversely, these operators can break the correlation between CP violation in the B_s and B_d systems and can induce significant new CP violation in ε_K.

5. SUPERSYMMETRY

Supersymmetric models generally provide new sources of flavor violation for both the quark and lepton sectors. The main new sources are the supersymmetry-breaking soft mass terms for squarks and sleptons and the trilinear couplings of a Higgs field with a squark-antisquark or slepton-antislepton pair. Let us focus on the squark sector. The new sources of flavor violation are most commonly analyzed in the basis in which the corresponding (down or up) quark mass matrix and the neutral gaugino vertices are diagonal. In this basis, the squark masses are not necessarily flavor diagonal, and they have the form

$$\tilde{q}^*_{Mi}\left(M^2_{\tilde{q}}\right)^{MN}_{ij}\tilde{q}_{Nj} = (\tilde{q}^*_{Li}\tilde{q}^*_{Rk})\begin{pmatrix}(M^2_{\tilde{q}})_{Lij} & A^q_{il}v_q \\ A^q_{jk}v_q & (M^2_{\tilde{q}})_{Rkl}\end{pmatrix}\begin{pmatrix}\tilde{q}_{Lj} \\ \tilde{q}_{Rl}\end{pmatrix}, \qquad 34.$$

where $M, N = L, R$ label chirality and $i, j, k, l = 1, 2, 3$ are generation indices. $(M^2_{\tilde{q}})_L$ and $(M^2_{\tilde{q}})_R$ are the supersymmetry-breaking squark masses squared. The A^q parameters enter in the trilinear scalar couplings $A^q_{ij}\phi_q\tilde{q}_{Li}\tilde{q}^*_{Rj}$, where ϕ_q $(q = u, d)$ is the q-type Higgs boson, and $v_q = \langle\phi_q\rangle$.

In this basis, flavor violation takes place through one or more squark mass insertions. Each mass insertion brings with it a factor of $(\delta^q_{ij})_{MN} \equiv (M^2_{\tilde{q}})^{MN}_{ij}/\tilde{m}^2_q$, where $\tilde{m}^2_q$ is a representative q-squark mass scale. Physical processes therefore constrain

$$\left[\left(\delta^q_{ij}\right)_{MN}\right]_{\text{eff}} \sim \max\left[\left(\delta^q_{ij}\right)_{MN}, \left(\delta^q_{ik}\right)_{MP}\left(\delta^q_{kj}\right)_{PN}, \ldots, (i \leftrightarrow j)\right]. \qquad 35.$$

For example,

$$\left[\left(\delta^d_{12}\right)_{LR}\right]_{\text{eff}} \sim \max\left[A^d_{12}v_d/\tilde{m}^2_d, \left(M^2_{\tilde{d}}\right)_{L1k}A^d_{k2}v_d/\tilde{m}^4_d, A^d_{1k}v_d\left(M^2_{\tilde{d}}\right)_{Rk2}/\tilde{m}^4_d, \ldots, (1 \leftrightarrow 2)\right]. \qquad 36.$$

Note that the contributions with two or more insertions may be less suppressed than those with only one.

In terms of mass-basis parameters, the $(\delta^q_{ij})_{MM}$s stand for a combination of mass splittings and mixing angles:

$$(\delta^q_{ij})_{MM} = \frac{1}{\tilde{m}^2_q} \sum_\alpha (K^q_M)_{i\alpha} (K^q_M)^*_{j\alpha} \Delta\tilde{m}^2_{q\alpha}, \qquad 37.$$

where K^q_M is the mixing matrix in the coupling of the gluino (and similarly for the bino and neutral wino) to $q_{Li} - \tilde{q}_{M\alpha}$. $\tilde{m}^2_q = \frac{1}{3}\sum_{\alpha=1}^3 \tilde{m}^2_{qM\alpha}$ is the average squark mass squared, and $\Delta\tilde{m}^2_{q\alpha} = \tilde{m}^2_{q\alpha} - \tilde{m}^2_q$. The situation simplifies considerably when two conditions in Equation 38 are satisfied (42), which means that a two-generation effective framework can be used (for simplicity, we omit the chirality index):

$$|K_{ik}K^*_{jk}| \ll |K_{ij}K^*_{jj}|, \quad |K_{ik}K^*_{jk}\Delta\tilde{m}^2_{q_k q_i}| \ll |K_{ij}K^*_{jj}\Delta\tilde{m}^2_{q_j q_i}|, \qquad 38.$$

where there is no summation over i, j, and k and where $\tilde{m}^2_{q_j q_i} = \tilde{m}^2_{q_j} - \tilde{m}^2_{q_i}$. Then the contribution of the intermediate $\tilde{q}_k$ can be neglected, and furthermore, to a good approximation, $K_{ii}K^*_{ji} + K_{ij}K^*_{jj} = 0$. For these cases, we obtain

$$(\delta^q_{ij})_{MM} = \frac{\Delta\tilde{m}^2_{q_j q_i}}{\tilde{m}^2_q} (K^q_M)_{ij} (K^q_M)^*_{jj}. \qquad 39.$$

It is also useful to employ, instead of $\tilde{m}_q$, the mass scale $\tilde{m}^q_{ij} = \frac{1}{2}(\tilde{m}_{q_i} + \tilde{m}_{q_j})$ (43). We also define

$$\langle\delta^q_{ij}\rangle = \sqrt{(\delta^q_{ij})_{LL}(\delta^q_{ij})_{RR}}. \qquad 40.$$

The new sources of flavor and *CP* violation contribute to FCNC processes via loop diagrams involving squarks and gluinos (or electroweak gauginos or higgsinos). If the scale of the soft supersymmetry breaking is below the teraelectronvolt scale, if the new flavor violation is of order one, and/or if the phases are of order one, then these contributions could be orders of magnitude above the experimental bounds. Requiring that the supersymmetric contributions not exceed the phenomenological constraints leads to constraints of the form $(\delta^q_{ij})_{MM} \ll 1$. Such constraints imply that either quasi-degeneracy ($\Delta\tilde{m}^2_{q_j q_i} \ll \tilde{m}^{q2}_{ij}$) or alignment ($|K^q_{ij}| \ll 1$), or a combination of the two mechanisms, is at work.

Table 4 presents the constraints obtained in References 44–47 as they appear in Reference 42. Wherever relevant, a phase suppression of order 0.3 in the mixing amplitude is allowed; that is, we quote the stronger of the bounds on $\mathrm{Re}(\delta^q_{ij})$ and $3\mathrm{Im}(\delta^q_{ij})$. The dependence of these bounds on the average squark mass $m_{\tilde{q}}$, the ratio $x \equiv m^2_{\tilde{g}}/m^2_{\tilde{q}}$, and the effect of arbitrary strong *CP*-violating phases can be found in Reference 42.

Table 4 The phenomenological upper bounds on $(\delta^q_{ij})_{MM}$ and $\langle\delta^q_{ij}\rangle$[a]

q	ij	$(\delta^q_{ij})_{MM}$	$\langle\delta^q_{ij}\rangle$
d	12	0.03	0.002
d	13	0.2	0.07
d	23	0.6	0.2
u	12	0.1	0.008

[a]Here $q = u, d$ and $M = L, R$. The constraints are given for $m_{\tilde{q}} = 1$ TeV and $x \equiv m^2_{\tilde{g}}/m^2_{\tilde{q}} = 1$. We assume that the phases could suppress the imaginary parts by a factor of ~0.3. The bound on $(\delta^d_{23})_{RR}$ is approximately three times weaker than that on $(\delta^d_{23})_{LL}$. The constraints on $(\delta^d_{12,13})_{MM}$, $(\delta^u_{12})_{MM}$, and $(\delta^d_{23})_{MM}$ are based on References 44–46, respectively.

Table 5 The phenomenological upper bounds on chirality mixing $(\delta^q_{ij})_{LR}$[a]

q	ij	$(\delta^q_{ij})_{LR}$
d	12	2×10^{-4}
d	13	0.08
d	23	0.01
d	11	4.7×10^{-6}
u	11	9.3×10^{-6}
u	12	0.02

[a]Here $q = u, d$. The constraints are given for $m_{\tilde{q}} = 1$ TeV and $x \equiv m^2_{\tilde{g}}/m^2_{\tilde{q}} = 1$. The constraints on $\delta^d_{12,13}$, δ^u_{12}, δ^d_{23}, and δ^q_{ii} are based on References 44–46 and 49, respectively. The relation between the neutron and quark electric dipole moments is assumed to be as in Reference 50.

For large tan β, some constraints are modified from those in **Table 4**. For instance, for tan $\beta = 30$ and $x = 1$ (42, 48), the effects of neutral Higgs exchange in B_s and B_d mixing give:

$$\langle\delta^d_{13}\rangle < 0.01 \cdot \left(\frac{M_{A^0}}{200\,\mathrm{GeV}}\right), \quad \langle\delta^d_{23}\rangle < 0.04 \cdot \left(\frac{M_{A^0}}{200\,\mathrm{GeV}}\right), \qquad 41.$$

where M_{A^0} denotes the pseudoscalar Higgs mass and the above bounds scale roughly as $(30/\tan\beta)^2$.

The experimental constraints on the $(\delta^q_{ij})_{LR}$ parameters in the quark-squark sector are presented in **Table 5**. The bounds are the same for $(\delta^q_{ij})_{LR}$ and $(\delta^q_{ij})_{RL}$, except for $(\delta^d_{12})_{MN}$, where the bound for $(\delta^d_{12})_{LR}$ is ten times weaker. Very strong constraints apply for the phase of $(\delta^q_{11})_{LR}$ from EDMs. For $x = 4$ and a phase smaller than 0.1, the EDM constraints on $(\delta^{u,d,\ell}_{11})_{LR}$ are weakened by a factor of approximately six.

Although the low-energy flavor measurements generally constrain only the combinations of the suppression factors from degeneracy and from alignment, such as in Equation 39, an interesting exception occurs when combining the measurements of $K^0 - \overline{K^0}$ and $D^0 - \overline{D^0}$ mixing to test the first two–generation squark doublets. Here, for masses below the teraelectronvolt scale, some level of degeneracy is unavoidable (16):

$$\frac{m_{\tilde{Q}_2} - m_{\tilde{Q}_1}}{m_{\tilde{Q}_2} + m_{\tilde{Q}_1}} \leq \begin{cases} 0.034 & \text{maximal phases} \\ 0.27 & \text{vanishing phases.} \end{cases} \qquad 42.$$

The strong constraints given in **Tables 4** and **5** can be satisfied if the mediation of supersymmetry breaking to the MSSM is MFV. Specifically, if at the scale of mediation the supersymmetry-breaking squark masses are universal, and if the A-terms vanish or are proportional to the Yukawa couplings, then the model is phenomenologically safe. Indeed, there are several known mechanisms of mediation that are MFV (e.g., 51), including gauge mediation (52–55), anomaly mediation (56, 57), and gaugino mediation (58). [The renormalization group flow in the MSSM with generic MFV soft-breaking terms at some high scale was recently discussed (41, 59).] However, we do not expect gravity mediation to be MFV, and it could provide subdominant yet observable flavor and *CP*-violating effects (60).

6. EXTRA DIMENSIONS

There are many models of extra dimensions, and the corresponding phenomenology, including the implications for flavor physics, changes from one extra-dimension framework to another. Yet, as in the supersymmetric case, one can classify the new sources of flavor violation as follows:

1. Bulk masses. If the SM fields propagate in the bulk of the extra dimensions, they can have bulk, vector-like masses. These mass terms are of particular importance to flavor physics because they induce fermion localization, which may yield hierarchies in the low-energy effective couplings. Furthermore, the bulk masses, which define the extra-dimension interaction basis, do not need to commute with the Yukawa matrices; therefore, they may induce contributions to FCNC processes, similar to the squark soft masses squared in supersymmetry.
2. Cutoff, UV physics. Because, generically, higher-dimensional field theories are nonrenormalizable, they rely on unspecified microscopic dynamics to provide UV completion of the models. Thus, they can be considered effective field theories, and the impact of the UV physics is expected to be captured by a set of operators suppressed by the framework-dependent cutoff scale. If we lack precise knowledge of the short-distance dynamics, the additional operators are expected to carry generic flavor structure and contribute to FCNC processes. This situation is somewhat similar to the case of gravity-mediated contributions to supersymmetry-breaking soft terms, which are generically expected to have an anarchic flavor structure and are suppressed by the Planck scale.
3. Brane-localized terms. The extra dimensions have to be compact, and they typically contain defects and boundaries of smaller dimensions (in order to yield, for example, a chiral, low-energy, four-dimensional theory). These special points may contain different microscopical degrees of freedom. Therefore, generically, one expects that a different and independent class of higher-dimension operators may be localized to this singular region in the extra-dimension manifold. (These operators are commonly denoted brane terms even though, in most cases, they have very little to do with string theory.) The brane-localized terms can, in principle, be of anarchic flavor structure and can provide new flavor and *CP*-violating sources. An important class of such operators consists of brane kinetic terms (BKTs), whose impact is somewhat similar to that of noncanonical kinetic terms, which generically arise in supersymmetric flavor models.

We now focus on flavor physics of five-dimensional models, with bulk SM fields, given that most of the literature focuses on this class. Furthermore, the new flavor structure that arises in five-dimensional models captures most of the known effects of extra-dimensional flavor models. Assuming a flat extra dimension, the energy range $\Lambda_{5\mathrm{D}}R$ [where $\Lambda_{5\mathrm{D}}$ is the five-dimensional effective cutoff scale and R is the extra dimension radius with the extra-dimension coordinate $y \in (0, \pi R)$] for which the five-dimensional effective field theory holds can be estimated as follows: Because gauge couplings in extra dimensional theories are dimensional—that is, $\alpha_{5\mathrm{D}}$ has mass dimension -1—a rough guess (which is confirmed, up to order-one corrections, by various methods of naïve dimensional analysis) would be $\Lambda_{5\mathrm{D}} \sim 4\pi/\alpha_{5\mathrm{D}}$ (61). Matching this five-dimensional gauge coupling to a four-dimensional coupling of the SM at leading order, $1/g^2 = \pi R/g_{5\mathrm{D}}^2$, we obtain

$$\Lambda_{5\mathrm{D}} R \sim \frac{4}{\alpha} \sim 30. \tag{43}$$

Generically, the mass of the lightest Kaluza-Klein (KK) states, M_{KK}, is $\mathcal{O}(R^{-1})$. If the extra-dimension theory is linked to the solution of the hierarchy problem and/or is directly accessible to near-future experiments, then $R^{-1} = \mathcal{O}(\mathrm{TeV})$. This implies an upper bound on the five-dimensional cutoff,

$$\Lambda_{5D} \lesssim 10^2\,\text{TeV} \ll \Lambda_K \sim 3 \times 10^5\,\text{TeV}, \qquad 44.$$

where Λ_K is the scale required to suppress the generic contributions to ε_K, as discussed above (see **Table 1**).

The above discussion ignores the possibility of splitting the fermions in the extra dimension. In split-fermion models, different bulk masses are assigned to different generations. Consequently, fermions are localized and separated in the bulk of the extra dimension in a manner that may successfully address the SM flavor puzzle (62). Separation in the extra dimension may suppress the contributions to ε_K from the higher-dimensional cutoff-induced operators. As shown in **Table 1**, the most dangerous operator is

$$O_K^4 = \frac{1}{\Lambda_{5D}^2}\,(\bar{s}_L d_R)(\bar{s}_R d_L)\,. \qquad 45.$$

This operator contains s and d fields of both chiralities. As a result, in a large class of split-fermion models, the overlap suppression would be similar to that accounting for the smallness of the down and strange four-dimensional Yukawa couplings. Integrating over the five-dimensional profiles of the four quarks may yield a suppression factor of $\mathcal{O}(m_d m_s/v^2) \sim 10^{-9}$. Together with the naïve scale suppression, $1/\Lambda_{5D}^2$, the coefficient of O_K^4 can be sufficiently suppressed to be consistent with the experimental bound.

In the absence of large BKTs, however, fermion localization generates order-one nonuniversal couplings to the gauge KK fields (63); for details, see, for instance, Reference 64. The case with large BKTs is similar to the warped case discussed below. The fact that the bulk masses are, generically, not aligned with the five-dimensional Yukawa couplings implies that KK gluon-exchange processes induce, among others, the following operator in the low-energy theory: $\left[(D_L)_{12}^2/(6M_{\text{KK}}^2)\right](\bar{s}_L d_L)^2$, where $(D_L)_{12} \sim \lambda$ is the left-handed down quark rotation matrix from the five-dimensional interaction basis to the mass basis. This structure provides only a mild suppression to the resulting operator. It implies that to satisfy the ε_K constraint, the KK and the inverse compactification scales must be above 10^3 TeV—beyond the direct reach of near-future experiments and too high to be linked to a solution of the hierarchy problem. This problem can be solved by tuning the five-dimensional flavor parameters and imposing appropriate five-dimensional flavor symmetries to make the tuning stable. Once the five-dimensional bulk masses are aligned with the five-dimensional Yukawa matrices, the KK gauge contributions vanish, and such a configuration is radiatively stable.

The warped extra-dimension [Randall-Sundrum (RS)] framework (65) provides a solution to the hierarchy problem. Moreover, with SM fermions propagating in the bulk, both the SM and NP flavor puzzles can be addressed. The light fermions could be localized away from the teraelectronvolt brane (66), where the Higgs is localized. Such a configuration can generate the observed Yukawa hierarchy and, at the same time, ensure that higher-dimensional operators are suppressed by a high-cutoff scale associated with the location of the light fermions in the extra dimension (67). Furthermore, because the KK states are localized near the teraelectronvolt brane, the couplings between the SM quarks and the gauge KK fields exhibit the hierarchical structure associated with SM masses and CKM mixings. This hierarchy in the couplings provides extra protection against nonstandard flavor-violating effects (68); such protection is termed the RS-GIM mechanism (69, 70). Interestingly, an analogous mechanism is at work in models with strong dynamics at the teraelectronvolt scale, with large anomalous dimension and partial compositeness (71). The link with strongly interacting models is indeed motivated by the AdS/CFT (anti–de Sitter space/conformal field theory) correspondence (72), which implies that the above five-dimensional framework is the dual description of four-dimensional composite Higgs models (73).

Concerning the quark zero modes, the flavor structure of the above models and the phenomenology can be captured by using the following simple rules (69, 74, 75). In the five-dimensional interaction basis, where the bulk masses $k\, C_x^{ij}$ are diagonal ($x = Q, U, D$; $i, j = 1, 2, 3$; k is the AdS curvature), the value f_{x^i} of the profile of the quark zero modes is given by

$$f_{x^i}^2 = (1 - 2c_{x^i})/(1 - \varepsilon^{1-2c_{x^i}}). \qquad 46.$$

Here c_{x^i} are the eigenvalues of the C_x matrices, $\varepsilon = \exp[-\xi]$, $\xi = \log[M_{\overline{\rm Pl}}/{\rm TeV}]$, and $M_{\overline{\rm Pl}}$ is the reduced Planck mass. If $c_{x^i} < 1/2$, then f_{x^i} is exponentially suppressed. Thus, order-one variations in the five-dimensional masses yield large hierarchies in the four-dimensional flavor parameters. We consider the cases in which the Higgs VEV either propagates in the bulk (76) or is localized on the IR brane. For a bulk Higgs case, the profile is given by $\tilde{v}(\beta, z) \simeq v\sqrt{k(1+\beta)}\bar{z}^{2+\beta}/\varepsilon$, where $\bar{z} \in (\varepsilon, 1)$ ($\bar{z} = 1$ on the IR brane) and $\beta \geq 0$. The $\beta = 0$ case describes a Higgs that is maximally spread into the bulk, saturating the AdS stability bound (77). The relevant part of the effective four-dimensional Lagrangian, which involves the zero modes and the first KK gauge states, can be approximated by (69, 74)

$$\mathcal{L}^{4D} \supset \left(Y_{5D}^{u,d}\right)_{ij} \phi^{u,d}\, \bar{Q}_i f_{Q_i} (U, D)_j\, f_{U_j,D_j} r_{00}^{\phi}(\beta, c_{Q_i}, c_{U_j,D_j}) + g_* G^1 x_i^\dagger x_i \left[f_{x^i}^2 r_{00}^g(c_{x^i}) - 1/\xi\right], \quad 47.$$

where $\phi^{u,d} = \tilde{\phi}, \phi$, g_* stands for a generic effective gauge coupling and summation over i and j is implied. The correction for the couplings from the case of fully IR-localized KK and Higgs states is given by the functions r_{00}^{ϕ} (74) and r_{00}^{g} (78, 79):

$$r_{00}^{\phi}(\beta, c_L, c_R) \approx \frac{\sqrt{2(1+\beta)}}{2+\beta-c_L-c_R} \quad \text{and} \quad r_{00}^{g}(c) \approx \frac{\sqrt{2}}{J_1(x_1)}\frac{0.7}{6-4c}\left(1+e^{c/2}\right), \qquad 48.$$

where $r_{00}^{\phi}(\beta, c_L, c_R) = 1$ for brane-localized Higgs and $x_1 \approx 2.4$ is the first root of the Bessel function, $J_0(x_1) = 0$.

In **Table 6** we present an example of a set of f_{x^i} values that, starting from anarchical five-dimensional Yukawa couplings, reproduce the correct hierarchy of the flavor parameters. We assume, for simplicity, an IR-localized Higgs. The values depend on two input parameters: (*a*) f_{U^3}, which has been determined assuming a maximally localized t_R ($c_{u^3} = -0.5$); and (*b*) $y_{5\rm D}$, the overall scale of the five-dimensional Yukawa couplings in units of k, which has been fixed to its maximal value, assuming three KK states. On general grounds, the value of $y_{5\rm D}$ is bounded from above, as a function of the number of KK levels, by the requirement that Yukawa interactions be perturbative below the cutoff of the theory, $\Lambda_{5\rm D} \sim N_{\rm KK} k$, and it is bounded from below to account for the large top mass. Thus, the following range for $y_{5\rm D}$ is obtained (e.g., 76, 80):

Table 6 Values of the f_{x^i} parameters that reproduce the observed quark masses and CKM mixing angles starting from anarchical five-dimensional Yukawa couplings[a]

Flavor	f_Q	f_U	f_D
1	$A\lambda^3 f_{Q^3} \sim 3\times10^{-3}$	$\frac{m_u}{m_t}\frac{f_{U^3}}{A\lambda^3} \sim 1\times10^{-3}$	$\frac{m_d}{m_b}\frac{f_{D^3}}{A\lambda^3} \sim 2\times10^{-3}$
2	$A\lambda^2 f_{Q^3} \sim 1\times10^{-2}$	$\frac{m_c}{m_t}\frac{f_{U^3}}{A\lambda^2} \sim 0.1$	$\frac{m_s}{m_b}\frac{f_{D^3}}{A\lambda^2} \sim 1\times10^{-2}$
3	$\frac{m_t}{v y_{5D} f_{U^3}} \sim 0.3$	$\sqrt{2}$	$\frac{m_b}{m_t} f_{U^3} \sim 2\times10^{-2}$

[a]The parameters are defined in Equation 46. We fix $f_{U^3} = \sqrt{2}$ and $y_{5D} = 2$.

Table 7 Most significant flavor constraints in the Randall-Sundrum framework[a]

	$M_G^{\min}$ (TeV)		$y_{5D}^{\min}$ or $f_{Q_3}^{\max}$	
Observable	**Infrared Higgs**	**$\beta = 0$**	**Infrared Higgs**	**$\beta = 0$**
$CPV\text{-}B_d^{LLLL}$	$12\, f_{Q_3}^2$	$12\, f_{Q_3}^2$	$f_{Q_3}^{\max} = 0.5$	$f_{Q_3}^{\max} = 0.5$
$CPV\text{-}B_d^{LLRR}$	$4.2/y_{5D}$	$2.4/y_{5D}$	$y_{5D}^{\min} = 1.4$	$y_{5D}^{\min} = 0.82$
$CPV\text{-}D^{LLLL}$	$0.73\, f_{Q_3}^2$	$0.73\, f_{Q_3}^2$	No bound	No bound
$CPV\text{-}D^{LLRR}$	$4.9/y_{5D}$	$2.4/y_{5D}$	$y_{5D}^{\min} = 1.6$	$y_{5D}^{\min} = 0.8$
ε_K^{LLLL}	$7.9\, f_{Q_3}^2$	$7.9\, f_{Q_3}^2$	$f_{Q_3}^{\max} = 0.62$	$f_{Q_3}^{\max} = 0.62$
ε_K^{LLRR}	$49/y_{5D}$	$24/y_{5D}$	Above 6.7[b]	$y_{5D}^{\min} = 8$

[a]The values of $y_{5D}^{\min}$ and $f_{Q_3}^{\max}$ correspond to $M_{KK} = 3$ TeV. The bounds are obtained assuming maximal CP-violating phases and $g_{s*} = 3$.

[b]Above bound means that for $M_{KK} = 3$ TeV, y_{5D} is outside the perturbative range.

$$\frac{1}{2} \lesssim y_{5D} \lesssim \frac{2\pi}{N_{KK}} \quad \text{for brane Higgs and} \quad \frac{1}{2} \lesssim y_{5D} \lesssim \frac{4\pi}{\sqrt{N_{KK}}} \quad \text{for bulk Higgs,} \qquad 49.$$

where we use the rescaling $y_{5D} \to y_{5D}\sqrt{1+\beta}$, which produces the correct $\beta \to \infty$ limit (81) and avoids subtleties in the $\beta = 0$ case.

With anarchical five-dimensional Yukawa matrices, an RS-residual little-CP problem remains (80): Too-large contributions to the neutron EDM (69) and sizable chirally enhanced contributions to ε_K (12, 78, 82–84) are predicted. The RS leading contribution to ε_K is generated by a tree-level KK-gluon exchange, which leads to an effective coupling for the chirality-flipping operator in Equation 45 of the type (78, 82–84)

$$\begin{aligned} C_4^K &\simeq \frac{g_{s*}^2}{M_{KK}^2} f_{Q_2} f_{Q_1} f_{d_2} f_{d_1} r_{00}^g(c_{Q_2}) r_{00}^g(c_{d_2}) \\ &\sim \frac{g_{s*}^2}{M_{KK}^2} \frac{2 m_d m_s}{(v y_{5D})^2} \frac{r_{00}^g(c_{Q_2}) r_{00}^g(c_{d_2})}{r_{00}^H(\beta, c_{Q_1}, c_{d_1}) r_{00}^H(\beta, c_{Q_2}, c_{d_2})}. \end{aligned} \qquad 50.$$

The final expression is independent of the f_{x^i}, so the bound in **Table 1** can be translated into constraints in the y_{5D}–M_{KK} plane. The analogous effects in the D and B systems yield numerically weaker bounds. Another class of contributions, which involves only left-handed quarks, is also important for constraining the f_Q–M_{KK} parameter space.

In **Table 7** we summarize the resulting constraints. For the purpose of a quantitative analysis, we set $g_{s*} = 3$, as obtained by matching to the four-dimensional coupling at one loop (76; for the impact of a smaller RS volume, see Reference 85). The constraints related to CP violation correspond to maximal phases and are subject to the requirement that the RS contributions be smaller than 30% (60%) of the SM contributions (9) in the B_d (K) system. The analytical expressions in the table have an accuracy of approximately 10% over the relevant range of parameters. Contributions from scalar exchange, either Higgs (81) or radion (86), are not included because they are more model dependent and are known to be weaker (87) in the brane-localized Higgs case.

Constraints from $\varepsilon'/\varepsilon_K$ have a different parameter dependence from that of the ε_K constraints. Specifically, for $\beta = 0$, the $\varepsilon'/\varepsilon_K$ constraint reads $M_G^{\min} = 1.2 y_{5D}$ TeV. Combining this constraint with the ε constraint yields $M_G^{\min} = 5.5$ TeV, with a corresponding $y_{5D}^{\min} = 4.5$ (74).

The constraints summarized in **Table 7** and the contributions to the neutron EDM, which generically require $M_{KK} > \mathcal{O}(20\,\text{TeV})$ (69), are a clear manifestation of the RS-residual little-CP problem. The problem can be amended by various alignment mechanisms (79, 80, 88). In this case, the bounds from the up sector, especially from *CP* violation in the *D* system (16, 47), become important. Constraints from $\Delta F = 1$ processes [in either the down sector (69, 84) or $t \to cZ$ (89)] are not included here because they are weaker, and furthermore, these contributions can be suppressed (84) due to the incorporation of a custodial symmetry (90).

7. FUTURE PROSPECTS

The NP flavor puzzle is why, and in what way, the flavor structure of teraelectronvolt-scale NP is nongeneric. Indeed, the flavor predictions of most NP models are not a consequence of their generic features but rather arise from the special structures that are imposed specifically to satisfy the existing severe flavor bounds. Therefore, flavor physics is a powerful indirect probe of NP. We hope that NP not far above the weak scale will be discovered at the LHC. A major issue will then be to understand its flavor structure. Although it is not easy to directly probe this flavor structure at high energy, much can be learned from low-energy flavor physics.

The precision with which we can probe high-scale physics in flavor physics experiments is often limited by theoretical uncertainties. Moreover, in the case of theoretically clean observables, the sensitivity to the NP scale increases slowly with the statistics of the experiment. Thus, the important questions in view of future experiments are the following:

1. What are the expected deviations from the SM predictions induced by NP at the teraelectronvolt scale?
2. Which observables are not limited by theoretical uncertainties?
3. In what case can we expect a substantial improvement on the experimental side?
4. What will the measurements teach us if deviations from the SM are (not) observed?

These questions have been analyzed in a series of recent works (e.g., 46, 49, 91–93), and the main conclusions can be summarized as follows:

1. The expected deviations from the SM predictions induced by NP at the teraelectronvolt scale with generic flavor structure have already been ruled out by many orders of magnitude. On general grounds, we can expect any size of deviation below the current bounds. In the most pessimistic frameworks, such as MFV, the typical size of the deviations is at the few-percent level in FCNC amplitudes.
2. The theoretical limitations are highly process dependent. In most multihadron final states, calculations of decay amplitudes are already limited by theoretical uncertainties. However, several channels involving leptons in the final state, and selected time-dependent asymmetries, have theoretical errors well below the current experimental sensitivity.
3. On the experimental side, there is a good prospect of improvement. As summarized in **Table 8**, one-order-of-magnitude improvements in several clean $B_{s,d}$, D, and K observables may be possible within a few years. Moreover, improvements of several orders of magnitudes are expected in top decays, which will be explored in detail for the first time at the LHC.
4. There is no doubt that new low-energy flavor data will be complementary with the high-p_T part of the LHC program. As illustrated above, the synergy of both data sets can teach us a great deal about NP at the teraelectronvolt scale.

Although some improvements can still be expected from running experiments, especially from CDF and DØ at Fermilab, a substantial step forward can be achieved only with new dedicated facilities. At the LHC, which has just begun to operate, the LHCb experiment is expected to collect approximately 10 fb^{-1} of data by 2015. Additionally, an LHCb upgrade with ten-times-higher luminosity is planned. At CERN, the NA62 experiment is expected to begin in 2012; it will

Table 8 Status and prospects of selected $B_{s,d}$, D, K, and t observables[a]

Observable	Standard Model prediction	Theory error[b]	Present result[c]	Future error[d]	Future facility
$\|V_{us}\|\ (K \to \pi \ell \nu)$	Input	0.5% → 0.1%$_{\text{Latt}}$	0.2246 ± 0.0012	0.1%	*K* factory
$\|V_{cb}\|\ (B \to X_c \ell \nu)$	Input	1%	$(41.54 \pm 0.73) \times 10^{-3}$	1%	Super-*B*
$\|V_{ub}\|\ (K \to \pi \ell \nu)$	Input	10% → 5%$_{\text{Latt}}$	$(3.38 \pm 0.36) \times 10^{-3}$	4%	Super-*B*
$\gamma\ (B \to DK)$	Input	<1°	$(70^{+27}_{-30})°$	3°	LHCb
$S_{B_d \to \psi K}$	$\sin(2\beta)$	$\lesssim 0.01$	0.671 ± 0.023	0.01	LHCb
$S_{B_s \to \psi \phi}$	0.036	$\lesssim 0.01$	$0.81^{+0.12}_{-0.32}$	0.01	LHCb
$S_{B_d \to \phi K}$	$\sin(2\beta)$	$\lesssim 0.05$	0.44 ± 0.18	0.1	LHCb
$S_{B_s \to \phi \phi}$	0.036	$\lesssim 0.05$	—	0.05	LHCb
$S_{B_d \to K^* \gamma}$	Few × 0.01	0.01	−0.16 ± 0.22	0.03	Super-*B*
$S_{B_s \to \phi \gamma}$	Few × 0.01	0.01	—	0.05	LHCb
A^d_{SL}	-5×10^{-4}	10^{-4}	$-(5.8 \pm 3.4) \times 10^{-3}$	10^{-3}	LHCb
A^s_{SL}	2×10^{-5}	$<10^{-5}$	$(1.6 \pm 8.5) \times 10^{-3}$	10^{-3}	LHCb
$A_{CP}(b \to s\gamma)$	<0.01	<0.01	−0.012 ± 0.028	0.005	Super-*B*
$\mathcal{B}(B \to \tau \nu)$	1×10^{-4}	20% → 5%$_{\text{Latt}}$	$(1.73 \pm 0.35) \times 10^{-4}$	5%	Super-*B*
$\mathcal{B}(B \to \mu \nu)$	4×10^{-7}	20% → 5%$_{\text{Latt}}$	$<1.3 \times 10^{-6}$	6%	Super-*B*
$\mathcal{B}(B_s \to \mu^+ \mu^-)$	3×10^{-9}	20% → 5%$_{\text{Latt}}$	$<5 \times 10^{-8}$	10%	LHCb
$\mathcal{B}(B_d \to \mu^+ \mu^-)$	1×10^{-10}	20% → 5%$_{\text{Latt}}$	$<1.5 \times 10^{-8}$	?	LHCb
$A_{\text{FB}}(B \to K^* \mu^+ \mu^-)_{q_0^2}$	0	0.05	(0.2 ± 0.2)	0.05	LHCb
$B \to K \nu \bar{\nu}$	4×10^{-6}	20% → 10%$_{\text{Latt}}$	$<1.4 \times 10^{-5}$	20%	Super-*B*
$\|q/p\|_{D\text{-missing}}$	1	$<10^{-3}$	$(0.86^{+0.18}_{-0.15})$	0.03	Super-*B*
ϕ_D	0	$<10^{-3}$	$(9.6^{+8.3}_{-9.5})°$	2°	Super-*B*
$\mathcal{B}(K^+ \to \pi^+ \nu \bar{\nu})$	8.5×10^{-11}	8%	$(1.73^{+1.15}_{-1.05}) \times 10^{-10}$	10%	*K* factory
$\mathcal{B}(K_L \to \pi^0 \nu \bar{\nu})$	2.6×10^{-11}	10%	$<2.6 \times 10^{-8}$	?	*K* factory
$R^{(e/\mu)}(K \to \pi \ell \nu)$	2.477×10^{-5}	0.04%	$(2.498 \pm 0.014) \times 10^{-5}$	0.1%	*K* factory
$\mathcal{B}(t \to c\, Z, \gamma)$	$\mathcal{O}(10^{-13})$	$\mathcal{O}(10^{-13})$	$<0.6 \times 10^{-2}$	$\mathcal{O}(10^{-5})$	LHC (100 fb^{-1})

[a]Based on information from References 46, 91, and 92.

[b]Latt refers to improvements in lattice quantum chromodynamics that are expected over the next 5 years.

[c]The bounds are at 90% CL.

[d]The errors refer to 10 fb^{-1} at LHCb, 50 ab^{-1} at Super-*B*, and 2 years at NA62 (the *K* factory). Errors followed by percentage signs are relative errors, whereas the others are absolute errors. A question mark indicates that a reliable estimate of future experimental prospects has not yet been found.

aim to collect $\mathcal{O}(100)$ events of the rare decay $K^+ \to \pi^+ \nu \bar{\nu}$ by 2014. Dedicated rare-*K* experiments are also planned at JParc and at Fermilab. The project to upgrade KEK-B into a Super-*B* factory, with a luminosity exceeding $5 \times 10^{35}\ \text{cm}^{-2}\ \text{s}^{-1}$, has just begun, and an even more ambitious Super-*B* project is currently under study in Italy.

In **Table 8** (17, 18, 46, 91, 92), we list possible improvements for a series of particularly significant $B_{s,d}$, D, K, and t observables, most of which are discussed above. The future improvements refer to 10 fb^{-1} at LHCb, 50 ab^{-1} at Super-*B*, and two years of nominal data taking at NA62. The table is not comprehensive, and the entries in the last two columns necessarily have significant uncertainties. However, it illustrates two important points: (*a*) There are still several clean observables from which we can expect significant improvements in the near future, and (*b*) progress in this field requires the combined efforts of different experimental facilities.

To conclude, we stress that the future of flavor physics is promising. The technology to collect much more flavor data exists, there are still several measurements that are not theory limited, and most well-motivated teraelectronvolt-scale extensions of the SM predict deviations within future experimental sensitivities and above the SM theoretical uncertainties. The combination of direct discoveries at the LHC and a pattern of deviations in flavor machines may help solve the NP flavor puzzle, indirectly probe physics at a scale much higher than the teraelectronvolt scale, and perhaps aid progress in understanding the SM flavor puzzle.

DISCLOSURE STATEMENT

The authors are not aware of any affiliations, memberships, funding, or financial holdings that might be perceived as affecting the objectivity of this review.

ACKNOWLEDGMENTS

We thank Kaustubh Agashe and Zoltan Ligeti for comments on the first version of this review. G.I. is supported by the European Union under contract MTRN-CT-2006-035482 Flavianet. Y.N. is supported by the Israel Science Foundation under grant number 377/07; by the German-Israeli Foundation for Scientific Research and Development; and by the United States–Israel Binational Science Foundation of Jerusalem, Israel. G.P. is supported by the Israel Science Foundation under grant number 1087/09, by Marie Curie grant number EU-FP7, by an International Reintegration Grant fellowship, and by a Peter and Patricia Gruber award.

LITERATURE CITED

1. Wolfenstein L. *Phys. Rev. Lett.* 51:1945 (1983)
2. Cabibbo N. *Phys. Rev. Lett.* 10:531 (1963)
3. Kobayashi M, Maskawa T. *Prog. Theor. Phys.* 49:652 (1973)
4. Buras AJ, Lautenbacher ME, Ostermaier G. *Phys. Rev. D* 50:3433 (1994)
5. Amsler C, et al. (Part. Data Group) *Phys. Lett.* B667:1 (2008)
6. Charles J, et al. (CKMfitter Group) *Eur. Phys. J. C* 41:1 (2005)
7. Nir Y. *Nucl. Phys. Proc. Suppl.* 117:111 (2003)
8. Ligeti Z. *Int. J. Mod. Phys. A* 20:5105 (2005)
9. Ligeti Z, Papucci M, Perez G. *Phys. Rev. Lett.* 97:101801 (2006); Agashe K, Papucci M, Perez G, Pirjol D. arXiv:hep-ph/0509117 (2005)
10. Silva JP, Wolfenstein L. *Phys. Rev. D* 55:5331 (1997)
11. Grossman Y, Nir Y, Worah MP. *Phys. Lett.* B407:307 (1997)
12. Bona M, et al. (UTfit Collab.) *J. High Energy Phys.* 0803:043 (2008)
13. Aaltonen T, et al. (CDF Collab.) *Phys. Rev. Lett.* 100:161802 (2008)
14. Abazov VM, et al. (DØ Collab.) *Phys. Rev. Lett.* 101:241801 (2008)
15. Bergmann S, et al. *Phys. Lett.* B 486:418 (2000)
16. Blum K, Grossman Y, Nir Y, Perez G. *Phys. Rev. Lett.* 102:211802 (2009)
17. Aaltonen T, et al. (CDF Collab.) *Phys. Rev. Lett.* 101:192002 (2008); CDF Collab. *Conf. Note 9496* (2008); Abazov VM, et al. (DØ Collab.) *Phys. Rev. Lett.* 100:192003 (2008)
18. Carvalho J, et al. (ATLAS Collab.) *Eur. Phys. J. C* 52:999 (2007); Aad G, et al. (ATLAS Collab.) arXiv:0901.0512 [hep-ex] (2009); de Roeck A, et al. *J. Phys. G* 34:995 (2006)
19. Coimbra RA, et al. *Phys. Rev. D* 79:014006 (2009); Nath P, et al. arXiv:1001.2693 [hep-ph] (2010)

20. Buchmuller W, Wyler D. *Nucl. Phys. B* 268:621 (1986); Aguilar-Saavedra JA. *Nucl. Phys. B* 812:181 (2009); Whisnant K, Yang JM, Young BL, Zhang X. *Phys. Rev. D* 56:467 (1997); Whisnant K, Yang JM, Young BL, Zhang X. *Phys. Rev. D* 56:467 (1997)
21. Fox PJ, et al. *Phys. Rev. D* 78:054008 (2008)
22. Gedalia O, Mannelli L, Perez G. arXiv:1002.0778 [hep-ph] (2010)
23. D'Ambrosio G, Giudice GF, Isidori G, Strumia A. *Nucl. Phys. B* 645:155 (2002)
24. Cirigliano V, Grinstein B, Isidori G, Wise MB. *Nucl. Phys. B* 728:121 (2005)
25. Davidson S, Palorini F. *Phys. Lett.* B642:72 (2006); Gavela MB, Hambye T, Hernandez D, Hernandez P. *J. High Energy Phys.* 0909:038 (2009)
26. Buras AJ, et al. *Phys. Lett.* B500:161 (2001)
27. Hurth T, Isidori G, Kamenik JF, Mescia F. *Nucl. Phys. B* 808:326 (2009)
28. Feldmann T, Mannel T. *J. High Energy Phys.* 0702:067 (2007)
29. Buras AJ. *Acta Phys. Polon. B* 34:5615 (2003)
30. Kagan AL, Perez G, Volansky T, Zupan J. *Phys. Rev. D* 80:076002 (2009)
31. Feldmann T, Mannel T. *Phys. Rev. Lett.* 100:171601 (2008)
32. Hall LJ, Rattazzi R, Sarid U. *Phys. Rev. D* 50:7048 (1994)
33. Blazek T, Raby S, Pokorski S. *Phys. Rev. D* 52:4151 (1995)
34. Isidori G, Retico A. *J. High Energy Phys.* 0111:001 (2001)
35. Hou WS. *Phys. Rev. D* 48:2342 (1993); Akeroyd AG, Recksiegel S. *J. Phys. G* 29:2311 (2003); Isidori G, Paradisi P. *Phys. Lett.* B639:499 (2006)
36. Carena M, Garcia D, Nierste U, Wagner CEM. *Nucl. Phys. B* 577:88 (2000); Carena M, Garcia D, Nierste U, Wagner CEM. *Phys. Lett.* B499:141 (2001); Degrassi G, Gambino P, Giudice GF. *J. High Energy Phys.* 0012:009 (2000)
37. Buras AJ, Chankowski PH, Rosiek J, Slawianowska L. *Nucl. Phys. B* 619:434 (2001); Buras AJ, Chankowski PH, Rosiek J, Slawianowska L. *Nucl. Phys. B* 659:3 (2003)
38. Hamzaoui C, Pospelov M, Toharia M. *Phys. Rev. D* 59:095005 (1999); Choudhury SR, Gaur N. *Phys. Lett.* B451:86 (1999); Babu KS, Kolda CF. *Phys. Rev. Lett.* 84:228 (2000)
39. Ellis JR, Lee JS, Pilaftsis S. *Phys. Rev. D* 76:115011 (2007)
40. Mercolli L, Smith C. *Nucl. Phys. B* 817:1 (2009)
41. Colangelo G, Nikolidakis E, Smith C. *Eur. Phys. J. C* 59:75 (2009)
42. Hiller G, Hochberg Y, Nir Y. *J. High Energy Phys.* 0903:115 (2009)
43. Raz G. *Phys. Rev. D* 66:037701 (2002)
44. Masiero A, Vempati SK, Vives O. arXiv:0711.2903 [hep-ph] (2007)
45. Ciuchini M, et al. *Phys. Lett.* B655:162 (2007)
46. Artuso M, et al. arXiv:0801.1833 [hep-ph] (2008)
47. Gedalia O, Grossman Y, Nir Y, Perez G. *Phys. Rev. D* 80:055024 (2009)
48. Foster J, Okumura Ki, Roszkowski L. *Phys. Lett.* B641:452 (2006); Isidori G, Retico A. *J. High Energy Phys.* 0209:063 (2002)
49. Raidal M, et al. *Eur. Phys. J. C* 57:13 (2008)
50. Gabbiani F, Gabrielli E, Masiero A, Silvestrini L. *Nucl. Phys. B* 477:321 (1996)
51. Shadmi Y, Shirman Y. *Rev. Mod. Phys.* 72:25 (2000)
52. Dine M, Nelson AE. *Phys. Rev. D* 48:1277 (1993)
53. Dine M, Nelson AE, Shirman Y. *Phys. Rev. D* 51:1362 (1995)
54. Dine M, Nelson AE, Nir Y, Shirman Y. *Phys. Rev. D* 53:2658 (1996)
55. Meade P, Seiberg N, Shih D. *Prog. Theor. Phys. Suppl.* 177:143 (2009)
56. Randall L, Sundrum R. *Nucl. Phys. B* 557:79 (1999)
57. Giudice GF, Luty MA, Murayama H, Rattazzi R. *J. High Energy Phys.* 9812:027 (1998)
58. Chacko Z, Luty MA, Nelson AE, Ponton E. *J. High Energy Phys.* 0001:003 (2000)
59. Paradisi P, Ratz M, Schieren R, Simonetto C. *Phys. Lett.* B668:202 (2008)
60. Feng JL, Lester CG, Nir Y, Shadmi Y. *Phys. Rev. D* 77:076002 (2008)
61. Kribs GD. arXiv:hep-ph/0605325 (2006)
62. Arkani-Hamed N, Schmaltz M. *Phys. Rev. D* 61:033005 (2000)

63. Delgado A, Pomarol A, Quiros M. *J. High Energy Phys.* 0001:030 (2000)
64. Agashe K. In *Proceedings of the Theoretical Advanced Study Institute in Elementary Particle Physics (TASI)*, ed. S Dawson, RN Mohapatra, p. 1. Singapore: World Sci. (2006)
65. Randall L, Sundrum R. *Phys. Rev. Lett.* 83:3370 (1999)
66. Grossman Y, Neubert M. *Phys. Lett.* B474:361 (2000)
67. Huber SJ, Shafi Q. *Phys. Lett.* B498:256 (2001); Gherghetta T, Pomarol A. *Nucl. Phys. B* 586:141 (2000)
68. Huber SJ. *Nucl. Phys. B* 666:269 (2003)
69. Agashe K, Perez G, Soni A. *Phys. Rev. Lett.* 93:201804 (2004); *Phys. Rev. D* 71:016002 (2005)
70. Burdman G. *Phys. Rev. D* 66:076003 (2002); Burdman G. *Phys. Lett.* B590:86 (2004)
71. Kaplan DB, Georgi H. *Phys. Lett.* B136:183 (1984); *Phys. Lett.* B145:216 (1984); Georgi H, Kaplan DB. *Phys. Lett.* B 145:216 (1984); Georgi H, Kaplan DB, Galison P. *Phys. Lett.* B143:152 (1984); Georgi H, Kaplan DB, Galison P. *Phys. Lett.* B143:152 (1984)
72. Maldacena JM. *Adv. Theor. Math. Phys.* 2:231 (1998); Aharony O, et al. *Phys. Rep.* 323:183 (2000); Witten E. *Adv. Theor. Math. Phys.* 2:253 (1998)
73. Arkani-Hamed N, Porrati M, Randall L. *J. High Energy Phys.* 0108:017 (2001); Rattazzi R, Zaffaroni A. *J. High Energy Phys.* 0104:021 (2001)
74. Gedalia O, Isidori G, Perez G. *Phys. Lett.* B682:200 (2009)
75. Contino R, Kramer T, Son M, Sundrum R. *J. High Energy Phys.* 0705:074 (2007)
76. Agashe K, Azatov A, Zhu L. *Phys. Rev. D* 79:056006 (2009)
77. Breitenlohner P, Freedman DZ. *Phys. Lett.* B115:197 (1982)
78. Csáki C, Falkowski A, Weiler A. *J. High Energy Phys.* 0809:008 (2008)
79. Csaki C, Curtin D. *Phys. Rev. D* 80:015027 (2009)
80. Csaki C, Perez G, Surujon Z, Weiler A. arXiv:0907.0474 [hep-ph] (2009)
81. Agashe K, Contino R. arXiv:0906.1542 [hep-ph] (2009); Azatov A, Toharia M, Zhu L. arXiv:0906.1990 [hep-ph] (2009)
82. Casagrande S, et al. *J. High Energy Phys.* 0810:094 (2008); Blanke M, et al. *J. High Energy Phys.* 0903:001 (2009)
83. Davidson S, Isidori G, Uhlig S. *Phys. Lett.* B663:73 (2008)
84. Blanke M, et al. *J. High Energy Phys.* 0903:108 (2009); Albrecht ME, et al. *J. High Energy Phys.* 0909:064 (2009); Buras AJ, Duling B, Gori S. *J. High Energy Phys.* 0909:076 (2009); Bauer M, Casagrande S, Haisch U, Neubert M. arXiv:0912.1625 [hep-ph] (2009)
85. Davoudiasl H, Perez G, Soni A. *Phys. Lett.* B 665:67 (2008)
86. Azatov A, Toharia M, Zhu L. *Phys. Rev. D* 80:031701 (2009)
87. Duling B. arXiv:0912.4208 [hep-ph] (2009)
88. Csaki C, Falkowski A, Weiler A. *Phys. Rev. D* 80:016001 (2009); Santiago J. *J. High Energy Phys.* 0812:046 (2008); Fitzpatrick AL, Randall L, Perez G. *Phys. Rev. Lett.* 100:171604 (2008)
89. Agashe K, Perez G, Soni A. *Phys. Rev. D* 75:015002 (2007)
90. Agashe K, Contino R, Da Rold L, Pomarol A. *Phys. Lett.* B641:62 (2006)
91. Grossman Y, Ligeti Z, Nir Y. *Prog. Theor. Phys.* 122:125 (2009)
92. Antonelli M, et al. arXiv:0907.5386 [hep-ph] (2009)
93. Buras AJ. arXiv:0910.1032 [hep-ph] (2009); Isidori G. arXiv:1001.3431 [hep-ph] (2010)

The Cold and Hot CNO Cycles

M. Wiescher,[1,2] J. Görres,[1,2] E. Uberseder,[1,2]
G. Imbriani,[3] and M. Pignatari[2,4]

[1]Department of Physics and [2]Joint Institute of Nuclear Astrophysics, University of Notre Dame, Notre Dame, Indiana 46556; email: mwiesche@nd.edu, jgoerres@nd.edu, eubersed@nd.edu

[3]Dipartimento di Scienze Fisiche, Università degli Studi di Napoli "Federico II," INFN Sezione di Napoli, Napoli I-80126, Italy; email: gianluca.imbriani@na.infn.it

[4]Department of Physics and Astronomy, University of Victoria, Victoria, British Columbia V8W 3P6, Canada, and TRIUMF, Vancouver, British Columbia V6T 2A3, Canada; email: mpignatari@gmail.com

Annu. Rev. Nucl. Part. Sci. 2010. 60:381–404

First published online as a Review in Advance on July 16, 2010

The *Annual Review of Nuclear and Particle Science* is online at nucl.annualreviews.org

This article's doi:
10.1146/annurev.nucl.012809.104505

0163-8998/10/1123-0381$20.00

Key Words

nuclear astrophysics, quiescent and explosive stellar hydrogen burning, radiative capture reactions

Abstract

New experimental methods and techniques, combined with the development of new theoretical tools, have opened new avenues to explore nuclear reactions of significance for nucleosynthesis at or near the actual temperatures of stellar burning. In particular, many reactions of the cold and hot CNO cycles have been investigated in recent years to provide a complete understanding of this critical hydrogen-burning mechanism in stars and stellar explosions. This has led to new interpretations of or new signatures for a number of critical hydrogen-burning environments and events. This article provides a summary of the most recent discoveries and results associated with CNO reactions, and it identifies existing shortcomings in the data as well as needs and opportunities for additional future experiments.

Contents

1. INTRODUCTION

Stellar hydrogen burning is characterized by two major reaction sequences: the *pp* chains and the carbon-nitrogen-oxygen (CNO) cycles. Whereas the *pp* chains represent a sequence of fusion processes of hydrogen nuclei forming ^4He, the CNO cycles are catalytic sequences of four proton capture reactions and two positron decays involving carbon, nitrogen, and oxygen nuclei with final α emission. In both cases, the generated energy per fusion corresponds to the ^4He-binding energy of 26.7 MeV.

The *pp* chains are triggered by the fusion of two protons that is based on the weak interaction. A subsequent radiative proton capture reaction forming ^3He, $^1H(p,\beta^+\nu)^2H(p,\gamma)^3He$, followed by the fusion reaction $^3He(^3He,2p)^4He$, completes the first *pp* chain. The second *pp* chain is triggered by the radiative capture process $^3He(\alpha,\gamma)^7Be$ followed by electron capture, proton capture, and α emission, $^7Be(e^-,\nu)^7Li(p,\alpha)^4He$. The alternative third *pp* chain is triggered by radiative proton capture on ^7Be with subsequent β-delayed α emission, $^7Be(p,\gamma)^8B(\beta^+\nu,\alpha)^4He$. The energy generation of the *pp* chains is limited by the $p+p$ fusion process, which is based on the weak interaction. Because of the extremely low cross section, *pp* chain–based hydrogen burning and the associated energy production are slow. The *pp* chains can operate on pure hydrogen fuel, which makes it the likely energy source for hydrogen burning in population III stars with primordial light element abundance distribution synthesized in the third minute of big bang nucleosynthesis. In 1936, von Weizsäcker (1) first suggested *pp* chains as a possible energy source in stars. In 1938, Bethe & Critchfield (2) derived the actual formalism, and in 1953, *pp* chains were identified as the primary energy source of our Sun (3).

The CNO cycle is a catalytic process that requires the presence of some initial carbon, nitrogen, and oxygen abundance in the stellar material. During the cyclic processing, the initial isotope abundances are changed and redistributed over the CNO mass range. Between 1937 and 1939, Bethe (4) and von Weizsäcker (5) independently proposed the first CNO cycle or CN cycle, $^{12}C(p,\gamma)^{13}N(\beta^+\nu)^{13}C(p,\gamma)^{14}N(p,\gamma)^{15}O(\beta^+\nu)^{15}N(p,\alpha)^{12}C$, as a possible mechanism for hydrogen burning in stars. In 1957, Burbidge et al. (6) proposed in their seminal paper on stellar nucleosynthesis a second CNO cycle, the NO cycle $^{16}O(p,\gamma)^{17}F(\beta^+\nu)^{17}O(p,\alpha)^{14}N(p,\gamma)^{15}O(\beta^+\nu)^{15}N(p,\gamma)^{16}O$, which is fueled mainly by the oxygen abundance in stellar material. A full NO cycle depends on the strength of the $^{15}N(p,\gamma)^{16}O$ radiative capture branch in the $^{15}N+p$ reaction channel. If the radiative capture branch is weak, the initial ^{16}O will eventually be distributed over the isotopes associated with the CN cycle (7). In an experiment in 1974, Rolfs & Rodney (8) found evidence for a third CNO cycle fueled by radiative proton capture on ^{17}O, as outlined by Dearborn & Schramm (9): $^{17}O(p,\gamma)^{18}F(\beta^+\nu)^{18}O(p,\alpha)^{15}N(p,\gamma)^{16}O(p,\gamma)^{17}F(\beta^+\nu)^{17}O$. Again, a full cycle depends on the strength of the radiative capture process that has to compete with the emission of an α particle from the ^{16}O and ^{18}F compound nuclei. Operating in a similar way, a possible fourth CNO cycle, $^{18}O(p,\gamma)^{19}F(p,\alpha)^{16}O(p,\gamma)^{17}F(\beta^+\nu)^{17}O(p,\gamma)^{18}F(\beta^+\nu)^{18}O$, would cause an even wider distribution of the initial fuel abundances. It is triggered by radiative capture on ^{18}O, which is not negligible compared with the $^{18}O(p,\alpha)^{15}N$ reaction (10, 11).

These competing (p,γ)–(p,α) reactions are branching points in the cycling process. Typically, the strong-interaction (p,α) branch is substantially stronger than the electromagnetic (p,γ) branch. However, in some cases, depending on the spin and energy of resonances, the latter can be of comparable strength (11), altering the reaction flow substantially in certain temperature regimes.

Present CNO hydrogen-burning simulations are all based on the concept of the intertwined tetra-cycle as shown in **Figure 1**. It is primarily a closed cyclic burning system dominated by the CN-cycle component, with no leakage toward higher masses. This is in particular due to the pronounced α cluster structure of the two critical compound nuclei ^{16}O and ^{20}Ne, which

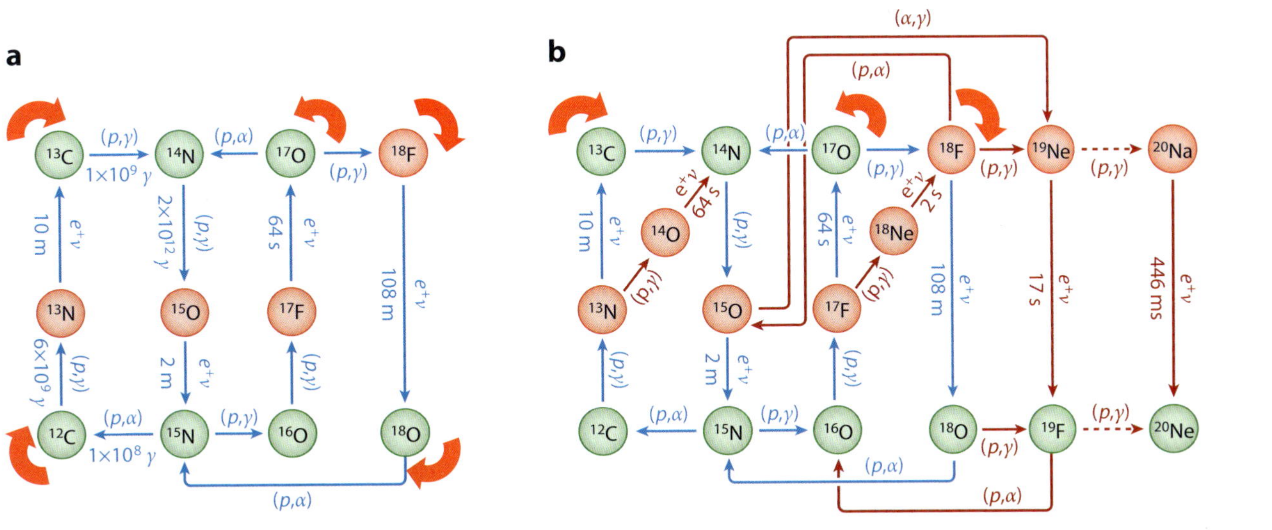

Figure 1

(*a*) The reaction flow in the three cold CNO cycles. (*b*) The hot CNO cycles, which represent a much more complex cyclic reaction pattern. Dashed lines indicate breakout reactions into the NeNa mass range, which occur at temperatures in excess of 0.4 GK.

enhance the strength of the $^{15}N(p,\alpha)^{12}C$ and $^{19}F(p,\alpha)^{16}O$ channels versus the competing radiative capture channels $^{15}N(p,\gamma)^{16}O$ and $^{19}F(p,\gamma)^{20}Ne$. With sufficient strength, these radiative capture reactions would provide a breakout link from the CNO cycles, passing the $Z = 8$ shell closure into the NeNa mass range at low temperatures (11).

The energy production rate in CNO burning depends on the timescale of the associated nuclear processes, which are determined by the various reaction and decay rates. These also determine the abundance distribution during CNO burning and in the CNO ashes. A detailed understanding of the nuclear reaction rates is key for a reliable analysis of stellar hydrogen burning with an appreciable CNO component.

Whereas the nuclear structure of these isotopes inhibits a breakout from the cold CNO cycles (12), high-temperature conditions provide alternative reaction pathways and proton capture reactions on radioactive isotopes such as ^{13}N, ^{17}F, and ^{18}F triggering the hot or β-limited CNO cycles (**Figure 1**) (13, 14). Energy production and nucleosynthesis in the cold CNO cycles depend critically on the reaction rates of the various proton capture processes, but in the hot CNO cycles, they are characterized by the β-decay rates of ^{14}O, ^{15}O, and ^{18}Ne (12). Breakout from the hot CNO cycles occurs at temperatures sufficiently high to allow α capture reaction on these isotopes, triggering a continuous reaction flow toward higher masses: the so-called αp process (15, 16).

In the following section, we provide a short overview of the role the CNO cycles play in a number of quiescent and explosive stellar-burning scenarios to stress the importance of a detailed understanding of the underlying nuclear physics. We discuss the nuclear physics aspects of the cold CNO cycles in Section 3. The nuclear physics questions associated with the description of the hot CNO cycles are presented in Section 4. Section 5 is dedicated to a discussion of possible CNO nuclear feeding and breakout processes that will alter the CNO abundance fuel. Finally, in Section 6, we discuss recent developments and plans to improve the experimental measurements of the most critical reactions in the cold and hot CNO cycles.

2. ASTROPHYSICAL ASPECTS OF CNO CYCLES

Here we describe the role of the CNO cycles in the stellar environment. We include the role of cold CNO burning as an energy source for massive main sequence stars and its impact on neutrino production in low-mass stars such as our Sun. Hot CNO cycles drive the thermonuclear runaway in classical nova explosions, whereas CNO fueling and breakout processes trigger the thermonuclear runaway on the surface of accreting neutron stars, a phenomenon that is observed as an X-ray burst. In **Figure 2**, we compare the CNO nucleosynthesis patterns calculated for constant density and temperature values typical for these scenarios. The temperature and density chosen for each of the simulations are $T = 2 \times 10^7$ K, $\rho = 10^2$ cm^{-3} (**Figure 2*c***); $T = 10^8$ K, $\rho = 10^3$ cm^{-3} (**Figure 2*b***); and $T = 4 \times 10^8$ K, $\rho = 10^4$ cm^{-3} (**Figure 2*a***). The timescales to reach equilibrium values for the CNO isotopes decrease drastically with increasing temperature, and the final abundance distribution shifts from primarily stable oxygen and nitrogen isotopes to a distribution of radioactive oxygen and neon isotopes.

2.1. CNO Cycles in Stellar Evolution

Quiescent hydrogen burning is the first burning stage in the evolution of stars. The ignition takes place through the *pp* chains when a sufficient temperature of approximately 6×10^6 K has been reached in the core of a gravitationally contracting protostar. The nuclear energy release builds sufficient internal pressure to stabilize the contraction, and the star maintains its equilibrium on the main sequence of the Hertzsprung-Russell diagram until the hydrogen fuel is burned.

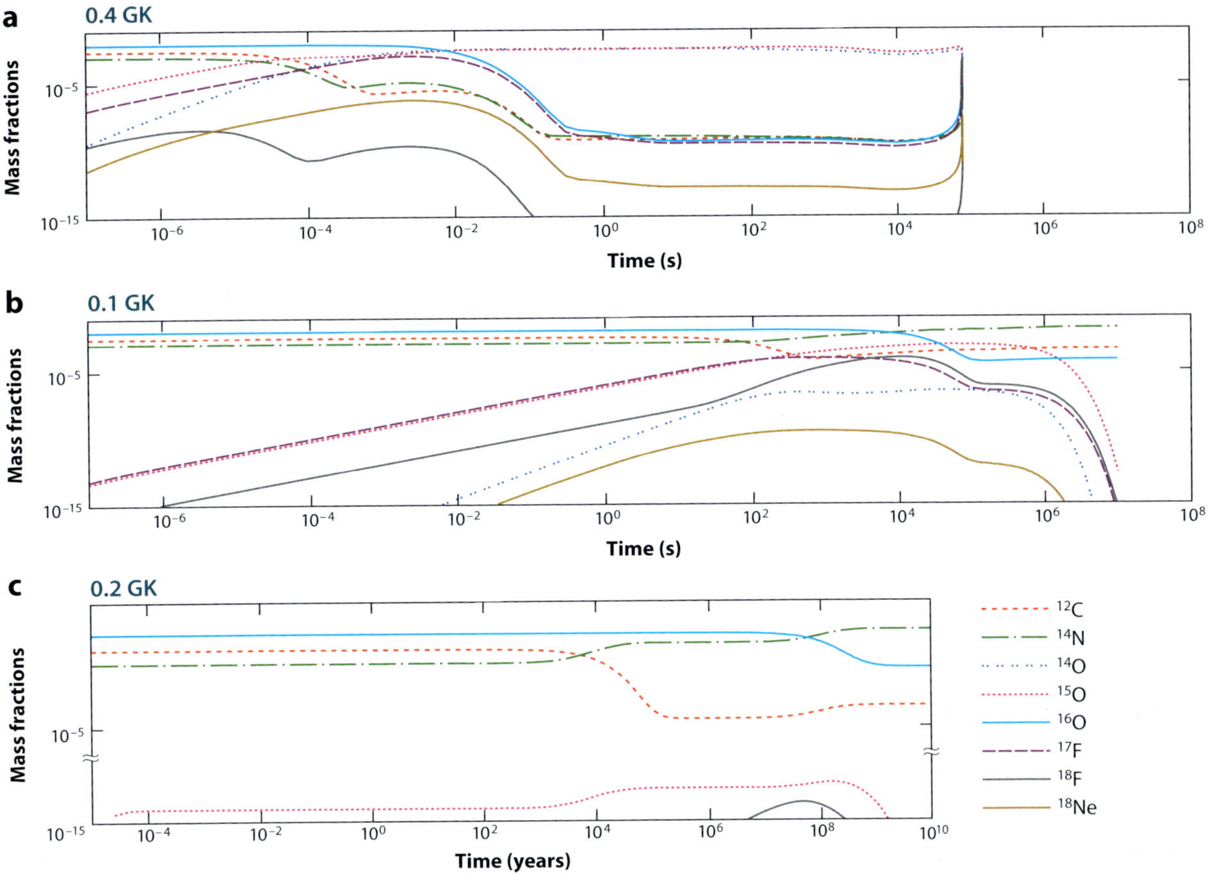

Figure 2

Evolution of selected CNO isotope abundances with time for different density and temperature conditions. These correspond to (*a*) peak conditions in novae explosions or the ignition conditions of X-ray bursts, to (*b*) conditions in the core of massive main sequence stars or in the hydrogen-burning shell of asymptotic giant branch stars, and to (*c*) conditions in the core of low-mass main sequence stars at the typical densities for these respective environments.

At ignition temperature, the *pp* chains are more efficient in energy production than the CNO cycles. However, the *p+p* fusion process has a relatively low Coulomb barrier and its overall energy production is only mildly temperature dependent: $\varepsilon_{pp} \propto T^4$. CNO-cycle processes are characterized by stronger Coulomb barriers and therefore ignite at a higher temperature, with a rapidly increasing energy production with temperature $\varepsilon_{CNO} \propto T^{18}$. At core temperatures of $T_C \approx 15 \times 10^6$ K, the hydrogen burning via the first CNO cycle (or CN cycle) becomes the most efficient energy source for hydrogen burning. Low-mass stars with $M \leq 1.5 \times M_{\mathrm{sun}}$ burn their hydrogen fuel primarily through the *pp* chains, with a small fraction from the CNO cycles. Massive stars with $M \geq 1.5 \times M_{\mathrm{sun}}$ maintain their stability primarily through CNO burning. The important difference between the two burning scenarios is the presence of a convective core in massive stars, because of the larger temperature gradient established by the CNO cycles.

The slowest reaction in the cycle, $^{14}\mathrm{N}(p,\gamma)^{15}\mathrm{O}$ for the cold CNO cycles, determines the timescale of the burning, which typically lasts a few million years depending on the core temperature of the star. During that time, most of the initial carbon in the core will be transformed into nitrogen. The detailed abundance distribution depends sensitively on the CNO reaction rates. The precise knowledge of these rates is therefore of crucial importance for the interpretation of new and more precise astronomical observations. In the following, we discuss a few examples of quiescent and explosive stellar environments where CNO burning plays a critical role and provides important signatures for the observer and modeler.

2.2. CNO Cycles in Low-Mass Stars

The determination of the ages for old globular clusters (GCs) strongly depends on the interplay of *pp* chains and CNO-cycle burning in the core of low-mass stars. GCs are the oldest components of Galactic halos. They represent a homogeneous and coeval group of old stars, having the same distance from us. The age of the oldest GCs is thought to coincide with the time of the star formation after the big bang and it can be derived by comparing the observed color-magnitude diagrams with theoretical isochrones obtained via a stellar evolution code. The age of GCs has been a relevant cosmological problem. For a long time, their age was estimated to be older than the age of the Universe obtained by means of a cosmological model (17, 18). During most of its life, a low-mass star burns hydrogen in the center via the *pp* chains. However, with the reduction of the hydrogen fuel, the energy production decreases and the core contracts gravitationally. This causes an increase of core temperature and density, and the CNO cycles become more efficient than the *pp* chains. Thus, the energy production in the cycle is directly correlated with the luminosity at the turnoff point from the main sequence in the color-magnitude diagram of the cluster as demonstrated in **Figure 3**. This is the critical parameter to fit in the comparison with the astronomical observation.

By contrast, the total lifetime is marginally affected by a change of the CNO rate, because it is determined mainly by the rate of the $p+p$ reaction. As a consequence, an increase of the CNO

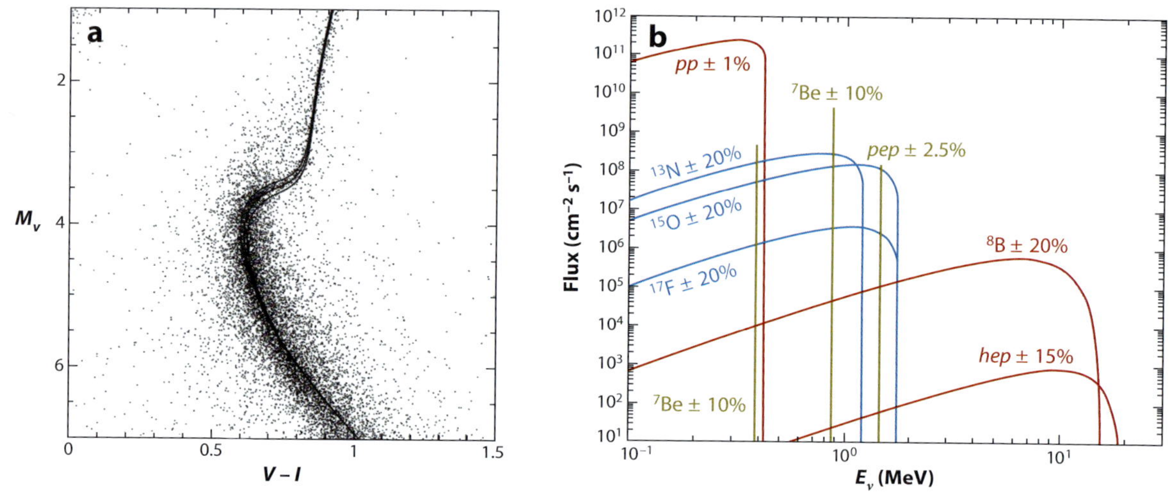

Figure 3

(*a*) The color-magnitude diagram of the globular cluster NGC 5904. The curves represent the best-fitting isochrones within the present uncertainties, namely 13, 14, and 15 Gy (see Reference 19 for details). (*b*) The predicted solar neutrino spectrum (26), including the CNO components associated with the decay of ^{13}N, ^{15}O, and ^{17}F (*blue*). The predictions are based on the presently adopted CNO reaction rates. The quoted uncertainties are based on new unpublished simulations performed by Oscar Straniero, Gianluca Imbriani, and Luciano Piersanti.

rate implies fainter turnoff points for a given age or younger ages for a given turnoff luminosity. Given the present reaction rates, the age of old GCs is approximately 14 ± 1 Gy (19). This age is in good agreement with 13.7 ± 0.2 Gy, which is the result of the WMAP experiment interpreted in the framework of the Λ cold dark matter model (19, 20).

2.3. CNO Cycles in Asymptotic Giant Branch Stars

The question of the origin of the element abundances at the surface of asymptotic giant branch (AGB) stars is determined by the interplay of CNO burning in the hydrogen-burning shell and deep-convection processes to the surface. Stars less massive than 8 M_{sun} during their evolution reach the AGB phase (21). An electron-degenerate CO core is surrounded by helium and hydrogen shells, where the hydrogen shell located just below the inner border of the convective envelope provides most of the energy required to balance the surface irradiation. Short-term thermonuclear runaways or thermal pulses localized in the helium shell recurrently make the helium-intershell zone unstable to convection, the external layers expand, and the hydrogen-burning shell temporarily loses efficiency. The entropy barrier created by the hydrogen-burning shell becomes weaker, and the convective envelope is allowed to penetrate the carbon-rich helium-intershell region (third dredge-up episode). As a result, elements previously synthesized in the internal layers, mainly carbon and *s*-process elements, are carried to the surface (21). The efficiency of the dredge-up, and consequently the amount of carbon and heavier elements carried on the surface, inversely scales with the rate of the CNO burning (22).

Different observations in low-mass red-giant branch and AGB stars ($M < 2M_{sun}$) seem to require extramixing processes, in which envelope material is mixed close to the hydrogen-burning shell and subsequently is partially processed by the CNO cycle (cool-bottom processing). For instance, in several presolar oxide grains recovered from meteorites, the measured oxygen isotopic ratios $^{17}O/^{16}O$ (higher than solar) and $^{18}O/^{16}O$ (lower than solar) were not explained by the first dredge-up, and extramixing during the red-giant phase would be required. Other isotopic ratios—for instance, $^{26}Al/^{27}Al$, C/N, and $^{12}C/^{13}C$ measured in grains—support such a conclusion. We stress that the physics mechanism that drives such extramixing is not yet clear, and several scenarios, including magnetic-induced mixing, have been proposed. In massive AGB stars ($4 < M/M_{sun} < 8$), a process known as hot-bottom burning is required to convert all the carbon dredged up in the envelope by the third dredge-up in nitrogen, which may eventually prevent massive AGB stars from having carbon-rich envelopes with $C/O > 1$ (e.g., 23 and references therein). After the AGB phase, along the post-AGB evolution track, the star may eventually experience a final thermal pulse event (very late thermal pulse). In such conditions, hydrogen left in a thin envelope is ingested in the helium-intershell region and burned at $T \leq 1.5 \times 10^8$ K, activating CNO cycles triggered mainly by the abundant ^{12}C that was produced by helium burning. Very late thermal-pulse events have been directly observed in great detail [Sakurai's object (24)], and the CNO reaction network has a crucial impact in nucleosynthesis. Similar hydrogen-ingestion events are more common at low metallicity, for instance, in helium-shell flash in AGB stars (23). For a review of cool-bottom processing, hot-bottom burning, and very late thermal pulse events, we refer the reader to Herwig 2005 (23 and references therein).

2.4. CNO Cycles in the Sun

For more than 30 years, the missing solar neutrino flux was considered one of the most interesting problems in physics (25). The need for physics beyond the standard electroweak model arose following (*a*) confirmation of the solar model, which was achieved using helioseismology to determine the interior sound speed (26), (*b*) new solar neutrino experiments [SAGE, GALLEX/GNO (27), and Super-Kamiokande (28)], and (*c*) the exclusion of nuclear solutions, such as an undiscovered low-energy resonance in the $^3He(^3He,2p)^4He$ (29). The simultaneous detection in the SNO experiment of ν_e from charged-current reactions (30) and of ν_e, ν_τ, and ν_μ from elastic scattering and neutral-current reactions showed definitively that electron neutrinos change flavor while traveling from the Sun's hot core, where they are produced as ν_e, to the Earth's surface, where they are detected. Additional knowledge about neutrino mass came from the KamLAND (31) reactor neutrino experiment and Borexino (32), which have recently measured low-energy solar neutrinos in real time.

The Sun is burning hydrogen mostly (~99%) by means of the *pp* chains; the remaining 1% is processed through the CNO cycles. Because of the weak interaction–based conversion of two protons into neutrons during the fusion process, two neutrinos are emitted for each helium nucleus produced. The emitted solar neutrinos are associated with five different weak-interaction processes in the *pp* chains and three in the CNO cycles. The flux and the energy distribution of neutrinos emitted from the Sun depend on the number and the type of the specific source reactions in the solar core, as shown in **Figure 3**.

After the initial solar neutrino problem was solved by introducing neutrino mass and mixing into the Standard Model, attention returned to one of the initial motivations for solar neutrino physics, that is, use of the neutrino flux as a probe of solar core hydrogen burning. The next generation of solar neutrino experiments (32), combined with the associated nuclear reaction cross sections, will allow researchers to make new measurements of solar properties, such as the core temperature and core metallicity, that can be compared to solar model predictions and assumptions. However, this comparison will be meaningful only if uncertainties in nuclear reaction rates, and thus in model neutrino flux predictions, are under control.

A particularly interesting case is the determination of the solar metallicity, where the $^{14}N(p,\gamma)^{15}O$ reaction is of crucial importance (33). The recent downward revision in the metal content of convective zones (34, 35) has generated interest in a check of the Standard Solar Model. Indeed, the new metal abundances significantly alter the agreement between predictions of the Standard Solar Model and helioseismology in the temperature region below the solar convective zone. CNO neutrinos play a key role in this study; in fact, their direct detection and identification will provide an independent approach to measure the core metallicity of the Sun. CNO neutrinos have an energy threshold of ~1 MeV, and their flux is approximately one order of magnitude lower than the ^7Be neutrino flux, which has comparable energy. A new generation of solar neutrino experiments, SNO+, and a second generation of Borexino aim to detect CNO neutrinos. By the time this is accomplished, the nuclear astrophysics community should have improved the present precision on the measurement of $^3He(\alpha,\gamma)^7Be$ and $^{14}N(p,\gamma)^{15}O$ cross sections.

2.5. Hot CNO Cycles in Novae

Classical novae contribute significantly to the production of light nuclei such as ^7Li, ^{13}C, ^{15}N, ^{17}O, ^{22}Na, and ^{26}Al (36, 37) and, consequently, to the production of γ lines, which has been observed from satellite-based γ-ray telescopes such as COMPTEL or INTEGRAL (38). Despite the relatively small amount of material that they eject during explosion (approximately $2 \times 10^{-5}\ M_{sun}$), the rate of nova explosion in the Galaxy is very high (approximately 30 years^{-1}). In addition, the overproduction factor for particular nuclei can be very high, such that novae may be a significant site for the production of these specific isotopes.

A nova outburst is a consequence of the accretion of hydrogen-rich material onto a white dwarf in a close binary system. The accreted material forms an envelope and ignites through the

CNO cycles. Because of the strong electron-degenerate conditions, the CNO burning triggers a thermonuclear runaway, which causes a rapid rise in temperature in the nuclear-burning region to values exceeding 10^8 K under all circumstances.

During the thermonuclear runaway, a major fraction of the CNO nuclei in the envelope is transformed into positron-unstable nuclei ^{14}O, ^{15}O, and ^{18}Ne, which serves both to constrain the nuclear energy generation on the dynamical timescale of the runaway and to yield distinctively nonsolar CNO isotopic abundance ratios. The decay energy of the ^{14}O, ^{15}O, ^{18}Ne, and ^{18}F daughters shapes the light curve of novae (39). The decay is also associated with the emission of 511-keV positron annihilation radiation, which can be detected by γ-ray observatories such as INTEGRAL (40).

2.6. CNO Breakout as X-Ray Burst Trigger

X-ray bursts are among the most fascinating of explosive astrophysical phenomena. The X-ray burster is characterized by a repeated sudden increase of X-ray emission within only a few seconds to a total luminosity of approximately 10^{40} erg. The recurrence time for single bursts can range from hours to days.

These bursts result from a thermonuclear explosion in the atmosphere of an accreting neutron star in a close binary system (41). When critical values for density and temperature are reached in the neutron-star atmosphere, the freshly accreted hydrogen and helium ignite, leading to a breakout from the slow, β-limited CNO cycles (12) via the ^{15}O$(\alpha,\gamma)^{19}$Ne to the rapid proton capture *rp* process (42). The *rp* process rapidly converts the light-element fuel into heavy elements from iron and nickel up to cadmium (43). The trigger conditions for the burst depend on the interplay between the triple-α fueling process and the efficiency of the ^{15}O$(\alpha,\gamma)^{19}$Ne breakout reaction from the hot CNO region.

In many cases, the burst frequency becomes irregular. Bursting may stop for months before starting up again. If fuel is supplied to the burning region at sufficient rates, the system never gets the chance to cool and build up a fuel reservoir, which results in steady-state burning. The stability of the system depends not only on the rate of accretion providing the overall nuclear fuel supply, but also on the nuclear trigger of the thermonuclear runaway. The balance depends on the ignition through the nuclear trigger reaction ^{15}O$(\alpha,\gamma)^{19}$Ne regulating the flow between the β-limited hot CNO cycle and the *rp* process (42, 44). This reaction is the gateway between the light nuclei and the heavy nuclei and between the nuclear reaction prior to the burst and the reaction during the burst. The impact of the uncertainties in the nuclear rates of breakout as well as other reactions on the overall nucleosynthesis pattern in X-ray bursts have recently been studied (45).

3. NUCLEAR PHYSICS ASPECTS OF CNO CYCLES

Nuclear reaction rates are the critical component for stellar-burning processes. The reaction rate $\langle \sigma \upsilon \rangle$ describes the probability of a nuclear interaction between two particles with an energy-dependent reaction cross section $\sigma(E)$ taking place in a stellar environment where the interacting particles have energies that follow a Maxwell-Boltzmann distribution. The reaction rate per particle pair is expressed by

$$\langle \sigma \upsilon \rangle = \left(\frac{8}{\pi\mu}\right)^{1/2} \frac{1}{(kT)^{3/2}} \int_0^\infty \sigma(E) E \times \exp\left(-\frac{E}{kT}\right) dE, \qquad 1.$$

where T is the temperature of the stellar environment, k is the Boltzmann factor, and μ refers to the reduced mass of the interacting particles. Traditionally, the resonant and the nonresonant reaction components in the cross section are treated separately by approximating the Maxwell-Boltzmann term and the S-factor, respectively, as constants (46). Modern mathematical techniques make solving the integral in Equation 1 possible by numerically taking all reaction components into account.

3.1. Experimental Goals and Techniques

Because CNO reactions are all charged particle–induced interactions, the overall energy dependency is determined by the tunnel probability through the Coulomb barrier. Consequently, cross sections decrease rapidly as center-of-mass energies are lowered. The cross section is therefore often expressed in terms of the so-called astrophysical S-factor $S(E)$ introduced by Salpeter in 1952 (47),

$$\sigma(E) = S(E) \times \frac{1}{E} \times \exp(-2\pi\eta), \text{ with } 2\pi\eta = \frac{2\pi \cdot z_1 z_2 \sqrt{\mu}}{\hbar \cdot \sqrt{2E}}, \qquad 2.$$

where the term $\exp(-2\pi\eta)$ approximates the Coulomb barrier between two charged particles with charges z_1 and z_2, orbital momentum $\ell = 0$, and $S(E)$ reflects the nuclear component in the reaction probability and higher-value orbital momentum–induced deviations. The most effective energy range for nuclear reactions depends on the stellar temperature T and the electrical charge of the interacting nuclei z_1 and z_2 and is known as the Gamow window with a mean energy of

$$E_o = 6.27 \times (z_1 z_2 \sqrt{\mu} kT)^{2/3} \text{ and a width of } \Delta E = 2.31 \times \sqrt{E_0 kT}. \qquad 3.$$

The main difficulty in determining reliable reaction rates for the CNO cycles is the uncertainty in the very low cross sections at the Gamow range. The cross-section contributions at these low energies depend critically on the reaction mechanism. In the case of CNO reactions, multiple components involve narrow and broad resonance and subthreshold-level contributions as well as direct nonresonant components with pronounced interference effects with each other. Because of the complexity of the reaction mechanisms, a direct calculation of the cross section is not possible, and the cross sections need to be determined experimentally. However, the experiments have not reached the actual Gamow energy range because of the low-subfemtobarn cross sections and the high background due to natural and cosmic ray–induced radioactivity in the detector environment. The presently used CNO rates are all based on the extrapolation of the existing low-energy experimental data (48). In many cases, the extrapolation is insufficiently accurate because of the difficulties in taking full account of the complexities of the reaction mechanisms. In the CNO reactions, extrapolations can deviate by more than an order of magnitude (49). Improved measurements at near-threshold energies and new experimental efforts are necessary to improve and expand the existing cross-section data toward lower energies.

A number of different approaches have been undertaken, and several new experimental techniques have been developed over the past decade. The direct measurement of cross sections has been severely handicapped by cosmic ray–induced background. A number of techniques have been developed to reduce the background component by improving detector configuration, event identification, and background-rejection techniques (50, 51). A major breakthrough was the development of the low-energy underground accelerator facility LUNA in the Gran Sasso laboratory (52). Because of the reduction in the cosmic-ray flux, the background in the detectors is substantially reduced, and for the first time, direct measurements of nuclear reactions near the Gamow range appear feasible. This advantage has been demonstrated in the $^{14}N(p,\gamma)^{15}O$ reaction, which was successfully measured down to energies of 70 keV at LUNA (53), reaching substantially lower limits than previous experiments (51). Complementing these direct experimental study techniques of CNO radiative capture and nuclear reactions are several new indirect approaches that selectively study critical reaction components. The indirect methods include the determination of the asymptotic normalization coefficient (ANC) by single-particle transfer reactions (54) and the Trojan-horse method by transfer reaction studies to the continuum (55). The primary aim of the ANC approach is to determine the strength of direct capture components in radiative capture reactions. This technique has been successfully applied to cases in which the direct capture component either contributes significantly to the total cross section such as in $^{16}O(p,\gamma)^{17}F$ (56) or is a critical parameter for a reliable extrapolation of existing direct measurement data as in $^{15}N(p,\gamma)^{16}O$ (57). The Trojan-horse method has been utilized for indirectly determining the strength of (p,α) reactions by observing virtual proton transfer with subsequent α emission as demonstrated in the case of $^{15}N(p,\alpha)^{12}C$ (58) and $^{18}O(p,\alpha)^{15}N$ (59). Both approaches represent powerful complimentary techniques to the direct measurement but depend on a number of approximations in the underlying theoretical models. Independently, they cannot determine total reaction rates with sufficient accuracy.

3.2. Reactions in the Cold CN Cycle

The CNO reactions are primarily radiative proton capture reactions with a variety of reaction components. The $^{12}C(p,\gamma)^{13}N$, $^{13}C(p,\gamma)^{14}N$, and $^{15}N(p,\gamma)^{16}O$ reactions are dominated by broad resonances controlling the reaction cross sections in the Gamow range; all three reactions also have an underlying direct capture component that causes interference patterns that influence the cross section at low energies. The $^{16}O(p,\gamma)^{17}F$ cross section, in contrast, is based on purely nonresonant direct capture transitions. The $^{14}N(p,\gamma)^{15}O$, $^{17}O(p,\gamma)^{18}F$, and $^{18}O(p,\gamma)^{19}F$ reaction cross sections are characterized by a multitude of resonances with pronounced direct capture components and complex interference patterns that need to be well understood for reliable extrapolation. Similar is the situation for the three important reactions $^{15}N(p,\alpha)^{12}C$, $^{17}O(p,\alpha)^{14}N$, and $^{18}O(p,\alpha)^{15}N$, which show broad resonances but also display evidence for an underlying direct transfer reaction component.

Great experimental progress has been made particularly in the low-energy study of $^{14}N(p,\gamma)^{15}O$, which is the slowest reaction in the CN cycle and therefore a key for understanding the timescale of the CNO cycles as well as the overall energy production and neutrino production associated with CNO burning. Detailed studies of the various γ transitions in the reactions have been performed at LUNA down to energies of 110 keV (60). These have been complemented by measurements of the total cross section down to 70 keV. These data allow improved extrapolation of the low-energy cross section into the Gamow range of stellar burning using R-matrix techniques (60), thereby removing the uncertainties discussed in previous extrapolation attempts (49). Yet, it can be observed that the range of uncertainty in higher-energy data (51, 61) affects the low-energy extrapolation. These discrepancies suggest the need for improved measurements in the high-energy range.

The $^{15}N(p,\alpha)^{12}C$ reaction has been extensively studied using both direct (62) and indirect reaction techniques (58), but there have been inconsistencies in the interpretation of the low-energy reaction cross section of the competing $^{15}N(p,\gamma)^{16}O$ cross section (63). Indirect studies suggest a significantly lower cross section value that would translate into a lower rate (57). This lower cross section seems confirmed by the results of the analysis of the $^{15}N(p,\gamma)^{16}O$ component in an earlier low-energy $^{14}N(p,\gamma)^{15}O$ measurement with natural nitrogen targets at the LUNA underground facility (64). Yet, the energy dependency of the proposed S-factor deviates from

R-matrix simulations, and it is not clear whether or to what extent the data are compromised by the large beam–induced $^{11}B(p,\gamma)^{12}C$ and $^{14}N(p,\gamma)^{15}O$ components. Additional experiments have been performed at LUNA and the University of Notre Dame to explore the energy dependency of the reaction cross section in more detail over a wide energy range (65). These studies confirm that the cross section is substantially lower than previously thought (63), which translates into a reduced leakage rate from the CN to the NO cycle in stellar burning.

The low-energy cross section of the $^{16}O(p,\gamma)^{17}F$ radiative capture reaction is characterized by two direct capture transitions to the ground state and the first excited state in ^{17}F. Detailed measurements have been performed for both transitions down to energies of 200 keV (66), and direct capture model simulations have been performed to extrapolate the cross section toward the Gamow range of stellar burning (67). Because of the absence of resonances, the cross section is low and the reaction is the slowest process in the NO cycle.

3.3. Reactions in the Cold NO Cycle

The second branching of the third CNO cycle is determined by the $^{17}O(p,\alpha)^{14}N$ and the $^{17}O(p,\gamma)^{18}F$ reaction rates. The $^{17}O(p,\alpha)^{14}N$ reaction has been measured down to energies of 400 keV (68). The radiative capture $^{17}O(p,\gamma)^{18}F$ has been explored down to energies of 300 keV (69), and a large number of transitions to the ground state and higher excited states have been identified. The cross sections of both processes are characterized by a number of broad and narrow resonances in the 500-to-800-keV energy range, tailing toward lower energies. Also contributing to the radiative capture channel is the nonresonant direct component (69), which may be severely affected by interference processes with resonances (70). In addition to the nonresonant direct reaction and resonant tail components, the low-energy range is also affected by two possible resonances at 70 keV and 193 keV, corresponding to the 5.673-MeV 1^- state and the 5.786-MeV 2^- state, respectively, in ^{18}F. These resonances determine the reaction rates for quiescent stellar hydrogen burning and explosive hydrogen burning in novae (71), respectively. The 70-keV resonance has been studied in the $^{17}O(p,\alpha)$ channel using direct (72, 73) and indirect (74) experimental techniques. The first study achieved only an upper limit value for the resonance strength (68). The results of the successful direct measurement (73) and the indirect study (74) carry large uncertainties and exhibit considerable discrepancies in their predictions. New experiments are necessary to remove the present uncertainties. The 193-keV resonance has been recently measured in the $^{17}O(p,\alpha)^{14}N$ reaction channel (75). The resulting resonance strength compares well with the results of independent work (76). The corresponding resonance in the $^{17}O(p,\gamma)^{18}F$ radiative capture channel was also measured (70) and is approximately three orders of magnitude weaker than the α-particle channel (75). A new low-energy study of the $^{17}O(p,\gamma)^{18}F$ reaction is presently proposed for the LUNA facility. First estimates of the reaction yield indicate that the measurement promises to be successful in the LUNA underground environment (G. Imbriani, unpublished results).

The main uncertainty in the present reaction rate determination is associated with the extrapolation of the experimental data into the Gamow range of quiescent or explosive stellar burning. The reliability of this procedure depends critically on an understanding of the reaction mechanism and of possible threshold structure effects to be included in the reaction model used for the extrapolation. Initial attempts relied on a simple Breit-Wigner formalism for treating resonances (77) and direct capture approximations within the framework of square-well or Wood-Saxon potential models (11, 69, 77). These approaches could not adequately account for the reaction's complex interferences. R-matrix methods (78) were used to describe multiple single-channel resonance systems for the first time in the case of $^{17}O(p,\alpha)^{14}N$ (68). That study clearly proved the usefulness of this approach to describe low-energy reaction patterns. However, the predictive power of the approach is limited and depends critically on knowledge of possible resonance contributions (79). A major limitation was that early studies were constrained to the multilevel analysis of single transitions in which each channel was analyzed separately (80). A parallel treatment of all reactions in a multichannel multilevel analysis approach offers a more comprehensive way of studying the different reaction contributions while reducing the uncertainties associated with this approach for extrapolating cross section data beyond the experimental range. The multichannel approach allows researchers to probe and identify the signatures of possible low-energy resonance or nonresonant contributions in all scattering and reaction channels. This is demonstrated in a recent analysis of CNO radiative capture reactions using a multichannel multilevel *R*-matrix approach to analyze the radiative capture, reaction, and elastic scattering channels in the $^{12}C + p$, $^{14}N + p$, $^{15}N + p$, and $^{16}O + p$ systems (81).

Figure 4 shows the *R*-matrix fits of selected channels to the experimental data and the *R*-matrix-based extrapolation of the reaction cross sections toward lower energies. The new results show appreciable differences compared with previous attempts to extrapolate existing data. The analysis

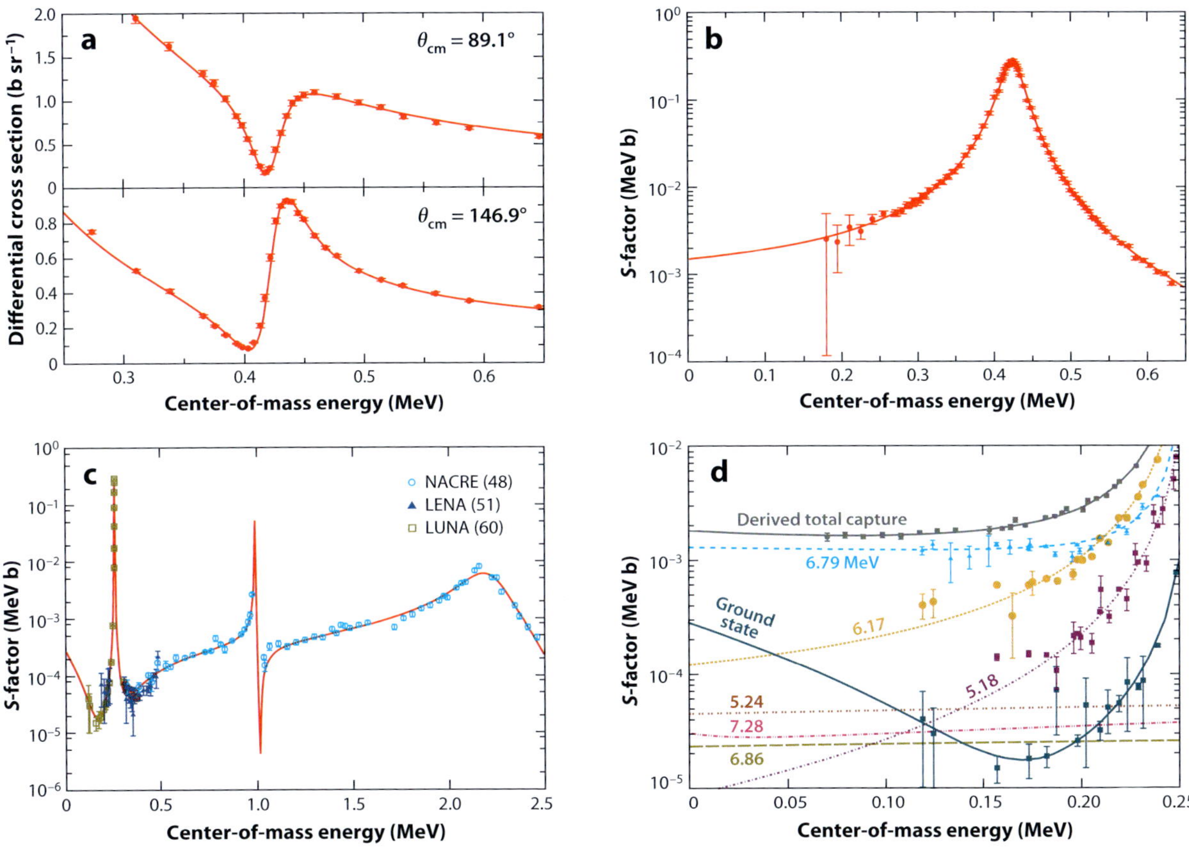

Figure 4

(*a*) *R*-matrix fit to the $^{12}C(p,p)^{12}C$ reaction. (*b*) R-matrix fit to the $^{12}C(p,\gamma)^{13}N$ reaction. The fit was performed in parallel with the scattering data. (*c*) *R*-matrix fit to the $^{14}N(p,\gamma_0)^{15}O$ reaction. (*d*) *R*-matrix-based low-energy extrapolation for the $^{14}N(p,\gamma)^{15}O$ reaction based on a fit to all available data concerning primary transitions. The solid gray line represents the sum of all fitted primary transitions and is shown in comparison with the derived total capture cross section (*gray dots*). The origin of the experimental data is discussed and summarized in Reference 81.

also demonstrates that the quality of R-matrix predictions and extrapolation for CNO reactions cannot rely on limited low-energy data alone but in many cases depends on the uncertainty in the experimental data over the entire excitation range.

4. THE HOT CNO CYCLES

The reaction rates of CNO processes increase exponentially with temperature. Therefore, the overall rate of converting four hydrogen nuclei to one helium nucleus and the associated energy release increase accordingly. The lifetime τ_i of a nucleus i in the CNO cycle with mass fraction X_i and mass number A_i at a density ρ is determined by the reaction rate $\langle\sigma\upsilon\rangle_i$ of the depleting proton capture and, if it is a radioactive species, by the β-decay rate λ_β:

$$\tau_i = \frac{1}{\lambda_\beta + \rho \times X_i/\mathrm{A}_i \times N_\mathrm{A}\langle\sigma\upsilon\rangle_i}. \qquad 4.$$

The total cycle time is the sum of all the lifetimes of the nuclei in the cycle $\tau = \sum_i \tau_i$. The cycle time and therefore the energy production are typically determined by the slowest reaction or decay process in the cycle. For example, the energy production rate in the CN cycle is determined by the slowest reaction, $^{14}\mathrm{N}(p,\gamma)^{15}\mathrm{O}$, whereas the energy production in the NO cycle is determined by the reaction rate of $^{16}\mathrm{O}(p,\gamma)^{17}\mathrm{F}$. The overall energy production ε_{CNO} in the CNO cycles depends on the total energy release in the cycle of $Q = 26.7$ MeV and can be approximated by

$$\varepsilon_{CNO} = \frac{\sum_i Q_i}{\tau} \approx \frac{4.5 \times 10^{25} \rho \times X_H Z_{CNO}}{T^{2/3}} \times e^{-15.228/T^{1/3}} (\mathrm{erg\,g^{-1}\,s^{-1}}), \qquad 5.$$

where ρ indicates density, X_H is the mass fraction of hydrogen, and Z_{CNO} is the mass fraction of the CNO material. With increasing temperature, the reaction rates eventually become larger than the typical β-decay rates λ_β:

$$\lambda_\beta \leq \rho \frac{X}{\mathrm{A}} N_\mathrm{A}\langle\sigma\upsilon\rangle. \qquad 6.$$

At these conditions, radioactive isotopes with a mass fraction X and mass number A are primarily depleted by radiative capture reactions instead of the β decay, and the "cold" cycle turns into a "hot" cycle in which the cycle time and the energy production rate depend on the β-decay rate only and are constant. The hot CN cycle is characterized by the fact that proton capture on $^{13}\mathrm{N}$ is faster than the competing β decay, leading to the sequence $^{12}\mathrm{C}(p,\gamma)^{13}\mathrm{N}(p,\gamma)^{14}\mathrm{O}(\beta^+\upsilon)^{14}\mathrm{N}(p,\gamma)^{15}\mathrm{O}(\beta^+\upsilon)^{15}\mathrm{N}(p,\alpha)^{12}\mathrm{C}$. Radiative proton capture on $^{14}\mathrm{O}$ and $^{15}\mathrm{O}$ is not possible because $^{15}\mathrm{F}$ and $^{16}\mathrm{F}$ are proton unbound. The burning is delayed by the decay of the $^{14}\mathrm{O}$ and $^{15}\mathrm{O}$ isotopes, which determines the cycle time of $\tau \approx 200$ s. For this reason, these isotopes are called waiting points. The energy production will be constant,

$$\varepsilon_{CNO} \approx (4.6 \times 10^{15}) \times Z_{CNO}(\mathrm{erg\ g^{-1}\ s^{-1}}), \qquad 7.$$

and there will be substantial enrichment of $^{14}\mathrm{O}$ and $^{15}\mathrm{O}$ isotopes.

Similarly the NO cycle has a hot mode: $^{16}\mathrm{O}(p,\gamma)^{17}\mathrm{F}(p,\gamma)^{18}\mathrm{Ne}(\beta^+\upsilon)^{18}\mathrm{F}(p,\alpha)^{15}\mathrm{O}(\beta^+\upsilon)^{15}\mathrm{N}(p,\gamma)^{16}\mathrm{O}$, with the two waiting-point nuclei $^{15}\mathrm{O}$ and $^{18}\mathrm{Ne}$. This hot NO cycle bypasses completely the $^{17}\mathrm{O}$ branching point of the third cold CNO cycle. Instead, a revised third cycle emerges through radiative capture on the $^{18}\mathrm{F}$ seed isotope: $^{18}\mathrm{F}(p,\gamma)^{19}\mathrm{Ne}(\beta^+\upsilon)^{19}\mathrm{F}(p,\alpha)^{16}\mathrm{O}(p,\gamma)^{17}\mathrm{F}(p,\gamma)^{18}\mathrm{Ne}(\beta^+\upsilon)^{18}\mathrm{F}$. To understand the transition from temperature-dependent cold to hot CNO burning, it is important to determine the reaction rates for the radiative proton capture reactions on the branching points for the transition from cold to hot cycles, $^{13}\mathrm{N}(p,\gamma)^{14}\mathrm{O}$ and $^{17}\mathrm{F}(p,\gamma)^{18}\mathrm{Ne}$, respectively.

4.1. Reactions of the Hot CNO Cycles

The $^{13}N(p,\gamma)^{14}O$ reaction cross section at low energies is determined by a broad $\Gamma = 38$ keV, 1^- resonance, and by the $s \rightarrow p$ direct capture to the 0^+ ground state in ^{14}O. The experimental study of this resonance was one of the first goals of radioactive-beam facilities and was completed successfully at Louvain-la-Neuve with radioactive ^{13}N beams (82). The results were confirmed by independent studies that used the Coulomb dissociation technique (83, 84). It is expected that the direct capture component interferes with the broad resonance, which introduces a large uncertainty in the low-energy cross-section extrapolation. Although no direct radioactive-beam measurement of the direct capture exists, there have been studies of the asymptotic attenuation coefficient, which can be used to derive the direct capture strength and the total reaction rate (85, 86). The direct capture component causes a substantial enhancement of the low-energy reaction cross section from the resonance tail (**Figure 5** shows an *R*-matrix prediction for the reaction cross section based on the available experimental data). This nearly doubles the reaction rate, which translates into a substantially lower ignition temperature for the hot CN cycle.

The reaction $^{17}F(p,\gamma)^{18}Ne$ triggers the hot NO cycle, and its cross section is characterized by direct capture $p \rightarrow s,d$ transitions to 2^+ bound states at 1.887 MeV and 3.616 MeV in ^{18}Ne (87). On the basis of ^{18}O-^{18}Ne mirror analysis, it was argued that a third resonance with a spin of 3^+ may contribute significantly to the reaction (88). The resonance level was indeed observed via inverse kinematics elastic scattering studies with ^{17}F beams at the HRIBF at the Oak Ridge National Laboratory (89). The experiment gave an excitation energy of 4.524 MeV, corresponding to a resonance energy of 600 keV in the center-of-mass system, which is higher than initially proposed (88) and reduces the impact of the level on the reaction rate. High-resolution $^{16}O(^{3}He,n)^{18}Ne$ transfer studies confirmed the results (90). A first direct measurement of the strength of the 3^+ $^{17}F(p,\gamma)$ resonance was recently completed using a radioactive ^{17}F beam (91). The results indicate that the reaction rate at temperatures below 0.4 GK is dominated by the direct capture component, whereas the 3^+ resonance at 600 keV dominates the rate at higher temperatures. **Figure 5** shows an *R*-matrix prediction for the reaction cross section, including the direct capture. However, the direct capture contribution is not experimentally confirmed but still relies on the theoretical predictions based on potential model calculations (88, 92). Although direct measurements of the direct capture will be difficult, indirect transfer measurements may be the best way to determine the ANC values of the two final states for direct capture to ^{18}Ne in order to derive the strengths of the transitions. Such a study would remove the remaining uncertainty in this reaction rate at nova temperatures (36).

A critical branching point in the hot CNO cycles are the competing reactions $^{18}F(p,\gamma)^{19}Ne$ and $^{18}F(p,\alpha)^{15}O$. The relative strengths of these reactions affect the abundance distribution in hot CNO burning and will determine the production of radioactive ^{18}F isotopes in nova explosions, which may affect the emission of 511-keV annihilation radiation (93). Considerable effort has been spent over the past decade to study these reactions experimentally using a variety of techniques. Transfer reaction studies demonstrated that the reaction rates will be determined by a number of resonant states in both reaction channels (94). The main goal over the past decade was to measure directly these resonances in the $^{18}F(p,\alpha)^{15}O$ and $^{18}F(p,\gamma)^{19}Ne$ channels. The first successful measurements with ^{18}F beams were performed at Louvain-la-Neuve (95) and Argonne National Laboratory (96). However, these studies suffered from limited ^{18}F beam intensity and provided only preliminary information about the contributions of strong resonances at higher energies. Substantially stronger ^{18}F beams were developed at the HRIBF, leading to a systematic study of the $^{18}F(p,\alpha)^{15}O$ reaction using indirect transfer methods (97), elastic scattering analysis (98–100), and direct experimental techniques (99, 101, 102). These studies provide information

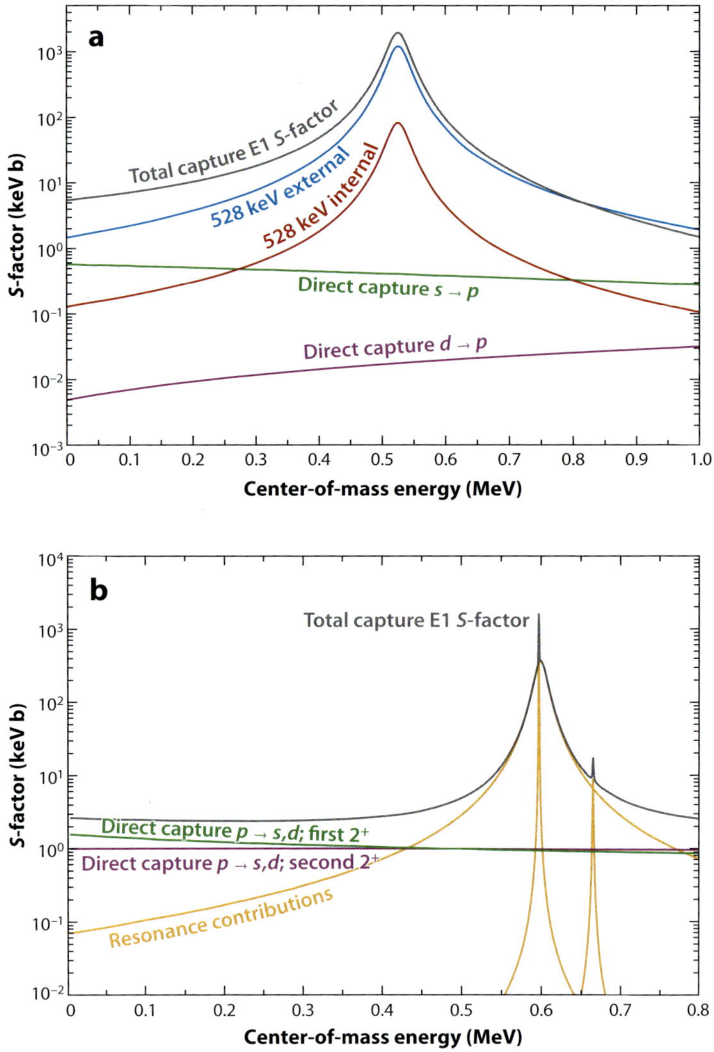

Figure 5

(*a*) The *S*-factor for the $^{13}N(p,\gamma)^{14}O$ reaction. The gray line indicates the total capture E1 *S*-factor. The direct capture components are represented by the purple and green lines for $d \rightarrow p$ and $s \rightarrow p$ transitions, respectively. The contributions from the $E_{cm} = 528$ keV resonance are shown explicitly for the internal and external regions. The external resonance contribution is entirely determined by the proton width of the resonance and the asymptotic normalization coefficient (ANC) of the ground state. The parameters were taken from Reference 86. (*b*) The calculated *S*-factor for the $^{17}F(p,\gamma)^{18}Ne$ reaction. The gray line represents the total capture *S*-factor. The orange lines indicate the resonance contributions to the reaction. The 1^- resonance at $E_{cm} = 597$ keV was calculated as an E1 decay to the ground state. The broad 3^+ resonance at $E_{cm} = 599.8$ keV and the 0^+ resonance at $E_{cm} = 665$ keV are assumed to be dominated by M1 decays to the first 2^+ state. The resonance parameters are taken from Reference 91. The green and purple lines are the E1 direct capture contributions arising from $p \rightarrow s,d$ transitions to the first and second 2^+ states, respectively. The ANCs for the bound states are calculated using the spectroscopic factors of Reference 69 for the analog states in ^{18}F.

about the strengths of the $3/2^-$ resonance at 330 keV and the 40-keV broad $3/2^+$ resonance at 665 keV, as well as its possible interference with lower-energy states. The low-energy tail of the 665-keV resonance was recently measured at Louvain-la-Neuve (103), and a careful *R*-matrix analysis was performed to investigate possible interference effects with the two $3/2^+$ low-energy resonances at 8 keV and 38 keV and to determine the *S*-factor in the Gamow range of nova burning (103, 104). A simultaneous inverse kinematics measurement of the $^{18}F(p,p)^{18}F$ elastic scattering and the $^{18}F(p,\alpha)^{15}O$ reaction was performed most recently at the ISAC facility at TRIUMF (105). Both reaction channels were systematically mapped over the range from 0.6-MeV to 11.6-MeV center-of-mass energy. This study confirmed the results of the previous measurements. A number of new resonances were also observed at higher energies; however, these resonances have only negligible impact on the reaction rate at stellar energies (104). This impressive amount of experimental effort reduced the uncertainties in previous predictions of the nuclear reaction rate. Still, some uncertainties remain. The main one is the measurement of the strength of the 330-keV resonance (101), which dominates the reaction rate at temperatures between 0.25 GK and 0.4 GK. More recent studies have focused on the higher-energy range, mapping the tails of broad resonances and other possible nonresonant contributions. More uncertainty is associated with the interference of the low-energy $3/2+$ resonance at 38 keV with the low-energy tail of the broad 655-keV resonance. The simulations suggest large uncertainties in the energy range near 200 keV, depending on the choice of interference pattern (102, 103). Additional experiments are necessary to remove these uncertainties.

Whereas the $^{18}F(p,\alpha)^{15}O$ reaction has been studied extensively, little attention has been given to the competing $^{18}F(p,\gamma)^{19}Ne$ radiative capture. It has been argued (56, 64) that this reaction is weaker by at least three to four orders of magnitude. A detailed analysis of the ^{19}Ne ^{19}F mirror structure and the properties of possible resonance states (106) provides important new information for an improved estimate of the $^{18}F(p,\gamma)^{19}Ne$ reaction rate. A study of the radiative capture at HRIBF or ISAC using recoil separator systems may be feasible for a more direct investigation of this radiative capture reaction (91, 107).

4.2. The $^{14}O(\alpha,p)^{17}F$ Reaction Branch

In the above sections, we concentrated on proton-induced reactions in the CNO cycles. However, higher temperatures also open new reaction possibilities, namely the $^{14}O(\alpha,p)^{17}F$ reaction. This reaction does not represent a "breakout" from the hot CNO cycles, as it is often mislabeled (108), but instead, it merges the hot CN and the hot NO cycles into a new cyclic CNO pattern: $^{14}O(\alpha,p)^{17}F(p,\gamma)^{18}Ne(\beta^+\upsilon)^{18}F(p,\alpha)^{15}O(\beta^+\upsilon)^{15}N(p,\alpha)^{12}C(p,\gamma)^{13}N(p,\gamma)^{14}O$. This cycle is slightly faster than the classical hot CNO cycles because it bypasses the slow ^{14}O β decay. This causes an increase in energy generation rate and a redistribution of the abundances by converting ^{14}O to ^{15}O. Nova explosions at such high temperatures should therefore show a higher $^{15}N/^{14}N$ ratio in their ejecta compared with classical novae, which are driven by the hot CNO cycles.

The reaction rate of $^{14}O(\alpha,p)^{17}F$ was originally thought (108) to be characterized by a broad 1^- resonance at 1.04 MeV, corresponding to an excited state in ^{18}Ne at 6.15 MeV and several narrow resonance contributions. However, microscopic multichannel calculations demonstrated that a strong nonresonant direct reaction component may enhance the rate considerably (109). The proposed rate depends critically on these contributions and pronounced interference effects between the broad resonance and a direct transfer component (110, 111). First attempts to study this reaction were performed using the time-reverse reaction $^{17}F(p,\alpha)^{14}O$ in inverse kinematics (112, 113). This study was complemented by the measurement of the inelastic scattering reaction $^{17}F(p,p\gamma)^{17}F$ for probing the $^{14}O(\alpha,p_1)$ reaction branch to the first excited state in ^{17}F (114, 115). A first direct measurement of $^{14}O(\alpha,p)^{17}F$ was performed using the CNS radioactive beam separator in the RIKEN accelerator facility (116). No conclusions were presented in terms of the reaction rate, but the presented data do not entirely agree with the cross-section predictions based on indirect studies. A direct measurement of the reaction is extremely important to understand the impact of this process in nova and X-ray burst reaction flow and nucleosynthesis analysis.

5. CNO FUELING AND BREAKOUT REACTIONS

The previous sections discuss the overall temperature dependency of the energy generation in CNO burning and the impact on the associated CNO nucleosynthesis. This simplified scenario is based on the fact that the CNO cycles are closed burning entities without significant feeding or loss through nuclear reaction processes and does not consider the possibility of changing the CNO fuel budget through convective processes in a stellar environment. The assumption of nuclear-confined CNO fuel conditions is also not satisfied toward higher temperatures, when new reaction sequences become strong enough to generate new CNO fuel from fusion reactions or to leak fuel through breakout processes (11).

5.1. CNO Fueling Reactions

Fueling the CNO region requires converting hydrogen and helium abundances from the mass $A = 12$ to the mass $A = 16$ region. This is a slow and difficult process because it requires bridging efficiently the mass $A = 5$ and $A = 8$ gaps. There are only two major charge particle–induced reaction processes that could generate a substantial production of fresh CNO material in stellar environments. The best known is the triple-α process, which fuses three ^4He into one ^{12}C nucleus (117); the second process is the hot *pp* chains, which represent a sequence of proton and α capture reactions bridging the mass $A = 5$ and $A = 8$ gaps (118).

The triple-α process is well understood in principle: The fusion of two ^4He nuclei forms the α cluster nucleus ^8Be, which is particle unbound by 92 keV and decays immediately back to two α particles. This relatively small difference in binding energy leads to an abundance equilibrium between ^4He and ^8Be, which depends on the temperature and density conditions in the stellar environment. A subsequent α capture on ^8Be forms ^{12}C through a 0^+ resonance cluster state at 7.654 MeV. The entire reaction rate depends critically on the properties of this resonance state, which decays by γ emission to the 2^+ first excited state at 4 MeV in ^{12}C and by pair production to the ground state of ^{12}C. The main uncertainties in deriving a reliable reaction rate are the errors in the partial widths for the decay into the pair production channel Γ_π and, to a lesser extent, in the γ partial width Γ_γ (119, 120). The presently adopted value for the pair production branch $\Gamma_\pi = 60.5 \pm 3.9$ μeV is based on inelastic proton scattering studies on ^{12}C (121, 122). A more recent inelastic electron scattering experiment yields a more precise but slightly lower value of $\Gamma_\pi = 52 \pm 1.4$ μeV (123). New measurements are necessary to remove the discrepancy and reduce the uncertainty in the pair emission branch to 5% (120).

The reaction rates for the triple-α process at higher and lower temperature regions is more complex. For low-temperature regions, the second α capture process is not resonant but depends critically on the cross section in the tails of higher-energy resonances (124), the strength of the E2 direct capture to the ground state, and the 2^+ first excited state in ^{12}C (125).

The presently adopted rate in the NACRE compilation (48) includes these nonresonant contributions in consideration of the rate, but it also takes into account possible contributions from higher-energy resonances that affect the rate at higher temperatures. Included in the rate is a possible 2^+ state at 9.2-MeV excitation energy. This state is predicted to have a significant influence on the reaction rate as resonance and through its low-energy tail. A recent experiment studying the three-particle breakup of ^{12}C states populated in the β decay of ^{12}B and ^{12}N does not confirm this proposed level. Instead, a strong, 3-MeV broad 0^+ resonance at 11 MeV was observed (126). Its tail reaches well into the lower-energy range, causing strong interference with the 0^+ state at 7.65 MeV. This further causes a substantial enhancement of the cross section at lower energies in the ^{8}Be(α,γ)^{12}C link, substantially enhancing the triple-α rate at lower temperatures.

The hot *pp* chains represent a number of different reaction sequences by which ^{4}He is transformed toward higher masses at high-temperature conditions primarily through α capture via the chain ^{3}He(α,γ)^{7}Be(α,γ)^{11}C(α,p)^{14}N and at high-density conditions primarily by the sequence ^{3}He(α,γ)^{7}Be(p,γ)^{8}B(p,γ)^{9}C(α,p)^{12}N(p,γ)^{13}O($\beta^+\upsilon$)^{13}N (118). Even though these reaction chains work efficiently, the main limitation is the limited supply of ^{3}He, ^{2}H seed material feeding the reaction chain. Therefore, the role of the hot *pp* chains in fueling the CNO cycles is negligible and cannot compete with the triple-α process. Despite the new results obtained in recent years for many reactions in the hot *pp* chains, a more detailed discussion of these results is beyond the scope of the present review.

5.2. CNO Breakout Reactions

CNO breakout reactions have been a matter of long debate (12). Initially only α capture reactions on CNO waiting-point nuclei such as ^{15}O(α,γ)^{19}Ne and ^{18}Ne(α,p)^{21}Na were included in the discussion (16, 127). More recently, two-proton capture reactions on these waiting-point nuclei ^{15}O($2p,\gamma$)^{17}Ne and ^{18}Ne($2p,\gamma$)^{20}Mg were considered as an alternative reaction branch, which could lead to breakout (128, 129). Taking into account the direct three-particle capture (129) increases the rate substantially compared with a pure sequential three-particle capture process (128). Recent simulations (J.L. Fisker & M. Wiescher, unpublished data) on the impact of these two-proton capture reactions show that they are negligible for the breakout process. Therefore, we do not consider them in the following evaluations.

The initial breakout branch from the CNO region is triggered by α capture on the highly enriched ^{15}O abundances in hot CNO cycle regimes. Theoretical predictions for the reaction rate of ^{15}O(α,γ)^{19}Ne predicted that the reaction rate is dominated by a 500-keV resonance, which corresponds to the $3/2^+$ state at 4.033 MeV in ^{19}Ne (130). Higher-energy states contribute only slightly; because of their high spin values, the resonance strengths are reduced as a result of the orbital momentum barrier. A direct measurement of the resonance strength is anticipated at future radioactive-beam facilities but is not yet possible because of beam-intensity limitations (131, 132). However, the resonance strengths can be derived from the α and γ partial widths of the resonance levels because the spin values are well known. The total width of the state Γ was measured using Doppler-shift attenuation methods (133–135). The resonance strengths depend critically on the α partial width of the resonance levels, and the initial assumptions (109) were tested by probing the mirror states in ^{19}F with the α transfer reaction (136). Complementary to that approach, numerous indirect measurements were made to probe the α decay of the unbound states in ^{19}Ne with increasingly sophisticated particle-detector systems (137–141). These efforts led to the determination of the relative α widths of most higher-energy states, but only upper limits were determined for the critical $3/2^+$ resonance level at 4.035 MeV. A new measurement yielded a value of $\Gamma_\alpha/\Gamma = 2.9 \pm 2.1 \times 10^{-4}$ within one standard deviation (142, 143). Although this is close to the initially predicted value (130), improved measurements are necessary to confirm the value independently and to reduce the large uncertainties of the result.

The second breakout reaction is ^{18}Ne(α,p)^{21}Na (128), which triggers the so-called αp process. Because of the higher Coulomb barrier involved, it is anticipated that this reaction takes place at higher temperatures than the $^{15}O(\alpha,\gamma)^{19}Ne$ reaction. Initially the information about the level structure of the compound nucleus ^{22}Mg above the α threshold was extremely limited, and only vague estimates of the reaction rate were possible. First attempts to measure the cross section directly at Louvain-la-Neuve (144, 145) were only partially successful because they covered higher-energy resonances that contribute little to the reaction rate at stellar energies.

More detailed information was generated by a $^{12}C(^{16}O,^{6}He)^{22}Mg$ transfer reaction study (146) that identified a number of potentially critical resonance states in ^{22}Mg. Independent transfer (147) and elastic scattering measurements (148) confirmed the observed resonance levels but suffered from limited energy resolution. More recently, however, researchers conducted a number of high-resolution $^{24}Mg(p,t)^{22}Mg$ transfer reaction studies (149, 150) that provide improved information about the excitation energies and, therefore, the resonance energies of the observed states. An ^{22}Ne–^{22}Mg mirror-level analysis and angular distribution measurements (150) provided information about the spin assignments of the $^{18}Ne(\alpha,p)^{21}Na$ resonance levels.

The main uncertainty in the reaction rates arises from the α and p partial widths, which determine the resonance strengths. The actual measurement of these parameters requires either transfer studies with radioactive ^{18}Ne and ^{21}Na beams to determine the ANC values, or a direct low-energy measurement of the reaction with high-intensity ^{18}Ne or, using the time-reverse approach, with ^{21}Na beams. More intense ^{18}Ne beams are presently under development at the TRIUMF ISAC radioactive-beam facility.

6. FUTURE OF CNO MEASUREMENTS AT DUSEL AND FRIB

There has been substantial progress over the past decade in our understanding of the CNO reactions. New measurements have been performed to map the cross sections of many critical CNO reactions with higher accuracy over a wide energy range. Particular efforts have been made to expand the experiments toward lower energies close to the Gamow window by using complementary experimental techniques to reduce the background by either passive shielding in an underground environment or active shielding by event identification and active background rejection. Shortcomings are typically associated with the limits on beam intensity, target technology, and detector efficiency. *R*-matrix analysis of these high-precision experimental data is a powerful tool for extrapolating the cross section into the stellar Gamow range so long as information about all reaction contributions is known and available.

Progress in the experimental studies of reactions associated with the hot CNO cycles and the breakout mechanism has been slow. There are experimental data for most reactions, but the uncertainties are still substantial. The main handicaps are in the difficulty of generating the necessary radioactive beams for these studies and in the limited intensity of the available radioactive beams. Complementary methods such as the ANC approach and the TH technique have been developed, but these techniques require high-precision reaction data to normalize their results through the use of *R*-matrix techniques.

Two major facility development projects in the United States—the Dakota Ion Accelerator for Nuclear Astrophysics (DIANA) and the Facility for Rare Ion Beams (FRIB)—are proposed to start operating around 2016, which will be essential for future progress in the field. It is anticipated that operation of these two facilities will help address and solve the remaining questions associated with the CNO mode of hydrogen burning.

6.1. Cold CNO Experiments Underground

The DIANA facility at DUSEL at the Homestake mine in South Dakota promises a unique environment for pursuing low-energy reaction studies with high-intensity beams. DIANA will be a dual-accelerator facility centered on a 400-kV high-intensity machine for low-energy proton- and α-beam studies and a 3-MV single-ended machine for the study of helium and heavy ion–induced nuclear reaction processes in stars. With the proposed high-beam intensities in the milliampere range, this facility will provide the opportunity for CNO experiments at substantially lower energies than presently available at the LUNA facility, reaching in many cases the Gamow range.

In Europe, two new proposals are being discussed: The underground Experimental Low-Energy Nuclear Astrophysics (ELENA) accelerator facility at the Boulby salt mine in the United Kingdom seeks to take advantage of the reduced level of neutron and natural activity in a salt environment. The disadvantage will be the reduced depth level compared with that of LUNA in the Gran Sasso laboratory. The second proposal is for an accelerator facility in an abandoned train tunnel in the Pyrenees mountains in Canfranc, Spain.

6.2. Hot CNO Experiments with Radioactive Beams

FRIB will be installed at Michigan State University, succeeding the present NSCL radioactive-beam facility. FRIB will provide a large number of radioactive beams for nuclear astrophysics–related experiments. The beams are being produced by fragmentation processes decelerated using gas or solid-stopper material techniques and then reaccelerated to the appropriate experimental stations. Reactions of relevance for the hot CNO cycles will be studied in inverse kinematics using recoil mass separator techniques developed at the Daresbury Recoil Separator at HRIBF and at the DRAGON separator at ISAC TRIUMF. New techniques are being developed to study (α,p) reactions in inverse kinematics. FRIB will provide superior beam intensities by up to several orders of magnitude compared with those of existing facilities. This will help to conclude the experimental study of reactions associated with the hot CNO cycles. Nuclear physics with radioactive beams has emerged as a powerful new direction in the field; new facilities have been built or are being built worldwide: RIKEN in Japan (already operating), SPIRAL II in France, and FAIR GSI in Germany will host an active research program in nuclear astrophysics. As these facilities become available, the nuclear physics community hopes to address the new and critical questions about stellar reaction cross sections and provide the final answer on the nuclear engine of stellar hydrogen burning.

DISCLOSURE STATEMENT

The authors are not aware of any affiliations, memberships, funding, or financial holdings that might be perceived as affecting the objectivity of this review.

ACKNOWLEDGMENTS

This work was supported by the Joint Institute for Nuclear Astrophysics under the National Science Foundation grant number PHY0822648. Also acknowledged is the support from the National Research Council of Canada (M.P.) and from the Istituto Nazionale di Fisica Nucleare of Italy (G.I.). The authors also want to express their gratitude to Oscar Straniero and Luciano Piersanti from the Osservatorio Astronomico di Collurania, Teramo, Italy, and to Richard E. Azuma from the University of Toronto, Canada, for agreeing to release some results prior to publication.

LITERATURE CITED

1. von Weizsäcker CF. *Phys. Z.* 38:176 (1937)
2. Bethe HA, Critchfield CL. *Phys. Rev.* 54:248, 862 (1938)

3. Salpeter EE. *Annu. Rev. Nucl. Sci.* 2:41 (1953)
4. Bethe HA. *Phys. Rev.* 55:103, 434 (1939)
5. von Weizsäcker CF. *Phys. Z.* 39:633 (1938)
6. Burbidge EM, Burbidge GR, Fowler WA, Hoyle F. *Rev. Mod. Phys.* 29:547 (1957)
7. Caughlan GR, Fowler WA. *Astrophys. J.* 136:453 (1962)
8. Rolfs CR, Rodney WS. *Astrophys. J. Lett.* 194:L63 (1974)
9. Dearborn D, Schramm DN. *Astrophys. J. Lett.* 194:L67 (1974)
10. Lorenz-Wirzba H, et al. *Nucl. Phys. A* 313:346 (1979)
11. Wiescher M, et al. *Nucl. Phys. A* 349:165 (1980)
12. Wiescher M, Görres J, Schatz H. *J. Phys. G* 25:R133 (1999)
13. Lazareff B, Audouze J, Starrfield S, Truran JW. *Astrophys. J.* 228:875 (1979)
14. Champagne A, Wiescher M. *Annu. Rev. Nucl. Part. Sci.* 42:39 (1992)
15. Wallace R, Woosley SE. *Astrophys. J. Suppl.* 45:389 (1982)
16. Van Wormer L, et al. *Astrophys. J.* 432:326 (1994)
17. Sandage A. *Astrophys. J.* 135:349 (1962)
18. Chaboyer B, Demarque P, Kernan PJ, Kraus LM. *Science* 271:975 (1996)
19. Imbriani G, et al. *Astron. Astrophys.* 420:625 (2004)
20. Gratton R, Sneden C, Carretta E. *Annu. Rev. Astron. Astrophys.* 42:385 (2004)
21. Busso M, Gallino R, Wasserburg GJ. *Annu. Rev. Astron. Astrophys.* 37:239 (1999)
22. Herwig F, Austin S, Lattanzio JC. *Phys. Rev. C* 73:025802 (2006)
23. Herwig F. *Annu. Rev. Astron. Astrophys.* 43:435 (2005)
24. Asplund M, et al. *Astron. Astrophys.* 343:507 (1999)
25. Davis R. *Phys. Rev. Lett.* 12:303 (1964)
26. Bahcall JN, Pinsonneault MH. *Phys. Rev. Lett.* 92:121301 (2004)
27. Abdurashitov JN, et al. *Phys. Rev. C* 80:015807 (2009)
28. Cravens JP, et al. *Phys. Rev. D* 78:032002 (2008)
29. Bonetti R, et al. *Phys. Rev. Lett.* 82:5205 (1999)
30. Aharmim B, et al. *Phys. Rev. C* 75:045502 (2007)
31. Abe S, et al. *Phys. Rev. Lett.* 100:221803 (2008)
32. Arpesella C, et al. *Phys. Rev. Lett.* 101:091302 (2008)
33. Haxton WC, Serenelli A. *Astrophys. J.* 687:678 (2008)
34. Asplund M, Grevesse N, Sauval AJ. *Am. Inst. Phys. Conf. Proc.* 336:25 (2005)
35. Piersanti L, Straniero O, Cristallo S. *Astron. Astrophys.* 462:1051P (2007)
36. Parete-Koon S, et al. *Astrophys. J.* 598:1239 (2003)
37. José J, Hernanz M, Iliadis C. *Nucl. Phys. A* 777:550 (2006)
38. Hernanz M, Gomez-Gomar J, Jose J, Coc A. *Astrophys. Lett. Comm.* 38:407 (1999)
39. Ness JU, et al. *Astrophys. J. Lett.* 594:L127 (2003)
40. Coc A, Hernanz M, Jose J, Thibaud JP. *Astron. Astrophys.* 357:561 (2000)
41. Woosley SR, Taam RE. *Nature* 263:101 (1976)
42. Schatz H, Aprahamian A, Görres J, Wiescher M. *Phys. Rep.* 294:167 (1998)
43. Woosley SE, et al. *Astrophys. J. Suppl.* 151:75 (2004)
44. Fisker JL, et al. *Astrophys. J.* 665:637 (2007)
45. Parikh A, Jose J, Moreno F, Iliadis C. *Astrophys. J. Suppl.* 178:110 (2008)
46. Fowler WA, Caughlan GR, Zimmerman BA. *Annu. Rev. Nucl. Part. Phys.* 13:69 (1975)
47. Salpeter E. *Phys. Rev.* 88:547 (1952)
48. Angulo C, et al. *Nucl. Phys. A* 656:3 (1999)
49. Angulo C, Descouvemont P. *Nucl. Phys. A* 690:755 (2001)
50. Couture A, et al. *Phys. Rev. C* 77:015802 (2008)
51. Runkle RC, et al. *Phys. Rev. Lett.* 94:082503 (2005)
52. Costantini H, et al. *Rep. Prog. Phys.* 72:086301 (2009)
53. Lemut A, et al. *Phys. Lett.* B634:483 (2006)
54. Mukhamedzhanov AM, Gagliardi CA, Tribble RE. *Phys. Rev. C* 63:024612 (2001)
55. Baur G. *Phys. Lett.* B178:135 (1986)

56. Gagliardi CA, et al. *Phys. Rev. C* 59:1149 (1999)
57. Mukhamedzhanov AM, et al. *Phys. Rev. C* 78:015804 (2008)
58. La Cognata M, et al. *Phys. Rev. C* 80:012801 (2009)
59. La Cognata M, et al. *Phys. Rev. Lett.* 101:152501 (2008)
60. Formicola A, et al. *Phys. Lett.* B591:61 (2004); Imbriani G, et al. *Eur. Phys. J. A* 25:455 (2005)
61. Marta M, et al. *Phys. Rev. C* 78:022802 (2008)
62. Redder A, et al. *Z. Phys. A* 305:325 (1982)
63. Rolfs CR, Rodney WS. *Nucl. Phys. A* 235:450 (1974)
64. Bemmerer D, et al. *J. Phys. G* 36:045202 (2009)
65. LeBlanc PJ. PhD thesis. Univ. Notre Dame, Ind. (2010)
66. Morlock R, et al. *Phys. Rev. Lett.* 79:3837 (1997)
67. Iliadis C, et al. *Phys. Rev. C* 77:045802 (2008)
68. Kieser WE, Azuma RE, Jackson KP. *Nucl. Phys. A* 331:155 (1979)
69. Rolfs C. *Nucl. Phys. A* 217:29 (1973)
70. Fox C, et al. *Phys. Rev. C* 71:055801 (2005)
71. Fox C, et al. *Phys. Rev. Lett.* 93:081102 (2004)
72. Berheide M, Rolfs CR, Schröder U, Trautvetter HP. *Z. Phys. A* 343:483 (1992)
73. Blackmon JC, et al. *Phys. Rev. Lett.* 74:2642 (1995)
74. Landre V, et al. *Phys. Rev. C* 40:1972 (1989)
75. Chafa A, et al. *Phys. Rev. C* 75:035810 (2007)
76. Newton JR, et al. *Phys. Rev. C* 75:055808 (2007)
77. Azuma RE, Rolfs CR. *Nucl. Phys. A* 227:291 (1974)
78. Lane A, Thomas R. *Rev. Mod. Phys.* 30:257 (1958)
79. Ugalde C, et al. *Phys. Rev. C* 77:035801 (2008)
80. Descouvemont P, Rauscher Y. *Nucl. Phys. A* 777:137 (2006)
81. Azuma RE, et al. *Phys. Rev. C*. In press (2010)
82. Decrock P, et al. *Phys. Rev. Lett.* 67:808 (1991)
83. Motobayashi T, et al. *Phys. Lett.* B264:259 (1951)
84. Kiener J, et al. *Nucl. Phys. A* 552:63 (1993)
85. Li ZH, et al. *Phys. Rev. C* 74:035801 (2006)
86. Guo B, Li ZH. *Chin. Phys. Lett.* 24:65 (2007)
87. Wiescher M, Kettner KU. *Astrophys. J.* 263:891 (1982)
88. Wiescher M, Görres J, Thielemann F-K. *Astrophys. J.* 326:384 (1988)
89. Bardayan DW, et al. *Phys. Rev. C* 62:055804 (2000)
90. Parpottas Y, et al. *Phys. Rev. C* 72:025802 (2005)
91. Chipps KA, et al. *Phys. Rev. Lett.* 102:152502 (2009)
92. García A, et al. *Phys. Rev. C* 43:2012 (1991)
93. Hernanz M, et al. *Astrophys. J. Lett.* 526:L97 (1999)
94. Utku S, et al. *Phys. Rev. C* 57:2731 (1998)
95. Coszach R, et al. *Phys. Lett.* B353:184 (1995)
96. Rehm KE, et al. *Phys. Rev. C* 53:1950 (1996)
97. Kozub RL, et al. *Phys. Rev. C* 71:032801 (2005)
98. Bardayan DW, et al. *Phys. Rev. C* 62:042802 (2000)
99. Bardayan DW, et al. *Phys. Rev. C* 63:065802 (2001)
100. Bardayan DW, et al. *Phys. Rev. C* 70:015804 (2004)
101. Bardayan DW, et al. *Phys. Rev. Lett.* 89:262501 (2002)
102. Chae KY, et al. *Phys. Rev. C* 74:012801 (2006)
103. De Séréville N, et al. *Phys. Rev. C* 79:015801 (2009)
104. Dalouzy JC, et al. *Phys. Rev. Lett.* 102:162503 (2009)
105. Murphy AS, et al. *Phys. Rev. C* 79:058801 (2009)
106. Nesaraja CD, et al., *Phys. Rev.* 75:055809 (2007)
107. D'Auria JM, et al. *Phys. Rev. C* 69:065803 (2004)
108. Wiescher M, et al., *Astrophys. J.* 316:162 (1987)

109. Funck C, Langanke K. *Nucl. Phys. A* 480:188 (1988)
110. Hahn KI, et al. *Phys. Rev.* 54:1999 (1996)
111. Bardayan DW, Smith MS. *Phys. Rev. C* 56:1647 (1997)
112. Harss B, et al. *Phys. Rev. C* 65:035803 (2002)
113. Blackmon JC, et al. *Nucl. Phys. A* 688:142c (2001)
114. Blackmon JC, et al. *Nucl. Phys. A* 718:127 (2003)
115. He JJ, et al. *Phys. Rev. C* 80:042801 (2009)
116. Notani M, et al. *Nucl. Phys. A* 746:113 (2004)
117. Hoyle F, Dunbar DNF, Wenzel WA, Whaling W. *Phys. Rev.* 92:1095 (1953)
118. Wiescher M, et al. *Astrophys. J.* 343:353 (1989)
119. Austin S. *Nucl. Phys. A* 758:375 (2005)
120. Buchmann LR, Barnes CA. *Nucl. Phys. A* 777:254 (2006)
121. Robertson RGH, Warner RA, Austin SM. *Phys. Rev. C* 15:1072 (1977)
122. Alburger DE. *Phys. Rev. C* 16:2394 (1977)
123. Crannell H, et al. *Nucl. Phys. A* 758:399 (2005)
124. Nomoto K, Thielemann FK, Miyaji S. *Astron. Astrophys.* 149:239 (1985)
125. Langanke K, Wiescher M, Thielemann FK. *Z. Phys. A* 324:147 (1986)
126. Fynbo HOU, et al. *Nature* 433:136 (2005)
127. Wallace RK, Woosley SE. *Astrophys. J. Suppl.* 45:389 (1981)
128. Görres J, Wiescher M, Thielemann FK. *Phys. Rev.* 51:392 (1995)
129. Grigorenko LV, Zhukov MV. *Phys. Rev. C* 72:015803 (2005)
130. Langanke K, Wiescher M, Fowler WA, Görres J. *Astrophys. J.* 301:629 (1886)
131. Fosshag E, et al. *Nucl. Instrum. Methods A* 480:124 (2002)
132. Vanderbist F, et al. *Eur. J. Phys.* 27:183 (2005)
133. Tan WP, et al. *Phys. Rev. C* 72:041302 (2005)
134. Kanungo R, et al. *Phys. Rev. C* 74:045803 (2006)
135. Mythili S, et al. *Phys. Rev. C* 77:035803 (2008)
136. Mao ZQ, Fortune HT, Lacaze AG. *Phys. Rev. Lett.* 74:3760 (1995); Mao ZQ, Fortune HT, Lacaze AG. *Phys. Rev. C* 53:1197 (1996)
137. Magnus PV, et al. *Nucl. Phys. A* 506:332 (1990)
138. Laird AM, et al. *Phys. Rev. C* 66:048801 (2002)
139. Davids B, et al. *Phys. Rev. C* 67:065808 (2003)
140. Rehm KE, et al. *Phys. Rev. C* 67:065809 (2003)
141. Visser DW, et al. *Phys. Rev. C* 69:048801 (2004)
142. Tan WP, et al. *Phys. Rev. Lett.* 98:242503 (2007)
143. Tan WP, et al. *Phys. Rev. C* 79:055805 (2009)
144. Bradfield-Smith W, et al. *Phys. Rev. C* 59:3402 (1999)
145. Groombridge D, et al. *Phys. Rev. C* 66:055802 (2002)
146. Chen AA, et al. *Phys. Rev. C* 63:065807 (2001)
147. Berg GPA, et al. *Nucl. Phys. A* 718:608 (2003)
148. He JJ, et al. *Phys. Rev. C* 80:015801 (2009)
149. Chae KY, et al. *Phys. Rev. C* 79:055804 (2009)
150. Matic A, et al. *Phys. Rev. C* 80:055804 (2009)

The Low-Energy Frontier of Particle Physics

Joerg Jaeckel[1] and Andreas Ringwald[2]

[1]Institute for Particle Physics Phenomenology, Durham University, Durham DH1 3LE, United Kingdom; email: joerg.jaeckel@durham.ac.uk

[2]Deutsches Elektronen-Synchrotron, D-22607 Hamburg, Germany; email: andreas.ringwald@desy.de

Annu. Rev. Nucl. Part. Sci. 2010. 60:405–37

First published online as a Review in Advance on July 19, 2010

The *Annual Review of Nuclear and Particle Science* is online at nucl.annualreviews.org

This article's doi:
10.1146/annurev.nucl.012809.104433

0163-8998/10/1123-0405$20.00

Key Words

theoretical and experimental tests, extensions of the Standard Model, axions, extra gauge bosons, hidden-matter particles

Abstract

Most embeddings of the Standard Model into a more unified theory, in particular those based on supergravity or superstrings, predict the existence of a hidden sector of particles that have only very weak interactions with visible-sector Standard Model particles. Some of these exotic particle candidates [for instance, axions, axion-like particles, and hidden $U(1)$ gauge bosons] may be very light, with masses in the subelectronvolt range, and may have very weak interactions with photons. Correspondingly, these very weakly interacting subelectronvolt particles (WISPs) may lead to observable effects in experiments (as well as in astrophysical and cosmological observations) searching for light shining through a wall, for changes in laser polarization, for nonlinear processes in large electromagnetic fields, and for deviations from Coulomb's law. We present the physics case and a status report of this emerging low-energy frontier of fundamental physics.

Contents

1. INTRODUCTION

We are entering an exciting time in particle physics: The Large Hadron Collider (LHC) is setting a new benchmark at the high-energy frontier and, through the collision of multiteraelectronvolt protons, is probing the structure of matter and space-time at an unprecedented level. There is much circumstantial evidence that the teraelectronvolt-scale physics exploited at the LHC will provide decisive insights into fundamental questions such as the origin of particle masses, the nature of dark matter in the universe, and the unification of all forces, including gravity. Indeed, most proposals to embed the Standard Model of particle physics into a more general, unified framework—notably those based on string theory or its low-energy incarnations, supergravity and supersymmetry—predict new heavy ($m \gtrsim 100$ GeV) particles that may be searched for at teraelectronvolt colliders. Some of these particles, such as neutralinos, are natural candidates for the constituents of cold dark matter (CDM) in the form of so-called weakly interacting massive particles (WIMPs).

However, there is also evidence of fundamental physics at the subelectronvolt scale. Indeed, atmospheric, reactor, and solar neutrino data strongly support the hypothesis that neutrinos have masses in the subelectronvolt range. Moreover, the vacuum energy density of the universe, as inferred from cosmological observations, points to the subelectronvolt range, $\rho_\Lambda \sim \text{meV}^4$. Many of the above-mentioned extensions of the Standard Model predict not only WIMPs, but also WISPs (very weakly interacting subelectronvolt particles). Prominent candidates for such particles are axions and axion-like particles (ALPs), and they often arise as Nambu-Goldstone bosons associated with the breakdown of global symmetries. Additional WISP candidates are massless or light extra, hidden $U(1)$ gauge bosons as well as light chiral fermions charged under this hidden $U(1)$. These particles are frequently encountered in string embeddings of the Standard Model. The latter may also contain light moduli fields and light gravitinos as WISP candidates.

Unlike for WIMPs, teraelectronvolt colliders are not the best means to search for WISPs. For this purpose, low-energy experiments exploiting lasers, microwave cavities, strong electromagnetic fields, torsion balances, and so forth seem to be superior. In this review we aim to present the physics case and a status report of this emerging low-energy frontier of fundamental physics.

The organization of this review is as follows. In Section 2 we argue that many extensions of the Standard Model predict new particles and phenomena at low energies. On the one hand, new light particles are suggested as solutions for puzzling experimental results, but on the other hand, they also appear to be a generic feature of underlying fundamental theories such as string theory. In Section 3 we turn to current constraints from astrophysics and cosmology. Moreover, we discuss a few interesting observations that could be explained by invoking WISPs. In Section 4 we describe how WISPs can be searched for in a variety of controlled laboratory experiments. We discuss the advantages of these experiments as well as the challenges they face. Finally, in Section 5 we summarize the current situation and discuss future prospects.

2. PHYSICS CASE FOR WISPs

2.1. Axions and Axion-Like Particles

The axion is a prime example of a WISP, a light particle very weakly coupled to the Standard Model. Below, we review how the axion arises from a solution to the strong-*CP* problem. In particular, we explain that both its mass and its coupling to two photons (which is crucial for many experiments) are small because the axion scale is large. Moreover, scalars and pseudoscalars with similar properties to the axion—so-called ALPs—arise naturally in string theory and are often invoked in cosmology to explain dark energy.

2.1.1. The strong-*CP* problem and axions. Quantum chromodynamics (QCD), the non-Abelian gauge theory describing strong interactions, allows for a *CP*-violating term in the Lagrangian,

$$\mathcal{L}_{\text{CP-viol.}} = \frac{\alpha_s}{4\pi}\theta \,\mathrm{tr}\, G_{\mu\nu}\tilde{G}^{\mu\nu} \equiv \frac{\alpha_s}{4\pi}\theta \frac{1}{2}\varepsilon^{\mu\nu\alpha\beta}\,\mathrm{tr}\, G_{\mu\nu}G_{\alpha\beta}, \qquad 1.$$

where G is the gluonic field strength. Similar to the strong coupling constant α_s, the fundamental parameter θ has to be determined experimentally. One of the most sensitive probes in this regard is the electric dipole moment of the neutron, which arises from the *CP*-violating term given in Equation 1 and should be of order

$$|d_n| \sim \frac{e}{m_n}\left(\frac{m_q}{m_n}\right)|\bar{\theta}| \sim 10^{-16}\,|\bar{\theta}|\; e\;\mathrm{cm}, \qquad 2.$$

where m_n (m_q) is the neutron (a light quark mass), e is the unit of electric charge, and

$$\bar{\theta} \equiv \theta + \arg\det M, \qquad 3.$$

where M is the quark mass matrix and $\bar{\theta}$ is the effective physical *CP*-violating parameter in the Standard Model. The current experimental upper bound on $|d_n| < 2.9 \times 10^{-26}$ e cm (1) places an extremely stringent limit on

$$|\bar{\theta}| \lesssim 10^{-10}. \qquad 4.$$

The strong-*CP* problem is the lack of an explanation for why the dimensionless parameter $\bar{\theta}$, a sum of two contributions of very different origins, is so unnaturally small.

The axion occurs in the course of a possible solution of this problem. In essence, Peccei & Quinn (2) proposed promoting θ to a dynamical field that can spontaneously relax to zero. The axion field a is introduced as a dynamical θ parameter, which has a shift symmetry of

$$a \rightarrow a + \text{constant}, \qquad 5.$$

which is broken only by the anomalous *CP*-violating terms. In other words, its low-energy effective Lagrangian is parameterized as

$$\mathcal{L}_a = \frac{1}{2}\partial_\mu a \partial^\mu a + \frac{\alpha_s}{4\pi f_a} a \,\mathrm{tr}\, G^{\mu\nu}\tilde{G}_{\mu\nu} + \frac{s\alpha}{8\pi f_a} aF^{\mu\nu}\tilde{F}_{\mu\nu} + \mathcal{L}_a^{\mathrm{int}}\left[\frac{\partial_\mu a}{f_a}; \psi\right], \qquad 6.$$

where s is a model-dependent parameter, F is the electromagnetic field strength, and ψ denotes generic Standard Model fields. The dimensionful axion decay constant f_a determines the strength of the interaction of the axion with the Standard Model particles. The θ term in the QCD Lagrangian can then be eliminated by being absorbed into the axion field: $a = \bar{a} - \bar{\theta} f_a$. Finally, the topological charge density $\propto \langle \mathrm{tr}\, G^{\mu\nu}\tilde{G}_{\mu\nu}\rangle \neq 0$, induced by topological fluctuations of the gluon fields such as QCD instantons, provides a nontrivial potential for the axion field $\bar{a}$, which is minimized at zero expectation value: $\langle\bar{a}\rangle = 0$. Thus, the $\bar{\theta}$ dependence is wiped out by the axion field, providing a natural explanation as to why the electric dipole moment of the neutron is so small. The nontrivial potential around $\langle\bar{a}\rangle = 0$ promotes the elementary particle excitation of the axion field, the axion, to a pseudo-Nambu-Goldstone boson (3), which we again denote by a, with a nonvanishing but parametrically small mass. This mass can be calculated via current algebra and expressed in terms of the light (u, d) quark masses, the pion mass m_π, and the pion decay constant f_π (3; also see Reference 1),

$$m_a = \frac{m_\pi f_\pi}{f_a}\frac{\sqrt{m_u m_d}}{m_u + m_d} \simeq 0.6\,\mathrm{meV} \times \left(\frac{10^{10}\,\mathrm{GeV}}{f_a}\right). \qquad 7.$$

For a large axion decay constant f_a, the axion is a prime example of a WISP (4): It is a very weakly interacting (see Equation 6) subelectronvolt mass particle. In particular, its coupling to photons, which for an axion, like for any other pseudoscalar, should be of the form

$$\mathcal{L}_{a\gamma\gamma} = -\frac{1}{4} g\, aF_{\mu\nu}\tilde{F}^{\mu\nu} = g\, a\, \vec{E}\cdot\vec{B}, \qquad 8.$$

which is very much suppressed (5):

$$g = \frac{\alpha}{2\pi f_a}\left(\frac{2}{3}\frac{m_u + 4m_d}{m_u + m_d} - s\right) \sim 10^{-13}\,\mathrm{GeV}^{-1}\left(\frac{10^{10}\,\mathrm{GeV}}{f_a}\right). \qquad 9.$$

Although expected to be small, the guaranteed coupling of axions to photons (Equation 9) may result, if axions exist, in observable consequences from processes involving large electromagnetic fields. Such fields often occur in astrophysical and cosmological environments (Section 3) and can be prepared in laboratory experiments (Section 4).

The proposal of an anomalous Peccei-Quinn shift symmetry is motivated to provide for a solution to the strong-*CP* problem. This concept has been generalized to other similar WISP candidates—namely ALPs—that may arise as (pseudo-) Nambu-Goldstone bosons from the breaking of other global symmetries such as, for example, family symmetries. However, in contrast to axions, for generic ALPs a nonzero coupling g to photons is not guaranteed. Moreover, for such ALPs a predictive relation between g and the mass is missing. Correspondingly, ALP searches, which exploit their interactions with photons, should try to cover the entire parameter space spanned by g and the mass of the ALP, not only the restricted parameter space (Equations 7 and 9) that is predicted for axions.

2.1.2. Axions and axion-like particles from string compactifications. The existence of axions and ALPs can also be strongly motivated from a top-down point of view. When the six extra spatial dimensions of string theory are compactified, axions and ALPs arise quite naturally as Kaluza-Klein zero modes of antisymmetric tensor fields, which are generically present in all string theories. Moreover, the (Chern-Simons) couplings of these form fields to the gauge fields, which are crucially determined by anomaly-cancellation conditions, result in the anomalous *CP*-violating couplings in the low-energy effective Lagrangian (Equation 6), which are necessary for the solution of the strong-*CP* problem. Thus, string compactifications suggest plenty of candidates for axions and axion-like WISPs (6–9). However, they do not really predict their presence, because there exist several known mechanisms by which they can be removed from the low-energy spectrum. Only those that escape such mechanisms are WISP candidates.

In the compactification of the weakly coupled heterotic string, a universal, model-independent axion appears as the Poincaré dual[1] of the (Neveu-Schwarz) antisymmetric tensor field $B_{\mu\nu}$, where μ and ν are tangent to 3+1-dimensional Minkowski space-time (6). The axion decay constant f_a is quite independent of the details of the compactification. To compute it, one considers the action of an $N = 1$ supergravity coupled to an $E_8 \times E_8$ pure gauge theory in 9+1 dimensions,

$$S_{\rm H} = \frac{2\pi M_s^8}{g_s^2}\int d^{10}x\sqrt{-g}R - \frac{M_s^6}{2\pi g_s^2}\int \frac{1}{4}{\rm tr}F \wedge {}^*F - \frac{2\pi M_s^4}{g_s^2}\int \frac{1}{2}H \wedge {}^*H + \ldots, \qquad 10.$$

which describes the dynamics of the massless bosonic excitations of the heterotic string in terms of the Ricci scalar R, the gauge-field strength F, and the field strength H of the two-form field B. By compactifying this theory on a six-dimensional manifold with volume V_6, the resulting effective action can be matched to its standard normalization in 3+1 dimensions:

$$S_{3+1} = \frac{M_P^2}{2}\int d^4x\sqrt{-g}R - \frac{1}{4g_{\rm YM}^2}\int d^4x\sqrt{-g}\,{\rm tr}F_{\mu\nu}F^{\mu\nu} - \frac{1}{f_a^2}\int \frac{1}{2}H \wedge {}^*H + \ldots, \qquad 11.$$

where

$$M_P^2 = (4\pi/g_s^2)M_s^8V_6; \quad g_{\rm YM}^2 = 4\pi g_s^2/(M_s^6V_6); \quad \text{and} \quad f_a^2 = g_s^2/(2\pi M_s^4V_6) \qquad 12.$$

express the reduced Planck mass $M_P = 2.4 \times 10^{18}$ GeV; the gauge coupling $g_{\rm YM}$; and the axion decay constant f_a in terms of the string coupling g_s, the string scale $M_s = 1/\ell_s$, and the volume V_6. By eliminating the volume V_6 and the string scale by means of the first two relations in Equation 12, one ends up with an axion decay constant of order of the Grand Unified Theory (GUT) scale (10),

$$f_a = \alpha_{\rm YM}M_P/(2\pi\sqrt{2}) \simeq 1.1 \times 10^{16}\,\text{GeV} \quad \text{for}\,\alpha_{\rm YM} = g_{\rm YM}^2/(4\pi) \sim 1/25. \qquad 13.$$

Model-dependent axions arise in the context of weakly coupled heterotic strings from massless excitations of the two-form B-field on the six-dimensional compact manifold (6). Correspondingly, their properties depend much more on the details of the compactification. Nevertheless, a recent exhaustive study (8) has elucidated that in this case too, the axion decay constant cannot be smaller than 10^{15} GeV. Similar conclusions have been drawn for the axions in strongly coupled heterotic string theory (8). These findings can be easily understood physically: It is the string scale M_s that primarily determines the axion decay constant (7). And in the heterotic case, this scale is large—for instance, $M_s = \sqrt{\alpha_{\rm YM}/(4\pi)}M_P$—for the weakly coupled heterotic string (Equation 12).

The size of f_a may be different in compactifications of type II string theories that give rise to intersecting brane worlds. In such theories, the Standard Model lives on a stack of D$(3+q)$-branes that are extended along the 3+1 noncompact dimensions and wrap q-dimensional, topologically nontrivial submanifolds in the compactification manifold while gravity propagates in the bulk, leading to a possibly smaller string scale at the expense of a larger compactification volume: $M_s \sim g_s M_P/\sqrt{V_6M_s^6}$ (**Figure 1**). In type II string theory, the axions arise from the massless

[1]To perform this dualization explicitly, one introduces the axion field as a Lagrange multiplier for the Bianchi identity for H, $dH = 1/(16\pi^2)({\rm tr}\,R \wedge R - {\rm tr}\,F \wedge F)$, and subsequently integrates over H (6, 8).

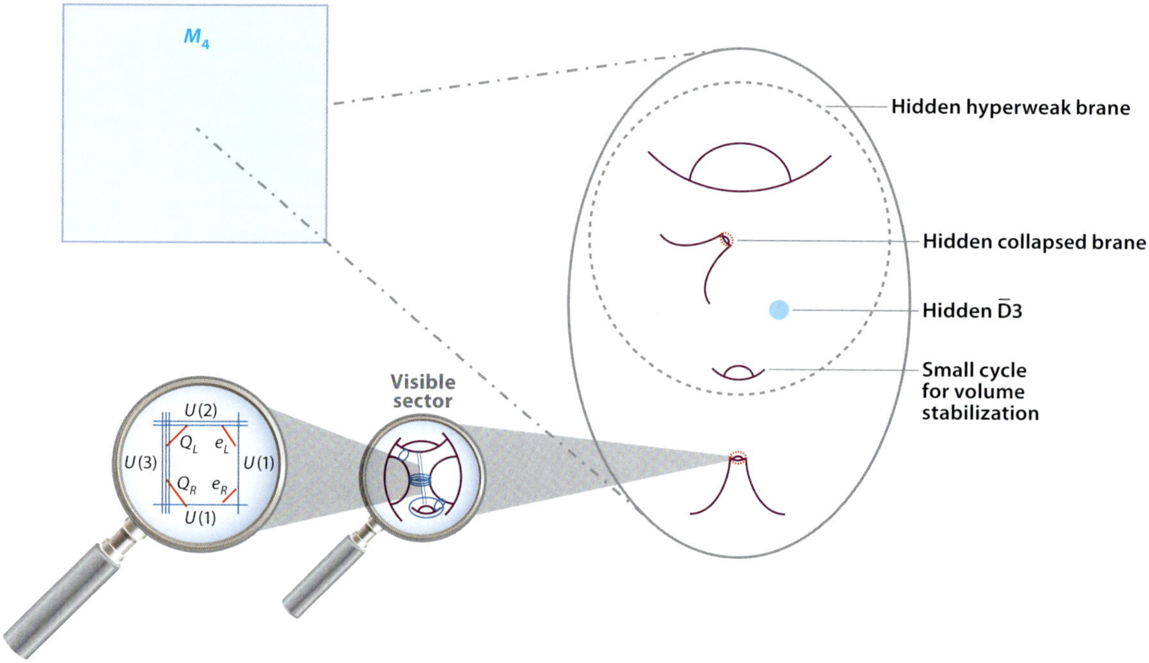

Figure 1

In compactifications of type II string theories, the Standard Model is locally realized by a stack of space-time-filling D-branes wrapping topologically nontrivial submanifolds in the compact dimensions. In general, there may also be hidden sectors localized at different places. They can arise from branes of different dimensions (D3 or D7 branes) that can be either of large extent or localized at singularities. Light visible and hidden-matter particles arise from strings located at intersection loci and stretched between brane stacks.

excitations of the (Ramond-Ramond) q-form gauge field C_q. The precise predictions depend on the particular embedding of the Standard Model, but generically one finds that the axion decay constant can be substantially lower than in the heteretic case, varying in the range (7)

$$f_a \sim \frac{M_P}{\sqrt{V_6 M_s^6}} \sim \frac{M_s}{g_s} \sim 10^{4 \div 17}\,\text{GeV}, \qquad 14.$$

which corresponds to a variation of the string scale between the teraelectronvolt and GUT scales.

2.1.3. Scalars and chameleons. Apart from possible light pseudoscalars, string compactifications generically also predict scalar particles—the dilaton and large numbers of moduli—that also appear massless at the compactification scale. Essentially massless scalar fields are also often invoked by cosmologists in the context of dark energy. A plausible explanation for the apparent acceleration of the cosmic expansion rate of the universe is provided by the presence of a spatially homogeneous scalar field rolling down a very flat potential (11).

Interactions between very light scalar fields and ordinary matter are strongly constrained by the nonobservation of so-called fifth force effects that lead to, for instance, violations of the equivalence principle (Section 4.4). Correspondingly, if such particles exist, then the forces mediated by them should be either much weaker than gravity or short range in the laboratory. The latter occurs in theories in which the mass of the scalar field depends effectively on the local density of matter—so-called chameleon field theories (12). Depending on the nonlinear-field self-interactions and on the interactions with the ambient matter, the chameleon may have a large mass in regions of high density (such as Earth) and a small mass in regions of low density (such as interstellar space). It is because such particles can hide so well from observations and experiments that they have been named chameleons.

2.2. Ultralight Hidden-Sector Particles

Similar to axions and ALPs, additional hidden-sector $U(1)$ gauge bosons are also a generic feature arising from string compactifications and, therefore, are well-motivated WISP candidates. In the standard compactification of the $E_8 \times E_8$ supergravity based on the heterotic string on a smooth (Calabi-Yau) manifold, the Standard Model gauge group is embedded in the first E_8 factor, whereas the second E_8 factor comprises a hidden gauge group that interacts with the first E_8 factor only gravitationally (13). The second E_8 factor may be broken in the course of compactification to products of non-Abelian and $U(1)$ gauge groups. The occurrence of hidden $U(1)$s can be exhaustively studied in toroidal orbifold compactifications of the heterotic string, which allow for a systematic scan of possible gauge-group factors and particles after compactification. Even if a realistic visible sector is required, numerous models may have possibly massless hidden $U(1)$s. In this "mini-landscape" of orbifold compactifications of the heterotic string (14), one encounters a breaking of the gauge symmetry to the Standard Model, a hidden-sector non-Abelian gauge symmetry, and typically at least one hidden $U(1)$, for example (15),

$$E_8 \times E_8 \rightarrow \underbrace{SU(3) \times SU(2) \times SU(1)}_{\text{Standard Model}} \times [SU(6) \times U(1)]. \tag{15.}$$

Compactifications of type II string theory also suggest the existence of hidden $U(1)$s, be they as Kaluza-Klein zero modes of the previously mentioned (Ramond-Ramond) form fields or as massless excitations of branes. As illustrated in **Figure 1**, type II compactifications generically involve space-time-filling hidden-sector branes that do not intersect with the Standard Model branes, often because of global consistency requirements.

Some of these hidden $U(1)$s may remain unbroken down to very small energy scales. In that case, their dominant interaction with the photon[2]—which is encoded in the low-energy effective Lagrangian

$$\mathcal{L} \supset -\frac{1}{4e^2}F_{\mu\nu}F^{\mu\nu} - \frac{1}{4g_h^2}X_{\mu\nu}X^{\mu\nu} + \frac{\chi}{2eg_h}F_{\mu\nu}X^{\mu\nu} + \frac{m_{\gamma'}^2}{2g_h^2}X_\mu X^\mu, \tag{16.}$$

where X_μ denotes the hidden $U(1)$ field with field strength $X_{\mu\nu}$ and gauge coupling g_h—occurs through kinetic mixing (17), with mixing parameter χ. Therefore, light hidden $U(1)$s (so-called hidden photons) are well-motivated WISP candidates, given that χ is expected to be small.

Kinetic mixing is generated at one loop by the exchange of heavy messengers that couple both to the visible $U(1)$ and to the hidden $U(1)$ gauge fields. In the context of compactifications of the heterotic string, the size of the mixing has been estimated as (18)

$$\chi \sim \frac{eg_h}{16\pi^2}C\frac{\Delta m}{M_P} \sim 10^{-5} \div 10^{-17} \quad \text{for} \quad C \gtrsim 10, \tag{17.}$$

where $\Delta m \sim 10^{5\div17}$ GeV is the possible range of mass splitting in the messenger sector once supersymmetry is broken.

[2]In principle, a hidden $U(1)$ gauge boson could also interact with the Standard Model particles via higher-dimensional operators (16).

A great variety of possible values of χ can also be accommodated in type II compactifications for the mixing between brane-localized hidden $U(1)$s and the visible $U(1)$. Here, kinetic mixing can be understood as originating from the exchange of closed strings through the bulk (19, 20). Generically, one finds that (21)

$$\chi \sim \frac{e g_h}{16\pi^2}, \tag{18.}$$

where the size of the hidden-sector gauge coupling g_h depends on the q-dimensional volume ($0 \leq q \leq 6$) of the cycle that the hidden brane wraps:

$$g_h^2 \simeq \frac{2\pi g_s}{V_q M_s^q} = 2\pi g_s \left(\frac{4\pi}{g_s^2} \frac{M_s^2}{M_P^2} \right)^{q/6}. \tag{19.}$$

This leads to a large range of possible values for the kinetic mixing,

$$10^{-12} \lesssim \chi \lesssim 10^{-3}, \tag{20.}$$

for the string scale varying between the teraelectronvolt scale and the GUT scale. Smaller values of kinetic mixing can be obtained in such setups in special cases where the one-loop contribution is canceled or vanishes. Moreover, exponentially suppressed values can be naturally obtained in flux compactifications with warped throats (20).

Masses for the hidden photons can arise via the standard Higgs mechanism but also via a Stueckelberg mechanism. In large-volume compactifications, small—even subelectronvolt—masses arise quite naturally (21). If the masses arise from a Stueckelberg mechanism, the mass and size of the kinetic mixing are typically linked through one scale—the string scale—and are therefore related to each other. Depending on the specific way in which the cycles wrap the singularities, one obtains expressions for the masses such as

$$(m_{\gamma'}^{\text{Stueck.}})^2 \simeq \frac{g_s}{2} \left(\frac{4\pi}{g_s^2} \frac{M_s^2}{M_P^2} \right)^z, \quad z = \frac{1}{3}, 1. \tag{21.}$$

For example, in the case of $z = 1$ we obtain, for a string scale of $M_s \sim 1$ TeV, $m_{\gamma'} \sim$meV and, from Equations 18 and 19, a mixing of $\chi \sim 10^{-12}$. The predictions for masses arising from the Higgs mechanism are less precise; however, they can also be tiny if the supersymmetry-breaking scale in the hidden sector is much smaller than in the visible sector.

Occasionally, there is also light hidden matter charged under the hidden $U(1)$s. After diagonalizing the gauge kinetic terms by a shift of $X \to X + \chi A$ and a multiplicative hypercharge renormalization, one observes that the hidden-sector particles acquire a minihypercharge $\varepsilon = \chi g_h$ (17). In particular, an eventual hidden Higgs particle may also be sought by exploiting its effective minihypercharge (22). Minicharged particles (MCPs) can arise from hidden-sector magnetic monopoles if the gauge fields mix via a nondiagonal θ term (23). Furthermore, MCPs can also arise from a nondiagonal mass term (24). Therefore, light MCPs are also very well motivated WISP candidates.

3. ASTROPHYSICAL AND COSMOLOGICAL CONSTRAINTS ON WISPs

As reviewed in Section 2, there exists a strong physics case for the possible existence of WISPs. Moreover, their possible masses and couplings span a very wide range in parameter space. Correspondingly, searches for signatures of WISPs must exploit a wide variety of observational and experimental techniques, ranging from cosmology and astrophysics to terrestrial laboratory experiments. At the present time, the strongest bounds on the existence of WISPs often come from stellar evolution and cosmology where, to the best of our knowledge, observations seem to agree with the standard budget of elementary particles. However, there are also some intriguing astronomical observations that are hard to explain with known physics and that could be interpreted as hinting at the existence of WISPs.

3.1. Bounds from Stellar Evolution

Production of WISPs in stars would substantially affect stellar evolution (25). WISPs are only rarely produced in the dense plasmas of stellar interiors, but they easily escape, thereby directly contributing to the total energy loss of the star. This mechanism differs from the standard energy loss of stars, which occurs mainly via photons from the stellar surface and neutrinos from the core. Therefore, the WISP luminosity is enhanced by a huge volume-to-surface factor, which leads to very strong constraints.

Stars evolve by fusing increasingly heavier nuclei in their cores. Heavier nuclei require hotter environments, and when a nuclear species is exhausted in the core, the latter slowly contracts and heats up until it reaches a new burning phase. WISP emission shortens normal burning phases because the energy-loss rate is higher than standard but the total energy is limited by the number of nuclei. However, WISP emission prolongs intermediate (red giant) phases because WISP cooling delays the point at which the appropriate temperature is reached during the core contraction. These effects have been used to constrain a variety of WISPs in different stellar environments (25, 26) for which information on evolutionary timescales is available. For standard QCD axions, the best constraints arise from white dwarf cooling (27, 28), both through the coupling to electrons and from the duration of the SN1987A neutrino burst (26) through the coupling to nucleons (**Figure 2*a***). The strongest limits for general ALPs with a two-photon coupling and MCPs arise from observations of horizontal branch (HB) stars in globular clusters (**Figures 2*b*, 3**) (33, 34). For very small masses, an even tighter limit on a two-photon coupling of ALPs can be obtained from the absence of a γ-ray burst in coincidence with a neutrino burst during the SN1987a explosion (35). The principle behind the latter bound is that ALPs would be produced in the supernova core from the Primakoff effect and reconverted into γ rays inside the galactic magnetic field.

The Sun is less sensitive than these other stars to axion or MCP emission, even though its properties are better known. Solar bounds have been obtained from studies of its lifetime, helioseismology, and the neutrino flux (30, 37), but although the data are more precise, the resulting constraints are weaker. However, as shown in **Figure 4**, this situation is different for hidden photons: The region in parameter space excluded by the solar lifetime (38) complements, in this case, the region excluded by the lifetime of HB stars (39).

The stellar bounds are very strong but also somewhat vulnerable: They can be considerably relaxed if the couplings to photons effectively depend on environmental conditions such as temperature and matter density (40), which definitely occurs in some models, such as the above-mentioned chameleons and the models presented in References 41 and 42.

3.2. Bounds from Big Bang Nucleosynthesis

Big bang nucleosynthesis (BBN) provides a unique probe of the early universe (for a recent review, see Reference 43). At temperatures below approximately 1 MeV, the weak reactions $p + e^- \leftrightarrow n + \nu_e$ in the primordial plasma became ineffective, fixing the n/p density ratio to $\sim$1/7. This freeze-out ratio depends crucially on the rate of cosmic expansion H, which in turn grows with increasing total energy density ρ of all particles in the primordial plasma: The larger ρ is, the sooner the pn freeze-out occurs and the closer n/p becomes to the high-temperature value of 1/2. After decoupling, during primordial nucleosynthesis, neutrons are confined mostly into ^{4}He

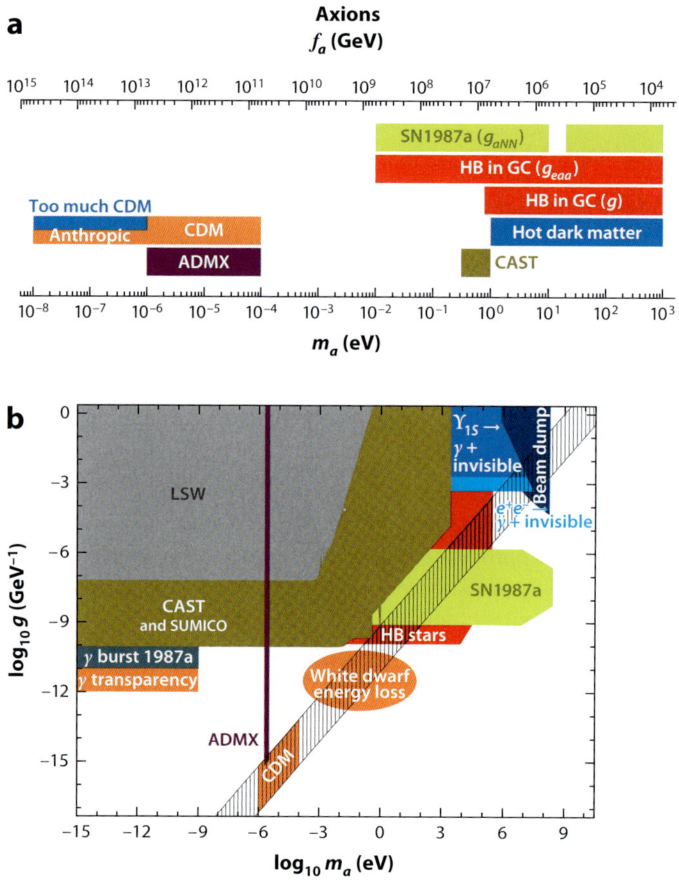

Figure 2

Summary of cosmological and astrophysical constraints for (*a*) axions (for the mass m_a or decay constant f_a) (26) and (*b*) axion-like-particles (ALPs; two-photon coupling g versus mass m_a of the ALP) (29–31). The mass region, in which the axion could be the cold dark matter (*orange regions labeled* CDM), can be extended toward smaller masses (larger $f_a \sim 10^{16}$ GeV) by anthropic reasoning. In the first plot the areas marked ADMX and CAST show the near-future search ranges. In the second plot the axion band is hatched. Other astrophysically interesting areas are colored orange. For comparison, we also show laboratory limits from the photon-regeneration experiments ADMX and LSW (32). Note that the limit from ADMX is valid only under the assumption that the local density of ALPs at Earth is given by the dark matter density. Abbreviations: GC, globular cluster; HB, horizontal branch; SN, supernova.

nuclei whose primordial abundance can then be measured. Such measurements lead to a bound on the nonstandard energy density ρ_x during BBN, which is usually expressed as the effective number of extra thermal neutrino species,

$$N_{\nu,x}^{\text{eff}} \equiv \frac{4}{7} \frac{30}{\pi^2 T^4} \rho_x. \qquad 22.$$

A recent determination of this number (44) resulted in

$$N_{\nu,x}^{\text{eff}} = -0.6^{+0.9}_{-0.8} \qquad 23.$$

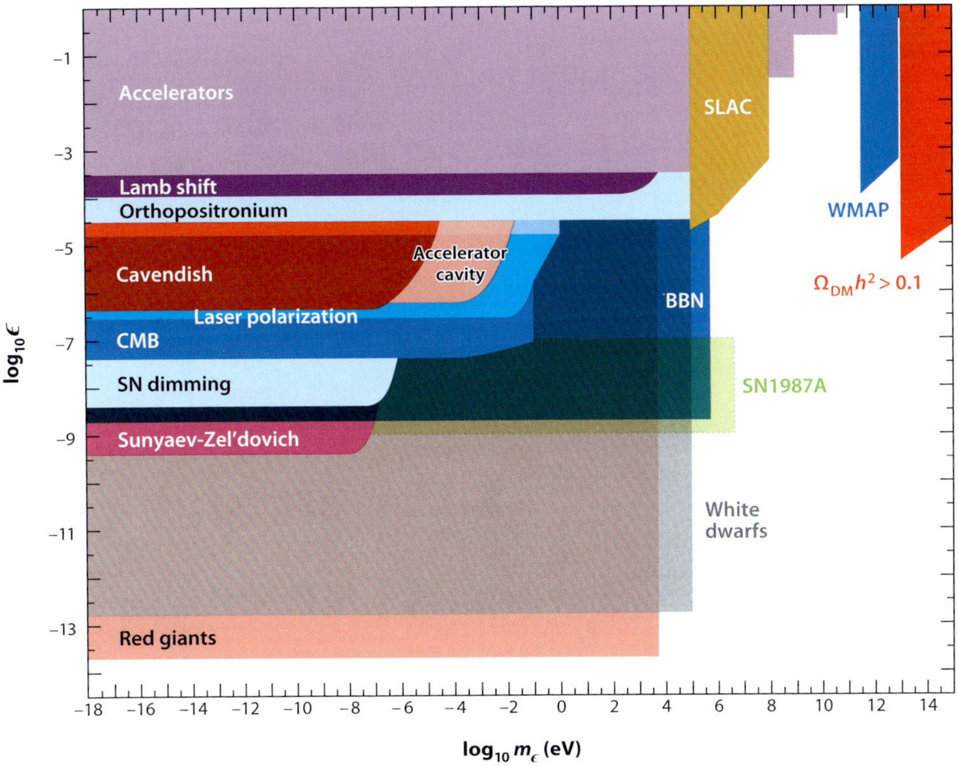

Figure 3

Summary of cosmological and astrophysical constraints for minicharged particles (fractional charge $\varepsilon = Q_\varepsilon/e$ versus mass m_ϵ) (21). [Additional interesting bounds may arise, for instance, by considering primordial black holes (36).] Also shown are the laboratory limits discussed in Section 4, along with the bounds from accelerator and fixed-target experiments (SLAC) at relatively large masses and couplings. Abbreviations: BBN, big bang nucleosynthesis; CMB, cosmic microwave background; SN, supernova.

for three standard neutrinos. Therefore, although an extra neutral spin-0 particle thermalized during BBN is allowed, this is not the case for other WISPs such as an MCP, for which

$$N^{\rm eff}_{\nu,\rm MCP} \geq 1, \qquad 24.$$

or a massive hidden photon, for which

$$N^{\rm eff}_{\nu,\gamma'} = 21/16. \qquad 25.$$

The interactions of MCPs and γ's with the standard bath should not allow thermalization before BBN. ψ MCPs are produced with a rate of $\Gamma(e^+e^- \to \psi\bar{\psi}) \sim \alpha^2\varepsilon^2 T/2$, whereas γ's are produced with a rate of $\Gamma(\gamma e^\pm \to \gamma' e^\pm) \sim \chi^2_{\rm eff}\Gamma_{\rm C}$, where $\Gamma_{\rm C}$ is the standard Compton scattering rate. Here $\chi_{\rm eff}$ is the effective $\gamma\gamma'$ mixing in the plasma, which for subelectronvolt γ' masses is $\chi_{\rm eff} \simeq \chi(m_{\gamma'}/\omega_P)^2$. The ratio of the γ' mass to the plasma frequency, $m_{\gamma'}/\omega_{\rm P}$, is extremely small before BBN and suppresses γ' production with respect to other WISPs. Correspondingly, a comparison with the expansion rate H reveals that MCPs with $\varepsilon < 2 \times 10^{-9}$ would be allowed (**Figure 3**) (45), but there are no significant bounds for hidden photons (42).

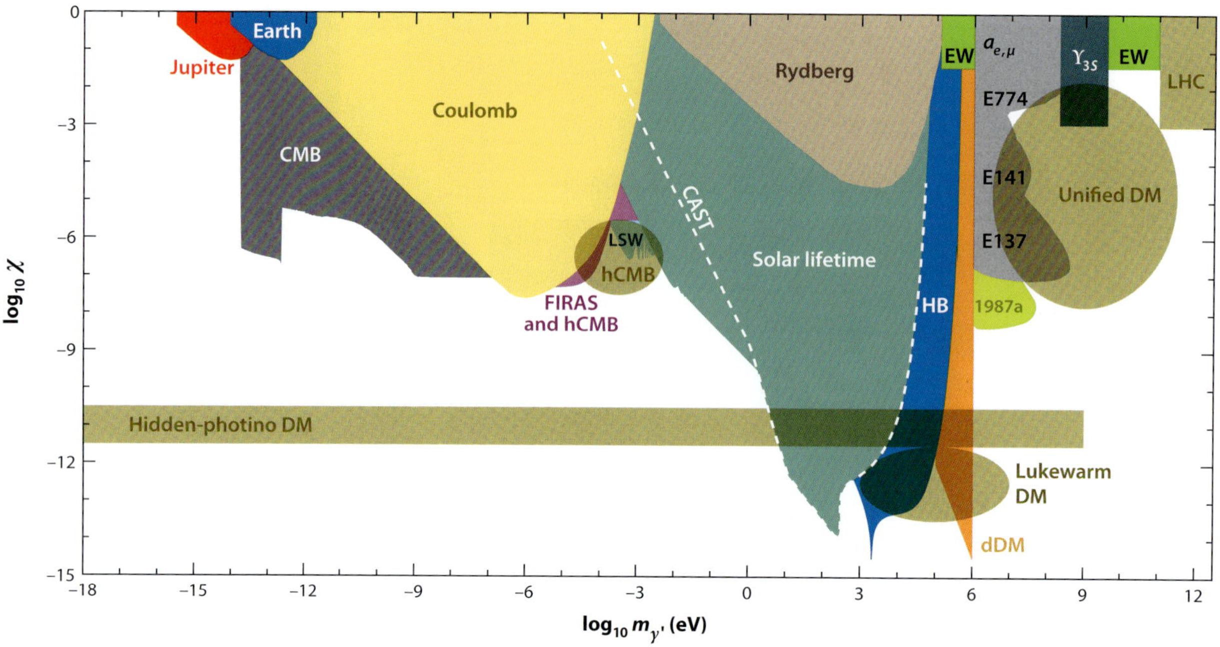

Figure 4

Summary of cosmological and astrophysical constraints for hidden photons (kinetic mixing χ versus mass $m_{\gamma'}$) (J. Redondo, private communication). We also show laboratory limits and, at higher masses, electroweak precision measurements (EW), bounds from upsilon decays (Υ_{3S}), and fixed-target experiments (three-digit numbers preceded by E). Especially interesting areas are colored light orange. Abbreviations: CMB, cosmic microwave background; DM, dark matter; dDM, decaying $\gamma' \to \gamma\gamma\gamma$ dark matter; HB, horizontal branch; hCMB, hidden cosmic microwave background.

3.3. Bounds from the Cosmic Microwave Background

The cosmic microwave background (CMB) features an almost perfect blackbody spectrum with $\mathcal{O}(10^{-5})$ angular anisotropies. It is released at a temperature $T \sim 0.1$ eV, but the reactions responsible for the blackbody shape freeze out much earlier, at $T \sim$ keV. Reactions such as $\gamma + \ldots \to$ WISP $+ \ldots$ would deplete photons in a frequency-dependent way, which can be constrained by the precise FIRAS spectrum measurements (46). This depletion can be used to constrain light MCPs and ALPs (47) as well as hidden photons (48). More generally (49), (resonant) production of hidden photons leads to distortions in the CMB spectrum (as measured by FIRAS), thereby strongly constraining their existence in a wide mass range (**Figure 4**); similar bounds can be obtained for ALPs, but they depend on the unknown strength of the intergalactic magnetic field (50). Similarly, CMB photons passing through the magnetic field of clusters in the presence of MCPs leads to a local distortion of the CMB spectrum in the direction of the cluster. Such distortions are constrained by measurements (51) of the so-called Sunyaev-Zel'dovich effect and lead to strong bounds on MCPs (52).[3] In addition, around $T \sim$ eV the primordial plasma is so sparse that WISPs would freely stream out of the density fluctuations, diminishing their contrast. Moreover,

[3]Analogously, light from distant supernovae passing through the (less well known) intergalactic magnetic field would be dimmed by MCP production, again constraining the existence of such particles (**Figure 3**) (53). Of course, the photon-axion conversion in the intergalactic medium would also contribute to supernova dimming (54), but the corresponding bound would be weaker than that from CMB measurements (50).

thermal WISPs contribute to the radiation-energy density, delaying the matter-radiation equality and reducing the contrast growth before decoupling. In this respect, such WISPs behave identically to standard neutrinos (55). Therefore, the extra contribution to the energy density ρ_x (and the couplings that would produce it) can again be constrained from the value of $N_\nu^{\rm eff}$ inferred from analysis of CMB anisotropies and other large-scale structure data (e.g., 44):

$$N_{\nu,x}^{\rm eff} = -0.1^{+2.0}_{-1.4}. \qquad 26.$$

This argument has been used to constrain axions (**Figure 2*a***) (56) and millielectronvolt γ's (**Figure 4**) (48). However, note that in the determination of the value of $N_{\nu,x}^{\rm eff}$ in Equation 26, Lyman-α forest data has been deliberately omitted. Lyman-α has systematically favored values of $N_{\nu,x}^{\rm eff}$ larger than zero (57), which may reveal the existence of a cosmic WISP relic density (Section 3.4.5). Alternatively, it may be due to an incorrect treatment of the bias parameter (58).

3.4. Possible Indirect Hints of WISPs

Other cosmological and astrophysical puzzles—interpreted in terms of WISPs—may indicate that the discovery of WISPs is just around the corner. Therefore, it is of great interest to search for these particles under controlled laboratory conditions.

3.4.1. Axions as cold dark matter. Importantly, axions may constitute the CDM in the universe. For very weak coupling, that is, for a large decay constant f_a, the ultralight axions are produced nonthermally in the early universe. At early times, at temperatures well above the QCD phase transition, the axion is effectively massless and the corresponding field can take any value, parameterized by the so-called misalignment angle θ_i. Later, as the temperature of the primordial plasma falls below the hadronic scale $T \lesssim$ GeV, the axion develops its mass m_a due to nonperturbative (topological instanton) effects. When the mass m_a becomes of order the Hubble expansion rate, the axion field begins to oscillate around its mean value $\langle a \rangle = 0$. These coherent and spatially uniform oscillations correspond to a coherent state of nonrelativistic axions, whose contribution to today's energy density, in terms of the critical energy density, can be estimated[4] to be (60)

$$\Omega_a h^2 = \kappa_a \left(\frac{f_a}{10^{12}\,\mathrm{GeV}} \right)^{1.175} \theta_i^2, \qquad 27.$$

where $0.5 \lesssim \kappa_a \lesssim$ few. Therefore, for generic values of the misalignment angle, $\theta_i = \mathcal{O}(1)$, the axion could be the main constituent of CDM in the universe, $\Omega_{\rm CDM} h^2 \sim 0.1$, if its decay constant is of order $f_a \sim 10^{12}$ GeV, which corresponds to a mass in the 10-μeV range. Larger values of f_a (**Figure 2*a***) are excluded because they would lead to an overclosure of the universe. However, if the initial θ_i is small, values of f_a near the GUT or Planck scale are still possible. This could be due to fine-tuning as well as anthropic selection (61).

Axion dark matter may indeed explain two additional puzzling observations: (*a*) an interesting alignment in the multipoles of the CMB anisotropies (62) and (*b*) the rotational curves of a number of galaxies, which provide evidence for additional structures in the galactic halo—so-called caustic rings (63). Both observations may be explained by Bose-Einstein condensation of dark matter axions (64).

[4]Note that this is a relatively crude estimate. In principle, the number of axions also depends on the order in which cosmological events take place, especially whether the breaking of the Peccei-Quinn symmetry occurred before or after inflation. In the latter case, for example, there are additional contributions from the formation and decay of cosmic strings and domain walls. Typically, these contributions are of similar order to the axion density produced in the misalignment mechanism (59).

Finally, in the context of dark matter, axions could also contribute a hot dark matter component. As for neutrinos, the fraction of this hot dark matter is proportional to the mass of the axion. Recent constraints on the size of a possible hot dark matter component provide a bound of $m_a \lesssim 1.2$ eV (**Figure 2*a***) (56).

3.4.2. Nonstandard energy loss in white dwarfs. Isern et al. (28) recently identified possible nonstandard energy loss in the white dwarf luminosity function. As the authors point out, such energy loss is compatible with the existence of axions with an axion-electron coupling,

$$g_{eea} \simeq 10^{-13}, \tag{28}$$

which suggests an axion decay constant and an axion mass[5] of

$$f_a \sim g_{eea} m_e \sim \text{few} \times 10^9\,\text{GeV} \quad \text{and} \quad m_a \sim \text{meV}, \tag{29}$$

respectively. In typical models, the corresponding coupling to photons is

$$g \sim \alpha / f_a \sim 10^{-12}\,\text{GeV}^{-1} \tag{30}$$

and, in other words, is very close to the stellar evolution bounds. But, apart from the obvious possibility that a more conventional explanation may be found for this nonstandard energy loss, we note that such energy loss could also be explained in terms of hidden photons coupling via nonrenormalizable operators to electrons (e.g., 65). Nevertheless, the values in Equations 29 and 30 should be considered benchmark values for future axion searches. Fortunately, as we discuss in the next section, this region of parameter space is not completely excluded from the possibilities to be checked in a controlled laboratory experiment based on photon regeneration.

3.4.3. Hints of cosmic-photon regeneration. Recent observations in teraelectronvolt γ astronomy may point to the existence of ALPs or at least may help provide sensible new constraints on their existence. Quite distant astrophysical sources have been observed in teraelectronvolt γ rays by H.E.S.S. and MAGIC, which was puzzling because the γ-ray absorption rate due to electron-positron pair production from extragalactic background light (EBL) was believed to be too strong to allow for their observation (66, 67). Clearly, a conventional explanation is that the EBL is less dense than expected and/or that the source spectra are harder than previously thought. Alternatively, such a high transparency of the universe may be explained by the conversion of γ rays into ALPs in the magnetic fields around the γ-ray sources. If so, these ALPs would travel unimpeded until they reach our galaxy, at which point they would reconvert into photons in the galactic magnetic fields (68). Alternatively, the conversion/reconversion could take place in the intergalactic magnetic fields (69; also see Reference 70 for a comprehensive bibliography and the current status). The intergalactic magnetic fields are not well known, but if they are assumed to be organized in randomly oriented patches, this would produce a characteristic scatter in luminosity relations. A powerful signature of cosmic-photon regeneration could therefore emerge if the reconstructed EBL, along different lines of sight toward different teraelectronvolt γ sources, were to display such a characteristic scatter (70). Similarly, photon-axion-photon conversion could allow super-GZK (Greisen-Zatsepin-Kuzmin) photons from ultrahigh-energy cosmic rays to travel large distances and arrive at Earth (71).

---

[5]For the mass of an ALP, everything below approximately 1 keV would be acceptable.

ALPs may also leave their imprint in the luminosity relations of active galactic nuclei. Mixing between photons and ALPs in the random magnetic fields in galaxy clusters may induce a characteristic scatter in the relations of X-ray versus optical luminosities of compact sources in these clusters. Evidence for such an effect was recently found in an analysis of luminosity relations of ~200 active galactic nuclei. This evidence hints at the existence of a very light ALP (72).

Furthermore, photon-ALP mixing could explain the puzzling origin of the (debated) correlation of the arrival directions of ultrahigh-energy cosmic rays and BL-Lac objects observed by AGASA and HiRes (73; however, see Reference 74). Moreover, the observed alignment of the polarization vectors of very distant quasars may also be explained by selective photon disappearance from photon-ALP oscillations (75). Finally, very similar ALPs have been invoked to solve some problematic aspects of the X-ray activity of the Sun, the longstanding corona problem, and the triggering of solar flares (76).

Intriguingly, in all the above-mentioned hints of cosmic-photon regeneration, the required ALP should have a very small mass (72),

$$m_a \ll 10^{-12}\,\text{eV} \div 10^{-9}\,\text{eV}, \qquad 31.$$

and a coupling in the range of

$$g \sim 10^{-12} \div 10^{-11}\,\text{GeV}^{-1}. \qquad 32.$$

Again, these values should be considered important benchmarks for future ALP laboratory searches. As we discuss below, photon-regeneration experiments can probe this region of mass and coupling.

3.4.4. Dark matter from hidden photons. Dark matter may also be directly related to hidden photons. Indeed there are several ways in which a hidden $U(1)$ may contribute to dark matter (**Figure 4**). First, in the $(\chi, m_{\gamma'}) \sim (10^{-12}, 0.1\,\text{MeV})$ region, the hidden photon itself could be a lukewarm dark matter candidate (39, 77). Second, for somewhat larger mixing angles in the $(\chi, m_{\gamma'}) \sim (10^{-11}, 100 \lesssim \text{GeV})$ region (**Figure 4**), the supersymmetric partner of the hidden photon (known as the hidden photino) is a promising dark matter candidate if its mass is in the 10–150-GeV range (78). Third, for $(\chi, m_{\gamma'}) \sim (10^{-23}, 0)$, the hidden photino, whose mass is in the teraelectronvolt range, could be a candidate for decaying dark matter and would thereby give rise to the above-mentioned excesses observed in galactic cosmic-ray positrons and electrons (79). Finally, in the $(\chi, m_{\gamma'}) \sim (10^{-4}, \text{GeV})$ region, the hidden photon plays an important role in models where the dark matter resides in the hidden sector (80). Such models aim to present a unified description of unexpected observations in astroparticle physics, notably the positron excess observed by the satellite experiment PAMELA (81) and the annual modulation signal seen by the direct dark matter search experiment DAMA (82). The massive hidden $U(1)$ can then mediate so-called dark forces. These values are also accessible to accelerator searches (83, 84) and have been motivated in various supersymmetric scenarios (84–86).

3.4.5. A hidden cosmic microwave background? As discussed in Section 3.3, hidden photons may contribute to the effective number of neutrinos. Indeed, such a contribution would lead to a higher number of effective neutrinos at the time of CMB and large-scale structure formation than at the time of BBN. Interestingly, some global cosmological analyses that account for precision cosmological data on the CMB and on the large-scale structure of the universe appear to require some extra radiation energy density from invisible particles apart from the three known neutrino species. The case for this scenario was strengthened by the recently released WMAP 7-year data (87). Hidden photons in the parameter region $(\chi, m_{\gamma'}) \sim (10^{-6}, 0.2\,\text{meV})$ (**Figure 4**) lead to a natural explanation for this finding (48). As an aside, note that these parameter values allow for interesting technological applications of hidden photons (88).

4. WISP SEARCHES WITH LOW-ENERGY PHOTONS

4.1. Photon-Regeneration Experiments

One of the most striking features of many new light bosons is that photon–light boson oscillations may occur in much the same way as those of the different neutrino species. Below we describe so-called light-shining-through-a-wall (LSW) experiments, which make direct use of this oscillation phenomenon. We also introduce the basic equations governing these oscillation phenomena.

4.1.1. Light shining through a wall: theory. One of the most striking consequences of the photon–light boson oscillations is the possibility of LSW (89–91). A schematic setup of an LSW experiment is shown in **Figure 5**. In such experiments, if an incoming photon can somehow be converted into a WISP, then the latter can transverse an opaque wall without being stopped. On the other side of the wall, the WISP can then reconvert into a photon.

This type of experiment is sensitive to a variety of WISPs (**Figure 6**). In particular, axions or ALPs can be sought by employing a magnetic field in the conversion regions, thereby facilitating the conversions of photons into axions via the two-photon interaction predicted for ALPs (**Figure 6*a***).

Such oscillation phenomena can be described via a nondiagonal mass term in the equation of motion that defines the photon A (with energy ω) and the new particle X (92) as

$$\left[\omega^2\mathbf{1} + \partial_z^2\mathbf{1} - \mathcal{M}^X\right]\begin{pmatrix} A \\ X \end{pmatrix} = 0, \qquad 33.$$

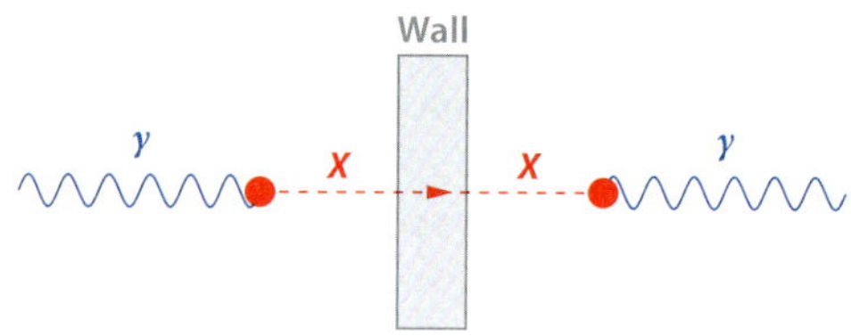

Figure 5

Schematic of a light-shining-through-a-wall experiment. An incoming photon γ is converted into a new particle X, which interacts only very weakly with the opaque wall. It passes through the wall and is subsequently reconverted into an ordinary photon that can then be detected.

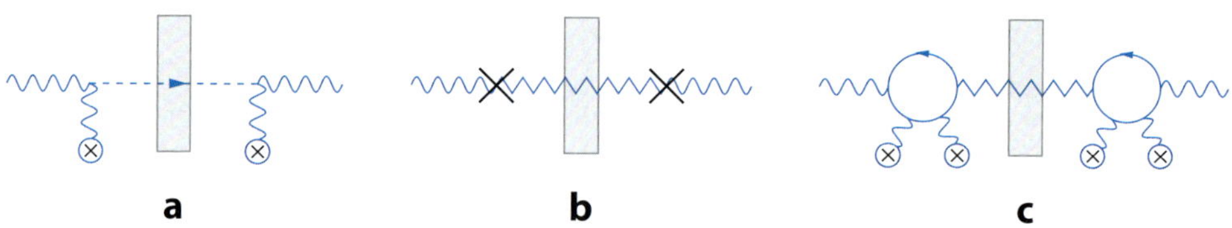

Figure 6

Explicit processes contributing to a potential signal in light-shining-through-a-wall experiments for various weakly interacting subelectronvolt particles. (*a*) Photon–axion-like particle oscillation. (*b*) Photon–hidden photon oscillation. (*c*) Photon–hidden photon oscillation facilitated by minicharged particles.

where we have suppressed the Lorentz structure. Indeed, for the types of particles discussed in Section 2, the equations of motion always separate into the two possible linear polarizations, but the mass matrices may differ between the two polarization directions.

The solutions to these equations of motion are of the form

$$v_1 = \exp[-\mathbf{i}(\omega t - k_1 z)] \begin{pmatrix} 1 \\ \delta \end{pmatrix}, \quad v_2 = \exp[-\mathbf{i}(\omega t - k_2 z)] \begin{pmatrix} -\delta \\ 1 \end{pmatrix}. \qquad 34.$$

If the off-diagonal entry in the mass matrix is small, then we can obtain simple analytical formulas for the mixing angle

$$\tan(2\,\delta) = 2\frac{\mathcal{M}^X_{12}}{\mathcal{M}^X_{11} - \mathcal{M}^X_{22}} \qquad 35.$$

as well as the wave numbers for the two mass eigenstates

$$k_1^2 = \omega^2 - \mathcal{M}^X_{11} \quad \text{and} \quad k_2^2 = \omega^2 - \mathcal{M}^X_{22}. \qquad 36.$$

By use of these formulas, we can easily find the transition amplitudes,

$$A(\gamma \to X) = \delta[\exp(\mathbf{i}k_1 z) - \exp(\mathbf{i}k_2 z)], \qquad 37.$$

and obtain

$$\begin{aligned} P(\gamma \to X, \ell) &= P(X \to \gamma, \ell) = |A(\gamma \to X)|^2 \\ &= |\delta|^2 \begin{pmatrix} \exp[-2\,\mathrm{Im}(k_1)\ell] + \exp[-2\,\mathrm{Im}(k_2)\ell] \\ \qquad -2\exp[-\mathrm{Im}(k_1 + k_2)\ell]\cos[\mathrm{Re}(k_1 + k_2)\ell] \end{pmatrix}. \end{aligned} \qquad 38.$$

In an LSW experiment, the photon must convert into a WISP and back. Therefore, the total probability that a photon will arrive at the detector is[6]

$$P(\gamma \to X, \ell_1)P(X \to \gamma, \ell_2). \qquad 39.$$

This probability can be enhanced by using optical cavities to reflect the light back and forth inside the production and regeneration regions (93, 94). One can easily imagine that, in the production region, the probability is enhanced by the number of passes toward the wall. On the regeneration side, such an enhancement is not as obvious. Nevertheless, a resonant cavity on the regeneration side would allow for an additional enhancement corresponding to the number of passes inside this cavity. The probability including these improvements is therefore

$$P_{\mathrm{LSW}} = N_1 N_2 P(\gamma \to X, \ell_1)P(X \to \gamma, \ell_2), \qquad 40.$$

where N_1 (N_2) is the effective number of passes in the respective cavity divided by a factor of two.

We now consider the three sets of particles discussed in Section 3: ALPs, massive hidden photons, and massless hidden photons with additional MCPs. ALPs couple to two photons. In this case, the photon-ALP oscillations have to be facilitated by the presence of a magnetic field (**Figure 5*a***), which provides for one of the two photons in the interaction. The presence of the magnetic field marks a preferred direction. Therefore, the matrix $\mathcal{M}$ can now depend on the polarization direction. Indeed, one obtains

$$\begin{aligned} \mathcal{M}^{a^-}_{\parallel} &= \begin{pmatrix} 0 & -gB\omega \\ -gB\omega & m_a^2 \end{pmatrix}, \quad & \mathcal{M}^{a^-}_{\perp} &= 0, \\ \mathcal{M}^{a^+}_{\parallel} &= 0, & \mathcal{M}^{a^+}_{\perp} &= \begin{pmatrix} 0 & -gB\omega \\ -gB\omega & m_a^2 \end{pmatrix}, \end{aligned} \qquad 41.$$

[6]Here, we neglect the typically very small absorption probability of the WISP X inside the wall.

where a^- indicates a pseudoscalar ALP coupling to $F\tilde{F}$ and a^+ represents a scalar coupling to F^2. Moreover, the subscripts $\parallel$ and $\perp$ indicate the polarization direction with respect to the magnetic field. Because the mass matrix is real, the expression for the probability in an LSW experiment (Equations 38 and 39) simplifies. If the polarization of the laser is at an angle θ with respect to the magnetic field, the probability for a pseudoscalar is

$$P_{\rm LSW} = 16\frac{[gB\omega\cos(\theta)]^4}{m_a^8}\sin^2\left(\frac{\ell_1 m_a^2}{4\omega}\right)\sin^2\left(\frac{\ell_2 m_a^2}{4\omega}\right). \qquad 42.$$

For a scalar, $\cos(\theta)$ must be replaced by $\sin(\theta)$. Moreover, for simplicity we have made the additional assumption that $\omega \gg m_a$.

The transition into hidden photons also occurs in the absence of a magnetic field. Accordingly, there is no preferred direction, and both polarizations have identical mass mixings:

$$\mathcal{M}^{\gamma'} = m_{\gamma'}^2\begin{pmatrix}\chi^2 & -\chi\\ -\chi & 1\end{pmatrix}. \qquad 43.$$

The corresponding transitions are depicted in **Figure 5*b***, and the transition probability is

$$P_{\rm LSW} = 16\chi^4\sin^2\left(\frac{\ell_1 m_{\gamma'}^2}{4\omega}\right)\sin^2\left(\frac{\ell_2 m_{\gamma'}^2}{4\omega}\right). \qquad 44.$$

Finally, for the combination of massless hidden photons and MCPs, we again need a magnetic field to allow for the transition

$$\mathcal{M}^{\gamma'+\rm MCP} = -2\omega^2 e_b^2\Delta N_i\begin{pmatrix}+\chi^2 & -\chi\\ -\chi & +1\end{pmatrix}. \qquad 45.$$

Here, $i = \parallel,\perp$ again indicate the polarization with respect to the magnetic field, and ΔN_i are the magnetic field–dependent, complex refractive indices describing the refraction and absorption due to the virtual and real production of MCPs (see Reference 95 for details).

4.1.2. Light shining through a wall: experiments. Most LSW experiments employ laser light in the optical regime. The typical lengths of the production and regeneration zones are of order a few meters. As one can see from Equations 42 and 44, an oscillation length of a few meters corresponds to particle masses in the millielectronvolt range. For ALPs, therefore, these experiments are sensitive for masses of up to a few millielectronvolts, and for hidden photons the sensitivity is usually optimal in this mass range. Various LSW experiments have been performed or are currently running (96, 97). The bounds are in the $g \leq$ few $\times$ 10^{-8} GeV^{-1} range for ALPs, and the kinetic mixing and minicharges probed are typically of the order of a few $\times$ 10^{-7} (**Figure 7**). Although these bounds are not yet competitive with those from astrophysical and cosmological realms—except for the case of hidden photons in the millielectronvolt mass range (**Figure 7*c***)—they are less model dependent (40–42). Moreover, so far only two of these experiments (BFRT and ALPS) have employed mirrors to enhance the transition probability on the production side.[7] The use of cavities on both sides (93, 94) could lead to significant improvements in the near future. However, doing so presents a technological challenge (99) because these cavities must be tuned to be resonant with each other (and, of course, a better-quality cavity reduces the bandwidth, making

[7]The ALPS Collaboration employed a Fabry-Pérot cavity, whereas the BFRT Collaboration used an optical delay line. If used only on the production side, the choice does not make a difference, but for resonant regeneration an optical cavity is needed.

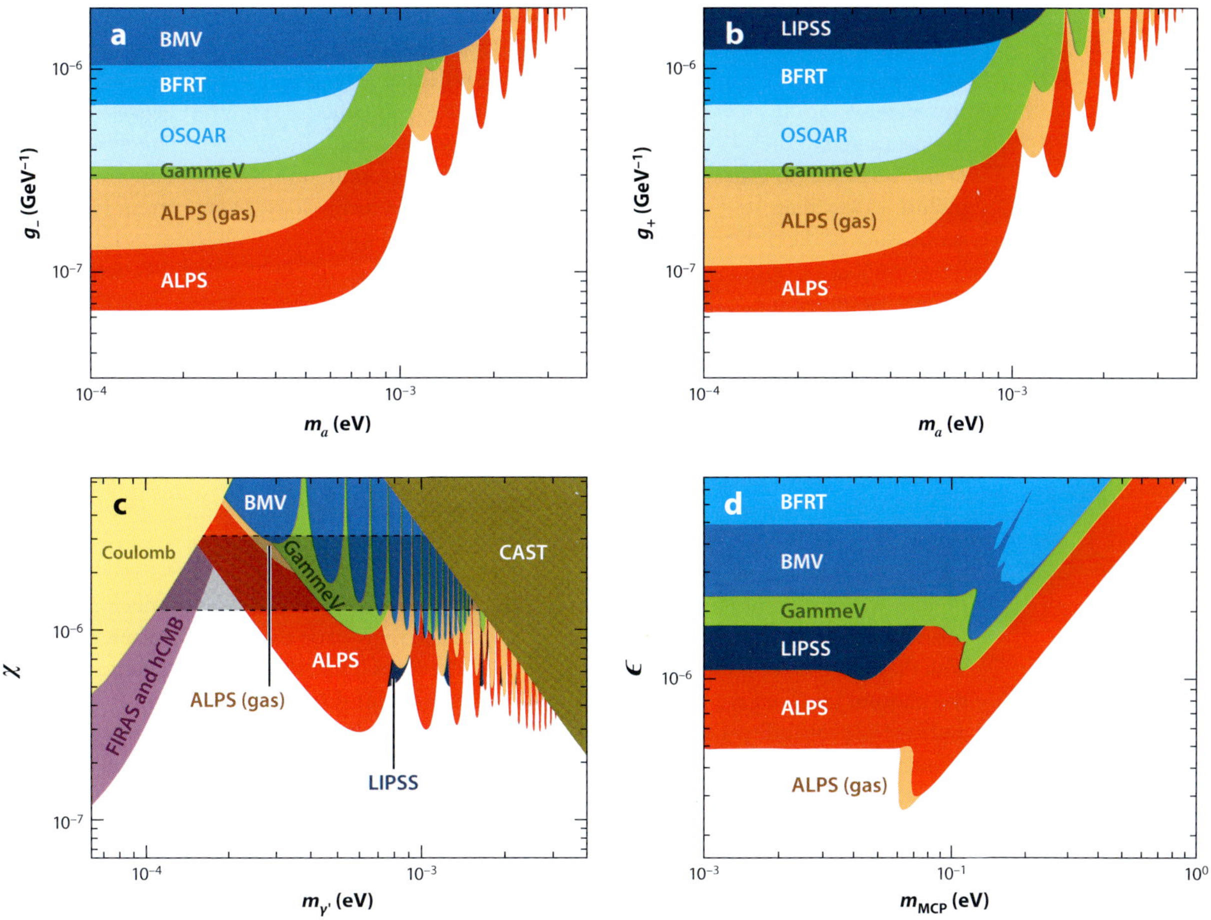

Figure 7

Sensitivities of light-shining-through-a-wall experiments. (*a*) Pseudoscalar and (*b*) scalar axion-like particles (ALPs). (*c*) Massive hidden photons and (*d*) massless hidden photons with an additional minicharged particle (MCP). In panel *c*, the band between the horizontal lines shows the region in which hidden photons could account for the apparent excess in the relic neutrino density recently reported by WMAP (87). Compilation from Reference 98. Abbreviation: hCMB, hidden cosmic microwave background.

this process even more difficult). Nevertheless, these methods may soon allow one to exceed, for the first time, the astrophysical bound on the ALP-photon coupling. They may even reach the ALP benchmark point (Equation 32)—$g \sim 10^{-11}$ GeV^{-1}—for a small enough ALP mass, and they could thereby test the ALP interpretation of reported hints at cosmic-photon regeneration (Section 3.4.3). To probe the axion benchmark point (Equation 29), however, one must be able to access higher masses. One possibility is to exploit alternating magnetic-field directions (91); another is to use phase-shift plates (100).

Higher masses could also be accessed by exploiting kiloelectronvolt photons from X-ray free-electron lasers or synchrotron radiation sources (101). However, at present the photon fluxes from these sources appear to be too small to be competitive with optical photons.

Further improvement in the sensitivity at low masses could also arise from experiments that employ electromagnetic waves in the radio-frequency range instead of laser light in the visible regime (102). The advantage of such microwave-shining-through-a-wall experiments is threefold. First, microwave cavities can be constructed such that light is effectively reflected back and forth inside the cavities up to 10^{11} times. This allows for a significant improvement over the current best optical cavities, in which light can be reflected only up to 10^5 times. Second, although still technologically challenging, tuning the cavities to resonance is simplified because extremely frequency-stable cavities exist in the microwave regime. Third, by using the phase information of the generator to distinguish between signal and background, one can detect microwaves with intensities less than 10^{-23} W with commercially available technology. With current technology, sensitivities of the order

$$g \sim 10^{-10}\,\mathrm{GeV}^{-1}, \quad \chi \sim 10^{-12}$$

appear achievable. The drawback is that due to the lower frequency and energy of microwaves, the sensitive range is typically restricted to particle masses of less than 0.1 meV. Several of these experiments are currently under construction or in the planning stages (103, 104), and one experiment has already obtained its first data (105).

Another variant of the LSW concept is the use of static magnetic fields (106). Such a setup consists of a highly sensitive magnetometer inside superconducting shielding, which is placed inside a strong (but subcritical) magnetic field. In ordinary electrodynamics, the magnetic field cannot permeate the superconductor, so no field should register on the magnetometer. However, photon–hidden-sector photon–photon oscillations would penetrate the superconductor, allowing the magnetic field to "leak" into the shielded volume and register on the magnetometer. Compared with the classic LSW setup, there are two crucial differences. First, the fields are (nearly) static, and the photons involved are virtual. Second, the magnetometer directly measures the field strength and not the probability. Correspondingly, the signal is suppressed only quadratically, $\propto \chi^2$, instead of quartically. For hidden-photon masses in the 0.002–200-meV range, the projected sensitivity for the mixing parameter lies in the 5×10^{-9}–10^{-6} regime.

4.1.3. Afterglow experiments. For chameleon particles (Section 2.1.3) and generally for particles whose mass increases with the local matter/energy density (40), LSW experiments do not always work, because the high density inside the wall (compared with that of the vacuum) increases the mass of the produced WISP and therefore creates a high potential barrier on which the particle is reflected.

However, this reflection can be employed in a slightly different type of experiment (107, 108). If the production zone is enclosed by suitably dense walls on all sides (transparent windows let the photons into the production zone, but they are typically too dense to let chameleons escape), then the chameleons produced are trapped inside the production zone and accumulate. After some time, the laser is switched off, and the accumulated chameleons can subsequently reconvert into photons, which then escape through the windows. Therefore, there is an afterglow after the laser is switched off (**Figure 8**).

Such an experiment has been performed by the GammeV Collaboration (109) and allows the exclusion of submillielectronvolt (vacuum mass) chameleons with couplings in the $g \sim (10^{-7}$–$10^{-6})\,\mathrm{GeV}^{-1}$ range. Improvements planned by the GammeV Collaboration (110) will further increase the sensitivity. Moreover, the ADMX Collaboration has performed a test experiment for chameleons with masses of approximately 2 μeV by driving the ADMX cavity with an external generator and looking for a microwave afterglow (104).

4.1.4. Helioscopes. As discussed in Section 3.1, WISPs can be produced inside the Sun, either through the same interactions that lead to the oscillation phenomena described above or via

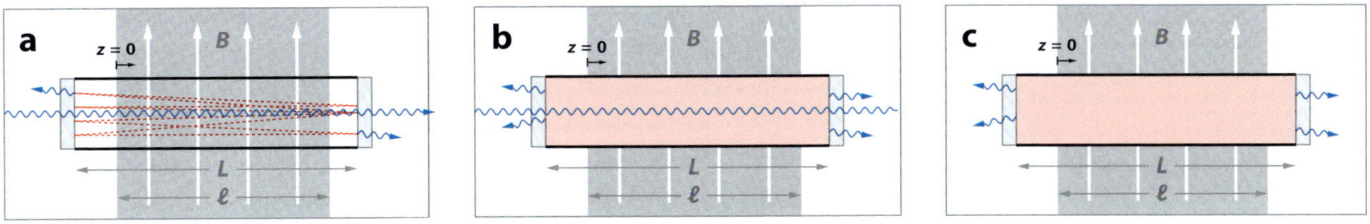

Figure 8

Illustration of an afterglow experiment searching for chameleon particles (107). (*a*) The vacuum tube is filled by means of a laser beam with chameleons via photon-chameleon conversion in a magnetic field. (*b*) An isotropic chameleon gas forms. (*c*) Afterglow from chameleon-photon conversion in a magnetic field.

additional derivative interactions with matter (e.g., axions). Due to the high number of photons inside the Sun and the high total number of interactions, these interactions would typically lead to the production of a huge number of WISPs. Helioscopes attempt to detect these WISPs on Earth (111, 112).

These helioscopes employ basically the same mechanism as an LSW experiment. However, the production side is replaced by WISP production inside the Sun; the regeneration side is the same as for a completely laboratory-based LSW experiment, with the detector pointing toward the Sun; and everything between the solar core and the regeneration side (the rest of the Sun, the atmosphere, the walls of the experimental hall, and so on) functions as the wall. The enormous number of interactions that take place inside the Sun leads to a large WISP flux, even if the coupling is tiny. This flux makes helioscopes an extremely powerful tool with which to search for WISPs. However, the same caveats discussed in Section 3.1 apply: If the production of WISPs inside the Sun is somehow suppressed, then helioscopes lose their sensitivity. The typical energy of photons inside the Sun is in the kiloelectronvolt range, which is also the typical energy of WISPs produced inside the Sun.

Currently two axion helioscopes—CAST and SUMICO—are running (29, 31). These two experiments employ large magnets; therefore, they are sensitive to ALPs as well as to hidden photons (with and without additional MCPs). Both experiments have performed searches for WISPs with kiloelectronvolt energies, which would yield X-ray photons inside the detector. CAST has also searched for WISPs in the electronvolt regime (113), yielding weaker but less model-dependent constraints on WISPs (40).

CAST has recently surpassed the HB constraints for ALPs with a two-photon coupling (**Figure 2*b***) (29), and the results have been used to limit a possible solar γ' flux (**Figure 4**) (38, 114). Additional improvements to the CAST apparatus are already in the planning stages (115). Further dedicated helioscopes, an add-on to SUMICO (116), and a stand-alone hidden-photon helioscope named SHIPS (117) are likely to increase the sensitivity for hidden photons toward smaller masses.

To search for solar axions, one could also use the magnetic field of Earth for regeneration and an X-ray satellite to look through the Earth at the solar core (118). Alternatively, one could replace the Sun with other cosmic objects as the source of axions (119).

4.1.5. Direct axion dark matter searches.

Finally, we turn to another type of photon-regeneration experiment that employs an external source of WISPs: axion dark matter searches. As the name suggests, these experiments focus primarily on axions.

As described in Section 3.4.1, axions were produced in the early universe and may form all or part of the dark matter. One could place the regeneration side of an LSW experiment (with a magnet) on Earth and wait for dark matter axions to enter the experiment and be converted into photons. Indeed, this is the basic idea behind axion dark matter experiments (111). In principle, such experiments (known as haloscope searches) are also sensitive to other types of WISPs, such as hidden photons. However, due to their different production mechanisms, these other WISPs may constitute only a small part of the dark matter (which itself puts a constraint on their existence). This dark matter is typically rather hot, which somewhat limits the sensitivity of haloscope searches for nonaxion WISPs. However, a final analysis of the results from haloscope searches has yet to be performed.

The ADMX (120) and CARRACK (121) Collaborations are currently operating axion haloscopes. Because the typical masses and energies expected for axion dark matter lie in the 1–100-μeV range, these experiments use microwave cavities for resonant regeneration (Section 4.1.2). Because dark matter axions are very cold—that is, their kinetic energy is very small compared to their mass—the energy of the regenerated photons is essentially given by the axion mass. To achieve resonant regeneration, the axion mass must therefore lie within the bandwidth of the microwave cavities. Because the axion mass is unknown, haloscope searches scan through a range of masses by changing the resonance frequency of the cavity. So far, ADMX has achieved enormous sensitivity but only for a relatively limited range of axion masses (**Figure 2**). Future improvements will significantly increase the scanning speed. After its Phase II upgrade is implemented, ADMX will attempt to scan the whole range from 10^{-6} eV to 10^{-5} eV. Note, however, that regeneration experiments for dark matter axions work only in the presence of an axion background. Usually, this background is assumed to be the size of the local dark matter density, and the ADMX bound given in **Figure 2** was obtained under this assumption.

As an alternative to a microwave cavity experiment, one could search for dark matter axions by using optical cavities filled with laser light. Photons absorbing dark matter axions would lead to the appearance of sidebands in the laser-light spectrum (122).

At very small masses [in the fine-tuning or anthropic region where the initial value of the axion field needs to be small (Section 3.4.1)], microwave cavities become too large to be practical. It may be possible to use resonant LC circuits instead (123), which would allow one to explore axion dark matter with decay constants up to the GUT scale: $f_a \sim 10^{16}$ GeV.

4.2. Laser Polarization Experiments

The experiments described in Section 4.1 search for regenerated photons as a signature of WISPs. An alternative is to search for the disappearance or phase modifications of photons, which would signify the real and virtual production of WISPs (95, 124, 125).

This idea is realized in laser polarization experiments, in which researchers shine (linearly) polarized laser light through a magnetic field and search for changes in polarization. The detectable changes are a rotation of the linearly polarized light, as well as a phase shift between different polarization directions, that modifies linear polarized light into elliptically polarized light. However, changes in the overall magnitude of the laser field are typically not detectable. As an example, the relevant processes for ALPs and MCPs are depicted in **Figure 9**.

To calculate the relevant changes in phase and amplitude, one can use the same equations of motion as in Section 4.1. Instead of the photon-to-WISP transition amplitude, the relevant quantity is now the photon-to-photon amplitude and the corresponding probability $P(\gamma \to \gamma)$.

For an observable effect to be obtained, the changes in the amplitude and phase of the photons must differ between the polarizations because changes in the overall strength of the laser field are not detectable. Therefore, in the presence of Lorentz symmetry, such experiments require a magnetic field that distinguishes a certain axis in order to generate an observable effect. For

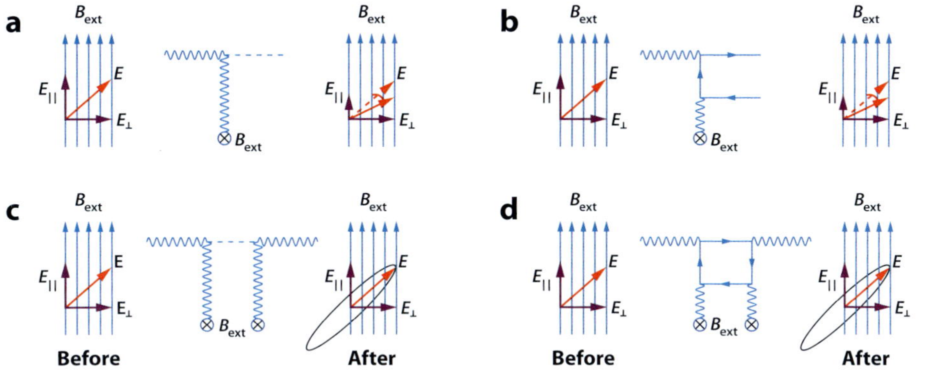

Figure 9

(*a,b*) Rotation and (*c,d*) ellipticity caused by the existence of (*a,c*) a light neutral spin-zero boson or (*b,d*) a light particle with a small electric charge. Figure adapted from Reference 126. Abbreviations: *B*, magnetic field; *E*, electric field.

the same reason, massive hidden photons with no additional matter and with a magnetic field and polarization-independent mass matrix (Equation 43) are not detectable in laser polarization experiments.

In contrast to LSW experiments, which have no Standard Model background,[8] two processes in quantum electrodynamics (QED) provide a background to laser polarization experiments. The first and most prominent is the QED contribution to the phase shift. The relevant diagram is the same as **Figure 9*c,d*** but with ordinary electrons in the loop. Current experiments are approximately two orders of magnitude away from the sensitivity required to measure this effect, and the next generation of experiments indeed aims to measure it. These backgrounds limit searches for ALPs via this process to approximately $g \gtrsim 10^{-7}\,\mathrm{GeV}^{-1}$ and for MCPs to $\varepsilon \gtrsim 10^{-7}$. The situation is much better for the rotation, where the Standard Model background arises from photon splitting and graviton and neutrino production. These backgrounds are much smaller and allow background-free discovery potential down to couplings of $g \lesssim 1/M_P$ and minicharges of $\varepsilon \lesssim 10^{-20}$ (128). Note, however, that spectacular experimental progress would be needed for such sensitivities to be achieved. Many of these experiments have already finished data taking, are running, or are under construction or in preparation (97, 129). Although their basic principle is the same, they differ in their use of pulsed or constant lasers and/or magnetic fields.

These polarization experiments typically measure only the difference in the speed of light between the two polarization directions. However, use of an interferometer, wherein one arm is placed in a magnetic field, allows one to measure the change in the speed of light for both polarizations individually. An interesting possibility is to equip one arm of a gravitational-wave detector with pulsed magnetic fields (130).

In an inhomogeneous magnetic field, axions would cause one polarization direction of the laser beam to be split into two components (similar to a Stern-Gerlach experiment). This feature could be exploited experimentally to search for axions (131).

[8]Actually, this is not quite true. There is a (tiny) background from the photon-graviton-photon process and an additional wall thickness–dependent one from neutrino tunneling (127).

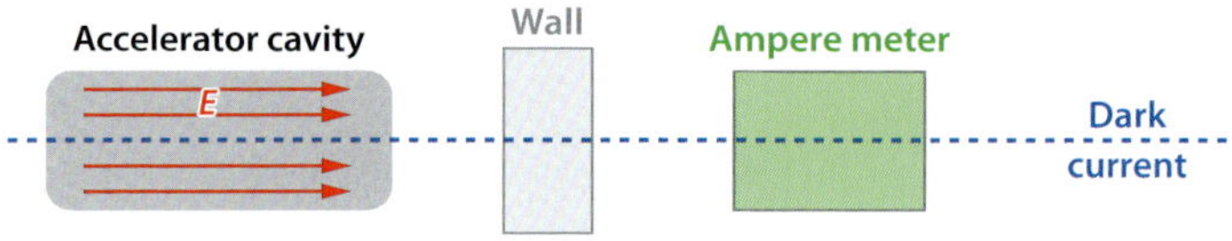

Figure 10

Schematic illustration of an accelerator cavity–dark current experiment that searches for minicharged particles. The dashed line represents the electric current produced by the acceleration of charged (anti)particles. Reprinted from Reference 133.

4.3. Experiments Using Strong Electromagnetic Fields

So far, we have described the propagation of light using linear equations of motion. All (possibly nonlinear) effects of background fields are included in effective mass matrices, which lead to small modifications of the propagation in vacuum. However, in strong electromagnetic fields, nonlinear quantum effects can lead to entirely new processes such as, for instance, Schwinger pair production (132). In the presence of light particles interacting with an electromagnetic field, such effects may set in at much smaller fields than in ordinary QED, where the critical field strength for such effects is set by the electron mass to be $\sim m_e^2/e$. These nonlinear effects can be used to search for WISPs. As an example, let us consider how one can search for MCPs by using Schwinger pair production.

The basic setup (133) is depicted in **Figure 10**. In a strong electric field, a vacuum pair of charged particles gains energy if the particles are separated by a distance along the lines of the electric field. If the electric field is strong enough (or if the distance is large enough) the energy gain can overcome the rest mass; in other words, the virtual particles can turn into real particles. This is the famous Schwinger pair–production mechanism (132). After the pair production, the electric field accelerates the particles and antiparticles, according to their charge, in opposite directions, which produces an electric current (**Figure 10**). If the current is made up of MCPs, then the individual particles have very small charges and interact only very weakly with ordinary matter. Therefore, they can pass through thick walls nearly unhindered. An electron current, however, would be stopped. After the particles pass through the wall, one can place an ampere meter nearby to detect the MCP current.

Typical accelerator cavities achieve field strengths of $\gtrsim 25\,\mathrm{MeV\,m^{-1}}$, and their sizes are typically of the order of tens of centimeters. Precision ampere meters can certainly measure currents down to the microampere range, and even smaller currents of order of picoamperes seem feasible. Using the Schwinger pair–production rate, we can then estimate the expected sensitivity for such an experiment to be

$$\varepsilon_{\text{sensitivity}} \sim 10^{-8} \div 10^{-6} \quad \text{for}\, m_\varepsilon \lesssim \mathrm{meV}. \qquad 46.$$

Therefore, such experiments have the potential for significant improvement over the current best laboratory bounds (124, 134, 135): $\varepsilon \lesssim$ few 10^{-7} (**Figure 3**).

4.4. Fifth Force Experiments

Some of the WISP candidates discussed herein, including scalar ALPs and hidden photons, mediate long-range forces between macroscopic bodies. Therefore, because of the cumulative effect of a macroscopic number of particles, experiments that test the inverse-square law of the Newton and Coulomb forces offer very sensitive probes of WISPs in a certain mass range (136).

In our discussion of fifth forces, we focus on WISPs that interact with photons. However, searches for non-Newtonian forces are also sensitive to a wide variety of WISPs that interact gravitationally or with very weak (effective) Yukawa interactions. Examples include the dilaton

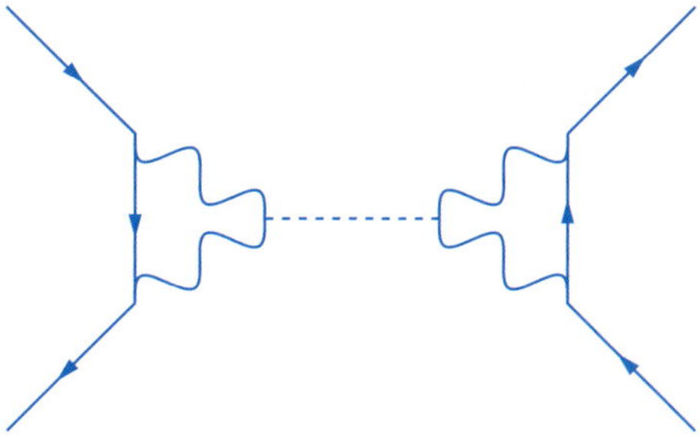

Figure 11

Exchange of a scalar axion-like particle coupled to two photons via Equation 47, between two protons, giving rise to a Yukawa-type non-Newtonian force between two neutral test bodies. Reprinted from Reference 139.

and other moduli, Kaluza-Klein gravitons, and the chameleons mentioned in Section 2.1.3. For a review, see Reference 136.

4.4.1. Bounds on axion-like particles from searches for non-Newtonian forces. Due to its pseudoscalar, (pseudo-)Goldstone boson nature, the exchange of a very light axion predominantly leads to a spin-dependent force that cannot be probed even by the most sensitive experiments that exploit unpolarized bodies. A spin-independent force is generated by the exchange of two axions,[9] which leads to a power-law correction to the spin-independent long-range force between neutral test bodies (138). However, the bounds inferred from corresponding torsion-balance-type experiments on f_a are not competitive with astrophysical bounds (139).

The situation is different for scalar ALPs, whose coupling to two photons occurs via the effective Lagrangian

$$\mathcal{L}_1 = \frac{g}{4}\phi F^{\mu\nu} F_{\mu\nu}. \qquad 47.$$

This coupling leads, via the radiative corrections shown in **Figure 11**, to a spin-independent non-Newtonian force between test bodies of the Yukawa type,

$$\propto (g m_p)^2 \exp(-m_\phi r). \qquad 48.$$

Given that such a force has not been observed in sensitive torsion-balance searches for Yukawa violations of the gravitational inverse-square law, one may place a very stringent limit (139, 140) of

$$g \lesssim 4 \times 10^{-17}\,\mathrm{GeV}^{-1} \qquad 49.$$

on the millielectronvolt mass range, along with even stronger constraints for smaller masses (**Figure 12**).

To test for pseudoscalar axions, one could also search for monopole-(spin-)dipole ($\sim\bar{\theta}^1$) and dipole-dipole interactions ($\sim\bar{\theta}^0$), where $\bar{\theta}$ is the remaining *CP*-violating angle due to the *CP* violation in the electroweak sector (136, 137; also see Reference 148 for recent data). However, the smallness of $\bar{\theta}$ limits the sensitivity of monopole-dipole searches, and dipole-dipole searches are experimentally challenging.

[9]Single-pseudoscalar exchange leads to long-range interactions between unpolarized test bodies only in the presence of *CP* violation. For axions, the resulting interaction potential is $\sim \bar{\theta}^2$ (136, 137). Although $\bar{\theta}$ is expected to be nonvanishing due to *CP* violation in the weak interactions, its value is very small (the bound on the electric dipole moment of the neutron yields $\bar{\theta} < 3 \times 10^{-10}$).

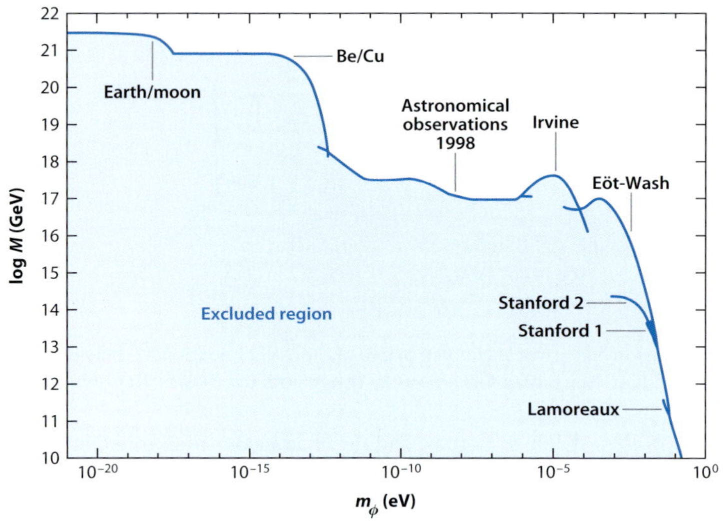

Figure 12

Constraints on the two-photon coupling $M = 1/g$ of a scalar axion-like particle versus its mass m_ϕ from searches for non-Newtonian forces exploiting torsion-balance techniques [Be/Cu (141), Irvine (142), Eöt-Wash (139, 143), Stanford (144)] and from Casimir force searches [Lamoreaux (145)]. Also shown are constraints from astrophysical observations [Earth/moon (146); also see Reference 147]. Compilation from Reference 140.

4.4.2. Bounds on WISPs from searches for non-Coulomb forces. Hidden photons (89, 149) and MCPs (135) also leave detectable imprints as modifications of Coulomb's law. For massive hidden photons, the nondiagonal mass term in the equations of motion leads to a modification of the potential between two charges:

$$V(r) = \frac{\alpha}{r}\left(1 + \chi^2 e^{-m_{\gamma'} r}\right). \qquad 50.$$

MCPs modify the potential at the one-loop level by the Uehling contribution. At large distances, the deviation is also exponential:

$$V(r) \approx \frac{\alpha}{r}\left[1 + \frac{\alpha\varepsilon^2}{4\sqrt{\pi}}\frac{\exp(-2mr)}{(mr)^{\frac{3}{2}}}\right] \quad \text{for} \quad mr \gg 1. \qquad 51.$$

The inverse-square law has been tested in the laboratory by Cavendish-type experiments that check for the absence of an electric field inside a charged conducting sphere (150). These experiments, originally performed to set limits on the photon mass (151)—which would also give an exponential deviation from the Coulomb law—and also to constrain the range and strength of an electrical fifth force (152), give a strong constraint on hidden photons, notably in the microelectronvolt regime (**Figure 4**). Similarly, they give the best laboratory constraints on MCPs in the submicroelectronvolt range. These results demonstrate that searches for deviations from Coulomb's law are a powerful way to search for WISPs. This provides a strong motivation to improve upon the last experiments performed nearly 40 years ago.

Static magnetic fields are also modified in the presence of hidden photons. In particular, the large-scale magnetic fields of planets can be used to test for hidden photons with very small masses (**Figure 4**) (151, 153).

5. SUMMARY AND OUTLOOK

For particle physicists, the term low-energy physics usually applies to physics below the current high-energy frontier, which today is approximately the electroweak scale ($\sim$100 GeV). Currently, three areas of low-energy physics are proving to be fruitful grounds in which to explore fundamental physics. First, there are experiments just below the electroweak scale, in the regime of roughly 1 to 100 GeV. This energy range is particularly interesting for high-precision flavor physics, but it could also provide insights into so-called dark forces, which have recently attracted a great deal of attention. Such insights could help resolve puzzling astrophysical observations (Section 3.4.4). Second, low-energy experiments and observations can be used to test fundamental symmetries, such as Lorentz symmetry, to an incredibly high precision (for a review, see Reference 154). Third, there may exist new particles with very small masses (possibly even subelectronvolt) but also very weak interactions with the known Standard Model particles. This last area has been the subject of this review.

Light weakly coupled particles with masses below 1 eV, so-called WISPs, are strongly motivated from both top-down and bottom-up considerations. A classic example of a WISP is the axion (Section 2.1.1). From a bottom-up point of view, the axion is predicted as a consequence of the Peccei-Quinn solution of the strong-*CP* problem. A new global symmetry is introduced, then spontaneously broken at a very high energy scale, $\sim f_a$. As a consequence, the axion—as a pseudo-Nambu-Goldstone boson of this symmetry—has both small mass, $m_a \sim$ meV (10^{10} GeV/f_a), and coupling to two photons, $g \sim 10^{-13}$ GeV^{-1} (10^{10} GeV/f_a). From a top-down point of view, in string theory particles with couplings similar to axions, namely ALPs, seem to be a natural consequence of the compactification of extra dimensions (Section 2.1.2). Again, their small couplings arise from the large energy scale involved: the string scale (these small couplings may also arise from the sizes of the extra dimensions). Searching for such particles is thereby a way to probe very high energy scales far beyond those attained by colliders such as LHC.

Axions are certainly not the only possible WISP candidates. String models often contain also additional $U(1)$ gauge groups and matter charged under them (Section 2.2). The corresponding extra, hidden photons can mix kinetically with the ordinary electromagnetic photon. Small mixing angles and correspondingly small interactions with ordinary matter arise from the high-energy scales involved and from the fact that the additional gauge factors are "far away" in the extra dimensions. Roughly speaking, probing small couplings allows us to look beyond our "immediate neighborhood" and probe the global structure of the compactification. For extra photons (and matter charged under them), small masses are possible but not necessary.

Astrophysics and cosmology already provide a powerful tool with which to constrain WISPs, and in many regions of parameter space, this tool sets the standard against which laboratory tests must measure up (Section 3). The rapidly increasing amount of available data will hopefully improve the situation further (e.g., 155).

Additionally, some astrophysical and cosmological observations provide suggestive hints for the existence of WISPs (Section 3.4). Importantly, both axions and hidden photons (or particles related to them) may contribute all or part of the dark matter. There also exist various astrophysical puzzles that can be solved by the presence of WISPs; these problems include, for instance, possible observations of high- and very high energy cosmic-photon regeneration, the observed alignment of the polarization vectors of very distant quasars, the energy loss of white dwarfs, and a slight excess in the effective number of neutrino species in CMB measurements. These interpretations can and should be tested in the laboratory.

One of the most important features of WISPs is that they have (very weak) interactions with photons. In combination with their small mass, this feature allows one to search for them in photon-regeneration experiments, laser polarization experiments, experiments that exploit strong electromagnetic fields, and fifth force experiments.

Currently, there are numerous laser LSW experiments (Section 4.1.2); this field is now entering a new generation. Important advances in laser technology are paving the way to exceed the sensitivity of current WISP helioscopes and to probe the above-mentioned explanations of astrophysical puzzles in terms of photon $\leftrightarrow$ WISP oscillations. Pioneering experiments that exploit high-quality microwave cavities, instead of laser light, for the generation and regeneration of WISPs are in the commissioning phase. Microwave-cavity searches for dark matter axions probe a complementary region in parameter space, compared with other photon-regeneration experiments, and within the current decade they should provide a definitive answer as to whether axions are the dominant part of CDM.

Laser polarization experiments (Section 4.2) will continue their quest to detect QED birefringence. At the same time, they will also considerably improve the bounds placed on WISPs and will help to develop optical techniques, such as cavities that are used in LSW experiments. Moreover, measurements that analyze the spectrum of laser light in an interferometer may be a promising way to search for dark matter axions.

Fifth force experiments (Section 4.4), especially searches for deviations from the gravitational inverse-square law, are already an established means of testing fundamental physics. Such experiments will continue to improve significantly, thereby providing new bounds on a variety of WISPs, especially light scalar fields. Other WISPs, such as hidden photons and MCPs, can be probed with high sensitivities by testing for deviations from Coulomb's law. Although tests of Coulomb's law are already at an impressive precision, the current best available bounds in the length range of meters are nearly 40 years old. Hopefully, with current technology considerable improvement will become possible.

The use of low-energy experiments with photons to search for WISPs may provide important information about fundamental particle physics that will complement results from high-energy colliders. Such experiments are already providing very strong bounds on light weakly interacting particles. Even more excitingly, the next few years will no doubt bring considerable advances and a huge discovery potential for new physics.

DISCLOSURE STATEMENT

The authors are not aware of any affiliations, memberships, funding, or financial holdings that might be perceived as affecting the objectivity of this review.

ACKNOWLEDGMENTS

The authors would like to thank Markus Ahlers, Holger Gies, Mark Goodsell, Axel Lindner, Alessandro Mirrizzi, Javier Redondo, and Christoph Weniger for interesting discussions, helpful suggestions, and joyful collaboration on the subjects discussed in this review.

LITERATURE CITED

1. Amsler C, et al. (Part. Data Group) *Phys. Lett.* B667:1 (2008)
2. Peccei RD, Quinn HR. *Phys. Rev. Lett.* 38:1440 (1977)
3. Weinberg S. *Phys. Rev. Lett.* 40:223 (1978); Wilczek F. *Phys. Rev. Lett.* 40:279 (1978)
4. Kim JE. *Phys. Rev. Lett.* 43:103 (1979); Dine M, Fischler W, Srednicki M. *Phys. Lett.* B104:199 (1981); Shifman MA, Vainshtein AI, Zakharov VI. *Nucl. Phys. B* 166:493 (1980); Zhitnitsky AR. *Sov. J. Nucl. Phys.* 31:260 (1980)

5. Bardeen WA, Tye SH. *Phys. Lett.* B74:229 (1978); Kaplan DB. *Nucl. Phys. B* 260:215 (1985); Srednicki M. *Nucl. Phys. B* 260:689 (1985)
6. Witten E. *Phys. Lett.* B149:351 (1984)
7. Conlon JP. *J. High Energy Phys.* 0605:078 (2006)
8. Svrcek P, Witten E. *J. High Energy Phys.* 0606:051 (2006)
9. Arvanitaki A, et al. arXiv:0905.4720 [hep-th] (2009)
10. Choi K, Kim JE. *Phys. Lett.* B154:393 (1985); Choi K, Kim JE. *Phys. Lett.* B165:71 (1985)
11. Wetterich C. *Nucl. Phys. B* 302:668 (1988); Ratra B, Peebles PJE. *Phys. Rev. D* 37:3406 (1988); Caldwell RR, Dave R, Steinhardt PJ. *Phys. Rev. Lett.* 80:1582 (1998)
12. Khoury J, Weltman A. *Phys. Rev. Lett.* 93:171104 (2004); Khoury J, Weltman A. *Phys. Rev. D* 69:044026 (2004); Brax P, et al. *Phys. Rev. D* 70:123518 (2004)
13. Candelas P, Horowitz GT, Strominger A, Witten E. *Nucl. Phys. B* 258:46 (1985); Witten E. *Nucl. Phys. B* 268:79 (1986)
14. Lebedev O, et al. *Phys. Lett.* B668:331 (2008)
15. Lebedev O, Ramos-Sanchez S. arXiv:0912.0477 [hep-ph] (2009)
16. Dobrescu BA. *Phys. Rev. Lett.* 94:151802 (2005)
17. Holdom B. *Phys. Lett.* B166:196 (1986)
18. Dienes KR, Kolda CF, March-Russell J. *Nucl. Phys. B* 492:104 (1997)
19. Lust D, Stieberger S. *Fortschr. Phys.* 55:427 (2007); Abel SA, Schofield BW. *Nucl. Phys. B* 685:150 (2004); Berg M, Haack M, Kors B. *Phys. Rev. D* 71:026005 (2005); Abel SA, Jaeckel J, Khoze VV, Ringwald A. *Phys. Lett.* B666:66 (2008)
20. Abel SA, et al. *J. High Energy Phys.* 0807:124 (2008)
21. Goodsell M, Jaeckel J, Redondo J, Ringwald A. *J. High Energy Phys.* 0911:027 (2009)
22. Ahlers M, Jaeckel J, Redondo J, Ringwald A. *Phys. Rev. D* 78:075005 (2008)
23. Brummer F, Jaeckel J, Khoze VV. *J. High Energy Phys.* 0906:037 (2009)
24. Kors B, Nath P. *Phys. Lett.* B586:366 (2004); Feldman D, Liu Z, Nath P. *Phys. Rev. D* 75:115001 (2007)
25. Raffelt GG. *Stars as Laboratories for Fundamental Physics: The Astrophysics of Neutrinos, Axions, and Other Weakly Interacting Particles*. Chicago: Univ. Chicago Press. 664 pp. (1996)
26. Raffelt GG. *Lect. Notes Phys.* 741:51 (2008)
27. Raffelt GG. *Phys. Lett.* B166:402 (1986)
28. Isern J, Garcia-Berro E, Torres S, Catalan S. arXiv:0806.2807 [astro-ph] (2008)
29. Andriamonje S, et al. (CAST Collab.) *J. Cosmol. Astropart. Phys.* 0704:010 (2007); Arik E, et al. (CAST Collab.) *J. Cosmol. Astropart. Phys.* 0902:008 (2009)
30. Schlattl H, Weiss A, Raffelt GG. *Astropart. Phys.* 10:353 (1999)
31. Inoue Y, et al. *Phys. Lett.* B668:93 (2008)
32. Redondo J. arXiv:0810.3200 [hep-ph] (2008)
33. Raffelt GG. *Phys. Rev. D* 33:897 (1986)
34. Raffelt GG, Dearborn DSP. *Phys. Rev. D* 36:2211 (1987)
35. Brockway JW, Carlson ED, Raffelt GG. *Phys. Lett.* B383:439 (1996); Grifols JA, Masso E, Toldra R. *Phys. Rev. Lett.* 77:2372 (1996)
36. Dai DC, Freese K, Stojkovic D. *J. Cosmol. Astropart. Phys.* 0906:023 (2009)
37. Gondolo P, Raffelt GG. arXiv:0807.2926 [astro-ph] (2008)
38. Redondo J. *J. Cosmol. Astropart. Phys.* 0807:008 (2008)
39. Redondo J, Postma M. *J. Cosmol. Astropart. Phys.* 0902:005 (2009)
40. Jaeckel J, et al. arXiv:hep-ph/0605313 (2006); Jaeckel J, et al. *Phys. Rev. D* 75:013004 (2007)
41. Masso E, Redondo J. *J. Cosmol. Astropart. Phys.* 0509:015 (2005); Mohapatra RN, Nasri S. *Phys. Rev. Lett.* 98:050402 (2007); Brax P, van de Bruck C, Davis AC. *Phys. Rev. Lett.* 99:121103 (2007); Jain P, Mandal S. *Int. J. Mod. Phys. D* 15:2095 (2006); Jain P, Stokes S. arXiv:hep-ph/0611006 (2006); Kim JE. *Phys. Rev. D* 76:051701 (2007)
42. Masso E, Redondo J. *Phys. Rev. Lett.* 97:151802 (2006); Redondo J. arXiv:0807.4329 [hep-ph] (2008)
43. Iocco F, et al. *Phys. Rep.* 472:1 (2009)
44. Simha V, Steigman G. *J. Cosmol. Astropart. Phys.* 0806:016 (2008)

45. Davidson S, Hannestad S, Raffelt G. *J. High Energy Phys.* 05:003 (2000); Berezhiani Z, Lepidi A. *Phys. Lett.* B 681:276 (2009)
46. Fixsen DJ, et al. *Astrophys. J.* 473:576 (1996)
47. Melchiorri A, Polosa A, Strumia A. *Phys. Lett.* B650:416 (2007)
48. Jaeckel J, Redondo J, Ringwald A. *Phys. Rev. Lett.* 101:131801 (2008)
49. Mirizzi A, Redondo J, Sigl G. *J. Cosmol. Astropart. Phys.* 0903:026 (2009)
50. Mirizzi A, Redondo J, Sigl G. *J. Cosmol. Astropart. Phys.* 0908:001 (2009)
51. Lawrence CR, Herbig T, Readhead ACS, Gulkis S. *Astrophys. J. Lett.* 449:L5 (1995); Bennett C, et al. (WMAP Collab.) *Astrophys. J. Suppl.* 148:97 (2003); de Petris M, et al. *Astrophys. J.* 574:L119 (2002)
52. Burrage C, Jaeckel J, Redondo J, Ringwald A. *J. Cosmol. Astropart. Phys.* 0911:002 (2009)
53. Ahlers M. *Phys. Rev. D* 80:023513 (2009)
54. Csaki C, Kaloper N, Terning J. *Phys. Rev. Lett.* 88:161302 (2002); Csaki C, Kaloper N, Terning J. *Phys. Lett.* B535:33 (2002)
55. Ichikawa K, Sekiguchi T, Takahashi T. *Phys. Rev. D* 78:083526 (2008)
56. Hannestad S, Mirizzi A, Raffelt G. *J. Cosmol. Astropart. Phys.* 0507:002 (2005); Melchiorri A, Mena O, Slosar A. *Phys. Rev. D* 76:041303 (2007); Hannestad S, Mirizzi A, Raffelt GG, Wong YYY. *J. Cosmol. Astropart. Phys.* 0708:015 (2007); Hannestad S, et al. arXiv:0910.5706 [hep-ph] (2009)
57. Seljak U, Slosar A, McDonald P. *J. Cosmol. Astropart. Phys.* 0610:014 (2006); Mangano G, et al. *J. Cosmol. Astropart. Phys.* 0703:006 (2007)
58. Hamann J, Hannestad S, Raffelt GG, Wong YYY. *J. Cosmol. Astropart. Phys.* 0708:021 (2007)
59. Sikivie P. *Lect. Notes Phys.* 741:19 (2008)
60. Preskill J, Wise MB, Wilczek F. *Phys. Lett.* B120:127 (1983); Abbott LF, Sikivie P. *Phys. Lett.* B120:133 (1983); Dine M, Fischler W. *Phys. Lett.* B120:137 (1983)
61. Hertzberg MP, Tegmark M, Wilczek F. *Phys. Rev. D* 78:083507 (2008)
62. Tegmark M, de Oliveira-Costa A, Hamilton A. *Phys. Rev. D* 68:123523 (2003); de Oliveira-Costa A, Tegmark M, Zaldarriaga M, Hamilton A. *Phys. Rev. D* 69:063516 (2004); Copi CJ, Huterer D, Schwarz DJ, Starkman GD. *Mon. Not. R. Astron. Soc.* 367:79 (2006)
63. Kinney WH, Sikivie P. *Phys. Rev. D* 61:087305 (2000); Sikivie P. *Phys. Lett.* B567:1 (2003)
64. Sikivie P. arXiv:0909.0949 [hep-ph] (2009)
65. Hoffmann S. *Phys. Lett.* B193:117 (1987)
66. Aharonian F, et al. (H.E.S.S. Collab.) *Nature* 440:1018 (2006)
67. Mazin D, Raue M. *Astron. Astrophys.* 471:439 (2007)
68. Hochmuth KA, Sigl G. *Phys. Rev. D* 76:123011 (2007); Hooper D, Serpico PD. *Phys. Rev. Lett.* 99:231102 (2007)
69. De Angelis A, Mansutti O, Roncadelli M. *Phys. Rev. D* 76:121301 (2007)
70. Mirizzi A, Montanino D. *J. Cosmol. Astropart. Phys.* 0912:004 (2009)
71. Csaki C, Kaloper N, Peloso M, Terning J. *J. Cosmol. Astropart. Phys.* 0305:005 (2003)
72. Burrage C, Davis AC, Shaw DJ. *Phys. Rev. Lett.* 102:201101 (2009)
73. Fairbairn M, Rashba T, Troitsky S. arXiv:0901.4085 [astro-ph] (2009)
74. Albuquerque IFM, Chou A. arXiv:1001.0972 [astro-ph] (2010)
75. Payez A, Cudell JR, Hutsemekers D. *AIP Conf. Proc.* 1038:211 (2008)
76. Zioutas K, Tsagri M, Semertzidis Y, Papaevangelou T. arXiv:0808.1545 [astro-ph] (2008); Zioutas K, Tsagri M, Papaevangelou T, Dafni T. *New J. Phys.* 11:105020 (2009)
77. Pospelov M, Ritz A, Voloshin MB. *Phys. Lett.* B662:53 (2008)
78. Ibarra A, Ringwald A, Weniger C. *J. Cosmol. Astropart. Phys.* 0901:003 (2009)
79. Shirai S, Takahashi F, Yanagida TT. arXiv:0902.4770 [hep-ph] (2009); Ibarra A, Ringwald A, Tran D, Weniger C. *J. Cosmol. Astropart. Phys.* 0908:017 (2009)
80. Arkani-Hamed N, Finkbeiner DP, Slatyer TR, Weiner N. *Phys. Rev. D* 79:015014 (2009)
81. Adriani O, et al. (PAMELA Collab.) *Nature* 458:607 (2009)
82. Bernabei R, et al. (DAMA Collab.) *Eur. Phys. J. C* 56:333 (2008)
83. Arkani-Hamed N, Weiner N. *J. High Energy Phys.* 0812:104 (2008); Essig R, Schuster P, Toro N. *Phys. Rev. D* 80:015003 (2009); Bjorken JD, Essig R, Schuster P, Toro N. *Phys. Rev. D* 80:075018 (2009)

84. Baumgart M, et al. *J. High Energy Phys.* 0904:014 (2009)
85. Chun EJ, Park JC. *J. Cosmol. Astropart. Phys.* 0902:026 (2009); Cheung C, Ruderman JT, Wang LT, Yavin I. *Phys. Rev. D* 80:035008 (2009); Morrissey DE, Poland D, Zurek KM. *J. High Energy Phys.* 0907:050 (2009); Cui Y, Morrissey DE, Poland D, Randall L. *J. High Energy Phys.* 0905:076 (2009)
86. Suematsu D. *J. High Energy Phys.* 0611:029 (2006); Katz A, Sundrum R. *J. High Energy Phys.* 0906:003 (2009); Batell B, Pospelov M, Ritz A. *Phys. Rev. D* 79:115008 (2009); Batell B, Pospelov M, Ritz A. *Phys. Rev. D* 80:095024 (2009); Dedes A, Giomataris I, Suxho K, Vergados JD. *Nucl. Phys. B* 826:148 (2010); Ruderman JT, Volansky T. arXiv:0908.1570 [hep-ph] (2009)
87. Komatsu E, et al. arXiv:1001.4538 [astro-ph] (2010)
88. Jaeckel J, Redondo J, Ringwald A. *Europhys. Lett.* 87:10010 (2009)
89. Okun LB. *Sov. Phys. JETP* 56:502 (1982)
90. Anselm AA. *Yad. Fiz.* 42:1480 (1985); Anselm AA. *Phys. Rev. D* 37:2001 (1988)
91. Van Bibber K, et al. *Phys. Rev. Lett.* 59:759 (1987)
92. Raffelt G, Stodolsky L. *Phys. Rev. D* 37:1237 (1988)
93. Hoogeveen F, Ziegenhagen T. *Nucl. Phys. B* 358:3 (1991)
94. Sikivie P, Tanner DB, van Bibber K. *Phys. Rev. Lett.* 98:172002 (2007)
95. Ahlers M, et al. *Phys. Rev. D* 76:115005 (2007)
96. Ruoso G, et al. (BFRT Collab.) *Z. Phys. C* 56:505 (1992); Robilliard C, et al. (BMV Collab.) *Phys. Rev. Lett.* 99:190403 (2007); Chou AS, et al. [GammeV (T-969) Collab.] *Phys. Rev. Lett.* 100:080402 (2008); Afanasev A, et al. (LIPSS Collab.) *Phys. Rev. Lett.* 101:120401 (2008); Fouche M, et al. (BMV Collab.) *Phys. Rev. D* 78:032013 (2008); Afanasev A, et al. (LIPSS Collab.) *Phys. Lett.* B679:317 (2009); Ehret K, et al. (ALPS Collab.) *Nucl. Instrum. Methods A* 612:83 (2009)
97. Cameron R, et al. (BFRT Collab.) *Phys. Rev. D* 47:3707 (1993)
98. Ehret K, et al. arXiv:1004.1313 [hep-ex] (2010)
99. Cantatore G. Presented at Patras Workshop Axions, WIMPs, WISPs, 5th, **http://axion-wimp2009.desy.de** (2009); Mueller G, Sikivie P, Tanner DB, van Bibber K. *Phys. Rev. D* 80:072004 (2009)
100. Jaeckel J, Ringwald A. *Phys. Lett.* B653:167 (2007)
101. Buchmuller W, Hoogeveen F. *Phys. Lett.* B237:278 (1990); Rabadan R, Ringwald A, Sigurdson K. *Phys. Rev. Lett.* 96:110407 (2006); Dias AG, Lugones G. *Phys. Lett.* B673:101 (2009)
102. Hoogeveen F. *Phys. Lett.* B288:195 (1992); Jaeckel J, Ringwald A. *Phys. Lett.* B659:509 (2008); Caspers F, Jaeckel J, Ringwald A. *J. Instrum.* 4:P11013 (2009)
103. Slocum P. Presented at Patras Workshop Axions, WIMPs, WISPs, 5th, **http://axion-wimp.desy.de** (2009); Williams P. Presented at Patras Workshop Axions, WIMPs, WISPs, 5th, **http://axion-wimp.desy.de** (2009)
104. Rybka G. Presented at AXIONS 2010 Workshop, Gainsville, Fla. **http://www.phys.ufl.edu/research/Axions2010/** (2010)
105. Povey R, Hartnett J, Tobar M. arXiv:1003.0964 (2010)
106. Jaeckel J, Redondo J. *Europhys. Lett.* 84:31002 (2008)
107. Ahlers M, et al. *Phys. Rev. D* 77:015018 (2008)
108. Gies H, Mota DF, Shaw DJ. *Phys. Rev. D* 77:025016 (2008)
109. Chou AS, et al. (GammeV Collab.) *Phys. Rev. Lett.* 102:030402 (2009)
110. Wester W. Presented at AXIONS 2010 Workshop, Gainsville, Fla. **http://www.phys.ufl.edu/research/Axions2010/** (2010); Upadhye A, Steffen JH, Weltman A. *Phys. Rev. D* 81:015013 (2010)
111. Sikivie P. *Phys. Rev. Lett.* 51:1415 (1983); Sikivie P. *Phys. Rev. Lett.* Erratum. 52:695 (1984)
112. van Bibber K, McIntyre PM, Morris DE, Raffelt GG. *Phys. Rev. D* 39:2089 (1989); Paschos EA, Zioutas K. *Phys. Lett.* B323:367 (1994)
113. Karuza M. Presented at Patras Workshop Axions, WIMPs, WISPs, 5th, **http://axion-wimp2009.desy.de** (2009)
114. Gninenko SN, Redondo J. *Phys. Lett.* B664:180 (2008)
115. Ferrer-Ribas E. Presented at Patras Workshop Axions, WIMPs, WISPs, 5th, **http://axion-wimp2009.desy.de** (2009)

116. Ohta R. Presented at Patras Workshop Axions, WIMPs, WISPs, 5th, **http://axion-wimp2009.desy.de** (2009)
117. Redondo J. Presented at PASCOS 2009, Hamburg, Ger. **http://pascos2009.desy.de/** (2009)
118. Davoudiasl H, Huber P. *Phys. Rev. Lett.* 97:141302 (2006); Davoudiasl H, Huber P. *J. Cosmol. Astropart. Phys.* 0808:026 (2008)
119. Fairbairn M, et al. *Eur. Phys. J. C* 52:899 (2007); Fairbairn M, Rashba T, Troitsky SV. *Phys. Rev. Lett.* 98:201801 (2007)
120. Duffy LD, et al. *Phys. Rev. D* 74:012006 (2006); Asztalos SJ, et al. (ADMX Collab.) arXiv:0910.5914 [astro-ph] (2009)
121. Tada M, et al. *Nucl. Phys. Proc. Suppl.* 72:164 (1999)
122. Melissinos AC. *Phys. Rev. Lett.* 102:202001 (2009)
123. Thomas S. Presented at AXIONS 2010 Workshop, Gainsville, Fla. **http://www.phys.ufl.edu/research/Axions2010/** (2010)
124. Gies H, Jaeckel J, Ringwald A. *Phys. Rev. Lett.* 97:140402 (2006)
125. Maiani L, Petronzio R, Zavattini E. *Phys. Lett.* B175:359 (1986)
126. Brandi F, et al. *Nucl. Instrum. Methods A* 461:329 (2001)
127. Gies H, Jaeckel J. *J. High Energy Phys.* 0908:063 (2009)
128. Ahlers M, Jaeckel J, Ringwald A. *Phys. Rev. D* 79:075017 (2009)
129. Zavattini E, et al. (PVLAS Collab.) *Phys. Rev. Lett.* 96:110406 (2006); Zavattini E, et al. (PVLAS Collab.) Erratum. *Phys. Rev. Lett.* 99:129901 (2007); Zavattini E, et al. (PVLAS Collab.) *Phys. Rev. D* 77:032006 (2008); Chen SJ, Mei HH, Ni WT. *Mod. Phys. Lett. A* 22:2815 (2007); Battesti R, et al. *Eur. Phys. J. D* 46:323 (2008); Pugnat P, et al. (OSQAR Collab.) *CERN-SPSC-2006-035*, *CERN-SPSC-P-331* (2006); Cantatore G, et al. arXiv:0809.4208 [hep-ex] (2008)
130. Boer D, Van Holten JW. arXiv:hep-ph/0204207 (2002); Zavattini G, Calloni E. *Eur. Phys. J. C* 62:459 (2009); Dobrich B, Gies H. *Europhys. Lett.* 87:21002 (2009); Dobrich B, Gies H. arXiv:0910.5692 [hep-ph] (2009)
131. Guendelman EI. *Phys. Lett.* B662:445 (2008); Redondo J. arXiv:1003.0410 [hep-ph] (2010)
132. Schwinger J. *Phys. Rev.* 82:664 (1951)
133. Gies H, Jaeckel J, Ringwald A. *Europhys. Lett.* 76:794 (2006)
134. Badertscher A, et al. *Phys. Rev. D* 75:032004 (2007); Ahlers M, et al. *Phys. Rev. D* 77:095001 (2008)
135. Jaeckel J. *Phys. Rev. Lett.* 103:080402 (2009)
136. Adelberger EG, et al. *Prog. Part. Nucl. Phys.* 62:102 (2009)
137. Moody JE, Wilczek F. *Phys. Rev. D* 30:130 (1984)
138. Ferrer F, Grifols JA. *Phys. Rev. D* 58:096006 (1998)
139. Adelberger EG, et al. *Phys. Rev. Lett.* 98:131104 (2007)
140. Dupays A, Masso E, Redondo J, Rizzo C. *Phys. Rev. Lett.* 98:131802 (2007)
141. Su Y, et al. *Phys. Rev. D* 50:3614 (1994)
142. Spero R, et al. *Phys. Rev. Lett.* 44:1645 (1980); Hoskins JK, Newman RD, Spero R, Schultz J. *Phys. Rev. D* 32:3084 (1985)
143. Kapner DJ, et al. *Phys. Rev. Lett.* 98:021101 (2007)
144. Chiaverini J, et al. *Phys. Rev. Lett.* 90:151101 (2003); Smullin SJ, et al. *Phys. Rev. D* 72:122001 (2005); Smullin SJ, et al. Erratum. *Phys. Rev. D* 72:129901 (2005)
145. Lamoreaux SK. *Phys. Rev. Lett.* 78:5 (1997); Lamoreaux SK. Erratum. *Phys. Rev. Lett.* 81:5475 (1998)
146. Dickey JO, et al. *Science* 265:482 (1994); Williams JG, Newhall XX, Dickey JO. *Phys. Rev. D* 53:6730 (1996)
147. Onofrio R. *New J. Phys.* 8:237 (2006)
148. Vasilakis G, Brown JM, Kornack TW, Romalis MV. *Phys. Rev. Lett.* 103:241801 (2009)
149. Popov V. *Turk. J. Phys.* 23:943 (1999)
150. Williams ER, Faller JE, Hill HA. *Phys. Rev. Lett.* 26:721 (1971)
151. Goldhaber AS, Nieto MM. *Rev. Mod. Phys.* 43:277 (1971)

152. Bartlett DF, Loegl S. *Phys. Rev. Lett.* 61:2285 (1988)
153. Goldhaber AS, Nieto MM. *Phys. Rev. Lett.* 21:567 (1968)
154. Kostelecky VA. arXiv:0802.0581 [gr-qc] (2008); Kostelecky VA, Mewes M. *Phys. Rev. D* 66:056005 (2002)
155. Chelouche D, Rabadan R, Pavlov S, Castejon F. *Astrophys. J. Suppl.* 180:1 (2009); Bassan N, Mirizzi A, Roncadelli M. arXiv:1001.5267 [astro-ph] (2010)

The Diffuse Supernova Neutrino Background

John F. Beacom

Departments of Physics and Astronomy and Center for Cosmology and Astro-Particle Physics, The Ohio State University, Columbus, Ohio 43210; email: beacom.7@osu.edu

Annu. Rev. Nucl. Part. Sci. 2010. 60:439–62

First published online as a Review in Advance on July 19, 2010

The *Annual Review of Nuclear and Particle Science* is online at nucl.annualreviews.org

This article's doi:
10.1146/annurev.nucl.010909.083331

0163-8998/10/1123-0439$20.00

Key Words

neutrino astronomy, massive stars, cosmic backgrounds

Abstract

The diffuse supernova neutrino background (DSNB) is the weak glow of megaelectronvolt neutrinos and antineutrinos from distant core-collapse supernovae. The DSNB has not been detected yet, but the Super-Kamiokande (SK) 2003 upper limit on the $\bar{\nu}_e$ flux is close to predictions, now quite precise, that are based on astrophysical data. If SK is modified with dissolved gadolinium to reduce detector backgrounds and increase the energy range for analysis, then it should detect the DSNB at a rate of a few events per year, providing a new probe of supernova neutrino emission and the cosmic core-collapse rate. If the DSNB is not detected, then new physics will be required. Neutrino astronomy, although uniquely powerful, has proven extremely difficult—only the Sun and the nearby Supernova 1987A have been detected to date—so the promise of detecting new sources soon is exciting indeed.

Contents

1. INTRODUCTION

The diffuse supernova neutrino background (DSNB) is the flux of neutrinos and antineutrinos emitted by all core-collapse supernovae in the causally reachable universe; it will appear isotropic and time independent in feasible observations. (Hereafter, neutrinos means neutrinos and antineutrinos, and supernovae means core-collapse supernovae, unless otherwise specified.) The DSNB results from the quasi-thermal megaelectronvolt neutrino spectrum emitted by newly formed neutron stars convolved with the rapid redshift evolution of the supernova rate. The cosmic energy density in neutrinos from core-collapse supernovae is comparable to that in photons from stars, $\sim$0.01 eV cm^{-3}, and is $\sim$10 times less than that of the cosmic microwave background. Some earlier literature referred to supernova relic neutrinos, but the term relic often caused confusion with the $\sim$10^{-4} eV neutrinos from the big bang.

Why is detection of the DSNB an important and relevant experimental goal?

1. Understanding supernovae is crucial to astrophysics and physics. Supernovae are central to the cosmic history of stellar birth and death, as well as to the production of chemical elements, neutron stars, black holes, cosmic rays, and gravitational waves. They are also integral to exploiting the extreme physical conditions of core collapse to probe the properties of neutrinos and hypothetical particles. New observations and syntheses are needed.
2. We cannot understand supernovae without detecting neutrinos. The optical supernova reveals little about the collapsing core, which is embedded in the stellar envelope; the total energy in photons is much less than that in neutrinos, leaves over a vastly longer time, and depends on the properties of the progenitor star. Neutrinos are emitted from the core, revealing its properties.
3. Detecting bursts of neutrinos from nearby supernovae is difficult. Prodigious emission and large detectors partially compensate the tiny detection cross section. For Milky Way supernovae (distance $D \sim 10$ kpc; 1 pc $= 3.1 \times 10^{18}$ cm), a neutrino burst would be detected easily, but the burst rate is only a few times per century. For supernovae in nearby galaxies ($D \sim 1$–10 Mpc), where the total burst rate is more than one per year, much larger detectors will be required.
4. The DSNB is a guaranteed steady source of supernova neutrinos. Although the probability of detecting even a single neutrino from a distant supernova is infinitesimally small, the number of supernovae per year is astronomically large, yielding a nonzero rate. The total rate and spectral shape of the DSNB are new probes of supernova neutrino emission and the cosmic core-collapse rate.

The DSNB has not been detected yet, but the discovery prospects are excellent. Super-Kamiokande (SK) is large enough to detect the DSNB at the level of a few events per year. These events are hidden by detector backgrounds, which are already low and could be substantially reduced with added gadolinium to detect neutrons, a proposed upgrade for which there is active research and development.

Detection of the DSNB will yield science that is difficult to obtain otherwise. It will measure the average supernova neutrino emission spectrum, and comparison to burst detections will probe the star-to-star variation. It will also be sensitive to the optically dark failed supernovae expected from high-mass stars.

This review is written for experimentalists, observers, and theorists in nuclear physics, particle physics, and astrophysics; it covers the many aspects of the DSNB. It is intended to be an accessible overview of the main themes needed to predict, detect, and interpret the DSNB. Details can be found in the references.

2. IMPOSSIBLE DREAMS OF NEUTRINO ASTRONOMY

Neutrino astronomy is a unique tool with which to answer some of the oldest and most important questions in nuclear and particle astrophysics (1): What happens when a massive star runs out of fuel? Are observed high-energy γ-ray sources powered by cosmic-ray protons or electrons? What is the origin of the highest-energy cosmic rays? Are there astrophysical neutrino sources unseen with electromagnetic observations? Are there new properties of neutrinos, or new particles, that can be revealed only with neutrino detectors?

The immense discovery potential of neutrino astronomy arises from the feeble interactions between neutrinos and matter, which allows neutrinos to escape from and thus to reveal the interiors of dense and distant astrophysical objects, undeflected, undegraded, and unobscured.

This powerful magic comes at a terrible price, as it is much less likely that the neutrinos will interact with a detector than with an entire star. As far as we know, neutrinos interact only via the weak nuclear force and gravity (but with only tiny masses).

In 1934, shortly after Pauli postulated the existence of the neutrino, Bethe & Peierls wrote: "If [there are no new forces]—one can conclude that there is no practically possible way of observing the neutrino" (2, p. 532). It took until 1956 for neutrinos to be detected, by Reines and Cowan and their colleagues (3), using a nuclear reactor source, for which Reines shared in the 1995 Nobel Prize in Physics (Cowan died in 1974). This detection allowed a first consideration of neutrino astronomy in principle.

Prescient remarks about this potential were made long ago (e.g., 4–8). As Bahcall wrote in 1964 (9, p. 300) about solar neutrinos, before their detection, "Only neutrinos, with their extremely small interaction cross sections, can enable us to see into the interior of a star and thus verify directly the hypothesis of nuclear energy generation in stars." The detection of neutrinos from the Sun and the nearby Supernova 1987A (SN1987A) shows that the required sensitivity has begun to be reached. Davis and Koshiba shared in the 2002 Nobel Prize in Physics for these first detections in neutrino astronomy, which led to a precise understanding of the solar fusion reactions, the discovery of neutrino mixing, and confirmation that Type II supernovae result from the deaths of massive stars.

No other astrophysical neutrinos have been detected, despite decades of effort. In Bahcall's 1989 book, *Neutrino Astrophysics* (10, pp. xxii–xxiii), which was almost completely about solar neutrinos, he wrote: "The title is more of an expression of hope than a description of the book's contents.... [T]he observational horizon of neutrino astrophysics may grow...perhaps in a time as short as one or two decades". Indeed, the first detections of distant neutrino sources now seem imminent, due to (*a*) more certain predictions based on new astronomical and laboratory data and (*b*) astounding improvements in detector sensitivity. Any detections will be of enormous scientific importance, certain to provide new information.

This review is about the exciting prospects for new results in extrasolar megaelectronvolt neutrino astronomy; as described below, these prospects build on a long history of work in experiment, observation, and theory. There are similarly encouraging prospects at the teraelectronvolt (10^6 MeV) scale (11, 12) and the exaelectronvolt (10^{12} MeV) scale (12, 13).

3. FRAMEWORK: DIFFUSE SUPERNOVA NEUTRINO BACKGROUND DETECTION SPECTRUM AT EARTH

Below, we discuss the framework to calculate the DSNB detection spectrum and summarize its three ingredients: the supernova neutrino emission (Section 4), the cosmic supernova rate (Section 5), and the detector capabilities (Section 6). We then return to the framework (Section 7) and discuss SK observations (Section 8), background reduction (Section 9), and conclusions (Section 10).

3.1. Simple Estimate of the Detection Rate

SN1987A was discovered optically in the Large Magellanic Cloud, a nearby satellite of the Milky Way, at a distance of approximately 50 kpc (14). Because SN1987A was so close, its neutrinos had probably interacted in the Kamiokande-II (Kam-II) and Irvine-Michigan-Brookhaven (IMB) water Cherenkov detectors, originally intended for proton decay searches. Kam-II reported 12 events and IMB reported 8 events; these events had energies of several tens of MeV and occurred in contemporaneous bursts of approximately 10 s, a few hours before the optical brightening

(15–18). The events were largely consistent with supernova neutrino interactions via the neutron inverse β-decay process, $\bar{\nu}_e + p \to e^+ + n$, where the positron is detected and where its energy is close to that of the neutrino.

Deferring many details, we can simply estimate the (steady) detection rate of DSNB neutrinos from the (burst) detection rate of SN1987A neutrinos as

$$\left[\frac{dN_\nu}{dt}\right]_{DSNB} \sim \left[\frac{dN_\nu}{dt}\right]_{87A} \left[\frac{N_{SN}\ M_{\rm det}}{4\pi D^2}\right]^{-1}_{87A} \left[\frac{N_{SN}\ M_{\rm det}}{4\pi D^2}\right]_{DSNB}. \qquad 1.$$

The detection rate in Kam-II during the burst was $\sim$1 s^{-1}. How are the inputs for the dimensionless product of conversion factors related? For SN1987A, $N_{SN} = 1$, whereas there are $N_{SN} \sim 100$ neutrino bursts in the universe starting within any 10-s interval. This accounts for the supernova rate of the Milky Way, $\sim 10^{-2}$ year^{-1}; the space density of comparable galaxies, $\sim 10^{-2}$ Mpc^{-3}; and the fact that the supernova rate per galaxy was $\sim$10 times higher in the past. The detector mass M_{det} of Kam-II was $\sim$10 times smaller than that of SK, which is 22.5 kton (22.5×10^9 g). The distance D of SN1987A was 0.050 Mpc, whereas for a typical supernova contributing the DSNB, at $z \sim 1$, it is $c/H_0 \sim 4000$ Mpc, $\sim 10^5$ times farther. The time-averaged DSNB detection rate in SK, noting changes in N_{SN}, M_{det}, and $1/D^2$, respectively, is

$$\left[\frac{dN_\nu}{dt}\right]_{DSNB} \sim (1\,{\rm s}^{-1}) \times 100 \times 10 \times 10^{-10} \sim 10^{-7}\,{\rm s}^{-1} \sim 3\,{\rm year}^{-1}, \qquad 2.$$

which is approximately the right number.

The Poisson expectation for the number of neutrinos detected per distant supernova is $\ll 1$, so nearly all supernovae yield 0 events, and a rare few unidentified supernovae yield only 1 event. Although the long-term average of the neutrino detection rate from Milky Way supernovae is $\sim$100 times larger than for the DSNB, it requires one to wait decades for large bursts from single identified supernovae. Future very large detectors may detect more frequent, but much smaller, bursts from identified supernovae in the nearest galaxies. Burst detection will be required to study the time profile and flavor dependence of supernova neutrino emission.

3.2. Line of Sight Integral for the Diffuse Supernova Neutrino Background Flux

The DSNB flux spectrum at Earth is calculated from the neutrino emission per supernova $\varphi(E_\nu)$ and the evolving core-collapse supernova rate $R_{SN}(z)$ as

$$\frac{d\phi}{dE_\nu}(E_\nu) = \int_0^\infty [(1+z)\varphi[E_\nu(1+z)]]\,[R_{SN}(z)]\left[\left|\frac{c\,dt}{dz}\right| dz\right]. \qquad 3.$$

This form arises as a standard line-of-sight integral for the radiation intensity (flux per solid angle) from a distribution of sources. Neglecting cosmological factors, the differential injection rate of neutrinos depends on the rate density (the number of neutrinos emitted per volume and time, i.e., the supernova neutrino emission multiplied by the supernova rate density) multiplied by the volume element $d\Omega r^2 dr$, whereas the fluxes fall off as $1/(4\pi r^2)$, canceling the r^2. The received intensity is then simply a line integral of the rate density. Because the sources are isotropic on the sky, the detection reaction has little directionality, and Earth is transparent to neutrinos, the intensity has been integrated over 4π to obtain the flux.

Equation 3 defines the neutrino number flux spectrum, in units of cm^{-2} s^{-1} MeV^{-1}. To obtain the energy flux spectrum, more familiar to photon astronomers, multiply through by E_ν. The first term in large brackets is the emission spectrum $\varphi(E_\nu)$, where a neutrino received at energy E_ν was emitted at a higher energy $E_\nu(1+z)$; the prefactor of $(1+z)$ on the spectrum accounts for the compression of the energy scale. The second term is the supernova rate density $R_{SN}(z)$.

The third term is the differential distance, where $|dt/dz|^{-1} = H_0(1+z)[\Omega_\Lambda + \Omega_m(1+z)^3]^{1/2}$; the cosmological parameters are $H_0 = 70$ km s^{-1} Mpc^{-1}, $\Omega_m = 0.3$, and $\Omega_\Lambda = 0.7$; and their uncertainties can be neglected.

Above, it is assumed that there is no variation in the supernova neutrino emission among different progenitor stars, which follows from the core-collapse criterion for massive stars and how supernova neutrino mixing effects work (both conditions are discussed in Section 4.2). It is also assumed that nothing except redshifting happens to the neutrinos between the supernovae and Earth because obscuration is always negligible, the neutrino mass eigenstates that propagate do not oscillate, and no exotic effects are expected [but such can be tested (19–22)]. These are realistic assumptions at the precision of the expected counts and energy resolution, and they can be generalized with appropriate averages.

3.3. Contemporary Inputs, Uncertainties, and Detectors

Throughout, we focus on the effective $\bar{\nu}_e$ spectrum at the surface of the supernovae, after any neutrino mixing effects in the star, because this is the same observable for both burst and DSNB detection. As described in Section 4, the neutrino emission in one flavor is characterized by its total energy (time-integrated luminosity) and average energy (3.15 T for a Fermi-Dirac spectrum, where T is the temperature in units of MeV). These are the parameters to be determined by a DSNB measurement because the supernova neutrino emission is now the most uncertain input and cannot be measured by astronomical techniques. Nominal values are $E_{\bar{\nu}_e,tot} \sim 5 \times 10^{52}$ erg (3×10^{58} MeV) and $E_{\bar{\nu}_e,avg} \sim 15$ MeV. If a black hole forms promptly, the luminosity from the collapsing core prior to being abruptly terminated can be higher than for neutron star formation, such that the larger time-integrated emission is larger (24).

The cosmic supernova rate is now well characterized, as shown in Section 5. For the supernova rate density, the local value is $\simeq (1.25 \pm 0.25) \times 10^{-4}$ Mpc^{-3} year^{-1}, and the rate grows by a factor of ~ 10 by $z = 1$. DSNB measurements are also sensitive to core collapses that are optically dark but neutrino bright.

Figure 1 shows the $\bar{\nu}_e$ flux spectrum; the SK 2003 limit is $\lesssim 1.2$ cm^{-2} s^{-1} for energies integrated above 19.3 MeV, ruling out spectra above about the middle of the $T = 8$ MeV band. Because the predictions are for generic thermal spectra, they could be used for other flavors as well. Equation 3 defines the flux spectrum of neutrinos. In Section 6, the event rate spectrum of positrons from the reaction $\bar{\nu}_e + p \rightarrow e^+ + n$ is introduced to allow comparison with the SK data.

3.4. Theoretical Predictions and Experimental Revolutions

Theoretical predictions of the DSNB flux go back more than 40 years, for instance to brief mentions by Zel'dovich & Guseinov (25) and Ruderman (8) and a calculation by Guseinov (26). First examples of detailed calculations, from the early 1980s, can be found in References 27–34; unfortunately, the true flux is not as large as in these predictions (but see the pessimistic lower limit noted in Reference 33). The variation between predictions, formerly orders of magnitude, decreased greatly beginning in the mid-1990s (e.g., 35–45). At roughly the end of that era, the DSNB was reviewed in 2004 by Ando & Sato (46). The SK 2003 upper limit on the DSNB was a key development, as it directly rules out models with fluxes more than a few times larger than present predictions.

It would be unfair to judge earlier predictions without adjusting their inputs to contemporary values. This seems difficult, as a wide variety of different formalisms, inputs, standards of precision,

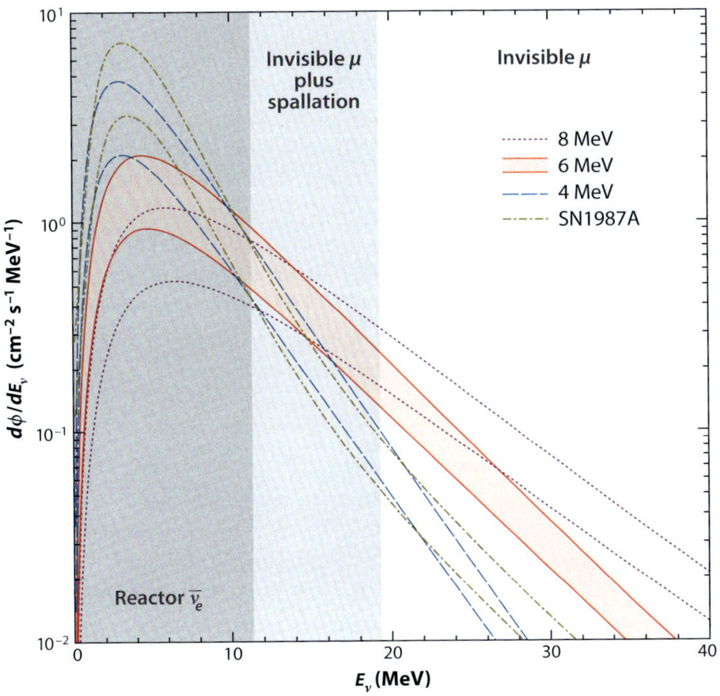

Figure 1

Predicted diffuse supernova neutrino background $\bar{\nu}_e$ flux spectrum. The spread between different labeled temperature bands indicates the uncertainty in the supernova neutrino emission; the widths of the bands indicate the uncertainty in the cosmic supernova rate. Gray shaded regions indicate the energy ranges in which noted detector backgrounds are important. Figure modified from Reference 23.

and methods of quoting results were used. However, this is easy in the present perspective, first noted in Reference 47, where the principal remaining unknown is the supernova neutrino emission, and the assumed values could be easily extracted from earlier papers. Most DSNB predictions did use reasonable supernova neutrino emission parameters (for the earliest papers, the existence of ν_τ and $\bar{\nu}_\tau$ was unknown), even before the detection of SN1987A. There were huge differences in the assumed supernova rate density because of observational uncertainties; these have been dramatically reduced by comprehensive star-formation and supernova rate measurements over the past decade. The results described in this review cover the range of reasonable allowed models for both the supernova neutrino emission and the supernova rate density.

4. FIRST INGREDIENT: SUPERNOVA NEUTRINO EMISSION

The detection of neutrinos from SN1987A established that some supernovae produce some neutrinos. Of course, our knowledge is much greater than that—there is strong evidence that core-collapse supernovae can be identified optically and that a powerful neutrino burst must accompany core collapse. To understand the properties of massive stars and how supernovae explode, it is necessary to confirm the ubiquity of neutrino emission from all types of supposed core-collapse supernovae and to refine our knowledge of this emission.

4.1. The Fates of Massive Stars

Massive stars race toward their doom. The lifetime of a star, which is mostly spent burning hydrogen in the core, is $\sim 30\,(8M_\odot/M)^{2.5} \times 10^6$ year, where $M_\odot = 2.0 \times 10^{33}$ g is the solar mass (48). The fusion reactions proceed in stages that depend on temperature, overcoming the Coulomb barriers for successively heavier elements only once the preceding elements are exhausted and the core contracts. Only stars with mass greater than approximately $8M_\odot$ become hot and dense enough in the core to burn past carbon and oxygen to iron (49, 50). A Chandrasekhar instability sets in once $\sim 1.4M_\odot$ of iron has been created, given that no further nuclear reactions can generate energy and electron degeneracy pressure provides insufficient support. Photonuclear reactions then endothermically destroy iron, leading to further collapse, which continues until the core reaches nuclear densities and cannot be compressed further. The subsequent bounce leads to the formation of an outgoing shock, which removes the envelope of the star and causes the optical supernova, which is bright for several months. What remains is a neutron star or black hole from the core and a shell remnant from the envelope.

Astronomical characterizations of supernovae are based on their optical properties (51). Supernovae must be distinguished by their spectra because their light curves alone are not usually enough. Many lines of evidence indicate that Types II, Ib, and Ic supernovae arise from the core collapse of massive stars and that they differ in the amount of progenitor envelope mass loss before explosion. Type II supernovae are several times more common than Types Ib and Ic, and the latter two seem to arise from ranges of successively larger progenitor masses. All should have comparable emission of all flavors of neutrinos and antineutrinos due to the mechanism of core collapse. Type Ia supernovae, which are also several times less common than Type II supernovae, seem to be powered by the runaway thermonuclear burning of a carbon and oxygen white dwarf in a binary system into iron; only relatively modest emission of low-energy electron neutrinos is expected. Detecting neutrinos will test the mapping between optical supernova types and the underlying physical mechanisms (52).

4.2. Core-Collapse Supernova Neutrino Emission

The basic picture of core collapse is well known (53). When the core collapses, there is a huge change in its gravitational potential energy. For a constant density profile, the change in this self-energy is

$$\Delta(P.E.) \simeq \left(\frac{3G_N M^2}{5R}\right)_{NS} - \left(\frac{3G_N M^2}{5R}\right)_{core} \simeq 3 \times 10^{53}\ \text{erg}, \qquad 4.$$

where the number follows from the mass and radius of a typical neutron star, $1.4\,M_\odot$ and 10 km, respectively (the subtracted term is negligible because the initial core radius is thousands of kilometers), and the change in gravitational potential energy is an incredible $\sim$10% of Mc^2 for the core (54–56). The mass of the collapsed core varies little with the mass of the progenitor star (57, 58). Because the protoneutron star is near nuclear density, no particles except neutrinos can escape, and core collapse is inevitably accompanied by a huge burst of neutrinos, carrying away nearly the whole change in gravitational potential energy. The neutrinos, which begin leaving the core at the instant of collapse, arrive at Earth before the light, which is delayed by at least the time it takes the shock to propagate through the envelope, typically hours. This time difference was observed for SN1987A and is a crucial confirmation of the basic picture.

It is thought that each of the six flavors of neutrinos and antineutrinos carries away a comparable fraction of the total energy. In numerical simulations of supernovae, this equipartition is only approximate, and the sense in which it might be violated in nature is not known (59–62). Most of the neutrino emission is in pairs—$\bar{\nu}_e + \nu_e$, $\bar{\nu}_\mu + \nu_\mu$, and $\bar{\nu}_\tau + \nu_\tau$—that are produced by weak interactions in the hot and dense matter in the core. The reaction $e^- + p \rightarrow \nu_e + n$ converts the nearly equal numbers of neutrons and protons from nuclei into the neutron-rich matter of the neutron star, producing an excess ν_e flux, but this is only a $\sim$10% effect.

The protoneutron star density is so high that even neutrinos do not readily escape. Instead, they diffuse out over several seconds and are thus emitted with a quasi-thermal spectrum characteristic of the surface of last scattering. If the protoneutron star were semitransparent, the neutrinos would leave on a timescale of $\sim$1 ms, with energies of $\sim$100 MeV. Taking opacity into account, and assuming blackbody emission, $L \sim E_{tot}/\tau \sim R^2 T^4$, the increase in the emission time by a factor of $\sim 10^4$ corresponds to the decrease in neutrino average energy by a factor of $\sim$10. A long emission timescale and moderate temperature were observed for SN1987A, confirming the high density of the protoneutron star.

The opacities of neutrino flavors with matter vary, leading to different neutrino temperatures, because the matter temperature of the protoneutron star falls with radius. The interactions of ν_μ, ν_τ, $\bar{\nu}_\mu$, and $\bar{\nu}_\tau$ occur only through the weak neutral current because their energies are too low to produce the corresponding charged leptons. For ν_e and $\bar{\nu}_e$, there are also interactions through the weak charged current, so the cross sections are larger; for ν_e, the coupling constants are larger, and there are more neutron than proton targets. Oft-quoted estimates for the temperatures are $T = 8$ MeV for ν_μ, ν_τ, $\bar{\nu}_\mu$, and $\bar{\nu}_\tau$; $T = 5$ MeV for $\bar{\nu}_e$; and $T = 4$ MeV for ν_e. More recent theoretical work indicates that the temperatures, and the splittings between them, are lower (64).

Neutrino mixing effects relate the initial neutrino spectra at the opaque emitting surface of the protoneutron star to those outside the supernova. Although the opacity is low outside the neutron star, the high density of matter and even other neutrinos significantly affect neutrino mixing (e.g., 65–68). The passage through high densities leaves the neutrinos in incoherent mass eigenstates that do not oscillate between the supernova and Earth. Mixing effects in Earth are suppressed by the isotropy of the sources and that of the detection cross section (69).

Neither the expected initial spectra nor the neutrino mixing effects are understood well enough, so we use the observable effective spectra. For $\bar{\nu}_e$, the effective time-integrated spectrum after mixing, assumed to be approximately of the Fermi-Dirac form, is (in units of MeV^{-1})

$$\varphi(E_\nu) = E_{\bar{\nu}_e,tot} \frac{120}{7\pi^4} \frac{E_\nu^2}{T^4} \frac{1}{(e^{E_\nu/T} + 1)}, \qquad 5.$$

where the total energy is $E_{\bar{\nu}_e,tot}$ and the average energy is $E_{\bar{\nu}_e,avg} = 3.15\,T$. These parameters are not fixed from theory, but rather will be determined from experiment and will test supernova simulations and neutrino mixing scenarios.

4.3. Comparison to Supernova 1987A

Although the SN1987A data were sparse, and the detectors were not optimized for that purpose, these data are the most direct evidence we have about supernova neutrino emission. A reconstruction of the effective $\bar{\nu}_e$ spectrum from the Kam-II and IMB data is shown in **Figure 2**; it is in reasonable agreement with expectations. Higher temperatures are favored by theoretical models, including neutrino mixing, as well as by observations of nucleosynthesis yields that depend on high-energy neutrinos (70–72). It remains to be seen whether SN1987A was different from an average supernova, and more data are needed, especially at high energies.

The SN1987A data are an important input for estimating the DSNB spectrum (40, 45), although care must be taken to note that the high-energy SN1987A neutrinos are the most relevant

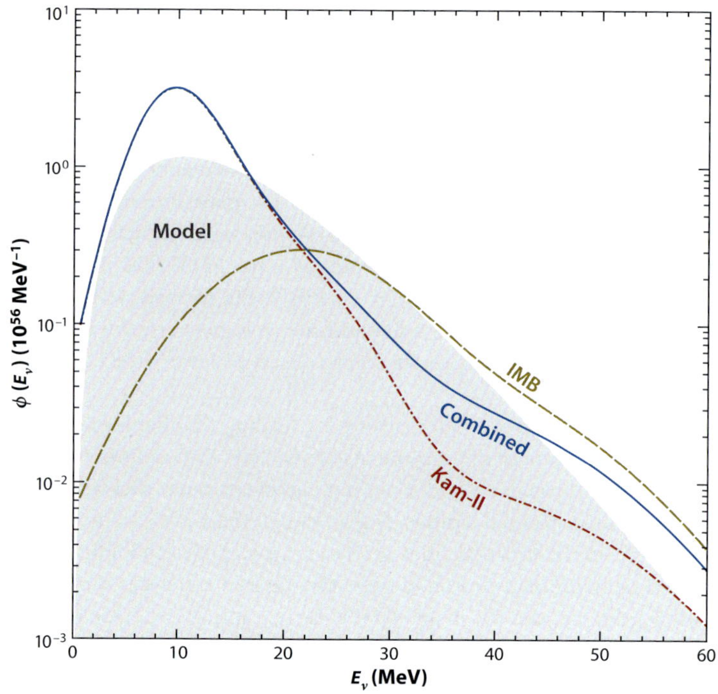

Figure 2

Time-integrated effective $\bar{\nu}_e$ spectrum from SN1987A. The colored curves indicate results reconstructed through use of only the Kam-II data, the IMB data, or their combination. The gray shaded region is a thermal spectrum with parameters $E_{\bar{\nu}_e,tot} \sim 5 \times 10^{52}$ erg and $E_{\bar{\nu}_e,avg} \sim 15$ MeV. Figure modified from Reference 63.

for predicting the DSNB (63). The SK 2003 search used a positron energy threshold of 18 MeV, and future searches may reach 10 MeV; redshift effects enhance the importance of high-energy emission.

5. SECOND INGREDIENT: COSMIC SUPERNOVA RATE

Supernovae are infrequent in the Milky Way, but not in the universe. Because the lifetimes of massive stars are very short on cosmological timescales, their cosmic birth and death rates are exactly equal. The cosmic star-formation rate has been measured precisely, by means of a great variety of techniques, principally based on the emission of massive stars. This measurement is enough to accurately give the rate of core collapses, and it is supplemented by direct measurements of the optical supernova rate, which have lower precision but are in good agreement.

5.1. Measured Star-Formation Rate

Star-formation rate measurements do not literally probe star formation, but begin with the total luminosity of the stars in a galaxy. If these stars were all of the same luminosity, so that their mass and lifetime were given by stellar evolution theory, then the star-formation rate would be determined by the total mass of these stars divided by their lifetime; this rate is the birth (and death) rate needed to keep the number of stars in equilibrium. The main star-formation rate indicators measure the emission from massive stars that are short lived on galactic timescales, so the equilibrium assumption should be good on average for these measurements.

A wide range of stellar masses are always present, complicating this procedure, but the degeneracies can be broken because stars of increasing mass have higher temperatures and much higher luminosities. Massive stars can be isolated by measurements of their continuum flux or of nebular emission lines from atomic recombinations in the surrounding gas they ionize. The distribution of stellar masses formed when star formation occurs is assumed to follow a universal initial mass function (IMF). For example, the conventional Salpeter IMF scales as $\psi(M) = dn/dM \propto M^{-2.35}$ for stellar masses M between $0.1M_\odot$ and $100M_\odot$ (73, 74); more realistic IMFs modify the low-mass end. With an IMF, calculated stellar and nebular emission spectra, and careful consideration of the effects of obscuration and reradiation, the star-formation rate can be determined from a galaxy luminosity spectrum (75). The high-mass star-formation rate is measured, but the formation rate of low-mass stars has to be deduced assuming an IMF. Doing so introduces a factor-two range in the normalization of the total star-formation rate due to the low-mass IMF uncertainties.

The comoving star-formation rate density, in units of $M_\odot$ Mpc^{-3} year^{-1}, is ~10 times larger at $z = 1$ than it is today at $z = 0$. Comoving coordinates are those for which the expansion of the universe has been removed; crudely, this defines the star-formation rate per average galaxy. Aside from the normalization uncertainty due to the choice of IMF, the precision of the combined data is at the tens-of-percent level for $z \lesssim 1$ (23, 76). From $z = 1$ to at least $z \sim 4$–5, the star-formation rate is nearly flat, with a precision that decreases with increasing redshift, becoming as poor as a factor of two, until a more uncertain decline at higher redshifts (e.g., 77–82).

These results, especially those in the $z \lesssim 1$–2 range relevant for the DSNB, are quite robust: (*a*) They have been measured by many different groups through the use of different instruments and techniques, and (*b*) extensive efforts have been made to calibrate the observables and their corrections for dust obscuration and incompleteness. The star-formation rate data are also in good agreement with measurements of the extragalactic background light that, analogous to the DSNB, is an integral measure of stellar emission over redshifts (23). These and other data favor an IMF of somewhat shallower slope than the Salpeter IMF.

5.2. Predicted Supernova Rate

To predict the core-collapse supernova rate density, $R_{SN}(z)$, in units of Mpc^{-3} year^{-1}, from the star-formation rate density, $R_{SF}(z)$, we use

$$R_{SN}(z) = R_{SF}(z)\frac{\int_8^{50} \psi(M)dM}{\int_{0.1}^{100} M\,\psi(M)dM} \simeq \frac{R_{SF}(z)}{143M_\odot}, \qquad 6.$$

where the upper integral gives the number of stars that lead to core collapse and the lower integral gives the total mass in stars; a Salpeter IMF has been used. Importantly, for $R_{SN}(z)$ the above-mentioned normalization uncertainty due to the choice of IMF goes away because nearly the same massive stars source the star-formation rate measurements and lead to core-collapse supernovae. For an IMF with a shallower slope, $R_{SF}(z)$ is smaller, but the ratio of integrals is larger, canceling in $R_{SN}(z)$ to the few-percent level (23).

In the upper integral, the choice of upper limit of integration is unimportant, as long as it is large; then the integral scales as $(8M_\odot/M_{lower})^{1.35}$ for a Salpeter IMF. Theoretical and indirect empirical work indicates that the minimum-mass star that leads to core collapse is $\sim 8M_\odot$. This hypothesis is now supported by direct observations. Archival imaging from the Hubble Space Telescope and other sources has allowed the identification of the stellar progenitors of some supernovae in nearby galaxies (e.g., 83–86). The luminosity of the star determines its mass, and a fit to progenitor data gives a minimum mass to lead to a core-collapse supernova of $(8.5 \pm 1.5)M_\odot$ (87). Also, the largest estimate so far of the maximum mass to instead lead to a white dwarf is $\sim 7M_\odot$ (88, 89).

Thus, there is a precise prediction of the cosmic core-collapse rate that is based on strong evidence from star-formation rate data and direct confirmation of the minimum stellar mass that leads to a core-collapse supernova. The star-formation rate data are based on observations of living massive stars, which do not differentiate whether these stars die as visible supernovae or as optically dark collapses. Because those outcomes have comparable neutrino emission, these data are enough to precisely predict the DSNB.

5.3. Measured Supernova Rate

As noted above, the relationship between star-formation and supernova rates is only a simple conversion with small uncertainties on the result. Because the redshift evolution of the star-formation rate is well measured, the evolution of the supernova rate is then precisely defined, and the conversion predicted above could be checked at a single redshift.

Measuring supernova rate densities is not easy, given that supernovae are infrequent in individual galaxies. Core-collapse supernovae (Types II, Ib, and Ic) are fainter and less uniform in their optical properties than are the thermonuclear supernovae (Type Ia) used as distance indicators, and they have received much less attention despite being more common.

At high redshift, repeated imaging of a volume can find supernovae largely independent of their host galaxies, thereby directly measuring the quantity of interest. At low redshift, galaxies from a catalog can be monitored. Because such monitoring finds the galaxies first and the supernovae second, it is subject to incompleteness due to faint galaxies, and one must carefully define the global volume and time monitored from those for each galaxy. Although such corrections have been made, catalog surveys tend to find lower supernova rates and less extreme supernovae than do volume surveys. Lastly, supernova rates estimated using supernovae that were not collected in a systematic way should be taken only as lower bounds.

The predicted supernova rate density is shown in **Figure 3**. There is very good agreement with the most reliable data, although the conversion was predicted absolutely and not fit. The uncertainty on the star-formation rate is taken as 20% on R_{SF} $(z = 0)$. The predicted supernova rate and its uncertainty are $R_{SN}(z = 0) = (1.25 \pm 0.25) \times 10^{-4}$ Mpc^{-3} year^{-1} (recall Section 3.1). At redshift zero, the lower limits on the rates are already quite high, matching the prediction well; the 10 Mpc was defined very conservatively and, along with the 28-Mpc point, is known to be incomplete (52, 87). Therefore, supernova rate measurements lower than this band must be less complete because the shape is fixed from the star-formation rate. At high redshift, the updated Dahlen et al. points are in excellent agreement (90, 91); these points were measured volumetrically and are larger than points that were not (92–94). After Reference 23 was published, the Dahlen et al. points decreased slightly but are still consistent with the band shown in **Figure 3** (T. Dahlen, private communication). That faint core-collapse supernovae were missed in surveys with rates below this band is supported by data on their rate relative to that of the brighter Type Ia supernovae. Note that the discovery rates of supernovae have dramatically increased with time and with advances in technology and interest.

The agreement between the predicted and measured supernova rates strongly supports this input for the DSNB calculation. Although a significant fraction ($\sim$50%) of dark collapses is allowed, it is not required ($\sim$10% is expected). These collapses may be intrinsically faint, simply obscured, or truly dark. In any case, their existence can only raise the level of the DSNB. In the near future,

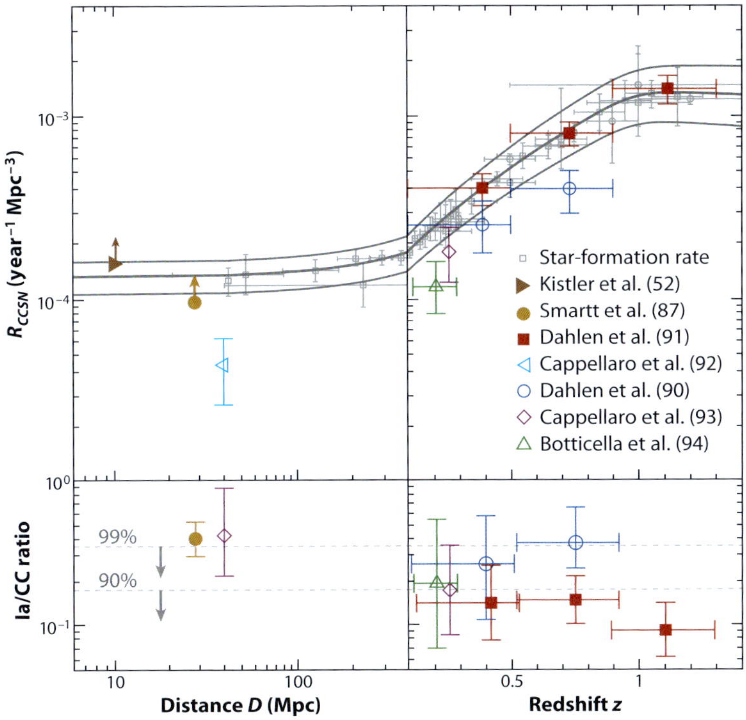

Figure 3

Predicted and measured cosmic core-collapse (CC) supernova rates. In the upper panel, the band and its width indicate the absolute prediction and its uncertainty from the star-formation rate data alone (*faint gray points indicate these data with the same conversion factor applied*). The most reliable supernova rate measurements are shown with solid symbols; those with open symbols must be less complete. In the lower panel, rate ratios are shown, which supports this hypothesis. Figure adapted from Reference 23.

supernova rate measurements will greatly improve, and their comparison with star-formation rate data will allow more precise tests of the fraction of dark collapses (23, 95–97).

6. THIRD INGREDIENT: NEUTRINO INTERACTIONS AND DETECTORS

The detection probability depends on the product of the DSNB flux and the effective area of the detector. Although supernovae make all flavors of neutrinos and antineutrinos, the best prospects are for $\bar{\nu}_e$ in SK, detected by $\bar{\nu}_e + p \rightarrow e^+ + n$, as SK has both large size and low background rates. Because SK (and Earth) is nearly transparent to neutrinos, the effective area is the neutrino-proton cross section multiplied by the number of target protons, independent of the shape or orientation of the detector. The properties of neutrinos have now been measured well, and exotic scenarios that would affect DSNB detection, such as large mixing with sterile species, are no longer favored.

6.1. Detection of Diffuse Supernova Neutrino Background Electron Antineutrinos

Electron antineutrinos are detected by the neutron inverse β decay reaction on free protons, $\bar{\nu}_e + p \rightarrow e^+ + n$. Here free protons means hydrogen nuclei and not the protons bound in heavier nuclei, for which nuclear binding effects suppress interactions at low energies; the atomic ionization state is irrelevant.

The total cross section for neutron inverse β decay is

$$\sigma(E_\nu) = \left[0.0952 \times 10^{-42}\,\text{cm}^2\,(E_\nu - 1.3\,\text{MeV})^2\right](1 - 7E_\nu/M_p), \qquad 7.$$

for neutrino energies above threshold, where $E_\nu > 1.8$ MeV; M_p is the proton mass (for $c = 1$). In the high-energy range, the rising cross section helps offset the falling flux spectrum. This expression, which is good to the several-percent level in the relevant energy range, neglects the small positron mass but does include the few-tens-of-percent recoil-order corrections, shown outside the square brackets. At the same orders, the kinematic relationship is

$$E_e = [(E_\nu - 1.3\,\text{MeV})](1 - E_\nu/M_p), \qquad 8.$$

where the isotropy of the neutrino angular distribution has been taken into account. The recoil-order correction to the kinematics can be neglected relative to the detector energy resolution. Full expressions for the cross section and kinematics can be found in References 98 and 99.

6.2. The Super-Kamiokande Detector

The fiducial volume of SK contains 22.5 kton of water and, hence, 1.5×10^{33} free protons (100). To reduce backgrounds, SK was built under 1 km of rock, is heavily shielded by separate outer and inner buffer regions, and was constructed to maintain extreme purity. SK is large enough to have a few DSNB events per year, and it already has low background rates. These facts are fortuitous, as the SK design was optimized not for the DSNB but rather for measurements of proton decay and atmospheric, accelerator, solar, and Milky Way supernova neutrinos.

In SK, relativistic charged particles are detected by the cones of optical Cherenkov light produced as they lose energy; the Cherenkov process is a small component of the energy-loss rate but is the only component that SK detects directly. The Cherenkov light travels through water to the photomultiplier tubes on the walls, which view a homogeneous volume of transparent water. From the patterns of received photons, the position, direction, energy, and identity of charged particles are determined. SK has excellent detection efficiency for the energies considered here.

7. FRAMEWORK REDUX: DIFFUSE SUPERNOVA NEUTRINO BACKGROUND DETECTION SPECTRUM AT EARTH

Now that we have reviewed the three ingredients—supernova neutrino emission, cosmic supernova rate, and detector capabilities—we return to the framework for the DSNB detection spectrum at Earth.

The event rate spectrum, in units of $\text{s}^{-1}\,\text{MeV}^{-1}$, where the detection cross section $\sigma(E_\nu)$, the number of proton targets N_p, and the shift between neutrino energy E_ν and positron energy $E_e \simeq E_\nu - 1.3$ MeV have been accounted for, is

$$\frac{dN_e}{dE_e}(E_e) = N_p\sigma(E_\nu)\int_0^\infty [(1+z)\varphi[E_\nu(1+z)]]\,[R_{SN}(z)]\left[\left|\frac{c\,dt}{dz}\right|dz\right]. \qquad 9.$$

For relevant energies, namely those above 10 MeV, the effective upper limit on the line of sight integral is $z \sim 1$–2, as beyond that limit the falling dt/dz is no longer compensated by a rising $R_{SN}(z)$ and the larger emission energies required for the same observed energy probe further into the high-energy tail of the emission spectrum.

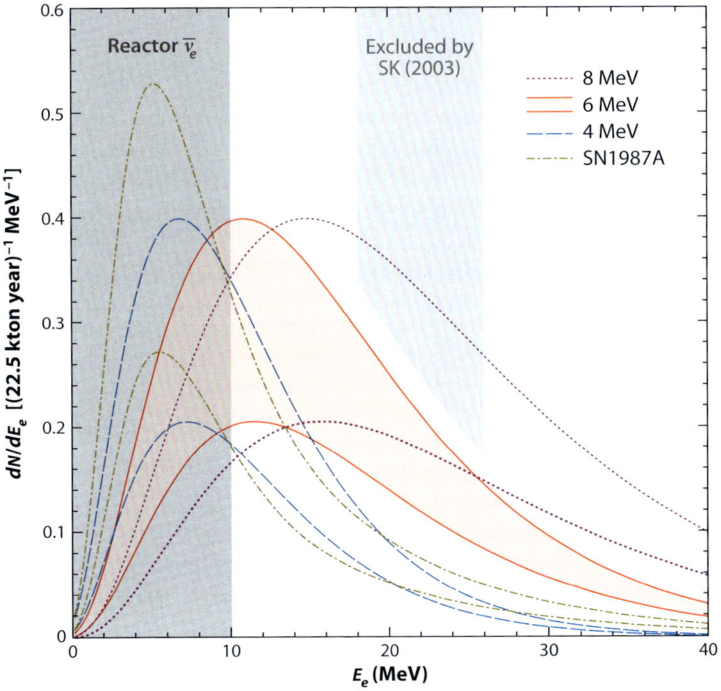

Figure 4

Predicted diffuse supernova neutrino background $\bar{\nu}_e$ event rate spectrum in positron energy. The labeled bands and their widths are as in **Figure 1**. Integrated event rates are tabulated in Reference 23. The 2003 exclusion from SK is shown; it is largely independent of the assumed temperature. The energy range of the irreducible reactor background is shaded; backgrounds at higher energies depend on whether gadolinium is added to SK. Figure modified from Reference 23.

In **Figure 4**, we show predictions for the DSNB detection spectrum, taking up-to-date inputs and their uncertainties into account. The range due to different emission parameters, especially at high energies, indicates the uncertainties to be reduced by measuring the DSNB. The uncertainty on the supernova rate is already modest and will soon be smaller. The detection reaction is neutron inverse β decay in SK, which is well understood. Because the expected number of DSNB events is at most a few per year, the Poisson statistical uncertainty will initially be significant. This predicted event rate spectrum can be directly compared with measured data, which are a combination of DSNB signal and detector backgrounds. The DSNB is especially sensitive to the important high-energy part of the emission spectrum. The predictions for different temperatures converge near 10 MeV, so a measurement there would probe $E_{\bar{\nu}_e,tot}$ alone; the falloff of the spectrum would then probe $E_{\bar{\nu}_e,avg}$ alone.

8. SUPER-KAMIOKANDE 2003 UPPER LIMIT ON THE DIFFUSE SUPERNOVA NEUTRINO BACKGROUND

It takes extraordinary efforts to operate a gigantic detector such as SK with low background rates at low energies. The SK DSNB limit (101) improved upon the Kam-II limit (102) by a factor of ~100, establishing the only experimental limit within range of contemporary predictions—an extremely important achievement.

8.1. Detector Backgrounds and Diffuse Supernova Neutrino Background Flux Limit

For the SK 2003 DSNB search, only visible positron or electron energies (they cannot be distinguished) above 18 MeV were used; above this energy, the detector background rates, due to atmospheric neutrinos, were relatively low. Nothing beyond expectations, including any surprise problems that would impede this or improved future searches, was found in this pioneering search.

To understand the atmospheric neutrino backgrounds, one must use the event rate spectrum in visible energy. The flux spectrum alone is misleading, as here the visible energy is quite different from the neutrino energy, unlike for the DSNB signal. The backgrounds, shown in **Figure 5** with visible energies of 18–82 MeV, arise from neutrinos with energies up to ~250 MeV that invisibly enter the detector and have charged-current interactions, mostly with oxygen nuclei, inside the fiducial volume.

Atmospheric ν_μ and $\bar{\nu}_\mu$ can produce nonrelativistic μ^- and μ^+, which do not produce Cherenkov light as they lose energy. When μ^- and μ^+ decay at rest, the electrons and positrons produced are relativistic, and their spectrum appears as the bump shown in **Figure 5**. This spectrum is the dominant background, but its shape is fixed and its normalization can be measured above the energy range of the DSNB. Atmospheric ν_e and $\bar{\nu}_e$ can produce e^- and e^+ with energies

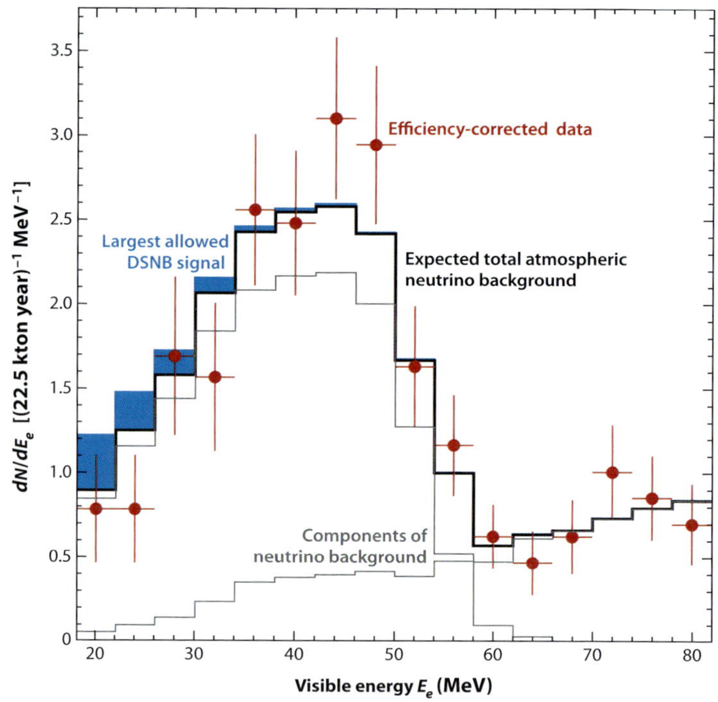

Figure 5

Results of the SK 2003 diffuse supernova neutrino background (DSNB) $\bar{\nu}_e$ search. The efficiency-corrected measured data are shown with red points and error bars. The expected total atmospheric neutrino background is represented by the thick black solid line, and its components are indicated by the thin gray solid lines. The largest allowed DSNB signal is shown by the shaded blue region added to the atmospheric neutrino background. Here dN/dE_e is shown, which can be directly compared in 1-MeV steps with **Figure 4**; to recover the dN results given in Reference 101, multiply the bin values by 4 MeV. Figure adapted from Reference 101.

well below the neutrino energy, due to nuclear effects; these produce the rise at high energies in **Figure 5**. A DSNB signal would have appeared as an excess at low energies (compare to **Figure 4**); none was observed, and a limit of $\phi(E_\nu > 19.3\,\text{MeV}) \lesssim 1.2\ \text{cm}^{-2}\ \text{s}^{-1}$ at the 90% confidence level was set (101).

8.2. Calculated Limits on Supernova Neutrino Emission

A DSNB limit expressed as a flux integrated above a certain energy, although a good starting point, is model dependent and makes comparisons difficult. One variable is not enough to characterize DSNB spectra, but the two natural variables of total energy $E_{\bar{\nu}_e,tot}$ and average energy $E_{\bar{\nu}_e,avg}$ for the $\bar{\nu}_e$ flavor are sufficient (47). These variables primarily parameterize the unknown effective supernova neutrino emission because the cosmic supernova rate is now known well enough. This reinterpretation of the SK DSNB limit, showing how the two emission parameters can compensate each other, is shown in **Figure 6**. Any change in the assumed normalization $R_{SN}\,(z = 0)$ corresponds to a simple change in $E_{\bar{\nu}_e,tot}$ alone, and uncertainties on the evolution $R_{SN}(z)/R_{SN}(z = 0)$ are small.

Excitingly, the sensitivity is near the parameters of supernova models (59–62) and the uncertain fits to the SN1987A data (103). A two-parameter limit on the effective $\bar{\nu}_e$ emission is simple for experimentalists to implement and corresponds to the minimal theory. If it is proven that there is

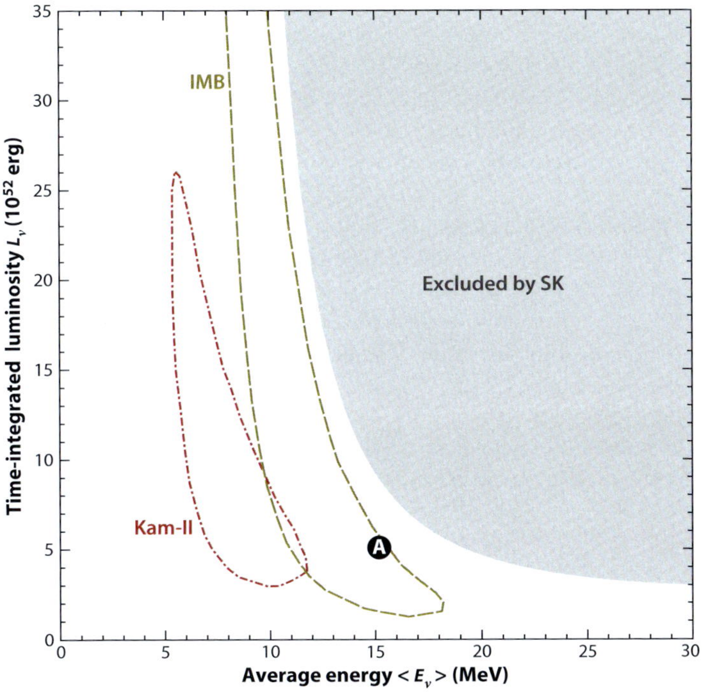

Figure 6

Constraints on the effective supernova $\bar{\nu}_e$ emission parameters. The gray excluded region follows from the null result of the SK 2003 diffuse supernova neutrino background search. Approximate regions from fits to the SN1987A data are shown (e.g., 103). Point A indicates the nominal parameter values. Figure modified from Reference 47.

large variety in supernova neutrino emission, large neutrino mixing effects, or exotic effects, then appropriate theoretical averages should be compared to this direct experimental limit.

9. PATHWAYS TO DISCOVERY

Progress on experimental sensitivity will have a decisive impact, as the expected signal is close to current limits. Astronomical observations and theoretical work will refine and enhance these prospects.

There are two striking features of the SK 2003 results. First, in the energy range used, the search is background limited, so the sensitivity will not improve linearly with exposure time, but naïvely only with the square root. Second, going to lower energies would give much more DSNB signal and much less atmospheric neutrino background; however, there is a large background rate due to the β decays of nuclei produced by spallation interactions of cosmic-ray muons. To detect the DSNB, SK therefore needs to reduce backgrounds and extend the search to lower energies. But how?

9.1. Proposal for Background Reduction in Super-Kamiokande

Because the DSNB detection reaction is $\bar{\nu}_e + p \rightarrow e^+ + n$, whereas the detector background reactions mostly do not produce neutrons, neutron detection is the key to separating signal from background. The neutron loses energy by elastic scattering and, at present, captures on a free proton, producing a 2.2-MeV γ ray that leads to an insufficient detection signal in SK.

The idea of positron-neutron coincidence detection goes back to the Reines and Cowan group (3), and it is well known how to implement such detection in oil-and-scintillator detectors, which have much higher light yields than do water Cherenkov detectors. For example, with the suspension of a small concentration of a gadolinium compound in the oil, neutrons capture on gadolinium due to its enormous cross section, producing an easily detectable 8-MeV γ-ray signal. The neutron is detected after the relatively long time of tens of microseconds after the positron, at nearly the same position.

However, until 2004 (104), it had not been shown that neutron detection with dissolved gadolinium in a water Cherenkov detector could be feasible. The key points are that soluble compounds exist—for instance, $GdCl_3$ and $Gd_2\ (SO_4)_3$—and that the γ rays would lead to a detectable signal in SK due to its high photomultiplier coverage. That these and many other technical concerns had positive answers was quite surprising. The ability to detect neutrons would give us unprecedented capabilities to separate signals and backgrounds in very large detectors. As described in Reference 104, doing so would reduce the atmospheric neutrino backgrounds and, more importantly, would dramatically reduce the large spallation radioactivity backgrounds, allowing SK to reduce its DSNB analysis threshold to $\sim$10 MeV, below which the reactor $\bar{\nu}_e$ background is overwhelming.

Figure 7 shows that there would be excellent prospects for the DSNB if SK had a $\sim$0.1% concentration of dissolved gadolinium. There is no known astrophysical source that can produce a comparable high-energy $\bar{\nu}_e$ flux, leaving a detection window for the DSNB. If an exotic process converts a small fraction of solar ν_e to $\bar{\nu}_e$, then the discovery of this new physics could partially get in the way of the DSNB, but such problems would be welcome (104, 105).

9.2. Research and Development for Gadolinium in Super-Kamiokande

Since 2004 (104), Vagins and other SK experimentalists (106–109) have been intensively researching the practicalities of introducing a dissolved gadolinium compound into SK. Our initial

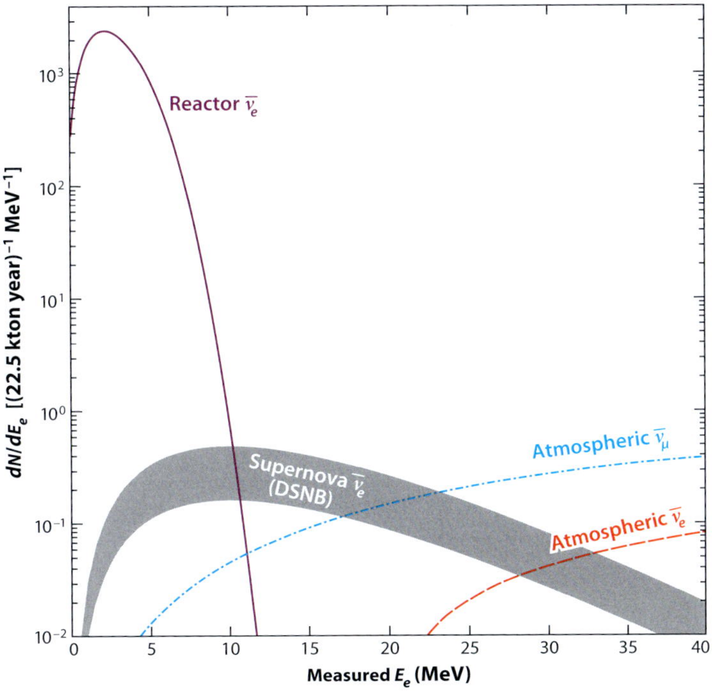

Figure 7

Expected detection rates in SK with dissolved gadolinium. The diffuse supernova neutrino background (DSNB) signal could be cleanly detected at a reasonable rate (see **Figure 4** for updated DSNB predictions and uncertainties). Figure modified from Reference 104.

investigations and the more extensive research that followed have uncovered no insurmountable problems regarding the infrastructure required to manage the gadolinium, the effects on the detector and other physics searches, or the positron-neutron coincidence detection. However, it has been determined that it is not advisable to pack a kilogram of white $GdCl_3$ powder in carry-on luggage for an international flight.

The progress on the comprehensive research and development program, and the seriousness with which the SK Collaboration is considering adding gadolinium, can be illustrated with two recent examples. First, a small container of gadolinium-loaded water and a neutron source were lowered into SK, and the detectable signal of neutron capture on gadolinium worked as expected (110; see also Reference 111). Second, a new 200-ton "miniature" (~1% scale) of SK with dissolved gadolinium is being built underground near SK to test all aspects of a gadolinium-loaded water Cherenkov detector (112).

9.3. Prospects for Other Detectors and Flavors

Oil-and-scintillator detectors the size of SK would have reduced atmospheric backgrounds and a reactor background that is relatively large at any location on Earth (113, 114). Ultimately, a high-statistics measurement of DSNB $\bar{\nu}_e$ will be needed to fully develop the science possibilities, probably with a water Cherenkov detector at least 10 times larger than SK (and hence with a comparable increase in DSNB signal rate, with the exact factor depending on the accessible energy range) (115, 116). The prospects for a high-statistics DSNB measurement have been explored (117–121).

Direct DSNB limits on other neutrino flavors are weaker than for $\bar{\nu}_e$ in SK. With all its data analyzed, the Sudbury Neutrino Observatory's sensitivity to DSNB ν_e will be only several times worse (122, 123), comparable to indirect limits (124), but limits for other flavors are worse by orders of magnitude (125, 126). Large liquid argon detectors are promising for DSNB ν_e (127, 128).

10. CONCLUSIONS

Neutrino astronomy can revolutionize astrophysics and physics due to its unique abilities to observe unseen processes and to probe subtle new particle interactions. Stellar fusion reactions and the core-collapse mechanism were revealed by the detection of neutrinos from the Sun and the nearby SN1987A. Solar neutrinos were crucial to the discovery of neutrino mixing, and supernova neutrinos probed exotic physics far beyond the reach of laboratory experiments.

However, the great promise of neutrino astronomy remains largely unfulfilled, as no other neutrino sources have yet been detected. Recall the courage of the pioneers of this field, who more than 40 years ago had neither adequate detectors nor accurate astrophysics to fortify their hopes. We have advantages of many orders of magnitude on each and can see the end of our wandering in the desert. To make progress on our understanding of supernovae and their role in astrophysics, as well as on our understanding of the neutrinos themselves, we must detect more supernova neutrinos.

Fortunately, first detection of the DSNB is within reach. Why are we so sure?

1. The framework for calculating the DSNB is solid. It is built on simple principles of cosmology that have been well tested with photons.
2. Supernova neutrino emission has been measured from SN1987A. Theory and indirect evidence suggest it should typically be larger.
3. The uncertainty on the cosmic supernova rate is modest. Observations of massive stars and supernovae are precise and quickly improving.
4. All aspects of DSNB signal detection are understood. Neutrino interactions and detector properties have each been well measured.

Therefore, even including uncertainties, there is a robust prediction that there must be a few DSNB neutrino events per year in SK. These events are hidden by backgrounds—what about the experimental prospects for revealing them?

1. SK has already shown sensitivity quite close to these predictions. Importantly, SK has controlled detector backgrounds nearly to the required level.
2. There are good prospects for improving the SK DSNB sensitivity. With gadolinium, SK would reduce its backgrounds and energy threshold.

Therefore, the DSNB could be discovered in SK. With sufficient detector background reduction, even a few signal events would be statistically significant and scientifically important. If the DSNB is not found, then there must be surprising new physics, which would require an upheaval in astrophysics or exotic new particle physics.

The enduring value of a DSNB measurement will come from its comparison to other observations and theory, which include the SN1987A data and numerical simulations of core collapse. The SK 2003 limit is already an important new result because it excludes neutrino emission as large as often assumed and is only a factor of two to four away from more conservative predictions. Significant developments in the theoretical predictions are also eagerly anticipated because, after more than 40 years of work, recent results have included increasingly realistic physics and accurate calculations, show explosions in some cases, and can integrate to several seconds in certain cases. These models will provide a rich context in which to interpret DSNB results and, eventually, to novel tests of neutrino properties and other new physics. In particular, a primary goal will be to understand the effective received spectra in terms of the initial spectra and the effects of neutrino mixing. Over the next several years, we will gain an improved and more comprehensive understanding of supernovae and their neutrinos as a result of the above improvements converging with unprecedented electromagnetic data, results from gravitational wave observatories, and—with a little luck—a Milky Way neutrino burst.

DISCLOSURE STATEMENT

The author is not aware of any affiliations, memberships, funding, or financial holdings that might be perceived as affecting the objectivity of this review.

ACKNOWLEDGMENTS

The author's research was supported by National Science Foundation CAREER grant PHY-0547102. I am grateful to my collaborators on the DSNB—Shin'ichiro Ando, Eli Dwek, Brian Fields, Andrew Hopkins, Shunsaku Horiuchi, Amy Lien, Louie Strigari, Mark Vagins, Terry Walker, Hasan Yüksel, and Pengjie Zhang—as well as to my additional collaborators on closely related topics—Dick Boyd, Will Farr, Joe Formaggio, Matt Kistler, Chris Kochanek, Tony Mezzacappa, José Prieto, Matthew Sharp, Kris Stanek, Yoichiro Suzuki, Todd Thompson, and Petr Vogel—for sharing their knowledge, enthusiasm, and wisdom. I especially thank Shin'ichiro Ando, Andrew Hopkins, Shunsaku Horiuchi, Todd Thompson, Mark Vagins, and Hasan Yüksel for helpful comments on the draft manuscript.

LITERATURE CITED

1. Freedman SJ, et al. arXiv:physics/0411216 (2004)
2. Bethe H, Peierls R. *Nature* 133:532 (1934)
3. Cowan CL, et al. *Science* 124:103 (1956)
4. Reines F, Cowan CL. *Nature* 178:446 (1956)
5. Marx G, Menyhard N. *Science* 131:299 (1960)
6. Pontecorvo B. *Sov. Phys. Usp.* 6:1 (1963)
7. Bahcall JN. *Science* 147:115 (1965)
8. Ruderman MA. *Rep. Prog. Phys.* 28:411 (1965)
9. Bahcall JN. *Phys. Rev. Lett.* 12:300 (1964)
10. Bahcall JN. *Neutrino Astrophysics*. Cambridge, UK: Cambridge Univ. Press. 584 pp. (1989)
11. Learned JG, Mannheim K. *Annu. Rev. Nucl. Part. Sci.* 50:679 (2000)
12. Anchordoqui LA, Montaruli T. arXiv:0912.1035 [astro-ph] (2009)
13. Beatty JJ, Westerhoff S. *Annu. Rev. Nucl. Part. Sci.* 59:319 (2009)
14. Arnett WD, Bahcall JN, Kirshner RP, Woosley SE. *Annu. Rev. Astron. Astrophys.* 27:629 (1989)
15. Hirata K, et al. *Phys. Rev. Lett.* 58:1490 (1987)
16. Hirata KS, et al. *Phys. Rev. D* 38:448 (1988)
17. Bionta RM, et al. *Phys. Rev. Lett.* 58:1494 (1987)
18. Bratton CB, et al. *Phys. Rev. D* 37:3361 (1988)
19. Ando S. *Phys. Lett.* B570:11 (2003)
20. Fogli GL, Lisi E, Mirizzi A, Montanino D. *Phys. Rev. D* 70:013001 (2004)
21. Goldberg H, Perez G, Sarcevic I. *J. High Energy Phys.* 11:023 (2006)

22. Baker J, Goldberg H, Perez G, Sarcevic I. *Phys. Rev. D* 76:063004 (2007)
23. Horiuchi S, Beacom JF, Dwek E. *Phys. Rev. D* 79:083013 (2009)
24. Nakazato K, Sumiyoshi K, Suzuki H, Yamada S. *Phys. Rev. D* 78:083014 (2008)
25. Zel'dovich YB, Guseinov OK. *Sov. Phys. Dokl.* 10:524 (1965)
26. Guseinov OK. *Sov. Astron.* 10:613 (1967)
27. Bisnovatyi-Kogan GS, Seidov ZF. *Sov. Astron.* 26:132 (1982)
28. Domogatskii GV. *Sov. Astron.* 28:30 (1984)
29. Krauss LM, Glashow SL, Schramm DN. *Nature* 310:191 (1984)
30. Bisnovatyi-Kogan GS, Seidov ZF. *Ann. N.Y. Acad. Sci.* 422:319 (1984)
31. Lagage PO. *Nature* 316:420 (1985)
32. Dar A. *Phys. Rev. Lett.* 55:1422 (1985)
33. Woosley SE, Wilson JR, Mayle R. *Astrophys. J.* 302:19 (1986)
34. Schramm DN, Mayle R, Wilson JR. *Nuovo Cim.* C9:443 (1986)
35. Totani T, Sato K, Yoshii Y. *Astrophys. J.* 460:303 (1996)
36. Malaney RA. *Astropart. Phys.* 7:125 (1997)
37. Hartmann DH, Woosley SE. *Astropart. Phys.* 7:137 (1997)
38. Kaplinghat M, Steigman G, Walker TP. *Phys. Rev. D* 62:043001 (2000)
39. Ando S, Sato K, Totani T. *Astropart. Phys.* 18:307 (2003)
40. Fukugita M, Kawasaki M. *Mon. Not. R. Astron. Soc.* 340:L7 (2003)
41. Strigari LE, Kaplinghat M, Steigman G, Walker TP. *J. Cosmol. Astropart. Phys.* 0403:007 (2004)
42. Iocco F, et al. *Astropart. Phys.* 23:303 (2005)
43. Strigari LE, Beacom JF, Walker TP, Zhang P. *J. Cosmol. Astropart. Phys.* 0504:017 (2005)
44. Daigne F, Olive KA, Sandick P, Vangioni E. *Phys. Rev. D* 72:103007 (2005)
45. Lunardini C. *Astropart. Phys.* 26:190 (2006)
46. Ando S, Sato K. *New J. Phys.* 6:170 (2004)
47. Yüksel H, Ando S, Beacom JF. *Phys. Rev. C* 74:015803 (2006)
48. Hansen CJ, Kawaler SD. *Stellar Interiors. Physical Principles, Structure, and Evolution*. Berlin: Springer. 445 pp. (1994)
49. Kennicutt RC. *Astrophys. J.* 277:361 (1984)
50. Heger A, et al. *Astrophys. J.* 591:288 (2003)
51. Filippenko AV. *Annu. Rev. Astron. Astrophys.* 35:309 (1997)
52. Kistler MD, et al. arXiv:0810.1959 [astro-ph] (2008)
53. Raffelt GG. *Stars as Laboratories for Fundamental Physics*. Chicago: Univ. Chicago Press. 664 pp. (1996)
54. Horowitz CJ, Piekarewicz J. *Phys. Rev. Lett.* 86:5647 (2001)
55. Page D, Reddy S. *Annu. Rev. Nucl. Part. Sci.* 56:327 (2006)
56. Lattimer JM, Prakash M. *Phys. Rep.* 442:109 (2007)
57. Zhang W, Woosley SE, Heger A. *Astrophys. J.* 679:639 (2008)
58. Fryer CL. *Astrophys. J.* 699:409 (2009)
59. Mezzacappa A. *Annu. Rev. Nucl. Part. Sci.* 55:467 (2005)
60. Ott CD, Burrows A, Dessart L, Livne E. *Astrophys. J.* 685:1069 (2008)
61. Janka HT, et al. *Phys. Rep.* 442:38 (2007)
62. Huedepohl L, et al. arXiv:0912.0260 [astro-ph] (2009)
63. Yüksel H, Beacom JF. *Phys. Rev. D* 76:083007 (2007)
64. Keil MT, Raffelt GG, Janka HT. *Astrophys. J.* 590:971 (2003)
65. Dighe AS, Smirnov AY. *Phys. Rev. D* 62:033007 (2000)
66. Fetter J, McLaughlin GC, Balantekin AB, Fuller GM. *Astropart. Phys.* 18:433 (2003)
67. Duan H, Fuller GM, Carlson J, Qian YZ. *Phys. Rev. Lett.* 97:241101 (2006)
68. Dasgupta B, Dighe A, Raffelt GG, Smirnov AY. *Phys. Rev. Lett.* 103:051105 (2009)
69. Ando S, Sato K. *Phys. Lett.* B559:113 (2003)
70. Woosley SE, Haxton WC. *Nature* 334:45 (1988)
71. Heger A, et al. *Phys. Lett.* B606:258 (2005)
72. Yoshida T, Kajino T, Hartmann DH. *Phys. Rev. Lett.* 94:231101 (2005)
73. Salpeter EE. *Astrophys. J.* 121:161 (1955)

74. Baldry IK, Glazebrook K. *Astrophys. J.* 593:258 (2003)
75. Kennicutt RC. *Annu. Rev. Astron. Astrophys.* 36:189 (1998)
76. Hopkins AM, Beacom JF. *Astrophys. J.* 651:142 (2006)
77. Yüksel H, Kistler MD, Beacom JF, Hopkins AM. *Astrophys. J.* 683:L5 (2008)
78. Kistler MD, et al. *Astrophys. J.* 705:L104 (2009)
79. Butler NR, Bloom JS, Poznanski D. *Astrophys. J.* 711:495 (2010)
80. Oesch PA, et al. *Astrophys. J.* 709:L16 (2010)
81. Bouwens RJ, et al. *Astrophys. J.* 705:936 (2009)
82. Yan H, et al. arXiv:0910.0077 [astro-ph] (2009)
83. Gal-Yam A, et al. *Astrophys. J.* 656:372 (2007)
84. Li WD, et al. *Astrophys. J.* 661:1013 (2007)
85. Prieto JL, et al. *Astrophys. J.* 681:L9 (2008)
86. Smartt SJ. *Annu. Rev. Astron. Astrophys.* 47:63 (2009)
87. Smartt SJ, Eldridge JJ, Crockett RM, Maund JR. *Mon. Not. R. Astron. Soc.* 395:1409 (2009)
88. Kalirai JS, et al. *Astrophys. J.* 676:594 (2008)
89. Williams KA, Bolte M, Koester D. *Astrophys. J.* 693:355 (2009)
90. Dahlen T, et al. *Astrophys. J.* 613:189 (2004)
91. Dahlen T, et al. Presented at Arcetri Supernova Rates Workshop, Florence. **http://www.arcetri.astro.it/filippo/snrate08/Abstracts_files/Dahlen.pdf** (Abstr.) (2008)
92. Cappellaro E, Evans R, Turatto M. *Astron. Astrophys.* 351:459 (1999)
93. Cappellaro E, et al. *Astron. Astrophys.* 430:83 (2005)
94. Botticella MT, et al. *Astron. Astrophys.* 479:49 (2008)
95. Kochanek CS, et al. *Astrophys. J.* 684:1336 (2008)
96. Lunardini C. *Phys. Rev. Lett.* 102:231101 (2009)
97. Lien A, Fields BD, Beacom JF. *Phys. Rev. D* 81:083001 (2010)
98. Vogel P, Beacom JF. *Phys. Rev. D* 60:053003 (1999)
99. Strumia A, Vissani F. *Phys. Lett.* B564:42 (2003)
100. Fukuda Y, et al. *Nucl. Instrum. Methods A* 501:418 (2003)
101. Malek M, et al. *Phys. Rev. Lett.* 90:061101 (2003)
102. Zhang W, et al. *Phys. Rev. Lett.* 61:385 (1988)
103. Jegerlehner B, Neubig F, Raffelt G. *Phys. Rev. D* 54:1194 (1996)
104. Beacom JF, Vagins MR. *Phys. Rev. Lett.* 93:171101 (2004)
105. Raffelt G, Rashba T. arXiv:0902.4832 [astro-ph] (2009)
106. Vagins MR, Svoboda RC. *Funded proposal to the US Dep. Energy Adv. Detect. Res. Prog.* **http://www.er.doe.gov/hep/hep_ADR/2003ADRWinners.shtml** (2003)
107. Vagins MR. *Funded proposal to the US Dep. Energy Adv. Detect. Res. Prog.* **http://www.er.doe.gov/hep/hep_ADR/2005ADRWinners.shtml** (2005)
108. Vagins MR. *Funded proposal to the US Dep. Energy Adv. Detect. Res. Prog.* **http://www.er.doe.gov/hep/hep_ADR/2007ADRWinners.shtml** (2007)
109. Nakahata M. *J. Phys. Conf. Ser.* 136:022042 (2008)
110. Watanabe H, et al. *Astropart. Phys.* 31:320 (2009)
111. Dazeley S, Bernstein A, Bowden NS, Svoboda R. *Nucl. Instrum. Methods A* 607:616 (2009)
112. Kibayashi A, et al. arXiv:0909.5528 [astro-ph] (2009)
113. Learned JG, Dye ST, Pakvasa S. arXiv:0810.4975 [hep-ex] (2008)
114. Würm M, et al. *Phys. Rev. D* 75:023007 (2007)
115. Bernstein A, et al. arXiv:0907.4183 [hep-ex] (2009)
116. Autiero D, et al. *J. Cosmol. Astropart. Phys.* 0711:011 (2007)
117. Ando S. *Astrophys. J.* 607:20 (2004)
118. Lunardini C. *Phys. Rev. D* 75:073022 (2007)
119. Volpe C, Welzel J. arXiv:0711.3237 [astro-ph] (2007)
120. Chakraborty S, Choubey S, Dasgupta B, Kar K. *J. Cosmol. Astropart. Phys.* 0809:013 (2008)
121. Galais S, Kneller J, Volpe C, Gava J. *Phys. Rev. D* 81:053002 (2009)
122. Beacom JF, Strigari LE. *Phys. Rev. C* 73:035807 (2006)

123. Aharmim B, et al. *Astrophys. J.* 653:1545 (2006)
124. Lunardini C. *Phys. Rev. D* 73:083009 (2006)
125. Aglietta M, et al. *Astropart. Phys.* 1:1 (1992)
126. Lunardini C, Peres OLG. *J. Cosmol. Astropart. Phys.* 0808:033 (2008)
127. Cocco AG, et al. *J. Cosmol. Astropart. Phys.* 0412:002 (2004)
128. Rubbia A. *J. Phys. Conf. Ser.* 171:012020 (2009)

The Color Glass Condensate

Francois Gelis,[1] Edmond Iancu,[1]
Jamal Jalilian-Marian,[2] and Raju Venugopalan[3]

[1]Institut de Physique Théorique, CEA/Saclay, F-91191 Gif-sur-Yvette Cedex, France; email: francois.gelis@cea.fr, edmond.iancu@cea.fr

[2]Department of Natural Sciences, Baruch College, New York, New York 10010 and City University of New York Graduate Center, New York, New York 10016; email: jamal.jalilian-marian@baruch.cuny.edu

[3]Physics Department, Brookhaven National Laboratory, Upton, New York 11973; email: raju@bnl.gov

Annu. Rev. Nucl. Part. Sci. 2010. 60:463–89

The *Annual Review of Nuclear and Particle Science* is online at nucl.annualreviews.org

This article's doi:
10.1146/annurev.nucl.010909.083629

0163-8998/10/1123-0463$20.00

Key Words

quantum chromodynamics, saturation, deep-inelastic scattering, glasma, multiparticle production, quark-gluon plasma, heavy-ion collisions

Abstract

We provide a broad overview of the theoretical status and phenomenological applications of the color glass condensate effective field theory, which describes universal properties of saturated gluons in hadron wave functions that are extracted from deep-inelastic scattering and hadron-hadron collision experiments at high energies.

Contents

1. HIGH-ENERGY QUANTUM CHROMODYNAMICS AND THE COLOR GLASS CONDENSATE

Only 5% of the mass of the universe is visible, but 99% of this visible matter is described by quantum chromodynamics (QCD), the theory of the strong interactions. QCD has been described as a nearly perfect theory, its only parameters being the masses of the current quarks (1). The vast bulk of visible matter therefore comprises emergent phenomena arising from the rich dynamics of the QCD vacuum and the interactions of the fundamental quark and gluon constituents of QCD. Although a wide range of experimental data and numerical lattice computations demonstrate that QCD is the correct theory of the strong interactions, outstanding questions remain about how a variety of striking phenomena arise in the theory, given that the complexity of the theory also makes it very difficult to solve.

Our focus in this review is on high-energy scattering in QCD. A traditional approach to these phenomena is to divide them into hard and soft scattering, which correspond, respectively, to large and small momentum exchanges in the scattering. In the former case, because of the asymptotic freedom of QCD, phenomena such as jets can be computed within a perturbative framework. In the latter case, because the momentum transfer is small, the so-called infrared slavery of QCD suggests that the coupling is large; therefore, the scattering is intrinsically nonperturbative and not amenable to first-principles analysis. This situation is problematic because soft dynamics constitute the bulk of QCD cross sections; in contrast, the perturbatively calculable hard cross sections are rare processes.[1]

We argue that the traditional separation of hard versus soft QCD dynamics is oversimplified because novel semihard scales generated dynamically at high energies allow one to

[1]Lattice QCD is not of much help here because it is best suited to computing static properties of the theory.

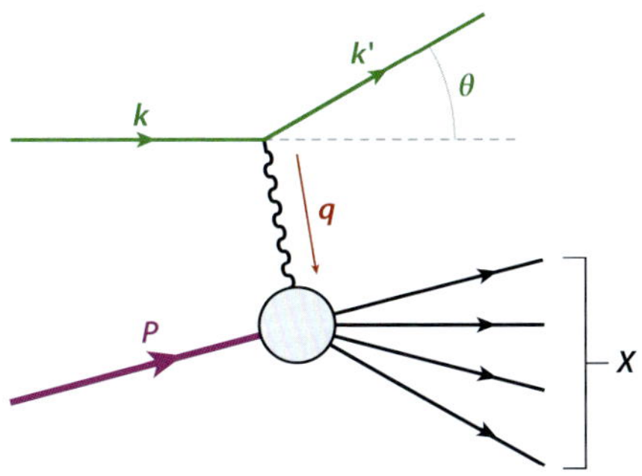

Figure 1

Kinematics of deep-inelastic scattering.

understand highly nonperturbative phenomena in QCD by using weak coupling methods. Specifically, consider inclusive cross sections in deep-inelastic scattering (DIS) experiments of electrons off hadrons. Inclusive cross sections in DIS (**Figure 1**) can be expressed in terms of the Lorentz invariants: (*a*) x, which corresponds at lowest order in perturbation theory to the longitudinal momentum fraction carried by a parton in the hadron; (*b*) the virtual photon momentum squared, $q^2 = -Q^2 < 0$, which is exchanged between the electron and the hadron; (*c*) the inelasticity y, the ratio of the photon energy to the electron energy in the hadron rest frame; and (*d*) the center-of-mass energy squared, s. These invariants satisfy the relation $x = Q^2/sy$. Rather than use the dichotomy of hard versus soft, we consider two asymptotic limits in DIS that better illustrate the QCD dynamics of high-energy hadron wave functions. The first, known as the Bjorken limit, corresponds to fixed x with $Q^2, s \to \infty$. The second is the Regge-Gribov limit of fixed Q^2, $x \to 0$ and $s \to \infty$.

In the Bjorken limit of QCD, one obtains the parton model wherein the hadron is viewed, in the infinite momentum frame (IMF), as a dilute collection of valence quarks and "wee" partons (in the terminology coined by Feynman)—gluons and sea quark pairs. Although the number of wee partons is large, the hadron is dilute (**Figure 2**) because the phase-space density is very small for $Q^2 \to \infty$. In this limit, to leading order (LO) in the coupling constant, the interaction of a probe with the hadron on the characteristic timescale $1/Q$ can be expressed as a hard interaction with an individual parton. In this impulse approximation, the interaction of the struck parton in the hadron with comoving partons is suppressed due to time dilation. The separation of hard and soft scattering (mentioned above) is valid here. Powerful tools such as the operator product expansion and factorization theorems extend to high orders in the coupling constant expansion, the hard-soft separation between process-dependent physics at the scale $1/Q$, and universal nonperturbative parton distribution functions (PDFs) at distance scales on order of the proton size. The evolution of the separation of hard and soft scales is given by renormalization group (RG) equations known as the DGLAP (Dokshitzer-Gribov-Lipatov-Altarelli-Parisi) equations (2).

The study of strong interactions in the Regge-Gribov limit predates QCD and underlies concepts such as soft Pomeron and Reggeon exchanges, whose dynamics—described by Reggeon field theory (RFT)—was believed to encompass many of the phenomena of multiparticle production. Albeit phenomenologically very suggestive (4), the dynamics of RFT is intrinsically nonperturbative and therefore not easily amenable to systematic computations. With the advent of high-energy colliders, hadron structure in the Regge-Gribov limit can be explored with $Q^2 \geq 1$ GeV^2. As strikingly demonstrated by the HERA DIS data (**Figure 2**), the gluon distribution $xG(x, Q^2)$ in a proton rises very fast with decreasing x at large, fixed Q^2—roughly as a power of $x^{-\lambda}$ with $\lambda \simeq 0.25$. In the

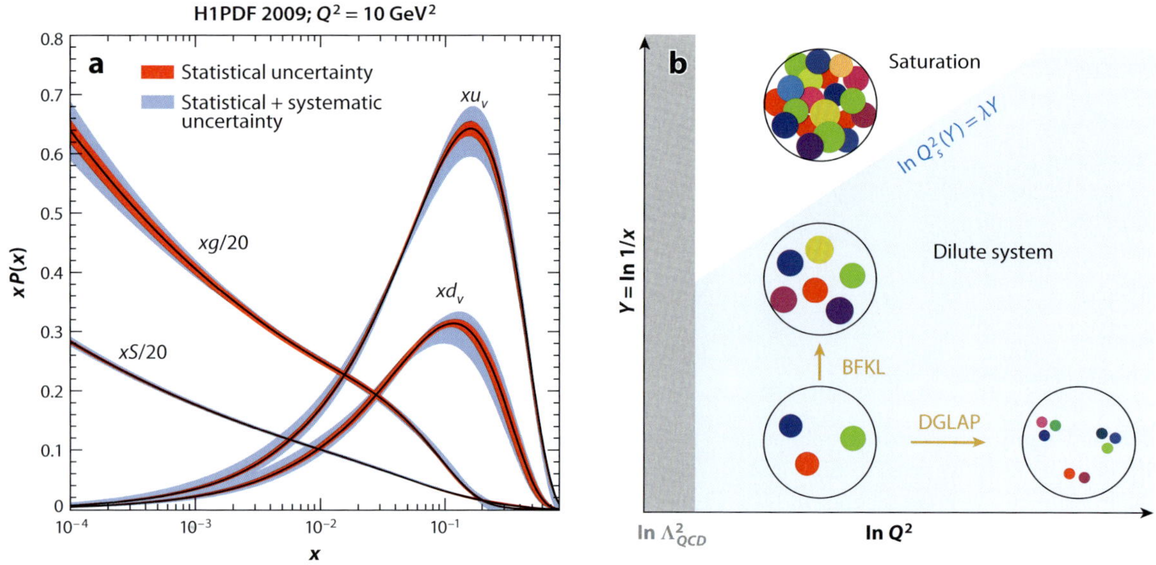

Figure 2

(*a*) The x evolution of the gluon, sea quark, and valence quark distributions for $Q^2 = 10\ \text{GeV}^2$, measured at HERA (3). (*b*) The phase diagram for quantum chromodynamics (QCD) evolution. Each colored dot represents a parton with transverse area $\delta S_\perp \sim 1/Q^2$ and longitudinal momentum $k^+ = xP^+$. Abbreviations: BFKL, Balitsky-Fadin-Kuraev-Lipatov; DGLAP, Dokshitzer-Gribov-Lipatov-Altarelli-Parisi.

IMF of the parton model, $xG(x, Q^2)$ is the number of gluons with a transverse area $\delta S_\perp \geq 1/Q^2$ and a fraction $k^+/P^+ \sim x$ of the proton longitudinal momentum.[2] In the Regge-Gribov limit, the rapid rise of the gluon distribution at small x is given by the BFKL (Balitsky-Fadin-Kuraev-Lipatov) equation (5), which we discuss below.

The stability of the theory formulated in the IMF requires that gluons have a maximal occupation number of order $1/\alpha_s$. This bound is saturated for gluon modes with transverse momenta $k_\perp \leq Q_s$, where $Q_s(x)$ is a semihard scale (i.e., the saturation scale) that increases as x decreases. In this novel saturation regime of QCD (**Figure 2*b***) (6), the proton becomes a dense many-body system of gluons. In addition to the strong-x dependence, the saturation scale Q_s has an A dependence because of the Lorentz contraction of the nuclear parton density in the probe rest frame. The dynamics of gluons in the saturation regime is nonperturbative, as is typical of strongly correlated systems. However, in a fundamental departure from RFT, this dynamics can be computed by use of weak coupling methods as a consequence of the large saturation scale that is dynamically generated by gluon interactions. Thus, instead of the hard-plus-soft paradigm of the Bjorken limit, one has a powerful new paradigm in the Regge-Gribov limit to compute the bulk of previously considered intractable scattering dynamics in hadrons and nuclei.

The color glass condensate (CGC) is the description of the properties of saturated gluons in the IMF in the Regge-Gribov limit. The effective degrees of freedom in this framework are color sources ρ at large x and gauge fields A^μ at small x. At high energies, because of time dilation, the former are frozen configurations on the natural timescales of the strong interactions and are distributed randomly from event to event. The latter are dynamical fields coupled to the static

[2] The light-cone coordinates are defined as $k^\pm = (k^0 \pm k^3)/\sqrt{2}$.

color sources. It is the stochastic nature of the sources, combined with the separation of timescales, that justifies the glass appellation. The condensate designation comes from the fact that in the IMF, saturated gluons have large occupation numbers $\mathcal{O}(1/\alpha_s)$, where typical momenta peak about a characteristic value of $k_\perp \sim Q_s$. The dynamical features of the CGC are captured by the Jalilian-Marian–Iancu–McLerran–Weigert–Leonidov–Kovner (JIMWLK) RG equation that describes how the statistical distribution $W[\rho]$ of the fast sources at a given x scale evolves when x decreases. The JIMWLK RG is Wilsonian in nature because weakly coupled fields integrated out at a given step in the evolution are interpreted as induced color charges that modify the statistical weight distribution.

The CGC framework is powerful because, given an initial nonperturbative distribution of sources at an initial scale x_0, it allows one to compute n-point gluon correlation functions and their evolution with x. In analogy to PDFs in the Bjorken limit, the distribution $W[\rho]$ captures the properties of saturated gluons. However, unlike PDFs, which in the IMF are parton densities, $W[\rho]$ carries much more information. Further, in contrast to the twist expansion of the Bjorken limit, which becomes extremely cumbersome beyond LO in $1/Q$, the CGC framework includes all twist correlations. Like the PDFs, the Ws are universal: The same distributions appear in computations of inclusive quantities in both lepton-nucleus and hadron-nucleus collisions.

An interesting aspect of the CGC is what it reveals about the collective dynamics of QCD at high parton densities. The CGC RG equations indicate that—at fixed impact parameter—a proton and a heavy nucleus become indistinguishable at high energy; the physics of saturated gluons is universal and independent of the details of the fragmentation region. This universal dynamics has a correspondence with reaction-diffusion processes in statistical physics; in particular, it may lie in a spin glass universality class. Understanding the nature of color screening and long-range order in this universal dynamics offers possibilities for progress in resolving fundamental QCD questions regarding properties such as confinement and chiral symmetry. A specific area of progress is in the mapping of the CGC degrees of freedom to the traditional language of Pomerons and the consequent prospects for understanding soft QCD dynamics. In nuclear collisions, CGC dynamics produces so-called glasma-field configurations at early times: strong longitudinal chromoelectric and chromomagnetic fields screened on transverse distance scales $1/Q_s$. These fields generate long-range rapidity correlations, sphaleron-like topological fluctuations characterized by a large Chern-Simons charge, and instabilities analogous to those seen in quantum electrodynamics plasmas.

The CGC framework is applicable to a wide variety of processes in $e+p$, $e+A$, $p+A$, and $A+A$ collisions. It provides an ab initio approach by which to study thermalization in heavy-ion collisions and the initial conditions for the evolution of a thermalized quark-gluon plasma (QGP). The interaction of hard probes with the glasma is little understood and is important for quantifying the transport properties of the QGP precursor. A further phenomenological application of the CGC is to the physics of ultrahigh-energy cosmic rays (7).

There are several reviews that discuss various aspects of the CGC and its applications in depth (8–10). Our aim is to provide a broad pedagogical overview of the current status of theory and phenomenology for nonexperts. Section 2 focuses on the theoretical status, and Section 3 discusses the CGC in the context of DIS, hadronic, and nuclear collisions. Section 4 presents the experimental results that probe the dynamics of saturated gluons. We end this review with a brief summary and outlook.

2. COLOR GLASS CONDENSATE: THEORETICAL STATUS

We begin with an elementary discussion of QCD bremsstrahlung in the Regge-Gribov limit and use this discussion to motivate the BFKL equation, discuss how this leads to gluon saturation, and describe its key features. We then present the theoretical status of the CGC effective theory, and finally we end with a discussion of more advanced topics that highlight the open issues.

2.1. Parton Evolution at High Energy

The structure of a hadron depends upon the scales resolved by an external probe. Quantum fields bound within the hadron can radiate other virtual quanta that, according to the uncertainty principle, live for only a short period of time. Therefore, what appears to be an isolated quantum at a given resolution scale reveals a complicated substructure when probed with finer space-time resolution. The structure of the hadron in the IMF is specified by the longitudinal (k^+) and transverse ($k_\perp$) momentum distributions of its quanta and by their correlations—the latter become increasingly important with increasing density.

In perturbative QCD, parton evolution proceeds via bremsstrahlung, which favors the emission of soft and collinear gluons. **Figure 3*a*** illustrates this elementary radiation process. When $x \ll 1$, to lowest order in α_s, the differential probability for this emission is given by

$$dP_{\text{Brem}} \simeq \frac{\alpha_s C_R}{\pi^2}\frac{d^2k_\perp}{k_\perp^2}\frac{dx}{x}, \qquad 1.$$

where C_R is the $SU(N_c)$ Casimir in the color representation of the emitter: N_c for a gluon and $(N_c^2-1)/2N_c$ for a quark. This formula demonstrates the collinear ($k_\perp \to 0$) and soft ($x \to 0$) singularities mentioned above, which produce a logarithmic enhancement of gluon emission at small $k_\perp$ and/or x. If the emitter at small x were a quark instead a gluon, there would be no small-x enhancement, whereas the collinear enhancement would be present for either emitter. This asymmetry is due to the spin-1 nature of the gluon.

A high-energy scattering such as that illustrated in **Figure 3*b*** probes the number of gluons with a given x and transverse momenta $k_\perp \le Q$. The gluon number (from Equation 1) is

$$x\frac{dN_g}{dx}(Q^2) = \frac{\alpha_s C_R}{\pi}\int_{\Lambda_{\text{QCD}}^2}^{Q^2}\frac{dk_\perp^2}{k_\perp^2} = \frac{\alpha_s C_R}{\pi}\ln\left(\frac{Q^2}{\Lambda_{\text{QCD}}^2}\right), \qquad 2.$$

where the cutoff Λ_{QCD} has been introduced as a crude way to account for confinement: When confined inside a hadron, a parton has a minimum virtuality of $\mathcal{O}(\Lambda_{\text{QCD}}^2)$.

In QCD, gluons can radiate softer gluons and thereby rapidly multiply (**Figure 3**). Each subsequent emission is α_s suppressed, but when the final value of x is small, such corrections

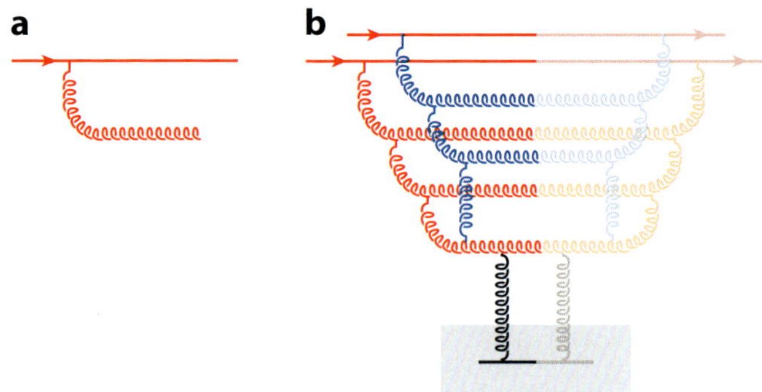

Figure 3

(*a*) Elementary bremsstrahlung radiation. (*b*) High-energy scattering with evolution and recombination. The faded sections denote partons in the complex conjugate amplitude.

become large. For a cascade with n intermediate gluons strongly ordered in x, one obtains

$$\alpha_s^n \int_x^1 \frac{\mathrm{d}x_n}{x_n} \int_{x_n}^1 \frac{\mathrm{d}x_{n-1}}{x_{n-1}} \cdots \int_{x_2}^1 \frac{\mathrm{d}x_1}{x_1} = \frac{1}{n!}\left(\alpha_s \ln \frac{1}{x}\right)^n. \qquad 3.$$

When $\alpha_s \ln(1/x) \gtrsim 1$, the correct result for the gluon distribution is obtained by summing contributions from all such ladders. The sum exponentiates, modifying Equation 2 into

$$x\frac{\mathrm{d}N_g}{\mathrm{d}x\mathrm{d}k_\perp^2} \sim \frac{\alpha_s C_R}{\pi}\frac{1}{k_\perp^2}\mathrm{e}^{\omega\alpha_s Y}, \quad Y \equiv \ln\frac{1}{x}, \qquad 4.$$

where ω is a number of order unity that is not fixed by this rough estimate. The variable Y is known as the rapidity.

To go beyond this simple power-counting argument, one must more accurately treat the kinematics of the ladder diagrams and include the associated virtual corrections. The result is the above-mentioned BFKL equation for the Y evolution of the unintegrated gluon distribution. The solution of this equation, which resums perturbative corrections $(\alpha_s Y)^n$ to all orders, confirms the exponential increase in Equation 4, albeit with a $k_\perp$-dependent exponent and modifications to the $k_\perp^{-2}$ spectrum of the emitted gluons.

An important property of the BFKL ladder is coherence in time. Because the lifetime of a parton in the IMF is $\Delta x^+ \sim k^+/k_\perp^2 \propto x$, the "slow" gluons at the lower end of the cascade have a much shorter lifetime than the preceding "fast" gluons. Therefore, for the purposes of small-x dynamics, fast gluons with $x' \gg x$ act as frozen color sources emitting gluons at the scale x. Because these sources may overlap in the transverse plane, their color charges add coherently, giving rise to a large color charge density. The average color charge density is zero by gauge symmetry, but fluctuations in the color charge density are nonzero and increase rapidly with $1/x$.

These considerations are at the heart of the CGC reformulation of BFKL evolution. However, in contrast to the original BFKL formulation, the CGC formalism (11–15) includes nonlinear effects that appear when the gluon density becomes large.

The quantity that controls gluon interactions in the IMF is their occupation number—the number of gluons of a given color per unit transverse phase space and per unit rapidity:

$$n(Y, k_\perp, b_\perp) \equiv \frac{(2\pi)^3}{2(N_c^2 - 1)}\frac{\mathrm{d}N_g}{dY\,\mathrm{d}^2\boldsymbol{k}_\perp \mathrm{d}^2\boldsymbol{b}_\perp}, \qquad 5.$$

where the impact parameter $\boldsymbol{b}_\perp$ is the gluon position in the transverse plane. If $n \ll 1$, the system is dilute, and gluon interactions become negligible. When $n \sim \mathcal{O}(1)$, gluons begin to overlap, but their interactions are suppressed by $\alpha_s \ll 1$. The interaction strength becomes of order unity when $n \sim \mathcal{O}(1/\alpha_s)$—it is then that nonlinear effects become important, leading to gluon saturation.

Gluon occupancy is further amplified if, instead of a proton, we consider a large nucleus with atomic number $A \gg 1$. The transverse area scales as $A^{2/3}$, and the gluon occupation number scales as $A^{1/3}$. Thus, for a large nucleus, saturation effects become important at larger values of x than for a proton.

More specifically, let us discuss a simplified version of the nonlinear evolution equation for the occupation number. Consider an elementary increment in rapidity: $Y \rightarrow Y + \mathrm{d}Y$. Each preexisting gluon in the hadron has a probability $\alpha_s \mathrm{d}Y$ of emitting an additional soft gluon—the average increase is $\mathrm{d}n \sim \alpha_s n \mathrm{d}Y$. Moreover, the emission vertex is nonlocal in $k_\perp$ because the transverse momentum of the parent gluon is shared among the two daughter gluons. At high energies, this nonlocality is well approximated as a diffusion in the logarithmic variable $t \equiv \ln(k_\perp^2)$. Finally, two preexisting gluons can merge and produce a single final gluon with rapidity $Y + \mathrm{d}Y$. This process is quadratic in $\alpha_s n$ and leads to a negative term in the evolution equation. Adding up these three effects, one obtains the evolution equation

$$\frac{\mathrm{d}n}{\mathrm{d}Y} \simeq \omega\alpha_s n + \chi\alpha_s \mathrm{d}_t^2 n - \beta\alpha_s^2 n^2, \qquad 6.$$

where ω, χ, and β are numbers of order unity. This equation mimics the BFKL equation (5), if one drops the quadratic term on the right-hand side, and mimics its nonlinear extensions that include saturation, if one keeps the quadratic term. In the next section, we further discuss the essential features of Equation 6.

2.2. Generic Features of Gluon Saturation

Although it is a toy model, Equation 6 captures essential features of saturation.[3] If n is not too large ($\alpha_s n \ll 1$), then one can neglect the quadratic term, and Equation 6 predicts an exponential growth in Y of the gluon occupation number. But when $\alpha_s n \sim 1$, the negative nonlinear term turns on and tames the growth. Equation 6 has a fixed point at $n = \omega/(\beta\alpha_s)$, where its right-hand side vanishes—the evolution stops when this value of n is reached, resulting in gluon saturation in the spirit of early works (6).

Equation 6 also reveals the emergence of a transverse momentum scale $Q_s(Y)$, which characterizes saturation. This scale is the $k_\perp$ at which gluon occupancy becomes of order $\mathcal{O}(1/\alpha_s)$. As a function of transverse momentum, the occupation number $n(Y, k_\perp)$ is $\mathcal{O}(1/\alpha_s)$ if $k_\perp \lesssim Q_s(Y)$, and it decreases rapidly above $Q_s(Y)$ [$n \propto 1/k_\perp^2$ for $k_\perp \gg Q_s(Y)$]. The shape of $n(Y, k_\perp)$ as a function of $k_\perp$ is known as the saturation front. Note that Q_s is the typical gluon transverse momentum at the rapidity Y; that is, Q_s increases with energy and becomes a semihard scale at sufficiently high energy, where $Q_s(Y) \gg \Lambda_{\mathrm{QCD}}$.

Equation 6 belongs to the generic class of reaction-diffusion processes (19). These are processes in which (*a*) an entity can hop to neighboring locations (diffusion term $\chi\alpha_s \partial_t^2 n$), (*b*) an entity can split into two identical entities (term $\omega\alpha_s n$), and (*c*) two entities can merge into a single one (term $-\beta\alpha_s^2 n^2$). In the limit of large occupation numbers, these processes admit the mean-field description of Equation 6, which in the context of statistical physics is known as the Fisher-Kolmogorov-Petrovsky-Piscounov (FKPP) equation (19). In QCD, the closest equation of this type is the Balitsky-Kovchegov (BK) equation (13). The correspondence between the BK and FKPP equations, first identified in Reference 20, clarifies the properties of saturation fronts in QCD in analogy with known properties of reaction-diffusion processes.

A crucial property is the emergence of traveling waves. The saturation front generated by this equation propagates without distortion at constant speed; one has $n(Y, t) = n(t - \lambda_s Y)$, where λ_s is a constant. This property has been verified in numerical studies of the BK equation (21, 22) and in analytic studies of the BFKL equation in the presence of a saturation boundary (23, 24). It provides a natural explanation of the geometric scaling phenomenon observed in the HERA data (25, 26); see Section 3.1. In QCD, a front moving with constant speed λ_s is equivalent to the saturation momentum increasing exponentially with Y,

$$Q_s^2(Y) \simeq Q_0^2 \mathrm{e}^{\lambda_s Y} \quad \text{with} \quad \lambda_s \approx 4.9\alpha_s, \qquad 7.$$

where Q_0 is some nonperturbative initial scale. For a large nucleus, Q_0^2 scales as $A^{1/3}$, as does $Q_s^2(Y)$ for any Y. This form of the saturation momentum is modified to $Q_s^2 = Q_0^2 e^{\sqrt{\lambda(Y-Y_0)}}$ when

[3]Our discussion oversimplifies the mechanism for gluon saturation. In the saturation regime, gluons form configurations that screen their color charge over transverse scales $\sim 1/Q_s$ (16, 17). Thus, gluons with momentum $k_\perp \lesssim Q_s$ are emitted from a quasi-neutral patch of color sources, and their occupation number grows only linearly in Y (16, 18)—much slower than the exponential growth in the region $k_\perp \gg Q_s$.

the running of the strong coupling is taken into account; see Reference 24 for a detailed study of higher-order effects on the energy dependence of Q_s.

2.3. The Color Glass Condensate

The CGC is an effective field theory (EFT) based on the separation of the degrees of freedom into fast frozen color sources and slow dynamical color fields (11). A RG equation—the JIMWLK equation (14, 15)—ensures the independence of physical quantities with respect to the cutoff that separates the two kinds of degrees of freedom.

The fast gluons with longitudinal momentum $k^+ > \Lambda^+$ are frozen by Lorentz time dilation in configurations specified by a color current $J^\mu_a \equiv \delta^{\mu+}\rho^a$, where $\rho^a(x^-, x_\perp)$ is the corresponding color charge density. However, slow gluons with $k^+ < \Lambda^+$ are described by the usual gauge fields A^μ of QCD. Because of the hierarchy in k^+ between these two types of degrees of freedom, they are coupled eikonally by the term $J_\mu A^\mu$. The fast gluons thus act as sources for the fields that represent the slow gluons. Although it is frozen for the duration of a given collision, the color source density ρ^a varies randomly event by event. The CGC provides a gauge-invariant distribution $W_{\Lambda^+}[\rho]$, which gives the probability of a configuration ρ. This weight functional encodes all the correlations of the color charge density at the cutoff scale Λ^+ that separates the fast and slow degrees of freedom. Given this statistical distribution, the expectation value of an operator at the scale Λ^+ is given by

$$\langle \mathcal{O} \rangle_{\Lambda^+} \equiv \int [D\rho] W_{\Lambda^+}[\rho] \mathcal{O}[\rho], \qquad 8.$$

where $\mathcal{O}[\rho]$ is the expectation value of the operator for a particular configuration ρ of the color sources.

The power counting of the CGC EFT is such that in the saturated regime the sources ρ are of order g^{-1}. Attaching an additional source to a given Feynman graph does not alter its order in g; the vertex at which this new source attaches to the graph is compensated by the g^{-1} of the source. Thus, computing an observable at a certain order in g^2 requires the resummation of all the contributions obtained by adding extra sources to the relevant graphs. The LO in g^2 is given by a sum of tree diagrams, which can be expressed in terms of classical solutions of the Yang-Mills equations. Moreover, for inclusive observables (27), these classical fields obey a simple boundary condition: They vanish when $t \to -\infty$.

Next-to-leading-order (NLO) computations in the CGC EFT involve a sum of one-loop diagrams embedded in the above classical field. To prevent double counting, momenta in loops are required to be below the cutoff Λ^+, which leads to a logarithmic dependence in Λ^+ of these loop corrections. These logarithms are large if Λ^+ is well above the typical longitudinal momentum scale of the observable considered, and they must be resummed.

For gluon correlations inside the hadron wave function, and also for sufficiently inclusive observables in a collision, the leading logarithms are universal and can be absorbed into a redefinition of the distribution $W_{\Lambda^+}[\rho]$ of the hard sources. The evolution of $W_{\Lambda^+}[\rho]$ with Λ^+ is governed by the functional JIMWLK equation

$$\frac{\partial W_{\Lambda^+}[\rho]}{\partial \ln(\Lambda^+)} = -\mathcal{H}\left[\rho, \frac{\delta}{\delta\rho}\right] W_{\Lambda^+}[\rho], \qquad 9.$$

where $\mathcal{H}$ is known as the JIMWLK Hamiltonian. This operator contains up to two derivatives $\partial/\partial\rho$ and arbitrary powers in ρ. Its explicit expression can be found in References 8, 14, and 15. We sketch the derivation of the JIMWLK equation in Section 3.1.

Numerical studies of JIMWLK evolution have been performed (22, 28). An analytic, albeit formal, solution to the JIMWLK equation has been constructed (29) in the form of a path integral. Alternatively, the evolution can be expressed as an infinite hierarchy of coupled nonlinear equations for n-point Wilson line correlators—often termed the Balitsky hierarchy (12). In this framework, the BK equation is a mean-field approximation of the JIMWLK evolution and is valid in the limit of a large number of colors $N_c \to \infty$. Numerical studies of the JIMWLK equation (22, 28) have found only small differences with the BK equation.

Finally, we comment on the initial condition for the JIMWLK equation, which is also important in understanding its derivation. The evolution should start at some cutoff value in the longitudinal momentum scale Λ_0^+ at which the saturation scale is already a (semi)hard scale, say $Q_{s0} \gtrsim 1$ GeV, for perturbation theory to be applicable. The gluon distribution at the starting scale is generally nonperturbative and requires a model. A physically motivated model for the gluon distribution in a large nucleus is the McLerran-Venugopalan (MV) model (11). In a large nucleus, there is a window in rapidity in which evolution effects are not large but x is still sufficiently small for a probe not to resolve the longitudinal extent of the nucleus. In this case, the probe "sees" a large number of color charges proportional to $A^{1/3}$. These charges add up to form a higher-dimensional representation of the gauge group, and they can therefore be treated as classical color distributions (11, 30). Further, the color charge distribution $W_{\Lambda_0^+}[\rho]$ is a Gaussian distribution[4] in ρ. The variance of this distribution—the color charge squared per unit area—is proportional to $A^{1/3}$ and provides a semihard scale that makes weak coupling computations feasible. In addition to its role in motivating the EFT and serving as the initial condition in JIMWLK evolution, the MV model allows for direct phenomenological studies in $p+A$ and $A+A$ collisions in regimes in which the values of x are not so small as to require evolution.

The dynamics of small-x gluons in QCD may be universal in more than one sense (33). A weak form of this universality is that their dynamics in both hadrons and nuclei is controlled only by the saturation scale with its particular dependence on energy and nuclear size. A stronger form of universality can be deduced from the solution of the running-coupling BK equation; the saturation scale for both hadrons and nuclei, at fixed impact parameter, becomes the same asymptotically with increasing energy (34). The strongest form of universality is that the RG flows in the saturation regime have a fixed point corresponding to universal critical exponents that describe the behavior of multiparton correlation functions. As discussed further below, the RG equations for high-energy QCD lie in a wide class of reaction-diffusion processes whose universal properties are remarkably close to those of spin glasses (35).

2.4. Advanced Theory Topics

Reaction-diffusion processes exhibit an extreme sensitivity to particle-number fluctuations (36–38) generated by gluon splittings, which produce correlations among pairs of gluons (39). This effect is of higher order in α_s and is linear in n because it results from the splitting of a single gluon. Conversely, producing two gluons without this splitting leads to a term that has one less power of α_s but is $\mathcal{O}(n^2)$. Thus, the splitting contribution is important in the dilute regime, where $n \lesssim \alpha_s$. Because, as mentioned previously, the dynamics of the saturation front is driven by the BFKL growth of its dilute tail, these fluctuations play an important role in the two-gluon density $\langle nn \rangle$ at high energy.

[4]There is a additional term, corresponding to the cubic Casimir, that is parametrically suppressed for large nuclei (31). This term generates Odderon excitations in the JIMWLK/BK evolution (32).

As shown in Equation 6, this two-gluon density enters in the nonlinear term, leading to saturation of the single-gluon density. Thus, gluon-number fluctuations in the dilute regime can strongly influence the approach toward saturation. For instance, as has been argued (36) for generic reaction-diffusion processes and, independently, in the QCD context (37), these fluctuations reduce the (average) speed of the saturation front. In addition to making the value of Q_s a fluctuating quantity, the fluctuations tend to wash out the geometric scaling property of the individual fronts (38, 39). Both effects are quantitatively important, as shown by explicit numerical simulations within various reaction-diffusion models [including those inspired by the QCD dynamics at high energy (40)].

There is presently no general theory that includes BFKL ladders, saturation, and fluctuations. [In terms of Feynman graphs, this corresponds to resumming Pomeron loop diagrams to all orders (39).] Attempts to construct such a theory (39, 41, 42) have led to incomplete formalisms that are difficult to exploit in phenomenological applications. Fortunately, the effect of these fluctuations can be considerably reduced by the running of the strong coupling (43), which tends to postpone their importance to unrealistically large rapidities. Thus, for practical applications at least up to LHC energies, a sufficient theory for the approach toward saturation is the LO mean-field evolution extended with running-coupling corrections. This theory developed significantly with the construction of the running-coupling version of the BK equation (44), and it has been successfully applied to studies of DIS phenomenology at HERA (45).

3. COLLISIONS IN THE COLOR GLASS CONDENSATE FRAMEWORK

The CGC is an effective theory for the wave function of a high-energy hadron or nucleus. In this section, we apply it, with particular emphasis on factorization, to DIS and hadronic collisions. In $A+A$ collisions, the formation of the glasma and its key features are emphasized.

3.1. The Color Glass Condensate and Deep-Inelastic Scattering at Small x

At small x, corresponding to large Ioffe times (46), DIS is characterized by the fluctuation of the virtual photon into a quark-antiquark pair which then scatters off the hadronic or nuclear target. The inclusive DIS cross section can be expressed as (47)

$$\sigma_{\gamma^* T} = \int_0^1 \mathrm{d}z \int \mathrm{d}^2\boldsymbol{r}_\perp |\psi(z, \boldsymbol{r}_\perp)|^2 \sigma_{\text{dipole}}(x, \boldsymbol{r}_\perp), \qquad 10.$$

where $\psi(z, \boldsymbol{r}_\perp)$ is the $q\bar{q}$ component of the wave function of the virtual photon (known from quantum electrodynamics) and $\sigma_{\text{dipole}}(x, \boldsymbol{r}_\perp)$ is the QCD dipole cross section for the quark-antiquark pair to scatter off the target. This process is shown in **Figure 4**, where we have assumed that the target moves in the $-z$ direction. In the LO CGC description of DIS, the target is described (**Figure 5**) as static sources with $k^- > \Lambda_0^-$. The field modes do not contribute at this order.

By employing the optical theorem, we can express $\sigma_{\text{dipole}}(x, \boldsymbol{r}_\perp)$ in terms of the forward scattering amplitude $T(\boldsymbol{x}_\perp, \boldsymbol{y}_\perp)$ of the $q\bar{q}$ pair at LO as

$$\sigma^{\text{LO}}_{\text{dipole}}(x, \boldsymbol{r}_\perp) = 2\int \mathrm{d}^2\mathbf{b} \int [D\rho] W_{\Lambda_0^-}[\rho] \boldsymbol{T}_{LO}\left(\boldsymbol{b} + \frac{\boldsymbol{r}_\perp}{2}, \boldsymbol{b} - \frac{\boldsymbol{r}_\perp}{2}\right), \qquad 11.$$

where, for a fixed configuration of the target color sources (48, 49),

$$\boldsymbol{T}_{\text{LO}}(\boldsymbol{x}_\perp, \boldsymbol{y}_\perp) = 1 - \frac{1}{N_c}\text{tr}[U(\boldsymbol{x}_\perp) U^\dagger(\boldsymbol{y}_\perp)]. \qquad 12.$$

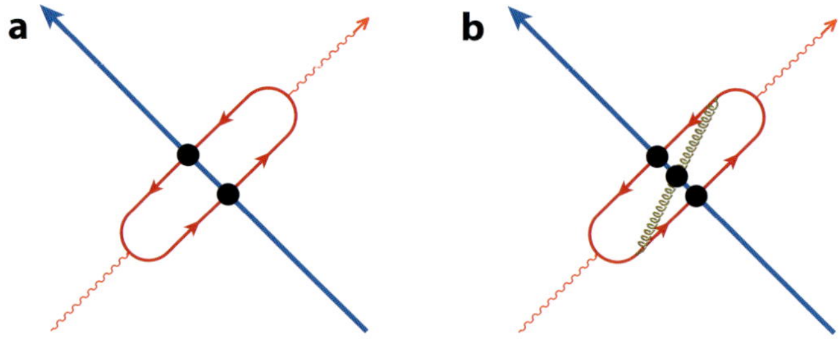

Figure 4

(*a*) Leading-order contribution to deep-inelastic scattering off the color glass condensate. (*b*) Next-to-leading-order contribution.

Here $U(\boldsymbol{x}_\perp)$ is a Wilson line that represents the interaction between a quark and the color fields of the target, defined as

$$U(\boldsymbol{x}_\perp) = T \exp ig \int^{1/xP^-} \mathrm{d}z^+ \mathcal{A}^-(z^+, \boldsymbol{x}_\perp). \qquad 13.$$

In this formula, $\mathcal{A}^-$ is the minus component of the gauge field generated (in Lorentz gauge) by the sources of the target; it is obtained by solving classical Yang-Mills equations with these sources. The upper bound xP^- (where P^- is the target longitudinal momentum and x is the kinematic variable defined in **Figure 1**) indicates that source modes with $k^- < xP^-$ do not contribute to this scattering amplitude. Thus, if the cutoff Λ_0^- of the CGC EFT is lower than xP^-, then $\boldsymbol{T}_{\mathrm{LO}}$ is independent of Λ_0^-.

However, when Λ_0^- is larger than xP^-, the dipole cross section is independent of x (because the CGC EFT does not have source modes near the upper bound xP^-) and depends on the unphysical parameter Λ_0^-. As we discuss below, this is related to the fact that Equation 11 is incomplete and receives large corrections from higher-order diagrams. Consider now the NLO contributions (one of them is shown in **Figure 4*b***) with gauge-field modes in the slice $\Lambda_1^- \le k^- \le \Lambda_0^-$ (**Figure 5**).

An explicit computation of the contribution of field modes in this slice gives

$$\delta \boldsymbol{T}_{\mathrm{NLO}}(\boldsymbol{x}_\perp, \boldsymbol{y}_\perp) = \ln\left(\frac{\Lambda_0^-}{\Lambda_1^-}\right) \mathcal{H}\, \boldsymbol{T}_{\mathrm{LO}}(\boldsymbol{x}_\perp, \boldsymbol{y}_\perp), \qquad 14.$$

where $\mathcal{H}$ is the JIMWLK Hamiltonian. All dependence on the cutoff scales is in the logarithmic prefactor alone. This Hamiltonian has two derivatives with respect to the classical field $\mathcal{A} \sim \mathcal{O}(1/g)$; $\mathcal{H}\boldsymbol{T}_{\mathrm{LO}}$ is of order $\alpha_s \boldsymbol{T}_{\mathrm{LO}}$ and therefore is clearly an NLO contribution.

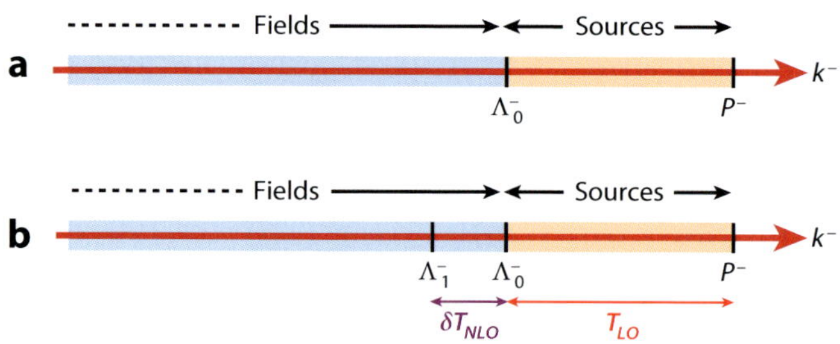

Figure 5

(*a*) Sources and fields in the color glass condensate effective theory. (*b*) Next-to-leading-order (NLO) correction from a layer of field modes just below the cutoff. Abbreviation: LO, leading order.

However, if the new scale Λ_1^- is such that $\alpha_s \ln(\Lambda_0^-/\Lambda_1^-) \sim 1$, this NLO term becomes comparable in magnitude to the LO contribution. By averaging the sum of LO and NLO contributions over the distribution of sources at the scale Λ_0^-, one obtains

$$\int [D\rho] W_{\Lambda_0^-}[\rho](\boldsymbol{T}_{\rm LO} + \delta \boldsymbol{T}_{\rm NLO}) = \int [D\rho] W_{\Lambda_1^-}[\rho] \boldsymbol{T}_{\rm LO}, \qquad 15.$$

where $W_{\Lambda_1^-} \equiv [1 + \ln(\Lambda_0^-/\Lambda_1^-)\mathcal{H}] W_{\Lambda_0^-}$. The NLO correction from quantum modes in the slice $\Lambda_1^- \leq k^- \leq \Lambda_0^-$ can thus be absorbed in the LO term, provided that we now use a CGC effective theory at Λ_1^- with the modified distribution of sources shown in Equation 15. In differential form, the evolution equation of the source distribution,

$$\frac{\mathrm{d}}{\mathrm{d}\ln(\Lambda^-)} W_{\Lambda^-} = -\mathcal{H}\, W_{\Lambda^-}, \qquad 16.$$

is the JIMWLK equation.

Repeating this elementary step, one progressively resums quantum fluctuations down to the scale $k^- \sim x P^-$. Thanks to Equation 15, the result of this resummation for the dipole cross section is formally identical to Equation 11, except that the source distribution is W_{xP^-} instead of $W_{\Lambda_0^-}$. Note that if one further lowers the cutoff below xP^-, the dipole cross section remains unchanged.

3.2. The Color Glass Condensate in *p+A* Collisions

Collisions between a dilute hadron projectile and a dense hadron target can be studied semianalytically in the CGC framework. The archetype of such collisions is a proton-nucleus collision. However, the dilute-dense treatment also applies to proton-proton collisions for measurements at forward rapidities where the wave function of one of the projectiles is probed at large x and that of the other at small x. These asymmetrical collisions can be treated by using conventional parton distributions for the proton and the CGC for the nucleus.

The simplest quantity to compute in this context is the single-particle-inclusive spectrum. In the CGC framework, this process arises from the scattering of a parton from the proton off the color field of the target nucleus (**Figure 6*a***). When the parton is a quark, the amplitude for the scattering is proportional to the Fourier transform of a Wilson line,

$$\mathcal{M}_{\rm LO} \propto \int \mathrm{d}^2\boldsymbol{x}_\perp e^{i\boldsymbol{p}_\perp\cdot\boldsymbol{x}_\perp} U(\boldsymbol{x}_\perp). \qquad 17.$$

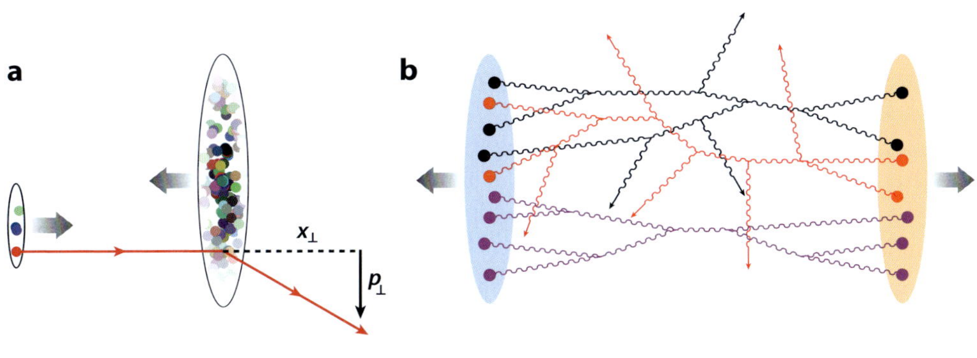

Figure 6

(*a*) Sketch of a proton-nucleus collision. (*b*) Example of a leading-order contribution in a nucleus-nucleus collision.

By squaring this amplitude, summing over the color of the quark in the final state, and averaging over the color of the incoming quark, we get

$$|\mathcal{M}|^2_{\rm LO} \propto \int d^2\boldsymbol{b}\, d^2\boldsymbol{r}_\perp\, e^{i\boldsymbol{p}_\perp\cdot\boldsymbol{r}_\perp}\, \boldsymbol{T}_{\rm LO}\left(\boldsymbol{b}+\frac{\boldsymbol{r}_\perp}{2}, \boldsymbol{b}-\frac{\boldsymbol{r}_\perp}{2}\right), \tag{18}$$

where $\boldsymbol{T}_{\rm LO}$ is the dipole scattering amplitude (see Equation 12) that already appeared in our discussion of DIS. Because the same quantity appears here, the treatment of NLO corrections we described in the DIS case is similarly applicable; one integrates out softer modes by lowering the cutoff of the CGC EFT, letting $W_{\Lambda^-}[\rho]$ evolve according to the JIMWLK equation. The scale to which one evolves the cutoff is $\Lambda^- = xP^-$ with $x = (p_\perp/\sqrt{s})\exp(-y)$, where $p_\perp$ is the transverse momentum of the scattered parton and y is its rapidity. Therefore, in the CGC framework, the cross section for this process is simply the Fourier transform of the dipole cross section $\sigma_{\rm dipole}$ $(x, \boldsymbol{r}_\perp)$. p+A collisions with final states that contain a photon or a lepton pair can be similarly expressed in terms of the same Fourier transform (50).

An additional contribution to the single-particle-inclusive spectrum in a p+A collision is due to an incoming gluon instead of a quark. The treatment is nearly identical to the incoming quark case, except that in Equation 17 one must replace the Wilson line in the fundamental representation by a Wilson line in the adjoint representation.

Similar calculations can be performed for processes with more complicated final states, such as the production of a quark-antiquark pair. Although this observable has a more complicated expression—it contains terms that are the product of four Wilson lines—its NLO corrections still comply with Equation 14, which ensures their factorization into the distribution of sources $W_{\Lambda^-}[\rho]$. For factorization to work, it is crucial to consider an observable that is sufficiently inclusive to allow the corresponding final state to be accompanied by an arbitrary number of gluons. Any restriction on associated gluon radiation does not permit NLO corrections to factorize simply in the distribution of sources.

A much more limited but widespread form of factorization is $k_\perp$ factorization (51), in terms of $k_\perp$-dependent, unintegrated quark and gluon distributions of the projectile and target. Within the CGC framework, these results can be reproduced for gluon (52, 53) and heavy quark (54) distributions at large transverse momenta $k_\perp \geq Q_s$; however, at smaller transverse momenta, $k_\perp$ factorization is broken even at LO (55–57).

3.3. Shattering Color Glass Condensates in A+A Collisions

Collisions between two nuclei (so-called dense-dense scattering) are complicated to handle on the surface. However, in the CGC framework, because the wave functions of the two nuclei are saturated, the collision can be treated as the collision of classical fields coupled to fast partons of each respective nucleus, described by the external current $J^\mu = \delta^{\mu+}\rho_1 + \delta^{\mu-}\rho_2$. The source densities of fast partons $\rho_{1,2}$ are both parametrically of order $1/g$, which implies that graphs involving multiple sources from both projectiles must be resummed (**Figure 6*b***).

At LO, inclusive observables[5] depend on the retarded classical color field $\mathcal{A}^\mu$, which solves the Yang-Mills equations $[\mathcal{D}_\mu, \mathcal{F}^{\mu\nu}] = J^\nu$ with the boundary condition $\lim_{x^0\to-\infty} \mathcal{A}^\mu = 0$. Among the observables to which this result applies is the expectation value of the energy-momentum tensor

[5]Exclusive observables may also be expressed in terms of solutions of the same Yang-Mills equations but with more complicated boundary conditions than for inclusive observables.

at early times after the collision. At LO,

$$T^{\mu\nu}_{\rm LO} = \frac{1}{4} g^{\mu\nu} \mathcal{F}^{\lambda\sigma} \mathcal{F}_{\lambda\sigma} - \mathcal{F}^{\mu\lambda} \mathcal{F}^{\nu}_{\lambda}, \quad 19.$$

where $\mathcal{F}^{\mu\nu}$ is the field strength of the classical field $\mathcal{A}^{\mu}$.

Although $A+A$ collisions are more complicated than $e+A$ or $p+A$ collisions, one can still factorize the leading higher-order corrections into the evolved distributions $W_{\Lambda^-}[\rho_1]$ and $W_{\Lambda^+}[\rho_2]$. At the heart of this factorization is a generalization of Equation 14 to the case in which the two projectiles are described in the CGC framework (58). When one integrates out the field modes in the slices $\Lambda_1^{\pm} \leq k^{\pm} \leq \Lambda_0^{\pm}$, the leading correction to the energy-momentum tensor is

$$\delta T^{\mu\nu}_{\rm NLO} = \left[\ln\left(\frac{\Lambda_0^-}{\Lambda_1^-}\right) \mathcal{H}_1 + \ln\left(\frac{\Lambda_0^+}{\Lambda_1^+}\right) \mathcal{H}_2 \right] T^{\mu\nu}_{\rm LO}, \quad 20.$$

where $\mathcal{H}_{1,2}$ are the JIMWLK Hamiltonians of the two nuclei, respectively. What is crucial here is the absence of mixing between the coefficients $\mathcal{H}_{1,2}$ of the logarithms of the two projectiles; they depend only on $\rho_{1,2}$, respectively, and not on the sources of the other projectile. Although the proof of this expression is somewhat involved, the absence of mixing is deeply rooted in causality. The central point is that because the duration of the collision (which scales as the inverse of the energy) is so brief, soft radiation must occur before the two nuclei have causal contact. Thus, logarithms associated with this radiation must have coefficients that do not mix the sources of the two projectiles.

Following the same procedure for Equation 20 as for the $e+A$ and $p+A$ cases, one obtains, for the energy-momentum tensor in an $A+A$ collision, the expression

$$\langle T^{\mu\nu} \rangle_{\rm LLog} = \int [D\rho_1 D\rho_2] W_1[\rho_1] W_2[\rho_2] T^{\mu\nu}_{\rm LO}. \quad 21.$$

This result can be generalized to multipoint correlations of the energy-momentum tensor:

$$\langle T^{\mu_1\nu_1}(x_1) \cdots T^{\mu_n\nu_n}(x_n) \rangle_{\rm LLog} = \int [D\rho_1 \, D\rho_2] W_1[\rho_1] W_2[\rho_2] \times T^{\mu_1\nu_1}_{\rm LO}(x_1) \cdots T^{\mu_n\nu_n}_{\rm LO}(x_n). \quad 22.$$

In this expression, all the correlations between the energy-momentum tensor at different points are from the distributions $W_{1,2}[\rho_{1,2}]$. Thus, the leading correlations are already built into the wave functions of the projectiles prior to the collision.

The expressions in Equations 21 and 22 are valid for proper times $\tau \sim 1/Q_s$ after the heavy-ion collision. The energy-momentum tensor for each configuration of sources $\rho_{1,2}$ is determined by solving classical Yang-Mills equations to compute the gauge fields $\mathcal{A}^{\rm cl.}_{\mu}[\rho_1, \rho_2]$ in the forward light cone with initial conditions determined by the classical CGC fields of each of the nuclei at $\tau = 0$ (55, 59–62). The corresponding nonequilibrium matter, with high occupation numbers $\sim 1/\alpha_s$, is known as the glasma (63). The glasma fields at early times are longitudinal chromoelectric and chromomagnetic fields that are screened at distances $1/Q_s$ in the transverse plane of the collision. As a consequence, the matter produced can be visualized (**Figure 7**) as comprising $R_A^2 Q_s^2$ color flux tubes of size $1/Q_s$, each of which produces $1/\alpha_s$ particles per unit rapidity. The flux-tube picture is supported by nonperturbative numerical solutions of the classical Yang-Mills equations (66). The glasma flux tubes generate n-particle long-range rapidity correlations (58, 64, 67, 68). These distributions are negative binomial distributions (69). They also carry topological charge (70); the resulting dynamical topological sphaleron transitions may result in observable metastable *CP*-violating domains (71).

How the glasma evolves into a thermalized QGP is not understood. An important ingredient is the role of instabilities (72). At early times, these instabilities arise at NLO from terms that break

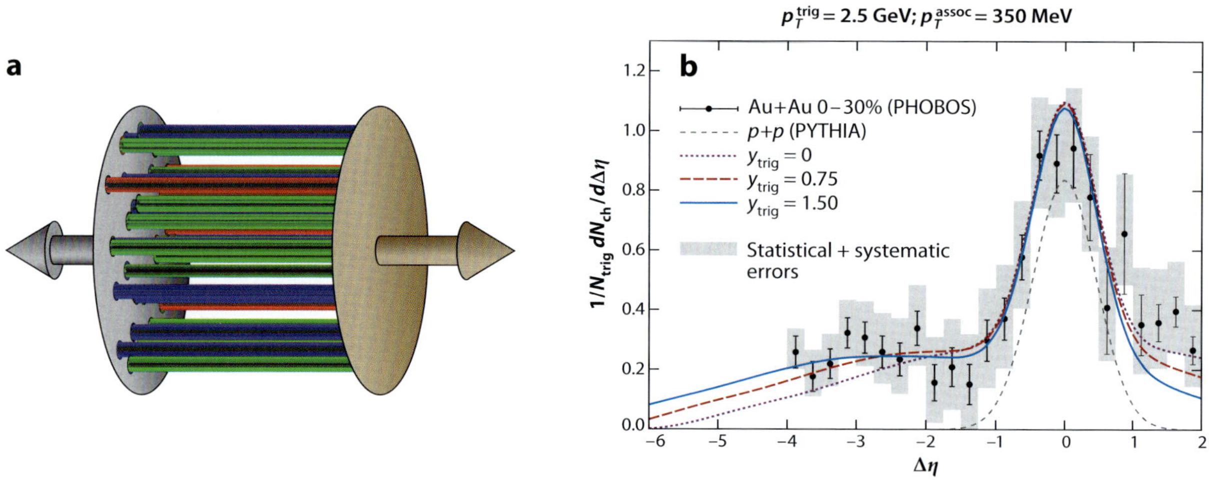

Figure 7

(*a*) Gauge-field configurations in the form of flux tubes of longitudinal chromoelectric and chromomagnetic fields screened on transverse scales $1/Q_s$. (*b*) Model comparison (64) to long-range rapidity correlations measured by the PHOBOS Collaboration (65).

the boost invariance of the LO term (10, 58). The modification to the evolution of the glasma is obtained by solving 3+1-dimensional Yang-Mills equations (73) for the (now) rapidity-dependent gauge fields convolved with a distribution that gives the spectrum of fluctuations (74). Although these effects may isotropize the system, early thermalization may also require collisions, whose role still needs to be clarified (75).

4. PHENOMENOLOGICAL APPLICATIONS OF THE COLOR GLASS CONDENSATE

In this section, we discuss the applications of the theoretical formalism outlined in Section 3 to analyze and predict a wide range of phenomena, including DIS in $e+p$ and $e+A$ collisions as well as the scattering of hadronic projectiles ranging from $p+p$ to $p+A$ to $A+A$ collisions. A unifying ingredient in many of the applications is the dipole cross section defined in Equation 11, albeit—as is apparent in the treatment of $A+A$ collisions—the fundamental ingredient is really the density matrix $W_Y\,[\rho]$. Because the JIMWLK equation (Equation 16) for this quantity is time consuming to solve,[6] many of the applications are in the context of models of the dipole cross section, which incorporate key features of saturation. These models provide an economical description of a wide range of data with only a few parameters. A good compromise between the full JIMWLK dynamics and models of the dipole cross section is the BK equation, which is a large-N_c realization of JIMWLK dynamics. With the recent availability of the NLO BK equation, global analyses of data are in order. Much of the following discussion is in the context of dipole models; improvements à la BK are highlighted wherever available.

[6]Other unknowns include higher-order corrections, initial conditions at low energy, and impact-parameter dependence of distributions.

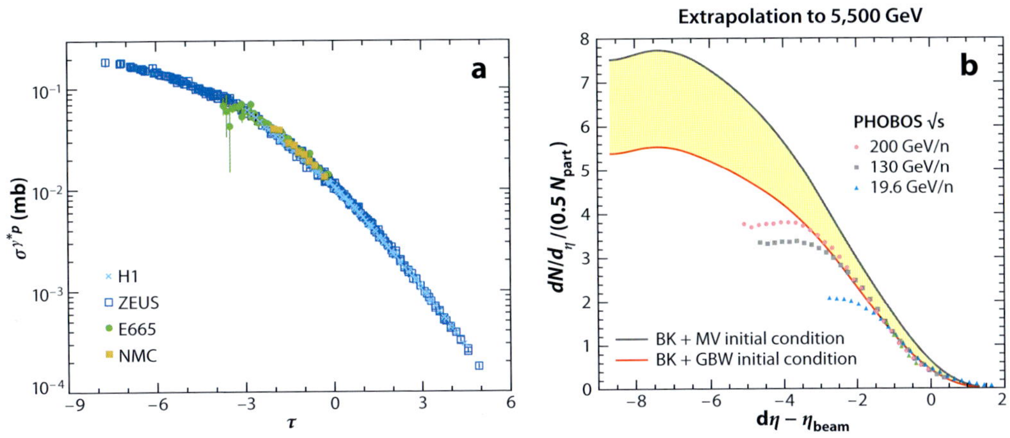

Figure 8

Geometrical scaling and limiting fragmentation. (*a*) $\sigma_{\gamma^* p}$ data at HERA for $x \leq 0.01$ and all Q^2 up to 450 GeV^2; τ is the scaling variable, $\tau \equiv Q^2/Q_s^2(x)$ (25, 26). (*b*) Particle multiplicities for several collision energies at RHIC (76), compared with the computation from Reference 77.

4.1. Deep-Inelastic Scattering in e+p and e+A Collisions

A remarkable observation (25) is that HERA data (25, 26) on the inclusive virtual photon-proton cross section for $x \leq 0.01$ scale as a function of the ratio $Q^2/Q_s^2(x)$ (**Figure 8*a***). This scaling is violated for larger values of x. Here $Q_s^2 = Q_0^2(x_0/x)^\lambda$ is the saturation scale, where $Q_0^2 = 1$ GeV^2, $x_0 = 3 \cdot 10^{-4}$, and $\lambda = 0.29$. This scaling is referred to as geometrical scaling because the survival probability of the color dipole into which the virtual photon fluctuates is close to unity or zero, depending on whether the ratio of the saturation radius ($\sim 1/Q_s$) to the size of the dipole ($\sim 1/Q$) is large or small, respectively. Recall that the saturation radius denotes the typical size of regions with strong color fields.

Geometric scaling has also been observed in inclusive diffraction, exclusive vector meson production, and deeply virtual Compton scattering (DVCS) data at HERA (78). The data also show violations of geometric scaling, which can be interpreted as consequences of BFKL diffusion (79), nonzero quark masses (80), and possibly DGLAP evolution (81). Note that the best scaling is obtained with a saturation scale that behaves like $Q_s^2(x) \propto x^{-0.3}$, a slower x dependence than predicted by the LO BK equation. This discrepancy is resolved by a resummed NLO computation of the saturation exponent (24), which indeed gives 0.3.

Although geometrical scaling is very suggestive of the presence of semihard dynamical scales in the proton, it is not conclusive in and of itself (82); more detailed comparisons to the data are essential. Despite their simplicity, saturation models (45, 79–81, 83–86) provide remarkably good descriptions of HERA data at small $x \leq 0.01$. The free parameters are fixed from fits to the total cross-section data alone; once they are fixed, the models predict a large variety of results, including the longitudinal (F_L), diffractive (F_2^D), and charm (F_2^c) structure functions; the virtual photon production of vector mesons (ρ, J/ψ); and DVCS. The most recent analysis of inclusive data (45) of DIS in e+p collisions is quite sophisticated: The energy dependence is given by the running-coupling BK equation, and the free parameters refer solely to the initial conditions and to the proton transverse area.

The phenomenon of hard diffraction in DIS is particularly sensitive to saturation. The simplest diffractive processes are events in which (*a*) the proton remains intact and (*b*) a large gap in rapidity with no particles extends between the rapidity of the proton and that of the fragmentation products of the virtual photon.

For small invariant masses, this process corresponds to elastic scattering of the $q\bar{q}$ dipole off the target; its cross section is evaluated as[7]

$$\left.\frac{d\sigma_{\rm diff}}{dt}\right|_{t=0} = \frac{1}{16\pi}\int_0^1 dz \int d^2\boldsymbol{r}_\perp |\psi(z,\boldsymbol{r}_\perp)|^2 \sigma^2_{\rm dipole}(x,\boldsymbol{r}_\perp). \qquad 23.$$

The dipole cross section in this expression, for small dipoles $r^2 \ll 1/Q_s^2$, is a color-singlet combination of two gluons that can be interpreted as a Pomeron exchange (87); for larger dipoles $r^2 \geq 1/Q_s^2$, the color-singlet exchange does not have this simple interpretation. The t distribution has the form $d\sigma_{\rm diff}/dt = \exp(-B_D|t|)d\sigma_{\rm diff}/dt|_{t=0}$, where B_D is the transverse area of the interaction region in the proton and is closely related to the transverse gluon radius in the proton, which is estimated to be 0.61 ± 0.04 fm (88). From this form of the diffractive cross section and Equation 23, we determine the total diffractive cross section to be

$$\sigma_D \sim \frac{B_D}{Q^2}\int_{Q^{-2}}^{Q_s^{-2}} \frac{dr^2}{r^4}(r^2 Q_s^2(x))^2 \sim B_D \frac{Q_s^2(x)}{Q^2}. \qquad 24.$$

Here we use the color transparency approximation $\sigma_{\rm dipole} \propto r^2 Q_s^2(x)$ for the dipole cross section. Unlike the inclusive cross section in Equation 10, which is dominated by small dipole sizes $\sim 1/Q$, the integrand of the diffractive cross section is dominated by larger size dipoles of size $\sim 1/Q_s$. By comparing Equation 24 with Equation 10, we deduce that in the saturation framework the ratio $\sigma_{\rm diff}/\sigma_{\rm tot}$ is approximately constant as a function of energy. The HERA data support this qualitative observation (**Figure 9*a***) and are in quantitative agreement with the detailed saturation model of Reference 84. Further, we note that excellent fits with $\chi^2 \sim 1$ are obtained in this saturation framework for exclusive vector meson production and DVCS (81) in addition to inclusive diffraction (**Figure 9*b***) (89). These exclusive processes can provide detailed information about the impact parameter dependence of the dipole cross section (90) and may even provide unique information about the partonic nature of short-range nuclear forces (88).

In contrast, if unitarization were due to soft physics [the prevailing viewpoint before the advent of saturation (91)], the natural cutoff for the integral would be $1/\Lambda_{\rm QCD}$. Diffraction would be nonperturbative even for hard Q, and the energy dependence of the diffractive cross section would be the square of the inclusive cross section, in disagreement with data.

Fits based on geometrical scaling of the $e{+}A$ fixed-target data with the functional form $Q_s^2 \propto A^\delta$ (where δ, naïvely, is 1/3) yield values of δ that range from 1/4 to 4/9 (92). However, the dipole formalism that describes $e{+}p$ data so successfully can be straightforwardly generalized to nuclei to construct the corresponding nuclear dipole cross section (85). When one compares this to results for the x, Q^2 dependence of inclusive $e{+}A$ data, one finds effectively that $Q^2_{s,A}(x) = Q^2_{s,p}(x)A^{1/3}$ (93). Nuclear diffractive distributions can be computed in saturation models (89, 94); predictions have also been made for semi-inclusive hadron production (95), exclusive vector meson production (96), and nuclear DVCS (97) in this framework.

In saturation models of nuclei, the small-x distributions in a nucleon are convolved with nuclear geometry to give the nuclear distributions. However, this process does not commute with the RG evolution in x of nuclear distributions determined at some initial scale. Initial computations of

[7]Forward diffraction [corresponding to $t = 0$, where $t = (P - P')^2$ is the momentum transfer squared between the incoming and outgoing protons] can be compared directly to inclusive DIS within a dipole model because it depends only on the dipole cross section.

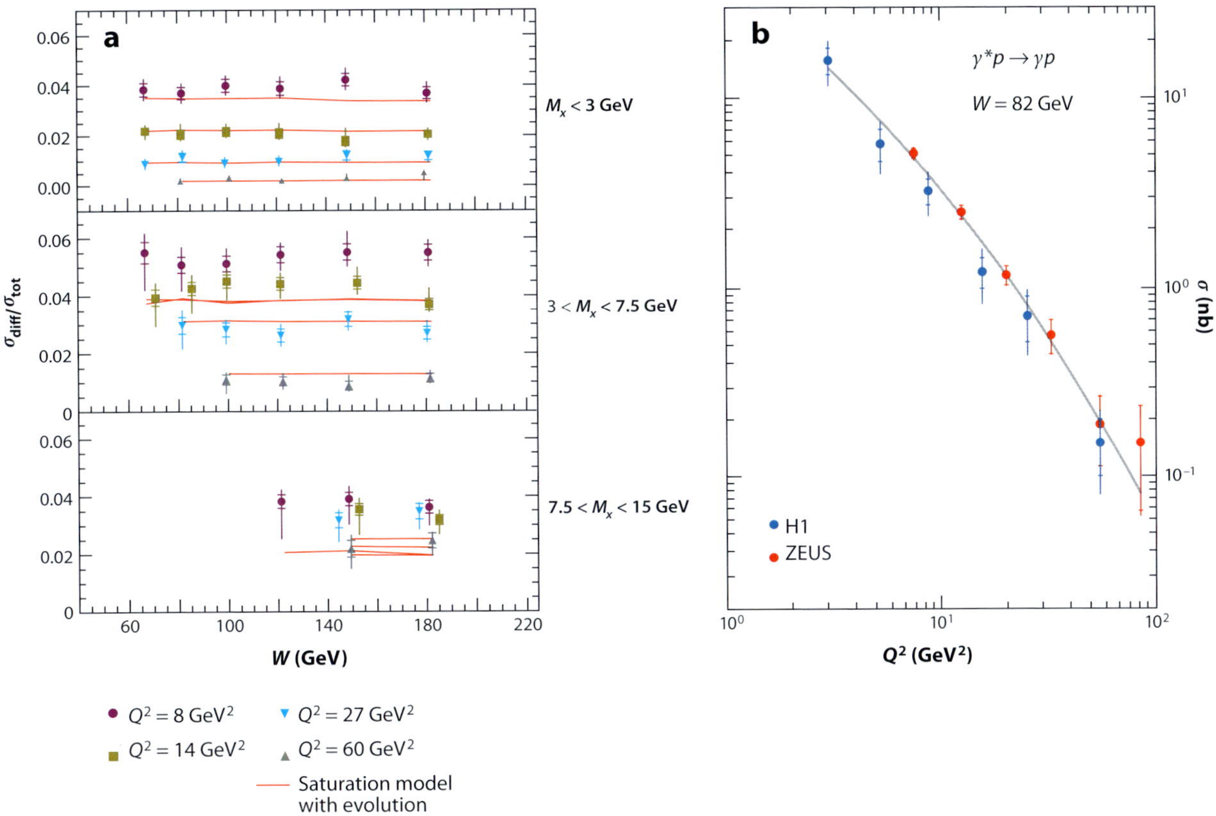

Figure 9

(*a*) ZEUS data for the ratio $\sigma_{\text{diff}}/\sigma_{\text{tot}}$, together with the corresponding prediction of the saturation model of Reference 84. (*b*) Deeply virtual Compton scattering measurement at HERA and comparison with results from Reference 81.

$e+A$–inclusive distributions in the NLO BK framework have been performed, and good agreement for existing fixed-target data has been obtained (64).

4.2. Particle Multiplicities in $d+A$ and $A+A$ Collisions

The CGC EFT is most reliable when at least one of the projectiles is dense in the sense discussed in Section 4.1. At RHIC, the world's first $d+A$ and $A+A$ collider, many features of the CGC are being tested. These include bulk features such as the rapidity and centrality dependence of particle multiplicities in $d+A$ and $A+A$ collisions, limiting fragmentation, particle spectra and correlations, and possibly more exotic features such as long-range rapidity correlations (known as ridge events) in $A+A$ collisions and local *CP* violation arising from topological fluctuations in $A+A$ collisions. In this section, we focus on particle multiplicities and discuss the other features in subsequent sections.

Limiting fragmentation is the well-known property of the strong interactions wherein the rapidity distribution in the fragmentation region becomes independent of the collision energy. When the beam energy is increased (**Figure 8*b***), $dN/d\eta'$ is the same at large η' for all energies, where $\eta' \equiv \eta - \eta_{\text{beam}}$. In the fragmentation region, large-x_1 modes are probed in one hadron or nucleus, and small-x_2 modes in the other. At small x_2, if gluonic matter is saturated, then parton distributions have a very weak dependence on x_2 or, equivalently, on $\eta + \eta_{\rm beam}$; cross sections depend only on x_1 or $\eta - \eta_{\rm beam}$. Deviations from limiting fragmentation at high energies are especially interesting because of a significant window in phase space wherein the RG evolution in x_2 to (or away from) the universal "black disk" of saturated gluon matter can be explored (77, 98). These constraints lead to specific predictions for the rapidity dependence of the multiplicity at the LHC in the BK RG framework (77, 99).

The CGC can predict only the distribution of initial gluons at a proper time $\tau \sim Q_s^{-1}$. In p+A or d+A collisions, this is not a significant limitation because only a few gluons are produced in the final state. The situation is much different in A+A collisions, in which the particle multiplicity is significantly larger and the system evolves from the nonequilibrium glasma to a QGP, the latter evolving subsequently as a fluid with low viscosity. For p+A collisions, the problem is solvable analytically (52, 53, 100). For A+A collisions, two kinds of calculations have been performed:

1. Exact numerical solutions of the Yang-Mills equations (55, 59, 61). Quantum evolution effects are not systematically included, and the ensemble of color charges is assumed to be Gaussian (MV model). This assumption may be reasonable at RHIC energies, where $x \sim 10^{-2}$. The inclusion of quantum effects from the wave functions (58, 69) is under control, but the problem of how to include quantum effects in the final state evolution is still unsolved (73, 74).
2. Approximate analytical calculations of the initial gluon spectrum (101). These calculations assume $k_\perp$ factorization, although this assumption is violated for momenta $p_\perp \le Q_s$. In this model, saturation effects are introduced via the unintegrated gluon distribution of the nuclei with the rapidity dependence of the gluon spectrum governed by that of the saturation scale $Q_s^2(x) \sim x^{-0.3}$. In the CGC framework, $\mathrm{d}N/\mathrm{d}\eta = Q_s^2 R_A^2/\alpha_s(Q_s)$; a unique feature is that the coupling runs as a function of Q_s (102), which is in agreement with the centrality dependence of RHIC data (103). Similar analyses of multiplicity distributions have been extended to the case of d+A collisions (104), and the results are in fair agreement with RHIC data.

4.3. Inclusive Spectra in d+A and A+A Collisions

Single- and double-inclusive spectra in d+A and A+A collisions have been computed in the CGC framework. The former, as discussed below, are in good agreement with measurements at RHIC. When compared with similar spectra in p+p collisions, they demonstrate interesting patterns of enhancement and suppression that provide considerable insight into multiple scattering and parton evolution in nuclei. Double-inclusive spectra in d+A collisions are significantly more sensitive to strong color-field dynamics; in A+A collisions, such spectra may explain the systematics of observed long-range ridge-like correlations.

4.3.1. $p_\perp$ dependence. One does not expect significant final-state interactions with on-shell partons in p+A collisions. Moreover, the inclusive gluon spectrum depends only on the Wilson line correlator $\langle U(0)U^\dagger(\boldsymbol{x}_\perp)\rangle$ (53, 100, 105), which can be determined from the dipole cross section used in DIS (106). The hadron spectrum in d+A collisions has been evaluated in this approach (107), and the results are in good agreement with RHIC data. Very recently, good agreement of single-inclusive distributions in d+A collisions were obtained in the NLO RG approach (108), consistent with its application in e+A collisions (64).

In A+A collisions, the strong final-state interactions probably alter the momentum distribution to be more isotropic at low momenta; the hard tail is modified by parton energy loss. Although these effects may in part be included in the glasma, the additional contributions at later stages are important.

4.3.2. Nuclear modification ratios. To quantify nuclear effects, $p_\perp$ spectra in A+A and d+A collisions can be compared with those in p+p collisions via the ratios

$$R_{AA} \equiv \frac{\left.\frac{dN}{dyd^2p_\perp}\right|_{AA}}{N_{\mathrm{coll}} \left.\frac{dN}{dyd^2p_\perp}\right|_{pp}} \text{ and } R_{dA} \equiv \frac{\left.\frac{dN}{dyd^2p_\perp}\right|_{dA}}{N_{\mathrm{coll}} \left.\frac{dN}{dyd^2p_\perp}\right|_{pp}}, \qquad 25.$$

where $N_{\mathrm{coll}} \sim A^{4/3}$ (A) is the number of binary collisions in A+A (d+A) collisions. At high $p_\perp$, one expects the ratios to scale with N_{coll}. In A+A collisions at RHIC, because no suppression is seen in d+A collisions at central rapidities, the large observed suppression in R_{AA} is rightly interpreted as a final-state effect, and a strong candidate is energy loss induced by the dense medium (109).

However, at small x, RG evolution à la BK predicts (110) that this ratio should be suppressed by saturation effects in the wave function of the nucleus. This can be studied in the clean environment of d+A collisions, where small x is probed by measuring the ratio at a positive rapidity in the direction of the deuterium nucleus. The ratio R_{dA} was measured by the BRAHMS Collaboration (111) up to rapidities of $\eta = 3.2$ and by the STAR Collaboration (112) at $\eta = 4$. At these large values of η, the value of R_{dA} is significantly below unity, consistent with the predicted trend. In addition, as anticipated, the suppression is greater for more central collisions. The onset of this forward suppression has also been studied semianalytically (100, 113), and more quantitative calculations have been performed (107, 114, 115) using models of the dipole cross section that have a realistic x dependence. More recently, such calculations were also performed in the NLO BK framework (108).

4.3.3. Two-hadron correlations in *d*+*A* and *A*+*A* collisions. Two-hadron correlations[8] are more sensitive in distinguishing between model predictions than are the single-inclusive results for which alternative explanations of the data are feasible. Striking preliminary results on these correlations in d+gold (Au) collisions have recently been presented by the STAR and PHENIX Collaborations. In the STAR measurements, back-to-back correlations of pairs have been studied in which (*a*) one particle in the pair is forward in rapidity and the other is at central rapidity and (*b*) both particles in the pair are at forward rapidities. The forward-central (FC) pairs probe $x \sim 10^{-2}$ in the Au nucleus, whereas the forward-forward (FF) pairs probe $x \sim 10^{-3}$ in the Au nucleus. A clear broadening of the backward hadron peak is seen in the transition from FC to FF, as well as with increasing centrality in the FF events. In the latter case, the distribution is so broad that no peak is visible! This particular effect was predicted in the CGC (117) and is in quantitative agreement with the STAR results (118). These results are also consistent with measurements by the PHENIX Collaboration on pair distributions in d+Au collisions. Related discussions can be found in Reference 119.

Although two-particle correlations in A+A collisions are typically greatly altered by final-state interactions, there is an important exception for particles widely separated in rapidity. Causality implies that such correlations are created at very early times (**Figure 10*a***). A simple estimate for ultrarelativistic particles whose space-time and momentum rapidities are strongly correlated gives

$$\tau_{\mathrm{max}} = \tau_{\text{freeze-out}} e^{-\frac{|\Delta\eta|}{2}}, \qquad 26.$$

for the latest time at which these particles could have been correlated. For a freeze-out time $\tau_{\text{freeze-out}} \approx 10$ fm/c and rapidity separations $\Delta\eta \geq 4$, one sees that these correlations must have

[8]Hadron-photon correlations have also been studied (116).

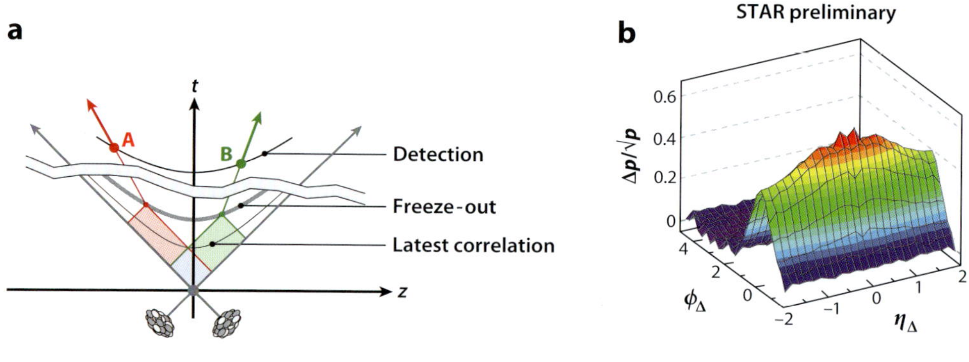

Figure 10

(*a*) Causal relations between two particles, A (*red dot*) and B (*green dot*), separated in rapidity. (*b*) Two-hadron correlation measured by the STAR Collaboration as a function of $\Delta\eta$ and $\Delta\phi$.

been generated before 1 fm/*c*. Thus, long-range rapidity correlations function as a chronometer of the evolution of the very strong glasma color fields produced early in the heavy-ion collision.

Striking long-range rapidity correlations have been seen in $A+A$ collisions in events with prominent jet-like structures. In such events, the spectrum of associated particles is collimated in the azimuthal separation $\Delta\varphi$ relative to the jet and shows a nearly constant amplitude in the strength of the pseudorapidity correlation $\Delta\eta$ up to $\Delta\eta \sim 1.5$ (120). These events are termed ridge events because they structurally resemble an extended mountain ridge in the $\Delta\eta - \Delta\varphi$ plane associated with a narrow jet peak. This collimated correlation persists up to $\Delta\eta \sim 4$ (65, 121). An important feature of ridge correlations is that the above-described structure is observed in two-particle correlations without a jet trigger (**Figure 10**) and persists without significant modification for the triggered events (122, 123). These events include all hadrons with momenta $p_\perp \geq 150$ MeV. In such events, a sharp rise in the amplitude of the ridge is observed (122, 124) in going from peripheral to central collisions.

The ridge structures seen in two-particle and three-particle (125) $A+A$ correlations can be explained as resulting from the transverse radial flow of the glasma flux tubes (see Section 3.3). The flux tubes are responsible for the long-range rapidity correlation in the ridge; the angular collimation occurs because particles produced isotropically in a given flux tube are collimated by the radial outward "Hubble" hydrodynamic flow of the flux tubes. Ideas about the angular collimation of particle distributions by flow were discussed in Reference 126. When combined with the long-range rapidity correlations provided by the glasma flux tubes, they provide a semiquantitative description of the ridge measurements (64, 67, 68, 127). More detailed studies are feasible with realistic hydrodynamic simulations (128). We note that the PHENIX Collaboration, which has also observed ridge-like structures (129), has presented preliminary data showing that such structures disappear as the $p_\perp$ of the trigger particle is increased (130). This result is in qualitative agreement with expectations in the glasma flux-tube picture. For a recent critical evaluation of this and alternative models, we refer the reader to Reference 131.

5. OUTLOOK: LHC AND FUTURE DEEP-INELASTIC SCATTERING COLLIDERS

At the LHC, very low values of x will be probed in $p+p$, $A+A$ (including the photoproduction-induced dynamics of peripheral $A+A$ collisions), and possibly $p+A$ collisions through a wide range of final states and diverse kinematic ranges. In nuclei, Q_s^2 will be large; estimates range from $\sim$2.6 to 4 GeV^2 in central collisions to $\sim$10 GeV^2 for $y = \pm 3$ units. The picture of strongly correlated (albeit weakly coupled) dynamics of glue in the CGC EFT outlined herein will be tested as never before. Detailed tests of BFKL dynamics and possibly the CGC are feasible in $p+p$ collisions by studying Mueller-Navelet jets (132) and forward jets, respectively. Diffractive final states, although always challenging to interpret, offer opportunities to understand how saturated gluons generate rapidity gaps. A strong test of CGC dynamics will become available in "Day 1" physics of bulk dynamics in $A+A$ collisions. More subtle tests of the RG flow of multiparton correlators will be conducted in a variety of final states ranging from quarkonia to electromagnetic probes to long-range rapidity correlations. In $p+A$ collisions at the LHC, many of the patterns observed in forward single-inclusive and dihadron correlation spectra will be present at central rapidities and will be much more dramatic at forward rapidities. Finally, the relative role of glasma dynamics relative to that of the QGP in $A+A$ collisions at the LHC is still unclear. If RHIC has reached the perfect hydrodynamic limit of maximal flow, what happens at the LHC?

Studies of the dynamics of gluon saturation and of the CGC, which will be complementary to those done at the LHC, can be performed at a future electron-ion collider (EIC).[9] First, with a wide lever arm in Q^2, the dynamics of saturation could be studied with precision in the regime where $Q^2 \sim Q_s^2 \gg \Lambda_{QCD}^2$, which is difficult to achieve at a hadron collider. Second, clarifying which aspects of the dynamics probed are universal and novel calls for an electron probe. The factorization theorems we have developed here suggest that the density matrices W are universal—how can we confirm this hypothesis and cleanly extract their rich dynamics? These issues are not merely academic and are strongly reflected in the structure of final states. For example, rapidity gaps are a large fraction of the cross section at HERA but represent a much smaller contribution in hadronic collisions at Fermilab, demonstrating a breakdown of factorization (133). The physics case for an EIC and electron probe has been outlined in Reference 134 and in a number of white papers and reports.

The prospects for unambiguous discovery and exploration of a novel many-body QCD regime of gluon saturation over the next decade are very good. The theoretical status of studies of this regime within the framework of the CGC EFT are increasingly robust, although many questions remain. The LHC and EIC experiments could provide definitive answers to these outstanding questions.

DISCLOSURE STATEMENT

The authors are not aware of any affiliations, memberships, funding, or financial holdings that might be perceived as affecting the objectivity of this review.

ACKNOWLEDGMENTS

F.G. and E.I. are supported in part by Agence Nationale de la Recherche via the program ANR-06-BLAN-0285-01. J.J.-M. is supported by the U.S. Department of Energy's Office of Nuclear Physics through grant number DE-FG02-09ER41620 and by the City University of New York through the PSC-CUNY Research Award Program, grant number 62625-40. R.V.'s research is supported by the U.S. Department of Energy under contract number DE-AC02-98CH10886.

[9]Current proposals include the EIC in the eRHIC version at BNL and the eLIC version at JLAB, the LHeC proposal at CERN, and an electron-ion collider at FAIR in GSI.

LITERATURE CITED

1. Wilczek F. *Mod. Phys. Lett. A* 21:701 (2006)
2. Altarelli G, Parisi G. *Nucl. Phys. B* 126:298 (1977); Gribov VN, Lipatov LN. *Sov. J. Nucl. Phys.* 15:438 (1972); Gribov VN, Lipatov LN. *Sov. J. Nucl. Phys.* 15:675 (1972); Dokshitzer Y. *Sov. Phys. JETP* 46:641 (1977)
3. Aaron FD, et al. (H1 Collab.) *Eur. Phys. J. C* 64:561 (2009)
4. Donnachie A, Landshoff PV. *Phys. Lett.* B437:408 (1998)
5. Lipatov LN. *Sov. J. Nucl. Phys.* 23:338 (1976); Kuraev EA, Lipatov LN, Fadin VS. *Sov. Phys. JETP* 45:199 (1977); Balitsky I, Lipatov LN. *Sov. J. Nucl. Phys.* 28:822 (1978)
6. Gribov LV, Levin EM, Ryskin MG. *Phys. Rep.* 100:1 (1983); Mueller AH, Qiu J-W. *Nucl. Phys. B* 268:427 (1986)
7. Stasto AM. *Int. J. Mod. Phys. A* 19:317 (2004); Drescher HJ, Dumitru A, Strikman M. *Phys. Rev. Lett.* 94:231801 (2005); Cooper-Sarkar A, Sarkar S. *J. High Energy Phys.* 0801:075 (2008); Henley EM, Jalilian-Marian J. *Phys. Rev. D* 73:094004 (2006)
8. Iancu E, Leonidov A, McLerran LD. arXiv:hep-ph/0202270 (2002); Iancu E, Venugopalan R. arXiv:hep-ph/0303204 (2003)
9. Jalilian-Marian J, Kovchegov Y. *Prog. Part. Nucl. Phys.* 56:104 (2006); Weigert H. *Prog. Part. Nucl. Phys.* 55:461 (2005); Gelis F, Venugopalan R. *Acta Phys. Polon. B* 37:3253 (2006)
10. Gelis F, Lappi T, Venugopalan R. *Int. J. Mod. Phys. E* 16:2595 (2007)
11. McLerran LD, Venugopalan R. *Phys. Rev. D* 49:2233 (1994); McLerran LD, Venugopalan R. *Phys. Rev. D* 49:3352 (1994); McLerran LD, Venugopalan R. *Phys. Rev. D* 50:2225 (1994)
12. Balitsky I. *Nucl. Phys. B* 463:99 (1996)
13. Kovchegov YV. *Phys. Rev. D* 61:074018 (2000)
14. Jalilian-Marian J, Kovner A, Leonidov A, Weigert H. *Nucl. Phys. B* 504:415 (1997); Jalilian-Marian J, Kovner A, Leonidov A, Weigert H. *Phys. Rev. D* 59:014014 (1999); Jalilian-Marian J, Kovner A, Leonidov A, Weigert H. *Phys. Rev. D* 59:034007 (1999), Jalilian-Marian J, Kovner A, Leonidov A, Weigert H. Erratum. *Phys. Rev. D* 59:099903 (1999); Jalilian-Marian J, Kovner A, Weigert H. *Phys. Rev. D* 59:014015 (1999); Kovner A, Milhano G, Weigert H. *Phys. Rev. D* 62:114005 (2000); Weigert H. *Nucl. Phys. A* 703:823 (2002)
15. Iancu E, Leonidov A, McLerran LD. *Nucl. Phys. A* 692:583 (2001); Iancu E, Leonidov A, McLerran LD. *Phys. Lett.* B510:133 (2001); Ferreiro E, Iancu E, Leonidov A, McLerran LD. *Nucl. Phys. A* 703:489 (2002)
16. Iancu E, McLerran LD. *Phys. Lett.* B510:145 (2001)
17. Mueller AH. *Nucl. Phys. B* 643:501 (2002); Iancu E, Itakura K, McLerran LD. *Nucl. Phys. A* 724:181 (2003)
18. Mueller AH. *Nucl. Phys. B* 558:285 (1999)
19. Van Saarloos W. *Phys. Rep.* 386:29 (2003)
20. Munier S, Peschanski R. *Phys. Rev. Lett.* 91:232001 (2003); Munier S, Peschanski R. *Phys. Rev. D* 69:034008 (2004)
21. Levin EM, Tuchin K. *Nucl. Phys. B* 573:833 (2000); Levin EM, Tuchin K. *Nucl. Phys. A* 691:779 (2001); Levin EM, Tuchin K. *Nucl. Phys. A* 693:787 (2001); Armesto N, Braun MA. *Eur. Phys. J. C* 20:517 (2001); Armesto N, Braun MA. *Eur. Phys. J. C* 22:351 (2001); Levin E, Lublinsky M. *Nucl. Phys. A* 696:833 (2001); Levin E, Lublinsky M. *Phys. Lett.* B521:233 (2001); Lublinsky M. *Eur. Phys. J. C* 21:513 (2001); Golec-Biernat K, Motyka L, Stasto AM. *Phys. Rev. D* 65:074037 (2002)
22. Rummukainen K, Weigert H. *Nucl. Phys. A* 739:183 (2004)
23. Iancu E, Itakura K, McLerran LD. *Nucl. Phys. A* 708:327 (2002); Mueller AH, Triantafyllopoulos DN. *Nucl. Phys. B* 640:331 (2002)
24. Triantafyllopoulos DN. *Nucl. Phys. B* 648:293 (2003)
25. Stasto AM, Golec-Biernat K, Kwiecinski J. *Phys. Rev. Lett.* 86:596 (2001)
26. Gelis F, Peschanski R, Schoeffel L, Soyez G. *Phys. Lett.* B647:376 (2007)
27. Gelis F, Venugopalan R. *Nucl. Phys. A* 776:135 (2006); Gelis F, Venugopalan R. *Nucl. Phys. A* 779:177 (2006)

28. Kovchegov YV, Kuokkanen J, Rummukainen K, Weigert H. *Nucl. Phys. A* 823:47 (2009)
29. Blaizot JP, Iancu E, Weigert H. *Nucl. Phys. A* 713:441 (2003)
30. Jeon S, Venugopalan R. *Phys. Rev. D* 70:105012 (2004)
31. Jeon S, Venugopalan R. *Phys. Rev. D* 71:125003 (2005)
32. Hatta Y, Iancu E, Itakura K, McLerran L. *Nucl. Phys. A* 760:172 (2005); Kovchegov YV, Szymanowski L, Wallon S. *Phys. Lett.* B586:267 (2004)
33. McLerran LD. *AIP Conf. Proc.* 490:42 (1999)
34. Mueller AH. *Nucl. Phys. A* 724:223 (2003)
35. Brunet E, Derrida B, Mueller AH, Munier S. *Europhys. Lett.* 76:1 (2006); Munier S. *Phys. Rep.* 473:1 (2009)
36. Brunet E, Derrida B. *Phys. Rev. E* 56:2597 (1997); Brunet E, Derrida B. *Comp. Phys. Commun.* 121/122:376 (1999); Brunet E, Derrida B. *J. Stat. Phys.* 103:269 (2001)
37. Mueller AH, Shoshi AI. *Nucl. Phys. B* 692:175 (2004)
38. Iancu E, Mueller AH, Munier S. *Phys. Lett.* B606:342 (2005)
39. Iancu E, Triantafyllopoulos DT. *Nucl. Phys. A* 756:419 (2005); Iancu E, Triantafyllopoulos DT. *Phys. Lett.* B610:253 (2005)
40. Iancu E, de Santana Amaral JT, Soyez G, Triantafyllopoulos DN. *Nucl. Phys. A* 786:131 (2007)
41. Mueller AH, Shoshi AI, Wong SMH. *Nucl. Phys. B* 715:440 (2005)
42. Kovner A, Lublinsky M. *Phys. Rev. D* 71:085004 (2005); Kovner A, Lublinsky M. *Phys. Rev. Lett.* 94:181603 (2005); Hatta Y, et al. *Nucl. Phys. A* 764:423 (2006); Balitsky I. *Phys. Rev. D* 72:074027 (2005)
43. Dumitru A, et al. *J. High Energy Phys.* 0708:062 (2007)
44. Balitsky I. *Phys. Rev. D* 75:014001 (2007); Balitsky I, Chirilli GA. *Phys. Rev. D* 77:014019 (2008); Gardi E, Kuokkanen J, Rummukainen K, Weigert H. *Nucl. Phys. A* 784:282 (2007); Kovchegov YV, Weigert H. *Nucl. Phys. A* 784:188 (2007)
45. Albacete JL, Armesto N, Milhano JG, Salgado CA. *Phys. Rev. D* 80:034031 (2009)
46. Gribov VN, Ioffe BL, Pomeranchuk IY. *Sov. J. Nucl. Phys.* 2:549 (1966); Ioffe BL. *Phys. Lett.* B30:123 (1969)
47. Nikolaev NN, Zakharov BG. *Phys. Lett.* B332:184 (1994)
48. McLerran LD, Venugopalan R. *Phys. Rev. D* 59:094002 (1999)
49. Venugopalan R. *Acta Phys. Polon. B* 30:3731 (1999)
50. Gelis F, Jalilian-Marian J. *Phys. Rev. D* 66:014021 (2002); Gelis F, Jalilian-Marian J. *Phys. Rev. D* 66:094014 (2002); Gelis F, Jalilian-Marian J. *Phys. Rev. D* 76:074015 (2007); Jalilian-Marian J. *Eur. Phys. J. C* 61:789 (2009); Kopeliovich BZ, Raufeisen J, Tarasov AV, Johnson MB. *Phys. Rev. C* 67:014903 (2003)
51. Collins JC, Ellis RK. *Nucl. Phys. B* 360:3 (1991); Catani S, Ciafaloni M, Hautmann F. *Nucl. Phys. B* 366:135 (1991); Levin EM, Ryskin MG, Shabelsky YM, Shuvaev AG. *Sov. J. Nucl. Phys.* 53:657 (1991)
52. Kovchegov YV, Mueller AH. *Nucl. Phys. B* 529:451 (1998)
53. Dumitru A, McLerran LD. *Nucl. Phys. A* 700:492 (2002)
54. Gelis F, Venugopalan R. *Phys. Rev. D* 69:014019 (2004)
55. Krasnitz A, Venugopalan R. *Phys. Rev. Lett.* 86:1717 (2001)
56. Blaizot JP, Gelis F, Venugopalan R. *Nucl. Phys. A* 743:57 (2004)
57. Nikolaev NN, Schafer W, Zakharov BG. *Phys. Rev. Lett.* 95:221803 (2005)
58. Gelis F, Lappi T, Venugopalan R. *Phys. Rev. D* 78:054019 (2008); Gelis F, Lappi T, Venugopalan R. *Phys. Rev. D* 78:054020 (2008); Gelis F, Lappi T, Venugopalan R. *Phys. Rev. D* 79:094017 (2009)
59. Krasnitz A, Venugopalan R. *Nucl. Phys. B* 557:237 (1999); Krasnitz A, Venugopalan R. *Phys. Rev. Lett.* 84:4309 (2000); Krasnitz A, Nara Y, Venugopalan R. *Phys. Rev. Lett.* 87:192302 (2001)
60. Krasnitz A, Nara Y, Venugopalan R. *Nucl. Phys. A* 727:427 (2003)
61. Lappi T. *Phys. Rev. C* 67:054903 (2003); Lappi T. *Phys. Lett.* B643:11 (2006)
62. Krasnitz A, Nara Y, Venugopalan R. *Phys. Lett.* B554:21 (2003); Krasnitz A, Nara Y, Venugopalan R. *Nucl. Phys. A* 717:268 (2003); Lappi T, Venugopalan R. *Phys. Rev. C* 74:054905 (2006)
63. Lappi T, McLerran LD. *Nucl. Phys. A* 772:200 (2006)

64. Dusling K, Gelis F, Lappi T, Venugopalan R. arXiv:0911.2720 [hep-ph] (2009)
65. Alver B, et al. (PHOBOS Collab.) arXiv:0903.2811 [nucl-ex] (2009)
66. Lappi T, Srednyak S, Venugopalan R. arXiv:0911.2068 [hep-ph] (2009)
67. Dumitru A, Gelis F, McLerran L, Venugopalan R. *Nucl. Phys. A* 810:91 (2008)
68. Dusling K, Fernandez-Fraile D, Venugopalan R. *Nucl. Phys. A* 828:161 (2009)
69. Gelis F, Lappi T, McLerran L. *Nucl. Phys. A* 828:149 (2009)
70. Kharzeev D, Krasnitz A, Venugopalan R. *Phys. Lett.* B545:298 (2002)
71. Kharzeev D, McLerran LD, Warringa HJ. *Nucl. Phys. A* 803:227 (2008)
72. Fujii H, Itakura K. *Nucl. Phys. A* 809:88 (2008); Fujii H, Itakura K, Iwazaki A. *Nucl. Phys. A* 828:178 (2009); Fukushima K. *Phys. Rev. C* 75:021902 (2007); Fukushima K. Erratum. arXiv:0711.2634 [hep-ph] (2007); Mrowczynski S. *Acta Phys. Polon. B* 39:1665 (2008)
73. Romatschke P, Venugopalan R. *Phys. Rev. Lett.* 96:062302 (2006); Romatschke P, Venugopalan R. *Phys. Rev. D* 74:045011 (2006)
74. Fukushima K, Gelis F, McLerran L. *Nucl. Phys. A* 786:107 (2007)
75. Xu Z, Greiner C. *Phys. Rev. C* 71:064901 (2005); Gelis F, Jeon S, Venugopalan R. *Nucl. Phys. A* 817:61 (2009)
76. Back BB, et al. (PHOBOS Collab.) *Phys. Rev. Lett.* 91:052303 (2003)
77. Gelis F, Stasto A, Venugopalan R. *Eur. Phys. J. C* 48:489 (2006)
78. Marquet C, Schoeffel L. *Phys. Lett.* B639:471 (2006)
79. Iancu E, Itakura K, Munier S. *Phys. Lett.* B590:199 (2004)
80. Soyez G. *Phys. Lett.* B655:32 (2007)
81. Kowalski H, Motyka L, Watt G. *Phys. Rev. D* 74:074016 (2006)
82. Caola F, Forte S. *Phys. Rev. Lett.* 101:022001 (2008)
83. Golec-Biernat K, Wüsthoff M. *Phys. Rev. D* 59:014017 (1999); Golec-Biernat K, Wüsthoff M. *Phys. Rev. D* 60:114023 (1999)
84. Bartels J, Golec-Biernat K, Kowalski H. *Phys. Rev. D* 66:014001 (2002)
85. Kowalski H, Teaney D. *Phys. Rev. D* 68:114005 (2003)
86. Forshaw JR, Sandapen R, Shaw G. *Phys. Rev. D* 69:094013 (2004); Forshaw JR, Sandapen R, Shaw G. *Phys. Lett.* B594:283 (2004); Forshaw JR, Sandapen R, Shaw G. *J. High Energy Phys.* 0412:052 (2004); Forshaw JR, Sandapen R, Shaw G. *J. High Energy Phys.* 0611:025 (2006)
87. Low FE. *Phys. Rev. D* 12:163 (1975); Nussinov S. *Phys. Rev. Lett.* 34:1286 (1975)
88. Caldwell A, Kowalski H. arXiv:0909.1254 [hep-ph] (2009)
89. Kowalski H, Lappi T, Marquet C, Venugopalan R. *Phys. Rev. C* 78:045201 (2008)
90. Munier S, Stasto AM, Mueller AH. *Nucl. Phys. B* 603:427 (2001); Rogers T, Guzey V, Strikman M, Zu X. *Phys. Rev. D* 69:074011 (2004)
91. Bjorken JD. arXiv:hep-ph/9601363 (1996)
92. Freund A, Rummukainen K, Weigert H, Schafer A. *Phys. Rev. Lett.* 90:222002 (2003); Armesto N, Salgado CA, Wiedemann UA. *Phys. Rev. Lett.* 94:022002 (2005)
93. Kowalski H, Lappi T, Venugopalan R. *Phys. Rev. Lett.* 100:022303 (2008)
94. Kugeratski MS, Goncalves VP, Navarra FS. *Eur. Phys. J. C* 46:413 (2006)
95. Marquet C, Xiao B-W, Yuan F. *Phys. Lett.* B682:207 (2009)
96. Goncalves VP, Kugeratski MS, Machado MVT, Navarra FS. *Phys. Rev. C* 80:025202 (2009); Dominguez F, Marquet C, Wu B. *Nucl. Phys. A* 823:99 (2009)
97. Machado MVT. arXiv:0905.4516 [hep-ph] (2009)
98. Jalilian-Marian J. *Phys. Rev. C* 70:027902 (2004)
99. Albacete JL. *Phys. Rev. Lett.* 99:262301 (2007)
100. Blaizot JP, Gelis F, Venugopalan R. *Nucl. Phys. A* 743:13 (2004)
101. Kharzeev D, Levin E. *Phys. Lett.* B523:79 (2001)
102. Mueller AH. *Nucl. Phys. B* 572:227 (2000)
103. Kharzeev D, Nardi M. *Phys. Lett.* B507:121 (2001)
104. Kharzeev D, Levin E, Nardi M. *Nucl. Phys. A* 730:448 (2004)
105. Dumitru A, Jalilian-Marian J. *Phys. Rev. Lett.* 89:022301 (2002); Dumitru A, Jalilian-Marian J. *Phys. Lett.* B547:15 (2002)

106. Gelis F, Jalilian-Marian J. *Phys. Rev. D* 67:074019 (2003)
107. Dumitru A, Hayashigaki A, Jalilian-Marian J. *Nucl. Phys. A* 765:464 (2006)
108. Albacete JL, Marquet C. arXiv:1001.1378 [hep-ph] (2010)
109. Bjorken JD. *FERMILAB-PUB-82-059-THY* (1982); Wang XN, Gyulassy M. *Phys. Rev. Lett.* 68:1480 (1992); Gyulassy M, Wang XN. *Nucl. Phys. B* 420:583 (1994); Baier R, Dokshitzer YL, Peigné S, Schiff D. *Phys. Lett.* B345:277 (1995); Baier R, et al. *Nucl. Phys. B* 483:291 (1997); Gyulassy M, Vitev I, Wang XN, Zhang BN. arXiv:nucl-th/0302077 (2003)
110. Kharzeev D, Levin E, McLerran LD. *Phys. Lett.* B561:93 (2003); Albacete JL, et al. *Phys. Rev. Lett.* 92:082001 (2004)
111. Arsene I, et al. (BRAHMS Collab.) *Phys. Rev. Lett.* 93:242303 (2004)
112. Adams J, et al. *Phys. Rev. Lett.* 97:152302 (2006)
113. Iancu E, Itakura K, Triantafyllopoulos D. *Nucl. Phys. A* 742:182 (2004)
114. Kharzeev D, Kovchegov Y, Tuchin K. *Phys. Rev. D* 68:094013 (2003); Kharzeev D, Kovchegov Y, Tuchin K. *Phys. Lett.* B599:23 (2004)
115. Goncalves VP, Kugeratski MS, Machado MVT, Navarra FS. *Phys. Lett.* B643:273 (2006)
116. Jalilian-Marian J. *Nucl. Phys. A* 770:210 (2006)
117. Marquet C. *Nucl. Phys. A* 796:41 (2007)
118. Marquet C. Presented at CATHIE-TECHQM Workshop, Brookhaven Natl. Lab. (2009); Tuchin K. arXiv:0912.5479 [hep-ph] (2009)
119. Kharzeev D, Levin E, McLerran LD. *Nucl. Phys. A* 748:627 (2005); Qiu JW, Vitev I. *Phys. Lett.* B632:507 (2006)
120. Adams J, et al. (STAR Collab.) *Phys. Rev. Lett.* 95:152301 (2005)
121. Molnar L. *J. Phys. G* 34:S593 (2007)
122. Adams J, et al. (STAR Collab.) *Phys. Rev. C* 73:064907 (2006)
123. Abelev BI, et al. (STAR Collab.) arXiv:0909.0191 [nucl-ex] (2009)
124. Daugherity M. (STAR Collab.) *J. Phys. G* 35:104090 (2008)
125. Netrakanti PK. (STAR Collab.) *J. Phys. G* 35:104010 (2008); Abelev BI, et al. (STAR Collab.) *Phys. Rev. Lett.* 102:052302 (2009)
126. Voloshin SA. *Phys. Lett.* B632:490 (2006); Shuryak EV. *Phys. Rev. C* 76:047901 (2007); Pruneau CA, Gavin S, Voloshin SA. *Nucl. Phys. A* 802:107 (2008)
127. Gavin S, McLerran L, Moschelli G. *Phys. Rev. C* 79:051902 (2009); Moschelli G, Gavin S, McLerran LD. *Eur. Phys. J. C* 62:277 (2009)
128. Andrade R, Grassi F, Hama Y, Qian WL. arXiv:0912.0703 [nucl-th] (2009); Osada T. arXiv:0911.2303 [nucl-th] (2009); Takahashi J, et al. Xu N, arXiv:0902.4870 [nucl-th] (2009)
129. Adare A, et al. (PHENIX Collab.) *Phys. Rev. C* 78:014901 (2008)
130. Chen J. (PHENIX Collab.) Presented at Winter Workshop Nucl. Dyn., 26th, Ochos Rios, Jam. (2010)
131. Nagle JL. *Nucl. Phys. A* 830:147C (2009)
132. Mueller AH, Navelet H. *Nucl. Phys. B* 282:727 (1987)
133. Alvero L, Collins JC, Terron J, Whitmore JJ. *Phys. Rev. D* 59:074022 (1999)
134. Deshpande A, Milner R, Venugopalan R, Vogelsang W. *Annu. Rev. Nucl. Part. Sci.* 55:165 (2005); Dainton JB, et al. *J. Instrum.* 1:P10001 (2006)

Supersymmetry Breaking and Gauge Mediation

Ryuichiro Kitano,[1] Hirosi Ooguri,[2,3] and Yutaka Ookouchi[4]

[1]Department of Physics, Tohoku University, Sendai 980-8578, Japan; email: kitano@tuhep.phys.tohoku.ac.jp

[2]Department of Theoretical Physics, California Institute of Technology, Pasadena, California 91125; email: ooguri@theory.caltech.edu

[3]Institute for the Physics and Mathematics of the Universe, University of Tokyo, Kashiwa, Chiba 277-8582, Japan

[4]Perimeter Institute for Theoretical Physics, Waterloo, Ontario N2L2Y5, Canada; email: yokochi@perimeterinstitute.ca

Annu. Rev. Nucl. Part. Sci. 2010. 60:491–511

First published online as a Review in Advance on July 19, 2010

The *Annual Review of Nuclear and Particle Science* is online at nucl.annualreviews.org

This article's doi: 10.1146/annurev.nucl.012809.104540

0163-8998/10/1123-0491$20.00

Key Words

model building, dynamical symmetry breaking, string theory

Abstract

We review recent works on supersymmetry breaking and gauge mediation. We survey our current understanding of dynamical supersymmetry-breaking mechanisms and describe new model-building tools that use duality, metastability, and stringy construction. We discuss phenomenological constraints and their solutions, paying particular attention to gaugino masses and electroweak symmetry breaking.

Contents

1. INTRODUCTION

Supersymmetry plays an important role in modern particle physics, even though there is no direct experimental evidence for its existence. It is a hypothetical space-time symmetry that transforms bosonic states into fermionic ones and vice versa; it is characterized by features such as vacuum stability ($E \geq 0$) and mild ultraviolet (UV) behavior of theories, that is, a restricted form of divergences. Many theorists consider that these features are indications that supersymmetry is an essential ingredient of the ultimate unified theory of elementary particles, which perhaps is string theory.

Supersymmetry has also provided examples of (partly) calculable strongly coupled theories; for example, there are strong/weak dualities among supersymmetric theories. Such dualities include those of the $\mathcal{N} = 4$ supersymmetric Yang-Mills theory (1–5) and the Seiberg-Witten theory (6), the Seiberg dualities (7, 8) in quantum chromodynamics (QCD)-like supersymmetric gauge theories, and the gauge/gravity correspondence (9). These dualities have improved our understanding of nonperturbative physics.

Particle theorists have applied supersymmetry to the Standard Model of particle physics. Their main motivation is to protect the Higgs potential against quantum corrections. Because of very weak UV divergences, one can naturally push the cutoff scale to an arbitrary high-energy scale. If the cutoff scale is postponed until, say, the Planck scale ($\sim 10^{18}$ GeV), then it offers a wonderful arena for model builders. One can build calculable field-theoretical models of high-scale or high-temperature phenomena such as inflation, baryogenesis, neutrino masses, fermion mass hierarchies, grand unification, and so on without worrying about the naturalness problems caused by large quantum corrections.

The successful unification of the coupling constants in the minimal supersymmetric extension of the Standard Model (MSSM) (10–12) has attracted many theorists to the supersymmetric world. Moreover, the lightest supersymmetric particle is a strong candidate for dark matter of the Universe. (See Reference 13 for a review.) Because of this beautiful and successful framework,

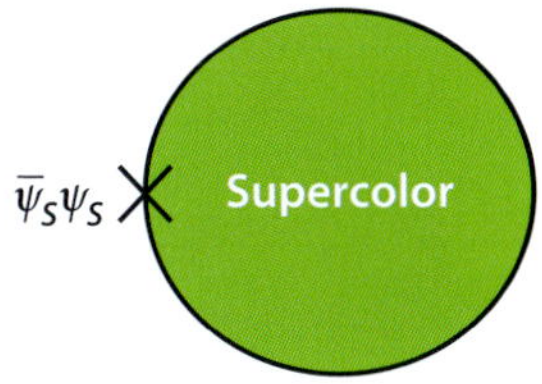

Figure 1

Dynamical supersymmetry breaking.

there have been many studies on supersymmetry searches at high-energy colliders such as LEP, Tevatron, LHC, and ILC, as well as through dark matter detections (14).

However, there is one missing link: the supersymmetry-breaking mechanism. We need mass splittings between bosons and fermions because of experimental constraints. Satisfying the experimental bounds themselves is not difficult. One can assume a supersymmetry-breaking sector that couples to the MSSM in such a way that large enough mass splittings are obtained. A nontrivial constraint on model building comes from consideration of the Higgs sector. Naturalness of electroweak symmetry breaking, together with the experimental bounds, requires that all the superpartners have masses of $\mathcal{O}(100 \text{ GeV})$.

In this article, we review recent developments toward a viable supersymmetric model of particle physics. We begin with a description of an early attempt at supersymmetric model building by Dine et al. (15), who discussed the concepts of dynamical supersymmetry breaking, (direct) gauge mediation, and their connection to the electroweak symmetry breaking. Even though the specific model presented in this paper is incomplete due to progress in supersymmetric gauge theories, the authors' scenario remains one of the most attractive and elegant. We then turn to recent progress in supersymmetry breaking and gauge mediation. Topics include the discovery of metastable vacua in the supersymmetric QCD (SQCD) by Intriligator, Seiberg, and Shih (ISS) (16), new formulations of gauge mediation and related topics, direct gauge-mediation models and gaugino masses, and connections to string theory. We conclude with problems associated with electroweak symmetry breaking.[1]

2. PROTOTYPE SCENARIO OF LOW-ENERGY SUPERSYMMETRY

In 1981, Witten (20) investigated whether supersymmetry breaking occurs dynamically. He proposed a natural framework for electroweak physics, known as supersymmetric technicolor, in which a technicolor force dynamically breaks both supersymmetry and the electroweak $SU(2) \times U(1)$. That same year, Dine et al. (15) and Dimopoulos & Raby (21) independently proposed concrete models along a similar line. Dine et al. (15) proposed a mechanism to generate gaugino and sfermion masses that is now known as direct gauge mediation. In this scenario, there is a QCD-like $SU(M)$ gauge theory that becomes strong at a scale Λ_{SC}, where SC stands for supercolor. It is assumed that supersymmetry is dynamically broken by a condensation of a pair of superquarks, ψ_S and $\bar{\psi}_S$, which are fermion components of the chiral superfields S and $\bar{S}$ (**Figure 1**). Because a condensation of the fermion pair, $\langle \bar{\psi}_S \psi_S \rangle \simeq \Lambda_{SC}^3$, is an F-component of the meson superfield ($M \sim S\bar{S}$), supersymmetry is spontaneously broken if such a condensation forms.

[1]There are numerous topics that we do not cover in this review, especially cosmological or astrophysical constraints on supersymmetric models. We refer the reader to, for instance, References 17–19 for constraints relevant for dynamical supersymmetry breaking and gauge mediation.

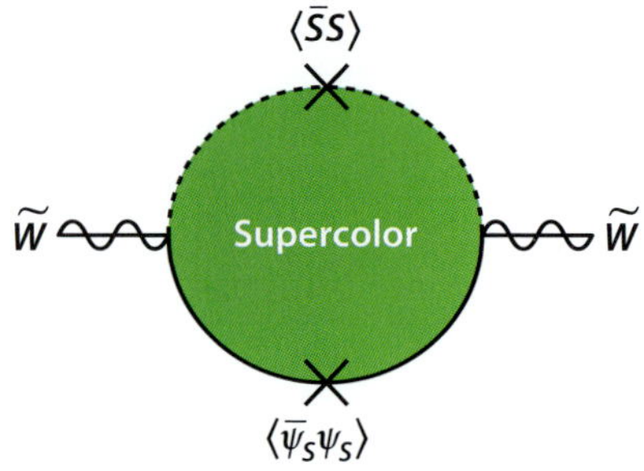

Figure 2

A diagram showing the contribution of gaugino masses.

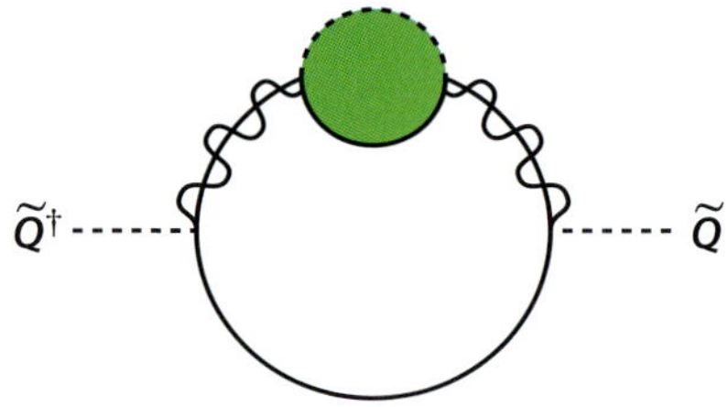

Figure 3

A diagram showing the contribution of sfermion masses.

Part of the global symmetry in the sector can be identified with the Standard Model gauge group by assigning quantum numbers to the superquarks. [In the original model, only the $SU(2) \times U(1)$ part was embedded.] The gauginos and sfermions can obtain masses through loop diagrams involving the Standard Model gauge interactions (**Figures 2** and **3**). To generate nonzero gaugino masses, one assumes that there is a boson-pair condensation,

$$\langle \bar{s}s \rangle \simeq \Lambda_{SC}^2, \qquad 1.$$

in addition to the fermion-pair condensation. Here $\bar{s}$ and s are the lowest components of $\bar{S}$ and S, respectively. The presence of both condensations ensures the breakdown of R-symmetry that is necessary for nonvanishing gaugino masses. To obtain $\mathcal{O}(100 \text{ GeV})$ gaugino and sfermion masses, one assumes that the dynamical scale Λ_{SC} is $\mathcal{O}(10 \text{ TeV})$ due to the loop factors ($\sim \alpha/4\pi$).

Finally, electroweak symmetry breaking can occur either through some other dynamics (technicolor) or through the standard Higgs mechanism. In both cases, one can introduce elementary Higgs fields to write down the Yukawa interactions while avoiding the naturalness issue, thanks to supersymmetry. A technicolor model has been proposed (15) in which the electroweak vacuum expectation value (VEV) and the Higgsino mass are generated through a strong dynamics at the $\mathcal{O}(300\,\text{GeV})$ energy scale, as illustrated in **Figures 4** and **5**. The authors (15) also discussed the possibility of using the usual Higgs mechanism.

This model has three essential ingredients: dynamical supersymmetry breaking with R-symmetry violation, gauge mediation, and electroweak symmetry breaking through supersymmetry breaking. In the following section, we review recent progress and our current understanding of these three components.

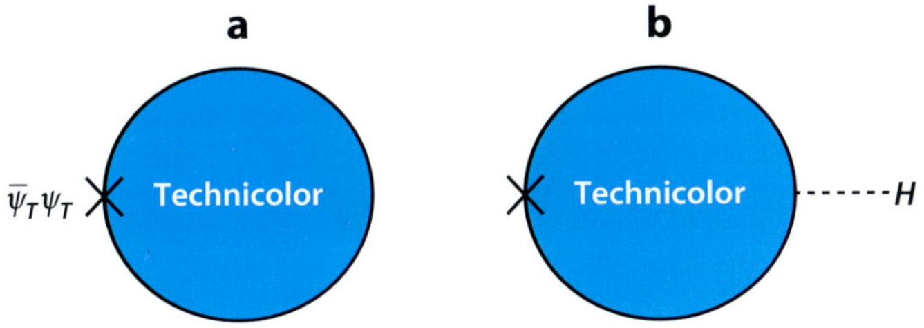

Figure 4

(*a*) Electroweak symmetry breaking. (*b*) The vacuum expectation value of the Higgs field.

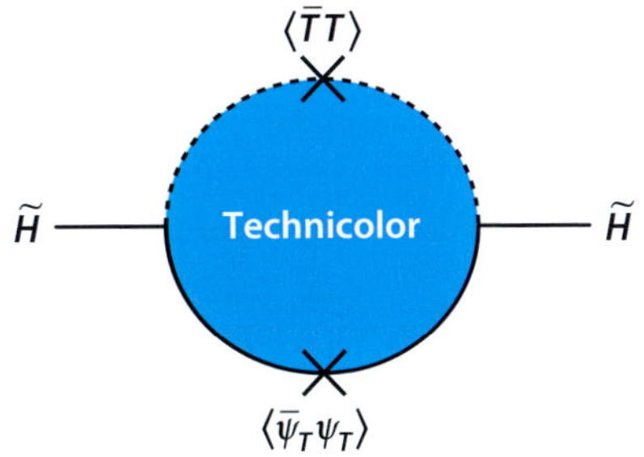

Figure 5

A diagram showing the contribution of the Higgsino mass.

3. SUPERSYMMETRY BREAKING

In our current understanding, the assumption (15) for supersymmetry breaking,

$$\langle \bar{\psi}_S \psi_S \rangle \simeq \Lambda_{\rm SC}^3, \qquad 2.$$

is not valid in SQCD theories. These theories have stable supersymmetric vacua (22, 23). Alternative possibilities have been considered, and there are successful models for dynamical supersymmetry breaking, for example chiral gauge theories (24–26) and theories with gauge-singlet fields (27, 28). See References 29 and 30 for more recent proposals.

Recently, dynamical supersymmetry breaking in SQCD theories has been revived by the work of ISS, who found a metastable supersymmetry breaking vacuum in $SU(N_c)$ supersymmetric gauge theories with massive N_f flavors for $N_c < N_f < 3N_c/2$ (16). The presence of the vacuum is established when $m \ll \Lambda$, where m and Λ are the quark mass and the dynamical scale, respectively. A nonvanishing fermion-pair condensation,

$$\langle \bar{\psi}_Q \psi_Q \rangle \sim m\Lambda^2, \qquad 3.$$

is obtained (rather than Equation 2), as illustrated in **Figure 6**.

Let us first discuss the linear sigma models for supersymmetry breaking, which are widely used (*a*) to establish the existence of a supersymmetry-breaking vacuum in strongly coupled theories (24–26, 31, 32) and (*b*) as effective theories to describe low-energy physics of a variety of supersymmetry-breaking models. We then introduce the ISS model and its connection to the linear sigma model. For a recent pedagogical review on supersymmetry breaking, see Reference 33.

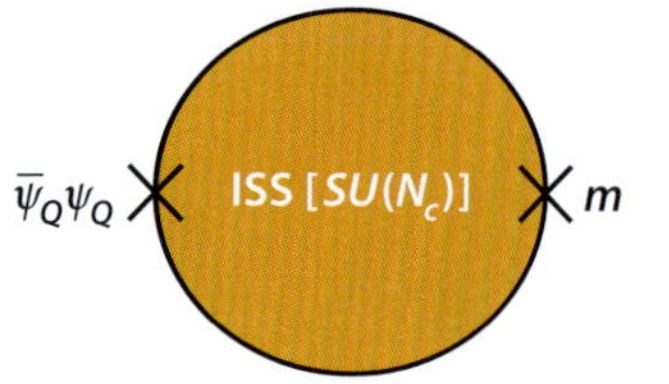

Figure 6

Dynamical supersymmetry breaking in the Intriligator-Seiberg-Shih (ISS) model.

3.1. Polonyi and Generalized O'Raifeartaigh Models

Supersymmetry is spontaneously broken when the F-component of a chiral superfield X acquires a VEV,

$$\langle F_X \rangle \neq 0, \qquad 4.$$

because $2\langle F_X \rangle = \langle \{iQ, X\}|_\theta \rangle$, which should vanish if $Q|0\rangle = 0$. It is easy to construct a linear sigma model to describe this phenomenon by using a single chiral superfield X. The superpotential is simply

$$W = \mu^2 X. \qquad 5.$$

The F-component of X acquires a VEV unless the Kähler potential is singular. This is the unique choice of the superpotential for a model with a single chiral superfield X. (There can be a small perturbation to the superpotential if one allows the vacuum to be metastable.) By making an appropriate shift of X, we can choose the stable point at $X = 0$. For this point to be stable, the Kähler potential expanded around it should be of the form

$$K = X^\dagger X - \frac{(X^\dagger X)^2}{\Lambda^2} + \cdots. \qquad 6.$$

Equations 5 and 6 define the linear sigma model for supersymmetry breaking: the Polonyi model (34).

The equation of motion for F_X and the potential minimization of X lead to

$$\langle F_X \rangle = \mu^2, \quad \text{where} \quad \langle X \rangle = 0. \qquad 7.$$

Therefore, supersymmetry is spontaneously broken. The fermion component ψ_X remains massless. This is the Goldstino fermion associated with the spontaneous supersymmetry breaking. The complex scalar field X obtains the following mass:

$$m_X^2 = \frac{4\mu^4}{\Lambda^2}. \qquad 8.$$

Up to $\mathcal{O}(1/\Lambda^2)$, there is no mass splitting between the scalar and the pseudoscalar parts due to an unbroken approximate R-symmetry with $R(X) = 2$.

Although it is a nonrenormalizable model, this model serves as the effective theory (if $\mu^2 \ll \Lambda^2$) for a large class of supersymmetry-breaking models. Conversely, one can establish the presence of a supersymmetry-breaking vacuum when a model reduces to the Polonyi model at a point of the field space by integrating out massive degrees of freedom. This technique has been used extensively (16, 24–26, 31, 32, 35–37).

One can also use the Polonyi model as a hidden sector, whose supersymmetry-breaking effects are communicated to the MSSM sector through interactions such as gravity or gauge interactions. Simple models of gauge mediation have been constructed (e.g., 38–41) by attaching the messenger fields to the above-described linear sigma model. These simple models belong to the indirect type of gauge mediation, rather than to the direct type depicted in **Figure 2** and discussed further below.

The O'Raifeartaigh model (42) and its generalization are alternative simple theories for supersymmetry breaking. (They reduce to the Polonyi model when the massive fields are much heavier than the size of supersymmetry breaking.) The generalized O'Raifeartaigh model is defined by a set of chiral superfields (Φ_i) and a superpotential up to cubic terms [the Wess-Zumino model (43)], where

$$\frac{\partial W}{\partial \Phi_i} = 0 \qquad 9.$$

cannot be solved simultaneously. The chiral superfields Φ_i have the canonical Kähler potential, $K = \Phi_i^\dagger \Phi_i$. These models have recently received renewed attention because many dynamical models (including the ISS model) reduce to it at low energy. One of the basic properties common to all the O'Raifeartaigh models is the existence of tree-level flat directions, known as pseudomoduli, that emanate from any local supersymmetry-breaking vacuum (44). Such pseudomoduli receive quantum corrections, and a potential along the flat direction is generated.

3.2. The Intriligator-Seiberg-Shih Model

Seiberg (8) showed that SQCD theories with quark mass m have stable supersymmetric vacua at

$$\langle M \rangle = \langle Q\bar{Q} \rangle \sim \Lambda^{(3N_c - N_f)/N_c} m^{(N_f - N_c)/N_c}, \qquad 10.$$

where N_c and N_f are numbers of color and flavor, respectively, which appeared to be inconsistent with dynamical supersymmetry breaking in these theories (22).

Recently, ISS showed that there can be a metastable supersymmetry-breaking vacuum far away from the supersymmetric vacua (16). The model they considered is a supersymmetric $SU(N_c)$ theory with N_f fundamental and antifundamental chiral superfields Q_i and $\bar{Q}_i$ ($i = 1, \ldots, N_f$). The superpotential is

$$W = m Q_i \bar{Q}_i, \qquad 11.$$

where the color and flavor indices are contracted. This is an asymptotically free theory and becomes strongly coupled in the infrared region. Therefore, the analysis in the electric description (i.e., the description in terms of quarks and gluons) is difficult near the origin of the meson space $M \sim Q\bar{Q}$. By the power of duality, the infrared physics of the model is described by the dual magnetic theory, which is weakly coupled for $N_c < N_f < 3N_c/2$ (7). Perturbative calculations in the magnetic theory are reliable near the origin of the meson-field space.

The dual gauge theory is $SU(N_f - N_c)$ with N_f dual quarks, q and $\bar{q}$, and the gauge-singlet meson fields M. The superpotential is

$$W_{\text{mag.}} = h\mu^2 M_{ii} - h q_i M_{ij} \bar{q}_j, \qquad 12.$$

where h is a coupling constant of $\mathcal{O}(1)$ and the term $h\mu^2 M$ corresponds to the quark mass terms in Equation 11. The coefficient $h\mu^2$ is naturally of $\mathcal{O}(m\Lambda)$. Again, i and j are the $SU(N_f)$ flavor indices. At the point where $\langle M_{ij} \rangle = 0$ and $\langle q_i \rangle = \langle \bar{q}_i \rangle \neq 0$ and the dual gauge group is completely broken, the flavor $SU(N_f)$ symmetry is broken down to $SU(N_f - N_c) \times SU(N_c)$. The effective superpotential at that point has the form of the O'Raifeartaigh model,[2]

$$W = h\,\mathrm{Tr}\left(\mu^2 X - X\rho\tilde{\rho} - \mu\rho\tilde{Z} - \mu Z\tilde{\rho}\right),$$

where X is the pseudomoduli field, which is a $N_c \times N_c$ part of the meson M; and ρ, $\tilde{\rho}$, Z, and $\tilde{Z}$ are massive fields. Once we integrate out these massive fields, the Coleman-Weinberg potential, which stabilizes the pseudomoduli at the origin, is generated (16). The F-component of the X field ($\sim \bar{\psi}_Q \psi_Q$) obtains a VEV triggered by an explicit breaking of chiral symmetry, as in **Figure 6**. Although the detailed structure is different, the dynamical supersymmetry breaking through a fermion-pair condensation (15, 20, 21) is revived in the metastable vacua of SQCD models. The tunneling rate into the true supersymmetric vacuum in Equation 10 is exponentially suppressed if $m \ll \Lambda$, because the supersymmetric vacuum is located far from the metastable vacuum, in comparison with the height of the potential along the meson direction $V^{1/4} \sim (m^2 M^2)^{1/4}$.

4. GAUGE MEDIATION

Inspired by the simplicity and genericity of metastable supersymmetry breaking, gauge-mediation model building has recently been invigorated.[3] The diagram in **Figure 2** has been reconsidered in light of our new knowledge of supersymmetry breaking, and it has been reformulated by use of current correlators. In this section, we review techniques to calculate the gaugino and sfermion masses.

4.1. Analytic Continuation into Superspace

One can perform explicit calculations of gaugino and sfermion masses when the supercolor boxes in **Figures 2** and **3** are loops of weakly coupled fields, known as messenger fields. Such models were first considered in 1982 (46–48) and revived a decade later (49–51). In the latter studies, explicit computations of gaugino and sfermion masses were performed. The highly predictive nature of gauge mediation was demonstrated, revealing that the gaugino and sfermion masses are functions only of their quantum numbers to a good approximation.

A powerful method has been developed to compute multiloop quantities from one-loop running data (52, 53). The results follow from the imposition of constraints due to holomorphy in a spurion, $X = X_0 + \theta^2 F$, on the effective action. One can use the method as the leading-order calculation in the F/X_0^2 expansion; higher-order terms in F arise from terms involving superderivatives, which are not considered. The gaugino masses are identified with the F-component of the holomorphic gauge coupling $\tau(X)$. The X dependence originates from the change of the β functions due to the decoupling of a messenger field whose mass is X_0 in the supersymmetric limit. The sfermion masses can be extracted from the wave function renormalizations $Z_Q(X, \bar{X})$ of the matter superfields Q. At leading order, Z_Q depends on X through the gauge couplings $\tau(X)$. By expanding the effective action in F, one can describe the gaugino and sfermion masses by one-loop quantities,

$$m_\lambda = -F \left.\frac{\partial \ln \tau(X)}{\partial X}\right|_{X=X_0} \simeq \Delta\beta_{g_a}^{(1)} \frac{F}{X_0},$$

[2]For simplicity we omit other pseudomoduli in the model. For details, see Reference 16.

[3]See Reference 45 for a review on earlier works on gauge mediation.

$$m_s^2 = -|F|^2 \left. \frac{\partial^2 \ln Z_Q(X, X^\dagger)}{\partial X \partial X^\dagger} \right|_{X=X_0} \simeq \gamma_s^{(1)} \Delta\beta_{g_a}^{(1)} \left| \frac{F}{X_0} \right|^2, \qquad 13.$$

where $\Delta\beta_{g_a}^{(1)}$ is a discontinuity of the coefficient of the one-loop β function at the threshold X_0, and $\gamma_s^{(1)}$ is the one-loop anomalous dimension of the matter fields. This method simplifies calculations of soft masses.

In addition to applications for gauge mediation, the technique of the analytic continuation into superspace can be used in computing the effective potential of the pseudomoduli in the supersymmetry-breaking sector. A leading-log effective potential for a pseudomoduli has been computed (54, 55). Even if the potential is generated only at higher loops, there is a regime in which the potential can be determined simply from a combination of one-loop running data. The results have been applied to survey pseudomoduli spaces for large classes of models (55).

4.2. General Gauge Mediation

The method described in Section 4.1 works only in cases in which the messenger fields are weakly coupled. Generally, to evaluate diagrams such as those shown in **Figures 2** and **3**, we need information about two-point functions of currents in the supersymmetry-breaking dynamics.

Recently, Meade et al. (56) provided the most general parameterization of the two-point functions by giving a model-independent definition of gauge mediation: In the MSSM gauge-coupling limit $\alpha_i \to 0$, the MSSM sector decouples from the hidden sector that breaks supersymmetry. Specifically, the MSSM gauge group becomes a global symmetry G of the hidden sector in this decoupling limit. All the information we need for gauge mediation is encoded in the currents and their correlation functions.

The conserved currents are real linear supermultiplets that satisfy $D^2\mathcal{J} = 0$. For a $U(1)$ symmetry, the current superfield can be written in components as

$$\mathcal{J} = J + i\theta j - i\bar{\theta}\bar{j} - \theta\sigma^\mu\bar{\theta} j_\mu + \cdots.$$

Two-point functions are constrained by the Lorentz invariance and the current conservation as follows:

$$\begin{aligned}
\langle J(p)J(-p)\rangle &= \tilde{C}_0(p^2/M^2),\\
\langle j_\alpha(p)\bar{j}_{\dot{\alpha}}(-p)\rangle &= -\sigma^\mu_{\alpha\dot{\alpha}} p_\mu \tilde{C}_{1/2}(p^2/M^2),\\
\langle j_\mu(p) j_\nu(-p)\rangle &= -(p^2\eta_{\mu\nu} - p_\mu p_\nu)\tilde{C}_1(p^2/M^2),\\
\langle j_\alpha(p) j_\beta(-p)\rangle &= \varepsilon_{\alpha\beta} M\tilde{B}(p^2/M^2).
\end{aligned}$$

Gaugino and scalar masses at leading order in the gauge coupling are governed by the two-point functions[4]

$$m_\lambda = g^2 M\tilde{B}(0) \quad \text{and} \quad m_s^2 = g^4 Y^2 A, \qquad 14.$$

where Y is the $U(1)$ charge of the sfermion and A is defined by

$$A = -\int \frac{d^4p}{(2\pi)^4} \frac{1}{p^2}[3\tilde{C}_1(p^2/M^2) - 4\tilde{C}_{1/2}(p^2/M^2) + \tilde{C}_0(p^2/M^2)].$$

[4]In general one-point functions, $\langle J\rangle$ is nonzero, which could lead to tachyonic sfermion. To forbid a contribution from the one-point function, the messenger $\mathbf{Z}_2$ parity was imposed $\mathcal{J} \to -\mathcal{J}$ (56).

Note that A is a linear combination of terms with different signs; thus, sfermion masses squared are not necessarily positive. Even under this generic situation, the scalar masses obey two sum rules,

$$\mathrm{Tr}(Y\, m_s^2) = 0 \quad \text{and} \quad \mathrm{Tr}[(B - L)m_s^2] = 0.$$

When generalizing the $U(1)$ theory to the full Standard Model gauge group $SU(3) \times SU(2) \times U(1)$, A and B are replaced by A_k and B_k ($k = 1, 2, 3$), corresponding to the three gauge-group factors. These are three real and complex numbers that provide nine independent parameters in general gauge mediation. Arbitrary phases of B_k typically lead to an unacceptable level of the electric dipole moment of an electron and a neutron. Therefore, it is plausible (*a*) to assume that the hidden sector is *CP* invariant and (*b*) to impose phenomenological constraints on the phases. Thus, we are left with six real parameters that span the parameter space of the general gauge mediation.

Are there simple models of weakly coupled messengers that cover the entire parameter space? This question was first raised by Carpenter et al. (57), who studied F-term supersymmetry-breaking models. The authors found models with the right number of parameters, but these models do not cover the entire parameter space. This question was investigated further by Buican et al. (58), who used D-term supersymmetry as suggested by Nakayama and colleagues (59). Buican et al. found models with weakly coupled messengers that span the whole parameter space. Although direct mediation models include cases with strongly coupled hidden sectors, the answer to the question posed at the beginning of this paragraph turns out to be positive.

The idea of analytic continuation into superspace (reviewed in Section 4.1) was extended to general gauge mediation for the small supersymmetry-breaking scale $F \ll M^2$ (60, 61). Distler & Robbins (60) and Intriligator & Sudano (61) showed identities for the two-point functions and reproduced the result shown in Equation 13 by exploiting the general mass formulae of Equation 14. For more works on general gauge mediation, see References 62–66.

5. DIRECT MEDIATION

With the technology of calculating or parameterizing the gaugino/sfermion masses, we can proceed to a discussion of model building with direct gauge mediation. Direct mediation is a type of gauge mediation in which messenger sectors are responsible for the supersymmetry breaking and (meta)stability of vacua[5] (**Figures 2** and **3**). In some cases, supersymmetry is restored if couplings to messengers are turned off. For a more precise definition of direct mediation, see References 57 and 71. Following two seminal publications (15, 21), this possibility of model building received a great deal of attention. See References 72–76 for some earlier works.

In this section, we discuss two major challenges in constructing phenomenologically viable models and their possible solutions: the Landau pole and the light gaugino mass problems. In a direct-type model, the messenger sector cannot be adjusted arbitrarily because it is closely tied to supersymmetry-breaking effects. Specifically, if there are a large number of messenger fields, the coupling constants of the Standard Model gauge interaction hit the Landau pole below the unification scale. One can see difficulties in, for example, the models of References 77, 78, and 79. In Reference 81, a model was proposed to alleviate this issue by using a nontrivial conformal fixed point. For more recent progress in this approach, see Reference 82.

[5]There is also semidirect mediation, in which messengers couple to the supersymmetry-breaking sector, but they are not relevant for the stability of a vacuum (67). Semidirect mediation includes mediator models (53, 68–70).

Another issue is how to generate large enough gaugino masses. Nonzero Majorana gaugino masses require that R-symmetry be broken.[6] However, simple O'Raifeartaigh-type models, including the ISS model, preserve an (approximate) R-symmetry at the supersymmetry-breaking vacuum. One needs a careful arrangement of particle content, superpotential, and coupling constants to achieve supersymmetry breaking without R-symmetry. Also, R-symmetry breaking is often not sufficient to guarantee large enough gaugino masses. Below we review these issues and discuss how to construct successful models.

5.1. Breaking of R-Symmetry

R-symmetry can be broken either spontaneously or explicitly. For a generalized O'Raifeartaigh model,[7] spontaneous R-symmetry breaking requires a field with an R-charge different from zero or two (87). Explicit examples of direct mediation with exotic R-charges have been constructed (89). Because such extraordinary gauge-mediation models do not necessarily preserve the approximate messenger parity (90), one-loop corrections generate tachyonic scalar masses in general. A way to suppress such one-loop contributions has been discussed (57).

Another possibility is to break an R-symmetry explicitly at the Lagrangian level, which has the added advantage of avoiding the unwanted R-axion. For a generic superpotential, stable supersymmetry-breaking vacua require an unbroken R-symmetry (23). However, explicit R-symmetry breaking is allowed if one permits metastable vacua. Specifically, if R-symmetry breaking is small, supersymmetric vacua are generated near the infinity in the field space, which guarantees the metastability of the supersymmetry-breaking vacuum. Because the R-symmetry is broken, there is no symmetry-related reason to prohibit the gaugino masses. However, generating large enough gaugino masses is still a challenge, as we explain below.

5.2. Anomalously Small Gaugino Mass

Let us consider the case in which the supersymmetry-breaking scale $\sqrt{F}$ is much smaller than the messenger scale M. The leading contribution to gaugino masses in the F/M^2 expansion is given by

$$m_\lambda = \frac{g^2}{(4\pi)^2} \cdot F_X \frac{\partial}{\partial X} \log[\det \mathcal{M}(X, M)], \qquad 15.$$

where $\mathcal{M}(M)$ is the fermion mass matrix of messengers and X is the pseudomoduli field with the nonzero F-component responsible for supersymmetry breaking (52). The right-hand side of this equation vanishes in a simple O'Raifeartaigh model even if the R-symmetry breaking is large (91). This problem has often been observed in models of direct gauge mediation (71, 77, 78, 92, 93). Therefore, researchers have had to consider models with $F/M^2 = \mathcal{O}(1)$, such as $\sqrt{F} \sim M \sim 10 - 100$ TeV, with which it is difficult to avoid the Landau pole problem. Moreover, it is interesting that the models with vanishing leading-order gaugino masses are severely constrained by a recent Tevatron bound on sparticle masses and a mass bound on a light gravitino (94).

---

[6] An alternative approach is to use Dirac gaugino mass (80–84), which requires extra fields in adjoint representation in the MSSM. Such models typically suffer from the Landau pole problem. Interestingly, R-symmetry does not have to be broken in such models. See References 85 and 86 for recent proposals.

[7] This statement does not hold for a model with gauge interactions (71) or for a model in which two-loop Coleman-Weinberg potential is dominant (88).

Recently, the reason for the small gaugino masses was explained on general grounds (95). Suppose that the low-energy effective theory near a given vacuum is described by the O'Raifeartaigh model with the superpotential

$$W = fX + \frac{1}{2}(\lambda_{ab}X + m_{ab})\phi_a\phi_b + \frac{1}{6}g_{abc}\phi_a\phi_b\phi_c. \quad 16.$$

By performing a unitary transformation of fields, one can rewrite the general O'Raifeartaigh model in this form (44). The fields ϕ_a are identified with messengers. The Kähler potential is canonical for all the chiral superfields. In this model, $\det(\lambda X + m)$ is a constant (independent of X) in a stable supersymmetry-breaking vacuum, that is, in the lowest-energy state (95). In the proof, it is assumed that there is no unstable point anywhere in the pseudomoduli space because otherwise there should be a lower-energy state.

This theorem means that the leading-order gaugino masses vanish at leading order in the F/M^2 expansion in this vacuum (from Equation 15, where $\mathcal{M} = \lambda X + m$), regardless of whether or how R-symmetry is broken. The fact that many dynamical supersymmetry-breaking models are effectively described by Equation 16 near the vacuum puts constraints on model building with direct gauge mediation. The metastability does not help qualitatively because the presence of a supersymmetric vacuum should be a small effect. Interestingly, anomalously small gaugino masses are related to global structure of the vacua of the theory and do not depend on details about how R-symmetry is broken.

One way to solve this gaugino mass problem is to choose our vacuum to be even more metastable, that is, to use an even higher energy state compared with the lowest supersymmetry-breaking vacuum so that the presence of an unstable point in the pseudomoduli space is allowed. In the next section we describe an example of such a model, which predates Reference 95.

5.3. Example of a Direct Mediation Model

As discussed above, a successful model for large enough gaugino masses requires R-symmetry breaking as well as a certain global structure of the pseudomoduli space. In this section, we review an example of a model with nonvanishing gaugino masses at leading order in F (96). It is a deformation of the ISS model described in Section 3: an $SU(N_c)$ gauge theory with N_f flavors with mass terms.

The idea is to embed the Standard Model gauge group into the global symmetry of the ISS model: $SU(N_f) \times U(1)_B$. To guarantee (meta)stability, to avoid the Landau pole below the unification scale, and to generate nonvanishing gaugino masses, it is assumed that (*a*) the quark masses are split into two groups, m_{light} for N_c flavors (Q_a, $a = 1, \ldots, N_c$) and m_{heavy} for $N_f - N_c$ flavors (Q_I, $I = 1, \ldots, N_f - N_c$); and (*b*) there is a quartic term, $(Q_I\bar{Q}_a)(Q_a\bar{Q}_I)$, in the superpotential. The quartic term breaks R-symmetry explicitly and makes the structure of the pseudomoduli space richer, helping to generate sizable gaugino masses.

The global symmetry is now $SU(N_f - N_c) \times SU(N_c) \times U(1)_B$. The Landau pole can be avoided when the Standard Model gauge group is embedded in the $SU(N_f - N_c)$ part and $m_{\text{heavy}} \gg m_{\text{light}}$.

In the magnetic description, the effective superpotential is given by

$$W = h\mu^2 \sum_{a=1}^{N_c} M_{aa} + hm^2 \sum_{I=1}^{N_f - N_c} M_{II} - h\,\mathrm{Tr}\, qM\bar{q} - hm_z \sum_{a,I} M_{Ia}M_{aI}, \quad 17.$$

where h is a parameter of $\mathcal{O}(1)$ and $\mu \ll m$ for $m_{\text{light}} \ll m_{\text{heavy}}$. The last term (the term with a coefficient hm_z) corresponds to the quartic term in the electric description. A part of the meson fields, M_{II}, M_{aI}, and M_{Ia}, and a part of the dual quarks, q_I and $\tilde{q}_I$, are charged under the Standard Model gauge group and thus are messenger fields. In addition to the perturbed ISS vacuum in which the meson fields M_{aa} acquire small VEVs that vanish as $m_z \to 0$, there are new supersymmetry-breaking vacua wherein the meson fields acquire large VEVs that go to infinity as $m_z \to 0$. The new vacua have lower energies than the perturbed ISS vacuum.

In the new vacua, gaugino masses vanish at leading order of F. We now understand why, thanks to the theorem of Reference 95. However, the original ISS vacuum has higher energy and thus evades the theorem. Near the origin of the meson space, the low-energy effective theory is the O'Raifeartaigh model with the superpotential

$$W = h\,\mathrm{Tr}\left(\mu^2 X - X\rho\tilde{\rho} - me^{-\theta}\rho\tilde{Z} - me^{\theta}Z\tilde{\rho} - m_z Z\tilde{Z}\right).$$

One-loop effects stabilize the potential, giving rise to the ISS vacuum. However, the tree-level scalar potential has an unstable region near the new vacua, $X \sim m^2/m_z$, so the ISS vacuum evades the theorem (95) and generates the gaugino masses

$$m_\lambda = \frac{g^2 N_c}{(4\pi)^2}\frac{h\mu^2}{m}\frac{m_z}{m} + O\left(\frac{m_z^2}{m^2}\right), \qquad 18.$$

where g is the coupling constant of the Standard Model gauge interaction. Scalar masses are also obtained by two-loop diagrams and can be adjusted to be the same order as the gaugino mass $\mathcal{O}(m_i) \simeq \mathcal{O}(m_\lambda)$ by setting $m_z \sim m/\sqrt{N_c}$.

6. STRINGY REALIZATION

Here we describe the realization of supersymmetry-breaking mechanisms and their mediations in string theory. There are two motivations for stringy constructions. One is to provide a unified framework for field-theory models in which to understand the nature of the parameters of these models. String theory has been used to develop powerful computational tools for field-theory effects and to provide geometric insights into supersymmetry-breaking phenomena. The second motivation is to better understand supersymmetry breaking in string theory. Over the past 25 years, much progress has been made in understanding of string theory in supersymmetric backgrounds, and many exact results on nonperturbative effects have been derived. It is desirable to extend these results to nonsupersymmetric situations.

Once we accept metastable vacua, numerous stringy realizations become possible. Because long-lived metastable vacua are ubiquitous in vector-like models such as SQCD, it is relatively easy to embed supersymmetry-breaking models in string theories. In the following sections, we review model building by intersecting branes, gauge/gravity dualities, and geometric transitions. For an interesting recent development of string phenomenology with F-theory, see Reference 97.

6.1. Brane Configuration

Open strings are collective coordinates of D-branes. On parallel Dp-branes, if we take the limit of string length $\ell_s \to 0$ while scaling the string coupling constant as $g_s = g_{\mathrm{YM}}^2 \ell_s^{3-p}$, the low-energy effective theory becomes the maximally supersymmetric Yang-Mills theory in $(p+1)$ dimensions with the gauge coupling given by g_{YM}. We can reduce the number of supercharges by considering configurations of intersecting branes. For example, if we have a pair of parallel NS5-branes, and if we suspend parallel D4-branes between them such that all the branes share four dimensions, then the low-energy effective theory is the $\mathcal{N}=2$ supersymmetric pure Yang-Mills theory in four

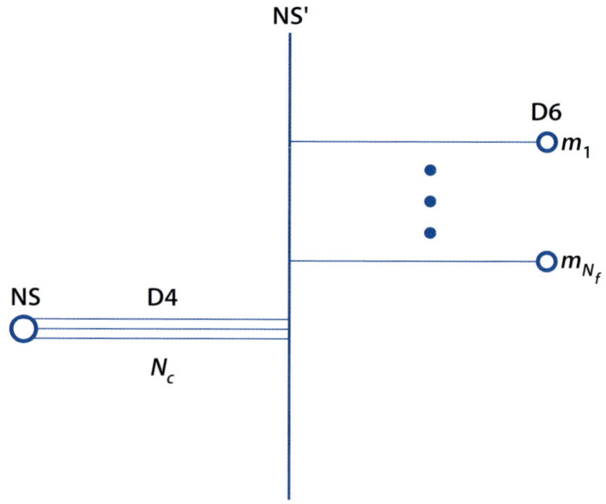

Figure 7

The electric brane configuration for $\mathcal{N} = 1$ supersymmetric gauge theory with massive flavors.

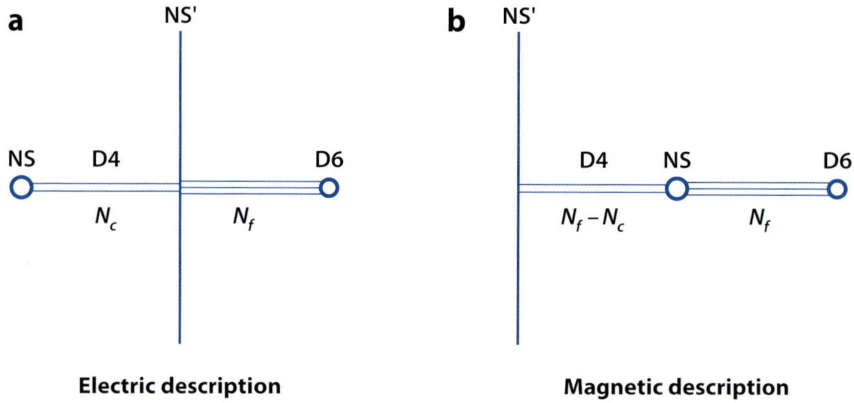

Figure 8

Supersymmetric brane configurations for $\mathcal{N} = 1$ supersymmetric gauge theory with massless flavors. (*a*) The electric description. (*b*) The magnetic description.

dimensions. Preserving supersymmetry requires that branes be in specific relative angles. Any other angle would break supersymmetry completely (see Reference 98 for a review).

Stringy description of the ISS model and its metastable vacua was initiated in References 99 and 100 and studied further in References 101 and 102. **Figure 7** shows the brane configuration for $\mathcal{N} = 1$ $SU(N_c)$ gauge theory with N_f fundamental chiral multiplets that have masses of $m_1, \ldots, m_{N_f}$—the ISS model. The metastable supersymmetry-breaking vacua of the ISS model can be found by going to the magnetic description. To go from the electric description to the magnetic dual, we exchange the location of the two NS5-branes. The number of branes changes during this process due to the Hanany-Witten mechanism. **Figure 8** shows the electric and magnetic branes' configurations when the masses are equal to zero. The gauge group in the magnetic description is $SU(N_f - N_c)$, as expected from the Seiberg duality. In both descriptions, the D4-branes are parallel and supersymmetry is preserved.

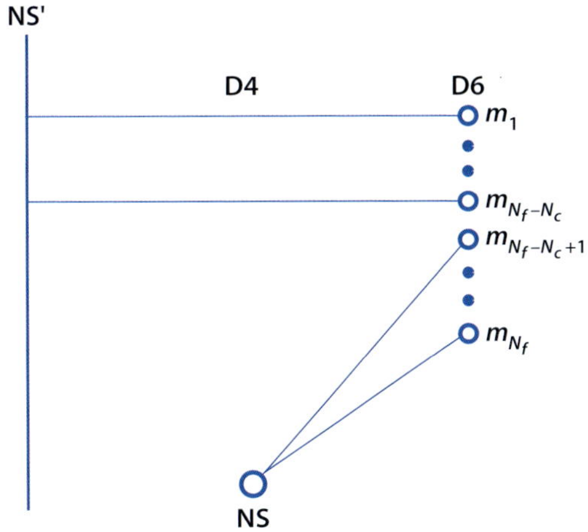

Figure 9

The brane configuration for the metastable supersymmetry-breaking vacuum.

Now, let us change the masses to nonzero. In the electric description, this can be done by moving the D6-branes up and down as in **Figure 7**. In the magnetic description, however, we cannot move the D6-branes in this way while keeping the D4-branes parallel to one another. We can keep $(N_f - N_c)$ D4-branes in parallel by reconnecting them at the NS5-branes, but the remaining N_c D4-branes have to be tilted, as shown in **Figure 9**. The resulting configuration breaks supersymmetry.

The brane configuration in this limit reproduces various features of vacuum structure of the ISS model such as global symmetries, pseudomoduli, and vacuum energy (99, 100). Thus, it is reasonable to identify this configuration with the supersymmetry-breaking metastable vacuum in the field theory.

Bena et al. (101) claimed that this brane configuration is not related to the original electric description and that it does not describe the ISS vacuum because of a certain asymptotic behavior of D-branes caused by brane bending at finite g_s. Giveon & Kutasov (102) proposed an alternative interpretation of brane bending involving string tachyons and argued that the brane configuration given in References 99 and 100 indeed describes the ISS vacuum. We refer the reader to these papers for details on the issue.

6.2. Holographic Gauge Mediation

Soft terms in gauge-mediation models are related to correlators of global symmetry currents in the hidden sector (56), as we reviewed in Section 4.2. Such relations are particularly useful for direct-type models in which hidden sectors are strongly coupled but the current correlators may be calculable. This idea has been applied to a class of strongly coupled hidden sectors to derive low-energy parameters of the models (63). Gauge/gravity duality in string theory can be useful because the current correlators in field theories are computed by Green's functions of the corresponding gauge fields in the gravity duals.

A well-studied example is the duality between the cascading $\mathcal{N} = 1$ $SU(N+M)\times SU(N)$ gauge theory and the warped deformed conifold geometry in type IIB string theory (103). Kachru et al. (104) constructed a metastable supersymmetry-breaking vacuum in this setup by putting anti-D3-branes at the tip of this conifold geometry. This vacuum can decay into a supersymmetric vacuum by quantum tunneling and brane/flux annihilation. Because this process occurs near the tip of the conifold, the UV description is not modified; we therefore expect that there would be a field-theory dual to describe the process. Both the supersymmetric vacuum and the metastable vacuum must have corresponding states in such a field theory. Recently, this construction was generalized (105, 106) to quiver gauge theories and string theory on quotients of the conifold, and the correspondence between metastable vacua in the field theories and in the gravity theories was clarified. Specifically, the model studied by Argurio et al. (106) gives a stringy realization of the direct gauge-mediation model constructed by Kitano et al. (96) and discussed in the previous section. Moreover, it provides a natural mechanism to generate small parameters in the model.

In the holographic gauge-mediation model[8] of Reference 112, strongly coupled messengers and hidden sectors are replaced by the supersymmetry-breaking solutions of type IIB supergravity (113), which takes into account the back-reaction of the antibranes. Although gaugino condensation breaks R-symmetry to $\mathbf{Z}_2$, the issue of anomalously small gaugino masses is not resolved in this model. McGuirk et al. (114) generalized this construction by using their supersymmetry-breaking solution presented in Reference 115.

6.3. Geometric Transition

Another way to construct low-energy gauge theories is to use D-branes wrapping cycles of Calabi-Yau manifolds. For example, consider a resolved conifold, and wrap its small two-sphere with N D5-branes. The low-energy effective theory is the $\mathcal{N} = 1$ pure Yang-Mills theory in four dimensions with gauge group $U(N)$. By a chain of dualities, one can relate this construction to an intersecting brane configuration of the type discussed above (116).

The geometric transition relates the resolved conifold to a deformed conifold with a small three-sphere. Specifically, the Veneziano-Yankielowicz superpotential of the $\mathcal{N} = 1$ Yang-Mills theory realized on the N D5-branes on the resolved conifold is equal to the Gukov-Vafa-Witten superpotential for the deformed conifold with N units of Ramond-Ramond fluxes through the three-sphere (117). The relation between the resolved and deformed conifolds is an example of the large-N duality and is closely related to the gauge/gravity duality described in References 103 and 118. It has been generalized to a variety of gauge theories and corresponding geometries, and their superpotentials have been computed exactly by using topological string theory (119).

Vafa (117) suggested that the geometric transition can be used even when both branes and antibranes are present. His proposal was made more precise by the authors of References 120–122, who conjectured that the large-N limit of brane/antibrane systems are Calabi-Yau manifolds with fluxes and that the physical potential can be computed using the method of topological string theory as in supersymmetric cases, except that some of the fluxes are negative. Field-theory descriptions of such metastable vacua have been given (122, 123) and their M-theory realizations explored (124, 125).

7. ELECTROWEAK SYMMETRY BREAKING

Electroweak symmetry breaking is the most relevant topic for particle physics. In the MSSM, electroweak symmetry breaking does not occur in the supersymmetric limit because supersymmetry guarantees the absence of the negative mass squared for the Higgs boson. Thus, there must be a coupling between the supersymmetry-breaking dynamics and the Higgs sector of the MSSM.

[8]For earlier related works in five-dimensional theory in a truncated AdS_5 background, see References 107–111.

The use of technicolor dynamics to break both supersymmetry and electroweak symmetry has been proposed (15, 20, 21). The Higgs fields obtain VEVs through a direct coupling to the dynamics and give masses to fermions in the Standard Model. However, in calculable models (e.g., 49–51), one usually assumes that the Higgs sector is separated from the supersymmetry-breaking dynamics and that some communication through messenger fields generates the Higgs potential and the Higgsino mass.

There are numerous difficulties with this scenario, and completely successful models have not been found. Here we explain these difficulties and the status of current research. The phenomenological requirements are (*a*) the Z boson mass, $m_Z = 91$ GeV; (*b*) the Higgs boson mass constraint, $m_h > 114$ GeV (126); (*c*) the Higgsino mass constraint, $m_{\tilde{H}} > 94$ GeV (127); (*d*) the gaugino mass constraint, $m_\lambda > O(100$ GeV) (127);[9] (*e*) the top quark mass, $m_t = 173$ GeV (131); and (*f*) the bound on the electric dipole moments of the neutron, $d_n < 3 \times 10^{-26} e$ cm (132).

As we explain below, it is not easy to naturally explain all of the above in supersymmetric models. This is generally known as the μ-problem. Most of the discussion below is not specific to gauge-mediation models, but the problems are particularly sharp in gauge mediation because it is designed to be calculable.

The Higgs potential in the MSSM is

$$V = (m_{H_u}^2 + \mu^2)|H_u|^2 + (m_{H_d}^2 + \mu^2)|H_d|^2 + (B\mu H_u H_d + \text{h.c.}) \\ + \frac{1}{8}(g_Y^2 + g_2^2)(|H_u|^2 - |H_d|^2)^2, \qquad 19.$$

where H_u and H_d are the neutral components of the Higgs fields in the MSSM and h.c. stands for the Hermitian conjugate. The part of the quadratic term μ^2 is the supersymmetric contribution to the Higgs potential. By definition, μ is the mass of the Higgsino. By supersymmetry, the quartic potential is related to the gauge couplings, g_Y and g_2, which control the Higgs boson mass. The other quadratic terms are soft supersymmetry-breaking terms that should arise from couplings to the supersymmetry-breaking sector. The B parameter is generally complex valued.

If the Higgs fields are only weakly coupled to the supersymmetry-breaking sector (e.g., 15), then the typical diagram from which to generate the Higgsino mass (the μ-term) is that shown in **Figure 5**, where the technicolor box is replaced by a messenger loop or a supersymmetry-breaking box. The problem is that if there is a one-loop diagram as in **Figure 5**, one-loop diagrams for generating $m_{H_u}^2$, $m_{H_d}^2$, and $B\mu$ should also be present. Because μ^2 is effectively two-loop valued, it suggests that the typical mass scale for the Higgs potential is larger than the Higgsino mass by a one-loop factor; that is, $m_Z \gg \mu$. Now we can see inconsistency between requirements *a* and *c* above.

One could evade the above problem by assuming that (*a*) the Higgs fields do not acquire VEVs from their potential and (*b*) there are other sources for the Z boson mass, such as technicolor models (15, 21). Small VEVs can be obtained through the diagram in **Figure 4*b***, and they are responsible for the fermion masses. However, because the top quark is rather heavy (requirement *e* above), such an assumption is not easily realized consistently.

Even if we somehow avoid the hierarchies in the parameters in the Higgs potential at the messenger-scale dynamics, we must also contend with the one-loop corrections to $m_{H_u}^2$ and $m_{H_d}^2$ that are proportional to the gaugino masses squared. Therefore, the gaugino masses cannot be very large compared with the Z boson mass. Together with requirement *d*, the gaugino masses are of $\mathcal{O}(100$ GeV). In gauge mediation, the gaugino masses are obtained at one-loop level. Therefore,

[9]One of the neutralinos can be very light (128–130).

the Higgs fields may also be weakly coupled to the supersymmetry-breaking sector. In that case, the hierarchy among $m^2_{H_u}$, $m^2_{H_d}$, $B\mu$, and μ^2, mentioned above, would be a generic consequence.

In the Higgs potential in Equation 19, there is a tree-level prediction for the Higgs boson mass: $m_h \leq m_Z$, which is clearly inconsistent with requirements *a* and *b*. This prediction requires a rather large quantum correction to the quartic couplings arising from supersymmetry breaking (133–135). In the MSSM, the largest contribution is from the top-stop loop diagrams. However, if there is a large quantum correction to the quartic coupling constant, there is also a large contribution to the quadratic term, especially to $m^2_{H_u}$. Such a large contribution requires that the parameters be fine-tuned to yield the correct size of the electroweak VEV, that is, the Z boson mass (136). See References 137 and 138 for recent discussions.

The degree of fine-tuning is somewhat milder when we have a sizable stop-stop Higgs coupling known as the A-term. However, a simple gauge-mediation model predicts that $A = 0$ at the messenger scale with which $\mathcal{O}(1\%)$ tuning is necessary (138).

One can try to build a model to generate the A-term. However, such a model requires that the A-term have the same phase as the gaugino masses, given that the relative phase is physical and observable. The constraint from requirement *f* means that the phase must be smaller than $\mathcal{O}(10^{-(2-3)})$. The same restriction applies for the B-parameter in the Higgs potential.

Several ways to address these problems have been considered. Use of semistrongly coupled models have been considered (39) to avoid large hierarchies among $m^2_{H_u}$, $m^2_{H_d}$, $B\mu$, and μ^2, where the *CP* phase is controlled by a symmetry. A perturbative model without the hierarchy in which the Higgs and the messenger sectors are extended has been presented (139). (Also, see References 51 and 140–142 for earlier discussions.) A possible way to obtain the correct size of m_Z even in the presence of the hierarchy has been pointed out (143); in this paper, an explicit example to realize the correct hierarchy pattern was presented. A mechanism to suppress $m^2_{H_u}$, $m^2_{H_d}$, and $B\mu$ relative to μ^2 by large renormalization effects has been discussed (144, 145). Finally, these discussions have been reformulated in the context of general gauge mediation (62). For another recent approach using more than one supersymmetry-breaking spurion, see Reference 146.

Natural electroweak symmetry breaking is one of the main motivations for the supersymmetric Standard Model. Supersymmetry breaking and its connection to the Higgs sector should be arranged so that electroweak symmetry breaking occurs naturally. We have encountered problems making this connection; however, the need for a special structure of the Higgs sector may hint at the actual underlying theory. Future or ongoing experiments, such as the LHC experiments, may help to reveal the role of supersymmetry in nature.

DISCLOSURE STATEMENT

The authors are not aware of any affiliations, memberships, funding, or financial holdings that might be perceived as affecting the objectivity of this review.

ACKNOWLEDGMENTS

We thank M. Graesser, Z. Komargodski, and D. Shih for comments on this manuscript. R.K. is supported in part by a Grant-in-Aid for Scientific Research, number 21840006, of the Japan Society for the Promotion of Science (JSPS). H.O. is supported in part by U.S. Department of Energy grant number DE-FG03-92-ER40701; the World Premier International Research Center Initiative of the Ministry of Education, Culture, Sports, Science and Technology of Japan; and a Grant-in-Aid for Scientific Research, number (C) 20540256, of JSPS. Y.O.'s research at the Perimeter Institute for Theoretical Physics is supported in part by the Government of Canada through the Natural Sciences and Engineering Research Council of Canada and by the Province of Ontario through the Ministry of Research and Innovation.

LITERATURE CITED

1. Montonen C, Olive DI. *Phys. Lett.* B72:117 (1977)
2. Witten E, Olive DI. *Phys. Lett.* B78:97 (1978)
3. Osborn H. *Phys. Lett.* B83:321 (1979)
4. Sen A. *Phys. Lett.* B329:217 (1994)
5. Vafa C, Witten E. *Nucl. Phys. B* 431:3 (1994)
6. Seiberg N, Witten E. *Nucl. Phys. B* 426:19 (1994)
7. Seiberg N. *Nucl. Phys. B* 435:129 (1995)
8. Seiberg N. *Phys. Rev. D* 49:6857 (1994)
9. Maldacena JM. *Adv. Theor. Math. Phys.* 2:231 (1998)
10. Dimopoulos S, Georgi H. *Nucl. Phys. B* 193:150 (1981)
11. Sakai N. *Z. Phys. C* 11:153 (1981)
12. Dimopoulos S, Raby S, Wilczek F. *Phys. Rev. D* 24:1681 (1981)
13. Jungman G, Kamionkowski M, Griest K. *Phys. Rep.* 267:195 (1996)
14. Amsler C, et al. (Part. Data Group) *Phys. Lett.* B667:1 (2008)
15. Dine M, Fischler W, Srednicki M. *Nucl. Phys. B* 189:575 (1981)
16. Intriligator KA, Seiberg N, Shih D. *J. High Energy Phys.* 04:021 (2006)
17. Moroi T, Murayama H, Yamaguchi M. *Phys. Lett.* B303:289 (1993)
18. Banks T, Kaplan DB, Nelson AE. *Phys. Rev. D* 49:779 (1994)
19. Bagger J, Poppitz E, Randall L. *Nucl. Phys. B* 426:3 (1994)
20. Witten E. *Nucl. Phys. B* 188:513 (1981)
21. Dimopoulos S, Raby S. *Nucl. Phys. B* 192:353 (1981)
22. Witten E. *Nucl. Phys. B* 202:253 (1982)
23. Nelson AE, Seiberg N. *Nucl. Phys. B* 416:46 (1994)
24. Affleck I, Dine M, Seiberg N. *Nucl. Phys. B* 241:493 (1984)
25. Affleck I, Dine M, Seiberg N. *Phys. Lett.* B137:187 (1984)
26. Affleck I, Dine M, Seiberg N. *Nucl. Phys. B* 256:557 (1985)
27. Izawa KI, Yanagida T. *Prog. Theor. Phys.* 95:829 (1996)
28. Intriligator KA, Thomas SD. *Nucl. Phys. B* 473:121 (1996)
29. Izawa KI, Takahashi F, Yanagida TT, Yonekura K. *Phys. Lett.* B677:195 (2009)
30. Izawa KI, Takahashi F, Yanagida TT, Yonekura K. *Phys. Rev. D* 80:085017 (2009)
31. Chacko Z, Luty MA, Ponton E. *J. High Energy Phys.* 12:016 (1998)
32. Murayama H. *Phys. Lett.* B355:187 (1995)
33. Intriligator KA, Seiberg N. *Class. Quantum Gravity* 24:S741 (2007)
34. Polonyi J. *Hung. Cent. Inst. Res. Rep. KFKI-77-93* (1977)
35. Ooguri H, Ookouchi Y, Park CS. *Adv. Theor. Math. Phys.* 12:405 (2008)
36. Marsano J, Ooguri H, Ookouchi Y, Park CS. *Nucl. Phys. B* 798:17 (2008)
37. Kitano R, Ookouchi Y. *Phys. Lett.* B675:80 (2009)
38. Kitano R. *Phys. Lett.* B641:203 (2006)
39. Ibe M, Kitano R. *J. High Energy Phys.* 08:016 (2007)
40. Murayama H, Nomura Y. *Phys. Rev. Lett.* 98:151803 (2007)
41. Murayama H, Nomura Y. *Phys. Rev. D* 75:095011 (2007)
42. O'Raifeartaigh L. *Nucl. Phys. B* 96:331 (1975)
43. Wess J, Zumino B. *Phys. Lett.* B49:52 (1974)
44. Ray S. *Phys. Lett.* B642:137 (2006)
45. Giudice GF, Rattazzi R. *Phys. Rep.* 322:419 (1999)
46. Dine M, Fischler W. *Phys. Lett.* B110:227 (1982)
47. Alvarez-Gaume L, Claudson M, Wise MB. *Nucl. Phys. B* 207:96 (1982)

48. Nappi CR, Ovrut BA. *Phys. Lett.* B113:175 (1982)
49. Dine M, Nelson AE. *Phys. Rev. D* 48:1277 (1993)
50. Dine M, Nelson AE, Shirman Y. *Phys. Rev. D* 51:1362 (1995)
51. Dine M, Nelson AE, Nir Y, Shirman Y. *Phys. Rev. D* 53:2658 (1996)
52. Giudice GF, Rattazzi R. *Nucl. Phys. B* 511:25 (1998)
53. Arkani-Hamed N, Giudice GF, Luty MA, Rattazzi R. *Phys. Rev. D* 58:115005 (1998)
54. Giveon A, Katz A, Komargodski Z, Shih D. *J. High Energy Phys.* 10:092 (2008)
55. Intriligator K, Shih D, Sudano M. *J. High Energy Phys.* 03:106 (2009)
56. Meade P, Seiberg N, Shih D. *Prog. Theor. Phys. Suppl.* 177:143 (2009)
57. Carpenter LM, Dine M, Festuccia G, Mason JD. *Phys. Rev. D* 79:035002 (2009)
58. Buican M, Meade P, Seiberg N, Shih D. *J. High Energy Phys.* 03:016 (2009)
59. Nakayama Y, Taki M, Watari T, Yanagida TT. *Phys. Lett.* B655:58 (2007)
60. Distler J, Robbins D. arXiv:0807.2006 [hep-ph] (2008)
61. Intriligator KA, Sudano M. *J. High Energy Phys.* 11:008 (2008)
62. Komargodski Z, Seiberg N. *J. High Energy Phys.* 03:072 (2009)
63. Ooguri H, Ookouchi Y, Park CS, Song J. *Nucl. Phys. B* 808:121 (2009)
64. Buican M, Komargodski Z. arXiv:0909.4824 [hep-ph] (2009)
65. Abel S, Dolan MJ, Jaeckel J, Khoze VV. *J. High Energy Phys.* 12:001 (2009)
66. Kobayashi T, Nakai Y, Takahashi R. arXiv:0910.3477 [hep-ph] (2009)
67. Seiberg N, Volansky T, Wecht B. *J. High Energy Phys.* 11:004 (2008)
68. Elvang H, Wecht B. *J. High Energy Phys.* 06:026 (2009)
69. Ibe M, Izawa KI, Nakai Y. *Phys. Rev. D* 80:035002 (2009)
70. Argurio R, Bertolini M, Ferretti G, Mariotti A. arXiv:0912.0743 [hep-ph] (2009)
71. Dine M, Mason JD. *Phys. Rev. D* 78:055013 (2008)
72. Poppitz E, Trivedi SP. *Phys. Rev. D* 55:5508 (1997)
73. Arkani-Hamed N, March-Russell J, Murayama H. *Nucl. Phys. B* 509:3 (1998)
74. Murayama H. *Phys. Rev. Lett.* 79:18 (1997)
75. Dimopoulos S, Dvali GR, Rattazzi R. *Phys. Lett.* B413:336 (1997)
76. Luty MA. *Phys. Lett.* B414:71 (1997)
77. Csaki C, Shirman Y, Terning J. *J. High Energy Phys.* 05:099 (2007)
78. Essig R, et al. *J. High Energy Phys.* 03:043 (2009)
79. Zur BK, Mazzucato L, Oz Y. *J. High Energy Phys.* 10:099 (2008)
80. Hall LJ, Randall L. *Nucl. Phys. B* 352:289 (1991)
81. Randall L, Rius N. *Phys. Lett.* B286:299 (1992)
82. Sato R, Yanagida TT, Yonekura K. arXiv:0910.3790 [hep-th] (2009)
83. Dine M, MacIntire D. *Phys. Rev. D* 46:2594 (1992)
84. Fox PJ, Nelson AE, Weiner N. *J. High Energy Phys.* 08:035 (2002)
85. Amigo SDL, Blechman AE, Fox PJ, Poppitz E. *J. High Energy Phys.* 01:018 (2009)
86. Benakli K, Goodsell MD. *Nucl. Phys. B* 816:185 (2009)
87. Shih D. *J. High Energy Phys.* 02:091 (2008)
88. Giveon A, Katz A, Komargodski Z. *J. High Energy Phys.* 06:003 (2008)
89. Cheung C, Fitzpatrick AL, Shih D. *J. High Energy Phys.* 07:054 (2008)
90. Dimopoulos S, Giudice GF. *Phys. Lett.* B393:72 (1997)
91. Izawa KI, Nomura Y, Tobe K, Yanagida T. *Phys. Rev. D* 56:2886 (1997)
92. Abel S, Durnford C, Jaeckel J, Khoze VV. *Phys. Lett.* B661:201 (2008)
93. Haba N, Maru N. *Phys. Rev. D* 76:115019 (2007)
94. Sato R, Yonekura K. arXiv:0912.2802 [hep-ph] (2009)
95. Komargodski Z, Shih D. *J. High Energy Phys.* 04:093 (2009)
96. Kitano R, Ooguri H, Ookouchi Y. *Phys. Rev. D* 75:045022 (2007)
97. Heckman JJ. arXiv:1001.0577 [hep-th] (2010)
98. Giveon A, Kutasov D. *Rev. Mod. Phys.* 71:983 (1999)
99. Ooguri H, Ookouchi Y. *Phys. Lett.* B641:323 (2006)
100. Franco S, Garcia-Etxebarria I, Uranga AM. *J. High Energy Phys.* 01:085 (2007)

101. Bena I, et al. *J. High Energy Phys.* 11:088 (2006)
102. Giveon A, Kutasov D. *Nucl. Phys. B* 778:129 (2007)
103. Klebanov IR, Strassler MJ. *J. High Energy Phys.* 08:052 (2000)
104. Kachru S, Pearson J, Verlinde HL. *J. High Energy Phys.* 06:021 (2002)
105. Argurio R, Bertolini M, Franco S, Kachru S. *J. High Energy Phys.* 01:083 (2007)
106. Argurio R, Bertolini M, Franco S, Kachru S. *J. High Energy Phys.* 06:017 (2007)
107. Gherghetta T, Pomarol A. *Nucl. Phys. B* 586:141 (2000)
108. Gherghetta T, Pomarol A. *Nucl. Phys. B* 602:3 (2001)
109. Goldberger WD, Nomura Y, Tucker-Smith D. *Phys. Rev. D* 67:075021 (2003)
110. Nomura Y, Tucker-Smith D. *Phys. Rev. D* 68:075003 (2003)
111. Nomura Y. arXiv:hep-ph/0410348 (2004)
112. Benini F, et al. *J. High Energy Phys.* 12:031 (2009)
113. DeWolfe O, Kachru S, Mulligan M. *Phys. Rev. D* 77:065011 (2008)
114. McGuirk P, Shiu G, Sumitomo Y. arXiv:0911.0019 [hep-th] (2009)
115. McGuirk P, Shiu G, Sumitomo Y. arXiv:0910.4581 [hep-th] (2009)
116. Ooguri H, Vafa C. *Nucl. Phys. B* 500:62 (1997)
117. Vafa C. *J. Math. Phys.* 42:2798 (2001)
118. Maldacena JM, Nunez C. *Phys. Rev. Lett.* 86:588 (2001)
119. Cachazo F, Intriligator KA, Vafa C. *Nucl. Phys. B* 603:3 (2001)
120. Aganagic M, Beem C, Seo J, Vafa C. *Nucl. Phys. B* 789:382 (2008)
121. Heckman JJ, Seo J, Vafa C. *J. High Energy Phys.* 07:073 (2007)
122. Aganagic M, Beem C, Freivogel B. *Nucl. Phys. B* 795:291 (2008)
123. Buican M, Malyshev D, Verlinde H. *J. High Energy Phys.* 06:108 (2008)
124. Marsano J, Papadodimas K, Shigemori M. *Nucl. Phys. B* 789:294 (2008)
125. Marsano J, Papadodimas K, Shigemori M. *Nucl. Phys. B* 804:19 (2008)
126. Barate R, et al. (LEP Work. Group) *Phys. Lett.* B565:61 (2003)
127. Abdallah J, et al. (DELPHI Collab.) *Eur. Phys. J. C* 31:421 (2003)
128. Hooper D, Plehn T. *Phys. Lett.* B562:18 (2003)
129. Belanger G, Boudjema F, Pukhov A, Rosier-Lees S. *J. High Energy Phys.* 0403:012 (2004)
130. Dreiner HK, et al. *Eur. Phys. J. C* 62:547 (2009)
131. Tevatron Electroweak Work. Group. arXiv:0903.2503 [hep-ex] (2009)
132. Baker CA, et al. *Phys. Rev. Lett.* 97:131801 (2006)
133. Okada Y, Yamaguchi M, Yanagida T. *Prog. Theor. Phys.* 85:1 (1991)
134. Ellis JR, Ridolfi G, Zwirner F. *Phys. Lett.* B257:83 (1991)
135. Haber HE, Hempfling R. *Phys. Rev. Lett.* 66:1815 (1991)
136. Barbieri R, Giudice GF. *Nucl. Phys. B* 306:63 (1988)
137. Chacko Z, Nomura Y, Tucker-Smith D. *Nucl. Phys. B* 725:207 (2005)
138. Kitano R, Nomura Y. *Phys. Rev. D* 73:095004 (2006)
139. Giudice GF, Kim HD, Rattazzi R. *Phys. Lett.* B660:545 (2008)
140. Dvali GR, Giudice GF, Pomarol A. *Nucl. Phys. B* 478:31 (1996)
141. Agashe K, Graesser M. *Nucl. Phys. B* 507:3 (1997)
142. de Gouvêa A, Friedland A, Murayama H. *Phys. Rev. D* 57:5676 (1998)
143. Csaki C, Falkowski A, Nomura Y, Volansky T. *Phys. Rev. Lett.* 102:111801 (2009)
144. Roy TS, Schmaltz M. *Phys. Rev. D* 77:095008 (2008)
145. Murayama H, Nomura Y, Poland D. *Phys. Rev. D* 77:015005 (2008)
146. Liu T, Wagner CEM. *J. High Energy Phys.* 06:073 (2008)

Fermilab's Intensity Frontier

André de Gouvêa[1] and Niki Saoulidou[2]

[1]Department of Physics and Astronomy, Northwestern University, Evanston, Illinois 60208; email: degouvea@northwestern.edu

[2]Particle Physics Division, Fermi National Accelerator Laboratory, Batavia, Illinois 60510; email: niki@fnal.gov

Annu. Rev. Nucl. Part. Sci. 2010. 60:513–38

First published online as a Review in Advance on July 19, 2010

The *Annual Review of Nuclear and Particle Science* is online at nucl.annualreviews.org

This article's doi:
10.1146/annurev-nucl-100809-131949

0163-8998/10/1123-0513$20.00

Key Words

rare processes, precision measurements, neutrinos, muons, kaons

Abstract

Particle physics experiments at the intensity frontier aim to probe nature through precision studies of the properties and interactions of its basic constituents, using intense particle beams and innovative detectors. We review the physics potential of several of these experiments, especially those that can be very effectively pursued at Fermilab in the near and intermediate future, assuming that a new intense proton source—Project X—will be available. We concentrate on flavor-violating phenomena that have been identified as the main particle physics drivers for Project X: the study of neutrino masses and mixing through long-baseline neutrino oscillations; searches for rare, flavor-violating muon processes; and precision measurements of kaon decays into neutrinos, $K \to \pi\nu\bar{\nu}$. We also comment on other opportunities, such as measurements of the anomalous magnetic moment of muon and neutrino-matter scattering.

Contents

1. INTRODUCTION

The search for a deeper understanding of nature at the smallest distance scales takes several distinct shapes and forms and involves research activities that range from observations of the universe when it was only 400,000 years old; to the study of particle collisions at teraelectronvolt center-of-mass energies; to searches for neutrino interactions at deep-underground, multikiloton-sized detectors. Recently (see **http://hepwww.physics.yale.edu/P5/**), these different types of activities have been broadly grouped into three distinct frontiers of fundamental science: the cosmic frontier, the energy frontier, and the intensity frontier.

The cosmic frontier consists of research efforts in which one attempts to learn about elementary particle physics from measurements of astrophysical and cosmological observables. These include studies of supernova explosions, cosmic-ray observatories, neutrino telescopes, precision measurements of the cosmic microwave background, and surveys of the large-scale structure of the universe. The energy frontier consists of research efforts that aim at directly probing the behavior of matter at the smallest accessible distance scales. This is the realm of high-energy particle accelerators including, most recently, LEP2 and the LHC at CERN, HERA at DESY, and the Tevatron at Fermilab. Future endeavors include teraelectronvolt-scale electron linear colliders and circular muon colliders.

This review concentrates on the intensity frontier and how it could be pursued at Fermilab in the near and intermediate future. The intensity frontier consists of research efforts that aim to probe nature through precision studies of the properties and fundamental interactions of its basic constituents. Many of these efforts—especially the ones pertinent to Fermilab—revolve around particle accelerators. Unlike energy frontier efforts, here the energy of the accelerator is not required to be as high as possible. Instead, the energy of the machine is defined, sometimes very precisely, by the physics question one is interested in addressing. It is the intensity and quality (purity, time and space profile, etc.) of the accelerated beam that determine the reach of intensity frontier experiments. Past, current, and future intensity frontier experiments include studies of neutrino oscillations; searches for rare muon, pion, and kaon processes; precision measurements of muon properties; heavy flavor (charm and bottom) factories; and the LEP1 experiments (where the energy was fixed at a special value: the Z boson mass).

Until the end of the Tevatron collider program, Fermilab will dedicate a large fraction of its resources to the energy frontier. Nonetheless, Fermilab plays a significant role in the intensity frontier that is expected to increase in the near and intermediate future, as will its growing investment in the cosmic frontier. Over the past several years, Fermilab has hosted several neutrino experiments, including MiniBooNE, SciBooNE, and MIPP, and is currently running the MiniBooNE, MINOS, and Minerνa experiments (which started taking data in 2009). These experiments are directly or indirectly related to measuring neutrino oscillations and understanding neutrino properties and interactions. A massive (15-kton) next-generation neutrino oscillation experiment—Noνa—is under construction and is expected to start taking data in 2012. Fermilab is also considering proposals to host a $\mu \to e$ conversion experiment, *Mu2e*, a next-generation experiment to measure the anomalous magnetic moment of the muon; and an experiment to measure kaon decays into neutrinos, $K^+ \to \pi^+ \nu\bar{\nu}$. Finally, Fermilab is considering building a new intense proton source—Project X—which will allow it to play a definitive role in intensity frontier activities by the end of the upcoming decade. Project X would allow one to reach the ultimate sensitivity in rare muon processes, realize a broadband-beam neutrino oscillation experiment with a >1,200-km baseline, improve our ability to see a nonzero electric dipole moment (EDM) for the muon, and improve our odds of not only discovering but studying some of the properties of new physics through rare kaon decays.

Here we review the physics potential of several intensity frontier experiments that could be pursued at Fermilab during the upcoming decade and beyond. We concentrate mostly on the physics reach of these experiments and briefly discuss more technical issues concerning the requirements for detectors and accelerators.

2. NEUTRINO OSCILLATIONS

2.1. Introduction

The existence of neutrinos was first proposed in 1930 by Pauli, who originally referred to them as neutrons. The name neutrino was introduced by Fermi shortly thereafter to differentiate between Pauli's neutrons and modern-day neutrons, which were discovered by Chadwick in 1932 (1). Electron antineutrinos were first observed in 1956, and muon neutrinos were first identified with the construction of the first neutrino beams in the 1960s. The third neutrino flavor, the tau neutrino, was directly observed just over a decade ago (2). For details on the history of neutrinos, see Reference 3. Neutrinos are the only known exclusively weakly interacting fundamental particles, which makes measuring their properties and interactions very challenging. As an example, it took more than 30 years since the first hint, and several experiments, to discover that neutrinos have mass and that, like the quarks, neutrinos mix.

Nonzero neutrino masses are the only concrete experimental evidence obtained so far for the existence of physics beyond the Standard Model. Over the past ten years, tremendous (experimental) progress has been made toward precisely measuring and better understanding neutrino mass-squared differences and mixing (4–11). For a summary of what we know about neutrino masses and mixing, see, for example, the most recent volume of the *Review of Particle Physics* (12; also see **http://pdg.lbl.gov/**).

Many open questions regarding neutrino masses and lepton mixing remain (see, for example, the review by Kayser in Reference 12). Note that we use the Particle Data Group parameterization for the neutrino mixing parameters.

1. What is the value of the third neutrino mixing angle, θ_{13}, for which there exists only an experimental upper bound, $\sin^2\theta_{13} < 0.05$ (16, 17)?
2. Do neutrinos violate *CP* invariance, and if so, by how much?
3. What is the neutrino mass hierarchy, or the sign of Δm^2_{31}?
4. What are the neutrino masses themselves? Neutrino oscillation experiments provide information only on mass-squared differences, Δm^2_{21} and Δm^2_{31}.
5. Are neutrinos Majorana or Dirac fermions, and is lepton number a good symmetry of nature?

These questions are important on their own, but they also provide necessary information with which to address more fundamental questions in particle physics. For instance, why is neutrino mixing so much different from quark mixing, and are the two mixing matrices related to one another? Is there an underlying symmetry that explains the pattern of lepton mixing? What is the mechanism behind neutrino masses? Why are neutrino masses so small compared to those of the quarks and charged leptons? Are the seeds for the matter-antimatter asymmetry of the universe contained in the mechanism behind neutrino masses? Finally, there remains the possibility that there are still more surprises to come in neutrino physics. One should keep in mind that neutrino properties are not well measured. During the next few decades, several neutrino experiments, driven by the well-defined questions posed above, will provide novel information and may teach us much more than we can anticipate.

The first three of the five questions posed above can be addressed with neutrino oscillation experiments. These make use of several distinct neutrino sources, including nuclear reactors, particle accelerators, so-called natural neutrinos from the Sun and other astronomical objects, and neutrinos produced in the atmosphere by cosmic rays.

The main goal of all next-generation (Phase I) experiments is to search for a nonzero value for θ_{13}. These can be grouped into two categories:

1. Reactor experiments, including Double Chooz (18), Daya Bay (19), and RENO (20). These disappearance experiments are searching for a deficit of $\bar{\nu}_e$, with respect to expectations. They can measure θ_{13} cleanly, free from possible degeneracies arising from the interplay of the different neutrino oscillation parameters. The three experiments mentioned above are under construction and will start taking data within the next 12 to 24 months.
2. Accelerator long-baseline off-axis experiments, including T2K (21) and NOνA (22). These appearance experiments are searching for an excess, with respect to expectations, of ν_e (or $\bar{\nu}_e$) in pion decay beams made up predominantly of ν_μ (or $\bar{\nu}_\mu$). These are, in principle, sensitive to other neutrino oscillation parameters in addition to θ_{13}. The T2K experiment started collecting data in 2009, and the NOνA experiment is under construction and expected to start taking data in 2012.

Phase I experiments have an ultimate reach down to $\sin^2(2\theta_{13}) \sim 0.01$, an order of magnitude beyond the current bound from the CHOOZ reactor experiment (16). In addition, the NOνA experiment, due to its very long baseline (810 km, compared with 295 km for T2K), can determine the neutrino mass hierarchy if θ_{13} is very close to the current experimental upper bound. No Phase I experiment has a significant ($>3\sigma$) discovery potential for *CP*-invariance violation.

The goals of third-generation (Phase II) experiments, currently in the planning stage, will depend on the results of Phase I experiments. If a nonzero θ_{13} is not revealed by Phase I experiments, Phase II experiments will aim at extending their reach for θ_{13} by approximately an order of magnitude. However, if θ_{13} effects are revealed by Phase I experiments, Phase II experiments will ultimately aim at (*a*) determining the neutrino mass hierarchy for, at least, all values of θ_{13} within the reach of Phase I experiments and (*b*) observing *CP*-violating effects, if they are present and large at least for all values of θ_{13} within the reach of Phase I experiments.

2.2. Brief Review of Neutrino Oscillations

Here, we are interested mostly in the oscillation of $\nu_\mu \leftrightarrow \nu_e$ (and the equivalent for antineutrinos). Furthermore, we concentrate on baselines L and neutrino energies E such that $\Delta m^2_{12} L/E \ll 1$. Taking into account that $\sin^2\theta_{13}$ is very small, and assuming that the electron number density N_e traversed by the neutrinos is a constant, one can write (e.g., 12)

$$P(\nu_\mu \to \nu_e) \simeq \sin^2(2\theta_{13}) \cdot T_1 - a \cdot \sin(2\theta_{13}) \cdot T_2 - a \cdot \sin(2\theta_{13}) \cdot T_3 + a^2 T_4, \tag{1}$$

where $a = \frac{\Delta m^2_{21}}{\Delta m^2_{31}}$, $x = \frac{2\sqrt{2} G_F N_e E}{\Delta m^2_{31}}$, $\Delta = \frac{\Delta m^2_{31} L}{4E}$ and

$$T_1 = \sin^2(\theta_{23}) \frac{\sin^2[(1-x)\Delta]}{(1-x)^2}; \tag{2}$$

$$T_2 = \sin(\delta_{CP}) \sin(2\theta_{12}) \sin(2\theta_{23}) \sin\Delta \frac{\sin(x\Delta)}{x} \frac{\sin[(1-x)\Delta]}{(1-x)}; \tag{3}$$

$$T_3 = \cos(\delta_{CP}) \sin(2\theta_{12}) \sin(2\theta_{23}) \sin\Delta \frac{\sin(x\Delta)}{x} \frac{\sin[(1-x)\Delta]}{(1-x)}; \tag{4}$$

and

$$T_4 = \cos^2(\theta_{23}) \sin^2(\theta_{12}) \frac{\sin^2(x\Delta)}{x^2}. \tag{5}$$

Here T_1 is the atmospheric term, T_4 is the solar term, and T_2 (T_3) is the *CP*-violating (*CP*-conserving) interference term. For antineutrinos, the matter effect parameter $x \leftrightarrow -x$. It is the dependency of $P(\nu_\mu \to \nu_e)$ on x that mostly provides sensitivity to the sign of Δm^2_{31}.

Measuring $P(\nu_\mu \to \nu_e)$ and extracting the unknown parameters θ_{13}, δ_{CP}, and the sign of Δm^2_{31} from Equation 1 are very nontrivial tasks, especially if we take into account that strong degeneracies are present between these parameters. For illustrative purposes, the degeneracy between "genuine" *CP* violation and "fake" *CP* violation arising from matter effects is depicted in **Figure 1**.

To achieve their goals, Phase II experiments require the following:

1. High statistics, given that the oscillation effects of interest are small. This implies more powerful neutrino beams and larger detectors than those used in Phase I.
2. Several measurements at different values of L or E to break oscillation parameter degeneracies.
3. Longer baselines than those of Phase I experiments to enhance matter effects and increase the potential for determining the neutrino mass hierarchy.

2.3. Possible Neutrino Beams from Fermilab and Project X

There are two main options for neutrino beams, each of which has different advantages and disadvantages. On-axis (or only slightly off-axis) wide-band beams (WBBs) allow one to use more of the available neutrino flux (higher event rates) and to study both the first and the second

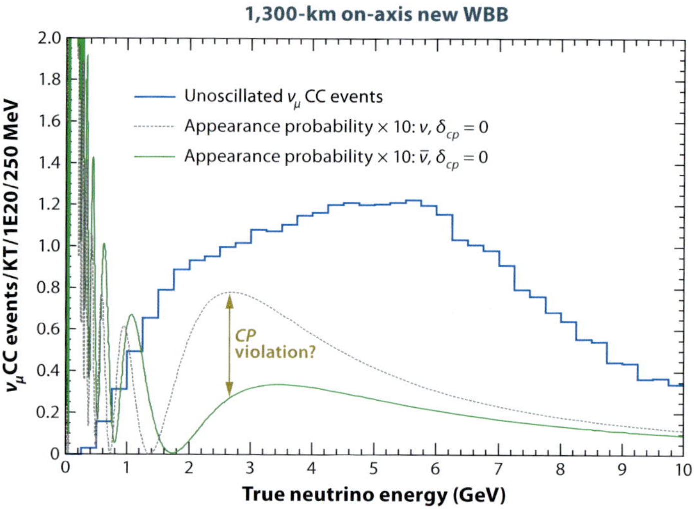

Figure 1

$P(\nu_\mu \to \nu_e)$ for neutrinos (*gray dotted line*) and antineutrinos (*green solid line*) superimposed to an unoscillated ν_μ spectrum (*blue histogram*). The difference between the neutrino and antineutrino oscillation probabilities arises from matter effects and not violation of the *CP* symmetry. Such strong degeneracies between "genuine" *CP* violation and "fake" *CP* violation make experimental measurements challenging. Abbreviations: CC, charged-current; WBB, wide-band beam.

oscillation maxima with one detector placed at a fixed location. Its main disadvantage is the higher neutral-current background rate to the ν_e-appearance signal, which results from the high-energy tail of the ν_μ spectrum. In the case of off-axis narrow-band beams (NBBs), the same neutral-current backgrounds are markedly smaller due to the suppression of the high-energy tail. Its main disadvantage is that it cannot probe a variety of E values with one detector at a fixed location. In this case, one needs to consider two detectors at two different locations (values of L), usually the first two oscillation maxima of the atmospheric oscillation length. This requirement, combined with the fact that the second oscillation maximum would require a larger off-axis angle (because it occurs at lower neutrino energies), strongly suppresses the event yield.

In **Figure 2** we illustrate the advantages and disadvantages discussed above, as well as the importance of being able to study both the first and second oscillation maxima: The second oscillation maximum (occurring at lower neutrino energies) does not suffer, at least not to the same extent as the first oscillation maximum, from the degeneracies between genuine and fake *CP* violation discussed above. Therefore, the ability to study the second oscillation maximum and compare the results obtained with those obtained in the first maximum is very important for the determination of both δ_{CP} and the neutrino mass hierarchy.

Fermilab is currently operating the 250-kW NuMI (Neutrinos at the Main Injector) beam, the most intense neutrino beam in the world. This beam is used by the MINOS (9) and Minerνa (23) experiments. There are approved near-future plans to upgrade the NuMI beam to 700 kW for the NOνA experiment. Over the course of the previous year, Fermilab developed a physics plan for the upcoming decade (24), which includes an upgrade to the accelerator complex—Project X. Project X (**Figure 3**) will be designed to provide approximately 2 MW of protons with energies ranging from 60 to 120 GeV. This feature will result in the ability to produce very high intensity conventional neutrino beams (protons on a target that yield pions that decay in flight, $\pi \to \mu\nu_\mu$, and can be sign-selected for a mostly ν_μ or $\bar{\nu}_\mu$ beam).

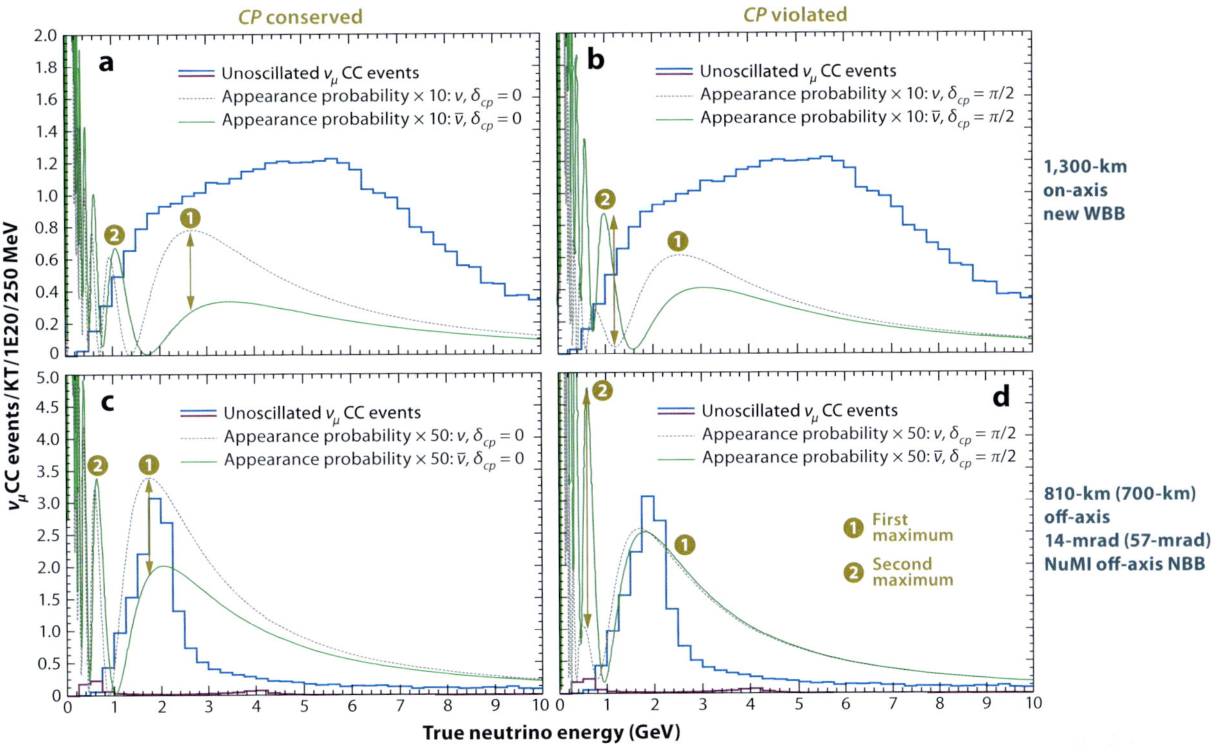

Figure 2

(*a*,*b*) Wide-band beams (WBBs). (*c*,*d*) Narrow-band beams (NBBs.) (*a*,*c*) *CP* conserved ($\delta_{CP} = 0$). (*b*,*d*) *CP* maximally violated ($\delta_{CP} = \pi/2$). Blue and purple histograms represent the unoscillated ν_μ spectrum, with $P(\nu_\mu \rightarrow \nu_e)$ (*dotted gray line*) and $P(\bar{\nu}_\mu \rightarrow \bar{\nu}_e)$ (*solid green line*) superimposed. The second oscillation maximum exhibits a difference in neutrino-antineutrino oscillation probabilities only when *CP* is violated (matter effects do not play a significant role), whereas the first oscillation maximum exhibits a difference between neutrino and antineutrino oscillation probabilities due to matter effects, even when *CP* is conserved. Information from both (first and second) maxima helps break these inherent degeneracies. In the WBB case, one detector is needed to study both first and second oscillation maxima, whereas in the NBB case two detectors at two different locations, one at the first oscillation maximum and another at the second, are necessary. Abbreviation: CC, charged-current.

2.4. Options for a New Far Detector

The NOνA experiment, which is under construction, is a 15-kton liquid scintillator detector that will use the NuMI beam in an off-axis NBB configuration. The off-axis angle is 14 mrad, and its baseline is 810 km. According to the brief discussion above, an optimal Phase II experiment will require a new detector, located further away from Fermilab. The Earth's geometry dictates that a new beam line will be required. A possible Phase II configuration makes use of the 1,300-km baseline from Fermilab to the Deep Underground Science Laboratory (DUSEL) in South Dakota and a new beam, preferably a WBB (see Section 2.3).

The two configurations (NOνA off axis and FNAL-to-DUSEL on axis), along with the advantages of the longer baseline (1,300 km versus 810 km), are illustrated in **Figure 4**. A longer baseline increases the matter effects and the potential to discover the neutrino mass hierarchy. In addition, the first and second oscillation maxima shift to higher energies, where both the neutrino fluxes and the cross sections are higher and the experimental determination of the type and energy of the interacting neutrinos is easier.

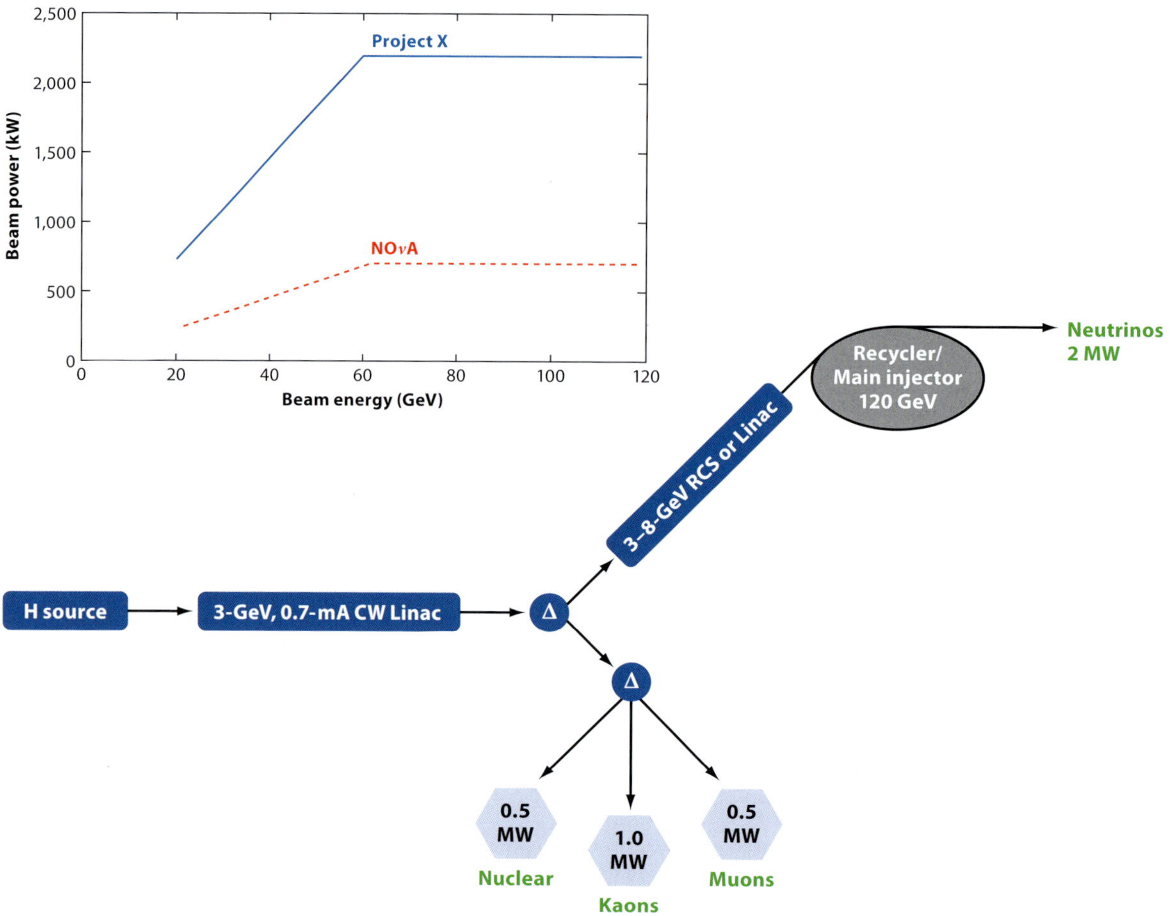

Figure 3

Project X beam power as a function of primary proton-beam energy. Reproduced from **http://www.fnal.gov/pub/directorate/steering/index.shtml**.

To be able to collect adequate statistics, the detectors of neutrino experiments have to be massive. The >1,000-km baselines usually require detector masses to be measured in tenths of megatons. Several ideas about how to construct massive neutrino detectors have been suggested over the past few years (25–32). The two main technologies under consideration are water Cherenkov (WC) and liquid argon (LAr) detectors. WC detectors, used in neutrino experiments, consist of a large volume of water instrumented with photomultiplier tubes. Charged particles created by neutrino interactions with the water produce Cherenkov light if their velocity exceeds that of light in water. Cherenkov light is emitted in a cone around the particle's trajectory and is detected by the photomultiplier tubes. Cosmic-ray muons passing through the detector also illuminate the photomultiplier tubes, so WC detectors must be located underground, where the muon flux is attenuated and where they are surrounded by an active muon veto. Electrons from electron-neutrino-induced charged-current interactions can be efficiently distinguished from muons because the electron-induced Cherenkov ring is very different from the muon-induced one, and hence they lead to a

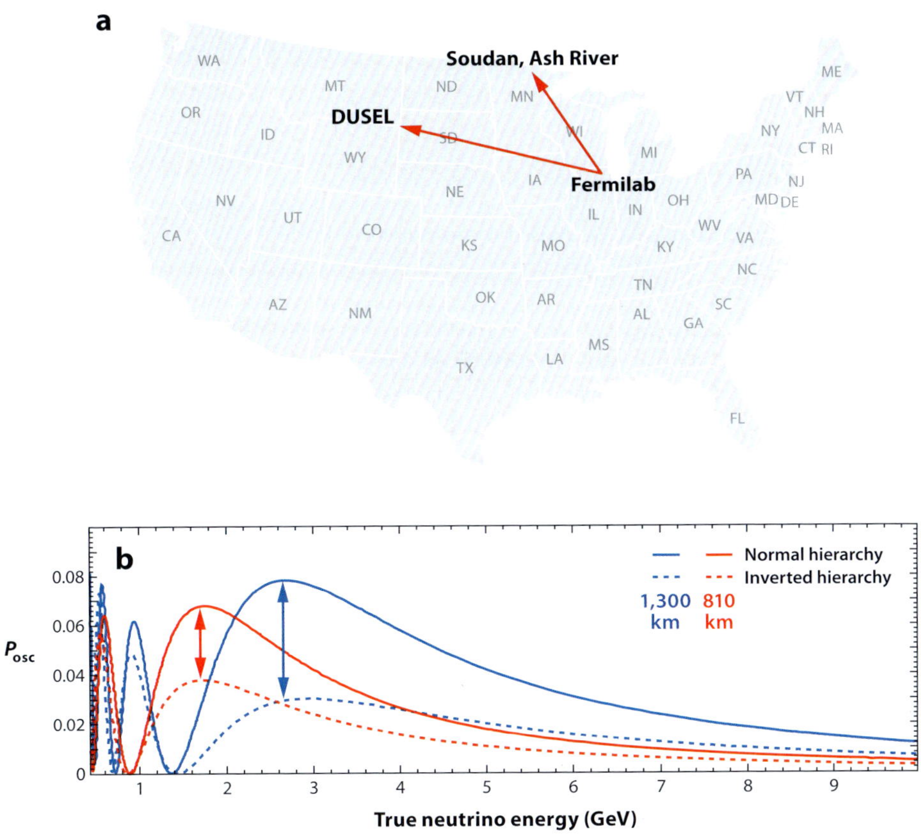

Figure 4

(*a*) Two long-baseline options in the United States. The first option is Fermilab to Soudan (MINOS) and Ash River (NOνA), with baselines of 735 km and 810 km, respectively, using the NuMI on-axis beam (MINOS) or the NuMI off-axis narrow-band beam (NOνA). The second option is Fermilab to DUSEL (Deep Underground Science and Engineering Laboratory), with a baseline of 1,300 km and a new wide-band beam. (*b*) $P(\nu_\mu \rightarrow \nu_e)$ for the normal (*continuous line*) and inverted (*dotted line*) neutrino mass hierarchies for two different baselines: 810 km (*red*) and 1,300 km (*blue*).

distinct hit pattern in the photomultiplier tubes. However, photons and electrons are very hard to distinguish in a WC detector.

LAr detectors are cryogenic vessels that contain the totally active LAr material and a time-projection chamber. The operation principle of a LAr time-projection chamber is that in very pure argon ionization tracks can drift (almost undistorted) in the presence of a uniform electric field over long distances. Imaging is provided by wire planes in different orientations placed at the end of the drift path. Such detectors have very high (millimeter-scale) spatial resolution and excellent tracking capability. Electrons from electron-neutrino-induced interactions are distinguished from muons, photons, and other particles with very high efficiency and purity due to the distinct event topologies and characteristics.

Both technologies have advantages and disadvantages. Next-generation WC detectors are a proven technology but would suffer from a relatively low efficiency (~20% on average, strongly energy dependent) and low background rejection (~1% on average, also strongly energy dependent). LAr detectors are not a proven technology for the mass scales of interest (~100 kton) and hence would require much more research and development efforts. However, LAr detectors are believed to have higher efficiency (~80%) and higher background-rejection capability (<1%) compared with WC detectors.

2.5. Physics Reach

If the right ingredients come into place—powerful neutrino beams coupled to large detectors at long baselines—Fermilab would be poised to host a world-leading staged program in neutrino physics for decades to come (24). The program would commence in the first half of the coming decade.

The first step of the staged program is the NOνA liquid scintillator experiment using the 700-kW NuMI off-axis NBB at a baseline of 810 km. A possible intermediate step, with interesting physics capabilities, could be an (technologically) upgraded detector consisting of ~5 kton of LAr, placed either in the NuMI beam or at DUSEL. The next step would be the construction—which would probably use a modular approach—of massive detectors (300 kton of WC and/or 100 kton of LAr) at DUSEL, in parallel with the construction of a new WBB from Fermilab to DUSEL. The initial beam power would be 700 kW. Finally, the construction of Project X would increase the neutrino beam power from 700 kW to 2 MW.

The physics capabilities of this staged program, in terms of discovery potentials for the parameters of interest [θ_{13}, δ_{CP}, and sign(Δm^2_{31})], are illustrated in **Figure 5**, which clearly shows the progressive increase in sensitivity. The ultimate reach fulfills the Phase II requirements listed above.

The next logical step would be to replace the pion decay beam with a neutrino factory (33): the ultimate neutrino beam. Another intriguing possibility, recently discussed in the literature, is to

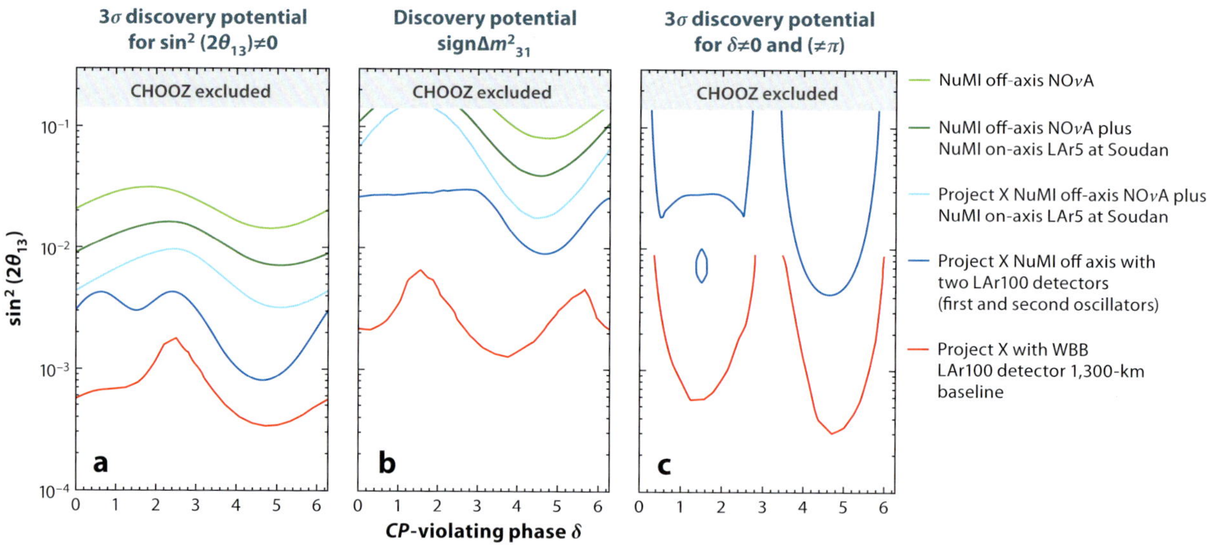

Figure 5

Fermilab staged plan. 3σ discovery potentials for (*a*) θ_{13}, (*b*) the neutrino mass hierarchy, and (*c*) *CP* violation in the $\delta_{CP} \times \sin^2 2\theta_{13}$ plane. From low to high, the discovery potentials are (*i*) NOνA with NuMI narrow-band beam (NBB) at 700 kW, (*ii*) NOνA plus 5-kton liquid argon (LAr) with NuMI NBB and wide-band beam (WBB) at 700 kW, (*iii*) NOνA plus 5-kton LAr with NuMI NBB and WBB at 2 MW, (*iv*) 50-kton LAr at first oscillation maximum plus 50-kton LAr at second oscillation maximum with NuMI NBB at 2 MW, and (*v*) 100-kton LAr (equivalent to a >300-kton water Cherenkov) at DUSEL with a new WBB at 2 MW.

replace the last step outlined above with a low-energy neutrino factory (34). We do not comment on this possibility further, except to state that, if it is accessible on a similar timescale, it should be seriously considered.

3. RARE MUON PROCESSES

3.1. Introduction

The existence of muons was first identified in cosmic-ray experiments in 1936. After much speculation, it was decided that muons have the same gauge quantum numbers as electrons and that they virtually always decay via the weak interaction: $\mu \to e\nu\bar{\nu}$. Early on, it was speculated that the muon was an excited electron and should, therefore, decay into an electron and a photon: $\mu \to e\gamma$.

Searches for $\mu \to e\gamma$ date back to the late 1940s. By the early 1960s, naïve estimates for the branching ratio for $\mu \to e\gamma$ from the physics responsible for $\mu \to e\nu\bar{\nu}$ were invalidated by experimental results. The fact that $\mu \to e\gamma$ did not occur with branching ratios above 10^{-6} was among the most pressing arguments supporting both the so-called two-neutrino hypothesis and the postulate that there are different lepton flavors accompanied by conserved lepton flavor numbers: electron number and muon number (later expanded to included tau number) (3). We now describe muon decay as $\mu^- \to e^-\bar{\nu}_e\nu_\mu$, and it has been experimentally established that the electron neutrino and the muon neutrino are different objects.

Over the last decade, however, it was discovered that lepton flavor numbers are not good quantum numbers—neutrinos oscillate! This finding implies that, barring accidental cancellations, all so-called charged-lepton flavor-violating (CLFV) processes are allowed and will occur at some order in perturbation theory. These processes include $\ell \to \ell'\gamma$, $\ell \to \ell'\ell''\bar{\ell}'''$, $\ell + X \to \ell' + X$, and $X \to \ell\bar{\ell}'$; where $\ell, \ell', \ldots = e, \mu,$ or τ; and X represents states that carry no lepton flavor number.

3.2. Standard Model Expectations?

Although we have determined that lepton flavor is violated, measurements of neutrino oscillation processes do not allow us to reliably estimate the rate for the various CLFV processes. The reason is that although neutrino oscillating phenomena depend only on neutrino masses and lepton mixing angles, the rates of the various CLFV processes may depend on the mechanism behind neutrino masses and lepton mixing. Different neutrino mass–generating Lagrangians (35) lead to very different rates for CLFV.

The massive neutrino contribution to CLFV, involving only active neutrinos, is absurdly small. For example (36),

$$B(\mu \to e\gamma) = \frac{3\alpha}{32\pi}\left|\sum_{i=2,3} U^*_{\mu i} U_{ei} \frac{\Delta m^2_{i1}}{M^2_W}\right| < 10^{-54}, \qquad 6.$$

where $U_{\alpha i}$ are elements of the neutrino mixing matrix, Δm^2_{ij} are the neutrino mass–squared differences, and M_W is the W boson mass. The above estimate applies to some neutrino mass models, including minimal scenarios with Dirac neutrinos. The reason for the tiny expected branching ratio is well known. CLFV processes are flavor-changing neutral-current processes and hence are subject to the Glashow-Illiopoulos-Maiani (GIM) mechanism (37).

In some neutrino mass–generating scenarios, the active neutrino contribution is severely subdominant. In the seesaw mechanism (38), for example, heavy neutrino contributions to CLFV are naïvely expected to be of order those of the light neutrinos, but there is no theorem that prevents them from being much, much larger. Current experimental constraints on the seesaw Lagrangian allow $B(\tau \rightarrow \mu\gamma)$ to be as large as 10^{-9}, $B(\mu \rightarrow e\gamma)$ to be as large as 4×10^{-13}, and normalized rates[1] for $\mu \rightarrow e$ conversion in ^{48}Ti to saturate the current experimental upper bound [4×10^{-12} at 90% CL (12, 13)] (39).

3.3. Model-Independent Analyses

Regardless of the mechanism that underlies neutrino masses, it is widely speculated that the rates for different CLFV processes should be just beneath current experimental upper bounds (see current bounds in References 12–15 and references therein). The reason is that we strongly suspect that there are new degrees of freedom with masses of approximately 1 TeV. Because lepton flavor numbers are not conserved, it is generically expected that virtual processes involving the new degrees of freedom mediate CLFV at some order in perturbation theory.

Some concrete new physics scenarios are briefly discussed in Section 3.4. In this section, we discuss two effective Lagrangians that mediate CLFV processes involving muons. Our goal is to illustrate (*a*) how sensitive searches for CLFV are to new physics, (*b*) whether they may reveal the nature of the new physics, and (*c*) how different CLFV channels compare with one another.

After electroweak symmetry breaking, CLFV is mediated by effective operators of dimension five and higher. Consider the following effective Lagrangian:[2]

$$\begin{aligned}\mathcal{L}_{\mathrm{CLFV}} = &\frac{m_\mu}{(\kappa+1)\Lambda^2}\bar{\mu}_R\sigma_{\mu\nu}e_L F^{\mu\nu}\\ &+\frac{\kappa}{(1+\kappa)\Lambda^2}\bar{\mu}_L\gamma_\mu e_L(\bar{u}_L\gamma^\mu u_L+\bar{d}_L\gamma^\mu d_L).\end{aligned} \qquad 7.$$

The subscripts L and R indicate the chirality of the different Standard Model fermion fields, $F^{\mu\nu}$ is the photon field strength, and m_μ is the muon mass. The coefficients of the two types of operators are parameterized by two constants: the mass dimension–one Λ parameter, which is meant to represent the effective mass scale of the new degrees of freedom, and the dimensionless parameter κ, which governs the relative size of the two different types of operators. The magnetic moment–type operator in the first line of Equation 7 directly mediates $\mu \rightarrow e\gamma$ and mediates $\mu \rightarrow eee$ and $\mu \rightarrow e$ conversion in nuclei at order α. The four-fermion operators in the second line of Equation 7, however, mediate $\mu \rightarrow e$ conversion at leading order and $\mu \rightarrow e\gamma$ at the one-loop level. For $\kappa \ll 1$, the dipole-type operator dominates CLFV phenomena, whereas for $\kappa \gg 1$, the four-fermion operators are dominant.

The sensitivity to Λ as a function of κ for $\mu \rightarrow e\gamma$ and $\mu \rightarrow e$ conversion efforts is depicted in **Figure 6*a***. For $\kappa \ll 1$, an experiment sensitive to $\mathrm{Br}(\mu \rightarrow e\gamma) > 10^{-13}$ would probe Λ values below 2,500 TeV, whereas for $\kappa \gg 1$, an experiment sensitive to Br($\mu \rightarrow e$ conversion in ^{48}Ti) > 10^{-16} would probe Λ values below 8,200 TeV.

A great deal of information can be extracted from **Figure 6*a***. First, CLFV already probes Λ values close to 1,000 TeV, whereas the currently running MEG experiment (41; also see **http://meg.web.psi.ch/**) will begin to probe $\Lambda \sim 10^3$ TeV and beyond. Second, a $\mu \rightarrow e$ conversion experiment is guaranteed to outperform a $\mu \rightarrow e\gamma$ experiment for any value of κ as

[1]In the case of the so-called $\mu \rightarrow e$ conversion in nuclei processes, one always refers to the rate for $\mu \rightarrow e$ conversion normalized to that for $\mu \rightarrow \nu_\mu$ conversion, which is often referred to as the capture rate. It is not uncommon to refer to this normalized rate as a branching ratio.

[2]The most general effective Lagrangian includes several other terms (40). The subsets included in Equations 7 and 8, however, are sufficient to illustrate all the issues discussed here.

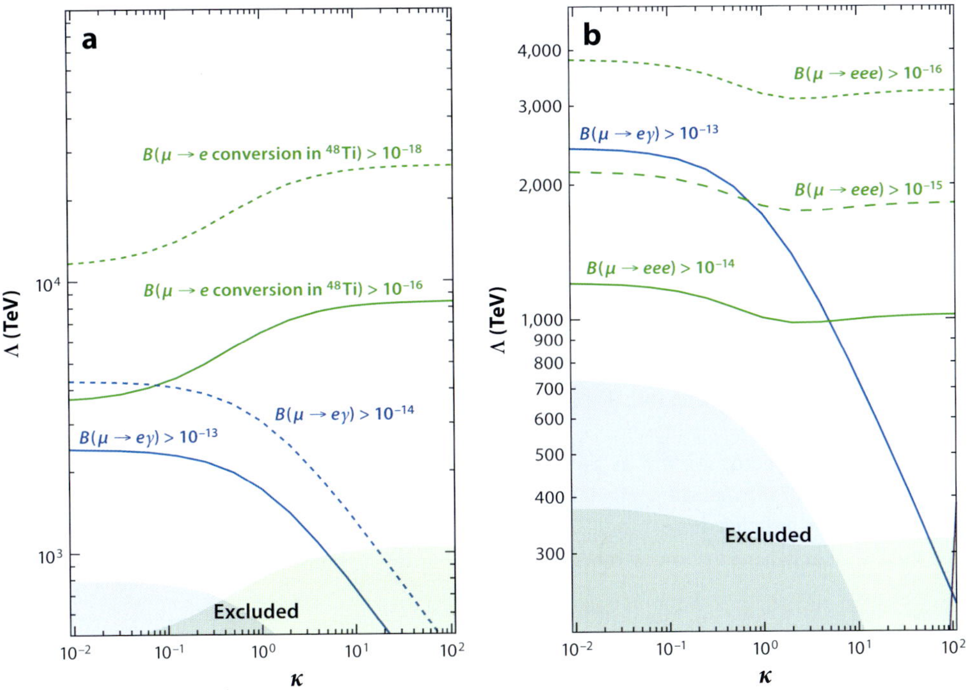

Figure 6

(*a*) Sensitivity of a $\mu \rightarrow e$ conversion in ^{48}Ti experiment that can probe normalized capture rates of 10^{-16} and 10^{-18}, and a $\mu \rightarrow e\gamma$ search that is sensitive to branching ratios of 10^{-13} and 10^{-14}, to the new physics scale Λ as a function of κ (Equation 7). (*b*) Sensitivity of a $\mu \rightarrow eee$ experiment that is sensitive to branching ratios 10^{-14}, 10^{-15}, and 10^{-16}, and a $\mu \rightarrow e\gamma$ search that is sensitive to a branching ratio of 10^{-13}, to the new physics scale Λ as a function of κ (Equation 8). Also depicted are the currently excluded regions of both parameter spaces.

long as it is a couple of orders of magnitude more sensitive. Because it is, experimentally, very hard to achieve sensitivity to branching ratios for $\mu \rightarrow e\gamma$ smaller than 10^{-14} (42), $\mu \rightarrow e$ conversion searches—which are not expected to hit any "wall" before normalized rates of (at least) approximately 10^{-18} (42)—seem to be the most effective way of pursuing CLFV after the ongoing MEG experiment is done at the Paul Scherrer Institute (PSI) (41).

We can ask what the consequences for CLFV are if the new physics is best captured by the following effective Lagrangian:

$$\mathcal{L}_{\rm CLFV} = \frac{m_\mu}{(\kappa+1)\Lambda^2}\bar{\mu}_R\sigma_{\mu\nu}e_L F^{\mu\nu} + \frac{\kappa}{(1+\kappa)\Lambda^2}\bar{\mu}_L\gamma_\mu e_L(\bar{e}\gamma^\mu e). \qquad 8.$$

Similar to the dimension-six operators in the second line of Equation 7, the dimension-six operator in the second line of Equation 8 mediates $\mu \rightarrow eee$ at tree level and $\mu \rightarrow e\gamma$ at one-loop level. Similar to Equation 7, the dimensionless parameter κ determines whether the dipole-like interaction or the four-fermion interaction is dominant when it comes to CLFV.

The sensitivity to Λ as a function of κ for $\mu \rightarrow e\gamma$ and $\mu \rightarrow eee$ efforts is depicted in **Figure 6*b***. Here, for $\kappa \gg 1$, an experiment sensitive to $\mathrm{Br}(\mu \rightarrow eee) > 10^{-15}$ will probe Λ values of up to 1,800 TeV. As in the example depicted in **Figure 6*a***, a $\mu \rightarrow eee$ experiment is guaranteed to outperform a $\mu \to e\gamma$ experiment for any value of κ, as long as it is a few hundred times more sensitive. Whether this can be realistically achieved is still under investigation (42).

The discussions above also illustrate another aspect of searches for CLFV violation. In the case of a positive signal, the amount of information regarding the new physics is limited. For example, a positive signal in a $\mu \to e$ conversion experiment does not allow one to measure either Λ or κ but only a function of the two. To learn more about the new physics, one needs to combine information involving the rate of a particular CLFV process with other observables. These include other CLFV observables (e.g., a positive signal in $\mu \to e\gamma$ and $\mu \to eee$ would allow one to measure both κ and Λ if Equation 8 describes CLFV), including those involving tau leptons (e.g., $\tau \to e\gamma$ and $\tau \to \mu\mu\mu$), studies of the electromagnetic properties of charged leptons (e.g., $g-2$, EDMs), precision studies of neutrino processes (including oscillations), and searches for new physics at the energy frontier. Valuable information can be obtained by observing $\mu \to e$ conversion in different nuclei (40, 43) or by studying the kinematic distribution of the final-state electrons in $\mu \to eee$ (see Reference 40 and references therein). Note that although searches for CLFV in the tau sector are not as powerful as those in the muon sector, they are expected to provide powerful information regarding the nature of new physics and its behavior with regard to the flavor sector. Furthermore, model-independently speaking, we cannot rule out the possibility that CLFV phenomena are much more likely to occur in the tau sector.

3.4. Different Scenarios and Examples

One can try to relate expectations for CLFV with our current understanding of the anomalous magnetic moment of the muon (discussed in detail in Section 5). In summary, the world's most precise measurement of the $g-2$ of the muon disagrees with the world's best Standard Model estimate at the 3σ level.

It is fair to speculate whether this is evidence for new physics. New, heavy physics contributions to the muon $g-2$ are captured by the following effective Lagrangian:[3]

$$\mathcal{L}_{g-2} \supset \frac{m_\mu}{\Lambda^2} \bar{\mu}_R \sigma_{\mu\nu} \mu_L F^{\mu\nu}. \qquad 9.$$

Current $g-2$ data point to $\Lambda \sim 8{,}000$ GeV. Equation 9, however, is very similar to Equations 7 and 8 in the limit $\kappa \ll 1$. In this case, we can relate the Λ in Equation 9 to that in Equations 7 and 8 in the following suggestive way: $(\Lambda_{\text{CLFV}})^{-2} = \theta_{e\mu}(\Lambda_{g-2})^{-2}$. Here the parameter $\theta_{e\mu}$ measures how flavor conserving the new physics is. For example, if $\theta_{e\mu} = 0$, then the new physics is strictly flavor conserving, whereas if the new physics is flavor indifferent, then $\theta_{e\mu} \sim 1$. In the latter case, negative searches for $\mu \to e\gamma$ already preclude a new physics interpretation to the muon $g-2$ results (they constrain $\Lambda > 750$ TeV!). However, if the muon $g-2$ discrepancy is really evidence for new physics, then searches for $\mu \to e\gamma$ reveal that the "amount" of flavor violation in the new physics sector is very small: $\theta_{e\mu} < 10^{-4}$. A more detailed comparison of these two probes of new physics can be found elsewhere (44).

A class of very well studied models that usually predict $\kappa \ll 1$ is that of supersymmetric versions of the Standard Model with R-parity conservation. Superpartner loops involving gauginos and sleptons lead to

$$\frac{1}{\Lambda^2} \sim \frac{g^2 e}{16\pi^2 M_{\text{SUSY}}^2} \theta_{e\mu}, \qquad 10.$$

[3]Another term, with the opposite chirality assignments, has been omitted.

where e and g are the electromagnetic and $SU(2)_L$ couplings, respectively, and $M_{\rm SUSY}$ is some effective superpartner mass. The flavor-violating parameter $\theta_{e\mu}$ can be expressed in terms of the off-diagonal slepton mass–squared matrix element $\Delta m^2_{\tilde{\mu}\tilde{e}}$,

$$\theta_{\mu e} \sim \frac{\Delta m^2_{\tilde{\mu}\tilde{e}}}{M^2_{\rm SUSY}}. \qquad 11.$$

Weak-scale superpartner masses are allowed only by CLFV data if $\theta_{\mu e}$ is small, which in this case translates into $\Delta m^2_{\tilde{\mu}\tilde{e}}$—much smaller than the diagonal slepton masses squared.

One way to guarantee that this is the case is to assume that the mechanism of supersymmetry breaking, which determines the slepton mass–squared matrix, is flavor blind so that, in the absence of Standard Model flavor effects, $\Delta m^2_{\tilde{\mu}\tilde{e}} \equiv 0$. Because flavor numbers are not conserved in the Standard Model, quantum-mechanical effects lead to nonzero, calculable $\Delta m^2_{\tilde{\mu}\tilde{e}}$. Not surprisingly, results will depend on the mechanism underlying neutrino masses. In the case of the seesaw mechanism, for example,

$$\Delta m^2_{\tilde{\alpha}\tilde{\beta}} \sim \frac{M^2_{\rm SUSY}}{8\pi^2} \sum_k (y)^*_{\alpha k}(y)_{\beta k} \ln \frac{M_{UV}}{M_{N_k}}, \qquad 12.$$

where y represents the neutrino Yukawa couplings; M_{N_k} represents the Majorana masses of the $k = 1, 2, 3$ right-handed neutrinos; and M_{UV} represents the ultraviolet scale at which the supersymmetry-breaking parameters originate.

Predictions for CLFV processes will depend, in this case, on the values of the neutrino Yukawa couplings. If these values are large, then the rates for several CLFV processes, especially $\mu \to e\gamma$ and $\mu \to e$ conversion in nuclei, would be expected to be large enough to be observed in current or next-generation experiments. For a concrete, detailed study see, for example, Reference 45.

Equation 12 reveals that, at least in principle, one can relate the rates for CLFV processes to a particular combination of neutrino Yukawa couplings. However, neutrino masses reveal, through the seesaw mechanism, a different combination of neutrino Yukawa couplings: $m_\nu \propto \sum_k y_{\alpha k} M_k^{-1} y_{\beta k}$. Hopefully, by combining neutrino masses, CLFV, and the measurement of superpartner masses and couplings at high-energy colliders, one can determine the neutrino Yukawa couplings and the right-handed neutrino masses. Doing so would not only nail down the physics behind neutrino masses but would also allow one to test whether the baryon asymmetry of the universe arises due to leptogenesis (46). For a detailed discussion see, for example, Reference 47.

Other new physics scenarios will lead to different values of κ, including $\kappa \sim 1$ and $\kappa \gg 1$. Supersymmetric models with trilinear R-parity violation serve as an excellent laboratory in which to explore different possibilities (48). In the presence of, say, large couplings λ' for the LQD superpotential term (see Reference 48 for details and notation), one realizes Equation 7 in the limit $\kappa \gg 1$, and

$$\frac{1}{\Lambda^2} \sim \frac{\lambda'^2}{M^2_{\rm SUSY}}. \qquad 13.$$

In such a scenario, the branching ratio for $\mu \to e$ conversion is much larger (several orders of magnitude) than those for $\mu \to e\gamma$ and $\mu \to eee$. The presence of other couplings may lead to scenarios in which $\kappa \sim 1$ in both Equation 7 and Equation 8, and yet another choice leads predominantly to Equation 8 in the limit $\kappa \gg 1$.

Large rates for CLFV are not restricted to supersymmetric scenarios. Any complete new physics scenario that addresses the gauge hierarchy problem will lead to CLFV rates that are much,

much larger than those associated with the naïve massive neutrino contribution (see Equation 6). Concrete examples, associated with different values of κ, are plentiful in the literature (e.g., 49–51).

3.5. Muon Flavor Violation at Fermilab

The MEG experiment at PSI (41), currently taking data, aims to be sensitive to $\mu^+ \to e^+\gamma$ decays if the branching ratio exceeds 10^{-13}. Upgrades to MEG are under consideration, and hopefully a sensitivity to branching ratios larger than 10^{-14} will be achieved. As discussed above, further sensitivity to this process is expected to be exceedingly challenging. For this reason, the community is dedicating a great deal of energy and resources to the exploration of $\mu \to e$ conversion processes in nuclei. It is expected that, over the coming decade, experiments sensitive to normalized conversion rates larger than 10^{-16} will be taking data.

One of these large efforts is currently being undertaken at Fermilab, and another is being actively pursued at J-PARC in Japan (52). The *Mu2e* proposal (53) aims to be sensitive to normalized rates larger than 6×10^{-17} at 90% CL for $\mu \to e$ conversion in aluminum. This sensitivity would represent a 10,000-fold improvement over the current upper bound. The *Mu2e* experiment will make use of 8-GeV protons from the Fermilab Booster and will take full advantage of the accelerator structure at Fermilab to stop a pure, intense muon beam on a thin target. The *Mu2e* apparatus design (**Figure 7**) is an evolution of the now-defunct MECO proposal at Brookhaven. One of the main features of the *Mu2e* design, advocated in the MECO proposal, is a pulsed beam with an extinction factor of better than 10^{-9}. This factor is required in order to reduce pion contaminants in the muon beam, which are one of the dominant sources of background.

Project X would allow upgrades to *Mu2e*, currently under study, to reach sensitivities better than the 10^{-18} level. The goals of such an upgrade will depend on the results obtained by the *Mu2e* experiment itself. If a positive signal is detected at the 10^{-16} level with the Project X–era upgrade, one could collect hundreds of $\mu \to e$ conversion events and perform a precision measurement of this rare phenomenon. If large event samples are expected, we may be able to study the $\mu \to e$ conversion process as a function of the target nucleus, which would reveal significantly more about the mechanism underlying the CLFV physics revealed at *Mu2e* (43). If no signal for CLFV is observed at *Mu2e*, the Project X–era upgrade would provide the ultimate sensitivity to CLFV; it would also be among the few opportunities we have to access physics that is well above the

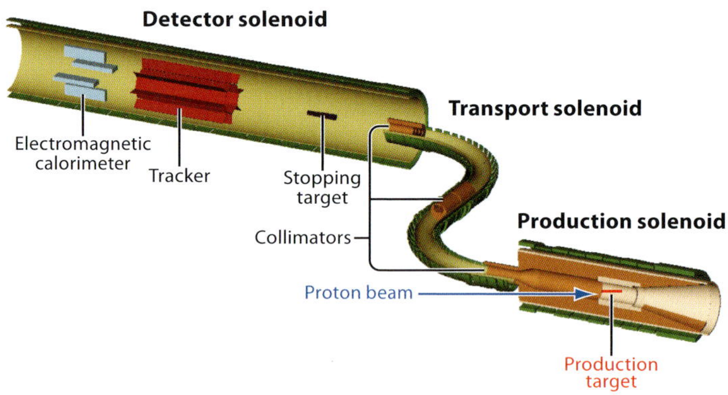

Figure 7

Envisioned layout of the *Mu2e* apparatus. The proton beam enters the production solenoid from the left, and muons are collected and transported via the transport solenoid to the detector solenoid. Reproduced from Reference 53.

teraelectronvolt scale and, hence, out of the reach of the LHC or next-generation lepton colliders (ILC, muon collider, etc).

In addition to the *Mu2e* program, initial studies of whether one could also pursue other searches for CLFV—including $\mu^+ \to e^+e^-e^+$—at Fermilab are currently under way. It appears (42) that sensitivities down to branching ratios of approximately 10^{-15} are achievable and that nontrivial improvements that go beyond access to very intense muon beams will be required to consider going beyond these limits. The main issue is that, for both $\mu^+ \to e^+e^-e^-$ and $\mu^+ \to e^+\gamma$, the sensitivity is ultimately limited by accidental backgrounds that grow with the beam intensity.

4. RARE KAON DECAYS

4.1. Introduction

Kaons were first discovered in the late 1940s in cosmic-ray experiments. Their large mass, combined with the fact that they were produced strongly but were abnormally long lived (meaning they decayed only weakly), led to the concept of strangeness.[4] Over the past 60 years, the study of kaon properties and interactions has played a pivotal role in shaping what we now refer to as the Standard Model. An incomplete list of kaon "accomplishments" includes, in modern terminology, the first observation of a flavor-changing charged current ($s \to u$ plus and off-shell W boson), which played a key role in defining the phenomenon of quark mixing; the $\theta \to \tau$ puzzle, which contributed definitively to the postulation that parity is violated in the weak interaction; the first observation of both a $\Delta F = 2$ (F = flavor) phenomenon and flavor oscillations ($K^0 \leftrightarrow \bar{K}^0$ mixing); and the first experimental observations of *CP* violation in fundamental physics.

Precision studies of kaon properties, including rare or forbidden decays, continue to challenge the standard paradigm and severely constrain weak-scale extensions of the Standard Model. For a recent comprehensive discussion, see Reference 54. It is well known, for example, that effective-operator analyses of $K^0 \leftrightarrow \bar{K}^0$ mixing yield some of the most stringent bounds (along with muon CLFV) on generic new physics (bounds on new physics scales of order $\Lambda > 1{,}000$ TeV) and that they are one of the driving forces behind the concept of minimal flavor violation (MFV) (55).

Here we concentrate on rare flavor-changing neutral-current decays of kaons into neutrino final states: $K^\pm \to \pi^\pm\nu\bar{\nu}$ and $K_L \to \pi^0\nu\bar{\nu}$. These are among the rarest observed particle decay modes (in the case of $K^\pm \to \pi^\pm\nu\bar{\nu}$) and, most importantly, are among the best-known (i.e., have the smallest theoretical uncertainty) exclusive hadronic processes, assuming the Standard Model is correct. Therefore, searches for $K \to \pi\nu\bar{\nu}$ remain among the most sensitive probes for new physics. This statement remains true in spite of the extraordinary results from the B factories, which have confirmed the Standard Model picture for the quark flavor sector and electroweak interactions.

4.2. Standard Model Expectations

In the Standard Model, the leading-order contribution to $K \to \pi\nu\bar{\nu}$ comes at the one-loop electroweak level and is GIM suppressed. The branching ratios can be written as (56–58)

$$B_{\rm SM}(K \to \pi\nu\bar{\nu}) = \frac{\tau_K M_K^5}{32\pi^3}(1+\Delta_{\rm EM})|f_+^{K_\pi}(0)|^2\mathcal{I}_\nu\left|\frac{g^4}{16\pi^2 M_W^2}Y_{+,L}\right|^2, \qquad 14.$$

[4] A similar phenomenon was observed with hyperons, and the concept of a new quantum number—strangeness—conserved by strong interactions but violated by the weak interactions also helped explain their lifetimes.

where we have written a combined expression for the charged kaon and the K_L decays. Here τ and M refer to the meson lifetime and mass, $f_+(0)$ is the $K \to \pi$ vector form factor at zero momentum transfer, $\mathcal{I}_\nu$ is the phase-space integral, $f_+(0)\mathcal{I}_\nu$ is best obtained from $K \to \pi\ell\nu$ decays [modulo V_{us} and isospin-breaking effects (57, 59)], and $\Delta_{\rm EM}$ (which is equal to zero for K_L) is a precisely known electromagnetic correction. All electroweak effects (short distance) are in the last term of Equation 14, where g is the $SU(2)_L$ gauge coupling, M_W is the W boson mass,[5]

$$Y_+ = V_{td}^* V_{ts} X(x_t) + V_{cd}^* V_{cs} [X(x_c) + |V_{us}|^2 \delta P_{c,u}], \tag{15.}$$

and

$$Y_L = \mathrm{Im}(Y_+) = \mathrm{Im}(V_{td}^* V_{ts}) X(x_t). \tag{16.}$$

Here V_{ij} are the elements of the Cabibbo-Kobayashi-Maskawa (CKM) mixing matrix; X is a loop function of $x_q = m_q/M_W$, where $q =$ charm, top; and $\delta P_{c,u}$ accounts for other effects that cannot be expressed concisely but is both small and under control [$\delta P_{c,u} = 0.04 \pm 0.02$ (60)]. In the near future, lattice computations should improve this result (61). Y_L vanishes in the limit of CP conservation (it is easy to see that it violates CP invariance explicitly), and the second equality in Equation 16 is approximate but numerically very accurate. An identical expression holds for $K_S \to \pi^0 \nu\bar{\nu}$, where $Y_S = \mathrm{Re}(Y_+)$. X is known at next-to-leading order in quantum chromodynamics (QCD) for the top quark and at next-to-next-to-leading order in the case of charm. For details, see Reference 62.

Current state-of-the-art theoretical and experimental results predict, within the Standard Model, that

$$B_{\rm SM}(K^+ \to \pi^+ \nu\bar{\nu}) = (8.51 \pm 0.72) \times 10^{-11} \tag{17.}$$

and

$$B_{\rm SM}(K_L \to \pi^0 \nu\bar{\nu}) = (2.54 \pm 0.48) \times 10^{-11}, \tag{18.}$$

where we have combined all sources of uncertainty in quadrature. The dominant contribution to the above uncertainties is from the current knowledge of the CKM matrix. In the near future (i.e., the next 5 years), these uncertainties will probably fall by a factor of approximately two (63). New lattice QCD results are expected to play a key role in reducing the uncertainty with which elements of the CKM matrix are determined. New results from the currently operational LHCb experiment will also help to reduce the uncertainties mentioned above.

4.3. Sensitivity to New Heavy Physics

New heavy physics contributes to $K \to \pi\nu\bar{\nu}$ via effective dimension-six operators. For example,

$$\mathcal{L}_{K\to\pi\nu\nu} \supset \frac{1}{\Lambda^2}(\bar{d}\gamma_\mu s) \sum_{\alpha=e,\mu\tau} (\nu_\alpha \gamma^\mu \nu_\alpha), \tag{19.}$$

which interferes with the Standard Model contribution.[6] Effectively, the new physics effect adds to the heavy W and Z contributions:

$$\frac{g^4}{16\pi^2 M_W^2} Y_{+,L} \to \frac{g^4}{16\pi^2 M_W^2} Y_{+,L} + \frac{c_{+,L}}{\Lambda^2}, \tag{20.}$$

[5] In practice, g and M_W are replaced by α evaluated at the Z boson mass, the Weinberg angle θ_W, and the Fermi constant G_F.

[6] As in the previous section, this is only a subset of the effective Lagrangian. Other dimension-six contributions include tensor-tensor and scalar-scalar interactions. These do not interfere with the Standard Model contribution, given the "wrong" helicity structure, but they can still manifest themselves in precision measurements of $K \to \pi\nu\bar{\nu}$.

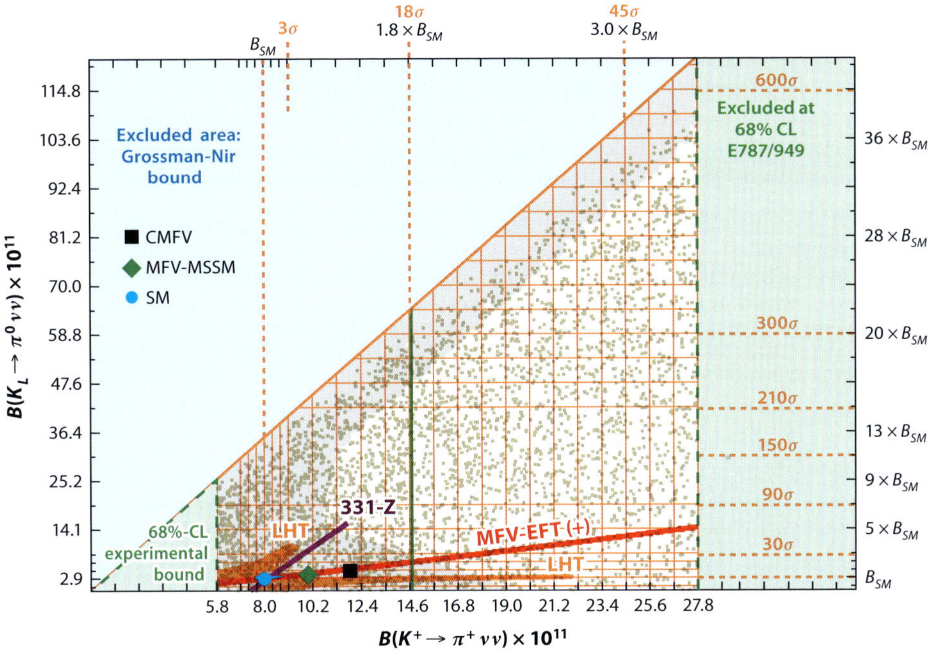

Figure 8

Predictions of different new physics models for the branching ratios for $K^+ \rightarrow \pi^+ \nu\bar{\nu}$ and $K_L \rightarrow \pi^0 \nu\bar{\nu}$. The dashed green lines indicate the 68%-CL limits obtained by the Brookhaven E787/949 experiments (64). The light brown dots depict allowed simultaneous values of the charged and neutral branching ratios in the Minimal Supersymmetric Standard Model (MSSM) with generic flavor structure. The gray shaded area depicts allowed simultaneous values of the charged and neutral branching ratios in models that allow for the existence of a fourth generation. The Grossman-Nir bound refers to an isospin analysis that forbids the light blue region (65). Abbreviations: CMFV, constrained minimal flavor violation; EFT, effective field theory; LHT, little-Higgs models with *T*-parity; MFV, minimal flavor violation; SM, Standard Model. Reprinted from Reference 68, courtesy of Federico Mescia.

where $c_{+,L}$ are calculable coefficients. Modestly precise measurements of the branching ratios for $K \rightarrow \pi\nu\bar{\nu}$ can probe very large Λ values. The Standard Model contribution is (*a*) proportional to $1/M_W^2$, (*b*) one-loop suppressed (factor $g^4/16\pi^2$), and (*c*) GIM suppressed (keep in mind that $V_{td}V_{ts} \sim \sin^5\theta_c \sim 10^{-4}$, where θ_c is the Cabibbo angle, so $Y_+ \sim 10^{-3}$). The combined measurement of $K^+ \rightarrow \pi^+\nu\bar{\nu}$ and $K_L \rightarrow \pi^0\nu\bar{\nu}$ is sensitive not only to the scale of new physics but also to its *CP* invariance–violating nature.

Figure 8 depicts predictions for $K^+ \rightarrow \pi^+\nu\bar{\nu}$ and $K_L \rightarrow \pi^0\nu\bar{\nu}$ in different manifestations of physics beyond the Standard Model, including different versions of the Minimal Supersymmetric Standard Model, little-Higgs models with *T*-parity (66, 67), and different models consistent with MFV. One can readily appreciate that the entire currently allowed region of the branching ratio space is peppered with predictions from different models for new physics. Several other concrete examples can be found in, for example, References 63, 68, and 69. These examples involve the "usual suspects," namely supersymmetric versions of the Standard Model, models with warped extra dimensions, little-Higgs models, and so on. Note that although in some cases the branching ratios for $K^+ \rightarrow \pi^+\nu\bar{\nu}$ and $K_L \rightarrow \pi^0\nu\bar{\nu}$ may be affected in a correlated way, in general one branching ratio can be affected much more than the other.

4.4. $K \rightarrow \pi\nu\bar{\nu}$ at Fermilab

$K^+ \rightarrow \pi^+\nu\bar{\nu}$ was first observed at Brookhaven (64, 70, 71), where seven candidate events were identified inside two distinct kinematic regions. These data allow one to extract $B(K^+ \rightarrow \pi^+\nu\bar{\nu}) = (17.3^{+11.5}_{-10.5}) \times 10^{-11}$ (71). Note that this more recent result is different, but only slightly, from that reported in Reference 64 and shown in **Figure 8**. The ongoing NA62 experiment at CERN aims to collect approximately 100 $K^+ \rightarrow \pi^+\nu\bar{\nu}$ decays with a signal-to-background ratio of approximately 1:10 (see **http://na62.web.cern.ch/NA62/**).

$K_L \rightarrow \pi^0\nu\bar{\nu}$ decays have never been observed. The current most stringent limit, which is still several orders of magnitude larger than the Standard Model expectation, comes from the E391a experiment at KEK: $B(K_L \rightarrow \pi^0\nu\bar{\nu}) < 6.7 \times 10^{-8}$ at 90% CL (72). The approved J-PARC E14 KOTO experiment (73) aims to observe approximately 100 $K_L \rightarrow \pi^0\nu\bar{\nu}$ decays, assuming the Standard Model expectation is correct.

Recently, Fermilab submitted proposal P996 (63) to study rare decays of stopped charged kaons, similar in spirit to the Brookhaven E949 experiment (see **http://www.phy.bnl.gov/e949**). The idea is to use the Tevatron accelerator as a 95% duty-factor "stretcher" (at 150 GeV) for a main injector beam (63). This would allow for an order-of-magnitude-greater kaon flux compared with the E949 flux. This, combined with a detector with ten times more acceptance than the E949 experiment, should allow one to accumulate several hundred $K^+ \rightarrow \pi^+\nu\bar{\nu}$ decays—approximately an order of magnitude larger than the goals of NA62 at CERN. The goals are to control the signal-to-noise ratio at better than 4:1 and to measure $B(K^+ \rightarrow \pi^+\nu\bar{\nu})$ at better than the 5% level. By the time this result would become available, the Standard Model uncertainty should be at a similar level, as discussed above. Such a measurement could dramatically change the scenario depicted in **Figure 8** and would be sensitive to new physics even if its contribution is only a fraction of the Standard Model contribution. Ultimately, any estimate of the impact of such a large statistical sample for $K^+ \rightarrow \pi^+\nu\bar{\nu}$ to our understanding of new physics will depend on the results of the NA62 experiment. More importantly, such a measurement can be performed with existing facilities and well-established detector technologies. In the Project X era, once access to more intense proton—and kaon—beams becomes available, one can envision reducing the uncertainty with which one can measure $B(K^+ \rightarrow \pi^+\nu\bar{\nu})$ by a factor of approximately two or three (74).

The much more challenging "nothing in, nothing out" search for $K_L \rightarrow \pi^0\nu\bar{\nu}$ could also be pursued at Fermilab in the Project X era. Estimates based on the Brookhaven KOPIO detector setup have been made, and it may be possible to collect approximately 1,000 $K_L \rightarrow \pi^0\nu\bar{\nu}$ decays, at the Standard Model branching ratio, while keeping the signal-to-noise ratio at 5:1 or better (74), which would be an improvement of an order of magnitude over the KOTO proposal at J-PARC. Similar to the P996 proposal, such a setup would allow a 5% measurement of $B(K_L \rightarrow \pi^0\nu\bar{\nu})$. As shown in **Figure 8**, such a measurement would be very sensitive to a panoply of new physics scenarios, especially once it is correlated with high-precision measurements of $B(K^+ \rightarrow \pi^+\nu\bar{\nu})$. The impact of a large statistical sample for $K_L \rightarrow \pi^0\nu\bar{\nu}$, in terms of discovering new physics or revealing some of its properties, will become more clear once results of the KOTO effort have been obtained.

5. OTHER OPPORTUNITIES

5.1. Muon $g - 2$ and Electric Dipole Moment

Over the past 30-plus years, precision measurements of the electromagnetic properties of the muon have provided among the most stringent bounds on quantum electrodynamics and electroweak interactions. The magnetic moment $\vec{\mu}$ of the muon is given by

$$\vec{\mu} = g_\mu \frac{e}{2m_\mu} \vec{s}, \tag{21.}$$

where e is the fundamental electric charge, m_μ and $\vec{s}$ are the muon mass and spin, and g_μ is the muon gyromagnetic factor. The anomalous magnetic moment, $a_\mu = (g_\mu - 2)/2$, is currently known at 0.54 ppm (75), whereas in the Standard Model it can be computed with similar precision and is known at 0.42 ppm. For recent reviews and overviews see, for example, References 76 and 77. Currently, there is a tantalizing 3σ discrepancy between theory and experiment:

$$a_\mu^{\text{exp}} - a_\mu^{\text{SM}} = (24.6 \pm 8.0) \pm 10^{-10}. \tag{22.}$$

The dominant contribution to the experimental uncertainty is statistical. The dominant contribution to the uncertainty in the estimation of the Standard Model expectation comes from the leading-order hadronic vacuum polarization. This is estimated mostly with the help of the optical theorem through measurements of electron-positron annihilation into hadrons at center-of-mass energies of approximately a few hundred to a few thousand megaelectronvolts. Results from several experiments are combined to make up the estimate discussed above (76, 77). More recently, data from the BaBar experiment (78), which slightly disagree with previous electron-positron scattering data, suggest that the deviation is smaller than the one summarized in Equation 22. See References 76–79 and 80 for more details. The same is true if one were to estimate the hadronic vacuum polarization contribution by using tau decay data. Clearly, next-generation experiments are required in order to probe whether this discrepancy is due to statistical fluctuations; the presence of new, heavy physics; or other, more subtle experimental issues.

Currently, a few proposals to improve on the precision of the measurement of a_μ, including a proposal to host an experiment at Fermilab (79), are under consideration.[7] The main goal of such an experiment would be to build on the success of the E821 experiment at Brookhaven and reach a precision of 0.14 ppm—a fourfold improvement over the precision reported in Reference 75—by accumulating a larger statistical sample (the results in Reference 75 are statistics limited) and at the same time reducing the systematic uncertainties by a factor of three (79). The Standard Model estimate for a_μ is expected to improve by a factor of two or three. Currently, the dominant contribution to the 0.42-ppm uncertainty mentioned above arises from estimates of the leading hadronic contribution to a_μ. These estimates are expected to improve significantly with upcoming new data on e^+e^- scattering and tau decays. If the current magnitude of the discrepancy shown in Equation 22 persists while the uncertainties improve as described above, the current 3σ hint would be upgraded to very robust 10σ evidence for new physics at the electroweak scale!

New heavy physics contributes to a_μ via the effective Lagrangian (Equation 9). The current discrepancy translates into $\Lambda \sim 8$ TeV. Assuming that the new physics contributes at the one-loop level, then

$$\left(\frac{1}{\Lambda}\right) \sim \frac{e}{16\pi^2} \left(\frac{Y}{M_{\text{new}}}\right)^2, \tag{23.}$$

where Y parameterizes the new couplings that allow the new physics to "talk" to the Standard Model. For $M_{\text{new}} \sim (\text{few} \times 100)$ GeV, the current discrepancy points to a large coupling $Y \sim 1$. Such scenarios are naturally realized in supersymmetric versions of the Standard Model, where the supersymmetry-breaking parameters are around the electroweak symmetry–breaking scale

[7]There are also discussions regarding hosting an experiment in Japan (81).

and the parameter tanβ is large. For more discussions on the muon anomalous magnetic moment and new physics at the electroweak scale, see Reference 82 and references therein.

Experimental setups capable of precision measurements of a_μ are often well suited for searches for a nonzero muon EDM. EDMs of fundamental fermions are sensitive not only to new physics around the electroweak scale but also to its *CP*-conserving character. In the Standard Model, the muon EDM is expected to be approximately 10^{-35} e·cm, whereas the current experimental upper bound is $d_\mu < 1.8 \times 10^{-19}$ e·cm at 95% CL (83). The Fermilab proposal to improve on a_μ (discussed above) should be sensitive to $d_\mu > 10^{-21}$ e·cm. Project X may enable third-generation searches for the muon EDM that could reach d_μ values larger than $d_\mu > 10^{-24}$ e cm. Although all these values are well below the Standard Model expectations, they are within the expectations for well-motivated scenarios of new physics at the electroweak scale. It is also expected that the sensitivity to the EDM of the electron and the neutron will improve significantly. If a positive result is found for any of these observables, we will need to collect as much information as possible about the *CP*-violating nature of the new physics, including, of course, d_μ.

5.2. Neutrino Scattering Physics and Short-Baseline Oscillations

The presence of intense neutrino beams allows for several different types of neutrino scattering experiments. Such experiments take advantage of the large neutrino fluxes to perform precision measurements of neutrino-matter interactions and to search for unexpected neutrino properties. These near detectors are of fundamental importance to the long-baseline program discussed above because they provide crucial information regarding neutrino cross sections and the energy and flavor composition of the neutrino beam. Concrete examples of measurements that are sensitive to new physics include searches for tau leptons in pion decay beams (which contain only muon-type and electron-type neutrinos) and precision measurements of neutrino-electron scattering (84).

An intriguing possibility involves using the Tevatron accelerator at Fermilab to shoot 800-GeV protons on a target and extract a high-energy pion decay neutrino beam. This type of beam was used by, for example, the NuTeV experiment during the 1990s. NuTeV's legacy includes some of the most precise measurements of neutrino interactions, along with the so-called NuTeV anomaly (85). The physics capabilities of a next-generation neutrino detector placed in a more powerful version of the NuTeV beam have been discussed in detail (86, 87).

6. CONCLUDING REMARKS

During the coming decades, progress in the understanding of nature at the smallest distance scales—what things are made of, and how these fundamental particles interact with one another—will require research efforts in several superficially disparate areas, from the LHC to gravitational-wave detectors to atomic physics experiments searching for *CP*-invariance violation. Fermilab is already strongly invested in experiments at the so-called intensity frontier and is poised to increase those investments and play a pivotal role in this very broad area of particle physics research. In this review we have given detailed descriptions of some of these experiments, concentrating on those that will profit most from the intense multimegawatt proton driver, Project X. Several other research opportunities at Fermilab after the end of the Tevatron collider run were not discussed at all. These include experiments that take advantage of Fermilab's unique antiproton source and a recent proposal to revisit the Tevatron-based fixed-target program at Fermilab to study, for example, charm quarks and tau-type neutrinos (88).

Concerning the exploration of the neutrino sector, Fermilab's current near- and longer-term physics program can provide world-class measurements and discoveries in both the oscillation and precision physics sectors. The currently available accelerator complex also places Fermilab in a position of unparallel strength in the study of muon interactions and properties and rare kaon decays. Searches for CLFV phenomena, especially $\mu^- \rightarrow e^-$ conversion in nuclei, can probe the flavor structure of new physics at the electroweak scale and will provide sensitivity to new degrees of freedom at energy scales above 1,000 TeV. The same may also be true of precise (better than 5–10%) third-generation measurements of charged and neutral $K \rightarrow \pi\nu\bar{\nu}$. Other measurements—including those that would improve our knowledge of the $g-2$ of the muon, and a high-statistics sample of neutrino-electron scattering events—may also provide nontrivial clues that would probably be difficult or impossible to collect otherwise.

Now that we are at the beginning of the LHC era, we may wonder whether it is wise to invest in the intensity frontier, rather than wait for LHC discoveries to illuminate the way forward. We believe, and have tried to illustrate herein, that the answer is an unqualified yes. Although LHC results may help shape our understanding of what intensity frontier experiments are revealing, the information the latter provide generally cannot be obtained at the energy frontier. For example, if there are new degrees of freedom at the electroweak scale, experiments in the intensity frontier will help reveal the flavor structure of this new physics and determine whether it obeys *CP* invariance. Results from the LHC may help reveal the origin of neutrino masses, but precision measurements of neutrino mixing parameters are required in order to establish any reliable connection. Furthermore, if there is new physics outside the reach of the LHC, only a handful of particle physics experiments can hope to show what is hiding beyond the energy frontier. Most, if not all, of these experiments lie within the intensity frontier.

There are several technological challenges shared between the intensity frontier today and the pursuit of particle physics tomorrow. Intense proton beam accelerators, the basis for the experiments discussed here, are also of fundamental importance for building third- (and fourth-) generation neutrino experiments—neutrino factories (33). Neutrino factories, however, can be interpreted as stepping stones toward the realization of a promising class of next-generation energy frontier machines—muon colliders.

APPENDIX: BRIEF OVERVIEW OF THE CURRENT AND NEAR-FUTURE ACCELERATOR COMPLEX AT FERMILAB AND PROJECT X

The current accelerator complex at Fermilab consists of seven different machines: the Linac (linear accelerator), the Booster, the Accumulator, the Debuncher, the Recycler, the Main Injector, and the Tevatron. Improvements that are currently under way are expected to provide 700 kW capability, at 120 GeV, for long-baseline neutrino experiments (e.g., NoνA). Project X is an upgrade to the initial stages of this complex that will result in more intense beams of protons. This intensity will be achieved by replacing the current Linac and Booster with a superconducting linac that, coupled with the Recycler and Main Injector, will provide high-intensity proton beams above 2 MW at beam energies between 8 and 120 GeV. This performance would be well beyond the capabilities of the present complex. For more details and information, see **http://www.fnal.gov/pub/projectx** (and links therein).

Project X is based on an 8-GeV superconducting H^- linac. The downstream 6 GeV would use superconducting cryomodules that are very similar to those being designed for the International Linear Collider (ILC). The Recycler would be used as a H^- stripper and accumulator ring. To take full advantage of the increased beam power made available by Project X, one would have to make changes to both the Recycler and the Main Injector. The Recycler would need a new H^- injection system, plus a new extraction line and new radio-frequency systems. The Main

Injector would require a new radio-frequency system. In addition, a neutrino target station that can withstand more than 2 MW of power would have to be constructed, as would a new neutrino beam line with different beam characteristics (i.e., a wide-band neutrino beam). The latter (target and neutrino beam line) are anticipated to be constructed as part of the LBNE (Long-Baseline Neutrino Experiment) project (for details, see **http://lbne.fnal.gov**). Some of these modifications are discussed in further detail in Section 2. With sufficient support, Project X could be constructed and operational by the end of this decade.

DISCLOSURE STATEMENT

The authors are not aware of any affiliations, memberships, funding, or financial holdings that might be perceived as affecting the objectivity of this review.

ACKNOWLEDGMENTS

The authors thank Andreas Kronfeld for useful discussions on rare kaon decays and Steve Holmes for comments on the manuscript and guidance regarding the current and near-future status of the Fermilab accelerator complex, including Project X.

LITERATURE CITED

1. Chadwick J. *Nature* 129:312 (1932)
2. Kodama K, et al. (DONUT Collab.) *Phys. Lett.* B504:218 (2001)
3. Franklin A. *Are There Really Neutrinos?—An Evidential History*. Cambridge, UK: Perseus (2001)
4. Hosaka J, et al. (Super-Kamiokande Collab.) *Phys. Rev. D* 73:112001 (2006)
5. Aharmim B, et al. (SNO Collab.) *Phys. Rev. C* 72:055502 (2005)
6. Araki T, et al. (KamLAND Collab.) *Phys. Rev. Lett.* 94:081801 (2005)
7. Ashie Y, et al. (Super-Kamiokande Collab.) *Phys. Rev. D* 71:112005 (2005)
8. Ahn MH, et al. (K2K Collab.) *Phys. Rev. D* 74:072003 (2006)
9. Michael DG, et al. (MINOS Collab.) *Phys. Rev. Lett.* 97:191801 (2006)
10. Ambrosio M, et al. (MACRO Collab.) *Eur. Phys. J. C* 36:323 (2004)
11. Allison WWM, et al. (Soudan-2 Collab.) *Phys. Rev. D* 72:052005 (2005)
12. Amsler C, et al. (Part. Data Group) *Phys. Lett.* B667:1 (2008)
13. Dolmen C, et al. (SINDRUM II Collab.) *Phys. Lett.* B317:631 (1993)
14. Bellgardt U, et al. (SINDRUM Collab.) *Nucl. Phys. B* 299:1 (1998)
15. Brooks ML, et al. (MEGA Collab.) *Phys. Rev. Lett.* 83:1521 (1999)
16. Apollonio M, et al. (CHOOZ Collab.) *Eur. Phys. J. C* 27:331 (2003)
17. Gonzalez-Garcia MC, Maltoni M, Salvado J. arXiv:1001.4524 [hep-ph] (2010)
18. Ardellier F, et al. (Double Chooz Collab.) arXiv:hep-ex/0606025 (2006)
19. Guo X, et al. (Daya Bay Collab.) arXiv:hep-ex/0701029 (2007)
20. Oh Y. (RENO Collab.) *Nucl. Phys. Proc. Suppl.* 188:109 (2009)
21. Itow Y, et al. (T2K Collab.) arXiv:hep-ex/0106019 (2001)
22. Ayres D, et al. NOνA Tech. Des. Rep. *Fermilab-design-2007-01* (2007)
23. Drakoulakos D, et al. (Minerνa Collab.) arXiv:hep-ex/0405002 (2004)
24. Beier E, et al. Fermilab Steer. Group Rep. *Fermilab-pub-07-672-DO, Fermilab-APC* (2007)
25. Nakamura K. *Front. Phys.* 35:359 (2000)
26. Hagiwara K, Okamura N, Senda KI. *Phys. Rev. D* 76:093002 (2007)
27. Jung CK. *AIP Conf. Proc.* 533:29 (2000)
28. Bartoszek L, et al. arXiv:hep-ex/0408121 (2004)
29. Diwan MV, et al. arXiv:hep-ex/0306053 (2003)
30. Cline DB, Sergiampietri F, Learned JG, McDonald K. *Nucl. Instrum. Methods A* 503:136 (2003)

31. Rubbia A. arXiv:hep-ph/0402110 (2004)
32. Barger V, et al. arXiv:0705.4396 [hep-ph] (2007)
33. Bandyopadhyay A, et al. (ISS Phys. Work. Group) *Rep. Prog. Phys.* 72:106201 (2009)
34. Bross A, et al. arXiv:0911.3776 [hep-ph] (2009)
35. Mohapatra RN, et al. *Rep. Prog. Phys.* 70:1757 (2007); Mohapatra RN, Smirnov A. *Annu. Rev. Nucl. Part. Sci.* 56:569 (2006); de Gouvêa A. *Mod. Phys. Lett. A* 19:2799 (2004)
36. Marciano WJ, Sanda AI. *Phys. Lett.* B67:303 (1977); Bilenky SM, Petcov ST, Pontecorvo B. *Phys. Lett.* B67:309 (1977); Cheng TP, Li LF. *Phys. Rev. D* 16:1425 (1977); Lee BW, Pakvasa S, Shrock RE, Sugawara H. *Phys. Rev. Lett.* 38:937 (1977); Lee BW, Pakvasa S, Shrock RE, Sugawara H. Erratum. *Phys. Rev. Lett.* 38:1230 (1977); Lee BW, Shrock RE. *Phys. Rev. D* 16:1444 (1977)
37. Glashow SL, Iliopoulos J, Maiani L. *Phys. Rev. D* 2:1285 (1970)
38. Minkowski P. *Phys. Lett.* B67:421 (1977); Gell-Mann M, Ramond P, Slansky R. In *Supergravity*, ed. D Freedman, P Van Niuwenhuizen, p. 315. Amsterdam: North Holland (1979); Yanagida T. In *Proceedings of the Workshop on Unified Theory and Baryon Number in the Universe*, ed. O Sawada, A Sugamoto, p. 95. Tsukuba, Jpn.: KEK (1979); Glashow SL. In *1979 Cargèse Lectures in Physics—Quarks and Leptons*, ed. M Lévy, et al., p. 707. New York: Plenum (1980); Mohapatra RN, Senjanović G. *Phys. Rev. Lett.* 44:912 (1980)
39. de Gouvêa A. arXiv:0706.1732 [hep-ph] (2007)
40. Kuno Y, Okada Y. *Rev. Mod. Phys.* 73:151 (2001)
41. Adam J, et al. (MEG Collab.) arXiv:0908.2594 [hep-ex] (2009)
42. Aysto J, et al. arXiv:hep-ph/0109217 (2001)
43. Cirigliano V, Kitano R, Okada Y, Tuzon P. *Phys. Rev. D* 80:013002 (2009)
44. Hisano J, Tobe K. *Phys. Lett.* B510:197 (2001)
45. Calibbi L, Faccia A, Masiero A, Vempati SK. *Phys. Rev. D* 74:116002 (2006)
46. Davidson S, Nardi E, Nir Y. *Phys. Rep.* 466:105 (2008)
47. Buckley MR, Murayama H. *Phys. Rev. Lett.* 97:231801 (2006)
48. de Gouvêa A, Lola S, Tobe K. *Phys. Rev. D* 63:035004 (2001)
49. de Gouvêa A, Giudice GF, Strumia A, Tobe K. *Nucl. Phys. B* 623:395 (2002)
50. Agashe K, Blechman AE, Petriello F. *Phys. Rev. D* 74:053011 (2006)
51. Blanke M, et al. *J. High Energy Phys.* 0705:013 (2007)
52. Cui YG, et al. (COMET Collab.) **http://ccdb4fs.kek.jp/tiff/2009/0924/0924011.pdf** (2009); Nishikawa K. Presented at PAC Meet., 9th, Ibasaki, Jpn. **http://j-parc.jp/NuclPart/Proposal_e.html** (2010)
53. Carey RM, et al. (*Mu2e* Collab.) *Fermilab-proposal-0973*. **http://lss.fnal.gov/archive/test-proposal/0000/fermilab-proposal-0973.shtml** (2008)
54. Antonelli M, et al. (FlaviaNet Work. Group Kaon Decays) arXiv:0801.1817 [hep-ph] (2008)
55. D'Ambrosio G, Giudice GF, Isidori G, Strumia A. *Nucl. Phys. B* 645:155 (2002)
56. Buchalla G, Buras AJ. *Nucl. Phys. B* 548:309 (1999)
57. Mescia F, Smith C. *Phys. Rev. D* 76:034017 (2007)
58. Brod J, Gorbahn M. *Phys. Rev. D* 78:034006 (2008)
59. Bijnens J, Ghorbani K. arXiv:0711.0148 [hep-ph] (2007)
60. Isidori G, Mescia F, Smith C. *Nucl. Phys. B* 718:319 (2005)
61. Isidori G, Martinelli G, Turchetti P. *Phys. Lett.* B633:75 (2006)
62. Buras AJ, Gorbahn M, Haisch U, Nierste U. *J. High Energy Phys.* 0611:002 (2006)
63. Comfort J, et al. *Fermilab-proposal-P996* (2009)
64. Anisimovsky VV, et al. (E949 Collab.) *Phys. Rev. Lett.* 93:031801 (2004)
65. Grossman Y, Nir Y. *Phys. Lett.* B398:163 (1997)
66. Blanke M, et al. *J. High Energy Phys.* 0701:066 (2007)
67. Blanke M, et al. arXiv:0906.5454 [hep-ph] (2009)
68. Isidori G. arXiv:0801.3039 [hep-ph] (2008)
69. Buras AJ. arXiv:0910.1032 [hep-ph] (2009)
70. Adler S, et al. (E949, E787 Collab.) *Phys. Rev. D* 77:052003 (2008)
71. Artamonov AV, et al. (E949 Collab.) *Phys. Rev. Lett.* 101:191802 (2008)

72. Ahn JK, et al. (E391a Collab.) *Phys. Rev. Lett.* 100:201802 (2008)
73. Nanjo H. *Proc. Sci.* KAON2009:047 (2009)
74. Bryman D. Presented at Workshop Phys. High-Intensity Proton Source, 4th, Batavia, Ill. **http://www.fnal.gov/directorate/Longrange/Steering_Public/workshop-physics-4th.html** (2009)
75. Bennett GW, et al. [Muon (g–2) Collab.] *Phys. Rev. D* 73:072003 (2006)
76. Prades J. arXiv:0909.2546 [hep-ph] (2009)
77. Jegerlehner F, Nyffeler A. *Phys. Rep.* 477:1 (2009)
78. Aubert B, et al. (BaBar Collab.) *Phys. Rev. Lett.* 103:231801 (2009)
79. Carey RM, et al. *Fermilab-proposal-0989*. **http://lss.fnal.gov/archive/test-proposal/0000/fermilab-proposal-0989.shtml** (2009)
80. Davier M, et al. *Eur. Phys. J. C* 66:1 (2010)
81. Imazato J. *Nucl. Phys. Proc. Suppl.* 129:81 (2004)
82. Czarnecki A, Marciano WJ. *Phys. Rev. D* 64:013014 (2001); Hertzog DW, et al. arXiv:0705.4617 [hep-ph] (2007)
83. Bennett GW, et al. [Muon (g–2) Collab.] *Phys. Rev. D* 80:052008 (2009)
84. de Gouvêa A, Jenkins J. *Phys. Rev. D* 74:033004 (2006)
85. Zeller GP, et al. (NuTeV Collab.) *Phys. Rev. Lett.* 88:091802 (2002); Zeller GP, et al. (NuTeV Collab.) Erratum. *Phys. Rev. Lett.* 90:239902 (2003)
86. Adams T, et al. (NuSOnG Collab.) *Int. J. Mod. Phys. A* 24:671 (2009)
87. Adams T, et al. (NuSOnG Collab.) *Int. J. Mod. Phys. A* 25:1 (2010)
88. Adams T, et al. arXiv:0905.3004 [hep-ex] (2009)

Big Bang Nucleosynthesis as a Probe of New Physics

Maxim Pospelov[1,2] and Josef Pradler[1]

[1]Perimeter Institute for Theoretical Physics, Waterloo, Ontario N2L 2Y5, Canada; email: jpradler@perimeterinstitute.ca

[2]Department of Physics and Astronomy, University of Victoria, Victoria, British Columbia V8P 1A1, Canada; email: pospelov@uvic.ca

Annu. Rev. Nucl. Part. Sci. 2010. 60:539–68

First published online as a Review in Advance on July 19, 2010

The *Annual Review of Nuclear and Particle Science* is online at nucl.annualreviews.org

This article's doi:
10.1146/annurev.nucl.012809.104521

0163-8998/10/1123-0539$20.00

Key Words

early universe, abundances of light elements, extensions of the Standard Model, dark matter

Abstract

Big bang nucleosynthesis (BBN), an epoch of primordial nuclear transformations in the expanding universe, has left an observable imprint in the abundances of light elements. Precision observations of such abundances, combined with high-accuracy predictions, provide a nontrivial test of the hot big bang and probe nonstandard cosmological and particle physics scenarios. We give an overview of BBN sensitivity to different classes of new physics: new particle or field degrees of freedom, time-varying couplings, decaying or annihilating massive particles leading to nonthermal processes, and catalysis of BBN by charged relics.

Contents

1. INTRODUCTION

The model of the hot expanding universe, once the subject of fierce scientific and philosophical debate, is now a well-established reality. Explosive progress in the field of cosmology over the past decade allowed for something that may well have been totally unanticipated in the previous decades: the high-precision determination of several cosmological parameters. This recent progress primarily clarifies the state of the universe at redshifts $z \lesssim$ few and exposes the relevant physics of the universe traced back to the epoch of the decoupling of the cosmic microwave background (CMB) at $z \lesssim 10^3$. At the same time, these recent advances also allow one to check the consistency with one of the building blocks of modern cosmology: big bang nucleosynthesis (BBN). The agreement, or rather the lack of qualitative disagreement between cosmology in the $z = 0$–10^3 interval and BBN at $z \sim 10^9$, is very important, as it leads to the conclusion that the very early universe was governed by the same physical laws of nature as the current universe and that it contained very similar if not identical particle content. Remarkably, the transformations and the synthesis of light elements in the expanding universe occur with the active participation of all the interactions known to date: strong, weak, electromagnetic, and gravitational. It is well known that even a mild modification of the standard conditions in the early universe, at the time of BBN or in the subsequent evolution, may lead to observable deviations in primordial abundances. Thus, employing the precise determinations of the primordial abundances allows one to set limits or constrain various scenarios with deviations from general relativity or from particle physics of the Standard Model (SM).

Primordial nucleosynthesis is not the earliest cosmological epoch. It must have been preceded by an era in which a mechanism capable of producing the observed baryon-antibaryon asymmetry of the universe (baryogenesis) was operative. Moreover, a period of generation of nearly scale-invariant cosmological perturbations—for which the leading candidate is inflation—had set the initial conditions from which the large-scale structure of the universe evolved. Neither the cosmological timing nor the main "players" in inflation or baryogenesis—such as the inflaton, the right-handed neutrino, and so on—are known with certainty. Although many successful models of this earliest epoch(s) exist, it is extremely difficult to directly test the associated physics in the laboratory. Thus, a multitude of baryogenesis and inflationary scenarios may be viable. In contrast, BBN relies on very well studied pieces of physics, such as spectra and reactions of light elements and their weak decays. It involves almost no free parameters and occurs in a well-understood sequence of events in the early universe. The consistency of BBN predictions with observations imposes an important calibration point on all new models of particle physics that requires them to achieve some degree of normalcy before $t = 1$ s. This requirement places powerful constraints on many extensions of the SM.

The purpose of this review is to reveal different physical mechanisms by which new physics can affect BBN and its light element–abundance predictions, which have been extensively discussed in the literature for as long as the BBN theory has existed. In fact, one of the first papers on the subject by Alpher et al. (1) points out that the nature of the neutrino species (Dirac versus Majorana) affects the neutron-to-proton freeze-out ratio, n/p, and therefore the primordial helium abundance. Since then, a great deal of research dedicated to nonstandard BBN scenarios has been performed, and a number of different ways in which new physics can change the outcome of BBN have been identified. Here, we present a nontechnical review of many aspects related to these interesting possibilities.

The developments over the past decade also confirmed the conclusion that the SM is not the ultimate theory. Important pieces of the puzzle come from cosmological observations suggesting that the energy balance of the modern universe is dominated by dark energy and dark matter. Is it possible that physics related to these mysterious substances interfered with the sequence of events that ultimately led to the fusion of the primordial elements? Although there is no compelling evidence that this occurred, some new physics models related to particle dark matter may reduce the tension between data and the standard BBN theory (SBBN) prediction for the lithium isotopes. Even though the resolution of this tension may be related to astrophysics, it is nevertheless intriguing to consider that the relics of the early universe may hint at deviations from the SM. Future developments in astrophysics, cosmology, and particle physics may help to clarify this question.

1.1. Universe at Redshift of a Billion: Basic Assumptions and Main Stages of Big Bang Nucleosynthesis

The primary activity of BBN took place in the era associated with photon temperatures between $T \simeq$ few MeV and $T \simeq 10$ keV in the cosmic time window $t \simeq (0.1 \div 10^4)$ s and may be considered as a transition from a neutron-proton statistical equilibrium with no other nuclear species to a universe with a significant presence of helium. BBN produced the bulk of ^{4}He and D as well as good fractions of ^{3}He and ^{7}Li observed in the current universe. All the other elements are believed to have been produced either by stars or by cosmic rays.

We begin by specifying the main assumptions on which SBBN theory rests.

1. It is assumed that the universe is spatially homogeneous and isotropic and that its energy density is completely dominated by radiation so that physical distances scale as $a(t) \propto t^{1/2}$. It is also assumed that the space-time geometry is flat with zero spatial curvature. The theory of general relativity relates the expansion rate with the time elapsed since the big bang and with the energy density of the ambient cosmological fluid ρ:

$$H \equiv \frac{\dot{a}}{a} = \sqrt{8\pi G_N \rho/3} \simeq \frac{1}{2t}, \qquad 1.$$

where the energy density of the radiation-dominated universe scales as $\rho \sim a^{-4} \sim T^4$. G_N denotes Newton's constant, $G_N = (8\pi M_{\rm Pl}^2)^{-1}$, where $M_{\rm Pl} \simeq 2.43 \times 10^{18}$ GeV is the reduced Planck mass. It is also assumed that the dark matter and dark energy components of the universe are well behaved, that is, that their contribution to the energy density at the time of BBN is negligible.

2. It is assumed that the initial temperature of the radiation-dominated epoch of the universe was well above the neutron-proton mass difference $\Delta m_{np} = 1.293$ MeV, so that the initial conditions for the nuclear reaction framework are well specified:

$$(n_n \simeq n_p)\big|_{T \gg \Delta m_{np}} = \frac{1}{2} n_b. \qquad 2.$$

The energy-momentum distribution of neutrons and protons is very close to being thermal. Moreover, a near-perfect spatial homogeneity for the distributions of neutrons and protons is assumed.

3. It is assumed that the particle content and their interactions are given by that of the SM and that by the time of BBN the baryon asymmetry was already present. The energy density is completely dominated by photons and the three species of SM neutrinos (as well as electrons and positrons before their annihilation), whereas neutrons and protons carry only a negligible fraction of the total energy density. The standard field content implies that there was an uneventful cosmological period between the time of BBN and the epoch of hydrogen recombination with electrons, which implies that the baryon-to-entropy density ratio measured at the period of CMB formation directly translates into the one at BBN:

$$\frac{n_b}{s}(t_{\rm BBN}) = \frac{n_b}{s}(t_{\rm CMB}). \qquad 3.$$

Indeed, the measurements of the Wilkinson Microwave Anisotropy Probe (WMAP) satellite have allowed us to pinpoint the baryon density for a standard ΛCDM cosmology—that is, a flat universe filled with baryons, cold dark matter, neutrinos, and a cosmological constant—to an accuracy of better than 3% (2). Expressed in terms of the baryon-to-photon ratio,

$$\eta_b(t_{\rm CMB}) = \frac{n_b}{n_\gamma}(t_{\rm CMB}) = (6.23 \pm 0.17) \times 10^{-10}, \qquad 4.$$

this number provides a measure of the nucleon content of the universe at BBN.

4. Finally, it is assumed that the properties of particles and nuclei (masses, couplings, scattering cross sections, and lifetimes) are identical between their current values and their values at $t_{\rm BBN}$. This condition follows from the assumption of a minimal field content, as hypothetical changing couplings and masses would necessitate new ultrasoft scalar fields.

In many theories with nonstandard cosmological or particle physics content, some of these assumptions can be violated. For example, late decays, that is, decays of metastable heavy particles during or after BBN, can lead to energy injection into the primordial plasma and thus to temporary but strong departures from thermal equilibrium for some species. The presence of the new relativistic degrees of freedom may affect the total energy density of the universe and, by modifying the Hubble expansion rate, may change the outcome of nuclear reactions. Before we consider such modifications in more detail, we remind the reader of the main components of SBBN and briefly review its current observational status.

SBBN theory, including the hits and misses of the original papers (3–5), is now well understood and described in detail in many textbooks and reviews (6–10). SBBN comprises the set of first-order differential Boltzmann equations on the abundances of the different species,

$$\frac{dY_i}{dt} = -H(T)T\frac{dY_i}{dT} = \sum(\Gamma_{ij}Y_j + \Gamma_{ikl}Y_kY_l + \ldots), \qquad 5.$$

where $Y_i = n_i/n_b$ are the time t (or temperature T)-dependent ratios between the number density n_i and the baryon number density n_b of light elements $i = p$, n, D, ^{4}He, and so on. The $\Gamma_{ij\ldots}$ represent generalized rates for element interconversion and decay that can be determined in experiments and/or inferred from theoretical calculations. $H(T)$ is the temperature-dependent Hubble expansion rate from Equation 1.

The full form of the (nonintegrated) Boltzmann equations should be given in terms of particle distribution functions over energy and momenta. However, in practice, the system of Equation 5, which assumes thermal distributions for nuclei, provides an excellent approximation because the frequent interactions with the numerous γs and $e^{\pm}$s in the plasma keep the light elements tightly coupled to the radiation field. The dependence of $H(T)$ on the temperature of the primordial plasma can be further specified:

$$H(T) = T^2 \left(\frac{8\pi^3 g_* G_N}{90} \right)^{1/2}, \quad \text{where } g_* = g_{boson} + \frac{7}{8} g_{fermion}, \qquad 6.$$

where the gs denote the excited relativistic degrees of freedom. This expression needs to be interpolated across the electron-positron annihilation epoch, in which the photon bath is heated with respect to the neutrino reservoir. The neutrinos maintain a quasi-thermal spectrum with temperature

$$T_\nu \simeq (4/11)^{1/3} T \qquad 7.$$

in the approximation of full neutrino decoupling at the time of electron-positron annihilation [with small calculable corrections (11)]. Following $e^{\pm}$ annihilation, the Hubble rate is given by $H(T) \simeq T_9^2/(2 \times 178\,\mathrm{s})$, where T_9 denotes the photon temperature T in units of 10^9 K.

A number of well-developed integration codes, pioneered by Wagoner et al. (12), allow one to solve the BBN system of equations numerically and to obtain the freeze-out values of the light elements. Also, good semianalytic results can be obtained (e.g., 13). The code that we use for this review is based on that of Reference 14. Here we incorporate some significant improvements and updates; physical constants, isotope masses, and conversion factors are determined from the evaluations (15, 16). For all important SBBN reactions (i.e., up to A = 7) we employ the results of Reference 17, with the exceptions of the $n(p,\gamma)$D and ^{3}He$(\alpha,\gamma)^7$Be reactions, for which we follow References 18 and 19, respectively. To arrive at an accurate prediction of ^{4}He, we numerically integrate all weak rates at each time step for which zero temperature radiative corrections, as well as relativistic Coulomb corrections (when applicable), are taken into account (20, 21). Finally, as assessed in Lopez & Turner (20), we apply a slight upward shift of 0.72% of the resulting ^{4}He abundance to account for more subtle, subleading corrections to the BBN reaction network. We then find very good agreement between our light element–abundance predictions and those presented (22) at the WMAP value (Equation 4) and with a neutron lifetime of $\tau_n = 885.7$ s.

Figure 1 shows the output of the SBBN reaction network. The evolution of all the main isotopes with A ≤ 7 is plotted as a function of time and temperature. Beginning from the earliest times, the following sequence of events occurs: (*a*) the chemical decoupling of neutrinos from the thermal bath, (*b*) the annihilation of electrons and positrons, (*c*) the freeze-out of neutrons and protons, (*d*) an "intermission" between the n/p freeze-out and the deuteron ignition at the end of the "bottleneck," (d) helium synthesis at $T_9 \simeq 0.85$ (70 keV), and finally (*e*) a follow-up stage in which the main nuclear reactions gradually drop out of equilibrium and the abundances of all light elements freeze out. Although some neutrons are still generated by residual deuterium (D) fusion below $T_9 \sim 0.1$ (10 keV), they are too few in number to cause any further change in the elemental abundances. Below we discuss the fusion of the light elements and compare their SBBN predictions with observations.

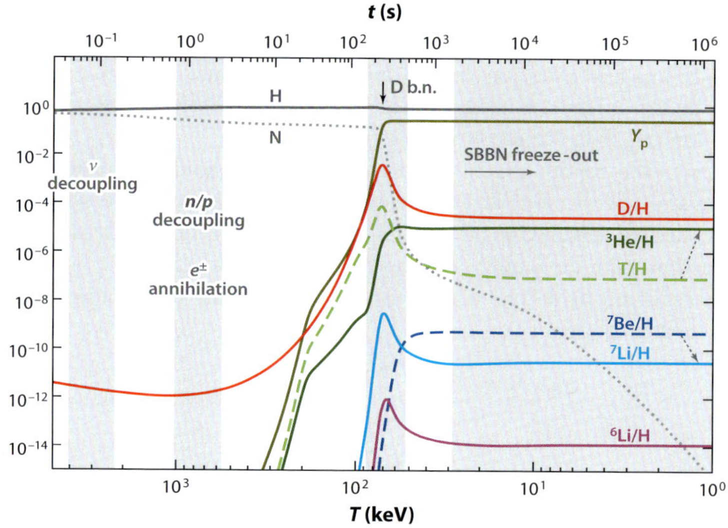

Figure 1

Time and temperature evolution of all standard big bang nucleosynthesis (SBBN)-relevant nuclear abundances. The vertical arrow indicates the moment at $T_9 \simeq 0.85$, when most of the He nuclei are synthesized. The gray vertical bands indicate main BBN stages. (*Left to right*) Neutrino decoupling, electron-positron annihilation and n/p freeze-out, D bottleneck, and freeze-out of all nuclear reactions. Protons (H) and neutrons (N) are given relative to n_b, whereas Y_p denotes the ^{4}He mass fraction. The freeze-out abundances are given by the horizontal lines on the right-hand side of the graph.

1.1.1. $\mathcal{O}(0.1)$ abundances: ^{4}He. The beauty of the SBBN prediction for ^{4}He lies in its simplicity: Only a few factors determine it. The rates for weak scattering processes that interconvert $n \leftrightarrow p$ at high plasma temperatures scale as $G_F^2 T^5$, where G_F is the Fermi constant. As the universe cools, these rates drop below the T^2-proportional Hubble rate $H(T)$ (Equation 6). The neutron-to-proton transitions slow down, and the ratio of their respective number densities cannot follow its chemical-equilibrium exponential dependence: $n/p|_{\rm eq} \simeq \exp(-\Delta m_{np}/T)$. Around $T \simeq 0.7$ MeV, this dependence freezes out to $n/p \simeq 1/6$ but continues to decrease slowly due to residual scattering and the β decays of neutrons. The formation of D during this intermission period is delayed by the process of photodissociation, which occurs efficiently because of the overwhelmingly large number of photons (Equation 4) with energies in excess of the deuteron-binding energy $E_d = 2.22$ MeV. Once the temperature drops to $T_9 \simeq 0.85$, the exponential Boltzmann suppression of such photons is sufficient to build a number density in D that is large enough to ignite other nuclear reactions. At these temperatures, the neutron-to-proton ratio drops to approximately 1/7, and very quickly, all neutrons are consumed and are incorporated into ^{4}He nuclei that have the highest binding energy per nucleon among all isotopes lighter than carbon. Thus, to a rather good accuracy,

$$Y_p \simeq \left. \frac{2n/p}{1+n/p} \right|_{T_9 \simeq 0.85}. \qquad 8.$$

The ^{4}He mass fraction Y_p is very weakly dependent on η_b as well as on the precise values for almost all nuclear reaction rates. Instead, Y_p is sensitive to the timing of major BBN events, such as the neutron-to-proton freeze-out and the end point of the D bottleneck. Consequently, the prediction for Y_p relies on such well-measured quantities as the Newton constant, the neutron-proton mass difference, the Fermi constant, the neutron lifetime, and the deuteron-binding energy.

The sensitivity of ^{4}He to all possible nonstandard scenarios that modify the timing of BBN events causes this isotope to serve as the BBN "chronometer" (9). SBBN predicts Y_p with an impressive precision (much better than 1%), so that the error bar is dominated by the uncertainty in the neutron lifetime.

The determination of primordial ^{4}He has been carried out by several groups (23, 24). Such determination is performed via observations of hydrogen- and helium-emission lines in extragalactic HII regions of low metallicity. Although the statistical error on Y_p—due to a large number of such observations—can be driven down to an $\mathcal{O}(10^{-3})$ level, the extraction of the primordial value is limited by systematic errors (25). Indeed, the latter errors are much larger than both the statistical error and the accuracy of the SBBN prediction. We quote the most conservative analysis by Olive & Skillman (25), together with a recent update on the SBBN prediction (22):

$$\text{SBBN: } Y_p = 0.2486 \pm 0.0002; \qquad 9.$$

$$\text{extrapolation to primordial value: } Y_p = 0.249 \pm 0.009. \qquad 10.$$

Another recent analysis (26) finds a somewhat smaller error bar, $Y_p \simeq 0.250 \pm 0.004$, and for the purpose of this review we assume that the observational range for Y_p is limited to the range between 0.24 and 0.26. Clearly this error range is one of the most important issues in the application of BBN to particle physics, as it translates directly into the tightness of constraints on many nonstandard scenarios. Further progress in understanding the error budget in the extraction of the primordial value of Y_p is needed.

1.1.2. $\mathcal{O}(10^{-5})$ abundances: D and ^{3}He. Rapid formation of D at $T_9 \sim 1$ is counterbalanced by the uptake of helium-producing reactions. Below $T_9 \sim 0.8$, the neutron supply for D quickly drops in abundance, producing a characteristic peak in the D/H ratio as a function of temperature (**Figure 1**). The near-complete burning of D then results in a rather small freeze-out value on the order of few $\times 10^{-5}$. Likewise, ^{3}He has a similar abundance to D. BBN predictions for both elements are very sensitive to nuclear reaction rates as well as to the η_b input. Nuclear reaction rates relevant for the formation of ^{3}He and D have been well measured to better than 10% accuracy. Therefore, by use of the current WMAP input, SBBN can make a fairly precise prediction for the abundances of these elements (22):

$$\text{SBBN: D/H} = 2.49 \pm 0.17 \times 10^{-5}; \qquad 11.$$

$$\text{SBBN: } ^3\text{He/H} = (1.00 \pm 0.07) \times 10^{-5}. \qquad 12.$$

D can be observed in the local universe as well as in highly redshifted clouds at low metallicities. The latter (27–31) is the preferred way of determining the primordial fraction of D. The observations of D absorption lines in quasar absorption systems (QALS) is challenging, given that only a tiny isotopic shift separates the D line from the main hydrogen line. Only in a few cases of QALS with sufficiently simple velocity structure has the D/H ratio been extracted. The scatter of those resulting D/H determinations poses a certain problem because it exceeds the naïvely averaged error bar. Artificially inflating the errors in order to account for this scatter produces the following result (31):

$$\text{QALS observations: D/H} = (2.82 \pm 0.21) \times 10^{-5}. \qquad 13.$$

This range agrees remarkably well with the SBBN prediction using the WMAP η_b-input. Despite the agreement one should be apprehensive of two potential problems. First, the galactic evolution of D presents astronomers with a number of puzzles and implies some significant degree of absorption of D onto dust grains. (For recent discussions, see, for instance, References 32 and 33.)

Do similar mechanisms exist at QALS, and if so, should they result in an upward correction to Equation 13 that would reflect the true primordial value of D/H? Second, the origin of scatter in the QALS data is not explained and could be a sign of underestimated systematic errors or additional depletion mechanisms. Until these issues are better understood, primordial values as high as 4×10^{-5} cannot be convincingly ruled out.

Unlike D, which can be depleted only in the course of the galactic evolution, both production and depletion of ^{3}He may occur; therefore, a direct interpretation of ^{3}He/H measurements as primordial is not possible. However, given the unidirectional evolution of D, one can conclude that the ^{3}He/D ratio will only grow, thereby providing us with a very important upper bound on the primordial ^{3}He/D ratio (34). For the purpose of this review, the primordial ratio can be constrained as (35)

$$\text{observational bound: } \frac{^3\text{He}}{\text{D}} < 1, \qquad 14.$$

which does not challenge SBBN but rather represents an important constraint on nonstandard BBN scenarios.

1.1.3. $\mathcal{O}(10^{-10})$ abundances: ^{7}Li. The formation of $A = 6, 7$ nuclei is suppressed in SBBN due to the absence of stable $A = 5$ elements. Tritium-α and ^{3}He-α fusion generates ^{7}Li and ^{7}Be nuclei but with rates that are much smaller than the Hubble expansion rate. Consequently, only a tiny number (few $\times 10^{-10}$) of ^{7}Li and ^{7}Be nuclei relative to hydrogen are generated. At a later stage of cosmological evolution, ^{7}Be is ultimately converted into ^{7}Li via electron capture. At the WMAP-measured value of the baryon-to-photon ratio η_b, more than 90% of the primordial lithium is produced in the form of ^{7}Be in the radiative capture process: $^4\text{He} + {}^3\text{He} \rightarrow {}^7\text{Be} + \gamma$. The output of ^{7}Be is almost linearly dependent on the corresponding S-factor for this reaction, which was recently remeasured by several groups (36–38). The current ~15% accuracy prediction for the combined ^{7}Be + ^{7}Li abundance of stands at (22)

$$\text{SBBN: } \frac{^7\text{Li}}{\text{H}} = 5.24^{+0.71}_{-0.67} \times 10^{-10}. \qquad 15.$$

The discrepancy between this prediction and observations is often referred to as the lithium problem.

Due to its very low abundance, lithium cannot be determined from observations of extragalactic absorption clouds. Instead, all observations of the ^{7}Li/H ratio must be performed in the atmospheres of low-metallicity galactic halo stars. The near constancy of ^{7}Li/H ratios at a low but wide range of metallicities and for some range of effective stellar temperature is known as the Spite plateau. For a long time, this plateau was thought to be an accurate representation of the primordial lithium abundance. A rather small scatter along the Spite plateau supports the interpretation of the measured ^{7}Li values as primordial. Although there have been numerous determinations of the ^{7}Li abundance on the Spite plateau, the most recent observations indicate some metallicity dependence of the ^{7}Li abundance (39, 40). Furthermore, scatter may not be negligible but rather may favor some amount of stellar ^{7}Li depletion (41). Currently, the observational status of the primordial lithium abundance is given by (42, 43)

$$\text{Spite plateau value: } \frac{^7\text{Li}}{\text{H}} = 1.23^{+0.34}_{-0.16} \times 10^{-10}, \qquad 16.$$

which is a factor of three to five lower than the SBBN-predicted amount of primordial lithium (Equation 15). It is important to keep in mind that measurements of lithium in globular clusters have resulted in somewhat higher abundances of $(2.19 \pm 0.28) \times 10^{-10}$ (43) (for other observational determinations of the ^{7}Li abundance consistent with References 42 and 43, see

References 44 and 45). The explanation of the discrepancy between Equations 15 and 16 could be of purely astrophysical origin. [A potential additional depletion of SBBN values would require some unorthodox modifications of secondary ^{7}Be-destroying reactions (46, 47), but so far this possibility has not found support from nuclear experiments (48).] Atmospheric ^{7}Li may have been partially depleted from atmospheres of population II stars due to additional settling mechanisms. Also, it has been suggested that stellar models that assume an approximately twofold suppression of ^{7}Li in such stars are consistent with observations of other abundances (49, 50). Finally, the lithium problem may point directly toward physics beyond the SBBN model—perhaps related to new physical processes at $T_9 \sim 0.5$. Unfortunately, given the controversial status of ^{7}Li, it is difficult to use this isotope for constraining models of new physics that modify its abundance. Instead, we point to certain interesting new physics possibilities that may be responsible for the reduction of ^{7}Li/H.

1.1.4. $\mathcal{O}(10^{-14})$ and less abundances: ^{6}Li and A $\geq$ 9 elements. ^{6}Li is the only stable A $=$ 6 element. The main SBBN reaction that produces it,

$$^4\text{He} + \text{D} \to {}^6\text{Li} + \gamma, \quad Q = 1.47\,\text{MeV}, \qquad 17.$$

is suppressed by four orders of magnitude relative to other radiative capture reactions, such as $^4\text{He} + T \rightarrow {}^7\text{Li} + \gamma$, and is suppressed by approximately seven to eight orders of magnitude relative to other photonless nuclear rates. The reason for the extra suppression is unique, as it arises from almost the same mass-to-charge ratio for ^{4}He and D, which inhibits the E1 transition, thereby making this radiative capture extremely inefficient. At the same time, the proton reaction that destroys ^{6}Li is very fast, which leads to an $\mathcal{O}(10^{-14})$-level prediction for primordial ^{6}Li. This is well below modern detection capabilities. Heavier elements with A $\geq$ 9, such as ^{9}Be, ^{10}B, and ^{11}B, are never produced in significant quantities in the SBBN framework because of the absence of stable A $=$ 8 nuclei, as ^{8}Be is underbound by 92 keV and decays to two α particles.

Detections of ^{9}Be, ^{10}B, and ^{11}B have been made for many stars at low metallicities and are not controversial. Observations of ^{9}Be (51, 52) are far above the $\mathcal{O}(10^{-18})$ SBBN prediction and exhibit a linear correlation with oxygen, clearly indicating its secondary (spallation) origin (53). The lowest level of detected ^{9}Be/H is at $\sim$few $\times$ 10^{-14}, which translates into a limit on the primordial fraction of 2×10^{-13} (54), assuming that there is no significant depletion of ^{9}Be in stellar atmospheres.

Because SBBN predicts very low ^{6}Li abundances, it is usually assumed that the origin of this isotope is not associated with BBN. Conversely, any level of ^{6}Li detection in stars at low metallicity in excess of $\mathcal{O}(10^{-12})$ is of interest for the test of both the SBBN framework and the galactic chemical-evolution models. Remarkably, the existence of a ^{6}Li plateau at an isotopic ratio of $^6\text{Li}/^7\text{Li} \simeq 0.05$ has been claimed from (2σ) detections in approximately 10 low-metallicity stars (45). This level of ^{6}Li/^{7}Li should be considered high, as it cannot be explained by cosmic-ray production at such low metallicities (55). However, in the thermal environment of stellar atmospheres the absorption lines of ^{7}Li and ^{6}Li are blended together so that observations become extremely challenging. Indeed, recent works (56) have claimed that additional convective motion in stellar atmospheres may lead to an ^{6}Li-unrelated distortion of ^{7}Li absorption lines, which could be confused with positive ^{6}Li detections. The analysis by Cayrel et al. (56) favors the latter explanation and sets a number of upper limits on ^{6}Li/^{7}Li in several stars of interest.

We now move to the main topic of our paper: modification of BBN by new physics. Instead of concentrating on very specific models of particle physics and/or gravitation, we review different classes of models that modify or abandon the main assumptions of SBBN listed in Section 1.

2. NEW PHYSICS AFFECTING THE TIMING OF MAIN BIG BANG NUCLEOSYNTHESIS EVENTS

As shown in **Figure 1**, most neutrons are incorporated into ^{4}He at a very specific moment: the end of the D bottleneck. The amount of helium and D that results has a strong dependence on the n/p ratio that, in turn, translates into a sensitivity to a much earlier event in the BBN history: the freeze-out of neutron-to-proton interconversion. Different physical processes may affect the timing of both events and thereby change the BBN output.

2.1. Big Bang Nucleosynthesis with New Degrees of Freedom

A classic application of BBN consists of constraining the number of the excited relativistic degrees of freedom. Traditionally, this procedure is formulated in terms of a constraint on the effective number of neutrino species, $N_{\nu,\mathrm{eff}}$. To be more general, we reformulate such a bound as a limit on the amount of so-called dark radiation, ρ_{dr}. By dark radiation we mean extra massless or nearly massless degrees of freedom that have the expected scaling as the universe expands: $\rho_{dr} \sim a^{-4}$. Furthermore, we assume that this additional relativistic component is totally passive and that it does not exchange energy with any SM species. This viewpoint can then be easily applied to the thermally decoupled extra bosonic or fermionic degrees of freedom. The addition of dark radiation to the cosmic-energy budget amplifies the Hubble expansion rate,

$$H_{\mathrm{SBBN}} \to H = H_{\mathrm{SBBN}}\sqrt{1+\rho_{dr}/\rho_{\mathrm{SM}}}, \qquad 18.$$

leading to an earlier freeze-out of the n/p ratio and thus to a larger helium mass fraction. In this formula, ρ_{SM} and H_{SBBN} are the SBBN energy density and the Hubble rate, respectively. **Figure 2** illustrates the changes to the abundances of helium, D, and lithium as a function of $\rho_{dr}/\rho_{\mathrm{SM}}$ and shows that even the somewhat generous assumption on the error bar of Y_p translates into a rather tight constraint on ρ_{dr}. One cannot achieve any significant degree of ^{7}Li suppression without violating other bounds. **Figure 2** formally extends to the region of negative ρ_{dr}. This may seem unphysical, but a negative ρ_{dr} effect can be mimicked by models with decreased neutrino temperatures due to, for instance, additional electromagnetic energy injection at approximately $T \sim 2$ MeV. Assuming that ^{4}He is limited to the range between 0.24 and 0.26, one arrives at the conclusion that

$$-0.06 < \frac{\rho_{dr}}{\rho_{SM}} < 0.14. \qquad 19.$$

Presented in a traditional way, this constraint corresponds to $2.6 \lesssim N_{\nu,\mathrm{eff}} \lesssim 4$.

The constraint on ρ_{dr} can be further specified to models at hand. It can be applied to limit the energy density carried by very light and thermally decoupled scalar fields. For example, if the dark energy sector at the time of BBN corresponds to a very light classical scalar field ϕ that tracks ρ_{SM} (57, 58), Equation 19 constrains the parameters of the potential $V(\phi)$. For an exponential potential of the form $V(\phi) = M^4 \exp\{-\lambda\phi/M_{Pl}\}$, the energy density carried by the tracker field is $\rho_\phi/\rho_{total} = 4/\lambda^2$, so that the magnitude of λ should be larger than approximately five so as to satisfy Equation 19.

The result of Equation 19 can also be developed further into constraints on the properties of the dark radiation fields and the temperature of their thermal decoupling from the SM field content. Suppose that the dark radiation sector contains N bosonic light neutral fields that are connected to the SM via some heavy Λ_n scale–suppressed operators of dimension $n > 4$. Then, the thermalization rate of dark radiation with the SM particles would typically scale as $\Gamma_{th} \sim T(T/\Lambda_n)^{2(n-4)}$. The temperature of thermal decoupling is given by the condition $H(T_d) \sim \Gamma_{th}$.

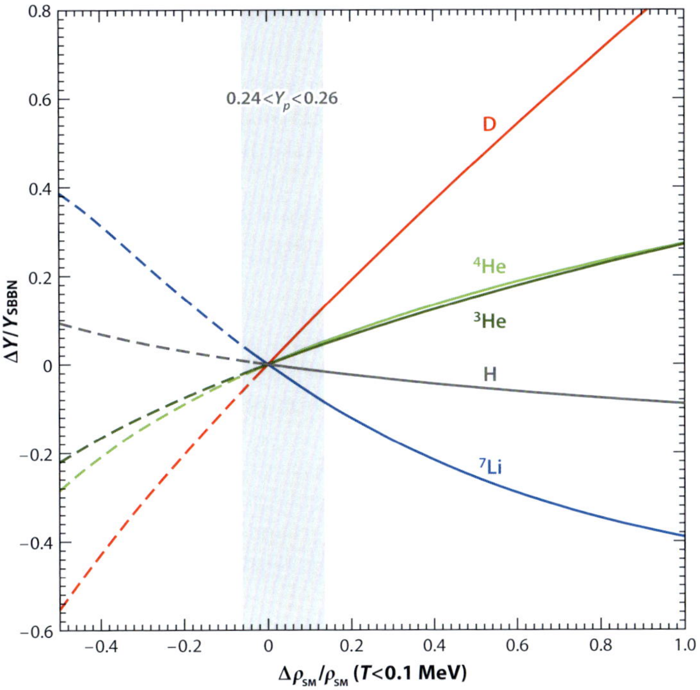

Figure 2

Change of nuclear abundances relative to their standard big bang nucleosynthesis (SBBN) values as a function of the dark radiation component $\rho_{dr}/\rho_{SM} = \Delta\rho_{SM}/\rho_{SM}$. The vertical band shows the allowed amount of dark radiation that keeps Y_p in the 0.24-to-0.26 window. Abbreviation: SM, Standard Model.

Therefore, for every given N that threatens to violate the bound of Equation 19, one can determine a minimal decoupling temperature, and for any given n, one can place a lower bound on the high-energy scale Λ_n. For example, if $N \geq 3$, then the thermal decoupling of these species must occur around the quantum chromodynamics hadronization epoch ($T \sim 200$ MeV) or earlier. By comparing the Hubble rate with the thermalization rates at that epoch, one arrives at

$$N \geq 3 \Rightarrow \Lambda_5 > 5 \times 10^8\,\text{GeV}; \quad \Lambda_6 > 5\,\text{TeV}. \qquad 20.$$

If the interaction is mediated by dimension-five operators, then the sensitivity to the Λ scale can extend very far—indeed, much beyond directly accessible energy scales in collider experiments. In exactly the same way, one can constrain properties and interactions of right-handed neutrinos, should they be light. For example, Equation 19 can be used to set limits on the strength of the coupling between left- and right-handed neutrino species. Finally, even if the right-handed neutrinos are heavy and not excited, BBN allows one to constrain nonstandard properties of the left-handed neutrino species. For example, a sizable magnetic moment or charge radius of neutrinos (that correspond to dimension-five and -six effective interactions with the external electromagnetic current) may be strong enough to prolong the thermal coupling of neutrinos to the electron-positron-photon fluid, thereby leading to essentially higher T_ν than predicted by the standard scenario (Equation 7). More discussions on the status of BBN with extra degrees of freedom can be found in earlier reviews (6–9).

2.2. Big Bang Nucleosynthesis with Sliding Couplings and Mass Scales

An alternative and more exotic way to affect the timing of main BBN events is realized in models that predict a time evolution of coupling constants and mass scales. A tractable version of such models represents a Brans-Dicke-type scalar field that couples to the SM operators $F^2_{\mu\nu}$ and $m_q \bar{q} q$, where m_q is a quark mass. The evolution of the scalar field creates the effect of changing m_q, electric charge, Λ_{QCD}, Higgs vacuum expectation value, and so forth. These changes, in turn, induce changes in the reaction rates, nuclear binding, and the position of resonances. Much effort has been devoted to calculations that make such connections explicit (59, 60). Several works have addressed the question of coupling variability in connection to BBN; for the most recent accounts, we refer the reader to, for instance, References 61–64.

Rather than delve into the intricacies of coupling and scale dependence of nuclear parameters, we limit ourselves to the following simplified framework. We assume that the primary effect of the change of couplings is on the value of the neutron-proton mass difference Δm_{np} and on the deuteron-binding energy E_d. We disregard other possible changes, that is, in the Coulomb barrier penetration, shifts in the position of nuclear resonances, and so on. Such an approximation is somewhat justified because both Δm_{np} and E_d enter in the exponents that control (*a*) the n/p ratio around its freeze-out time and (*b*) the D abundance around the end of the bottleneck. By using this approach, we are certain to capture the main effects related to the change of couplings because Δm_{np} influences the n/p freeze-out and the value of E_d dictates the end of the D bottleneck. Moreover, the deuteron-binding energy is among the most sensitive nuclear parameters to the variation of m_q/Λ_{QCD} due to the strong cancellation of m_q-dependent against m_q-independent components in E_d (62).

Figure 3 shows contours of D, ^{4}He, and ^{7}Li abundances in the Δm_{np}–E_d plane when both variations are implemented simultaneously. The abundances of elements are very sensitive to the variations of these quantities; helium in particular is extremely sensitive to Δm_{np}. Moreover, among all elements, lithium appears to be the most sensitive to the variation of the deuteron-binding energy because for smaller values of E_d, the end of the D bottleneck is delayed and the ^{3}He$(\alpha,\gamma)^7$Be reaction is even less efficient than in SBBN. The possibility of suppressing the lithium abundance while maintaining the agreement with helium and D was noted in recent papers (62–64). A rather strong sensitivity of E_d to the m_q/Λ_{QCD} parameter ensures that even a small variation of the latter leads to a drastic change in lithium abundance.

3. NONEQUILIBRIUM BIG BANG NUCLEOSYNTHESIS

One of the main assumptions of standard BBN is that all reactions occur between partners that are perfectly thermalized with the photon background fluid. However, even the standard nuclear processes that occur during BBN lead to a nonthermal energy component, which is released when mass is converted into nuclear binding energy. It is easy to estimate how much of this energy is injected. Because the total binding energy of helium is 28 MeV, the total energy release is just under 2 MeV per nucleon. Although such an energy injection is noticeable when it occurs later in the history of the universe, the release of such energy at $T_9 = 0.85$ does not lead to a change in SBBN predictions (for an analysis of some SBBN nonequilibrium reactions, see Reference 65).

More drastically, nonstandard decaying or annihilating particles can also increase the energy release per nucleon. Indeed, if this occurs predominantly after nucleosynthesis is complete, it can lead to strong departures from the observed pattern of primordial abundances. Therefore, BBN provides us with a significant sensitivity to this class of models, even when the total energy density

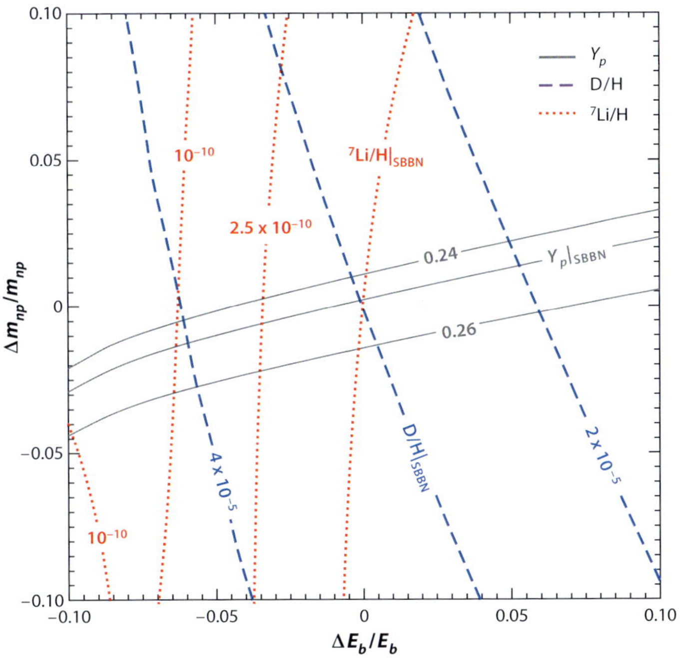

Figure 3

Contours of Y_p, D/H, and ^{7}Li/H are plotted in the parameter space of variable neutron-proton mass difference Δm_{np} and deuteron-binding energy E_d, normalized to their current values. The 5–10% downward change in E_b can significantly reduce the ^{7}Li abundance.

stored in the decaying and annihilating species is completely subdominant to the energy density of the universe. This issue has taken on a great deal of importance due to its potential connection to particle dark matter, in which weakly interacting massive particles (WIMPs) can either source nonequilibrium BBN via their annihilation or be produced in the decays of some metastable parent particles.

BBN scenarios with additional energy injection have received plenty of attention since their inception (66–73; more recent and more detailed treatments can be found in References 74–78). By accounting for the qualitative differences in the abundance signatures, one may distinguish between electromagnetic decays to γs, $e^{\pm}$, and possibly other leptons and decays to hadronic final states that lead to extra energetic nucleons. Whereas electromagnetic decays have a significant impact on BBN only at late times ($\tau \gtrsim 10^5$ s), after all reactions are effectively frozen, the hadronic decays may have observable consequences even if they occur as early as a few seconds.

3.1. Big Bang Nucleosynthesis with Electromagnetic and Hadronic Energy Injection

When the decaying particle produces mostly electromagnetic radiation, the treatment of nonequilibrium BBN is relatively simple. Because the density of the early universe is quite significant, decaying particles quickly yield electromagnetic showers. As a result, a potentially very large energy release per decaying particle [i.e., $\mathcal{O}$(1 TeV)] can be transferred to a large number of $\mathcal{O}$(10 MeV)-energy photons, some of which may have a chance of interacting with and/or disintegrating light nuclei before their energy is further dispersed and thermalized.

The main physical process that regulates the maximal energy of particles in the shower is the pair production in the scattering of energetic γs on thermal photons; that is, $\gamma + \gamma_T \to e^- + e^+$. This leads to a so-called zeroth-generation differential photon spectrum in the form of a broken power law (79),

$$p_\gamma(E_\gamma) = \begin{cases} K_0(E_\gamma/E_{\rm low})^{-1.5} & \text{for} \quad E_\gamma < E_{\rm low} \\ K_0(E_\gamma/E_{\rm low})^{-2.0} & \text{for} \quad E_{\rm low} < E_\gamma < E_C\,, \\ 0 & \text{for} \quad E_\gamma > E_C \end{cases} \qquad 21.$$

where the power break occurs at $E_{\rm low} \simeq m_e^2/(80T)$ and the spectrum is cut off at the threshold of pair production: $E_C \simeq m_e^2/22T$ (73). The overall normalization constant K_0 of the spectrum is determined by requiring that the primary injected (electromagnetic) energy E_0 be carried by the photon cascade; that is, $E_0 = \int dE_\gamma E_\gamma p_\gamma$.

The ansatz (Equation 21) for the spectrum immediately tells us the temperature and time of injection that allow for the photodisintegration of a certain element. The highest temperature, $T_{\rm ph}$ (one-to-one related to the earliest cosmological time), at which photodisintegration can occur can be determined by equating E_C to the nuclear binding energies E_b against photodissociation:

$$T_{\rm ph} \simeq \begin{cases} 7\,{\rm keV} & \text{for} \quad {}^7{\rm Be} + \gamma \to {}^3{\rm He} + {}^4{\rm He} \quad (E_b = 1.59\,{\rm MeV}) \\ 5\,{\rm keV} & \text{for} \quad {\rm D} + \gamma \to n + p \quad (E_b = 2.22\,{\rm MeV}). \\ 0.6\,{\rm keV} & \text{for} \quad {}^4{\rm He} + \gamma \to {}^3{\rm He(T)} + n(p) \quad (E_b \simeq 20\,{\rm MeV}) \end{cases} \qquad 22.$$

Once the temperature is so low that the photodisintegration of ^{4}He can occur, net production of D and ^{3}He becomes possible. An amount of ^{6}Li can be produced either via photodissociation of ^{7}Be and ^{7}Li or through secondary interactions of the products of ^{4}He destruction. This important mechanism is discussed in some detail below.

The photons in the cascade (Equation 21) undergo further degradation via the (slower) processes of Compton scattering, pair production on nuclei, and elastic $\gamma\gamma$ scattering so that the total number of energetic photons is given by competition between the injection (decay or annihilation) rate $\Gamma_{\rm inj}$ and the total energy-loss rate $\Gamma_\gamma(E_\gamma)$. The energy spectrum can then be obtained in the form of a quasi-static equilibrium solution (82),

$$f_\gamma^{\rm qse} = n_X \frac{\Gamma_{\rm inj} p_\gamma(E_\gamma)}{\Gamma_\gamma(E_\gamma)}, \qquad 23.$$

where n_X is the time-dependent number density of the decaying and annihilating particles, $\Gamma_{\rm inj} = \tau_X^{-1}$ for decays, and $\Gamma_{\rm inj} = \frac{1}{2} n_X \langle \sigma_{\rm ann} v \rangle$ for self-annihilation. Depending on the temperature of the primordial plasma at the time of energy injection, both production and destruction of elements may occur. One can incorporate these possibilities into an additional set of Boltzmann equations, which include the nonthermal photon population and abundances of nuclei, denoted here as T, A, and P ($A_T > A_A > A_P$):

$$\begin{aligned} -HT\frac{dY_A}{dT} &= \sum_T Y_T \int_0^\infty dE_\gamma f_\gamma^{\rm qse}(E_\gamma)\sigma_{\gamma+T\to A}(E_\gamma) \\ &- Y_A \sum_P \int_0^\infty dE_\gamma f_\gamma^{\rm qse}(E_\gamma)\sigma_{\gamma+A\to P}(E_\gamma). \end{aligned} \qquad 24.$$

The solution to this set of equations constrains the amount and timing of deposited electromagnetic energy. **Figure 4** shows the results of a sample calculation for the model with a decaying particle with a lifetime of $\tau_X = 10^8$ s and an initial energy density relative to baryons of $m_X n_X/(m_p n_b) = m_X Y_X/m_p \simeq 1$; there is an extra assumption that half of the rest mass of the species, $m_X = 1$ TeV, is released in the form of electromagnetic energy into the thermal bath. One can see the significant

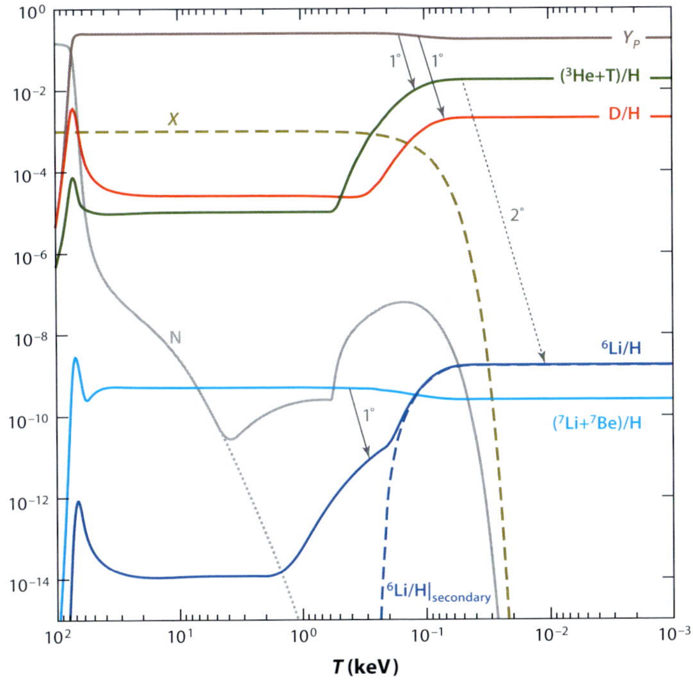

Figure 4

Consequences of late decays of a heavy 1-TeV mass particle X that releases half of its rest mass in the form of electromagnetic energy. The threshold of ^{4}He disintegration is clearly visible below 1 keV. Primary abundance flows are indicated by solid arrows, whereas the dashed arrow indicates the secondary transformation of A = 3 nuclei into ^{6}Li. The model is excluded by the overproduction of D, ^{3}He, and ^{6}Li.

increase in ^{3}He, D, and ^{6}Li abundances. The model considered in this graph is in stark conflict with observations and is therefore excluded.

It is intriguing to investigate whether the injection of electromagnetic energy may reduce the abundance of ^{7}Li + ^{7}Be. A dedicated study (80) found that such a reduction is generally difficult to achieve because either D is also destroyed or ^{3}He/D is overproduced beyond observationally acceptable levels. Although the overall conclusion is negative—that is, the electromagnetic destruction of ^{7}Li is problematic—we stress that "just-so" solutions can always be found. For example, consider a relatively light but abundant neutral particle with mass m in the narrow range between ~3.5 and 4.5 MeV. ^{7}Be could indeed be destroyed while D, ^{3}He, and ^{6}Li remain unaffected because the maximum released energy, $m/2$, lies below their photodisintegration threshold.

The treatment of energy injection via hadronic channels is significantly more complicated because additional effects must be taken into account. Partons emitted in a decay or annihilation are quickly hadronized, and the highly energetic fragmentation products [mostly pions but also (anti)nucleons] are released into the plasma. Only long-lived mesons—namely charged pions ($\pi^{\pm}$) and kaons ($K^{\pm}$, K_L^0)—with lifetimes of $\mathcal{O}(10^{-8}$ s) have a chance to interact with background nuclei before decaying. A very early hadronic energy injection (300 keV $\lesssim T \lesssim$ 1 MeV) allows the π mesons to participate in charge-exchange reactions: $\pi^- + p \rightarrow \pi^0 + n$. Such reactions lead to a higher n/p freeze-out value, thereby increasing the helium mass fraction Y_p. Similar effects can be caused by antinucleons, which tend to annihilate predominantly on protons. The most important difference with respect to electromagnetic energy releases comes from the injection of energetic nucleons after the formation of ^{4}He. The propagation of energetic nucleons through the primordial plasma causes spallation processes on ^{4}He. Reactions caused by energetic nucleons—$n + {}^4\text{He} \to T + p + n$, $n + {}^4\text{He} \to D + p + 2n$, and others—in turn generate energetic $A = 3$ elements that can participate in endothermic nuclear reactions that are forbidden in SBBN.

What determines the efficiency of spallation processes and of subsequent secondary nonthermal reactions is the rate at which the energetic charged nuclei and neutrons are stopped by the primordial plasma. The dominant thermalization processes for charged nuclei are Coulomb interactions with $e^{\pm}$ and Thomson scattering off thermal photons. Below the e^+e^- annihilation threshold, the stopping power of the plasma rapidly drops with temperature, reaching its minimum at approximately $T \simeq 20$ keV. The energy loss of the neutrons initially occurs via its magnetic-moment interaction with electrons and positrons. At later stages ($T \lesssim 80$ keV), neutrons lose their energy dominantly by scattering on protons and ^{4}He.

The resulting constraints on energy injection are described in great detail elsewhere (74–78). BBN results are most sensitive to the energy injection after 10^8 s, where constraints as strong as ~1 MeV per nucleon result from the secondary ^{6}Li production and do not generally depend on the branching ratio to hadrons. Constraints on electromagnetic energy injection depend, to a large extent, on whether ^{4}He can be photodissociated, and they become much weaker if energy is released when the temperature is above 0.6 keV. Hadronic decays allow one to probe much earlier times. For fully hadronic decays to energetic $q\bar{q}$ pairs, a similar sensitivity (1 MeV per nucleon) applies to injection at much earlier times, specifically $t \sim 10^4$ s. Perhaps one of the most interesting features of energy injection with nucleons in the final state is the effect of extra neutrons at $T \sim 40$ keV, which may lead to an important depletion of the total lithium abundance.

3.1.1. Neutron excess at 40 keV and suppression of ^{7}Li. An interesting aspect of nonthermal BBN with hadronic decays and annihilations is the possibility of alleviating the tension between the Spite plateau value and the predicted abundance of ^{7}Li—that is, to solve the ^{7}Li problem. In SBBN, the ^{7}Be abundance is controlled by the $^3\text{He} + {}^4\text{He} \to {}^7\text{Be} + \gamma$ reaction and by the combination of the two reactions that destroy it:

$$n + {}^7\text{Be} \to p + {}^7\text{Li} \text{ and } p + {}^7\text{Li} \to {}^4\text{He} + {}^4\text{He}. \qquad 25.$$

The increase of neutrons in a narrow temperature interval, 60 keV $\gtrsim T \gtrsim$ 30 keV (in other words, during or just after ^{7}Be synthesis) may amplify the efficiency of this destruction. The injection of $\gtrsim 10^{-5}$ neutrons per baryon, regardless of the microscopic cause of such an injection, may enhance $^7\text{Be} \to {}^7\text{Li}$ and lead to an overall depletion of $^7\text{Be} + {}^7\text{Li}$ (69, 74). When the temperature drops below 30 keV, the ^{7}Li-burning reaction drops out of equilibrium, the lithium destruction stops, and the supply of extra neutrons has no further effects on the overall $^7\text{Be} + {}^7\text{Li}$ abundance. At the same time, most of the extra neutrons injected around the time of BBN are removed by the radiative recombination with protons, leading to the generation of extra D. Therefore, one should expect that this mechanism of depleting ^{7}Be is tightly constrained by the abundance of D. The temperature evolution of elemental abundances in the presence of an extra source of thermal neutrons is illustrated in **Figure 5**.

Elevated neutron concentrations can be caused by decaying or annihilating particles that contain a significant number of baryons in the final state. This presence is amplified when the energetic particles lose their energy in collisions with the primordial plasma. Many details of such processes depend on concrete particle physics realizations: particle masses, abundances, lifetimes, branching ratios to nucleons, and so forth—all play important roles. In **Figure 6**, we both emulate the process of energy injection by assuming very rapid thermalization and study the effect of extra neutrons on $^7\text{Li} + {}^7\text{Be}$ and D abundances by varying the amount of injected neutrons while

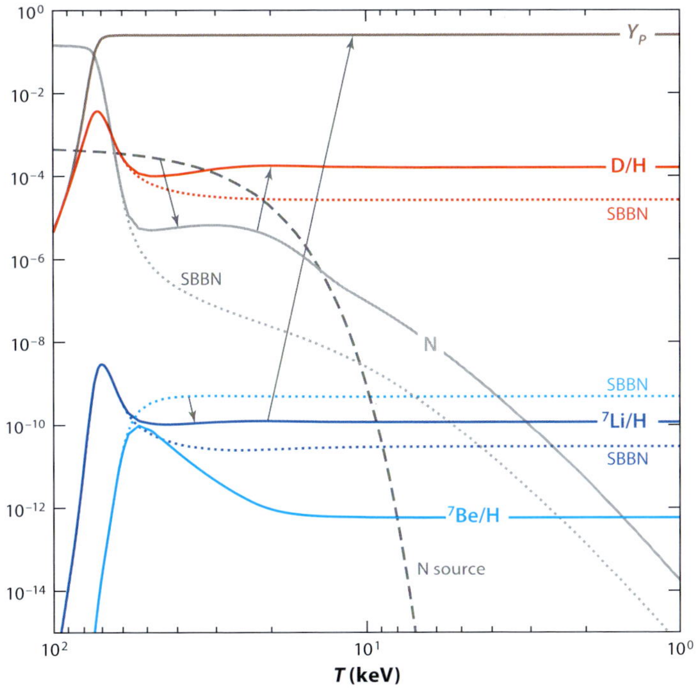

Figure 5

Effects on element abundances in response to an elevated (thermal) neutron (N) content, sourced by a decaying species, at early times. Although neutrons are mostly incorporated into D, they also transfer ^{7}Be nuclei into the ^{7}Li reservoir, from which they are more susceptible to proton burning. This mechanism suppresses the overall outcome of the A = 7 elements. Abbreviation: SBBN, standard big bang nucleosynthesis.

assuming different time (or temperature) patterns for such injections. Specifically, we include cases of decaying particles (with different lifetimes) and annihilating WIMPs, with or without possible Sommerfeld or resonant enhancement of annihilation (81). If not carefully tuned, the injection of neutrons that reduces the total ^{7}Li abundance by a factor of two would overpredict the D/H ratio. Only a relatively small subset of models, typically those whose decaying particles have lifetimes below 1,000 s, are consistent with D/H, and provide enough suppression for ^{7}Li. However, the injection of energetic neutrons, ignored in this treatment, is also important because of secondary and tertiary reactions induced by the neutrons. These reactions lead to an increase of rare light elements such as ^{6}Li and ^{9}Be. This method of correcting the lithium problem is subject to additional constraints from ^{6}Li and ^{9}Be.

3.1.2. Production of ^{6}Li, ^{9}Be, and ^{10}B in secondary collisions. As stated in Section 1, the abundances of the elements predicted by SBBN are extremely low. The path to them is heavily hindered as it is guarded by (*a*) the breaks at A = 5 and A = 8; (*b*) the low efficiency of the SBBN reaction (Equation 17); and (*c*) the efficient destruction of these elements by protons, which stops only at $T \sim 10$ keV, when the destruction rates fall below the Hubble expansion rate. Nonequilibrium BBN provides an important mechanism for circumventing at least some of that suppression through access to endothermic reactions that have much higher cross sections (68).

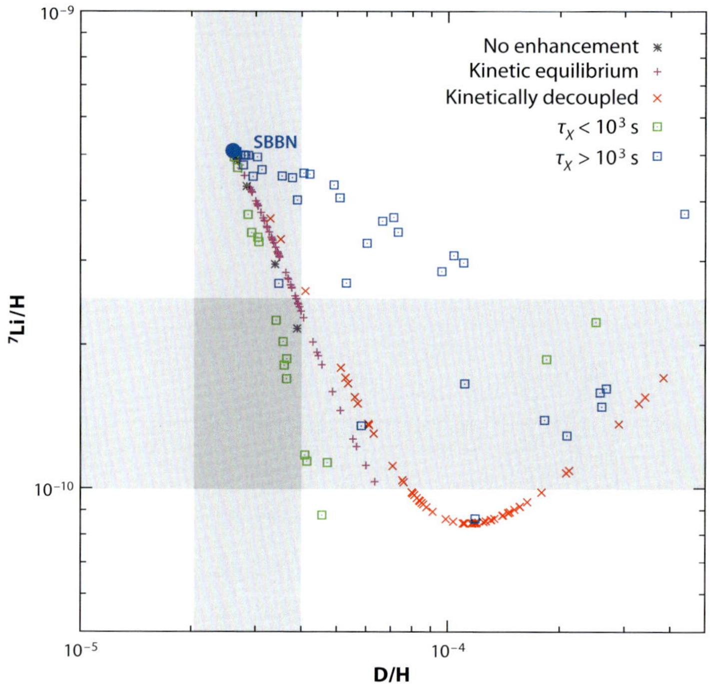

Figure 6

^{7}Li versus D abundance in the standard big bang nucleosynthesis (SBBN) scenario with extra neutrons. The overall number of injected neutrons, as well as the temporal patterns of injection, is varied. Patterns consistent with particle decays and annihilation, both with and without Sommerfeld enhancement, are considered.

Energetic mass-three nuclei (^{3}He and tritium), produced via electromagnetic or hadronic energy injection (i.e., via spallation or photodisintegration), collide with α particles from the thermal bath. Such secondary collisions produce ^{6}Li via the following set of reactions:

$$\begin{aligned} &{}^3\mathrm{H} + {}^4\mathrm{He} \to {}^6\mathrm{Li} + n, && Q = -4.78\,\mathrm{MeV}; \\ &{}^3\mathrm{He} + {}^4\mathrm{He} \to {}^6\mathrm{Li} + p, && Q = -4.02\,\mathrm{MeV}; \quad \text{and} \\ &{}^3\mathrm{H} + {}^4\mathrm{He} \to {}^6\mathrm{He} + p, && Q = -7.5\,\mathrm{MeV}; \end{aligned} \qquad 26.$$

where ^{6}He subsequently β decays to ^{6}Li. For energies of projectiles that are approximately 10 MeV, the cross sections for these nonthermal processes are on the order of 100 mb—10^7 times larger than the SBBN cross section for the production of ^{6}Li. The overall efficiency of producing ^{6}Li from energetic ^{3}He and tritium reaches its maximum at $\mathcal{O}(10^{-4} - 10^{-3})$ below $T = 10$ keV; therefore, even one energetic nucleus of tritium per one million protons can have a very significant impact on ^{6}Li. This enhancement figure underscores the sensitivity of ^{6}Li to nonthermal BBN and makes such sensitivity an important probe of energy-injection mechanisms in the early universe. **Figure 7** illustrates the importance of accounting for secondary reactions of the type shown in Equation 26 when tracking the ^{6}Li output in scenarios with electromagnetic energy release at late times. In our treatment of photodissociation processes, we primarily follow the approach of Reference 82.

The doubtful observational status of ^{6}Li and the possibility of its stellar depletion over and above the depletion of ^{7}Li should stimulate studies of ^{9}Be and boron production via the

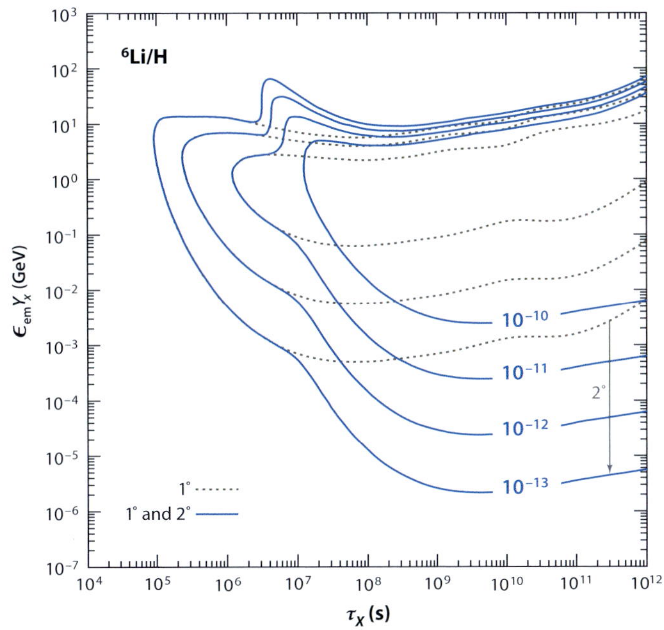

Figure 7

Contour lines of ^{6}Li/H in the plane of X-lifetime τ_X and total injected (electromagnetic) energy per baryon: $\varepsilon_{\rm em} Y_X$. Only below ~10 keV ($t > 10^4$ s) are ^{7}Li and ^{7}Be dissociated by energetic photons producing primary ^{6}Li (*dotted lines*). Additionally, ^{6}Li is fused via the processes of Equation 26. The importance of this secondary production at late times is represented by the vertical arrow, which indicates the significant elevation of total ^{6}Li/H (*solid lines*) with respect to the primary yield.

nonthermal nucleosynthesis. So far, this subject has escaped the attention of the groups working on nonequilibrium BBN. Note that tertiary processes may be responsible for the nonequilibrium ^{9}Be production. The ^{6}He and ^{6}Li nuclei emerging from secondary reactions (Equation 26) can collide further with particles from the bath:

$$\begin{aligned} &{}^6\text{Li} + {}^4\text{He} \to {}^9\text{Be} + p, \quad Q = -2.13\,\text{MeV}; \\ &{}^6\text{He} + {}^4\text{He} \to {}^9\text{Be} + n, \quad Q = 0.60\,\text{MeV}. \end{aligned} \qquad 27.$$

The nuclei can also generate ^{9}Be from A = 6 nuclei with similar efficiencies to the A = 3 → A = 6 transition. The actual output of ^{9}Be relative to ^{6}Li is enhanced beyond a crude estimate of $\mathcal{O}(10^{-4} - 10^{-3})$ if, for example, the energy injection occurs both at very early times and above $T = 12$ keV, where ^{6}Li is more rapidly destroyed by proton reactions than ^{9}Be is. Given the absence of a primordial ^{9}Be plateau down to the level of 10^{-13}, it is desirable to investigate the tertiary production mechanisms of beryllium and boron in more detail (M. Pospelov & J. Pradler, manuscript in preparation).

3.2. Energy Injection by Weakly Interacting Massive Particles

Decays of heavy relics may represent the simplest possible way to achieve energy injection. However, without specifying the model and the physical mechanisms that lead to the production of unstable particles and their subsequent decay, one cannot estimate the amount or the timing of the energy injection. In contrast, WIMPs represent a somewhat more restricted framework in which the leading mechanisms regulating WIMP abundance at BBN time is their self-annihilation at earlier times. Assuming a weak-scale mass for relic WIMPs, as well as a typical range for the annihilation cross section, one can deduce these WIMPs' abundance as a function of mass and annihilation rate. The most restrictive framework is that of stable neutral WIMPs that form the dominant component of cold dark matter. Although most of the WIMPs annihilate when the temperature drops below the WIMP mass, a residual annihilation persists even at BBN times (83). The question of WIMP annihilation at early times, during BBN, is important in view of the attempts to detect signatures of WIMP annihilation in our own galaxy. Indeed, a broad similarity among WIMP velocities inside the galaxy and during BBN tells us that the kinematics of WIMP collisions must be very similar.

The most straightforward case to consider is a single species of WIMP dark matter X with annihilation rate $(\sigma v)_0 \simeq 3 \times 10^{-26}\ \mathrm{cm}^3\ \mathrm{s}^{-1}$, averaged over the thermal bath at the freeze-out. Such an annihilation rate yields a total energy density of WIMPs that is close to what is required for cold dark matter by observations. This result provides us with a convenient normalization point for the annihilation rate at arbitrary WIMP velocities in the center-of-mass frame:

$$\sigma(v)v = S(v)(\sigma v)_0. \qquad 28.$$

In the simplest case, in which the annihilation is mediated by short-distance forces and proceeds in the s-wave, the velocity dependence is trivial: $S(v) = 1$. The fraction of WIMPs that is still annihilating at the time of BBN during one Hubble time H^{-1} is given by

$$f_X(t_{\mathrm{BBN}}) \simeq \left. \frac{\langle \sigma(v)v \rangle n_X}{H} \right|_{t=t_{\mathrm{BBN}}}, \qquad 29.$$

where $\langle \cdots \rangle$ denotes a thermal average that can be calculated once the velocity dependence of $S(v)$ and the thermal distribution of WIMPs are specified.

We are interested in finding the fraction of annihilating WIMP particles within a Hubble time at $T < 10$ keV, specifically, at a temperature scale below which ^{6}Li is no longer susceptible to nuclear burning in the $^6\mathrm{Li}\,(p, \alpha)\,^3\mathrm{He}$ reaction. In general, this fraction can be expressed via the freeze-out temperature $T_f \sim 0.05\, m_X$ and the number of effective degrees of freedom at an arbitrary temperature $T < T_f$ (10):

$$f_X(T) \simeq \left[\frac{g(T)}{g(T_f)} \right]^{1/2} \frac{T}{T_f} \langle S(v) \rangle_T. \qquad 30.$$

There are several generic options for the temperature scaling of the S-factor in Equation 30. In the simplest case of $S(v) = 1$, one can use Equation 30 to deduce that for a WIMP of mass $M_X = 100$ GeV ($T_F \sim 5$ GeV), only a small fraction ($f_X \approx 6 \times 10^{-7}$) of the X particles has the chance to annihilate at $T \simeq 10$ keV and below. This corresponds to an injection of approximately 10^{-15} GeV per photon. Nevertheless, even this tiny fraction may be sufficient to produce a ^{6}Li abundance up to the 10^{-12} level, provided that hadronic annihilation channels dominate. Lighter WIMPs may have a more pronounced effect on BBN (74). Short-range mediated annihilation in the p-wave gives $S(v) \sim v^2$ and negligible energy release at BBN.

Much-enhanced annihilation rates occur in models where interactions are mediated by an attractive Coulomb-like force and/or have near-threshold resonances. Both mechanisms of enhancing the galactic annihilation have been widely discussed (e.g., 81) in an attempt to link some cosmic-ray anomalies—such as the elevated positron fraction $e^+/(e^- + e^+)$ observed by the PAMELA satellite experiment (84)—to dark matter annihilation. For example, an attractive $-\alpha'/r$ potential in the WIMP sector leads to a significant enhancement of annihilation at low velocities via a well-known Sommerfeld factor, $\sigma v \sim (\pi\alpha'/v)[1 - \exp(-\pi\alpha'/v)]^{-1}$, when

$v \lesssim \pi\alpha'$ and $S(v) \simeq \pi\alpha'/v$. This enhancement, in turn, generates a $\sim T^{-1/2}$ scaling of $\langle S(v)\rangle_T$ in Equation 30 when the WIMPs are still in kinetic equilibrium with the plasma. Once the WIMPs' interactions with the thermalized plasma species have ceased, $\langle S(v)\rangle_T$ falls even more rapidly as $\sim T^{-1}$. Therefore, the fraction of WIMPs annihilated in a Hubble time scales with photon temperature as follows:

$$\text{Sommerfeld enhancement}: \begin{cases} f_X(T) \sim T^{1/2} & \text{kinetic equilibrium} \\ f_X(T) \sim \text{constant} & \text{kinetically decoupled.} \end{cases}$$

Similarly, the presence of narrow resonances just above the XX-annihilation threshold may drastically boost the annihilation at low energies. For a narrow resonance at some energy E_R above the di-WIMP threshold, the pattern of energy injection has a sharp cutoff:

$$\text{Resonant enhancement}: \begin{cases} f_X(T) \sim \exp(-E_R/T) & \text{kinetic equilibrium} \\ f_X(T) \sim \exp(-E_R T_{kd}/T^2) & \text{kinetically decoupled,} \end{cases}$$

where T_{kd} is the temperature of WIMP kinetic decoupling. Notably, the last pattern (resonant annihilation, kinetically decoupled WIMPs) has the same time dependence as the energy injection of particle decays, $\exp(-t/\tau_X)$, on account of the $t \sim 1/T^2$ relation. If the annihilation cross sections $\langle\sigma v\rangle(t_{\rm BBN})$ are enhanced by some large factor $[\sim\mathcal{O}(10^3)]$ relative to the fiducial WIMP rate $(\sigma v)_0$, then WIMP physics can have a significant impact on the primordial abundance of ^{6}Li. However, a large class of models that predict the WIMP enhancement of the PAMELA positron signal produce light particles (e.g., photons, electrons, and light mesons) via annihilation, which has a much weaker effect on ^{6}Li.

A case of independent interest is the decay of a WIMP-like state with a stable non-SM particle in the final state. The initial abundance of so-called parent WIMPs is given by its annihilation at the freeze-out, whereas their decays may source the dark matter energy density. If the amount of energy released in such decays is on the order of the parent WIMP mass, then in such a scenario an amount of energy comparable to the dark matter energy density could be released into the primordial plasma. If so, BBN with energy injection sets a rather strict constraint: The lifetime of the parent particle must be shorter than $\sim 10^7$ s if the decays are fully electromagnetic and shorter than 10^4 s if there is a significant hadronic component among the decay products. In the next section we explain that the sensitivity to the lifetime of such parent WIMP particles can be even stronger if they carry negative electric charge.

4. CATALYZED BIG BANG NUCLEOSYNTHESIS

In this section we discuss changes to BBN in a scenario where particles from a new physics sector participate in thermal nuclear reactions before decaying. Such an unusual situation may arise if the particles are charged under the electromagnetic or strong group of the SM (85). The idea of particle physics catalysis of nuclear reactions dates from the 1950s, when muon-catalyzed fusion became a subject of active theoretical and experimental research in nuclear physics. More recently, interest in the possibility of nuclear catalysis by hypothetical negatively charged particles that live long enough to participate in nuclear reactions at the time of BBN has intensified (54, 77, 85–95; see also References 96–98 for earlier work). The essence of the idea is very simple: A negatively charged massive particle, which we term X^-, gets into a bound state with the nucleus of mass m_N and charge Z, forming a large compound nucleus with charge $Z-1$, mass $M_X + m_N$, and binding energy in the $\mathcal{O}(0.1-1)$-MeV range. Once the bound state is formed, the Coulomb barrier is reduced, signaling higher reactivity between the compound nucleus and other nuclei. Even more importantly, new reaction channels may open up and preclude SBBN-suppressed production mechanisms (e.g., Equation 17) (85), thereby enabling the synthesis of very rare isotopes such as ^{6}Li and ^{9}Be.

Interest in catalyzed BBN (CBBN) is partly motivated by a possible connection to dark matter. Even though dark matter should not be charged, it is possible that it comes with a relatively long-lived charged counterpart. An example of this kind is weak-scale supersymmetry (SUSY), where the lightest supersymmetric particle is the gravitino and the next-to-lightest supersymmetric particle is a charged slepton. In that case the decay of the next-to-lightest supersymmetric particle is greatly delayed by the smallness of the gravitino-lepton-slepton coupling $\sim M_{\rm Pl}^{-1}$. A similar example is a nearly degenerate stau-neutralino system; in this case, the longevity of the charged stau against the decay to the dark matter neutralino is ensured as long as the mass splitting of the stau-neutralino system is below 100 MeV (99). In these two examples, both the gravitino and neutralino represent viable dark matter candidates. A very important aspect of CBBN is that the abundance of charged particles before they begin to decay is given by their annihilation rate at freeze-out. In most models, these particles' abundance is easy to calculate. If no special mechanisms are introduced to boost the annihilation rate, then the abundance of charged particles per nucleon is not small [$Y_X = n_X/n_b \sim (0.001-0.1)m_X/\text{TeV}$]. Moreover, the charged states accompanying dark matter are often required in order to facilitate the WIMP-annihilation process, which results in a cosmologically acceptable dark matter abundance—a phenomenon known as coannihilation.

4.1. Catalysis by Stable Charged Particles

In this section we describe BBN catalysis by use of negatively charged heavy relics X^-; we do not attempt to place this phenomenon in a specific particle physics model framework, and we ignore the effects of energy injection due to the X^- decay. We describe an elementary particle, so typical collider bounds would require $m_{X^-} \gtrsim 100$ GeV.

4.1.1. Properties of the bound states. For light nuclei that participate in BBN, we assume that the reduced mass of the nucleus-X^- system is well approximated by the nuclear mass, so that the binding energy is given by $Z^2\alpha^2 m_A/2$ when the Bohr orbit is larger than the nuclear radius. However, this is a poor approximation for all nuclei heavier than A $=$ 4, and the effect of the finite nuclear charge radius must be taken into account. In **Table 1** we give the binding energies and the recombination temperatures for each bound state. The recombination temperature is the temperature at which the photodissociation rate of bound states becomes smaller than the Hubble expansion rate. Below these temperatures, bound states are practically stable. The most important benchmark temperatures for CBBN are $T \sim 30$, 8, and 0.5 keV for the respective bound states (^{7}BeX^-), (^{4}HeX^-), and (pX^-) when they can be formed without efficient suppression by the photodissociation processes. Importantly, these properties of the bound states are generic

Table 1 Properties of the bound states

Bound state	a_0 (fm)[a]	$\lvert E_b \rvert$ (keV)[b]	T_0 (keV)[c]
(p X^-)	29	25	0.6
(^{4}He X^-)	3.63	346	8.2
(^{7}Be X^-)	1.03	1350	32
(^{8}Be X^-)	0.91	1430	34

[a]Bohr radius $a_0 = 1/(Z\alpha m_N)$.
[b]Binding energies E_b calculated for realistic charge radii.
[c]Photodissociation decoupling temperatures T_0.

to any CBBN realization when X^- does not participate in strong interactions; in other words, they are determined entirely by the charge of X^- and the electromagnetic properties of the nuclei and therefore are applicable to SUSY and non-SUSY models alike. Also, the (^{8}BeX^-) compound nucleus is stable, which may allow the synthesis of A > 8 elements in CBBN.

4.1.2. Catalysis at 30 keV: suppression of ^{7}Be. While the universe cools to temperatures of 30 keV, the abundances of D, ^{3}He, ^{4}He, ^{7}Be, and ^{7}Li are already close to their freeze-out values, although several nuclear processes remain faster than the Hubble rate. At such temperatures, a negatively charged relic can get into bound states with ^{7}Be and form a (^{7}BeX^-) composite object through the photorecombination process. Once a composite object is formed, new destruction mechanisms for ^{7}Be appear. For models with weak currents connecting nearly mass-degenerate X^- and X^0 states, a very fast internal conversion followed by the p-destruction of ^{7}Li becomes possible:

$$(^7\mathrm{Be}X^-) \to {}^7\mathrm{Li} + X^0; \quad {}^7\mathrm{Li} + p \to 2\alpha. \qquad 31.$$

Alternatively, ^{7}Li can be destroyed via recombination with X^- and subsequent conversion to unstable ^{7}He (90). When a $X^- \to X^0$ weak transition is not allowed, the destruction of ^{7}Be can be achieved via the following chain:

$$(^7\mathrm{Be}X^-) + p \to (^8\mathrm{B}X^-) + \gamma; \quad (^8\mathrm{B}X^-) \to (^8\mathrm{Be}X^-) + e^+\nu, \qquad 32.$$

which is greatly enhanced by atomic resonances in the (^{7}BeX^-) system (89).

The rates for both mechanisms may be faster than the Hubble rate and may therefore lead to a sizable suppression of ^{7}Be abundance if (^{7}BeX^-) bound states are forming efficiently. In other words, (^{7}BeX^-) serves as a bottleneck for the CBBN depletion of ^{7}Be. The recombination rate per ^{7}Be nucleus leading to (^{7}BeX^-) is given by the product of the recombination cross section and the concentration of X^- particles. It is easy to show that for $Y_X \lesssim 0.01$, the recombination rate is too slow to lead to a significant depletion of ^{7}Be. Detailed calculations of the recombination rate and numerical analyses of CBBN at 30 keV (89, 94) find that a suppression of ^{7}Be by a factor of two is possible (*a*) for $Y_X \geq 0.1$, if the mechanism in Equation 32 alone is operative, and (*b*) for $Y_X \geq 0.02$, if the internal conversion of Equation 31 is allowed. Even if $Y_X \sim \mathcal{O}(0.1-1)$, X^0 could still be compatible with dark matter as long as m_{X^0} is suppressed relative to m_{X^-}.

4.1.3. Catalysis at 8 keV: enhancement of ^{6}Li and ^{9}Be. As the universe continues to cool below 10 keV, the efficient formation of (^{4}HeX^-) bound states becomes possible. If we make the reasonable assumption that $Y_X \lesssim Y_{^4\mathrm{He}}$, then the rate of formation of bound states per X^- particle is given by the recombination cross section and the concentration of the helium nuclei. A numerical analysis of recombination reveals that at $T \simeq 5$ keV, approximately 50% of the available X^- particles are in bound states with ^{4}He (85).

As soon as (^{4}He X^-) is formed, new reaction channels open up. In particular, a photonless thermal production of ^{6}Li becomes possible:

$$(^4\mathrm{He}X^-) + \mathrm{D} \to {}^6\mathrm{Li} + X^-; \quad Q \simeq 1.13\,\mathrm{MeV}. \qquad 33.$$

This rate exceeds the SBBN production rate by approximately six orders of magnitude. The production of ^{9}Be may also be catalyzed, possibly by many orders of magnitude relative to the SBBN values, with the following thermal nuclear chain (93):

$$(^4\mathrm{He}X^-) + {}^4\mathrm{He} \to (^8\mathrm{Be}X^-) + \gamma; \quad (^8\mathrm{Be}X^-) + n \to {}^9\mathrm{Be} + X^-. \qquad 34.$$

Both reactions at these energies are dominated by resonant contributions, although the efficiency of the second process in Equation 34 is not fully understood. A recent calculation (95) places

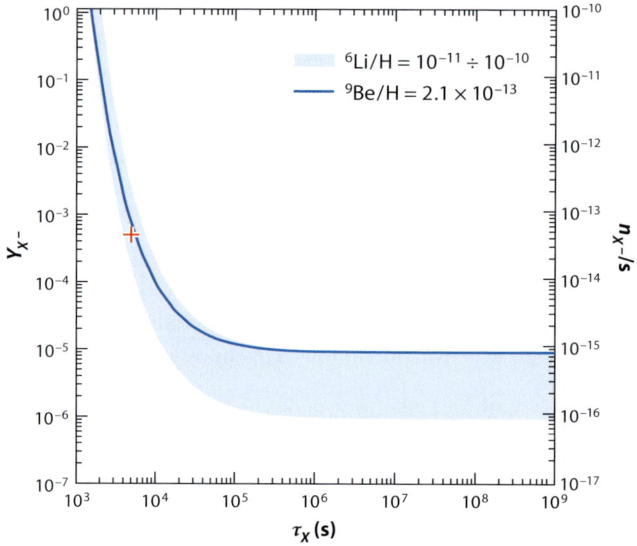

Figure 8

Catalyzed big bang nucleosynthesis constraints on the abundance versus lifetime of X^-. The red cross corresponds to a point in the parameter space; the temporal development of ^{6}Li and ^{9}Be for this point is shown in **Figure 9** (54).

the resonance found in the (^{9}BeX^-) system (93) below the threshold. However, because this is a reaction with a neutral particle, even the subthreshold resonance may give a strong enhancement of the neutron capture.

Current estimates and calculations of the CBBN rates are used to determine the generic constraints on the lifetimes and abundances of charged particles. The essence of these limits is depicted in **Figures 8** and **9**, which show that for typical X^- abundances the lifetime of the charged particles would have to be limited by a few thousand seconds [unless a special mechanism for suppressing Y_X is found (e.g., 100, 101)]. This is the most important fact to be learned from the catalysis of BBN by charged particles. Although to lowest order nonthermal BBN is sensitive to the energy density of decaying particles, the CBBN processes are controlled by the number density of X^-, which underscores the complementary character of these constraints. In some models in which both catalysis and nonequilibrium nucleosynthesis occur, catalysis dominates the nonthermal production of ^{6}Li for all particles with hadronic branching ratio $B_h \lesssim 10^{-2}$ (92), whereas ^{7}Li destruction is usually dominated by neutron injection effects unless $B_h \lesssim 10^{-4}$.

4.1.4. Catalysis below 1 keV? Here we discuss the possibility of (pX^-) catalysis of nuclear reactions (91, 97). Although it is conceivable that the absence of the Coulomb barrier for this compound nucleus may lead to significant changes in SBBN/CBBN predictions, in practice (pX^-)-related mechanisms are of only secondary importance in most cases. The large radius and shallow binding of this system lead to a fast charge-exchange reaction on helium: $(pX^-) + {}^4\text{He} \rightarrow ({}^4\text{He}X^-) + p$. This reaction reduces the abundance of (pX^-) below 10^{-6} relative to hydrogen as long as $Y_{X^-} \lesssim Y_{^4\text{He}}$, making further reactions inconsequential for any observable element (54). In the less likely case in which $Y_{X^-} \gtrsim Y_{^4\text{He}}$, significant late-time processing due to (pX^-) bound states may still occur. Nevertheless, such late-time BBN typically leads to observationally unacceptable final BBN yields.

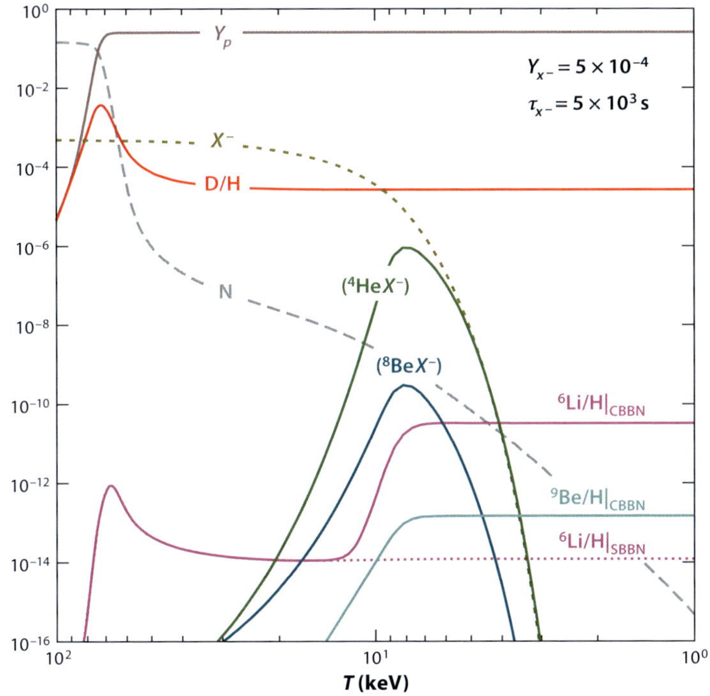

Figure 9

The temperature evolution of the bound-state abundances (^{4}HeX^-) and (^{8}BeX^-), as well as the synthesis of ^{6}Li and ^{9}Be at $T \sim 8$ keV. The initial abundance of negatively charged particles is $Y_X = 5 \times 10^{-4}$, and the lifetime is 5,000 s. Abbreviations: CBBN, catalyzed big bang nucleosynthesis; SBBN, standard big bang nucleosynthesis.

Unlike in the SBBN case, and even in nonequilibrium nucleosynthesis that utilizes mostly measured nuclear reaction rates, CBBN rates cannot be measured in the laboratory; significant nuclear theory input for the calculation of the reaction rates is required. However, because the X^- participates only in electromagnetic interactions, such calculations are feasible, and dedicated nuclear theory studies (88, 95) for certain important CBBN processes (such as those of Equations 32 and 33) have already been performed.

4.2. Catalysis by Strongly Interacting Relics

Catalysis by strongly interacting relics, X_s, is another generic possibility that has recently been addressed (102, 103). The main difficulty of this scenario lies in determining the properties of the composite nuclei that contain X_s. For some types of X_s, some progress can be made. Specifically, if X_s represents an isospin doublet similar to (n, p), one could draw some conclusions about the binding energies with nucleons by utilizing our extensive theoretical knowledge of two-nucleon systems. For example, a heavy scalar quark (squark), would attract one light quark and become a spin-1/2 "mesino" with similar interaction properties as normal nucleons. In other examples involving a particle in the adjoint representation, such as a gluino, it is almost impossible to determine whether it would bind to a single nucleon and, if so, with what energy. Regardless of the type of strongly interacting relic, the relic's abundance is expected to be suppressed by more than five orders of magnitude compared with X^- due to a very efficient annihilation channel following hadronization (104).

Detailed investigations of mesino-nucleon bound systems (103) offer insight into X_s-CBBN. By making fairly generous assumptions about the uncertainties in the interaction strength, the binding energy of the mesino-nucleon system would typically be between 5 and 40 MeV, which is much larger than the deuteron-binding energy E_d. A more tightly bound nucleon-meson system is expected because the shallow binding of the deuteron is to some extent accidental and because going from a nucleon-nucleon system to a nucleon-mesino system decreases the kinetic energy and results in a stronger binding. The main cosmological consequence of a tightly bound mesino-nucleon system is the formation of such bound states long before the end of the SBBN D bottleneck and, hence, at times when neutrons are abundant. For example, a ~20-MeV binding would correspond to a new bottleneck at $T \sim 1$ MeV, thereby opening up an interesting new possibility for nucleosynthesis in the neutron-rich environment around each X_s particle. Unfortunately, a detailed study of such a nucleosynthetic path is currently beyond theoretical capabilities, mostly due to the rather uncertain spectrum of light mesino–containing nuclei. Although some researchers have claimed that nucleosynthesis proceeds to form the $A \geq 6$ elements (102), it appears equally plausible that nucleosynthesis around strongly interacting particles stops at a "compound helium"—a bound state consisting of a mesino particle and three nucleons. Nucleosynthesis of heavier elements around a mesino may be inhibited by (p, α) reactions, as well as by an expected weak binding in a system containing five particles (one mesino and four nucleons) (103). Much more theoretical work will be required before we fully understand the primordial catalysis caused by strongly interacting particles.

5. CONCLUSIONS

BBN, a very short period in the history of the universe, is an important reference point going back to cosmological times of a few seconds. In this review, we have shown how new physics can modify the synthesis of light elements and therefore be probed by contrasting the observations of helium and D (as well as lithium, beryllium, and boron) with the BBN predictions. All three generic methods—modifications to the timing of main BBN events via extra degrees of freedom or space-time-dependent couplings and mass scales, nonequilibrium nucleosynthesis triggered by energy injection via the annihilation or decay of heavy particles, and particle catalysis of BBN reactions—are very important to many extensions of the SM. In some models, the sensitivity of BBN to new physics is exceptional and indeed exceeds all other direct and indirect probes. A representative example is the nontrivial limit on the lifetime of a metastable charged particle (e.g., a stau decaying to a gravitino) that follows from the considerations of the ^{6}Li abundance, a limit that currently cannot be constrained other than in the BBN.

The amount of lithium predicted by SBBN is of the same order of magnitude as the value indicated by the Spite plateau. However, a detailed comparison with the standard cosmology and particle physics input shows that SBBN overpredicts the abundance of ^{7}Li by a factor of three to five. This discrepancy has become more concrete, given that the baryon-to-photon ratio and the main SBBN reactions determining ^{7}Li abundance are known to an accuracy of better than 10%. We have suggested possible solutions to the problem that may reside in subtle astrophysical effects or in new cosmological and/or particle physics ingredients. Here we summarize these (and some additional) options.

- The most economical solution to the lithium discrepancy would be an astrophysical mechanism that depletes lithium from the photic zone in the atmospheres of population II stars.

Several ideas as to the nature of this mechanism have been proposed in the literature, and hints at traces of such depletion in the scatter of points along the Spite plateau have been found (45). Although some amount of depletion is certainly possible, it is premature to concede that it can account for the discrepancy of a factor of approximately three. Clearly this issue remains one of the most important astrophysical issues in the cosmology of the early universe.

- Nuclear physics is unlikely to be responsible for the solution to the lithium problem, as all main reactions participating in creation and destruction of lithium are well known. Perhaps the last subject worthy of detailed investigation is a possible resonant enhancement of the ^{7}Be+D reaction (47), which could contribute to the depletion of ^{7}Be, albeit at the very end of the most optimistic range for the parameters of such a resonance. This issue can be directly clarified through experiments.
- Given that lithium is observed in the Milky Way, whereas the measurements of D and especially η_b are global, one may speculate that baryons and possibly dark matter are distributed nonuniformly and that the 30–50% downward fluctuation of η_b at the local patch of the universe leads to a local value of lithium that is close to the Spite plateau value (see, for instance, Reference 105 for a recent discussion). Of course, the variations in baryon number by a factor of a few, if persistent at all scales, would create very strong isocurvature-type features in the CMB polarization maps. However, it is also possible that models with anticorrelated baryon–dark matter fluctuations with a very blue perturbation spectrum may survive all observational constraints.
- Particle physics models with unstable or annihilating particles may have an important effect on the lithium abundance. Although the photodisintegration of lithium by a factor of three is difficult to reconcile with the D abundance and/or the ^{3}He/D ratio (unless the mass of a decaying particle is carefully chosen), the energy injection with some yield of nucleons in the final state may fare much better. Indeed, the injection of neutrons (regardless of the details of the particle physics mechanism) before the freeze-out of $^7\mathrm{Li}(p, \alpha)\alpha$ reaction can reduce the ^{7}Be and therefore the ^{7}Li abundance. A necessary consequence of this scenario is a somewhat elevated D/H abundance and probably elevated concentrations of other rare isotopes of lithium (and possibly beryllium and boron).
- The catalytic suppression of ^{7}Li by negatively charged particles relies on the fact that ($^7\mathrm{Be}X^-$) is the deepest bound state among all X^--bound nuclei, which can form at temperatures of approximately 30 keV. The most efficient suppression of ^{7}Be happens in models allowing for weak transitions to ^{7}Li at these temperatures. The need for a large number of X^- at 30 keV also implies the presence of X^- at $T \sim 10$ keV, which would lead to the catalyzed synthesis of ^{6}Li and ^{9}Be.
- The last and perhaps most exotic option is the possibility of coupling constants and mass scales changing over time. Indeed, a $\sim$5% change in the D binding energy can create a very strong downward shift of the ^{7}Li abundance while remaining consistent with helium and D abundances.

We do not yet know how to resolve the lithium problem; only continuing progress in particle physics, cosmology, and astrophysics will help to clarify this intriguing discrepancy.

Finally, we discuss some anticipated future developments that are not necessarily directly related to BBN but rather are related to questions discussed in our review.

- CMB physics will continue to deliver very important results. Of particular interest to this review is the increased sensitivity on small angular scales (high-l multipoles) to anisotropies in which the effects of extra dark radiation states or certain types of isocurvature fluctuations are the most pronounced. Future data from the Planck satellite experiment will allow us to probe the dark radiation energy density with an accuracy of $\Delta N_{\nu,\mathrm{eff}} \sim 0.5$ or better; results thus obtained will thereby provide an independent check of the constraints resulting from ^{4}He abundances.

- The Large Hadron Collider is expected to deliver its first physics results in the near future. The teraelectronvolt energy frontier will be effectively probed, which may lead to the discovery of stable or massive metastable particles. If such particles were neutral, their signature would constitute missing energy, which is difficult but not impossible to disentangle from SM processes. However, if a metastable particle were charged under a strong or electromagnetic gauge group, the signatures would be much more spectacular, with long charged tracks that would allow us to identify the mass of such objects. If charged metastable states are discovered at the LHC, the theories of nonequilibrium and catalyzed nucleosynthesis would be further reinforced.
- A massive expansion of the dark matter search program, both in space and in underground laboratories, may point to the existence of WIMPs in the near future. If WIMPs turn out to be relatively light (in the 10–50-GeV window), then the energy injection in the early universe could lead to altered ^{7}Li and ^{6}Li abundances. Heavier WIMPs with enhanced annihilation cross sections devised to fit the PAMELA positron anomaly (84) may also lead to changes in lithium abundances, and therefore any future development that supports or disfavors the WIMP interpretation of the PAMELA anomaly is of interest for the BBN predictions.

Further theoretical progress in non-SBBN scenarios is expected. Given the doubtful status of ^{6}Li observations and its extreme fragility, it is highly desirable to perform calculations of beryllium and boron abundances generated in BBN scenarios with energy injection. Doing so may place further constraints on annihilating/decaying WIMP scenarios. Additionally, more detailed calculations of some CBBN rates, with a proper many-body nuclear physics input, are needed.

DISCLOSURE STATEMENT

The authors are not aware of any affiliations, memberships, funding, or financial holdings that might be perceived as affecting the objectivity of this review.

LITERATURE CITED

1. Alpher RA, Follin JW, Herman RC. *Phys. Rev.* 92:1347 (1953)
2. Dunkley J, et al. (WMAP Collab.) *Astrophys. J. Suppl.* 180:306 (2009)
3. Alpher RA, Bethe H, Gamow G. *Phys. Rev. Lett.* 73:803 (1948)
4. Alpher R, Herman RC. *Rev. Mod. Phys.* 22:153 (1950)
5. Hayashi C. *Progr. Theor. Phys.* 5:224 (1950)
6. Malaney RA, Mathews GJ. *Phys. Rep.* 229:145 (1993)
7. Sarkar S. *Rep. Prog. Phys.* 59:1493 (1996)
8. Iocco F, et al. arXiv:0809.0631 [astro-ph] (2008)
9. Steigman G. *Annu. Rev. Nucl. Part. Sci.* 57:463 (2007)
10. Jedamzik K, Pospelov M. *New J. Phys.* 11:105028 (2009)
11. Dolgov AD, Fukugita M. *Phys. Rev. D* 46:5378 (1992)
12. Wagoner RV, Fowler WA, Hoyle F. *Astrophys. J.* 148:3 (1967)
13. Mukhanov VF. *Int. J. Theor. Phys.* 43:669 (2004)
14. Kawano L. *Fermilab-pub-92-004-A* (1992)
15. Audi G, Wapstra AH, Thibault C. *Nucl. Phys. A* 729:337 (2002)
16. Mohr PJ, Taylor BN. *Rev. Mod. Phys.* 77:1 (2005)
17. Descouvemont P, et al. *At. Data Nucl. Data Tables* 88:203 (2004)

18. Ando S, Cyburt RH, Hong SW, Hyun CH. *Phys. Rev. C* 74:025809 (2006)
19. Cyburt RH, Davids B. *Phys. Rev. C* 78:064614 (2008)
20. Lopez RE, Turner MS. *Phys. Rev. D* 59:103502 (1999)
21. Wilkinson DH. *Nucl. Phys. A* 377:474 (1982)
22. Cyburt RH, Fields BD, Olive KA. *J. Cosmol. Astropart. Phys.* 0811:12 (2008)
23. Peimbert M, Luridiana V, Peimbert A. *Astrophys. J.* 666:636 (2007)
24. Izotov YI, Thuan TX, Stasinska G. *Astrophys. J.* 662:15 (2007)
25. Olive KA, Skillman ED. *Astrophys. J.* 617:29 (2004)
26. Fukugita M, Kawasaki M. *Astrophys. J.* 646:691 (2006)
27. Burles S, Tytler D. *Astrophys. J.* 499:699 (1998)
28. Levshakov SA, Dessauges-Zavadsky M, D'Odorico S, Molaro P. *Astrophys. J.* 565:696 (2007)
29. Crighton NHM, Webb JK, Ortiz-Gill A, Fernandez-Soto A. *Mon. Not. R. Astron. Soc.* 355:1042 (2004)
30. O'Meara JM, et al. *Astrophys. J.* 649:L61 (2006)
31. Pettini M, et al. *Mon. Not. R. Astron. Soc.* 391:1499 (2008)
32. Linsky JL, et al. *Astrophys. J.* 647:1106 (2006)
33. Prodanovic T, Steigman G, Fields BD. arXiv:0910.4961 [astro-ph] (2009)
34. Sigl G, Jedamzik K, Schramm DN, Berezinsky VS. *Phys. Rev. D* 52:6682 (1995)
35. Geiss J, Gloeckler G. *Space Sci. Rev.* 130:5 (2007)
36. Nara Singh BS, Hass M, Nir-El Y, Haquin G. *Phys. Rev. Lett.* 93:262503 (2004)
37. Gyurky G, et al. *Phys. Rev. C* 75:035805 (2007)
38. Brown TAD, et al. *Phys. Rev. C* 76:055801 (2007)
39. Aoki W, et al. *Astrophys. J.* 698:1803 (2009)
40. Sbordone L. arXiv:1003.4510 [astro-ph] (2010)
41. Melendez J, et al. arXiv:0912.2949 [astro-ph] (2009)
42. Ryan SG, Norris JE, Beers TC. *Astrophys. J.* 523:654 (1999)
43. Bonifacio P, et al. *Astron. Astrophys.* 390:91 (2002)
44. Pasquini L, Molaro P. *Astron. Astrophys.* 322:109 (1997)
45. Asplund M, et al. *Astrophys. J.* 644:229 (2006)
46. Coc A, et al. *Astrophys. J.* 600:544 (2004)
47. Cyburt RH, Pospelov M. arXiv:0906.4373 [astro-ph] (2009)
48. Angulo C, et al. *Astrophys. J.* 630:L105 (2005)
49. Richard O, Michaud G, Richer J. *Astrophys. J.* 619:538 (2005)
50. Korn A, et al. *Nature* 442:657 (2006)
51. Primas F, Asplund M, Nissen PE, Hill V. *Astron. Astrophys.* 364:42 (2000)
52. Boesgaard AM, Novicki MC. *Astrophys. J.* 641:1122 (2006)
53. Fields BD, Olive KA, Vangioni-Flam E. *Astrophys. J.* 623:1083 (2005)
54. Pospelov M, Pradler J, Steffen FD. *J. Cosmol. Astropart. Phys.* 0811:020 (2008)
55. Prantzos N. *Astron. Astrophys.* 448:665 (2005)
56. Cayrel R, et al. *Astron. Astrophys.* 473:37 (2007)
57. Ratra B, Peebles PJE. *Phys. Rev. D* 37:3406 (1988)
58. Wetterich C. *Nucl. Phys. B* 302:668 (1988)
59. Flambaum VV, Shuryak EV. *Phys. Rev. D* 65:103503 (2002)
60. Flambaum VV, Wiringa RB. *Phys. Rev. C* 76:054002 (2007)
61. Nollett KM, Lopez RE. *Phys. Rev. D* 66:063507 (2002)
62. Dmitriev VF, Flambaum VV, Webb JK. *Phys. Rev. D* 69:063506 (2004)
63. Coc A, et al. *Phys. Rev. D* 76:023511 (2007)
64. Dent T, Stern S, Wetterich C. *Phys. Rev. D* 76:063513 (2007)
65. Voronchev VT, Nakao Y, Nakamura M. *J. Cosmol. Astropart. Phys.* 0805:010 (2008)
66. Ellis JR, Nanopoulos DV, Sarkar S. *Nucl. Phys. B* 259:175 (1985)
67. Levitan YL, Sobol IM, Khlopov MY, Chechetkin VM. *Sov. J. Nucl. Phys.* 47:109 (1988)
68. Dimopoulos S, Esmailzadeh R, Hall LJ, Starkman GD. *Astrophys. J.* 330:545 (1988)
69. Reno MH, Seckel D. *Phys. Rev. D* 37:3441 (1988)
70. Dimopoulos S, Esmailzadeh R, Hall LJ, Starkman GD. *Nucl. Phys. B* 311:699 (1989)

71. Ellis JR, et al. *Nucl. Phys. B* 373:399 (1992)
72. Khlopov MY, Levitan YL, Sedelnikov EV, Sobol IM. *Phys. Atom. Nucl.* 57:1393 (1994)
73. Kawasaki M, Moroi T. *Astrophys. J.* 452:506 (1995)
74. Jedamzik K. *Phys. Rev. D* 70:063524 (2004)
75. Kawasaki M, Kohri K, Moroi T. *Phys. Lett.* B625:7 (2005)
76. Kawasaki M, Kohri K, Moroi T. *Phys. Rev. D* 71:083502 (2005)
77. Cyburt RH, et al. *J. Cosmol. Astropart. Phys.* 0611:014 (2006)
78. Cyburt RH, et al. *J. Cosmol. Astropart. Phys.* 0910:021 (2009)
79. Protheroe RJ, Stanev T, Berezinsky VS. *Phys. Rev. D* 51:4134 (1995)
80. Ellis JR, Olive KA, Vangioni E. *Phys. Lett.* B619:30 (2005)
81. Pospelov M, Ritz A. *Phys. Lett.* B671:391 (2009)
82. Cyburt RH, Ellis JR, Fields BD, Olive KA. *Phys. Rev. D* 67:103521 (2003)
83. Jedamzik K. *Phys. Rev. D* 70:083510 (2004)
84. Adriani O, et al. *Nature* 458:607 (2008)
85. Pospelov M. *Phys. Rev. Lett.* 98:231301 (2007)
86. Kohri K, Takayama F. *Phys. Rev. D* 76:063507 (2007)
87. Kaplinghat M, Rajaraman A. *Phys. Rev. D* 74:103004 (2006)
88. Hamaguchi K, et al. *Phys. Lett.* B650:268 (2007)
89. Bird C, Koopmans K, Pospelov M. *Phys. Rev. D* 78:083010 (2008)
90. Jittoh T, et al. *Phys. Rev. D* 76:125023 (2007)
91. Jedamzik K. *Phys. Rev. D* 77:063524 (2008)
92. Jedamzik K. *J. Cosmol. Astropart. Phys.* 0803:008 (2008)
93. Pospelov M. arXiv:0712.0647 [hep-ph] (2007)
94. Kusakabe M, et al. *Astrophys. J.* 680:846 (2008)
95. Kamimura M, Kino Y, Hiyama E. *Prog. Theor. Phys.* 121:1059 (2009)
96. De Rujula A, Glashow SL, Sarid U. *Nucl. Phys. B* 333:173 (1990)
97. Dimopoulos S, Eichler D, Esmailzadeh R, Starkman GD. *Phys. Rev. D* 41:2388 (1990)
98. Rafelski J, Sawicki M, Gajda M, Harley D. *Phys. Rev. A* 44:4345 (1991)
99. Jittoh T, Sato J, Shimomura T, Yamanaka M. *Phys. Rev. D* 73:055009 (2006)
100. Ratz M, Schmidt-Hoberg K, Winkler MW. *J. Cosmol. Astropart. Phys.* 0810:026 (2008)
101. Pradler J, Steffen FD. *Nucl. Phys. B* 809:318 (2009)
102. Kusakabe M, Kajino T, Yoshida T, Mathews GJ. *Phys. Rev. D* 80:103501 (2009)
103. Sharman J. MSc. thesis. Univ. Victoria, Victoria, BC, Can. (2010)
104. Kang J, Luty MA, Nasri S. *J. High Energy Phys.* 09:086 (2008)
105. Holder GP, Nollett KM, van Engelen A. arXiv:0907.3919 [astro-ph] (2009)

Collective Neutrino Oscillations

Huaiyu Duan,[1] George M. Fuller,[2] and Yong-Zhong Qian[3]

[1]Theoretical Division, Los Alamos National Laboratory, Los Alamos, New Mexico 87545; email: Huaiyu.Duan@MailAPS.org

[2]Department of Physics, University of California, San Diego, La Jolla, California 92093; email: gfuller@ucsd.edu

[3]School of Physics and Astronomy, University of Minnesota, Minneapolis, Minnesota 55455; email: qian@physics.umn.edu

Annu. Rev. Nucl. Part. Sci. 2010. 60:569–94

First published online as a Review in Advance on July 20, 2010

The *Annual Review of Nuclear and Particle Science* is online at nucl.annualreviews.org

This article's doi:
10.1146/annurev.nucl.012809.104524

0163-8998/10/1123-0569$20.00

Key Words

neutrino masses and mixings, neutrino spectral swaps/splits, supernova neutrino signals

Abstract

We review the rich phenomena associated with neutrino flavor transformation in the presence of neutrino self-coupling. Our exposition centers on three collective neutrino oscillation scenarios: (*a*) a simple bipolar neutrino system that initially consists of monoenergetic ν_e and $\bar{\nu}_e$, (*b*) a homogeneous and isotropic neutrino gas with multiple neutrino/antineutrino species and continuous energy spectra, and (*c*) a generic neutrino gas in an anisotropic environment. We use each of these scenarios to illustrate key facets of collective neutrino oscillations. We discuss the implications of collective neutrino flavor oscillations for core-collapse supernova physics and for the prospects of obtaining and/or constraining fundamental neutrino properties, such as the neutrino mass hierarchy and θ_{13} from a future observed supernova neutrino signal.

Contents

1. INTRODUCTION

1.1. Neutrino Mixing and Astrophysics

Neutrinos and the phenomena associated with neutrino flavor transformation stand at the nexus of two exciting recent developments: the success of experimental neutrino physics and the tremendous growth of astronomy and astrophysics. The former enterprise has provided key insights into neutrino mass and vacuum mixing and promises more (1), and the latter has provided fundamental cosmological parameters and is revealing how structure and elemental abundances emerge and evolve in the universe (2–5). Moreover, there is feedback between these subjects. For example, observations of large-scale structure and the cosmic microwave background radiation currently provide our best limits on the neutrino rest masses (6, 7). Both the early universe and the massive star core-collapse and supernova explosion environments can be dominated by neutrinos. Neutrino flavor transformation in each of these environments may give insights into astrophysics and even into fundamental neutrino properties. Obtaining these insights requires confident modeling of neutrino flavor evolution in environments where neutrino-neutrino interactions produce vexing nonlinearity. The neutrino mass and mixing data already gathered from the experiments make a compelling case that we must solve this problem.

Experiments and observations have established that the neutrino energy (mass) states $|\nu_i\rangle$ ($i = 1, 2, 3$) are not coincident with the weak interaction states $|\nu_\alpha\rangle$ ($\alpha = e, \mu, \tau$). The relation between these bases is given by $|\nu_\alpha\rangle = \sum_i U^*_{\alpha i}|\nu_i\rangle$ (8), where the Maki-Nakagawa-Sakata (MNS) matrix elements $U_{\alpha i}$ are parameterized by three vacuum mixing angles (θ_{12}, θ_{23}, and θ_{13}) and a *CP*-violating phase (δ). Two of these, $\sin^2\theta_{12} \approx 0.31$ and $\sin^2\theta_{23} \approx 0.5$, are measured outright, whereas there is a firm upper limit on the third, $\sin^2\theta_{13} < 0.04$ at 2σ (9). Observations of solar neutrinos show flavor conversion in the $\nu_e \rightleftharpoons \nu_{\mu/\tau}$ channel with characteristic mass-squared splitting $\Delta m^2_{\odot} \approx 7.6 \times 10^{-5}\ \mathrm{eV}^2$. Atmospheric neutrino measurements show near-maximal vacuum mixing in the $\nu_\mu \rightleftharpoons \nu_\tau$ channel with corresponding mass-squared splitting $\Delta m^2_{\mathrm{atm}} \approx 2.4 \times 10^{-3}\ \mathrm{eV}^2$. However, experiments do not reveal the absolute neutrino rest masses m_i ($i = 1, 2, 3$), nor do they show whether these neutrino mass eigenvalues are ordered in the normal mass hierarchy ($m_3 > m_2 > m_1$) or in the inverted mass hierarchy ($m_2 > m_1 > m_3$).

Future terrestrial neutrino experiments (10–15) will be directed primarily toward measuring θ_{13} and, if θ_{13} is big enough, the neutrino mass hierarchy and possibly even the *CP*-violating phase δ. Planned reactor and long-baseline experiments may be able to measure θ_{13} if it satisfies $\sin^2\theta_{13} > 10^{-4}$. This limit ultimately is set by constraints on the neutrino flux and detector mass. To find significantly larger neutrino fluxes we must turn to cosmic sources, such as core-collapse supernovae (16–23).

Stars whose masses are greater than $\sim 8\ M_{\odot}$ end their lives in gravitational collapse and the production of a neutron star or, if they are massive enough, a black-hole remnant (24–27). Neutrinos play a role in nearly every aspect of the evolution of these core-collapse supernovae, from dominating lepton number and entropy loss from the epoch of core carbon/oxygen burning onward, to providing the bulk of energy and lepton number transport during collapse itself. Neutrinos may provide the heating necessary to engender convection (28–35) and, for instance, the standing accretion shock instability, which may create an explosion (36, 37).

A key point is that gravitational collapse causes an appreciable fraction of the rest mass of the Chandrasekhar mass ($\sim 1.4 M_{\odot}$) core to appear as seas of trapped neutrinos, which subsequently diffuse out of the core on timescales of seconds (38–40). At core bounce, the energy in the neutrino seas trapped in the core is $\sim 10^{52}$ ergs, but by ~ 10 s after core bounce an energy of some 10^{53} ergs, or $\sim 10\%$ of the rest mass of the core, has been radiated away as neutrinos of all kinds. The emergent neutrino and antineutrino energy spectra and fluxes at the neutrino sphere vary with time post–core bounce, but there are epochs in which the energy spectra and/or the luminosities in the various neutrino flavors differ. Moreover, charged-current, flavor-specific neutrino interaction processes such as $\nu_e + n \rightleftharpoons p + e^-$ and $\bar{\nu}_e + p \rightleftharpoons n + e^+$ are important both for energy and electron lepton number deposition as well as for determining neutrino transport physics (41–43). It is therefore interesting and necessary to assess whether interconversion of neutrino flavors in the supernova environment affects explosion physics and neutrino-heated nucleosynthesis and how such flavor transformation might affect a supernova neutrino burst signature in a terrestrial detector.

1.2. Brief History of Collective Neutrino Oscillations

Early studies of neutrino flavor transformation centered on solar neutrinos, especially after it was recognized first by Wolfenstein (44), and then by Mikheyev & Smirnov (45), that the medium through which the neutrino propagates could alter the effective neutrino mass and mixing properties. Almost immediately after the Mikheyev-Smirnov-Wolfenstein (MSW) mechanism was understood, it was realized that the forward coherent scattering of neutrinos with other neutrinos, somewhat misleadingly termed neutrino self-coupling or neutrino self-interaction, could generate a similar effect (46, 47). Subsequently, the effects of neutrino self-interaction were investigated independently in the core-collapse supernova and the early universe scenarios. The studies of the supernova environment first focused on MSW-like effects (41, 46, 48–51). However, the neutrino self-interaction potential is very different from the matter potential in that it can have nonvanishing off-diagonal elements in the interaction basis (52, 53). Studies using the complete neutrino self-interaction potential showed that neutrinos could experience self-maintained coherent oscillations or collective oscillations in lepton-degenerate early universe scenarios (54–64). By collective oscillations we mean a significant fraction of neutrinos oscillating coherently with respect to one another.

Early research on MSW-like evolution, where the flavor off-diagonal potentials were minimal, as well as the first paper (65) to point out that collective effects could occur in supernovae, focused on relatively large neutrino mass–squared differences. However, neutrinos and antineutrinos can be transformed collectively and simultaneously even for the small, measured neutrino mass–squared differences (66, 67). Additionally, ordinary matter does not necessarily suppress collective neutrino oscillations, at least in homogeneous and isotropic environments (68). As a consequence, collective neutrino oscillations may occur deep in the envelope of a supernova (69, 70). These studies showed that the neutrino energy spectra would be modified differently for the normal and inverted neutrino mass hierarchies (**Figure 1**).

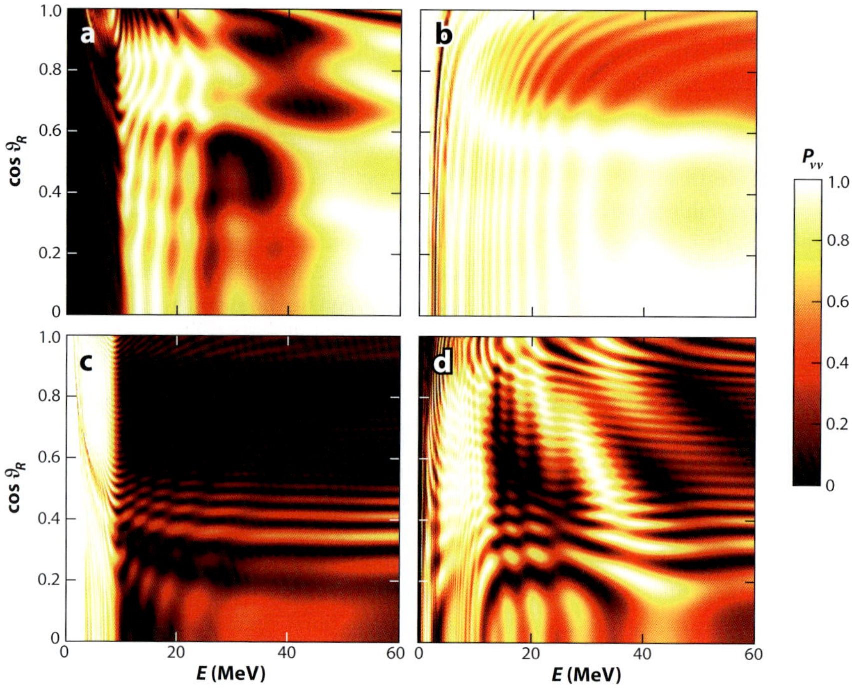

Figure 1

Survival probabilities $P_{\nu\nu}$ for (*a*,*c*) neutrinos and (*b*,*d*) antineutrinos as functions of both neutrino energy E and emission angle ϑ_R in a numerical calculation using the neutrino bulb model (see **Figure 3**) and the two-flavor mixing scheme. The most prominent features are (*a*,*c*) the approximately angle-independent step-like changes in the neutrino survival probabilities—these are the spectral swaps/splits first shown in Reference 70. The energy spectra of ν_e and ν_μ with energy below (above) $E \simeq 9$ MeV in this calculation are almost completely swapped in panel *a* (*c*), which employs the normal (inverted) neutrino mass hierarchy. Spectral swaps/splits are the result of a collective neutrino oscillation mode known as the precession mode. The vertical fringes (the energy-dependent features in the figure) are the result of Mikheyev-Smirnov-Wolfenstein flavor transformation, which is energy dependent. The horizontal fringes (the angle-dependent features) are the result of the kinematic decoherence of bipolar neutrino oscillations (72). The movie version of these calculations is available in Reference 87. Figure adapted from figure 3 of Reference 69. Copyright 2006 by the American Physical Society.

Over the past few years, many papers have been written on the collective neutrino oscillation phenomenon and its physical effects (71–111). Following this literature can be bewildering. For example, synchronized and bipolar neutrino oscillations are frequently perceived as the two most important collective neutrino oscillation modes in supernovae. However, numerical studies suggest that supernova neutrinos are probably never synchronized because of the nonvanishing matter density (76), and the bipolar oscillation is not a collective neutrino oscillation mode in anisotropic environments such as supernovae (72).

1.3. Goal and Organization of This Review

The purpose of this review is to provide a relatively short but in-depth exposition of the properties of collective neutrino oscillations that are reasonably well understood. Specifically, we elucidate the physics behind striking features such as the swaps of supernova neutrino energy spectra caused by collective oscillations, as shown in **Figure 1**. Our strategy is to illustrate key facets of collective neutrino oscillations by describing three increasingly complex models.

The rest of the review is organized as follows. In Section 2 we try to make the connection between the so-called wave function language and the spin language, which are commonly used for studying noncollective and collective neutrino oscillations, respectively. In Section 3 we discuss the flavor evolution of a homogeneous and isotropic neutrino gas that consists initially of monoenergetic, pure ν_e and $\bar{\nu}_e$. This simple model can be solved analytically and offers important insights into the behavior of more complicated systems. In Section 4 we discuss homogeneous and isotropic gases that consist of neutrinos with continuous energy spectra. We explore adiabatic solutions to the neutrino flavor evolution equations and demonstrate how the spectral swap/split phenomenon can be explained by one of these solutions. In Section 5 we discuss some of the important neutrino oscillation properties that are unique to anisotropic environments and explain why the spectral swap/split phenomenon may occur in these environments despite the anisotropy in neutrino fields. In Section 6 we apply the current understanding of the collective neutrino oscillation phenomenon to supernova environments. In Section 7 we give a summary and identify several issues in collective neutrino oscillations that remain to be understood.

2. NEUTRINO MIXING IN DENSE NEUTRINO GASES

2.1. Equations of Motion

Here we focus on two-flavor neutrino mixing scenarios, namely $\alpha = e, \mu$ and $i = 1, 2$, where $|\nu_\mu\rangle$ is a linear combination of the physical $|\nu_\mu\rangle$ and $|\nu_\tau\rangle$. We do this for pedagogical purposes, although our choice is physically justifiable because the physical ν_μ and ν_τ are nearly maximally mixed in vacuum and experience nearly identical interactions in the supernova environment (112, 113). We discuss collective neutrino oscillations with the full three-flavor mixing machinery in Section 4.3.

We consider only coherent neutrino flavor evolution, in which the effects of neutrino inelastic scattering and associated quantum decoherence can be neglected. This is generally applicable in the region well above the neutron star in supernova models. [Solution of the complete problem of neutrino flavor evolution with both elastic and inelastic neutrino scattering would necessitate the use of the full quantum kinetic equations (7, 43, 114–116).] We also assume that neutrinos are relativistic and that general relativistic effects can be ignored. With these assumptions, a mean-field Schrödinger-like equation,

$$\mathrm{i}\frac{\mathrm{d}}{\mathrm{d}x}\psi = \mathsf{H}\psi, \qquad 1.$$

is taken to describe flavor evolution along a neutrino world line (117). Here $\hbar = c = 1$, x is the distance along the world line of the neutrino; ψ is the neutrino flavor wave function, taken to be a vector in (neutrino) flavor space; and

$$\mathsf{H} = \frac{\Delta m^2}{2E}\mathsf{B} + \lambda\mathsf{L} + \mathsf{H}_{\nu\nu} = \frac{\Delta m^2}{2E}\mathsf{B} + \sqrt{2}G_{\mathrm{F}}n_e\mathsf{L} + \sqrt{2}G_{\mathrm{F}}\int \mathrm{d}^3\mathbf{p}'(1 - \hat{\mathbf{p}}\cdot\hat{\mathbf{p}}')(\rho_{\mathbf{p}'} - \bar{\rho}_{\mathbf{p}'}) \qquad 2.$$

is the Hamiltonian. (See References 118–122 for discussions of the applicability of the one-particle effective approximation assumed in Equation 1.) The wave function for an antineutrino also obeys Equation 1, but with the replacements $\lambda \to -\lambda$ and $\mathsf{H}_{\nu\nu} \to -\mathsf{H}^*_{\nu\nu}$ in Equation 2. The flavor space is spanned by the neutrino interaction basis or, equivalently, by the mass basis. We use the word flavor in a more general sense, and we avoid using the phrase flavor state, which in the literature can mean either the interaction state $|\nu_\alpha\rangle$ or the flavor quantum state $|\psi\rangle$. (We use symbols in sans serif font, such as H, to denote basis-dependent matrices in flavor space.)

The first term in Equation 2 induces neutrino flavor transformation because in the interaction basis,

$$\mathsf{B} = \mathsf{U}\left(\frac{1}{2}\mathrm{diag}[-1, 1]\right)\mathsf{U}^\dagger = \frac{1}{2}\begin{bmatrix} -\cos 2\theta_{\mathrm{v}} & \sin 2\theta_{\mathrm{v}} \\ \sin 2\theta_{\mathrm{v}} & \cos 2\theta_{\mathrm{v}} \end{bmatrix} \qquad 3.$$

is nondiagonal. In Equation 3, U is the MNS matrix and θ_{v} is the so-called vacuum mixing angle within the range $(0, \pi/4]$. In Equation 2, Δm^2 is the mass-squared difference appropriate for $|\nu_2\rangle$ and $|\nu_1\rangle$, and E is the energy of the neutrino. Here $\Delta m^2 > 0$ and $\Delta m^2 < 0$ correspond to the normal neutrino mass hierarchy (NH) and the inverted neutrino mass hierarchy (IH), respectively, as discussed above for the full three-flavor mixing case. Although this formalism is completely general, one of the most interesting supernova cases is where $|\Delta m^2| \simeq \Delta m^2_{\mathrm{atm}}$ and $\theta_{\mathrm{v}} \simeq \theta_{13} \ll 1$.

The second term in Equation 2 arises from coherent neutrino-electron forward exchange scattering (44). In this term, which is referred to as the matter term, G_{F} is the Fermi constant, n_e is the net electron number density, and $\mathsf{L} = \mathrm{diag}[1, 0]$ in the interaction basis. Note that adding or subtracting a multiple of the identity matrix to or from H (e.g., the contribution of neutral-current neutrino-electron scattering) gives only an overall phase to ψ and, therefore, does not affect neutrino oscillations.

The last term in Equation 2 stems from coherent neutral-current neutrino-neutrino forward exchange scattering (46–48, 52), where $\hat{\mathbf{p}}$ and $\hat{\mathbf{p}}'$ are the unit vectors for the propagation directions of the test neutrino and the background neutrino or antineutrino, respectively. In the interaction basis, at location $\mathbf{x}$ and at time t, the (flavor) density matrices for neutrinos and antineutrinos with momentum $\mathbf{p}'$ and with our assumptions can be written as

$$[\rho_{\mathbf{p}'}(t, \mathbf{x})]_{\alpha\beta} = \sum_{\nu'} n_{\nu',\mathbf{p}'}(t, \mathbf{x})\langle\nu_\alpha|\psi_{\nu',\mathbf{p}'}(t, \mathbf{x})\rangle\langle\psi_{\nu',\mathbf{p}'}(t, \mathbf{x})|\nu_\beta\rangle \qquad 4a.$$

and

$$[\bar{\rho}_{\mathbf{p}'}(t, \mathbf{x})]_{\beta\alpha} = \sum_{\bar{\nu}'} n_{\bar{\nu}',\mathbf{p}'}(t, \mathbf{x})\langle\bar{\nu}_\alpha|\psi_{\bar{\nu}',\mathbf{p}'}(t, \mathbf{x})\rangle\langle\psi_{\bar{\nu}',\mathbf{p}'}(t, \mathbf{x})|\bar{\nu}_\beta\rangle, \qquad 4b.$$

respectively, where $|\psi_{\nu'(\bar{\nu}'),\mathbf{p}'}\rangle$ is the state of a neutrino ν' (antineutrino $\bar{\nu}'$) with momentum $\mathbf{p}'$ and $n_{\nu',\mathbf{p}'}$ ($n_{\bar{\nu}',\mathbf{p}'}$) is the corresponding number density of the neutrino (antineutrino). Note that the order of the indices on the matrix representation of $\bar{\rho}$ in Equation 4*b* follows the convention in Reference 114. The advantage of this definition is that $\bar{\rho}$ transforms in the same way as does ρ when transforming from the interaction basis to the mass basis or vice versa.

2.2. Neutrino Flavor Polarization Vector

It becomes more difficult to analyze neutrino oscillations using the wave function formalism when $\mathsf{H}_{\nu\nu}$ is significant because $\mathsf{H}_{\nu\nu}$ is a sum of density matrices that involve bilinear forms of the wave functions. The density matrices contain all the physical information about the neutrino mixing problem. The diagonal elements of the density matrices give the number densities of neutrinos in the weak interaction states or the mass states, depending on the basis used, and the off-diagonal elements of the density matrices contain the neutrino mixing information. For simplicity let us first take a working example, a homogeneous and isotropic neutrino gas whose flavor content can vary with time t. Homogeneity and isotropy imply that the factor $(1-\hat{\mathbf{p}}\cdot\hat{\mathbf{p}}')$ in Equation 2 averages to one. Therefore, the neutrino propagation direction does not matter, and $\rho_{\mathbf{p}}(t,\mathbf{x}) \to \rho_E(t)$ and $\bar{\rho}_{\mathbf{p}}(t,\mathbf{x}) \to \bar{\rho}_E(t)$. These density matrices obey the equations of motion (EoMs)

$$\mathrm{i}\dot{\rho}_E = \left[\frac{\Delta m^2}{2E}\mathsf{B} + \lambda\mathsf{L} + \sqrt{2}G_\mathrm{F}\int_0^\infty \mathrm{d}E'(\rho_{E'} - \bar{\rho}_{E'}),\, \rho_E\right] \qquad \text{5a.}$$

and

$$\mathrm{i}\dot{\bar{\rho}}_E = \left[-\frac{\Delta m^2}{2E}\mathsf{B} + \lambda\mathsf{L} + \sqrt{2}G_\mathrm{F}\int_0^\infty \mathrm{d}E'(\rho_{E'} - \bar{\rho}_{E'}),\, \bar{\rho}_E\right]. \qquad \text{5b.}$$

Because ρ_E and $\bar{\rho}_E$ are 2×2 Hermitian matrices, they can be mapped into vectors in a three-dimensional Euclidean space, which we also refer to as flavor space. We define the components of the (neutrino flavor) polarization vector $\vec{P}_\omega$ to be

$$P_{\omega,a} = \left(\frac{1}{n_\nu}\right)\left(\frac{|\Delta m^2|}{2\omega^2}\right)\times\begin{cases}\mathrm{Tr}(\rho_E\sigma_a) & \text{for neutrinos,}\\ -\mathrm{Tr}(\bar{\rho}_E\sigma_a) & \text{for antineutrinos,}\end{cases} \qquad \text{6.}$$

where the (angular) vacuum oscillation frequency is $\omega = \frac{\Delta m^2}{2E}$ for neutrinos and $\frac{\Delta m^2}{(-2E)}$ for antineutrinos, and σ_a ($a = 1, 2, 3$) are the Pauli matrices. (We use symbols with the vector hat, such as $\vec{P}$, to denote vectors in flavor space and symbols in the bold font, such as $\mathbf{p}$, to denote vectors in physical three-dimensional coordinate space.) Note that $\vec{P}_\omega$ can be normalized by an arbitrary factor. For example, if $\vec{P}_\omega$ is normalized to unity, then $\rho_E \propto (1+\frac{1}{2}\vec{P}_\omega\cdot\vec{\sigma})$ for neutrinos and $\bar{\rho}_E \propto (1-\frac{1}{2}\vec{P}_\omega\cdot\vec{\sigma})$ for antineutrinos. In Equation 6 we defined $\vec{P}_\omega$ to be normalized by n_ν, the initial total number density of a certain neutrino species ν, which is chosen to be $\bar{\nu}_e$ in the rest of this review. However, we usually take $\mu = \sqrt{2}G_\mathrm{F}n_\nu$ as a measure of the strength of neutrino self-interaction. Therefore, in some cases it is more appropriate to normalize $\vec{P}_\omega$ by the number density of other neutrino species, for instance, when $n_{\bar{\nu}_e}$ is negligible. If all neutrinos and antineutrinos are in the interaction states, then $\vec{P}_\omega \propto (n_{\nu_e,\omega} - n_{\nu_\mu,\omega})\hat{e}_3^{(\mathrm{I})}$ for neutrinos and $\vec{P}_\omega \propto -(n_{\bar{\nu}_e,\omega} - n_{\bar{\nu}_\mu,\omega})\hat{e}_3^{(\mathrm{I})}$ for antineutrinos, where $n_{\nu,\omega}$ ($\nu = \nu_e, \nu_\mu, \bar{\nu}_e, \bar{\nu}_\mu$) is the corresponding number density in the neutrino or antineutrino mode ω, and $\hat{e}_3^{(\mathrm{I})}$ is one of the interaction-basis vectors.

Because the Pauli matrices are traceless, the trace of the density matrix is not contained in the polarization vector. According to Equation 5, the traces of the density matrices do not change with time. (This corresponds to one of our assumptions that neutrinos are not created or annihilated.) These terms can be easily reintroduced in, for example, the calculation of the neutrino energy spectra. Use of Equations 5 and 6 shows that

$$\dot{\vec{P}}_\omega = (\omega\vec{B} + \lambda\vec{L} + \mu\vec{D})\times\vec{P}_\omega, \qquad \text{7.}$$

where vectors $\vec{B} = \mathrm{Tr}(\mathsf{B}\,\vec{\sigma})$ and $\vec{L} = \mathrm{Tr}(\mathsf{L}\,\vec{\sigma})$ are parallel to the (vacuum) mass and interaction-basis vectors $\hat{e}_3^{(\mathrm{V})}$ and $\hat{e}_3^{(\mathrm{I})}$, respectively (**Figure 2**), and $\vec{D} = \int_{-\infty}^{\infty}\vec{P}_\omega\,\mathrm{d}\omega$ is the total polarization vector. In the absence of neutrino self-coupling, $\vec{P}_\omega$ can be considered a "magnetic spin." In this analogy, the "magnetic spin" is coupled to two "magnetic fields," $\vec{B}$ and $\vec{L}$, with gyromagnetic

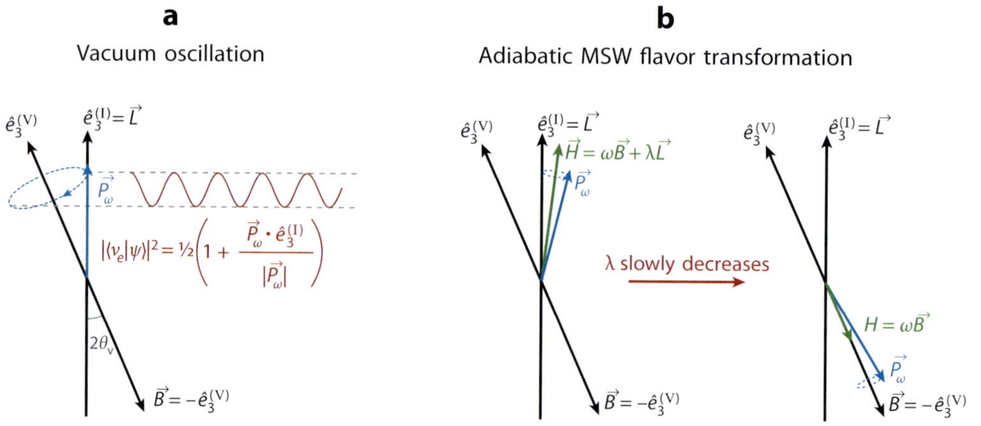

Figure 2

Geometric pictures for (*a*) vacuum oscillation and for (*b*) Mikheyev-Smirnov-Wolfenstein (MSW) flavor transformation. The flavor Hilbert space spanned by $|\nu_\alpha\rangle$ ($\alpha = e, \mu$) or $|\nu_i\rangle$ ($i = 1, 2$) can be mapped onto the flavor Euclidean space that is spanned by $\hat{e}_a^{(I)}$ or $\hat{e}_a^{(V)}$ ($a = 1, 2, 3$). The interaction-basis vectors $\hat{e}_{1,3}^{(I)}$ can be obtained by rotating the (vacuum) mass-basis vectors $\hat{e}_{1,3}^{(V)}$ by $2\theta_v$ about $\hat{e}_2^{(I)} = \hat{e}_2^{(V)}$. (*a*) The polarization vector $\vec{P}_\omega$, which describes a neutrino initially in $|\nu_e\rangle$, and its precession about $\vec{B}$. The projection of this precession motion onto the $\hat{e}_3^{(I)}$ axis represents the flavor oscillation of the neutrino. (*b*) The precession of $\vec{P}_\omega$ in the presence of ordinary matter. If the matter density varies slowly, the angle between $\vec{P}_\omega$ and $\vec{H} = \omega\vec{B} + \lambda\vec{L}$ remains constant. This represents adiabatic MSW flavor transformation of the neutrino. Also see Reference 123 for a more detailed discussion on this geometric interpretation.

ratios of $-\omega$ and $-\lambda$, respectively. (Note that a real magnetic spin $\mathbf{s}$ with the gyromagnetic ratio γ in the presence of magnetic field $\mathbf{B}$ obeys the EoM $\dot{\mathbf{s}} = -\gamma\mathbf{B} \times \mathbf{s}$.) Equivalently, $\vec{P}_\omega$ behaves as a "magnetic spin" coupled to the total "magnetic field" $\vec{H} = \omega\vec{B} + \lambda\vec{L}$ with the "gyromagnetic ratio" of -1. This picture allows geometric interpretations for both vacuum oscillations and MSW flavor transformation (**Figure 2**).

We give a few comments about the polarization vector notation before we consider neutrino self-coupling. First, the polarization vector notation is fully equivalent to the neutrino flavor isospin notation (68), where the neutrino flavor isospin for a neutrino or antineutrino is $\vec{s}_\omega = \frac{1}{2}\frac{\vec{P}_\omega}{|\vec{P}_\omega|}$. Second, in Equation 6 $\vec{P}_\omega$ is defined with a minus sign for the antineutrino. Although the physics is not changed by the choice of notation, our definition of $\vec{P}_\omega$ is convenient for analyzing collective neutrino oscillations when both neutrinos and antineutrinos are present (see Section 2.3). This notation has been adopted in the most recent literature.

2.3. Synchronized Neutrino Oscillations, Corotating Frames, and Matter Effects

The neutrino self-coupling is represented as the coupling between polarization vectors in Equation 7. Let us consider a homogeneous and isotropic neutrino gas with $\lambda = 0$ and μ constant. Use of Equation 7 shows that the so-called total energy of the "magnetic spins,"

$$\mathcal{E} = \int_{-\infty}^{\infty} \omega(\vec{P}_\omega \cdot \vec{B})\mathrm{d}\omega + \frac{\mu}{2}\vec{D}^2, \qquad 8.$$

is constant in this case (68). Here the first term is the total energy of coupling between the so-called "magnetic field" and "spins," and the second term is the total spin-spin coupling energy. If μ is large, then $|\vec{D}| \simeq \sqrt{2\mathcal{E}/\mu}$ is approximately constant. This implies that a dense neutrino gas can experience self-maintained coherent oscillations (58). For example, if a dense neutrino gas consists initially of neutrinos of the same flavor (so that all $\vec{P}_\omega$ are initially aligned), then the flavor evolution of these neutrinos is coherent (i.e., all $\vec{P}_\omega$ remain aligned) even when they have different energies. This phenomenon has been termed synchronized neutrino oscillation because all neutrinos (and antineutrinos) in such a system oscillate collectively with (angular) frequency $\Omega_{\rm sync}$. The synchronized oscillation frequency is an average of all ωs (62):

$$\Omega_{\rm sync} = |\vec{D}|^{-2} \int_{-\infty}^{\infty} (\vec{D} \cdot \vec{P}_\omega)\omega \mathrm{d}\omega. \quad 9.$$

Consider another neutrino system that is similar to the synchronized system discussed above except that the oscillation frequency ω of each neutrino or antineutrino is shifted by a common value ω_0. Any polarization vector, say $\vec{P}_\omega$, in this system should move in a way similar to $\vec{P}_{\omega-\omega_0}$ in the synchronized system, except for an extra precession about $\vec{B}$ with frequency ω_0. In other words, this system behaves just like the synchronized system in a reference frame that rotates about $\vec{B}$ with frequency ω_0. Indeed, in this noninertial corotating frame (68), each polarization vector is coupled to a nonphysical field $-\omega_0 \vec{B}$, and Equation 7 (with $\lambda = 0$) becomes

$$\dot{\vec{P}}_\omega = [(\omega - \omega_0)\vec{B} + \mu \vec{D}] \times \vec{P}_\omega. \quad 10.$$

Therefore, this neutrino system experiences synchronized flavor transformation, just like the one discussed above, but with $\Omega_{\rm sync}$ shifted by ω_0. Because the oscillation frequency ω can be shifted to any value by using an appropriate corotating frame, the criterion for synchronization should not be $\mu \gg |\langle\omega\rangle|$ but rather $\mu \gg \Delta\omega$ (68), where $\langle\omega\rangle$ and $\Delta\omega$ are the average value and the spread in ω for the neutrino system, respectively. Note that the reversal of the direction of $\vec{P}_\omega$ for the antineutrino (Equation 6) is important when the corotating frame is used. If this reversal is not incorporated into the definition of $\vec{P}_\omega$, then the direction of $\vec{P}_\omega$ also changes if the sign of the corresponding vacuum oscillation frequency changes on transformation to the corotating frame.

Note that if $\vec{B}$ is parallel to $\vec{L}$, the matter effect can be completely removed by transforming to an appropriate corotating frame. For a general case, on transforming to the corotating frame Equation 7 becomes

$$\dot{\vec{P}}_\omega = (\omega\vec{B} + \mu \vec{D}) \times \vec{P}_\omega \quad \text{and} \quad \dot{\vec{B}} = -\lambda \vec{L} \times \vec{B}. \quad 11.$$

The matter effect does not disappear here but rather causes $\vec{B}$ to rotate in the corotating frame. If the matter density is large ($\lambda \gg |\omega|$), however, the fast-rotating $\vec{B}$ in Equation 11 can be replaced by $(\vec{B} \cdot \vec{L})\vec{L}$ for collective neutrino oscillations (68). In other words, for collective neutrino oscillations and in the presence of large matter density, matter effects may be ignored and the effective neutrino mixing parameters become $\theta_{\rm eff} \simeq 0$ and $\Delta m^2_{\rm eff} = \Delta m^2 \cos 2\theta_{\rm v}$.

2.4. Solving for Supernova Neutrino Flavor Evolution

Although collective neutrino oscillations may occur in any environment in which neutrino fluxes are significant, recent studies of this phenomenon have focused on the core-collapse supernova environment. The supernova environment is far more complex than the early universe, in part because of its inhomogeneity and anisotropy. This complexity is enhanced because neutrinos of different flavors and energies and propagating in different directions are coupled by neutrino self-interaction (50). Full simulations of neutrino oscillations with neutrino self-interaction in a general supernova environment are beyond current numerical capabilities. Here we briefly discuss two schemes commonly used in investigating collective neutrino oscillations. Both of these schemes employ the neutrino bulb model, in which the supernova environment is spherically symmetric

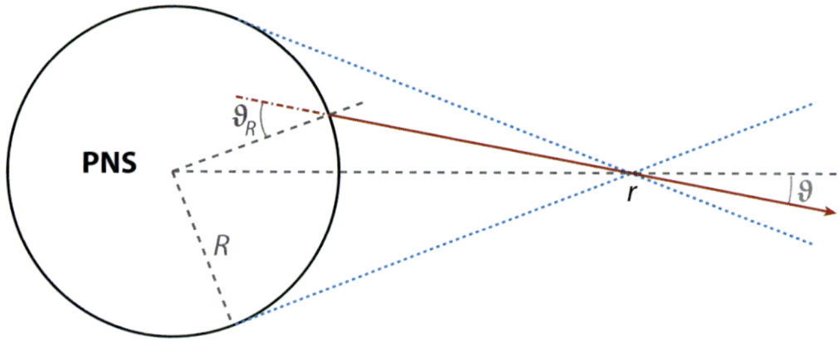

Figure 3

The geometric layout of the neutrino bulb model. In this model all neutrinos are emitted half-isotropically from the surface (neutrino sphere) of the protoneutron star (PNS), which has radius R. Spherical symmetry and isotropic emission on the neutrino sphere imply that all neutrinos with the same initial flavor, energy, and emission angle ϑ_R have identical flavor evolution histories. The neutrino polarization vector $\vec{P}_{\omega,\vartheta_R}(r)$ is uniquely determined by ω, ϑ (or ϑ_R), and r. Here ϑ is the angle at radius r between the neutrino trajectory direction and the radial direction. Figure adapted from figure 1 of Reference 70. Copyright 2006 by the American Physical Society.

around the center of the protoneutron star (PNS) (**Figure 3**). The polarization vectors in this model obey the EoM

$$\cos\vartheta \frac{\mathrm{d}}{\mathrm{d}r}\vec{P}_{\omega,\vartheta_R}(r) = [\omega\vec{B} + \lambda(r)\vec{L} + \vec{H}_{\nu\nu,\vartheta_R}(r)] \times \vec{P}_{\omega,\vartheta_R}(r), \qquad 12.$$

where

$$\vec{H}_{\nu\nu,\vartheta_R}(r) = \sqrt{2}G_{\mathrm{F}}n_{\bar{\nu}_e}(R)\int_{-\infty}^{\infty}\mathrm{d}\omega'\int_{\cos\vartheta_{\max}}^{1}\mathrm{d}(\cos\vartheta')(1-\cos\vartheta\,\cos\vartheta')\vec{P}_{\omega',\vartheta'_R}(r). \qquad 13.$$

Note that $\vartheta = \arcsin(R\sin\vartheta_R/r)$ and has the maximum value $\vartheta_{\max} = \arcsin(R/r)$ in the neutrino bulb model (**Figure 3**). Also note that here we choose to normalize $\vec{P}_\omega$ by $n_{\bar{\nu}_e}(R) = L_{\bar{\nu}_e}/(2\pi R^2\langle E_{\bar{\nu}_e}\rangle)$, the total number density of $\bar{\nu}_e$ at the neutrino sphere, where $L_{\bar{\nu}_e}$ and $\langle E_{\bar{\nu}_e}\rangle$ are the energy luminosity and the spectrum-averaged energy of $\bar{\nu}_e$ at the neutrino sphere, respectively. For the neutronization-burst epoch where the ν_e flux is much larger than the fluxes of all other neutrino species, $\vec{P}_\omega$ should be normalized by $n_{\nu_e}(R)$ instead. In typical numerical simulations, R is in the range of 10 to 60 km, and the luminosities and the average energies of neutrinos are in the ranges of 10^{50} to 10^{53} erg s^{-1} and 10 to 30 MeV, respectively.

Equation 12 can be solved numerically without any further assumptions. This is known as the multiangle scheme. The other scheme is the so-called single-angle scheme (50). In this latter scheme it is assumed that $\vec{P}_{\omega,\vartheta_R}(r) = \vec{P}_\omega(r)$ is the same for different neutrino trajectories. There are several variants of the single-angle scheme that lead to qualitatively similar results. In one of the variants, $\vec{P}_\omega(r)$ is computed along the radial direction ($\vartheta_R = 0$), and Equation 12 becomes

$$\frac{\mathrm{d}}{\mathrm{d}r}\vec{P}_\omega(r) = [\omega\vec{B} + \lambda(r)\vec{L} + \sqrt{2}G_{\mathrm{F}}n_{\bar{\nu}_e}\mathcal{D}(r/R)\vec{D}] \times \vec{P}_\omega(r), \qquad 14.$$

where the geometric factor $\mathcal{D}(r/R) = \frac{1}{2}[1-\sqrt{1-(\frac{R}{r})^2}]^2$ partially accounts for the angle effect and geometric dilution of the neutrino fluxes in the neutrino bulb model. Comparing Equations 7 and 14 shows that the flavor evolution of neutrinos in the single-angle scheme is equivalent to that of a homogeneous and isotropic neutrino gas expanding with "time" r. In this analogy, the strength of the neutrino self-coupling is $\mu(r) = \sqrt{2}G_{\mathrm{F}}n_{\bar{\nu}_e}(R)\mathcal{D}(r/R)$. The radial direction is a rather special direction in the neutrino bulb model. In another variant of the single-angle scheme, it is assumed that all neutrinos are emitted with $\vartheta_R = \pi/4$ (75). Alternatively, $\vec{H}_{\nu\nu,\vartheta}$ can be averaged over neutrino trajectories (70, 90). Each of these variants also leads to

Equation 14 when $r \gg R$. However, in this limit $\mathcal{D}(r/R) \to \frac{1}{4}\left(\frac{R}{r}\right)^4$ for these variants instead of $\frac{1}{8}\left(\frac{R}{r}\right)^4$ in the single-angle scheme that employs the radial trajectory. These variants can improve the agreement between the results of the single-angle and multiangle calculations (90). Although the single-angle scheme is frequently used for its simplicity, it must be emphasized that it misses some of the important properties (e.g., anisotropy) of the neutrino bulb model and can lead to incorrect results when such properties play important roles in collective neutrino oscillations (see Section 5).

3. SIMPLE BIPOLAR NEUTRINO SYSTEMS

3.1. Bipolar Systems and the Flavor Pendulum

To illustrate another important example of collective neutrino oscillations, let us consider a homogeneous and isotropic gas that initially consists of monoenergetic ν_e and $\bar{\nu}_e$. This neutrino system is represented by two polarization vectors, $\vec{P}_\omega$ and $\vec{P}_{-\omega}$, for the neutrino and the antineutrino, respectively. We assume that $\lambda = 0$, that μ is fixed, and that $|\vec{P}_\omega| = (1+\varepsilon)|\vec{P}_{-\omega}|$, where ε is the fractional excess of neutrinos over antineutrinos. At $t = 0$, $\vec{P}_\omega$ points in the direction of $\hat{e}_3^{(\mathrm{I})}$, which is tilted away from $\hat{e}_3^{(\mathrm{V})}$ by $2\theta_\mathrm{v}$, and $\vec{P}_{-\omega}$ points in the direction opposite to $\vec{P}_\omega$. Neutrino systems that are represented by two nearly oppositely directed polarization vector groups are termed bipolar systems. Using the corotating frame technique, the discussion in this section easily can be applied to, for instance, a gas consisting initially of ν_e and ν_μ with energies $E_{\nu_e} \neq E_{\nu_\mu}$.

A peculiar case is one in which $\varepsilon = 0$ and $\vec{P}_\omega$ is initially aligned with $\hat{e}_3^{(\mathrm{V})} = -\vec{B}$ (i.e., $\theta_\mathrm{v} = 0$). On the one hand, use of energy conservation (Equation 8) demonstrates (68) that if $\omega > 0$, neither $\vec{P}_\omega$ nor $\vec{P}_{-\omega}$ can move, and so the initial configuration of the system is absolutely stable. On the other hand, if $\omega < 0$ and $\mu \gg |\omega|$, $\vec{P}_\omega$ and $\vec{P}_{-\omega}$ can nearly swap their directions (but with a slight bend toward each other). This implies that the initial configuration of the system is unstable. Therefore, the $\nu_e - \bar{\nu}_e$ system may experience insignificant flavor oscillations when $\Delta m^2 > 0$ and $\theta_\mathrm{v} \ll 1$. However, this system can experience significant flavor oscillations when $\Delta m^2 < 0$ and $\theta_\mathrm{v} \ll 1$. Such collective neutrino oscillations are known as bipolar oscillations.

The EoMs of the simple bipolar system,

$$\dot{\vec{P}}_\omega = (\omega\vec{B} + \mu\vec{D}) \times \vec{P}_\omega, \quad \text{and} \quad \dot{\vec{P}}_{-\omega} = (-\omega\vec{B} + \mu\vec{D}) \times \vec{P}_{-\omega}, \tag{15}$$

have been solved analytically (58, 59). Instead of presenting this solution, let us rewrite Equation 15 as (71)

$$\dot{\vec{D}} = \mu^{-1}\vec{q} \times \vec{g} \quad \text{and} \quad \vec{D} = \mu^{-1}\vec{q} \times \dot{\vec{q}} + \sigma_\mathrm{s}\vec{q}, \tag{16}$$

where $\vec{q} = \vec{Q}/|\vec{Q}| = (\vec{P}_\omega - \vec{P}_{-\omega} - \frac{\omega}{\mu}\vec{B})/|\vec{Q}|$, $\vec{g} = -\mu\omega|\vec{Q}|\vec{B}$, and $\sigma_\mathrm{s} = \vec{q}\cdot\vec{D}$ is constant. Equation 16 describes the motion of a fictitious gyroscopic pendulum, or flavor pendulum, with total angular momentum $\vec{D}$ in a uniform gravitational field in which the acceleration of gravity is $\vec{g}$. The pendulum consists of a massless rod with a point particle of mass μ^{-1} and spin σ_s attached to the end of the rod at position $\vec{q}$. The flavor pendulum that represents a symmetric bipolar system ($\varepsilon = 0$) has no internal spin. The stable and unstable configurations of the system discussed above correspond to the lowest and highest positions, respectively, that the pendulum can reach.

Generally, the flavor pendulum can experience two kinds of motion: a precession about the $\vec{B}$ axis and a nutation around the average precession track. The nutation motion corresponds to bipolar neutrino oscillations. However, like a child's top, the flavor pendulum can "defy gravity" and precess almost uniformly if μ is large enough. Specifically, the flavor pendulum can become a so-called sleeping top and does not fall from its highest position if (71, 73)

$$\mu > \mu_{\rm cr} \equiv \frac{2|\omega|}{(\sqrt{1+\varepsilon}-1)^2}. \qquad 17.$$

This precession behavior of the flavor pendulum in the large μ limit represents synchronized oscillations of the bipolar system with $\Omega_{\rm sync} = (1 + 2\varepsilon^{-1})\omega$.

3.2. Bipolar Systems with Slowly Decreasing Neutrino Density

Let us now focus on the IH case with $\theta_{\rm v} \ll 1$. Significant neutrino oscillations can occur in this case. If the neutrino density decreases, the mass μ^{-1} of the pendulum becomes larger. The swing amplitude of the pendulum that represents a symmetric bipolar system decreases as μ decreases. The maximum swing amplitude of this pendulum can be found in the adiabatic limit where μ changes slowly (73). In this case, $\vec{P}_\omega$ and $\vec{P}_{-\omega}$ become aligned and antialigned, respectively, with $\vec{B}$ (i.e., $\nu_e \to \nu_2$ and $\bar{\nu}_e \to \bar{\nu}_2$) as μ decreases toward 0.

The asymmetric bipolar system is more interesting. Neutrino oscillations in this case are synchronized, and the flavor pendulum precesses uniformly in the limit $\mu \gg |\omega|$. If μ decreases very slowly, the flavor pendulum can still experience a nearly pure precession motion for any given μ. In this case, $\vec{P}_\omega$ and $\vec{P}_{-\omega}$ lie in the same plane with $\vec{B}$, and their directions can be readily solved for (**Figure 4**) (73, 79). Assuming $\varepsilon > 0$, $\vec{P}_{-\omega}$ becomes antialigned with $\vec{B}$ (i.e., $\bar{\nu}_e \to \bar{\nu}_2$) as $\mu \to 0$. Meanwhile the direction of $\vec{P}_\omega$ can be determined from the constancy of $\vec{D} \cdot \vec{B}$ (71; also see Equation 16, where $\vec{g} \propto \vec{B}$).

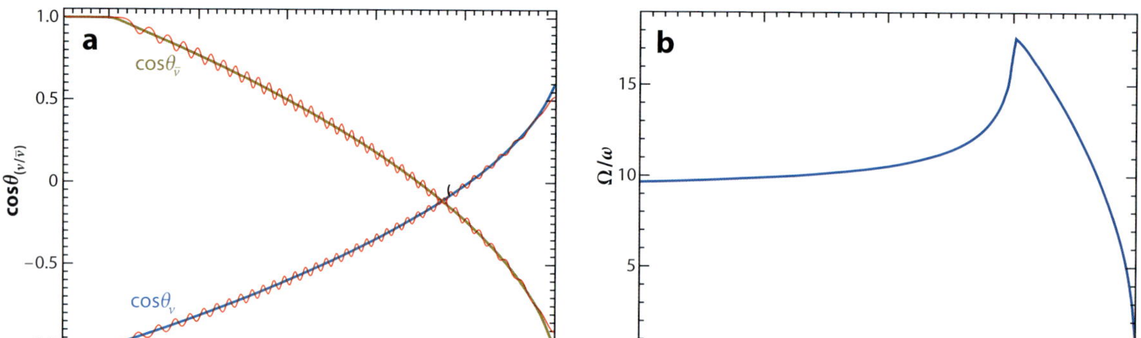

Figure 4

Evolution with decreasing μ for a simple bipolar system that consists initially of monoenergetic ν_e and $\bar{\nu}_e$. The neutrino excess is $\varepsilon = 0.25$, and the mixing parameters are $\theta_{\rm v} = 0.01$ and $\Delta m^2 < 0$. (*a*) The configuration of the polarization vectors, where θ_ν ($\theta_{\bar{\nu}}$) is the angle between $\vec{P}_\omega$ ($\vec{P}_{-\omega}$) and $\vec{B} = -\hat{e}_3^{(\rm V)}$. The thick colored lines represent an assumed pure precession motion for the flavor pendulum, and the thin wavy red lines represent a case where μ decreases slowly and linearly with time. At $\mu \gtrsim \mu_{\rm cr}$ the flavor pendulum is a so-called sleeping top, and there are no flavor oscillations. When $\mu < \mu_{\rm cr}$, the flavor pendulum experiences precession as well as nutation around the average precession track. (*b*) The precession frequency Ω of the flavor pendulum, assuming that it is executing pure precession. In this case, the value Ω approaches $\Omega_{\rm sync}$ and ω in the limits $\mu \to \infty$ and $\mu \to 0$, respectively. The analytical solution for the pure precession motion of the polarization vectors can be found in Reference 79. This figure is adapted from figures 3*a* and 4 in Reference 73. Copyright 2007 by the American Physical Society.

4. ISOTROPIC AND HOMOGENEOUS NEUTRINO GASES

4.1. Static Solutions for the Neutrino Flavor Evolution Equation

Here we consider a homogeneous and isotropic gas that consists of neutrinos with continuous energy spectra. Let us first seek static solutions to the EoM for the polarization vectors (Equation 7) for the case in which both the matter density and the neutrino densities are constant. One possibility is that the solution is stationary and that the $\vec{P}_\omega$s do not evolve with time t. This solution is only possible if the "spin" $\vec{P}_\omega$ is parallel to the corresponding total "magnetic field" $\vec{H}_\omega = \omega\vec{B} + \lambda\vec{L} + \mu\vec{D}$. For this case, and noting that $\vec{H}_\omega$ does not depend on $\vec{P}_\omega$, we can obtain the following equations for the total polarization vector $\vec{D}$ (76):

$$D_1 = (\lambda \sin 2\theta_{\rm v} + \mu D_1) \int_{-\infty}^{\infty} \frac{\epsilon_\omega |\vec{P}_\omega|}{|\vec{H}_\omega|} \mathrm{d}\omega, \tag{18a}$$

$$D_2 = \mu D_2 \int_{-\infty}^{\infty} \frac{\epsilon_\omega |\vec{P}_\omega|}{|\vec{H}_\omega|} \mathrm{d}\omega, \tag{18b}$$

and

$$D_3 = \int_{-\infty}^{\infty} (-\omega + \lambda \cos 2\theta_{\rm v} + \mu D_3) \frac{\epsilon_\omega |\vec{P}_\omega|}{|\vec{H}_\omega|} \mathrm{d}\omega, \tag{18c}$$

where $D_a = \vec{D} \cdot \hat{e}_a^{(\rm V)}$ $(a = 1, 2, 3)$ and $\epsilon_\omega = +1$ (-1) if $\vec{P}_\omega$ is aligned (antialigned) with $\vec{H}_\omega$. The stationary solution to the EoM for $\vec{P}_\omega$ can be found from Equations 18*a*–*c* and the alignment condition. We note that Equations 18*a* and 18*b* generally imply that $D_2 = 0$ if $\lambda \neq 0$, which in turn implies that $\vec{P}_\omega \cdot \hat{e}_2^{(\rm V)} = 0$ for any ω.

Equations 18*a* and 18*b* become equivalent when $\lambda = 0$, and $\vec{D}$ is underconstrained by Equations 18*a*–*c* in this case because when $\lambda = 0$, the EoM for $\vec{P}_\omega$ possesses a rotational symmetry about $\vec{B}$. In other words, if $\{\vec{P}_\omega(t)|\forall\omega\}$ (meaning the polarization vectors for all frequencies ω at a given time t) solves the EoM, then $\{\vec{P}'_\omega(t)|\forall\omega\}$ also solves the EoM, in which $\vec{P}'_\omega(t)$ is obtained from $\vec{P}_\omega(t)$ by rotation about $\vec{B}$ by an arbitrary angle ϕ. Here ϕ is independent of ω and t. This is a generalization of the rotational symmetry of the flavor pendulum about the "gravity" vector $\vec{g} \propto \vec{B}$. Such a continuous symmetry in the EoMs generally implies the existence of collective motion and a conservation law. For example, a translational symmetry of a group of particles along some direction implies the possibility of the collective motion of the particles in that direction and the conservation of the total momentum of the particles in the same direction. As discussed in Section 3, for the simple bipolar system this rotational symmetry implies the possibility of a pure precession of the flavor pendulum in which both polarization vectors of the system precess with the same frequency Ω. Therefore, it is natural to seek a static solution to the EoM in which $\lambda = 0$, and μ is a constant, and all the $\vec{P}_\omega$s precess about $\vec{B}$ with the same frequency Ω. In this case, all the $\vec{P}_\omega$s are stationary in a corotating frame, which rotates about $\vec{B}$ with frequency Ω. In this corotating frame, $\vec{P}_\omega$ is either aligned or antialigned with $\vec{H}_\omega = (\omega - \Omega)\vec{B} + \mu\vec{D}$, and the components of $\vec{D}$ in the corotating frame can be found from equations similar to Equations 18*a*–*c*. These equations can be recast as the following two simple sum rules (74):

$$1 = \int_{-\infty}^{\infty} \frac{\epsilon_\omega |\vec{P}_\omega|}{\sqrt{[(\Omega - \omega)/\mu + D_3]^2 + D_\perp^2}} \mathrm{d}\omega \tag{19}$$

and

$$\Omega = \int_{-\infty}^{\infty} \frac{\epsilon_\omega \omega |\vec{P}_\omega|}{\sqrt{[(\Omega - \omega)/\mu + D_3]^2 + D_\perp^2}} \mathrm{d}\omega, \tag{20}$$

where $D_\perp$ is the component of $\vec{D}$ that is perpendicular to $\hat{e}_3^{(V)}$. The rotational symmetry of the system about $\vec{B}$ implies that $D_3 = -\vec{D} \cdot \vec{B}$ is constant.

4.2. Adiabatic Solutions and the Spectral Swap/Split

If λ and μ vary slowly with time, adiabatic solutions that correspond to the static solutions discussed above can be obtained. We term them the (adiabatic, ν-enhanced) MSW solution and the (adiabatic) precession solution. **Figure 5** shows these two adiabatic solutions for a single-angle scheme, together with the corresponding numerical solutions for Equation 14. In these calculations it is assumed that (*a*) the most abundant neutrino species at the neutrino sphere are ν_e and $\bar{\nu}_e$ and (*b*) $\theta_v \ll 1$. For the precession solution the matter field is removed with an appropriate corotating frame. The flavor pendulum model provides insight into how the numerical solution evolves from the MSW solution toward the precession solution. In the IH case, a large matter density essentially keeps the flavor pendulum (with $\varepsilon > 0$) near its highest position. This configuration becomes unstable when $\mu < \mu_{cr}$. In the NH case, the ν-enhanced MSW flavor transformation raises the flavor pendulum to near its highest position, and the configuration again becomes unstable when $\mu < \mu_{cr}$.

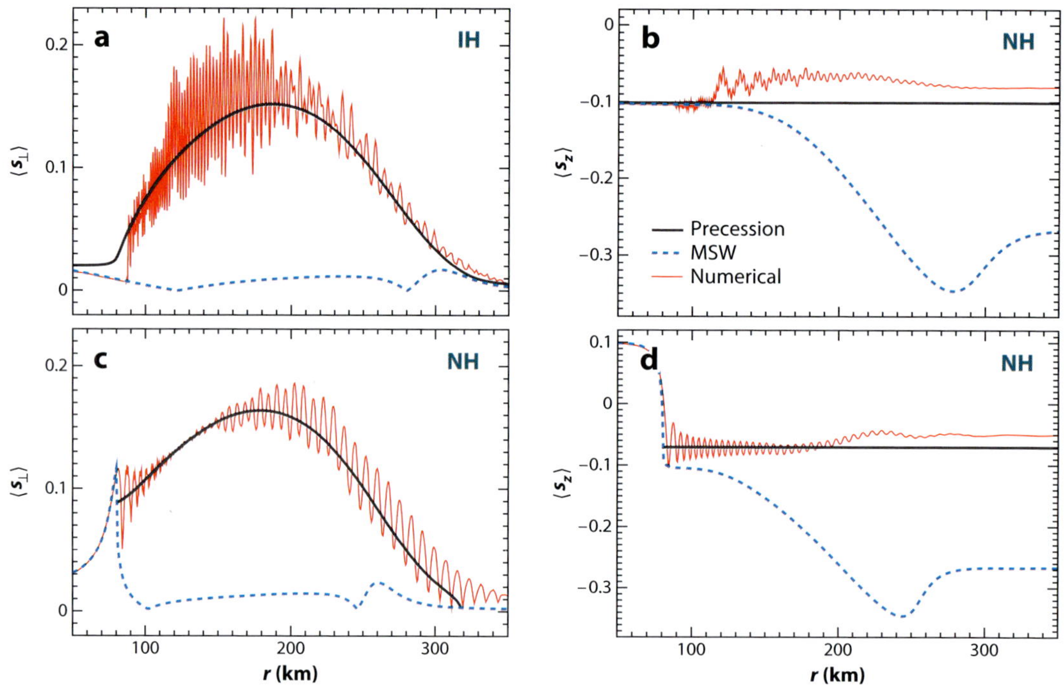

Figure 5

Comparison of the numerical solution, the ν-enhanced Mikheyev-Smirnov-Wolfenstein (MSW) solution, and the precession solution in a single-angle scheme, where $\theta_v = 0.1$, $\langle s_\perp \rangle \propto D_\perp$, and $\langle s_z \rangle \propto D_3$. The numerical solution initially follows the MSW solution. (*a*) In the inverted neutrino mass hierarchy (IH) case, the MSW solution becomes unstable at $r \simeq 88$ km, and thereafter the numerical solution shows oscillations around the precession solution. (*b*) In the normal neutrino mass hierarchy (NH) case, the numerical solution follows the MSW solution through the resonance (where $\langle s_z \rangle \simeq 0$) before it shifts to follow the precession solution track. Figure adapted from figure 1 of Reference 76. Copyright 2007 by the American Physical Society.

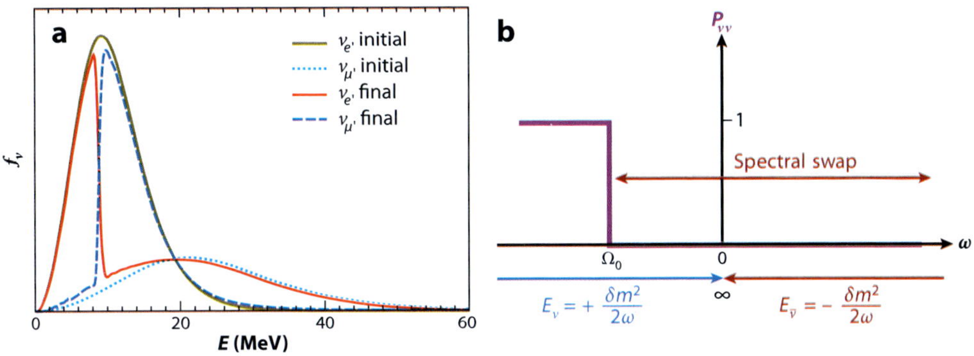

Figure 6

Illustration of the stepwise spectral swap phenomenon in the two-flavor mixing case with $\Delta m^2 < 0$ (inverted neutrino mass hierarchy) that was discovered by Duan et al. (70). (*a*) The stepwise swapping of ν_e and ν_μ energy spectra about $E_s \simeq 9$ MeV in a single-angle scheme. The spectra of $\bar{\nu}_e$ and $\bar{\nu}_\mu$ are nearly fully swapped in this calculation. (*b*) The corresponding survival probability $P_{\nu\nu}$, which is a step-like function of ω. For the normal neutrino mass hierarchy case, the step-like structure of $P_{\nu\nu}(\omega)$ is pushed rightward to $\Omega_0 > 0$. Because $E_s = |\frac{\Delta m^2}{2\Omega_0}|$ splits a neutrino spectrum into two parts with different flavors, this phenomenon is also sometimes termed spectral split. Reprinted with permission from Reference 124. Copyright 2009, American Institute of Physics.

Just as in the conventional adiabatic MSW flavor transformation case (**Figure 2**), in the adiabatic precession solution $\vec{P}_\omega$ follows $\tilde{\vec{H}}_\omega$, whose direction (and magnitude) changes as μ decreases. This induces neutrino flavor transformation. Specifically, as $\mu \to 0$, $\tilde{\vec{H}}_\omega \to (\omega - \Omega_0)\vec{B}$, where $\Omega_0 = \Omega(\mu = 0)$. This means that the adiabatic collective precession mode converts the initial ν_e into the mass state $|\nu_1\rangle$ or $|\nu_2\rangle$, depending on whether ω is smaller or larger than Ω_0 (70). This phenomenon, known as the stepwise spectral swap or spectral split, is most dramatic when $\theta_v \ll 1$ (**Figure 6**). The swap/split energy $E_s = |\frac{\Delta m^2}{2\Omega_0}|$ can be determined from the constancy of $\vec{D} \cdot \vec{B}$ (74).

4.3. Precession Solution in the Three-Flavor Mixing Scenario

The neutrino polarization vector defined in Equation 6 can be easily generalized to the three-flavor mixing scenario by replacing the Pauli matrices with the Gell-Mann matrices Λ_a ($a = 1, 2, \ldots, 8$) (83, 123). However, because an eight-dimensional polarization vector, or Bloch vector, cannot be as easily visualized as its three-dimensional counterpart, we discuss the collective precession mode by using the matrix formalism. To this end, we define the polarization matrix $\mathsf{P}_\omega = \frac{1}{2}\sum_{a=1}^{8}(P_{\omega,a}\Lambda_a)$, where $P_{\omega,a}$ is the ath component of the Bloch vector $\vec{P}_\omega$. We note that here the definition of the Bloch vector $\vec{P}_\omega$ follows the same sign convention for antineutrinos as in Equation 6. The polarization matrix obeys the EoM

$$\mathrm{i}\dot{\mathsf{P}}_\omega = [\omega_{\mathrm{L}}\mathsf{B}_{\mathrm{L}} + \omega_{\mathrm{H}}\mathsf{B}_{\mathrm{H}} + \mu\mathsf{D}, \mathsf{P}_\omega], \qquad 21.$$

where $\mathsf{D} = \int_{-\infty}^{\infty}\mathsf{P}_\omega \mathrm{d}\omega$ is the total polarization matrix. In Equation 21, $\omega_{\mathrm{L}} = \pm\frac{\delta m^2}{2E}$ and $\mathsf{B}_{\mathrm{L}} = -\frac{1}{2}\Lambda_3$ (in the mass basis) correspond to the small mass splitting, which we define as $\delta m^2 \equiv m_2^2 - m_1^2 \simeq \Delta m_\odot^2$. Also in Equation 21, $\omega_{\mathrm{H}} = \omega = \pm\frac{\Delta m^2}{2E}$ and $\mathsf{B}_{\mathrm{H}} = -\frac{1}{\sqrt{3}}\Lambda_8$ (in the mass basis) correspond to the large mass splitting, which we define as $\Delta m^2 = m_3^2 - \frac{1}{2}(m_1^2 + m_2^2) \simeq \pm\Delta m_{\mathrm{atm}}^2$. For simplicity we have ignored the matter field because it can be removed by employing the corotating frame technique (84).

The static precession solution (with constant μ) can be obtained by assuming that $\{\tilde{\mathsf{P}}_\omega, \Omega_{\rm L}, \Omega_{\rm H} \mid \forall\omega\}$ solves the equation

$$[\tilde{\mathsf{H}}_\omega, \tilde{\mathsf{P}}_\omega] = [(\omega_{\rm L} - \Omega_{\rm L})\mathsf{B}_{\rm L} + (\omega_{\rm H} - \Omega_{\rm H})\mathsf{B}_{\rm H} + \mu\tilde{\mathsf{D}}, \tilde{\mathsf{P}}_\omega] = 0, \qquad 22.$$

where $\tilde{\mathsf{P}}_\omega$, $\Omega_{\rm L}$, and $\Omega_{\rm H}$ are constant and $\tilde{\mathsf{D}} = \int_{-\infty}^{\infty} \tilde{\mathsf{P}}_\omega \, d\omega$. In the two-flavor mixing scenario ($\mathsf{B}_{\rm L} = 0$), Equation 22 corresponds to the condition that the "spin" $\vec{\tilde{P}}_\omega$ is stationary in a corotating frame and parallel to the total "magnetic field" $\vec{\tilde{H}}_\omega$ in this reference frame. Equation 22 is known as the precession ansatz (84) because it implies a static precession solution to the EoM (Equation 21). This solution, $\{\mathsf{P}_\omega(t), \Omega_{\rm L}, \Omega_{\rm H} \mid \forall\omega\}$, can be written as

$$\mathsf{P}_\omega(t) = \exp[-\mathrm{i}(\Omega_{\rm L}\mathsf{B}_{\rm L} + \Omega_{\rm H}\mathsf{B}_{\rm H})t]\tilde{\mathsf{P}}_\omega \exp[\mathrm{i}(\Omega_{\rm L}\mathsf{B}_{\rm L} + \Omega_{\rm H}\mathsf{B}_{\rm H})t]. \qquad 23.$$

Because $\tilde{\mathsf{P}}_\omega$ and $\tilde{\mathsf{H}}_\omega$ commute, they can be simultaneously diagonalized by a unitary matrix X. Therefore, we have $\mathsf{X}\tilde{\mathsf{H}}_\omega\mathsf{X}^\dagger = \mathrm{diag}[\tilde{h}_{\omega,1}, \tilde{h}_{\omega,2}, \tilde{h}_{\omega,3}]$, where $\tilde{h}_{\omega,1} < \tilde{h}_{\omega,2} < \tilde{h}_{\omega,3}$, and $\mathsf{X}\tilde{\mathsf{P}}_\omega\mathsf{X}^\dagger = \mathrm{diag}[\tilde{p}_{\omega,1}, \tilde{p}_{\omega,2}, -(\tilde{p}_{\omega,1} + \tilde{p}_{\omega,2})]$. When μ varies slowly with t, the adiabatic precession solution can be obtained from the static precession solutions (with different μ) by using the adiabatic ansatz (84)

$$\frac{\partial}{\partial\mu}\tilde{p}_{\omega,1} = \frac{\partial}{\partial\mu}\tilde{p}_{\omega,2} = 0. \qquad 24.$$

In the two-flavor mixing scenario, the adiabatic ansatz corresponds to the assumption that $\vec{\tilde{P}}_\omega$ remains parallel to $\vec{\tilde{H}}_\omega$. If neutrinos follow the adiabatic precession solution, then as $\mu \to 0$, $\tilde{\mathsf{P}}_\omega$ becomes a diagonal matrix in the mass basis (see Equation 22). The diagonal elements of $\mathsf{P}_\omega|_{\mu=0} = \tilde{\mathsf{P}}_\omega|_{\mu=0}$ are $\tilde{p}_{\omega,1}$, $\tilde{p}_{\omega,2}$, and $-(\tilde{p}_{\omega,1} + \tilde{p}_{\omega,2})$, and these elements have the same order of appearance as the diagonal elements of $\mathsf{H}_\omega|_{\mu=0} = (\omega_{\rm L} - \Omega_{\rm L})\mathsf{B}_{\rm L} + (\omega_{\rm H} - \Omega_{\rm H})\mathsf{B}_{\rm H}$. For example, in the mass basis we have $\mathsf{P}_\omega|_{\mu=0} = \mathrm{diag}[\tilde{p}_{\omega,2}, \tilde{p}_{\omega,1}, -(\tilde{p}_{\omega,1} + \tilde{p}_{\omega,2})]$ if $\tilde{\mathsf{H}}_\omega|_{\mu=0} = \mathrm{diag}[\tilde{h}_{\omega,2}, \tilde{h}_{\omega,1}, \tilde{h}_{\omega,3}]$, where $\tilde{h}_{\omega,1} < \tilde{h}_{\omega,2} < \tilde{h}_{\omega,3}$. The fact that $\mathsf{P}_\omega|_{\mu=0}$ is diagonal in the mass basis implies that there can be multiple spectral swaps/splits in the final neutrino energy spectra. Because $\alpha = \frac{\delta m^2}{|\Delta m^2|} \ll 1$, these spectral swaps/splits form hierarchically and appear at different neutrino densities (**Figure 7**).

5. ANISOTROPIC AND/OR INHOMOGENEOUS NEUTRINO GASES

5.1. Kinematic Decoherence of Collective Neutrino Oscillations

The oscillations of neutrinos with different momenta can become out of phase (i.e., collective oscillations can break down), which is sometimes referred to as kinematic decoherence. Of course, this is not to be confused with quantum decoherence, which can be induced by any neutrino scattering process that changes neutrino momentum. In a homogeneous and isotropic neutrino gas, the condition for kinematic decoherence is $\Delta\omega \gg \mu$. In this limit, the coupling among "spins" is not strong enough to maintain a collective motion, and $\vec{P}_\omega$ precesses about $\vec{B}$ with vacuum oscillation frequency ω. In Section 2.3 we showed that, when $\Delta\omega \ll \mu$, synchronized neutrino oscillations do not decohere kinematically, and $|\vec{D}|$ is approximately constant because of energy conservation. Numerical simulations suggest that bipolar neutrino oscillations are also stable (55, 57), although as yet we know of no conservation law that could explain this phenomenon.

If collective neutrino oscillations also exist in an anisotropic environment, then the wave fronts of the oscillation waves of the neutrinos must coincide with one another. In the neutrino bulb

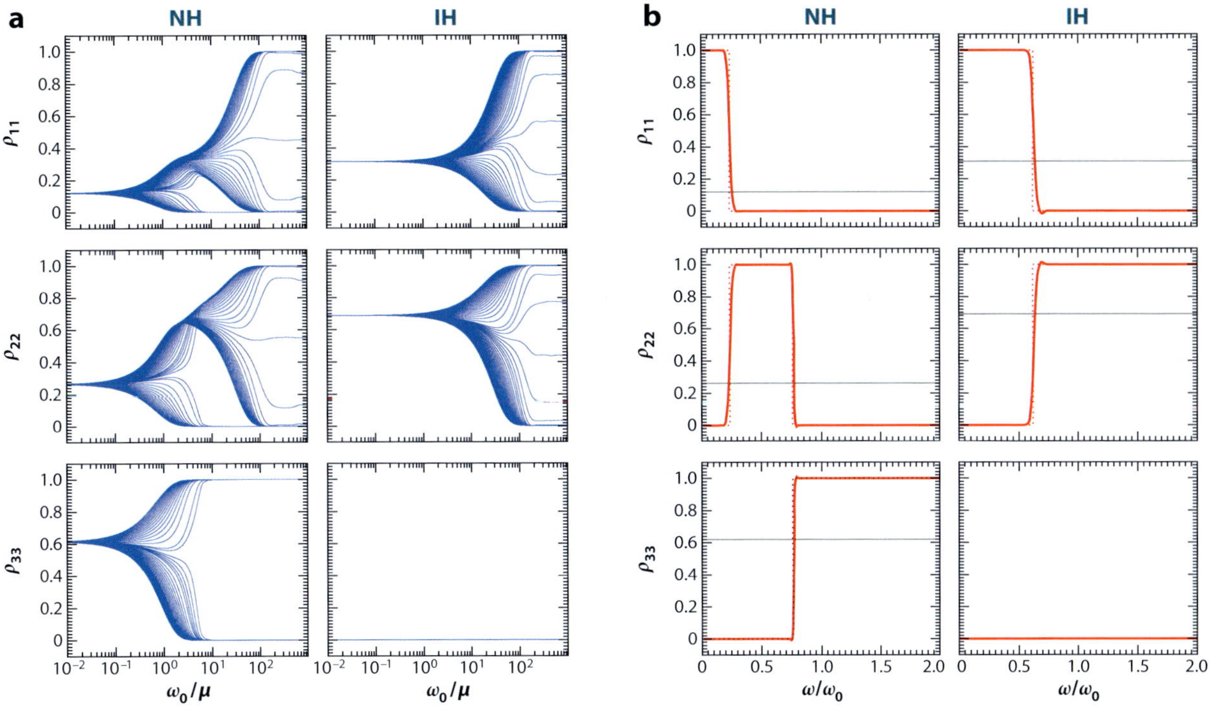

Figure 7

The evolution of the diagonal elements of the neutrino density matrices in the mass basis during the hierarchical formation of spectral swaps/splits in a neutrino gas. Initially P_ω, and the density matrix ρ_ω, is the same across the whole energy spectrum $\omega \in [0, 2\omega_0]$ (*a, thin blue lines*). When $\mu \gg \omega_0$, all neutrinos are in the heaviest eigenstate (i.e., the state with the largest eigenvalue) of $\tilde{\mathsf{H}}_\omega \simeq \mu\tilde{\mathsf{D}}$. For the normal neutrino mass hierarchy (NH) case, two spectral swaps/splits form hierarchically at $\mu \sim \omega_0$ and $\mu \sim \alpha\omega_0$, respectively. Curves in panel *a* give ρ_ω with various values of ω. At $\omega_0 \gg \mu \gg \alpha\omega_0$, the heaviest eigenstate of $\tilde{\mathsf{H}} \simeq (\omega - \Omega_{\mathrm{H},0})\mathsf{B}_\mathrm{H} + \mu\tilde{\mathsf{D}}$ is $|\nu_3\rangle$ for neutrinos with $\omega > \Omega_{\mathrm{H},0}$. Subsequently these neutrinos no longer participate in collective oscillations. At $\mu \ll \alpha\omega_0$, the remaining neutrinos are in $|\nu_2\rangle$ or $|\nu_1\rangle$, depending on whether $\omega_\mathrm{L} = \alpha\omega$ is larger or smaller than $\Omega_{\mathrm{L},0}$. In either case, the state corresponds to the heaviest eigenstate of $\tilde{\mathsf{H}}_\omega \simeq (\omega_\mathrm{L} - \Omega_{\mathrm{L},0})\mathsf{B}_\mathrm{L}$ (ignoring the decoupled state $|\nu_3\rangle$). This leads to two spectral swaps/splits in the final neutrino energy spectra (*b, thick red lines in the NH column*). In the inverted neutrino mass hierarchy (IH) case, the ρ_{33} are initially set to zero for all ω. As a result, $|\nu_3\rangle$ is completely decoupled from the equation of motion, and no spectral swap/split forms at $\mu \sim \omega_0$. The dotted red lines in panel *b* are computed by using the constancy of $\mathrm{Tr}(\mathsf{D}\cdot\Lambda_3)$ and $\mathrm{Tr}(\mathsf{D}\cdot\Lambda_8)$ and by assuming that the spectral swaps/splits are infinitely sharp. Figure reprinted with permission from Reference 85. Copyright 2008 by the American Physical Society.

model these wave fronts are spheres that cocenter with the PNS. However, in this model a neutrino propagating along a nonradial trajectory would travel a longer distance than would a radially propagating neutrino (50). In other words, a nonradially propagating neutrino appears to have a larger oscillation frequency $\omega' = \omega/\cos\vartheta$ along r (see Equation 12). This frequency would enlarge $\Delta\omega$ and would require stronger neutrino self-coupling to maintain collective oscillations among neutrinos propagating along different trajectories. Note that the "magnetic field" $\vec{H}_{\nu\nu,\vartheta}(r)$ generated by other "spins" is trajectory dependent, which adds another potential source of kinematic decoherence.

Raffelt & Sigl (72) pointed out that a neutrino gas with an initially symmetric bipolar configuration (i.e., $\vec{P}_\nu = -\vec{P}_{\bar{\nu}}$) can experience quick kinematic decoherence even in the presence of a small anisotropy, and in this case, both $|\vec{P}_\nu|$ and $|\vec{P}_{\bar{\nu}}|$ evolve toward zero. Later it was found that an asymmetric bipolar neutrino system [i.e., $|\vec{P}_\nu| = (1+\varepsilon)|\vec{P}_{\bar{\nu}}|$ with $\varepsilon \neq 0$] may or may not experience kinematic decoherence, depending on the value of ε (75). For typical choices of other parameters in a neutrino bulb model, kinematic decoherence is suppressed if $\varepsilon \gtrsim 0.3$.

In the above discussion, we have ignored ordinary matter. In the presence of a large matter density, $\vec{P}_\omega$ tends to precess around $-\vec{L}$ with frequency $\omega' = -\lambda + \omega \cos 2\theta_v$. In the isotropic environment, $\omega' \to \omega \cos 2\theta_v$ in the corotating frame that rotates about $\vec{L}$ with frequency λ. In this case, therefore, when neutrinos experience collective oscillations, ordinary matter can be ignored. In an anisotropic environment, such as the neutrino bulb model, we have a frequency of $\omega' = (-\lambda + \omega \cos 2\theta_v)/\cos\vartheta$ along r. Clearly, it is not possible to remove the matter effect for all neutrino trajectories, which implies that collective neutrino oscillations would not exist in a region where the net number density of electrons is much larger than that of neutrinos (91).

5.2. Precession Mode in the Anisotropic Environment

In Section 3 we showed that with $\varepsilon \neq 0$ the flavor pendulum possesses an internal spin. The existence of this internal spin makes it possible for the flavor pendulum to experience a precession motion and a nutation motion simultaneously. These two kinds of motion of the flavor pendulum correspond to the precession mode and the bipolar mode of neutrino oscillations. The findings in References 72 and 75 suggest that bipolar neutrino oscillations, which are the most prominent when ε is small, become noncollective in the anisotropic environment. These findings also suggest that the precession mode, which becomes important for cases with sufficiently large ε, can remain collective in the anisotropic environment. This hypothesis is partially supported by the fact that neutrino oscillations calculated in the single-angle scheme and in the multiangle scheme possess common features, such as the spectral swap/split phenomenon, which—as explained with the single-angle supernova scheme or in homogeneous and isotropic environments—results from neutrino oscillations in the collective precession mode.

The EoM of the polarization vectors in a stationary environment can be written as (114, 116, 125)

$$\hat{\mathbf{p}} \cdot \nabla \vec{P}_{\omega,\hat{\mathbf{p}}}(\mathbf{x}) = [\omega \vec{B} + \vec{H}_{\nu\nu,\hat{\mathbf{p}}}(\mathbf{x})] \times \vec{P}_{\omega,\hat{\mathbf{p}}}(\mathbf{x}), \tag{25.}$$

where $\lambda = 0$ for now, all the polarization vectors are normalized by $n_\nu(\mathbf{x}_0)$, the number density of neutrino species ν at location $\mathbf{x}_0$, and

$$\vec{H}_{\nu\nu,\hat{\mathbf{p}}}(x) = \sqrt{2} G_F n_\nu(\mathbf{x}_0) \int_{-\infty}^{\infty} \mathrm{d}\omega' \int \mathrm{d}^2\hat{\mathbf{p}}'(1 - \hat{\mathbf{p}} \cdot \hat{\mathbf{p}}') \vec{P}_{\omega',\hat{\mathbf{p}}'}(\mathbf{x}). \tag{26.}$$

As in the isotropic neutrino gas case, Equation 25 also exhibits rotational symmetry about $\vec{B}$. If $\{\vec{P}_{\omega,\hat{\mathbf{p}}}(\mathbf{x}) \,|\, \forall\omega, \hat{\mathbf{p}}\}$ solves the EoM, then $\{\vec{P}'_{\omega,\hat{\mathbf{p}}}(\mathbf{x}) \,|\, \forall\omega, \hat{\mathbf{p}}\}$ also solves the EoM; in the latter, $\vec{P}'_{\omega,\hat{\mathbf{p}}}(\mathbf{x})$ is obtained from $\vec{P}_{\omega,\hat{\mathbf{p}}}(\mathbf{x})$ by rotation about $\vec{B}$ by an arbitrary angle ϕ, and ϕ is independent of ω, $\hat{\mathbf{p}}$, and $\mathbf{x}$. This symmetry leads to the conservation law (93)

$$\nabla \cdot \left[\int_{-\infty}^{\infty} \mathrm{d}\omega \int \mathrm{d}^2\hat{\mathbf{p}} (\vec{P}_{\omega,\hat{\mathbf{p}}} \cdot \vec{B}) \hat{\mathbf{p}} \right] = 0. \tag{27.}$$

The rotational symmetry of the EoM about $\vec{B}$ can lead to a collective precession mode for neutrino oscillations (**Figure 8**), even in an anisotropic environment.

Let **K** be the wave vector of the collective neutrino oscillation wave in a stationary environment. If $\vec{P}_{\omega,\hat{\mathbf{p}}}$ experiences pure precession, then this vector must precess about $\vec{B}$ with frequency $\hat{\mathbf{p}} \cdot \mathbf{K}$ (in flavor space) as the corresponding neutrino propagates along its world line (in coordinate space). Similar to the isotropic neutrino gas case, this means that $\vec{P}_{\omega,\hat{\mathbf{p}}}$ is parallel to $\vec{H}_{\omega,\hat{\mathbf{p}}} = (\omega - \hat{\mathbf{p}} \cdot \mathbf{K})\vec{B} + \vec{H}_{\nu\nu,\hat{\mathbf{p}}}$. Therefore the corresponding neutrino or antineutrino must be in a mass

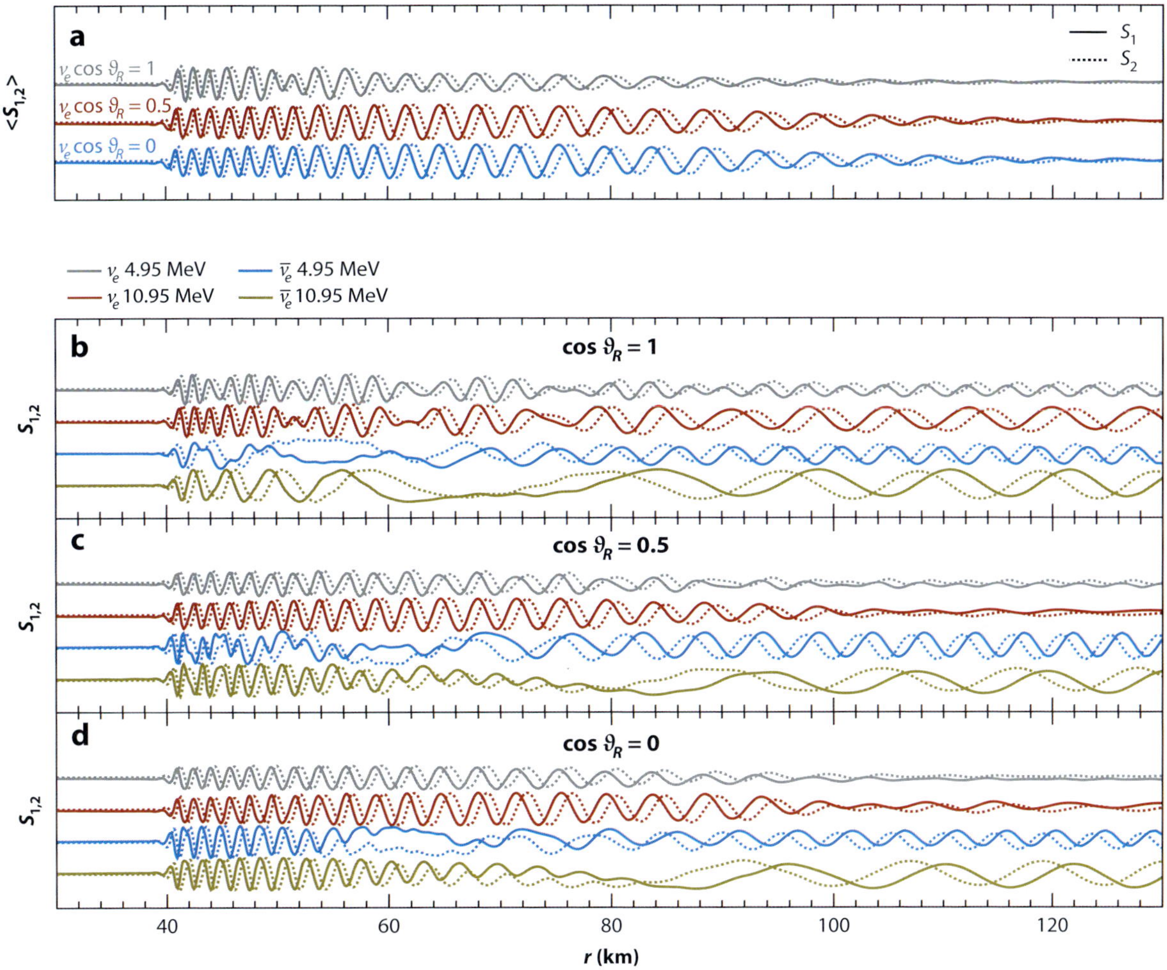

Figure 8

The collective precession of $\vec{P}_{\omega,\vartheta_R}$ in a neutrino bulb model with $\theta_v \ll 1$ and the inverted mass hierarchy ($\Delta m^2 < 0$). (*b–d*) For most neutrinos the "spin components" $s_1 \propto \vec{P}_{\omega,\vartheta_R} \cdot \hat{e}_1^{(V)}$ and $s_2 \propto \vec{P}_{\omega,\vartheta_R} \cdot \hat{e}_2^{(V)}$ oscillate in phase right after collective neutrino oscillations begin (for this case, at $r \simeq 40$ km). In-phase oscillations of s_1 and s_2 and the constant relative phase of these components imply the collective precession of the polarization vectors. The neutrinos or antineutrinos that drop out of the collective precession mode at smaller radii are those with vacuum oscillation frequencies that are farther from the collective oscillation frequency and those propagating along trajectories with larger $\cos\vartheta_R$. (*a*) The energy-averaged components $\langle s_1 \rangle$ and $\langle s_2 \rangle$. The collective precession mode stands out when noncollective neutrino oscillations average to zero. Figure adapted from figure 1 of Reference 93. Copyright 2009 by the Institute of Physics.

eigenstate if (93)

$$|\omega - \hat{\mathbf{p}} \cdot \mathbf{K}| \gg |\vec{H}_{\nu\nu,\hat{\mathbf{p}}}|. \qquad 28.$$

Equation 28 gives the criterion for when neutrinos or antineutrinos drop out of the collective precession mode and begin to oscillate incoherently with respect to other neutrinos. The neutrinos or antineutrinos that drop out in this way have oscillation frequencies ω that are so different from the collective oscillation frequency that neutrino self-interaction is not strong enough to maintain the corresponding polarization vectors in collective precession.

With the replacements $\omega \to (\omega \cos 2\theta_v - \lambda)$ and $\vec{B} \to -\vec{L}$, the above discussion also applies to the case with a large λ. Including these replacements, Equation 28 indicates that collective precession is indeed suppressed when the number density of electrons is much larger than that of neutrinos.

6. COLLECTIVE NEUTRINO OSCILLATIONS IN SUPERNOVAE

6.1. Neutrino Oscillation Regimes

With the picture for neutrino oscillations developed above, we can utilize the strength of neutrino self-interaction to sketch out neutrino oscillation regimes (**Figure 9**). We focus the following discussion on the two-flavor mixing scenario with $|\Delta m^2| \simeq \Delta m^2_{\rm atm}$ and $\theta_v \simeq \theta_{13}$. We designate $R^-_{\rm coll}$ as the radius closest to the PNS where collective oscillations set in. Likewise $R^+_{\rm coll}$ is the outer radius where collective oscillations cease. In both the NH and IH cases, $R^+_{\rm coll}$ can be estimated from the condition

$$\sqrt{2}G_{\rm F}n_{\bar{\nu}_e}(R^+_{\rm coll}) \simeq \frac{\Delta m^2_{\rm atm}}{\langle E_{\bar{\nu}_e}\rangle}, \qquad 29.$$

where $n_{\bar{\nu}_e}(r)$ is the total number density of the antineutrinos at r that are initially $\bar{\nu}_e$ at the neutrino sphere and $\langle E_{\bar{\nu}_e}\rangle$ is the average energy of these antineutrinos. Here we use $\bar{\nu}_e$ as the representative neutrino species and $\frac{\Delta m^2_{\rm atm}}{\langle E_{\bar{\nu}_e}\rangle}$ as an estimate of the frequency spread of the neutrino spectrum.

In the NH case, assuming a fully synchronized neutrino system, $R^-_{\rm coll}$ is approximately where (65)

$$\sqrt{2}G_{\rm F}n_e(R^-_{\rm coll}) \simeq \Omega_{\rm sync}. \qquad 30.$$

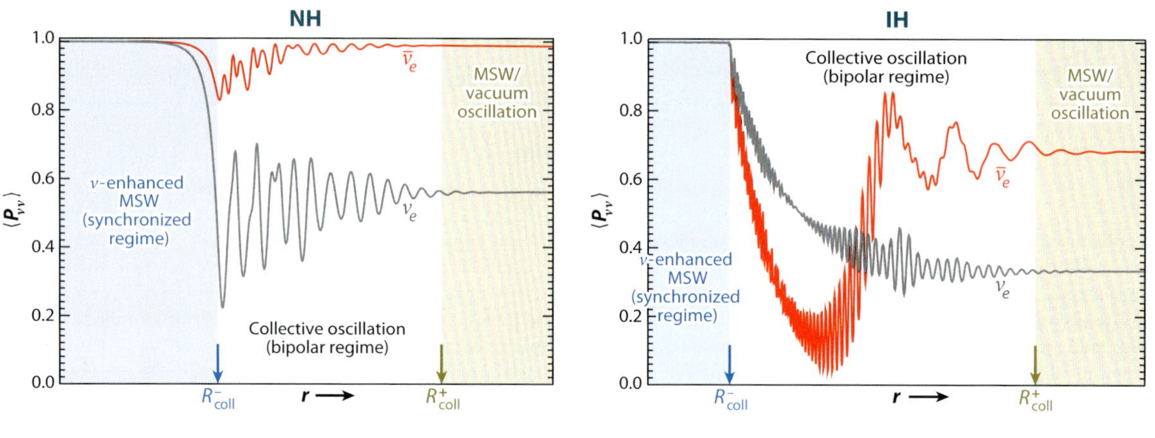

Figure 9

Schematic plots of the neutrino oscillation regimes in the core-collapse supernova environment in the two-flavor mixing scenario with $|\Delta m^2| \simeq \Delta m^2_{\rm atm}$. Near the protoneutron star (PNS), and in the synchronized regime ($r \lesssim R^-_{\rm coll}$), collective neutrino oscillations are suppressed by either the large matter density or the large neutrino fluxes themselves. In this regime neutrinos can still experience ν-enhanced Mikheyev-Smirnov-Wolfenstein (MSW) flavor transformation. Far away from the PNS ($r \gtrsim R^+_{\rm coll}$), where the neutrino fluxes are negligible, neutrinos experience either vacuum oscillations or conventional MSW flavor transformation, depending on the matter density and the energy of the neutrino. If $R^+_{\rm coll} > R^-_{\rm coll}$, then there is a window ($R^-_{\rm coll} \lesssim r \lesssim R^+_{\rm coll}$, known as the bipolar regime) where neutrinos experience collective neutrino oscillations and where spectral swaps/splits develop. The curves show the energy-averaged neutrino survival probabilities for electron neutrinos and antineutrinos in single-angle calculations that are chosen to be representative of some late-time supernova conditions.

[In the large neutrino flux limit, the behavior of a neutrino system experiencing ν-enhanced MSW flavor transformation is the same as that of a synchronized neutrino system (76).] We note that for an iron core–collapse supernova at early times and in the NH case, $R^{+}_{\text{coll}} \leq R^{-}_{\text{coll}}$, meaning that the collective neutrino oscillation regime does not exist. At later times in these models a collective neutrino oscillation regime can appear when the matter density becomes relatively small. In contrast, collective neutrino oscillations can occur at early epochs in an O-Ne-Mg core–collapse supernova (80) because the progenitors of these supernovae have relatively lower masses (8–$12M_{\odot}$), and therefore, after core bounce, the PNS has a dilute, lower-density envelope (126, 127).

In the IH case, the value of R^{-}_{coll} is the larger of the values extracted from the following two conditions:

$$n_{\bar{\nu}_e}(R^{-}_{\text{coll}}) \simeq \frac{1}{(\sqrt{1+\varepsilon}-1)^2}\frac{\Delta m^2_{\text{atm}}}{\sqrt{2}G_{\text{F}}\langle E_{\bar{\nu}_e}\rangle} \quad \text{and} \quad n_{\bar{\nu}_e}(R^{-}_{\text{coll}}) \simeq n_e(R^{-}_{\text{coll}}). \qquad 31.$$

The first condition in Equation 31 is based on the flavor pendulum model (see Equation 17), and the second condition accords with the discussion in Section 5.1. In the very early epochs of an iron core–collapse supernova, the collective neutrino oscillation regime may not exist in the IH case because of the presence of a large matter density.

6.2. Effects of Collective Neutrino Oscillations

The supernova neutrino energy spectra can be dramatically modified by collective neutrino oscillations. Perhaps the most prominent feature to arise from collective neutrino oscillations is the spectral swap/split (e.g., **Figure 1**). Obtaining this feature can depend on supernova conditions. For example, the calculations used to produce **Figure 1** would give no spectral swap/split if the matter density were too large (78, 91).

When ν_e and $\bar{\nu}_e$ are the most abundant neutrino species at the neutrino sphere, some qualitative aspects of collective neutrino oscillations in supernovae can be understood by using the flavor pendulum model. In the IH case, the initial configuration of the flavor pendulum is near its highest position and is unstable when the neutrino flux is below some critical value. (See the discussion in Section 4.2.) This implies that collective neutrino oscillations and their effects on neutrino energy spectra are relatively insensitive to the matter density or the exact value of θ_{13}. In the NH case, however, the flavor pendulum would be near its stable configuration unless MSW flavor transformation displaces it from this position. As a result, collective neutrino oscillations and their effects depend on the efficiency of the ν-enhanced MSW flavor transformation. In turn, the efficiency of this flavor transformation is sensitive to both the matter profile and the value of θ_{13}. This general picture is confirmed by multiangle calculations (77). The results shown in Reference 77 suggest that it may be possible to resolve the neutrino mass hierarchy with an observed supernova neutrino signal, even if the absolute neutrino masses and/or θ_{13} are too small to be measured in the laboratory.

Neutrino signals detected at very early times after core bounce may be important probes of neutrino mixing. For iron core–collapse supernovae, the neutrino spectra at very early times are modified by the conventional MSW mechanism and can be easily calculated (128). For O-Ne-Mg core–collapse supernovae, however, the matter density is so low above the PNS that even at these very early times, where the ν_e luminosity can be very large ($\sim 10^{53}$ erg s^{-1}), collective neutrino oscillations can create step-like features (swaps/splits) in the observed neutrino energy spectra (80). If these features are observed in a galactic supernova neutrino burst, then they could serve as diagnostics of the neutrino mass hierarchy and, in the NH case, even provide a measure of θ_{13}.

However, the spectral swap/split phenomenon is sensitive to the neutrino luminosities and energy spectra at the neutrino sphere. Presently there are rather large uncertainties in these quantities, especially for the late-time supernova environment. Also, very large neutrino fluxes do not necessarily imply a large neutrino oscillation effect; indeed, in the IH case a larger neutrino luminosity pushes R^{-}_{coll} to a larger radius.

Of course, collective neutrino oscillations are not the only way that neutrinos may experience flavor transformation in supernovae. As in the Sun, in supernovae neutrinos can also experience the conventional MSW flavor transformation. However, the matter profile in supernovae may not be smooth (e.g., because of shocks or turbulence) near MSW resonance regions, which can produce some interesting phenomena (101, 129–135).

In principle, the alteration of supernova neutrino energy spectra by collective neutrino oscillations and/or other processes of neutrino flavor transformation could affect supernova dynamics, shock reheating, and nucleosynthesis in neutrino-heated ejecta (41, 48, 50, 108, 136). However, modeling of this physics in realistic supernova conditions is at present primitive.

7. SUMMARY AND OPEN ISSUES

In this review, we have discussed collective neutrino oscillations in a simple bipolar neutrino system, in homogeneous and isotropic neutrino gases, and in anisotropic neutrino gases. The simple bipolar neutrino system (described by the flavor pendulum) is the simplest of these scenarios and is solvable analytically. It can be used to understand many qualitative features of collective neutrino oscillations in supernovae. The single-angle scheme essentially treats supernova neutrinos as a homogeneous and isotropic gas, and adiabatic neutrino flavor transformation in such a gas can be used to understand the spectral swap/split phenomenon in the supernova environment. An anisotropic neutrino gas can possess unique characteristics (e.g., suppression of collective flavor oscillations by large matter density), which makes it an important target for study because realistic physical environments such as core-collapse supernovae can be highly anisotropic.

We have covered some of the basic properties of collective neutrino oscillations, with emphasis on the two-flavor mixing case. There is much about the collective neutrino oscillation phenomenon that remains to be understood. For example, in Section 4.1 we skipped over the adiabaticity condition when we discussed adiabatic neutrino flavor transformation. Adiabaticity criteria developed so far (79) are difficult to use in practice.

There are other open issues in our current understanding of collective neutrino oscillations in supernovae. For example, there can be multiple spectral swaps/splits in the final neutrino energy spectra (**Figure 10**) (78, 103, 105). The appearance of these features, however, depends on the luminosities and energy spectra of the different neutrino species at the neutrino sphere. This finding cannot be explained by a grand collective precession mode in which all neutrinos and antineutrinos participate. In a more recent paper (110), it was reported that qualitatively different results could appear in some collective neutrino oscillation scenarios, depending on whether two-flavor mixing or full three-flavor mixing is used. All of these discoveries point to the need for a systematic study of supernova neutrino oscillations with neutrino self-interaction and full three-flavor mixing at various supernova epochs in which neutrino luminosities and energy spectra can be different. Also, numerical simulations of supernova explosions have clearly shown that realistic supernova environments may be highly anisotropic and inhomogeneous (28, 30–37, 137). It would be interesting to see how density inhomogeneities arising from, for instance, shock waves and turbulence might affect collective neutrino oscillations. It remains a towering numerical challenge to integrate fully hydrodynamic, three-dimensional supernova models with calculations of neutrino oscillations that include neutrino self-interaction. However, such

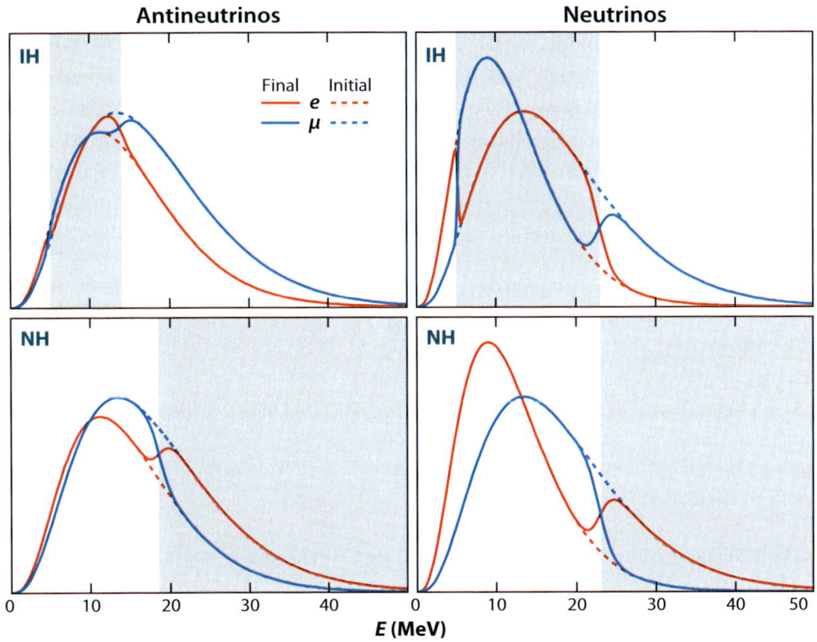

Figure 10

Multiple spectral swaps/splits in a two-flavor single-angle calculation with $|\Delta m^2| \simeq \Delta m^2_{\rm atm}$ and $\theta_{\rm v} \ll 1$. The solid and dashed curves represent the final and initial energy spectra, respectively. The red and blue curves are for e and μ flavors, respectively. The shaded regions mark the energy ranges in which spectral swaps occur. Unlike previous numerical calculations, this calculation assumes that ν_μ and $\bar{\nu}_\mu$, rather than ν_e and $\bar{\nu}_e$, are the most abundant neutrino species at the neutrino sphere. Abbreviations: IH, inverted neutrino mass hierarchy; NH, normal neutrino mass hierarchy. Reprinted with permission from Reference 103. Copyright 2009 by the American Physical Society.

self-consistent integration may be necessary to study the interplay between supernova physics and neutrino oscillations. Specifically, a full understanding of neutrino oscillations in supernovae may hold the key to deciphering the neutrino signal from a future galactic supernova. The stakes are high. Deciphering a supernova neutrino signal could provide important insights into supernova astrophysics. It could also provide key insights into fundamental neutrino properties, and such insights could be complementary to those sought by future neutrino experiments.

DISCLOSURE STATEMENT

The authors are not aware of any affiliations, memberships, funding, or financial holdings that might be perceived as affecting the objectivity of this review.

ACKNOWLEDGMENTS

We thank J. Carlson, J. Cherry, A. Friedland, W. Haxton, D. Kaplan, C. Kishimoto, A. Kusenko, A. Mezzacappa, and S. Reddy for useful conversations. We especially thank G. Raffelt for reading the manuscript and providing valuable comments. Our work was supported in part by the Institutional Computing Program at Los Alamos National Laboratory (LANL) and the National

Energy Research Scientific Computing Center, which is supported by the U.S. Department of Energy (DOE), Office of Science, under contract DE-AC02-05CH11231. Our work was also supported by DOE grant DE-FG02-00ER41132 at the Institute for Nuclear Theory; DOE grant DE-FG02-87ER40328 at the University of Minnesota; National Science Foundation grant PHY-06-53626 at the University of California, San Diego; and an Institute of Geophysics and Planetary Physics/LANL minigrant. The research of H.D. is supported by the LANL Laboratory Directed Research and Development program through the director's postdoctoral fellowship at LANL.

LITERATURE CITED

1. Camilleri L, Lisi E, Wilkerson JF. *Annu. Rev. Nucl. Part. Sci.* 58:343 (2008)
2. Timmes FX, Woosley SE, Weaver TA. *Astrophys. J. Suppl.* 98:617 (1995)
3. Abel T, Bryan GL, Norman ML. *Science* 295:93 (2002)
4. Samtleben D, Staggs S, Winstein B. *Annu. Rev. Nucl. Part. Sci.* 57:245 (2007)
5. Steigman G. *Annu. Rev. Nucl. Part. Sci.* 57:463 (2007)
6. Hannestad S. *Annu. Rev. Nucl. Part. Sci.* 56:137 (2006)
7. Fuller GM, Kishimoto CT. *Phys. Rev. Lett.* 102:201303 (2009)
8. Amsler C, et al. (Part. Data Group) *Phys. Lett.* B667:1 (2008)
9. Schwetz T, Tortola MA, Valle JWF. *New J. Phys.* 10:113011 (2008)
10. Mena O, Parke SJ. *Phys. Rev. D* 70:093011 (2004)
11. Guo X, et al. (Daya Bay Collab.) arXiv:hep-ex/0701029 (2007)
12. Bandyopadhyay A, et al. (ISS Phys. Work. Group) *Rep. Prog. Phys.* 72:106201 (2009)
13. Ray RE. *J. Phys. Conf. Ser.* 136:022019 (2008)
14. Terri R. (T2K Collab.) *Nucl. Phys. Proc. Suppl.* 189:277 (2009)
15. Adamson P, et al. (MINOS Collab.) *Phys. Rev. Lett.* 103:261802 (2009)
16. Fuller GM, Haxton WC, McLaughlin GC. *Phys. Rev. D* 59:085005 (1999)
17. Lunardini C, Smirnov AY. *J. Cosmol. Astropart. Phys.* 0306:009 (2003)
18. Beacom JF, Vagins MR. *Phys. Rev. Lett.* 93:171101 (2004)
19. Cadonati L, Calaprice FP, Chen MC. *Astropart. Phys.* 16:361 (2002)
20. Ahrens J, et al. (AMANDA Collab.) *Astropart. Phys.* 16:345 (2002)
21. Sharp MK, Beacom JF, Formaggio JA. *Phys. Rev. D* 66:013012 (2002)
22. Ikeda M, et al. (Super-Kamiokande Collab.) *Astrophys. J.* 669:519 (2007)
23. Autiero D, et al. *J. Cosmol. Astropart. Phys.* 0711:011 (2007)
24. Colgate SA, Grasberger WH, White RH. *Astron. J.* 66:280 (1961)
25. Woosley SE, Heger A, Weaver TA. *Rev. Mod. Phys.* 74:1015 (2002)
26. Woosley S, Janka T. *Nat. Phys.* 1:147 (2005)
27. Janka H-T, et al. *Phys. Rep.* 442:38 (2007)
28. Herant M, et al. *Astrophys. J.* 435:339 (1994)
29. Burrows A, Hayes J, Fryxell BA. *Astrophys. J.* 450:830 (1995)
30. Buras R, Rampp M, Janka HT, Kifonidis K. *Astron. Astrophys.* 447:1049 (2006)
31. Bruenn SW, et al. *J. Phys. Conf. Ser.* 46:393 (2006)
32. Ott CD, Burrows A, Dessart L, Livne E. *Astrophys. J.* 685:1069 (2008)
33. Hammer NJ, Janka HT, Mueller E. *Astrophys. J.* 714:1371 (2010)
34. Murphy JW, Burrows A. *Astrophys. J.* 688:1159 (2008)
35. Bruenn SW, et al. *J. Phys. Conf. Ser.* 180:012018 (2009)
36. Blondin JM, Mezzacappa A. *Astrophys. J.* 642:401 (2006)
37. Blondin JM, Mezzacappa A. *Nature* 445:58 (2007)
38. Bethe HA, Wilson JR. *Astrophys. J.* 295:14 (1985)
39. Bethe HA. *Rev. Mod. Phys.* 62:801 (1990)
40. Arnett D. *Supernovae and Nucleosynthesis*. Princeton: Princeton Univ. Press (1996)
41. Qian Y-Z, et al. *Phys. Rev. Lett.* 71:1965 (1993)
42. Horowitz CJ. *Phys. Rev. D* 65:043001 (2002)

43. Prakash M, Lattimer JM, Sawyer RF, Volkas RR. *Annu. Rev. Nucl. Part. Sci.* 51:295 (2001)
44. Wolfenstein L. *Phys. Rev. D* 17:2369 (1978)
45. Mikheyev SP, Smirnov AY. *Yad. Fiz.* 42:1441 (1985)
46. Fuller GM, Mayle RW, Wilson JR, Schramm DN. *Astrophys. J.* 322:795 (1987)
47. Nötzold D, Raffelt GG. *Nucl. Phys. B* 307:924 (1988)
48. Fuller GM, Mayle RW, Meyer BS, Wilson JR. *Astrophys. J.* 389:517 (1992)
49. Fuller GM. *Phys. Rep.* 227:149 (1993)
50. Qian Y-Z, Fuller GM. *Phys. Rev. D* 51:1479 (1995)
51. Qian Y-Z, Fuller GM. *Phys. Rev. D* 52:656 (1995)
52. Pantaleone JT. *Phys. Rev. D* 46:510 (1992)
53. Pantaleone JT. *Phys. Lett.* B287:128 (1992)
54. Samuel S. *Phys. Rev. D* 48:1462 (1993)
55. Kostelecký VA, Pantaleone JT, Samuel S. *Phys. Lett.* B315:46 (1993)
56. Kostelecký VA, Samuel S. *Phys. Lett.* B318:127 (1993)
57. Kostelecký VA, Samuel S. *Phys. Rev. D* 49:1740 (1994)
58. Kostelecký VA, Samuel S. *Phys. Rev. D* 52:621 (1995)
59. Samuel S. *Phys. Rev. D* 53:5382 (1996)
60. Kostelecký VA, Samuel S. *Phys. Lett.* B385:159 (1996)
61. Pantaleone JT. *Phys. Rev. D* 58:073002 (1998)
62. Pastor S, Raffelt GG, Semikoz DV. *Phys. Rev. D* 65:053011 (2002)
63. Dolgov AD, et al. *Nucl. Phys. B* 632:363 (2002)
64. Abazajian KN, Beacom JF, Bell NF. *Phys. Rev. D* 66:013008 (2002)
65. Pastor S, Raffelt GG. *Phys. Rev. Lett.* 89:191101 (2002)
66. Balantekin AB, Yüksel H. *New J. Phys.* 7:51 (2005)
67. Fuller GM, Qian Y-Z. *Phys. Rev. D* 73:023004 (2006)
68. Duan H, Fuller GM, Qian Y-Z. *Phys. Rev. D* 74:123004 (2006)
69. Duan H, Fuller GM, Carlson J, Qian Y-Z. *Phys. Rev. Lett.* 97:241101 (2006)
70. Duan H, Fuller GM, Carlson J, Qian Y-Z. *Phys. Rev. D* 74:105014 (2006)
71. Hannestad S, Raffelt GG, Sigl G, Wong YYY. *Phys. Rev. D* 74:105010 (2006)
72. Raffelt GG, Sigl G. *Phys. Rev. D* 75:083002 (2007)
73. Duan H, Fuller GM, Carlson J, Qian Y-Z. *Phys. Rev. D* 75:125005 (2007)
74. Raffelt GG, Smirnov AY. *Phys. Rev. D* 76:081301(R) (2007)
75. Esteban-Pretel A, et al. *Phys. Rev. D* 76:125018 (2007)
76. Duan H, Fuller GM, Qian Y-Z. *Phys. Rev. D* 76:085013 (2007)
77. Duan H, Fuller GM, Carlson J, Qian Y-Z. *Phys. Rev. Lett.* 99:241802 (2007)
78. Fogli GL, Lisi E, Marrone A, Mirizzi A. *J. Cosmol. Astropart. Phys.* 0712:010 (2007)
79. Raffelt GG, Smirnov AY. *Phys. Rev. D* 76:125008 (2007)
80. Duan H, Fuller GM, Carlson J, Qian Y-Z. *Phys. Rev. Lett.* 100:021101 (2008)
81. Esteban-Pretel A, et al. *Phys. Rev. D* 77:065024 (2008)
82. Lunardini C, Mueller B, Janka HT. *Phys. Rev. D* 78:023016 (2008)
83. Dasgupta B, Dighe A. *Phys. Rev. D* 77:113002 (2008)
84. Duan H, Fuller GM, Qian Y-Z. *Phys. Rev. D* 77:085016 (2008)
85. Dasgupta B, Dighe A, Mirizzi A, Raffelt GG. *Phys. Rev. D* 77:113007 (2008)
86. Dasgupta B, Dighe A, Mirizzi A. *Phys. Rev. Lett.* 101:171801 (2008)
87. Duan H, Fuller GM, Carlson J. *Comput. Sci. Disc.* 1:015007 (2008)
88. Sawyer RF. *Phys. Rev. D* 79:105003 (2009)
89. Chakraborty S, Choubey S, Dasgupta B, Kar K. *J. Cosmol. Astropart. Phys.* 0809:013 (2008)
90. Dasgupta B, Dighe A, Mirizzi A, Raffelt GG. *Phys. Rev. D* 78:033014 (2008)
91. Esteban-Pretel A, et al. *Phys. Rev. D* 78:085012 (2008)
92. Gava J, Volpe C. *Phys. Rev. D* 78:083007 (2008)
93. Duan H, Fuller GM, Qian Y-Z. *J. Phys. G* 36:105003 (2009)
94. Raffelt GG. *Phys. Rev. D* 78:125015 (2008)
95. Blennow M, Mirizzi A, Serpico PD. *Phys. Rev. D* 78:113004 (2008)

96. Fogli G, Lisi E, Marrone A, Tamborra I. *J. Cosmol. Astropart. Phys.* 0904:030 (2009)
97. Lunardini C, Peres OLG. *J. Cosmol. Astropart. Phys.* 0808:033 (2008)
98. Guo X-H, Huang M-Y, Young B-L. *Phys. Rev. D* 79:113007 (2009)
99. Minakata H, Nunokawa H, Tomas R, Valle JWF. *J. Cosmol. Astropart. Phys.* 0812:006 (2008)
100. Liao W. arXiv:0904.0075 [hep-ph] (2009)
101. Duan H, Kneller JP. *J. Phys. G* 36:113201 (2009)
102. Liao W. arXiv:0904.2855 [hep-ph] (2009)
103. Dasgupta B, Dighe A, Raffelt GG, Smirnov AY. *Phys. Rev. Lett.* 103:051105 (2009)
104. Galais S, Kneller J, Volpe C, Gava J. *Phys. Rev. D* 81:053002 (2010)
105. Fogli G, Lisi E, Marrone A, Tamborra I. *J. Cosmol. Astropart. Phys.* 0910:002 (2009)
106. Esteban-Pretel A, Tomas R, Valle JWF. *Phys. Rev. D* 81:063003 (2010)
107. Gava J, Kneller J, Volpe C, McLaughlin GC. *Phys. Rev. Lett.* 103:071101 (2009)
108. Chakraborty S, Choubey S, Goswami S, Kar K. arXiv:0911.1218 [hep-ph] (2009)
109. Lazauskas R, Lunardini C, Volpe C. *J. Cosmol. Astropart. Phys.* 0904:029 (2009)
110. Friedland A. *Phys. Rev. Lett.* 104:191102 (2010)
111. Gava J, Jean-Louis CC. arXiv:0912.5206 [hep-ph] (2009)
112. Balantekin AB, Fuller GM. *Phys. Lett.* B471:195 (1999)
113. Caldwell DO, Fuller GM, Qian Y-Z. *Phys. Rev. D* 61:123005 (2000)
114. Sigl G, Raffelt GG. *Nucl. Phys. B* 406:423 (1993)
115. McKellar BHJ, Thomson MJ. *Phys. Rev. D* 49:2710 (1994)
116. Strack P, Burrows A. *Phys. Rev. D* 71:093004 (2005)
117. Halprin A. *Phys. Rev. D* 34:3462 (1986)
118. Bell NF, Rawlinson AA, Sawyer RF. *Phys. Lett.* B573:86 (2003)
119. Friedland A, Lunardini C. *J. High Energy Phys.* 10:043 (2003)
120. Sawyer RF. *Phys. Rev. D* 72:045003 (2005)
121. Friedland A, McKellar BHJ, Okuniewicz I. *Phys. Rev. D* 73:093002 (2006)
122. Balantekin AB, Pehlivan Y. *J. Phys. G* 34:47 (2007)
123. Kim CW, Kim J, Sze WK. *Phys. Rev. D* 37:1072 (1988)
124. Duan H. *AIP Conf. Proc.* 1182:36 (2009)
125. Cardall CY. *Phys. Rev. D* 78:085017 (2008)
126. Nomoto K. *Astrophys. J.* 277:791 (1984)
127. Nomoto K. *Astrophys. J.* 322:206 (1987)
128. Kachelriess M, et al. *Phys. Rev. D* 71:063003 (2005)
129. Sawyer RF. *Phys. Rev. D* 42:3908 (1990)
130. Loreti FN, Qian Y-Z, Fuller GM, Balantekin AB. *Phys. Rev. D* 52:6664 (1995)
131. Schirato RC, Fuller GM. arXiv:astro-ph/0205390 (2002)
132. Fogli GL, Lisi E, Montanino D, Mirizzi A. *Phys. Rev. D* 68:033005 (2003)
133. Kneller JP, McLaughlin GC. *Phys. Rev. D* 73:056003 (2006)
134. Friedland A, Gruzinov A. arXiv:astro-ph/0607244 (2006)
135. Dasgupta B, Dighe A. *Phys. Rev. D* 75:093002 (2007)
136. Sigl G. *Phys. Rev. D* 51:4035 (1995)
137. Fryer CL, Warren MS. *Astrophys. J.* 574:L65 (2002)

Triggering on Heavy Flavors at Hadron Colliders

Luciano Ristori[1] and Giovanni Punzi[2]

[1]Istituto Nazionale di Fisica Nucleare, I-56127 Pisa, Italy; email: luciano.ristori@pi.infn.it

[2]Department of Physics, University of Pisa, I-56126 Pisa, Italy; email: giovanni.punzi@pi.infn.it

Annu. Rev. Nucl. Part. Sci. 2010. 60:595–614

First published online as a Review in Advance on July 20, 2010

The *Annual Review of Nuclear and Particle Science* is online at nucl.annualreviews.org

This article's doi:
10.1146/annurev.nucl.012809.104501

0163-8998/10/1123-0595$20.00

Key Words

trigger, electronics, particle physics, track reconstruction, pattern recognition, *b* physics

Abstract

For a number of interesting processes in the sector of heavy flavors, the quality of measurements made at hadron colliders is very similar to the quality achieved at e^+e^- colliders (known as B factories). The key to performing such measurements in a hadron environment is the ability to select rare processes from background in real time, that is, to trigger on them. Two distinctive features of heavy-flavor decays have been used for this purpose: the presence of leptons in the final state and secondary vertices produced by the relatively long lifetime. The selection of events based on long lifetime, although technically very challenging, is the most inclusive of all such techniques, allowing access to the widest range of channels. The focus of this review is on the innovative concepts that permitted the reconstruction of tracks produced in hadron collisions with sufficient speed and accuracy for use at trigger level to detect heavy-flavor decays.

Contents

1. INTRODUCTION

The study of hadrons that contain a *b* quark is very effective in testing our understanding of the fundamental interactions and the validity of the Standard Model (1–3) and, possibly, in detecting hints of new physics. Measurements of quantities such as cross sections, branching ratios, lifetimes, masses of new states, *CP* asymmetries, and neutral meson oscillation frequencies have improved our understanding of the Standard Model, constrained the elements of the CKM (Cabibbo-Kobayashi-Maskawa) quark-mixing matrix (4, 5), and provided guidance to the planning of future machines and experiments. Numerous measurements have been performed in experiments at the *B* factories (the BaBar and Belle Collaborations) and at the Tevatron (the CDF and DØ Collaborations) (6; also see **http://pdg.lbl.gov**).

1.1. *B* Production at Hadron Colliders

Due to large cross sections, hadronic collisions at high energy are the most effective way to produce large quantities of heavy-flavored hadrons. The largest rates are currently provided by the Tevatron, which produces of order 10^{11} *b*-flavored particles per year. However, large quantities of other particles are produced as well, and *b* production represents only a fraction (of order 10^{-3}) of the total. This large background was the reason for the design of dedicated e^+e^- colliders for the selective production of *B* particles through the decay of $b\bar{b}$ resonances [$\Upsilon(4s)$, $\Upsilon(5s)$]. These colliders were named *B* factories not because of the large production rate, which is many times lower than at hadron machines, but because of the specificity of the production of *b*-flavored particles with a much more favorable signal-to-background ratio.

Physics results obtained in recent years have shown that, for a number of interesting processes in the heavy-flavor sector, the quality of measurements performed at hadron colliders can be very similar to those performed at the B factories in terms of both precision and the number of events collected. An advantage of hadron colliders is that, due to their production mechanism, more types of heavy-flavored particles can be studied, including B_s, B_c, and B baryons. However, the key to performing such measurements is the ability to select such processes from the background in real time—that is, to trigger on those processes. Triggering on heavy-flavor events is the central topic of this review.

1.2. The Trigger

The collision rate at hadron colliders is much higher than the recording capability of experiments. The only way to exploit abundant heavy-flavor production is to select collisions in real time by recording only a small fraction for further processing and immediately discarding the data that contain only uninteresting, light-flavored processes.

The Tevatron, for example, produces approximately 2.5 million collisions per second at each interaction region. Two beams of protons and antiprotons are accelerated up to ~1,000 GeV in opposite directions; they then collide at two designated points along the circumference. Dozens of particles are created from vacuum in each collision. The detectors installed around the interaction regions record all the signals generated by all these particles. Approximately 500 Kbytes are needed to store all the information for each collision.

With current technology, the required storage space makes it impossible to permanently store the data from each and every collision. A typical large hard drive (~1 Tb) would fill up in less than 1 s, so it is necessary to be selective and keep only the events in which we are interested. If, for example, we wish to study collisions in which b quarks are produced, we would retain the data from approximately one in a thousand collisions.

All the signals from the different detector elements must be processed by complex computer algorithms before a clear picture of what occurred in each collision can emerge. This processing takes typically of order of 1 s for a modern central processing unit (CPU) per collision. It would take millions of CPUs working in parallel to process 2.5 million collisions per second and send only the data of interest to permanent storage. Present hadron collider experiments (CDF and DØ) use computer farms comprising ~1,000 CPUs to process events in real time; therefore, the event rate must be reduced by more than three orders of magnitude before full reconstruction can be afforded. Such selection is performed by a machine that can interpret the data from a collision, possibly with some approximation, in much less time than any commercial CPU needs. We term such a device a trigger.

The trigger is a very important component of any experiment at a hadron collider such as the Tevatron or the LHC. Its quality is primarily what limits the phenomena that can be observed and the measurements that can be made. Events that do not pass the trigger requirements are lost forever.

2. TRIGGERING ON HEAVY FLAVORS

Two features of heavy-flavor decays have been exploited to distinguish them from the background of light particles: (*a*) the presence of leptons and (*b*) the relatively long lifetime of their ground states, which can decay only through weak interaction. Although detection of leptons in the final state is a powerful tool in selecting event samples with a high heavy-flavor content, it limits the processes that can be studied to a small subset of the total. Conversely, the long lifetime of states that decay by weak interaction creates so-called secondary vertices at the point in space where they decay, irrespective of the specific decay mode, and allows the collection of very inclusive samples of heavy flavors. For this reason, triggering on secondary vertices, although technically very difficult, has provided a large fraction of the flavor physics results from hadron colliders.

2.1. Leptons

The weak decay of b and c quarks proceeds through the production of a virtual W boson, which can decay into a charged lepton-neutrino pair. This decay leads to the production of either an electron or a muon, with a $\sim$10% probability of each (production of tau leptons is less probable due to their high mass). These leptons can be used to tag the presence of heavy flavors because they stand out as a signature of their weak decay. Production of leptons from other sources is limited and easily recognizable (e.g., the Drell-Yan process), and those from weak decays of the lighter s quark are produced with significantly lower transverse momenta. However, the presence of a neutrino, which escapes detection, implies that it is not possible to completely reconstruct the decay with the momenta of all the particles produced in the final state. This limitation is more important in hadron collisions than at e^+e^- colliders, where knowledge of the initial total energy helps constrain the kinematics of the decay. Leptons are also used to recognize heavy-flavor decays through the $J/\psi \rightarrow e^+e^-$ and $J/\psi \rightarrow \mu^+\mu^-$ decays. Some interesting B hadron decays produce a J/ψ particle in the final state; thus, requiring the presence of a pair of leptons with the correct invariant mass is a good way to select them at trigger level. In hadron collisions at high energies, a sizable percentage ($\sim$20%) of all the produced J/ψ particles comes from B decays. Other important B decays have nonresonant pairs of leptons in the final state and can also be selected using this method.

2.2. Secondary Vertices

Detecting the presence of a secondary vertex in a particular event is not a trivial task: It requires one to reconstruct tracks and to recognize that some of these tracks do not originate from the collision point. When a long-lived particle decays after traveling some distance from the point where it was created, the trajectories of the decay products generally do not point back to the collision point. The distance of closest approach of the extrapolated trajectory to the collision point, known as the impact parameter, provides the key signal for the detection of secondary vertices. An impact parameter that is significantly different from zero is the primary clue that the track originates from a secondary vertex and that it is a good candidate for the product of the decay of a long-lived particle.

A typical situation is shown in **Figure 1**. The primary vertex is where the primary collision took place. Most of the particles produced in the collision originate from there. If a long-lived particle is created in the primary collision, it will travel some distance L (decay path) before decaying, and will generate secondary tracks originating from a secondary vertex. If we extrapolate back the trajectory of a secondary track, we find that, in general, it will miss the primary vertex by some distance d (impact parameter),

$$d = L \sin\theta,$$

where θ is the angle between the direction of the secondary track and the direction of the parent particle, and

$$L = \beta\gamma ct,$$

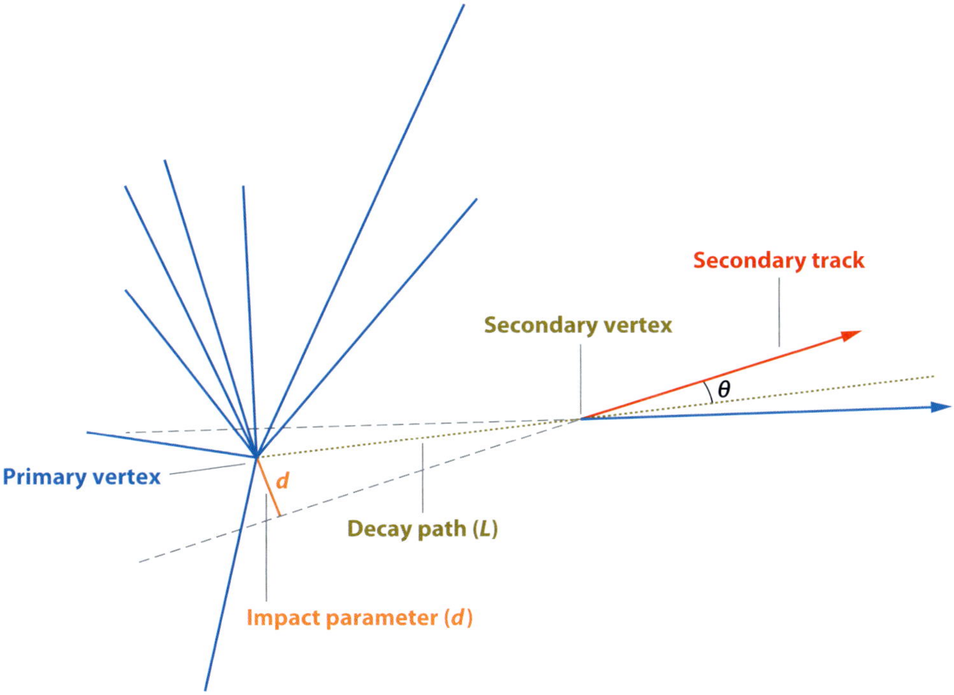

Figure 1

When a long-lived particle decays after traveling some distance, the trajectories of the decay products do not point back to the collision point. The distance of closest approach of the extrapolated trajectory to the collision point is known as the impact parameter.

where β and γ are the usual parameters of the Lorentz transformation from the rest frame of the parent particle to the laboratory, t is the proper decay time, and c is the speed of light.

If the particles involved in the process are all relativistic, then on average, θ is of the order of $1/\gamma$, and for $\gamma \gg 1$, the impact parameter is

$$\langle d \rangle \simeq \beta\gamma c\tau(1/\gamma) = \beta c\tau \simeq c\tau,$$

where τ is the lifetime of the parent particle. Note that, with these approximations, the average size of the impact parameter is independent from the Lorentz boost γ.

For particles containing the b quark, $c\tau$ is of the order of 450 μm, and for particles containing the c quark, $c\tau$ can be as small as 123 μm. These values set the scale for the precision of the impact-parameter measurement needed to detect secondary vertices from heavy-flavor decay to the order of a few tens of micrometers.

The first experiment that attempted to exploit the presence of secondary vertices to trigger on heavy-flavor decays for a real measurement was probably WE82 at CERN (7). This experiment searched for b and c production in pion-nucleon collisions at an energy of 350 GeV in the laboratory frame. WE82 used six planes of detectors to reconstruct tracks in real time, scanning for secondary vertices in a total time of 35 μs. It selected b and c quarks with a rejection factor of order 10 against lighter quarks by requiring three primary tracks accompanied by two nonprimary tracks. This device probably achieved the maximum possible performance with the technology available at that time; however, given the ~1-MHz interaction rate and the initial signal-to-background ratio for heavy flavors of only $\mathcal{O}(10^{-6})$, the device was only usable in conjunction with other triggers, and it could only provide modest help.

A much more complex and ambitious impact-parameter trigger was designed for the BTeV experiment (8–10). The detector, which was never built, would have used an extensive telescope comprising 30 layers of silicon detectors and a complex tracking algorithm, supported by a large array of computers. The trigger was designed to work in three dimensions, with an input rate of 7.6 MHz. The trigger was intended to build tracks beginning from pairs of hits in adjacent detectors, then add a third hit and trigger on *B* events by using two detached tracks with large enough transverse momentum in the same arm of the detector.

The DØ experiment can perform track reconstruction and measure impact parameters in real time (11); however, the general organization of the DØ trigger and its timing prevent the implementation of a pure impact-parameter trigger. Impact-parameter information can be used in DØ only as a complement to other requirements, such as the presence of leptons or a high-P_T jet.

The LHCb experiment plans to trigger on secondary vertices by using an array of standard processors to reconstruct tracks in real time (12). Currently, CDF is the only experiment to have successfully operated a pure impact-parameter trigger that yielded numerous physics results. Our review is devoted to the description of the CDF trigger's design principles and some of the innovative underlying ideas.

3. SUPERFAST TRACK RECONSTRUCTION

The task of reconstructing the trajectories of charged particles (tracks) in the detector can be broken down into two steps. The first step is track finding; the second step is track fitting. Track finding involves choosing, from among all the hits present in the detector, those hits caused by the same particle. Doing so may be difficult if many particles are present in the same event and if noise hits are also present, as is normally the case. Track fitting involves extracting track parameters (e.g., the particle momentum and impact parameter) from the coordinates of the hits. These steps are usually executed multiple times to find and fit all the tracks in each event, and they require sophisticated algorithms that typically consume several seconds of CPU time per event. To use this information in the trigger, we must find a way to perform a similar task within approximately a few microseconds—that is, 100,000 to 1 million times faster than a commercial CPU.

A possible solution to this problem emerged toward the end of the 1980s, when very large scale integration (VLSI) technology, which brought us microprocessors and the home computer, became affordable not only for industry giants but also for universities and small research projects. Some universities taught classes in which students designed VLSI circuits and had them fabricated (15). At that time it was realized that VLSI technology could be used to build a machine that could reconstruct tracks in a very short time.

3.1. Pattern Recognition by Brute Force

A time-consuming task in track reconstruction is pattern recognition: One needs both to try many different combinations of hits to find those that satisfy all geometrical constraints and to discard false hits arising from electronic noise or uncorrelated particles. All those different combinations are usually tested sequentially, and because there are so many of them, a great deal of time is needed to do so. But what if we could test these combinations in parallel, all at the same time? Use of VLSI technology makes it possible to integrate thousands of logic elements on a small silicon

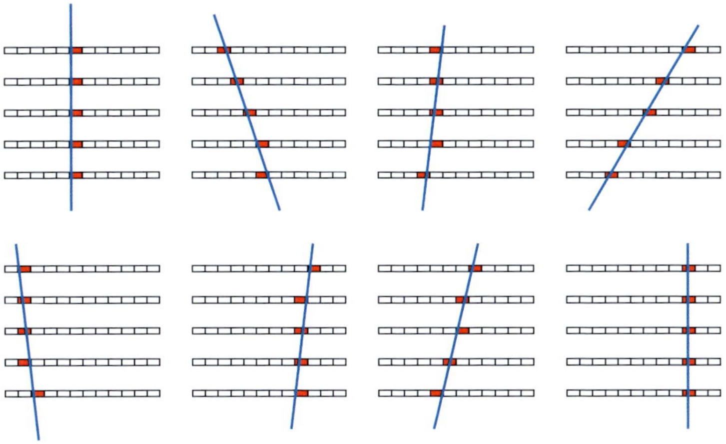

Figure 2

An ideal detector with five sensitive layers, each of which is divided into a number of segments. A charged track crossing the five layers fires one and only one segment per layer, producing a pattern of hits. If we let the track parameters span a certain volume of the phase space, a corresponding finite set of distinct patterns is generated.

chip and implement pattern recognition as a massively parallel algorithm. These ideas led to the invention of a special VLSI system: the associative memory device (16).

We use a simple example to illustrate the principle of operation of the associative memory device. **Figure 2** shows an ideal detector with five sensitive layers, each of which is divided into a number of segments. These segments could be, for example, five parallel layers of microstrip silicon detectors, as seen from along the strip direction. For the sake of simplicity, we assume that a charged track crossing these five layers fires one and only one strip per layer, thereby producing a pattern of hits. If we assume that each strip is uniquely identified by a coordinate, then each pattern consists of five coordinates (one per layer). If we change the track parameters enough, the resulting pattern changes. If we let the track span a certain volume of the parameter space, a corresponding, finite set of distinct patterns is generated. We term such sets pattern banks. **Figure 2** shows several tracks, each of which contributes a different pattern to the pattern bank.

Figure 3 shows a typical pattern-recognition problem consisting of so-called noise hits mixed together with hits from a real track. We need to implement an algorithm that is able to tell the difference between the scenarios shown in the two panels: In **Figure 3*a*** there is a combination of hits, one per detector layer, that corresponds to a pattern stored in the pattern bank and could therefore have been produced by a track; in **Figure 3*b*** no such combination exists. As discussed above, we usually solve this problem by trial and error by sifting through all possible combinations of hits, one per layer, until we find a combination that is compatible with having been generated by a single track. However, we can also proceed in the opposite way by scanning the pattern bank until we find a pattern whose coordinates correspond to a hit present in the detector. The advantage of the second approach is that, as we discuss in the following section, it lends itself easily to a massively parallel implementation.

Imagine that we have generated a pattern bank as described above. Now imagine that instead of storing all the patterns in normal random-access memory (RAM) (17–19), we store them in a device that is similar to a RAM but for which each storage element has additional intelligence built in: In addition to storing information, it can ascertain whether all of its coordinates (one per layer) correspond to a detector segment that was actually fired. If this method is used, the process

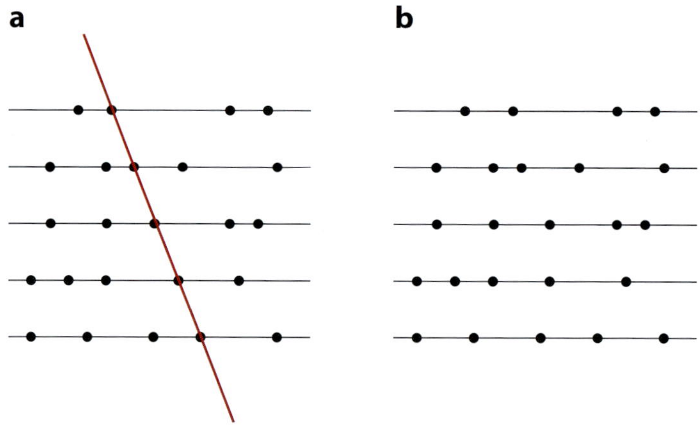

Figure 3

A typical pattern-recognition problem, consisting of noise hits superimposed on a track. The problem is to implement an algorithm that can tell the difference between the two cases shown in panels *a* and *b*, namely that in panel *a* we can find a combination of hits, one per detector layer, that can be produced by a single track, whereas in panel *b* no such combination exists.

can go much faster: We simply feed all the coordinates of all the fired segments into this device, and each pattern can determine whether there is a match. The time needed to perform pattern recognition is therefore equal to the time needed to extract the coordinates of all the hits from the detector and is independent of the size of the pattern bank.

3.2. The Associative Memory Device

The associative memory is based on the general concept of content-addressable memory (CAM) (21). In a CAM device, data are stored in the same way as in a RAM device (17–19) but are retrieved differently. To retrieve data from a RAM, one provides an address and obtains the data stored at that address. To retrieve data from a CAM, one provides the data and obtains the address at which that data is stored. CAM's typical applications are, for example, in cell phone and computer network address routing, where it is important to perform table lookups in the shortest possible time.

The schematics of a single CAM bit implemented as a CMOS (complementary metal-oxide semiconductor) device is shown in **Figure 4**. For more information on CAM architecture, see References 21 and 22.

Figure 5 represents the internal structure of the associative memory device that was designed for the CDF experiment and successfully used for many years (20). In the following discussion, we refer specifically to this implementation.

The associative memory device for the CDF experiment was designed for six detector layers. Each pattern is composed of six coordinates (one per layer), and each coordinate is encoded with 12 bits. Each coordinate can be accessed for reading or writing using the address bus to select one specific pattern and the layer bus to select one specific detector layer. Data are transmitted over the data bus and routed to or from the appropriate detector layer by a multiplexer driven by the layer bus.

The same data bus and layer bus used for writing are also used to feed hits from the detector to the associative memory device. One hit is encoded as a three-bit layer number (0–5) and as a 12-bit coordinate (0–4,095). The coordinate of each hit is transmitted to the correct layer and compared

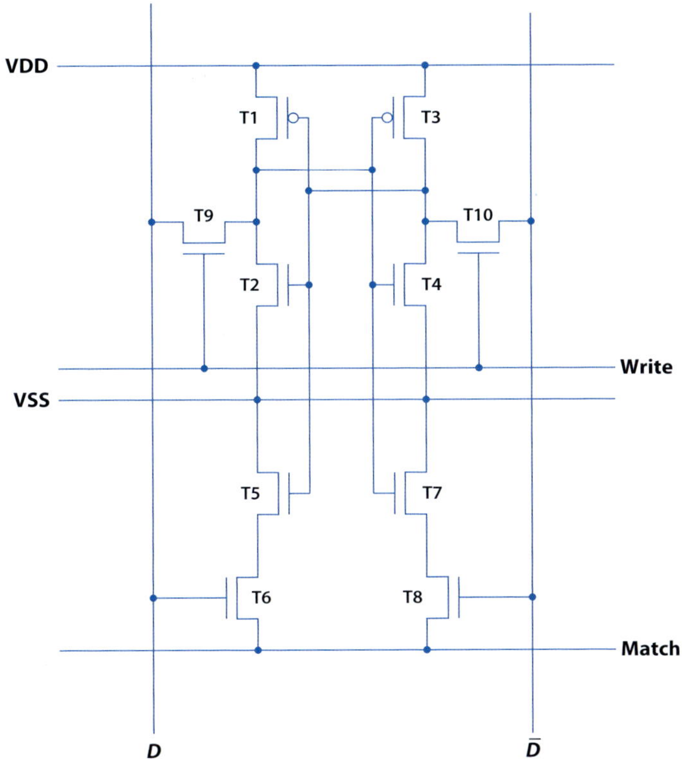

Figure 4

The schematics of a complementary metal-oxide semiconductor implementation of one bit of content-addressable memory (CAM). VSS and VDD refer to the source and drain polarization voltages, respectively. Transistors T1–T4 form the positive feedback loop that creates the two-state system that holds one bit of information. Transistors T9 and T10 are activated by the Write line to write the memory with the content of the data lines (D and $\bar{D}$). Transistors T5–T8 form a comparator that pulls the Match line down if the content of the memory does not match the data on the data lines. By stringing together a number of CAM bits, one can build CAM words of any length.

to the corresponding memory content for all patterns in parallel. Every stored coordinate that matches the data on the bus causes the hit bit to be set. This process is repeated for all the hit coordinates in one event.

After all the coordinates have been fed, the number of hit bits is counted in parallel for all the patterns, and the match bit is set only if the count reaches six. This means that all the coordinates in each of the six detector layers have been detected and that they therefore correspond to a possible trajectory of a single track as stored in the pattern bank.

Thereafter, all the match bits are fed into a priority encoder (23), which drives the address bus in output mode. The address of the first match bit is thereby extracted from the associative memory device. The same match bit is then cleared, causing the address of the second match bit to be encoded on the address bus. The process is repeated until all the match bits have been cleared, effectively causing the list of the addresses of all matching patterns to be output sequentially. Interestingly, this process differs from RAMs in which the address bus is always used as an input: Here, the address bus is used as an input bus when the memory is written and as an output bus when the memory is read.

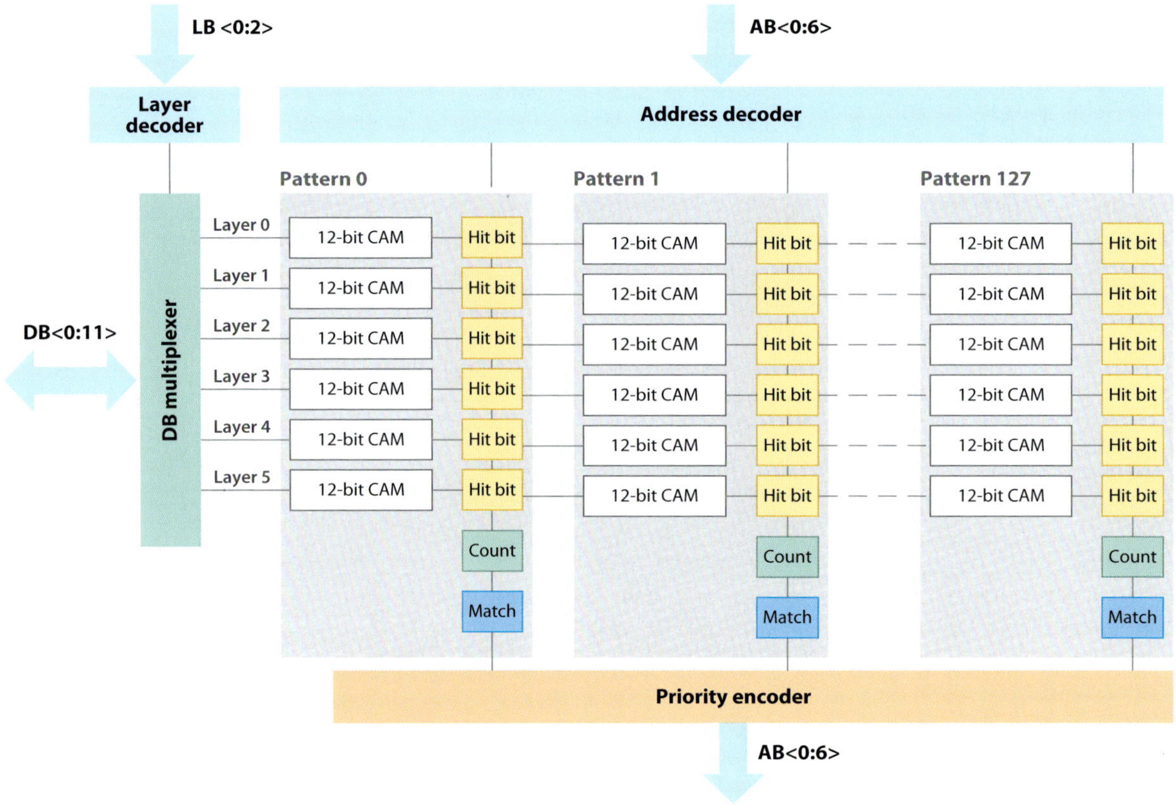

Figure 5

The internal structure of the associative memory device. Abbreviations: AB, address bus; DB, data bus; LB, layer bus.

It is often desirable to recognize patterns that have only a partial match with data so that we can find a track even when one or more hits have been lost due to detector inefficiencies. In most tracking detectors, allowing for some missing hits along a track still ensures sufficient rejection against random combinations and also increases the efficiency of finding all the tracks. For this reason, the associative memory device was designed with the option of having the match bit set when the hit count reaches a programmable threshold, which can be set to any number from one to six. If the threshold is set to six, only the patterns matching on all six detector layers are output; if the threshold is set to five, all the patterns matching on five or six detector layers are output. At CDF, requiring at least five hits out of six yielded sufficient rejection against background and a significantly larger track-finding efficiency than did requiring six out of six.

3.3. Generating the Pattern Bank

We now turn to the problem of constructing the pattern bank to be loaded in the associative memory device. The most direct and conceptually simple strategy is to use a Monte Carlo simulation method. Doing so involves generating tracks that are randomly extracted from a distribution in parameter space, similar to what we would expect in real experimental conditions, by simulating the trajectory of each track through the detector layers to determine the coordinate of the crossing point in each layer. All possible different hit patterns generated by simulated tracks are stored and used to fill the pattern bank. Unfortunately, in practical cases, the size of the pattern bank thereby created may be unmanageably large. The size of the pattern bank grows with a power of the number of segments in each detector layer. The power is typically the number of degrees of freedom of the track-parameter space. In the simple case of straight tracks (**Figure 2**), the power is two, and in the case of nonnegligible curvature due to a magnetic field, it becomes three. For a silicon detector a few centimeters wide and with a spatial resolution of order 10 μm, the number of segments would be of order of a few thousand and would require millions or even billions of patterns to be stored in the associative memory. These numbers are entirely unrealistic, considering that the maximum density achieved in the VLSI implementation of the associative memory device for the CDF experiment was 128 patterns on a 1-cm^2 silicon chip (20). The solution to this problem is to reduce the number of segments into which we subdivide the detector by increasing the segment size. This is done by grouping a number of strips in a silicon detector into a single logic segment and treating them, as far as the associative memory device is concerned, as a single detector element.

Using segment sizes ranging from 125 to 250 μm, CDF created a pattern bank of the order of 10^5, which was implemented through the use of the custom VLSI technology available in the 1990s. The increase in segment size, however, caused a significant loss of resolution, which had to be recovered. In the following section, we explain how to use the output of the associative memory to recover full detector resolution and extract the parameters of each track with high precision.

4. LINEARIZED TRACK FINDING AND FITTING

To find secondary vertices, we must extract the parameters of each track given the position of the hits. In particular, we need to measure the impact parameter of each track with a precision of a few tens of micrometers. In off-line analyses, this task is performed by fitting the analytical expression of the trajectory of a particle to the position of the hits in the detector, using a minimization algorithm that typically works by successive approximations. For our purposes, such an approach is not an option, given the time constraints imposed by the trigger requirements.

The solution to this problem was to substitute all the functions involved in the calculations with linear approximations, thereby allowing a much simpler minimization that could be performed analytically. For each track parameter, this procedure yielded a simple scalar product of the hit coordinates with an array of coefficients that depended only on the geometry of the detector and which could be computed in advance once and for all. In fact, the idea of separating pattern recognition from the precision measurement of the track parameters is one of the key points that led to the success of CDF's silicon vertex trigger (SVT) because it allows both for easier pattern recognition with coarser resolution and easier determination of track parameters, which become limited to the much smaller space within the boundaries of a single pattern.

4.1. Linearizing the Constraints

Each candidate track is represented as a list of n hit coordinates, one per detector layer, and can be thought of as a point $\boldsymbol{x}$ within a set $\mathcal{C} \subset R^n$, limited in extension by the physical boundaries on the hit coordinates. Clearly, not every point in $\mathcal{C}$ is equally likely to represent a track: Only those with coordinates aligned along a track path are reasonable candidates. The set of all acceptable candidates is therefore some subset $\mathcal{T} \subset \mathcal{C}$, to be precisely defined by some criterion (such as a χ^2 cut) that determines the efficiency of the set, which is defined as the probability that a real track crossing the detector produces a candidate belonging to $\mathcal{T}$. The process of deciding whether a given candidate track belongs to the subset $\mathcal{T}$ is termed track finding.

Suppose each track can be labeled by a number $m < n$ of real parameters $\boldsymbol{p}$ (e.g., P_T, ϕ_0, and d). In the ideal case of perfect resolution, the hit coordinates are uniquely determined by the parameters $\boldsymbol{p}$, and the set $\mathcal{T}$ reduces to an m-dimensional surface contained in $\mathcal{C}$, described by n parametric equations,

$$\boldsymbol{x} = \boldsymbol{x}(p_1, \cdots p_m), \tag{1.}$$

which can also be cast in implicit form by eliminating the parameters. Doing so yields $n - m$ constraint equations,

$$f_i(\boldsymbol{x}) = 0; \quad i = 1, \cdots n - m. \tag{2.}$$

The constraint functions $\{f_i\}$ can be determined, either analytically or numerically, from our knowledge of geometry. These functions formally solve the problem of track finding: For each candidate track, one evaluates the $f_i(\boldsymbol{x})$ and accepts the track if and only if all f_is are zero.

This holds exactly only for infinite resolution. The effect of finite resolution is to make the f_is random variables that take values slightly different from zero, thereby adding some finite thickness to the m-dimensional surface $\mathcal{T}$. The measure of this effect is given by the covariance matrix F_{ij} of the f_is, which can be calculated at first order from the covariance matrix $\boldsymbol{S}$ of the coordinates $\boldsymbol{x}$:

$$F_{ij} \simeq \sum_{k\ell} \frac{\partial f_i}{\partial x_k} \frac{\partial f_j}{\partial x_\ell} S_{k\ell}. \tag{3.}$$

The covariance matrix, being symmetric, can be made diagonal through a rotation in the space of the f_is. Furthermore, the new coordinates $\tilde{f}_i$ can be scaled to unit variance so that a χ^2 forms as $\chi^2 = \sum_i \tilde{f}_i^2$, where $\langle \tilde{f}_i^2 \rangle = 1$.

The values of the constraint functions $\tilde{f}_i(\boldsymbol{x})$ must be computed for each candidate track. Generally, they are complicated functions, so we approximate them with linear functions by expanding about some convenient point $\boldsymbol{x}_0$:

$$\tilde{f}_i \simeq \frac{\partial \tilde{f}_i}{\partial \boldsymbol{x}} \cdot (\boldsymbol{x} - \boldsymbol{x}_0) = \boldsymbol{v}_i \cdot \boldsymbol{x} + c_i. \tag{4.}$$

Geometrically, this amounts to approximating $\mathcal{T}$ with its tangent hyperplane in $\boldsymbol{x}_0$, and the $\boldsymbol{v}_i$s are the vectors orthogonal to the hypersurface in $\boldsymbol{x}_0$. Clearly, the approximation works well when $\mathcal{T}$ is nearly flat. In general, to obtain a sufficient precision, it is necessary to segment $\mathcal{T}$ into several smaller regions and perform the expansion around the central point $\boldsymbol{x}_0$ of each of them.

All constants can be calculated analytically from our knowledge of the detector geometry, but there is a numerical method that is simpler and more intuitive and that also allows easy handling of misalignments. This technique is known as principal components analysis (24).

This method begins from a sample of vectors $\boldsymbol{x}$ belonging to $\mathcal{T}$ generated either by Monte Carlo or from real data. The covariance matrix $\boldsymbol{M}$ of $\boldsymbol{x}$ can be estimated as

$$M_{ij} \simeq \langle (x_i - \langle x_i \rangle)(x_j - \langle x_j \rangle) \rangle. \tag{5.}$$

If we diagonalize $\boldsymbol{M}$, we generally find (*a*) $n - m$ eigenvectors with small eigenvalues, which form a base in the linear subspace orthogonal to the surface, and (*b*) m eigenvectors with larger eigenvalues, which form a base in the linear subspace tangent to the surface. The small eigenvalues correspond to the thickness of the surface caused by finite detector resolution, whereas the large eigenvalues correspond to the variance of the coordinates caused by spanning the track-parameter space.

We can now take the coordinates with respect to the eigenvectors orthogonal to the surface, scale them to unit variance, and identify them with the $\tilde{f}_i$s. This procedure is much more convenient than any analytical calculation because it is simpler and can be performed directly on samples of real data, thereby automatically incorporating corrections for possible detector misalignments that are already implicitly contained in $\boldsymbol{M}$.

4.2. Linearized Parameter Estimation

Now we come to the problem of estimating the parameters of the tracks we have found. We can invert (at least locally) Equation 1 to obtain the parameters as a function of the hit coordinates:

$$p_i = p_i(\boldsymbol{x}). \qquad 6.$$

By applying the same reasoning that we applied to the constraints in the previous section, we can use a linear approximation of Equation 6:

$$p_i \simeq \boldsymbol{w}_i \cdot (\boldsymbol{x} - \boldsymbol{x}_0) + p_i(\boldsymbol{x}_0) = \boldsymbol{w}_i \cdot \boldsymbol{x} + q_i. \qquad 7.$$

The optimal values of the constants $(\boldsymbol{w}_i, q_i)$ can be found simply by optimizing the resolution, that is, by minimizing the average quadratic deviation from the true values $\tilde{p}_i$:

$$\frac{\partial \langle [\tilde{p}_i - p_i(\boldsymbol{x})]^2 \rangle}{\partial \boldsymbol{w}_i} = 0. \qquad 8.$$

These constants can be readily obtained from a sample of tracks with known parameters. This is an important difference with respect to the case of the constraints discussed above, where we could obtain the constants from a sample of real tracks. In this case, we can use only Monte Carlo, and we are forced to rely on the accuracy of the detector geometry used in the simulation. We can easily determine the q_is from real data by using our knowledge of the average values $\langle \tilde{p}_i \rangle$ of each parameter for the track sample.

4.3. Indexing Parameters with Road Number

The strategy for track finding and fitting outlined above is based on the assumption that the linear approximations we are using are good enough to ensure that the performance in terms of background rejection and parameter resolution is not significantly degraded. Clearly, this strategy can be beautifully matched to the pattern recognition performed by the associative memory device (Section 3). Each track candidate found by the associative memory device is typically bound within a very small volume of the track-parameter space, and hit-coordinates range in a very small interval. Therefore, we can calculate in advance a different set of constants for the linear approximation for each road and index them with the road number, that is, the address in associative memory at which the corresponding pattern is stored. Given a track candidate, by using the road number we can immediately retrieve the constants relative to that road and calculate the χ^2 and, possibly, the track parameters. In practical cases, the associative memory device parameter-space segmentation may be much finer than needed for the linear approximations to be good, and the same set of constants may be used for multiple road numbers. For the CDF implementation, nearly 400,000 roads were needed in the associative memory device, but only 72 different sets of geometrical coefficients were sufficient to cover the whole detector.

5. THE SILICON VERTEX TRIGGER

The principles discussed in Section 4 are very general and could be applied to different situations and different experimental conditions. Below, we describe in some detail how these principles were applied to the design of the SVT for the CDF experiment (25–31).

5.1. The Tracking System of CDF

A more detailed description of the CDF upgraded detector can be found in References 32–34. The trigger system uses information from the silicon vertex detector and the central drift chamber, known respectively as SVXII (35) and COT (36). All these components are immersed in a longitudinal magnetic field produced by a superconducting solenoid.

Five layers of SVXII are used for triggering. Their distances from the beam range from 2.5 to 10.6 cm. Each layer of the silicon detector is subdivided into 12 ϕ sectors and six segments along the beam direction (i.e., the z coordinate). The ensemble of the five layers at the same ϕ and same z is termed a wedge. There are a total of $12 \times 6 = 72$ wedges in SVXII. For each charged track produced in the η interval between -1 and $+1$, the COT provides approximately 90 points, distributed more or less uniformly along the trajectory and ranging from 40 to 140 cm away from the beam. The signals from the COT are processed by the extremely fast tracker (XFT) trigger processor (37), which finds tracks and computes their ϕ and P_T in a time of approximately 5 μs. The precision of the XFT is far from sufficient to detect the impact parameters of displaced tracks from heavy-flavor decays, so the tracks found by the XFT are fed to the SVT, where they are combined with information from the silicon detector, first to measure impact parameters with a precision of a few tens of micrometers and second to make them available in time for a trigger decision ($\sim$20 μs).

5.2. Silicon Vertex Trigger Architecture

The SVT uses five concentric SVXII layers (0–4), with strips parallel to the beam (z coordinate), and two track parameters as calculated by the XFT (ϕ and P_T). We assign five layers of the associative memory to the five silicon layers. Hit coordinates are segmented into 250-μm bins. The two track parameters output by the XFT (ϕ and P_T) are mapped to one parameter ϕ_0 (angle ϕ at the origin of the track), segmented into 20-mrad bins, and assigned to the sixth layer in the associative memory device. Later, the two original parameters of each track are used in the fit, together with full-resolution hit coordinates from the five silicon layers.

For each event, the SVT reads out data from the SVXII, combines this information with information from the XFT, finds all tracks with $P_T > 2$ GeV/c, and reconstructs them with an impact-parameter accuracy of the order of 30 μm. All these steps must be performed within approximately 20 μs. **Figure 6** shows the SVT's main physical components and their interconnections. The SVT consists of more than 100 large electronic boards of 10 different types.

The SVT is implemented as a data-driven architecture. Data communication between functional blocks is obtained through point-to-point high-speed links. Various tasks are divided among functional blocks. Each functional block begins working on data as soon as it obtains them and outputs results as soon as they are ready: There is no attempt to synchronize the operation of different blocks by any other means than the flow of data. The communication protocol has been designed for maximum simplicity and speed, and it is uniform throughout the system.

Each SVXII wedge is read out through an independent data path. Sparsified[1] data from each readout line are fed into one hit finder, whose task is to find pulse-height clusters and compute the coordinate of the centroid of each cluster. There are a total of 72 hit finders, one per SVXII wedge. Hit coordinates from the six hit finders from the six wedges at the same ϕ and tracks from XFT (the COT track processor) are merged into a single data stream and fed to both the associative memory device and the hit buffer. The task of the associative memory device is to perform the

[1] Sparsification is the action of eliminating from the output data stream all the channels containing no relevant information (empty channels).

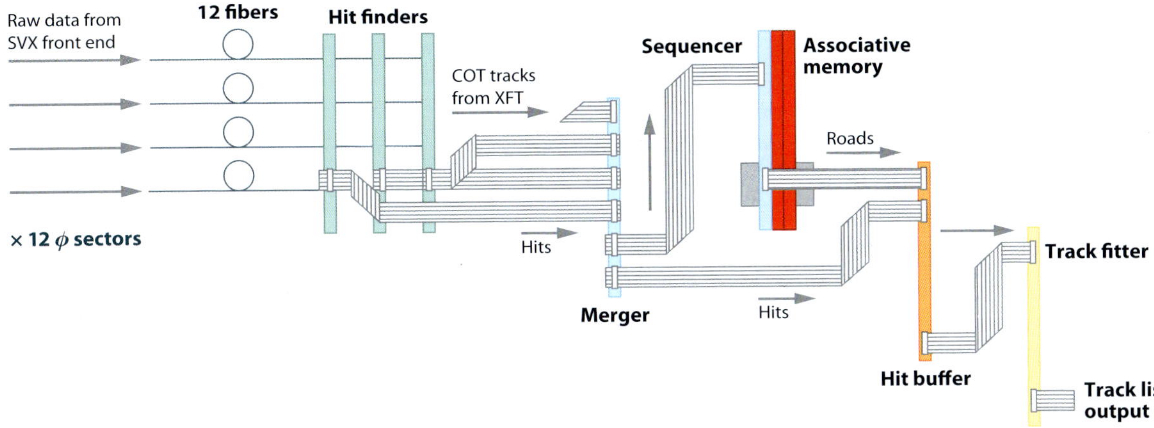

Figure 6

The physical components of the silicon vertex trigger and how they are interconnected. Abbreviations: COT, central drift chamber; SVX, silicon vertex detector; XFT, extremely fast tracker.

first stage of the pattern-recognition process using 250-μm bins on SVXII layers. These coarse-resolution track candidates are termed roads. Roads are transmitted from the associative memory device to the hit buffer, which associates each road with the proper hit coordinates and XFT track and transmits them to the track fitters. Each track fitter receives one or more roads, each with the corresponding full-resolution hits from the hit buffer, and reconstructs one or more tracks within each road by using linearized constraints and linearized parameter estimation (discussed above). Each processor outputs the parameters (P_T, ϕ, d, and z) of every track it finds. All the data from the 12 track fitters are merged into a single track list that is output by SVT and used by the overall trigger logic to build different physics triggers.

5.3. The Beam Spot

The SVT approach to tagging nonprimary tracks is to look at the impact parameter relative to the center of the beam spot. We abandon any attempt to find the position of the primary vertex of the event because (*a*) finding the primary collision vertex is difficult and (*b*) the results are often poor for B events, in which the track multiplicity is low. Moreover, at the typical luminosity for Run II, we have multiple primary vertices in a single event. The Tevatron beam spot has a transverse sigma of ~30 μm, and for B physics we are interested in tracks with an impact parameter of at least ≈100 μm. Therefore, use of the beam spot center instead of the primary vertex results in only a modest degradation in performance.

Because SVT measures only transverse coordinates, it observes only the transverse projection of the luminous region of the Tevatron. For this reason, we need to ensure that the detector axis is parallel to the beam; otherwise, the beam spot will appear to SVT as an elongated region, impairing its ability to recognize secondary tracks. Also, the need for measuring impact parameters with such a high precision imposes very strict requirements on the mechanical tolerances of the detector. The entire mechanical structure of the SVXII detector was built so that, after installation, all detector elements would be at their nominal relative positions within 20 μm (35).

For trigger purposes, therefore, we need to measure the impact parameter of all tracks with respect to the center of the beam spot projected on the plane transverse to the beam. To do so,

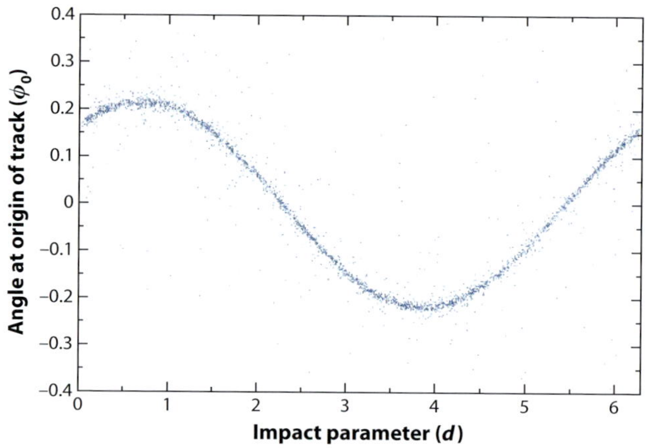

Figure 7

A typical d-ϕ_0 plot. The sine wave shape is created by tracks originating from a beam spot that is slightly displaced from the center of the coordinate system.

we need to know the exact position of the beam at the interaction point inside the CDF detector. This position is not constant in time; rather, it changes slightly whenever protons and antiprotons are injected into the machine. The position also drifts slowly during data taking. These changes can be of the order of hundreds of micrometers, and they cannot be neglected without affecting the performance of SVT. It follows that we need to measure the position of the beam spot in real time and use it to compute the impact parameters of all the tracks. The SVT strategy is to have two independent processes. One method finds all tracks in each event and computes their impact parameter with respect to the center of the SVX detector, which is assumed to be the origin of the coordinate system. This process is performed as described above. Another method takes all the tracks found by the SVT in all events and plots their impact parameter d versus ϕ_0. A typical d-ϕ_0 plot is shown in **Figure 7**. If the beam spot were located exactly in the center of the coordinate system, all points would lie on the horizontal axis because all the tracks would have zero impact parameter. If the beam spot were slightly displaced from the center, then a nonzero impact parameter would be measured with respect to the origin of the coordinate system, and a sine wave shape would be obtained as a function of ϕ_0:

$$d = x_0 \sin(\phi_0) - y_0 \cos(\phi_0),$$

where x_0 and y_0 are the coordinates of the beam spot that can be obtained by fitting the above expression to the d-ϕ_0 plot.

Given the very high rate at which the SVT processes events, statistics of several hundred thousand tracks can be collected in a matter of seconds, and a very precise determination of the position of the beam can be obtained and used to correct the impact parameter of all the tracks in real time. **Figure 8** shows the distribution of the impact parameters with respect to the actual position of the beam spot of a sample of tracks reconstructed by SVT.

5.4. Trigger Paths

The track list output by SVT is used by the CDF trigger processors to look for decays of long-lived particles. The selection criteria are typically based on the requirement of at least two tracks in the same event with a transverse momentum greater than 2 GeV/c and an impact parameter

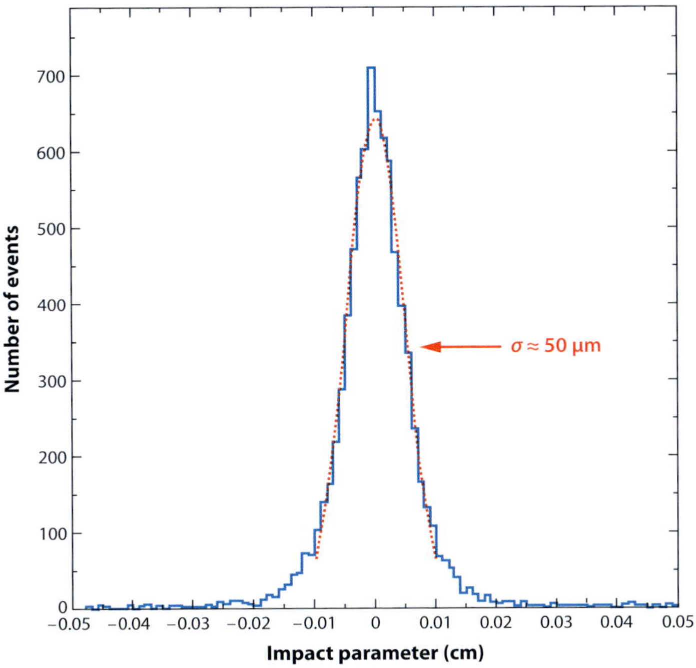

Figure 8

The distribution of the impact parameters of a sample of tracks as reconstructed by the silicon vertex trigger (SVT). A Gaussian with $\sigma \approx 50$ μm fits the core of the distribution very well. The width of the distribution is interpreted as the convolution of the actual transverse size of the beam spot ($\approx$25 μm) with the impact-parameter resolution of the SVT ($\approx$43 μm).

greater than 100 μm, but the exact value of these cuts is optimized as a function of instantaneous luminosity and can range up to 2.5 GeV/*c* and 120 μm, respectively. A number of distinct trigger paths, with slightly different cuts, are provided in parallel and are enabled, disabled, or prescaled[2] according to the evolution of the instantaneous luminosity in the duration of a Tevatron store. In this way, we strive to optimize the number of signal events that are permanently recorded and available for physics analyses (38).

6. PHYSICS RESULTS

The SVT was designed and built by CDF between 1990 and 2000. It was installed in the CDF control room in the summer of 2000 and began operation in October of the same year. The first publication of the Tevatron Run II results (a measurement of the mass of the D_s) was based on data collected by the SVT (39).

The SVT has been operating without interruption since 2000, and it has allowed CDF to perform important measurements that would otherwise have been impossible. Examples of such measurements are those of the B_s oscillation frequency (40) and the direct *CP* asymmetry in the charmless decay of B_d, B_s, and Λ_b (41–43). The SVT has also provided CDF with the world's largest sample of D^0 mesons, paving the way to precision measurements of D^0 oscillations and

[2]Prescaling is the action of sampling only a fraction of the events selected by a specific trigger path.

Figure 9

The likelihood plot for the B_s oscillation frequency, as obtained by CDF (40). The contribution of hadronic decays, collected by the silicon vertex trigger, is shown in red, and the contribution from lepton triggers is shown in blue.

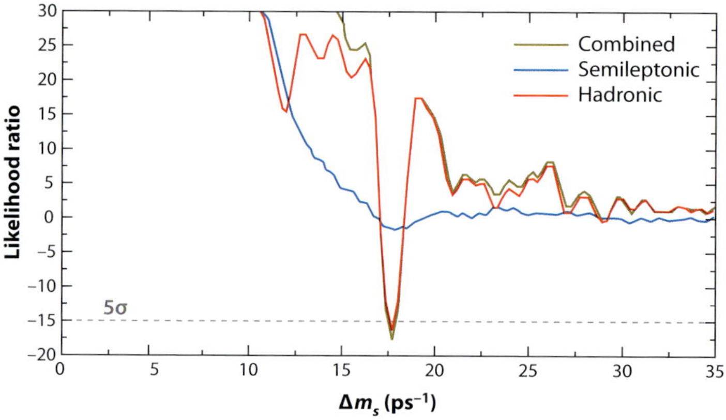

direct *CP* violation (44). The range of possible measurements using the SVT data sample is still being explored.

The measurement of B_s oscillations is a good example of the importance of triggering on displaced vertices rather than on the presence of leptons. The proper time of the B_s decay is obtained from the measurement of its momentum and the path length traveled before decay, but in all semileptonic decays a neutrino in the final state escapes detection and precludes a precise momentum measurement, which in turn severely affects the precision of the determination of the proper time. **Figure 9** shows the likelihood plot for the B_s oscillation frequency as obtained by CDF (40). **Figure 10** shows the proper time resolution for different decay modes of the B_s.

Figure 10

The proper time resolution as a function of proper time, as obtained by CDF for the measurement of the B_s oscillation frequency (40). Plots for different decay modes of the B_s are shown.

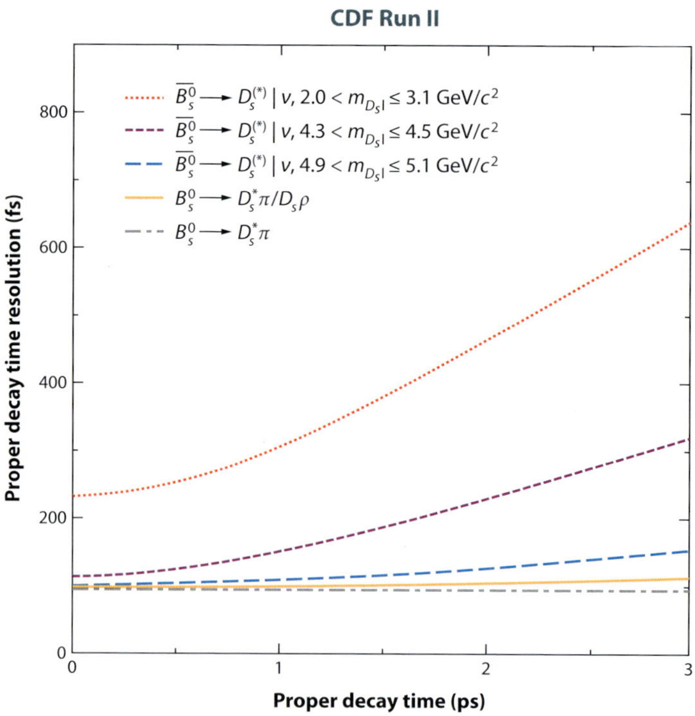

The advantage of hadronic decay modes over semileptonic decay modes is evident. Without the SVT, we would still be waiting for an observation of the B_s oscillations.

7. CONCLUSIONS

The performance of SVT at CDF shows that by using a sophisticated trigger to measure impact parameters in real time, hadron colliders can achieve results in the field of *c* and *b* physics that were difficult to imagine in the past and that are competitive with, or in some cases superior to, results obtained at dedicated e^+e^- colliders.

The experience of designing, building, running, and doing physics with the SVT at CDF has generated ideas about how to use the same principles for the next generation of hadron colliders, namely the LHC. Although nothing similar to the SVT has been included in the baseline design of current LHC detectors, ATLAS is now considering the addition of a hardware-based fast track finder (13, 14) as part of its upgrades.

The complexity of track-finding algorithms, which require a great deal of computer power, will be an important factor in determining the quality of the physics results from the LHC and other future high-energy physics experiments. Innovative techniques to efficiently solve pattern-recognition and track-fitting problems on the basis of the latest electronic technologies could significantly increase these experiments' physics reach, and they certainly represent an important and fascinating challenge for future generations of experimental particle physicists.

DISCLOSURE STATEMENT

The authors are not aware of any affiliations, memberships, funding, or financial holdings that might be perceived as affecting the objectivity of this review.

LITERATURE CITED

1. Weinberg S. *Phys. Rev. Lett.* 19:1264 (1967)
2. Salam A. In *Elementary Particle Theory*, ed. N Svartholm, pp. 367–77. Stockholm: Almquist & Wiksells (1969)
3. Glashow SL, Iliopoulos J, Maiani L. *Phys. Rev. D* 2:1285 (1970)
4. Kobayashi M, Maskawa T. *Prog. Theor. Phys.* 49:652 (1973)
5. Cabibbo N. *Phys. Rev. Lett.* 10:531 (1963)
6. Amsler C, et al. (Part. Data Group) *Phys. Lett.* B667:1 (2008)
7. Hurst RB, et al. (WA82 Collab.) *Nucl. Phys. Proc. Suppl.* 16:302 (1990)
8. Kaplan DM. (BTeV Collab.) *Proc. APS/DPF/DPB Summer Study Future Part. Phys.* 1:E704 (2001)
9. Christian CD, et al. *Nucl. Instrum. Methods A* 473:152 (2001)
10. Zimmermann S, et al. *Nucl. Instrum. Methods A* 465:224 (2000)
11. Caron S. (DØ Collab.) *Nucl. Phys. Proc. Suppl.* 156:143 (2006)
12. Augusto Alves A, et al. (LHCb Collab.) *J. Instrum.* 3:S08005 (2008)
13. Annovi A, et al. *IEEE Trans. Nucl. Sci.* 48:575 (2001)
14. Brubaker E, et al. *IEEE Trans. Nucl. Sci.* 55:145 (2008)
15. Mead C, Conway L. In *Introduction to VLSI Systems*, p. v. New York: Addison-Wesley (1980)
16. Dell'Orso M, Ristori L. *Nucl. Instrum. Methods A* 278:436 (1989)
17. Weste NHE, Eshraghian K. In *Principles of CMOS VLSI Design: A System Perspective*, pp. 564–80. New York: Addison-Wesley. 2nd ed. (1993)
18. Taub H, Schilling D. In *Digital Integrated Electronics*, pp. 420–31. New York: McGraw-Hill (1977)
19. Millman J, Grabel A. In *Microelectronics*, pp. 352–66. New York: McGraw-Hill. 2nd ed. (1987)
20. Morsani F, et al. *Nucl. Instrum. Methods A* 315:446 (1992)

21. Weste NHE, Eshraghian K. In *Principles of CMOS VLSI Design: A System Perspective*, pp. 589–90. New York: Addison-Wesley. 2nd ed. (1993)
22. Pagiamtzis K, Sheikholeslami A. *IEEE J. Solid-State Circuits* 41:712 (2006)
23. Millman J, Grabel A. In *Microelectronics*, pp. 286–91. New York: McGraw-Hill. 2nd ed. (1987)
24. Pearson K. *Philos. Mag.* 2:559 (1901); Jolliffe IT. *Principal Component Analysis*. New York: Springer. 2nd ed. (2002)
25. Bardi A, et al. *Nucl. Instrum. Methods A* 409:658 (1998)
26. Ashmanskas W, et al. *Nucl. Instrum. Methods A* 447:218 (2000)
27. Ashmanskas W, et al. *Nucl. Instrum. Methods A* 477:451 (2002)
28. Bardi A, et al. *Nucl. Instrum. Methods A* 485:178 (2002)
29. Ashmanskas W, et al. *Nucl. Instrum. Methods A* 501:201 (2003)
30. Ashmanskas W, et al. *Nucl. Instrum. Methods A* 518:532 (2004)
31. Adelman J, et al. *Nucl. Instrum. Methods A* 569:111 (2006)
32. Acosta D, et al. *Phys. Rev. D* 71:032001 (2005)
33. Acosta D, et al. *Phys. Rev. D* 71:052003 (2005)
34. Abulencia A, et al. *J. Phys. G* 34:2457 (2007)
35. Sill A, et al. *Nucl. Instrum. Methods A* 447:1 (2000)
36. Affolder T, et al. *Nucl. Instrum. Methods A* 526:249 (2004)
37. Thomson EJ, et al. *IEEE Trans. Nucl. Sci.* 49:1063 (2002)
38. Anikeev K, et al. arXiv:hep-ph/0201071 (2002)
39. Acosta DE, et al. *Phys. Rev. D* 68:072004 (2003)
40. Abulencia A, et al. *Phys. Rev. Lett.* 97:242003 (2006)
41. Abulencia A, et al. *Phys. Rev. Lett.* 97:211802 (2006)
42. Aaltonen T, et al. *Phys. Rev. Lett.* 103:031801 (2009)
43. Morello M, et al. *Nucl. Phys. Proc. Suppl.* 170:39 (2007)
44. Aaltonen T, et al. (CDF Collab.) *Phys. Rev. Lett.* 100:121802 (2008)

Advances in Calorimetry

James E. Brau,[1] John A. Jaros,[2] and Hong Ma[3]

[1]Department of Physics, University of Oregon, Eugene, Oregon 97403-1274; email: jimbrau@uoregon.edu

[2]Particle Physics and Astrophysics, SLAC National Accelerator Laboratory, Menlo Park, California 94025-7015; email: john@slac.stanford.edu

[3]Physics Department, Brookhaven National Laboratory, Upton, New York 11973-5000; email: hma@bnl.gov

Annu. Rev. Nucl. Part. Sci. 2010. 60:615–44

The *Annual Review of Nuclear and Particle Science* is online at nucl.annualreviews.org

This article's doi:
10.1146/annurev.nucl.012809.104449

0163-8998/10/1123-0615$20.00

Key Words

calorimeters, energy measurement, particle identification, compensation

Abstract

We review recent advances in calorimetry for high-energy physics. We provide an overview of the fundamentals of calorimetry, then survey calorimeters used in recent experiments, which represent the current state of the art. We conclude with descriptions of several research and development efforts and discuss future directions of the field.

Contents

1. INTRODUCTION

Calorimetry in particle and nuclear physics refers to techniques for measuring particle energies by absorbing them in matter. These are well-established techniques that serve as a critical component of experiments that explore the energy frontier, flavor physics, and neutrinos at accelerators, and experiments that address astrophysics from Earth and in space. This review concentrates on recent advances in calorimetry and emphasizes high-energy applications, typically for energies in excess of 1 GeV. We summarize calorimetry fundamentals, survey recent major applications, and discuss new developments that are likely to influence future experiments.

2. BRIEF OVERVIEW OF CALORIMETRY FUNDAMENTALS

Calorimetry relies heavily on an understanding of the interactions of particles in matter, and it makes use of several techniques for detecting these interactions. Two general classes of interactions occur: (*a*) those initiated by electrons, positrons, and photons, which produce cascades of these electromagnetic (EM) particles; and (*b*) those initiated by strongly interacting particles, which result in hadron showers. The showers are detected through observation of ionization or light generated by the particles in the shower. This section describes these aspects of calorimetry and concludes with a brief overview of the requirements for detector performance imposed by the physics goals of modern experiments.

Accurate calorimeter simulation codes have played an essential role in the development of calorimeters (1). They are used extensively in the design stage of experiments, as well as in data analysis, where extraction of the most accurate physics results depend on reliable simulations. Cascades of EM particles are well defined (2) and very accurately simulated. However, hadron shower development is complicated by nuclear breakup, which includes the loss of energy to nuclear binding energy and the production of neutrons. Many current calorimeter studies rely on the GEANT4 simulation toolkit (3), which supports several different hadronic physics models.

2.1. Key Properties of Electromagnetic Showers

A high-energy electron, positron, or photon entering a medium initiates an EM cascade of electrons, positrons, and photons (2). Bremsstrahlung emission of a photon by the entering electron or positron, or electron-positron pair production by the photon, initiates the cascade and is followed by bremsstrahlung and pair production. Above 1 GeV, the cross sections for these processes are nearly energy independent. The loss of energy through bremsstrahlung by high-energy electrons is characterized by the radiation length X_0,

$$dE/dx|_{radiation} = -E/X_0, \tag{1}$$

where E is the energy of the electron. For a medium of atomic number Z and atomic mass A, the radiation length is given by (4)

$$X_0 \approx \frac{716.4\,\mathrm{g\,cm^{-2}\,A}}{\mathrm{Z(Z+1)}\ln(287/\sqrt{\mathrm{Z}})}. \tag{2}$$

Typical values are $X_0 = 6.37$ g cm^{-2} for lead, 12.86 g cm^{-2} for copper, 13.84 g cm^{-2} for iron, and 21.82 g cm^{-2} for silicon. The conversion length for a high-energy photon to materialize as an electron-positron pair is $\frac{9}{7}X_0$, which leads to the cascade. Surviving high-energy photons follow an exponential attenuation with this conversion length.

Within the cascade, electrons and positrons deposit energy by ionizing and exciting atoms in the medium and emit Cherenkov radiation while relativistic. A good approximation (5) to the average longitudinal profile ($t = x/X_0$) with fit parameters a and b is

$$\frac{dE}{dt} = E_0 b \frac{(bt)^{a-1}e^{-bt}}{\Gamma(a)}. \tag{3}$$

Approximate values for the fit parameters are $a \approx 1 + 0.5[\log(E/E_c) - \alpha]$ and $b \approx 0.5$, where $\alpha = 0.5$ for e^+, e^- and $\alpha = -0.5$ for γ.

The longitudinal maximum of the cascade, $t_{\max} \approx \log(E/E_c) - \alpha$, occurs when the average shower-particle energy falls below the production threshold (6) [characterized (4) by the critical energy $E_c \approx 610$ MeV/(Z + 1.24)] and particle multiplication ceases. Fluctuations in the shower profile and length are dominated by the early high-energy processes that initiate the cascade. The total track length of electrons and positrons in the shower is proportional to the energy of the incident particle. Consequently, measurements of shower properties that are proportional to the total track length (such as ionization and Cherenkov or scintillation light) are a good measure of energy.

The transverse profile of the energy deposition is characterized by the Molière radius [$R_M = 0.0348X_0(\mathrm{Z} + 1.24)$], with approximately 90% of the energy contained within the cylinder of radius R_M. The achievable energy, position, and angular resolution for EM showers are limited by the measurement process rather than by the physics of the shower development. Generally, total-absorption measurements in homogeneous media produce the best energy resolution, and highly granular, maximal density produces the best position and angular resolution. Refer to Reference 4 for a more complete description and additional references.

2.2. Key Properties of Hadronic Showers

Hadron calorimetry is essential to modern particle physics experiments in which measurements of hadronic jets are required. In contrast to electrons and photons, hadrons can interact with matter through the strong interaction, leading to various effects that make measurements of hadronic showers more complex than EM showers. The scale of hadronic interactions is characterized by

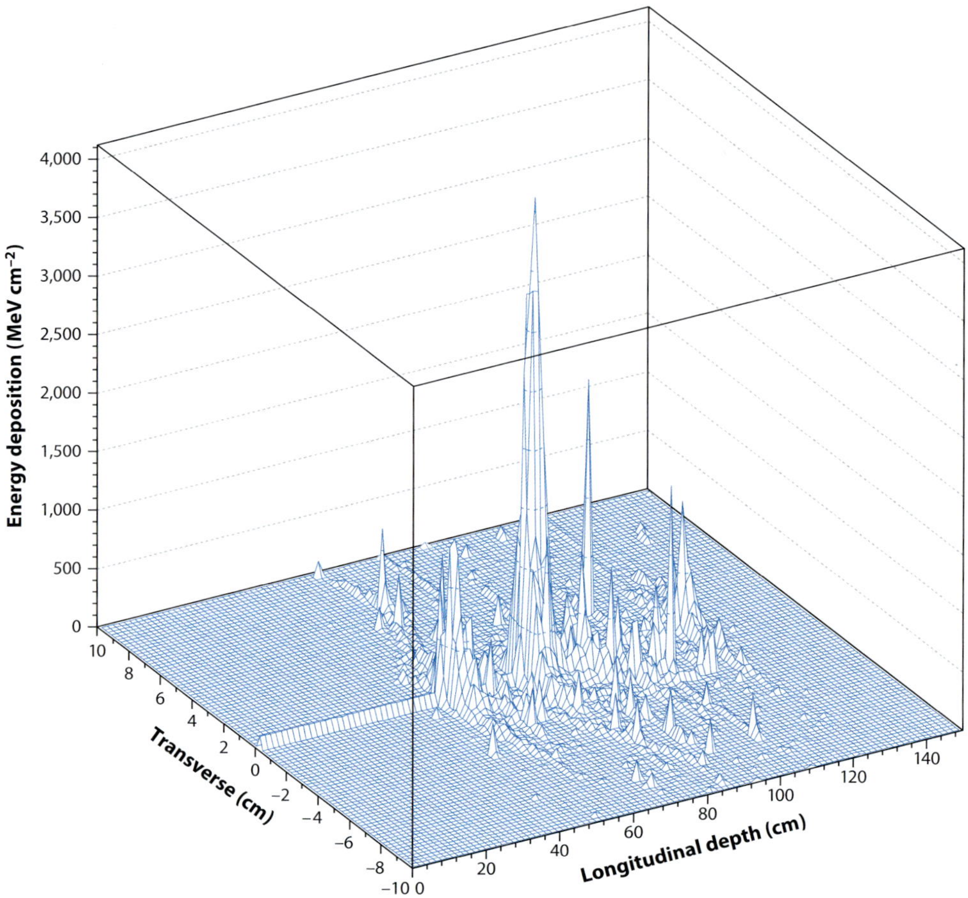

Figure 1

The energy deposition density for a typical hadron shower in lead, by a 100-GeV pion, in longitudinal and transverse dimensions (7).

the interaction length: $\lambda = 37.8A^{0.312}$ g cm^{-2}. On average, a hadron travels an interaction length before undergoing a nuclear interaction, which may create many other hadrons, including π^0s, as well as nuclear fragments. Photons from π^0 decays initiate EM showers, whereas charged hadrons deposit energy by ionization and both neutral and charged hadrons undergo subsequent nuclear interactions. A hadron shower develops in matter through the cascade of these secondary particles, with lumpy energy deposits in the medium, as illustrated by a 100-GeV pion shower in lead in **Figure 1** (7). Containing 95% (99%) of a 100-GeV pion shower in iron requires a calorimeter depth of 7.5 (12) interaction lengths (4).

A fraction of the secondary particles produced by the strong interaction are π^0s, which constitute approximately one-third of all pions. These particles subsequently decay into photons, giving rise to the EM fraction of the hadronic shower. The rest of the hadron's energy can be dissipated as ionization by charged particles, recoil and excitation of nuclei, and evaporation of nucleons through spallation reactions. **Figure 2** shows the average fraction of the various components of the non-EM part of the hadron shower. A hadron shower releases a significant amount of energy that is not detected by the usual detection methods, such as that expended in overcoming nuclear

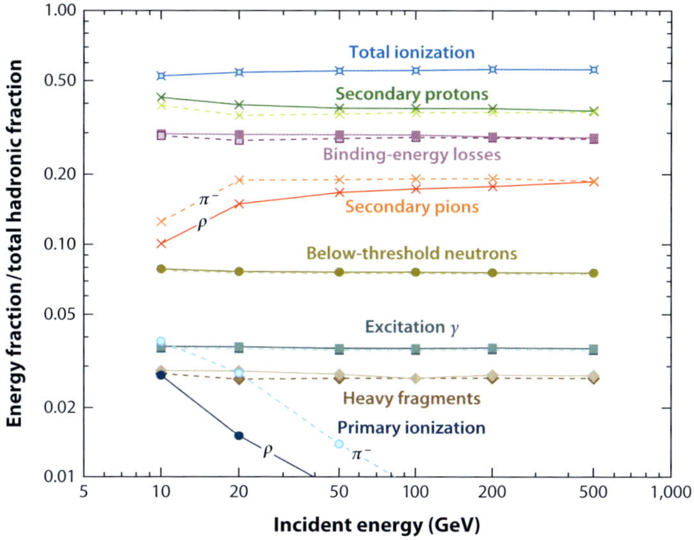

Figure 2

Energy loss of the non-π^0 components of the hadron shower initiated by protons (*solid lines*) and negative pions (*dashed lines*) in iron, in a fraction of the total non-π^0 energy, estimated by CALOR simulation (8).

binding energy, and some of the visible energy is deposited on long timescales. Neutrons play an important part in the total observable energy. They lose kinetic energy by elastic nuclear collisions and are eventually thermalized and captured by nuclei, giving rise to photons from nuclear photoemission on the timescale of microseconds.

Let $\langle f_{em}\rangle$ be the EM fraction of the energy, $\langle f_h\rangle = 1 - \langle f_{em}\rangle$ the purely hadronic fraction, e the calorimeter response to an EM shower per unit energy, and h the calorimeter response to a hadronic shower per unit energy. The ratio of the calorimeter's response to pions and electrons of the same incident energy can then be written as

$$\pi/e = \langle f_{em}\rangle + \langle f_h\rangle h/e = 1 - (1 - h/e)\langle f_h\rangle. \qquad 4.$$

As the incident hadron energy increases, the number of hadronic interactions increases and $\langle f_h\rangle$ decreases, which is fairly well described by the power law (8, 9)

$$\langle f_h\rangle \sim (E/E_0)^{m-1}, \qquad 5.$$

where E_0 is roughly the threshold for inelastic interactions (~1 GeV for incident pions) and m is experimentally determined to be in the range between 0.80 and 0.87. **Figure 3** shows the energy dependence of $\langle f_{em}\rangle$ and its standard deviation as simulated by FLUKA (9). The fractional fluctuation is large at low energy (48% at 10 GeV) and is still quite appreciable at the energy of high-energy colliders (28% at 100 GeV).

If $h/e \neq 1$, the hadronic calorimeter is intrinsically nonlinear, and its response is subject to the fluctuations of the EM fraction event by event. Several ways to improve hadron calorimeter energy resolution, based on minimizing these fluctuations, are as follows.

1. Compensation (10). For a calorimeter with $h/e = 1$, the nonlinearity is removed (11). This has been achieved in a scintillator sampling calorimeter as well as in a uranium liquid argon (LAr) calorimeter.
2. Weighting. Because EM showers tend to have higher energy density in the detector, measurements of the energy density in a highly segmented detector allow energy clusters to be

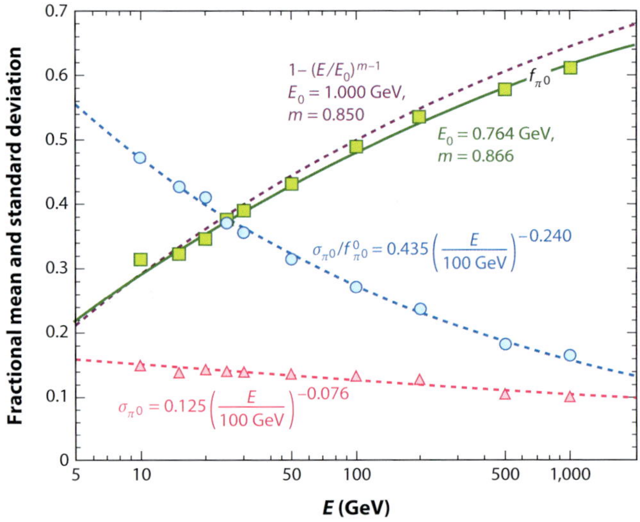

Figure 3

Energy dependence of the mean π^0 fraction of hadron showers, $f^0_{\pi^0} \sim \langle f_{em} \rangle$, its standard deviation (σ_{π^0}) of f_{π^0} distribution, and the standard deviation relative to the mean (σ_{π^0}/f_{π^0}) for negative pions incident on a lead calorimeter, simulated with FLUKA (9).

classified as EM or hadronic. The resolution can thus be improved by properly weighting the EM and hadronic clusters (12).

3. Dual readout. Techniques have recently been developed to measure the EM and hadron components of the shower separately. By summing these components with the right weights, the response to the hadronic showers can be optimized (13).

In high-impact collisions, quarks and gluons materialize as jets after fragmentation. The energy resolution of interest is usually the jet-energy resolution. Although improving the jet-energy resolution has traditionally relied on improving the hadronic response of calorimeters, particle-flow calorimetry (Section 4.3) has been proposed to optimize the resolution by separately measuring the neutral and charged components of the jets through the use of calorimeter clusters and tracks reconstructed in front of the calorimeter.

2.3. Signal-Detection Techniques

Most of the energy deposited in a calorimeter by high-energy electrons, positrons, photons, and charged and neutral hadrons is ultimately transferred to charged particles, which lose energy via ionization and excitation, as described by the Bethe-Bloch equation (4). Solid-state detectors, noble-liquid calorimeters, and gas detectors detect and measure this ionization signal directly. In scintillation materials, a fraction of the ionization energy is transferred to atomic or molecular excitations. Photosensors detect the light emitted in deexcitation. Only relativistic particles produce Cherenkov radiation, which constitutes a very small fraction of the total deposited energy but can be detected in suitable media by photosensors. All three signals are proportional to the energy deposited and have been exploited in calorimeters for high-energy physics experiments. Details can be found in standard references (4, 14).

The size of the ionization signal depends on the characteristics of the detecting medium. In solid-state detectors, such as silicon, a minimum ionizing charged particle (mip) generates a signal of approximately 22,000 electrons in a 300-μm-thick detector, thanks to the small 3.6-eV band gap. The signal is generated on a timescale of tens of nanoseconds and can be detected after electronic amplification. The noble liquids, with ionization potentials greater than 20 eV, generate comparatively fewer electrons per centimeter but still yield of order 50,000 electrons cm^{-1}, enough for detection with electronic amplification. Recombination of some of the ionization signal results in excitation and subsequent fluorescence. The electron drift velocity in LAr saturates around 1 cm $\mu\mathrm{s}^{-1}$, so collection times in LAr are much longer than in silicon. Gas detectors generally sample a very small fraction of the total energy deposited in a calorimeter. A mip generates only approximately 100 electrons cm^{-1} passing through argon gas at standard temperature and pressure, so amplification is needed. Proportional chambers, gas electron multipliers (GEMs), micromesh gaseous structure detectors (MicroMegas), and resistive plate chambers (RPCs) use regions of high electric fields to produce electron avalanches, generating signals that are easily detectable with modern electronics. The intrinsic resolution of the ionization signal in moderately thick detectors, such as scintillator or LAr cells, is worse than expected from simple Gaussian errors because of the possibility of infrequent large energy-transfer collisions. Even so, practical considerations such as amplifier noise, sampling statistics, and calibration systematics usually dominate the resolution. In thin-silicon and gas detectors, Landau fluctuations in the original ionization signal, gain fluctuations, and the small sampling fraction limit the achievable resolution.

Energy loss to excitation of atomic and molecular levels can give rise to photon emission in the visible region, which is readily detectable by photomultiplier tubes (PMTs), photodiodes, avalanche photodiodes, or silicon photomultipliers (SiPMs). In a plastic scintillator, only a few percent of the energy deposited results in photon emission, yielding roughly 1 visible photon per 100 eV deposited. Although even this yield produces roughly 20,000 photons cm^{-1} in a typical plastic scintillator, the detected signal is much smaller because not all the light is collected, and the quantum efficiency of the photon sensor is often well below 100%. In noble liquids and inorganic crystals, however, a much larger fraction of the initial energy loss results in excitation, and the corresponding light yields can be an order of magnitude larger. In the case of total-absorption calorimeters exploiting inorganic crystals or noble liquids, the number of detected photons per mip is large enough to ensure high resolution. The pulse length of scintillators ranges from a few nanoseconds in fast scintillating plastics to hundreds of nanoseconds or more in some crystals.

Although very little of the energy deposited by a high-energy EM shower appears in Cherenkov radiation, the large numbers of relativistic electrons and positrons produce a useful Cherenkov signal in suitably transparent media, such as plastic, quartz, and lead glass. The signal size in plastic, for example, is only $\sim$500 photons cm^{-1} per electron, well below that of a scintillator. The time dispersion of the Cherenkov signal is minimal, so subnanosecond timing is possible in suitable detectors. Because typical hadron showers have (excepting their EM components) comparatively few relativistic particles and thus comparatively small signals, Cherenkov detectors are generally employed for EM calorimetry.

2.4. Physics Requirements on Calorimeter Performance

EM calorimeters are needed to detect and measure photon energies and angles, identify π^0s, and identify electrons and measure their energy. Hadronic calorimeters are needed to detect and measure neutral hadrons and, in conjunction with EM calorimeters, to trigger on and measure high-energy jets and missing energy and to reconstruct di- and multijet final states.

Calorimeters play a critical role in particle identification and can enhance electron-energy measurements. Electron identification requires good transverse segmentation in EM calorimeters to isolate the energy deposited by electrons from other particles, as well as good longitudinal segmentation to distinguish EM shower profiles from other energy depositions. Good segmentation also permits photon position and angle measurements. Photons are identified as EM showers without an associated charged track. Moderate energy resolution [$\approx 15\%/\sqrt{E(\text{GeV})}$] is adequate for good electron identification. Really superb EM energy resolution is needed for high-energy spectroscopy; a good signal-to-noise ratio in π^0 reconstruction; and distinguishing rare processes from backgrounds, for example, Higgs $\rightarrow \gamma\gamma$. Excellent EM energy resolution is also needed for the precision measurement of electron energy because radiative processes may degrade the momentum measurement in the tracker. The deposition of isolated energy in the hadron calorimeter identifies the presence of a neutral hadron, whether K_L or neutron. With sufficient segmentation, a hadronic calorimeter also identifies penetrating mips as muon candidates.

High-energy parton-parton collisions are identified in hadron interactions by the presence of high-p_T jets in the final state; many new particle searches rely on missing-energy signatures. Therefore, both high-E_T and high–missing E_T signatures are critical for triggering on and analyzing events at hadron colliders. Moderate jet-energy resolutions [$\approx 60\%/\sqrt{E(\text{GeV})}$] have been adequate for this purpose and for identifying and reconstructing multijet final states, such as those from top. To maximize the signals of these particles against combinatoric backgrounds, or to optimize the search for new particles like the Higgs, requires even better dijet mass resolution, which depends on jet-energy resolution as $\Delta M/M = (\Delta E_{jet}/E_{jet})/\sqrt{2}$. Setting the desired mass resolution to be equal to the natural width of the Z^0 requires $\Delta E_{jet}/E_{jet} \approx 3-4\%$. This goal is beyond the current state of the art, but it motivates the new methods described below. Gluon radiation, imperfect jet finding, noise from underlying events, and missing low-momentum particles degrade mass resolution. Improved jet-energy resolution leads directly to improved accuracy in measuring the energy of reconstructed particles and improved resolution in measuring missing energy. Good missing-energy resolution also requires full solid-angle coverage and hermeticity.

3. CURRENT STATE OF THE ART OF CALORIMETRY

High-energy calorimeters have used several technologies. Such calorimeters must operate over a wide range of energies, all the way up to that of the Large Hadron Collider (LHC), $E_{cms} = 14\,\text{TeV}$, measuring with precision the energy and position of single particles and jets of particles. At lower energies, high-energy calorimeters must reconstruct π^0 decays, measure exclusive channels in bottom and charm decays, and detect radiative decays of heavy vector mesons. In this section, we survey current EM calorimeters, sampling hadron calorimeters, and total-absorption hadron calorimeters. See the tables for summaries of key parameters of representative high-energy calorimeters, including calorimeters used in neutrino experiments (**Table 1**), calorimeters used in hadron colliders and *ep* colliders (**Table 2**), calorimeters used in e^+e^- flavor factories (**Table 3**), and calorimeters used in high-energy e^+e^- experiments (**Table 4**).

3.1. Electromagnetic Calorimeters

The current generation of e^+e^- flavor factories—KEKB, PEP-II, CESR, BEPC III, DAFNE, and VEPP-4M—have general-purpose detectors that utilize several techniques for EM calorimetry: cesium iodide crystals doped with tellurium [CsI(Tl)], liquid krypton (LKr), and Shashlyk scintillating fibers. Special-purpose experiments have used liquid xenon (LXe). Silicon-tungsten (Si/W) EM calorimeters were first used at LEP and SLC for precision luminosity monitoring. At the LHC, the high-rate, high-radiation environment demands fast radiation-hard techniques, including LAr, lead tungstate ($PbWO_4$), and Shashlyk calorimeters. Examples of EM calorimeters based on crystals, scintillating fibers, cryogenic liquids, and silicon are discussed below.

Table 1 Parameters of representative calorimeters used in neutrino experiments[a]

Parameter	Collaboration					
	CDHS	**CHARM II**	**NuTeV**	**MINOS**	**T2K**	**NOνA**
E range (GeV)	<400	<400	<400	<120	0.2–5	<10
Physics goal	ν_μ interactions	νe scattering	Electroweak	ν_μ disappearance	$\nu_\mu \rightarrow \nu_e$	$\nu_\mu \rightarrow \nu_e$
Dimensions (m)	3.75 (D) × 21 (L)	4 (D) × 4 (H) × 35.7 (L)	3 (D) × 3 × 15.7 (L)	8 (D) × 31 (L)	39 (D) × 42 (H)	15.7 (D) × 15.7 (H) × 78 (L)
Total mass (kton)	1.25	0.69	0.69	5.4	50	15
Technology	Fe-scintillator sandwich with DC	Glass/streamer tubes with strips and scintillators	Scintillator/steel with DC	Fe-scintillator sandwich	Water Cherenkov/ PMTs	Liquid scintillator (70% of mass)/APDs
Energy resolution	$190\%/\sqrt{E}$(hadronic)	$27.5\%/\sqrt{E}+3.0\%$ (EM)	$50\%/\sqrt{E} \oplus 4\%$ (EM)	$55\%\sqrt{E}$(hadronic)	—	~6% (EM)
	FWHM	$23.3\% \sqrt{E} + 8.4\%$ (hadronic)	$86\%/\sqrt{E} \oplus 2\%$ (hadronic)	—	—	(0.5–3 GeV)
First year of operation	1976	1986	1996	2005	2010	Under construction
Reference(s)	15, 16	17	18	19, 20	21	22

[a]Abbreviations: APD, avalanche photodiode; D, diameter; DC, drift chambers; EM, electromagnetic; FWHM, full width at half maximum; H, height; L, length; PMT, photomultiplier tube.

Table 2 Parameters of calorimeters used in hadron colliders and *ep* colliders[a,b]

	Collaboration					
Parameter	**DØ**	**CDF**	**ATLAS**	**CMS**	**H1**	**ZEUS**
Accelerator	Tevatron ($p\bar{p}$)	Tevatron ($p\bar{p}$)	LHC (pp)	LHC (pp)	HERA (ep)	HERA (ep)
E range (GeV)	800	800	3,000	3,000	200	200
Barrel R_{in} (m)	0.75	1.7	1.0	1.3	0.9	1.2
Barrel R_{out} (m)	2.5	3.6	3.8	3.0	2.6	2.9
EM technology	U/LAr	Pb/scintillator	Pb/LAr	$PbWO_4$	Pb/LAr	U/scintillator
Hadron technology	U/LAr	Fe/scintillator	Barrel: Fe/scintillator Endcap: Cu/LAr Forward: W/Cu/LAr	Barrel: brass/scintillator Endcap: brass/scintillator Forward: Fe/quartz	Fe/LAr	U/scintillator
EM segmentation $\Delta\eta \times \Delta\phi$ or lateral size	0.1×0.1	0.1×0.26 to 0.09×0.087	Presampler: 0.025×0.1 Strips: 0.0031×0.1 Main: 0.25×0.25 Back: 0.5×0.25	Si preshower: 1.9 mm $\times$ 61 mm Main: 0.017×0.017	10–100 cm^2	4.9×23.3 cm^2 to $(5–10) \times 20$ cm^2
Hadronic segmentation $\Delta\eta \times \Delta\phi$ or lateral size	0.1×0.1	0.1×0.26	0.1×0.1 to 0.2×0.2	0.087×0.087 to 0.175×0.175	50–1,000 cm^2	24.4×27.1 cm^2
EM resolution a (GeV$^{-1/2}$)	15.7%	13.5%	10–12%	2.8–3.6%	11%	18%
Hadronic resolution a (GeV$^{-1/2}$)	41%	80%	50%	85%	50%	35%
First year of operation	1994	1988	2009	2009	1992	1992
Reference(s)	23, 24	25	26	27, 28	12, 29, 30	31–33

[a]The energy resolution is parameterized as $\sigma_E/E = \sqrt{[a/\sqrt{E(\mathrm{GeV})}]^2 + [b/E(\mathrm{GeV})]^2 + c^2}$. Only the sampling term a is listed.
[b]Abbreviations: EM, electromagnetic; LAr, liquid argon.

Table 3 Parameters of calorimeters used in e^+e^- flavor factories[a]

Parameter	Collaboration					
	CLEO II	KLOE	BELLE	BaBar	BES III	KEDR
E range (GeV)	0.015–5.0	0.020–0.28	0.20–4.0	0.020–4.0	0.020–2	0.015–5.0
Physics goal(s)	Tau/charm	Kaon; *CP* violation	Bottom; *CP* violation	Bottom; *CP* violation	Tau/charm	Tau/charm
Barrel dimensions (m)	$R_{in} = 1.02$; L = 2.50	$R_{in} = 2.0$; L = 4.3	$R_{in} = 1.25$; L = 2.98	$R_{in} = 0.92$; L = 2.93	$R_{in} = 0.94$; L = 2.75	$R_{in} = 0.75$; L = 2.58
EM technology	CsI(Tl)	Pb/scinillating fiber	CsI(Tl)	CsI(Tl)	CsI(Tl)	LKr/CsI
Segmentation ($\theta \times \phi$)	0.05×0.05	0.04×0.04	0.044×0.044	0.051×0.051	0.055×0.055	0.13×0.13
Energy resolution (%)	$0.35/E^{3/4} + 1.9 - 0.1E$	$5.7/\sqrt{E}$	$1.53/E^{1/4} \oplus 1.18$	$2.3/E^{1/4} \oplus 1.35$	$2.5/\sqrt{E}$	3.0 (1.8 GeV); 6.0 (0.1 GeV)
Angular resolution (mrad)	$2.8/\sqrt{E} + 1.9$	$6/\sqrt{E}$	$0.2 \oplus 2.7/\sqrt{E} \oplus 1.4/E^{.25}$	$4.16/\sqrt{E}$	$6.4/\sqrt{E}$	≈ 1.3
First year of operation	1991	2000	1999	1999	2008	2004
Reference(s)	34	35	36	37, 38	39	40

[a]Abbreviations: EM, electromagnetic; L, length; LKr, liquid krypton.

Table 4 Parameters of calorimeters used in high-energy e^+e^- experiments[a]

Parameter	Collaboration				
	ALEPH	DELPHI	L3	OPAL	SLD
E range (GeV)	0.2–100	0.2–100	0.1–100	0.1–100	0.2–50
Physics goal	Electroweak	Electroweak	Electroweak	Electroweak	Electroweak
EM barrel dimensions (m)	R_{in} = 1.85 L = 4.3	R_{in} = 2.1 L = 4.0	R_{in} = 0.50 L = 2.0	R_{in} = 2.45 L = 7.0	R_{in} = 1.77 L = 6.2
Hadronic barrel dimensions (m)	R_{in} = 3.0 L = 7.24	R_{in} = 3.25 L = 7.0	R_{in} = 0.9 L = 1.6	R_{in} = 3.39 L = 10.0	R_{in} = 3.25 L = 7.4
EM technology	Pb/PWC	Pb/TPC	BGO	Pb glass	Pb/LAr
Hadronic technology	Fe/lim streamer	Fe/lim streamer	U/PWC	Fe/lim streamer	Pb/LAr; Fe/lim streamer
EM segmentation $\theta \times \phi$ (mrad)	16 × 16	1.9 × 17.5	40 × 40	40 × 40	33 × 36
Hadronic segmentation $\theta \times \phi$ (mrad)	65 × 65	52 × 65	44 × 44	150 × 150	72 × 66
EM resolution (%)	$18/\sqrt{E}$	7 (45 GeV)	5 (0.1 GeV)/ 1 (5 GeV)	$6.3/\sqrt{E} \oplus 0.2$	$15/\sqrt{E}$
Hadronic resolution (%)	$84/\sqrt{E}$	$120/\sqrt{E}$	$55/\sqrt{E}$	$120/\sqrt{E}$	$\approx 60/\sqrt{E}$
EM angular resolution (mrad)	$3.7/\sqrt{E}$	$36/\sqrt{E} \oplus 2.5(\theta)$; $97/\sqrt{E} \oplus 10(\phi)$	—	4.5	5–10
First year of operation	1989	1989	1989	1989	1991
Reference(s)	41	42	43, 44	45	46–47

[a]Abbreviations: BGO, bismuth germinate; EM, electromagnetic; L, length; LAr, liquid argon; lim, limited; PWC, proportional wire chamber; TPC, time-projection chamber.

CsI(Tl) is used for EM calorimetry in the flavor factories (34, 36, 37, 39). CsI(Tl) is slow (1,000 ns), dense, and not very radiation hard (10 Gy), but it has excellent light yield ($\approx$15,000 photons MeV^{-1}) and excellent resolution [$\approx 2\%/E\ (\mathrm{GeV})^{1/4}$] and is well matched to the low-rate, low-radiation e^+e^- environment. CsI(Tl) detects 20-MeV photons efficiently, measures high-energy photons with excellent precision, and can identify and reconstruct states with multiple π^0 decays. It is read out with photodiodes and preamplifiers and is calibrated with light flashers, sources, and physics signals. Typical crystals have dimensions of $5 \times 5\ \mathrm{cm}^2$ and a depth of 16 X_0. Modern detectors employ ~5,000 crystals, which cover most of the solid angle.

$PbWO_4$ provides crystal EM calorimetry for CMS at the LHC (48). High resolution is required to search for the Higgs $\rightarrow \gamma\gamma$ decay amid copious backgrounds. $PbWO_4$ is very dense (X_0 = 0.89 cm), fast (25 ns), and radiation hard (10^5 Gy), but it has rather low light yield ($\approx$200 photons MeV^{-1}). Radiation hardness is achieved by doping with yttrium. Avalanche photodiodes are used as photodetectors in regions of moderate radiation but high magnetic field. Vacuum phototriodes are used in the endcaps, where the transverse B field is low and the dose is higher. Additional amplification is provided on the detector, and a dynamic range of nearly 5,000 is accommodated. A light-monitoring system is incorporated to correct for light-output changes with radiation dose. Excellent energy resolution [$\approx 3\%/\sqrt{E(\mathrm{GeV})} \oplus 0.3\%$] is obtained. Some of the crystals have been calibrated in test beams, and all are cross-calibrated with cosmic muons. The detector uses nearly 76,000 crystals.

Recent scintillating-fiber EM calorimeters are based on the Shashlyk readout scheme (49) and are installed in the ALICE (50) and LHCb (51) detectors at the LHC, in PHENIX at RHIC, at Novosibirsk, and in PANDA (52) at GSI. A recent example is shown in **Figure 4**. In this scheme, stacks of alternating scintillator and lead absorber plates are read out with multiple wavelength-shifting fibers that are threaded through the entire stack, perpendicular to the plates, and then bundled behind the stack and coupled to photosensors. Despite the transverse nonuniformity resulting from the readout fibers, the calorimeter's response is uniform to the percent level across the plates. The scheme naturally accommodates tower geometry, good transverse segmentation ($\approx 2 \times 2$ cm^2), large numbers of layers ($\sim$50 or more), and a good sampling fraction. Shashlyk calorimeters can have very good energy resolutions [$\approx 3\%/\sqrt{E(\mathrm{GeV})} \oplus 1\%$] (53). Plastic scintillators are fast and reasonably radiation hard, and such calorimeters are often used in triggering. In fixed-target or forward geometries, traditional PMTs are used for readout. In the ALICE detector, where the calorimeter lies within the solenoid, an avalanche photodiode serves as photon sensor. Monitoring gain is accomplished by threading a clear plastic fiber through the stack to the photosensor and pulsing it with a light-emitting diode.

LAr has long been used in sampling EM calorimetry; its current application at the LHC derives from its excellent radiation hardness, good energy resolution, and high rate capability. The ATLAS accordion calorimeter (26), shown in **Figure 5**, uses a novel arrangement of lead absorber and collection electrodes, which form alternating zigzag layers, arranged radially. The LAr gap between the layers is only 2 mm thick, limiting the collection time to $\sim$400 ns. The ionization signal is shaped to generate a fast signal that is suitable for the LHC's 40-MHz crossing time. The design naturally accommodates tower geometry and longitudinal and fine transverse segmentation within the cryogenic volume. The calorimeter accommodates a huge dynamic range (30 MeV to 3 TeV) by using three parallel channels with different gains. It delivers an energy resolution of $10\%/\sqrt{E(\mathrm{GeV})} \oplus 0.7\%$ and has good linearity. Excellent channel-to-channel calibration is accomplished by injecting precisely known currents to the circuit near the electrodes.

LKr, with its shorter radiation length, has been used for total-absorption EM calorimetry and offers very good energy and spatial resolution in both fixed-target and colliding-beam applications. Recently, it has been incorporated in the KEDR detector at BINP (40). A small fraction of Kr^{85}, which is radioactive, is a natural contaminant; its decays roughly double the nominal electronic noise when shaping times of a few microseconds are employed. The calorimeter has 27 tons of LKr in a large cryostat. High-voltage and readout electrodes of copper-clad G10 are spaced roughly 2 cm apart throughout the liquid. Both 10×10 cm^2 pads and 5-mm-wide strips are read out in three radial tower segments. The calorimeter delivers 3% energy resolution and roughly 1-mm spatial resolution for photons or mips.

The MEG experiment at PSI has developed a special-purpose total-absorption LXe calorimeter to search for $\mu \rightarrow e\gamma$ (54). LXe is a very fast scintillator with excellent light yield in the ultraviolet (UV) region. The inner wall of the LXe cryostat has roughly 850 UV-sensitive phototubes to record the time, energy, and position of the photon from the muon decay. It achieves superb timing resolution of 0.15 ns, good energy resolution (4.8 MeV), and moderate spatial resolution (15 mm).

Si/W calorimeters were first employed as luminosity monitors for the SLD and LEP experiments, in which they improved the precision of luminosity measurements to a fraction of a percent. Silicon technology offers high segmentation, stable calibrations, and good resolution. From its inception, SLD (55) operated with this technology, and LEP experiments replaced their initial luminosity monitors with silicon upgrades (56). ALEPH (57), OPAL (58), and DELPHI (59) employed silicon technology, which is now being considered for full-scale EM calorimeters for future experiments (Section 4.3).

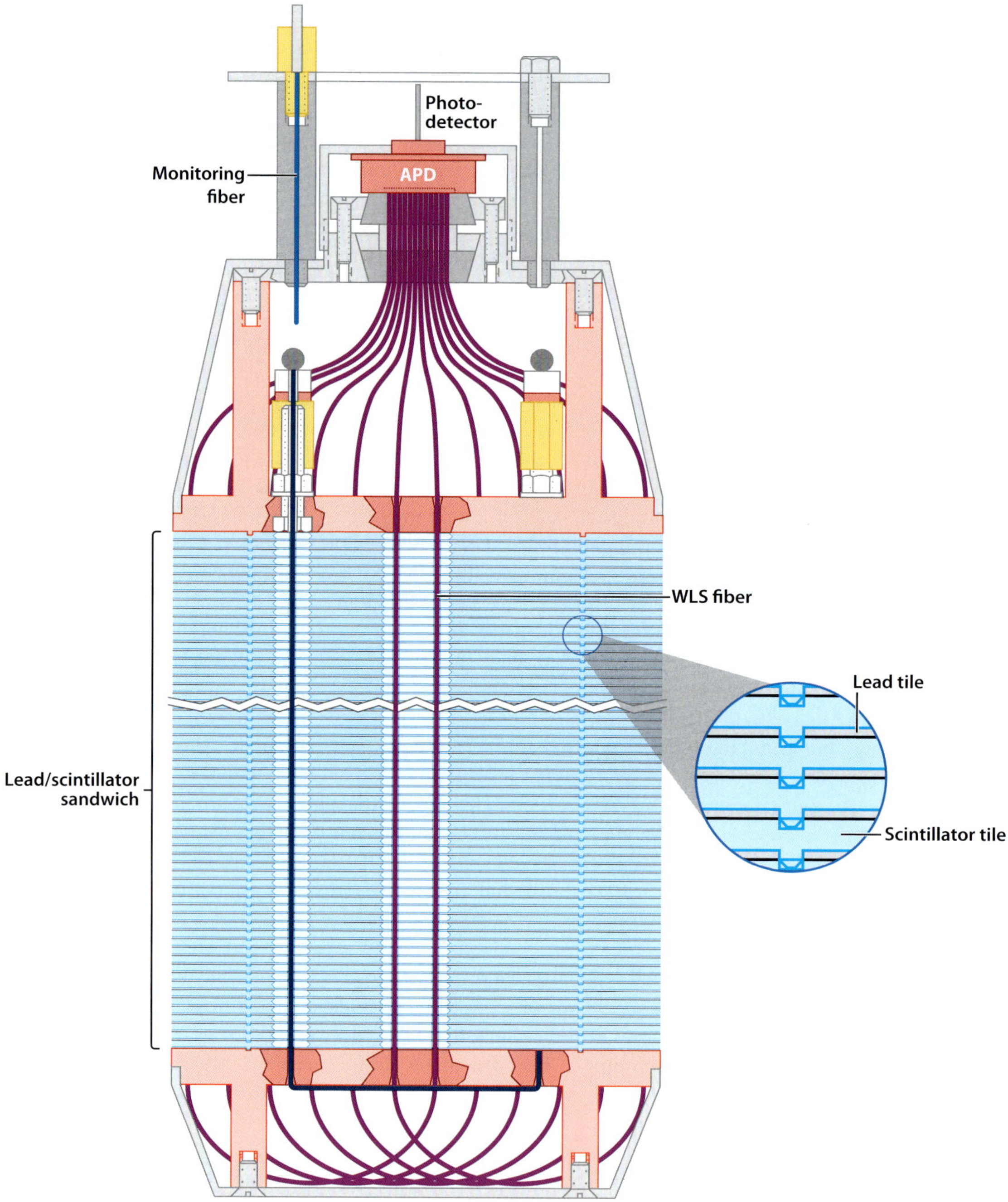

Figure 4

The Shashlyk module design (49). Abbreviations: APD, avalanche photodiode; WLS, wavelength shifter.

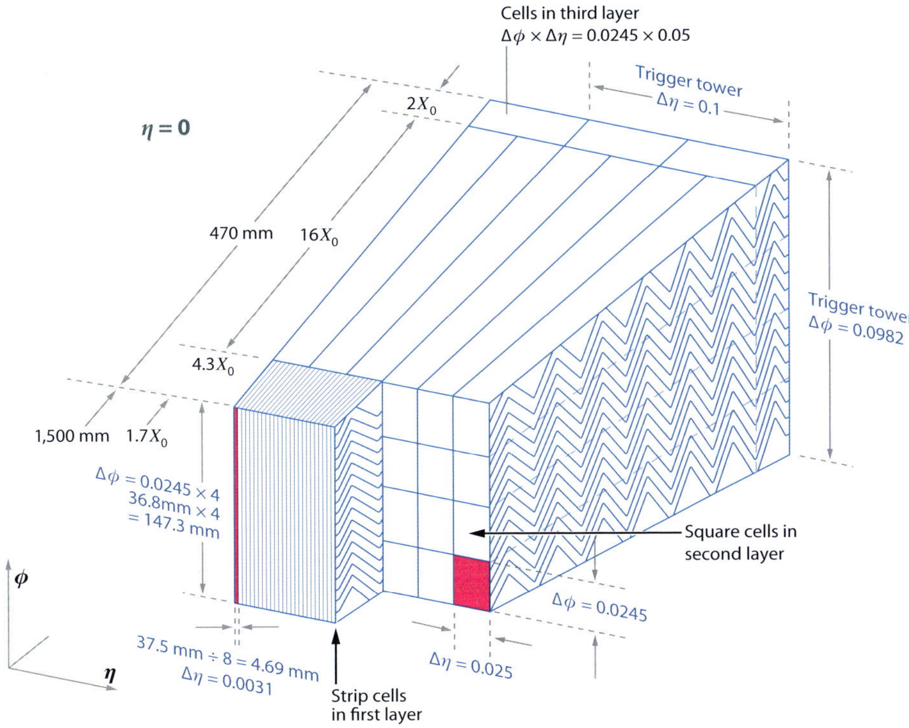

Figure 5

The ATLAS accordion electromagnetic calorimeter, with different layers clearly visible. Also shown is the granularity in pseudorapidity (η) and azimuth (ϕ) of the cells in each of the three layers and in the trigger towers (26).

3.2. Hadronic Sampling Calorimeters

With advances in our understanding of hadron showers (60–63), hadron calorimeters have achieved remarkable performance. In sampling calorimeters, the detector samples a fraction of the energy deposited; the rest is deposited in the absorber. A few notable examples are described below. They illustrate how several different techniques can reach effective compensation.

As discussed in Section 2.2, hadronic energy resolution is optimized if the calorimeter has $h/e \sim 1$. Such compensating calorimeters have been built in recent decades. The calorimeter for the ZEUS detector is a uranium-scintillator sampling calorimeter (31, 32), optimized for the measurement of hadronic jets. The compensation is achieved by suppressing the EM shower response with a high-Z absorber and by adjusting the thickness of scintillator to enhance the response to neutrons. On the basis of a previous experience with a compensating uranium-scintillator calorimeter (64), the absorber thickness for both the EM and the hadronic part of the calorimeter modules was chosen to be 1 X_0 (3.3 mm). The scintillator thickness of 2.6 mm is optimal for achieving $h/e = 1$. This particular choice of parameters yields a pion energy resolution of $\sigma_E/E = 35\%/\sqrt{E(\mathrm{GeV})}$ and a resolution for EM showers of $\sigma_E/E = 18\%/\sqrt{E(\mathrm{GeV})}$. The uniform EM and hadronic configuration results in a somewhat coarser sampling for EM showers.

The DØ calorimeter, designed at approximately the same time, uses a uranium absorber with LAr as the active medium to achieve compensation. The EM sections use thin plates (3 or 4 mm in the central calorimeter and endcap calorimeter, respectively) made from nearly pure depleted uranium. The fine hadronic sections are made from 6-mm-thick uranium-niobium (2%) alloy. The coarse hadronic modules contain relatively thick (46.5-mm) plates of copper (in the central calorimeter) or stainless steel (endcap calorimeter). In Tevatron Run I, the beam bunch-crossing time was 3.5 μs. With an electronics integration time of 2.2 μs, the calorimeter is nearly compensating, with $e/\pi < 1.05$ for energy above 30 GeV (65). The hadronic response is enhanced by the long integration time, during which signals from nuclear excitation and thermal neutrons are collected. For Tevatron Run II, the calorimeter electronics for the upgraded DØ calorimeters were redesigned to optimize the signal-to-noise ratio (including pileup effects) for a much shorter bunch-crossing time (132 ns at the time of upgrade design; 396 ns in real operation). This resulted in a reduced response to the hadronic particles. A correction factor, ranging from 0.65 to 0.8 for 35-GeV to 300-GeV jets, corrects the energy scale (66).

The ATLAS calorimeter is an example of a noncompensating calorimeter (26). It consists of a lead-LAr accordion EM calorimeter for $|\eta|$ up to 3.2, a steel-scintillator hadronic calorimeter in the barrel, and a copper-LAr calorimeter in the endcap. In addition, a forward calorimeter with novel tube geometry extends the η coverage to 4.9. The hadronic calibration scheme is based on the technique developed by the H1 experiment (12, 67). Because of the highly segmented readout (a total of 180,000 channels and up to eight longitudinal layers), the energies measured in the calorimeter at the EM scale are first formed into three-dimensional topological clusters, where neighboring energy deposits are grouped together based on their significance above measured noise. The clusters are then classified as EM like or hadron like based on cluster-shape variables. The hadron-like clusters receive weights based on their energy density. The weights are derived from detailed Monte Carlo simulations (68).

In a hadron collider experiment, the final jet-energy scale is determined in situ using physics processes, after taking into account various detector and physics effects, which include multiple interactions, energy loss in dead material, η-dependent response correction, and underlying-event and out-of-cone corrections. The CDF (69) and DØ (65, 66) experiments at the Tevatron have demonstrated these techniques, achieving uncertainties of the jet-energy scale of approximately 1 to 3%.

3.3. Massive Calorimeters

Calorimeters whose active sensitive medium either is the target material itself or is embedded within the target material have been used to measure the total energy of the particles produced in high-energy neutrino interactions. Due to the small neutrino cross sections, these calorimeters must be massive. Early massive neutrino detectors were sampling calorimeters, such as the 1.25-kton CDHS calorimeter (15), the 0.69-kton CHARM II calorimeter (17), the 0.69-kton CCFR/NuTeV calorimeter (18), and the 5.4-kton MINOS calorimeter (19). These calorimeters were limited in energy resolution due to their small sampling fractions and the internal mismatch in response to EM and hadronic showers.

A recent example of a fully active detector, used in the MiniBooNE (70) neutrino experiment at Fermilab, is an 800-ton, pure–mineral oil (CH_2) Cherenkov detector read out with 1,280 PMTs. The 15-kton NOνA detector (22) is under construction in northern Minnesota, where it will study the high-energy neutrino beam from Fermilab. This "totally active" detector is built of polyvinyl chloride extrusions, filled with liquid scintillator, to minimize the mass of inactive material. The detector consists of 357,120 total channels.

In Japan, the high-energy neutrino beam from KEK, as well as the new beam from JPARC, has been detected by the Super-Kamiokande (Super-K) detector (21), which is a 50-kton ultra-pure water Cherenkov system contained by a cylindrical stainless-steel tank that is 42 m tall and

39 m in diameter. Cherenkov light is detected by 11,146 PMTs. Water Cherenkov detectors were developed to search for proton decay, but they also achieve a low-energy (approximately megaelectronvolt-range) sensitivity for neutrinos. The transparent dielectric medium of large mass provides an active target for neutrino interactions. The large Super-Kamiokande detector was preceded by the 8-kton IMB (71) and 3-kton Kamiokande detectors—which serendipitously detected low-energy Supernova 1987A neutrinos. For future experiments, megaton detectors are under consideration to increase the high-energy neutrino-event rate. To reach the highest possible target and active volume, to increase rate and energy reach, and to detect extraterrestrial sources, calorimetry using the atmosphere, sea water, and polar ice has been developed [for example, Auger (72) and IceCube (73)].

4. RESEARCH AND DEVELOPMENT AND FUTURE DIRECTIONS

Although calorimetry has matured into a very effective and widely applied technique, advances continue, driven by physics requiring higher resolution, faster response, more radiation hardness, and more massive detectors. The availability and understanding of a variety of new crystals promise improved performance. Total-absorption techniques, including water Cherenkov and cryogenic-liquid detectors, are under development for ultramassive calorimeters. The advent of optimized particle-flow detectors aims to capitalize on the synergy of tracking and calorimetry to improve jet-energy resolution. The simultaneous detection of different components of the hadron shower (dual readout) also has potential for improved jet-energy resolution.

4.1. New Crystals for Calorimetry

Inorganic crystals have been used in EM calorimetry for many years, providing exceptional resolution and overall performance due to the improved sensitivity of total-absorption techniques. Crystals with high light yield, fast response, high densities (4–8 gm cm^{-3}), and high effective atomic number are preferred. Examples of past applications are given in Section 3.1. Crystals that can provide new calorimeter-performance levels continue to be developed.

There has been recent interest in developing LSO:Ce or LYSO:Ce crystals for future calorimetry applications (74, 75). LYSO:Ce has been proposed for use as the forward calorimeter material at a super-*B* factory detector due to its high level of radiation hardness, in addition to other attractive features. Similar considerations make it an attractive candidate for use at the super LHC (SLHC) (76). With a density of 7.4 gm cm^{-3}, a radiation length of 1.14 cm, an extremely high light yield [85% of NaI(Tl)], and a 40-ns decay time, LYSO:Ce has the potential to make a very dense, compact, high-performance EM calorimeter.

The family of yttrium and lutetium perovskites has attracted interest due to its broad range of energy linearity and potentially extremely good intrinsic energy resolution. Such crystals include YAP ($YAlO_3$:Ce) and LuAP ($LuAlO_3$:Ce) (77). CeF_3 and PbF_2 also show good potential for excellent energy resolution. Of particular importance to dual-readout applications of any crystal (Section 4.4) is good UV transmission, which is needed in order to efficiently detect Cherenkov light. PbF_2 is particularly strong in this property.

4.2. Total-Absorption Calorimetry

Massive total-absorption detectors are motivated by high-energy physics and astrophysics opportunities addressing fundamental questions. The last unmeasured neutrino mixing angle, θ_{13}, and the search for neutrino-sector *CP* violation motivate significantly increasing detector mass.

Improved proton decay sensitivity and supernova neutrino detection require massive detectors in the 100-kton–1-Mton class. Two leading technologies for such detectors are the water Cherenkov technique and the cryogenic-liquid time-projection chamber (TPC).

4.2.1. Water Cherenkov detectors. Detection of accelerator-generated neutrinos (as well as extraterrestrial or atmospheric neutrinos) is limited by the small neutrino cross section. Massive absorbers are needed to produce sizable event rates. Water detectors can be relatively easily scaled up in size. Neutrinos interacting in water produce charged particles, some of which (particularly electrons and positrons) in turn generate Cherenkov radiation. The water Cherenkov detector is an excellent calorimeter for measuring the energy of the electron produced in electron-neutrino charged-current interactions, as it has a demonstrated resolution of 20% for 10-MeV electrons that improves with energy. The technique does not produce a very good energy measurement of hadronic products because most of the resulting energy loss occurs for particles below the Cherenkov radiation threshold.

The 50-kton Super-K successfully expanded the technique that had been proven feasible by the smaller IMB and Kamiokande I detectors. The next generation of detectors for future long-baseline neutrino experiments is being planned (78). Trade-offs between the required physics reach and costs motivate megaton-scale detectors (79). The configuration of the near megaton detector is an important consideration. Several configurations are under study by various groups. The Hyper-Kamiokande detector is envisioned as a 500-m-long tank with a cross-sectional area of 2,500 m^2. Because Cherenkov light attenuates, the tank should be subdivided into 50-m-long compartments. The large area that requires coverage of photodetection has motivated the consideration of photodetectors larger than the 20-inch PMTs used by Super-K, but safety considerations have now limited the size under study to 20 inches. Instrumenting with a density of two 20-inch PMTs per square meter (the same as Super-K) would require approximately 200,000 PMTs. Photodetector research and development (R&D) is proceeding on several fronts, including higher quantum efficiency, flat and thin devices, and operation in high magnetic fields.

A promising large new photodetector under development (80) is the Hybrid Avalanche Photodetector (**Figure 6**), which may offer easier construction due to its simpler structure, and therefore would offer simplified mass production and lower manufacturing cost as well as excellent quantum efficiency. A 13-inch version has been successfully developed and tested (81). The internal surface of the Hybrid Avalanche Photodetector glass envelope is coated with a photocathode and a light reflector. The very high voltage (approximately +20 kV) used for the acceleration of electrons toward an avalanche diode results in a high gain ($\sim10^5$). The time resolution, 190 ps, is better than that of a dynode structure, such as 1.4 ns for the 13-inch PMTs. The elimination of a costly dynode structure, in favor of an avalanche diode, reduces cost.

4.2.2. Cryogenic-liquid time-projection chamber. Since it was proposed in 1977 (82), the LAr TPC has been a promising technique for measuring neutrino interactions. It provides three-dimensional bubble chamber–like images of particle tracks, with millimeter resolution. Over the past 20 years, the ICARUS Collaboration (83) has designed and constructed a series of detectors, culminating in the current ~300-ton detector, which has been operated successfully. **Figure 7** shows EM and hadron showers in the ICARUS detector.

Currently, the high-energy physics and astrophysics communities are pushing for the development of LAr TPCs on a much larger (~100-kton) scale to address some of the most fundamental questions in physics. Compared with the large water Cherenkov detectors, the LAr TPC has several advantages: It is sensitive to all charged particles; its fine resolution allows very high

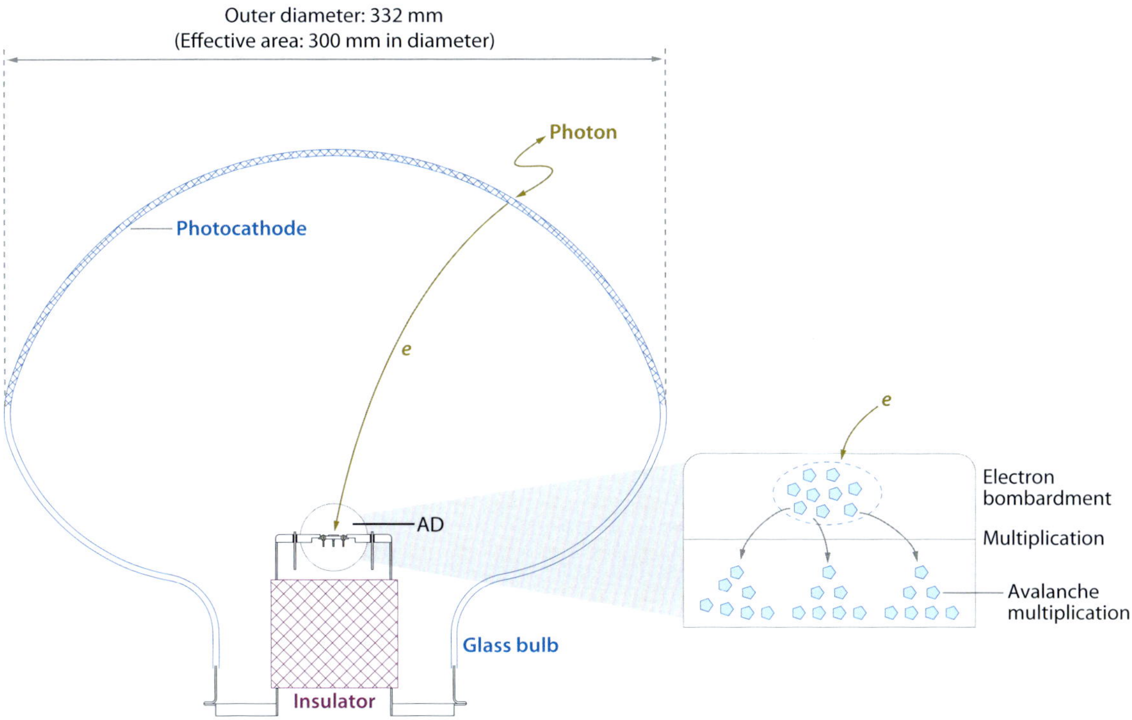

Figure 6

Outline of the large-aperture Hybrid Avalanche Photodetector and its principle of operation (80). Abbreviation: AD, avalanche diode.

efficiency for electron-neutrino interactions; and it demonstrates sensitivity to $p \rightarrow \bar{\nu}K^+$ decays by identifying the final-state kaon through *dE/dx* and kaon decay.

In a LAr TPC, ionization electrons liberated by passing charged particles are drifted in a strong electric field in ultrapure LAr to the edge of the detector. A series of wire planes then record the passing charge in stereo views and measure the arrival time. A three-dimensional picture of the interaction can be reconstructed from the timing signals and the positions of the wires. The collected charges, after correcting for losses during drift, provide precise energy measurements of the particles that are contained in the detector.

Scintillation light emission (UV at $\lambda = 128$ nm) provides a complementary measure of energy deposition in LAr. The relative yield between ionization and scintillation depends on the electric field applied in the active LAr volume. At the nominal field of 500 V cm^{-1} for a large LAr TPC, both processes are active. The detector can be instrumented with photomultipliers and wavelength shifters for detecting the prompt light.

A large LAr TPC detector poses many technical challenges, including the following:

1. LAr purification. To drift electrons over $\mathcal{O}(1)$ m, the contamination of electronegative molecules must be reduced to the level of 0.1 ppb. Maintaining such purity in a large vessel will be a challenge.
2. Readout of a large number of signals. For a large detector, cost and system complexity rule out carrying $\sim10^5$–10^6 signals out of the large cryogenic volume with cables. Cold electronics in LAr, with signal multiplexing and optical transmission, is being developed. A novel readout technique, the LAr large electron multiplier TPC (84), uses both liquid and

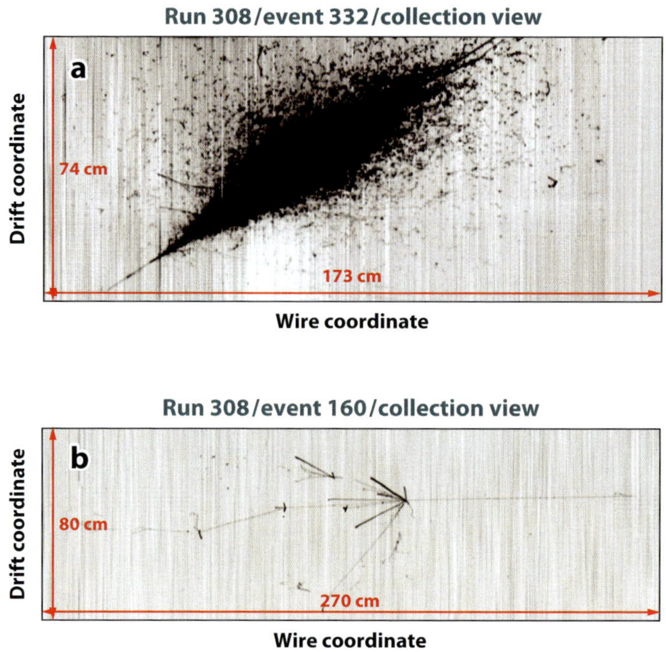

Figure 7

(*a*) An electromagnetic shower and (*b*) a hadron shower in the ICARUS detector (83).

vapor phases in pure argon to allow signal amplification in the gas phase after the electrons are drifted to the surface.

3. Engineering and safety issues of large cryogenic vessels deep underground. To achieve all the physics goals, a LAr TPC at a depth of 4,300 meters water equivalent is desired (85), although a less deep location can be considered for some of the physics goals.
4. Better knowledge of the LAr properties, including transverse and longitudinal diffusion, recombination, optical absorption, and effects of nuclear Fermi motion on low-energy neutrino scattering.

A few experiments and R&D efforts are ongoing or planned for the near future. ArgoNeut (86) is a small-scale (175-liter) LAr TPC in the NuMI beam at FNAL that aims to study neutrino interactions in LAr at low energy. MicroBooNE (87) is a proposed LAr TPC detector that is planned to run in the on-axis Booster and off-axis NuMI beam on the surface at Fermilab. MODULAr (88) is the proposed continuation of the ICARUS experiment, with a 20-kton fiducial volume of LAr for the CERN/Gran Sasso neutrino complex. A long-baseline neutrino experiment at DUSEL with the neutrino beam from FNAL (see **http://lbne.fnal.gov/**) uses a proposed LAr TPC detector, initially of 5 kton and ultimately of $\sim$100 kton. GLACIA (89) represents a new concept of using a single container (similar to those used for liquid–natural gas storage) for 100 kton of LAr, together with the large electron multiplier TPC readout.

4.3. Particle-Flow Calorimetry

Particle-flow calorimetry is a new paradigm for high-resolution calorimetry; it is motivated in part by the need to distinguish the dijet invariant masses of W and Z bosons on an event-by-event basis at a future e^+e^- linear collider. The particle-flow algorithm (PFA) concept (91) evolved from

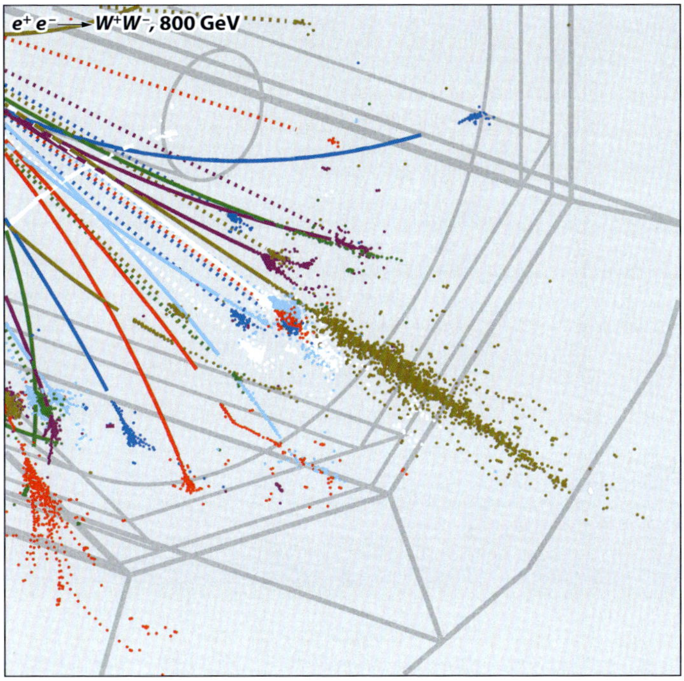

Figure 8

Simulation of part of an $e^+e^- \rightarrow W^+W^-$ event, in which one of the *W*s decays into jets (90).

the idea of so-called energy-flow calorimetry (92), which was developed at LEP to improve upon the jet-energy resolutions that purely calorimetric measurements could provide. The idea is to measure the energy of all the particles in a jet, using the tracker for the charged particles, the EM calorimeter for prompt photons, and the EM and hadronic calorimeters to capture the neutral hadrons. Because most of the energy in an event is in charged particles that are measured with high resolution, and most of the remaining energy is in photons that are typically measured with good resolution, the resolution on the total jet energy is improved significantly compared with that obtained with the calorimeters alone. In principle, the jet-energy resolution could be improved from the LEP standard of roughly $60\%/\sqrt{E(\mathrm{GeV})}$ to approximately $20\%/\sqrt{E(\mathrm{GeV})}$—more than enough to distinguish *W* and *Z* decays. The key to achieving such resolution is isolating the energy deposited in the calorimeters by charged particles from that deposited by photons and neutral hadrons. Photons and charged and neutral hadrons shower in passing through the calorimeters, and their showers may overlap with one another. In practice, it is the confusion in assigning energy depositions to one source or another that limits the jet-energy resolution. Elaborate pattern-recognition codes, dubbed PFAs, have been developed over the past few years to optimize such assignments, as have imaging calorimeters, which incorporate new detector technologies that permit very high transverse and longitudinal segmentation. **Figure 8** shows a full Monte Carlo simulation of the energy depositions in a highly segmented calorimeter and suggests that the pattern-recognition problems are tractable. The PFAs and the new particle-flow calorimetry techniques are described below.

At the level of detailed full Monte Carlo simulations, PFAs have demonstrated excellent jet-energy resolution. For example, PandoraPFA (93) has been developed in the context of detector

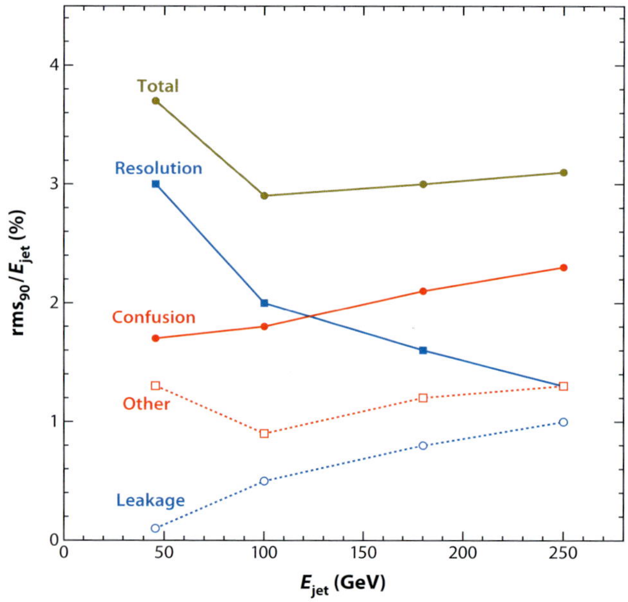

Figure 9

The contributions to the jet-energy resolution obtained with PandoraPFA as a function of energy. The total is (approximately) the quadrature sum of the components (93). Abbreviations: PFA, particle-flow algorithm; rms, root mean square.

studies for the International Linear Collider (ILC) and has now demonstrated jet-energy resolutions, σ_E/E, that range between 3 and 4% for jets from 45 to 500 GeV. PFAs typically have non-Gaussian resolution functions, with a narrow core but some additional tail, and the practice in the field is to quote the root mean square of the part of the distribution that includes 90% of events (known as rms90). As shown in **Figure 9**, the intrinsic calorimeter resolution dominates the jet-energy resolution at the lowest jet energies; at the highest energies, as the jets become more collimated and showers overlap, the confusion term dominates. The result is that the net resolution is relatively constant over the whole energy range, in contrast to that of a typical calorimeter ($\sim 1/\sqrt{E}$). The Iowa PFA (94) is another algorithm that has been developed for linear collider detector studies. It demonstrates similar if slightly lower performance over a more restricted range of energies than that of Pandora, but it confirms that PFAs work. Both codes are being improved and extended to treat higher energies. PandoraPFA has been applied to a detector modified for Compact Linear Collider energies (4-T field and 8-λ hadronic calorimeter) and has achieved a jet-energy resolution of 3.5% at 500 GeV. This resolution is significantly better than the purely calorimetric energy resolution with the same detector, demonstrating the efficacy of PFAs at even high energies.

Evaluating PFA performance in simulation depends on hadronic shower codes, which are imperfect (95). Fortunately, PandoraPFA has demonstrated rather stable performance for several popular shower codes, so shower uncertainties may not be a problem. The real proof of PFAs awaits experimental test. Given that it is difficult to generate well-calibrated hadronic jets outside of collider experiments, experimental proofs are coming piecemeal. Studies of combined EM and hadronic PFA calorimeters by the CALICE Collaboration are under way; these studies include the response to single hadrons and comparisons of observed shower properties to simulations. The

PFA approach is being adapted to the LHC experiments, which have calorimeters with coarser segmentation, with good results (96).

Developing EM and hadronic calorimeters that are well suited for particle-flow calorimetry has required new detector technologies capable of supporting a high degree of segmentation. Silicon sensors, scintillating tiles, and a variety of gas detectors are suitable.

Si/W sampling calorimeters have been successfully used as luminosity monitors at SLD and LEP (56). More recently, they have been proposed as fully imaging particle-flow EM calorimeters for ILC detectors. They offer good energy resolution, excellent spatial and angular resolution, unmatched segmentation, and stable calibrations. The proposed systems are huge, constituting hundreds of square meters of silicon sensors, and are therefore costly. The CALICE Collaboration has developed a prototype Si/W EM calorimeter based on silicon-diode sensors with 1-cm^2 pixels (97). It delivers good energy resolution [$\sigma_E/E = 16.6\%/\sqrt{E(\mathrm{GeV})} \oplus 1\%$], few-millimeter spatial resolution, $\approx$50-mrad angular resolution, and excellent linearity. An Oregon/SLAC collaboration (98) is developing a Si/W calorimeter suitable for an ILC detector. It integrates a readout ASIC (99) directly onto a hexagonal silicon sensor with 1,024 14-mm^2 pixels, using bump bonding and double-metal layers for signal routing on the sensor. Doing so minimizes the thickness of the detector layer, which preserves a small Molière radius to limit transverse shower size. To date, prototype sensors and readout chips have been produced and tested. Both CALICE and Oregon/SLAC are developing full detector prototypes to address system issues. A variation on this theme is being developed by a Rutherford group (100) using a monolithic active pixel sensor, which incorporates $50 \times 50\ \mu\mathrm{m}^2$ pixels and readout in a single CMOS (complementary metal-oxide semiconductor) chip. The monolithic active pixel sensor detector works as a digital calorimeter by counting the number of "hit" pixels. In principle, this is an excellent measure of the number of particles in the shower and could deliver superior energy resolution and the ultimate in spatial segmentation.

Scintillating tiles and strips can also provide the high segmentation required of a particle-flow calorimeter. They have become a realistic possibility with the advent of multipixel photon counters (MPPCs) and SiPMs, which can provide readout of individual tiles or strips. MPPCs and SiPMs (101) are photodiode detectors with areas of 10 to 100 mm^2 and $\sim$1,000 pixels, which are operated in Geiger mode. Each pixel is photosensitive, so the number of hit pixels counts the number of incident photons. A prototype EM calorimeter has been tested by the CALICE Collaboration (102). Cross talk, nonuniformity of response, temperature dependence, and MPPC saturation had to be dealt with, but the device was reasonably linear and performed well in an electron beam, with $\sigma_E/E = 14\%/\sqrt{E(\mathrm{GeV})} \oplus 3\%$. CALICE has also developed a highly segmented sampling scintillating-tile hadron calorimeter, the AHCAL, with SiPM readout (103). Corrections for SiPM saturation and gain, temperature dependence, and voltage dependence are required. Beam tests show reasonably good linearity, EM resolution of $22\%/\sqrt{E(\mathrm{GeV})}$, single-hadron resolution of approximately $60\%/\sqrt{E(\mathrm{GeV})}$, and good agreement with simulations. Data from the AHCAL are being used to test PFAs.

Gas counters are relatively inexpensive and can be highly segmented (1×1 cm^2 pixels), so they are a natural candidate for particle-flow hadron calorimeters. Studies with RPCs (104) are the most advanced, but GEMs (105) and MicroMegas (106) are also being developed. All of these detectors operate as digital calorimeters, where the number of hit pixels measures the energy. Glass RPCs have been used in beam tests with electronics based on the DCAL chip (107); they have a noise rate of 4×10^{-5} Hz per pad and >90% efficiency. The average multiplicity is 1.5 hits per mip. Their response to electrons and pions agrees with simulation and demonstrates the feasibility of a gas-based digital calorimeter. A full 40-layer, 1×1 m^2 calorimeter is under construction to test the full calorimetric response.

4.4. Dual-Readout Calorimetry

The principle of hadron calorimetry is to measure the energy deposited in the calorimeter and to use the measurement to infer the energy of the incident particle (or jet) of interest. Traditionally, hadron calorimeters detect signals related to the total energy deposited in the active medium of the calorimeter, thereby estimating the energy of the shower-initiating particle. Signals may come from, for instance, generated scintillation light or ionization. Techniques such as compensation (10) and energy weighting (12) have been invented to balance the response of the active medium to the components of the shower, principally EM for π^0s generated within the hadronic shower, and charged hadrons. Each of these components results in different efficiencies for transfer of the generated energy. An alternative to equally summing the generated signals is to measure them separately and to apply the appropriate factors (based on relative efficiencies) to determine the particle's energy. Two components dominate, so by separately detecting the EM components (from π^0s) and the non-EM components, a balanced sum can be made (108). Without this balanced sum, the energy resolution is limited by the fluctuations in the fractions of energy that are EM and hadronic, as well as by their differing conversion efficiencies of energy to signal. Detector R&D motivated by this possibility includes early work with scintillating crystals (109).

The DREAM Collaboration has realized this dual-readout approach on the scale of a full-sized calorimeter; they used a copper calorimeter structure, employing embedded extruded rods filled with fibers that measure both components (13). Because EM particles (electrons and positrons) are still relativistic at low energies, their energy deposition is well characterized by Cherenkov light, whereas heavy particles generate Cherenkov light only at higher energy. Therefore, both components can be measured by detecting the Cherenkov light and scintillation light separately. The DREAM calorimeter uses scintillating fibers and quartz fibers (which are sensitive to the Cherenkov light). The calorimeter contains 5,580 2-m-long rods, of which 5,130 are equipped with fibers. The fiber-containing holes are 2.5 mm in diameter. The full calorimeter consists of 19 longitudinally unsegmented hexagonal towers that are approximately 80 mm apex to apex and 200 cm deep. A central tower is surrounded by two hexagonal rings, an inner ring of six towers, and an outer ring of 12 towers.

The quartz fibers are sensitive to relativistic particles, whereas the scintillating fibers detect all charged particles. Two signals, Q and S (which stand for quartz and scintillator, respectively), are normalized to electrons, so for each event a particle of energy E gives average signals of

$$Q = E[f_{em} + h/e_Q(1 - f_{em})] \qquad 6.$$

and

$$S = E[f_{em} + h/e_S(1 - f_{em})], \qquad 7.$$

where f_{em} is the EM shower fraction and $h/e_{Q,S}$ is the ratio of hadron to electron response in each set of fibers. **Figure 10** shows the measured correlation between Cherenkov and scintillator signals in the DREAM calorimeter for 100 GeV π^-. The DREAM calorimeter has achieved improved resolution by combining these two signals (i.e., the Q/S method). With this combined analysis, the resolution measured by either method alone is improved for simulated jets of particles produced by interacting particles to $64\%/\sqrt{E(\mathrm{GeV})} \oplus 0.6\%$. The small constant term (0.6%) means that the resolution continues to improve to very high energy. Unfortunately, the calorimeter suffers from limited lateral extent so that significant particle leakage occurs, and the ultimate performance of the technique cannot be measured without making assumptions.

The DREAM Collaboration plans a triple-readout calorimeter in which the neutron signal is measured for each event using the time response of the scintillation signal. The DREAM

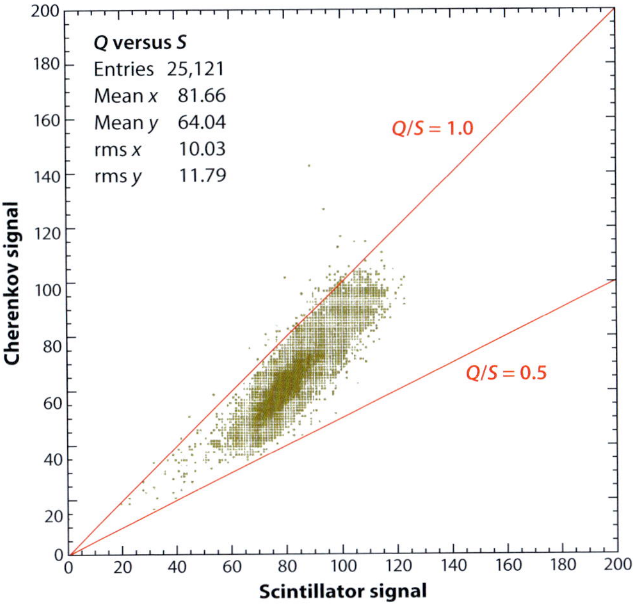

Figure 10

Cherenkov signals versus scintillator signals for 100-GeV π^- in the DREAM calorimeter (13). Abbreviations: rms, root mean square; Q, quartz; S, scintillator.

Collaboration argues they may achieve hadronic calorimeter resolution approaching that of the EM calorimeter.

Another approach in dual-readout calorimetry which is based on a totally active scintillating-crystal hadron calorimeter has been proposed and is being studied (110). This homogeneous hadron calorimeter concept proposes to separately detect the Cherenkov and scintillation light generated in a large array of scintillating crystals through wavelength filters and photodiodes. Several crystals have been evaluated for this application, and at present scintillating PbF_2 crystals are favored (111).

4.5. Calorimetry in High-Rate/High-Radiation Environments

With the anticipated development of higher-energy and higher-luminosity accelerators in the coming decades, calorimeters will have to work in higher–radiation level environments and endure higher interaction rates. Ongoing R&D is addressing both upgrades for existing calorimeters and the design of new calorimeters at future accelerators. Current R&D efforts for the SLHC and the forward calorimeters for a high-energy e^+e^- linear collider are discussed below.

4.5.1. Calorimetry at the super LHC. The design luminosity of the LHC is 10^{34} cm^{-2} s^{-1} at 14 TeV. In the next decade, the LHC luminosity is expected to be upgraded beyond the nominal luminosity (112). The total integrated luminosity is expected to exceed 3,000 fb^{-1} by the end of LHC operation. Because the calorimeters at the current CMS and ATLAS experiments were designed to tolerate approximately 1,000 fb^{-1} and an instantaneous luminosity of 10^{34} cm^{-2} s^{-1}, some components in the calorimeter systems will have to be upgraded or replaced.

The ATLAS EM calorimetry is expected to be robust against the increased radiation backgrounds, but the current readout will not survive beyond 700 fb^{-1}. ATLAS is evaluating a novel SiGe Bi-CMOS technology for the upgrade of the LAr front-end readout followed by a completely new readout electronics chain, including on-detector and back-end electronics. The new readout is entirely digital, which will allow a more sophisticated fast trigger (76). The front-end electronics on the detector is required to withstand an exposure of up to 300 krad of ionizing radiation and a neutron fluence of 10^{13} n cm^{-2}.

The ATLAS endcap hadronic calorimeter uses highly integrated amplifiers and summing chips, which operate at LAr temperature inside the cryostat. More radiation-hard technologies are being investigated to ensure the operation at SLHC at a neutron fluence of up to 2×10^{15} n cm^{-2} (113).

A concern for the forward calorimeter is the degradation of the LAr signal at high beam intensity due to charge buildup in the LAr itself (114). The critical ionization rate, at which the rate of ion-pair creation per unit volume is equal to the rate at which positive ions are removed from the gap, is proportional to $E^2\mu_+ a^{-2}$, where E is the electric field, μ_+ is the positive ion mobility, and a is the gap size. The charge-buildup effect in the ATLAS forward calorimeter is mitigated by having very thin (250-μm) gaps. An R&D effort is currently being carried out to study the performance of small test modules at IHEP/Protovino at intensities above those expected for ATLAS at SLHC (113). The resulting ionization is significantly higher than the critical density. The expected energy deposition at nominal luminosity is approximately 45 W in each of the forward calorimeters (115). An alternative to modifying the ATLAS forward calorimeter to reduce the charge-buildup effect is to place a warm calorimeter in front of it to reduce its ionization load. A diamond detector with a copper or tungsten absorber is being considered.

On the basis of measurements of crystal degradation with radiation, it is expected that the CMS barrel EM calorimeter ($|\eta| < 1.4$) will survive well into the SLHC era. Current efforts are focused instead on understanding the performance degradation of the endcap EM calorimeter, in which radiation levels will be significantly higher.

The upgrade of the CMS hadronic calorimeter (116) involves replacing the current hybrid photodetectors with SiPMs, which have lower noise and higher gain. Researchers have also proposed introducing longitudinal segmentation in the hadron endcap calorimeter to meet the challenges due to high occupancy and to replace the scintillators in the endcap with quartz plates to increase the radiation hardness.

4.5.2. Forward calorimetry for beamstrahlung pairs at the ILC. One of the most challenging detectors for linear collider experiments is the forward calorimeter. When intense beams of electrons and positrons collide at the center of the detector, the beam-beam interactions cause the emission of beamstrahlung photons. Although most of these photons leave the detector along the beam line, many are converted to electron-positron pairs. Some of the electrons and positrons are deflected into larger polar angles to become a major detector background, whereas others travel down the beam line, where they intercept the forward calorimeters. The current ILC detector design includes two calorimeters in the very forward region, BeamCal and LumiCal, to provide fast and precise measurements of the beam properties and luminosity and to ensure detector hermeticity (118).

The BeamCal covers polar angles down to 5 mrad. Such coverage is required in order to identify high-energy electrons from hard e^+e^- scattering on top of the background of beamstrahlung pairs, which can deposit several teraelectronvolts per bunch crossing in the forward region. Fine granularity and a small Molière radius are required. The BeamCal is also expected to be used for beam tuning by measuring the flow of the energies from the beamstrahlung pairs as the beams are adjusted. **Figure 11** shows the density of energy deposited by beamstrahlung remnants for one bunch crossing in the $r - \phi$ plane for two different beam configurations (117).

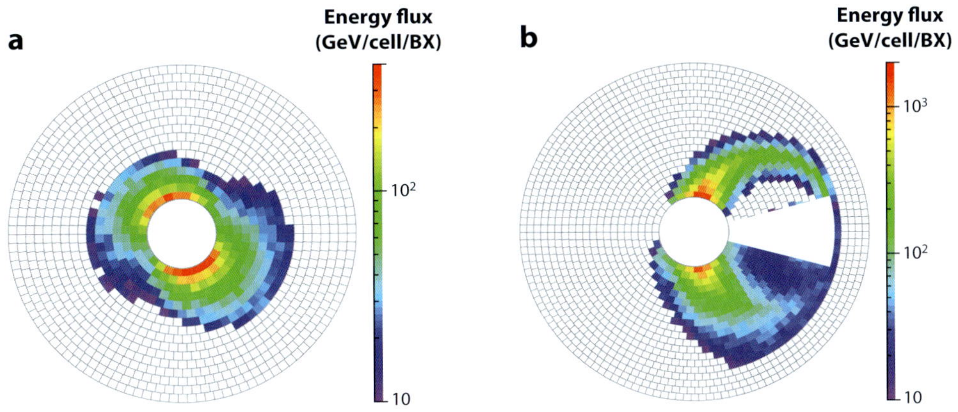

Figure 11

The energy density of beamstrahlung remnants per bunch crossing as a function of position in the $r - \phi$ plane at (*a*) a 2-mrad crossing angle and (*b*) a 20-mrad crossing angle with detector-integrated dipole configuration (117).

The expected radiation dose is up to 10 MGy per year for the nominal accelerator parameters at a center-of-mass energy of 500 GeV. The technology being considered is a tungsten sampling calorimeter, with sensors that are stable under the radiation load and have good linearity over a dynamic range of approximately 10^4. The chemical vapor deposition diamond sensor is the primary candidate (119), although gallium arsenide and silicon sensors are also being considered. The performance of several diamond sensors as a function of the absorbed dose has been measured in a 10-MeV electron beam for doses up to 7 MGy. A linearity better than 30% for a particle density up to a few 10^6 cm^{-2} (10 ns)$^{-1}$ was observed. The change in charge-collection distance after high-radiation dosage was also measured. An R&D program for the forward calorimeters for ILC experiments is being carried out by the ILC FCAL Collaboration.

5. CONCLUSION

Calorimetry is a critical technology in modern high-energy and nuclear physics experimentation. It is a field marked by innovation and advances on many fronts. Physics needs continue to push technological developments. Accelerator advances demand higher levels of performance in radiation hardness, resolution, and speed. Neutrino physics requires more massive and sensitive detectors, and cosmic-ray physics needs improved understanding of ultrahigh-energy hadron shower development. The prospects for future advances in calorimetry are bright.

DISCLOSURE STATEMENT

The authors are not aware of any affiliations, memberships, funding, or financial holdings that might be perceived as affecting the objectivity of this review.

ACKNOWLEDGMENTS

We gratefully acknowledge contributions from discussions with our collaborators and colleagues, many of whom are cited below. Three were particularly helpful in the final review of our paper: Martin Breidenbach, Howard Gordon, and Ren-yuan Zhu. Our work was supported by the U.S. Department of Energy (grant DE-FG02-96ER40969 at Oregon and contracts DE-AC02-76SF00515 at SLAC and DE-AC02-98CH1-886 at BNL) and the U.S. National Science Foundation. J.E.B. expresses gratitude for research support from Philip H. Knight.

LITERATURE CITED

1. Brau JE. *Nucl. Instrum. Methods A* 312:483 (1992)
2. Nelson W, Hirayama H, Rogers D. *SLAC Rep. No. 0265.* **http://www.slac.stanford.edu/pubs/slacreports/slac-r-265.html** (1985)
3. Agostinelli S, et al. *Nucl. Instrum. Methods A* 506:250 (2003)
4. Amsler C, et al. *Phys. Lett.* B667:1 (2008)
5. Longo E, et al. *Nucl. Instrum. Methods* 128:283 (1975)
6. Fabjan CW, Ludlam T. *Annu. Rev. Nucl. Part. Sci.* 32:335 (1982)
7. Huston J. Appendix H. In *The Fundamental Particles and Their Interactions*, ed. WB Rolnick, p. 425. Reading, UK: Addison-Wesley (1994)
8. Gabriel T, et al. *Nucl. Instrum. Methods A* 338:336 (1994)
9. Groom D. *Nucl. Instrum. Methods A* 572:633 (2007); Groom D. Erratum. *Nucl. Instrum. Methods A* 593:628 (2008)
10. Fabjan CW, et al. *Nucl. Instrum. Methods* 141:61 (1977)
11. Wigmans R. *Annu. Rev. Nucl. Part. Sci.* 41:133 (1991)
12. Abt I, et al. (H1 Collab.) *Nucl. Instrum. Methods A* 386:348 (1997)
13. Akchurin N, et al. *Nucl. Instrum. Methods A* 537:537 (2005)
14. Leo WR. *Techniques for Nuclear and Particle Physics Experiments: A How-To Approach*. Berlin: Springer. 368 pp. (1987)
15. Holder M, et al. *Nucl. Instrum. Methods* 148:235 (1978)
16. Holder M, et al. *Nucl. Instrum. Methods* 151:69 (1978)
17. De Winter K, et al. (CHARM-II Collab.) *Nucl. Instrum. Methods A* 278:670 (1989)
18. Harris DA, et al. (NuTeV Collab.) *Nucl. Instrum. Methods A* 447:377 (2000)
19. Adamson P, et al. (MINOS Collab.) *IEEE Trans. Nucl. Sci.* 49:861 (2002)
20. Adamson P, et al. (MINOS Collab.) *Phys. Rev. D* 73:072002 (2006)
21. Fukuda Y, et al. *Nucl. Instrum. Methods A* 501:418 (2003)
22. Feldman GJ. In *Neutrino Oscillations*, ed. JA Thomas, PL Vahle, p. 217. Singapore: World Sci. (2008)
23. Abachi S, et al. (DØ Collab.) *Nucl. Instrum. Methods A* 338:185 (1994)
24. Abazov V, et al. (DØ Collab.) *Nucl. Instrum. Methods A* 565:463 (2006)
25. Abe F, et al. (CDF Collab.) *Nucl. Instrum. Methods A* 271:387 (1988)
26. Aad G, et al. (ATLAS Collab.) *J. Instrum.* 3:S08003 (2008)
27. Chatrchyan S, et al. (CMS Collab.) *J. Instrum.* 3:S08004 (2008)
28. Bayatian GL, et al. (CMS Collab.) *CERN-LHCC-2006-001* (2006)
29. Abt I, et al. (H1 Collab.) *Nucl. Instrum. Methods A* 386:310 (1997)
30. Andrieu B, et al. *Nucl. Instrum. Methods A* 336:460 (1993)
31. Derrick M, et al. *Nucl. Instrum. Methods A* 309:77 (1991)
32. Andresen A, et al. *Nucl. Instrum. Methods A* 309:101 (1991)
33. Bernstein A, et al. *Nucl. Instrum. Methods A* 336:23 (1993)
34. Kubota Y, et al. (CLEO Collab.) *Nucl. Instrum. Methods A* 320:66 (1992)
35. Ambrosino F, et al. *Nucl. Instrum. Methods A* 598:239 (2009)
36. Abashian A, et al. (Belle Collab.) *Nucl. Instrum. Methods A* 479:117 (2002)
37. Aubert B, et al. (BaBar Collab.) *Nucl. Instrum. Methods A* 479:1 (2002)
38. Kocian M. (BaBar Collab.) Presented at Int. Conf. Calorim. High Energy Phys., 10th, Pasadena, Calif. (2002)
39. Zhou L, et al. *J. Phys. Conf. Ser.* 160:012009 (2009)
40. Anashin VV, et al. *Nucl. Instrum. Methods A* 478:420 (2002)
41. Decamp D, et al. (ALEPH Collab.) *Nucl. Instrum. Methods A* 294:121 (1990)

42. Aarnio PA, et al. (DELPHI Collab.) *Nucl. Instrum. Methods A* 303:233 (1991)
43. Sumner RL. (L3 Collab.) *Nucl. Instrum. Methods A* 265:252 (1988)
44. Adriani O, et al. (L3 Collab.) *Nucl. Instrum. Methods A* 302:53 (1991)
45. Ahmet K, et al. (OPAL Collab.) *Nucl. Instrum. Methods A* 305:275 (1991)
46. Axen D, et al. *Nucl. Instrum. Methods A* 328:472 (1993)
47. Benvenuti AC, et al. *Nucl. Instrum. Methods A* 276:94 (1989)
48. Lustermann W. *J. Phys. Conf. Ser.* 160:012044 (2009)
49. Atoian GS, et al. *Nucl. Instrum. Methods A* 584:291 (2008)
50. Atoian GS, et al. *Nucl. Instrum. Methods A* 320:144 (1992)
51. Cortese P, et al. (ALICE Collab.) *CERN-LHCC-2008-014* (2008)
52. Machikhiliyan I. (LHCb Collab.) *J. Phys. Conf. Ser.* 160:012047 (2009)
53. Semenov PA, et al. (PANDA Collab.) *Nucl. Instrum. Methods A* 598:224 (2009)
54. Gallucci G. (MEG Collab.) *J. Phys. Conf. Ser.* 160:012011 (2009)
55. Berridge SC, et al. *IEEE Trans. Nucl. Sci.* 39:1242 (1992)
56. Strom DM. In *Proceedings of the Third International Conference on Calorimetry in High Energy Physics*, ed. J Siegrist, p. 357. Singapore: World Sci. (1993)
57. Bederede D, et al. *Nucl. Instrum. Methods A* 365:117 (1995)
58. Abbiendi G, et al. (OPAL Collab.) *Eur. Phys. J. C* 14:373 (2000)
59. Almehed S, et al. *Nucl. Instrum. Methods A* 305:320 (1991)
60. Brau J, Gabriel TA. *Nucl. Instrum. Methods A* 238:489 (1985)
61. Brückmann H, Kowalski H. *ZEUS Intern. Note, 86/026 DESY* (1986)
62. Wigmans R. *Nucl. Instrum. Methods A* 259:389 (1987)
63. Brau JE, Gabriel TA. *Nucl. Instrum. Methods A* 275:190 (1989)
64. Akesson T, et al. *Nucl. Instrum. Methods A* 241:17 (1985)
65. Abbott B, et al. *Nucl. Instrum. Methods A* 424:352 (1999)
66. Hegeman J. *J. Phys. Conf. Ser.* 160:012024 (2009)
67. Issever C, Borras K, Wegener D. *Nucl. Instrum. Methods A* 545:803 (2005)
68. Barillari T, et al. *ATLAS Public Note, ATL-LARG-PUB-2009-001* (2009)
69. Bhatti A, et al. *Nucl. Instrum. Methods A* 566:375 (2006)
70. Aguilar-Arevalo AA, et al. (MiniBooNE Collab.) *Nucl. Instrum. Methods A* 599:28 (2009)
71. Becker-Szendy R, et al. *Nucl. Instrum. Methods A* 324:363 (1993)
72. Abraham J, et al. (Auger Collab.) *Nucl. Instrum. Methods A* 523:50 (2004)
73. Karle A, et al. (IceCube Collab.) *Nucl. Instrum. Methods A* 604:S46 (2009)
74. Chen J, Mao R, Zhang L, Zhu R. *IEEE Trans. Nucl. Sci.* 54:718 (2007)
75. Chen J, Mao R, Zhang L, Zhu R. *IEEE Trans. Nucl. Sci.* 54:1319 (2007)
76. Strässner A. *ATL-LARG-PROC-2009-016*. **http://cdsweb.cern.ch/record/1213928/files/ATL-LARG-PROC-2009-016.pdf** (2009)
77. Kuntner C, et al. *Nucl. Instrum. Methods A* 493:131 (2002)
78. Nakamura K. *Int. J. Mod. Phys. A* 18:4053 (2003)
79. Barger V. arXiv:0705.4396 [hep-ph] (2007)
80. Kawai Y, et al. *Nucl. Instrum. Methods A* 579:42 (2007)
81. Kawai Y, et al. *IEEE Nucl. Sci. Symp. Conf. Rec.* 40:3 (2009)
82. Rubbia C. *CERN-EP-INT-77-8* (1977)
83. Amerio S, et al. (ICARUS Collab.) *Nucl. Instrum. Methods A* 527:329 (2004)
84. Badertscher A, et al. arXiv:0811.3384 [physics.ins-det] (2008)
85. Bernstein A, et al. arXiv:0907.4183 [hep-ex] (2009)
86. Soderberg M. (ArgoNeuT Collab.) arXiv:0910.3433 [physics.ins-det] (2009)
87. Chen H, et al. (MicroBooNE Collab.) *Fermilab proposal 0974* (2007)
88. Baibussinov B, et al. *Astropart. Phys.* 29:174 (2008)
89. Rubbia A. *J. Phys. Conf. Ser.* 171:012020 (2009)
90. Videau H. *Proc. Int. Conf. Linear Collid.* 1:105 (2004)
91. Brient JC. Presented at Int. Linear Collid. Workshop. 5th, Fermilab, Batavia, Ill. (2000)
92. Buskulic D, et al. (ALEPH Collab.) *Nucl. Instrum. Methods A* 360:481 (1995)

93. Thomson MA. *Nucl. Instrum. Methods A* 611:25 (2009)
94. Charles MJ, Mallik U, Kim TJ. arXiv:0901.4532 [physics-ins.det] (2009)
95. Apostolakis J, et al. *J. Phys. Conf. Ser.* 160:012073 (2009)
96. CMS Collab. *CMS Phys. Anal. Summ. CMS-PAS-PFT-09-001* (2009)
97. Adloff C, et al. (CALICE Collab.) *J. Phys. Conf. Ser.* 160:012065 (2009)
98. Strom DM, et al. *IEEE Trans. Nucl. Sci.* 52:868 (2005)
99. Freytag D, et al. *SLAC-PUB-13462* (2008)
100. Ballin JA, et al. arXiv:0901.4457 [physics-ins.det] (2009)
101. Buzhan P, et al. *Nucl. Instrum. Methods A* 504:48 (2003)
102. Jeans D. (CALICE Collab.) *J. Phys. Conf. Ser.* 160:012026 (2009)
103. Garutti E. (CALICE Collab.) *J. Phys. Conf. Ser.* 160:012077 (2009)
104. Repond J. *J. Phys. Conf. Ser.* 160:012066 (2009)
105. Yu J. *Proc. Int. Linear Collid. Workshop* 2:567 (2007)
106. Adloff C, Blaha J, Espargiliere A, Karyotakis Y. arXiv:0901.4927 [physics-ins.det] (2009)
107. Hoff JR, et al. Presented at IEEE Nucl. Sci. Symp. Med. Imaging Conf., El Conquistador, P.R. (2005)
108. Mockett PM. *Stanford 1983: Proc. Dyn. Spectrosc. High Energy* 1:335 (1983)
109. Theodosiou GE, et al. *IEEE Trans. Nucl. Sci.* 31:57 (1984)
110. Para A. Presented at Tohoku Linear Collid. Workshop, Tohoku, Jpn. (2009)
111. Zhu R. Presented at Int. Linear Collid. Workshop, Chicago (2009)
112. Mangano ML. arXiv:0910.0030 [hep-ph] (2009)
113. Schacht P. *ATL-LARG-PROC-2009-013*. **http://cdsweb.cern.ch/record/1212930/files/** (2009)
114. Rutherfoord JP. *Nucl. Instrum. Methods A* 482:156 (2002)
115. Artamonov A., et al. (ATLAS Collab.) *J. Instrum.* 3:P02010 (2008)
116. Dugad SR. Presented at Int. Conf. Technol. Instrum. Part. Phys., 1st, Tsukuba, Ibaraki, Jpn. *CERN-CMS-CR-2009-057* (2009)
117. Bambade P, Drugakov V, Lohmann W. *Pramana* 69:1123 (2007)
118. Behnke T, et al. arXiv:0712.2356 [physics-ins.det] (2007)
119. Schuwalow S. *Nucl. Instrum. Methods A* 598:258 (2009)

Radiative and Electroweak Penguin Decays of B Mesons

Tobias Hurth[1] and Mikihiko Nakao[2]

[1]Institute for Physics, Johannes Gutenberg University, D-55099 Mainz, Germany; email: Tobias.Hurth@cern.ch

[2]High Energy Accelerator Research Organization (KEK) and the Graduate University for Advanced Studies (Sokendai), Tsukuba 305-0801, Japan; email: mikihiko.nakao@kek.jp

Annu. Rev. Nucl. Part. Sci. 2010. 60:645–77

The *Annual Review of Nuclear and Particle Science* is online at nucl.annualreviews.org

This article's doi:
10.1146/annurev.nucl.012809.104424

0163-8998/10/1123-0645$20.00

Key Words

B factory, flavor physics, flavor-changing neutral current, inclusive decays, exclusive decays

Abstract

The huge data sets collected at the two B factories, Belle and BaBar, have made it possible to explore the radiative penguin process $b \to s\gamma$, the electroweak penguin process $b \to s\ell^+\ell^-$, and the suppressed radiative process $b \to d\gamma$ in detail—all in exclusive channels and inclusive measurements. Theoretical tools have also advanced to meet or surpass the experimental precision, especially in inclusive calculations and the various ratios of exclusive channels. In this article, we review the theoretical and experimental progress over the past decade in the radiative and electroweak penguin decays of B mesons.

Contents

1. INTRODUCTION

The B meson system, which is a bound state that consists of a b quark and a light antiquark, provides an ideal laboratory for the precise study of the Standard Model (SM) of particle physics and thus facilitates the search for new physics (NP). Because the b quark mass is much larger than the typical scale of the strong interaction, the otherwise troublesome long-distance strong interactions are generally less important and are under better control than in other, lighter meson systems. Radiative penguin[1] decays of the B meson with the emission of a photon (γ) and electroweak penguin decays with the emission of a lepton pair ($\ell^+\ell^-$, $\ell = e, \mu$) are of particular interest in this respect. These processes reveal the flavor-changing neutral current (FCNC), that is, the transition of a b quark with an electric charge of $-1/3$ into an s or a d quark of the same charge. In the SM, the FCNC occurs only via virtual-loop diagrams (**Figure 1**). Additional NP contributions to these decay rates are not necessarily suppressed with respect to the SM contribution. Examples of such NP scenarios include those in which the SM particles in the loop diagram are replaced by hypothetical new particles at a high mass scale; so far, they have not been directly accessible in collider experiments. Radiative and electroweak penguin decays are highly sensitive to NP because

[1]The name penguin decays was first introduced in Reference 1 as the result of a bet. A more detailed account of the name can be found in Reference 2.

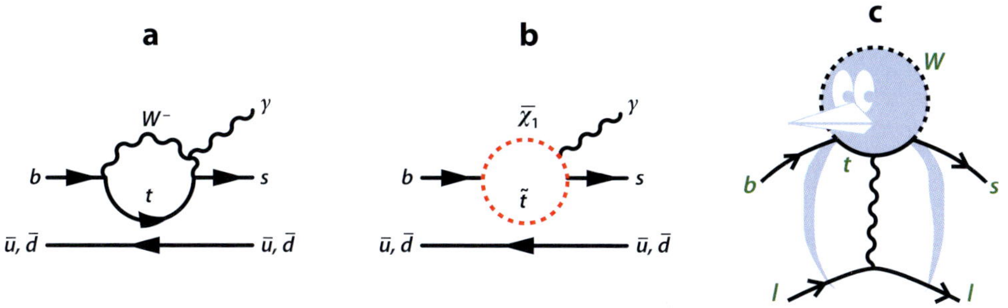

Figure 1

Examples of radiative penguin decay diagrams in (*a*) the Standard Model and (*b*) beyond. (*c*) A penguin.

they are theoretically well understood and have been extensively measured at the *B* factories. The search for such NP effects complements the search for new particles produced at collider experiments.

The first generation of the *B* factories at KEK (i.e., the Belle experiment at the KEKB e^+e^- collider; see **http://belle.kek.jp/**) and at SLAC (i.e., the BaBar experiment at the PEP-II e^+e^- collider; see **http://www.slac.stanford.edu/BFROOT/**) have collected huge samples of *B* meson decays and have thereby established the SM picture of *CP* violation and other flavor-changing processes in the quark sector. These processes are governed by a single 3×3 unitarity matrix referred to as the Cabibbo-Kobayashi-Maskawa (CKM) matrix (3, 4). The CKM matrix can be illustrated by a unitarity triangle in the complex plane that is overconstrained by measurements from the *B* factories, the Tevatron *B* physics programs [namely the CDF (see **http://www-cdf.fnal.gov/physics/new/bottom/bottom.html**) and DØ (see **http://www-d0.fnal.gov/Run2Physics/WWW/results/b.htm**) experiments], and earlier kaon decay experiments. In other words, none of the current measurements of *B* meson decays have observed an unambiguous sign of NP (5, 6). Although this experimental result is an impressive success of the CKM theory within the SM, there is still room for sizable new effects from new flavor structures, given that FCNC processes have been tested only up to the 10% level.

The nonexistence of large NP effects in the FCNC processes hints at the famous flavor problem, namely why FCNCs are suppressed. This problem must be solved in any viable NP model. Either the mass scale of the new degrees of freedom is very high or the new flavor-violating couplings are small for reasons that remain to be found. For example, assuming generic new flavor-violating couplings, the present data on $K - \bar{K}$ mixing imply a very high NP scale of order 10^3–10^4 TeV, depending on whether the new contributions enter at loop level or at tree level. In contrast, theoretical considerations on the Higgs sector, which is responsible for the mass generation of the fundamental particles in the SM, call for NP at order 1 TeV. As a consequence, any NP below the 1-TeV scale must have a nongeneric flavor structure. The present measurements of *B* decays, especially of FCNC processes, already significantly restrict the parameter space of NP models. For further considerations on NP, the reader is referred to another article in this volume (7) and to Reference 8.

Quark-level FCNC processes such as $b \to s\gamma$, to which NP may contribute, cannot be directly measured because the strong interaction forms hadrons from the underlying quarks. Instead, the experimentally measured and theoretically calculated process is a *B* meson decay into a photon plus an inclusive hadronic final-state X_s, which includes all the hadron combinations that carry the strange quantum number $s = +1$ of the s quark.[2] Exclusive final states with one or a few specific hadrons in the final state (e.g., $B \to K^*\gamma$) have less predictive power theoretically; however, because the measurements are easier and better defined, there are other useful observables beyond branching fractions, in particular the *CP*, forward-backward, isospin, and polarization asymmetries. In the future, a large overconstrained set of measurements of these observables will allow us to detect specific patterns and to distinguish between various NP scenarios.

This review covers progress in radiative and electroweak decays over the past decade, during which a huge number of B factory results were accumulated and significant progress in various theoretical aspects was achieved. The pioneering work that led to the first observation of the $b \to s\gamma$ process by CLEO (9, 10) was discussed in an earlier volume of this journal (11).

Our review is organized as follows. In Section 2, we describe theoretical tools for radiative and electroweak penguin decays, and in Section 3 we describe experimental techniques. We give theoretical predictions in Section 4 and summarize the measurements of radiative and electroweak penguin decays in Section 5. Finally, we briefly discuss future prospects in Section 6.

2. THEORETICAL FRAMEWORK

Inclusive B decays are theoretically clean because they are dominated by partonic (perturbatively calculable) contributions. Nonperturbative corrections are in general rather small (12–14). This result can be derived with the help of the heavy mass expansion (HME) of the inclusive decay rates in inverse powers of the b quark mass. Up-to-date predictions of exclusive B decays are based on the quantum chromodynamics (QCD)-improved factorization (QCDF) and soft collinear effective theory (SCET) methods. In general, exclusive modes have larger nonperturbative QCD corrections than do inclusive modes.

2.1. Electroweak Effective Hamiltonian

Rare B decays are governed by an interplay between the weak and strong interactions. The QCD corrections that arise from hard gluon exchange bring in large logarithms of the form $\alpha_s^n(m_b)\log^m(m_b/M)$, where $M = m_t$ or $M = m_W$ and $m \leq n$ (with $n = 0, 1, 2, \ldots$). These large logarithms are a natural feature in any process in which two different mass scales are present. To obtain a reasonable result, one must resum at least the leading-log (LL) series, $n = m$, with the help of renormalization-group techniques. Working to next-to-leading-log (NLL) or next-to-next-to-leading-log (NNLL) precision means that one resums all the terms with $n = m + 1$ or $n = m + 2$. A suitable framework in which to achieve the necessary resummations of the large logarithms is an effective low-energy theory with five quarks; this framework is obtained by integrating out the heavy particles, which in the SM are the electroweak bosons and the top quark.

This effective field theory approach serves as a theoretical framework for both inclusive and exclusive modes. The standard method of the operator product expansion (OPE) (15, 16) allows for the separation of the B meson decay amplitude into two distinct parts: the long-distance

[2]In this review, we use the following notations and conventions: We denote the inclusive decay as $B \to X_s\gamma$ when charge conjugation is implied, or as $\bar{B} \to X_s\gamma$ and $B \to X_{\bar{s}}\gamma$ to reflect the quark charges of the underlying processes $b \to s\gamma$ and $\bar{b} \to \bar{s}\gamma$, respectively, when *CP* and angular asymmetries are concerned. Here, B denotes either an isospin- and *CP*-averaged state of B°, $\bar{B}^0$, B^+, and B^- mesons or an isospin-averaged state of B° and B^+ (in the latter case, $\bar{B}$ denotes $\bar{B}^0$ and B^-). Expressions are constructed similarly for inclusive $X_d\gamma$ and $X_s\ell^+\ell^-$ final states and for isospin-averaged exclusive final states. In the literature, the notation $\bar{B} \to X_s\gamma$ is also commonly used for the case that includes charge conjugation.

contributions contained in the operator matrix elements and the short-distance physics described by the so-called Wilson coefficients.

The electroweak effective Hamiltonian (17–19) can be written as

$$\mathcal{H}_{\rm eff} = \frac{4G_F}{\sqrt{2}} \sum C_i(\mu, M)\, \mathcal{O}_i(\mu), \qquad 1.$$

where $\mathcal{O}_i(\mu)$ are the relevant operators and $C_i(\mu, M)$ are the corresponding Wilson coefficients. As the heavy fields are integrated out, the complete top and W mass dependency is contained in the Wilson coefficients. Clearly, only within the observable $\mathcal{H}_{\rm eff}$ does the scale dependency cancel out. G_F denotes the Fermi coupling constant.

From the μ independence of the effective Hamiltonian, one can derive a renormalization-group equation (RGE) for the Wilson coefficients $C_i(\mu)$,

$$\mu \frac{d}{d\mu} C_i(\mu) = \gamma_{ji}\, C_j(\mu), \qquad 2.$$

where the matrix γ is the anomalous dimension matrix of the operators $\mathcal{O}_i$, which describe the anomalous scaling of the operators with respect to the scaling at the classical level. At leading order, the solution is given by

$$\tilde{C}_i(\mu) = \left[\frac{\alpha_s(\mu_W)}{\alpha_s(\mu)}\right]^{\frac{\tilde{\gamma}_{ii}^0}{2\beta_0}} \tilde{C}_i(\mu_W) = \left[\frac{1}{1+\beta_0 \frac{\alpha_s(\mu)}{4\pi} \ln \frac{\mu_W^2}{\mu^2}}\right]^{\frac{\tilde{\gamma}_{ii}^0}{2\beta_0}} \tilde{C}_i(\mu_W), \qquad 3.$$

where $\mu d/d\mu\, \alpha_s = -2\beta_0 \alpha_s^2/(4\pi)$, and β_0 and $\tilde{\gamma}_{ii}^0$ correspond to the leading anomalous dimensions of the coupling constant and of the operators, respectively. The tilde indicates that the diagonalized anomalous dimension matrix is used.

Although the Wilson coefficients $C_i(\mu)$ enter both inclusive and exclusive processes and can be calculated with perturbative methods, the calculational approaches to the matrix elements of the operators differ between the two cases. Within inclusive modes, one can use the quark-hadron duality to derive a well-defined HME of the decay rates in powers of Λ/m_b (20–25). In particular, the decay width of the $B \to X_s\gamma$ is well approximated by the partonic decay rate, which can be calculated in renormalization group–improved perturbation theory (26, 27):

$$\Gamma(B \to X_s\gamma) = \Gamma\left(b \to X_s^{\rm parton}\gamma\right) + \Delta^{\rm nonpert.}. \qquad 4.$$

In exclusive processes, however, one cannot rely on the quark-hadron duality, so one must estimate the matrix elements between meson states. A promising approach is the QCDF method, which has been systematized for nonleptonic decays in the heavy quark limit (28–30). In addition, a more general quantum field theoretical framework for QCDF, known as SCET, has been proposed (31–36). This method allows for a perturbative calculation of QCD corrections to naïve factorization and is the basis for the up-to-date predictions for exclusive rare B decays. However, within this approach, a general quantitative method to estimate the important Λ/m_b corrections to the heavy quark limit is missing.

2.2. Perturbative Corrections to Inclusive Decays

Within inclusive B decay modes, short-distance QCD effects are very important. For example, in the $B \to X_s\gamma$ decay these effects lead to a rate enhancement by a factor of greater than two. Such effects are induced by hard gluon exchanges between the quark lines of the one-loop electroweak diagrams. The corresponding large logarithms have to be summed as discussed above.

The effective electroweak Hamiltonian that is relevant to $b \to s/d\,\gamma$ and $b \to s/d\,\ell^+\ell^-$ transitions in the SM reads

$$\mathcal{H}_{\mathrm{eff}} = -\frac{4G_F}{\sqrt{2}}\left[\lambda_q^t \sum_{i=1}^{10} C_i \mathcal{O}_i + \lambda_q^u \sum_{i=1}^{2} C_i(\mathcal{O}_i - \mathcal{O}_i^u)\right], \qquad 5.$$

where the explicit CKM factors are $\lambda_q^t = V_{tb}V_{tq}^*$ and $\lambda_q^u = V_{ub}V_{uq}^*$. The unitarity relations $\lambda_q^c = -\lambda_q^t - \lambda_q^u$ have already been used. The dimension-six operators are

$$\begin{aligned}
\mathcal{O}_1 &= (\bar{s}_L\gamma_\mu T^a c_L)(\bar{c}_L\gamma^\mu T^a b_L), & \mathcal{O}_2 &= (\bar{s}_L\gamma_\mu c_L)(\bar{c}_L\gamma^\mu b_L),\\
\mathcal{O}_1^u &= (\bar{s}_L\gamma_\mu T^a u_L)(\bar{u}_L\gamma^\mu T^a b_L), & \mathcal{O}_2^u &= (\bar{s}_L\gamma_\mu u_L)(\bar{u}_L\gamma^\mu b_L),\\
\mathcal{O}_3 &= (\bar{s}_L\gamma_\mu b_L)\sum_q(\bar{q}\gamma^\mu q), & \mathcal{O}_4 &= (\bar{s}_L\gamma_\mu T^a b_L)\sum_q(\bar{q}\gamma^\mu T^a q),\\
\mathcal{O}_5 &= (\bar{s}_L\Gamma b_L)\sum_q(\bar{q}\Gamma' q), & \mathcal{O}_6 &= (\bar{s}_L\Gamma T^a b_L)\sum_q(\bar{q}\Gamma' T^a q),\\
\mathcal{O}_7 &= e/16\pi^2 m_b(\bar{s}_L\sigma^{\mu\nu}b_R)F_{\mu\nu}, & \mathcal{O}_8 &= g_s/16\pi^2 m_b(\bar{s}_L\sigma^{\mu\nu}T^a b_R)G^a_{\mu\nu},\\
\mathcal{O}_9 &= e^2/16\pi^2(\bar{s}_L\gamma_\mu b_L)\sum_\ell(\bar{\ell}\gamma^\mu\ell), & \mathcal{O}_{10} &= e^2/16\pi^2(\bar{s}_L\gamma_\mu b_L)\sum_\ell(\bar{\ell}\gamma^\mu\gamma_5\ell),
\end{aligned} \qquad 6.$$

where $\Gamma = \gamma_\mu\gamma_\nu\gamma_\lambda$ and $\Gamma\prime = \gamma^\mu\gamma^\nu\gamma^\lambda$. The subscripts L and R refer to left- and right-handed components, respectively, of the fermion fields. In $b \to s$ transitions, the contributions proportional to λ_s^u are rather small, whereas in $b \to d$ decays, where λ_d^u is of the same order as λ_d^t, these contributions play an important role in *CP* and isospin asymmetries. The semileptonic operators $\mathcal{O}_9$ and $\mathcal{O}_{10}$ occur only in the semileptonic $b \to s/d\,\ell^+\ell^-$ modes.

Among the four-quark operators, only the effective couplings for $i = 1, 2$ are large at the low scale $\mu = m_b$ [$C_{1,2}(m_b) \sim 1$], whereas the couplings of the other four-quark operators have almost negligible values. However, the dipole operators [$C_7(m_b) \sim -0.3,\ C_8(m_b) \sim -0.15$] and the semileptonic operators [$C_9(m_b) \sim 4,\ C_{10}(m_b) \sim -4$] also play a significant role.

There are three principal calculational steps that lead to the LL (NNLL) result within the effective field theory approach:

1. The full SM theory must be matched with the effective theory at the scale $\mu = \mu_W$, where μ_W denotes a scale of order m_W or m_t. The Wilson coefficients $C_i(\mu_W)$ pick up only small QCD corrections, which can be calculated within fixed-order perturbation theory. In the LL (NNLL) program, the matching has to be worked out at the $O(\alpha_s^0)$ [$O(\alpha_s^2)$] level.
2. The evolution of these Wilson coefficients from $\mu = \mu_W$ down to $\mu = \mu_b$ must then be performed with the help of the renormalization group, where μ_b is of the order of m_b. Because the matrix elements of the operators evaluated at the low scale μ_b are free of large logarithms, the latter are contained in resummed form in the Wilson coefficients. For the LL (NNLL) calculation, this RGE step has to be performed using the anomalous-dimension matrix up to order α_s^1 (α_s^3).
3. To LL (NNLL) precision, the corrections to the matrix elements of the operators $\langle s\gamma|\mathcal{O}_i(\mu)|b\rangle$ at the scale $\mu = \mu_b$ must be calculated to order α_s^0 (α_s^2) precision. The calculation also includes bremsstrahlung corrections.

2.2.1. $B \to X_s\gamma$. The error of the LL prediction of the $B \to X_s\gamma$ branching fraction (37–40) is dominated by a large renormalization-scale dependency at the $\pm 25\%$ level, which indicates the importance of the NLL series. By convention, the dependency on the renormalization scale μ_b is obtained by the variation $m_b/2 < \mu_b < 2m_b$. The three calculational steps of the NLL enterprise—Step 1 (41, 42), Step 2 (43, 44), and Step 3 (45–49)—have been performed by many

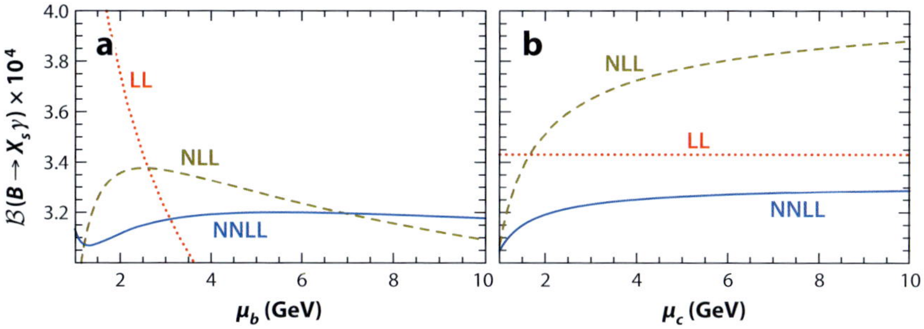

Figure 2

Renormalization-scale dependency of $\mathcal{B}(B \to X_s\gamma)$ in units of 10^4 at leading log (LL) (*dotted red lines*), next-to-leading log (NLL) (*dashed dark yellow lines*), and next-to-next-to-leading log (NNLL) (*solid blue lines*). The plots describe the dependency on (*a*) the low-energy scale μ_b and (*b*) the charm mass renormalization scale μ_c. From Reference 54.

different groups and have been independently checked. The resulting NLL prediction had a small dependency on the scale μ_b as well as on the matching scale μ_0 below 5%. But, as first observed in Reference 50, there was a large charm mass scheme dependency because the charm loop vanishes at the LL level and the significant charm dependency begins only at the NLL level. By varying m_c/m_b in the conservative range $0.18 \leq m_c/m_b \leq 0.31$, which covers both the pole mass value (with its numerical error) and the running mass value $\bar{m}_c(\mu_c)$ with $\mu_c \in [m_c, m_b]$, one finds an uncertainty of almost 10% (51, 52). This uncertainty is the dominant error in the NLL prediction. The renormalization scheme for m_c is an NNLL issue, and a complete NNLL calculation reduces this large uncertainty by at least a factor of two (53). This finding motivated the NNLL calculation of the $B \to X_s\gamma$ branching fraction.

Following a global effort, such an NNLL calculation was recently performed and led to the first NNLL prediction of the $B \to X_s\gamma$ branching fraction (54). This result is based on various highly nontrivial perturbative calculations (55–65). Some perturbative NNLL corrections have not yet been included in the present NNLL estimate, but they are expected to be smaller than the current perturbative uncertainty of 3% (59, 66–69).

In the present NNLL prediction (54), the reduction of the renormalization-scale dependency at the NNLL is shown in **Figure 2**. The most important effect occurs for the charm mass $\overline{MS}$ renormalization scale μ_c, which has been the main source of uncertainty at the NLL. The current uncertainty of $\pm 3\%$ due to higher-order $[\mathcal{O}(\alpha_s^3)]$ effects can be estimated via the NNLL curves in **Figure 2**. The reduction factor of the perturbative error is a factor of greater than three. The central value of the NNLL prediction is based on the choices $\mu_b = 2.5$ GeV and $\mu_c = 1.5$ GeV.

At NNLL QCD accuracy, subdominant, perturbative, electroweak two-loop corrections are also relevant and have been calculated to be -3.6% (70–73). They are included in the present NNLL prediction.

2.2.2. $B \to X_s\ell^+\ell^-$. Compared with the $B \to X_s\gamma$ decay, the inclusive $B \to X_s\ell^+\ell^-$ decay presents a complementary and more complex test of the SM, given that different perturbative electroweak contributions add to the decay rate. This inclusive mode is also dominated by perturbative contributions, if one eliminates $c\bar{c}$ resonances with the help of kinematic cuts. In the so-called perturbative q^2 windows below and above the resonances, namely in the low–dilepton mass region $1\,\text{GeV}^2 < q^2 = m_{\ell\ell}^2 < 6\,\text{GeV}^2$ and in the high–dilepton mass region where $q^2 > 14.4\ \text{GeV}^2$, theoretical predictions for the invariant mass spectrum are dominated by the perturbative contributions. A theoretical precision of order 10% is possible.

The complete NLL contributions have been presented (40, 74). For the NNLL calculation, many components were taken over from the NLL calculation of the $B \to X_s\gamma$ mode. The additional components for the NNLL QCD precision have been calculated (44, 56, 57, 75–84).

More recently, electromagnetic corrections were calculated: NLL quantum electrodynamics (QED) two-loop corrections to the Wilson coefficients are of $O(2\%)$ (84). Also, in the QED one-loop corrections to matrix elements, large collinear logarithms of the form $\log(m_b^2/m_\ell^2)$ survive integration if only a restricted part of the dilepton mass spectrum is considered. These collinear logarithms add another contribution of order +2% in the low-q^2 region for $B \to X_s\mu^+\mu^-$ (85). For the high-q^2 region, one finds −8% (86).

2.2.3. $B \to X_d\gamma$ and $B \to X_d\ell^+\ell^-$. The perturbative QCD corrections in the inclusive decays $B \to X_d\gamma$ (51, 52, 87) and $B \to X_d\ell^+\ell^-$ (88, 89) can be treated completely analogously to those in the corresponding $b \to s$ modes. The effective Hamiltonian is the same in these processes, up to the obvious replacement of the s quark field by the d quark field. However, because $\lambda_u = V_{ub}V_{ud}^*$ for $b \to d\gamma$ is not small with respect to $\lambda_t = V_{tb}V_{td}^*$ and $\lambda_c = V_{cb}V_{cd}^*$, one must also account for the operators proportional to λ_u, namely $\mathcal{O}_{1,2}^u$ in Equation 5. The matching conditions $C_i(m_W)$ and the solutions of the RGEs, which yield $C_i(\mu_b)$, coincide with those needed for the corresponding $b \to s$ processes (87, 88).

2.3. Hadronic Power Corrections to Inclusive Modes

The inclusive modes $B \to X_s\gamma$ and $B \to X_s\ell^+\ell^-$ are dominated by the partonic contributions. Indeed, if only the leading operator in the effective Hamiltonian ($\mathcal{O}_7$ for $B \to X_s\gamma$, $\mathcal{O}_9$ for $B \to X_s\ell^+\ell^-$) is considered, the HME makes it possible to calculate the inclusive decay rates of a hadron containing a heavy quark, especially a b quark (20–25). The optical theorem relates the inclusive decay rate of a hadron H_b to the imaginary part of the forward scattering amplitude

$$\Gamma(H_b \to X) = \frac{1}{2m_{H_b}} \Im\langle H_b|\mathbf{T}|H_b\rangle, \qquad 7.$$

where the transition operator $\mathbf{T}$ is given by $\mathbf{T} = i\int d^4x\, T[\mathcal{H}_{\text{eff}}(x)\mathcal{H}_{\text{eff}}(0)]$. The insertion of a complete set of states, $|X\rangle\langle X|$, leads to the standard formula for the decay rate:

$$\Gamma(H_b \to X) = \frac{1}{2m_{H_b}} \sum_X (2\pi)^4\delta^4(p_i - p_f)|\langle X|\mathcal{H}_{\text{eff}}|H_b\rangle|^2. \qquad 8.$$

It is then possible to construct an OPE of the operator $\mathbf{T}$, which is expressed as a series of local operators that are suppressed by powers of the b quark mass and written in terms of the b quark field:

$$T[\mathcal{H}_{\text{eff}}\mathcal{H}_{\text{eff}}] \overset{OPE}{=} \frac{1}{m_b}\left(\mathcal{O}_0 + \frac{1}{m_b}\mathcal{O}_1 + \frac{1}{m_b^2}\mathcal{O}_2 + \ldots\right). \qquad 9.$$

This construction is based on the parton-hadron duality: The sum is performed over all exclusive final states, and the energy release in the decay is large with respect to the QCD scale, $\Lambda \ll m_b$. With the help of the heavy quark effective theory (HQET) (90, 91), namely the new, heavy quark spin-flavor symmetries that arise in the heavy quark limit $m_b \to \infty$, the hadronic matrix elements within the OPE $\langle H_b|\mathcal{O}_i|H_b\rangle$ can be further simplified. In this well-defined expansion, the free quark model is the first term in the constructed expansion in powers of $1/m_b$ and therefore is the dominant contribution. In the applications to inclusive rare B decays, one finds no correction of order Λ/m_b to the free quark–model approximation. The corrections to the partonic decay rate begin with $1/m_b^2$ only, which implies a rather small numerical impact of the nonperturbative corrections on the decay rate of inclusive modes. However, there are additional subtleties to consider if other than the leading operators are taken into account (see below).

2.3.1. $B \to X_s\gamma$. These techniques can be used directly in the decay $B \to X_s\gamma$ to single out nonperturbative corrections to the branching fraction: If one neglects perturbative QCD corrections and assumes that the decay $B \to X_s\gamma$ is due to the leading electromagnetic dipole operator $\mathcal{O}_7$ alone, then the photon would always be emitted directly from the hard process of the b quark decay. One has to consider the time-ordered product $T\,\mathcal{O}_7^+(x)\mathcal{O}_7(0)$. Using the OPE for $T\,\mathcal{O}_7^+(x)\mathcal{O}_7(0)$ and HQET methods, as discussed above, the decay width $\Gamma(B \to X_s\gamma)$ reads (modulo higher terms in the $1/m_b$ expansion) (26, 27)

$$\Gamma_{B\to X_s\gamma}^{(\mathcal{O}_7,\mathcal{O}_7)} = \frac{\alpha_{\rm em} G_F^2 m_b^5}{32\pi^4}|V_{tb}V_{ts}|^2\, C_7^2(m_b)\left(1+\frac{\delta_{\rm rad}^{NP}}{m_b^2}\right), \quad \delta_{\rm rad}^{NP} = \frac{1}{2}\lambda_1 - \frac{9}{2}\lambda_2, \qquad 10.$$

where λ_1 and λ_2 are the HQET parameters for the kinetic energy and the chromomagnetic energy, respectively. If the $B \to X_s\gamma$ decay width is normalized by the (charmless) semileptonic one, the nonperturbative corrections of order $1/m_b^2$ cancel out within the ratio $\mathcal{B}(B \to X_s\gamma)/\mathcal{B}(B \to X_u\ell\nu)$.

However, as noted in Reference 92, there is no OPE for the inclusive decay $B \to X_s\gamma$, if one considers operators beyond the leading electromagnetic dipole operator $\mathcal{O}_7$. Voloshin (93) has identified a contribution to the total decay rate in the interference of the $b \to s\gamma$ amplitude due to the electromagnetic dipole operator $\mathcal{O}_7$ and the charming penguin amplitude due to the current-current operator $\mathcal{O}_2$. This resolved photon contribution contains subprocesses in which the photon couples to light partons instead of connecting directly to the effective weak interaction vertex (94). If one treats the charm quark as heavy, then it is possible to expand the contribution in local operators. The first term in this expansion may be the dominating one (92, 95, 96). This nonperturbative correction is suppressed by λ_2/m_c^2 and is estimated to be of order 3% compared with the leading-order (perturbative) contribution to the decay rate $\Gamma_{b\to s\gamma}$, which arises from the electromagnetic dipole operator $\mathcal{O}_7$:

$$\frac{\Delta\Gamma_{B\to X_s\gamma}^{(\mathcal{O}_2,\mathcal{O}_7)}}{\Gamma_{b\to s\gamma}^{LL}} = -\frac{1}{9}\frac{C_2}{C_7}\frac{\lambda_2}{m_c^2} \simeq +0.03. \qquad 11.$$

However, if the charm mass is assumed to scale as $m_c^2 \sim \Lambda m_b$, then the charm penguin contribution must be described by the matrix element of a nonlocal operator (92, 95–97).

Recently, another example of such nonlocal matrix elements within the power-suppressed contributions to the decay $B \to X_s\gamma$ was identified (97)—specifically, in the interference of the $b \to s\gamma$ transition amplitude mediated by the electromagnetic dipole operator $\mathcal{O}_7$, where the $b \to sg$ amplitude is mediated by the chromomagnetic dipole operator $\mathcal{O}_8$, followed by the fragmentation of the gluon into an energetic photon and a soft quark-antiquark pair. A naïve dimensional estimate of these power corrections leads to

$$\frac{\Delta\Gamma_{B\to X_s\gamma}^{(\mathcal{O}_7,\mathcal{O}_8)}}{\Gamma_{b\to s\gamma}^{LL}} \sim \pi\alpha_s\frac{C_8}{C_7}\frac{\Lambda}{m_b}, \qquad 12.$$

whereas an estimate using the vacuum insertion method for the nonlocal matrix elements indicates an effect of −3%.

Power corrections to the high-energy part of the $B \to X_s\gamma$ photon spectrum can be parameterized systematically in terms of subleading shape functions. For the interference of the

$\mathcal{O}_7$–$\mathcal{O}_7$ pair, these nonlocal operators reduce to local operators, if one considers the total decay rate (98–100), whereas other resolved photon contributions to the total decay rate—such as the previously analyzed $\mathcal{O}_7$–$\mathcal{O}_8$ interference term (97)—cannot be described by a local OPE. A recent systematic analysis (94) of all resolved photon contributions related to other operators in the weak Hamiltonian establishes this breakdown of the local OPE within the hadronic power corrections as a generic result. Clearly, estimating such nonlocal matrix elements is very difficult, and an irreducible theoretical uncertainty of $\pm(4$–$5)\%$ for the total *CP*-averaged decay rate, defined with a photon-energy cut of $E_\gamma = 1.6$ GeV, remains (94). This result strongly indicates that the theoretical efforts for the $B \to X_s\gamma$ mode have reached the nonperturbative boundaries. The complete effect of power corrections on *CP* asymmetries has not yet been estimated.

2.3.2. $B \to X_s\ell^+\ell^-$. Hadronic power corrections in the decay $B \to X_s\ell^+\ell^-$ that scale with $1/m_b^2$, $1/m_b^3$ (26, 27, 101–104), and $1/m_c^2$ (96) have also been considered. They can be calculated analogously to those in the decay $B \to X_s\gamma$. However, a systematic analysis of hadronic power corrections, including all relevant operators, has yet to be performed. Thus, an additional uncertainty of $\pm 5\%$ should be added to all theoretical predictions for this mode on the basis of a simple dimensional estimate.

In the high-q^2 region, one encounters the breakdown of the HME at the end point of the dilepton mass spectrum: Whereas the partonic contribution vanishes, the $1/m_b^2$ and $1/m_b^3$ corrections tend toward a nonzero value. In contrast to the end-point region of the photon-energy spectrum in the $B \to X_s\gamma$ decay, no partial all-order resummation into a shape function is possible. However, for an integrated high-q^2 spectrum, an effective expansion is found in inverse powers of $m_b^{\text{eff}} = m_b \times (1 - \sqrt{s_{\text{min}}})$ rather than m_b (105, 106). The expansion converges less rapidly, depending on the lower–dilepton mass cut $s_{\text{min}} = q_{\text{min}}^2/m_b^2$ (77).

The large theoretical uncertainties could be significantly reduced by normalizing the $B \to X_s\ell^+\ell^-$ decay rate to the semileptonic $B \to X_u\ell\bar{\nu}$ decay rate with the same q^2 cut (104):

$$\mathcal{R}(s_0) = \int_{\hat{s}_0}^{1} d\hat{s}\, \frac{d\Gamma(B \to X_s\ell^+\ell^-)}{d\hat{s}} \Bigg/ \int_{\hat{s}_0}^{1} d\hat{s}\, \frac{d\Gamma(B^0 \to X_u\ell\nu)}{d\hat{s}}. \qquad 13.$$

For example, the uncertainty due to the dominating $1/m_b^3$ term would be reduced from 19% to 9% (85).

2.3.3. $B \to X_d\gamma$. The nonperturbative contributions in the decay $B \to X_d\gamma$ can be treated analogously to those in the decay $B \to X_s\gamma$. The power corrections that scale as $1/m_b^2$ (in addition to the CKM factors) are the same for the two modes. Also, the systematic analysis of resolved contributions given in Reference 94 can be applied to this case. However, the long-distance contributions from the intermediate u quark in the penguin loops are critical. They are suppressed in the $B \to X_s\gamma$ mode by the CKM-matrix elements. In $B \to X_d\gamma$, there is no CKM suppression, and one must account for the nonperturbative contributions that arise from the operator $\mathcal{O}_1^u$. The contribution due to the $\mathcal{O}_1^u$–$\mathcal{O}_7$ interference scales with Λ/m_b (102). However, this interference contribution vanishes in the total *CP*-averaged rate of $B \to X_s\gamma$ at order Λ/m_b (94). This result applies to the total rate of $B \to X_d\gamma$ as well. Other interference terms, namely the double-resolved contributions $\mathcal{O}_1^u$–$\mathcal{O}_8$ and $\mathcal{O}_1^u$–$\mathcal{O}_1^u$, arise first at order $1/m_b^2$, as they can also be deduced from the results presented in Reference 94. Thus, there is no power correction due to the operator $\mathcal{O}_1^u$ in the total rate of $B \to X_d\gamma$ at order Λ/m_b, which implies that the *CP*-averaged decay rate of $B \to X_d\gamma$ is as theoretically clean as the decay rate of $B \to X_s\gamma$.

2.3.4. $B \to X_d\ell^+\ell^-$. In the case of $B \to X_d\ell^+\ell^-$, long-distance contributions due to u quark loops can be avoided in the low-q^2 window $1\,GeV^2 < q^2 < 6\,GeV^2$. The ρ and ω resonances are given below, and the $c\bar{c}$ (J/ψ, ψ') resonances are above this window (88). The effect of their respective tails can be taken into account within the Krüger-Sehgal (KS) approach (see Section 2.5) (107, 108). In this low-q^2 region, one can then treat the nonperturbative power corrections analogously to those in the decay $B \to X_s\ell^+\ell^-$, and one can expect a similar theoretical accuracy in this q^2 window.

2.4. Nonperturbative Corrections due to Kinematical Cuts

There are additional subtleties in inclusive modes. Kinematical cuts induce additional sensitivities to nonperturbative physics.

2.4.1. $B \to X_s\gamma$. In the measurement of the inclusive mode $B \to X_s\gamma$, one needs cuts in the photon-energy spectrum to suppress the background from other B decays (**Figure 3**).

These shape-function effects were taken into account in the experimental analysis, and the corresponding theoretical uncertainties due to this model dependency are reflected in the extrapolation error of the experimental results (see Section 5.1). The extrapolation is done from the experimental energy cut values down to 1.6 GeV by use of three different theoretical schemes (71, 110–112) for averaging.

Again constraining the analysis to the leading operator $\mathcal{O}_7$, a cut around 1.6 GeV might not guarantee that a theoretical description in terms of a local OPE is sufficient because of the sensitivity to the scale $\Delta = m_b - 2E_\gamma$ (113). A multiscale OPE with three short-distance scales—m_b, $\sqrt{m_b\Delta}$, and Δ—has been proposed to connect the shape function and the local OPE region. Recently, such additional perturbative cutoff-related effects have been calculated to NNLL precision by the use of SCET methods (114–116). Such perturbative effects due to the additional scale are negligible for 1.0 GeV but are of order 3% at 1.6 GeV (114). The size of these effects at 1.6 GeV is similar to the 3% higher-order uncertainty in the present NNLL prediction. However, the numerical

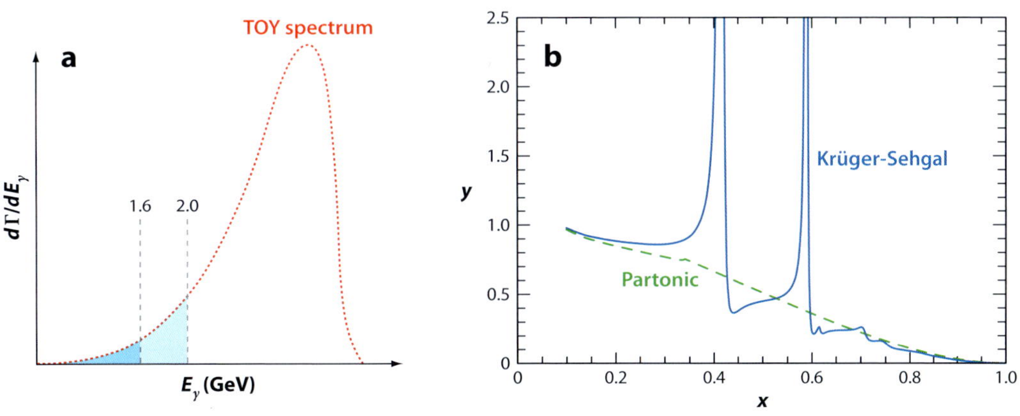

Figure 3

Spectra in inclusive modes. (*a*) Cut in the photon-energy spectrum in $B \to X_s\gamma$. (*b*) Differential $B \to X_s\ell^+\ell^-$ branching fraction as a function of $s = q^2/m_b^2 \equiv m_{\ell^+\ell^-}^2/m_b^2$, including the effect of charm resonances in the Krüger-Sehgal method (*solid blue line*). For comparison, the dashed green curve shows the same quantity obtained within a purely partonic calculation at next-to-leading-log precision. Modified from Reference 109.

consistency of the SCET analysis has recently been questioned (117). Far away from the end point ($E_0 = 1.6$ GeV), the logarithmic and nonlogarithmic terms cancel; the same result was presented in Reference 118. Within the resummation of the cutoff-enhanced logarithms, this feature leads to an overestimate of the $O(\alpha_s^3)$ terms (117). Further work is needed to clarify this issue.

There is an alternative approach to the cut effects in the photon-energy spectrum that is based on dressed gluon exponentiation and on the incorporation of Sudakov and renormalon resummations (118, 119). The greater predictive power of this approach is related in part to the assumption that nonperturbative power corrections associated with the shape function follow the pattern of ambiguities present in the perturbative calculation (120). In the future, these additional perturbative cut effects could be analyzed and combined together with those already included in the experimental average.

2.4.2. $B \to X_s \ell^+ \ell^-$. In the inclusive decay $B \to X_s \ell^+ \ell^-$, the hadronic and dilepton-invariant masses are independent kinematical quantities. A hadronic invariant mass cut is imposed in the experiments (see Section 5.7). The high–dilepton mass region is not affected by this cut, but in the low–dilepton mass region the kinematics with a jet-like X_s and $m_X^2 \leq m_b \Lambda$ implies the relevance of the shape function. A recent SCET analysis shows that, through the use of the universality of the shape function, a 10–30% reduction in the dilepton mass spectrum can be accurately computed. Nevertheless, the effects of subleading shape functions lead to an additional uncertainty of 5% (121, 122). A more recent analysis (123) estimates the uncertainties due to subleading shape functions more conservatively. By scanning over a range of models of these functions, one finds corrections in the rates relative to the leading-order result to be between -10% and $+10\%$, with equally large uncertainties. In the future, it may be possible to decrease such uncertainties significantly by constraining both the leading and subleading shape functions using the combined $B \to X_s \gamma$, $B \to X_u \ell \bar{\nu}$, and $B \to X_s \ell^+ \ell^-$ data (123).

2.5. Charmonium Resonance Contributions

One must also consider the on-shell $c\bar{c}$ resonances, which have to be taken out. Whereas in the decay $B \to X_s \gamma$ the intermediate ψ background, namely $B \to \psi X_s$ followed by $\psi \to X' \gamma$, is suppressed for the high-energy cut E_γ and can be subtracted from the $B \to X_s \gamma$ decay rate, the $c\bar{c}$ resonances show up as large peaks in the dilepton-invariant mass spectrum in the decay $B \to X_s \ell^+ \ell^-$.

As discussed in Section 2.2, these resonances can be removed by making appropriate kinematic cuts in the invariant mass spectrum. However, nonperturbative contributions away from the resonances within the perturbative windows are also important. In the KS approach (107, 108) one absorbs factorizable long-distance charm rescattering effects (in which the $B \to X_s c\bar{c}$ transition can be factorized into the product of $\bar{s}b$ and $c\bar{c}$ color-singlet currents) into the matrix element of the leading semileptonic operator $\mathcal{O}_9$. Following the inclusion of nonperturbative corrections scaling with $1/m_c^2$, the KS approach avoids double counting. For the integrated branching fractions, one finds an increase of (1–2)% in the low-q^2 region due to the KS effect, whereas in the high-q^2 region the increase is well below the uncertainty due to the $1/m_b$ corrections. As shown in **Figure 3**, the integrated branching fraction is dominated by this resonance background, which exceeds the nonresonant charm-loop contribution by two orders of magnitude. This feature should not be misinterpreted as a striking failure of global parton-hadron duality (109), which postulates that the sum over the hadronic final states, including resonances, should be well approximated by a quark-level calculation (124). Crucially, the charm resonance contributions to the decay $B \to X_s \ell^+ \ell^-$ are expressed in terms of a phase-space integral over the absolute square of a correlator. For such a quantity, global quark-hadron duality is not expected to hold. Nevertheless, local quark-hadron duality (which, of course, also implies global duality) may be reestablished by resumming Coulomb-like interactions (109).

2.6. Soft Collinear Effective Theory for Exclusive Decays

The Wilson coefficients of the weak effective Hamiltonian are process independent and therefore can be used directly in the description of exclusive modes. It is the computation of the hadronic matrix elements between meson states that is difficult in the case of exclusive modes and that limits the theoretical precision. The naïve approach consists of writing the amplitude $A \simeq C_i(\mu_b)\langle \mathcal{O}_i(\mu_b)\rangle$ and parameterizing $\langle \mathcal{O}_i(\mu_b)\rangle$ in terms of form factors. A substantial improvement can be obtained by using the QCDF method (28–30) and its field theoretical formulation, the SCET method (31–36). These methods form the basis of the up-to-date predictions of exclusive B decays. Within this framework one can show that, even if the form factors were known with infinite precision, the description of exclusive decays would be incomplete due to the existence of so-called nonfactorizable strong interaction effects that do not correspond to form factors.

The QCDF and SCET methods were first systematized for nonleptonic decays in the heavy quark limit. In contrast to the HQET, SCET does not correspond to a local operator expansion. Whereas HQET is applicable to B decays if the energy transfer to light hadrons is small, for example to $B \to D$ transitions at small recoil to the D meson, HQET is not applicable if some of the outgoing, light particles have momenta of order m_b. If so, one faces a multiscale problem that can be tackled within SCET. In this case, there are three relevant scales: (*a*) $\Lambda = \text{few} \times \Lambda_{\text{QCD}}$, the soft scale set by the typical energies and momenta of the light degrees of freedom in the hadronic bound states; (*b*) m_b, the hard scale set by both the heavy b quark mass and the energy of the final-state hadron in the B meson rest frame; and (*c*) the hard collinear scale $\mu_{\text{hc}} = \sqrt{m_b \Lambda}$, which appears through interactions between the soft and energetic modes in the initial and final states. The dynamics of hard and hard collinear modes can be described perturbatively in the heavy quark limit $m_b \to \infty$. Thus, SCET describes B decays to light hadrons with energies much larger than their masses, assuming that their constituents have momenta collinear to the hadron momentum.

2.6.1. $B \to K^*\gamma$ and $B \to \rho\gamma$.

The application of the QCDF formalism to radiative and semileptonic decays was first proposed in Reference 125. For $B \to K^*\gamma$, or more generally for $B \to V\gamma$, where V is a light vector meson, the QCDF formula for the hadronic matrix element of each operator of the effective Hamiltonian in the heavy quark limit and, to all orders in α_s, reads

$$\langle V\gamma|\mathcal{O}_i|B\rangle = T_i^I F^{B\to V_\perp} + \int_0^\infty \frac{d\omega}{\omega}\phi_B(\omega)\int_0^1 du\, \phi_{V_\perp}(u) T_i^{II}(\omega, u). \qquad 14.$$

This formula separates the process-independent nonperturbative quantities (*a*)$F^{B\to V_\perp}$, a form factor evaluated at maximum recoil ($q^2 = 0$); and (*b*) ϕ_B and $\phi_{V_\perp}$, the light-cone distribution amplitudes (LCDAs) for the heavy and light mesons, respectively, from the perturbatively calculable quantities T^I and T^{II}. The latter correspond to vertex and spectator corrections, respectively, and have been calculated to $O(\alpha_s^1)$ (126–129). More recently, some α_s^2 terms were also presented (130).

Light-cone wave functions of pseudoscalar and vector mesons that enter the factorization formula have been studied in detail through the use of light-cone QCD sum rules (131–134). However, not much is known about the B meson LCDA, whose first negative moment enters the factorized amplitude at $O(\alpha_s)$. Because this moment also enters the factorized expression for the $B \to \gamma$ form factor, it might be possible to extract its value from measurements of decays such as $B \to \gamma e\nu$, if the power corrections are under control.

The QCDF formula also includes an important simplification in the form factor description. The $B \to V$ form factors at large recoil have been analyzed in SCET (135–137) and are independent of the Dirac structure of the current in the heavy quark limit; as a consequence, all $B \to V_\perp$ form factors reduce to a single form factor up to factorizable corrections in the heavy quark and large energy limits (125, 138).

Field theoretical methods such as SCET make it possible to reach a deeper understanding of the QCDF approach. The various momentum regions are represented by different fields in the effective field theory. The hard scattering kernels T^I and T^{II} can be shown to be Wilson coefficients of effective field operators. Using SCET, one can prove the factorization formula to all orders in α_s and to leading order in Λ/m_b (139). QCD is matched on SCET in a two-step procedure that separates the hard scale $\mu \sim m_b$ and then the hard collinear scale $\mu \sim \sqrt{\Lambda m_b}$ from the hadronic scale Λ. The vertex correction term T^I involves the hard scale, whereas the spectator scattering term T^{II} involves both the hard collinear and hard scales. This is why large logarithms have to be resummed (139), which can be done most efficiently in SCET.

In principle, the field theoretical framework of SCET allows one to go beyond the leading-order result in Λ/m_b (140). However, a breakdown of factorization is expected at that order (30). For example, in the analysis of $B \to K^*\gamma$ decays at subleading order, an infrared divergence is encountered in the matrix element of $\mathcal{O}_8$ (141). In general, power corrections involve convolutions, which turn out to be divergent. Currently, no solution to this well-analyzed problem of end-point divergencies within power corrections is available (136, 142, 143). Thus, within the QCDF/SCET approach, a general, quantitative method to estimate the important Λ/m_b corrections to the heavy quark limit is missing, significantly limiting the precision in phenomenological applications.

Nevertheless, some very specific power corrections are still computable and are often numerically important. Indeed, this is the case for the annihilation and weak exchange amplitudes in $B \to \rho\gamma$. The annihilation contributions also represent the leading contribution to isospin asymmetries (141). All these corrections are included in recent analyses of radiative exclusive decays (144–146). Moreover, the method of light-cone QCD sum rules can help provide estimates of such unknown subleading terms. For example, power corrections for the indirect *CP* asymmetries in $B \to V\gamma$ decays have been analyzed in this manner (147).

2.6.2. $B \to K^{(*)}\ell^+\ell^-$. There is also a factorization formula for exclusive semileptonic B decays, such as $B \to K^*\ell^+\ell^-$, that are analogous to the one for the radiative decay $B \to K^*\gamma$ (126, 146). The simplification due to form factor relations is even more drastic. The hadronic form factors can be expanded in the small ratios Λ/m_b and Λ/E, where E is the energy of the light meson. If we neglect corrections of order $1/m_b$ and α_s, the seven a priori independent $B \to K^*$ form factors reduce to two universal form factors, $\xi_\perp$ and $\xi_\parallel$ (125, 138). This reduction makes it possible to design interesting ratios of observables in which any soft form factor dependency cancels out for all dilepton masses q^2 at leading order in α_s and Λ/m_b (148).

The theoretical simplifications of the QCDF/SCET approach are restricted to the kinematic region in which the energy of the K^* is of the order of the heavy quark mass; that is, $q^2 \ll m_B^2$. Moreover, the influences of very light resonances below 1 GeV^2 question the QCDF results in this region. In addition, the longitudinal amplitude in the QCDF/SCET approach generates a logarithmic divergence in the limit $q^2 \to 0$, which indicates problems in the theoretical description below 1 GeV^2 (126). Thus, the factorization formula applies well in the dilepton mass range $1\,\text{GeV}^2 < q^2 < 6\,\text{GeV}^2$.

Clearly, the QCDF and SCET methods are also applicable to phenomenologically important semileptonic decays such as $B \to K\ell^+\ell^-$ (126, 149), $B \to \rho\ell^+\ell^-$ (146), and $B_s \to \phi\ell^+\ell^-$. The decay mode into a pseudoscalar is analogous to the decay mode into a longitudinal vector meson.

3. EXPERIMENTAL TECHNIQUES

The $\Upsilon(4S)$ resonance produced by the e^+e^- collision at the B factories provides a clean sample of $B^0\bar{B}^0$ and B^+B^- meson pairs as well as strong kinematical constraints that are otherwise unavailable, particularly at hadron colliders. The main background is from continuum light quark pair production ($e^+e^- \rightarrow q\bar{q}$; $q = u, d, s, c$), which has a cross section only three times larger than that of $B\bar{B}$ production. Radiative and electroweak penguin B decays are efficiently measured at the B factories thanks to their clear signatures: a high-energy photon and a lepton pair, respectively.

3.1. Exclusive B Reconstruction

A B meson decaying into an exclusive final state is reconstructed by measuring all long-lived decay products ($\pi^\pm$, $K^\pm$, $e^\pm$, $\mu^\pm$, and γ), selecting intermediate states of certain invariant masses, and calculating two standard variables: the beam-energy constrained mass $M_{\rm bc} = \sqrt{s/4 - |p_B^*|^2}$ (also referred to as the beam-energy substituted mass, $M_{\rm ES}$) and the energy difference $\Delta E = E_B^* - \sqrt{s}/2$. Here, $\sqrt{s}/2$ is the beam energy, and p_B^* and E_B^* are the momentum and energy, respectively, of the reconstructed B meson candidate in the $\Upsilon(4S)$ rest frame.

$M_{\rm bc}$ has a peak at the B meson mass and ΔE has a peak at zero (**Figure 4**). The resolution of $M_{\rm bc}$ is significantly better than that of ΔE, as the former is dominated by the spread of the beam energy, whereas the latter is dominated by the detector resolution. The ΔE variable is sensitive to misreconstructed background events, whereas $M_{\rm bc}$ has little separation power for them. When a kaon is misidentified as a pion, ΔE shifts by approximately 50 MeV, and when a low-momentum pion is added or missed, ΔE shifts by more than the pion mass. An exclusive $b \rightarrow d\gamma$ final state is thus separated from a similar $b \rightarrow s\gamma$ state with ΔE, but this separation is marginal due to the photon-energy resolution of the electromagnetic calorimeter. Therefore, pion-to-kaon separation is crucial for measurement of the suppressed $b \rightarrow d\gamma$ processes.

Background events due to random combinations of particles are also reduced by correct identification of particle species. For the $b \rightarrow s\ell^+\ell^-$ processes, electrons and muons are almost completely separated from the more abundant hadrons. In addition, various techniques based on the event topology can be applied to suppress the background from continuum $q\bar{q}$ events.

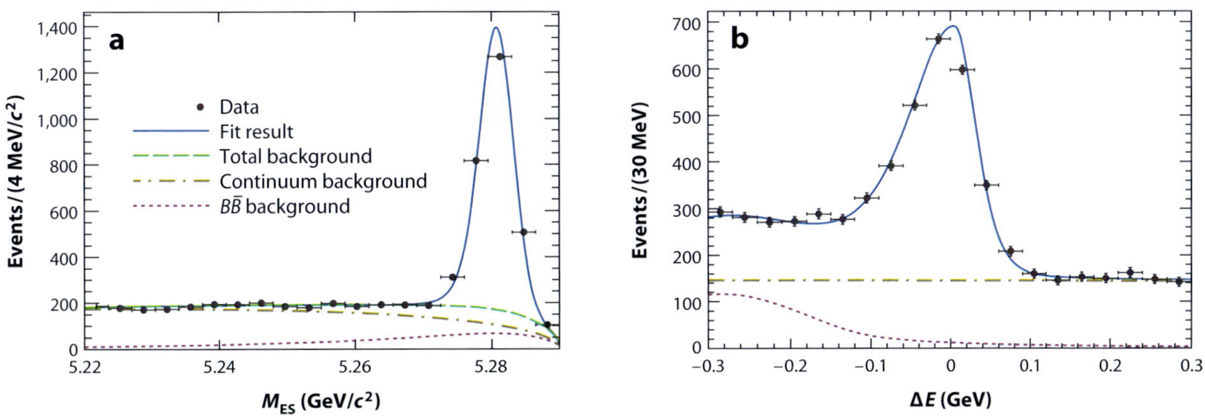

Figure 4

Examples of $M_{\rm ES}$ ($M_{\rm bc}$) and ΔE for $B^0 \rightarrow K^{*0}\gamma$ by BaBar. From Reference 150.

3.2. Inclusive Measurement Techniques

A fully inclusive measurement of $B \to X_s\gamma$, in which the system recoiling against the emitted photon is not reconstructed, has been performed at the B factories thanks to the clean environment. The dominant background photon sources are (*a*) the copiously produced $\pi^0 \to \gamma\gamma$ decays; (*b*) $\eta \to \gamma\gamma$, to a lesser extent; and (*c*) other secondary and initial-state radiation photons in continuum $q\bar{q}$ events. These contributions can be safely subtracted because they are measured in events taken 60 or 40 MeV below the $\Upsilon(4S)$ resonance (i.e., off resonance). Here, small corrections due to the center-of-mass energy difference are applied to the production cross section and the reconstruction efficiency. To avoid sacrificing the B decay sample for other studies, the size of the off-resonance data sample is only ~10% of the on-resonance sample from both Belle and BaBar and is the dominant source of statistical and systematic errors (CLEO collected one-third of the sample as off resonance). The second severe background source arises from similar secondary photons from B decays. These contributions are subtracted from the expected photon spectrum on the basis of measured π^0 and η spectra from B decays, various control samples, or Monte Carlo simulation.

An alternative technique is to measure as many exclusive modes as possible and then calculate their sum (i.e., the sum-of-exclusive method). Exclusive branching fractions measured to date do not saturate the inclusive process, but one can still infer the total branching fraction by estimating the fraction of unmeasured modes of typically ~45% (or ~30% if K_L^0 modes are accounted for by corresponding K_S^0 modes) using simulated hadronization processes. In the simulation, a light quark pair is generated according to the SM mass spectrum, and final-state hadrons are produced by the PYTHIA program (151). This method also provides direct information about the B meson. For example, the B meson momentum defines the B meson rest frame, and charge and flavor information allows *CP*- and isospin-asymmetry measurements. So far, the sum-of-exclusive method is the only way to perform inclusive measurements of $B \to X_s\ell^+\ell^-$ and $B \to X_d\gamma$ decays.

Another, potentially definitive method is the so-called B-reco technique, in which the other B meson is fully reconstructed, thereby allowing the target B decay to be measured in a very clean environment. The efficiency is as low as a fraction of a percent and will be more important in future experiments.

4. PRESENT THEORETICAL PREDICTIONS

Theoretical predictions have significantly improved in the past decade along with the development of the theoretical tools. There have also been improvements in the relevant experimental input quantities, as discussed below.

4.1. Inclusive Penguin Decays

The inclusive radiative and electroweak penguin modes offer theoretically clean observables because nonperturbative corrections are small and well under control. This assessment also applies to the branching ratio of $B \to X_d\gamma$ mode, as discussed in Section 2.3.

4.1.1. Inclusive $B \to X_s\gamma$. The stringent bounds obtained from $B \to X_s\gamma$ on various nonstandard scenarios are a clear example of the importance of clean FCNC observables for discriminating NP models. The branching fraction for $B \to X_q\gamma$ ($q = s, d$) can be parameterized as

$$\mathcal{B}(B \to X_q\gamma)_{E_\gamma > E_0} = \mathcal{B}(B \to X_c e\bar{\nu})_{\text{exp}} \frac{6\alpha_{\text{em}}}{\pi C} \left| \frac{V_{tq}^* V_{tb}}{V_{cb}} \right|^2 [P(E_0) + N(E_0)], \qquad 15.$$

where $\alpha_{em} = \alpha_{\rm em}^{\rm on-shell}$ (70); $C = |V_{ub}|^2/|V_{cb}|^2 \times \Gamma[B \to X_c e\bar{\nu}]/\Gamma[B \to X_u e\bar{\nu}]$; and $P(E_0)$ and $N(E_0)$ denote the perturbative and nonperturbative contributions, respectively. The latter are normalized to the charmless semileptonic rate to separate the charm dependency.

The first NNLL prediction, which is based on the perturbative calculations discussed in Section 2.2 and on the analyses of nonperturbative corrections presented in Sections 2.3 and 2.4, for a photon-energy cut $E_\gamma > 1.6$ GeV (54), reads

$$\mathcal{B}(B \to X_s\gamma)_{\rm NNLL} = (3.15 \pm 0.23) \times 10^{-4}. \qquad 16.$$

The overall uncertainty consists of nonperturbative (5%), parametric (3%), perturbative (scale) (3%), and m_c-interpolation ambiguities (3%), which are added in quadrature. An additional scheme dependency in the determination of the prefactor C has been found (152); it is within the perturbative uncertainty of 3% (117).

Thus, the SM prediction and the experimental average (see Section 5.1) are consistent at the 1.2σ level. This finding implies very stringent constraints on NP models, such as (*a*) the bound on the charged Higgs mass in the two–Higgs doublet model (153, 154) ($M_{H^+} > 295$ GeV at 95% CL) (54) and (*b*) the bound on the inverse compactification radius of the minimal universal extra-dimension model ($1/R > 600$ GeV at 95% CL) (155). In both cases, the bounds are much stronger than those derived from other measurements. Constraints within various supersymmetric extensions have been analyzed (156–164; for overviews, see References 12 and 165). Bounds on the little Higgs model with T-parity have also been presented (166). Finally, model-independent analyses in the effective field theory approach without (167) and with (168, 169) the assumption of minimal flavor violation also show the strong constraining power of the $B \to X_s\gamma$ branching fraction.

4.1.2. Inclusive $B \to X_d\gamma$. The theoretical predictions for the branching fraction $\mathcal{B}(B \to X_d\gamma)$ for photon energies $E_\gamma > 1.6$ GeV read (51, 52)

$$\mathcal{B}(B \to X_d\gamma) = \left(1.38\ {}^{+0.14}_{-0.21}\Big|_{\frac{m_c}{m_b}} \pm 0.15_{\rm CKM} \pm 0.09_{\rm param.} \pm 0.05_{\rm scale}\right) \times 10^{-5} \qquad 17.$$

and

$$\frac{\mathcal{B}(B \to X_d\gamma)}{\mathcal{B}(B \to X_s\gamma)} = \left(3.82\ {}^{+0.11}_{-0.18}\Big|_{\frac{m_c}{m_b}} \pm 0.42_{\rm CKM} \pm 0.08_{\rm param.} \pm 0.15_{\rm scale}\right) \times 10^{-2}. \qquad 18.$$

These predictions are of NLL order and are fully consistent with previous results (87). A good part of the uncertainties cancel out in the ratio. The errors are dominated by CKM uncertainties, and thus the measurement of $\mathcal{B}(B \to X_d\gamma)$ constrains the CKM parameters. This measurement is also of specific interest with respect to NP because its CKM suppression by the factor $|V_{td}|^2/|V_{ts}|^2$ in the SM may not hold in extended models.

4.1.3. Direct *CP* asymmetry. Other important observables are the direct *CP* asymmetries ($q = s, d$), whose sign is always defined in terms of $b - \bar{b}$, or

$$\mathcal{A}_{CP}(\bar{B} \to X_q\gamma) \equiv \frac{\Gamma(\bar{B} \to X_q\gamma) - \Gamma(B \to X_{\bar{q}}\gamma)}{\Gamma(\bar{B} \to X_q\gamma) + \Gamma(B \to X_{\bar{q}}\gamma)}. \qquad 19.$$

As first noted in Reference 170, the SM predictions are almost independent from the photon-energy cutoff and, for $E_\gamma > 1.6$ GeV, read (51, 52)

$$\mathcal{A}_{CP}(\bar{B} \to X_s\gamma) = \left(0.44\ {}^{+0.15}_{-0.10}\Big|_{\frac{m_c}{m_b}} \pm 0.03_{\rm CKM}\ {}^{+0.19}_{-0.09}\Big|_{\rm scale}\right) \times 10^{-2} \qquad 20.$$

and

$$\mathcal{A}_{CP}(\bar{B} \to X_d\gamma) = \left(-10.2 \,{}^{+2.4}_{-3.7}\Big|_{\frac{m_c}{m_b}} \pm 1.0_{\text{CKM}} \,{}^{+2.1}_{-4.4}\Big|_{\text{scale}}\right) \times 10^{-2}. \quad 21.$$

The two CP asymmetries are connected by the relative CKM factor $\lambda^2\,[(1-\rho)^2+\eta^2]$. The small SM prediction for the CP asymmetry in the decay $B \to X_s\gamma$ is a result of three suppression factors: (*a*) α_s, to have a strong phase; (*b*) CKM suppression of order λ^2; and (*c*) GIM suppression of order $(m_c/m_b)^2$, which reflects that in the limit $m_c = m_u$, any CP asymmetry in the SM would vanish.

On the basis of CKM unitarity, one can derive the following U-spin relation between the unnormalized CP asymmetries (171):

$$\left[\Gamma(\bar{B} \to X_s\gamma) - \Gamma(B \to X_{\bar{s}}\gamma)\right] + \left[\Gamma(\bar{B} \to X_d\gamma) - \Gamma(B \to X_{\bar{d}}\gamma)\right] = 0. \quad 22.$$

U-spin-breaking effects can be estimated within the HME (even beyond the partonic level), so one arrives at the following prediction for the total (or untagged) $\bar{B} \to X_{s+d}\gamma$ asymmetry (172, 173):

$$|\Delta\mathcal{B}(\bar{B} \to X_s\gamma) + \Delta\mathcal{B}(\bar{B} \to X_d\gamma)| \sim 1 \cdot 10^{-9}. \quad 23.$$

Because this null test is based on the CKM unitarity, it represents a clear test for new CP phases beyond the CKM phase (172, 173). NP sensitivities of direct CP asymmetries have been analyzed (51, 170).

4.1.4. Inclusive $B \to X_s\ell^+\ell^-$. The decay $B \to X_s\ell^+\ell^-$ is particularly attractive because it offers several kinematic observables. The angular decomposition of the decay rate provides three independent observables, H_T, H_A, and H_L, from which one can extract the short-distance electroweak Wilson coefficients that test for NP (174):

$$\frac{d^3\Gamma}{dq^2\,dz} = \frac{3}{8}\left[(1+z^2)H_T(q^2) + 2(1-z^2)H_L(q^2) + 2zH_A(q^2)\right]. \quad 24.$$

Here $z = \cos\theta_\ell$, θ_ℓ is the angle between the negatively charged lepton and the $\bar{B}$ meson in the center-of-mass frame of the dilepton system, and q^2 is the dilepton mass squared. H_A is equivalent to the forward-backward asymmetry, and the dilepton mass spectrum is given by $H_T + H_L$. The observables depend on the Wilson coefficients C_7, C_9, and C_{10} in the SM. The present measurements of the $B \to X_s\ell^+\ell^-$ already favor the SM sign of the coefficient C_7, which is undetermined by the $B \to X_s\gamma$ mode (175).

As discussed above, these observables are dominated by perturbative contributions in the perturbative low- and high-q^2 windows, which are below ($1\,\text{GeV}^2 < q^2 < 6\,\text{GeV}^2$) and above ($q^2 > 14.4\,\text{GeV}^2$) the $c\bar{c}$ resonances, respectively. The present predictions are based on the perturbative calculations to NNLL precision in QCD and to NLL precision in QED (see Section 2.2). For the branching fraction in the low-q^2 region, one arrives at (85)

$$\mathcal{B}(B \to X_s\ell^+\ell^-)_{\text{low}} = \begin{cases} (1.59 \pm 0.11) \times 10^{-6} & (\ell = \mu) \\ (1.64 \pm 0.11) \times 10^{-6} & (\ell = e) \end{cases}, \quad 25.$$

and for the high-q^2 region, one arrives at (86)

$$\mathcal{B}(B \to X_s\ell^+\ell^-)_{high} = \begin{cases} 2.40 \times 10^{-7} \times (1^{+0.29}_{-0.26}) & (\ell = \mu) \\ 2.09 \times 10^{-7} \times (1^{+0.32}_{-0.30}) & (\ell = e) \end{cases}. \quad 26.$$

As suggested in Reference 104, normalizing the $B \to X_s\ell^+\ell^-$ decay rate in the high-q^2 region to the semileptonic $B \to X_u\ell\bar{\nu}$ decay rate with the same q^2 cut (Equation 13) significantly reduces the nonperturbative uncertainties (86):

$$\mathcal{R}\left(\hat{q}_0^2 = 14.4\, GeV^2\right) = \begin{cases} 2.29 \times 10^{-3} \times (1 \pm 0.13) & (\ell = \mu) \\ 1.94 \times 10^{-3} \times (1 \pm 0.16) & (\ell = e) \end{cases}. \quad 27.$$

The value of q_0^2 for which the forward-backward asymmetry vanishes,

$$\left(q_0^2\right)[X_s \ell^+ \ell^-] = \begin{cases} (3.50 \pm 0.12)\,\mathrm{GeV}^2 & (\ell = \mu) \\ (3.38 \pm 0.11)\,\mathrm{GeV}^2 & (\ell = e) \end{cases}, \quad 28.$$

is one of the most precise predictions in flavor physics and also determines the relative sign and magnitude of the coefficients C_7 and C_9 (86). However, unknown subleading nonperturbative corrections of order $O(\alpha_s \Lambda / m_b)$, which are estimated to give an additional uncertainty of order 5%, have to be added in all observables of the $B \to X_s \ell^+ \ell^-$ mode (see Section 2.3).

In all predictions, it is assumed that there is no cut in the hadronic mass region (see Section 2.4). Furthermore, after including the NLL QED matrix elements, the electron and muon channels receive different contributions due to terms involving $\ln(m_b^2/m_\ell^2)$ (see Section 2.2). This is the only source of the difference between these two channels. All collinear photons are assumed to be included in the X_s system, and the dilepton-invariant mass does not contain any photons; in other words, $q^2 = (p_{\ell^+} + p_{\ell^-})^2$. Present experimental settings at the B factories are different, and therefore the theoretical predictions have to be modified (83).

This difference between the settings also means that deviations from the SM prediction ($R_{X_s}^{\mathrm{SM}} = 1$) in the muon-electron ratio

$$R_{X_s} = \Gamma(B \to X_s \mu^+ \mu^-)_{[q_a^2, q_b^2]} / \Gamma(B \to X_s e^+ e^-)_{[q_a^2, q_b^2]} \quad 29.$$

can result from a different treatment of collinear photons in the two modes. This ratio is interesting because it is sensitive to the neutral Higgs boson of two–Higgs doublet models at large $\tan\beta$ (176, 177), which is also valid in corresponding ratios $R_{K^{(*)}}$ of exclusive modes: In the SM, one finds $R_K = 1$, as well as $R_{K^*} = 0.75$ when integrated over all q^2, including $M_{e^+e^-} < 2m_\mu$.

4.2. Exclusive Penguin Decays

The exclusive penguin modes offer a larger variety of experimentally accessible observables than do the inclusive ones, but the nonperturbative uncertainties in the theoretical predictions are in general sizable.

4.2.1. $B \to K^*\gamma$ and $B \to \rho\gamma$. The large hadronic uncertainties, which arise from the nonperturbative input of the QCDF formula and from our limited knowledge of power corrections, do not allow precise predictions of the branching fractions of exclusive modes. However, within ratios of exclusive modes such as asymmetries, parts of the uncertainties cancel out, and one may hope for higher precision.

The ratio $R_{\mathrm{th}}(\rho\gamma/K^*\gamma)$ [and similarly $R_{\mathrm{th}}(\omega\gamma/K^*\gamma)$] is given by (144–147)

$$R_{\mathrm{th}}(\rho\gamma/K^*\gamma) = \frac{\mathcal{B}_{\mathrm{th}}(B \to \rho\gamma)}{\mathcal{B}_{\mathrm{th}}(B \to K^*\gamma)} = S_\rho \left|\frac{V_{td}}{V_{ts}}\right|^2 \frac{\left(M_B^2 - m_\rho^2\right)^3}{\left(M_B^2 - m_{K^*}^2\right)^3} \zeta^2 \left[1 + \Delta R(\rho/K^*)\right], \quad 30.$$

where m_ρ is the mass of the ρ meson; ζ is the ratio of the transition form factors; $\zeta = \bar{T}_1^{\rho}(0)/\bar{T}_1^{K^*}(0)$; and $S_\rho = 1$ and 1/2 for the $\rho^\pm$ and ρ^0 mesons, respectively. The quantity $(1+\Delta R)$ entails the explicit $O(\alpha_s)$ corrections as well as the power-suppressed contributions. These functions also depend on CKM parameters, namely $\phi_2 \equiv \alpha = \arg(-V_{cb}V_{cb}^*/V_{td}V_{tb}^*)$ and $R_{ut} = |V_{ud}V_{ub}^*/V_{td}V_{tb}^*|$, and one finds numerically (146) that

$$\Delta R(\rho^{\pm}/K^{*\pm}) = \left\{1 - 2R_{ut}\cos\phi_2\left[0.24^{+0.18}_{-0.18}\right] + R^2_{ut}\left[0.07^{+0.12}_{-0.07}\right]\right\} \qquad 31.$$

and

$$\Delta R(\rho^{0}/K^{*0}) = \left\{1 - 2R_{ut}\cos\phi_2\left[-0.06^{+0.06}_{-0.06}\right] + R^2_{ut}\left[0.02^{+0.02}_{-0.01}\right]\right\}. \qquad 32.$$

These results are consistent with the predictions given in References 144, 145, and 147. Obviously, the neutral mode is better suited for the determination of $|V_{td}/V_{ts}|$ than is the charged mode, in which the function ΔR is dominated by the weak annihilation contribution, which leads to a larger error. The most recent determination of the ratio $\zeta = \bar{T}_1^{\rho}(0)/\bar{T}_1^{K^*}(0)$ within the light-cone QCD sum rule approach (178), $1/\zeta = 1.17 \pm 0.09$, leads to the determination of $|V_{td}/V_{ts}|$ via Equation 30 (see Section 5.4). However, the experimental data on the branching fractions of $B \to K^*\gamma$ and $B \to \rho\gamma$ calls for a larger error on ζ, if one assumes no large power corrections beyond the known annihilation terms (146).

4.2.2. Isospin asymmetry in radiative decays. Another important observable is the isospin-breaking ratio given by

$$\Delta_{0+}(B \to K^*\gamma) = \frac{\Gamma(B^0 \to K^{*0}\gamma) - \Gamma(B^+ \to K^{*+}\gamma)}{\Gamma(B^0 \to K^{*0}\gamma) + \Gamma(B^+ \to K^{*+}\gamma)}, \qquad 33.$$

where the partial decay rates are *CP* averaged. In the SM, spectator-dependent effects enter only at the order Λ/m_b, whereas isospin breaking in the form factors is expected to be a negligible effect. Therefore, the SM prediction is as small as $O(5\%)$ (141, 144–147). The ratio is especially sensitive to NP effects in the penguin sector, specifically the ratio of the two effective couplings C_6/C_7. The analogous isospin ratio in the ρ sector strongly depends on CKM parameters (146):

$$\Delta(\rho\gamma) = \frac{\Gamma(B^+ \to \rho^+\gamma)}{2\Gamma(B^0 \to \rho^0\gamma)} - 1 = \left(-4.6^{+5.4}_{-4.2}\Big|_{\text{CKM}} - 5.6^{+5.8}\Big|_{\text{had}}\right) \times 10^{-2}. \qquad 34.$$

The hadronic error is mainly due to the weak annihilation contribution, to which a 50% error is assigned.

4.2.3. *CP* asymmetries in radiative decays. In the *CP* asymmetries, the uncertainties due to form factors cancel out to a large extent. But both the scale dependency and the dependency on the charm quark of the next-to-leading-order predictions are rather large because the *CP* asymmetries arise at $O(\alpha_s)$ only. Although the direct *CP* asymmetry in $B \to K^*\gamma$ is doubly Cabibbo suppressed and expected to be very small within the QCDF/SCET approach, one finds $O(-10\%)$ predictions for the direct *CP* asymmetries in the $B \to \rho\gamma$ mode (127, 144, 146). Because the weak annihilation contribution does not contribute significantly here, the neutral and charged modes are of similar sizes (146):

$$\mathcal{A}_{CP}(\bar{B}^0 \to \rho^0\gamma) = \left(-10.4^{+1.6}_{-2.4}\big|_{\text{CKM}} - 3.6^{+3.0}\big|_{\text{had}}\right)\% \qquad 35.$$

and

$$\mathcal{A}_{CP}(B^- \to \rho^-\gamma) = \left(-10.7^{+1.5}_{-2.0}\Big|_{\text{CKM}} - 3.7^{+2.6}\Big|_{\text{had}}\right)\%. \qquad 36.$$

Finally, we reiterate that all predictions of exclusive observables within the QCDF/SCET approach may receive further uncertainties due to the unknown power corrections. This possibility might be especially important in the case of *CP* asymmetries.

The time-dependent *CP* asymmetry is given by two parameters, $\mathcal{S}_{CP}$ and $\mathcal{A}_{CP}$:

$$\mathcal{A}_{CP}(B \to f; \Delta t) = \mathcal{S}_{CP}\sin(\Delta m \Delta t) + \mathcal{A}_{CP}\cos(\Delta m \Delta t), \qquad 37.$$

where $\mathcal{A}_{CP}$ represents the size of the direct *CP* asymmetry discussed above.[3] In hadronic decay modes such as $B \to \mathcal{J}/\psi K_S^0$, a large value of $\mathcal{S}_{CP}$, due to the angle $\phi_1 \equiv \beta = -\arg(V_{td}V_{tb}^*/V_{ud}V_{ub}^*)$ of the unitarity triangle, has been established, and a similarly large *CP* asymmetry is expected for hadronic penguin decays. This asymmetry is suppressed in radiative penguin decays because the photon helicities are opposite between those from B^0 and $\bar{B}^0$ decays under the left-handed current of SM weak decays, and they do not interfere in the limit of massless quarks. This finding implies a suppression factor of m_s/m_b in the leading contribution to $\mathcal{S}_{CP}$ that is induced by the electromagnetic dipole operator $\mathcal{O}_7$:

$$\mathcal{S}_{CP}^{\mathrm{SM}} = -\sin 2\phi_1 \frac{m_s}{m_b}\left[2 + O(\alpha_s)\right] + \mathcal{S}^{\mathrm{SM},s\gamma g}. \qquad 38.$$

However, there are also additional contributions, $\mathcal{S}^{\mathrm{SM},s\gamma g}$, induced by the process $b \to s\gamma g$ via operators other than $\mathcal{O}_7$ (179, 180). These corrections are not helicity suppressed but rather are power suppressed. A conservative dimensional estimate of the contribution from a nonlocal SCET operator series leads to $|\mathcal{S}^{\mathrm{SM},s\gamma g}| \approx 0.06$ (179, 180), whereas within a QCD sum rule calculation, the contribution due to soft gluon emission is estimated to be $\mathcal{S}^{\mathrm{SM},s\gamma g} = -0.005 \pm 0.01$ (147, 181), which leads to $\mathcal{S}_{CP}^{\mathrm{SM}} = -0.022 \pm 0.015^{+0}_{-0.01}$.[4] The QCD sum rule estimates of power corrections, namely long-distance contributions that arise from photon and soft gluon emission from quark loops (147), lead to analogous results for the other radiative decay modes, such as $B \to \rho\gamma$ (147). If a large value of $\mathcal{S}_{CP}$ beyond the SM prediction is observed, it will signal a new, right-handed current beyond the SM.

4.2.4. $B \to K^*\ell^+\ell^-$. The isospin asymmetry in the mode $B \to K^*\ell^+\ell^-$, as in the radiative mode, is a subleading Λ/m_b effect, but the dominant isospin-breaking effects can be calculated perturbatively, whereas other Λ/m_b corrections can simply be estimated. Thus, the exact uncertainty is difficult to estimate due to unknown power corrections, but the observable may still be useful in the NP search because of its high sensitivity to specific Wilson coefficients (182).

The decay $\bar{B}^0 \to \bar{K}^{*0}\ell^+\ell^-$ (with $\bar{K}^{*0} \to K^-\pi^+$ on the mass shell) is completely described by four independent kinematic variables: the lepton pair–invariant mass squared, q^2, and the three angles θ_ℓ, θ_K, and ϕ (for their precise definitions, see References 148 and 183). Summing over the spins of the final particles, one can write the differential decay distribution as (184, 185)

$$\frac{d^4\Gamma_{\bar{B}}}{dq^2\, d\theta_\ell\, d\theta_K\, d\phi} = \frac{9}{32\pi} I(q^2, \theta_\ell, \theta_K, \phi) \sin\theta_\ell \sin\theta_K. \qquad 39.$$

By integrating two of the angles, one finds

$$\frac{d\Gamma'}{d\theta_K} = \frac{3\Gamma'}{4}\sin\theta_K\left(2F_L\cos^2\theta_K + (1 - F_L)\sin^2\theta_K\right) \qquad 40.$$

and

$$\frac{d\Gamma'}{d\theta_\ell} = \Gamma'\left(\frac{3}{4}F_L\sin^2\theta_\ell + \frac{3}{8}(1 - F_L)(1 + \cos^2\theta_\ell) + A_{FB}\cos\theta_\ell\right)\sin\theta_\ell. \qquad 41.$$

The observables appear linearly in the expressions so the fits can be performed on data binned in q^2. The fraction of the longitudinal polarization F_L from the kaon angular distribution and the

[3]The symbol $\mathcal{C}_{CP} = -\mathcal{A}_{CP}$ is also often used.

[4]This does not necessarily contradict a larger, time-dependent *CP* asymmetry of approximately 10% within the inclusive mode found in Reference 179, because the SCET estimate (179, 180) shows that the expansion parameter is Λ/Q. Here Q is the kinetic energy of the hadronic part. There is no contribution at leading order. Thus, the effect is expected to be larger for larger invariant hadronic mass. The K^* mode must have the smallest effect, below the average 10%.

forward-backward asymmetry A_{FB} from the lepton angular distribution are accessible this way. The latter observable is defined as follows (θ_ℓ is defined below Equation 24):

$$A_{FB}(q^2) \equiv \frac{1}{d\Gamma/dq^2}\left(\int_0^1 d(\cos\theta_\ell)\frac{d^2\Gamma}{dq^2 d\cos\theta_\ell} - \int_{-1}^0 d(\cos\theta_\ell)\frac{d^2\Gamma}{dq^2 d\cos\theta_\ell}\right). \qquad 42.$$

The hadronic uncertainties of these two differential observables are large. However, the value of the dilepton-invariant mass q_0^2, for which the differential forward-backward asymmetry vanishes, can be predicted in quite a clean way. In the QCDF approach at leading order in Λ/m_b, the value of q_0^2 is free from hadronic uncertainties at order α_s^0. A dependency on the soft form factor and on the light-cone wave functions of the B and K^* mesons appears only at order α_s^1. At next-to-leading order, one finds (146)

$$q_0^2[K^{*0}\ell^+\ell^-] = 4.36^{+0.33}_{-0.31}\,\text{GeV}^2 \quad \text{and} \quad q_0^2[K^{*+}\ell^+\ell^-] = 4.15^{+0.27}_{-0.27}\,\text{GeV}^2. \qquad 43.$$

The small difference is due to isospin-breaking power corrections. However, an uncertainty due to unknown power corrections should be re-added to the theoretical error bars. The zero is highly sensitive to the ratio of the two Wilson coefficients C_7 and C_9. Thus, such a measurement would have a huge phenomenological impact.

In the near future, a full angular analysis based on the fourfold differential decay rate in Equation 39 will become possible. Such rich information would allow for the design of observables with specific NP sensitivity and reduced hadronic uncertainties (148, 183). These observables would be constructed in such a way that the soft form factor dependency would cancel out at leading order for all dilepton masses, and they would have much higher sensitivity to new right-handed currents than would observables that are already accessible via the projection fits (148, 183, 185). In these optimized observables, the unknown Λ/m_b corrections would be the source of the largest uncertainty. Further detailed NP analyses of such angular observables have been presented (186, 187). A full angular analysis provides high sensitivity to various Wilson coefficients, but the sensitivity to new weak phases is restricted (183, 188).

5. PRESENT EXPERIMENTAL RESULTS

The huge samples of B meson decays collected by Belle and BaBar have made it possible to fully explore the radiative penguin decays $b \to s\gamma$ and $b \to d\gamma$, as well as the electroweak penguin decays $b \to s\ell^+\ell^-$.

5.1. Inclusive $B \to X_s\gamma$ Branching Fraction

An experimental challenge is how to lower the minimum photon energy to 1.6 GeV (see Section 2.4). Before the construction of the B factories, a minimum photon energy of 2.0 GeV was required in the measurement by CLEO (189). BaBar has a minimum photon-energy requirement of 1.9 GeV, based on 89 million $B\bar{B}$ pairs (190), whereas Belle first reported the result of 1.8 GeV, based on 152 million $B\bar{B}$ pairs (191). Belle recently lowered the limit to 1.7 GeV by using 657 million $B\bar{B}$ pairs (**Figure 5**) (192). Belle measured the branching fraction to yield $\mathcal{B}(B \to X_s\gamma) = (345 \pm 15 \pm 40) \times 10^{-6}$ for $E_\gamma > 1.7$ GeV, whereas BaBar measured it to yield $(367 \pm 29 \pm 34 \pm 29) \times 10^{-6}$ for $E_\gamma > 1.9$ GeV.

The sum-of-exclusive technique and the B-reco technique have been used by BaBar with a minimum photon-energy requirement of 1.9 GeV (193, 194), which corresponds to a maximum recoil mass requirement of 2.8 GeV. Belle also made a sum-of-exclusive measurement using a very early data set (195).

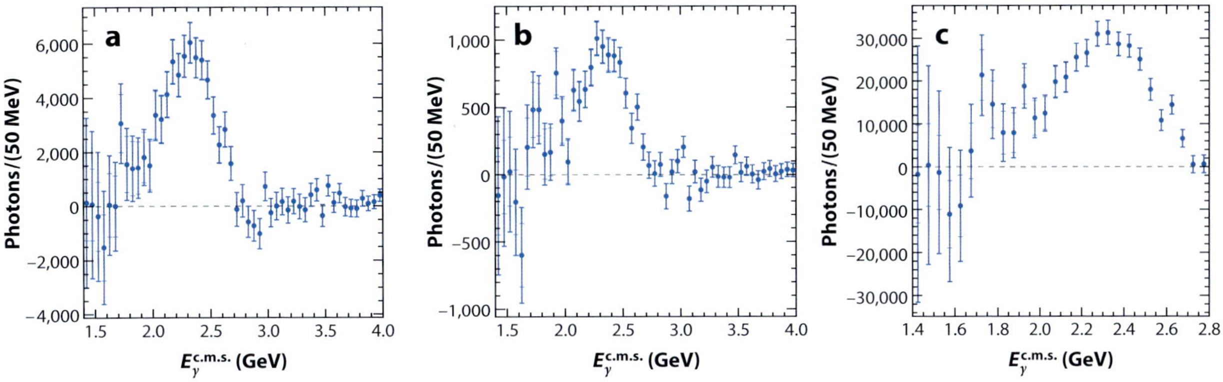

Figure 5

Photon-energy spectrum for $B \to X_s\gamma$, as measured by Belle (*a*) without a lepton tag, (*b*) with a lepton tag, and (*c*) their average from Reference 192.

To calculate the average branching fraction based on the same phase space and to compare it with theory predictions, the Heavy Flavor Averaging Group (HFAG) (196) has made an extrapolation of the branching fraction to the same minimum photon energy of 1.6 GeV (112). The extrapolation factor is 0.985 ± 0.004 (0.894 ± 0.016) for 1.7 (2.0) GeV. The average thereby obtained is

$$\mathcal{B}(B \to X_s\gamma) = (352 \pm 23 \pm 9) \times 10^{-6}, \qquad 44.$$

where the first error is statistical and systematic combined and the second is due to the extrapolation. The result is in agreement with the SM prediction given in Equation 16, and it provides stringent constraints on NP, as discussed in Section 4.1.

5.2. Exclusive Measurements of $b \to s\gamma$ Processes

The recoil system of $B \to X_s\gamma$ below 1.1 GeV is dominated by the K^* resonance, as the spin-0 state is forbidden. Above 1.1 GeV, X_s is a mixture of various resonant and nonresonant states and therefore can usually be modeled as a continuum spectrum in the inclusive $B \to X_s\gamma$ analysis. The K^* signal makes it possible to use the channel for various studies, as discussed below. The $B \to K^*\gamma$ branching fractions have been measured precisely by Belle (197) and BaBar (150) and have been averaged by HFAG to yield

$$\begin{aligned}\mathcal{B}(B^0 \to K^{*0}\gamma) &= (43.3 \pm 1.5) \times 10^{-6}\\ \mathcal{B}(B^+ \to K^{*+}\gamma) &= (42.1 \pm 1.8) \times 10^{-6},\end{aligned} \qquad 45.$$

which corresponds to approximately 12% of the total $B \to X_s\gamma$ branching fraction. The SM predictions for the branching fraction have been calculated by many groups. They are consistent with the measured values but have very large errors of 30% to 50%, which arise mainly from the uncertainty of the $B \to K^*$ form factor (126, 127).

The resonant structure of the high-mass X_s system has also been explored. So far, $B \to K_2^*(1430)\gamma$ (198, 199) and $B \to K_1(1270)\gamma$ (200) have been measured, but other decay channels such as $B \to K_1(1400)\gamma$ seem to have small branching fractions and have not yet been observed. In addition, many multibody final states have been measured, such as $B \to K\pi\pi\gamma$ (198, 201) (including $B \to K^*\pi\gamma$ and $B \to K\rho\gamma$), $B \to K\phi\gamma$ (202, 203), $B \to K\eta\gamma$ (204, 205), $B \to K\eta'\gamma$ (206), and $B \to \Lambda\bar{p}\gamma$ (207).

5.3. *CP* and Isospin Asymmetries in $b \to s\gamma$ Processes

The measurement of the inclusive direct *CP* asymmetry (Equation 19) was performed by use of the sum-of-exclusive method to tag the flavor of the B candidate. For B° ($\bar{B}^0$), only the self-tagging modes with a K^+ (K^-) were used. The measured asymmetry was corrected for a small dilution due to the doubly misidentified pair of a charged kaon and a pion. The results, based on 152 and 383 million $B\bar{B}$ samples by Belle and BaBar, are $0.002 \pm 0.050 \pm 0.030$ (208) and $-0.011 \pm 0.030 \pm 0.014$ (209), respectively, and have been averaged by HFAG to yield

$$\mathcal{A}_{CP}(\bar{B} \to X_s\gamma) = -0.012 \pm 0.028. \qquad 46.$$

This measurement is consistent with null asymmetry, and the size of the error is still much larger than the SM precision (Equation 20). Assuming the systematic error can be reduced along with the statistical error, a data set two orders of magnitude larger would be more sensitive to NP, although still insufficient to measure the small $\mathcal{A}_{CP}$ predicted by the SM.

In the exclusive $B \to K^*\gamma$ channel, the direct *CP* asymmetry is also small but has less-understood theoretical uncertainties. Experimentally, it can be measured more precisely. The current HFAG average is

$$\mathcal{A}_{CP}(B \to K^*\gamma) = -0.003 \pm 0.017, \qquad 47.$$

which is also consistent with null asymmetry.

In the fully inclusive measurement, flavor information is not available for the signal side, but it can be obtained from the charge of the lepton in the event that the other B decays into a semileptonic final state. In that case, it is not possible to discriminate $B \to X_d\gamma$ from $B \to X_s\gamma$, and the measured asymmetry corresponds to a combined one. This combined asymmetry (note the different normalization in comparison with Equation 23) has been measured by BaBar to be

$$\mathcal{A}_{CP}(B \to X_{s+d}\gamma) = -0.110 \pm 0.115 \pm 0.017, \qquad 48.$$

which is consistent with null asymmetry but has a much larger error than do the other two asymmetry measurements.

The measurement of the isospin asymmetry (Equation 33) is another way to utilize the reconstructed $B \to K^*\gamma$ events. Here, the measured branching fractions are corrected by the lifetime ratio: $\tau_{B^+}/\tau_{B^0} = 1.071 \pm 0.009$ (210). Usually, B decay branching fractions are quoted based on the assumption of $\mathcal{B}(\Upsilon(4S) \to B^+B^-) = \mathcal{B}(\Upsilon(4S) \to B^0\bar{B}^0) = 0.5$, but the isospin-asymmetry measurement is already precise enough to be affected by the difference in these branching fractions. Belle measures $\Delta_{0+}(B \to K^*\gamma) = 0.012 \pm 0.044 \pm 0.026$ without this correction, whereas BaBar measures $\Delta_{0+}(B \to K^*\gamma) = 0.066 \pm 0.021 \pm 0.022$ using $\mathcal{B}(\Upsilon(4S) \to B^+B^-) = 0.516 \pm 0.006$ and $\mathcal{B}(\Upsilon(4S) \to B^0\bar{B}^0) = 0.484 \pm 0.006$ (210). After scaling the Belle result and including the CLEO result using the aforementioned lifetime and production ratios, the naïve world average is

$$\Delta_{0+}(B \to K^*\gamma) = 0.062 \pm 0.027. \qquad 49.$$

This average is in agreement with the SM expectation.

A similar isospin asymmetry can be also measured for the inclusive $B \to X_s\gamma$ decay by use of the sum-of-exclusive method. The result by BaBar is

$$\Delta_{0+}(B \to X_s\gamma) = -0.006 \pm 0.058 \pm 0.009 \pm 0.024, \qquad 50.$$

which is consistent with null asymmetry but not yet as precise as that for $B \to K^*\gamma$.

The measurement of the time-dependent *CP* asymmetry (Equation 37) for $b \to s\gamma$ faces two experimental challenges. First, the modes and statistics that can be used for time-dependent *CP*-asymmetry measurements are rather limited. Although the $B^0 \to K^{*0}\gamma$ branching fraction is not very small, only one-ninth of the events that decay into the $K_S^0(\to \pi^+\pi^-)\pi^0\gamma$ final state can be used. The B meson decay vertex position has to be extrapolated from the displaced $K_S^0 \to \pi^+\pi^-$ vertex and the K_S^0 momentum vector. Therefore, the K_S^0 decays inside the vertex detector volume (55% in Belle, 68% in BaBar) and the resulting vertex resolution must be somewhat degraded.

Because any $B \to P^0 Q^0 \gamma$ final states [where P^0 and Q^0 are CP eigenstates (211)] can be used, the $B \to K_S^0\pi^0\gamma$ events with $M_{K_S^0\pi^0}$ up to 1.8 GeV, including $K_2^*(1430)$, were measured by both Belle (212) and BaBar (213). They have been averaged by HFAG to be

$$\mathcal{S}_{CP}\left(B \to K_S^0\pi^0\gamma\right) = -0.15 \pm 0.20, \tag{51}$$

in which the $B^0 \to K^{*0}\gamma$ contribution gives $\mathcal{S}_{CP}(B \to K^*\gamma) = -0.16 \pm 0.22$.

For additional channels, BaBar has measured $\mathcal{S}_{CP}(B \to K_S^0\eta\gamma) = -0.18^{+0.49}_{-0.46} \pm 0.12$ (205), and Belle has measured $\mathcal{S}_{CP}(B \to K_S^0\rho^0\gamma) = 0.11 \pm 0.33^{+0.05}_{-0.09}$ (214). The latter is slightly diluted by the $B \to K^{*+}\pi^-\gamma$ events, by a factor which was measured to be $0.83^{+0.19}_{-0.03}$ but is free from the restriction of K_S^0 vertexing and has a statistical error comparable in size to that of the $B \to K_S^0\pi^0\gamma$ mode. Currently, all results are compatible with null asymmetry with errors that are still not small enough to provide nontrivial constraints on right-handed currents, but this observable will be one of the best ways to search for NP in future experiments.

5.4. Measurements of $b \to d\gamma$ Processes

Three exclusive decay modes are considered to be the easiest modes to study the $b \to d\gamma$ process: $B^+ \to \rho^+\gamma$, $B^0 \to \rho^0\gamma$, and $B^0 \to \omega\gamma$. Although these modes have been searched for since the beginning of Belle and BaBar, only in the later stage of these experiments were measurements of the $B \to \rho\gamma$ modes established. This is partly because of the large $B \to K^*\gamma$ background and partly due to the huge continuum background, which is more severe for modes without a kaon in the final state. Therefore, large statistics and good particle identification are essential; Belle has the advantage in the former, whereas BaBar leads in the latter.

To gain statistics, these three modes have been combined by assuming their naïve quark contents through the use of $\Gamma(B^+ \to \rho^+\gamma) = 2\Gamma(B^0 \to \rho^0\gamma) = 2\Gamma(B^0 \to \omega\gamma)$. In the latest measurements by both Belle (215) and BaBar (216), the $B^0 \to \rho^0\gamma$ mode was measured with more than 5σ significance and $B^+ \to \rho^+\gamma$ with more than 3σ significance, whereas $B^0 \to \omega\gamma$ remains unestablished with significance less than 3σ. Using the symbol $B \to (\rho, \omega)\gamma$ for the combined results that are adjusted for the $B^+ \to \rho^+\gamma$ mode, the averaged branching fraction by HFAG becomes

$$\mathcal{B}[B \to (\rho,\omega)\gamma] = \left(1.30^{+0.18}_{-0.19}\right) \times 10^{-6}. \tag{52}$$

The results are consistent with the SM predictions. However, these predictions are affected by form factor uncertainties and do not have effective prediction power for NP.

A more effective way to use these results is to combine them with the $B \to K^*\gamma$ measurements to determine $|V_{td}/V_{ts}|$. Using Equation 30, Belle and BaBar reported the value of $|V_{td}/V_{ts}|$ to be $0.195^{+0.020}_{-0.019} \pm 0.015$ and $0.233^{+0.025+0.022}_{-0.024-0.021}$, respectively, where the errors are experimental and theoretical. These results from the penguin diagrams are in agreement with the determination from the box diagrams using the ratio of the B^0 and B_s^0 mixing parameters $\Delta m_d/\Delta m_s$, where Δm_d is measured at the B factories and Δm_s at the Tevatron. The results are also in agreement with the more indirect determination by the unitarity triangle fit from other observables. This is a nontrivial test of the CKM scheme. However, although the experimental errors are still larger than the theoretical errors, the size of the theoretical error is unlikely to be reduced.

A possible way to improve this situation utilizes the inclusive $B \to X_d\gamma$ measurement with the sum-of-exclusive method. BaBar has reconstructed the $X_d\gamma$ system in seven final states ($\pi^+\pi^-\gamma$, $\pi^+\pi^0\gamma, \pi^+\pi^-\pi^+\gamma, \pi^+\pi^-\pi^0\gamma, \pi^+\pi^-\pi^+\pi^-\gamma, \pi^+\pi^-\pi^+\pi^0\gamma, \pi^+\eta\gamma$) (217) in the mass range $0.6 < M_{X_d} < 1.8$ GeV, which covers approximately 50% of the total branching fraction. To reduce the uncertainty due to missing modes and phase space, $X_s\gamma$ modes were also measured in the corresponding seven final states in the same mass range, where the first π^+ was replaced with K^+. The ratio of the two inclusive branching fractions is

$$\frac{\mathcal{B}(B \to X_d\gamma)}{\mathcal{B}(B \to X_s\gamma)} = 0.033 \pm 0.013 \pm 0.009, \qquad 53.$$

which is converted to $|V_{td}/V_{ts}| = 0.177 \pm 0.043 \pm 0.001$, where the theory error does not include the effect due to the limited mass range. This result is also in good agreement with other determinations.

The direct *CP* asymmetry for $B \to \rho\gamma$ can be as large as $\simeq$10% in the SM, whereas the time-dependent *CP* asymmetry is doubly suppressed due to the photon helicity and the cancellation of the CKM element V_{td}. The latter appears in the mixing and in the $b \to d$ penguin decay. However, the $\rho^0 \to \pi^+\pi^-$ decay provides clear vertex information for $B^0 \to \rho^0\gamma$. These *CP* asymmetries have been measured by Belle (215, 218) as

$$\begin{aligned} \mathcal{A}_{CP}(B^+ \to \rho^+\gamma) &= -0.11 \pm 0.32 \pm 0.09, \\ \mathcal{A}_{CP}(B^0 \to \rho^0\gamma) &= -0.44 \pm 0.49 \pm 0.14, \quad \text{and} \\ \mathcal{S}_{CP}(B^0 \to \rho^0\gamma) &= -0.83 \pm 0.65 \pm 0.18. \end{aligned} \qquad 54.$$

So far the results are consistent with null asymmetry. A nonzero direct *CP* violation may be measured earlier in $B \to \rho\gamma$ than in $B \to K^*\gamma$.

The isospin asymmetry (Equation 34) is also expected to be as large as $\simeq$5% in $B \to \rho\gamma$. Belle measures $\Delta(\rho\gamma) = -0.48^{+0.21+0.08}_{-0.19-0.09}$ (215), and BaBar measures $\Delta(\rho\gamma) = -0.43^{+0.25}_{-0.22} \pm 0.10$ (216); both measurements show a large isospin asymmetry. The average by HFAG is

$$\Delta(\rho\gamma) = -0.46^{+0.17}_{-0.16}. \qquad 55.$$

A significant nonzero isospin asymmetry could indicate NP.

5.5. Exclusive $B \to K^{(*)}\ell^+\ell^-$ Branching Fraction

Despite their small branching fractions, the exclusive decay channels $B \to K^{(*)}\ell^+\ell^-$ have been measured efficiently with small background at Belle and BaBar, given that their final states are the same as those of $B \to J/\psi K^{(*)}$ for which the *B* factories were designed. Here, $K^{(*)}$ is one of K^+, K^0_S, K^{*+}, and K^{*0}, and $\ell^+\ell^-$ is either e^+e^- or $\mu^+\mu^-$.

Electrons are identified by their energy deposit through an electromagnetic shower in the calorimeter. The minimum momentum is required to be greater than 0.4 GeV by Belle or 0.5 GeV by BaBar. The momentum of the bremsstrahlung photons that may be emitted by the electrons are added to their momenta if they are found near the electron direction. Muons have to reach and penetrate into the outer muon detectors, and the minimum momentum is required to be 0.7 GeV by Belle or 1.0 GeV by BaBar. The dilepton mass regions around J/ψ and $\psi(2S)$ are vetoed.

The branching fractions, averaged over the lepton and kaon flavors and integrated over the dilepton masses, assuming the SM distribution over the vetoed J/ψ and $\psi(2S)$, were measured by Belle (219) and BaBar (220) and have been averaged by HFAG as

$$\begin{aligned} \mathcal{B}(B \to K\ell^+\ell^-) &= (4.5 \pm 0.4) \times 10^{-7} \quad \text{and} \\ \mathcal{B}(B \to K^*\ell^+\ell^-) &= (10.8^{+1.2}_{-1.1}) \times 10^{-7}. \end{aligned} \qquad 56.$$

The results are consistent with SM expectations. At present, the irreducible form factor uncertainty in the SM calculations prevents these results from placing meaningful constraints on NP.

A small subset of these combinations, $B^+ \to K^+\mu^+\mu^-$ and $B^0 \to K^{*0}\mu^+\mu^-$, can be efficiently measured at hadron colliders. CDF has reported the most precise measurements of these modes (221).

5.6. $B \to K^{(*)}\ell^+\ell^-$ Asymmetries and Angular Distributions

The direct *CP* and isospin asymmetries in $B \to K^{(*)}\ell^+\ell^-$ are also useful in the search for NP. The direct *CP* asymmetries are consistent with null values, $\mathcal{A}_{CP}(B \to K^+\ell^+\ell^-) = -0.01 \pm 0.09$ and $\mathcal{A}_{CP}(B \to K^*\ell^+\ell^-) = -0.07 \pm 0.08$, as averaged by HFAG. However, nonzero, negative isospin asymmetries in the small-q^2 region of $B \to K\ell^+\ell^-$ (3.2 σ) and $B \to K^*\ell^+\ell^-$ (2.7σ) have been reported by BaBar (3.9σ when combined). The corresponding isospin asymmetries by Belle are 1.4σ and 1.8σ from zero and are consistent with both BaBar's results and null asymmetry. The isospin asymmetry combined for $B \to K^{(*)}\ell^+\ell^-$ and averaged by HFAG is

$$A_I^{K^{(*)}} = -0.45 \pm 0.10. \qquad 57.$$

The SM prediction is essentially zero at this level of statistics (see Section 4.2).

Muon to electron ratios in $B \to K^{(*)}\ell^+\ell^-$ (Equation 29) have also been measured by both Belle and BaBar. The results are consistent with the SM, and their naïve averages are $R_K = 1.02 \pm 0.18$ and $R_{K^*} = 0.88 \pm 0.17$.

The four-body decay configuration of $B \to K^*\ell^+\ell^- \to K\pi\ell^+\ell^-$ allows extraction of further information from the angular distributions of the final-state particles. The most interesting observables are the fraction of longitudinal polarization F_L from the kaon angular distribution (Equation 40) and the forward-backward asymmetry A_{FB} from the lepton angular distribution (Equation 41). Belle has measured F_L and A_{FB} in six bins of q^2 (219), whereas BaBar has done so in two bins (222). Current statistics are not enough to tell whether there is a zero-crossing point at low q^2, although the results favor the case with no crossing, for which the sign of the Wilson coefficient C_7 is flipped. Both results have positive A_{FB} for high q^2 (**Figure 6**), which sets nontrivial constraints on the Wilson coefficients. CDF has also measured F_L and A_{FB} in the same six bins as Belle for $B^0 \to K^{*0}\mu^+\mu^-$ events (221). The results are in agreement with Belle and BaBar.

5.7. Inclusive $B \to X_s\ell^+\ell^-$ Branching Fraction

The inclusive $B \to X_s\ell^+\ell^-$ branching fraction has been measured by Belle and BaBar using the sum-of-exclusive technique. The X_s system includes final states with one kaon and up to four (two) pions that have masses up to 2.0 (1.8) GeV for the result by Belle (BaBar). Belle recently announced a preliminary result based on 657 million $B\bar{B}$ (223), and BaBar's result is based on 89 million $B\bar{B}$ (224). In Belle's new analysis, partial branching fractions are measured in bins of the X_s mass, and then the total branching fraction is calculated as their sum. This method reduces the large systematic error observed in previous studies that arose from the strong-X_s mass dependency of the efficiency and the unknown fractions of exclusive channels $B \to K^{(*)}\ell^+\ell^-$. The measurement is still dominated by the statistical error and will be more precise in the future. Belle and BaBar reported branching fractions as $\mathcal{B}(B \to X_s\ell^+\ell^-) = (3.33 \pm 0.80^{+0.19}_{-0.24}) \times 10^{-6}$ and $\mathcal{B}(B \to X_s\ell^+\ell^-) = (5.6 \pm 1.5 \pm 0.6 \pm 1.1) \times 10^{-6}$, respectively, which were averaged by HFAG as

$$\mathcal{B}(B \to X_s\ell^+\ell^-) = (3.66^{+0.76}_{-0.77}) \times 10^{-6}, \qquad 58.$$

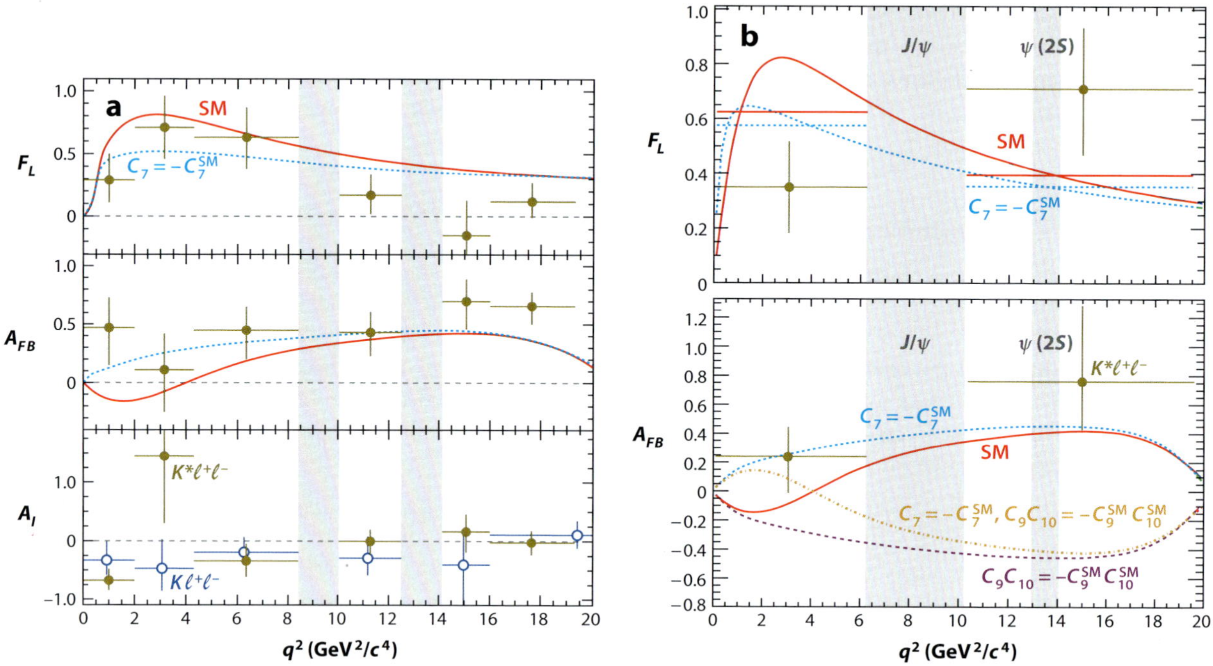

Figure 6

Longitudinal polarization fraction, forward-backward asymmetry, and isospin asymmetry of $B \to K^*\ell^+\ell^-$ by (*a*) Belle (219) and (*b*) BaBar (222). The solid line shows the Standard Model (SM) predictions, and the other curves represent non-SM extreme cases.

integrated over the entire subset of phase space with $q^2 > 0.2$ GeV, including the vetoed J/ψ and $\psi(2S)$ regions. The results are in good agreement with the SM prediction. They strongly disfavor the case with the flipped sign of C_7 (175).

6. OUTLOOK

Remarkably, the B factories have measured all the observables within the radiative and electroweak penguin decays at values that are consistent with SM predictions. These measurements rule out $O(1)$ corrections to the SM and identify the CKM theory as the dominant effect for flavor violation as well as for CP violation. The success of the simple CKM theory of CP violation was honored with the Nobel Prize in Physics in 2008. Theoretical tools and precision have significantly advanced during the past decade, and we are ready either to challenge the SM if a clear deviation is found or to discriminate different NP scenarios if direct evidence is found at the LHC.

Also, the future offers great experimental opportunities in flavor physics. LHCb has finally started taking data and promises to overwhelm many B factory results, and ATLAS and CMS will also contribute to flavor physics. In radiative and electroweak penguin decays, the most promising measurements will be the angular analysis of $B^0 \to K^{*0}\mu^+\mu^-$ and the analysis of time-dependent CP asymmetry in $B_s \to \phi\gamma$; the latter measurement cannot be performed at the B factories due to the fast B_s oscillation. However, the theoretically clean inclusive modes and many modes involving neutral particles such as the π^0 can be pursued only at the e^+e^- B factories. Two proposed super-B factories, Belle II at KEK and SuperB in Italy, would accumulate two-orders-of-magnitude-larger data samples. Such data would push experimental precision to its limit.

Theoretical and experimental techniques are ready for such large data samples. The results provided by LHCb and the next-generation e^+e^- and B factories are eagerly awaited, as they may be the key to identifying physics beyond the SM.

DISCLOSURE STATEMENT

The authors are not aware of any affiliations, memberships, funding, or financial holdings that might be perceived as affecting the objectivity of this review.

ACKNOWLEDGMENTS

We thank Christoph Greub, Colin Jessop, and Kurtis Nishimura for their careful reading of the manuscript, and Thorsten Feldmann, Matthias Neubert, and Gil Paz for comments. T.H. thanks the CERN Theory Group for its hospitality during his visits to CERN.

LITERATURE CITED

1. Ellis JR, Gaillard MK, Nanopoulos DV, Rudaz S. *Nucl. Phys. B* 131:285 (1977); Ellis JR, Gaillard MK, Nanopoulos DV, Rudaz S. Erratum. *Nucl. Phys. B* 132:541 (1978)
2. Shifman MA. arXiv:hep-ph/9510397 (1995)
3. Kobayashi M, Maskawa T. *Prog. Theor. Phys.* 49:652 (1973)
4. Cabibbo N. *Phys. Rev. Lett.* 10:531 (1963)
5. Artuso M, et al. *Eur. Phys. J. C* 57:309 (2008)
6. Antonelli M, et al. arXiv:0907.5386 [hep-ph] (2009)
7. Isidori G, Nir Y, Perez G. *Annu. Rev. Nucl. Part. Sci.* 60: in press (2010)
8. Buras AJ. arXiv:0910.1032 [hep-ph] (2009)
9. Ammer R, et al. (CLEO Collab.) *Phys. Rev. Lett.* 71:674 (1993)
10. Alam MS, et al. (CLEO Collab.) *Phys. Rev. Lett.* 74:2885 (1995)
11. Lingel K, Skwarnicki T, Smith JG. *Annu. Rev. Nucl. Part. Sci.* 48:253 (1998)
12. Hurth T. *Rev. Mod. Phys.* 75:1159 (2003)
13. Hurth T, Lunghi E. Presented at Workshop CKM Unit. Triangle, 2nd, Durham, UK (2003)
14. Hurth T. *Int. J. Mod. Phys. A* 22:1781 (2007)
15. Wilson KG. *Phys. Rev.* 179:1499 (1969)
16. Wilson KG. *Phys. Rev. D* 3:1818 (1971)
17. Gaillard MK, Lee BW. *Phys. Rev. Lett.* 33:108 (1974)
18. Altarelli G, Maiani L. *Phys. Lett.* B52:351 (1974)
19. Witten E. *Nucl. Phys. B* 122:109 (1977)
20. Chay J, Georgi H, Grinstein B. *Phys. Lett.* B247:399 (1990)
21. Bigi II, Uraltsev NG, Vainshtein AI. *Phys. Lett.* B293:430 (1992); Bigi II, Uraltsev NG, Vainshtein AI. Erratum. *Phys. Lett.* B297:477 (1993)
22. Bigi II, et al. arXiv:hep-ph/9212227 (1992)
23. Bigi II, Shifman MA, Uraltsev N. *Annu. Rev. Nucl. Part. Sci.* 47:591 (1997)
24. Manohar AV, Wise MB. *Phys. Rev. D* 49:1310 (1994)
25. Manohar AV, Wise MB. *Camb. Monogr. Part. Phys. Nucl. Phys. Cosmol.* 10:1 (2000)
26. Falk AF, Luke ME, Savage MJ. *Phys. Rev. D* 49:3367 (1994)
27. Ali A, Hiller G, Handoko LT, Morozumi T. *Phys. Rev. D* 55:4105 (1997)
28. Beneke M, Buchalla G, Neubert M, Sachrajda CT. *Phys. Rev. Lett.* 83:1914 (1999)
29. Beneke M, Buchalla G, Neubert M, Sachrajda CT. *Nucl. Phys. B* 591:313 (2000)
30. Beneke M, Buchalla G, Neubert M, Sachrajda CT. *Nucl. Phys. B* 606:245 (2001)
31. Bauer CW, Fleming S, Luke ME. *Phys. Rev. D* 63:014006 (2000)
32. Bauer CW, Fleming S, Pirjol D, Stewart IW. *Phys. Rev. D* 63:114020 (2001)

33. Bauer CW, Stewart IW. *Phys. Lett.* B516:134 (2001)
34. Bauer CW, Pirjol D, Stewart IW. *Phys. Rev. D* 65:054022 (2002)
35. Beneke M, Chapovsky AP, Diehl M, Feldmann T. *Nucl. Phys. B* 643:431 (2002)
36. Hill RJ, Neubert M. *Nucl. Phys. B* 657:229 (2003)
37. Ciuchini M, et al. *Phys. Lett.* B316:127 (1993)
38. Ciuchini M, Franco E, Martinelli G, Reina L. *Nucl. Phys. B* 415:403 (1994)
39. Cella G, Curci G, Ricciardi G, Vicere A. *Nucl. Phys. B* 431:417 (1994)
40. Misiak M. *Nucl. Phys. B* 393:23 (1993); Misiak M. Erratum. *Nucl. Phys. B* 439:461 (1995)
41. Adel K, Yao YP. *Phys. Rev. D* 49:4945 (1994)
42. Greub C, Hurth T. *Phys. Rev. D* 56:2934 (1997)
43. Chetyrkin KG, Misiak M, Munz M. *Phys. Lett.* B400:206 (1997); Chetyrkin KG, Misiak M, Munz M. Erratum. *Phys. Lett.* B425:414 (1998)
44. Gambino P, Gorbahn M, Haisch U. *Nucl. Phys. B* 673:238 (2003)
45. Ali A, Greub C. *Z. Phys. C* 49:431 (1991)
46. Greub C, Hurth T, Wyler D. *Phys. Rev. D* 54:3350 (1996)
47. Pott N. *Phys. Rev. D* 54:938 (1996)
48. Buras AJ, Czarnecki A, Misiak M, Urban J. *Nucl. Phys. B* 611:488 (2001)
49. Buras AJ, Czarnecki A, Misiak M, Urban J. *Nucl. Phys. B* 631:219 (2002)
50. Gambino P, Misiak M. *Nucl. Phys. B* 611:338 (2001)
51. Hurth T, Lunghi E, Porod W. *Nucl. Phys. B* 704:56 (2005)
52. Hurth T, Lunghi E, Porod W. *Eur. Phys. J. C* 33:382 (2004)
53. Asatrian HM, et al. *Phys. Lett.* B619:322 (2005)
54. Misiak M, et al. *Phys. Rev. Lett.* 98:022002 (2007)
55. Misiak M, Steinhauser M. *Nucl. Phys. B* 683:277 (2004)
56. Bobeth C, Misiak M, Urban J. *Nucl. Phys. B* 574:291 (2000)
57. Gorbahn M, Haisch U. *Nucl. Phys. B* 713:291 (2005)
58. Gorbahn M, Haisch U, Misiak M. *Phys. Rev. Lett.* 95:102004 (2005)
59. Czakon M, Haisch U, Misiak M. *J. High Energy Phys.* 0703:008 (2007)
60. Melnikov K, Mitov A. *Phys. Lett.* B620:69 (2005)
61. Blokland I, et al. *Phys. Rev. D* 72:033014 (2005)
62. Asatrian HM, et al. *Nucl. Phys. B* 749:325 (2006)
63. Asatrian HM, et al. *Nucl. Phys. B* 762:212 (2007)
64. Bieri K, Greub C, Steinhauser M. *Phys. Rev. D* 67:114019 (2003)
65. Misiak M, Steinhauser M. *Nucl. Phys. B* 764:62 (2007)
66. Boughezal R, Czakon M, Schutzmeier T. *J. High Energy Phys.* 0709:072 (2007)
67. Ligeti Z, Luke ME, Manohar AV, Wise MB. *Phys. Rev. D* 60:034019 (1999)
68. Asatrian HM, Ewerth T, Gabrielyan H, Greub C. *Phys. Lett.* B647:173 (2007)
69. Ewerth T. *Phys. Lett.* B669:167 (2008)
70. Czarnecki A, Marciano WJ. *Phys. Rev. Lett.* 81:277 (1998)
71. Kagan AL, Neubert M. *Eur. Phys. J. C* 7:5 (1999)
72. Baranowski K, Misiak M. *Phys. Lett.* B483:410 (2000)
73. Gambino P, Haisch U. *J. High Energy Phys.* 0110:020 (2001)
74. Buras AJ, Munz M. *Phys. Rev. D* 52:186 (1995)
75. Asatryan HH, Asatrian HM, Greub C, Walker M. *Phys. Rev. D* 65:074004 (2002)
76. Asatryan HH, Asatrian HM, Greub C, Walker M. *Phys. Rev. D* 66:034009 (2002)
77. Ghinculov A, Hurth T, Isidori G, Yao YP. *Nucl. Phys. B* 685:351 (2004)
78. Ghinculov A, Hurth T, Isidori G, Yao YP. *Eur. Phys. J. C* 33:s288 (2004)
79. Asatrian HM, Bieri K, Greub C, Hovhannisyan A. *Phys. Rev. D* 66:094013 (2002)
80. Asatrian HM, Asatryan HH, Hovhannisyan A, Poghosyan V. *Mod. Phys. Lett. A* 19:603 (2004)
81. Ghinculov A, Hurth T, Isidori G, Yao YP. *Nucl. Phys. B* 648:254 (2003)
82. Greub C, Pilipp V, Schupbach C. *J. High Energy Phys.* 0812:040 (2008)
83. Huber T, Hurth T, Lunghi E. arXiv:0807.1940 [hep-ph] (2008)
84. Bobeth C, Gambino P, Gorbahn M, Haisch U. *J. High Energy Phys.* 0404:071 (2004)

85. Huber T, Lunghi E, Misiak M, Wyler D. *Nucl. Phys. B* 740:105 (2006)
86. Huber T, Hurth T, Lunghi E. *Nucl. Phys. B* 802:40 (2008)
87. Ali A, Asatrian H, Greub C. *Phys. Lett.* B429:87 (1998)
88. Asatrian HM, Bieri K, Greub C, Walker M. *Phys. Rev. D* 69:074007 (2004)
89. Seidel D. *Phys. Rev. D* 70:094038 (2004)
90. Isgur N, Wise MB. *Adv. Ser. Dir. High Energy Phys.* 10:549 (1992)
91. Neubert M. *Phys. Rep.* 245:259 (1994)
92. Ligeti Z, Randall L, Wise MB. *Phys. Lett.* B402:178 (1997)
93. Voloshin MB. *Phys. Lett.* B397:275 (1997)
94. Benzke M, Lee SJ, Neubert M, Paz G. arXiv:1003.5012 [hep-ph] (2010)
95. Grant AK, Morgan AG, Nussinov S, Peccei RD. *Phys. Rev. D* 56:3151 (1997)
96. Buchalla G, Isidori G, Rey SJ. *Nucl. Phys. B* 511:594 (1998)
97. Lee SJ, Neubert M, Paz G. *Phys. Rev. D* 75:114005 (2007)
98. Lee KSM, Stewart IW. *Nucl. Phys. B* 721:325 (2005)
99. Bosch SW, Neubert M, Paz G. *J. High Energy Phys.* 0411:073 (2004)
100. Beneke M, Campanario F, Mannel T, Pecjak BD. *J. High Energy Phys.* 0506:071 (2005)
101. Chen JW, Rupak G, Savage MJ. *Phys. Lett.* B410:285 (1997)
102. Buchalla G, Isidori G. *Nucl. Phys. B* 525:333 (1998)
103. Bauer CW, Burrell CN. *Phys. Rev. D* 62:114028 (2000)
104. Ligeti Z, Tackmann FJ. *Phys. Lett.* B653:404 (2007)
105. Neubert M. *J. High Energy Phys.* 0007:022 (2000)
106. Bauer CW, Ligeti Z, Luke ME. *Phys. Rev. D* 64:113004 (2001)
107. Krüger F, Sehgal LM. *Phys. Lett.* B380:199 (1996)
108. Krüger F, Sehgal LM. *Phys. Rev. D* 55:2799 (1997)
109. Beneke M, Buchalla G, Neubert M, Sachrajda CT. *Eur. Phys. J. C* 61:439 (2009)
110. Benson D, Bigi II, Uraltsev NG. *Nucl. Phys. B* 710:371 (2005)
111. Bosch SW, Lange BO, Neubert M, Paz G. *Nucl. Phys. B* 699:335 (2004)
112. Buchmuller O, Flacher H. *Phys. Rev. D* 73:073008 (2006)
113. Neubert M. *Eur. Phys. J. C* 40:165 (2005)
114. Becher T, Neubert M. *Phys. Rev. Lett.* 98:022003 (2007)
115. Becher T, Neubert M. *Phys. Lett.* B633:739 (2006)
116. Becher T, Neubert M. *Phys. Lett.* B637:251 (2006)
117. Misiak M. arXiv:0808.3134 [hep-ph] (2008)
118. Andersen JR, Gardi E. *J. High Energy Phys.* 0701:029 (2007)
119. Andersen JR, Gardi E. *J. High Energy Phys.* 0506:030 (2005)
120. Gardi E. arXiv:hep-ph/0606080 (2006)
121. Lee KSM, Stewart IW. *Phys. Rev. D* 74:014005 (2006)
122. Lee KSM, Ligeti Z, Stewart IW, Tackmann FJ. *Phys. Rev. D* 74:011501 (2006)
123. Lee KSM, Tackmann FJ. *Phys. Rev. D* 79:114021 (2009)
124. Poggio EC, Quinn HR, Weinberg S. *Phys. Rev. D* 13:1958 (1976)
125. Beneke M, Feldmann T. *Nucl. Phys. B* 592:3 (2001)
126. Beneke M, Feldmann T, Seidel D. *Nucl. Phys. B* 612:25 (2001)
127. Bosch SW, Buchalla G. *Nucl. Phys. B* 621:459 (2002)
128. Ali A, Parkhomenko AY. *Eur. Phys. J. C* 23:89 (2002)
129. Descotes-Genon S, Sachrajda CT. *Nucl. Phys. B* 693:103 (2004)
130. Ali A, Pecjak BD, Greub C. *Eur. Phys. J. C* 55:577 (2008)
131. Braun VM, Filyanov IE. *Z. Phys. C* 44:157 (1989)
132. Braun VM, Filyanov IE. *Z. Phys. C* 48:239 (1990)
133. Ball P, Braun VM, Koike Y, Tanaka K. *Nucl. Phys. B* 529:323 (1998)
134. Ball P, Braun VM. *Nucl. Phys. B* 543:201 (1999)
135. Bauer CW, Pirjol D, Stewart IW. *Phys. Rev. D* 67:071502 (2003)
136. Beneke M, Feldmann T. *Nucl. Phys. B* 685:249 (2004)
137. Lange BO, Neubert M. *Nucl. Phys. B* 690:249 (2004)

138. Charles J, et al. *Phys. Rev. D* 60:014001 (1999)
139. Becher T, Hill RJ, Neubert M. *Phys. Rev. D* 72:094017 (2005)
140. Feldmann T, Hurth T. *J. High Energy Phys.* 0411:037 (2004)
141. Kagan AL, Neubert M. *Phys. Lett.* B539:227 (2002)
142. Becher T, Hill RJ, Neubert M. *Phys. Rev. D* 69:054017 (2004)
143. Arnesen CM, Ligeti Z, Rothstein IZ, Stewart IW. *Phys. Rev. D* 77:054006 (2008)
144. Ali A, Lunghi E, Parkhomenko AY. *Phys. Lett.* B595:323 (2004)
145. Bosch SW, Buchalla G. *J. High Energy Phys.* 0501:035 (2005)
146. Beneke M, Feldmann T, Seidel D. *Eur. Phys. J. C* 41:173 (2005)
147. Ball P, Jones GW, Zwicky R. *Phys. Rev. D* 75:054004 (2007)
148. Egede U, et al. *J. High Energy Phys.* 0811:032 (2008)
149. Bobeth C, Hiller G, Piranishvili G. *J. High Energy Phys.* 0712:040 (2007)
150. Aubert B, et al. (BaBar Collab.) *Phys. Rev. Lett.* 103:211802 (2009)
151. Sjostrand T, Mrenna S, Skands P. *J. High Energy Phys.* 0605:026 (2006)
152. Gambino P, Giordano P. *Phys. Lett.* B669:69 (2008)
153. Ciuchini M, Degrassi G, Gambino P, Giudice GF. *Nucl. Phys. B* 527:21 (1998)
154. Borzumati F, Greub C. *Phys. Rev. D* 58:074004 (1998)
155. Haisch U, Weiler A. *Phys. Rev. D* 76:034014 (2007)
156. Bertolini S, Borzumati F, Masiero A, Ridolfi G. *Nucl. Phys. B* 353:591 (1991)
157. Degrassi G, Gambino P, Giudice GF. *J. High Energy Phys.* 0012:009 (2000)
158. Carena MS, Garcia D, Nierste U, Wagner CEM. *Phys. Lett.* B499:141 (2001)
159. Degrassi G, Gambino P, Slavich P. *Phys. Lett.* B635:335 (2006)
160. Borzumati F, Greub C, Hurth T, Wyler D. *Phys. Rev. D* 62:075005 (2000)
161. Besmer T, Greub C, Hurth T. *Nucl. Phys. B* 609:359 (2001)
162. Ciuchini M, Franco E, Masiero A, Silvestrini L. *Phys. Rev. D* 67:075016 (2003); Ciuchini M, Franco E, Masiero A, Silvestrini L. Erratum. *Phys. Rev. D* 68:079901 (2003)
163. Ciuchini M, et al. *Phys. Rev. Lett.* 92:071801 (2004)
164. Altmannshofer W, Guadagnoli D, Raby S, Straub DM. *Phys. Lett.* B668:385 (2008)
165. Altmannshofer W, et al. *Nucl. Phys. B* 830:17 (2010)
166. Blanke M, et al. *Acta Phys. Polon. B* 41:657 (2010)
167. Ali A, Lunghi E, Greub C, Hiller G. *Phys. Rev. D* 66:034002 (2002)
168. D'Ambrosio G, Giudice GF, Isidori G, Strumia A. *Nucl. Phys. B* 645:155 (2002)
169. Hurth T, Isidori G, Kamenik JF, Mescia F. *Nucl. Phys. B* 808:326 (2009)
170. Kagan AL, Neubert M. *Phys. Rev. D* 58:094012 (1998)
171. Soares JM. *Nucl. Phys. B* 367:575 (1991)
172. Hurth T, Mannel T. *Phys. Lett.* B511:196 (2001)
173. Hurth T, Mannel T. *AIP Conf. Proc.* 602:212 (2001)
174. Lee KSM, Ligeti Z, Stewart IW, Tackmann FJ. *Phys. Rev. D* 75:034016 (2007)
175. Gambino P, Haisch U, Misiak M. *Phys. Rev. Lett.* 94:061803 (2005)
176. Yan QS, Huang CS, Liao W, Zhu SH. *Phys. Rev. D* 62:094023 (2000)
177. Hiller G, Krüger F. *Phys. Rev. D* 69:074020 (2004)
178. Ball P, et al. *Phys. Rev. D* 0604:046 (2006)
179. Grinstein B, Grossman Y, Ligeti Z, Pirjol D. *Phys. Rev. D* 71:011504 (2005)
180. Grinstein B, Pirjol D. *Phys. Rev. D* 73:014013 (2006)
181. Ball P, Zwicky R. *Phys. Lett.* B642:478 (2006)
182. Feldmann T, et al. *Phys. Rev. D* 0301:074 (2003)
183. Egede U, et al. arXiv:1005.0571 [hep-ph] (2010)
184. Krüger F, Sehgal LM, Sinha N, Sinha R. *Phys. Rev. D* 61:114028 (2000); Krüger F, Sehgal LM, Sinha N, Sinha R. Erratum. *Phys. Rev. D* 63:019901 (2001)
185. Krüger F, Matias J. *Phys. Rev. D* 71:094009 (2005)
186. Altmannshofer W, et al. *J. High Energy Phys.* 0901:019 (2009)
187. Bobeth C, Hiller G, Piranishvili G. *J. High Energy Phys.* 0807:106 (2008)
188. Egede U, et al. arXiv:0912.1349 [hep-ph] (2009)

189. Chen S, et al. (CLEO Collab.) *Phys. Rev. Lett.* 87:251807 (2001)
190. Aubert B, et al. (BaBar Collab.) *Phys. Rev. Lett.* 97:171803 (2006)
191. Koppenburg P, et al. (Belle Collab.) *Phys. Rev. Lett.* 93:061803 (2004)
192. Limosani A, et al. (Belle Collab.) *Phys. Rev. Lett.* 103:241801 (2009)
193. Aubert B, et al. (BaBar Collab.) *Phys. Rev. D* 72:052004 (2005)
194. Aubert B, et al. (BaBar Collab.) *Phys. Rev. D* 77:051103R (2008)
195. Abe K, et al. (Belle Collab.) *Phys. Lett.* B511:151 (2001)
196. Barberio E, et al. arXiv:0808.1297 [hep-ex]. **http://www.slac.stanford.edu/xorg/hfag** (2009)
197. Nakao M, et al. (Belle Collab.) *Phys. Rev. D* 69:112001 (2004)
198. Nishida S, et al. (Belle Collab.) *Phys. Rev. Lett.* 89:231801 (2002)
199. Aubert B, et al. (BaBar Collab.) *Phys. Rev. D* 70:091105R (2004)
200. Yang H, et al. (Belle Collab.) *Phys. Rev. Lett.* 94:111802 (2005)
201. Aubert B, et al. (BaBar Collab.) *Phys. Rev. Lett.* 98:211804 (2007)
202. Drutskoy A, et al. (Belle Collab.) *Phys. Rev. Lett.* 92:051801 (2004)
203. Aubert B, et al. (BaBar Collab.) *Phys. Rev. D* 75:051102 (2007)
204. Nishida S, et al. (Belle Collab.) *Phys. Lett.* B610:23 (2005)
205. Aubert B, et al. (BaBar Collab.) *Phys. Rev. D* 79:011102 (2009)
206. Wedd R, et al. (Belle Collab.) arXiv:0810.0804 [hep-ex] (2008)
207. Lee YJ, et al. (Belle Collab.) *Phys. Rev. Lett.* 95:061802 (2005)
208. Nishida S, et al. (Belle Collab.) *Phys. Rev. Lett.* 93:031803 (2004)
209. Aubert B, et al. (BaBar Collab.) *Phys. Rev. Lett.* 101:171804 (2008)
210. Amsler C, et al. (Part. Data Group) *Phys. Lett.* B667:1 (2008)
211. Atwood D, Gershon T, Hazumi M, Soni A. *Phys. Rev. D* 71:076003 (2005)
212. Ushiroda Y, et al. (Belle Collab.) *Phys. Rev. D* 74:111104R (2006)
213. Aubert B, et al. (BaBar Collab.) *Phys. Rev. D* 78:071102R (2008)
214. Li J, et al. (Belle Collab.) *Phys. Rev. Lett.* 101:251601 (2008)
215. Taniguchi N, et al. (Belle Collab.) *Phys. Rev. Lett.* 101:111801 (2008)
216. Aubert B, et al. (BaBar Collab.) *Phys. Rev. D* 78:112001 (2008)
217. Aubert B, et al. (BaBar Collab.) *Phys. Rev. Lett.* 102:161803 (2009)
218. Ushiroda Y, et al. (Belle Collab.) *Phys. Rev. Lett.* 100:021602 (2008)
219. Wei JT, et al. (Belle Collab.) *Phys. Rev. Lett.* 103:171801 (2009)
220. Aubert B, et al. (BaBar Collab.) *Phys. Rev. Lett.* 102:091803 (2009)
221. CDF Collab. *CDF public note 10047* (2009)
222. Aubert B, et al. (BaBar Collab.) *Phys. Rev. D* 79P:031102R (2009)
223. Iijima T. Presented at XXIV Int. Symp. Lepton Photon Interact., 24th, Hamburg, Ger. (2009)
224. Aubert B, et al. (BaBar Collab.) *Phys. Rev. Lett.* 93:081802 (2004)

Cumulative Indexes

Contributing Authors, Volumes 51–60

Chapter Titles, Volumes 51–60

Prefatory Chapters

Accelerators

Instrumentation and Techniques

Nuclear Applications

Nuclear Experiment

Particle Interactions at High Energies

Particle Spectroscopy

Weak and Electromagnetic Interactions

Special Topics